ENCYCLOPEDIA OF PHYSICS

EDITED BY

S. FLÜGGE

VOLUME III/2

PRINCIPLES OF THERMODYNAMICS AND STATISTICS

WITH 25 FIGURES

SPRINGER-VERLAG

BERLIN HEIDELBERG GMBH

1959

HANDBUCH DER PHYSIK

HERAUSGEGEBEN VON
S. FLÜGGE

BAND III/2

PRINZIPIEN DER THERMODYNAMIK UND STATISTIK

MIT 25 FIGUREN

SPRINGER-VERLAG
BERLIN HEIDELBERG GMBH
1959

by Springer-Verlag Berlin Heidelberg 1959
Ursprünglich erschienen bei Springer-Verlag OHG. Berlin · Gottingen · Heidelberg 1959
Softcover reprint of the hardcover 1st edition 1959

ISBN 978-3-642-45913-9   ISBN 978-3-642-45912-2 (eBook)
DOI 10.1007/ 978-3-642-45912-2

Inhaltsverzeichnis.

Thermodynamics, Classical and Statistical.

By

E. A. GUGGENHEIM.

With 4 Figures.

A. Fundamentals of classical thermodynamics.

1. Scope of thermodynamics. The most important conception in thermodynamics is temperature. The essential properties of temperature will be described below. Anticipating this we may define thermodynamics as that part of physics concerned with the dependence on temperature of any equilibrium property. This definition may be illustrated by a simple example. Consider the distribution of two immiscible liquids such as mercury and water in a gravitational field. The equilibrium distribution is that in which the heavier liquid, mercury, occupies the part of accessible space where the gravitational potential is lowest and the lighter liquid, water, occupies the part of the remaining accessible space where the gravitational potential is lowest. This equilibrium distribution is, apart from the effect of thermal expansion which we neglect, independent of temperature. Consequently the problem does not involve thermodynamics, but only hydrostatics. Now consider by contrast the distribution in a gravitational field of two completely miscible fluids such as bromine and carbon disulphide. The relative proportions of the two substances will vary from place to place, the proportion of the heavier liquid, bromine, being greatest at the lowest gravitational potential and conversely. The precise relation between the composition and the gravitational potential depends on the temperature, assumed uniform, of the mixture. Clearly this is a problem in thermodynamics, not merely hydrostatics.

We shall now mention a few other typical examples to show that thermodynamics has a bearing on most branches of physics, including elasticity, hydrodynamics, electrostatics and electrodynamics. In the relation, known as Hooke's law, of proportionality between tension and extension the coefficient of proportionality will in general be temperature dependent. In so far as its variation with temperature is relevant thermodynamics is involved. To study the temperature dependence of the compressibility of a fluid, that of the dielectric coefficient of a dielectric, that of the permeability of a paramagnetic material, that of the electromotive force of a cell and in fact the temperature dependence of any equilibrium property thermodynamics is needed.

2. Energy and heat. First law. Leaving temperature for the moment, we must now say something about energy. The conception of energy arose first in mechanics and was extended to electrostatics and electrodynamics. When these branches of physics are idealized so as to exclude friction, viscosity, hysteresis, temperature gradients, temperature dependence of properties and related phenomena the fundamental property of energy can be described in two alternative ways.

I. When several systems interact in any way with one another, the whole set of systems being isolated from the rest of the universe, the sum of the energies of the several systems remains constant.

II. When a single system interacts with the rest of the universe (its surroundings) the increase of the energy of this system is equal to the work done on the system by the rest of the universe (its surroundings).

Under the idealized conditions mentioned above these two descriptions are equivalent, but when these restrictions are removed the two descriptions are no longer equivalent and we have to make a choice between them. Of the alternatives we choose I and with this choice the extended conception of energy is termed total energy and is denoted by U. The formulation I is then a statement of the conservation of total energy. It is based indirectly on the classical experiments of Joule.

Let us now consider in greater detail the interaction between a pair of systems, supposed isolated from the rest of the universe. Using subscripts A, B to relate to the two systems we have

$$dU_A + dU_B = 0 \tag{2.1}$$

or

$$dU_A = - dU_B \tag{2.2}$$

but in general this is not equal to the work w_{BA} done by B on A. In other words there can be exchange of energy between A and B of a kind other than work. Such an exchange of energy is that determined by a temperature difference and is called heat. If then we denote the heat flow from B to A by q_{BA}, we have the following relations

$$dU_A = w_{BA} + q_{BA}, \tag{2.3}$$

$$dU_B = w_{AB} + q_{AB}, \tag{2.4}$$

$$w_{AB} + w_{BA} = 0, \tag{2.5}$$

$$q_{AB} + q_{BA} = 0. \tag{2.6}$$

This set of relations together constitutes the first law of thermodynamics.

The sign of q is determined by the temperature difference between A and B, and the universal convention is to define the sign of a temperature difference in such a way that heat always flows from the higher to the lower temperature.

The above analysis of the most general interaction between two systems can immediately be extended to the most general interaction between a given system and the rest of the universe. If we denote by U the total energy of the system, by q the heat flow from the surroundings to the system and by w the work done on the system we have

$$dU = q + w. \tag{2.7}$$

When two systems are separated by a boundary or wall which allows heat to flow freely from the one to the other, the two systems are said to be in thermal contact. When on the contrary the boundary or wall is of such a nature as to prevent any flow of heat the wall is said to be insulating. When a system is completely surrounded by an insulating boundary, the system is said to be thermally insulated and any process taking place in the system is called adiabatic. We thus have

$$q = 0, \quad dU = w, \quad \text{adiabatic process.} \tag{2.8}$$

We have already mentioned that if two systems have different temperatures, then over and above any work done by the one system on the other there can be a further exchange of energy, called a flow of heat, but only in one direction namely from the higher to the lower temperature. It is a corollary of this that if the two systems have the same temperature, there will be no flow of heat between them. They are then said to be in thermal equilibrium.

3. Thermostats and thermometers. Consider now two systems in thermal contact, one very much smaller than the other, for example a short thin metallic wire immersed in a large quantity of water. If the quantity of water is large enough (or the wire small enough), then in the process of attaining thermal equilibrium the change in the physical state of the water will be entirely negligible compared with that of the wire. This situation is described differently according as we are primarily interested in the small system or in the large one.

If we are primarily interested in the small system, the wire, then we regard the water as a means of controlling the temperature of the wire and we refer to the water as a temperature bath or thermostat.

If on the other hand we are primarily interested in the large system, the water, we regard the wire as an instrument for recording the temperature of the water and we refer to the wire as a thermometer. This recording of temperature can be rendered quantitative by measuring some property of the thermometer, such as its electrical resistance, which varies with temperature.

4. Absolute temperature. The choice of thermometers is very wide especially as there is a choice both of the substance constituting the thermometer and of the property measured. Consequently there is a wide, effectively infinite, choice of temperature scales. There is however one particular scale which has outstandingly simple characteristics which can be described in a manner independent of the properties of any particular substance or class of substances. This temperature is called absolute temperature. It was first defined by Kelvin and is denoted by T. It is the only scale that we shall use. It will be defined by its properties, especially its relation to entropy. The question how T can best be measured must necessarily be postponed to the next chapter.

5. Phase. Extensive and intensive properties. At this stage it is convenient to define phase, extensive property and intensive property. A phase is a system or part of a system which is completely homogeneous. An extensive property is any property, such as mass, whose value for the whole system is equal to the sum of its values for the separate phases. Its value for a phase of given nature is proportional to the amount of the phase. Important examples of extensive properties are volume and total energy. An intensive property is any property, such as density, whose value is constant throughout a phase. Its value in a phase of given nature is independent of the amount of the phase. Important examples of intensive properties are pressure and temperature.

6. Natural and reversible processes. We must now consider a classification of processes due to Planck. All the independent infinitesimal processes that might conceivably take place may be divided into three types: natural processes, unnatural processes and reversible processes.

Natural processes are all such as actually do occur in nature; they proceed in a direction towards equilibrium.

An unnatural process is one in a direction away from equilibrium; such a process never occurs in nature.

As a limiting case between natural and unnatural processes we have reversible processes, which consist of the passage in either direction through a continuous series of equilibrium states. Reversible processes do not actually occur in nature, but in whichever direction we contemplate a reversible process we can by a small change in the conditions produce a natural process differing as little as we choose from the reversible process contemplated.

We shall illustrate the three types by examples. Consider a system consisting of a liquid together with its vapour at a pressure P. Let the equilibrium vapour pressure of the liquid be p. Consider now the process of the evaporation of a small quantity of the liquid. If $P < p$, this is a natural process and will in fact take place. If on the other hand $P > p$, the process contemplated is unnatural and cannot take place; in fact the contrary process of condensation will take place. If $P = p$ then the process contemplated and its converse are reversible, for by slightly decreasing or increasing P we can make either occur. The last case may be described in an alternative manner as follows. If $P = p - \delta$, where $\delta > 0$, then the process of evaporation is a natural one. Now suppose δ gradually decreased. In the limit $\delta \to 0$, the process becomes reversible.

We have defined a reversible process as a hypothetical passage through equilibrium states. If we have a system interacting with its surroundings either through the performance of work or through the flow of heat, we shall use the term reversible process only if there is throughout the process equilibrium between the system and its surroundings. If we wish to refer to the hypothetical passage of the system through a sequence of internal equilibrium states, without necessarily being in equilibrium with its surroundings we shall refer to a reversible change. We shall illustrate this distinction by examples.

Consider a system consisting of a liquid and its vapour in mutual equilibrium in a cylinder closed by a piston opposed by a pressure equal to the equilibrium vapour pressure corresponding to the temperature of the system. Suppose now that there is a flow of heat through the walls of the cylinder, with a consequent evaporation of liquid and work done on the piston at constant temperature and pressure. The change in the system is a reversible change, but the whole process is a reversible process only if the medium surrounding the cylinder is at the same temperature as the liquid and vapour; otherwise the flow of heat through the walls of the cylinder is not reversible and so the process as a whole is not reversible, though the change in the system within the cylinder is reversible.

As a second example consider a flow of heat from one system in complete internal equilibrium to another system in complete internal equilibrium. Provided both systems remain in internal equilibrium then the change which each system undergoes is a reversible change, but the whole process of heat flow is not a reversible process unless the two systems are at the same temperature.

If a system is in complete equilibrium, any conceivable infinitesimal change in it must be reversible. For a natural process is an approach towards equilibrium, and as the system is already in equilibrium the change cannot be a natural one. Nor can it be an unnatural one, for in that case the opposite infinitesimal change would be a natural one, and this would contradict the supposition that the system is already in equilibrium. The only remaining possibility is that, if the system is in complete equilibrium, any conceivable infinitesimal change must be reversible.

7. Entropy. Second law. We are now ready to introduce the quantity, invented by Clausius, called entropy, the enumeration of whose properties constitutes the second law of thermodynamics. The entropy S is an extensive property of a system which can change in two distinct ways namely by interaction with the surroundings and by changes taking place inside the system. Symbolically we may write this as

$$dS = d_e S + d_i S \qquad (7.1)$$

where dS denotes the increase of the entropy of the system, $d_e S$ denotes the part of this increase due to interaction with the surroundings and $d_i S$ denotes the part of this increase due to changes taking place inside the system.

The entropy increase $d_e S$ due to interaction with the surroundings is related to the heat q flowing into the system from its surroundings by

$$d_e S = \frac{q}{T} \tag{7.2}$$

where T is the absolute temperature. If temperature were measured on any other scale this formula would have to be modified.

The entropy increase $d_i S$ due to changes taking place inside the system is positive for all natural changes, is zero for all reversible changes and is never negative. In symbols

$$d_i S > 0 \qquad \text{natural change,} \tag{7.3}$$

$$d_i S = 0 \qquad \text{reversible change.} \tag{7.4}$$

Consequently, by addition of the relations for $d_e S$ and $d_i S$,

$$dS > \frac{q}{T} \qquad \text{natural change,} \tag{7.5}$$

$$dS = \frac{q}{T} \qquad \text{reversible change.} \tag{7.6}$$

If a system is not at a uniform temperature the formulae involving T cannot be applied to the whole system, but can still be applied to any part of the system which is at a uniform temperature.

It is to be noted that in this formulation of the second law of thermodynamics, the absolute temperature T and the entropy S are introduced simultaneously as two fundamental quantities. In a more mathematical formulation T^{-1} appears as an integrating factor leading to the complete differential dS. Instead of taking this formulation as an axiom it can be deduced logically from various physical assumptions.

It was deduced by CLAUSIUS from the assumption: a mere transfer of heat from a colder to a hotter body, unaccompanied by some other process, is not possible. It was deduced by KELVIN from the assumption: it is impossible that the only result of a series of changes undergone by a system of bodies should consist in taking from a body a certain quantity of heat and transforming it entirely into work. It was deduced by CARATHÉODORY[1] from the more general assumption: it is impossible to reach all states in the neighbourhood of any arbitrary initial state by an adiabatic process.

8. Temperature scale. The fundamental relation defining entropy and absolute temperature remains unaltered if S is replaced by cS and T by $c^{-1}T$ where c is any constant. Thus to complete the definitions it is necessary to fix the value of T at some identifiable temperature or in other words to fix the unit of temperature called the degree. This is done by specifying that at the triple point of water, namely the condition of temperature and pressure at which ice, liquid water and water vapour are in mutual equilibrium, the absolute temperature is exactly 273.16 degrees. The absolute temperature, with the size of the degree

---

[1] For a simple clear account of this derivation see R. EISENSCHITZ: Sci. Progr. **170**, 246 (1955).

thus fixed, is called the Kelvin thermodynamic scale and is denoted by °K. Thus the temperature of the triple point of water is by definition 273.16 °K.

The freezing point of water saturated with air at a pressure of one atmosphere, called the normal freezing point (nfp) of water, is 273.15 °K. The boiling point of pure water at a pressure of one atmosphere, called the normal boiling point (nbp) of water, is 373.15 °K. Thus the nbp of water exceeds its nfp by just 100 degrees. This is of course no accident but was the determining factor in defining the temperature of the triple point of water.

For many practical purposes it is considered convenient to use a temperature scale such that the temperatures most usually met in the laboratory and in ordinary life shall be represented by numbers between 0 and 100. For this purpose the scale used in scientific work is the Celsius scale (sometimes loosely called the centigrade scale) and denoted by the letter C. The temperature on the Celsius scale is defined as the excess of the temperature on the Kelvin scale over that of the nfp of water also on the Kelvin scale. Thus the relation between the two scales is

$$x \,°K = (x - 273.15) \,°C , \tag{8.1}$$

$$y \,°C = (y + 273.15) \,°K , \tag{8.2}$$

$$\text{nfp of water} = 273.15 \,°K = 0 \,°C , \tag{8.3}$$

$$\text{nbp of water} = 373.15 \,°K = 100 \,°C . \tag{8.4}$$

In both the Kelvin scale and the Celsius scale the unit is the centigrade degree. There are other scales based on other kinds of degrees, but these are little used in scientific work and need not be mentioned further.

9. Thermal equilibrium. According to the fundamental property of temperature the condition for thermal equilibrium between two parts of a system is equality of temperature. We can now verify that this condition is consistent with the general condition of equilibrium, namely that any infinitesimal change should be reversible. We consider a thermally insulated system consisting of two parts α and β at uniform temperatures T^α and T^β respectively. The entropy S of the whole system is equal to the sum of S^α the entropy of α and S^β the entropy of β: that is to say

$$S = S^\alpha + S^\beta . \tag{9.1}$$

Consider now the flow of an infinitesimal positive quantity q of heat from α to β. Then we have

$$dS^\alpha = d_e S^\alpha = - \frac{q}{T^\alpha} , \tag{9.2}$$

$$dS^\beta = d_e S^\beta = \frac{q}{T^\beta} . \tag{9.3}$$

The increase in the entropy of the whole system due to this flow of heat is therefore

$$dS = d S^\alpha + d S^\beta = - \frac{q}{T^\alpha} + \frac{q}{T^\beta}$$

$$= q \left(\frac{1}{T^\beta} - \frac{1}{T^\alpha} \right) . \tag{9.4}$$

The condition that the flow of positive heat q from α to β should be a natural process is that the consequent change of entropy of the whole system should be an increase, that is to say that the last expression should be positive. Thus heat will flow naturally from α to β provided

$$T^\alpha > T^\beta. \tag{9.5}$$

We have thus reached the conclusion that according to our definition of the absolute temperature T, heat flows naturally from a higher to a lower temperature.

In the special case $T^\alpha = T^\beta$ we have

$$dS = dS^\alpha + dS^\beta = 0, \tag{9.6}$$

and there is no entropy increase associated with the flow of heat. The two parts of the system are in thermal equilibrium and so the flow of heat contemplated is reversible.

10. Single closed phase. We shall now develop some general thermodynamic relations applicable to a single phase of fixed content assumed to be in internal equilibrium so that any infinitesimal change is reversible. We then have according to (7.4) and (7.2)

$$d_i S = 0, \tag{10.1}$$

$$dS = d_e S = \frac{q}{T}, \tag{10.2}$$

$$q = T\,dS. \tag{10.3}$$

We shall further assume that the only work is that done by a uniform external pressure P producing a volume decrease $-dV$ so that

$$w = -P\,dV. \tag{10.4}$$

Substituting into the relation

$$dU = q + w \tag{10.5}$$

we obtain

$$dU = T\,dS - P\,dV. \tag{10.6}$$

We may usefully compare and contrast this formula with that for a hypothetical hydrostatic system without any thermal properties; the difference appears in the presence of the thermal (non-mechanical) term $T\,dS$. It is clear that a single phase in virtue of its thermal properties has two degrees of freedom, whereas a hypothetical hydrostatic system without thermal properties would have only one.

11. Free energy. If we regard the state of the phase as defined by the two independent variables S and V then formula (10.6) shows how the total energy U depends on these variables. However the choice of S and V as independent variables is by no means the most convenient. A much more convenient choice of independent variables is T and V. The change to these can be attained by means of a Legendre transformation. We define a quantity F by

$$F = U - T\,S. \tag{11.1}$$

Differentiating this and using formula (10.6) for dU we obtain

$$dF = -S\,dT - P\,dV. \tag{11.2}$$

This quantity F was first used by Gibbs who denoted it by ψ and called it the force function for constant temperature. Prior to Gibbs the related function $-F/T$

had been used by Massieu. Helmholtz independently of Gibbs defined the quantity, denoted it by F and called it free energy the name which we still use.

The equation relating dU to dS, dV and the equation relating dF to dT, dV are called the fundamental equations for the variables S, V and T, V respectively. U is called a characteristic function for the variables S, V and F a characteristic function for the variables T, V.

It is evident that F, like U and S, is an extensive property so that a system of several phases has a free energy equal to the sum of the free energies of the several phases. Moreover the definition of F is not restricted to the situation where the only work is that done by the external pressure. We shall now study the most important properties of the free energy F in a system of uniform but not necessarily constant temperature. From the definition of F,

$$F = U - TS, \tag{11.3}$$

we derive

$$dF = dU - S\,dT - T\,dS. \tag{11.4}$$

Comparing this with the previous relations

$$dU = q + w \tag{11.5}$$

$$T\,dS = T\,d_e S + T\,d_i S = q + T\,d_i S, \tag{11.6}$$

we obtain

$$dF = -S\,dT - T\,d_i S + w. \tag{11.7}$$

Since $d_i S$ is positive for natural processes, zero for reversible processes and never negative we have

$$dF = -S\,dT + w, \quad \text{reversible}, \tag{11.8}$$

$$dF < -S\,dT + w, \quad \text{natural}, \tag{11.9}$$

and in particular at constant temperature, that is to say for isothermal changes,

$$dF = w, \quad \text{reversible isothermal}, \tag{11.10}$$

$$dF < w, \quad \text{natural isothermal}. \tag{11.11}$$

We here refer to the two alternative statements I and II concerning energy in a hypothetical system without any thermal properties and recall that in thermodynamics the total energy is the extension of energy described by statement I. We now see that the extension to thermodynamics of energy described by statement II is the free energy F subject to the restriction to reversible isothermal processes. This is the explanation of Gibbs' name for F force function for constant temperature whereas U is the force function for constant entropy.

12. Other characteristic functions. Returning now to a phase of fixed content when the only work is that done by a uniform external pressure P, we can obtain fundamental equations for other pairs of variables by other Legendre transformations. Defining

$$H = U + PV, \tag{12.1}$$

$$G = F + PV = U - TS + PV \tag{12.2}$$

differentiating and using the previous formulae (10.6) for dU and (11.2) for dF we obtain

$$dH = T\,dS + V\,dP, \tag{12.3}$$

$$dG = -S\,dT + V\,dP. \tag{12.4}$$

The characteristic function for the variables S, P denoted by H was introduced by GIBBS with the symbol χ and the name heat function for constant pressure. It has several names of which the commonest is now enthalpy. The characteristic function for the variables T, P was introduced by GIBBS with the symbol ζ but without any name. In the past it has been called the thermodynamic potential but is now usually called the Gibbs function. Some American chemists call G, not F, free energy. It is evident that H and G are extensive properties so that a system of several phases has an enthalpy and a Gibbs function equal to the sum of those of the several phases.

13. Notation. The notation adopted here is that recommended by the International Union of Pure and Applied Physics. It is used almost universally in Europe, but not always in America. The usage of various authors is shown in the following table.

Table 1.

Preferred Names	Entropy	Energy	Enthalpy	Free energy or Helmholtz function	Gibbs function
Present notation	S	U	H	F	G
MASSIEU (1869)		U	U'	$-T\psi$	$-T\psi'$
GIBBS (1876)	η	ε	χ	ψ	ζ
HELMHOLTZ (1882)	S	U		F	
DUHEM (1886)	S	U		F	φ
LORENTZ (1921, 1927)	η	ε		Ψ	ζ
PLANCK (1932 edition)	S	U		F	
LEWIS and RANDALL (1923)	S	E	H	A	F
PARTINGTON (1924)	S	U	H	F	Z
SCHOTTKY (1929)	S	U	H	F	G
FOWLER (1936)	S	E	H	F	G
BRÖNSTED (1936)	S	E	H	F	G
DE DONDER and VAN RYSSELBERGHE (1936)	S	E	H	F	G
FOWLER and GUGGENHEIM (1939)	S	E	H	F	G
MACDOUGALL (1939)	S	E	H	A	F
MACINNES (1939)	S	U	H	F	Z
SLATER (1939)	S	U	H	A	G
MAYER and MAYER (1940)	S	E	H	A	F
ZEMANSKY (1943)	S	U	H	A	G
PRIGOGINE and DEFAY (1944)	S	E	H	F	G
DE BOER (1946)	S	U	W	F	G
EVERETT (1954)	S	U	H	F	G
HAASE (1956)	S	U	H	F	G
MÜNSTER (1956)	S	U	H	F	G

14. Energy and enthalpy. We shall now describe some of the most important properties of H, comparing and contrasting them with those of U. We consider a system of one or several phases at uniform but not necessarily constant temperature and pressure. We have then

$$q = dU - w = dU + P\,dV \tag{14.1}$$

and consequently

$$q = dH - V\,dP. \tag{14.2}$$

From these we derive immediately the general condition for constant volume calorimetry

$$q = dU, \qquad V \text{ constant} \tag{14.3}$$

and that for constant pressure calorimetry

$$q = dH, \qquad P \text{ constant.} \tag{14.4}$$

The relation $q = dU$ also holds for a volume change against zero pressure as when a tap separating a fluid from a vacuum is opened. In JOULE's classical experiment a gas was alowed to expand suddenly into a vacuum. The conditions of this experiment were

$$w = 0, \quad q = 0, \quad dU = 0, \quad \text{free expansion.} \tag{14.5}$$

Consequently if we denote the initial conditions by the subscript i and the final conditions by the subscript f, we have

$$U(T_i, V_i) = U(T_f, V_f), \quad \text{free expansion.} \tag{14.6}$$

JOULE measured the temperature before and after expansion and could not detect any change. This was in fact due to the insensitivity of his equipment. In so far as the temperature change was negligible it would follow that

$$U(T, V_i) = U(T, V_f) \tag{14.7}$$

or in other words that at constant temperature the total energy of a gas is independent of its volume. This is in fact not true for a real gas, but we shall see later that it is a property of what is called an ideal gas.

The relation $q = dH$ also holds for the condition of the experiment called throttling first carried out by JOULE and KELVIN. In this experiment a stream of gas in a thermally insulated container is forced through a plug, the pressure being greater on the near side than on the far side. Consider now the whole system in a steady state such that in each unit of time a certain mass of gas is pushed in at a pressure P_1 and during the same time an equal mass of gas streams away at a pressure P_2. We use the subscript 1 to denote the state of the gas being pushed in and the subscript 2 to denote that of the gas streaming away. Then during the time considered a mass of gas of pressure P_2, volume V_2, temperature T_2 and energy U_2 is displaced by an equal mass of pressure P_1, volume V_1, temperature T_1 and energy U_1. During this time the work done on the system is

$$w = P_1 V_1 - P_2 V_2. \tag{14.8}$$

Since the system is supposed thermally insulated $q = 0$ and so

$$U_2 - U_1 = w = P_1 V_1 - P_2 V_2 \tag{14.9}$$

and consequently from the definition (12.1) of H

$$H_1 = H_2, \quad \text{insulated throttling} \tag{14.10}$$

or in differential form

$$dH = 0, \quad \text{insulated throttling.} \tag{14.11}$$

15. Gibbs function. We turn now to consider the most important property of G. We consider a system of uniform, but not necessarily constant, temperature and pressure. From the definition (12.2) of G we have

$$dG = dF + P\,dV + V\,dP. \tag{15.1}$$

Comparing this with our previous relation (11.7) for dF

$$dF = -S\,dT - T\,d_i S + w$$
$$= -S\,dT - T\,d_i S - P\,dV \tag{15.2}$$

we deduce

$$dG = -S\,dT + V\,dP - T\,d_i S \tag{15.3}$$

and in particular at constant T, P

$$dG = - T d_i S \qquad \text{constant } T, P \tag{15.4}$$

and consequently since $d_i S \geq 0$

$$dG = 0, \qquad \text{reversible, constant } T, P, \tag{15.5}$$

$$dG < 0, \qquad \text{natural, constant } T, P. \tag{15.6}$$

Since the general condition of equilibrium is that any infinitesimal change should be reversible, we deduce the condition that G should be stationary at constant T, P.

Incidentally the formula for dG having a term $- T d_i S$ on the right does not contradict the fundamental equation (12.4)

$$dG = - S d T + V d P \tag{15.7}$$

since this was for a single phase in complete equilibrium so that necessarily $d_i S$ vanishes.

The important equilibrium condition $dG = 0$ can be instructively derived in a slightly different way as follows. We have the following relations

$$q = dH, \qquad\qquad\qquad\qquad\qquad\qquad \text{constant } P, \tag{15.8}$$

$$q = T d S, \qquad\qquad\qquad\qquad\qquad\qquad \text{reversible}, \tag{15.9}$$

$$dH = T d S, \qquad\qquad\qquad\qquad\qquad\qquad \text{reversible, constant } P, \tag{15.10}$$

$$dG = d(H - T S) = dH - T d S = 0, \quad \text{reversible, constant } T, P, \tag{15.11}$$

$$dG = 0, \qquad\qquad\qquad\qquad\qquad\qquad \text{equilibrium, constant } T, P. \tag{15.12}$$

16. Equilibrium conditions. At this stage it is useful to collect together some equivalent conditions for distinguishing between a natural process, which is an approach towards equilibrium, and a reversible process, which is a condition of equilibrium. In each case the former is expressed by an inequality, the latter by an equality. Starting from (7.3) and (7.4)

$$d_i S \geq 0 \tag{16.1}$$

we can derive all the following equivalent statements

$$T d S \geq q, \tag{16.2}$$

$$dS \geq 0, \qquad dU = w, \qquad\qquad\qquad \text{adiabatic}, \tag{16.3}$$

$$dS \geq 0, \qquad dU = 0, \qquad dV = 0, \qquad \text{adiabatic, constant volume}, \tag{16.4}$$

$$dU \leq 0, \qquad dS = 0, \qquad dV = 0, \tag{16.5}$$

$$dF \leq 0, \qquad dT = 0, \qquad dV = 0, \tag{16.6}$$

$$dH \leq 0, \qquad dS = 0, \qquad dP = 0, \tag{16.7}$$

$$dG \leq 0, \qquad dT = 0, \qquad dP = 0. \tag{16.8}$$

Whereas these equalities and inequalities are all equivalent, the most useful are those relating to F and G.

17. Stability. In thermodynamics as in mechanics, unstable equilibrium is a figment not met in practice. Consequently the last four equilibrium conditions state that U or F or H or G must always be a minimum, never a maximum, for given values of the independent variables S, V or T, V or S, P or T, P respectively. It is useful to classify equilibria, all stable, into absolutely stable and metastable. In the former the minimum is the lowest existing one, while in the latter there exists another lower minimum.

18. Phase equilibrium. We shall later consider conditions for equilibrium between two or more phases each containing several chemical substances, but we are already in a position to study the equilibrium condition between two phases consisting of a single substance. Let us denote the two phases by the superscripts α and β. We use the symbol n to denote the quantity, measured either in grams or moles or any other unit, of substance. We shall also here and elsewhere use small letters to denote the ratio of an extensive property to the quantity of the phase. In this notation we have for the two phase system

$$G = n^\alpha g^\alpha + n^\beta g^\beta , \qquad (18.1)$$

$$S = n^\alpha s^\alpha + n^\beta s^\beta , \qquad (18.2)$$

$$V = n^\alpha v^\alpha + n^\beta v^\beta . \qquad (18.3)$$

Let us now consider the transference of a quantity dn of substance from the phase α to the phase β at constant temperature and constant pressure. Since then each phase remains unchanged in all respects except in its quantity, it is evident that g^α and g^β remain constant. The equilibrium condition (16.8)

$$dG = 0, \qquad \text{equilibrium, constant } T, P \qquad (18.4)$$

thus becomes

$$(g^\beta - g^\alpha)\, dn = 0, \qquad \text{equilibrium, constant } T, P \qquad (18.5)$$

and consequently

$$g^\alpha = g^\beta, \qquad \text{equilibrium.} \qquad (18.6)$$

We may therefore reformulate the condition for two phase equilibrium of a single substance as follows. The Gibbs function per unit quantity of substance must have the same value in two phases in mutual equilibrium.

We can now study the relation between temperature and pressure at two phase equilibrium of a single substance. Applying the fundamental equation (12.4) to unit quantity of each phase we obtain

$$dg^\alpha = - s^\alpha dT + v^\alpha dP , \qquad (18.7)$$

$$dg^\beta = - s^\beta dT + v^\beta dP . \qquad (18.8)$$

If now we vary the temperature T and the pressure P so as to maintain equilibrium between the two phases we must have

$$dg^\alpha = dg^\beta \qquad (18.9)$$

and consequently

$$- s^\alpha dT + v^\alpha dP = - s^\beta dT + v^\beta dP \qquad (18.10)$$

or

$$\frac{dT}{dP} = \frac{v^\beta - v^\alpha}{s^\beta - s^\alpha} . \qquad (18.11)$$

Since the equilibrium condition (18.6) can be written as

$$h^\alpha - T s^\alpha = h^\beta - T s^\beta \qquad (18.12)$$

we have

$$T(s^\beta - s^\alpha) = h^\beta - h^\alpha \qquad (18.13)$$

and so formula (18.11) can be rewritten as

$$\frac{dT}{dP} = \frac{T(v^\beta - v^\alpha)}{h^\beta - h^\alpha} \qquad (18.14)$$

called the Clapeyron equation. We see that increase of pressure leads to an increase of the equilibrium temperature when the phase having the greater specific enthalpy also has the greater volume and conversely. In particular since the liquid phase always has a higher enthalpy than the solid, increase of pressure leads to an increase of the melting temperature when the solid has a higher density than the liquid. This is the usual case, but in the system water+ice the liquid phase has the higher density and consequently dT/dP is negative.

19. Use of fundamental equations. By means of a fundamental equation all the thermodynamic quantities can be expressed in terms of the chosen characteristic function and its derivatives with respect to the corresponding variables. For example if we choose as independent variables T, P we deduce from (12.4)

$$S = -\frac{\partial G}{\partial T}, \tag{19.1}$$

$$H = G - T\frac{\partial G}{\partial T}, \qquad \text{Gibbs-Helmholtz relation}, \tag{19.2}$$

$$V = \frac{\partial G}{\partial P}, \tag{19.3}$$

$$F = G - P\frac{\partial G}{\partial P}, \tag{19.4}$$

$$U = G - T\frac{\partial G}{\partial T} - P\frac{\partial G}{\partial P}, \tag{19.5}$$

$$\frac{\partial S}{\partial T} = \frac{1}{T}\frac{\partial H}{\partial T} = -\frac{\partial^2 G}{\partial T^2}, \tag{19.6}$$

$$-\frac{\partial S}{\partial P} = \frac{\partial V}{\partial T} = \frac{\partial^2 G}{\partial T \partial P}, \qquad \text{Maxwell relation}, \tag{19.7}$$

$$\frac{\partial H}{\partial P} = V - T\frac{\partial V}{\partial T} = \frac{\partial G}{\partial \Gamma} - T\frac{\partial^2 G}{\partial T \partial P}, \tag{19.8}$$

$$\frac{\partial V}{\partial P} = \frac{\partial^2 G}{\partial P^2}. \tag{19.9}$$

The T, P system is the most useful, except for gases when the T, V system is equally useful. In the latter system we deduce from (11.2)

$$S = -\frac{\partial F}{\partial T}, \tag{19.10}$$

$$U = F - T\frac{\partial F}{\partial T}, \qquad \text{Gibbs-Helmholtz relation}, \tag{19.11}$$

$$P = -\frac{\partial F}{\partial V}, \tag{19.12}$$

$$G = F - P\frac{\partial F}{\partial P}, \tag{19.13}$$

$$H = F - T\frac{\partial F}{\partial T} - P\frac{\partial F}{\partial P}, \tag{19.14}$$

$$\frac{\partial S}{\partial T} = \frac{1}{T}\frac{\partial U}{\partial T} = -\frac{\partial^2 F}{\partial T^2}, \tag{19.15}$$

$$\frac{\partial S}{\partial V} = \frac{\partial P}{\partial T} = -\frac{\partial^2 F}{\partial T \partial V}, \qquad \text{Maxwell relation}, \tag{19.16}$$

$$\frac{\partial U}{\partial V} = -P + T\frac{\partial P}{\partial T} = \frac{\partial F}{\partial V} - T\frac{\partial^2 F}{\partial T \partial V}. \tag{19.17}$$

20. Open phases. Chemical potentials. Our fundamental equations were derived, as stated, for a single phase in internal equilibrium and of fixed content. Such a phase is called a closed phase. We shall now extend these equations to a phase of variable content, called an open phase. For this purpose we introduce as further independent variables the quantities n_i of the several chemical species i. The quantity may be measured as a mass, but we are not obliged to use the same mass as unit for different species. It is in fact more usual and more convenien' to measure n_i in moles, but we postpone the thermodynamic definition of a mole to the next chapter.

We now extend formula (10.6)

$$dU = T\,dS - P\,dV \qquad \text{constant } n_i \tag{20.1}$$

to

$$dU = T\,dS - P\,dV + \sum_i \mu_i\,dn_i \tag{20.2}$$

where

$$\mu_i = \left(\frac{\partial U}{\partial n_i}\right)_{S,V}. \tag{20.3}$$

These quantities μ_i were first used by Gibbs who called them potentials of the several chemical substances. The name now used is chemical potential. From the definitions (11.1), (12.1) and (12.2) of F, H and G we immediately deduce

$$dF = -S\,dT - P\,dV + \sum_i \mu_i\,dn_i, \tag{20.4}$$

$$dH = \quad T\,dS + V\,dP + \sum_i \mu_i\,dn_i, \tag{20.5}$$

$$dG = -S\,dT + V\,dP + \sum_i \mu_i\,dn_i \tag{20.6}$$

so that

$$\mu_i = \left(\frac{\partial U}{\partial n_i}\right)_{S,V} = \left(\frac{\partial F}{\partial n_i}\right)_{T,V} = \left(\frac{\partial H}{\partial n_i}\right)_{S,P} = \left(\frac{\partial G}{\partial n_i}\right)_{T,P}. \tag{20.7}$$

We have thus extended the four fundamental equations to open phases. O these equations (20.6) is the most useful. From it we deduce that, with T, P, ν as independent variables

$$\frac{\partial \mu_i}{\partial T} = -\frac{\partial S}{\partial n_i}, \tag{20.8}$$

$$\frac{\partial \mu_i}{\partial P} = \frac{\partial V}{\partial n_i}. \tag{20.9}$$

21. Partial quantities. In the system with T, P, n_i as independent variable it is useful to define a quantity v_i by

$$v_i = \frac{\partial V}{\partial n_i}. \tag{21.1}$$

This quantity v_i, in contrast to V, is an intensive property. It is called th partial volume of the substance i. We similarly define other partial quantitie derived from other extensive properties, in particular

$$u_i = \frac{\partial U}{\partial n_i}, \tag{21.2}$$

$$f_i = \frac{\partial F}{\partial n_i}, \tag{21.3}$$

$$h_i = \frac{\partial H}{\partial n_i}, \tag{21.4}$$

$$g_i = \frac{\partial G}{\partial n_i}, \tag{21.5}$$

$$s_i = \frac{\partial S}{\partial n_i}, \tag{21.6}$$

$$v_i = \frac{\partial V}{\partial n_i}. \tag{21.7}$$

We notice from the fourth fundamental equation (20.6) that

$$g_i = \mu_i. \tag{21.8}$$

Corresponding to any relation between extensive properties there is a derived relation between the partial quantities. In particular

$$f_i = u_i - T s_i, \tag{21.9}$$

$$h_i = u_i + P v_i, \tag{21.10}$$

$$\mu_i = g_i = f_i + P v_i = u_i - T s_i + P v_i, \tag{21.11}$$

$$\frac{\partial \mu_i}{\partial T} = \frac{\partial g_i}{\partial T} = - s_i, \tag{21.12}$$

$$\mu_i - T \frac{\partial \mu_i}{\partial T} = g_i - T \frac{\partial g_i}{\partial T} = h_i, \tag{21.13}$$

$$\frac{\partial \mu_i}{\partial P} = \frac{\partial g_i}{\partial P} = v_i. \tag{21.14}$$

Since at given T, P each extensive property such as V is a homogeneous function of the first degree in n_i we have by Euler's theorem

$$V = \sum_i n_i \frac{\partial V}{\partial n_i} = \sum_i n_i v_i \tag{21.15}$$

and similarly

$$U = \sum_i n_i u_i, \tag{21.16}$$

$$F = \sum_i n_i f_i, \tag{21.17}$$

$$H = \sum_i n_i h_i, \tag{21.18}$$

$$G = \sum_i n_i g_i, \tag{21.19}$$

$$S = \sum_i n_i s_i. \tag{21.20}$$

In the case of a mixture the specific quantity v is related to the partial quantities v_i by

$$v \sum_i n_i = \sum_i n_i v_i \tag{21.21}$$

and there are of course similar relations for u_i, f_i, h_i, g_i, s_i. In the special case of a phase composed of a single substance each partial property becomes a specific property equal to the extensive property per unit quantity of the phase. We have already used these quantities in Sect. 18 for our discussion of a two phase equilibrium of a single substance.

22. Gibbs-Duhem relation. From the relation (21.8)

$$g_i = \mu_i \tag{22.1}$$

we deduce

$$G = \sum_i n_i \mu_i. \tag{22.2}$$

Differentiating we have

$$dG = \sum_i \mu_i d n_i + \sum_i n_i d \mu_i, \tag{22.3}$$

and comparing with the fourth fundamental equation (20.6)

$$dG = -SdT + VdP + \sum_i \mu_i dn_i \qquad (22.4)$$

we obtain

$$SdT - VdP + \sum_i n_i d\mu_i = 0. \qquad (22.5)$$

This relation is known as the Gibbs-Duhem relation. It is used specially in the form

$$\sum_i n_i d\mu_i = 0 \qquad \text{constant } T, P. \qquad (22.6)$$

23. Composition. We are often more interested in the relative composition of a phase than in the quantity of the phase. It is then convenient to use fractions x_i (mole fractions if n_i is measured in moles, mass fractions if n_i is measured in grams or other units of mass) defined by

$$x_i = \frac{n_i}{\sum_k n_k}. \qquad (23.1)$$

By definition the mole fractions (or mass fractions) satisfy the identity

$$\sum_i x_i = 1. \qquad (23.2)$$

We can rewrite the Gibbs-Duhem relation (22.6) in terms of mole fractions (or mass fractions) as

$$\sum_i x_i d\mu_i = 0 \qquad \text{constant } T, P. \qquad (23.3)$$

We can likewise rewrite (21.21) in terms of mole fractions (or mass fractions) as

$$v = \sum_i x_i v_i. \qquad (23.4)$$

Similarly

$$g = \sum_i x_i g_i = \sum_i x_i \mu_i. \qquad (23.5)$$

24. Degrees of freedom of single phase. A single phase containing c chemically distinct substances is completely described by $c+2$ independent variables which we may take as T, P, n_i. If however we are not concerned with the quantity of the phase but only with its intensive properties, then it is sufficiently described by $c+1$ independent variables and a single phase is accordingly said to have $c+1$ degrees of freedom. We may take as the $c+2$ variables defining such a phase T, P, x_i subject to the identity

$$\sum_i x_i = 1. \qquad (24.1)$$

Alternatively we may use the $c+2$ variables T, P, μ_i subject to the Gibbs-Duhem relation

$$SdT - VdP + \sum_i x_i d\mu_i = 0. \qquad (24.2)$$

25. Physicochemical equilibrium. Affinity. We shall now show that the equilibrium conditions previously obtained can be expressed in a compact and convenient form in terms of the chemical potentials μ_i. We shall use the variables

T, P, n_i but the conclusion reached is independent of this choice. We recall the general condition (16.8)

$$dG \leq 0 \qquad \text{constant } T, P \tag{25.1}$$

where the inequality relates to an approach towards equilibrium and the equality to existing equilibrium. Using superscripts to denote phases we have

$$dG^\alpha = - S^\alpha dT^\alpha + V^\alpha dP^\alpha + \sum_i \mu_i^\alpha dn_i^\alpha \tag{25.2}$$

and so for a system of several phases at uniform, but not necessarily constant, temperature and pressure

$$dG = - \sum_\alpha S^\alpha dT + \sum_\alpha V^\alpha dP + \sum_\alpha \sum_i \mu_i^\alpha dn_i^\alpha. \tag{25.3}$$

In particular at constant temperature and pressure

$$dG = \sum_\alpha \sum_i \mu_i^\alpha dn_i^\alpha \qquad \text{constant } T, P. \tag{25.4}$$

The general condition relating to equilibrium thus becomes

$$\sum_\alpha \sum_i \mu_i^\alpha dn_i^\alpha \leq 0. \tag{25.5}$$

We now consider a physicochemical process described by the formula

$$\nu_A A + \cdots \to \nu_B B + \cdots \tag{25.6}$$

meaning that ν_A units of A in a given phase and the like change into ν_B units of B in a given phase and the like. We may note that if quantities are measured in moles then all the ν's can be taken as small integers called stoichiometric numbers. When this change takes place to an infinitesimal extent we may write

$$- \frac{dn_A}{\nu_A} = \cdots = \frac{dn_B}{\nu_B} = \cdots = d\xi \tag{25.7}$$

where ξ is called the extent of reaction. The condition relating to equilibrium thus becomes

$$A \leq 0 \tag{25.8}$$

where A is defined by

$$A = \sum_A \nu_A \mu_A - \sum_B \nu_B \mu_B \tag{25.9}$$

and is called the affinity of the process. With this definition a process will take place only in the direction in which the affinity is positive and at equilibrium the affinity vanishes.

The simplest conceivable kind of process is the transfer of a single species i from a given phase α to a given phase β. For this process the affinity is

$$A = \mu_i^\alpha - \mu_i^\beta. \tag{25.10}$$

We thus see that each chemical species i tends to move from a phase of higher μ_i to a phase of lower μ_i. Hence the name chemical potential for μ. In particular μ_i has the same value in any two phases in equilibrium.

26. Gibbs' phase rule. Let us now consider a system consisting of c independent chemical substances or components in p phases. The state of the system, apart from the quantities of the several phases, may then be completely described by the $c+2$ quantities T, P, μ_i all of which have uniform values throughout the system. But these quantities are not independent since their variations are restricted by Gibbs-Duhem relations (22.5)

$$S^\alpha\, dT - V^\alpha\, dP + \sum_i n_i^\alpha\, d\mu_i = 0 \tag{26.1}$$

of which there is one relating to each phase, that is p in all. Consequently the number of degrees of freedom of a system of c components and p phases is $c - p + 2$. This statement is known as Gibbs' phase rule.

27. Temperature dependence of affinity. The affinity A of a process may be regarded as a measure of the tendency for the process to take place. This tendency will depend on the temperature and it is therefore of interest to consider how A depends on the temperature. We have, using the independent variables T, P, n_i,

$$\begin{aligned}
\frac{\partial A}{\partial T} &= \sum_A \nu_A \frac{\partial \mu_A}{\partial T} - \sum_B \nu_B \frac{\partial \mu_B}{\partial T} \\
&= -\sum_A \nu_A\, s_A + \sum_B \nu_B\, s_B
\end{aligned} \tag{27.1}$$

and consequently

$$\begin{aligned}
A - T \frac{\partial A}{\partial T} &= \sum_A \nu_A\,(\mu_A + T\, s_A) - \sum_B \nu_B\,(\mu_B + T\, s_B) \\
&= \sum_A \nu_A\, h_A - \sum_B \nu_B\, h_B \\
&= -\Delta H
\end{aligned} \tag{27.2}$$

where the operator symbol Δ denotes the increase of a property when the process takes place to the extent that ξ increases by unity. It is often convenient to use the above relation in the form

$$\frac{d(A/T)}{d(1/T)} = -\Delta H. \tag{27.3}$$

28. Conventional zeros of U and S. It has already been mentioned that only changes in U and in S, not their absolute values, have any physical meaning. We have only to take care always to use a consistent convention. If chemical reactions were excluded we could choose arbitrary zeros of U and S for each chemical substance. But since chemical reactions do take place we are entitled only to assign arbitrary values to each independent substance and it is natural to choose as the independent substances the elements. The following conventions are usual and almost universal.

The zero of U is fixed by the statement that the value of $H = U + PV$ is zero for each element in its stable state at 298.15 °K, that is 25 °C, and a pressure of one atmosphere.

The zero of S is fixed by the statement that S is zero for each element in its stable crystalline state at 0 °K.

29. Internal stability. We shall now discuss some of the conditions necessary for a phase to be internally stable. We shall begin with a thermal condition and for this purpose we consider a single phase of fixed composition. Let its entropy be S its volume V and its total energy U. Imagine one half of this phase to change so as to have an entropy $\frac{1}{2}(S + dS)$ and volume $\frac{1}{2}V$ while the other half changes so as to have an entropy $\frac{1}{2}(S - dS)$ and volume $\frac{1}{2}V$. According to Taylor's expansion the total energy of the first half becomes

$$\frac{1}{2}\left\{ U + \frac{\partial U}{\partial S}\, dS + \frac{1}{2}\frac{\partial^2 U}{\partial S^2}\,(dS)^2 \right\} \tag{29.1}$$

when small quantities of third and higher orders are neglected: all partial differentiations are at constant V. The total energy of the second half becomes similarly

$$\frac{1}{2}\left\{V - \frac{\partial U}{\partial S}\,dS + \frac{1}{2}\,\frac{\partial^2 U}{\partial S^2}\,(dS)^2\right\}. \tag{29.2}$$

Hence by addition the total energy of the whole system has increased by the second order small quantity

$$\frac{1}{2}\,\frac{\partial^2 U}{\partial S^2}\,(dS)^2 \tag{29.3}$$

while the total entropy and volume remain unchanged. Now a condition for a system to be in stable equilibrium is that for given values of entropy and volume, the total energy should be a minimum. If then the original state of the system was stable, the change considered must lead to an increase of the energy. Hence we obtain as a necessary condition for stable equilibrium

$$\left(\frac{\partial^2 U}{\partial S^2}\right)_V > 0. \tag{29.4}$$

Since by (10.6)

$$\left(\frac{\partial U}{\partial S}\right)_V = T \tag{29.5}$$

the condition (29.4) for stability may be rewritten as

$$\left(\frac{\partial S}{\partial T}\right)_V > 0. \tag{29.6}$$

We shall now derive a hydrostatic condition of stability and for this purpose we consider a single phase of fixed composition of temperature T, volume V and free energy F. Suppose half the mass to change so as to have a volume $\frac{1}{2}(V + dV)$ and the other half to change so as to have a volume $\frac{1}{2}(V - dV)$, the temperature remaining uniform and unchanged. Then by an argument analogous to the previous one we find that the free energy of the system has increased by the second order small quantity

$$\frac{1}{2}\left(\frac{\partial^2 F}{\partial V^2}\right)_T (dV)^2 \tag{29.7}$$

while the temperature and total volume are unchanged. Now a condition for a system to be in stable equilibrium is that for given values of temperature and volume the free energy should be a minimum. Hence we obtain as a necessary condition for stability

$$\left(\frac{\partial^2 F}{\partial V^2}\right)_T > 0. \tag{29.8}$$

Since by (11.2)

$$\left(\frac{\partial F}{\partial V}\right)_T = -P \tag{29.9}$$

we may rewrite the condition (29.8) for stability as

$$\left(\frac{\partial V}{\partial P}\right)_T < 0. \tag{29.10}$$

We shall now obtain a physicochemical condition for stability. We consider a phase at constant uniform temperature and pressure. Let n_i denote the quantity of a particular substance i in this phase and let G denote its Gibbs function. Imagine this phase to split into two halves one containing a quantity $\frac{1}{2}(n_i + dn_i)$ and the other a quantity $\frac{1}{2}(n_i - dn_i)$ of the substance i, the temperature, pressure and distribution of the other substances remaining unchanged. According to Taylor's expansion the Gibbs function of the whole system has increased by the second order small quantity

$$\frac{1}{2}\left(\frac{\partial^2 G}{\partial n_i^2}\right)_{T,P}(dn_i)^2. \tag{29.11}$$

But a condition for stability is that G should be a minimum at a given temperature and given pressure. We thus obtain the stability condition

$$\left(\frac{\partial^2 G}{\partial n_i^2}\right)_{T,P} > 0 \tag{29.12}$$

which by use of (22.4) may be rewritten as

$$\left(\frac{\partial \mu_i}{\partial n_i}\right)_{T,P} > 0. \tag{29.13}$$

If we consider a more general change in composition involving two or more kinds of substance, then by similar reasoning we find that for stability the quantity

$$\frac{1}{2}\frac{\partial^2 G}{\partial n_i^2}(dn_i)^2 + \frac{\partial^2 G}{\partial n_i\,\partial n_k}dn_i\,dn_k +$$

$$+ \frac{1}{2}\frac{\partial^2 G}{\partial n_k^2}(dn_k)^2 \tag{29.14}$$

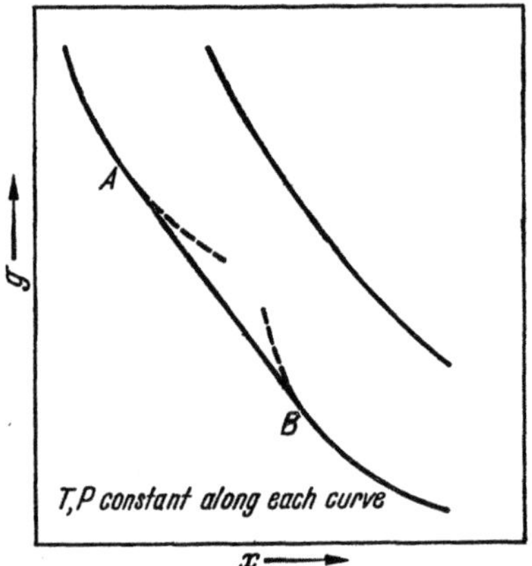

Fig. 1. Stable and metastable isotherms[1].

must be positive for all small variations dn_i, dn_k. This leads to the further condition

$$\frac{\partial^2 G}{\partial n_i^2}\frac{\partial^2 G}{\partial n_k^2} \geqq \left(\frac{\partial^2 G}{\partial n_i\,\partial n_k}\right)^2. \tag{29.15}$$

In the simple case of a binary phase a more convenient form of the stability conditions is obtained by considering changes in g the Gibbs function of unit quantity regarded as a function of T, P and x, the fractional amount of either component. By reasoning similar to that used above we find that for stability the quantity

$$\left(\frac{\partial^2 g}{\partial x^2}\right)_{T,P}(dx)^2 \tag{29.16}$$

must be positive for all variations dx. This leads to the condition

$$\left(\frac{\partial^2 g}{\partial x^2}\right)_{T,P} > 0. \tag{29.17}$$

The last condition is illuminated by Fig. 1 in which g is plotted against x at given T, P. If now we imagine a phase of composition x to split into two, one of greater and the other of smaller x, the new value of g is then given by a point on the straight line joining the two points representing the two new phases. If this point lies above the one representing the original phase, the system will revert to its original state which is stable. In the contrary case the original phase is not stable. It is then clear from the diagram that while the upper curve represents phases all stable, the phases represented by the broken parts of the lower curve between A and B are metastable with respect to a mixture of phases represented by A and B. If the two broken parts of the curve were joined there

<hr>

[1] Figs. 1—4 after E. A. GUGGENHEIM: Thermodynamics, North Holland Publishing Company, Amsterdam 1950.

would necessarily be another part of the curve where $\partial^2 g/\partial x^2$ is negative and this would represent phases completely unstable.

From these necessary conditions for stability other weaker conditions can be derived. For example we have

$$\left(\frac{\partial S}{\partial T}\right)_P = \left(\frac{\partial S}{\partial T}\right)_V + \left(\frac{\partial S}{\partial V}\right)_T\left(\frac{\partial V}{\partial T}\right)_P$$

$$= \left(\frac{\partial S}{\partial T}\right)_V + \left(\frac{\partial P}{\partial T}\right)_V\left(\frac{\partial V}{\partial T}\right)_P$$

$$= \left(\frac{\partial S}{\partial T}\right)_V - \left\{\left(\frac{\partial V}{\partial T}\right)_P\right\}^2\left(\frac{\partial P}{\partial V}\right)_T. \tag{29.18}$$

Since we have already obtained the necessary conditions for stability

$$\left(\frac{\partial S}{\partial T}\right)_V > 0, \qquad -\left(\frac{\partial P}{\partial T}\right)_V > 0, \tag{29.19}$$

it follows a fortiori that

$$\left(\frac{\partial S}{\partial T}\right)_P > 0 \tag{29.20}$$

regardless of the sign of $(\partial V/\partial T)_P$.

30. Third law. We conclude this chapter with a brief mention of the third law of thermodynamics. We consider an isothermal change represented formally by

$$\alpha \to \beta, \qquad \text{constant } T \tag{30.1}$$

where all phases involved are internally stable but not necessarily in equilibrium with one another. The change considered may be a purely physical one such as a change of pressure or change of an external magnetic field, or a physicochemical change such as an allotropic change. Let ΔS denote the increase in the entropy of the system when this change takes place isothermally. The third law states

$$\Delta S \to 0 \quad \text{as} \quad T \to 0. \tag{30.2}$$

The simplest example is a change of pressure. We then have

$$\left(\frac{\partial S}{\partial P}\right)_T \to 0 \quad \text{as} \quad T \to 0. \tag{30.3}$$

Using Maxwell's relation (19.7)

$$\left(\frac{\partial S}{\partial P}\right)_T = -\left(\frac{\partial V}{\partial T}\right)_P \tag{30.4}$$

we deduce

$$\left(\frac{\partial V}{\partial T}\right)_P \to 0 \quad \text{as} \quad T \to 0 \tag{30.5}$$

so that the coefficient of thermal expansion tends to zero as $T \to 0$.

As a second example consider the fusion of a solid under equilibrium conditions. If we denote the entropy increase on fusion by $\Delta_f S$ and the volume increase by $\Delta_f V$ we have, provided $\Delta_f V$ does not vanish,

$$\frac{\Delta_f S}{\Delta_f V} \to 0 \quad \text{as} \quad T \to 0. \tag{30.6}$$

But by Clapeyron's relation (18.11)

$$\frac{dP}{dT} = \frac{\Delta_f S}{\Delta_f V}, \tag{30.7}$$

where P and T relate to the two phase equilibrium between solid and liquid, it follows that

$$\frac{dP}{dT} \to 0 \quad \text{as} \quad T \to 0. \tag{30.8}$$

This relation has been verified for helium the only substance of which a liquid phase can persist down to sufficiently low temperatures to permit an unambiguous extrapolation to the limit $T \to 0$.

31. Unattainability of absolute zero. The third law is closely related to the following principle. It is impossible by any procedure, no matter how idealized, to reduce the temperature of any system to the absolute zero in a finite number of operations.

Let us consider a process such as change of volume, change of external field or allotropic change, denoted formally by

$$\alpha \to \beta. \tag{31.1}$$

We shall use superscripts α, β to denote properties of the system in the states α, β respectively. Then the entropy of the system in either state depends on the temperature according to the formulae

$$S^\alpha = S^{0\alpha} + \int_0^T \frac{\partial S^\alpha}{\partial T}\, dT, \tag{31.2}$$

$$S^\beta = S^{0\beta} + \int_0^T \frac{\partial S^\beta}{\partial T}\, dT \tag{31.3}$$

where $S^{0\alpha}$, $S^{0\beta}$ are the limiting values of S^α, S^β respectively as $T \to 0$. It is known from quantum theory that both integrals converge. Suppose now that we start with the system in the state α at the temperature T' and that we can make the process $\alpha \to \beta$ take place adiabatically. Let the final temperature when the system has reached the state β be T''. We shall now consider the possibility or impossibility of T'' being zero. From the second law of thermodynamics we know that for an adiabatic process the entropy increases if there is any irreversible (natural) change and remains constant if the change is completely reversible. It follows that the chances of attaining as low a final T as possible are most favourable when the change is completely reversible. We shall therefore consider only such changes. For the reversible adiabatic change

$$\alpha,\ T' \to \beta,\ T'' \tag{31.4}$$

we have then

$$S^{0\alpha} + \int_0^{T'} \frac{\partial S^\alpha}{\partial T}\, dT = S^{0\beta} + \int_0^{T''} \frac{\partial S^\beta}{\partial T}\, dT. \tag{31.5}$$

If T'' is to be zero we must then have

$$S^{0\beta} - S^{0\alpha} = \int_0^{T'} \frac{\partial S^\alpha}{\partial T}\, dT. \tag{31.6}$$

If now $S^{0\beta} > S^{0\alpha}$ it will always be possible, owing to the stability condition $\partial S/\partial T > 0$, to choose an initial T' satisfying this equation. By making the process $\alpha \to \beta$ take place adiabatically from this initial T' it will be possible to reach $T'' = 0$. From the premise of the unattainability of $T = 0$ we conclude

$$S^{0\beta} \leq S^{0\alpha}. \tag{31.7}$$

Similarly we can show that if $S^{0\alpha} > S^{0\beta}$ we could reach $T = 0$ by making the reverse process $\beta \to \alpha$ take place adiabatically from a suitably chosen initial temperature. From the premise of unattainability of $T = 0$ we conclude

$$S^{0\alpha} \leq S^{0\beta}. \tag{31.8}$$

From the two inequalities (31.7) and (31.8) we deduce

$$S^{0\alpha} = S^{0\beta} \tag{31.9}$$

which can also be written

$$\Delta S \to 0 \quad \text{as} \quad T \to 0 \tag{31.10}$$

in agreement with the third law.

We can show conversely that given $\Delta S \to 0$ as $T \to 0$, neither the process $\alpha \to \beta$ nor the reverse process $\beta \to \alpha$ can be used to reach $T = 0$. For with this assumption we now have for the adiabatic process

$$\alpha, \, T' \to \beta, \, T'' \tag{31.11}$$

the initial temperature T' and the final temperature T'' related by

$$\int_0^{T'} \frac{\partial S^\alpha}{\partial T} \, dT = \int_0^{T''} \frac{\partial S^\beta}{\partial T} \, dT. \tag{31.12}$$

Hence to reach $T'' = 0$ we should require

$$\int_0^{T'} \frac{\partial S^\alpha}{\partial T} \, dT = 0 \tag{31.13}$$

but, since $\partial S^\alpha / \partial T > 0$ always, this is impossible for any non-zero T'. Hence the process $\alpha \to \beta$ can not be used to reach $T = 0$. The proof for the reverse process $\beta \to \alpha$ is exactly analogous.

B. Applications of classical thermodynamics.

32. Determination of absolute temperature. It was mentioned at the beginning of the previous chapter that temperature is the most important conception in thermodynamics. In fact thermodynamics is entirely concerned with how properties depend on temperature. We have throughout used the scale of temperature called absolute temperature and it is high time we discussed how to measure it on this scale. The key to this problem is the Gibbs-Helmholtz relation (19.11)

$$U = F - T \frac{\partial F}{\partial T}, \tag{32.1}$$

or rather the derived relation

$$\Delta U = \Delta F - T \frac{\partial \Delta F}{\partial T} \tag{32.2}$$

where the operator symbol Δ denotes the increase of a quantity in a process defined with respect to initial and final volumes independent of the temperature. Suppose that ΔF can be expressed as a power series in T

$$\Delta F = a_0 + a_1 T + \sum_{r>1} a_r T^r + \sum_{s>0} a_s T^{-s}. \tag{32.3}$$

Then from the Gibbs-Helmholtz relation (32.1) we deduce

$$\Delta U = a_0 - \sum_{r>1} (r-1) a_r T^r + \sum_{s>0} (s+1) a_s T^{-s}. \tag{32.4}$$

Of especial interest are the terms in a_0 and a_1. The term a_0 independent of temperature is the same in F and U as would be expected since for temperature-independent properties there is no need to distinguish between F and U. In striking contrast to this there is no term in ΔU corresponding to the term in F which is directly proportional to T. If therefore we can find some isothermal process in which U does not change then the change in F will be directly proportional to T and this would lead to a simple way of determining T. This is in fact the case for the compression of a gas in the limit of great dilution. For this reason a gas affords a convenient thermometric substance for the determination of absolute temperature. To show this it is more convenient to use P than V as independent variable. We accordingly use the other Gibbs-Helmholtz relation (19.2)

$$H = G - T \frac{\partial G}{\partial T} \tag{32.5}$$

where the partial differentiation is now at constant pressure. Differentiating with respect to pressure P we have

$$\frac{\partial H}{\partial P} = V - T \frac{\partial V}{\partial T} \tag{32.6}$$

or, if we use lower case letters to denote values per unit quantity of substance (specific quantities),

$$\frac{\partial h}{\partial P} = v - T \frac{\partial v}{\partial T}. \tag{32.7}$$

This relation applies to any single phase. We now consider the characteristics of a gas. In Chap. E we shall derive theoretically the properties of an ideal gas from a model consisting of non-interacting particles, but we are now concerned not with the theory of an ideal gas, but with the empirical properties of a real gas. For our present purpose the most important of these properties are the following.

1. At constant temperature the product Pv varies only slowly with pressure and tends to a finite value as P tends to zero. We may therefore write

$$Pv = A + BP + C'P + \cdots, \qquad \text{constant } T. \tag{32.8}$$

The reason for writing C' rather than C is that we shall require to use C for a related quantity. In this formula A, B, C' are independent of P but of course depend on T.

2. The adiabatic throttling experiment leads to the relation

$$h(\vartheta_1, P_1) = h(\vartheta_2, P_2) \tag{32.9}$$

where ϑ denotes temperature measured on an arbitrary scale and the subscripts 1 and 2 refer respectively to conditions before and after throttling. By a purely calorimetric experiment at constant pressure P_2 we can measure

$$h(\vartheta_1, P_2) - h(\vartheta_2, P_2) \tag{32.10}$$

and in view of the previous relation this is equal to

$$h(\vartheta_1, P_2) - h(\vartheta_1, P_1). \tag{32.11}$$

We thus obtain experimental information of the dependence of h on the pressure at constant temperature, that is to say of $(\partial h/\partial P)_T$. It is found that $(\partial h/\partial P)_T$ remains finite as $P \to 0$. This empirical fact is sufficient for our purpose.

We now return to the relation (32.8) which we rewrite as

$$v = \frac{A}{P} + B + C' P + \cdots. \tag{32.12}$$

Substituting this into formula (32.7) we obtain

$$\frac{\partial h}{\partial P} = \left(A - T \frac{dA}{dT}\right) \frac{1}{P} + \left(B - T \frac{dB}{dT}\right) + \left(C' - T \frac{dC'}{dT}\right) P + \cdots \tag{32.13}$$

from which it would appear that $\partial h/\partial P \to \infty$ like P^{-1} as $P \to 0$ in contradiction with experiment. The apparent contradiction is resolved only if

$$A - T \frac{dA}{dT} = 0 \tag{32.14}$$

which is equivalent to

$$A \propto T. \tag{32.15}$$

We accordingly rewrite (32.8) as

$$P v = RT + B P + C' P^2 + \cdots \tag{32.16}$$

where R is independent of temperature and pressure. If then Pv is measured at several pressures and the values extrapolated to zero pressure, this limiting value is directly proportional to the absolute temperature. This provides the basis for the precise determination of absolute temperature by gas thermometry.

33. Gas constant. The value of R depends on the unit of quantity to which v relates. We can conveniently choose the unit quantity of each pure substance, so that R has the same value for all substances. It is customary to do this with the quantity of oxygen gas taken as exactly 32 grams. The quantity thus defined is independent of molecular theory. It will be shown in Chap. E that one mole of every substance contains the same number of molecules in confirmation of Avogadro's hypothesis. R is called the gas constant and has the value

$$R = 8.3147 \text{ J deg}^{-1} \text{ mole}^{-1} \tag{33.1}$$

$$= 0.082057 \text{ l atm deg}^{-1} \text{ mole}^{-1}. \tag{33.2}$$

From now onwards the mole will generally be used as the unit of quantity unless the contrary is stated.

34. Absolute activity. Having defined the gas constant R we find it convenient to define a new quantity λ_i called absolute activity related to the chemical potential μ_i per mole by

$$\mu_i = RT \ln \lambda_i. \tag{34.1}$$

There is no need to use both μ_i and λ_i, but the latter is often the more convenient especially in statistical mechanics. We may note that the temperature dependence of λ_i takes the simple form

$$\frac{\partial \ln \lambda_i}{\partial T} = -\frac{h_i}{RT^2}. \tag{34.2}$$

35. Thermodynamic functions of single gas. From the formula for v in a gas we can derive formulae for the other thermodynamic quantities. We obtain

$$\mu = \int v\, dP = \mu^\dagger + RT \ln \frac{P}{P^\dagger} + BP + \frac{1}{2}C'P^2 + \cdots \tag{35.1}$$

where $P^\dagger$ denotes some standard pressure, usually one atmosphere, and $\mu^\dagger$ depends on T and $P^\dagger$ but is independent of P. We also have

$$s = -\frac{\partial \mu}{\partial T} = -\frac{d\mu^\dagger}{dT} - R \ln \frac{P}{P^\dagger} - P\frac{dB}{dT} - \frac{1}{2}\frac{dC'}{dT}P^2 - \cdots, \tag{35.2}$$

$$h = \mu - T\frac{d\mu^\dagger}{dT} + \left(B - T\frac{dB}{dT}\right)P + \frac{1}{2}\left(C' - T\frac{dC'}{dT}\right)P^2 + \cdots, \tag{35.3}$$

$$f = \mu^\dagger - RT + RT \ln \frac{P}{P^\dagger} - \frac{1}{2}C'P^2 - \cdots, \tag{35.4}$$

$$u = \mu^\dagger - T\frac{d\mu^\dagger}{dT} - RT - T\frac{dB}{dT}P - \frac{1}{2}\left(C' + T\frac{dC'}{dT}\right)P^2 + \cdots. \tag{35.5}$$

The Joule-Thomson coefficient which measures the ratio of the temperature drop to the pressure drop in the throttling experiment is

$$\left(\frac{\partial T}{\partial P}\right)_h = -\frac{(\partial h/\partial P)_T}{(\partial h/\partial T)_P} = \frac{(B - T\, dB/dT) + (C' - T\, dC'/dT)\, P + \cdots}{(\partial h/\partial T)_P}. \tag{35.6}$$

At low pressures we may replace this by

$$\left(\frac{\partial T}{\partial P}\right)_h = \frac{B - T\, dB/dT}{(\partial h/\partial T)_P} \tag{35.7}$$

and the inversion temperature where $(\partial T/\partial P)_h$ changes sign is given by

$$\frac{dB}{dT} = \frac{B}{T}, \qquad \text{inversion point.} \tag{35.8}$$

The dependence of B on the temperature is a wide subject for the application of statistical mechanics. It can here be only briefly mentioned. At high temperatures B is positive and nearly independent of the temperature. As the temperature decreases, so B decreases, changes sign and continues to decrease more and more rapidly. The temperature at which B is zero, being the temperature at which Boyle's law is most nearly obeyed, is called the Boyle temperature. The simplest type of formula able to represent the variation of B over a wide range of temperature is

$$B = b - \frac{a}{T} - \frac{c}{T^2}, \qquad a, b, c \text{ constants.} \tag{35.9}$$

The simpler formula with c omitted (van der Waals) and that with a omitted (Berthelot) can be made to fit only over much shorter ranges of temperature.

The formula (32.16) expressing Pv as a power series in P can be transformed into a power series in v^{-1}

$$Pv = RT\left(1 + \frac{B}{v} + \frac{C}{v^2} + \cdots\right) \tag{35.10}$$

where

$$C = B^2 + RT\,C'. \tag{35.11}$$

The coefficients $B, C, \ldots$ in this expansion are called second, third, $\ldots$ virial coefficients. Whereas the expansion in powers of P is sometimes more convenient in thermodynamic calculations, the expansion in powers of v^{-1} is much more important in the molecular theory of real gases.

36. Ideal gas. Except in precision work or work at high pressures, it is often permissible to neglect terms in $B, C', \ldots$ or $B, C, \ldots$. When this is done we speak of the gas as an ideal gas. It is sometimes stated that the properties of a real gas approach those of an ideal gas as $P \to 0$. Such a statement is true for some properties, but untrue for others. In particular

$$Pv - (Pv)^{\text{ideal}} \to 0, \qquad\qquad P \to 0, \tag{36.1}$$

$$v - v^{\text{ideal}} \to B \neq 0, \qquad P \to 0, \tag{36.2}$$

$$\frac{v}{P} - \left(\frac{v}{P}\right)^{\text{ideal}} \to \frac{B}{P} \to \infty, \qquad P \to 0. \tag{36.3}$$

For convenience of reference we collect together the formulae for the thermodynamic properties of an ideal gas containing n moles

$$G = n\mu = n\mu^\dagger + n\,RT \ln \frac{P}{P^\dagger}, \tag{36.4}$$

$$S = -n\frac{d\mu^\dagger}{dT} - n\,R \ln \frac{P}{P^\dagger}, \tag{36.5}$$

$$H = n\left(\mu^\dagger - T\frac{d\mu^\dagger}{dT}\right), \tag{36.6}$$

$$PV = n\,RT, \tag{36.7}$$

$$F = G - PV = n\left(\mu^\dagger - RT\right) + n\,RT \ln \frac{n\,RT}{P^\dagger V}, \tag{36.8}$$

$$U = n\left(\mu^\dagger - T\frac{d\mu^\dagger}{dT} - RT\right). \tag{36.9}$$

37. Fugacity. When a gas is not ideal it is sometimes found convenient to introduce a quantity p^* called the fugacity defined by

$$RT \ln \frac{p^*}{P} = \int_0^P \left(v - \frac{RT}{P}\right) dP$$

$$= BP + \frac{1}{2} C'P^2 + \cdots. \tag{37.1}$$

We then have

$$\mu = \mu^\dagger + RT \ln \frac{p^*}{P^\dagger} \tag{37.2}$$

which formally resembles the relation (36.4) for an ideal gas

$$\mu = \mu^\dagger + RT \ln \frac{P}{P^\dagger}. \tag{37.3}$$

The simplification attained by the introduction of the fugacity is one of appearance or elegance, but leads to nothing quantitive unless we express the fugacity in terms of the pressure and we are then back where we started. Considerable care is required in the use of fugacity to avoid mistakes. For example the formula for the entropy is

$$S = - n\frac{d\mu^\dagger}{dT} - n\,R\ln\frac{p^*}{p^\dagger} - n\,RT\left(\frac{\partial\ln p^*}{\partial T}\right)_P \tag{37.4}$$

where the danger is that of overlooking the last term.

38. Extensible spring. We now leave gases for the time being and turn to various systems in which the work takes other forms than that due to a uniform external pressure. As our first example we consider the extension of a spring. If we denote the length of the spring by l and the tension by X, we have for the work

$$w = X\,dl \tag{38.1}$$

and so by (11.8) for the free energy

$$dF = -S\,dT + X\,dl. \tag{38.2}$$

We can express this in the integrated form

$$F = F^0 + \int_{l^0}^{l} X\,dl \tag{38.3}$$

where F^0 denotes the value of F and l^0 the value of l when the spring is unstretched. The integral has to be evaluated at constant temperature. If the spring obeys Hooke's law, we have

$$X = \frac{k(l - l^0)}{l^0} \tag{38.4}$$

where k is independent of l but in general depends on T. Using this expression for X and evaluating the integral we obtain

$$F - F^0 = \frac{k(l - l^0)^2}{2l^0}. \tag{38.5}$$

By using the Gibbs-Helmholtz relation

$$U = F - T\left(\frac{\partial F}{\partial T}\right)_l, \tag{38.6}$$

we obtain for the total energy

$$U - U^0 = \left(k - T\frac{dk}{dT}\right)\frac{(l - l^0)^2}{2l^0} \tag{38.7}$$

where U^0 is the value of U for the unstretched spring. For the heat $-q$ which must be removed to keep the temperature constant while the spring is stretched we have

$$-q = T\frac{dk}{dT}\frac{(l - l^0)^2}{2l^0}. \tag{38.8}$$

39. Parallel plate condenser. As our next example we choose a parallel plate condenser. We denote the area of each plate by A, the distance between the plates by l, the charge on the plates by $\pm Q$ and the potential difference between the plates by E. We have

$$w = E\,dQ \tag{39.1}$$

and consequently by (11.8)

$$dF = -S\,dT + E\,dQ. \tag{39.2}$$

When we neglect edge effects we have the relation between E and Q, in the rational system,

$$E = \frac{Q\,l}{\varepsilon\,A} \qquad (39.3)$$

where ε is the permittivity which we assume to be at constant temperature independent of Q. Substituting this value for E into the formula for dF and integrating at constant temperature we obtain

$$F - F^0 = \frac{Q^2\,l}{2\,\varepsilon\,A} \qquad (39.4)$$

where F^0 denotes the value of F for the uncharged condenser. By using the Gibbs-Helmholtz relation

$$U = F - T\left(\frac{\partial F}{\partial T}\right)_Q \qquad (39.5)$$

we obtain for the total energy

$$U - U_0 = \left(\frac{1}{\varepsilon} + \frac{1}{\varepsilon^2}\frac{d\varepsilon}{dT}\right)\frac{Q^2\,l}{2\,A}\,. \qquad (39.6)$$

40. Electrochemical cell. As our next example we choose an electrochemical cell. The essential characteristic of a cell is that a chemical process can take place in it in such a manner that this chemical process is necessarily accompanied by a transfer of electric charge from one terminal to the other without building up a charge anywhere inside the cell. Moreover the charge which flows from the one terminal to the other is directly proportional to the change in the extent of chemical reaction. It would be out of place to discuss here the chemistry of cells in general. We shall merely quote one example and we choose the most used cell, the ordinary lead accumulator. When one faraday (product of Avogadro's number and proton charge) flows through this cell the chemical change is given by the formula

$$\tfrac{1}{2}\,\mathrm{Pb(s)} + \tfrac{1}{2}\,\mathrm{PbO_2(s)} + \mathrm{H_2SO_4(l)} \rightarrow \mathrm{PbSO_4(s)} + \mathrm{H_2O(l)} \qquad (40.1)$$

where (s) denotes a solid phase and (l) a liquid phase. The decrease in the Gibbs function when this process takes place will be denoted by $-\Delta G$. Its value will depend on, amongst other things, the relative proportions of sulphuric acid and water in the liquid phase, but we shall not be concerned with this except to recall the practical consequence that the running down of the cell is accompanied by a decrease in the ratio of sulphuric acid to water and consequently a decrease in the density of the cell liquid.

We now suppose the two terminals of the cell to be put into contact respectively with two points of a potentiometer bridge so placed that the electric potential of the right contact exceeds that of the left contact by an amount E'. Then in general an electric current will flow through the cell and between the two points of contact with the potentiometer bridge. If either of the points of contact is moved along the bridge the current will increase or decrease and it will change sign when E' has a certain value E. When E' is slightly less than E there will be a flow of current from left to right in the cell and from right to left in the potentiometer bridge; this flow of current will be accompanied by a well defined chemical change in the cell. When E' is slightly greater than E there will be a flow of current in the opposite direction and the accompanying chemical change in the cell will also be reversed. When E' is equal to E there will be no flow of current and no chemical change, but by a small shift in the point of contact between cell terminal and potentiometer bridge a small current can be made to flow in either direction. This is a typical and a particularly realistic example of a reversible process. The value E of E' at which the current changes sign is the electromotive force of the cell.

We now stipulate that $E' = E$ so that the electromotive force of the cell is balanced against the potential difference in the potentiometer bridge and we consider the flow of one faraday through the cell, the temperature being maintained constant throughout and the pressure on every phase being kept constant. Then since this process is reversible and isothermal the work done on the cell is equal to the increase in free energy, that is to say

$$w = \Delta F. \tag{40.2}$$

In the present case w consists of two distinct parts, namely

a) the work $-P \Delta V$ done by the external pressure when the volume change is ΔV,

b) the electrical work $-\mathsf{F} E$ done by the potentiometer on the cell in transferring one faraday F through a potential difference E.

We therefore have

$$-\mathsf{F} E = \Delta F + P \Delta V = \Delta G. \tag{40.3}$$

We thus see that the electrical work obtainable from a reversible isothermal process at constant pressure is equal to the decrease in the Gibbs function. Alternatively we may write

$$\mathsf{F} E = A \tag{40.4}$$

where A denotes the affinity of the chemical reaction associated with the flow of one faraday.

Using the Gibbs-Helmholtz relation (19.2)

$$H = G - T \frac{\partial G}{\partial T} \tag{40.5}$$

we obtain for the enthalpy increase ΔH of the chemical reaction associated with the flow of one faraday

$$\Delta H = -\mathsf{F} \left(E - T \frac{dE}{dT} \right). \tag{40.6}$$

This relation is in fact the one obtained by Helmholtz which has led to the name Gibbs-Helmholtz relation for the more general formula previously obtained by Gibbs.

41. Paramagnetic substance. As our next example we choose a small specimen of a paramagnetic substance in a magnetic field. It is convenient to describe the magnetic field by the value of the magnetic induction $\boldsymbol{B}$ before the specimen was introduced. We denote this by $\boldsymbol{B}_e$. For the sake of simplicity we assume that $\boldsymbol{B}_e$ is uniform. If we denote the magnetic moment of the specimen by m then the work w is related to $\boldsymbol{B}_e$ and m by the formula

$$w = \boldsymbol{B}_e \, dm \tag{41.1}$$

and consequently by (11.8)

$$dF = -S \, dT + \boldsymbol{B}_e \, dm. \tag{41.2}$$

It is usually convenient to choose $\boldsymbol{B}_e$, rather than m, as independent variable. We accordingly use a Legendre transformation

$$\varLambda = F - \boldsymbol{B}_e \, m \tag{41.3}$$

and obtain

$$d\varLambda = -S \, dT - m \, d\boldsymbol{B}_e. \tag{41.4}$$

The relation between m and $\boldsymbol{B}_e$ depends both on the magnetic permeability of the specimen and on its shape. In the simplest case of a spherical specimen of volume v the magnetization M is uniform and parallel to $\boldsymbol{B}_e$. It is given by

$$M = \frac{3(\mu - \mu_0)}{\mu + 2\mu_0} \frac{\boldsymbol{B}_e}{\mu_0} \tag{41.5}$$

where μ_0 is the permeability of empty space and consequently

$$m = \frac{3(\mu - \mu_0)}{\mu + 2\mu_0} \frac{\boldsymbol{B}_e v}{\mu_0} . \tag{41.6}$$

Substituting this into the formula for $d\boldsymbol{H}$ and integrating we obtain

$$\boldsymbol{H} - F^0 = \frac{3(\mu - \mu_0)}{(\mu + 2\mu_0)} \frac{\boldsymbol{B}_e^2 v}{2\mu_0} \tag{41.7}$$

where F^0 is the value of F or of $\boldsymbol{H}$ when $\boldsymbol{B}_e$ vanishes. Differentiating with respect to T we obtain

$$S - S^0 = - \frac{\boldsymbol{B}_e^2 v}{2\mu_0} \frac{9\, d\mu/dT}{(\mu + 2\mu_0)^2} \tag{41.8}$$

where S^0 is the value of S when $\boldsymbol{B}_e$ vanishes.

42. Entropy changes. We now return to systems in which the only work is that done by the external pressure and shall now describe how the entropy, or rather an entropy change, of such a system is determined experimentally. For the moment we confine ourselves to a single substance. We use as independent variables temperature and pressure, and we begin by considering changes at constant pressure. We then have for unit quantity of substance

$$d h = T\, d s, \qquad \text{constant } P \tag{42.1}$$

and consequently

$$s_2 - s_1 = \int_1^2 \frac{d h}{T} , \qquad \text{constant } P \tag{42.2}$$

where 1 and 2 relate to an initial and a final state. There are two kinds of change to which this formula is applicable, namely temperature changes and phase changes. For a temperature change we write the formula as

$$s_2 - s_1 = \int_1^2 \frac{\partial h}{\partial T} \frac{d T}{T} , \qquad \text{constant } P . \tag{42.3}$$

The quantity $(\partial h/\partial T)_P$ which can be determined calorimetrically is called the heat capacity of the substance at constant pressure. For a phase change the relation becomes

$$s_2 - s_1 = \frac{h_2 - h_1}{T} , \qquad \text{phase equilibrium} \tag{42.4}$$

where T is the temperature at which the two phases are in mutual equilibrium at the given pressure while $h_2 - h_1$, the heat absorbed in this change, is directly measurable by calorimetry.

By use of the above relations and straightforward calorimetry it is thus possible to determine the entropy difference between any two states of a single substance at the same pressure. To complete our description of the determination of entropy changes we have to describe the determination of an entropy change associated with an isothermal pressure change. For this purpose we use Maxwell's relation (19.7)

$$\frac{\partial s}{\partial P} = -\frac{\partial v}{\partial T} = -\alpha v, \quad \text{constant } T \tag{42.5}$$

where α denotes the coefficient of (cubic) thermal expansion and consequently

$$s_2 - s_1 = -\int_1^2 \alpha v \, dP, \quad \text{constant } T. \tag{42.6}$$

Since α is a measurable quantity the integral can be evaluated.

From values of entropy changes determined as described together with values of enthalpy changes determined directly by calorimetry, changes in the Gibbs function are then calculated from $g = h - Ts$. A different set of relations can be developed for the independent variables T, V but these relations are less useful and will not be discussed here.

43. Gas and liquid states. We turn now to a discussion of the relationship between the gaseous and liquid states of a single substance. The $P-v$ isotherms of a pure substance divide into two classes according as the temperature lies above or below a critical temperature T_c. Examples of each class for carbon dioxide are shown in Fig. 2.

When the molar volume v is sufficiently large both classes approximate to the rectangular hyperbola $Pv = RT$ of a perfect gas. As the molar volume diminishes, the forms of the two classes are quite different. At temperatures greater than the critical, there is a smooth regular variation along the whole isotherm, which can be expressed mathematically by saying that it is a single analytic curve or expressed physically by saying that throughout the isotherm there is a single gaseous phase. At temperatures below the critical, on the other hand, the isotherm consists of three analytically distinct parts separated by discontinuities of the slope. The middle portion is a straight line parallel to the v axis. These parts represent respectively the pure vapour, the saturated vapour in equilibrium with the liquid and the pure liquid. The isothermal curve for the critical temperature T_c is the borderline between the two classes of isotherms. In this isotherm the straight portion is reduced to a single point of horizontal inflexion.

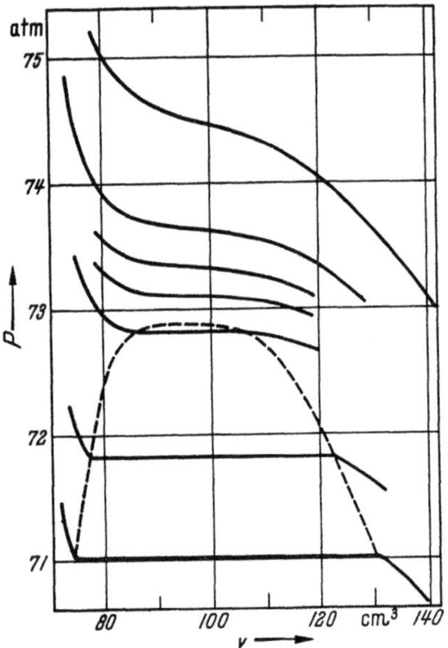

Fig. 2. Isotherms of carbon dioxide.

The broken curve is on the left the locus of the points representing the liquid phase under the pressure of its vapour and on the right the locus of the points representing the saturated vapour. As the temperature increases the molar volume of the liquid at the pressure of its vapour increases, while the molar volume of the saturated vapour decreases. At the critical temperature the isotherm has a point of horizontal inflexion at the top of the broken curve where the liquid and vapour phases cease to be distinguishable. The state represented by this point is called the critical state: the pressure and molar volume in the critical state are called the critical pressure P_c and the critical volume v_c respectively.

To recapitulate, above the critical temperature the substance can exist in only one state, the gaseous. Below the critical temperature it can exist in two states, the liquid with a molar volume less than the critical volume and the gas with a molar volume greater than the critical volume. The equilibrium pressure between the two phases, liquid and vapour can have values up to but not exceeding the critical pressure.

44. Continuity of state. The relation between P and v for a single substance below the critical temperature is shown schematically in Fig. 3. The portion KL represents the liquid state, the portion VW the gaseous state, and the straight portion LV the two phase system liquid-vapour.

At the given temperature the substance can be brought from the liquid state to the gaseous state, or conversely, only by a change during part of which two separate phases will be present. By introducing temperature alterations, however, it is possible to bring the substance from the gaseous state represented by W to the liquid state represented by K by a continuous change throughout which there is never more than one phase present. It is only necessary to raise the temperature above the critical temperature, keeping the volume sufficiently greater than the critical volume, then compress the fluid to a volume below the critical volume, keeping the temperature

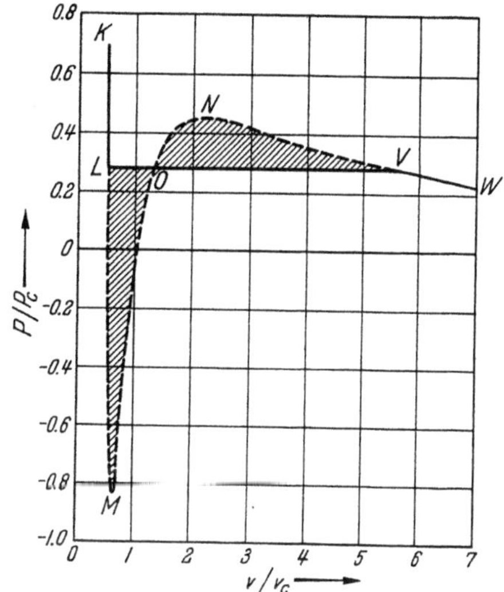

Fig. 3. Continuity between liquid and gas phases.

above the critical temperature, and finally cool the liquid to its original temperature, keeping the volume sufficiently below the critical volume. This possibility of continuity between the liquid and gaseous states was first realized by JAMES THOMSON, and he suggested that the portions KL and VW of the isotherm are actually parts of one smooth curve, such as $KLMONVW$. In point of fact, states corresponding to the portion VN are realizable as supersaturated vapour, and under certain circumstances the same may be true of the portion LM representing superheated liquid. Each of these portions represents states stable with respect to infinitesimal variations, but metastable relative to the two phase system liquid + saturated vapour. The portion of the curve MON, on the other hand, represents states absolutely unstable since here $\partial v/\partial P$ is negative.

Although the states represented by points on the curve $LMONV$ are either metastable or absolutely unstable, still they are equilibrium states. It follows that the sequence of states represented by the curve $LMONV$ corresponds to a reversible process. The change in the chemical potential μ of the fluid in passing through this sequence of states is given by

$$\mu^G - \mu^L = \int \frac{\partial \mu}{\partial P} dP = \int v \, dP \qquad (44.1)$$

where the integrals are to be evaluated along the curve $LMONV$. But, since the two states represented by L and G can exist in equilibrium with each other, we have

$$\mu^G = \mu^L \tag{44.2}$$

and consequently

$$\int v\,dP = 0 \tag{44.3}$$

where the integral is to be evaluated along the curve $LMONV$. The geometrical significance of this is that the two shaded areas LMO and ONV are of equal area. This is known as Maxwell's rule of equal areas.

It is instructive to consider continuity of state in terms of the free energy F. Imagine F to be plotted as vertical coordinate against T and V as horizontal cartesian coordinates. The resulting locus is a curved surface. Consider now cross-sections of this surface by planes $T = $ const. Examples of these are shown diagrammatically in Fig. 4 and since by (19.12)

$$\left(\frac{\partial F}{\partial V}\right)_T = -P, \tag{44.4}$$

the slope of each curve at each point is equal to $-P$.

In the upper curve we see that as V increases, the negative slope steadily decreases numerically and so P decreases steadily. This is typical of any temperature above the critical.

In the lower curve we see that there are two portions $K'L'$ and $V'W'$ in which the negative slope decreases steadily as V increases and these are joined by a straight line $L'V'$ touching $K'L'$ at L' and touching $V'W'$ at V'.

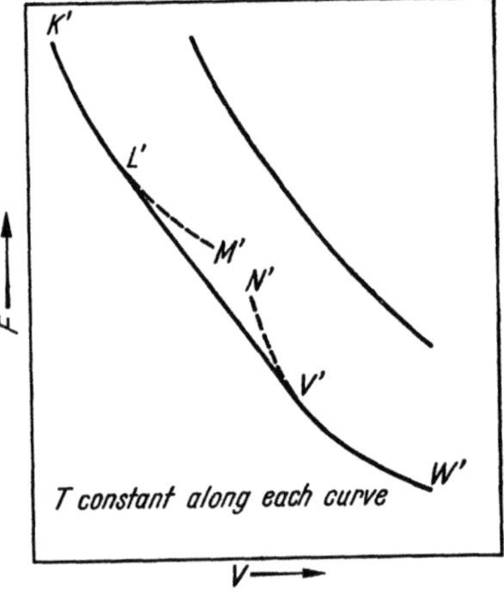

Fig. 4. Stable and metastable isotherms.

These three portions correspond to liquid, to gas and to a two phase liquid-vapour system. This is typical of a temperature below the critical. The broken portion of curve $L'M'$ represents superheated liquid and the broken portion $V'N'$ represents supersaturated vapour. We see immediately that all states represented by these portions of curve are metastable, for any point on either of them lies above a point of the same volume V on the straight line $L'V'$. This means that the free energy of the superheated liquid or supersaturated vapour is greater than that of a system of the same volume consisting of a mixture of liquid L' and vapour V'.

The portions of curve $L'M'$ and $N'V'$ have curvature concave upwards so that

$$\frac{\partial P}{\partial V} = -\frac{\partial^2 F}{\partial V^2} < 0 \tag{44.5}$$

and consequently they represent states internally stable, though metastable with respect to a two phase mixture. If however we wish to unite these two portions into a single smooth curve, the middle portion would necessarily have a curvature concave downwards. This would correspond to a positive value of $\partial P/\partial V$ and so to unstable states and such states are never realizable. It may therefore be argued that no physical significance could be attached to this part of the F-T-V surface. Nevertheless, if the realizable parts $K'L'M'$ and $N'V'W'$ of the surface could be represented by the same analytical function, it would be reasonable from a mathematical point of view to consider the complete surface.

Having constructed such a surface and considering a section corresponding to a particular temperature below the critical, we could then plot $P = -\partial F/\partial V$ against V and so construct a continuous curve such as that shown in the P-v diagram. From this construction it follows of necessity that in Fig. 3 the area below the broken curve $LMONV$ and the area below the straight line LV are both equal to the height of L' above V' in Fig. 4. Consequently these two areas are equal. From this it follows immediately that the two shaded areas are equal as already proved. Since the portion MON of the curve cannot be realized experimentally, instead of saying that the two phase equilibrium is determined by the condition of equality of the two shaded areas, it is perhaps more correct to say that L and V being known the connecting portion of the curve must be sketched in such a manner as to make the two shaded areas equal; otherwise it would be nonsensical, for then $-P$ would not be the slope of any conceivable curve on the F-V diagram.

45. Liquid-vapour equilibrium. We now return to the discussion of phase equilibrium in a single substance, in particular equilibrium between liquid and vapour. Denoting the two phases by the superscripts α and β we have the equilibrium condition

$$\mu^\alpha = \mu^\beta. \tag{45.1}$$

If we vary temperature and pressure in such a way as to maintain equilibrium between the two phases, we have

$$\frac{\partial \mu^\alpha}{\partial T}\,dT + \frac{\partial \mu^\alpha}{\partial P}\,dP = \frac{\partial \mu^\beta}{\partial T}\,dT + \frac{\partial \mu^\beta}{\partial P}\,dP \tag{45.2}$$

or by use of (21.12) and (21.14)

$$-s^\alpha\,dT + v^\alpha\,dP = -s^\beta\,dT + v^\beta\,dP \tag{45.3}$$

and consequently

$$\frac{dT}{dP} = \frac{v^\beta - v^\alpha}{s^\beta - s^\alpha} = \frac{T(v^\beta - v^\alpha)}{h^\beta - h^\alpha} \tag{45.4}$$

which is the Clapeyron relation already met in the previous chapter.

When one of the phases β is a gas it is often sufficiently accurate to treat this phase as an ideal gas, replacing v^β by RT/P, and at the same time to neglect v^α compared with v^β. We then have

$$\frac{dT}{dP} = \frac{RT^2}{P\,\Delta_e h} \tag{45.5}$$

where $\Delta_e h = h^\beta - h^\alpha$ denotes the molar heat of evaporation. We can rewrite (45.5) as

$$\frac{d\ln P}{d(1/T)} = -\frac{\Delta_e h}{R} \tag{45.6}$$

known as the Clapeyron-Clausius relation. If then $\ln P$ is plotted against T^{-1} the slope of the curve at each point is equal to $-\Delta_e h/R$. In practice the plot of $\ln P$ against T^{-1} is found to be nearly straight even as far as the critical temperature although there is no known reason for this.

The Clapeyron-Clausius relation incidentally provides a method, rarely mentioned, for determining the molar mass in the vapour. For by measuring the saturated vapour pressure at several temperatures we can use the relation to determine the heat of evaporation per mole. We can also make a direct calorimetric measurement of the heat of evaporation per unit mass. By comparison we obtain the ratio of the mole to the unit of mass, that is to say the molar mass.

46. Two-phase equilibrium. Let us now return to the Clapeyron relation for equilibrium between any two phases. Using Δ to denote the difference between the value of a quantity in the two phases we may write the relation (18.14) as

$$\frac{dP}{dT} = \frac{\Delta s}{\Delta v} = \frac{\Delta h}{T \Delta v}. \tag{46.1}$$

We shall now study the temperature dependence under equilibrium conditions of the quantities Δs and Δh occurring in this formula. For Δs we have, using T and P as variables,

$$\begin{aligned}
\frac{d \Delta s}{dT} &= \frac{\partial \Delta s}{\partial T} + \frac{dP}{dT} \frac{\partial \Delta s}{\partial P} \\
&= \Delta \frac{\partial s}{\partial T} - \frac{\Delta s}{\Delta v} \frac{\partial \Delta v}{\partial T} \\
&= \Delta \frac{\partial s}{\partial T} - \Delta s \frac{\partial \ln \Delta v}{\partial T}.
\end{aligned} \tag{46.2}$$

From this we derive for Δh, still using T and P as variables

$$\begin{aligned}
\frac{d \Delta h}{dT} &= \frac{d(T \Delta s)}{dT} = \Delta s + T \frac{d \Delta s}{dT} \\
&= \Delta s + T \Delta \frac{\partial s}{\partial T} - T \Delta s \frac{\partial \ln \Delta v}{\partial T} \\
&= \frac{\Delta h}{T} + \Delta \frac{\partial h}{\partial T} - \Delta h \frac{\partial \ln \Delta v}{\partial T}
\end{aligned} \tag{46.3}$$

a formula due to Planck.

47. Ideal gas mixture. We now turn to systems containing more than one chemical substance, beginning with a single gas phase. For the sake of brevity and simplicity we shall neglect the second and higher virial coefficients treating the gas as ideal. In Chap. E we shall derive the properties of an ideal gaseous mixture from a model of particles between which the interaction energy is negligible, but we here use a purely thermodynamic definition of an ideal gaseous mixture. We define a gaseous mixture as ideal when the free energy at given temperature and given volume is the sum of those of the component gases each at the given temperature and given volume and treated as ideal. According to this definition combined with (36.8) the free energy F of an ideal gaseous mixture containing n_i moles of the species i in a volume V has the form

$$F = \sum_i n_i (\mu_i^\dagger - RT) + RT \sum_i n_i \ln \frac{n_i RT}{P^\dagger V} \tag{47.1}$$

where $P^\dagger$ is a chosen standard pressure, usually one atmosphere, while $\mu_i^\dagger$ depends on T and on $P^\dagger$ but is independent of V and the n_i's. We immediately derive for the pressure

$$P = -\frac{\partial F}{\partial V} = \sum_i n_i \frac{RT}{V} \tag{47.2}$$

from which we see that the pressure is the sum of the pressures which each gas would produce if present alone at the same temperature and volume.

Using the last relation to change the variable from V to P we obtain for the Gibbs function

$$G = F + PV = \sum_i n_i \mu_i^\dagger + RT \sum_i n_i \ln \frac{P}{P^\dagger} + RT \sum_i n_i \ln \frac{n_i}{\sum_k n_k}. \qquad (47.3)$$

We immediately deduce for the chemical potential of the species i

$$\mu_i = \frac{\partial G}{\partial n_i} = \mu_i^\dagger + RT \ln \frac{P}{P^\dagger} + RT \ln \frac{n_i}{\sum_k n_k}$$

$$= \mu_i^\dagger + RT \ln \frac{P}{P^\dagger} + RT \ln x_i \qquad (47.4)$$

where x_i is the mole fraction of the species i. If we define the partial pressure p_i of the species i by

$$p_i = x_i P \qquad (47.5)$$

we can write formula (47.4) as

$$\mu_i = \mu_i^\dagger + RT \ln \frac{p_i}{P^\dagger}. \qquad (47.6)$$

We see that the chemical potential of each species i is thus determined by its own partial pressure and is independent of the partial pressures of the other species present.

The corresponding relation for the absolute activity λ_i of the species i has the form

$$\lambda_i = \lambda_i^\dagger \, p_i / P^\dagger \qquad (47.7)$$

where $\lambda_i^\dagger / P^\dagger$ depends only on the temperature.

From the formula for G we deduce for the entropy S and the enthalpy H

$$S = -\frac{\partial G}{\partial T} = -\sum_i n_i \frac{d\mu_i^\dagger}{dT} - R \sum_i n_i \ln \frac{P}{P^\dagger} - R \sum_i n_i \ln \frac{n_i}{\sum_k n_k}, \qquad (47.8)$$

$$H = G + TS = \sum_i n_i \left(\mu_i^\dagger - T \frac{d\mu_i^\dagger}{dT} \right). \qquad (47.9)$$

We could also obtain an equivalent formula for S from formula (47.1), namely

$$S = -\left(\frac{\partial F}{\partial T} \right)_{V,\, n_i} = -\sum_i n_i \frac{d\mu_i^\dagger}{dT} - R \sum_i n_i \ln \frac{n_i RT}{P^\dagger V}$$

$$= -\sum_i n_i \frac{d\mu_i^\dagger}{dT} - R \sum_i n_i \ln \frac{p_i}{P^\dagger}. \qquad (47.10)$$

We notice that the entropy of the mixture is equal to the sum of those of the unmixed gases each at the same T, V or each at the same T, p_i. On the other hand at given T, P we see that the entropy exceeds the sum of those of the unmixed gases each at the same T, P by the positive quantity

$$- R \sum_i n_i \ln \frac{n_i}{\sum_k n_k} \qquad (47.11)$$

called the entropy of mixing. A safer name would be entropy of mixing at given total pressure.

48. Liquid mixture. We turn now to liquid mixtures. This is a vast subject and we shall have to restrict our discussion to the most important and simplest aspects. We choose as independent variables T, P, n_i so that the characteristic function is the Gibbs function G or g per mole. We have

$$\frac{\partial g}{\partial P} = v \tag{48.1}$$

and since for a liquid at ordinary pressures $Pv \ll RT$ we may usually ignore terms of the order Pv. This means that we regard G and consequently also the μ_i's as independent of the pressure. We now define the Gibbs function of mixing $\Delta_m G$ as the amount by which the Gibbs function of the mixture exceeds those of the unmixed pure liquids at the same temperature. We thus write

$$G = \sum_i n_i \mu_i^0 + \Delta_m G \tag{48.2}$$

where μ_i^0 denotes the chemical potential of the pure liquid i at the same temperature. All the thermodynamic properties of the mixture are determined by the form of $\Delta_m G$, that is to say its dependence on temperature and more especially on composition.

49. Ideal mixture. Excess functions. It will be shown in Chap. E that in a mixture of isotopes, whether solid, liquid or gas

$$\Delta_m G = RT \sum_i n_i \ln \frac{n_i}{\sum_k n_k} \tag{49.1}$$

which we recognize has the same form as in an ideal gaseous mixture. The simplest possible behaviour of a liquid mixture would be that corresponding to this formula for $\Delta_m G$ and such a liquid mixture is called ideal. But, except for mixtures of isotopes, no liquid mixture is accurately ideal and only a few known mixtures are approximately ideal. Nevertheless the ideal mixture provides a useful standard with which to compare the behaviour of real mixtures. We accordingly write

$$\Delta_m G = RT \sum_i n_i \ln \frac{n_i}{\sum_k n_k} + G^E \tag{49.2}$$

where G^E called the excess Gibbs function is a measure of the deviation from ideality. We have now to consider the form of G^E, but for the sake of brevity we shall restrict ourselves to mixtures of only two substances, that is to say binary mixtures.

Denoting the two substances by the subscripts 1 and 2 we have

$$\Delta_m G = RT\, n_1 \ln \frac{n_1}{n_1 + n_2} + RT\, n_2 \ln \frac{n_2}{n_1 + n_2} + G^E. \tag{49.3}$$

We may expect to be able to express G^E empirically as a polynomial expression in n_1 and n_2 which must be homogeneous of the first degree. Since moreover G^E must vanish when $n_1 = 0$ and when $n_2 = 0$ we may expect the leading term to be proportional to $n_1 n_2/(n_1 + n_2)$. This being so the most convenient polynomial expression for G^E is

$$G^E = \frac{n_1 n_2}{n_1 + n_2} \left\{ A_0 + A_1 \frac{n_2 - n_1}{n_1 + n_2} + A_2 \left(\frac{n_2 - n_1}{n_1 + n_2} \right)^2 \right\} \tag{49.4}$$

where the coefficients A_0, A_1, A_2 depend only on the temperature. Whatever may be the basis for choosing this form for G^E, it is an empirical fact that almost all liquid mixtures yet studied, excluding electrolytes, can be fitted by such a formula without going beyond A_2, most of them without going beyond A_1 and quite a number without going beyond A_0. All the thermodynamic properties of a mixture can be expressed in terms of the properties of the pure substances and the coefficients A_0, A_1, A_2 and their temperature derivatives.

50. Simple mixture. For the sake of brevity we shall confine our discussion from here onwards to the simple case where in (49.4) the terms in A_1, A_2 may be neglected. We accordingly write, dropping the subscript from A_0,

$$\Delta_m G = RT\, n_1 \ln \frac{n_1}{n_1 + n_2} + RT\, n_2 \ln \frac{n_2}{n_1 + n_2} + A\, \frac{n_1 n_2}{n_1 + n_2} \tag{50.1}$$

and shall call mixtures conforming to this formula simple mixtures. The detailed description of the behaviour of these simple mixtures serves as a semi-quantitative description of many other mixtures. We immediately deduce for the enthalpy of mixing and the entropy of mixing

$$\Delta_m H = \Delta_m G - T\, \frac{\partial \Delta_m G}{\partial T} = \left(A - T\, \frac{dA}{dT}\right) \frac{n_1 n_2}{n_1 + n_2}, \tag{50.2}$$

$$\Delta_m S = - \frac{\partial \Delta_m G}{\partial T} = - R\, n_1 \ln \frac{n_1}{n_1 + n_2} - R\, n_2 \ln \frac{n_2}{n_1 + n_2} - \frac{\partial A}{\partial T} \frac{n_1 n_2}{n_1 + n_2}. \tag{50.3}$$

We deduce for the chemical potentials

$$\mu_1 - \mu_1^0 = \frac{\partial \Delta_m G}{\partial n_1} = RT \ln \frac{n_1}{n_1 + n_2} + A\left(\frac{n_2}{n_1 + n_2}\right)^2, \tag{50.4}$$

$$\mu_2 - \mu_2^0 = \frac{\partial \Delta_m G}{\partial n_2} = RT \ln \frac{n_2}{n_1 + n_2} + A\left(\frac{n_1}{n_1 + n_2}\right)^2. \tag{50.5}$$

It is convenient to denote the mole fractions of 1 and 2 by $1 - x$ and x respectively. With this notation we have

$$\mu_1 = \mu_1^0 + RT \ln (1 - x) + A\, x^2, \tag{50.6}$$

$$\mu_2 = \mu_2^0 + RT \ln x + A\, (1 - x)^2, \tag{50.7}$$

or in terms of absolute activities

$$\frac{\lambda_1}{\lambda_1^0} = (1 - x) \exp\{A\, x^2\}, \tag{50.8}$$

$$\frac{\lambda_2}{\lambda_2^0} = x \exp\{A(1 - x)^2\}. \tag{50.9}$$

The ratios λ/λ^0 are called relative activities and are often denoted by a.

We now consider the equilibrium between a simple liquid mixture and the vapour phase treated as an ideal gaseous mixture. The condition for equilibrium between the two phases is that μ_1 and μ_2, or λ_1 and λ_2, should have the same value in both phases. We recall that for an ideal gas phase by (47.7)

$$\lambda_1 = \lambda_1^\ddagger\, p_1/P^\dagger, \qquad \lambda_2 = \lambda_2^\ddagger\, p_2/P^\dagger \tag{50.10}$$

where $\lambda_1^\ddagger$, $\lambda_2^\ddagger$ are constant at constant temperature. It follows from (50.8) to (50.10)

$$\frac{p_1}{p_1^0} = (1 - x) \exp\{A\, x^2\}, \tag{50.11}$$

$$\frac{p_2}{p_2^0} = x \exp\{A(1 - x)^2\} \tag{50.12}$$

where p_1^0, p_2^0 are the values of the saturated vapour pressures of the two pure liquids at the given temperature.

In the limit $x \ll 1$ these relations simplify to

$$\frac{p_1}{p_1^0} = 1 - x \qquad (50.13)$$

known as Raoult's law and

$$p_2 = x\, p_2^0\, e^A \quad \text{or} \quad p_2 \propto x \qquad (50.14)$$

known as Henry's law.

51. Azeotropic mixture. The total vapour pressure of the simple liquid mixture is according to (50.11) and (50.12)

$$P = p_1 + p_2 = p_1^0 (1 - x) \exp\{A\, x^2\} + p_2^0\, x \exp\{A(1 - x)^2\}. \qquad (51.1)$$

When we differentiate P with respect to x we obtain

$$\frac{dP}{dx} = [- p_1^0 \exp\{A\, x^2\} + p_2^0 \exp\{A(1 - x)^2\}]\,\{1 - 2A\, x\,(1 - x)\} \qquad (51.2)$$

which vanishes when

$$\frac{p_1^0}{p_2^0} = \exp\{A(1 - 2x)\}. \qquad (51.3)$$

We may without loss of generality assume that $p_1^0 \geq p_2^0$. If then A is positive, this equation has a real root in the range $0 \leq x \leq 1$ provided

$$e^A > \frac{p_1^0}{p_2^0} \geq 1 \qquad (51.4)$$

and P will have a maximum at this value of x. If on the other hand A is negative P will have a minimum provided

$$e^{-A} > \frac{p_1^0}{p_0^2} \geq 1. \qquad (51.5)$$

It is interesting to notice that the condition for P to be stationary can be rewritten as

$$\frac{p_1}{1 - x} = \frac{p_2}{x}. \qquad (51.6)$$

If we denote mole fractions in the vapour phase by $1 - y$ and y we have

$$p_1 = (1 - y)\, P, \qquad p_2 = y\, P, \qquad (51.7)$$

and the condition at the stationary value of P is

$$x = y, \qquad (51.8)$$

that is to say the vapour and liquid have the same relative composition. A liquid of such composition can be completely evaporated at constant temperature and pressure without any change of composition and is called azeotropic.

Azeotropic mixtures need not belong to the class of simple mixtures. We shall therefore discuss their behaviour quite generally. Any liquid binary phase by itself has three degrees of freedom which we can take as given by the three independent variables, T, P, x. Alternatively if we use the four variables T, P, μ_1, μ_2 these are not independent but are connected by the Gibbs-Duhem relation, which may be written as

$$s^L\, dT^L - v^L\, dP^L + (1 - x)\, d\mu_1^L + x\, d\mu_2^L = 0 \qquad (51.9)$$

in which the superscript L denotes the liquid phase. Similarly a gaseous binary phase by itself has three degrees of freedom and can be described by the four variables T, P, μ_1, μ_2 subject to the Gibbs-Duhem relation

$$s^G\, dT^G - v^G\, dP^G + (1 - y)\, d\mu_1^G + y\, d\mu_2^G = 0 \tag{51.10}$$

in which the superscript G denotes the gas phase. We now consider a liquid phase and a gas phase in mutual equilibrium. The two phases have common values of T, P, μ_1, μ_2 so that the two Gibbs-Duhem relations become

$$s^L\, dT - v^L\, dP + (1 - x)\, d\mu_1 + x\, d\mu_2 = 0\,, \tag{51.11}$$

$$s^G\, dT - v^G\, dP + (1 - y)\, d\mu_1 + y\, d\mu_2 = 0\,. \tag{51.12}$$

By subtraction we obtain

$$(s^G - s^L)\, dT - (v^G - v^L)\, dP + (x - y)\, (d\mu_1 - d\mu_2) = 0\,. \tag{51.13}$$

We now impose the condition for the mixture to be azeotropic

$$x = y \tag{51.14}$$

and (51.13) reduces to

$$(s^G - s^L)\, dT - (v^G - v^L)\, dP = 0\,. \tag{51.15}$$

From this it follows immediately that, at a given temperature, the pressure has a stationary value and that, at a given pressure, the temperature has a stationary value. It also follows that if temperature and pressure are varied simultaneously so as to conserve the azeotropic condition, we have

$$\frac{dT}{dP} = \frac{v^G - v^L}{s^G - s^L}\,, \qquad \text{azeotropy} \tag{51.16}$$

of the same form as the Clapeyron relation (18.11) for a single substance. This is not surprising since, as long as the condition of azeotropy is preserved, the mixture is behaving as a single substance.

52. Equilibrium between simple mixture and pure solid. We now return to simple mixtures and shall consider the condition for equilibrium between such a mixture and a solid phase consisting of the pure substance 1. We denote the temperature of this equilibrium by T and the freezing point of the pure liquid 1 by T^0. We also use the superscript 0 to denote other properties of the pure liquid 1 and the superscript s to denote properties of the pure solid phase 1. Symbols without superscripts relate to the liquid mixture. We then have the following relations. The two phase equilibrium condition between the liquid mixture and the solid is

$$\mu_1(T) = \mu_1^s(T) \tag{52.1}$$

which it is more useful to write as

$$\frac{\mu_1(T)}{T} = \frac{\mu_1^s(T)}{T}\,. \tag{52.2}$$

Similarly the two phase equilibrium condition between the pure liquid and the solid is

$$\frac{\mu_1^0(T^0)}{T^0} = \frac{\mu_1^s(T^0)}{T^0}\,. \tag{52.3}$$

The relation between μ_1/T in the mixture and in the pure liquid at the same temperature T is according to (50.6)

$$\frac{\mu_1(T)}{T} - \frac{\mu_1^0(T)}{T} = R \ln (1 - x) + \frac{A}{T} x^2. \tag{52.4}$$

The relations between values of μ_1/T at the temperatures T and T^0 for the pure liquid and the solid are according to (21.13)

$$\frac{\mu_1^0(T^0)}{T^0} - \frac{\mu_1^0(T)}{T} = - \int\limits_T^{T^0} \frac{h_1^0}{T^2}\, dT, \tag{52.5}$$

$$\frac{\mu_1^s(T^0)}{T^0} - \frac{\mu_1^s(T)}{T} = - \int\limits_T^{T^0} \frac{h_1^s}{T^2}\, dT. \tag{52.6}$$

By addition and subtraction of these relations we obtain

$$R \ln (1 - x) + \frac{A}{T} x^2 = - \int\limits_T^{T^0} \frac{h_1^0 - h_1^s}{T^2}\, dT$$

$$= - \int\limits_T^{T^0} \frac{\Delta_f h^0}{T^2}\, dT \tag{52.7}$$

where $\Delta_f h^0$ denotes the heat of fusion of the pure substance 1. We can abbreviate this formula to

$$- \ln (1 - x) - \frac{A}{RT} x^2 = \frac{[\Delta_f h^0]}{R} \left(\frac{1}{T} - \frac{1}{T^0} \right) \tag{52.8}$$

where [] denotes averaging over the range of reciprocal temperature from $1/T$ to $1/T^0$. When $x \ll 1$ this formulas may be replaced by the approximation

$$x = \frac{\Delta_f h^0}{RT^{02}} (T^0 - T) \tag{52.9}$$

which is essentially the same as a relation due to van't Hoff.

53. Incomplete miscibility. The stability condition for a binary mixture not to split into two phases is according to (29.17)

$$\frac{\partial^2 g}{\partial x^2} > 0. \tag{53.1}$$

For a binary simple mixture we deduce from (50.1)

$$g = \frac{G}{n_1 + n_2}$$

$$= (1 - x)\,\mu_1^0 + x\,\mu_2^0 + RT (1 - x) \ln (1 - x) + RT\, x \ln x + A\, x (1 - x) \tag{53.2}$$

so that

$$\frac{\partial g}{\partial x} = \mu_2^0 - \mu_1^0 + RT \ln \frac{x}{1 - x} + A (1 - 2x), \tag{53.3}$$

$$\frac{\partial^2 g}{\partial x^2} = \frac{RT}{x(1 - x)} - 2A, \tag{53.4}$$

and so $\partial^2 g/\partial x^2$ is positive infinite at $x = 0$ and $x = 1$ and has a minimum value $2(2RT - A)$ at $x = \tfrac{1}{2}$. Thus the condition for complete miscibility, according to the formulae of simple mixtures is that $A/RT < 2$ and the critical condition occurs when $A/RT = 2$.

Pairs of liquids are known, for example water and nicotine, which are completely miscible above a certain critical temperature and below another critical temperature, but are incompletely miscible in the intermediate temperature range. It is interesting to note that even simple mixtures can behave in this way when A is a quadratic function of the temperature provided the three coefficients in the quadratic expression have suitable signs and magnitudes. To be precise if A has the quadratic form

$$\frac{A}{R} = 2T + \frac{t^2 - (T - T_0)^2}{\Theta} \tag{53.5}$$

where Θ, T_0 and t are positive constants and $t < T_0$, then it is clear that $A/RT = 2$ when $T = T_0 - t$ or $T = T_0 + t$. It can also be verified that $A/RT > 2$ when $T_0 - t < T < T_0 + t$ and that $A/RT < 2$ when $T > T_0 + t$ or $T < T_0 - t$. Consequently the temperature range of incomplete miscibility extends from $T_0 - t$ to $T_0 + t$.

Incidentally the converse behaviour would occur, that is complete miscibility only between the two critical temperatures, if Θ were negative. Such behaviour is unknown.

54. Duhem-Margules relation. This completes our discussion of simple binary mixtures. We now return to the most general kind of liquid mixture and recall the Gibbs-Duhem relation (23.3)

$$\sum_i x_i \, d\mu_i = 0, \qquad \text{constant } T, P \tag{54.1}$$

or in so far as we neglect the dependence on pressure of the properties of a liquid

$$\sum_i x_i \, d\mu_i = 0, \qquad \text{constant } T. \tag{54.2}$$

Now the value of each μ_i must be the same in the liquid phase as in the gas phase in equilibrium with it. If then we treat the gas phase as ideal we have according to (47.6)

$$\mu_i = \mu_i^\dagger + RT \ln \frac{p_i}{p^\dagger} \tag{54.3}$$

so that

$$d\mu_i = RT \, d \ln p_i, \qquad \text{constant } T. \tag{54.4}$$

Substituting this into (54.2) we have

$$\sum_i x_i \, d \ln p_i = 0, \qquad \text{constant } T \tag{54.5}$$

which is known as the Duhem-Margules relation. For a binary mixture this reduces to

$$(1 - x) \, d \ln p_1 + x \, d \ln p_2 = 0 \qquad \text{constant } T. \tag{54.6}$$

It is readily verified that this relation is satisfied by the previous formulae for a simple mixture which can be rewritten as

$$\ln p_1 = \ln p_1^0 + \ln (1 - x) + A x^2/RT, \tag{54.7}$$

$$\ln p_2 = \ln p_2^0 + \ln x + A(1 - x)^2/RT. \tag{54.8}$$

55. Liquid-vapour interface. We have hitherto assumed that every system consists of one or more completely homogeneous phases bounded by sharply defined geometrical surfaces. This is an over-simplification because the interface between any two phases will rather be a thin layer across which the physical properties vary continuously from those of the interior of one phase to those of the interior of the other. We shall now consider the thermodynamic properties of these surfaces between two phases.

For the sake of brevity and simplicity we shall restrict our considerations to a plane boundary between a liquid and a vapour. We shall throughout assume that the liquid, the vapour and the boundary are in mutual equilibrium. Between the homogeneous liquid phase denoted by the superscript L and the homogeneous gas phase denoted by the superscript G we postulate a surface phase denoted by σ. We assume that the boundary between the liquid and the surface phase is a plane, which we call $L\sigma$, and that the boundary between the gas phase and the surface phase is a plane, which we call $G\sigma$. We assume that these planes $L\sigma$ and $G\sigma$ are parallel to each other and so orientated that the surface phase is homogeneous in directions parallel to these planes. The planes $L\sigma$ and $G\sigma$ must be sufficiently far apart for the liquid phase to be completely homogeneous as far as $L\sigma$ and for the gas phase to be completely homogeneous as far as $G\sigma$. Subject to this restriction the placing of these planes is arbitrary but the physical deductions from the formulae obtained are invariant with respect to displacement of these planes. It will usually be possible and convenient to suppose the distance apart of the planes, that is to say the thickness of the surface phase, to be between 10 Å and 100 Å.

Since the surface phase σ thus defined is a material system with a well defined volume and material content, its thermodynamic properties require no special definition. We may speak of its temperature, free energy, composition and so on as for a homogeneous bulk phase. The only functions that call for special comment are the pressure and the surface tension. In any homogeneous bulk phase the force across unit area is equal in all directions and is called the pressure. But in σ the force across unit area is not the same in all directions. In the direction normal to the interface the force per unit area in the surface phase is the same as in the liquid and gas phases and we denote this by P, but in the direction parallel to the interface the force is less. If we denote the volume of the surface phase by V^σ and its area by A then the work w done on the surface phase when it is deformed has the form

$$w = - P\, dV^\sigma + \gamma\, dA \tag{55.1}$$

where γ is called the surface tension. This expression takes the place of $- P\, dV$ for a homogeneous phase. We consequently have for variations of the free energy of the surface phase

$$dF^\sigma = - S^\sigma\, dT - P\, dV^\sigma + \gamma\, dA + \sum_i \mu_i\, dn_i^\sigma. \tag{55.2}$$

There is no need to add superscripts to T, P, μ_i because these have the same values in the surface as in the liquid and gas.

Since at given temperature F^σ must be a homogeneous function of the first degree of the extensive variables V^σ, A, n_i^σ we have by Euler's theorem

$$F^\sigma = V^\sigma \frac{\partial F^\sigma}{\partial V^\sigma} + A\, \frac{\partial F^\sigma}{\partial A} + \sum_i n_i^\sigma \frac{\partial F^\sigma}{\partial n_i^\sigma}$$

$$= - P V^\sigma + \gamma A + \sum_i n_i^\sigma \mu_i. \tag{55.3}$$

In analogy with the definition (12.2) of the Gibbs function for a bulk phase α

$$G^\alpha = F^\alpha + P V^\alpha \tag{55.4}$$

we now define the Gibbs function G^σ of the surface phase by

$$G^\sigma = F^\sigma + P V^\sigma - \gamma A . \tag{55.5}$$

We immediately deduce

$$dG^\sigma = - S^\sigma dT + V^\sigma dP - A\, d\gamma + \sum_i \mu_i d n_i^\sigma , \tag{55.6}$$

$$G^\sigma = \sum_i \mu_i n_i^\sigma , \tag{55.7}$$

$$S^\sigma dT - V^\sigma dP + A\, d\gamma + \sum_i n_i^\sigma d\mu_i = 0 . \tag{55.8}$$

These equations differ from those for bulk phases only in the terms $A\, d\gamma$. The last equation is the analogue of the Gibbs-Duhem relation (22.5). If we divide through by A we obtain the more convenient form

$$- d\gamma = s_u^\sigma dT - \tau\, dP + \sum_i \Gamma_i d\mu_i \tag{55.9}$$

where s_u^σ is the entropy per unit area in the surface phase, τ is the thickness of the surface phase and Γ_i is the number of moles of the substance i in unit area of the surface σ. We call Γ_i the surface concentration of i.

Formula (55.9) expresses variations of γ in terms of variations of T, P and the μ_i's, but these are not independent variables. If the number of independently variable chemical substances is c, then according to Gibbs' phase rule the liquid-vapour system has c degrees of freedom. It is usually most convenient to choose as independent variables the temperature T and $c-1$ mole fractions x_2, x_3, ..., x_c in the liquid phase. For the sake of brevity we shall from here onwards restrict ourselves to binary systems. We denote the mole fractions of the species 1 and 2 in the liquid phase by $1-x$ and x respectively. We choose as independent variables T and x.

We have the relations

$$- d\gamma = s_u^\sigma dT - \tau\, dP + \Gamma_1 d\mu_1 + \Gamma_2 d\mu_2 , \tag{55.10}$$

$$d\mu_1 = - S_1^L dT + V_1^L dP + \frac{\partial \mu_1}{\partial x}\, dx , \tag{55.11}$$

$$d\mu_2 = - S_2^L dT + V_2^L dP + \frac{\partial \mu_2}{\partial x}\, dx . \tag{55.12}$$

In these formulae P is equal to the total vapour pressure of the liquid. As explained in our earlier discussion of liquids the terms $V^L dP$ are then negligible and so is the term $\tau\, dP$. We may therefore without loss of accuracy omit these terms and it was in order to make this possible that we chose as independent variable a mole fraction x in the liquid phase rather than a mole fraction in the gas phase. The question sometimes raised "how does the surface tension depend on the pressure?" is meaningless since the usual, or any useful, set of independent variables does not include the pressure.

Neglecting the terms in dP and substituting for $d\mu_1$, $d\mu_2$ from (55.11), (55.12) into (55.10) we have, dropping the superscript L

$$- d\gamma = (s_u^\sigma - \Gamma_1 s_1 - \Gamma_2 s_2)\, dT + \left(\Gamma_1 \frac{\partial \mu_1}{\partial x} + \Gamma_2 \frac{\partial \mu_2}{\partial x}\right) dx$$

$$= (s_u^\sigma - \Gamma_1 s_1 - \Gamma_2 s_2)\, dT + \left(\Gamma_2 - \frac{x\,\Gamma_1}{1-x}\right) \frac{\partial \mu_2}{\partial x}\, dx \tag{55.13}$$

by virtue of the Gibbs-Duhem relation.

Whereas the values of s_u^σ, Γ_1 and Γ_2 depend on the position assigned to the plane $L\sigma$, it can readily be verified that when this plane is shifted the quantities

$$s_u^\sigma - \Gamma_1 s_1 - \Gamma_2 s_2, \qquad (55.14)$$

$$\Gamma_2 - \frac{x\,\Gamma_1}{1-x} \qquad (55.15)$$

remain invariant. With regard to shift of the other plane $G\sigma$ little need be said in the present connection. For our approximation of neglecting terms in $P\tau$, as we are doing, is equivalent to assuming that the quantity of matter per unit volume in the gas phase is negligible compared with that in unit volume of the surface layer. Consequently if we shifted the geometrical surface $G\sigma$ away from the liquid even to the extent of doubling the value of τ, the change in the physical content of the surface layer would be negligible and consequently the values of Γ_1, Γ_2, S^σ would not be appreciably affected.

56. Dependence of surface tension on temperature. If we now consider variations of temperature at constant composition of the liquid, we deduce from (55.13)

$$-\frac{d\gamma}{dT} = s_u^\sigma - \Gamma_1 s_1 - \Gamma_2 s_2, \qquad \text{constant } x. \qquad (56.1)$$

The right-hand side of this formula is the entropy of unit area of surface less the entropy of the same material contained in the liquid phase.

This relation involving entropies can be transformed to one involving total energies. Since we are neglecting terms in Pv we may write with sufficient accuracy

$$\mu_1 = u_1 - T s_1, \qquad (56.2)$$

$$\mu_2 = u_2 - T s_2 \qquad (56.3)$$

and using the subscript u to denote unit area of surface

$$u_u^\sigma - T s_u^\sigma - \gamma = f^\sigma - \gamma = \Gamma_1 \mu_1 + \Gamma_2 \mu_2. \qquad (56.4)$$

Eliminating s_u^σ, s_1, s_2 from these relations we obtain

$$\gamma - T \frac{d\gamma}{dT} = u_u^\sigma - \Gamma_1 u_1 - \Gamma_2 u_2, \qquad \text{constant } x. \qquad (56.5)$$

The right-hand side of this formula is the energy of unit area of surface less the energy of the same material contained in the liquid phase.

These formulae can of course be applied to the pure substance 1 by setting $\Gamma_2 = 0$.

57. Dependence of surface tension on composition. Turning now to variations of composition at constant temperature we deduce from (55.13)

$$-\frac{d\gamma}{dx} = \Gamma_1 \frac{\partial \mu_1}{\partial x} + \Gamma_2 \frac{\partial \mu_2}{\partial x}, \qquad \text{constant } T. \qquad (57.1)$$

This relation is known as Gibbs' adsorption isotherm. Since each μ has the same value as in the vapour phase, we may, if we treat the vapour as an ideal gas, as a consequence of (47.6) write

$$d\mu_1 = RT\, d\ln p_1, \qquad \text{constant } T, \qquad (57.2)$$

$$d\mu_2 = RT\, d\ln p_2, \qquad \text{constant } T, \qquad (57.3)$$

where p_1, p_2 are the partial pressures of the saturated vapour. Substituting (57.2) and (57.3) into (57.1) we obtain

$$-\frac{1}{RT}\frac{d\gamma}{dx} = \Gamma_1 \frac{\partial \ln p_1}{\partial x} + \Gamma_2 \frac{\partial \ln p_2}{\partial x}, \qquad \text{constant } T. \qquad (57.4)$$

By using the Duhem-Margules relation (54.6)

$$(1-x)\frac{\partial \ln p_1}{\partial x} + x\frac{\partial \ln p_2}{\partial x} = 0, \qquad \text{constant } T \qquad (57.5)$$

we may eliminate either p_1 or p_2 from (57.4) thus obtaining

$$-\frac{1}{RT}\frac{d\gamma}{dx} = -\frac{I}{x}\frac{\partial \ln p_1}{\partial x} = \frac{I}{1-x}\frac{\partial \ln p_2}{\partial x}, \qquad \text{constant } T \qquad (57.6)$$

where I is defined as

$$I = (1-x)\,\Gamma_2 - x\,\Gamma_1. \qquad (57.7)$$

Thus from measurements of γ and p_1 or of γ and p_2 over a range of values of x at constant T values of I can be computed. Separate values of Γ_1 and Γ_2 cannot be determined nor would such values be physically significant, since they would depend on the precise position arbitrarily assigned to the plane $L\sigma$. The value of I by contrast is invariant with respect to shifts of this plane. We may usefully regard I as a measure of the relative adsorption of the two species.

In the particular case of a simple mixture we have by (50.11) and (50.12)

$$\frac{\partial \ln p_1}{\partial x} = -\frac{1}{1-x} + \frac{2A}{RT}\,x, \qquad \text{constant } T, \qquad (57.8)$$

$$\frac{\partial \ln p_2}{\partial x} = \frac{1}{x} - \frac{2A}{RT}\,(1-x), \qquad \text{constant } T, \qquad (57.9)$$

and so according to (57.6)

$$-\frac{d\gamma}{dx} = \left\{\frac{RT}{x(1-x)} - 2A\right\}I, \qquad \text{constant } T. \qquad (57.10)$$

C. Fundamentals of statistical mechanics.

58. Scope of statistical mechanics[1]. The science of statistical mechanics has the special function of providing reasonable methods for treating the behaviour of mechanical systems under circumstances such that our knowledge of the condition of the system is less than the maximal knowledge which would be theoretically possible. The principles of ordinary mechanics, whether classical or quantal, enable us to make reasonable predictions about the future state of a mechanical system from a precise knowledge of its initial state. On the other hand the principles of statistical mechanics enable us to make reasonable predictions about the future condition of a system, which may be expected to hold on the average, from an incomplete knowledge of its initial state.

Since our actual contacts with the physical world are such that we never do have the maximum knowledge of systems regarded as theoretically allowable, the idea of the precise state of a system is in any case an abstract limiting concept.

---

[1] The present approach follows closely that of TOLMAN. The author is grateful to the Clarendon Press for permission to borrow freely from the text of "The Principles of Statistical Mechanics" by TOLMAN, parts of which have been taken over almost verbatim in Sects. 58 and 59.

This is specially obvious and significant for systems having a large number of degrees of freedom. Hence the methods of ordinary mechanics really apply to somewhat highly idealized situations and the methods of statistical mechanics provide a significant supplement in the direction of decreased abstraction and closer correspondence between theoretical methods and actual experience.

The statistical mechanical procedure consists in abandoning any attempt to follow the changes taking place in a particular system and in studying instead the behaviour of a collection or *ensemble of systems* of similar structure to the system of actual interest distributed over a range of states. From a knowledge of the *average behaviour* of the systems in a *representative ensemble*, appropriately chosen so as to correspond to the partial knowledge that we do have about the system of interest, we can then make predictions of what may be expected on the average for the particular system which concerns us.

It may seem surprising that the methods of statistical mechanics should be so simple and effective as they actually prove to be. The substitution of a whole ensemble of systems of similar structure in place of a single system would not of itself lead to increased simplicity of treatment. The simplification is due to the circumstance that the average behaviour of a suitably chosen collection of systems may be much easier to treat than the precise behaviour of a single complicated system or a single system in a complicated environment.

59. Basic assumption. For the development of statistical mechanics in its modern quantal form it is necessary to introduce a fundamental assumption which may be formulated thus. The observable equilibrium properties of any system are calculated by averaging over all states of the representative ensemble of systems giving the same weight to every distinct state of the ensemble which is accessible under the given conditions. According to the laws of quantum mechanics the ensemble, assumed to be enclosed in a finite volume, can exist only in discrete states of definite discrete energies. To each of these energies belongs a definite number of linearly independent eigenfunctions of the ensemble. Each such distinct eigenfunction represents a distinct state of the ensemble. The averaging required is over all these states without favour provided only that under the given conditions the ensemble can occupy the state. As is so often the case in theoretical developments, this hypothesis is not itself suitable for direct empirical test. It has however the indirect check that is provided by the extensive agreement found between the deductive results obtained from it and the observational results obtained by experiment. Moreover it is the simplest hypothesis, if not the only simple hypothesis, consistent with the laws of quantum mechanics. It is a consequence of these laws that the probability w_{ik} per unit time for a system known to be in the state i to change into the state k is equal to the probability w_{ki} per unit time for the system known to be in the state k to change into the state i. In view of this symmetrical relationship between pairs of states any hypothesis for weighting unequally would seem on the face of it unreasonable. Owing to the need for conservation of energy the states i and k referred to must of course be of equal energy. Consequently in a change involving emission or absorption of radiation the radiation must be included in the system.

60. Classical approximation. Liouville's theorem. It would take up too much space to discuss classical systems fully, but a few words about classical systems regarded as approximations to quantal systems may be useful. A classical mechanical system of f degrees of freedom may be described by f generalized coordinates $q_1, \ldots q_f$ and their f conjugate momenta $p_1, \ldots p_f$. An ensemble composed of N such systems may correspondingly be described by Nf generalized coordinates and their Nf conjugate momenta. It is convenient to have a quasi-geometrical language for describing the condition of each system in the ensemble and for describing the condition of the ensemble as a whole. For this purpose, corresponding to any system of f degrees of freedom we can construct a conceptual Euclidean space of $2f$ dimensions with $2f$ rectangular axes one for each of the coordinates $q_1, \ldots q_f$ and one for each of the momenta $p_1, \ldots p_f$ whose values would describe the condition of the system. Using the terminology of GIBBS we speak of such a conceptual space as a phase space for the kind of system under consideration. Since each quantum state of a system with f degrees of freedom corresponds to an element of classical phase space of extension h^f our fundamental assumption reduces in the classical limit to assigning to any element of phase space a weight proportional to its extension. Consequently an ensemble of total energy between E and $E + dE$ uniformly distributed over its possible quantum states may in the classical limit be represented by points uniformly distributed over that part of phase space corresponding to the prescribed energy interval.

It is desirable to verify that such a uniform distribution of representative points in phase space will persist. This persistence, known as Liouville's theorem, can be proved by classical mechanics as follows.

Consider the density ϱ of representative points in phase space at the point $q_1, \ldots q_f, p_1, \ldots p_f$. The number dN of representative points in the element of phase space $dq_1 \ldots dq_f \, dp_1 \ldots dp_f$ is given by

$$dN = \varrho \, dq_1 \ldots dq_f \, dp_1 \ldots dp_f. \tag{60.1}$$

This number will in general be changing. If we consider the two planes perpendicular to the q_1 axis located at q_1 and $q_1 + dq_1$ the number of representative points entering through the first of these surfaces per unit time is

$$\varrho \, \dot{q}_1 \, dq_2 \ldots dq_f \, dp_1 \ldots dp_f. \tag{60.2}$$

The corresponding number of representative points leaving through the opposite plane is

$$\left(\varrho \, \dot{q}_1 + \varrho \, \frac{\partial \dot{q}_1}{\partial q_1} \, dq_1 + \frac{\partial \varrho}{\partial q_1} \, \dot{q}_1 \, dq_1 \right) dq_2 \ldots dq_f \, dp_1 \ldots dp_f. \tag{60.3}$$

The consequent increase in dN per unit time is

$$- \left(\varrho \, \frac{\partial \dot{q}_1}{\partial q_1} + \frac{\partial \varrho}{\partial q_1} \, \dot{q}_1 \right) dq_1 \ldots dq_f \, dp_1 \ldots dp_f \tag{60.4}$$

and the consequent increase in ϱ is

$$- \left(\varrho \, \frac{\partial \dot{q}_1}{\partial q_1} + \frac{\partial \varrho}{\partial q_1} \, \dot{q}_1 \right). \tag{60.5}$$

Summing all contributions from the motion in all directions of the representative points we obtain

$$\left(\frac{\partial \varrho}{\partial t} \right)_{q,p} = - \sum_{i=1}^{f} \left[\varrho \left(\frac{\partial \dot{q}_i}{\partial q_i} + \frac{\partial \dot{p}_i}{\partial p_i} \right) + \frac{\partial \varrho}{\partial q_i} \, \dot{q}_i + \frac{\partial \varrho}{\partial p_i} \, \dot{p}_i \right]. \tag{60.6}$$

In view of the canonical equations of motion

$$\dot{q}_i = \frac{\partial H}{\partial p_i}, \qquad \dot{p}_i = - \frac{\partial H}{\partial q_i} \tag{60.7}$$

where $H(q_i, p_i)$ is the Hamiltonian, we have

$$\frac{\partial \dot{q}_i}{\partial q_i} + \frac{\partial \dot{p}_i}{\partial p_i} = \frac{\partial}{\partial q_i} \frac{\partial H}{\partial p_i} - \frac{\partial}{\partial p_i} \frac{\partial H}{\partial q_i} = 0 \tag{60.8}$$

and formula (60.6) reduces to

$$\left(\frac{\partial \varrho}{\partial t}\right)_{q, p} + \sum_{i=1}^{f} \left(\frac{\partial \varrho}{\partial q_i} \dot{q}_i + \frac{\partial \varrho}{\partial p_i} \dot{p}_i\right) = 0. \tag{60.9}$$

This may be written in the alternative form

$$\frac{d\varrho}{dt} = 0 \tag{60.10}$$

where d/dt relates to the change of density in the neighbourhood of any selected representative point moving through phase space. It follows that any distribution which is at any time uniform over all parts of phase space corresponding to equal energy will remain so. This is Liouville's theorem which played an important part in the discovery of the correct averaging rule in classical theory. The averaging rule in quantum theory is much simpler and does not require Liouville's theorem.

61. Thermal equilibrium. We shall now apply the general principles of statistical mechanics as enunciated above to study the equilibrium properties of a pair of different systems denoted by A and B respectively. There is no restriction on the nature or size of either system. Possible examples are (i) a crystal of copper, (ii) a two phase liquid-vapour system composed of water and air, (iii) a single molecule of CO_2, (iv) a harmonic oscillator of given frequency. In accordance with the procedure already described instead of considering a single system A and a single system B we shall consider an ensemble composed of a large number N_A of similar systems A and a large number N_B of similar systems B. We regard the individual systems as labelled. We assume that the systems can exchange energy but that the interaction between the systems is so weak that we may with sufficient accuracy describe each system as being in a definite state with definite energy. We may then specify the state of the ensemble by specifying the state of each A system and of each B system. The first question which we shall study is this. Given that the total energy of the whole ensemble has a value E what is the average value of the fraction of all the A systems in a particular state p? The answer is clearly equal to $\overline{N}_p/N_A$ where $\overline{N}_p$ denotes the number of systems A in the state p averaged over all states of the ensemble of total energy E.

We suppose the states of an A system to be numbered and we denote such a state in general by r and its energy by E_r. Similarly we denote a state of a B system in general by s and its energy by E_s. Consider now the sum $\sum_r z^{E_r}$ which contains one term corresponding to each state r of a system A; and similarly the sum $\sum_s z^{E_s}$. Now consider the multiple product

$$\left(\sum z^{E_r}\right)^{N_A} \left(\sum_s z^{E_s}\right)^{N_B}. \tag{61.1}$$

When this product is expanded there will be one term corresponding to each possible state of the ensemble described by the states of the N_A constituent systems A and N_B constituent systems B. Moreover the power to which z is raised in any term is equal to the sum of the energies of the constituent systems and consequently equal to the total energy of the ensemble. If then we pick out all the terms in which z is raised to the power E, these terms correspond to all the accessible states of the ensemble consistent with the condition that its overall energy has the prescribed value E. The number of these terms is equal to the number of accessible states of the ensemble consistent with an overall energy E. We may therefore say that this number C of accessible states of the ensemble is equal to the coefficient of z^E in the above multiple product. We now choose a unit of energy small enough so that the ratio of each E_r and each E_s to this unit may with sufficient accuracy be regarded as integral. If we now regard z as a complex variable then by Cauchy's theorem we have

$$C = \frac{1}{2\pi i} \int \left(\sum_r z^{E_r}\right)^{N_A} \left(\sum_s z^{E_s}\right)^{N_B} z^{-E-1}\, dz \qquad (61.2)$$

where the integration is round any contour within the circle of convergence of the integrand circulating once counterclockwise round the origin.

By similar reasoning the number of ensemble states consistent with an overall energy E in which a chosen A system is in the particular state p of energy E_p is equal to the coefficient of z^E in the multiple product

$$z^{E_p}\left(\sum_r z^{E_r}\right)^{N_A-1}\left(\sum_s z^{E_s}\right)^{N_B} . \qquad (61.3)$$

Consequently using Cauchy's theorem we have

$$\frac{C\,\overline{N_p}}{N_A} = \frac{1}{2\pi i} \int z^{E_p}\left(\sum_r z^{E_r}\right)^{N_A-1}\left(\sum_s z^{E_s}\right)^{N_B} z^{-E-1} dz . \qquad (61.4)$$

We recall that the numbers N_A, N_B of the systems A, B respectively constituting the ensemble have to be large numbers, in fact the larger the better. Consequently we require to study the behaviour of these contour integrals in the limit as $N_A, N_B \to \infty$. This can be done rigorously by the method of steepest descents. We shall merely outline the procedure.

Consider the factor multiplying dz/z on the positive real axis. It tends to infinity as $z \to 0$ and as $z \to 1$. Somewhere between, say at $z = \vartheta$, there is a unique minimum. For the contour of integration take the circle of radius ϑ and centre the origin, that is $z = \vartheta e^{i\alpha}$. Then for values of z on this circle, if N_A, N_B and E are large, the modulus of this factor has a strong maximum at $z = \vartheta$, $\alpha = 0$. Owing also to the fact that the differential coefficient of this factor vanishes at $\alpha = 0$, the complex terms there are trivial and the whole effective contribution to the integral comes from near this point. This argument is unaffected by extra factors in the integrand which are not raised to a high power such as N_A, N_B or E. Consequently we may, if we like, take such factors outside the sign of integration if in them we replace z by ϑ. Thus we may replace formula (61.4) by

$$\frac{C\,\overline{N_p}}{N_A} = \frac{\vartheta^{E_p}}{2\pi i \sum_r \vartheta^{E_r}} \int \frac{\left(\sum_r z^{E_r}\right)^{N_A}\left(\sum_s z^{E_s}\right)^{N_B} dz}{z^{E+1}} = \frac{C\,\vartheta^{E_p}}{\sum_r \vartheta^{E_r}} . \qquad (61.5)$$

Consequently, the average value of the fraction of systems A in the state p is given by

$$\frac{\overline{N_p}}{N_A} = \frac{\vartheta^{E_p}}{\sum_r \vartheta^{E_r}} . \qquad (61.6)$$

By precisely the same argument we see that the fraction of B systems in a chosen state t of energy E_t is given by

$$\frac{\overline{N_t}}{N_B} = \frac{\vartheta^{E_t}}{\sum\limits_{s} \vartheta^{E_s}}. \tag{61.7}$$

In both these formulae ϑ is defined as the value of z at which the remaining factor multiplying dz/z in the integrand has a minimum on the real axis. Consequently ϑ is determined by

$$N_A \frac{\sum\limits_{r} E_r \vartheta^{E_r}}{\sum\limits_{r} \vartheta^{E_r}} + N_B \frac{\sum\limits_{s} E_s \vartheta^{E_s}}{\sum\limits_{s} \vartheta^{E_s}} = E. \tag{61.8}$$

By comparing this with formulae (61.6) and (61.7) we see that

$$\sum\limits_{r} \overline{N_r} E_r + \sum\limits_{s} \overline{N_s} E_s = E \tag{61.9}$$

which confirms that our formulae for the average numbers N_r, N_s are consistent, as they must be, with the requirement that the energy of the ensemble is equal to the sum of the energies of all the constituent systems.

62. Canonical ensemble. Absolute temperature. The most striking feature of ϑ is that the average distribution over states of the system A and that of the system B are both determined by the same value of ϑ. It is moreover obvious that if instead of two systems A, B we had considered several systems A, B, C, D, ... the average distribution of each system would be determined by the same value of ϑ. It is evident that ϑ has the qualitative property of temperature, namely that the several systems at equilibrium have the same temperature. We may further regard ϑ as a measure of the temperature of the whole ensemble and we may regard the ensemble as a temperature bath or thermostat. We may now usefully change our point of view and regard the overall energy E as determined by the temperature ϑ instead of the converse. In other words we may take ϑ instead of E as one of the prescribed properties of the systems A, B, C, D, ... being considered. To recapitulate, if temperature is measured on the scale here denoted by ϑ, then the average fraction of systems A in the state p of energy E_p is equal to $\vartheta^{E_p}/\sum\limits_{r} \vartheta^{E_r}$ regardless of how many other systems B, C, D, ... may also be in the same temperature bath. An ensemble of systems thus distributed over the accessible states is called a *canonical ensemble*. The canonical ensemble is the statistical representation of a temperature bath in classical thermodynamics.

The temperature scale ϑ is related to the more usual temperature scale T by the identity

$$\vartheta = \exp\left(-1/kT\right) \tag{62.1}$$

where k is a universal arbitrarily chosen constant. This temperature T will later be identified with Kelvin's thermodynamic temperature usually called absolute temperature. It is evident that the value assigned to the constant k determines the size of the temperature unit called the degree. In order that T should have the value $273.16°$ K at the triple point of water, in accordance with international convention, the value that must be assigned to k is 1.3803×10^{-16} erg deg^{-1}. The average fraction of systems A in the state p of energy E_p is equal to $\exp\left(-E_p/kT\right)/\sum\limits_{r} \exp\left(-E_r/kT\right)$.

63. Partition function. The sum Q_A defined by

$$Q_A \equiv \sum_r \vartheta^{E_r} \equiv \sum_r e^{E_r/kT} \tag{63.1}$$

which will occur frequently in later discussions is called the *partition function* (in German: *Zustandsumme*) of the system A.

From formula (61.6) for the average value of the fraction of systems A in the state p it follows that the average value of the energy of a system A is

$$\begin{aligned}
\overline{E}_A &= \frac{1}{N_A} \sum_r \overline{N}_r E_r = \frac{\sum\limits_r E_r \exp\left(-E_r/kT\right)}{\sum\limits_r \exp\left(-E_r/kT\right)} \\
&= kT^2 \frac{\partial}{\partial T}\left\{\ln \sum_r \exp\left(-E_r/kT\right)\right\} \\
&= kT^2 \frac{\partial \ln Q}{\partial T}.
\end{aligned} \tag{63.2}$$

64. Strains and stresses. Work. Up to this point it has been tacitly assumed that the energy E_r of the particular quantum state labelled r is an invariable constant. This is not so since in general E_r will depend on certain macroscopic coordinates in particular those which determine the boundary of the system A being considered. A set of such macroscopic coordinates that are independently variable may be considered as defining the independent strains of the system. Such coordinates of strain will be denoted by y_i. In general a change in y_i implies a change in E_r so that we may write

$$dE_r = \sum_i \frac{\partial E_r}{\partial y_i} dy_i \tag{64.1}$$

and $\partial E_r/\partial y_i$ is the stress corresponding to the strain determined by the coordinate y_i when the system A is in the state r. For the sake of brevity and simplicity we shall now restrict ourselves to the variation of only one of the coordinates y_i and shall denote this coordinate simply as y. We accordingly write

$$dE_r = \frac{\partial E_r}{\partial y} dy = Y_r \, dy \tag{64.2}$$

where we have written Y_r for $\partial E_r/\partial y$.

As already stated Y_r is the stress or generalized force corresponding to the generalized coordinate y when the system is in the quantum state r. We have now to consider the question what is the average value Y_A of this stress when the system A is maintained in a temperature bath. In accordance with our standard procedure instead of a single system A we consider a canonical ensemble of systems A. The average value $\overline{Y}_A$ is then obtained by multiplying Y_r by the average fraction of systems in the ensemble in the state r and summing over all states r. We thus obtain using (61.6)

$$\begin{aligned}
\overline{Y}_A &= \sum_r \frac{\overline{N}_r}{N_A} \frac{\partial E_r}{\partial y} \\
&= \sum_r \frac{\partial E_r}{\partial y} \frac{\exp\left(-E_r/kT\right)}{\sum\limits_s \exp\left(-E_s/kT\right)}.
\end{aligned} \tag{64.3}$$

This may be rewritten more compactly as

$$\overline{Y}_A = - kT \frac{\partial}{\partial y} \left\{ \ln \sum_r \exp\left(- E_r/kT\right) \right\}$$

$$= - kT \frac{\partial \ln Q_A}{\partial y} \tag{64.4}$$

where Q_A is the partition function of the system A at the temperature T.

65. Free energy. It is expedient to define a function $F_A(T, y)$ by

$$F_A(T, y) = - kT \ln Q_A. \tag{65.1}$$

We may now rewrite formula (64.4) for $\overline{Y}_A$ as

$$\overline{Y}_A = \frac{\partial F}{\partial y} \tag{65.2}$$

and formula (63.2) for $\overline{E}_A$ as

$$\overline{E}_A = - T^2 \frac{\partial (F_A/T)}{\partial T} = F_A - T \frac{\partial F_A}{\partial T}. \tag{65.3}$$

We may usefully compare this with the Gibbs-Helmholtz formula (19.11) of classical thermodynamics

$$U = F - T \frac{\partial F}{\partial T}. \tag{65.4}$$

Since it is eminently reasonable to identify the measured force Y with the statistical average force $\overline{Y}_A$ and the thermodynamic total energy U with the statistical average energy $\overline{E}_A$, it appears that the statistically defined function $F_A(T, y)$ has the properties of the thermodynamic free energy F. Incidentally we have confirmed that the absolute temperature of classical thermodynamics is the same as the temperature defined statistically by the relation $\vartheta = \exp(-1/kT)$.

66. Thermodynamic relations. Now that we have correlated the statistically defined average quantities with their classical thermodynamic counterparts we may drop the distinction between the two sets of quantities. We may at the same time drop the subscript A for the system A.

To recapitulate we have from (65.2) and (65.4)

$$\frac{\partial F}{\partial y_i} = Y_i \qquad \frac{\partial F}{\partial T} = \frac{F - U}{T} \tag{66.1}$$

and consequently

$$dF = \frac{F - U}{T} dT + \sum_i Y_i \, dy_i. \tag{66.2}$$

It is important at this stage to emphasize that the statistical definitions of T, F, U, Y_i do not require the system A to be macroscopic. Even if A is a single molecule in a temperature bath of temperature T confined within a boundary defined by coordinates y_i, then F, U and Y_i are functions of T, y_i completely defined by averages.

67. Macroscopic systems. It has already been stressed that all the formulae obtained hitherto, including those defining the thermodynamic functions are valid for any system large or small, macroscopic or molecular, in a temperature bath with a specified temperature. We now turn to study some striking differences between systems having a small number of degrees of freedom and systems having a large number, in particular macroscopic systems in which the number of degrees of freedom usually exceeds 10^{20}. These differences can be brought to light by studying in detail the properties of systems composed of one or several or many harmonic oscillators. A characteristic property of a harmonic oscillator is that the energies of the quantum states are evenly spaced. Owing to this property the detailed algebraic formulae are simpler than for other systems, but the most important conclusions reached are of general validity.

It is convenient here to take as the zero of energy that of the lowest state of the harmonic oscillator and to denote the separation between consecutive energy levels by ε. The energies of the states are then defined by $E_r = r\varepsilon$ with $r = 0, 1, 2, \ldots$ The partition function, which we shall denote by Q_1, is according to (63.1)

$$Q_1 = \sum_{r=0}^{\infty} e^{-r\varepsilon/kT} = (1 - e^{-\varepsilon/kT})^{-1}. \tag{67.1}$$

The average fraction of time that the oscillator has the energy $r\varepsilon$ is proportional to $\exp(-r\varepsilon/kT)$ which decreases monotonically as r increases.

We now turn to consider a system consisting of a pair of similar but distinguishable harmonic oscillators. The distinguishability might be due to different positions or to different directions of oscillation. We shall denote the partition function of this pair of oscillators by Q_2. Since the state of the pair of oscillators is conveniently described by specifying the state of each oscillator and since the energy of the pair is equal to the sum of the individual energies, it follows that Q_2 is given by

$$Q_2 = \left(\sum_{r=0}^{\infty} e^{-r\varepsilon/kT} \right)^2 = (1 - e^{-\varepsilon/kT})^{-2}$$
$$= 1 + 2 e^{-\varepsilon/kT} + 3 e^{-2\varepsilon/kT} + \cdots$$
$$= \sum_{r=0}^{\infty} (r + 1) e^{-r\varepsilon/kT}. \tag{67.2}$$

We can immediately extend this reasoning to a system composed of $N-1$ harmonic oscillators and we shall denote the partition function by Q_{N-1}. The choice of $N-1$ rather than N is purely one of convenience the formulae being slightly simpler. We obtain

$$Q_{N-1} = \left(\sum_{r=0}^{\infty} e^{-r\varepsilon/kT} \right)^{N-1} = (1 - e^{-\varepsilon/kT})^{-(N-1)}$$
$$= \sum_{r=0}^{\infty} \frac{(N+r)!}{N!\,r!} e^{-r\varepsilon/kT}. \tag{67.3}$$

The factor $(N+r)!/N!\,r!$ is the number of states of the system having the same energy $r\varepsilon$. This situation is described by stating that the energy level $r\varepsilon$ is $(N+r)!/N!\,r!$ *fold degenerate*. If we consider the fraction of time that the system has an energy $r\varepsilon$, in contrast to the fraction of time the system spends in one particular state of energy $r\varepsilon$, we see that it is proportional to q_r defined by

$$q_r = \frac{(N+r)!}{N!\,r!} e^{-r\varepsilon/kT}. \tag{67.4}$$

We note that whereas $e^{-r\varepsilon/kT}$ is a monotonically decreasing function of r, the factor $(N+r)!/N!\,r!$ is a monotonically increasing function. Consequently their product q_r has a unique maximum. We shall now study this maximum in the case that $N \gg 1$. It can be verified that in the neighbourhood of the maximum

$r \gg 1$. We may therefore use Stirling's approximation for the factorials. Differentiating with respect to r and denoting the value of r at the maximum by m we deduce from (67.4)

$$\frac{1}{q_r}\frac{\partial q_r}{\partial r} = \ln\frac{N+r}{r} - \frac{\varepsilon}{kT} = 0\,, \qquad r = m \qquad (67.5)$$

and consequently

$$\frac{N}{r} = e^{\varepsilon/kT} - 1\,, \qquad r = m \qquad (67.6)$$

thus verifying that m is of the same order of magnitude as N. Substituting the found value of m for r in formula (67.4) for q_r and using Stirling's theorem we obtain

$$\begin{aligned}
\ln q_m &= N\ln\frac{N+m}{N} + m\ln\frac{N+m}{m} - \frac{m\,\varepsilon}{kT}\\
&= N\ln\frac{N+m}{N}\\
&= -N\ln(1 - e^{-\varepsilon/kT})\\
&= \ln\{(1 - e^{-\varepsilon/kT})^{-N}\}\\
&= \ln Q_{N-1}\,.
\end{aligned} \qquad (67.7)$$

We have thus reached the remarkable conclusion that provided N is large enough to allow the use of Stirling's approximation then the logarithm of the partition function Q_{N-1} may with sufficient accuracy be replaced by the logarithm of its maximum term q_m. This is not the same thing as saying that Q_{N-1} is equal to its maximum term. A more precise analysis shows that provided $N \gg 1$

$$\frac{Q_{N-1}}{q_m} = \alpha\,N^{\frac{1}{2}} \qquad (67.8)$$

where α is nearly independent of N. Taking logarithms we have

$$\ln\frac{Q_{N-1}}{q_m} = \frac{1}{2}\ln N + \ln\alpha \qquad (67.9)$$

which when $N \gg 1$ is negligible compared with $\ln q_m$ which we have seen is of the order of magnitude of N. This is so important and fundamental for our subsequent discussion that consideration of a numerical example may be worthwhile. We shall take $N = 10^{20}$ a rather small value for a macroscopic system. The value of ε/kT is unimportant. For simplicity we shall take $e^{-\varepsilon/kT} = \frac{1}{2}$. We have then

$$\tfrac{1}{2}\ln N = 23 \qquad (67.10)$$

and even if we assign to α the improbably high value of 10^6 we still have

$$\tfrac{1}{2}\ln N + \ln\alpha = 37 \qquad (67.11)$$

which is utterly trivial compared with

$$\ln q_m = N\ln 2 = 10^{20}\ln 2 = 7\times 10^{19}\,. \qquad (67.12)$$

This confirms that with trivial inaccuracy we may replace $\ln Q$ by $\ln q_m$.

This striking conclusion is nowise peculiar to a system composed of harmonic oscillators, chosen as illustrative for the sake of algebraic simplicity but is common to all systems having a large number of degrees of freedom and especially macroscopic systems. If the partition function of a macroscopic system is expressed as

$$Q = \sum_{E_r} q_r = \sum_r g_r\, e^{-E_r/kT} \tag{67.13}$$

and q_r has its maximum value at $r=m$ then quite generally $\ln q_m$ is of order of magnitude N while $\ln (Q/q_m)$ is of order of magnitude $\ln N$. Consequently $\ln (Q/q_m)$ is entirely negligible compared with $\ln q_m$ and so without sensible inaccuracy $\ln Q$ may be replaced by $\ln q_m$.

The free energy F of the system is then according to (65.1) given by

$$F = - kT \ln Q = - kT \ln q_m = - kT \ln g_m + E_m \tag{67.14}$$

and the total energy U according to (65.4) by

$$U = F - T\,\frac{\partial F}{\partial T} = E_m \tag{67.15}$$

which is physically obvious since the system spends effectively all its time in states with energy differing inappreciably from E_m.

For a system in a temperature bath of given temperature T, if and only if the system is macroscopic, its instantaneous energy E never deviates appreciably from the thermodynamic total energy U. Hence for a macroscopic system we may, for given values of the strain coordinates y_i, regard the energy $U=E$ as determining the temperature instead of the temperature determining the (average) energy $\overline{E}=U$.

68. Entropy. We shall now give some alternative definitions of entropy and shall show that for a macroscopic system they are equivalent. For a system, having only one or a small number of degrees of freedom in a temperature bath of temperature T the only satisfactory definition S_T of the entropy is

$$S_T = \frac{U-F}{T} = -\frac{\partial F}{\partial T} = \frac{\partial}{\partial T}\,(kT \ln Q)$$
$$= k \ln Q + kT\,\frac{\partial \ln Q}{\partial T}. \tag{68.1}$$

This definition also holds good for a macroscopic system at a temperature T. An apparently quite different definition S_U of the entropy of a macroscopic system of specified energy U is

$$S_U = k \ln g_m \tag{68.2}$$

where g_m is the number of states of energy $E_m = U$ or the degree of degeneracy of the energy level E_m. But since the system is assumed to be macroscopic we have with negligible inaccuracy

$$S_T = \frac{\partial}{\partial T}\,(kT \ln Q) = \frac{\partial}{\partial T}\,(kT \ln q_m)$$
$$= \frac{\partial}{\partial T}\,(kT \ln g_m - E_m) = k \ln g_m = S_U. \tag{68.3}$$

There are yet other definitions of entropy but they are less important. Provided the system is macroscopic they are all essentially equivalent.

69. Heat. This is an appropriate place to derive the fundamental property of entropy, namely its relation to heat absorbed. We shall denote heat absorbed by the system by q and work done on the system by w. Conservation of energy

requires that
$$dU = q + w. \tag{69.1}$$
The work w is given by
$$w = \sum_i Y_i \, dy_i = \sum_i \frac{\partial F}{\partial y_i} \, dy_i = dF - \frac{\partial F}{\partial T} \, dT$$
$$= d(U - TS) + S \, dT = dU - T \, dS \tag{69.2}$$
and consequently
$$q = dU - w = T \, dS \tag{69.3}$$
in agreement with formula (7.6) of classical thermodynamics.

70. Approach to equilibrium. Up to this point it has been tacitly assumed that the system under consideration was in equilibrium. We must now consider briefly what happens when a system out of equilibrium moves into equilibrium. We shall not discuss this exhaustively, but shall concentrate our attention on two important examples.

We first consider two separate systems A and B in the same temperature bath at temperature T but otherwise quite independent. If their partition functions under these conditions are Q_A and Q_B, then the partition function of the pair regarded as a single system is under the same conditions $Q_A Q_B$. Now suppose that the systems are allowed to interact in any way, their temperature and total volume being kept constant. For example the boundary separating the two systems might move so that one system contracts and the other expands; or the material contents of the two systems might mix; or there might be a chemical reaction between the two systems. If such an interaction, or indeed any other kind of interaction, occurs then new states become accessible to the pair of systems which were previously inaccessible and so the partition function Q_{AB} of the pair of systems will contain new terms in addition to those of $Q_A Q_B$. Consequently
$$Q_{AB} > Q_A Q_B. \tag{70.1}$$
Using a similar notation for the initial and final free energies, we have by (65.1)
$$F_A = - kT \ln Q_A , \tag{70.2}$$
$$F_B = - kT \ln Q_B , \tag{70.3}$$
$$F_{AB} = - kT \ln Q_{AB} , \tag{70.4}$$
$$F_{AB} < F_A + F_B \tag{70.5}$$
and we see that any isothermal interaction results in a decrease of the free energy in agreement with classical thermodynamics. Strictly speaking there is no exception to the last inequality, but in effect the decrease in the free energy will be inappreciable in the special case that the two systems A and B happen to be in equilibrium in the initial condition. When they are allowed to interact the new states which then become accessible to the pair of systems are rarely occupied states while the frequently occupied states of the combined system at equilibrium are the same states as those occupied before the interaction took place.

The other example which we shall consider is that of two macroscopic systems A and B initially isolated and then brought together in such a manner as to be able to exchange energy without any change involving the performance of work. Initially the system A has a temperature T_A, a total energy U_A and an entropy S_A and the system B has a temperature T_B, a total energy U_B and a temperature T_B.

Finally the pair of systems have the same temperature T_{AB}, a total energy U_{AB} and a total entropy S_{AB}. Then owing to conservation of energy we have

$$U_{AB} = U_A + U_B \qquad (70.6)$$

and it will now be proved that

$$S_{AB} > S_A + S_B. \qquad (70.7)$$

Since the systems are macroscopic we have by (68.2)

$$S_A = k \ln g_A \qquad (70.8)$$

where g_A denotes the number of states of energy U_A accessible to the isolated system A. Similarly and with a similar notation

$$S_B = k \ln g_B. \qquad (70.9)$$

The entropy of the pair of systems is initially

$$S_A + S_B = k \ln g_A + k \ln g_B = k \ln (g_A \, g_B). \qquad (70.10)$$

The final entropy is likewise

$$S_{AB} = k \ln g_{AB} \qquad (70.11)$$

where g_{AB} denotes the number of accessible states for the pair of systems of total energy $U_{AB} = U_A + U_B$. But this set of states evidently includes all the states such that the energy of A is U_A and the energy of B is U_B as well as numerous other states of total energy $U_A + U_B$. It is thus evident that

$$g_{AB} > g_A \, g_B. \qquad (70.12)$$

From (70.10) to (70.12) it follows immediately that

$$S_{AB} > S_A + S_B \qquad (70.13)$$

in accordance with classical thermodynamics.

Strictly there is no exception to this inequality, but actually the increase in entropy is inappreciable in the special case that the initial temperature T_A, T_B happen to be equal. In this case the highly populated states in the initial and final conditions are identical.

From these two examples we see that there is nothing mysterious in the increasing property of entropy at constant energy or in the decreasing property of free energy at constant temperature. These properties are merely manifestations of increase in the number of accessible states.

D. Statistical mechanics of molecules.

71. System composed of molecules. In the previous chapter we have used statistical mechanics to study the average behaviour of a system and to express this behaviour in terms of the accessible states of the system and the energy values of these states. We considered both macroscopic and molecular systems. We shall now study in greater detail the behaviour of a macroscopic system composed of a large number of molecules. The word molecule will be used to include atoms, ions, electrons, or any other units which are essentially independent so that the state of the whole system may be described by a specification of the states of all the individual molecules. The molecules can interchange energy but the interaction between them is assumed to be so weak that the energy of the whole system may be taken to be the sum of the energies of the individual molecules.

72. Fermi-Dirac and Bose-Einstein statistics. We recall the law of quantum theory that the allowed eigenfunctions of the system must be symmetrical (antisymmetrical) with respect to all identical molecules composed of an even (odd) number of fundamental particles namely protons, neutrons and electrons.

Let us now consider a system composed of N_A identical even molecules A and N_B identical odd molecules B all free to move in the same enclosure. Let N_r denote the number of molecules A in a quantum state r of energy E_r and let N_s denote the number of molecules B in a quantum state s of energy E_s. Then there exists one quantum state of the system for each distinct set of values of N_r and N_s subject to the pair of restrictions

$$\sum_r N_r = N_A , \tag{72.1}$$

$$\sum_s N_s = N_B \tag{72.2}$$

and also the restriction expressing Pauli's exclusion principle for the odd molecules only, namely

$$\text{every } N_s = 0 \quad \text{or} \quad 1 . \tag{72.3}$$

If we specify that the energy of the system is E we have the further restriction

$$\sum_r N_r E_r + \sum_s N_s E_s = E . \tag{72.4}$$

Consider now the product of sums

$$\prod_r \{1 + x\, z^{E_r} + (x\, z^{E_r})^2 + (x\, z^{E_r})^3 + \cdots\} = \prod_r (1 - x\, z^{E_r})^{-1} . \tag{72.5}$$

When this product is expanded there will be one term corresponding to each conceivable set of values of N_r. The power to which x is raised represents $\sum_r N_r$ and the power to which z is raised represents $\sum_r N_r E_r$. Consider also the product

$$\prod_s (1 + y\, z^{E_s}) . \tag{72.6}$$

When this product is expanded there will be one term corresponding to each conceivable set of values of N_s, subject to the restriction that every N_s is either zero or unity. The power to which y is raised represents $\sum_s N_s$ and the power to which z is raised represents $\sum_s N_s E_s$. Finally consider the continued product

$$\prod_r (1 - x\, z^{E_r})^{-1} \prod_s (1 + y\, z^{E_s}) . \tag{72.7}$$

When this product is expanded there will be one term corresponding to each conceivable set of values of N_r and each conceivable set of values of N_s subject to the restriction that every N_s is either zero or unity. The power to which x is raised represents $\sum_r N_r$, the power to which y is raised represents $\sum_s N_s$ and the power to which z is raised represents $\sum_r N_r E_r + \sum_s N_s E_s$. Consequently the number C of states of the system, composed of N_A molecules A and N_B molecules B, having a total energy E is equal to the coefficient of

$$x^{N_A}\, y^{N_B}\, z^{E} \tag{72.8}$$

in the expansion of the product. We now regard x, y, z as complex variables. Using Cauchy's theorem three times over we have for the number C of states of the system having a total energy E

$$C = \left(\frac{1}{2\pi i}\right)^3 \iiint \frac{dx\,dy\,dz}{x^{N_A+1}\,y^{N_B+1}\,z^{E+1}} \prod_r (1 - x\,z^{E_r})^{-1} \prod_s (1 + y\,z^{E_s}) \qquad (72.9)$$

where each integration is round any contour within the circle of convergence of the integrand circulating once counterclockwise round the origin.

The product

$$\prod_r (1 - x\,z^{E_r})^{-1} \prod_s (1 + y\,z^{E_s}) \qquad (72.10)$$

contains a factor

$$(1 - x\,z^{E_p})^{-1} = 1 + x\,z^{E_p} + (x\,z^{E_p})^2 + (x\,z^{E_p})^3 + \cdots \qquad (72.11)$$

and no other factor countaining E_p. If we replace this factor by

$$x\,z^{E_p} + 2\,(x\,z^{E_p})^2 + 3\,(x\,z^{E_p})^3 + \cdots = x\,z^{E_p}\,(1 - x\,z^{E_p})^{-2} \qquad (72.12)$$

we obtain

$$x\,z^{E_p}\,(1 - x\,z^{E_p})^{-1} \prod_r (1 - x\,z^{E_r})^{-1} \prod_s (1 + y\,z^{E_s}). \qquad (72.13)$$

We have by this change multiplied by N_p every term representing the occupation of the state p by a number N_p of molecules A. If then we pick the coefficient of

$$x^{N_A}\,y^{N_B}\,z^{E} \qquad (72.14)$$

in this new product it will be equal to $C\,\overline{N_p}$ where $\overline{N_p}$ denotes the average value of N_p when the whole system has an energy E. Using Cauchy's theorem as before we obtain

$$C\,\overline{N_p} = \left(\frac{1}{2\pi i}\right)^3 \iiint \frac{dx\,dy\,dz}{x^{N_A+1}\,y^{N_B+1}\,z^{E+1}} \left\{ x\,z^{E_p}\,(1 - x\,z^{E_p})^{-1} \right\} \times$$
$$\times \prod_r (1 - x\,z^{E_r})^{-1} \prod_s (1 + y\,z^{E_s}). \qquad (72.15)$$

By precisely similar reasoning if N_t' denotes the number of molecules B in a particular state t then the average value $\overline{N_t'}$ when the whole system has an energy E is given by

$$C\,\overline{N_t'} = \left(\frac{1}{2\pi i}\right)^3 \iiint \frac{dx\,dy\,dz}{x^{N_A+1}\,y^{N_B+1}\,z^{E+1}} \left\{ y\,z^{E_t}\,(1 + y\,z^{E_t})^{-1} \right\} \times$$
$$\times \prod_r (1 - x\,z^{E_r})^{-1} \prod_s (1 + y\,z^{E_s}). \qquad (72.16)$$

Since the system is macroscopic N_A, N_B, E are large and we may therefore use the method of steepest descents as in the previous chapter. The factor multiplying $dx\,dy\,dz/xyz$ in the integral for C has for real values of x, y, z a unique maximum at a point $\lambda_A, \lambda_B, \vartheta$. Further the contours of integration can be chosen to pass through this point so that on the contours the integrand has a single dominant maximum at this point whose neighbourhood contributes the whole effective value of the integral. Such a point is called a col or saddle-point. If then we substitute $\lambda_A, \lambda_B, \vartheta$ for x, y, z in any extra factors in the integrand, we may take these factors outside the sign of integration. It is evident as it was in the previous chapter that ϑ is a measure of temperature and we again replace ϑ by $e^{-\frac{1}{kT}}$. We thus obtain from (72.9), (72.15) and (72.16)

$$\overline{N}_p = \frac{\lambda_A \, e^{-E_p/kT}}{1 - \lambda_A \, e^{-E_p/kT}}, \tag{72.18}$$

$$\overline{N}_t' = \frac{\lambda_B \, e^{-E_t/kT}}{1 + \lambda_B \, e^{-E_t/kT}} \tag{72.19}$$

or alternatively

$$\frac{\overline{N}_p}{1 + \overline{N}_p} = \lambda_A \, e^{-E_p/kT}, \quad \text{symmetrical eigenfunctions}, \tag{72.20}$$

$$\frac{\overline{N}_t'}{1 - \overline{N}_t'} = \lambda_B \, e^{-E_t'/kT}, \quad \text{antisymmetrical eigenfunctions}. \tag{72.21}$$

These two distribution laws are known by the names Bose-Einstein and Fermi-Dirac respectively.

The conditions determining the saddle-point λ_A, λ_B, $\vartheta = e^{-1/kT}$ are

$$\sum_r \frac{\lambda_A \, e^{-E_r/kT}}{1 - \lambda_A \, e^{-E_r/kT}} = N_A, \tag{72.22}$$

$$\sum_s \frac{\lambda_B \, e^{-E_s/kT}}{1 + \lambda_B \, e^{-E_s/kT}} = N_B, \tag{72.23}$$

$$\sum_r \frac{\lambda_A E_r \, e^{-E_r/kT}}{1 - \lambda_A \, e^{-E_r/kT}} + \sum_s \frac{\lambda_B E_s \, e^{-E_s/kT}}{1 + \lambda_B \, e^{-E_s/kT}} = E. \tag{72.24}$$

From (72.18) and (72.19) we see that these are equivalent to the necessary equalities

$$\sum_r \overline{N}_r = N_A, \tag{72.25}$$

$$\sum_s \overline{N}_s = N_B, \tag{72.26}$$

$$\sum_r \overline{N}_r E_r + \sum_s \overline{N}_s E_s = E. \tag{72.27}$$

It is evident that if there are more than two kinds of molecules each kind will obey either the Bose-Einstein or the Fermi-Dirac distribution according as a molecule is composed of an even or an odd number of primary particles.

We may abbreviate the two distribution laws (72.20) and (72.21) into the single formula

$$\frac{\overline{N}_p}{1 \pm \overline{N}_p} = \lambda \, e^{-E_p/kT} \tag{72.28}$$

where the plus sign relates to even molecules and the minus sign to odd molecules. This formula with the appropriate sign holds for any one kind of molecule, whatever other molecules may be present in the same enclosure provided, of course, that all interaction energies between molecules are neglected.

73. Chemical equilibrium. This distribution law can readily be extended to systems in which chemical changes can take place. We shall illustrate this by the simple case where there are three kinds of molecules A, B and AB. We denote the stoichiometric number of A molecules, whether free or bound to B, by N' and the stoichiometric number of B molecules by N''. We denote the number of free molecules A by N_A, the number of free molecules B by N_B and the number of molecules AB by N_{AB}. Initially we imagine N_A, N_B and N_{AB} to be fixed. The system then behaves just like those previously considered and we have distribution laws for the three kinds of molecules of the standard form expressed in terms of λ_A, λ_B and λ_{AB} respectively. Now suppose that by the introduction of a catalyst we enable the chemical change represented by

$$A + B \rightleftharpoons AB \tag{73.1}$$

to take place. The numbers N_A, N_B and N_{AB} are then no longer given constants but are only subject to the restrictions

$$N_A + N_{AB} = N', \tag{73.2}$$

$$N_B + N_{AB} = N''. \tag{73.3}$$

If then we express the total number of accessible states of the system for given values of N', N'' and the total energy E as a triple complex integral we obtain

$$C = \left(\frac{1}{2\pi i}\right)^3 \iiint \frac{dx\,dy\,dz}{x^{N'+1}\,y^{N''+1}\,z^{E+1}} \prod_r \left(1 \mp x\,z^{E_r}\right)^{\mp 1} \times$$

$$\times \prod_s \left(1 \mp y\,z^{E_s}\right)^{\mp 1} \prod_t \left(1 \mp x\,y\,z^{E_t}\right)^{\mp 1} \tag{73.4}$$

where the subscripts r, s, t relate to states of the molecules A, B, AB respectively. The noticeable feature of the integrand is that the last factor relating to AB molecules contains xy where the factor relating to A contains x and the factor relating to B contains y. When we apply the method of steepest descents we obtain the usual type of distribution for the A and B molecules while for the AB molecules we obtain a similar formula except that it contains $\lambda_A \lambda_B$ instead of λ_{AB}. Thus the distribution laws for the system at chemical equilibrium are obtained from those at fixed values of N_A, N_B, N_{AB} by merely substituting $\lambda_A \lambda_B$ for λ_{AB}. In other words the condition for equilibrium with respect to the reaction

$$A + B \rightleftharpoons AB \tag{73.5}$$

can be expressed by the relation

$$\lambda_{AB} = \lambda_A \lambda_B \tag{73.6}$$

or alternatively

$$\ln \lambda_A + \ln \lambda_B = \ln \lambda_{AB}. \tag{73.7}$$

This result can immediately be extended to more general chemical reactions. Thus the condition for chemical equilibrium with respect to the reaction

$$\nu_A A + \nu_B B + \cdots \rightleftharpoons \nu_L L + \nu_M M + \cdots \tag{73.8}$$

where ν_A, ν_B, ... are small integers (or simple fractions) is

$$\nu_A \ln \lambda_A + \nu_B \ln \lambda_B + \cdots = \nu_L \ln \lambda_L + \nu_M \ln \lambda_M + \cdots . \tag{73.9}$$

We now compare this equilibrium condition with that of classical thermodynamics which according to (25.8) and (25.9) may be expressed as

$$\nu_A \mu_A + \nu_B \mu_B + \cdots = \nu_L \mu_L + \nu_M \mu_M + \cdots . \tag{73.10}$$

It appears that $\ln \lambda$ is closely related to μ, in fact differing from μ only by a factor the same for every molecular species. We shall now establish this relation between $\ln \lambda$ and μ.

74. Grand partition function. Absolute activities. For the sake of brevity we again consider a system of only two kinds of molecules A and B. Formula (72.9) for the total number C of accessible states may be generalized to

$$C = \left(\frac{1}{2\pi i}\right)^3 \iiint \frac{dx\,dy\,dz}{x^{N_A+1}\,y^{N_B+1}\,z^{E+1}} \prod_r (1 \mp x\,z^{E_r})^{\mp 1} \prod_s (1 \mp y\,z^{E_s})^{\mp 1}. \qquad (74.1)$$

Let us denote by $\varXi$ the value at the col of the continued product in the integrand, namely

$$\varXi = \prod_r (1 \mp \lambda_A\,e^{-E_r/kT})^{\mp 1} \prod_s (1 \mp \lambda_B\,e^{-E_s/kT})^{\mp 1}. \qquad (74.2)$$

We suppose that the energies E_r, E_s depend in general on geometrical variables described by generalized coordinates y_i but for the sake of brevity we shall refer explicitly to only one such coordinate y. As previously we use Y_r to denote the corresponding generalized force $\partial E_r/\partial y$ and similarly Y_s. We accordingly regard $\varXi$ as a function of the variables λ_A, λ_B, y, T. We shall now study the dependence of $\ln \varXi$ on each of these variables. We find the conditions at the col

$$\lambda_A \frac{\partial \ln \varXi}{\partial \lambda_A} = \sum_r \frac{\lambda_A\,e^{-E_r/kT}}{1 \mp \lambda_A\,e^{-E_r/kT}} = \sum_r \overline{N}_r = N_A, \qquad (74.3)$$

$$\lambda_B \frac{\partial \ln \varXi}{\partial \lambda_B} = \sum_s \frac{\lambda_B\,e^{-E_s/kT}}{1 \mp \lambda_B\,e^{-E_s/kT}} = \sum_s \overline{N}_s = N_B, \qquad (74.4)$$

$$- kT \frac{\partial \ln \varXi}{\partial y} = \sum_r \frac{\dfrac{\partial E_r}{\partial y}\lambda_A\,e^{-E_r/kT}}{1 \mp \lambda_A\,e^{-E_r/kT}} + \sum_s \frac{\dfrac{\partial E_s}{\partial y}\lambda_B\,e^{-E_s/kT}}{1 \mp \lambda_B\,e^{-E_s/kT}}$$

$$= \sum_r \overline{N}_r \frac{\partial E_r}{\partial y} + \sum_s \overline{N}_s \frac{\partial E_s}{\partial y} = \sum_r \overline{N}_r Y_r + \sum_s \overline{N}_s Y_s = Y, \qquad (74.5)$$

$$kT^2 \frac{\partial \ln \varXi}{\partial T} = \sum_r \frac{E_r\,\lambda_A\,e^{-E_r/kT}}{1 \mp \lambda_A\,e^{-E_r/kT}} + \sum_s \frac{E_s\,\lambda_B\,e^{-E_s/kT}}{1 \mp \lambda_B\,e^{-E_s/kT}}$$

$$= \sum_r \overline{N}_r E_r + \sum_s \overline{N}_s E_s = U. \qquad (74.6)$$

Consequently for the most general variation of $\ln \varXi$ we have

$$d \ln \varXi = \frac{U}{kT^2}\,dT - \frac{Y}{kT}\,dy + N_A\,d \ln \lambda_A + N_B\,d \ln \lambda_B \qquad (74.7)$$

which we can rewrite as

$$d\,(kT \ln \varXi) = \left(\frac{U}{T} + k \ln \varXi\right) dT - Y\,dy + N_A\,kT\,d \ln \lambda_A + N_B\,kT\,d \ln \lambda_B. \qquad (74.8)$$

We shall relate this to a formula of classical thermodynamics. We start from the classical thermodynamic relations

$$U - TS - Yy = n_A \mu_A + n_B \mu_B, \qquad (74.9)$$

$$0 = S\,dT + y\,dY + n_A\,d\mu_A + n_B\,d\mu_B \qquad (74.10)$$

which is a generalization of (22.5). We can rewrite formula (74.10) as

$$0 = \left(S + \frac{n_A \mu_A + n_B \mu_B}{T}\right) dT + y\,dY + n_A\,T\,d\left(\frac{\mu_A}{T}\right) + n_B\,T\,d\left(\frac{\mu_B}{T}\right)$$

$$= \left(\frac{U}{T} - \frac{Yy}{T}\right) dT + y\,dY + n_A\,T\,d\left(\frac{\mu_A}{T}\right) + n_B\,T\,d\left(\frac{\mu_B}{T}\right). \qquad (74.11)$$

Subtracting $d(Yy)$ from both sides we have

$$-d(Yy) = \left(\frac{U}{T} - \frac{Yy}{T}\right) dT - Y\, dy + n_A\, T\, d\left(\frac{\mu_A}{T}\right) + n_B\, T\, d\left(\frac{\mu_B}{T}\right). \qquad (74.12)$$

This formula is directly comparable with the statistical relation (74.8). We see that they become equivalent if we set

$$kT \ln \varXi = -Yy, \qquad (74.13)$$

$$\mu_A = \frac{N_A}{n_A}\, kT \ln \lambda_A = L\, kT \ln \lambda_A = RT \ln \lambda_A, \qquad (74.14)$$

$$\mu_B = \frac{N_B}{n_B}\, kT \ln \lambda_B = L\, kT \ln \lambda_B = RT \ln \lambda_B \qquad (74.15)$$

where L is Avogadro's number. We have thus confirmed, as anticipated that $\ln \lambda$ differs from μ only by a factor common to all molecular species and we see that this factor is RT. The name for $\varXi$ is *grand partition function* and the name for λ_A is the *absolute activity* of the species A.

75. Boltzmann's distribution law. We have studied the properties of the absolute activities λ and especially their connection with the conditions for chemical equilibrium in a manner covering both even molecules obeying the Bose-Einstein distribution law and odd molecules obeying the Fermi-Dirac distribution law. We shall now explain that the distinction between these two kinds of distribution is in fact nearly always unimportant. From the distribution laws (72.20) and (72.21) written in the form

$$\frac{\overline{N}_p}{1 \pm \overline{N}_p} = \lambda_A\, e^{-E_p/kT} \qquad (75.1)$$

it is evident that both reduce to

$$\overline{N}_p = \lambda_A\, e^{-E_p/kT}, \qquad (75.2)$$

provided

$$\overline{N}_p \ll 1, \qquad (75.3)$$

and this inequality will hold whenever N_A is small compared with the number of molecular states of energy less than kT. The physical explanation is simply that under these circumstances the eventuality of any state being occupied by more than one molecule is so remote that the question whether this eventuality is allowed or disallowed (Pauli's exclusion principle) becomes unimportant. This condition holds for all applications of statistical mechanics to gases under conditions normally met and in fact to practically all applications to material molecular systems with one important exception, namely to the system of conducting ("free") electrons in a metal. The reason for this exception is that the mass of the electron is about 4×10^3 times smaller than that of the lightest gaseous molecule, and consequently the number of molecular states in a given volume of phase space is about $(4 \times 10^3)^3$ or 6×10^{10} fewer. Except in the discussion of electrons in a metal we shall henceforth assume the simpler distribution law

$$\overline{N}_p = \lambda_A\, e^{E_p/kT} \qquad (75.4)$$

called Boltzmann's distribution law.

Since of necessity

$$\sum_r \overline{N}_r = N_A \qquad (75.5)$$

it follows that

$$\lambda_A = \frac{N_A}{\sum_r e^{-E_r/kT}}. \qquad (75.6)$$

We observe that $\sum_r e^{-E_r/kT}$ is structurally similar to the partition function of a system. We call this sum the partition function of the molecule A and denote[1] it by Q_A.

We have then

$$Q_A = \sum_r e^{-E_r/kT}, \tag{75.7}$$

$$\lambda_A = \frac{N_A}{Q_A}, \tag{75.8}$$

$$\frac{\overline{N}_p}{\overline{N}_A} = \frac{e^{-E_p/kT}}{Q_A}. \tag{75.9}$$

76. System of localized molecules. We have derived Boltzmann's distribution law as an approximation, nearly always valid, to either the Bose-Einstein or the Fermi-Dirac distribution law for molecules confined in the same enclosure. We shall now describe different conditions when Boltzmann's distribution law is accurate. This holds when each molecule is attached to a site on a lattice. We first suppose that the molecules cannot change places. Then the molecules are distinguishable by the sites to which they are attached and consequently the eigenfunctions of the system, instead of having to be symmetrical or antisymmetrical with respect to the molecules, are simply products of the eigenfunctions of the several molecules. For a system of N_A molecules A of energy E_A the number C of states is equal to the coefficient of z^{E_A} in

$$\left(\sum_r z^{E_r}\right)^{N_A}. \tag{76.1}$$

Consequently

$$C = \frac{1}{2\pi i} \int \frac{dz}{z^{E_A+1}} \left(\sum_r z^{E_r}\right)^{N_A} \tag{76.2}$$

and the average number $\overline{N}_p$ of molecules in the state p is given by

$$C\frac{\overline{N}_p}{N_A} = \frac{1}{2\pi i} \int \frac{dz}{z^{E_A+1}} \frac{z^{E_p}}{\sum_r z^{E_r}} \left(\sum_r z^{E_r}\right)^{N_A} \tag{76.3}$$

where both integrations are round a contour within the circle of convergence circulating once counterclockwise round the origin. Using the method of steepest descents as previously we obtain from (76.2) and (76.3) after identifying the col $z = \vartheta$ with $e^{-1/kT}$,

$$\frac{\overline{N}_p}{N_A} = \frac{e^{-E_p/kT}}{\sum_r e^{-E_r/kT}} = \frac{e^{-E_p/kT}}{Q_A} \tag{76.4}$$

which is Boltzmann's distribution law.

In the above argument it was assumed that each molecule is tied to a particular site. We shall now study briefly the effect of making the opposite assumption that the molecules are free to change places subject to the restriction that there is only one molecule on each site. Then since the molecules can be distributed in $N_A!$ ways there would be an increase by this factor of the number of states were it not for the restriction of symmetry or antisymmetry with respect to the now identical molecules. This restriction reduces the number of states by a factor $N_A!$ so that the number of states is in fact just the same whether the molecules can change places or not.

---

[1] In the previous chapter Q_A denoted the partition function of a *system* A. Here Q_A denotes the partition function of a *molecule* A. It should always be clear from the context whether A denotes a system or a molecule.

77. Classical statistics. Free energy. Whenever the molecules obey Boltzmann's distribution law, either accurately or to an adequate approximation, the system is said to be subject to *classical statistics*.

We shall now obtain a formula for the free energy of a system obeying classical statistics. From (74.2) and (74.13) we have

$$Yy = -kT \sum_r \ln\left(1 \mp \lambda_A e^{-E_r/kT}\right)^{\mp 1} - kT \sum_s \ln\left(1 \mp \lambda_B e^{-E_s/kT}\right)^{\mp 1} \qquad (77.1)$$

where for the sake of brevity we have included only two kinds of molecules A and B. We use the thermodynamic relation

$$\begin{aligned} F &= n_A \mu_A + n_B \mu_B + Yy \\ &= N_B kT \ln \lambda_A + N_B kT \ln \lambda_B + Yy \end{aligned} \qquad (77.2)$$

and using (77.1) we obtain

$$\begin{aligned} F = N_A kT \ln \lambda_A &- kT \sum_r \ln\left(1 \mp \lambda_A e^{-E_r/kT}\right)^{\mp 1} + \\ &+ N_B kT \ln \lambda_B - kT \sum_s \ln\left(1 \mp \lambda_B e^{-E_s/kT}\right)^{\mp 1}. \end{aligned} \qquad (77.3)$$

The condition for Boltzmann's distribution law to be a sufficient approximation is

$$\lambda_A e^{-E_r/kT} \ll 1, \quad \text{all } r, \qquad (77.4)$$

$$\lambda_B e^{-E_s/kT} \ll 1, \quad \text{all } s. \qquad (77.5)$$

We may therefore without sensible loss of accuracy replace formula (77.3) by

$$\begin{aligned} F = N_A kT \ln \lambda_A &- kT \lambda_A \sum_r e^{-E_r/kT} \\ &+ N_B kT \ln \lambda_B - kT \lambda_B \sum_s e^{-E_s/kT}. \end{aligned} \qquad (77.6)$$

Using (75.6)

$$\lambda_A = N_A \Big/ \sum_r e^{-E_r/kT} = N_A/Q_A, \qquad (77.7)$$

$$\lambda_B = N_B \Big/ \sum_s e^{-E_s/kT} = N_B/Q_B. \qquad (77.8)$$

in (77.6) we obtain

$$\begin{aligned} F = &- N_A kT \left(\ln Q_A - \ln N_A + 1\right) \\ &- N_B kT \left(\ln Q_B - \ln N_B + 1\right). \end{aligned} \qquad (77.9)$$

Since N_A, N_B are large we may use Stirling's approximation

$$\ln N_A! = N_A \left(\ln N_A - 1\right), \quad \ln N_B! = N_B \left(\ln N_B - 1\right) \qquad (77.10)$$

and consequently may rewrite (77.9) as

$$F = -kT \ln \frac{Q_A^{N_A} Q_B^{N_B}}{N_A! \, N_B!}. \qquad (77.11)$$

According to (65.1) the general relation between the free energy F and the partition function of the system, which we now denote by Q_S, is

$$F = -kT \ln Q_S. \qquad (77.12)$$

The necessary and sufficient condition for these two formulae to be consistent is

$$Q_S = \frac{Q_A^{N_A} Q_B^{N_B}}{N_A! \, N_B!}, \qquad (77.13)$$

and this condition is satisfied when the system obeys classical statistics. For Q_A contains one term for each state of a single molecule and so $Q_A^{N_A}$ contains one term for each combination of states of the N_A molecules and the factor $1/N_A!$ takes account of the requirement for symmetry or antisymmetry in the N_A identical molecules. Similar remarks apply to the factors relating to the molecules B.

We return now to a system of several kinds of molecules in a common enclosure obeying classical statistics. The free energy F is given by the generalization of (77.9), namely

$$F = \sum_A N_A \, kT \, (-\ln Q_A + \ln N_A - 1),\qquad(77.14)$$

and consequently the chemical potential μ_A and absolute activity λ_A are given by

$$\frac{\mu_A}{L} = kT \ln \lambda_A = \frac{\partial F}{\partial N_A} = - kT \ln \frac{Q_A}{N_A}.\qquad(77.15)$$

The free energy of a system of molecules A located on lattice sites is given by

$$F = - N_A \, kT \ln Q_A \qquad(77.16)$$

and consequently

$$\frac{\mu_A}{L} = kT \ln \lambda_A = \frac{\partial F}{\partial N_A} = - kT \ln Q_A.\qquad(77.17)$$

The difference between formulae (77.15) and (77.17) for μ_A is more apparent than real, because in the former the partition function Q_A relates to the whole enclosure while in the latter it relates to a single site. We could in fact rewrite formula (77.17) as

$$\frac{\mu_A}{L} = kT \ln \lambda_A = - kT \ln \frac{Q_A'}{N_A}\qquad(77.18)$$

where Q_A' is the partition function obtained by summing over all states of N_A sites instead of only those on a single site.

It is now clear that all the thermodynamic functions, in particular F and μ_A can be expressed in terms of the molecular partition functions Q_A. The remainder of our task consists then in studying these molecular partition functions. We begin by discussing some general properties.

78. Factorization of partition function. By suitable choice of coordinates Schrödinger's equation for a molecule becomes separable in some of or all the variables. The overall eigenfunctions are then products of eigenfunctions relating to sets of degrees of freedom, or possibly individual degrees of freedom, and the overall energies are sums of energies in these sets of degrees of freedom. In particular the three degrees of freedom describing the motion of the molecular centre of mass are separable from the remaining degrees of freedom which for the sake of brevity we shall call internal degrees of freedom. Thus the energy of any molecular state is of the form

$$E_{rs} = E_r^{\mathrm{tr}} + E_s^{\mathrm{int}}\qquad(78.1)$$

where the superscripts tr and int denote translational and internal respectively. It follows immediately that the molecular partition function Q is of the form

$$\begin{aligned}
Q &= \sum_r \sum_s \exp\left(- E_{rs}/kT\right)\\
&= \sum_r \sum_s \exp\left(- [E_r^{\mathrm{tr}} + E_s^{\mathrm{int}}]/kT\right)\\
&= \sum_r \exp\left(- E_r^{\mathrm{tr}}/kT\right) \sum_s \exp\left(- E_s^{\mathrm{int}}/kT\right)\\
&= Q_{\mathrm{tr}} \, Q_{\mathrm{int}}\qquad(78.2)
\end{aligned}$$

where
$$Q_{\text{tr}} = \sum_r \exp\left(-E_r^{\text{tr}}/kT\right), \tag{78.3}$$

$$Q_{\text{int}} = \sum_s \exp\left(-E_s^{\text{int}}/kT\right). \tag{78.4}$$

In other words the overall partition function factorizes into partition functions relating to the two sets of degrees of freedom. It is evident from this example that whenever there is separation of variables the partition function can be factorized. It will accordingly be sufficient and convenient to consider in detail partition functions relating to particular degrees of freedom instead of overall partition functions.

79. Degenerate energy levels. At this stage it is convenient to introduce a small change in our notation. For any molecular partition function Q, whether relating to a single degree of freedom or to several or to all, instead of writing as hitherto
$$Q = \sum_r \exp\left(-E_r/kT\right), \tag{79.1}$$

we shall group together all states of equal energy and write
$$Q = \sum_{E_i} g_i \exp\left(-E_i/kT\right) \tag{79.2}$$

where g_i denotes the number of states having equal energy E_i and is called the *degree of degeneracy* of the energy level E_i. It is precisely analogous to the quantity, denoted similarly in the previous chapter, namely, the degree of degeneracy of the energy level of a (macroscopic) system. In a partition function relating to a single degree of freedom all the g_i's are usually unity. An important exception is the degree of freedom relating to electron spin. In a partition function relating to a few degrees of freedom the g_i's are simple integers. In a partition function relating to many degrees of freedom, such as those of a macroscopic system, the g_i's can become very large numbers of the order of magnitude $N!$, where N denotes the number of molecules in the system, as we have already seen in the previous chapter.

80. Energy zero. It is usually convenient to take as the energy zero in each degree of freedom, or set of degrees of freedom, the energy of the lowest accessible states. This will always be done unless the contrary is stated. These lowest states will be denoted by the subscript 0.

81. Unexcited degrees of freedom. The general behaviour of a partition function for a set of degrees of freedom depends markedly on how the separation between the energy levels compares with kT. There are two important extreme cases.

We label the energy levels $E_0, E_1, E_2, \ldots$ with $E_0 = 0$ by convention. When
$$E_1 \gg kT, \tag{81.1}$$

so that a fortiori
$$E_i \gg kT, \quad i > 1, \tag{81.2}$$

the partition function reduces effectively to its first term g_0, which as already mentioned is usually unity except for degrees of freedom relating to spin. Under these circumstances effectively all the molecules will be in the lowest energy state, that is to say according to our convention will have no energy, in these degrees of freedom. Such degrees of freedom are called *unexcited*. The apparent misbehaviour of these degrees of freedom was one of the mysteries, perhaps the greatest mystery, of classical theory.

82. Classical degrees of freedom. The opposite extreme case occurs when the spacing of the energy levels, or at least the lower ones, is small compared with kT. The sum may then without significant loss of accuracy be replaced by an integral. In the simplest case of a single degree of freedom described by a single coordinate q with its conjugate momentum p formula (79.2) becomes

$$Q = \int dq \int dp \, g(E) \exp(-E/kT)$$
$$= \frac{1}{h} \int dq \int dp \exp(-E/kT) \tag{82.1}$$

since there is on the average one quantum state per area h in the phase plane. The corresponding formula for f degrees of freedom is

$$Q = \frac{1}{h^f} \int dq_1 \dots dq_f \int dp_1 \dots dp_f \exp(-E/kT). \tag{82.2}$$

Degrees of freedom behaving in this manner could be satisfactorily accounted for according to classical theory and are therefore called *classical degrees of freedom*.

There is no a priori reason to expect most degrees of freedom to be either unexcited or classical, but it so happens, as we shall see in the next chapter, that many of the most important molecular degrees of freedom are in fact one or the other, and moreover that almost all the remaining important degrees of freedom are vibrations which are nearly harmonic.

E. Statistical thermodynamics.

83. Scope of statistical thermodynamics. The name statistical thermodynamics is used for the combined application of classical thermodynamics and statistical mechanics to obtain results more detailed than can be obtained from the former alone and to obtain them more briefly than is possible from the latter alone. This will be made clear by various examples.

84. Ideal monatomic gas. The first system to be considered is an ideal gas composed of a single kind of monatomic molecules. The epithet ideal implies that the energy of the intermolecular interactions is negligible. The system is thus regarded as composed of molecules able to exchange energy by collisions but otherwise essentially independent. As explained in the previous chapter all the equilibrium properties of such a system are determined by the partition function of a molecule. The molecular degrees of freedom are separable into the following sets. First the intranuclear degrees of freedom which are always unexcited and will not be further discussed. Second the motion of the electrons relative to the nucleus; these electronic degrees of freedom are almost always unexcited, important exceptions being Cl and NO. The partition function Q_{el} for the electronic degrees of freedom then reduces to g_0 the degree of degeneracy of the lowest electronic level. This is 1 for the inert elements, 1 for mercury, 2 for the alkali metals and 4 for the halogen atoms I and Br. Third the translational motion of the centre of mass which coincides with the nucleus. This translational motion is itself separable into motion in three mutually orthogonal directions. Consequently the translational partition function Q_{tr} factorizes as

$$Q_{tr} = Q_x Q_y Q_z \tag{84.1}$$

where the subscripts x, y, z refer to the three directions of motion. We shall now consider motion in the x direction which we assume to be confined between two parallel walls distant a apart. The energy E, purely kinetic, is related to the momentum p_x by

$$E_x = \frac{p_x^2}{2m} \tag{84.2}$$

and the quantum states are determined by

$$2a\,|p_x| = n\,h \qquad n = 1, 2, 3, \ldots \tag{84.3}$$

where n is an integral quantum number. The energy values are thus

$$E_x = \frac{n^2 h^2}{8m a^2} \tag{84.4}$$

and so the partition function is

$$Q_x = \sum_{n=1}^{\infty} \exp\left(- n^2 h^2/8m a^2 kT\right). \tag{84.5}$$

We may rewrite this more briefly as

$$Q_x = \sum_{n=1}^{\infty} \exp\left(- n^2 \Theta_x/T\right) \tag{84.6}$$

where Θ_x is a characteristic temperature defined by

$$\Theta_x = \frac{h^2}{8m a^2 k}. \tag{84.7}$$

If as an example we take helium atoms enclosed between walls one centimetre apart, we have

$$\Theta_x = \frac{6.62 \times 10^{-27} \times 6.62 \times 10^{-27}}{8 \times 4 \times 1.66 \times 10^{-24} \times 1.0 \times 1.0 \times 1.38 \times 10^{-16}} \ \mathrm{deg}$$

$$= 6.0 \times 10^{-15}\ \mathrm{deg}. \tag{84.8}$$

If we had chosen a heavier atom, Θ_x would have been still smaller. Even if we reduced the distance apart of the walls to 10^{-2} cm, the value of Θ_x would still be only 6×10^{-11} deg. It is thus clear that with a tremendous margin of safety we may treat this degree of freedom as classical. We accordingly replace the sum in (84.6) by an integral obtaining

$$Q_x = \int_0^{\infty} \exp\left(- n^2 \Theta_x/T\right) dn = \frac{\pi^{\frac{1}{2}}}{2}\left(\frac{T}{\Theta_x}\right)^{\frac{1}{2}}$$

$$= (2\pi m kT)^{\frac{1}{2}}\, a/h. \tag{84.9}$$

Once we know that this degree of freedom is classical we can, if we prefer, use a more classical approach. We need only recall that there is one state per area h in the phase plane. We thus obtain

$$Q_x = \frac{1}{h}\int_0^{a} dx \int_{-\infty}^{+\infty} dp_x \exp\left(- p_x^2/2m kT\right)$$

$$= \frac{a}{h}\int_{-\infty}^{+\infty} dp_x \exp\left(- p_x^2/2m kT\right)$$

$$= (2\pi m kT)^{\frac{1}{2}}\, a/h \tag{84.10}$$

in agreement with (84.9).

We obtain precisely similar formulae for motion in the y and z directions. If then the gas is contained in a box of sides a, b, c and of volume V we have

$$Q_{tr} = Q_x\, Q_y\, Q_z = (2\pi m\, kT)^{\frac{3}{2}}\, a\, b\, c/h^3$$
$$= (2\pi m\, kT)^{\frac{3}{2}}\, V/h^3. \tag{84.11}$$

This result is in fact independent of the shape of the box.

The free energy of a gas of N molecules in a box of volume V is according to (77.9)

$$F = N E_0 + N\, kT\,(-\ln Q + \ln N - 1)$$
$$= N E_0 + N\, kT\,(-\ln Q_{tr} - \ln Q_{int} + \ln N - 1) \tag{84.12}$$

where E_0 is the energy of the molecule in its lowest state. The term $N E_0$ is required because inside each partition function the energy is measured from that of the lowest state. As long as we are concerned only with changes of temperature and pressure of a permanent gas the term $N E_0$ is an arbitrary additive constant without physical significance, but if later we are concerned with the condensation of the gas to a liquid or a crystal or with a chemical reaction in which the molecules are destroyed then the term $N E_0$ may be required in order to preserve consistency in the choice of energy zero.

Substituting (84.11) into (84.12) we obtain

$$F = N E_0 + \frac{3}{2} N\, kT \ln (h^2/2\pi\, m\, kT) - N\, kT \ln g_0 + N\, kT \ln \frac{N}{V} - N\, kT. \tag{84.13}$$

Q_{int} has in accordance with Sect. 81 been replaced by g_0, the degree of degeneracy of the lowest electronic energy level. For the chemical potential μ and absolute activity λ we deduce by (20.7) and (34.1)

$$\frac{\mu}{L} = kT \ln \lambda = \frac{\partial F}{\partial N}$$
$$= E_0 + \frac{3}{2} kT \ln (h^2/2\pi\, m\, kT) - kT \ln g_0 + kT \ln \frac{N}{V} \tag{84.14}$$

which could have been obtained directly from the relation (77.7) $\lambda = N/Q$.

For the entropy S and total energy U we obtain using (19.10) and (19.11)

$$\frac{S}{n R} = \frac{S}{N k} = -\frac{1}{N k}\frac{\partial F}{\partial T}$$
$$= -\frac{3}{2} \ln \frac{h^2}{2\pi\, m\, kT} + \ln g_0 - \ln \frac{N}{V} + \frac{5}{2}, \tag{84.15}$$

$$U = F - T \frac{\partial F}{\partial T} = N \left(E_0 + \frac{3}{2} kT\right). \tag{84.16}$$

If we regard the volume V as an example of a strain coordinate y, then the corresponding generalized force Y is equal and opposite to the pressure P. Consequently

$$P = -\frac{\partial F}{\partial V} = \frac{N\, kT}{V} \tag{84.17}$$

which is the well known equation of state of a perfect gas. Alternatively this proves that the temperature scale of a perfect gas is the same as the thermodynamic scale.

By use of (84.17) the entropy may be expressed in terms of P instead of N/V. Thus

$$\frac{S}{n R} = -\frac{3}{2} \ln \frac{h^2}{2\pi\, m} + \ln g_0 - \ln P + \frac{5}{2} + \frac{5}{2} \ln kT. \tag{84.18}$$

When numerical values are inserted we obtain

$$\frac{S}{n\,R} = \frac{5}{2}\ln\frac{T}{1.59_3\,\mathrm{deg}} + \ln\frac{M}{\mathrm{g\ mole^{-1}}} - \ln\frac{P}{\mathrm{atm}} + \ln g_0 \qquad (84.19)$$

where $M = L\,m$ denotes molar mass.

85. Maxwell's distribution law and related formulae. The derivation of formula (84.17) is typical of the procedure of statistical thermodynamics. We shall now give an alternative derivation using only statistical mechanics without thermodynamics.

We begin with the distribution law for molecular translation in the x direction. The fraction of molecules having momentum of magnitude

$$|p_x| = \frac{n\,h}{2a} \qquad (85.1)$$

is

$$\frac{\exp\left(-p_x^2/2m\,kT\right)}{\sum \exp\left(-p_x^2/2m\,kT\right)}. \qquad (85.2)$$

Since this degree of freedom is classical and since the quantum states are evenly spaced with respect to $|p_x|$ we may replace the above statement by the statement that the fraction of molecules having a momentum between p_x and $p_x + dp_x$ in the x direction is equal to

$$\frac{\exp\left(-p_x^2/2m\,kT\right)dp_x}{\int\limits_{-\infty}^{+\infty} \exp\left(-p_x^2/2m\,kT\right)dp_x} = (2\pi m\,kT)^{-\frac{1}{2}}\exp\left(-p_x^2/2m\,kT\right)dp_x. \qquad (85.3)$$

This may be expressed in terms of velocity $u_x = p_x/m$ by the statement that the fraction of molecules having a velocity component in the x direction between u_x and $u_x + du_x$ is

$$\frac{\exp\left(-m\,u_x^2/2\,kT\right)du_x}{\int\limits_{-\infty}^{+\infty} \exp\left(-m\,u_x^2/2\,kT\right)du_x} = (2\pi kT/m)^{\frac{1}{2}}\exp\left(-m\,u_x^2/2\,kT\right)du_x. \qquad (85.4)$$

This is known as Maxwells's distribution law.

We can immediately derive the average value of the kinetic energy in the x direction of a molecule. We have

$$\frac{1}{2}m\,\overline{u_x^2} = \frac{\displaystyle\int\limits_{-\infty}^{+\infty}\tfrac{1}{2}m\,u_x^2\exp\left(-m\,u_x^2/2\,kT\right)du_x}{\displaystyle\int\limits_{-\infty}^{+\infty}\exp\left(-m\,u_x^2/2\,kT\right)du_x}. \qquad (85.5)$$

Integrating the numerator by parts we obtain

$$\frac{1}{2}m\,\overline{u_x^2} = \frac{\displaystyle\tfrac{1}{2}kT\int\limits_{-\infty}^{+\infty}\exp\left(-m\,u_x^2/2\,kT\right)du_x}{\displaystyle\int\limits_{-\infty}^{+\infty}\exp\left(-m\,u_x^2/2\,kT\right)du_x} = \frac{1}{2}kT. \qquad (85.6)$$

We are now ready to derive a formula for the pressure by equating it to the transfer of momentum per unit time to unit area of the wall of the container. We consider an element of wall normal to the x direction of area A. It is evident from symmetry that molecular motion in the y and z directions can affect neither the number of collisions per unit time nor the transfer of momentum in a collision. We shall accordingly ignore motion in the y and z directions and consider only motion in the x direction. Let us begin by considering only molecules having given velocity u_x towards the element of wall and we denote the number of such molecules per unit volume by $C(u_x)$. The number of such molecules striking the element of area A in time t is $C(u_x)\,u_x A\,t$. The momentum normal to the wall of such a molecule is $m u_x$ before and $-m u_x$ after hitting the wall. The transfer of momentum to the wall is thus $2 m u_x$. The total transfer of momentum to the element of wall due to these molecules is thus $C(u_x)\,2 m u_x^2 A\,t$. If we denote by C the total number of molecules per unit volume, then the total transfer of momentum to the element of wall in the time t due to all the molecules is $\frac{1}{2} C\,2 m \overline{u_x^2} A\,t$ where the factor $\frac{1}{2}$ takes account of the fact that at any moment half the molecules are moving towards the wall and half away from it. Consequently the pressure being equal to the transfer of momentum per unit time per unit area of wall is given by

$$P = C\,m\overline{u_x^2} = C\,kT = \frac{N}{V}\,kT \qquad (85.7)$$

in agreement with the formula obtained by use of thermodynamics.

86. Rotation of linear molecules. We shall now turn our attention to gases formed of linear molecules including in particular all diatomic molecules. Such molecules have all the separable degrees of freedom of a monatomic molecule and in addition the following degrees of freedom all separable at least to a degree of accuracy adequate for ordinary purposes: two rotational degrees of freedom and $3a-5$ vibrational degrees of freedom where a denotes the number of atoms in the molecule. We must now consider the several kinds of degrees of freedom in turn.

We begin with the rotational degrees of freedom. In general a rigid body has three rotational degrees of freedom, but a linear molecule has only two because any motion having angular momentum about the molecular axis is excluded. This is not because such motion is impossible but simply because all motion of this kind is included either in the electronic degrees of freedom or in the nuclear spin degrees of freedom. All such degrees of freedom are, with the one exception of nitric oxide, unexcited. The rotational states of two degrees of freedom are defined by the two quantum numbers j and m such that the total angular momentum p_ϑ has the values

$$p_\vartheta = \{j(j+1)\}^{\frac{1}{2}}\,h/2\pi, \qquad j = 0, 1, 2, \ldots \qquad (86.1)$$

and the component of this angular momentum about a chosen axis fixed in space has the values

$$m\,h/2\pi \qquad m = -j, -j+1, \ldots 0 \ldots j-1, j. \qquad (86.2)$$

The rotational energy E_{rot} is given by

$$E_{\mathrm{rot}} = p_\vartheta^2/2I = j(j+1)\,h^2/8\pi^2 I \qquad (86.3)$$

where I is the principle moment of inertia. Since the energy is independent of m and there are $2j+1$ values of m for a given j, each energy level is $2j+1$ fold degenerate. The partition function Q_{rot} is consequently

$$Q_{\mathrm{rot}} = \sum_{j=0}^{\infty} (2j+1)\exp\{-j(j+1)\,h^2/8\pi^2 I\,kT\}. \qquad (86.4)$$

It is convenient to introduce a rotational characteristic temperature $\Theta_{\rm rot}$ defined by

$$\Theta_{\rm rot} = h^2/8\pi^2 I k. \tag{86.5}$$

The partition function then becomes

$$Q_{\rm rot} = \sum_{j=0}^{\infty} (2j+1)\exp\{-j(j+1)\,\Theta_{\rm rot}/T\}. \tag{86.6}$$

The value of I and so of $\Theta_{\rm rot}$ for various molecules can be determined spectroscopically. The value of $\Theta_{\rm rot}$ for H_2 is $85°$, for D_2 $42.5°$, for HCl $15.0°$ and it is smaller for other linear molecules. Consequently for all linear molecules, other than H_2 and its isotopes, at all temperatures at which measurements can be made on the gas the rotational degrees of freedom are effectively classical. We accordingly replace the sum by an integral and setting $j(j+1)=y$ we obtain in place of (86.6)

$$Q_{\rm rot} = \int_0^{\infty} dy \exp\left(-y\,\Theta_{\rm rot}/T\right) = \frac{T}{\Theta_{\rm rot}}. \tag{86.7}$$

This formula can also be obtained by classical reasoning. If we use polar coordinates ϑ, φ and their conjugate momenta p_ϑ, p_φ the rotational energy $E_{\rm rot}$ is given by

$$E_{\rm rot} = (p_\vartheta^2 + p_\varphi^2\,\mathrm{cosec}^2\,\vartheta)/2I. \tag{86.8}$$

Remembering that there is one quantum state per volume h^2 in four dimensional phase space, we obtain for the partition function

$$Q_{\rm rot} = \frac{1}{h^2} \int_0^{\pi} d\vartheta \int_0^{2\pi} d\varphi \int_{-\infty}^{\infty} dp_\vartheta \int_{-\infty}^{\infty} dp_\varphi \exp\left(-\frac{p_\vartheta^2 + p_\varphi^2\,\mathrm{cosec}^2\,\vartheta}{2IkT}\right). \tag{86.9}$$

Integrating with respect to p_ϑ and p_φ we obtain

$$\left. \begin{aligned}
Q_{\rm rot} &= \frac{2\pi I kT}{h^2} \int_0^{\pi} \sin\vartheta\,d\vartheta \int_0^{2\pi} d\varphi \\
&= \frac{2\pi I kT}{h^2}\,4\pi = \frac{T}{\Theta_{\rm rot}}
\end{aligned} \right\} \tag{86.10}$$

in agreement with formula (86.7).

87. Symmetry number. This completes the investigation of $Q_{\rm rot}$ for molecules having no centre of symmetry, but if the molecule has a centre of symmetry the argument requires modification. By semi-classical reasoning when a symmetrical molecule is rotated through two, not four, right angles, it is in a configuration identical with its initial configuration. In order not to count each distinct configuration twice over the above double integral with respect to ϑ and φ should be halved. We accordingly replace formula (86.10) by

$$Q_{\rm rot} = \frac{T}{\sigma\,\Theta_{\rm rot}} \tag{87.1}$$

where σ called the symmetry number is 2 for a symmetrical molecule and 1 for an unsymmetrical molecule.

88. Ortho and para states. The same conclusion is reached in quantum theory but the argument is more sophisticated. Let us denote by R either half of a symmetrical linear molecule, not counting the central atom if any. Thus in OCO by R we mean O while in HCCH by R we mean CH. According as R contains an even or odd number of protons and neutrons the overall eigenfunction must be symmetrical or antisymmetrical with respect to the two R's. Now the translational, electronic and vibrational eigenfunctions are all symmetrical, but the nuclear and rotational eigenfunctions may be either. If the resultant nuclear spin of R is such that there are ϱ distinct eigenfunctions for the nuclei of R, then for the pair of R's in the same molecule there are ϱ^2 distinct eigenfunctions. Of these $\frac{1}{2}\varrho(\varrho+1)$ are symmetrical and molecules in these states are called ortho while $\frac{1}{2}\varrho(\varrho-1)$ are antisymmetrical and molecules in these states are called para. It follows from the symmetry rule that if R contains an even (odd) number of protons and neutrons then ortho molecules may occupy only rotational states symmetrical (antisymmetrical) with respect to the R's and para molecules may occupy only states antisymmetrical (symmetrical) with respect to the R's. In every case a given rotational state is accessible either to ortho molecules or to para molecules but never to both. Now the eigenfunctions for the rotational motion are surface harmonics. Those with even values of j are symmetrical and those with odd values of j are antisymmetrical. Consequently alternate energy levels are occupied by ortho and para molecules. Provided the rotational degrees of freedom are classical, that is to say always except for H_2 and its isotopes, the effect of omitting alternate energy levels from Q_{rot} will be simply to halve the value of the integral. Consequently the final result of the symmetry requirements is to reduce Q_{rot} by a factor $\frac{1}{2}$ in agreement with the previous semi-classical argument.

Provided we include the symmetry number in the rotational partition function we may regard the overall nuclear partition function for a linear molecule as equal to the product of those of the constituent atoms. We need then pay no further attention to the nuclear degrees of freedom.

89. Electronic degrees of freedom. The electronic degrees of freedom of linear molecules, with the exception of nitric oxide, are unexcited and the partition function reduces to g_0 the degree of degeneracy of the lowest energy level. For almost all linear molecules considered to be chemically saturated g_0 is 1 but for O_2 it is 3. The molecule NO has a ground state $^2\Pi_{\frac{1}{2}}$ and an excited state $^2\Pi_{\frac{3}{2}}$ each doubly degenerate. If we denote their energy separation by $k\Theta_{\mathrm{el}}$, then the electronic partition function has the form

$$Q_{\mathrm{el}} = 2\left\{1 + \exp\left(-\Theta_{\mathrm{el}}/T\right)\right\} \tag{89.1}$$

and Θ_{el} has the value $178°$.

90. Internal vibrations. There remain for discussion only the vibrational degrees of freedom. A linear molecule composed of a atoms has, apart from electronic and intranuclear degrees of freedom, in all $3a$ degrees of freedom. Of these 3 are translational, 2 are rotational and the remaining $3a-5$ are vibrational. The molecule has then $3a-5$ normal modes of vibration. Each of these can with sufficient accuracy usually be treated as a harmonic vibration with a characteristic frequency.

The quantization of a harmonic oscillator leads to energy values, when the energy zero is taken as a state of rest,

$$E_{\mathrm{vib}} = (v + \tfrac{1}{2})\,h\nu, \quad v = 0, 1, 2, \ldots \tag{90.1}$$

where v is the quantum number and ν the characteristic frequency. If in accordance with our usual practice we take as our energy zero that of the lowest vibrational state, we have instead

$$E_{\text{vib}} = v\,h\,\nu, \qquad v = 0, 1, 2, \dots . \tag{90.2}$$

The partition function for this vibration is

$$Q_{\text{vib}} = \sum_{v=0}^{\infty} \exp\left(- v\,h\,\nu/kT\right). \tag{90.3}$$

It is convenient to introduce a vibrational characteristic temperature Θ_{vib} defined by

$$\Theta_{\text{vib}} = h\,\nu/k. \tag{90.4}$$

We have then

$$Q_{\text{vib}} = \sum_{v=0}^{\infty} e^{-v\,\Theta_{\text{vib}}/T} = (1 - e^{-\Theta_{\text{vib}}/T})^{-1}. \tag{90.5}$$

The values of Θ_{vib} for various molecules have been determined from their infra-red and Raman spectra.

91. Thermodynamic functions for linear molecules. We can now write down the extra terms in the free energy of a gas composed of linear molecules over and above those for a gas composed of monatomic molecules, namely

$$F_{\text{rot}} + F_{\text{vib}} = N\,kT \left\{ \ln \frac{\sigma\,\Theta_{\text{rot}}}{T} + \sum_{i=1}^{a-5} \ln\left(1 - e^{-\Theta_i/T}\right) \right\} \tag{91.1}$$

where Θ_i denotes a vibrational characteristic temperature. Since this is independent of the volume, the pressure is given by the same formula as for a gas of monatomic molecules.

For the extra terms in the chemical potential, the total energy and the entropy we derive from (91.1) using (20.7), (19.11) and (19.10)

$$\mu_{\text{rot}} + \mu_{\text{vib}} = L\left(\frac{\partial F_{\text{rot}}}{\partial N} + \frac{\partial F_{\text{vib}}}{\partial N}\right)$$

$$= RT \left\{ \ln \frac{\sigma\,\Theta_{\text{rot}}}{T} + \sum_{i=1}^{a-5} \ln\left(1 - e^{-\Theta_i/T}\right) \right\}, \tag{91.2}$$

$$\frac{U_{\text{rot}} + U_{\text{vib}}}{n} = RT + \sum_{i=1}^{a-5} \frac{R\,\Theta_i}{e^{\Theta_i/T} - 1}, \tag{91.3}$$

$$\frac{S_{\text{rot}} + S_{\text{vib}}}{n\,R} = -\ln \frac{\sigma\,\Theta_{\text{rot}}}{T} - \sum_{i=1}^{a-5} \ln\left(1 - e^{-\Theta_i/T}\right) +$$

$$+ \sum_{i=1}^{a-5} \frac{\Theta_i/T}{e^{\Theta_i/T} - 1}. \tag{91.4}$$

92. Hydrogen. It has been mentioned that the molecule H_2 has the exceptionally high rotational characteristic temperature $\Theta_{\text{rot}} = 85°$. As a result of this its rotational degrees of freedom are just classical at room temperature but cease to be so at lower temperatures and they in fact become effectively unexcited within the range of experiment. Hydrogen consequently has quite exceptional properties which will now be described. Since the hydrogen nucleus is a single proton the overall eigenfunction of the hydrogen molecule must be antisymmetric with respect to the two nuclei. The proton has a spin $\frac{1}{2}h/2\pi$ and its component in a chosen direction can be $\frac{1}{2}h/2\pi$ or $-\frac{1}{2}h/2\pi$. Hence there are two spin eigenfunctions which we denote by σ_+ and σ_-. For a pair of protons labelled 1 and 2 there are three symmetrical spin eigenfunctions, namely $\sigma_+(1)\,\sigma_+(2)$ and $\sigma_-(1)\,\sigma_-(2)$ and $\sigma_+(1)\,\sigma_-(2)+\sigma_-(1)\,\sigma_+(2)$ and one antisymmetrical eigenfunction namely $\sigma_+(1)\,\sigma_-(2)-\sigma_-(1)\,\sigma_+(2)$. That is to say there are three ortho states and one para state of spin. In order that the overall eigenfunction shall be antisymmetrical the ortho molecules may occupy only odd rotational states and the para molecules only even rotational states. Consequently the partition function for nuclear spin and molecular rotations combined has the form

$$Q_{\text{spin, rot}} = Q_p + 3\,Q_o \qquad (92.1)$$

where

$$Q_p = 1 + 5\,e^{-6\,\Theta_{\text{rot}}/T} + 9\,e^{-20\,\Theta_{\text{rot}}/T} + \cdots, \qquad (92.2)$$

$$Q_o = 3\,e^{-2\,\Theta_{\text{rot}}/T} + 7\,e^{-12\,\Theta_{\text{rot}}/T} + 11\,e^{-30\,\Theta_{\text{rot}}/T} + \cdots. \qquad (92.3)$$

At temperatures above $300°$ K these sums may be replaced by integrals and effectively

$$Q_p = Q_o = \frac{1}{2}\,\frac{T}{\Theta_{\text{rot}}} \qquad (92.4)$$

so that

$$Q_{\text{spin, rot}} = 4\,\frac{T}{2\Theta_{\text{rot}}}. \qquad (92.5)$$

The factor 4 is the normal spin factor 2^2 and the remaining factor is the usual classical rotational partition function with the symmetry number 2 in the denominator.

In order to study the behaviour of hydrogen at lower temperatures, when it is no longer classical, one further important fact must be mentioned. Under normal condition there is no mechanism for reversing nuclear spins and consequently interchange between para and ortho forms is too slow to be observed. When hydrogen is cooled it therefore behaves as a metastable mixture of one quarter para hydrogen with a rotational partition function Q_p and three quarters ortho hydrogen with a rotational partition function Q_o. The contribution of the rotational degrees of freedom to the total energy is thus given by

$$U_{\text{rot}} = \tfrac{1}{4}\,N\,k\,\Theta_{\text{rot}}\,\{5\times 6\ e^{-6\,\Theta_{\text{rot}}/T} + 9\times 20\ e^{-20\,\Theta_{\text{rot}}/T} + \cdots\} +$$
$$+ \tfrac{3}{4}\,N\,k\,\Theta_{\text{rot}}\,\{3\times 2\ e^{-2\,\Theta_{\text{rot}}/T} + 7\times 12\ e^{-12\,\Theta_{\text{rot}}/T} + \cdots\}. \qquad (92.6)$$

By the use of charcoal as a catalyst the ortho-para conversion can be accelerated so that it becomes possible to prepare pure para hydrogen having a rotational partition function Q_p.

The molecule D_2 has $\Theta_{\text{rot}} = 42.5°$ K. The principles determining its behaviour are the same as for H_2 with several differences of detail. The nucleus of D consists of a proton and a neutron. The overall eigenfunction must therefore be symmetrical. Thus the ortho molecules occupy the even rotational states and the para molecules occupy the odd rotational states. The deuterium nucleus has a spin $h/2\pi$ of which the component in a chosen direction can take the values $-h/2\pi$ or 0 or $h/2\pi$. Thus a deuterium nucleus has three spin eigenfunctions. For a pair of deuterium nuclei there are consequently six symmetrical and three antisymmetrical spin eigenfunctions. Hence deuterium behaves under normal conditions as a metastable mixture of six ninths (or two thirds) ortho with a rotational partition function

$$Q_o = 1 + 5\,e^{-6\,\Theta_{rot}/T} + 9\,e^{-20\,\Theta_{rot}/T} + \cdots \tag{92.7}$$

and three ninths (or one third) para with a rotational partition function

$$Q_p = 3\,e^{-2\,\Theta_{rot}/T} + 7\,e^{-12\,\Theta_{rot}/T} + 11\,e^{-30\,\Theta_{rot}/T} + \cdots . \tag{92.8}$$

The contribution of the rotational degree of freedom to the total energy is thus given by

$$U_{rot} = \tfrac{6}{9} N k\,\Theta_{rot}\{5\times 6\,e^{-6\,\Theta_{rot}/T} + 9\times 20\,e^{-20\,\Theta_{rot}/T} + \cdots\} +$$
$$+ \tfrac{3}{9} N k\,\Theta_{rot}\{3\times 2\,e^{-2\,\Theta_{rot}/T} + 7\times 12\,e^{-12\,\Theta_{rot}/T} + \cdots\} . \tag{92.9}$$

By use of charcoal as a catalyst the para-ortho conversion can be accelerated and pure ortho deuterium obtained.

It is thus possible to obtain pure para hydrogen and pure ortho deuterium. It is impossible to obtain hydrogen containing more than three quarters ortho molecules or deuterium containing less than two thirds ortho molecules. There is excellent agreement between theory and experiment on the heat capacities of ordinary hydrogen, para hydrogen, ordinary deuterium and ortho deuterium derived from the above formulae for the rotational contributions to the energy.

93. Degrees of freedom of non-linear molecules. We turn now to non-linear molecules. The essential difference from linear molecules is that they have three rotational degrees of freedom and only $3a-6$, instead of $3a-5$, vibrational degrees of freedom. We begin by considering molecules having the shape of a symmetrical top since the symmetry makes their behaviour comparatively simple. We denote the two equal transverse moments of inertia by I_A and the axial moment of inertia by I_C. Since there are three rotational degrees of freedom there are three quantum numbers j, λ, m with the following physical meanings. The overall angular momentum is

$$\{j\,(j+1)\}^{\frac{1}{2}}\,h/2\pi, \quad j = 0, 1, 2, \ldots . \tag{93.1}$$

The angular momentum about the molecular axis of symmetry is

$$\lambda\,h/2\pi, \quad \lambda = -j, -j+1, \ldots 0, \ldots j-1, j. \tag{93.2}$$

The component of angular momentum about a chosen axis fixed in space is

$$m\,h/2\pi, \quad m = -j, -j+1, \ldots 0, \ldots j-1, j.$$

The rotational energy is

$$E_{rot} = \frac{1}{2I_A}\{j\,(j+1) - \lambda^2\}\left(\frac{h}{2\pi}\right)^2 + \frac{\lambda^2}{2I_C}\left(\frac{h}{2\pi}\right)^2 \tag{93.3}$$

and is thus independent of m. It is convenient to introduce two rotational characteristic temperatures Θ_A, Θ_C defined by

$$\Theta_A = h^2/8\pi^2 I_A\,k, \tag{93.4}$$

$$\Theta_C = h^2/8\pi^2 I_C\,k. \tag{93.5}$$

We can then rewrite the rotational energy as

$$E_{rot} = j\,(j+1)\,k\,\Theta_A + \lambda^2\,k\,(\Theta_C - \Theta_A). \tag{93.6}$$

Since there are $2j+1$ values of m for each value of j, each energy level is $2j+1$ fold degenerate. The rotational partition function is thus

$$Q_{\text{rot}} = \sum_{j=0}^{\infty} \sum_{\lambda=-j}^{j} (2j+1) \exp\{-j(j+1)\,\Theta_A/T\} \exp\{-\lambda^2\,(\Theta_C-\Theta_A)/T\}. \quad (93.7)$$

Both Θ_A and Θ_C are always small compared with T under conditions where measurements can be made on gases so that the rotational degrees of freedom are classical. Under these conditions it can be shown[1] by use of the Euler-Maclaurin formula that without significant loss of accuracy the formula for Q_{rot} reduces to

$$Q_{\text{rot}} = \pi^{\frac{1}{2}} \frac{T}{\Theta_A} \left(\frac{T}{\Theta_C}\right)^{\frac{1}{2}} \quad (93.8)$$

which can also be derived[2] by a purely classical treatment.

If a molecule has three unequal moments of inertia I_A, I_B, I_C the quantization of its motion is too complicated to be described here, but provided the motion is effectively classical, a condition always satisfied in practice, the resultant formula for the rotational partition function is closely similar to that for symmetrical top molecules. It is

$$Q_{\text{rot}} = \pi^{\frac{1}{2}} \left(\frac{T^3}{\Theta_A\,\Theta_B\,\Theta_C}\right)^{\frac{1}{2}} \quad (93.9)$$

where Θ_A, Θ_C are defined as previously and Θ_B similarly by

$$\Theta_B = 8\pi^2\,I_B/h^2\,k. \quad (93.10)$$

Just as for linear molecules these formulae have to be modified to take account of symmetry requirements with respect to identical nuclei. The resultant corrected formula is

$$Q_{\text{rot}} = \frac{\pi^{\frac{1}{2}}}{\sigma} \left(\frac{T^3}{\Theta_A\,\Theta_B\,\Theta_C}\right)^{\frac{1}{2}} \quad (93.11)$$

where the symmetry number σ is equal to the number of indistinguishable orientations of the molecule. For example σ is 1 for the scalene triangle NOCl, 2 for the isosceles triangle OH_2, 3 for the symmetrical triangular pyramid NH_3, 4 for the rectangle C_2H_4, 6 for the equilateral triangle BF_3, 12 for the regular tetrahedron CH_4 and 12 for the regular hexagon C_6H_6.

As regards the vibrational degrees of freedom of a non-linear molecule, these differ only in number but not in behaviour from those of a linear molecule.

94. Thermodynamic functions for non-linear molecules. We can now write down the extra terms in the thermodynamic functions of a gas of non-linear molecules over and above those of a gas of monatomic molecules. We obtain using (77.12), (19.11), (19.10) and (20.7)

$$\frac{F_{\text{rot}} + F_{\text{vib}}}{n} = \frac{1}{2}\,RT \ln \frac{\sigma^2\,\Theta_A\,\Theta_B\,\Theta_C}{\pi T^3} + RT \sum_{i=1}^{a-6} \ln\left(1 - e^{-\Theta_i/T}\right), \quad (94.1)$$

$$\frac{U_{\text{rot}} + U_{\text{vib}}}{n} = \frac{3}{2}\,RT + R \sum_{i=1}^{a-6} \frac{\Theta_i}{e^{\Theta_i/T} - 1}, \quad (94.2)$$

$$S_{\text{rot}} + S_{\text{vib}} = \frac{(U_{\text{rot}} + U_{\text{vib}}) - (F_{\text{rot}} + F_{\text{vib}})}{T}, \quad (94.3)$$

$$\mu_{\text{rot}} + \mu_{\text{vib}} = \frac{1}{2}\,RT \ln \frac{\sigma^2\,\Theta_A\,\Theta_B\,\Theta_C}{\pi\,T^3} + RT \sum_{i=1}^{a-6} \ln\left(1 - e^{-\Theta_i/T}\right). \quad (94.4)$$

[1] I. E. VINEY: Proc. Cambridge Phil. Soc. **29**, 142 (1933).
[2] R. H. FOWLER: Statistical Mechanics, p. 62. Cambridge 1936.

The values of Θ_A, Θ_C for symmetrical top molecules can be obtained from spectroscopic measurements. For molecules with three unequal moments of inertia these can be calculated from interatomic distances determined by electron diffraction or estimated by analogy with other simpler molecules. The values of Θ_i are determined from infrared and Raman spectra.

95. Ideal gaseous mixtures. We now turn from single ideal gases to ideal gaseous mixtures. Since the word ideal implies neglect of the interaction energy between molecules, it is evident that if several ideal gases are present in the same container the free energy is the sum of those of the individual gases when present alone in the same container. If we denote the several species by $A, B, \ldots$ we have

$$F(T, V, N_A, N_B, \ldots) = F(T, V, N_A) + F(T, V, N_B) + \cdots \qquad (95.1)$$

and consequently

$$U(T, V, N_A, N_B, \ldots) = U(T, V, N_A) + U(T, V, N_B) + \cdots, \qquad (95.2)$$

$$S(T, V, N_A, N_B, \ldots) = S(T, V, N_A) + S(T, V, N_B) + \cdots, \qquad (95.3)$$

$$P(T, V, N_A, N_B, \ldots) = P(T, V, N_A) + P(T, V, N_B) + \cdots. \qquad (95.4)$$

Formula (95.4) expresses Dalton's law of partial pressures.

We now denote the total number of molecules by N,

$$N = N_A + N_B + \cdots, \qquad (95.5)$$

the total number of molecules per unit volume by C,

$$C = N/V, \qquad (95.6)$$

and the molecular fractions by $x_A, x_B, \ldots,$

$$x_A = N_A/(N_A + N_B + \cdots), \qquad (95.7)$$

und we change from the set of variables $V, N_A, N_B, \ldots$ to the set $N, C, x_A, x_B, \ldots$ object to the identity

$$x_A + x_B + \cdots = 1. \qquad (95.8)$$

Then each term in F of the form $kTN_A \ln(N_A/V)$ transforms to $NkTx_A \ln x_A C$ and consequently

$$F(T, N, C, x_A, x_B, \ldots) = x_A F(T, N, C, x_A = 1) +$$
$$+ x_B F(T, N, C, x_B = 1) + \cdots + NkT(x_A \ln x_A + x_B \ln x_B + \cdots). \qquad (95.9)$$

If we compare the mixture with the single gases, not in the same volume V, but at the same concentration C then the free energy of the mixture exceeds that of the unmixed gases by the negative quantity

$$NkT(x_A \ln x_A + x_B \ln x_B + \cdots) \qquad (95.10)$$

called the free energy of mixing. Correspondingly the entropy of the mixture exceeds that of the unmixed gases at the same concentration C, not in the same volume V, by the quantity

$$-Nk(x_A \ln x_A + x_B \ln x_B + \cdots) \qquad (95.11)$$

called the entropy of mixing. The corresponding total energy of mixing is zero.

Since the relation

$$P = C R T \tag{95.12}$$

holds for a mixture as well as for a single ideal gas, the above statements concerning free energy of mixing and entropy of mixing remain true if for same concentration C we read same pressure P. Thus

$$F(T, N, P, x_A, x_B, \ldots) = x_A F(T, N, P, x_A = 1) +$$
$$+ x_B F(T, N, P, x_B = 1) + \cdots + N k T (x_A \ln x_A + x_B \ln x_B + \cdots). \tag{95.13}$$

When, on the other hand, we use partial pressures $p_A = x_A P$ we obtain

$$F(T, N, p_A, p_B, \ldots) = x_A F(T, N, p_A, 0, \ldots) + x_B F(T, N, 0, p_B, \ldots) + \cdots \tag{95.14}$$

so that with these variables the free energy of mixing is zero.

96. Simple crystals. We now turn from gases to crystals and we begin with those crystals, in particular metals, constituted of a single kind of atom and without any polyatomic molecular structure. The only degrees of freedom, apart from the unexcited intranuclear and electronic degrees of freedom, are the translational degrees of freedom of the nuclei. The translational motion of the nuclei consists of vibrations about their equilibrium positions on the crystal lattice. These vibrations are the ones excited by a sound wave and are therefore called the acoustic degrees of freedom. The atoms do not, of course, vibrate independently of one another but the vibrations can be resolved into independent normal modes, each of which may with sufficient accuracy be treated as a harmonic vibration. Just as a non-linear molecule composed of a atoms has $3a - 6$ normal vibrational modes, so a crystal composed of N atoms has $3N - 6$ normal vibrational modes and this number differs insignificantly from $3N$. This number is so large that we may safely regard the distribution of frequencies as continuous. We accordingly denote by $\varrho(\nu)\, d\nu$ the number of frequencies between ν and $\nu + d\nu$ and we have then

$$\int_0^\infty \varrho(\nu)\, d\nu = 3N \tag{96.1}$$

or if there is a finite maximum frequency ν_m

$$\int_0^{\nu_m} \varrho(\nu)\, d\nu = 3N. \tag{96.2}$$

Since the partition function of a single harmonic oscillator is $(1 - e^{-h\nu/kT})^{-1}$, when the energy of the lowest state is taken as zero, the partition function Q_{ac} for the acoustic modes in the crystal is given by

$$\ln Q_{ac} = - \int_0^{\nu_m} \ln (1 - e^{-h\nu/kT})\, \varrho(\nu)\, d\nu. \tag{96.3}$$

If however we take as zero energy a hypothetical state with all the atoms at rest in their equilibrium positions on the lattice we have instead

$$\ln Q_{ac} = - \int_0^{\nu_m} \left\{ \frac{h\nu}{2kT} + \ln (1 - e^{-h\nu/kT}) \right\} \varrho(\nu)\, d\nu. \tag{96.4}$$

The contribution of these degrees of freedom to the free energy of the crystal is according to (77.12)

$$F_{ac} = - kT \ln Q_{ac} = \int_0^{\nu_m} \{ \tfrac{1}{2} h\nu + kT \ln (1 - e^{-h\nu/kT}) \} \varrho(\nu)\, d\nu \tag{96.5}$$

and consequently the contribution to the total energy is

$$U_{ac} = F_{ac} - T \frac{\partial F_{ac}}{\partial T}$$

$$= \int\limits_0^{\nu_m} \left(\frac{1}{2} h\nu + \frac{h\nu}{e^{h\nu/kT} - 1} \right) \varrho(\nu)\, d\nu. \tag{96.6}$$

We shall now study the behaviour of the energy at high temperatures, that is to say when $kT > h\nu_m$ and a fortiori $kT > h\nu$ for $\nu < \nu_m$. We expand in powers of $h\nu/kT$ and neglecting terms of order $h^2\nu^2/kT$ obtain

$$U_{ac} = kT \int\limits_0^{\nu_m} \varrho(\nu)\, d\nu = 3NkT \tag{96.7}$$

in agreement with the classical empirical law of Dulong and Petit.

At low temperatures the detailed behaviour of the crystal depends on the spectrum described by $\varrho(\nu)$. The precise form of $\varrho(\nu)$ depends on the interatomic interactions and is too complicated to have much practical use. It is therefore useful to consider simple approximations to $\varrho(\nu)$. The simplest possible approximation, proposed by EINSTEIN in 1907, is to replace the whole spectrum by a single average frequency. If we denote this frequency by ν_E, we have

$$F_{ac} = 3N \left\{ \tfrac{1}{2} h\nu_E + kT \ln(1 - e^{-h\nu_E/kT}) \right\}. \tag{96.8}$$

It is convenient to define a characteristic temperature Θ_E by $k\Theta_E = h\nu_E$. We then have

$$F_{ac} = 3N \left\{ \tfrac{1}{2} k\Theta_E + kT \ln(1 - e^{-\Theta_E/T}) \right\} \tag{96.9}$$

and consequently

$$U_{ac} = 3Nk\Theta_E \left\{ \frac{1}{2} + \frac{1}{e^{\Theta_E/T} - 1} \right\}, \tag{96.10}$$

$$\frac{L}{N} \frac{\partial U_{ac}}{\partial T} = 3R \left\{ \frac{\Theta_E/2T}{\sinh(\Theta_E/2T)} \right\}^2. \tag{96.11}$$

At high temperatures this tends to $3R$ in agreement with the empirical law of Dulong and Petit. At lower temperatures by suitable choice of the value of Θ_E good agreement with experiment can be obtained for temperatures down to about $T = \tfrac{1}{2}\Theta_E$ but at lower temperatures the heat capacity decreases much more slowly than predicted by the above formula. This is because according to Einstein's approximation all the modes become unexcited together whereas in fact the higher frequencies die out at higher temperatures than the lower frequencies.

Much better agreement with experiment at the lowest temperatures without loss of agreement at the higher temperatures is obtained by assuming for $g(\nu)$ a form which is more nearly correct for the lowest frequencies. This approximation was proposed by DEBYE in 1912. The lowest frequencies have the longest wave lengths and these wave lengths will be large compared with the spacing between the atoms. Consequently the distribution of the lowest frequencies will be the same as in a continuum. It follows from purely geometrical considerations that in a three dimensional continuum $g(\nu) \propto \nu^2$. Debye's approximation consists in assuming this proportionality to hold up to a maximum frequency ν_m at which there is a sharp cut-off of the spectrum. We accordingly set

$$\varrho(\nu)\, d\nu = \frac{9N}{\nu_m^3} \nu^2\, d\nu \tag{96.12}$$

where the factor 9 ensures that the total number of modes is $3N$. We now have

$$F_{\text{ac}} = \frac{9N}{\nu_m^3} \int_0^{\nu_m} \left\{ \frac{1}{2}\, h\nu + kT \ln\left(1 - e^{-h\nu/kT}\right) \right\} \nu^2\, d\nu . \qquad (96.13)$$

It is convenient to introduce a characteristic temperature Θ_{D} defined by $k\Theta_{\text{D}} = h\nu_m$. At the same time we write u for $h\nu/kT$. We have then

$$F_{\text{ac}} = \frac{9}{8}\, Nk\Theta_{\text{D}} + \frac{9NkT^4}{\Theta_{\text{D}}^3} \int_0^{\Theta_{\text{D}}/T} \ln\left(1 - e^{-u}\right) u^2\, du . \qquad (96.14)$$

Integration by parts gives

$$F_{\text{ac}} = \frac{9}{8} Nk\Theta_{\text{D}} + 3NkT \ln\left(1 - e^{-\Theta_{\text{D}}/T}\right) - \frac{3NkT^4}{\Theta_{\text{D}}^3} \int_0^{\Theta_{\text{D}}/T} \frac{u^3}{e^u - 1}\, du \qquad (96.15)$$

and for the total energy we have

$$U_{\text{ac}} = \frac{9}{8} Nk\Theta_{\text{D}} + \frac{9NkT^4}{\Theta_{\text{D}}^3} \int_0^{\Theta_{\text{D}}/T} \frac{u^3}{e^u - 1}\, du \qquad (96.16)$$

the remaining terms cancelling. The integral occurring in the last two formulae has to be evaluated numerically. This has been done and values have been tabulated[1] as functions of T/Θ_{D}. It is found that for values of T/Θ_{D} greater than $\frac{1}{3}$ the values of F_{ac}, U_{ac} and $S_{\text{ac}} = (U_{\text{ac}} - F_{\text{ac}})/T$ according to these formulae of Debye can be closely approximated by the simpler formulae of Einstein when we take $\Theta_{\text{E}} = 0.72\,\Theta_{\text{D}}$. It is obvious that Θ_{E} must be less than Θ_{D} because the former represents an average frequency, the latter a maximum frequency. At lower temperatures the behaviour predicted by Debye's formulae is quite different from that predicted by Einstein's and is usually a good approximation to the behaviour found experimentally.

It is interesting to consider further Debye's formulae in the limit of low temperatures when $T \ll \Theta_{\text{D}}$. For this purpose we return to formula (96.14) and replace the upper limit of integration Θ_{D}/T by ∞ and obtain

$$-\int_0^{\infty} \ln\left(1 - e^{-u}\right) u^2\, du = \int_0^{\infty} \sum_{n=1}^{\infty} \frac{u^2}{n}\, e^{-nu}\, du$$

$$= \sum_{n=1}^{\infty} \frac{1}{n^4} \int_0^{\infty} y^2 e^{-y}\, dy = 2 \sum_{n=1}^{\infty} \frac{1}{n^4} = \frac{\pi^4}{45} . \qquad (96.17)$$

Substituting this into (96.14) we obtain

$$F_{\text{ac}} = \frac{9}{8} Nk\Theta_{\text{D}} - \frac{\pi^4}{5} \frac{NkT^4}{\Theta_{\text{D}}^3} , \qquad T \ll \Theta_{\text{D}} . \qquad (96.18)$$

The terms neglected in this approximation are of order $e^{-\Theta_{\text{D}}/T}$ or less. The corresponding formulae for the total energy and entropy are

$$U_{\text{ac}} = \frac{9}{8} Nk\Theta_{\text{D}} + \frac{3\pi^4}{5} \frac{NkT^4}{\Theta_{\text{D}}^3} , \qquad (96.19)$$

$$S_{\text{ac}} = \frac{4\pi^4}{5} \frac{NkT^3}{\Theta_{\text{D}}^3} \qquad (96.20)$$

---

[1] J. A. Beattie: J. Math. Phys. (Mass. Inst. Tech.) **6**, 1 (1926/27).

so that

$$\frac{L}{N}\frac{\partial U_{\text{ac}}}{\partial T} = \frac{L}{N}\,T\,\frac{\partial S_{\text{ac}}}{\partial T} = \frac{12\pi^4}{5}\,R\,\frac{T^3}{\Theta_{\text{D}}^3} \qquad (96.21)$$

which is known as Debye's cube law.

97. Other crystals. We shall now consider briefly crystals having a more complicated structure. We begin with crystals composed of well defined linear molecules such as I_2. In a crystal composed of N iodine molecules, apart from the unexcited intranuclear and electronic degrees of freedom, there are $6N$ degrees of freedom. It is an adequate approximation to regard these as $3N$ translational degrees of freedom of the centres of mass of the I_2 molecules, $2N$ orientational degrees of freedom of the I_2 molecules and N internal vibrations of the I_2 molecules. The $3N$ translational degrees of freedom can be treated by Debye's approximation as in an atomic crystal. The N internal vibrations can be treated like the internal vibrations of gaseous molecules. The $2N$ orientational degrees of freedom consist of libration (wobble) about an equilibrium orientation. Accurate quantitative formulae for these librations are not available but it is safe to assume that at the lowest temperatures these librations will behave as harmonic vibrations.

In a crystal consisting of N nonlinear molecules such as CCl_4 each composed of a atoms there are $3N$ translational acoustic modes, $3N$ librational degrees of freedom and $(3a-6)N$ internal vibrational modes. The librational degrees of freedom are too complicated to be treated quantitatively but at the lowest temperatures they presumably become harmonic vibrations.

We could make either of two extreme assumptions about the behaviour of the librational modes. The first alternative assumption is that each molecule librates independently of the others. The second alternative assumption is that the librational frequencies form a continuous spectrum similar to that of the translational acoustic modes. According to the former assumption the librational modes will die out faster than the acoustic modes as the temperature is decreased and will become unexcited when the acoustic modes are obeying Debye's cube law. According to the latter assumption the librational modes as well as the translational acoustic modes will obey Debye's cube law. According to either assumption the cube law will hold at low temperatures for the complete set of degrees of freedom that are not unexcited. On the basis of this rather thin argument it is customary to extrapolate experimental measurements to $T=0$ according to the formulae

$$U(T) - U(0) = \tfrac{1}{4}Na\,T^4, \qquad (97.1)$$

$$S(T) - S(0) = \tfrac{1}{3}Na\,T^3, \qquad (97.2)$$

$$\frac{\partial U}{\partial T} = T\,\frac{\partial S}{\partial T} = Na\,T^3. \qquad (97.3)$$

It is moreover customary to extrapolate according to these same formulae for crystals containing hydrogen bonds such as H_2O and also ionic crystals such as $CaCO_3$. The error, if any, is the less the shorter the range of extrapolation.

In view of the above considerations, and others which cannot be discussed here, the thermodynamic functions of a crystal, in contrast to those of a gas, can never be calculated theoretically as accurately as they can be measured experimentally. Consequently Debye's formulae or improved variants of them should be used only for the purpose of extrapolation to $T=0$.

98. Ideal mixed crystal. We shall now discuss briefly mixed crystals composed of several kinds of molecules able to replace one another in the crystal structure. We shall however confine our attention to molecules so similar that an interchange in position of two different molecules does not sensibly affect the potential energy nor the acoustic spectrum of the crystal. Such a mixed crystal is called ideal. The most important application of the conception of an ideal mixed crystal is to one composed of isotopic molecules. Let us initially consider only two kinds of molecules A and B. The partition function of a crystal composed of N_A molecules of A is of the form

$$Q_A = (e^{-E_{A0}/kT} q_A)^{N_A} \tag{98.1}$$

where E_{A0} relates to the energy per molecule when every molecule is in its lowest state and q_A takes care of the acoustic and internal degrees of freedom. Similarly for a crystal composed of N_B molecules B

$$Q_B = (e^{-E_{B0}/kT} q_B)^{N_B}. \tag{98.2}$$

We turn now to an ideal crystal composed of N_A molecules A and N_B molecules B. According to the definition of ideal the energy when all the molecules are in their lowest states is $N_A E_{A0} + N_B E_{B0}$ and the contribution to the partition function of the acoustic and internal degrees of freedom is $q_A^{N_A} q_B^{N_B}$. There are moreover $(N_A + N_B)!/N_A! N_B!$ ways of arranging the two kinds of molecules on the lattice. Consequently the partition function of the mixed crystal is

$$Q = \frac{(N_A + N_B)!}{N_A! N_B!} (e^{-E_{A0}/kT} q_A)^{N_A} (e^{-E_{B0}/kT} q_B)^{N_B}. \tag{98.3}$$

Owing to (77.12) the free energy of the mixed crystal exceeds that of the two pure crystals out of which it could be formed by

$$- kT \ln Q + kT \ln Q_A + kT \ln Q_B$$
$$= - kT \ln \frac{(N_A + N_B)!}{N_A! N_B!} \tag{98.4}$$

or, when we use Stirling's approximation for the factorials,

$$kT \left(N_A \ln \frac{N_A}{N_A + N_B} + N_B \ln \frac{N_B}{N_A + N_B} \right). \tag{98.5}$$

If we denote the total number of molecules by N and the molecular fractions by x_A, x_B this can be rwritten as

$$N kT (x_A \ln x_A + x_B \ln x_B). \tag{98.6}$$

This can be extended to any number of kinds of molecules. Thus the free energy of mixing is the negative quantity

$$N kT(x_A \ln x_A + x_B \ln x_B + \cdots) \tag{98.7}$$

and the corresponding entropy of mixing is the positive quantity

$$- N k (x_A \ln x_A + x_B \ln x_B + \cdots). \tag{98.8}$$

We notice that the expressions for the free energy of mixing and entropy of mixing in an ideal mixed crystal are precisely the same as in an ideal gaseous mixture at constant pressure.

99. Relation of gas to crystal. We now want to correlate the equilibrium properties of a crystal with those of a gas of the same composition. Before doing so we must devote some attention to energy zeros. As already mentioned we usually define the partition function of each degree of freedom relative to an energy zero equal to the lowest state in that degree of freedom. Consequently the free energy of a system of N molecules contains, as well as terms of the form $-kT \ln Q$, a term NE_0 where E_0 is the energy of the system when all the molecules are in their lowest states. It is hardly necessary to mention that absolute values of free energy, total energy and entropy have no physical meaning; only changes in these quantities are significant. In particular E_0 by itself has no physical meaning. But when we consider the difference between the free energy of a gas and the free energy of a crystal composed of the same N molecules this will contain a term $N(E_0^G - E_0^C)$ where the superscripts G, C denote gas, crystal respectively. This same term occurs in the total energy but cancels out from the entropy. Since a priori we known nothing about the value of $E_0^G - E_0^C$ it is most convenient to use the entropy in correlating the equilibrium properties of the crystal with those of a gas of the same composition. We shall accordingly devote some attention to the entropy difference $S^G(T) - S^C(T')$ where T is a temperature convenient for the study of the gas and T' a temperature convenient for the study of the crystal. For T it is usual to choose either the standard temperature $T = 298.15$ °K equal to 25 °C or when this is not convenient the normal boiling point of the substance. For the crystal we have seen that the lower the temperature the safer it is to make theoretical predictions. It is therefore convenient and usual to consider the entropy difference $S^G(T) - S^C(T' \to 0)$ where the extrapolation to $T' = 0$ has been performed by use of Debye's cube law. This difference can usefully be discussed theoretically and can also be determined calorimetrically. It thus forms a useful link between theory and experiment.

Initially we shall disregard isotopic composition and shall later return to consider what effects the presence of isotopes can have. We begin by making a list of the degrees of freedom. These are

1. Intranuclear, especially nuclear spin,
2. Translational (in the crystal acoustic),
3. Rotational in gas or orientational in crystal,
4. Internal vibrational,
5. Electronic.

Of these the intranuclear make the same contribution to S^G as to S^C, provided the symmetry number is included in the gaseous rotational partition function and they may therefore be disregarded. The contributions of the remaining degrees of freedom to the gas have been discussed in detail and they can be computed provided the characteristic temperatures have been determined spectroscopically or otherwise. By comparison with the calorimetrically determined difference $S^G(T) - S^C(T' \to 0)$ we thus obtain the contribution of these degrees of freedom to $S^C(T' \to 0)$. It is found that this contribution is almost always zero. We shall now discuss the interpretation of this.

100. Crystal at absolute zero. If the partition function for any set of degrees of freedom is

$$Q = g_0 + g_1 e^{-E_1/kT} + g_2 e^{-E_2/kT} + \cdots \tag{100.1}$$

the contribution of these degrees of freedom to the molecular entropy is

$$\frac{\partial (kT \ln Q)}{\partial T} = k \ln Q + \frac{g_1 E_1 e^{-E_1/kT} + \cdots)}{T(g_0 + g_1 e^{-E_1/kT} + \cdots)} \tag{100.2}$$

which tends to $k \ln g_0$ as T tends to zero. The experimental fact that as $T \to 0$ the contribution of several sets of degrees of freedom to the entropy of the crystal tends to zero may therefore be interpreted to mean that the lowest energy level of each of these sets of degrees of freedom is nondegenerate ($g_0 = 1$). This means that every molecule in the crystal has only one state of lowest energy. This is what we should obviously expect unless there were any degeneracy due to alternative orientations of equal energy. This type of degeneracy cannot occur in the acoustic modes since in the crystal lattice there is clearly only one equilibrium position for each atom. Nor can it occur in the internal vibrational modes since each mode has a single state of lowest energy specified by $v = 0$ where v is its vibrational quantum number. The complete absence of degeneracy in the electronic degrees of freedom is not so obvious. We have seen that almost all chemically saturated gaseous molecules are in an unexcited electronic state which is nondegenerate and there is no reason to expect any electronic degeneracy in the crystal. But we must also consider odd molecules such as NO and Na which in the gaseous state have an electronic degeneracy $g_0 = 2$. If we use g_0 to denote the electronic degeneracy in the gaseous molecule, then in a crystal of N molecules owing to the interaction between them the $g_0 N$ electronic states are not of equal energy and the electrons fill the lowest of these states and they can do this in only one way. Thus the interaction between the molecules removes the electronic degeneracy. Whether this happens by the occurrence of electronic energy bands or by chemical combination of odd molecules to form dimers is for the present discussion irrelevant. There remains for discussion the orientational degrees of freedom in the crystals. The interpretation of the absence of degeneracy is simply that each molecule has only one possible equilibrium orientation. If we denote the number of possible equilibrium orientations of each molecule by o then the theoretical interpretation of the calorimetric measurements of $S^G - S^C$ is that $o = 1$ usually but that there are a few exceptions. Before studying the implication of this it will be useful to introduce a digression on the subject of isotopes.

101. Isotopes. The presence of isotopes can have three effects. The first effect is that in the formulae for the gaseous thermodynamic functions $\ln M$ must be replaced by the suitably weighted sum $\sum_i x_i \ln M_i$ where x_i is the molecular fraction of the particular isotope i having a molar mass M_i. Similarly each term of the form $\ln \Theta$ where Θ is either a rotational or a vibrational characteristic temperature must be replaced by the suitably weighted average $\sum_i x_i \ln \Theta_i$ where Θ_i is the characteristic temperature of the isotope i. This effect is trivial.

The second effect is the presence of the terms called entropy of mixing $-Nk \sum_i x_i \ln x_i$. We have already pointed out that this same expression is applicable both to the crystal and to the gas. Hence as long as the solid and the gas have the same isotopic composition the entropy of mixing makes zero contributions to $S^G - S^C$ and so may be disregarded.

The third effect to be considered is that associated with differences in symmetry. Let us consider the particular example of Cl_2. There are three kinds of molecules $^{35}Cl^{35}Cl$, $^{37}Cl^{37}Cl$ and $^{35}Cl^{37}Cl$. For the molecules $^{35}Cl^{35}Cl$ and $^{37}Cl^{37}Cl$ the symmetry number σ is 2 while for the molecule $^{35}Cl^{37}Cl$ the value of σ is 1. In the crystal on the other hand o for $^{35}Cl^{37}Cl$ will have the value 2 because each molecule can be reversed to give a physically distinct state of the crystal of effectively the same energy, whereas for $^{35}Cl^{35}Cl$ and $^{37}Cl^{37}Cl$ there are not two distinguishable equilibrium orientations and so $o=1$. Now S^G contains a term $-k\ln\sigma$ and S^C contains a term $k\ln o$ and so S^G-S^C contains a term $-k\ln\sigma o$. Consequently S^G-S^C depends on σ and o only through the combination σo. We notice that this product σo has the same value 2 for all three types of molecules. Disregard of the isotopic composition means assigning to $^{35}Cl^{37}Cl$ a fictitious value of $\sigma=2$ instead of $\sigma=1$ and to o a fictitious value $o=1$ instead of $o=2$. In the experimentally measurable quantity S^G-S^C the two errors cancel.

102. Exceptional crystals. We have already mentioned that for almost all crystals $o=1$ and we have discussed the physical significance of this. We have also mentioned that exceptions exist and we shall now discuss why these occur. The simplest known exceptions are CO and NNO. These linear molecules have in common the property of such low polarity that in the crystal they simulate the molecule $^{35}Cl^{37}Cl$. This means that instead of a regular orientation of the molecules there is randomness between two directions. It is true that however small the asymmetry of the molecular field, the difference of energy between regular and random orientations must become small compared with NkT as $T\to 0$, but at such low temperatures the molecules have not sufficient energy to reorientate. In other words as $T\to 0$ the crystal with random orientations becomes metastable, but this is the only crystalline state having a real existance and it is more important than a hypothetical crystal with regular orientations. The following table relating to σ and o summarises the position and shows that the behaviour of CO resembles that of the symmetrical molecule Cl_2 more than that of the unsymmetrical molecule HCl.

	$^{35}Cl^{35}Cl$	$^{35}Cl^{37}Cl$	ClCl isotopes ignored	CO	HCl
σ	2	1	2	1	1
o	1	2	1	2	1
σo	2	2	2	2	1

A slightly more complicated example is NO. It is probable that in the crystal the unit is the square dimer $\begin{smallmatrix}ON\\NO\end{smallmatrix}$ and that there is randomness between the two orientations $\begin{smallmatrix}NO\\ON\end{smallmatrix}$ and $\begin{smallmatrix}ON\\NO\end{smallmatrix}$. This leads consequently to an entropy term $k\ln 2$ per dimer or $\tfrac{1}{2}k\ln 2$ per monomer. A yet more complicated example is that of H_2O. The O atoms form a perfectly regular lattice and between each pair of neighbouring O atoms there is an H atom. Each H atom is closer to one of its neighbour O atoms than to the other and each O atom has two of its four neighbouring H atoms close to it. Subject to these restrictions there is randomness in the arrangement of the H atoms. It can be shown that this randomness leads to $o=4\times\tfrac{6}{16}=\tfrac{3}{2}$.

All the above mentioned theoretical values of o are in agreement with experiment within the experimental accuracy but this is a distorted way of describing the situation. What in fact happened is that these few exceptions to the usual rule $o=1$ were discovered by experiment and the above mentioned theoretical explanations were proposed and are generally accepted. We must be prepared for other exceptions but they are likely to be few. It is a priori conceivable that OCS might have $o=2$ but experiment shows that in fact $o=1$. This means that the O and S atoms differ too much in size for a random orientation to occur, but without experiment we could not be sure.

The position may be summarized as follows. For nearly all existing crystals $o=1$. There are a few exceptions with $o>1$ and these are understood to be metastable. The statement $o=1$ for all stable crystals is true but not very useful because only experiment can detect which crystals are metastable.

103. Conventional entropy values. We have found it convenient to discuss the value of o defined as the number of distinct orientations per molecule in the crystal in the limit $T\to0$ and the value of o has been deduced from the experimental value of $S^G(T)-S^C(T\to0)$. Let us now consider an alternative approach. Although only entropy differences, never absolute entropies, are measurable the contribution of specified degrees of freedom to the entropy has a definite value. Let us then define the *conventional entropy* of a gas as the sum of the contributions to the entropy of the following degrees of freedom

> translational,
> rotational,
> internal vibrational,
> electronic,

but omitting the contributions from

> intranuclear, especially spin,
> isotopic composition.

According to this convention we can compute $S^G(T)$. Substituting this value into the experimental quantity $S^G(T)-S^C(T\to0)$ we obtain a value of $S^C(T\to0)$ which we call the *conventional entropy* of the crystal as $T\to0$. If now we write

$$S^C(T\to0)=Nk\ln o \qquad (103.1)$$

this gives us a new definition of o. According to this new definition the statements $o=1$ and $S^C(T\to0)=0$ are equivalent. So are the statements $o>1$ and $S^C(T\to0)>0$. The two definitions of o are in fact equivalent for all the substances which we have mentioned and for all substances for which para-ortho separation may be ignored, but not for H_2 and D_2. The conventional value of $S^C(T\to0)$ for these two substances depends on the extent to which para-ortho equilibrium has been established. We shall here consider only the situation when this equilibrium has been completely established. Then H_2 is in the pure para form. The contribution to the entropy of all the degrees of freedom included in the conventional definition is zero. Moreover the spin factor in the partition function, which is $2^2=4$ in the gas at high temperatures, is reduced to 1 in the crystal as $T\to0$. Thus the contribution of nuclear spin to

$$S^G(T)-S^C(T\to0) \qquad (103.2)$$

is

$$Nk(\ln4-\ln1)=Nk\ln4. \qquad (103.3)$$

When conventionally the spin contribution is ignored in S^G, we obtain conventionally

$$S^C(T\to0)=-Nk\ln4. \qquad (103.4)$$

Thus in the limit $T\to0$ the conventional entropy of para hydrogen is negative.

The situation for deuterium is similar. At high temperatures the contribution of nuclear spin to the entropy is $Nk\ln\varrho^2=Nk\ln9$ and at low temperatures for the pure ortho form it is $Nk\ln\{\tfrac{1}{2}\varrho(\varrho+1)\}=Nk\ln6$. Hence conventionally

$$S^C(T\to0)=Nk(\ln6-\ln9)=-Nk\ln\tfrac{3}{2} \qquad (103.5)$$

which is negative.

104. Gaseous chemical equilibrium. Our next application of statistical thermodynamics is to chemical equilibrium. We recall that the equilibrium condition for the process

$$v_A A + v_B B + \cdots \rightleftharpoons v_L L + v_M M + \cdots \tag{104.1}$$

is according to (25.8) and (25.9)

$$A = (v_A \mu_A + v_B \mu_B + \cdots) - (v_L \mu_L + v_M \mu_M + \cdots) = 0 \tag{104.2}$$

or alternatively by use of (34.1)

$$\lambda_A^{v_A} \lambda_B^{v_B} \cdots = \lambda_L^{v_L} \lambda_M^{v_M} \cdots . \tag{104.3}$$

Further progress consists in expressing the μ's or λ's in terms of temperature, pressure, composition and molecular characteristics. We shall initially confine our attention to a reaction between perfect gases. For any component gas A of a perfect gaseous mixture we may express λ_A in the form (47.7) namely

$$\lambda_A = \lambda_A^\dagger \, p_A / P^\dagger \tag{104.4}$$

where $p_A = x_A P$ denotes the partial pressure of A in the mixture and $\lambda_A^\dagger$ depends on the temperature and the nature of the gas A but is independent of the pressure and of the other gases in the mixture. Substitution of (104.4) into (104.3) gives

$$\frac{p_L^{v_L} p_M^{v_M} \cdots}{p_A^{v_A} p_B^{v_B} \cdots} = K(T) \tag{104.5}$$

where $K(T)$, called the equilibrium constant, is given by

$$\ln K = v_A \ln \lambda_A^\dagger + v_B \ln \lambda_B^\dagger + \cdots - v_L \ln \lambda_L^\dagger - v_M \ln \lambda_M^\dagger - \cdots -$$
$$- (v_A + v_B + \cdots - v_L - v_M - \cdots) \ln P^\dagger \tag{104.6}$$

which we may abbreviate to

$$\ln K = - \varDelta \sum v \ln \frac{\lambda^\dagger}{P^\dagger} . \tag{104.7}$$

Thus the evaluation of K reduces to that of the individual $\lambda^\dagger$'s. For a gas of monatomic molecules we obtain by substituting $p = NkT/V$ into (84.14)

$$\ln \lambda^\dagger = \frac{E_0}{kT} + \frac{3}{2} \ln (h^2/2\pi m) - \frac{5}{2} \ln kT + \ln P^\dagger - \ln g_0 . \tag{104.8}$$

For polyatomic linear molecules we obtain from (91.2)

$$\ln \lambda^\dagger = \frac{E_0}{kT} + \frac{3}{2} \ln (h^2/2\pi m) - \frac{5}{2} \ln kT + \ln P^\dagger - \ln g_0 +$$
$$+ \ln \frac{\sigma \Theta_{\mathrm{rot}}}{T} + \sum_{i=1}^{a-5} \ln (1 - e^{-\Theta_i/T}) \tag{104.9}$$

and for nonlinear molecules from (94.4)

$$\ln \lambda^\dagger = \frac{E_0}{kT} + \frac{3}{2} \ln (h^2/2\pi m) - \frac{5}{2} \ln kT + \ln P^\dagger - \ln g_0 +$$
$$+ \frac{1}{2} \ln \frac{\sigma^2 \Theta_A \Theta_B \Theta_C}{\pi T^3} + \sum_{i=1}^{a-6} \ln (1 - e^{-\Theta_i/T}) . \tag{104.10}$$

For the purpose of numerical calculation it is convenient to rewrite

$$\frac{3}{2}\ln(h^2/2\pi m) - \frac{5}{2}\ln kT = -\frac{3}{2}\ln\frac{M}{\text{g mole}^{-1}} - \frac{5}{2}\ln\frac{T}{4.33_1\,°\text{K}} \qquad (104.11)$$

where $M = Lm$ is the molar mass or conventional "molecular weight". Provided the information concerning sizes, shapes and characteristic frequencies of the molecules has been determined spectroscopically, all the terms in $\ln \lambda^\dagger$ can be computed except E_0. Consequently $\ln K$ can be computed except for a term

$$-\frac{\nu_L E_{L0} + \nu_M E_{M0} + \cdots - \nu_A E_{A0} - \nu_B E_{B0} - \cdots}{kT} \qquad (104.12)$$

which we shall study further. For this purpose we consider the temperature dependence of K. Using (34.2) we deduce

$$RT^2\frac{\partial \ln K}{\partial T} = \nu_L h_L + \nu_M h_M + \cdots - \nu_A h_A - \nu_B h_B - \cdots = \Delta H, \qquad (104.13)$$

where ΔH is the heat of reaction and the individual $h_A, \ldots$ are given by

$$-h_A = RT^2\frac{\partial \ln \lambda_A}{\partial T} = RT^2\frac{\partial \ln \lambda_A^\dagger}{\partial T}. \qquad (104.14)$$

For a gas of monatomic molecules we deduce from (104.8)

$$h_A = LE_{A0} + \tfrac{5}{2}RT, \qquad (104.15)$$

for one of linear molecules from (104.9)

$$h_A = LE_{A0} + \frac{7}{2}RT + \sum_{i=1}^{a-5}\frac{R\Theta_i}{e^{\Theta_i/T}-1}, \qquad (104.16)$$

and for one of nonlinear molecules from (104.10)

$$h_A = LE_{A0} + 4RT + \sum_{i=1}^{a-6}\frac{R\Theta_i}{e^{\Theta_i/T}-1}. \qquad (104.17)$$

We see then that the heat of reaction ΔH can be computed at any temperature except for the term

$$L(\nu_L E_{L0} + \nu_M E_{M0} + \cdots - \nu_A E_{A0} - \nu_B E_{B0} - \cdots). \qquad (104.18)$$

Thus both K and ΔH can be computed at any temperature apart from the single unknown quantity

$$\Delta \Sigma \nu E_0 = \nu_L E_{L0} + \nu_M E_{M0} + \cdots - \nu_A E_{A0} - \nu_B E_{B0} - \cdots. \qquad (104.19)$$

Consequently a measurement of the heat of reaction ΔH at any one temperature may be used to determine the value of $\Delta \Sigma \nu E_0$ and then this may be used to compute K at any temperature and ΔH at any other temperature.

105. Chemical equilibrium involving crystals. These considerations can be extended to equilibria involving crystal phases as well as ideal gases. We shall merely indicate briefly how this is done. For a crystal in contrast to a gas we may with sufficient accuracy regard λ as independent of the pressure so that formally $\lambda_A(T, P) = \lambda_A^\dagger(T)$. Consequently we may write

$$\frac{p_L^{\nu_L} p_M^{\nu_M} \cdots}{p_A^{\nu_A} p_B^{\nu_B} \cdots} = K(T), \qquad (105.1)$$

provided the product on the left contains p factors relating only to the gaseous species and we still have

$$\ln K = \nu_A \ln \lambda_A^{\dagger} + \nu_B \ln \lambda_B^{\dagger} + \cdots - \nu_L \ln \lambda_L^{\dagger} - \nu_M \ln \lambda_M^{\dagger} - \cdots \tag{105.2}$$

where the terms on the right will be of two kinds, those relating to gaseous species and those relating to crystalline species. For the gaseous species each $\ln \lambda^{\dagger} - E_0/kT$ is computable as before. For each crystalline species, we ignore the dependence of λ on pressure, and we have

$$\frac{d \ln \lambda_A}{dT} = - \frac{h_A}{RT^2}. \tag{105.3}$$

The integrated form of this relation is

$$\ln \lambda_A = \ln \lambda_A (T \to 0) - \int_0^T \frac{h_A}{RT^2} dT. \tag{105.4}$$

Calorimetric measurements of the temperature dependence of the heat function, that is to say of the heat capacity, down to low temperatures together with an extrapolation to $T=0$ by Debye's formula enable us to evaluate the integral on the right. The first term on the right of (105.4) may be written

$$\ln \lambda_A (T \to 0) = \frac{E_{A0}}{kT} - \frac{s_A (T \to 0)}{R} \tag{105.5}$$

where $s_A (T \to 0)$ is the conventional molar entropy at $T=0$. Apart from the few exceptions due to metastability already discussed this is zero and formula (105.5) reduces to

$$\ln \lambda_A (T \to 0) = \frac{E_{A0}}{kT}. \tag{105.6}$$

We can also write down a formula for the heat of reaction ΔH which will contain gaseous terms of the same form as previously and crystal terms of the form

$$h_A = L E_{A0} + \{h_A (T) - h_A (0)\} \tag{105.7}$$

where $\{h_A (T) - h_A (0)\}$ is a calorimetrically measurable quantity. It is clear that once again the single unknown quantity

$$\Delta \Sigma \nu E_0 = \nu_L E_{L0} + \nu_M E_{M0} + \cdots - \nu_A E_{A0} - \nu_B E_{B0} - \cdots \tag{105.8}$$

occurs both in the formula for $\ln K$ and in that for ΔH.

It follows that spectroscopic determination of the required molecular characteristics of the gaseous molecules together with calorimetric determination of the temperature dependence of H for each crystalline species, down to sufficiently low temperatures to allow extrapolation to $T=0$ would enable us to calculate both K and ΔH at any temperature were it not for the one unknown quantity $\Delta \Sigma \nu E_0$. Consequently a measurement of the heat of reaction ΔH at a single temperature enables us to evaluate the single unknown $\Delta \Sigma \nu E_0$ and so compute K at any temperature and ΔH at any other temperature.

F. Degenerate ideal gases and radiation.

106. Degenerate ideal gas. In Chap. D it was explained that Boltzmann statistics is nearly always an adequate approximation to either Fermi-Dirac statistics or Bose-Einstein statistics. The condition for this is that $\lambda \exp (E_0/kT) \ll 1$ or, if as usual we take E_0 as our energy zero, simply $\lambda \ll 1$. This inequality is satisfied by all ordinary gases under the conditions in which measurements can be made. We shall now consider what happens when the condition $\lambda \ll 1$ is not satisfied and in particular the opposite extremes. These are, as we shall verify, $\lambda \to \infty$ in Fermi-Dirac statistics and $\lambda \to 1$ in Bose-Einstein statistics.

We shall confine our discussion almost entirely to a gas of structureless particles and we continue to neglect any interaction between the particles. When the condition $\lambda \ll 1$ no longer holds, and especially when λ is comparable to unity or greater, such a gas is called a degenerate ideal gas.

We begin by examining the distribution of quantum states of particles of mass m enclosed in a rectangular box of sides with lengths a, b, c. The quantum states are described by three quantum numbers n_x, n_y, n_z and their energies relative to a state of rest as zero are

$$E = \frac{h^2}{8m}\left(\frac{n_x^2}{a^2} + \frac{n_y^2}{b^2} + \frac{n_z^2}{c^2}\right), \quad n_x, n_y, n_z = 1, 2, 3, \dots . \tag{106.1}$$

Since, for a given value of E,

$$\frac{n_x^2}{a^2} + \frac{n_y^2}{b^2} + \frac{n_z^2}{c^2} = \frac{8mE}{h^2} \tag{106.2}$$

the number of states of (kinetic) energy less than or equal to E is equal to the volume of an octant of the ellipsoid defined by the last equation in n_x, n_y, n_z space. This ellipsoid has semi-axes $(8mEa^2/h^2)^{\frac{1}{2}}$, $(8mEb^2/h^2)^{\frac{1}{2}}$ and $(8mEc^2/h^2)^{\frac{1}{2}}$. The volume of the octant is therefore

$$\frac{1}{8}\frac{4\pi}{3}\left(\frac{8mE}{h^2}\right)^{\frac{3}{2}} a\,b\,c = \frac{4\pi}{3}\left(\frac{2mE}{h^2}\right)^{\frac{3}{2}} V \tag{106.3}$$

where V denotes the volume of the box. It can be shown that this number is independent of the shape of the box provided it is reasonably large. By differentiation we obtain for the number of states in the energy range E to $E + dE$

$$g(E)\,dE = 2\pi\left(\frac{2m}{h^2}\right)^{\frac{3}{2}} V E^{\frac{1}{2}}\,dE . \tag{106.4}$$

We have derived this formula with E measured from a zero which is not the lowest quantum state but the state of rest. Under the conditions where we are justified in replacing the discrete distribution of energies by a continuum the difference is quite inappreciable. We shall, as usual, take the energy of the lowest state as zero and still use this formula.

We recall that the condition for replacing the discrete distribution of energies by a continuum is a valid approximation provided the spacing of the states is small compared with kT. When this condition is satisfied we call the degree of freedom classical. In the present case the condition is that

$$\frac{h^2}{m\,a^2} \ll kT \tag{106.5}$$

or

$$\frac{h}{(m\,kT)^{\frac{1}{2}}} \ll V^{\frac{1}{3}} . \tag{106.6}$$

This inequality states that the de Broglie wave length of particles having a kinetic energy kT is small compared with the linear dimensions of the container. We again emphasize that this condition is always satisfied, even for particles as light as electrons, in any macroscopic container. This inequality must be sharply contrasted with the condition for the validity of Boltzmann statistics namely

$$\frac{h}{(m\,kT)^{\frac{1}{2}}} \ll \left(\frac{V}{N}\right)^{\frac{1}{3}} \tag{106.7}$$

which states that the de Broglie wave length of particles having a kinetic energy kT is small compared with the average spacing between neighbouring particles. This condition is also satisfied for all ordinary gases under experimental conditions but it is not at all satisfied by the conduction electrons in a metal even at quite high temperatures. For example for electrons at 10^3 °K we have

$$\frac{h}{(m\,kT)^{\frac{1}{2}}} = \frac{6.6\times 10^{-27}\times 10^8}{(0.9\times 10^{-27}\times 1.4\times 10^{-16}\times 10^3)^{\frac{1}{2}}}\,\text{Å} = 60\,\text{Å} \tag{106.8}$$

which is much greater than the atomic spacing in a metal.

We now construct the grand partition functions $\varXi$,

$$\ln \varXi = \int_0^\infty g(E)\ln\left(1 + \lambda\,e^{-E/kT}\right)dE, \qquad \text{Fermi-Dirac}, \tag{106.9}$$

$$\ln \varXi = \int_0^\infty -g(E)\ln\left(1 - \lambda\,e^{-E/kT}\right)dE, \qquad \text{Bose-Einstein}. \tag{106.10}$$

Considerable use will be made of the last formula in our discussion of a Bose-Einstein gas and for this purpose formula (106.4) for the distribution function $g(E)$ will be used. But the behaviour of a Fermi-Dirac gas can be investigated without direct reference to $\varXi$.

107. Fermi-Dirac gas. We now consider specifically a Fermi-Dirac gas and we begin by recalling the distribution law (72.19) for the number N_r of particles in the state r. We rewrite this as

$$N_r = \frac{1}{\lambda^{-1}\,e^{E_r/kT} + 1} \tag{107.1}$$

and we now restrict our discussion to the case $\lambda \gg 1$ which is just the opposite of that of Boltzmann statistics. Qualitatively this distribution law means that the particles, like the electrons in an atom, will crowd into the states of low energy, one particle in each, leaving the states of higher energy empty. Between the effectively filled energy levels and the effectively empty levels there will be one energy level, which we denote by E_F, at which on the average exactly half the states are occupied. The absolute activity λ and chemical potential μ are then related to E_F by the simple formulae

$$\lambda = \exp\left(E_F/kT\right), \tag{107.2}$$

$$\mu/L = E_F. \tag{107.3}$$

E_F is called the Fermi level. The distribution law (107.1) may be rewritten in terms of E_F

$$N_r = \frac{1}{e^{(E_r - E_F)/kT} + 1}. \tag{107.4}$$

It follows that

$$N_r = 1, \qquad E_F - E_r \gg kT, \tag{107.5}$$

$$N_r = 0, \qquad E_r - E_F \gg kT. \tag{107.6}$$

Moreover since we are supposing that $\lambda \gg 1$ it follows that $kT \gg E_F$ and so the intermediate range of energies where $|E_r - E_F|$ is not much greater than kT

is only a small fraction of E_F. Hence as far as the average energy and other average properties are concerned we may replace the exact distribution law by the approximation

$$N_r = 1, \quad E_r < E_F, \tag{107.7}$$

$$N_r = 0, \quad E_r > E_F. \tag{107.8}$$

From this we deduce using (106.4)

$$N = \sum_r N_r = \int_0^{E_F} g(E)\, dE = \int_0^{E_F} 2\pi \left(\frac{2m}{h^2}\right)^{\frac{3}{2}} V E^{\frac{1}{2}}\, dE$$

$$= \frac{4\pi}{3} \left(\frac{2m}{h^2}\right)^{\frac{3}{2}} V E_F^{\frac{3}{2}} \tag{107.9}$$

and consequently

$$E_F = \frac{h^2}{2m} \left(\frac{3N}{4\pi V}\right)^{\frac{2}{3}} \tag{107.10}$$

from which we see that E_F is independent of T but proportional to $(N/V)^{\frac{2}{3}}$. It follows that $\lambda \to \infty$ as $T \to 0$.

The total energy U is determined by

$$\frac{U}{N} = \frac{\int_0^{E_F} g(E)\, E\, dE}{\int_0^{E_F} g(E)\, dE} = \frac{\int_0^{E_F} E^{\frac{3}{2}}\, dE}{\int_0^{E_F} E^{\frac{1}{2}}\, dE} = \frac{3}{5} E_F \tag{107.11}$$

which is thus independent of the temperature. Consequently we have

$$\frac{F}{N} = \frac{U}{N} = \frac{3}{5} E_F. \tag{107.12}$$

The pressure is determined by

$$PV = -V \frac{\partial F}{\partial V} = -\frac{3}{5} NV \frac{\partial E_F}{\partial V} = \frac{2}{5} N E_F = \frac{2}{3} U. \tag{107.13}$$

For the Gibbs function G we have

$$\frac{G}{N} = \frac{F}{N} + \frac{PV}{N} = E_F \tag{107.14}$$

confirming the thermodynamic relation

$$\frac{G}{N} = \frac{\mu}{L} = kT \ln \lambda. \tag{107.15}$$

108. Higher approximation. The accuracy of the approximation which we have investigated is sufficient for many purposes. A more detailed investigation[1] shows that more accurate formulae are

$$\frac{F}{N} = \frac{3}{5} E^* - \frac{\pi^2}{4} \frac{k^2 T^2}{E^*}, \tag{108.1}$$

$$\frac{U}{N} = \frac{3}{5} E^* + \frac{\pi^2}{4} \frac{k^2 T^2}{E^*}, \tag{108.2}$$

$$\frac{PV}{N} = \frac{2}{5} E^* + \frac{\pi^2}{6} \frac{k^2 T^2}{E^*} = \frac{2}{3} \frac{U}{N}, \tag{108.3}$$

$$\frac{G}{N} = \frac{\mu}{L} = kT \ln \lambda = E^* - \frac{\pi^2}{12} \frac{k^2 T^2}{E^*} \tag{108.4}$$

---

[1] A. Sommerfeld: Z. Physik **47**, 1 (1928).

where E^* is defined by

$$E^* = \frac{h^2}{2m}\left(\frac{3N}{4\pi V}\right)^{\frac{2}{3}}.$$

(108.5)

In the approximation which neglects the terms in $k^2 T^2/E^*$ there is no difference between E^* and E_F.

109. Conducting electrons. The non-degenerate ideal Fermi-Dirac gas which we have discussed provides the simplest model of the collection of conducting electrons in a metal. It would take us too far afield to examine this application and we shall confine ourselves to two comments. Whereas we have been considering a system of structureless particles, an electron is not structureless since it has a spin. Consequently the number of states of given energy is doubled. The effect of this is to introduce an extra factor $2^{-\frac{2}{3}}$ in E_F and E^*, but the relations between the thermodynamic functions and E_F or E^* are unaffected. The other modification of the theory for application to the conducting electrons in a metal concerns the potential energy. We have throughout assumed that the potential energy is uniform and have ignored it. When we consider electrons in a metal it may be that the assumption of a uniform potential $-\chi$ is a reasonable approximation but this potential due to the atomic cores (nuclei and inner shells of electrons) will certainly not be independent of the volume. Hence the formula for the free energy must be modified to

$$\frac{F}{N} = -\chi + \frac{3}{5} E_F$$

(109.1)

and the resulting formula for the pressure is

$$\frac{PV}{N} = V\frac{\partial \chi}{\partial V} + \frac{2}{5} E_F.$$

(109.2)

In fact the pressure is negligible and the equilibrium volume is determined by the condition

$$V\frac{\partial \chi}{\partial V} + \frac{2}{5} E_F = 0.$$

(109.3)

110. Generalization to s dimensions. Incidentally the behaviour of a Fermi-Dirac degenerate ideal gas, by contrast with a Bose-Einstein degenerate ideal gas, is nowise strikingly dependent on the number s of dimensions. The formulae obtained in the first approximation are in s dimensions

$$g(E) \propto E^{\frac{1}{2}s-1},$$

(110.1)

$$E_F^{\frac{1}{2}s} \propto \frac{N}{V} \text{ independent of } T,$$

(110.2)

$$\frac{F}{N} = \frac{U}{N} = \frac{s}{s+2} E_F,$$

(110.3)

$$\frac{PV}{N} = \frac{2}{s+2} E_F.$$

(110.4)

111. Bose-Einstein gas above condensation temperature. We now turn to consider a Bose-Einstein ideal gas. We begin with the grand partition function $\varXi$.

We have, using (106.4),

$$\ln \Xi = \int_0^\infty - \ln (1 - \lambda\, e^{-E/kT})\, g(E)\, dE$$

$$= 2\pi \left(\frac{2m}{h^2}\right)^{\frac{3}{2}} V \int_0^\infty - E^{\frac{1}{2}} \ln (1 - \lambda\, e^{-E/kT})\, dE$$

$$= 2\pi \left(\frac{2m}{h^2}\right)^{\frac{3}{2}} V \int_0^\infty E^{\frac{1}{2}} \sum_{j=1}^\infty \lambda^j j^{-1}\, e^{-jE/kT}\, dE$$

$$= 2\pi \left(\frac{2m\,kT}{h^2}\right)^{\frac{3}{2}} V \sum_{j=1}^\infty \lambda^j j^{-\frac{5}{2}} \int_0^\infty y^{\frac{1}{2}}\, e^{-y}\, dy$$

$$= \left(\frac{2\pi\, m\,kT}{h^2}\right)^{\frac{3}{2}} V \sum_{j=1}^\infty \lambda^j j^{-\frac{5}{2}}. \tag{111.1}$$

From this we immediately derive using the relations of Sect. 74

$$\frac{PV}{kT} = \ln \Xi = \left(\frac{2\pi\, m\,kT}{h^2}\right)^{\frac{3}{2}} V \sum_{j=1}^\infty \lambda^j j^{-\frac{5}{2}}, \tag{111.2}$$

$$N = \lambda \frac{\partial \ln \Xi}{\partial \lambda} = \left(\frac{2\pi\, m\,kT}{h^2}\right)^{\frac{3}{2}} V \sum_{j=1}^\infty \lambda^j j^{-\frac{3}{2}}, \tag{111.3}$$

$$\frac{U}{kT} = T \frac{\partial \ln \Xi}{\partial T} = \frac{3}{2} \left(\frac{2\pi\, m\,kT}{h^2}\right)^{\frac{3}{2}} V \sum_{j=1}^\infty \lambda^j j^{-\frac{5}{2}} \tag{111.4}$$

and consequently

$$\frac{PV}{NkT} = \frac{U}{\frac{3}{2} N kT} = \frac{\sum_{j=1}^\infty \lambda^j j^{-\frac{5}{2}}}{\sum_{j=1}^\infty \lambda^j j^{-\frac{3}{2}}}. \tag{111.5}$$

When $\lambda \ll 1$ we may replace each $\sum_{j=1}^\infty$ by λ and the formulae reduce to the familiar relations previously obtained by using Boltzmann statistics.

We shall now examine the behaviour of the gas when λ is not small compared with unity. It is convenient to define a temperature T_0 by

$$\frac{N}{V} \left(\frac{h^2}{2\pi\, m\,kT_0}\right)^{\frac{3}{2}} = \sum_{j=1}^\infty j^{-\frac{3}{2}} = \zeta\left(\frac{3}{2}\right) = 2.612 \tag{111.6}$$

so that T_0 is proportional to $(N/V)^{-\frac{2}{3}}$ and is independent of T. We shall initially confine ourselves to temperatures greater than T_0. We have then for given N/V

$$\sum_{j=1}^\infty \lambda^j j^{-\frac{3}{2}} = \left(\frac{T_0}{T}\right)^{\frac{3}{2}} \sum_{j=1}^\infty j^{-\frac{3}{2}} = \left(\frac{T_0}{T}\right)^{\frac{3}{2}} \zeta\left(\frac{3}{2}\right) \tag{111.7}$$

from which it is clear that as T decreases towards T_0 the value of λ increases monotonically towards 1. Our formulae for the thermodynamic functions express these in terms of T, V, λ. We want to obtain other formulae in which the thermodynamic functions are expressed in terms of T, V, N.

It is convenient to introduce the abbreviation x defined by

$$x = \frac{N}{V}\left(\frac{h^2}{2\pi m kT}\right)^{\frac{3}{2}}$$

(111.8)

equivalent to

$$x = \sum_{j=1}^{\infty} \lambda^j j^{-\frac{3}{2}} = \left(\frac{T_0}{T}\right)^{\frac{3}{2}} \sum_{j=1}^{\infty} j^{-\frac{3}{2}} = \left(\frac{T_0}{T}\right)^{\frac{3}{2}} \zeta\left(\frac{3}{2}\right).$$

(111.9)

We may then rewrite (111.7) as

$$x = \lambda\left(1 + a_1\lambda + a_2\lambda^2 + a_3\lambda^3 + \cdots\right),$$

(111.10)

$$a_1 = 2^{-\frac{3}{2}}, \quad a_2 = 3^{-\frac{3}{2}}, \quad a_3 = 4^{-\frac{3}{2}}\ldots.$$

(111.11)

Inverting this expansion we obtain

$$\lambda = x\left(1 - b_1 x - b_2 x^2 - b_3 x^3 + \cdots\right),$$

(111.12)

$$b_1 = 2^{-\frac{3}{2}} = 0.3535_5,$$

(111.13)

$$b_2 = 3^{-\frac{3}{2}} - 2^{-2} = -0.0575_5,$$

(111.14)

$$b_3 = 2^{-3} - 5.6^{-\frac{3}{2}} + 5.2^{-\frac{3}{2}} = 0.0057_6$$

(111.15)

which according to (111.9) is equivalent to

$$\lambda = x\left\{1 - c_1\left(\frac{T_0}{T}\right)^{\frac{3}{2}} - c_2\left(\frac{T_0}{T}\right)^3 - c_3\left(\frac{T_0}{T}\right)^{\frac{9}{2}} - \cdots\right\},$$

(111.16)

$$c_1 = 2.612\, b_1 = 0.9235,$$

(111.17)

$$c_2 = (2.612)^2\, b_2 = -0.3926,$$

(111.18)

$$c_3 = (2.612)^3\, b_3 = 0.1027.$$

(111.19)

From this we obtain immediately

$$\frac{G}{NkT} = \frac{\mu}{RT} = \ln\lambda = \ln x - d_1\left(\frac{T_0}{T}\right)^{\frac{3}{2}} - d_2\left(\frac{T_0}{T}\right)^3 - d_3\left(\frac{T_0}{T}\right)^{\frac{9}{2}} - \cdots,$$

(111.20)

$$d_1 = c_1 = 0.9235,$$

(111.21)

$$d_2 = c_2 + \tfrac{1}{2}c_1^2 = 0.0338,$$

(111.22)

$$d_3 = c_3 + c_1 c_2 + \tfrac{1}{3}c_1^3 = 0.0026.$$

(111.23)

The power series in $(T_0/T)^{\frac{3}{2}}$ expresses the deviation from the formulae of Boltzmann statistics. We can now derive formulae for all the other thermodynamic functions by various routes. The simplest procedure is to write down the leading terms given by Boltzmann statistics and then to adjust the coefficients of the power series so as to satisfy the required thermodynamic relations. We obtain

$$\frac{G}{NkT} = \frac{\mu}{RT} = \ln\lambda$$

$$= \frac{3}{2}\ln\frac{h^2}{2\pi m kT} + \ln\frac{N}{V} - d_1\left(\frac{T_0}{T}\right)^{\frac{3}{2}} - d_2\left(\frac{T_0}{T}\right)^3 - d_3\left(\frac{T_0}{T}\right)^{\frac{9}{2}} - \cdots,$$

(111.24)

$$\frac{F}{NkT} = \frac{3}{2}\ln\frac{h^2}{2\pi m kT} + \ln\frac{N}{V} - 1 - \frac{1}{2}d_1\left(\frac{T_0}{T}\right)^{\frac{3}{2}} - \frac{1}{3}d_2\left(\frac{T_0}{T}\right)^3 - \frac{1}{4}d_3\left(\frac{T_0}{T}\right)^{\frac{9}{2}} - \cdots,$$

(111.25)

$$\frac{PV}{NkT} = \frac{U}{\frac{3}{2}NkT} = \frac{H}{\frac{5}{2}NkT} = 1 - \frac{1}{2}d_1\left(\frac{T_0}{T}\right)^{\frac{3}{2}} - \frac{2}{3}d_2\left(\frac{T_0}{T}\right)^3 - \frac{3}{4}d_3\left(\frac{T_0}{T}\right)^{\frac{9}{2}} - \cdots,$$

(111.26)

$$\frac{S}{Nk} = \frac{5}{2} - \frac{3}{2}\ln\frac{h^2}{2\pi m kT} - \ln\frac{N}{V} - \frac{1}{4}d_1\left(\frac{T_0}{T}\right)^{\frac{3}{2}} - \frac{2}{3}d_2\left(\frac{T_0}{T}\right)^3 - \frac{7}{8}d_3\left(\frac{T_0}{T}\right)^{\frac{9}{2}} - \cdots,$$

(111.27)

$$\frac{1}{\frac{3}{2}Nk}\frac{dU}{dT} = 1 + \frac{1}{4}d_1\left(\frac{T_0}{T}\right)^{\frac{3}{2}} + \frac{4}{3}d_2\left(\frac{T_0}{T}\right)^3 + \frac{21}{8}d_3\left(\frac{T_0}{T}\right)^{\frac{9}{2}} - \cdots.$$

(111.28)

It is noteworthy that these series not only converge but do so quite rapidly even when $T = T_0$. For example we have when $T = T_0$, from (111.5)

$$\frac{PV}{NkT_0} = \frac{U}{\frac{3}{2}NkT_0} = \frac{H}{\frac{5}{2}NkT_0} = \frac{\sum_{j=1}^{\infty} j^{-\frac{5}{2}}}{\sum_{j=1}^{\infty} j^{-\frac{3}{2}}} = \frac{\zeta\left(\frac{5}{2}\right)}{\zeta\left(\frac{3}{2}\right)} = \frac{1.341}{2.612} = 0.5134 \qquad (111.29)$$

while from (111.26)

$$\frac{PV}{NkT_0} = 1 - \frac{1}{2}d_1 - \frac{2}{3}d_2 - \frac{3}{4}d_3 = 1 - 0.4618 - 0.0225 - 0.0020 = 0.5137. \qquad (111.30)$$

At the same time as $T \to T_0$

$$\frac{G}{NkT_0} = \frac{\mu}{RT_0} = \ln \lambda \to 0, \qquad (111.31)$$

$$\frac{F}{NkT_0} = \frac{G}{NkT_0} - \frac{PV}{NkT_0} \to -\frac{\zeta\left(\frac{5}{2}\right)}{\zeta\left(\frac{3}{2}\right)} = -0.5134, \qquad (111.32)$$

whereas breaking off the power series in (111.25) at $(T_0/T)^{\frac{3}{2}}$ we have using (111.6) and (111.8)

$$\begin{aligned}
\frac{F}{NkT} &= \ln x - 1 - \frac{1}{2}d_1 - \frac{1}{3}d_2 - \frac{1}{4}d_3 - \cdots \\
&= \ln\left\{\zeta\left(\frac{3}{2}\right)\right\} - 1 - \frac{1}{2}d_1 - \frac{1}{3}d_2 - \frac{1}{4}d_3 - \cdots \\
&= -0.5136.
\end{aligned} \qquad (111.33)$$

112. Bose-Einstein gas below condensation temperature. This completes our investigation of a Bose-Einstein ideal gas at temperatures above T_0. We shall now investigate the more interesting behaviour at temperatures below T_0. We begin with the distribution law (72.18) which we rewrite as

$$N_r = \frac{1}{\lambda^{-1} e^{E_r/kT} - 1}. \qquad (112.1)$$

In particular, if we take the lowest energy as zero,

$$N_0 = \frac{1}{\lambda^{-1} - 1} = \frac{\lambda}{1 - \lambda} \qquad (112.2)$$

from which it is physically obvious that λ can never exceed 1. We have already seen that as $T \to T_0$ from above $\lambda \to 1$. From the last formula it also appears that when $\lambda = 1$ the value of N_0 becomes infinite. This remark need not disturb us when we remember that all our statistical formulae for molecular distributions have been derived on the assumption that N is large and effectively infinite. Our formulae relate to the limiting conditions $N \to \infty$, $V \to \infty$ while N/V remains finite so that the mathematical statement $N_0 \to \infty$ merely implies the physical statement that the ratio N_0/N is finite in contrast to the ratio N_r/N for $r \neq 0$ which tends to zero as $N \to \infty$ because in this limit the states form a continuum. We are concerned with the two limiting operations

$$N \to \infty \qquad V \to \infty \qquad N/V \text{ finite,} \qquad (112.3)$$

and

$$\lambda \to 1. \qquad (112.4)$$

The thorough mathematical investigation[1] of the combined result of these two limiting operations is difficult but the conclusion eventually reached is simple. It is that when $T \leq T_0$ the set of all states other than $r = 0$ may with sufficient accuracy be replaced by the continuum with our previous formula for $g(E)$, while the ground state $r = 0$ must be considered separately. We shall not attempt to reproduce this mathematical investigation, but before making use of the conclusion reached we shall show that it is physically plausible.

We suppose the gas contained in a rectangular box with sides of lengths a, b, c in the x, y, z directions. We may without loss of generality assume that the longest side is a. Then the first excited state is $n_x = 2$, $n_y = 1$, $n_z = 1$ and the energy of this state, when the energy of the ground state is taken as zero, is

$$E_1 = 3\,\frac{h^2}{8\,m\,a^2}\,. \tag{112.5}$$

We recall that T_0 is defined according to (111.6) by

$$k T_0 = \frac{h^2}{2\pi\,m}\left(\frac{N}{2.612\,a\,b\,c}\right)^{\frac{2}{3}}. \tag{112.6}$$

It follows that E_1/kT_0 is of the order of magnitude $N^{-\frac{2}{3}}$, say 10^{-16}. Let us now write $\lambda^{-1} = 1 + \delta$ where $\delta \ll 1$. We have then

$$N_r = \frac{1}{(1 + \delta)\,e^{E_r/kT} - 1}\,. \tag{112.7}$$

In particular

$$N_1 = \frac{1}{(1 + \delta)\,e^{E_1/kT} - 1}\,, \tag{112.8}$$

$$N_0 = \delta^{-1}. \tag{112.9}$$

We now further suppose that δ is much smaller than $N^{-\frac{2}{3}}$ but not small compared with N^{-1}. Evidently then we may replace the above formula for N_1 by

$$N_1 = \frac{1}{e^{E_1/kT} - 1} \tag{112.10}$$

and a fortiori

$$N_r = \frac{1}{e^{E_r/kT} - 1}\,, \qquad r \neq 0, \tag{112.11}$$

but we must not write $\delta = 0$ in

$$N_0 = \delta^{-1}\,. \tag{112.12}$$

Under these conditions we may regard the distribution law for all states other than the ground state as given by formulae similar to those for $T > T_0$ with $\lambda = 1$. We have then

$$N - N_0 = \left(\frac{2\pi\,m\,kT}{h^2}\right)^{\frac{3}{2}} V\,\zeta\!\left(\frac{3}{2}\right), \tag{112.13}$$

$$\frac{U}{kT} = \frac{3}{2}\left(\frac{2\pi\,m\,kT}{h^2}\right)^{\frac{3}{2}} V\,\zeta\!\left(\frac{5}{2}\right) \tag{112.14}$$

and we have shown that these formulae are good approximations provided $1 - \lambda \ll E_1/kT_0$. But it is in fact true, though less obviously, that these formulae are valid even when $1 - \lambda$ is comparable with E_1/kT_0. This is because in the

---

[1] S.R. DE GROOT and C.A. TEN SELDAM: Physica, The Hague **15**, 671 (1949).

integrals

$$\int_0^\infty g(E)\, e^{-E/kT}\, dE,\tag{112.15}$$

$$\int_0^\infty g(E)\, E\, e^{-E/kT}\, dE\tag{112.16}$$

which lead to the above relations, the integrands have maxima at values of E comparable with kT and the contributions from values of E comparable with $E_1 \ll kT_0$ are negligible when T is comparable with T_0 and even when T is much smaller than T_0. In fact these formulae for $N-N_0$ and U/kT are valid from $T=T_0$ down to temperatures where U differs inappreciably from zero. The final conclusion of a deep mathematical analysis is that we may with accuracy adequate for all practical purposes use (112.14) throughout the temperature range from $T=T_0$ down to $T=0$ and so write

$$\frac{U}{kT} = \frac{3}{2}\left(\frac{2\pi m kT}{h^2}\right)^{\frac{3}{2}} V \zeta\left(\frac{5}{2}\right), \qquad 0 < T \leq T_0.\tag{112.17}$$

113. Einstein condensation. The physical significance of formula (112.17) is remarkable. It means that the total energy of a system of N molecules in a volume V at a temperature $T \leq T_0$ is independent of N. If at given $T \leq T_0$ and given V we increase N then all the extra molecules go into the ground state and so contribute nothing to the energy. In this respect there is an analogy with a saturated vapour in equilibrium with a liquid phase. If at constant T, V more molecules are added they increase the amount of liquid without affecting the pressure of the vapour. Because of this analogy the phenomenon occurring in a Bose-Einstein ideal gas when $T \leq T_0$ is called Einstein condensation since the situation was first described by Einstein.

Formula (112.17) for U/kT together with $\lambda = 1$ leads immediately to formulae for all the other thermodynamic functions. These formulae are

$$-\frac{F}{kT} = \frac{PV}{kT} = \frac{U}{\frac{3}{2}kT} = \frac{S}{\frac{5}{2}k} = \left(\frac{2\pi m kT}{h^2}\right)^{\frac{3}{2}} V \zeta\left(\frac{5}{2}\right) = (N - N_0)\frac{\zeta(\frac{5}{2})}{\zeta(\frac{3}{2})},\tag{113.1}$$

$$\frac{G}{kT} = 0, \quad \mu = 0, \quad \lambda = 1,\tag{113.2}$$

$$\frac{1}{k}\frac{dU}{dT} = \frac{15}{4}\left(\frac{2\pi m kT}{h^2}\right)^{\frac{3}{2}} V \zeta\left(\frac{5}{2}\right).\tag{113.3}$$

As $T \to T_0$ we have

$$\left(\frac{2\pi m kT}{h^2}\right)^{\frac{3}{2}} V \zeta\left(\frac{5}{2}\right) \to \left(\frac{2\pi m kT_0}{h^2}\right)^{\frac{3}{2}} V \zeta\left(\frac{5}{2}\right) = \frac{\zeta(\frac{5}{2})}{\zeta(\frac{3}{2})} = 0.5134.\tag{113.4}$$

We have thus obtained two sets of formulae for the several thermodynamic functions, one set for the condition $\lambda < 1$ and the other for the condition $\lambda = 1$. We shall now combine these into a single set of formulae valid under all conditions. For this purpose we introduce $N_\emptyset$ defined as the number of molecules not in the ground state, so that

$$N = N_0 + N_\emptyset.\tag{113.5}$$

We have then according to (112.13) and (112.17)

$$N_\emptyset = \left(\frac{2\pi m kT}{h^2}\right)^{\frac{3}{2}} V \sum_{j=1}^\infty \lambda^j j^{-\frac{3}{2}},\tag{113.6}$$

$$U = \frac{3}{2} kT \left(\frac{2\pi m kT}{h^2}\right)^{\frac{3}{2}} V \sum_{j=1}^{\infty} \lambda^j j^{-\frac{5}{2}}, \tag{113.7}$$

$$\frac{U}{N_{\emptyset}} = \frac{3}{2} kT \frac{\sum\limits_{j=1}^{\infty} \lambda^j j^{-\frac{5}{2}}}{\sum\limits_{j=1}^{\infty} \lambda^j j^{-\frac{3}{2}}}. \tag{113.8}$$

We stress that these relations are valid in both the region

$$\lambda < 1, \qquad N/N_{\emptyset} = 1 \tag{113.9}$$

and the region

$$\lambda = 1, \qquad N/N_{\emptyset} > 1. \tag{113.10}$$

These formulas may be differentiated any number of times with respect to T at $\lambda = 1$ without producing any discontinuity between the two regions.

114. Variation of T at constant N, V. The formulae of Sect. 113 constitute the mathematically simplest description of the thermodynamics of a Bose-Einstein ideal gas. This is however not always the most convenient form. In particular it may be desirable to consider the behaviour of a system at constant V, N as T is varied. We shall show that at $T = T_0$ or $x = \zeta(\frac{3}{2})$ while U and dU/dT are continuous, there is a discontinuity in d^2U/dT^2.

From here onwards it is assumed that V and N are constant while T may be varied. Under these conditions we have by (111.8) and (111.9)

$$x = \frac{N}{V}\left(\frac{h^2}{2\pi m kT}\right)^{\frac{3}{2}} = \zeta\left(\frac{3}{2}\right)\left(\frac{T_0}{T}\right)^{\frac{3}{2}}, \tag{114.1}$$

$$\frac{dx}{dT} = -\frac{3}{2}\frac{x}{T}, \tag{114.2}$$

$$\frac{d^2x}{dT^2} = \frac{15}{4}\frac{x}{T^2}. \tag{114.3}$$

As $\lambda \to 1$, that is as $T \to T_0$, then $x \to \zeta(\frac{3}{2})$. We have also under all conditions by comparing (113.6) with (114.1)

$$\frac{N_{\emptyset}}{N} = x^{-1} \sum_{j=1}^{\infty} \lambda^j j^{-\frac{3}{2}}, \tag{114.4}$$

and in particular when $\lambda < 1$ so that $N_{\emptyset}/N = 1$

$$\sum_{j=1}^{\infty} \lambda^j j^{-\frac{3}{2}} = x. \tag{114.5}$$

Differentiating (114.5) with respect to x, we obtain

$$\frac{d\lambda}{dx} = \frac{\lambda}{\sum\limits_{j=1}^{\infty} \lambda^j j^{-\frac{1}{2}}} \to 0 \quad \text{as} \quad \lambda \to 1. \tag{114.6}$$

It follows immediately that $d\lambda/dT \to 0$ as $\lambda \to 1$, that is to say as $T \to T_0$, and consequently

$$\frac{dU}{dT} = \left(\frac{\partial U}{\partial T}\right)_{\lambda} + \left(\frac{\partial U}{\partial \lambda}\right)_{T}\frac{d\lambda}{dT} \to \left(\frac{\partial U}{\partial T}\right)_{\lambda=1} \quad \text{as} \quad T \to T_0 \tag{114.7}$$

so that dU/dT is continuous at $T = T_0$.

We now return to formula (114.6) for $d\lambda/dx$ and differentiate again with respect to x. We obtain

$$\frac{d^2\lambda}{dx^2} = \frac{\lambda}{\left(\sum\limits_{j=1}^{\infty} \lambda^j j^{-\frac{1}{2}}\right)^2} - \frac{\sum\limits_{j=1}^{\infty} \lambda^j j^{\frac{1}{2}}}{\left(\sum\limits_{j=1}^{\infty} \lambda^j j^{-\frac{1}{2}}\right)^3}. \tag{114.8}$$

When $\lambda \to 1$ both numerator and denominator of the second term diverge, but it can be shown[1] that their ratio tends to $(2\pi)^{-1}$. The first term clearly tends to zero. Hence

$$\frac{d^2\lambda}{dx^2} \to -\frac{1}{2\pi} \quad \text{as} \quad \lambda \to 1 \quad \text{that is as} \quad T \to T_0 + 0. \tag{114.9}$$

We conclude that as $T \to T_0 + 0$, since $d\lambda/dT = 0$,

$$\frac{d^2\lambda}{dT^2} \to \frac{d^2\lambda}{dx^2}\left(\frac{dx}{dT}\right)^2 = -\frac{1}{2\pi}\left\{\frac{3}{2}\frac{\zeta(\frac{3}{2})}{T_0}\right\}^2 = -\frac{9}{8\pi}\left\{\frac{\zeta(\frac{3}{2})}{T_0}\right\}^2. \tag{114.10}$$

There is a consequent discontinuity in $d^2 U/dT^2$, namely

$$\left(\frac{d^2 U}{dT^2}\right)_{T=T_0+0} - \left(\frac{d^2 U}{dT^2}\right)_{T=T_0-0} = \left(\frac{\partial U}{\partial\lambda}\right)_{T=T_0+0}\left(\frac{d^2\lambda}{dT^2}\right)_{T=T_0+0}. \tag{114.11}$$

But we have

$$\left(\frac{\partial U}{\partial\lambda}\right)_{T=T_0+0} = \frac{\zeta(\frac{3}{2})}{\zeta(\frac{5}{2})}\, U_{T=T_0} = \frac{3}{2} N k T_0 \tag{114.12}$$

and consequently

$$\left(\frac{d^2 U}{dT^2}\right)_{T=T_0+0} - \left(\frac{d^2 U}{dT^2}\right)_{T=T_0-0} = -\frac{27}{16\pi}\left\{\zeta\left(\frac{3}{2}\right)\right\}^2\frac{Nk}{T_0} = -3.66\frac{Nk}{T_0}. \tag{114.13}$$

We may summarize these results for a system of given V, N at $T = T_0$

$$\lambda \text{ is continuous and equal to 1;} \tag{114.14}$$

$$\frac{d\lambda}{dT} \text{ is continuous and equal to 0;} \tag{114.15}$$

$$\frac{d^2\lambda}{dT^2} \text{ has a discontinuity from } \frac{9}{8\pi}\left\{\zeta\left(\frac{3}{2}\right)\right\}^2 \text{ to 0;} \tag{114.16}$$

$$\frac{dU}{dT} \text{ is continuous and equal to } \frac{15}{4}\frac{\zeta(\frac{5}{2})}{\zeta(\frac{3}{2})} N k = 1.93\, N k; \tag{114.17}$$

$$\frac{d^2 U}{dT^2} \text{ has a discontinuity of magnitude } \frac{27}{16\pi}\left\{\zeta\left(\frac{3}{2}\right)\right\}^2\frac{Nk}{T_0} = 3.66\frac{Nk}{T_0}. \tag{114.18}$$

Since ^{4}He atoms obey Bose-Einstein statistics it has been suggested that the Einstein condensation has some bearing on the transition from helium I to helium II in the liquid phase, but this suggestion is still controversial and we shall not pursue it.

115. Extension to s dimensions. We shall now consider briefly to what extent the Einstein condensation is dependent on space being 3-dimensional[2]. In the general case of s dimensions we should have when $\lambda < 1$ the relation

$$\sum_{j=1}^{\infty} \lambda^j j^{-\frac{1}{2}s} = \frac{N}{V}\left(\frac{h^2}{2\pi m k T}\right)^{\frac{1}{2}s} = x \tag{115.1}$$

---

[1] R. H. Fowler and H. Jones: Proc. Cambridge Phil. Soc. **34**, 576 (1938).
[2] S. R. de Groot and C. A. ten Seldam: Physica, The Hague **15**, 671 (1949).

of which our previous relation (111.9),

$$x = \frac{N}{V}\left(\frac{h^2}{2\pi m kT}\right)^{\frac{3}{2}} = \sum_{j=1}^{\infty} \lambda^j j^{-\frac{3}{2}} \tag{115.2}$$

was the special case $s=3$. In the 2-dimensional case we have

$$\sum_{j=1}^{\infty} \lambda^j j^{-1} = \frac{N}{V}\frac{h^2}{2\pi m kT} = x \tag{115.3}$$

and since the series $\sum_{j=1}^{\infty} j^{-1}$ diverges it is clear that $\lambda \to 1$ when $T \to 0$ so that there is no condensation temperature $T_0 > 0$. On the other hand for the 4-dimensional case we have

$$\sum_{j=1}^{\infty} \lambda^j j^{-2} = \frac{N}{V}\left(\frac{h^2}{2\pi m kT}\right)^2 = x \tag{115.4}$$

and $\lambda \to 1$ at the temperature T_0 defined by

$$\frac{N}{V}\left(\frac{h^2}{2\pi m kT_0}\right)^2 = \sum_{j=1}^{\infty} j^{-2} = \zeta(2) = \frac{\pi^2}{6}. \tag{115.5}$$

To investigate the behaviour at $T=T_0$ we begin with the relation

$$\sum_{j=1}^{\infty} \lambda^j j^{-2} = x \tag{115.6}$$

and differentiate with respect to x obtaining

$$\frac{d\lambda}{dx} = \frac{\lambda}{\sum_{j=1}^{\infty} \lambda^j j^{-1}} = -\frac{\lambda}{\ln(1-\lambda)} \to 0 \quad \text{as} \quad \lambda \to 1 \tag{115.7}$$

as in the 3-dimensional case.

Differentiating again with respect to x we obtain

$$\frac{d^2\lambda}{dx^2} = \frac{\lambda}{\{\ln(1-\lambda)\}^2} + \frac{\lambda(1-\lambda)^{-1}}{\{\ln(1-\lambda)\}^3} \to \infty \quad \text{as} \quad \lambda \to 1. \tag{115.8}$$

As a consequence of this for given values of V, N as $T \to T_0 + 0$ we find that $\dfrac{d^2 U}{dT^2} \to \infty$.

In space of more than 4 dimensions the behaviour is again different. We have

$$\frac{d\lambda}{dx} = \frac{\lambda}{\sum_{j=1}^{\infty} \lambda^j j^{-\frac{1}{2}s+1}} \to \frac{1}{\zeta(\frac{1}{2}s-1)} \neq 0 \quad \text{when} \quad s > 4 \tag{115.9}$$

so that for given V, N there is at $T=T_0$ a discontinuity in $d\lambda/dT$ and consequently a discontinuity in dU/dT.

From this brief discussion we see that in 2 dimensions there is no Einstein condensation, in 3 dimensions dU/dT is continuous but there is a finite discontinuity in $d^2 U/dT^2$, in 4 dimensions there is an infinite discontinuity in $d^2 U/dT^2$ and in more than 4 dimensions there is a finite discontinuity in dU/dT.

116. Radiation, enumeration of modes. We shall next derive the equilibrium properties of radiation and their dependence on temperature. We begin by enumerating the vibrational modes of radiation in an enclosure. This is an essentially geometrical problem which closely resembles the problem of enumerating the quantum states of a Bose-Einstein gas in an enclosure. For the sake of simplicity we suppose the enclosure to be a rectangular box with sides of lengths a, b, c parallel to the x, y, z axes respectively. Then for a wave travelling parallel to the x-axis the half wave length $\frac{1}{2}\lambda$ must be an integral submultiple of the length a of the box, that is to say

$$\frac{2}{\lambda} = \frac{l}{a}, \qquad l \text{ integral.} \tag{116.1}$$

Similar conditions hold for waves parallel to the y and z axes. For a wave travelling in any direction it can be shown that the boundary conditions of the box lead to the general condition on the wave length

$$\left(\frac{2}{\lambda}\right)^2 = \frac{l^2}{a^2} + \frac{m^2}{b^2} + \frac{n^2}{c^2} \qquad l, m, n \text{ integral.} \tag{116.2}$$

Hence the number of possible vibrational modes having wave lengths greater than a given λ is equal to the number of positive integers l, m, n satisfying

$$\frac{l^2}{a^2} + \frac{m^2}{b^2} + \frac{n^2}{c^2} \leqq \left(\frac{2}{\lambda}\right)^2. \tag{116.3}$$

This is effectively equal to the number of points x, y, z, with positive integral coordinates inside the ellipsoid

$$\frac{x^2}{a^2} + \frac{y^2}{b^2} + \frac{z^2}{c^2} = \left(\frac{2}{\lambda}\right)^2 \tag{116.4}$$

and therefore to the volume of an octant of this ellipsoid, namely

$$\frac{1}{8} \frac{4\pi}{3} \frac{a\,b\,c}{(\frac{1}{2}\lambda)^3} = \frac{4\pi}{3} \frac{V}{\lambda^3} \tag{116.5}$$

where V is the volume of the box. It can be shown that this number is independent of the shape of the box.

We have up to this point overlooked the fact that a light wave of given wave length and given direction can be resolved into two waves polarized in mutually perpendicular directions. This fact leads to a doubling of the number of independent vibrational modes. Consequently the total number of these modes having wave lengths greater than λ is

$$\frac{8\pi}{3} \frac{V}{\lambda^3}. \tag{116.6}$$

It follows that the number of modes having frequencies less than ν is

$$\frac{8\pi}{3} \frac{V\nu^3}{c^3} \tag{116.7}$$

where c is the speed of light. Finally by differentiation with respect to ν we obtain for the number of vibrational modes with frequencies between ν and $\nu + d\nu$

$$g(\nu)\, d\nu = \frac{8\pi V}{c^3} \nu^2\, d\nu. \tag{116.8}$$

These modes, like the particles composing an ideal gas, can exchange energy with one another but are otherwise independent. We shall see that they contribute additively to the free energy and the total energy.

117. Radiation, energy per mode. There are several possible ways of determining the contribution of each vibrational mode to the thermodynamic functions. These differ in the form of the fundamental assumption concerning the nature of light. Here we shall assume only that the possible values of the energy in a particular vibrational mode i of frequency ν_i are integral multiples of $h\nu_i$. We introduce this assumption into the general approach to the study of energy distribution described at the beginning of Chap. 4.

We accordingly return to the system of N_A identical even molecules and N_B identical odd molecules in the same enclosure but we now recognize that the enclosure also contains radiation. By reasoning exactly analogous to that used previously, we obtain for the number C of states of the system having a total energy E a multiple integral similar to that obtained previously but containing for each vibrational mode i an extra factor

$$1 + z^{h\nu_i} + z^{2h\nu_i} + z^{3h\nu_i} + \cdots = (1 - z^{h\nu_i})^{-1}. \tag{117.1}$$

We notice that there is no new symbol analogous to x and y because for radiation there is no restriction analogous to conservation of the number of particles.

Again by reasoning analogous to that used in Sect. 72, if we denote by $\overline{E}_i$ the average energy in the vibrational mode i, then $C\overline{E}_i$ is given by a multiple integral differing from the new integral for C by replacing the factor $(1 - z^{h\nu_i})^{-1}$ by the factor

$$0 + h\nu_i z^{h\nu_i} + 2h\nu_i z^{2h\nu_i} + 3h\nu_i z^{3h\nu_i} + \cdots = h\nu_i z^{h\nu_i} (1 - z^{h\nu_i})^{-2}. \tag{117.2}$$

Exactly as previously we apply the method of steepest descents and after identifying the value of z at the col with $e^{-1/kT}$, we obtain

$$\begin{aligned}
\overline{E}_i &= \frac{h\nu_i e^{-h\nu_i/kT}(1 - e^{-h\nu_i/kT})^{-2}}{(1 - e^{-h\nu_i/kT})^{-1}} \\
&= \frac{h\nu_i}{e^{h\nu_i/kT} - 1}
\end{aligned} \tag{117.3}$$

which we may rewrite as

$$\frac{\overline{E}_i}{h\nu_i} = \frac{1}{e^{h\nu_i/kT} - 1}. \tag{117.4}$$

If we interpret $\overline{E}_i/h\nu_i$ as the average number N_i of photons in the vibrational mode i each having energy $h\nu_i$ we see that the photons obey Bose-Einstein statistics with $\lambda = 1$

$$\overline{N}_i = \frac{1}{\lambda^{-1} e^{h\nu_i/kT} - 1}, \qquad \lambda = 1. \tag{117.5}$$

The physical significance of $\lambda = 1$ is that the number of photons in the mode i is unrestricted. There is some analogy with a Bose-Einstein ideal gas at a temperature below the condensation temperature T_0, but we shall not pursue this analogy.

118. Thermodynamic functions for radiation. We now combine formula (116.8) for the number of vibrational modes in a given frequency range with formula (117.3) for the average energy in each mode. We thus obtain for the energy in the frequency range ν to $\nu + d\nu$

$$\frac{8\pi V}{c^3} \frac{h\nu^3 d\nu}{e^{h\nu/kT} - 1} \tag{118.1}$$

and for the total energy U

$$U = \frac{8\pi V h}{c^3} \int_0^\infty \frac{\nu^3}{e^{h\nu/kT} - 1} d\nu. \tag{118.2}$$

The corresponding formula for the free energy F is

$$F = \frac{8\pi V kT}{c^3} \int_0^\infty \nu^2 \ln\left(1 - e^{-h\nu/kT}\right) d\nu \tag{118.3}$$

which can be verified as satisfying the Gibbs-Helmholtz relation $U = F - T\, \partial F/\partial T$. The pressure P of the radiation is given by

$$PV = -V\frac{\partial F}{\partial V} = -F = -\frac{8\pi V kT}{c^3} \int_0^\infty \nu^2 \ln\left(1 - e^{h\nu/kT}\right) d\nu. \tag{118.4}$$

Formula (118.4) could have been obtained more directly, if less convincingly, by writing down the grand partition function $\varXi$ for photons obeying Bose-Einstein statistics with $\lambda = 1$. We recall that the contribution to $\ln \varXi$ of particles of a given kind in a given state of energy E_r is

$$\sum_r \ln\left(1 - \lambda\, e^{-E_r/kT}\right). \tag{118.5}$$

Now a photon in a given vibrational mode i has only one energy $h\nu_i$ and so, when we put $\lambda = 1$, the contribution of such photons to $\ln \varXi$ is simply

$$-\ln\left(1 - e^{-h\nu_i/kT}\right). \tag{118.6}$$

Summing the contributions of all vibrational modes we obtain

$$\ln \varXi = -\frac{8\pi V}{c^3} \int_0^\infty \nu^2 \ln\left(1 - e^{-h\nu/kT}\right) d\nu \tag{118.7}$$

and consequently by (74.13)

$$PV = kT \ln \varXi = -\frac{8\pi V kT}{c^3} \int_0^\infty \nu^2 \ln\left(1 - e^{-h\nu/kT}\right) d\nu \tag{118.8}$$

in agreement with formula (118.4) derived by classical thermodynamics from formula (118.2) for U.

119. Stefan-Boltzmann law. The above formulae show clearly the contribution of each frequency range to the several thermodynamic functions. If we are interested only in the overall contribution we can obtain more compact formulae by evaluating the integrals. We begin with the free energy. We deduce from (118.3)

$$F = -\frac{8\pi V k^4 T^4}{c^3 h^3} I \tag{119.1}$$

where I is the integral defined by

$$\begin{aligned}
I &= -\int_0^\infty \xi^2 \ln\left(1 - e^{-\xi}\right) d\xi \\
&= \int_0^\infty \sum_{j=1}^\infty \xi^2 j^{-1} e^{-j\xi}\, d\xi \\
&= \sum_{j=1}^\infty j^{-4} \int_0^\infty \eta^2 e^{-\eta}\, d\eta \\
&= 2\sum_{j=1}^\infty j^{-4} = 2\zeta(4) = \frac{\pi^4}{45}.
\end{aligned} \tag{119.2}$$

Substituting (119.2) into (119.1) we obtain

$$F = -\tfrac{1}{3}\,a\,T^4 V \tag{119.3}$$

where

$$a = \frac{8\pi^5 k^4}{15 c^3 h^3} = 7.561 \times 10^{-15} \text{ erg cm}^{-3} \text{ deg}^{-4}. \tag{119.4}$$

For the other thermodynamic functions we deduce immediately from (119.3)

$$S = -\frac{\partial F}{\partial T} = \frac{4}{3}\,a\,T^3 V, \tag{119.5}$$

$$U = F + TS = a\,T^4 V, \tag{119.6}$$

$$P = -\frac{\partial F}{\partial V} = \frac{1}{3}\,aT^4 = \frac{1}{3}\frac{U}{V}, \tag{119.7}$$

$$G = F + PV = 0. \tag{119.8}$$

The last relation confirms that $\ln \lambda = 0$ or $\lambda = 1$.

The proportionality between U/V and T^4 is known as the Stefan-Boltzmann law.

We have chosen the definition of the universal constant a so that $a\,T^4$ is the equilibrium value of the radiation per unit volume in an enclosure. If a small hole is made in an isothermal enclosure then it can be shown by simple geometrical considerations that the radiation emitted through the hole per unit area and per unit time is σT^4 where σ is given by

$$\sigma = \tfrac{1}{4}\,a\,c = 5.667 \times 10^{-5} \text{ erg cm}^{-2} \text{ sec}^{-1} \text{ deg}^{-4}. \tag{119.9}$$

This constant σ is called the Stefan-Boltzmann constant.

G. External fields.

120. Ideal gas in gravitational field. In this chapter we shall discuss the behaviour of systems in external fields confining ourselves to important illustrative examples. We begin with the simple problem of the isothermal distribution of an ideal gas in a gravitational field. This is a classical problem in which the only degrees of freedom involved are translational. Moreover the translational kinetic energy is separable from the potential energy. Consequently equal elements of volume in real space are proportional to equal elements of the six-dimensional phase space of a molecule. If we denote the gravitational potential by Φ, then the number of molecules of a species i and mass m_i per unit volume C_i is related to the gravitational potential energy of a molecule $m_i \Phi$ by

$$C_i \propto \exp\left(- m_i\,\Phi/kT\right) \tag{120.1}$$

and this relation holds for each species independently of what other species are present. This can be expressed in the differential form

$$kT \operatorname{grad} C_i = - m_i\,C_i \operatorname{grad} \Phi. \tag{120.2}$$

If we sum over all species using the equation of a perfect gas,

$$kT \sum_i C_i = P \tag{120.3}$$

and the formula for the density ϱ,

$$\varrho = \sum_i m_i\,C_i, \tag{120.4}$$

we obtain

$$\mathrm{grad}\, P = - \varrho\, \mathrm{grad}\, \Phi. \tag{120.5}$$

This relation can be derived from pure hydrostatics, but the distribution law for each species can not.

121. Molecules in electrostatic field. We turn now to electrostatic fields. It would seem that formulae similar to those for a gravitational field would be applicable to ions in an electrostatic field. This is formally true, but in practice the electrostatic interaction between ions, in contrast to the gravitational attraction between molecules, will usually not be negligible. The electrostatic field acting on the ions is then not entirely an external one. The problem becomes complicated and depends significantly on the geometrical conditions. We shall not consider it further. We shall rather study the polarization and orientation of molecules in a uniform electric field.

We assume that the energy E_i of a molecule in a given quantum state i in an external electric field $\boldsymbol{E}$ is of the form

$$E_i = E_i^0 - a_i\, \boldsymbol{E} - \tfrac{1}{2} b_i\, \boldsymbol{E}^2, \qquad E_i^0,\, a_i,\, b_i\ \text{constant}, \tag{121.1}$$

and that higher powers of $\boldsymbol{E}$ are negligible. The component of the electric dipole moment of a molecule parallel to the field, when the molecule is in a given state i, is by definition

$$\mu_i = a_i + b_i\, \boldsymbol{E} = - \frac{\partial E_i}{\partial \boldsymbol{E}}. \tag{121.2}$$

If we now construct the molecular partition function

$$\begin{aligned}
Q &= \sum_i \exp\left(- E_i/kT\right) \\
&= \sum_i \exp\left\{- (E_i^0 - a_i\, \boldsymbol{E} - \tfrac{1}{2} b_i\, \boldsymbol{E}^2)/kT\right\}
\end{aligned} \tag{121.3}$$

we see that $\bar{\mu}$ the average value of the component of the electric moment of a molecule parallel to the field, at a temperature T is

$$\bar{\mu} = \frac{- \sum_i (\partial E_i/\partial \boldsymbol{E}) \exp\left\{- (E_i^0 - a_i\, \boldsymbol{E} - \tfrac{1}{2} b_i\, \boldsymbol{E}^2)/kT\right\}}{\sum_i \exp\left\{- (E_i^0 - a_i\, \boldsymbol{E} - \tfrac{1}{2} b_i\, \boldsymbol{E}^2)/kT\right\}} = kT\, \frac{\partial \ln Q}{\partial \boldsymbol{E}}. \tag{121.4}$$

It is usually possible to divide the quantum states of a molecule with sufficient accuracy into two classes: those, including the lowest energy state, with energy steps among themselves small compared with kT and the remainder separated from those of the first class by energy steps large compared with kT. The latter class may evidently be ignored. In the former class it is usually a good approximation to assume that a_i is the product of a constant μ characteristic of the molecule and an orientational factor determined by an orientational quantum number, which is the quantal analogue of $\cos \vartheta$ where ϑ is the angle between the external field $\boldsymbol{E}$ and a fixed axis in the molecule. At the same time it may be assumed that b_i either has the same value for all low energy quantum states or may be replaced by an average value for all these quantum states. Subject to these conditions the problem may be treated classically. We need then make no further mention of quantum states. Instead we postulate an energy of the form

$$E = E^0 - \boldsymbol{E}\, \mu \cos \vartheta - \tfrac{1}{2} \gamma\, \boldsymbol{E}^2 \tag{121.5}$$

where E^0 is the constant energy at zero field, μ is the permanent electric dipole moment of the molecule and γ is the molecular polarizability. The conditions required to justify this classical treatment are satisfied by typical rigid molecules, but they are not satisfied by molecules such as $CH_2Cl \cdot CH_2Cl$ having internal rotational degrees of freedom.

In the classical treatment which we shall adopt, the only degrees of freedom involved are the rotational ones and moreover the kinetic energy may be separated from the orientational potential energy and cancelled out. Equal elements of phase space are then proportional to equal elements of the solid angle $\sin \vartheta \, d\vartheta \, d\varphi$ where ϑ is the angle between the moment and the field while φ is the azimuthal angle. The molecular partition function Q_{or} for the orientational degrees of freedom therefore becomes after omitting irrelevant constant factors

$$Q_{\mathrm{or}} = \int_0^\pi \exp\left\{\left(\boldsymbol{E}\,\mu \cos \vartheta + \frac{1}{2}\gamma \,\boldsymbol{E}^2\right)\middle/ kT\right\} \sin \vartheta \, d\vartheta$$
$$= 2 \exp\left(\frac{\frac{1}{2}\gamma \boldsymbol{E}^2}{kT}\right) \frac{kT}{\mu \boldsymbol{E}} \sinh \frac{\mu \boldsymbol{E}}{kT} . \tag{121.6}$$

The average component $\bar{\mu}$ of the molecular electric moment along the field becomes

$$\bar{\mu} = kT \frac{\partial \ln Q_{\mathrm{or}}}{\partial T} = \mu\, L\left(\frac{\mu \boldsymbol{E}}{kT}\right) + \gamma \,\boldsymbol{E} \tag{121.7}$$

where $L(x)$ called Langevin's function is defined by

$$L(x) = \coth x - \frac{1}{x} . \tag{121.8}$$

For all ordinary fields at ordinary temperatures $\mu\,\boldsymbol{E} \ll kT$ and we may use the approximation

$$L(x) = \tfrac{1}{3} x \qquad x \ll 1 \tag{121.9}$$

and we therefore have

$$\bar{\mu} = \frac{\mu^2 \,\boldsymbol{E}}{3kT} + \gamma \,\boldsymbol{E} . \tag{121.10}$$

In the above discussion we have tacitly ignored any interaction between molecules. Consequently the treatment applies only to an ideal gas. If C denotes the number of molecules per unit volume, ε the permittivity of the gas and ε_0 the permittivity of empty space, we have

$$C\bar{\mu} = (\varepsilon - \varepsilon_0)\, \boldsymbol{E} \tag{121.11}$$

and consequently

$$\frac{\varepsilon}{\varepsilon_0} - 1 = C\left(\frac{\mu^2}{3\,\varepsilon_0\, kT} + \frac{\gamma}{\varepsilon_0}\right). \tag{121.12}$$

Hence by measuring the dielectric constant $\varepsilon/\varepsilon_0$ over a range of temperatures at constant concentration C, the values of both μ and γ can be determined.

122. Paramagnetic atoms in magnetic field. We now leave electric fields and turn to the behaviour of a paramagnetic substance in a magnetic field. For this problem, in contrast to the electrical one, classical theory is quite inadequate. We consider a system containing paramagnetic atoms or ions sufficiently dilute so that mutual interactions are negligible. Each atom or ion can exist in quantum states specified by the five quantum numbers n, L, S, J and M. The energy is determined mostly by n, to a less extent by L, to a much smaller extent by S and J, and is in the absence of an external field independent of M. The quantum number L is a measure of the orbital angular momentum which has the value $\{L(L+1)\}^{\frac{1}{2}} h/2\pi$. The quantum number S is a measure of the spin angular momentum which has the value $\{S(S+1)\}^{\frac{1}{2}} h/2\pi$. The quantum number J measures the resultant total angular momentum which has the value $\{J(J+1)\}^{\frac{1}{2}} \times h/2\pi$. The quantum number M measures the component of resultant angular momentum in the direction of a magnetic field. M can have any of the $2J+1$ values $-J, -J+1, \ldots J-1, J$.

In the absence of an external field the energy of the atom or ion is determined by n, L, S, J and its ground level will be specified by definite values of these four quantum numbers. We shall assume that the first excited state, usually defined by the same values of n, L, S and a different value of J, is separated from the ground level by an amount large compared with kT. This condition is usually satisfied, but there are exceptions which we shall not discuss here. Under these conditions the energies determined by the $2J+1$ values of M in a magnetic field of induction $\boldsymbol{B}$ are given by

$$E_M = E^0 - M g \beta \boldsymbol{B}, \quad M = -J, \ldots J \tag{122.1}$$

where E^0 is the energy when the field vanishes and β is Bohr's magneton defined by

$$\beta = e\,h/4\pi\,m_e = 9.273 \times 10^{-24} \text{ A m}^2 \tag{122.2}$$

where $-e$ is the charge and m_e the mass of the electron. g is called Landé's splitting factor and is defined by

$$g = 1 + \frac{J(J+1) + S(S+1) - L(L+1)}{2J(J+1)}. \tag{122.3}$$

We shall not here discuss the physical significance of g. We only point out that for a singlet state we have

$$S = 0, \quad J = L, \quad g = 1, \quad \text{singlet state} \tag{122.4}$$

and for an S state

$$L = 0, \quad J = S, \quad g = 2, \quad S \text{ state.} \tag{122.5}$$

In all other cases g has a value intermediate between 1 and 2.

We now construct the orientational partition function Q_{or} from which we omit all factors independent of M and $\boldsymbol{B}$. We have

$$\begin{aligned}
Q_{\text{or}} &= \sum_{M=-J}^{+J} \exp\{- M g \beta \boldsymbol{B}/kT\} \\
&= \frac{\sinh\{(J + \tfrac{1}{2}) g \beta \boldsymbol{B}/kT\}}{\{\sinh \tfrac{1}{2} g \beta \boldsymbol{B}/kT\}}.
\end{aligned} \tag{122.6}$$

The average value m of the atomic magnetic moment in the direction of the field is related to the partition function by a formula analogous to formula (121.4) for the average electric moment $\bar{\mu}$ in an electric field. Thus

$$\begin{aligned}
\frac{m}{\beta} &= \frac{kT}{\beta} \frac{\partial \ln Q_{\text{or}}}{\partial \boldsymbol{B}} \\
&= (J + \tfrac{1}{2}) g \coth \frac{(J + \tfrac{1}{2}) g \beta \boldsymbol{B}}{kT} - g \coth \frac{g \beta \boldsymbol{B}}{2kT}.
\end{aligned} \tag{122.7}$$

The right hand side regarded as a function of $g\beta\boldsymbol{B}/kT$ is called Brillouin's function.

In ordinarily obtainable magnetic fields at ordinary temperatures $\beta B \ll kT$ and the last formula may then be replaced by the approximation

$$m = \frac{J(J+1)\,g^2\,\beta^2\,B}{3\,kT}. \tag{122.8}$$

The rational volume susceptibility χ (a dimensionless quantity) is then given by

$$\chi = \frac{C\,\mu_0\,m}{B} = \frac{C\,\mu_0\,J(J+1)\,g^2\,\beta^2}{3\,kT} \tag{122.9}$$

where C is the number of paramagnetic atoms or ions per unit volume and μ_0 is the permeability of empty space

$$\mu_0 = 4\pi \cdot 10^{-7}\,\mathrm{JA^{-2}\,m^{-1}}. \tag{122.10}$$

The inverse proportionality of χ to the temperature is known as Curie's law.

123. Diamagnetism. We conclude with a few remarks about diamagnetism without discussing its origin. It is sufficient to know that the diamagnetism of an atom or ion is independent of M and is therefore independent of the temperature. Moreover the diamagnetism of a given paramagnetic atom or ion is smaller than its paramagnetism by a factor of the order

$$\frac{kT}{\text{electronic kinetic energy}} \tag{123.1}$$

which at ordinary temperatures is about 10^{-3}. Consequently for any given paramagnetic atom or ion, the diamagnetic contribution of this atom or ion will be negligible compared with its paramagnetic contribution, but it does not follow that we may neglect the total contribution of all the diamagnetic atoms or ions also present in the substance. A simple example will make this clear. In potassium ferric alum $KFe(SO_4)_2 \cdot 12H_2O$ the paramagnetism is entirely due to the Fe^{3+} ion. To obtain an accurate experimental value of this paramagnetic ionic susceptibility, we should correct the observed molecular susceptibility of $KFe(SO_4)_2 \cdot 12H_2O$ by adding the numerical value of the molecular susceptibility of the diamagnetic substance ordinary alum $KAl(SO_4)_2 \cdot 12H_2O$. In this procedure we have justifiably ignored the diamagnetic contribution of the Fe^{3+} ion, or rather the difference between this and that of the Al^{3+} ion, but we have not ignored the appreciable diamagnetic contributions of the 47 other atoms in the molecule. Whenever a comparison is made between a theoretical and an experimental value of a paramagnetic susceptibility, it is to be presumed that the latter has been corrected in this way for the diamagnetic contribution.

H. Historical notes.

124. Introduction. Our exposition of classical thermodynamics and statistical mechanics has been planned so as to describe these subjects as we understand them today with emphasis on the most convincing and most powerful methods of approach. In this exposition there was no room for dealing with any of the topics historically. This in no way implies any suggestion that the history of the subject is unimportant. The history of these subjects is on the contrary both important and interesting but it is something different from the subjects themselves. It is the object of this chapter to give a brief historical summary in the form of notes on the most striking advances in our knowledge of both classical thermodynamics and statistical mechanics.

125. Classical thermodynamics. 1824. CARNOT: Réflexions sur la puissance motrice du feu, Bachelier, Paris, 1824. Reprinted in 1912 by Hermann, Paris and in 1953 by Blanchard, Paris, together with some notes discovered after Carnot's death in 1832 and communicated to the Académie des Sciences in 1878 by Carnot's brother.

In his book Carnot presented the first study of reversible cycles and derived his well known theorem stated on p. 38:

«La puissance motrice de la chaleur est indépendante des agens mis en œuvre pour la réaliser; sa quantité est fixée uniquement par les températures des corps entre lesquels se fait en dernier résultat le transport du calorique.»

This may be expressed in modern terminology as follows. If any system is taken round a reversible cycle in which heat is absorbed only at one temperature ϑ_1 and is rejected only at one temperature ϑ_2, the ratio of the work done w to the heat q_1 absorbed at the temperature ϑ_1 depends only on ϑ_1, ϑ_2 and is independent of the nature of the system.

In this formulation no mention is made of the heat q_2 rejected at the temperature ϑ_2. There are phrases in Carnot's book which suggest that $q_1 = q_2$ in contradiction with the principle of conservation of energy.

1832. Whatever Carnot may have thought in 1824, it is clear from the posthumously published note that before he died in 1832 he understood this principle. Thus on p. 134 of the reprint of Carnot's writings we read:

«La chaleur n'est autre chose que la puissance motrice ou plutôt que le mouvement qui a changé de forme, c'est un mouvement. Partout où il y a destruction de puissance motrice dans les particules des corps, il y a en même temps production de chaleur en quantité précisément proportionnelle à la quantité de puissance motrice détruite; réciproquement, partout où il y a destruction de chaleur, il y a production de puissance motrice.

On peut donc poser en thèse générale que la puissance motrice est en quantité invariable dans la nature, qu'elle n'est jamais à proprement parler ni produite, ni détruite. A la vérité elle change de forme c'est-à-dire qu'elle produit tantôt un genre de mouvement, tantôt un autre, mais elle n'est jamais anéantie.»

1834. Clapeyron: J. école polytechnique **14** (1834). Reprinted in Ann. Phys., Lpz. **59**, 446, 566 (1843). (In this volume the numbering of the pages is faulty; there are two pages with each of the numbers 449—464.)

In this article Clapeyron applied Carnot's theorem to the study of equilibrium. On p. 568 of the reprint he gave the formula

$$k = \left(1 - \frac{\delta}{\varrho}\right) \frac{dp}{dt} \, C$$

where k denotes the latent heat of evaporation per unit volume of a liquid, ϱ denotes the density of this liquid, δ denotes the density of the saturated vapour, p denotes the saturation vapour pressure at the temperature t and C is a function of the temperature independent of the nature of the substance. Clapeyron wrote: „Diese Gleichung sagt, daß die latente Wärme, welche gleiche Volume Dampf von verschiedenen Flüssigkeiten bei derselben Temperatur und unter dem entsprechenden Druck enthalten, proportional ist dem Coëfficienten dp/dt des Drucks in Bezug zur Temperatur." The quantity C in the above formula was later identified by Clausius with the thermodynamic absolute temperature.

1840—1845. Joule's experiments over the years 1840—1845, [Phil. Mag. **27**, 205 (1845)], showed that the energy required to raise the temperature of a pound of water by 1° Fahrenheit (that is to say the so-called British thermal unit) is always equal to 817 ft lbf; the accurate value is 780 ft lbf.

1847. This provided a firm basis for the explicit formulation of the principle of conservation energy by Helmholtz: [Über die Erhaltung der Kraft. Physik. Ges. Berlin (1847)].

1848. Thomson (later Lord Kelvin) [Proc. Cambridge Phil. Soc. **1**, 69 (1848)] gave the thermodynamic definition of absolute temperature.

On p. 69 he wrote "The characteristic property of the scale which I now propose is, that all degrees have the same value; that is, that a unit of heat descending from a body A at the temperature T^0 of this scale, to a body B at the temperature $(T-1)^0$, would give out the same mechanical effect, whatever be the number T. This may justly be termed an absolute scale, since its characteristic is quite independent of the physical properties of any specific substance."

1850. Clausius [Ann. Phys., Lpz. **79**, 368, 500 (1850)] clarified the significance of conservation of energy in Carnot's theorem.

1851. Thomson (later Lord Kelvin) [Trans. Roy. Soc. Edinb. **20**, 261 (1853)] formulated the first and second laws of thermodynamics together.

On p. 264 he wrote "The whole theory of the motive power of heat is founded on the two following propositions, due respectively to Joule and to Carnot and Clausius.

Prop. I (Joule). When equal quantities of mechanical effect are produced by any means whatever, from purely thermal sources, or lost in purely thermal effects, equal quantities of heat are put out of existence, or are generated.

Prop. II (Carnot and Clausius). If an engine be such that, when it is worked backwards, the physical and mechanical agencies in every part of its motions are all reversed; it produces as much mechanical effect as can be produced by any thermo-dynamic engine, with the same temperatures of source and refrigerator, from a given quantity of heat."

On p. 265 he wrote "The demonstration of the second proposition is founded on the following axiom:

"It is impossible, by means of inanimate material agency, to derive mechanical effect from any portion of matter by cooling it below the temperature of the coldest of the surrounding objects."

1854. Clausius [Ann. Phys., Lpz. **93**, 481 (1854)] gave an exposition of the first and second laws of thermodynamics. A few typical quotations follow.

On p. 488: ,,Es kann nie Wärme aus einem kälteren in einen wärmeren Körper übergehen, wenn nicht gleichzeitig eine andere damit zusammenhängende Aenderung eintritt."

On p. 497: ,,Nennt man zwei Verwandlungen, welche sich, ohne dazu eine sonstige bleibende Veränderung zu erfordern, gegenseitig ersetzen können, aequivalent, so hat die Entstehung der Wärmemenge Q von der Temperatur t aus Arbeit den Aequivalenzwerth

$$\frac{Q}{T},$$

und der Uebergang der Wärmemenge Q von der Temperatur t_1 zur Temperatur t_2 den Aequivalenzwerth

$$Q\left(\frac{1}{T_2} - \frac{1}{T_1}\right),$$

worin T eine von der Art des Processes, durch welchen die Verwandlung geschieht, unabhängige Temperaturfunction ist."

On p. 500: ,,Demnach gilt für alle umkehrbaren Kreisprocesse als analytischer Ausdruck des zweiten Hauptsatzes der mechanischen Wärmetheorie die Gleichung:

$$\int \frac{dQ}{T} = 0."$$

On p. 504: ,,Die algebraische Summe aller in einem Kreisprocesse vorkommenden Verwandlungen kann nur positiv seyn."

On p. 505 it is assumed ,,dass ein permanentes Gas, wenn es sich bei constanter Temperatur ausdehnt, nur so viel Wärme verschluckt, wie zu der dabei gethanen äusseren Arbeit verbraucht wird", and from this it is proved that the thermodynamic scale of temperature coincides with the perfect gas scale.

1865. Clausius [Ann. Phys., Lpz. **125**, 353 (1865)] gave a revised version of thermodynamics.

On p. 390 he wrote ,,Sucht man für S einen bezeichnenden Namen, so könnte man, ähnlich wie von der Grösse U gesagt ist, sie sey der Wärme- und Werkinhalt des Körpers, von der Grösse S sagen, sie sey der Verwandlungsinhalt des Körpers. Da ich es aber für besser halte, die Namen derartiger für die Wissenschaft wichtiger Grössen aus den alten Sprachen zu entnehmen, damit sie unverändert in allen neuen Sprachen angewandt werden können, so schlage ich vor, die Grösse S nach dem griechischen Worte ἡ τροπή, die Verwandlung, die Entropie des Körpers zu nennen. Das Wort Entropie habe ich absichtlich dem Worte Energie möglichst ähnlich gebildet, denn die beiden Grössen, welche durch diese Worte benannt werden sollen, sind ihren physikalischen Bedeutungen nach einander so nahe verwandt, dass eine gewisse Gleichartigkeit in der Benennung mir zweckmässig zu seyn scheint."

This paper closes on p. 400 with the well known words:
,,Die Energie der Welt ist constant.
Die Entropie der Welt strebt einem Maximum zu."

1869. Massieu [C. R. Acad. Sci., Paris **69**, 858 (1869)] introduced the two characteristic functions now denoted by $-F/T$ and $-G/T$.

1873. Horstmann [Ann. Chem. u. Pharm. **170**, 192 (1873)] gave a thermodynamic derivation of the equilibrium conditions for the dissociation of $CaCO_3$ and of PCl_5, including their temperature dependence, treating the vapour as a perfect gas. An analogous treatment is applicable to any equilibrium involving only perfect gases and solids.

1875. Gibbs [Trans. Connecticut Acad. **3**, 108, 343 (1875)] in a paper entitled "On the equilibrium of heterogeneous systems" developed the whole of chemical thermodynamics. In this paper he introduced the functions now denoted by F, H, G and μ.

1882. Helmholtz [Sitzgsber. Akad. Wiss. Berlin **1**, 22 (1882)] independently of Gibbs introduced the free energy denoted by F and obtained formulae relating to the electromotive force of chemical cells.

On p. 31 we meet the equation

$$U = F - \vartheta \frac{\partial F}{\partial \vartheta}$$

now called the Gibbs-Helmholtz relation.

On p. 33 Helmholtz wrote: „Die Function F fällt, wie wir gesehen haben, für isotherme Veränderungen mit dem Werthe der potentiellen Energie für die unbeschränkt verwandelbaren Arbeitswerthe zusammen. Ich schlage deshalb vor, diese Grösse die freie Energie des Körpersystems zu nennen.“

1886. Duhem [Le potentiel thermodynamique et ses applications, p. 33, (1886)] derived the relation now called the Gibbs-Duhem relation.

1887. Planck [Ann. Phys., Lpz. **30**, 563 (1887)] classified processes into two classes:

„1. in solche, welche sich vollständig rückgängig machen lassen — reversible, umkehrbare, neutrale Processe —,

2. in solche, bei denen dies nicht möglich ist — ich habe sie natürliche Processe genannt“, the former being characterized by an overall conservation of entropy, the latter by creation of entropy.

1906. Nernst [Nachr. kgl. Ges. Wiss. Göttingen, Math.-phys. Kl. 1 (1906)] enunciated his theorem. On p. 5 he writes: „Obwohl, wie oben erwähnt, die Grössen A und U im Allgemeinen nicht einander gleich sind, so ist es doch sehr auffällig, dass, wenigstens bei nicht zu hohen Temperaturen, in der Regel der Unterschied beider Grössen innerhalb mässiger Grenzen bleibt. Freilich sind bei diesem Vergleich Gase und verdünnte Lösungen auszuschalten, weil bei diesen bekanntlich Q, nicht aber A von der Konzentration unabhängig ist. Schon lange war mir nun in dieser Hinsicht aufgefallen, dass bei galvanischen Kombinationen, bei welchen in der Gleichung des stromliefernden chemischen Prozesses nur feste Körper und sehr konzentrierte Lösungen vorkommen, die Unterschiede zwischen A und U auffällig klein sind; ferner sei auch an das Verhalten der sogenannten idealen konzentrierten Lösungen erinnert. So drängte sich die Annahme auf, dass in solchen Fällen in der nächsten Nähe des absoluten Nullpunktes ein völliges Zusammenfallen beider Grössen stattfindet, und es würde als Grenzgesetz

$$\lim \frac{dA}{dT} = \lim \frac{dQ}{dT} \text{ für } T = 0$$

sich ergeben.“

1909. Carathéodory [Math. Ann. **67**, 355 (1909)] showed that the existence and properties of entropy can be derived from the axiom, stated on p. 363, „In jeder beliebigen Umgebung eines willkürlich vorgeschriebenen Anfangszustandes gibt es Zustände, die durch adiabatische Zustandsänderungen nicht beliebig approximiert werden können.“

126. Statistical mechanics. 1860. Maxwell [Phil. Mag. **19**, 22 (1860)] derived his distribution law from the assumption of the independence of the velocity components in three perpendicular directions.

1865. Loschmidt [Sitzgsber. Akad. Wiss. Wien **52** (2), 395 (1865)] estimated molecular diameters as 10^{-7} cm.

1870. Clausius [Phil. Mag. **40**, 122 (1870)] derived his virial theorem: The mean vis viva of the system is equal to its virial.

1871. Boltzmann [Sitzgsber. Akad. Wiss. Wien. **63** (2), 397 (1871)] extended Maxwell's distribution law to other degrees of freedom and showed that the distribution if once established would persist.

On p. 414 he wrote „Wenn zu einer beliebigen Zeit t die Anzahl der Moleküle der Gasart G in der Volumeinheit, deren Zustand zwischen den Grenzen

$$\xi_1 \quad \text{und} \quad \xi_1 + d\xi_1 \ldots w_r \quad \text{und} \quad w_r + dw_r$$

liegt, gleich ist:

$$dN = A e^{-h\varphi} d\xi_1 d\eta_1 \ldots d\zeta_{r-1} du_1 dv_1 \ldots dw_r \tag{21}$$

und die Zustandsvertheilung unter den Molekülen der übrigen Gasarten durch eine analoge Formel gegeben ist, so wird dieselbe weder durch die Bewegung der Atome in den Molekülen, noch durch die Zusammenstösse der Moleküle verändert, sie erhält sich also unverändert durch beliebige Zeit. Dabei ist φ die Summe der Kraftfunction und der gesammten lebendigen

Kraft des Moleküls, h eine für alle Gasarten gleiche, A aber eine für die verschiedenen Gasarten verschiedene Constante. Die Constante h kann so gewählt werden, daß die gesammte im Gase enthaltene lebendige Kraft, folglich auch seine Temperatur jeden beliebigen endlichen Werth erhält. Die Constanten A dagegen bestimmen die Dichte und das Mischungsverhältnis der Gase. Ist N die Anzahl der Moleküle der Gasart G in der Volumeinheit, so hat man

$$A = \frac{N}{\iint \ldots e^{-h\varphi} \, d\xi_1 \, d\eta_1 \ldots dw_r} \, ; \tag{22}$$

denn dN über alle möglichen Werthe der darin enthaltenen Variabeln integrirt, muss N liefern. Die durch die Formel (21) dargestellte Zustandsvertheilung erfüllt also alle Anforderungen, denen die wirkliche Zustandsvertheilung unter den Gasmolekülen zu genügen hat, und stellt man sie einmal unter den Gasmolekülen her, so wird sie durch die Zusammenstösse nicht mehr alterirt. Der Beweis dieses Satzes scheint mir vollkommen streng und von jeder Voraussetzung frei zu sein, die nicht auf mathematische Gewissheit, sondern blos auf grössere oder geringere Wahrscheinlichkeit Anspruch machen könnte. Dagegen ist mir der Beweis, dass diese Zustandsvertheilung die einzige ist, die durch die Zusammenstösse nicht verändert wird, bis jetzt noch nicht gelungen. Es ist jedoch dieser Umstand, dass ein und dasselbe Gas bei gleicher Temperatur und gleicher Dichte mehrerer Zustände fähig ist und dass es blos von den Anfangsbedingungen abhängt, welchen derselben es annimmt, a priori unwahrscheinlich und wird auch durch keinerlei Erfahrungen bestätigt.“

1871. BOLTZMANN [Sitzgsber. Akad. Wiss. Wien. **63** (2), 712 (1871)] forged the first link between statistical mechanics and thermodynamics by showing that the properties of entropy follow from the distribution law.

On p. 719 he wrote „Wir wollen nun zu dem Beweise schreiten, dass das Differential der zugeführten Wärme δQ, dividirt durch die mittlere lebendige Kraft $T = \dfrac{m}{2}\,\overline{c^2}$ eines Atoms, immer ein vollständiges Differential ist, und zugleich die Grösse, deren Differential sie ist, analytisch bestimmen.“

1872. BOLTZMANN [Sitzgsber. Akad. Wiss. Wien. **66**, 275 (1872)] showed that a certain function E of the distribution of the molecules in phase space can never increase and that E has its minimum value when the distribution is that of the Maxwell-Boltzmann law.

On p. 359 he wrote „Wenn die Zustandsvertheilung zu Anfang der Zeit keine gleichförmige war, so enthält die Function f auch die Coordinaten x, y, z derjenigen Stelle des Gases, für welche sie die Geschwindigkeitsvertheilung darstellt. Alsdann tritt an die Stelle von E ein etwas allgemeinerer Ausdruck. Wenn

$$f(t, x, y, z, \xi_1, \eta_1 \ldots w_r) \, dx \, dy \, dz \, d\xi_1 \ldots dw_r$$

die Zahl der Moleküle ist, welche sich zur Zeit t im Volumelemente $dx \, dy \, dz$ mit den Coordinaten x, y, z befinden, und für welche die den Zustand bestimmenden Variablen zwischen den Grenzen

$$\xi_1 \text{ und } \xi_1 + d\xi_1 \ldots w_r \text{ und } w_r + dw_r$$

liegen, so kann die Größe

$$E = \iint \ldots f \log f \cdot dx \, dy \, dz \, d\xi_1 \ldots dw_r$$

nicht zunehmen.“

Later BOLTZMANN used the symbol H instead of E, whence the name “Boltzmann's H-theorem”.

1877. BOLTZMANN [Sitzgsber. Akad. Wiss. Wien. **76** (2), 373 (1877)] introduced the idea that entropy is related to probability.

On p. 374 he wrote „Es ist also damit ausgesprochen, dass man den Zustand des Wärmegleichgewichtes dadurch berechnen kann, dass man die Wahrscheinlichkeit der verschiedenen möglichen Zustände des Systems aufsucht. Der Anfangszustand wird in den meisten Fällen ein sehr unwahrscheinlicher sein, von ihm wird das System immer wahrscheinlicheren Zuständen zueilen, bis es endlich den wahrscheinlichsten, d.h. den des Wärmegleichgewichtes, erreicht hat. Wenden wir dies auf den zweiten Hauptsatz an, so können wir diejenige Grösse, welche man gewöhnlich als die Entropie zu bezeichnen pflegt, mit der Wahrscheinlichkeit des betreffenden Zustandes identificiren.“

1900. PLANCK [Verh. dtsch. phys. Ges. **2**, 237 (1901); Ann. Phys. **4**, 553 (1901)] invented the quantum theory of radiating oscillators.

1902. GIBBS [Elementary principles in statistical mechanics developed with special reference to the rational foundation of thermodynamics, 1902] invented the canonical ensemble as a representation of a system in a temperature bath.

1907. Einstein [Ann. Phys. **22**, 180 (1907)] applied quantum theory to the thermal properties of solids.

1912. Debye [Ann. Phys. **39**, 789 (1912)] gave an improved quantal treatment of solids.

1921. Planck [Ann. Phys. **66**, 369 (1921)] introduced the use of Zustandsummen (partition functions).

1921. Ehrenfest and Trkal [Proc. Amst. Acad. Sci. **23** (1), 162 (1921)] introduced symmetry numbers into the treatment of polyatomic molecules.

1922. Darwin and Fowler [Phil. Mag. **44**, 450 (1922)] applied the method of steepest descents to the derivation of the fundamental laws of statistical mechanics.

1924. Bose [Z. Physik **26**, 178 (1924)] invented new statistics for photons.

1924. Einstein [Sitzgsber. Akad. Wiss. Berlin 261 (1924)] extended Bose statistics to molecules with non-zero rest mass.

1924. Einstein [Sitzgsber. Akad. Wiss. Berlin 3 (1924)] gave the theory of "condensation" in a Bose-Einstein gas.

1925. Pauli [Z. Physik **31**, 776 (1925)] formulated the exclusion principle.

1925. Heisenberg [Z. Physik **33**, 879 (1925)] formulated quantum mechanics.

1926. Fermi [Z. Physik **36**, 902 (1926)] invented new statistics for identical particles vibrating about a common fixed point.

1926. Dirac [Proc. Roy. Soc. Lond., Ser. A **112**, 661 (1926)] independently invented statistics for identical free particles in a box.

1927. Heisenberg [Z. Physik **41**, 252 (1927)] formulated the law that the eigenfunctions of a system are antisymmetrical with respect to all the electrons in the system.

1927. Dennison [Proc. Roy. Soc. Lond., Ser. A **115**, 483 (1927)] gave the theory of para and ortho molecules.

Axiomatik der Thermodynamik*.

Von

G. FALK und H. JUNG.

Mit 5 Figuren.

Einleitung. Mit der Axiomatik der Thermodynamik ist unlösbar der Name CARATHÉODORYs verbunden. Man wird daher von einer axiomatischen Behandlung der Thermodynamik in erster Linie eine Auseinandersetzung mit CARATHÉODORYs „Untersuchungen über die Grundlagen der Thermodynamik"[1] erwarten. Diese Erwartung scheint von der hier präsentierten Theorie nicht erfüllt zu werden, da sie sich im Gesamtaufbau wie in Details von der CARATHÉODORYs wesentlich unterscheidet. Dennoch bildete die Durcharbeitung der Carathéodoryschen Abhandlung nicht nur den Ausgangspunkt, sondern schlechthin die Basis unserer Untersuchungen, und wenn die schließliche Form des axiomatischen Aufbaues von der CARATHÉODORYs abwich, so war dies unter anderem durch die Sachlage bedingt, daß die Thermodynamik nicht nur die Theorie „mechanisch-thermischer" Systeme ist, sondern die einer *allgemeinen Art von Gleichgewichten*, ganz zu schweigen von den Aspekten der *statistischen Mechanik*, die außer acht zu lassen, heute ebenfalls nicht mehr möglich ist.

Neben CARATHÉODORYs Arbeit stellen vor allem die klassischen Abhandlungen von GIBBS[2] eine unvergleichliche Einführung in die logische Struktur der Thermodynamik dar, wenn auch in diesen Schriften nicht explizite von Axiomatik die Rede ist. Die Gibbsschen Abhandlungen sind daher nicht ohne Einfluß auf vorliegende Axiomatik geblieben. Andererseits glauben wir sagen zu können, daß die hier entwickelte Theorie auch manche Eigenheit der Gibbsschen Darstellung leichter verständlich machen wird[3].

Eine wesentliche Schwierigkeit beim Aufbau der Theorie war eine klare Erkenntnis der Rolle von Entropie und Temperatur. Man ist versucht, den bisherigen axiomatischen Darstellungen eine Überbewertung des Temperatur-Begriffes nachzusagen. So wichtig nämlich die Temperatur für alle praktischen Anwendungen der Thermodynamik auch ist (und in dieser Hinsicht kann ihre prädominante Rolle kaum übertrieben werden), so leicht verhindert die große Vertrautheit mit ihr eine klare Einsicht in ihre logische Funktion innerhalb der Theorie. Hinzu kommt, daß die Thermodynamik, wie ihr Name sagt, als Theorie *thermischer* Systeme, das heißt von Systemen mit Temperatur, entwickelt wurde und daß ihre Formulierung deshalb auf thermische Systeme zugeschnitten ist. Damit schien ihr Geltungsbereich abgesteckt und auf thermische Systeme beschränkt. Merkwürdigerweise änderte sich die Situation auch nicht als die

---

* KURT REIDEMEISTER zum 65. Geburtstag.
Vorliegender Artikel ist eine Originalabhandlung. Er enthält keinen allgemeinen Bericht über die Literatur der Axiomatik der Thermodynamik.
[1] C. CARATHÉODORY: Math. Ann. **67**, 355 (1909).
[2] J. W. GIBBS: Scientific papers Vol. I. New York 1906.
[3] Dies gilt vor allem für die ersten Kapitel der Abhandlung "On the equilibrium of heterogeneous substances".

Entwicklung der statistischen Mechanik die primäre Stellung der Entropie offensichtlich machte. In Wahrheit geben nicht Temperatur und Wärmemenge die Grundlagen zur Konstruktion der Entropie ab, sondern die *adiabatischen* und *quasistatischen Prozesse* (Ziff. 8 bis 10). Eines der logischen Probleme besteht somit darin, die Begriffe „adiabatisch" und „quasistatisch" festzulegen. Wenn dieses Problem beim klassischen Aufbau der Thermodynamik unter Verwendung der Begriffe Temperatur und Wärme scheinbar nicht auftritt, so deshalb, weil die Begriffe adiabatisch und quasistatisch mehr oder weniger stillschweigend benutzt werden. Carathéodory, der zum ersten Male die fundamentale Rolle der adiabatischen Isolation hervorhebt, vermeidet bei der Konstruktion der Entropie zwar die Wärme als Grundbegriff, nicht aber die (empirische) Temperatur[1], die sich, wie gesagt, bei konsequenter Analyse des physikalischen Sachverhaltes ebenfalls als entbehrlich erweist. Die Einführung der *quasistatischen* Prozesse verkoppelt Carathéodory mit Forderungen über Eigenschaften spezieller Systeme, die er „einfach" nennt. Ob hierin eine physikalisch wesentliche Beschränkung liegt oder nicht, sei dahingestellt. Dieser Sachverhalt zeigt bereits, daß Carathéodorys *Unerreichbarkeits-Axiom ohne die begleitenden Axiome über einfache Systeme* nicht schlechthin äquivalent genannt werden kann mit den klassischen Formulierungen des zweiten Hauptsatzes, welche den Begriff der quasistatisch übertragenen Wärme und damit den des quasistatischen Prozesses enthalten.

Die Grundidee der vorliegenden Darstellung der Thermodynamik kann folgendermaßen skizziert werden: Die Konstruktion thermodynamischer Standard-Variablen beruht auf zwei Operationen, die wir als *System-Zusammensetzung* und *System-Reduktion*[2] bezeichnen. Die System-Zusammensetzung wird generell in der Physik zur Konstruktion von *metrischen* oder „extensiven" Variablen verwendet (wozu auch Energie und Entropie gehören). Sie ist keineswegs für die Thermodynamik charakteristisch. Die System-Reduktion hingegen ist die operative Quelle der für die Thermodynamik kennzeichnenden Variablen. Sie läßt sich entweder als eine Auswahl unter den (allgemein physikalischen) metrischen Variablen oder als Definition von *Kontakt-Variablen* bzw. „intensiven" Variablen formulieren.

Hervorgehoben sei in diesem Zusammenhang noch die Behandlung des Stetigkeits-Begriffes. Die Stetigkeit, oder besser, die *Kontinuierlichkeit* von physikalischen Variablen wird im folgenden niemals als ein „primärer" Begriff bei der Definition der Variablen benutzt, sondern stets als physikalisch irrelevante und lediglich mathematisch zweckmäßige Zusatzforderung, die im Prinzip ebenso gut fortgelassen werden könnte, ohne daß mehr bewirkt würde als eine Schwerfälligkeit in der Formulierung. Dadurch treten die *kombinatorischen* Züge der physikalischen Größen in den Vordergrund. Dieses Verfahren spiegelt direkt die physikalische Wirklichkeit wider, denn empirische Sachverhalte haben stets die Form kombinatorischer Aussagen. Im folgenden wird daher die Stetigkeit physikalischer Variablen erst nach Klärung der kombinatorischen Sachlage eingeführt. Dies erleichtert die logische Kontrollierbarkeit der Theorie. Um Mißverständnissen vorzubeugen, sei noch einmal betont: Es ist nicht beabsichtigt, Kontinuierlichkeit, Stetigkeit und Differenzierbarkeit als *Beschreibungsmittel* auszuschalten, sondern deutlich zu machen, welche Tragweite man von diesen Begriffen in einer physikalischen Theorie erwarten kann. Wesentlich ist daher nicht die Vermeidung der Begriffe Kontinuum und Stetigkeit in der Beschreibung physikalischer Sachverhalte schlechthin, sondern nur bei der *Formulierung der*

---

[1] Für eine nähere Analyse der Carathéodoryschen Darstellung vgl. den Anhang.

[2] Mit „System-Reduktion" ist die Herstellung eines inneren Gleichgewichtes gemeint. Siehe Ziff. 14.

Grundlagen. Überall dort, wo die Begriffe durch Grenzprozesse eingeführt werden können, sind sie in dem hier gemeinten Sinn „harmlos", denn unter Grenzprozessen versteht man die Erweiterung bzw. Abschließung eines Objektbereiches, der selbst ohne Verwendung der Grenzprozesse, d.h. kombinatorisch erklärbar sein muß[1].

Die Aufgabe, eine hochentwickelte Theorie zu axiomatisieren, kann unter Umständen recht kompliziert sein, auch wenn die endgültige Form wieder einfach aussieht. Um die dabei zu erwartenden Schwierigkeiten abzuschätzen, stelle man sich z.B. die Aufgabe, die vertraute Variable Energie durch ihre Eigenschaften zu kennzeichnen, indem man keinen Begriff benutzt, der sich seinerseits aus der Energie ableitet, wie Arbeit, Wärme und dergleichen[2]. Die Feststellung etwa, sie sei eine Zustands-Funktion, besagt hinsichtlich ihrer Kennzeichnung so gut wie nichts. Jede derartige Aufgabe erfordert zunächst eine Analyse des Vorhandenen und möglicherweise die Untersuchung einer beträchtlichen Zahl von Einzelbeispielen; erst danach kann man an die Synthese einer axiomatischen Kennzeichnung herangehen. Obwohl die eleganteste Form der axiomatischen Darstellung die sein mag, Herkunft und Motivierung der Axiome zu unterdrücken, haben wir uns bemüht, die zu den Axiomen führenden analysierenden Betrachtungen in den wichtigsten Punkten darzulegen. Diese Darstellungsweise entspricht zwar nicht der mathematischen Tradition, ist aber angesichts der Rolle der Axiomatik in der Physik vorzuziehen.

Zur Rechtfertigung der axiomatischen Behandlung einer physikalischen Theorie möchten wir vor allem zwei Gründe anführen: Einmal die logische Durchdringung einer Theorie als wissenschaftlichen Selbstzweck und zum anderen die durch Erkenntnis der logischen Struktur ermöglichte *systematische Suche nach Erweiterungen der Theorie* als pragmatischen Zweck. Wir hoffen, zu beiden Aufgaben beizutragen.

Die Literatur-Zitate sind nach Gründen der Zweckmäßigkeit ausgewählt worden. Sie erheben keinen Anspruch auf Vollständigkeit.

A. Grundbegriffe und formale Beschreibungsmittel.

1. Zustände und Übergänge. Die Thermodynamik handelt von Systemen und ihren Zuständen sowie von Übergängen zwischen diesen Zuständen. Bei einer axiomatischen Beschreibung spielt der Zustand die Rolle eines (nicht näher zu erklärenden) Grundobjektes der Theorie: Es wird nur vorausgesetzt, daß von *Zuständen* z als unterscheidbaren Objekten gesprochen werden kann. Jedes „System" besitzt eine wohldefinierte Menge $\mathfrak{Z}$ von Zuständen[3]. *Übergänge* $z \to z'$ sind geordnete Zustandspaare (z, z') mit $z \neq z'$. Es ist zweckmäßig, Zustandsmengen durch Punktmengen zu repräsentieren und Übergänge $z \to z'$ durch gerichtete Strecken, die jeweils von z nach z' verlaufen. Es sei ausdrücklich

---

[1] Ein Beispiel, in dem die Stetigkeit nicht in dem gemeinten Sinn, sondern „primär" verwendet wird, bildet die Forderung, daß eine Zustandsmenge ein *n-dimensionaler Raum* sei, ohne ein bestimmtes Koordinatensystem zu charakterisieren. Denn der Begriff der *Dimension* macht in einer Weise von der Stetigkeit Gebrauch, die typisch nicht-kombinatorischer Art ist. Dies ist deshalb interessant, weil der Begriff der Dimension oder der „Anzahl der Freiheitsgrade" wesentlich in der Physik benutzt wird. Die Lösung dieses Dilemmas liegt darin, daß bei der Fundierung einer physikalischen Theorie der Zustands-Raum nicht invariant definiert wird, sondern mittels ausgezeichneter Variablen oder geometrisch gesprochen: Man definiert erst ein Koordinatensystem und mit seiner Hilfe einen Raum. Um diesen Sachverhalt auszudrücken, nennen wir in Kap. B, II die fraglichen ausgezeichneten Variablen *thermodynamische Koordinaten*.

[2] Siehe Ziff. 5.

[3] Mengen werden durch deutsche Buchstaben $\mathfrak{Z}, \mathfrak{M}, \ldots$ bezeichnet.

betont, daß wir den Zustandsmengen zunächst keine Kontinuums-Eigenschaften zuschreiben. Entsprechend dieser kombinatorischen Ausgangssituation empfiehlt es sich, zur Veranschaulichung von Zustandsmengen diskrete Punktmengen zu benutzen.

Um ein System aus einem Zustand z in einen Zustand z' zu bringen, bedarf es irgendwelcher „Einwirkungen von außen". Ohne zunächst nach der physikalischen Charakterisierung verschiedener Typen solcher Einwirkungen zu fragen, läßt sich sagen, daß von jedem Typ nur ganz bestimmte Übergänge erzeugt werden können. Jedes Zustandspaar (z, z') repräsentiert also in bezug auf einen Einwirkungs-Typ einen *möglichen* oder *nicht möglichen* Übergang. Diese Alternative beschreiben wir durch eine Funktion $F(z, z')$, die für alle Paare (z, z') erklärt ist und den Wert $+1$ annimmt, wenn der Übergang $z \to z'$ möglich oder $z = z'$ ist und -1, wenn $z \to z'$ unmöglich ist. Eine solche Funktion nennen wir eine *Relation* (auf $\mathfrak{Z}$). Jeder Typ von Einwirkungen auf das System definiert also eine Relation. Wir können aber auch umgekehrt sagen: Jede Relation auf $\mathfrak{Z}$ definiert eine Einwirkung auf das System. Damit ist erklärt, was im Sinn der Theorie eine „System-Einwirkung" ist.

Alle in der Thermodynamik betrachteten System-Einwirkungen haben die wesentliche Eigenschaft, daß ihre Relationen *transitiv* sind, d.h. wenn $z \to z'$ und $z' \to z''$ unter einer Einwirkung möglich sind, so ist auch $z \to z''$ möglich. In Symbolen: Wenn $F(z, z') = +1$ und $F(z', z'') = +1$, so ist auch $F(z, z'') = +1$. Wir bezeichnen solche transitiven Relationen als *Übergangs-Relationen*.

2. Vollständige, zerfällende und verzweigungslose Übergangs-Relationen. Ein einzelner Übergang $z \to z'$ soll in bezug auf eine Übergangs-Relation F *reversibel* heißen, wenn $F(z, z') = F(z', z) = +1$, d.h. wenn unter der durch F definierten Einwirkung sowohl $z \to z'$ als auch $z' \to z$ möglich sind. Dementsprechend definieren wir: Eine Teilmenge von $\mathfrak{Z}$ heißt in bezug auf F *reversible Zustandsmenge*, falls jeder Übergang in ihr reversibel ist. Schließlich nennen wir ein $z \to z'$ *irreversibel*, wenn $F(z, z') = -F(z', z) = +1$ gilt.

Jede auf einer Zustandsmenge $\mathfrak{Z}$ erklärte Übergangs-Relation F erlaubt nun eine Einteilung von $\mathfrak{Z}$ in *maximale reversible Zustandsklassen* $\mathfrak{M}$. Die Konstruktion dieser Klassen ist ganz elementar durchführbar: Zu einem Zustand z sucht man ein z', für das sowohl $z \to z'$ als auch $z' \to z$ möglich sind. Von z' ausgehend sucht man dann ein z'', für das wieder $z' \to z''$ und $z'' \to z'$ möglich sind; dann sind wegen der Transitivität der Übergangs-Relation auch $z \to z''$ und $z'' \to z$ möglich. Fortsetzung des Verfahrens liefert eine maximale Menge $\mathfrak{M}$ von Zuständen, die alle untereinander reversibel verbindbar sind. Beginnt man die Konstruktion statt mit z mit irgendeinem anderen Zustand von $\mathfrak{M}$, so erhält man offensichtlich dieselbe Menge $\mathfrak{M}$. Jeder Zustand definiert also eindeutig eine maximale reversible Menge und es kann nicht vorkommen, daß ein und derselbe Zustand zu zwei verschiedenen dieser Mengen gehört: Die Mengen $\mathfrak{M}$ schöpfen die ganze Zustandsmenge $\mathfrak{Z}$ aus, ohne sich gegenseitig zu überdecken, sie bilden also eine *Einteilung von $\mathfrak{Z}$ in Klassen*.

Sind nun $\mathfrak{M}$ und $\mathfrak{M}'$ zwei verschiedene derartige Klassen und z sowie z' je ein Zustand aus ihnen, so gibt es für die beiden Relations-Werte $F(z, z')$ und $F(z', z)$ nur folgende drei Kombinations-Möglichkeiten:

$$F(z, z') = -F(z', z) = \begin{cases} +1, & \text{d.h. } z \to z' \\ -1, & \text{d.h. } z' \to z \end{cases} \text{ irreversibel,}$$

$$F(z, z') = F(z', z) = -1, \qquad \begin{cases} \text{d.h. } z \to z' \\ \text{und } z' \to z \end{cases} \text{ unmöglich.}$$

Die reversible Kombination $F(z, z') = F(z', z) = +1$ scheidet wegen der Voraussetzung $\mathfrak{M} \neq \mathfrak{M}'$ aus. Offensichtlich ändern nun $F(z, z')$ und $F(z', z)$ ihre Werte nicht, wenn man z beliebig in $\mathfrak{M}$ und z' beliebig in $\mathfrak{M}'$ variiert. Die Übergangs-Relation zwischen den Zuständen hat also *eine Relation $\mathscr{F}$ zwischen den Klassen* zur Folge. Für diese Relation gilt natürlich $\mathscr{F}(\mathfrak{M}, \mathfrak{M}) = +1$. Außerdem bestehen für $\mathfrak{M} \neq \mathfrak{M}'$ die Kombinations-Möglichkeiten

$$a) \quad \mathscr{F}(\mathfrak{M}, \mathfrak{M}') = - \mathscr{F}(\mathfrak{M}', \mathfrak{M}) = \left\{ \begin{matrix} +1, \\ -1, \end{matrix} \right\}$$
$$b) \quad \mathscr{F}(\mathfrak{M}, \mathfrak{M}') = + \mathscr{F}(\mathfrak{M}', \mathfrak{M}) = -1 . \quad (2.1)$$

Wesentlich hierbei ist, daß wir mit Kenntnis der Relation $\mathscr{F}$ zwischen den Klassen gleichzeitig die Übergangs-Relation F zwischen den Zuständen besitzen; denn liegen z, z' innerhalb derselben Klasse, so gilt gemäß Definition der Klassen $F(z, z') = F(z', z) = +1$, liegen sie in verschiedenen Klassen, so gibt (2.1) ihren Relations-Wert. Beachtet man schließlich, daß die Relation $\mathscr{F}$ transitiv ist, so haben wir

Satz 1: Jede Übergangs-Relation $F(z, z')$ auf $\mathfrak{Z}$ ist äquivalent einer Einteilung von $\mathfrak{Z}$ in Klassen $\mathfrak{M}$ und einer Übergangs-Relation $\mathscr{F}(\mathfrak{M}, \mathfrak{M}')$, die für $\mathfrak{M} \neq \mathfrak{M}'$ keine reversiblen Lösungen besitzt, d.h. bei der die Kombination $\mathscr{F}(\mathfrak{M}, \mathfrak{M}') = \mathscr{F}(\mathfrak{M}', \mathfrak{M}) = +1$ nicht vorkommt.

Wir treffen nun die

Definition: Eine Übergangs-Relation heißt *vollständig*, wenn bei ihr die Werte-Kombination $F(z, z') = F(z', z) = -1$ nicht vorkommt.

In bezug auf eine System-Einwirkung mit vollständiger Übergangs-Relation sind also zwei beliebige Zustände z, z' stets in mindestens einer Richtung durch einen Übergang verbindbar. Die Vollständigkeit von $F(z, z')$ überträgt sich auf $\mathscr{F}(\mathfrak{M}, \mathfrak{M}')$ und schließt (2.1 b) aus. Damit ergibt sich für die Klassen-Relation $\mathscr{F}$ als kennzeichnende Eigenschaft neben der Transitivität die *Antisymmetrie* $\mathscr{F}(\mathfrak{M}, \mathfrak{M}') = - \mathscr{F}(\mathfrak{M}', \mathfrak{M})$. Eine antisymmetrische transitive Relation ist aber nichts anderes als eine Ordnungs-Relation[1] und somit haben wir

Satz 2: Die vollständigen Übergangs-Relationen auf einer Menge entsprechen umkehrbar eindeutig den *angeordneten Klassen-Einteilungen* dieser Menge.

Wir erklären einen weiteren Typ von Übergangs-Relationen durch folgende

Definition: Eine Übergangs-Relation heißt *zerfällend* (oder „symmetrisch"), wenn bei ihr die Kombination $F(z, z') = -F(z', z)$ nicht vorkommt.

In diesem Fall existieren also keine (bezüglich F) irreversiblen Übergänge, sondern nur reversible und beidseitig unmögliche. Die Zerfällungs-Eigenschaft überträgt sich auf die Klassen-Relation $\mathscr{F}(\mathfrak{M}, \mathfrak{M}')$ und schließt (2.1 a) aus. Damit besitzt die Klassen-Relation neben der Transitivität die kennzeichnende Eigenschaft (2.1 b), so daß die Klassen durch keine Übergänge miteinander verbunden sind. Dementsprechend definiert eine zerfällende Übergangs-Relation nur eine Klassen-Einteilung von $\mathfrak{Z}$, aber *keine Anordnung* der Klassen, und die Kenntnis dieser Klassen genügt, um die Relation $F(z, z')$ vollständig festzulegen („Äquivalenz-Relation").

Schließlich wollen wir noch eine dritte Art von Übergangs-Relationen F einführen. Wir bilden dazu wieder alle bezüglich F maximalen reversiblen Klassen $\mathfrak{M}$.

---

[1] Beschreibt man nämlich $a < b$ bzw. $b < a$ durch $F(a, b) = +1$ bzw. $F(a, b) = -1$, so folgt aus $F(a, b) = +1$ stets $F(b, a) = -1$ und aus $F(a, b) = +1$ sowie $F(b, c) = +1$ stets $F(a, c) = +1$.

Für diese reversiblen Klassen gibt es nur die beiden Möglichkeiten (2.1a) und (2.1b), d.h. zwei Klassen sind entweder einseitig oder gar nicht miteinander verbindbar. Ein beliebig herausgegriffenes $\mathfrak{M}$ definiert also entsprechend Fig. 1a die Gesamtheit aller seiner „Vorgänger" und „Nachfolger", d.h. die Menge aller $\mathfrak{M}'$, $\mathfrak{M}''$, ..., $\mathfrak{M}^{(i)}$, ..., die mit $\mathfrak{M}$ in einer der beiden Anordnungen $\mathfrak{M}$, $\mathfrak{M}^{(i)}$ oder $\mathfrak{M}^{(i)}$, $\mathfrak{M}$ den Relations-Wert $\mathscr{F} = +1$ haben. Die in Fig. 1a angedeutete „Verzweigung" der Menge aller $\mathfrak{M}^{(i)}$ wollen wir aber gerade ausschließen, so daß jedes $\mathfrak{M}$ genau eine „Kette" von Klassen $\mathfrak{M}^{(i)}$ festlegt, wie in Fig. 1b dargestellt ist. Dementsprechend treffen wir die

Definition: Eine Übergangs-Relation $\mathscr{F}$ heißt *verzweigungslos*, wenn zwei beliebige Klassen $\mathfrak{M}'$, $\mathfrak{M}''$, die Nachfolger oder Vorgänger in bezug auf ein und dieselbe Klasse $\mathfrak{M}$ sind, auch gegenseitig im Verhältnis von Nachfolger und Vorgänger stehen.

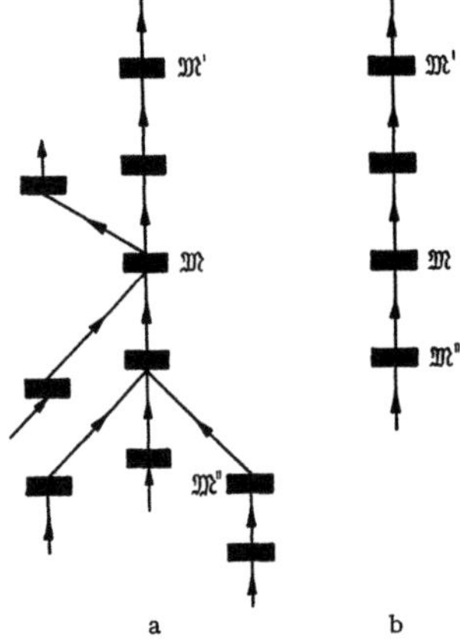

Fig. 1a und b. Zum Begriff der verzweigungslosen Übergangs-Relation.

Verzweigungslose Übergangs-Relationen sind also festgelegt durch die Eigenschaft: Aus $\mathscr{F}(\mathfrak{M}, \mathfrak{M}') = -\mathscr{F}(\mathfrak{M}', \mathfrak{M})$ und $\mathscr{F}(\mathfrak{M}, \mathfrak{M}'') = -\mathscr{F}(\mathfrak{M}'', \mathfrak{M})$ folgt stets $F(\mathfrak{M}', \mathfrak{M}'') = -\mathscr{F}(\mathfrak{M}'', \mathfrak{M}')$. Infolgedessen bestimmt jede reversible Klasse $\mathfrak{M}$ eindeutig *eine Klassen-Kette* $\mathfrak{C}$ (von Klassen $\mathfrak{M}$), und zwischen den verschiedenen $\mathfrak{C}$ gibt es keine Übergänge. In den Klassen-Ketten geschrieben lautet $\mathscr{F}$ daher $\mathscr{F}^*(\mathfrak{C}, \mathfrak{C}') = \mathscr{F}^*(\mathfrak{C}', \mathfrak{C}) = -1$ für $\mathfrak{C} \neq \mathfrak{C}'$. Wählt man nun in jeder Kette genau eine reversible Klasse aus, und bezeichnet man eine Gesamtheit solcher Klassen als *Ketten-Querschnitt*, dann gilt: Zwischen den Zuständen eines Ketten-Querschnitts ist $\mathscr{F}$ *zerfällend*. Hingegen ist $\mathscr{F}$ zwischen den Zuständen einer einzelnen Kette *vollständig*. Vollständige und zerfällende Übergangs-Relationen sind also Grenzfälle der verzweigungslosen Übergangs-Relation, nämlich die erste für den Fall, daß es nur *eine* Klassen-Kette gibt, die zweite für den Fall, daß jede Kette nur aus einer reversiblen Klasse besteht.

Unter den Übergangs-Relationen sind die genannten Typen die wichtigsten. Dies ist unmittelbar einsichtig, wenn man sich vergegenwärtigt, daß in den einfachsten physikalischen Beispielen die vollständige Übergangs-Relation für das Verhalten adiabatisch abgeschlossener Systeme charakteristisch ist, während die zerfällende den energetischen Abschluß beschreibt. Im allgemeinen sind sowohl adiabatische als auch energetische Isolation *verzweigungslose* Übergangs-Relationen: Die adiabatische Isolation deshalb, weil es Variablen gibt, die sich unter ihr nicht verändern lassen (wie z.B. die Massen-Variable), die energetische Isolation, weil sie gleichzeitig auch eine adiabatische Isolation ist. Diese Bemerkung sei hier nur als vorläufiger Hinweis aufgefaßt.

3. Zustands-Funktionen und Variablen. Als Zustands-Funktion $f(z)$ bezeichnet man bekanntlich jede eindeutige Zuordnung von Zahlen f zu den Zuständen z von $\mathfrak{Z}$. Wir betrachten nur Zuordnungen von reellen Zahlen, d.h. reellwertige Zustands-Funktionen. Eine Zustands-Funktion $f(z)$ teilt nun $\mathfrak{Z}$ in elementfremde Klassen $\mathfrak{M}$, wobei eine Klasse alle diejenigen Zustände z enthält, für die $f(z)$ denselben Wert hat. Umgekehrt gibt jede Klassen-Einteilung von $\mathfrak{Z}$ Anlaß zu *mehreren* Zustands-Funktionen, nämlich allen Funktionen $f(z)$, die auf jeder der Klassen konstante Werte haben, aber auf verschiedenen

Klassen verschiedene Werte. Wir sagen: *Die Zustands-Funktionen auf $\mathfrak{Z}$ sind Darstellungen der Klassen-Einteilungen von $\mathfrak{Z}$.* Zwei Funktionen f, g, welche dieselbe Klassen-Einteilung darstellen, haben offenbar die Eigenschaft, daß sie durch eine umkehrbar eindeutige Abbildung zwischen ihren Werte-Bereichen auseinander hervorgehen: Solche Funktionen sollen (uneigentlich) *äquivalent* heißen.

Bisher haben wir die Tatsache außer acht gelassen, daß die reellen Zahlen *angeordnet* sind. Die obige Definition der Äquivalenz von Zustands-Funktionen nimmt auf die Anordnung keine Rücksicht, was z.B. daraus ersichtlich ist, daß sie sich wörtlich auf komplex-wertige Zustands-Funktionen ausdehnen ließe, das heißt auf einen Zahlbereich, der keine Anordnung besitzt. Will man auch die Eigenschaft der Anordnung der Klassen durch Zustands-Funktionen ausdrücken, so müssen die Werte der Zustands-Funktionen diese Anordnung wiedergeben. Die Äquivalenz von Zustands-Funktionen muß dann eingeschränkt werden auf eine Äquivalenz unter Erhaltung der Anordnung der Funktions-Werte, das heißt die Zustands-Funktionen sind definiert bis auf eine allgemeine „Skalen-Transformation". Eine solche Zustands-Funktion, deren Werte-Anordnung invariante Bedeutung besitzt, wird aber in der Physik als Variable eines Systems bezeichnet. Wir treffen daher die

Definition: Jede *angeordnete* Klassen-Einteilung $\{\mathfrak{M}\}$ der Zustände von $\mathfrak{Z}$ ist eine *Variable* auf $\mathfrak{Z}$. Jede Zustands-Funktion $f(\mathfrak{M})$, deren Werte-Anordnung die Anordnung der Klassen wiedergibt, heißt eine *Darstellung dieser Variablen.*

Da eine System-Einwirkung, die durch eine vollständige Übergangs-Relation beschrieben wird, eine angeordnete Klassen-Einteilung erzeugt, ergibt sich sofort der

Satz 3: Jede System-Einwirkung mit vollständiger Übergangs-Relation definiert eine Variable.

Die durch eine *zerfallende* Übergangs-Relation (Äquivalenz-Relation) definierte Klassen-Einteilung von $\mathfrak{Z}$ ist nicht angeordnet und definiert somit *keine* Variable. Jede Darstellung $f(z)$ einer solchen Relation induziert zwar eine Anordnung der Klassen, aber es gibt andere Darstellungen $g(z)$, welche beliebige andere Anordnungen erzeugen, und es ist ohne Hinzunahme weiterer Informationen nicht möglich, eine dieser willkürlichen Anordnungen gegenüber anderen als „richtig" auszuzeichnen.

4. Kontinuierliche Variablen. Zustandsraum. Wenn der Werte-Bereich einer Variablen-Darstellung $f(z)$ ein Intervall der reellen Zahlengeraden vollständig bedeckt, dann heißt die Variable *kontinuierlich*[1]. Die zu einer kontinuierlichen Variablen gehörige Klassen-Einteilung $\{\mathfrak{M}\}$ von $\mathfrak{Z}$ bildet eine einparametrig-zusammenhängende Schar $\mathfrak{M}(f)$ mit den Werten f der darstellenden Funktion als Parameter. Ist das darstellende Intervall offen, das heißt besitzt die Variable keine erste und keine letzte Klasse in $\mathfrak{Z}$, so kann man zu einer solchen Darstellung übergehen, deren Werte-Bereich die ganze Zahlengerade ist; denn ein beliebiges offenes Intervall reeller Zahlen läßt sich bekanntlich eindeutig und stetig (sogar differenzierbar) auf die ganze Zahlengerade transformieren. Unabhängig von den Darstellungen ist die wesentliche Eigenschaft kontinuierlicher

---

[1] Die Kontinuums-Eigenschaft einer Variablen ist bekanntlich keiner experimentellen Nachprüfung zugänglich; denn eine physikalisch „kontinuierliche" Variable kann im mathematischen Sinne höchstens als abzählbar-dicht bezeichnet werden. Der Grund für die Einführung des mathematischen Kontinuums liegt in der dadurch geschaffenen Möglichkeit, den Kalkül der Analysis ausnutzen zu können. Siehe auch S. 131.

Variablen die, daß es in bezug auf die zugehörige Übergangs-Relation zu je zwei Klassen $\mathfrak{M}'$, $\mathfrak{M}''$ stets eine dritte Klasse $\mathfrak{M}$ gibt, die zwischen $\mathfrak{M}'$ und $\mathfrak{M}''$ liegt.

In $\mathfrak{Z}$ sei nun ein Satz von n kontinuierlichen Variablen x_1, x_2, ..., x_n vorgegeben, deren darstellende Funktionen $x_i(z)$ die ganze reelle Zahlengerade durchlaufen mögen. Dann bestimmt jeder Zustand z aus $\mathfrak{Z}$ eindeutig einen Bildpunkt $x_1(z)$, $x_2(z)$, ..., $x_n(z)$ im n-dimensionalen Zahlenraum (Raum der reellen Zahlen-n-tupel), und jeder Punkt des Zahlenraumes hat mindestens ein Original z in $\mathfrak{Z}$. Hat jeder Bild-Punkt sogar nur ein einziges z als Original, so heißt $\mathfrak{Z}$ ein *n-dimensionaler Zustandsraum* $\mathfrak{Z}_n$ mit den *Koordinaten* $x_1(z)$, $x_2(z)$, ..., $x_n(z)$.

5. Metrische Variablen. Variablen auf einer Zustandsmenge $\mathfrak{Z}$ haben wir definiert als geordnete Klassen-Einteilungen $\{\mathfrak{M}\}$ von $\mathfrak{Z}$. Jede ordnungstreue Belegung der Klassen mit reellen Zahlen x nannten wir eine Darstellung $x(\mathfrak{M})$ der betreffenden Variablen. Die Darstellungen von Variablen sind demgemäß bis auf beliebige ordnungstreue Skalen-Transformationen erklärt. Bei kontinuierlichen Variablen wird man trivialerweise verlangen, daß neben der Anordnung auch die Kontinuität der Klassenschar $\mathfrak{M}(x)$ vom Scharparameter dargestellt werden soll. Dann sind die Funktionen $x(\mathfrak{M})$ sogar bis auf ordnungstreue und stetige (bzw. differenzierbare) Skalen-Transformationen erklärt. Die sog. empirischen Temperaturskalen bieten hierfür ein geläufiges Beispiel. Andererseits ist man von vielen in der Physik auftretenden Variablen gewohnt, daß ihre Darstellungen nur *lineare Skalen-Transformationen* zulassen: Maßstabsdilatation und Nullpunktsverschiebung. Solche Variablen bzw. ihre Darstellungen wollen wir *metrische Variablen* nennen[1].

In Ziff. 1 bis 4 wurde gezeigt, wie der allgemeine Begriff einer Variablen sich auf bestimmte Eigenschaften von Relationen reduzieren läßt. Ein solches Vorgehen entsprach der Absicht, den Variablenbegriff von seiner elementarsten Seite her zu verstehen: Die rein *kombinatorischen Merkmale* der System-Einwirkungen und gewisse *Existenz-Aussagen*[2] bildeten den alleinigen Ausgangspunkt. Es handelt sich nunmehr darum, auch den Begriff der metrischen Variablen auf Eigenschaften von Relationen zurückzuführen. In Ziff. 5β geben wir am Beispiel der Energie eine entsprechende Analyse. Dabei wird sich herausstellen, daß die Existenz einer metrischen Variablen und ihre finite *Konstruierbarkeit* immer dann gesichert ist, wenn es gelingt, eine *Wechselwirkung* aufzuweisen, die einem Erhaltungssatz äquivalent ist[3].

α) *Das zusammengesetzte System.* Zur Formulierung des Begriffes der Wechselwirkung zwischen zwei Systemen $\mathfrak{Z}_1$ und $\mathfrak{Z}_2$ ist es zweckmäßig, *das aus $\mathfrak{Z}_1$ und $\mathfrak{Z}_2$ zusammengesetzte System* $\mathfrak{Z} = [\mathfrak{Z}_1, \mathfrak{Z}_2]$ einzuführen, dessen Zustände durch die Zustandspaare $[z_1, z_2]$ repräsentiert werden, wobei z_1 und z_2 die Zustände von $\mathfrak{Z}_1$ und $\mathfrak{Z}_2$ durchlaufen. Der konkrete Anlaß zu dieser Festsetzung ist folgender: Wenn man zwei Systeme mit den Zustandsmengen $\mathfrak{Z}_1$, $\mathfrak{Z}_2$ nicht unmittelbar aneinander koppelt, dann lassen sich die Zustände z_1, z_2 unabhängig voneinander realisieren. Dementsprechend ist es zunächst irrelevant, ob man von den gleichzeitigen Zuständen z_1, z_2 der beiden betrachteten Systeme spricht oder von den

---

[1] Metrische Variablen lassen sich natürlich auch nicht-metrisch darstellen. Dann gehört aber zu jeder Variablen-Darstellung $x(z)$ noch eine metrische Dichte $\varrho(z) \neq$ const mit der Eigenschaft $\varrho \cdot dx =$ invariant.

[2] Zum Beispiel von der Art, daß ein System $\mathfrak{Z}$ in bezug auf eine vollständige Übergangs-Relation F zu zwei Klassen $\mathfrak{M}'$, $\mathfrak{M}''$ stets eine dritte Klasse $\mathfrak{M}$ enthält, die zwischen $\mathfrak{M}'$, $\mathfrak{M}''$ liegt.

[3] „Finite Konstruierbarkeit" soll heißen, daß mit Hilfe der Wechselwirkung eine Vorschrift abgeleitet werden kann, die es erlaubt, für jeden vorgelegten Zustand z den zugehörigen Variablen-Wert $x(z)$ in *endlich* vielen Schritten zu konstruieren. Eine derartige Vorschrift ist natürlich nichts anderes als das formale Abbild der konkreten Meßmethode.

Zuständen $z = [z_1, z_2]$ des zusammengesetzten Systems $\mathfrak{Z} = [\mathfrak{Z}_1, \mathfrak{Z}_2]$. Wesentlich ist, daß man die Einwirkung auf die beiden Systeme $\mathfrak{Z}_1, \mathfrak{Z}_2$ in einer solchen Weise spezialisieren kann, daß zwar für jedes der Teilsysteme $\mathfrak{Z}_1, \mathfrak{Z}_2$ jeder einzelne Übergang $z_1 \rightarrow z_1'$ bzw. $z_2 \rightarrow z_2'$ möglich ist, diese Übergänge sich aber nicht in beliebiger Kombination gleichzeitig realisieren lassen. Offenbar kennzeichnet dies gerade den *kombinatorischen Sachverhalt einer Wechselwirkung*. Zwei gleichzeitig realisierbare Übergänge $z_1 \rightarrow z_1'$, $z_2 \rightarrow z_2'$ bedeuten nun im Hinblick auf die Zustandsmenge $\mathfrak{Z} = [\mathfrak{Z}_1, \mathfrak{Z}_2]$ einen möglichen Übergang $z \rightarrow z'$ mit $z = [z_1, z_2]$ und $z' = [z_1', z_2']$. Wir können also festsetzen: Jede Art von Wechselwirkung zwischen $\mathfrak{Z}_1$ und $\mathfrak{Z}_2$, möglicherweise unter zusätzlicher äußerer Einwirkung[1] auf das Gesamtsystem, repräsentiert eine Übergangs-Relation $F([z_1, z_2], [z_1', z_2']) = \pm 1$ auf dem zusammengesetzten System.

Es sei noch darauf hingewiesen, daß der Begriff *System* bisher nicht definiert wurde. Wir haben uns lediglich auf *Zustandsmengen* berufen, die den Systemen eindeutig zugeordnet sind. Insbesondere wurde der Begriff *System-Zusammensetzung* einfach durch die (mengentheoretische) Produktbildung $\mathfrak{Z} = [\mathfrak{Z}_1, \mathfrak{Z}_2]$ erklärt. Dies hat einige Konsequenzen im Hinblick auf den gewohnten Sprachgebrauch des Wortes System. Aus zwei Systemen $\mathfrak{Z}_1$ und $\mathfrak{Z}_2$ lassen sich danach nämlich, je nach Art der Wechselwirkung zwischen ihnen, verschiedene „Gesamtsysteme" aufbauen, die in ihren Eigenschaften durchaus nicht übereinstimmen und die nicht einmal durch die Menge aller Zustandspaare $[z_1, z_2]$ repräsentiert zu werden brauchen. Die Wechselwirkung „thermisches Gleichgewicht" zwischen $\mathfrak{Z}_1$ und $\mathfrak{Z}_2$ zum Beispiel hätte zur Folge, daß die Zustände des Gesamtsystems nur von einer Teilmenge der Paare $[z_1, z_2]$ repräsentiert werden, nämlich von derjenigen, bei der z_1 und z_2 jeweils Zustände gleicher Temperatur sind. Andererseits würde eine Wechselwirkung mit Stoff-Austausch, das heißt das Verändern von Variablen, die bei adiabatischer Isolation der Einzelsysteme unveränderlich sind und daher zur Charakterisierung der Zustände von $\mathfrak{Z}_1$ und $\mathfrak{Z}_2$ oft ganz außer acht gelassen werden, dazu führen, daß das Gesamtsystem Zustände besitzt, die unter den Paaren $[z_1, z_2]$ gar nicht vorkommen. Wir werden erst später eine Definition des Begriffes „thermodynamisches System" geben; hier sei aber schon darauf hingewiesen, daß der diskutierte Fall des Stoff-Austausches eine definitorische Alternative anbietet: *Entweder* betrachten wir auch in diesem Fall das Gesamtsystem als zusammengesetzt, dann müssen in physikalischen Systemen (wie $\mathfrak{Z}_1, \mathfrak{Z}_2$) stets auch solche Variablen berücksichtigt werden, die zwar bei adiabatischer Isolation nicht variabel sind, dagegen bei irgendeiner der in Betracht kommenden Wechselwirkungen sichtbar werden können[2], *oder* wir bezeichnen das Gesamtsystem als nicht-zusammengesetzt. Im zweiten Fall müßten wir konsequenterweise zwischen „reduziblen" und „irreduziblen" physikalischen Systemen unterscheiden je nachdem, ob sie sich als aus Teilsystemen zusammengesetzt auffassen lassen oder nicht. Die genannte Alternative ist beim Aufbau der Thermodynamik insofern von Bedeutung, als je nach

---

[1] Die Unterscheidung der Begriffe „Wechselwirkung" und „System-Einwirkung" kommt gewissen Vorstellungs-Gewohnheiten entgegen. Es wäre konsequenter, diesen Unterschied nicht zu machen und nur von den Wechselwirkungen eines Systems oder nur von den Einwirkungen auf ein System zu sprechen.

[2] Man beachte, daß die Variablen eines Systems wesentlich bestimmt sind durch die Gesamtheit aller anderen Systeme, mit denen Wechselwirkung zugelassen wird. Der logische Ausgangspunkt zur Festlegung der Variablen ist daher nicht das isolierte Einzelsystem, sondern das Einzelsystem als Mitglied einer vorgegebenen Menge von Systemen bzw. als Teilsystem eines übergeordneten (aus allen betrachteten Systemen) zusammengesetzten Systems. Diese Beschreibung entspricht genau dem realen physikalischen Sachverhalt. Das übergeordnete Gesamtsystem wird dabei als isoliertes System betrachtet. Manchmal spielt in thermodynamischen Überlegungen die „Welt" die Rolle dieses isolierten Gesamtsystems.

der Festlegung des Begriffes System die adiabatische Isolation einmal eine verzweigungslose, zum anderen dagegen eine vollständige Übergangs-Relation ist.

β) Orientierende Betrachtungen. Wir wollen nunmehr zeigen, welche Eigenschaften eine Wechselwirkung besitzen muß, damit sie sich durch einen Erhaltungssatz darstellen läßt. Dazu gehen wir von einem geläufigen Sachverhalt aus: Zwei Systeme $\mathfrak{Z}_1$ und $\mathfrak{Z}_2$ mit den Energie-Funktionen $U_1(z_1)$ und $U_2(z_2)$ mögen miteinander wechselwirken und zwar so, daß bei jedem einzelnen Wechselwirkungs-Prozeß die Gesamtenergie unverändert bleibt. Betrachtet man nun alle Zustands-quadrupel $([z_1, z_2], [z_1', z_2'])$ als Übergänge des aus $\mathfrak{Z}_1$ und $\mathfrak{Z}_2$ zusammengesetzten Systems, so ist für jedes beliebig herausgegriffene Quadrupel die Bedingung der Energie-Erhaltung $U_1 + U_2 = U_1' + U_2'$ entweder erfüllt oder nicht erfüllt. Diese Alternative läßt sich als Relation $F([z_1, z_2], [z_1', z_2']) = \pm 1$ schreiben. Da F für ein vorgegebenes Zustandsquadrupel seinen Wert nicht verändert, wenn die einzelnen Zustände durch energetisch gleichwertige ersetzt werden, ist es zweckmäßig, statt F die entsprechende Relation über den Energie-Klassen[1] $\mathfrak{E}_1, \mathfrak{E}_1', \dots$ von $\mathfrak{Z}_1$ und $\mathfrak{E}_2, \mathfrak{E}_2' \dots$ von $\mathfrak{Z}_2$ zu benutzen: $\mathscr{F}([\mathfrak{E}_1, \mathfrak{E}_2], [\mathfrak{E}_1', \mathfrak{E}_2']) = \pm 1$. Das oben genannte Problem der Kennzeichnung von Wechselwirkungen, die auf einen Erhaltungssatz führen, besteht also darin, diejenigen Eigenschaften der Klassen-Relation $\mathscr{F}$ aufzufinden, welche der Darstellbarkeit der Wechselwirkung durch einen Erhaltungssatz äquivalent sind.

Wir betrachten dazu folgende Operation: An dem System $\mathfrak{Z}_1$ seien zwei beliebige Energie-Klassen $\mathfrak{E}_1^0$ und $\mathfrak{E}_1'$ markiert[2]. Dann trete $\mathfrak{Z}_1$ mit dem System $\mathfrak{Z}_2$ in Wechselwirkung derart, daß das Gesamtsystem energetisch abgeschlossen bleibt und das System $\mathfrak{Z}_1$ den Übergang $\mathfrak{E}_1^0 \to \mathfrak{E}_1'$ erfährt. Dabei muß auch $\mathfrak{Z}_2$ seine Energieklasse wechseln: $\mathfrak{E}_2^0 \to {}'\mathfrak{E}_2$. Offenbar ist ${}'\mathfrak{E}_2$ sogar eindeutig bestimmt, wenn $\mathfrak{E}_2^0$ vorgegeben wird, und zwar unabhängig davon, ob die Energie-Klassen stets von denselben Zuständen repräsentiert werden oder nicht. Bringt man, nach Lösung der Wechselwirkung zwischen $\mathfrak{Z}_1$ und $\mathfrak{Z}_2$, das erste System wieder in seine anfängliche Energie-Klasse $\mathfrak{E}_1^0$, das zweite hingegen nicht, so wird durch Wiederholung der Operation in Anwendung auf ${}'\mathfrak{E}_2$ (als Anfangszustand von $\mathfrak{Z}_2$) eine neue Energie-Klasse ${}''\mathfrak{E}_2$ gewonnen. Iteration des Verfahrens liefert eine Folge $\mathfrak{E}_2^0 \to {}'\mathfrak{E}_2 \to {}''\mathfrak{E}_2 \to \cdots$, und die Umkehrung des erzeugenden Überganges ergibt eine zweite Folge $\mathfrak{E}_2^0 \to \mathfrak{E}_2' \to \mathfrak{E}_2'' \to \cdots$. Jedes $\mathfrak{E}_2$ als Anfangszustand bestimmt in dieser Weise eine absteigende und eine aufsteigende Folge von Energie-Klassen von $\mathfrak{Z}_2$. Die beiden von $\mathfrak{E}_2^0$ ausgehenden Folgen können als eine einzige betrachtet werden, denn jede von ihnen läßt sich auch „rückwärts" durchlaufen und mündet dabei in die andere. Diese Eigenschaft läßt sich natürlich nicht aus obiger Konstruktion entnehmen, sondern impliziert eine zusätzliche Forderung. Weiterhin besagt die Erfahrung, daß obige Konstruktion „äquidistanter" Energie-Klassen $\mathfrak{E}_2$ nicht abhängig ist von dem speziell gewählten erzeugenden Klassen-Paar $\mathfrak{E}_1^0$, $\mathfrak{E}_1'$; denn zwei Energie-Klassen irgendeines Systems, die durch Kopplung mit den Übergängen $\mathfrak{E}_2^0 \to \mathfrak{E}_2'$ oder $\mathfrak{E}_2' \to \mathfrak{E}_2^0$ ineinander überführt werden können, sind als Substituenten für $\mathfrak{E}_1^0$ und $\mathfrak{E}_1'$ brauchbar, da sie dieselbe Folge von $\mathfrak{E}_2$-Klassen liefern. Zwei Klassen $\mathfrak{E}_2^0$, $\mathfrak{E}_2'$ definieren

---

[1] Als „Energie-Klassen" eines Systems bezeichnen wir die Zustandsklassen gleicher Energie. Diese Klassen lassen sich unabhängig von der Zustands-Funktion durch den Begriff der energetischen Isolation definieren. Vergleiche Ziff. 11.

[2] Ein rechts (links) angebrachter Strich-Index soll darauf hinweisen, daß die betreffende Energie-Klasse mehr (weniger) Energie besitzt als die willkürlich ausgezeichnete, mit 0 indizierte Klasse. Diese Festsetzung ist natürlich nur möglich durch Vorgriff auf die noch zu beweisende Anordnung der Energie-Klassen. Sie darf infolgedessen bei keiner Schluß-folgerung benutzt werden.

also ohne Bezugnahme auf einen erzeugenden Übergang (wie $\mathfrak{E}_1^0 \to \mathfrak{E}_1'$) eindeutig eine Folge von Energie-Klassen in $\mathfrak{Z}_2$, und zwei beliebige, benachbarte Klassen dieser Folge definieren wiederum dieselbe Folge. Gibt man weiterhin in allen anderen Zustandsmengen je eine Klasse $\mathfrak{E}^0$ vor, dann läßt sich mit Hilfe des Überganges $\mathfrak{E}_2' \to \mathfrak{E}_2^0$ je eine Klasse $\mathfrak{E}'$ erzeugen und somit eine Folge von Energie-Klassen festlegen. Außerdem gilt: Jeder beliebige Übergang zwischen zwei benachbarten Klassen einer Folge läßt sich durch jeden anderen derartigen Übergang erzeugen, falls beide Übergänge entgegengesetzte Richtung haben. Dies trifft sogar dann zu, wenn die Übergänge eine gleich große Anzahl von Klassen ,,überspringen''. Beliebige äquidistante Klassenpaare stehen also in der Beziehung gegenseitiger Erzeugbarkeit.

Alle diese Eigenschaften der energetischen Wechselwirkung lassen sich wie folgt zusammenfassen[1]:

a) Jeder Übergang $\mathfrak{E}_1 \to \mathfrak{E}_1'$ erzeugt aus einer Klasse $\mathfrak{E}_2$ *eindeutig* eine Klasse $\mathfrak{E}_2'$.

b) Wenn zwei Übergänge sich gegenseitig erzeugen, so gilt dasselbe für die beiden entgegengesetzt orientierten Übergänge, das heißt die Relation $\mathscr{J}$ ist *symmetrisch* in den Klassenpaaren $[\mathfrak{E}_1, \mathfrak{E}_2]$.

c) Wenn drei Übergänge in der Beziehung stehen, daß der erste und zweite sowie der entgegengesetzte zweite und dritte sich gegenseitig durch Wechselwirkung hervorbringen, dann erzeugen sich stets auch der erste und dritte gegenseitig.

d) Sind $\mathfrak{E}_1 \to \mathfrak{E}_1' \to \mathfrak{E}_1''$ und $\mathfrak{E}_2 \to \mathfrak{E}_2' \to \mathfrak{E}_2''$ sich durch Wechselwirkung gegenseitig erzeugende konsekutive Übergänge, so erzeugen sich auch die Übergänge $\mathfrak{E}_1 \to \mathfrak{E}_1''$ und $\mathfrak{E}_2 \to \mathfrak{E}_2''$ gegenseitig, das heißt die Relation ist *transitiv* in den Klassenpaaren $[\mathfrak{E}_1, \mathfrak{E}_2]$.

Man erkennt aus b) und d), daß $\mathscr{J}$ eine *zerfällende Übergangs-Relation* zwischen den Klassenpaaren ist, also erzeugt $\mathscr{J}$ auf der Zustandsmenge $\mathfrak{Z} = [\mathfrak{Z}_1, \mathfrak{Z}_2]$ eines zusammengesetzten Systems eine Klassen-Einteilung $\{\mathfrak{E}\}$. Da $\mathscr{J}$ die Wechselwirkung der Teilsysteme $\mathfrak{Z}_1, \mathfrak{Z}_2$ unter *energetischer Isolation* des Gesamtsystems beschreiben soll, ist diese Klassen-Einteilung $\{\mathfrak{E}\}$ nichts anderes als die Einteilung von $\mathfrak{Z}$ in energetisch gleichwertige Zustände. Außerdem ist unmittelbar zu erkennen, daß jede Klasse $\mathfrak{E}$ die Vereinigungsmenge einer bestimmten Gesamtheit von Klassenpaaren $[\mathfrak{E}_1, \mathfrak{E}_2]$ ist: $\mathfrak{E} = [\mathfrak{E}_1^0, \mathfrak{E}_2^0] + [\mathfrak{E}_1', \mathfrak{E}_2'] + [\mathfrak{E}_1'', \mathfrak{E}_2''] + \cdots$, was sich unter Benutzung der Energie-Funktionen ausdrückt als $U = U_1^0 + U_2^0 = U_1' + U_2' = U_1'' + U_2'' = \cdots = \mathrm{const}$. Weiterhin besagt die Eigenschaft a), daß innerhalb einer beliebigen Klasse $\mathfrak{E}$ jede der Klassen $\mathfrak{E}_1, \mathfrak{E}_2$ vorkommt, und daß dabei jede Klasse $\mathfrak{E}_1^{(i)}$ mit genau einer Klasse $\mathfrak{E}_2^{(j)}$ verbunden ist und umgekehrt. Schließlich gestattet c) den Nachweis, daß je zwei Klassen $\mathfrak{E}_2^0, \mathfrak{E}_2'$ eindeutig eine Folge von Klassen $\mathfrak{E}_2$ festlegen.

Es ist offensichtlich, in welcher Weise die erläuterte Konstruktion der Klassen-Folgen für die Konstruktion einer Variablen, das heißt einer *angeordneten* Klassen-Einteilung benutzt werden kann: Zunächst liegt nur eine Klassen-Einteilung in $\mathfrak{Z}_k$ ohne Anordnung vor. Eine Wechselwirkung mit den Eigenschaften a) bis d) gibt sodann die Möglichkeit, mit irgend zwei beliebigen Klassen $\mathfrak{E}_k^0, \mathfrak{E}_k'$ eine *Folge*, das heißt eine angeordnete Menge von Klassen $\mathfrak{E}_k^{(i)}$ zu erzeugen. Wenn es gelingt, diese Folge so zu erweitern, daß *alle* Klassen $\mathfrak{E}_k$ aus $\mathfrak{Z}_k$ in ihr vorkommen, dann ist in eindeutiger Weise eine *Variable* auf $\mathfrak{Z}_k$ definiert. Dazu genügt es, die beiden folgenden Forderungen zu stellen:

---

[1] Die oben getroffene Vereinbarung über rechts bzw. links gestellte Indices soll der Einfachheit halber jetzt wieder außer acht bleiben.

e) Zu jedem Übergang $\mathfrak{E}_k^0 \to \mathfrak{E}_k'$ gibt es einen „halbierenden" Übergang, der durch zweimalig iterierte Wechselwirkung aus der Energie-Klasse $\mathfrak{E}_k^0$ die Klasse $\mathfrak{E}_k'$ erzeugt: $\mathfrak{E}_k^0 \to \mathfrak{E}_k^{\frac{1}{2}}$, $\mathfrak{E}_k^{\frac{1}{2}} \to \mathfrak{E}_k'$ und alle derartigen Halbierungs-Übergänge erzeugen ein und dieselbe Klasse $\mathfrak{E}_k^{\frac{1}{2}}$.

f) Jede vorgegebene Energie-Klasse $\mathfrak{E}_k$ kann durch endlich viele Halbierungs-Schritte in eine beliebige Klassen-Folge aufgenommen werden.

Zusammen mit den Bedingungen a) bis d) liefert e) die Aussage, daß ein zu $\mathfrak{E}_k^0 \to \mathfrak{E}_k'$ gehöriger Halbierungs-Übergang *alle* Übergänge zwischen benachbarten Klassen der aus $\mathfrak{E}_k^0 \to \mathfrak{E}_k'$ konstruierten Folge halbiert. Offensichtlich überträgt sich dieser Halbierungs-Prozeß auch auf entsprechende Klassen-Folgen anderer Systeme. Mit anderen Worten, die Eigenschaft e) sichert ein eindeutiges Verfahren, jeden Teilschritt der Klassen-Folgen in zwei Schritte zu zerlegen, und die Iteration dieser Zerlegung ermöglicht eine beliebige Verfeinerung aller Klassen-Folgen. Belegt man nun eine Folge ..., $''\mathfrak{E}_k \to '\mathfrak{E}_k \to \mathfrak{E}_k^0 \to \mathfrak{E}_k' \to \mathfrak{E}_k'' \to \cdots$ der Reihe nach mit den ganzen Zahlen ..., $-2, -1, 0, +1, +2, \ldots$ und benutzt man in entsprechender Weise Dualbrüche zur Kennzeichnung der Halbierungs-Schritte, so bewirkt die Eigenschaft e), daß zu jeder rationalen Zahl U_k eine Klasse $\mathfrak{E}_k$ gehört[1], und Eigenschaft f) zieht nach sich, daß auch umgekehrt jede Klasse $\mathfrak{E}_k$ eine rationale Zahl U_k erhält. Wir sagen, die Energie-Klassen bilden eine *kontinuierliche*[2] *metrische Variable*, für die eine darstellende Funktion $U_k(\mathfrak{E}_k)$ eindeutig festgelegt ist durch die Zuordnung der Werte 0 und 1 zu den Klassen $\mathfrak{E}_k^0$ und $\mathfrak{E}_k'$. Die Auszeichnung von je einer Energie-Klasse $\mathfrak{E}^0$ in allen anderen Zustandsmengen liefert dann einen Satz von Energie-Funktionen $U_1, U_2, \ldots$, mit deren Hilfe sich die energetische Wechselwirkungs-Relation als Erhaltungssatz darstellen läßt: $\mathscr{J}([\mathfrak{E}_k, \mathfrak{E}_l], [\mathfrak{E}_k', \mathfrak{E}_l']) = +1$ gilt genau dann, wenn

$$U_k(\mathfrak{E}_k) + U_l(\mathfrak{E}_l) = U_k(\mathfrak{E}_k') + U_l(\mathfrak{E}_l').\tag{5.1}$$

Die Zustands-Funktion $U_k(\mathfrak{E}_k) + U_l(\mathfrak{E}_l) = U^*$ kennzeichnet also mit jedem ihrer Werte U^* genau eine Energie-Klasse $\mathfrak{E}(U^*)$ des zusammengesetzten Systems $\mathfrak{Z} = [\mathfrak{Z}_k, \mathfrak{Z}_l]$. Damit ist aber keineswegs gesagt, daß der Parameter $U^*(\mathfrak{E})$ eine metrische Darstellung der Energie-Variablen U des zusammengesetzten Systems ist, denn U ist nicht definiert durch $U_k + U_l$, sondern durch die Wechselwirkungen zwischen $\mathfrak{Z} = [\mathfrak{Z}_k, \mathfrak{Z}_l]$ und den anderen Systemen $\mathfrak{Z}_i$. In formaler Hinsicht ist darüber hinaus noch nicht einmal gesichert, daß die Zustandsmenge $\mathfrak{Z} = [\mathfrak{Z}_k, \mathfrak{Z}_l]$ wieder ein System repräsentiert, dessen energetische Wechselwirkung mit anderen Systemen wieder die Eigenschaften a) bis f) hat. Wir fordern deshalb:

g) Je zwei Systeme mit den Zustandsmengen $\mathfrak{Z}_k, \mathfrak{Z}_l$ definieren eindeutig ein drittes System mit der Zustandsmenge $\mathfrak{Z} = [\mathfrak{Z}_k, \mathfrak{Z}_l]$. Die Energie-Klassen $\mathfrak{E}$ von $\mathfrak{Z}$ stimmen überein mit den Zustandsklassen $U_k(\mathfrak{E}_k) + U_l(\mathfrak{E}_l) = \text{const}$, das heißt die Zustands-Funktion $U(\mathfrak{E})$ hängt umkehrbar eindeutig von $U_k + U_l$ ab.

Wenn nun $U^*(\mathfrak{E}) = U_1 + U_2$ ein „falscher" Parameter wäre, der auch nach Addition einer geeigneten Konstanten nicht mit $U(\mathfrak{E})$ übereinstimmt, dann würde im allgemeinen eine Energie-Differenz $\Delta U = U(\mathfrak{E}) - U(\mathfrak{E}')$ von der zugehörigen Differenz ΔU^* verschieden sein. Es sei zum Beispiel $0 < \Delta U^* < \Delta U$, dann ergibt sich die Möglichkeit, durch folgenden Kreis-Prozeß Energie zu erzeugen: 1. Übertragung der Energie ΔU vom System $\mathfrak{Z} = [\mathfrak{Z}_1, \mathfrak{Z}_2]$ an ein System $\mathfrak{Z}_3$. Hierbei ändert sich U^* um $-\Delta U^* = -\Delta(U_1 + U_2)$. 2. Zerlegung von $\mathfrak{Z}$ in $\mathfrak{Z}_1, \mathfrak{Z}_2$ und

---

[1] Wir sehen davon ab, daß die Folgen der Energie-Klassen nach „unten" abbrechen.
[2] Siehe Bemerkung S. 131.

Rückführung der Systeme $\mathfrak{Z}_1$, $\mathfrak{Z}_2$ auf ihre Anfangsenergie durch je eine Wechselwirkung mit $\mathfrak{Z}_3$. Dazu muß $\mathfrak{Z}_3$ insgesamt die Energie $\varDelta U_1 + \varDelta U_2 = \varDelta U^*$ abgeben, so daß seine Energie im Endzustand wegen $\varDelta U - \varDelta U^* > 0$ über dem Anfangswert liegt. Um diese Möglichkeit auszuschließen, muß gefordert werden, daß $U^*(\mathfrak{E}) = U_1 + U_2$ eine metrische Darstellung der Energie-Variablen U des zusammengesetzten Systems ist. Die Wechselwirkungs-Eigenschaft welche uns zu dieser Aussage geführt hat, läßt sich unter Beachtung der genannten Prozesse 1, 2 wie folgt formulieren:

h) Wenn der Übergang $[\mathfrak{E}_1, \mathfrak{E}_2] \to [\mathfrak{E}_1', \mathfrak{E}_2']$ eines zusammengesetzten Systems $\mathfrak{Z} = [\mathfrak{Z}_1, \mathfrak{Z}_2]$ von einem Übergang $\mathfrak{E}_3 \to \mathfrak{E}_3'$ erzeugt wird, dann erzeugen die beiden Übergänge $[\mathfrak{E}_1, \mathfrak{E}_2] \to [\mathfrak{E}_1, \mathfrak{E}_2']$, $[\mathfrak{E}_1, \mathfrak{E}_2'] \to [\mathfrak{E}_1', \mathfrak{E}_2']$ eine Zerlegung von $\mathfrak{E}_3 \to \mathfrak{E}_3'$ in zwei Teil-Schritte $\mathfrak{E}_3 \to \mathfrak{E}_3^* \to \mathfrak{E}_3'$.

Der hier am Beispiel der Energie diskutierte Zusammenhang zwischen einer metrischen Variablen und einer zugehörigen Wechselwirkung mit Erhaltungssatz ist *genereller* Natur. *Die üblicherweise als extensiv bezeichneten Variablen sind Größen von solcher Art.* Zu ihnen zählt bekanntlich auch die Entropie S[1].

Zum Schluß noch eine Bemerkung: In Ziff. 4 wurde der Begriff einer *kontinuierlichen* Variablen in gewohntem mathematischen Sinn verwendet, nämlich zur Kennzeichnung einer geordneten Menge (von Klassen), die sich isomorph auf die reellen Zahlen abbilden läßt. Hingegen haben wir auf S. 130 das Wort kontinuierlich benutzt, obwohl die fragliche Klassen-Einteilung nur die Struktur der rationalen Zahlen besitzt. Diese formale Inkonsequenz rechtfertigt sich dadurch, daß die im allgemeinen Sprachgebrauch als kontinuierlich bezeichneten *physikalischen* Variablen in Wirklichkeit nur *rational* sein können, denn ihre konkrete Aufweisung ist stets an finite Meßprozesse gebunden. Andererseits ist es zweckmäßig, jede „physikalisch kontinuierliche" Variable durch Hinzunahme von Grenzprozessen zu einer „mathematisch kontinuierlichen" Variablen zu erweitern; das ermöglicht, wie in Ziff. 4 erwähnt, die Anwendung des analytischen Kalküls, ohne den physikalischen Inhalt zu berühren. Bei der Längenmessung, die auch nichts anderes ist als die Konstruktion einer metrischen Variablen, ist dieses Problem von alters her bekannt: Betrachtet man das iterierte Anlegen und Teilen eines Maßstabes längs einer Geraden g als konkrete Methode zur Fixierung der Punkte auf g, so ist klar, daß nach Maßgabe des jeweils verwendeten Meßaufwandes jede beliebige Strecke auf g durch eine *endliche* Anzahl von Punkten repräsentiert wird, und diese Punkte besitzen rationale Zahlen als Koordinaten. Damit ist Eigenschaft f) erfüllt. Die Annahme, daß jede Strecke als *physikalisches* Objekt einer ständig verschärften Zerlegung in (endlich viele) Punkte keine prinzipielle Grenze setzt, kommt in Eigenschaft e) zum Ausdruck. Ohne Zweifel ist damit der unmittelbare Sachverhalt der Punkt-Mannigfaltigkeit auf einer Geraden vollständig erfaßt. Bekanntlich ergibt sich aber in formaler Hinsicht die Notwendigkeit, irrationale Punkte einzuführen, sobald die Punkte einer Geraden g durch zusätzliche Konstruktionen fixiert werden, die *außerhalb* g verlaufen. Hier ist nicht der Ort, auf diese Fragen näher einzugehen. Wir werden vielmehr den Begriff der kontinuierlichen Variablen ohne weitere Rechtfertigung in gewohnter Weise benutzen.

γ) Die formale Beschreibung. Die Bedingungen für die Existenz einer metrischen Variablen sollen nun in abstrakter Weise zusammengefaßt werden. Es sei $\mathfrak{G}$ eine Gesamtheit von Mengen $\mathfrak{Z}_1, \mathfrak{Z}_2, \ldots$, die mit jedem Paar $\mathfrak{Z}_i, \mathfrak{Z}_k$ auch

---

[1] In Ziff. 10 wird gezeigt, daß die reversibel-adiabatischen Prozesse an zusammengesetzten Systemen eine Wechselwirkung mit Erhaltungssatz repräsentieren und somit die Existenz einer metrischen Variablen S sichern.

deren Produkt-Menge $[\mathfrak{Z}_i, \mathfrak{Z}_k]$ enthält. Dann bezeichnen wir jedes $\mathfrak{Z}_k$ als *System mit einer metrischen Variablen* X, wenn folgendes der Fall ist:

(i) Jedes $\mathfrak{Z}_k$ besitzt eine Klassen-Einteilung $\{\mathfrak{R}_k\}$ seiner Elemente.

(ii) Auf den Klassenquadrupeln $[\mathfrak{R}_k, \mathfrak{R}_l]$, $[\mathfrak{R}'_k, \mathfrak{R}'_l]$ je zweier Systeme $\mathfrak{Z}_k$, $\mathfrak{Z}_l$ ist eine zweiwertige Funktion $\mathscr{J}([\mathfrak{R}_k, \mathfrak{R}_l], [\mathfrak{R}'_k, \mathfrak{R}'_l]) = \pm 1$ erklärt, welche in bezug auf die Klassenpaare $[\mathfrak{R}_k, \mathfrak{R}_l]$ die Struktur einer *zerfällenden Übergangs-Relation* besitzt.

(iii) $\mathscr{J}([\mathfrak{R}_k, \mathfrak{R}_l], [\mathfrak{R}'_k, \mathfrak{X}]) = +1$ sowohl als $\mathscr{J}([\mathfrak{R}_k, \mathfrak{R}_l], [\mathfrak{X}, \mathfrak{R}'_l]) = +1$ sind stets nach $\mathfrak{X}$ auflösbar und zwar eindeutig.

(iv) Mit $\mathscr{J}([\mathfrak{R}_k, \mathfrak{R}_l], [\mathfrak{R}'_k, \mathfrak{R}'_l]) = +1$ und $\mathscr{J}([\mathfrak{R}'_l, \mathfrak{R}_m], [\mathfrak{R}_l, \mathfrak{R}'_m]) = +1$ gilt auch $\mathscr{J}([\mathfrak{R}_k, \mathfrak{R}_m], [\mathfrak{R}'_k, \mathfrak{R}'_m]) = +1$.

(v) Wenn $\mathfrak{R}$, $\mathfrak{R}'$ zwei beliebige Klassen eines Systems $\mathfrak{Z}$ sind, dann lassen sich die beiden Gleichungen $\mathscr{J}([\mathfrak{R}, \mathfrak{Y}], [\mathfrak{X}, \mathfrak{W}]) = +1$ und $\mathscr{J}([\mathfrak{X}, \mathfrak{Y}], [\mathfrak{R}', \mathfrak{W}]) = +1$ stets *eindeutig* nach $\mathfrak{X}$ auflösen; $\mathfrak{X}$ definiert die *Halbierung* des Klassenpaares $[\mathfrak{R}, \mathfrak{R}']$.

Satz 1: Es sei $\mathfrak{R}^0_1$, $\mathfrak{R}'_1$; $\mathfrak{R}^0_2$, $\mathfrak{R}^0_3$, ... eine beliebige Auswahl von Klassen, dann gehört dazu in jedem System $\mathfrak{Z}_k$ eine Klassen-Folge ... $\mathfrak{R}^0_k \to \mathfrak{R}'_k \to \mathfrak{R}''_k$ ... mit der Eigenschaft, daß zwei entgegengesetzt orientierte, äquidistante Klassen-Paare den Relations-Wert $\mathscr{J} = +1$ haben[1].

(vi) Sind $\mathfrak{R}^0$, $\mathfrak{R}'$, $\mathfrak{R}$ drei beliebige Klassen eines Systems, dann kann durch endlich viele Halbierungs-Schritte die aus $\mathfrak{R}^0$, $\mathfrak{R}'$ erzeugte Folge so weit verfeinert werden, daß $\mathfrak{R}$ in ihr enthalten ist.

Satz 2: Die Klassen-Folgen von Satz 1 lassen sich mit Hilfe des Halbierungs-Prozesses zu einparametrigen Klassen-Scharen erweitern, in denen jede Klasse vorkommt. Gleichzeitig ist damit ein finites Verfahren gegeben, das die Klassen-Scharen umkehrbar eindeutig und ordnungstreu auf die rationalen Zahlen abbildet: $X = X(\mathfrak{R})$, $\mathfrak{R} = \mathfrak{R}(X)$.

Satz 3: Benutzt man für jedes System $\mathfrak{Z}$ den in Satz 2 definierten Parameter $X(\mathfrak{R})$ zur Darstellung der Klassen $\mathfrak{R}$; so läßt sich die Wechselwirkungs-Relation $\mathscr{J}$ als Erhaltungssatz schreiben: $\mathscr{J}([\mathfrak{R}_k, \mathfrak{R}_l], [\mathfrak{R}'_k, \mathfrak{R}'_l]) = +1$ genau dann, wenn $X_k + X_l = X'_k + X'_l$.

(vii) Die gemäß (i) auf $\mathfrak{Z}$ definierten Klassen sind identisch mit den Zerfällungs-Klassen von (ii).

Im Hinblick auf eine spätere Anwendung formulieren wir noch den mit (ii), (iii), (vii) äquivalenten

Satz 4: Jede Klasse $\mathfrak{R}$ eines zusammengesetzten Systems $\mathfrak{Z} = [\mathfrak{Z}_k, \mathfrak{Z}_l]$ ist die Vereinigungsmenge einer bestimmten Gesamtheit von Klassen-Paaren $[\mathfrak{R}_k, \mathfrak{R}_l]$, wobei zu jedem $\mathfrak{R}_k$ aus $\mathfrak{Z}_k$ genau ein $\mathfrak{R}_l$ gehört und umgekehrt. Unter Verwendung von Satz 3 geht diese Aussage unmittelbar hervor aus: $X_k(\mathfrak{R}_k) + X_l(\mathfrak{R}_l) = \text{const}$.

(viii) Es sei $[\mathfrak{R}_k, \mathfrak{R}_l] \to [\mathfrak{R}'_k, \mathfrak{R}'_l]$ ein Übergang $\mathfrak{R} \to \mathfrak{R}'$ an einem zusammengesetzten System $\mathfrak{Z} = [\mathfrak{Z}_k, \mathfrak{Z}_l]$, dann legt $\mathscr{J}([\mathfrak{R}, \mathfrak{R}_m], [\mathfrak{R}', \mathfrak{R}'_m]) = +1$ eindeutig ein $\mathfrak{R}^*_m$ fest, derart daß $\mathscr{J}([\mathfrak{R}_k, \mathfrak{R}_m], [\mathfrak{R}'_k, \mathfrak{R}^*_m]) = +1$ sowie $\mathscr{J}([\mathfrak{R}_l, \mathfrak{R}^*_m], [\mathfrak{R}'_l, \mathfrak{R}'_m]) = +1$.

---

[1] Genau genommen müßte noch der Möglichkeit Rechnung getragen werden, daß die Klassen-Folgen einseitig oder auch zweiseitig abbrechen. Dann ist die Eigenschaft (iii) offensichtlich nicht unbeschränkt gültig. Da ein solcher Sachverhalt die wesentlichen Struktur-Merkmale der Relation nicht verändert, verzichten wir auf seine explizite Formulierung.

Damit ergibt sich

Satz 5: Wenn auf jedem $\mathfrak{Z}_k$ von $\mathfrak{G}$ eine Funktion $X_k(z_k)$ erklärt ist derart, daß sich diese Funktionen gegenüber der Produktbildung $[\mathfrak{Z}_i, \mathfrak{Z}_n]$ additiv verhalten, dann repräsentiert die Alternative $X_i(z_i) + X_k(z_k) =$ oder $\neq X_i(z_i') + X_k(z_k')$ eine Relation mit den Eigenschaften (i) bis (viii).

6. Durch Kontakt-Relationen definierte Klassen-Einteilungen. Es sollen schließlich noch diejenigen Klassen-Einteilungen erwähnt werden, die durch Kontakt-Gleichgewichte definiert sind. Dazu betrachten wir die geläufige Konstruktion der Zustandsklassen gleicher Temperatur durch die Eigenschaften des thermischen Gleichgewichtes. Stehen zwei Systeme $\mathfrak{Z}_1$ und $\mathfrak{Z}_2$ im thermischen Gleichgewicht und wird $\mathfrak{Z}_1$ in einem bestimmten Zustand z_1 festgehalten, so kann $\mathfrak{Z}_2$ nicht jeden Zustand annehmen, sondern nur die einer bestimmten Untermenge $\mathfrak{N}_2$ von $\mathfrak{Z}_2$. Diese Untermenge ist dem Zustand z_1 von $\mathfrak{Z}_1$ eindeutig zugeordnet; wir bezeichnen sie deshalb mit $\mathfrak{N}_2(z_1)$ und nennen sie die zu z_1 gehörige *Gleichgewichts-Klasse* in $\mathfrak{Z}_2$. Jedem Zustand z_1 von $\mathfrak{Z}_1$ ist also eindeutig eine Gleichgewichts-Klasse $\mathfrak{N}_2(z_1)$ in $\mathfrak{Z}_2$ zugeordnet und ebenso jedem Zustand z_2 von $\mathfrak{Z}_2$ eine Gleichgewichts-Klasse $\mathfrak{N}_1(z_2)$ in $\mathfrak{Z}_1$. Aus dieser Zuordnung wird nun eine umkehrbar-eindeutige Zuordnung der *Klassen $\mathfrak{N}_1$ und $\mathfrak{N}_2$* zueinander durch die Eigenschaft der Transitivität des thermischen Gleichgewichtes: Befindet sich $\mathfrak{Z}_1$ mit $\mathfrak{Z}_2$ im thermischen Gleichgewicht und $\mathfrak{Z}_2$ mit $\mathfrak{Z}_3$, so auch $\mathfrak{Z}_1$ mit $\mathfrak{Z}_3$.

Die behauptete eindeutige Zuordnung der Klassen $\mathfrak{N}_1$ und $\mathfrak{N}_2$ läßt sich leicht einsehen. Zwei beliebigen Zuständen z_1, z_3 von $\mathfrak{Z}_1, \mathfrak{Z}_3$ seien in $\mathfrak{Z}_2$ die Gleichgewichts-Klassen $\mathfrak{N}_2(z_1)$ und $\mathfrak{N}_2(z_3)$ zugeordnet. Dann müssen die Mengen $\mathfrak{N}_2(z_1)$ und $\mathfrak{N}_2(z_3)$ entweder elementfremd oder identisch sein, und letzteres ist genau dann der Fall, wenn sich z_1 und z_3 im thermischen Gleichgewicht befinden. Enthalten nämlich $\mathfrak{N}_2(z_1)$ und $\mathfrak{N}_2(z_3)$ einen gemeinsamen Zustand z_2, so ist, da z_2 und z_1 sowohl als z_2 und z_3 miteinander im thermischen Gleichgewicht stehen, auch z_1 mit z_3 im Gleichgewicht. Daraus folgt aber, daß sich z_1 auch mit allen Zuständen von $\mathfrak{N}_2(z_3)$ im thermischen Gleichgewicht befindet. Es muß also $\mathfrak{N}_2(z_3)$ ganz in $\mathfrak{N}_2(z_1)$ enthalten oder sogar $\mathfrak{N}_2(z_3) = \mathfrak{N}_2(z_1)$ sein, in Zeichen: $\mathfrak{N}_2(z_3) \subseteq \mathfrak{N}_2(z_1)$. Ebenso schließt man unter Vertauschung der Rollen von z_1 und z_3 auf $\mathfrak{N}_2(z_1) \subseteq \mathfrak{N}_2(z_3)$, was zusammen mit der vorigen Feststellung bereits die Behauptung $\mathfrak{N}_2(z_1) = \mathfrak{N}_2(z_3)$ liefert. Sind nun z_2 und z_2' zwei Zustände aus $\mathfrak{N}_2(z_1)$, so gilt auch $\mathfrak{N}_1(z_2) = \mathfrak{N}_1(z_2')$. Denn da sich z_2 mit z_3 im thermischen Gleichgewicht befindet, ist nach dem eben bewiesenen Hilfssatz $\mathfrak{N}_1(z_2) = \mathfrak{N}_1(z_3)$. Ebenso gilt aber $\mathfrak{N}_1(z_2') = \mathfrak{N}_1(z_3)$. Somit haben alle Zustände von $\mathfrak{N}_2(z_1)$ dieselbe Gleichgewichts-Klasse $\mathfrak{N}_1$ in $\mathfrak{Z}_1$ und umgekehrt alle Zustände von $\mathfrak{N}_1$ dieselbe Gleichgewichts-Klasse $\mathfrak{N}_2$ in $\mathfrak{Z}_2$. Damit ist die umkehrbar-eindeutige Zuordnung zwischen den Gleichgewichts-Klassen der Systeme bewiesen.

Offensichtlich lassen sich dieselben Überlegungen auch auf andere Kontakt-Gleichgewichte[1] anwenden, was besonders deutlich wird, wenn man sie in abstrakter Weise beschreibt: Ein Kontakt-Gleichgewicht zwischen zwei Systemen $\mathfrak{Z}_i, \mathfrak{Z}_j$ repräsentiert eine Relation $F(z_i, z_j) = \pm 1$, wobei F den Wert $+1$ genau dann annimmt, wenn z_i, z_j im Gleichgewicht stehen. Jede solche *Kontakt-Relation* besitzt die Eigenschaften:

1. Zu jedem $z_i \in \mathfrak{Z}_i$ existiert ein $z_j \in \mathfrak{Z}_j$ mit $F(z_i, z_j) = +1$.
2. Aus $F(z_i, z_j) = +1$ und $F(z_j, z_k) = +1$ folgt $F(z_i, z_k) = +1$.

Allein aus diesen Eigenschaften erhält man, wie oben gezeigt, den

Satz: Eine Kontakt-Relation F erzeugt in jedem System (welches das betreffende Kontakt-Gleichgewicht zuläßt) eine Klassen-Einteilung und zwischen den

---

[1] Zum Beispiel Druck-Gleichgewicht und chemische Gleichgewichte.

Klassen verschiedener Systeme eine umkehrbare eindeutige Zuordnung derart, daß $F(z_i, z_j) = +1$ genau dann, wenn z_i, z_j in zugeordneten Klassen liegen.

Man beachte, daß die durch eine Kontakt-Relation in einem System $\mathfrak{Z}$ definierte Klassen-Einteilung *keine* Variable von $\mathfrak{Z}$ ist, denn die Klassen besitzen keine ausgezeichnete Anordnung. Der Begriff des Gleichgewichtes bezieht sich eben nur auf die Frage, ob verschiedene Zustände miteinander im Gleichgewicht sind oder nicht, im Fall des thermischen Gleichgewichtes also, ob sie gleiche oder verschiedene Temperaturen haben, jedoch nicht, welcher Zustand die „höhere" Temperatur hat. Die fundamentale Eigenschaft der Gleichgewicht-Klassen, eine größer-kleiner-Relation zu besitzen, ist in den bisherigen Betrachtungen nicht enthalten. Sie kann nur aus zusätzlichen physikalischen Informationen gewonnen werden. Wir werden sehen, daß die Verknüpfung von Energie U und Entropie S bei bestimmten Prozessen die thermischen Gleichgewichts-Klassen anordnet, und mit Zahl-Werten versieht. Die Verhältnisse liegen bei allen Arten von Kontakt-Gleichgewichten analog; lediglich das maßgebende metrische Variablenpaar (im Fall des thermischen Kontaktes also U und S) ist von Fall zu Fall ein anderes.

Schließlich sei noch erwähnt, daß die Gleichgewichts-Klassen *eine* wesentliche Eigenschaft physikalischer Variablen besitzen: Jede in *einem* System vorgegebene Anordnung der Klassen überträgt sich wegen der oben erläuterten umkehrbar eindeutigen Zuordnung auf die Klassen aller anderen Systeme, welche das betreffende Kontakt-Gleichgewicht erlauben.

B. Die Struktur der Thermodynamik.

I. Entropie und Energie.

7. Zum Begriff des physikalischen Systems. Wir haben bei Einführung des Begriffes der Zustandsmenge gesagt, daß jedes System eine Zustandsmenge $\mathfrak{Z}$ „besitzt". In den weiteren Formulierungen haben wir dann der Kürze halber zwar oftmals nur vom „System $\mathfrak{Z}$" gesprochen, meistens aber die Zustandsmenge des Systems gemeint. Es ist klar, daß ein System mehr ist als eine Menge, da seine Variablen sicherlich zu seinen charakteristischen Merkmalen zählen. Die späteren Entwicklungen werden deutlich machen, daß die Beschreibung eines Systems genauer in folgender Weise geschieht: Es gibt zwei universelle Variablen, Energie und Entropie, mit deren Hilfe die Systeme eingeteilt werden in solche mit gleichartigen metrischen Variablen, die dann als *charakteristische Koordinaten* fungieren. Energie und Entropie als Funktionen dieser Koordinaten definieren das einzelne physikalische System. Diese Behauptung möge hier als Hinweis auf den im folgenden einzuschlagenden Weg dienen.

Wenn von nun ab der Ausdruck „das System $\mathfrak{Z}$" gebraucht wird, so ist damit die Zustandsmenge $\mathfrak{Z}$ einschließlich der charakteristischen Variablen gemeint. Wenn die Menge $\mathfrak{Z}$ allein gemeint ist, werden die Redewendungen „Zustandsmenge $\mathfrak{Z}$" oder „Zustandsmenge des Systems $\mathfrak{Z}$" verwendet.

Die *thermodynamischen Systeme* sind durch bestimmte Eigenschaften ihrer charakteristischen Koordinaten gekennzeichnet, und eine Axiomatik der Thermodynamik hat im wesentlichen die Kennzeichnung dieser *thermodynamischen Koordinaten* zur Aufgabe.

8. Adiabatische Isolation. Empirische Entropie. Eine der grundlegenden thermodynamischen bzw. allgemein physikalischen System-Einwirkungen ist durch die Gesamtheit derjenigen Zustandsübergänge gegeben, die sich bei adiabatischer Isolation des Systems realisieren lassen. Diese Gesamtheit ist im allgemeinen einer verzweigungslosen Übergangs-Relation äquivalent. Wir formulieren daher

Axiom 1: Jedes thermodynamische System $\mathfrak{Z}$ besitzt eine System-Einwirkung *adiabatische Isolation* mit der Struktur einer verzweigungslosen Übergangs-Relation.

Die adiabatische Isolation zerlegt also die Zustandsmenge eines thermodynamischen Systems in Klassen-Ketten. Jede der Ketten repräsentiert eine Variable, die wir als *empirische Entropie* σ bezeichnen[1]. Demnach gehören zu einem System so viele empirische Entropien, wie es adiabatische Klassen-Ketten besitzt.

Wir wollen, wo es zweckmäßig ist, Axiom 1 stets dahingehend ergänzt denken, daß die empirischen Entropien σ *kontinuierlich* sind.

Es ist klar, daß sich jede Darstellung einer empirischen Entropie bei Übergängen unter adiabatischer Isolation nur einseitig ändern kann. Entweder kann sie niemals abnehmen oder niemals zunehmen, je nachdem ob man aufeinanderfolgenden Entropie-Klassen steigende oder fallende Zahlen-Werte von $\sigma(z)$ zuordnet. Wir entscheiden uns für die erste der beiden Möglichkeiten und sagen deshalb: Die adiabatische Isolation repräsentiert eine „Wachstums-Größe".

An dieser Stelle scheint es angebracht, auf die in Ziff. 5a erwähnte Alternative in der Definition eines physikalischen Systems zurückzukommen. Das Beispiel der Massen-Variable zeigt, daß es Variablen geben kann, welche durch die adiabatische Isolation eines Systems gehemmt werden und dadurch die Übergangs-Relation zerfällen. Gegenüber einer vollständigen Übergangs-Relation, die unmittelbar eine Variable definiert, bedeutet dies natürlich eine Komplikation. Daher könnte man, wie in Ziff. 5a erwähnt, versucht sein, durch eine geeignete Definition des System-Begriffes diese Schwierigkeit zu vermeiden (indem man nur adiabatisch bewegliche Variablen aufnimmt), aber damit verschiebt man das Problem nur auf die System-Zusammensetzung. Die Unzweckmäßigkeit dieses Vorgehens ist erwiesen, seit GIBBS die Vorteile des anderen Weges offenbar machte, solche *adiabatisch-gehemmten* Variablen in den System-Begriff aufzunehmen.

Wir betrachten also die adiabatische Übergangs-Relation als verzweigungslos. Dann ist es unter adiabatischer Isolation des Systems möglich, jede Klassen-Kette in einer Richtung zu durchlaufen, jedoch unmöglich, von einer Kette auf eine andere zu gelangen. Dies kann erst geschehen, wenn die adiabatische Isolation aufgehoben wird, also zum Beispiel ein geeignetes zweites System $\mathfrak{Z}_2$ mit dem ersten $\mathfrak{Z}_1$ zusammengesetzt und das Gesamt-System $[\mathfrak{Z}_1, \mathfrak{Z}_2]$ wieder adiabatisch isoliert wird. Dann bestehen folgende zwei Möglichkeiten: Entweder gibt es in $\mathfrak{Z}_1$ Übergänge zwischen den Klassen-Ketten, die in $\mathfrak{Z}_2$ von Übergängen innerhalb *einer* Klassen-Kette begleitet sind und umgekehrt, oder jeder Übergang, der in $\mathfrak{Z}_1$ die Klassen-Ketten wechselt, ist in $\mathfrak{Z}_2$ mit einem Übergang derselben Art verbunden und umgekehrt. Betrachtet man nur adiabatisch-reversible Übergänge des Gesamtsystems $[\mathfrak{Z}_1, \mathfrak{Z}_2]$, so werden im ersten Fall die Entropie-Klassen verschiedener Klassen-Ketten umkehrbar-eindeutig einander zugeordnet und damit eine empirische Entropie für das ganze System $\mathfrak{Z}_1$ erklärt. In diesem Fall läßt sich, dank der Existenz von $\mathfrak{Z}_2$, die adiabatische Übergangs-Relation des Systems $\mathfrak{Z}_1$ von einer verzweigungslosen zu einer vollständigen

---

[1] Der Begriff der empirischen Entropie als durch die adiabatische Isolation definierte Variable wurde unabhängig von H. A. BUCHDAHL [Z. Physik **152**, 425 (1958)] eingeführt. Zum ersten Male findet er sich in der Dissertation von H. JUNG, Aachen 1957.

erweitern[1]. Im zweiten Fall dagegen gibt es keine derartige erzwungene Zuordnung zwischen den Entropie-Klassen verschiedener Ketten, so daß man prinzipiell die Freiheit hat, zwischen den Entropie-Klassen verschiedener Ketten jede beliebige Zuordnung zu stiften, welche die Anordnung erhält. Dies gilt auch für die weiter unten eingeführte metrische Entropie, wobei allerdings die Freiheit der Zuordnung auf je eine Entropie-Klasse einer Kette eingeschränkt wird. Formal betrachtet *verhalten sich die verschiedenen Klassen-Ketten eines Systems ebenso zueinander wie verschiedene Systeme mit je einer empirischen Entropie.* Daher wird im folgenden implizite auch diese Frage miterledigt.

9. Die empirische Entropie zusammengesetzter Systeme. Im folgenden wollen wir voraussetzen, daß ein thermodynamisches System *eine* empirische Entropie besitzt. Unser Interesse gilt sodann dem Zusammenhang zwischen der empirischen Entropie eines zusammengesetzten Systems $\mathfrak{Z} = [\mathfrak{Z}_1, \mathfrak{Z}_2]$ und den empirischen Entropien seiner Komponenten $\mathfrak{Z}_1, \mathfrak{Z}_2$.

$\alpha)$ *Ein Beispiel.* Wir betrachten zunächst ein von T. Ehrenfest-Afanassjewa angegebenes Beispiel[2], dessen wesentliche Züge dann in abstrakterer Formulierung wiederholt und vervollständigt werden sollen.

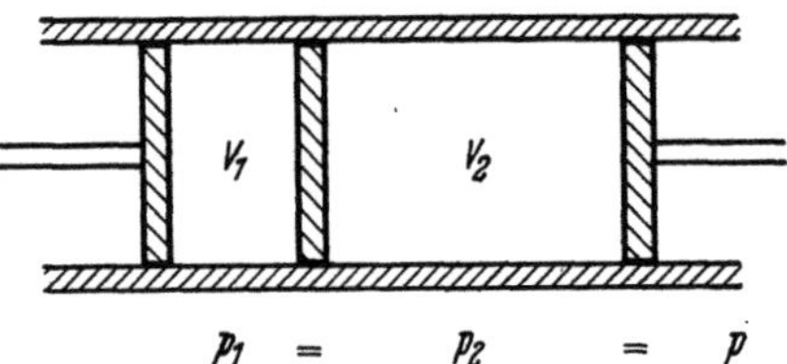

Fig. 2. Zur empirischen Entropie zweier Gase in Druck-Kopplung und thermischem Gleichgewicht.

Zwei mit Gas gefüllte und gegen Wärme-Austausch isolierte Zylinder mit Kolben stehen über eine wärme-undurchlässige, bewegliche Wand in Druck-Kopplung (Fig. 2). Die Zustände des Gesamtsystems seien über den Koordinaten V_1, V_2, p aufgetragen. Hinreichend langsame Bewegung der Kolben liefert dann zu jedem beliebigen Ausgangszustand eine reversibel durchlaufbare maximale Zustandsklasse, die sich im Koordinatenraum als Kurve $\mathfrak{C}$ darstellt. Sobald man einen plötzlichen Schlag auf die Kolben ausübt, springt der Zustand von einer Kurve $\mathfrak{C}'$ auf eine andere $\mathfrak{C}''$, und dieser Übergang $\mathfrak{C}' \to \mathfrak{C}''$ läßt sich bei Aufrechterhalten der Wärme-Isolation durch kein Experiment rückgängig machen. Die zugehörige Übergangs-Relation ist nicht vollständig; denn es gibt Kurvenpaare $\mathfrak{C}^*$, $\mathfrak{C}^{**}$, die unter den angegebenen Bedingungen der Wärme-Isolation weder in der einen noch in der anderen Anordnung miteinander verbindbar sind. Dies sieht man leicht folgendermaßen ein: Die Teilsysteme, welche die Entropien σ_1 und σ_2 haben mögen, erfahren bei den betrachteten Prozessen je für sich adiabatische Zustandsübergänge, für die also $\sigma_1'' \geqq \sigma_1'$ und $\sigma_2'' \geqq \sigma_2'$ gilt. Zu jeder Kurve $\mathfrak{C}$ gehört nun ein σ_1 und ein σ_2. Wählt man also $\mathfrak{C}^*$ und $\mathfrak{C}^{**}$ so, daß $\sigma_1^{**} < \sigma_1^*$ und $\sigma_2^* < \sigma_2^{**}$, dann ist offenbar $\mathfrak{C}^* \to \mathfrak{C}^{**}$ sowohl als $\mathfrak{C}^{**} \to \mathfrak{C}^*$ unmöglich. Die Wärme-Isolation definiert demnach für das System der Fig. 2 keine vollständige Übergangs-Relation und damit auch keine empirische Entropie.

Es ist klar, daß das System diese Eigenschaft dadurch erhält, daß wir die Möglichkeit des thermischen Kontaktes im Inneren des Systems ignoriert haben. Erlaubt man nämlich für die druck-koppelnde, wärme-undurchlässige Wand Eingriffe, die ihre wärme-isolierende Eigenschaft zeitweilig auf-

---

[1] Dies zeigt, wie sehr eine „Isolation" von der Gesamtheit der Systeme abhängt, die mit dem betrachteten System wechselwirken können. In der traditionellen Form der Thermodynamik spiegelt sich das in der ausgiebigen Benutzung von Kreis-Prozessen wider.

[2] T. Ehrenfest-Afanassjewa: Zur Axiomatisierung des zweiten Hauptsatzes der Thermodynamik. Z. Physik **33**, 933 (1925).

heben, so stellt man fest, daß es im allgemeinen[1] auf jeder Kurve $\mathfrak{C}$ einen Zustand z_0 gibt, der sich nicht verändert, wenn die Wärme-Isolation zwischen den Teilsystemen beseitigt wird (Temperaturdifferenz $\vartheta_1 - \vartheta_2 = 0$). Bringt man nun das System in einen solchen Zustand z_0 und führt unter Aufrechterhalten des Wärme-Kontaktes mit den Kolben langsame Bewegungen aus, so entsteht eine reversibel durchlaufbare Kurve $\mathfrak{A}$, die eine Schar von $\mathfrak{C}$-Kurven zu einer Fläche $\mathfrak{S}$ verheftet. Auf $\mathfrak{S}$ sind zwei beliebige Zustände z', z'' stets durch einen Kurvenzug $\mathfrak{C}' - \mathfrak{A} - \mathfrak{C}''$ reversibel verbindbar. Die Gesamtheit der Flächen (adiabatisch-reversible Klassen) $\mathfrak{S}$ bildet hingegen eine Schar, zwischen deren Mitgliedern nur irreversible Übergänge möglich sind. Man erhält damit eine empirische Entropie für das Gesamtsystem.

β) Allgemeine Analyse der Konstruktion der empirischen Entropie zusammengesetzter Systeme. Die adiabatische Isolation eines zusammengesetzten Systems $\mathfrak{Z} = [\mathfrak{Z}_1, \mathfrak{Z}_2]$ zerlegt die Zustandsmenge $\mathfrak{Z}$ in maximale adiabatisch-reversible Klassen $\mathfrak{S}$. Um auch die *Anordnung* der Klassen-Einteilung $\{\mathfrak{S}\}$ zu erkennen, machen wir folgendes Gedankenexperiment: $\mathfrak{Z}$ möge sich in einem beliebigen Anfangszustand $z^0 = [z_1^0, z_2^0]$ befinden. Die adiabatischen Zustands-Veränderungen der *Teil*systeme erlauben dann zwei Entropie-Klassen $\mathfrak{S}_1$, $\mathfrak{S}_2$ *unabhängig* voneinander zu durchlaufen. Alle in diesem Klassenpaar enthaltenen Zustände $z = [z_1, z_2]$ sind also adiabatisch-reversibel miteinander verbindbar, das heißt die Gesamtheit der Zustände $[\mathfrak{S}_1, \mathfrak{S}_2]$ ist auf jeden Fall in der zu z^0 gehörigen Entropie-Klasse $\mathfrak{S}^0$ enthalten. Zur vollständigen Bestimmung von $\mathfrak{S}^0$ müssen wir aber auch die Zustands-Veränderungen berücksichtigen, bei denen die adiabatische Isolation *zwischen* den Teilsystemen aufgehoben ist. Geschieht dies z. B. mittels eines thermischen Kontaktes, so wird im allgemeinen eine irreversible Zustands-Veränderung, ein thermischer Ausgleichs-Prozeß, einsetzen. Es existiert aber in $\mathfrak{S}^0$ mindestens ein Zustand, bei dem dies nicht so ist. Von einem derartigen Zustand ausgehend, ist es dann möglich, unter Aufrechterhalten des thermischen Kontaktes die Klassen $\mathfrak{S}_1$ und $\mathfrak{S}_2$ adiabatisch-reversibel zu verlassen. Man gelangt so in irgendwelche neuen Klassen $\mathfrak{S}_1'$, $\mathfrak{S}_2'$ die sich wieder unabhängig voneinander durchlaufen lassen, sobald der thermische Kontakt beseitigt wird. Fortsetzung des Verfahrens führt schließlich auf die maximale adiabatisch-reversible Klasse $\mathfrak{S}^0$. Wir denken uns jetzt alle $\mathfrak{S}$ von $\mathfrak{Z}$ ermittelt und fragen, ob zwei beliebige $\mathfrak{S}$, $\mathfrak{S}'$ stets in (genau) einer Anordnung adiabatisch-irreversibel verbindbar sind. Dies ist tatsächlich der Fall, denn wenn z, z' zwei Zustände aus $\mathfrak{S}$, $\mathfrak{S}'$ sind, dann muß eine der folgenden vier Kombinationen vorliegen: 1. $\sigma_1 > \sigma_1'$, $\sigma_2 > \sigma_2'$; 2. $\sigma_1 < \sigma_1'$, $\sigma_2 < \sigma_2'$; 3. $\sigma_1 > \sigma_1'$, $\sigma_2 < \sigma_2'$; 4. $\sigma_1 < \sigma_1'$, $\sigma_2 > \sigma_2'$. Für 1 bzw. 2 erkennt man unmittelbar, daß sich der Übergang $\mathfrak{S}' \to \mathfrak{S}$ bzw. $\mathfrak{S} \to \mathfrak{S}'$ durch adiabatische Einwirkung auf die *Teil*systeme erzwingen läßt. Im Fall 3 denken wir uns mit Hilfe des thermischen Kontaktes zwischen beiden Systemen einen adiabatisch-reversiblen Prozeß von $\mathfrak{Z}$ durchlaufen, der das System $\mathfrak{Z}_1$ von $\mathfrak{S}_1$ in $\mathfrak{S}_1'$ überführt[2]. Dabei muß $\mathfrak{Z}_2$ *eindeutig* in ein $\mathfrak{S}_2^*$ übergehen, denn wenn es in bezug auf den genannten Prozeß zwei verschiedene $\mathfrak{S}_2^*$, $\mathfrak{S}_2^{**}$ gäbe (es sei $\sigma_2^* < \sigma_2^{**}$), dann wäre an dem Gesamtsystem $\mathfrak{Z}$ die Übergangs-Folge $[\mathfrak{S}_1', \mathfrak{S}_2^*] \to [\mathfrak{S}_1, \mathfrak{S}_2] \to [\mathfrak{S}_1', \mathfrak{S}_2^{**}]$ adiabatisch-reversibel,

---

[1] Das heißt, wenn die beiden Gase nicht gerade dieselben thermischen Druck-Koeffizienten haben, also z. B. identisch sind; dann erleiden nämlich beide Teilsysteme bei reversibler Volumen-Änderung stets die gleiche Temperatur-Änderung, so daß die Temperaturen niemals einander gleich werden können, wenn sie es nicht zu Anfang schon waren. Bei einem solchen System muß man für die Zwischenwand auch „Druck-Entkopplung" zulassen, um es zu einem System mit empirischer Entropie zu machen.

[2] Die Möglichkeit eines solchen Prozesses kommt in der klassischen Formulierung der Thermodynamik dadurch zum Ausdruck, daß man behauptet, zwei beliebige Zustände eines Systems $\mathfrak{Z}_1$ können durch Arbeits- und Wärme-Zufuhr reversibel verbunden werden.

und dies würde bedeuten, daß der Übergang zwar am System $\mathfrak{Z}_2$ adiabatisch-irreversibel ist, daß er aber mit Hilfe der thermischen Kopplung von $\mathfrak{Z}_2$ an die „Umgebung" $\mathfrak{Z}_1$ unter adiabatischem Abschluß des Gesamtsystems rückgängig gemacht werden könnte, ohne in der Umgebung eine adiabatisch-irreversible Zustands-Veränderung zu hinterlassen. Eine solche Möglichkeit schließen wir aber aus, und deshalb muß entweder $\sigma_2^* > \sigma_2'$ oder $\sigma_2^* < \sigma_2'$ gelten, das heißt im Fall 3 ist entweder $\mathfrak{S}' \to \mathfrak{S}$ oder $\mathfrak{S} \to \mathfrak{S}'$ möglich. 4 erledigt sich in analoger Weise. Damit ist für die Klassen-Einteilung $\{\mathfrak{S}\}$ von $\mathfrak{Z}$ die Anordnung nachgewiesen.

Fragt man rückblickend unter Außerachtlassung der einzelnen Annahmen nach einer knappen Beschreibung des Resultates der Analyse, so ergibt sich: Repräsentieren $\mathfrak{Z}_1$ und $\mathfrak{Z}_2$ zwei thermodynamische Systeme, so bestimmen diese eindeutig ein zusammengesetztes System $\mathfrak{Z} = [\mathfrak{Z}_1, \mathfrak{Z}_2]$ wenigstens insoweit als es die Klassen-Einteilung der empirischen Entropie angeht. Es ist wichtig, sich den Sinn dieser Aussage unter Berücksichtigung des oben diskutierten Begriffes System klarzumachen. Zunächst ist $\mathfrak{Z}$ nur eine *Menge*, die in definierter Weise aus den Mengen $\mathfrak{Z}_1$ und $\mathfrak{Z}_2$ gebildet wird. $\mathfrak{Z}_1$ und $\mathfrak{Z}_2$ repräsentieren aber thermodynamische *Systeme*, das heißt außer den Zustandsmengen sind auch ihre charakteristischen Klassen-Einteilungen als gegeben zu betrachten. Dann, so sagen die obigen Untersuchungen, ist auch das durch $\mathfrak{Z}$ repräsentierte System insofern charakterisiert als seine empirische Entropie durch die empirischen Entropien der Teilsysteme festgelegt ist. Die thermische Wechselwirkung hatte lediglich die Funktion eines Konstruktionsmittels. Das Resultat läßt sich jedoch ohne Bezugnahme auf bestimmte Konstruktionsweisen formulieren. Dies ist für die axio-matische Beschreibung wesentlich.

γ) Axiomatische Formulierung. Die hinsichtlich der empirischen Entropie gewonnene Aussage, nämlich daß zwei Systeme $\mathfrak{Z}_1$ und $\mathfrak{Z}_2$ mit je einer empirischen Entropie eindeutig die angeordnete Klassen-Einteilung der empirischen Entropie auf der Zustandsmenge $\mathfrak{Z} = [\mathfrak{Z}_1, \mathfrak{Z}_2]$ festlegen, läßt eine Verallgemeinerung als naheliegend erscheinen: Was über die empirische Entropie gesagt wird, gilt auch für andere charakteristische Variablen. Erweitert man die Betrachtung auf die Zusammensetzung von drei oder mehr Systemen, so ist klar, daß die Reihenfolge der Verknüpfung physikalisch bedeutungslos ist. Formal läßt sich dies als Assoziativität der Zusammensetzung beschreiben: $[\mathfrak{Z}_1, [\mathfrak{Z}_2, \mathfrak{Z}_3]] = [[\mathfrak{Z}_1, \mathfrak{Z}_2], \mathfrak{Z}_3] = [\mathfrak{Z}_1, \mathfrak{Z}_2, \mathfrak{Z}_3]$. Wir formulieren daher

Axiom 2: Zwei oder mehr thermodynamische Systeme $\mathfrak{Z}_1, \mathfrak{Z}_2, \ldots$ bestimmen eindeutig ein neues thermodynamisches System $\mathfrak{Z} = [\mathfrak{Z}_1, \mathfrak{Z}_2, \ldots]$.

Wie der Zusammenhang zwischen den Variablen des Systems $\mathfrak{Z}$ und denen der Systeme $\mathfrak{Z}_1, \mathfrak{Z}_2$ im einzelnen aussieht, wird erst durch weitere Axiome geregelt. Für die empirische Entropie, als einer der fraglichen Variablen, liefern die obigen Betrachtungen

Axiom 3: Zwischen den Entropie-Klassen $\mathfrak{S}$ des Systems $\mathfrak{Z} = [\mathfrak{Z}_1, \mathfrak{Z}_2]$ und den Entropie-Klassen $\mathfrak{S}_1, \mathfrak{S}_2$ von $\mathfrak{Z}_1, \mathfrak{Z}_2$ besteht folgender Zusammenhang:

a) Jede Entropie-Klasse $\mathfrak{S}$ ist die Vereinigungsmenge einer bestimmten Gesamtheit von Entropie-Klassen-Paaren $[\mathfrak{S}_1, \mathfrak{S}_2]$, symbolisch: $\mathfrak{S} = \sum [\mathfrak{S}_1, \mathfrak{S}_2]$.

b) In jeder Entropie-Klasse $\mathfrak{S}$ kommt jede Entropie-Klasse von $\mathfrak{Z}_1$ und $\mathfrak{Z}_2$ vor.

c) Sind $\mathfrak{S}_1' > \mathfrak{S}_1$ zwei beliebige Entropie-Klassen des Systems $\mathfrak{Z}_1$, und ist $\mathfrak{S}_2$ eine beliebige Entropie-Klasse des Systems $\mathfrak{Z}_2$, so gilt für die beiden zugehörigen Entropie-Klassen $\mathfrak{S}'$ und $\mathfrak{S}$ von $\mathfrak{Z}$ die Relation $\mathfrak{S}' > \mathfrak{S}$.

Offensichtlich äquivalent mit Axiom 3a ist die Aussage, daß alle Zustände $z = [z_1, z_2]$, die in einem Entropie-Klassen-Paar $[\mathfrak{S}_1, \mathfrak{S}_2]$ liegen, ein und derselben Entropie-Klasse $\mathfrak{S}$ angehören. Oder auch: Die adiabatische Übergangs-Relation $F(z, z') = \pm 1$ des Systems $[\mathfrak{Z}_1, \mathfrak{Z}_2]$ kann auf die Form $\mathscr{J}([\mathfrak{S}_1, \mathfrak{S}_2], [\mathfrak{S}_1', \mathfrak{S}_2']) = \pm 1$ gebracht werden. Ebenfalls äquivalent ist

Satz 1: Die empirische Entropie σ des Systems $\mathfrak{Z}$ ist allein eine Funktion von σ_1 und σ_2.

Nehmen wir an, daß die empirischen Entropien der betrachteten Systeme kontinuierliche Variablen sind, so ist Axiom 3b äquivalent mit

Satz 2: Die Gleichungen $\sigma = \sigma(x, \sigma_2)$ und $\sigma = \sigma(\sigma_1, y)$ besitzen bei beliebigen Werten von σ, σ_1, σ_2 stets Lösungen x, y.

Berücksichtigt man schließlich noch die Axiome 2 und 3c, so ergibt sich der wichtige

Satz 3: Die empirische Entropie $\sigma = \sigma(\sigma_1, \sigma_2, \ldots)$ eines zusammengesetzten Systems $\mathfrak{Z} = [\mathfrak{Z}_1, \mathfrak{Z}_2, \ldots]$ ist eine stetige, monoton-wachsende Funktion der empirischen Entropien der Teilsysteme. — Die nach Satz 2 existierenden Lösungen x und y sind demnach sogar eindeutig.

An dieser Stelle erscheint eine kurze Erläuterung über die *formale* Rangordnung der ersten drei Axiome angebracht: Axiom 1 führt den Grundbegriff der adiabatischen Übergangs-Relation *eines* Systems ein, Axiom 2 den Grundbegriff der Verknüpfung *zweier* Systeme. Das Axiom 3 schließlich macht eine Aussage über den Zusammenhang jener beiden Grundbegriffe, denn es bezieht sich auf die Tatsache, daß die Verknüpfung zweier beliebiger Systeme $\mathfrak{Z}_1$, $\mathfrak{Z}_2$ zu dem dritten System $\mathfrak{Z} = [\mathfrak{Z}_1, \mathfrak{Z}_2]$ eine ganz bestimmte Verknüpfungs-Struktur der drei zugehörigen adiabatischen Übergangs-Relationen zur Folge hat. Diese Struktur kommt am sinnfälligsten in Satz 3 zum Ausdruck.

Hinsichtlich der *physikalischen* Rangordnung wird man geneigt sein, alle die Axiome als trivial anzusprechen, die entweder an einem sehr geläufigen konkreten Sachverhalt abgelesen sind oder im Rahmen der klassischen Darstellung der Thermodynamik mehr oder weniger selbstverständliche Voraussetzungen bilden. Dann bleibt Axiom 3c als einzige nicht-triviale Aussage, die wegen ihrer besonderen Bedeutung noch einmal auf folgende Weise formuliert werden soll: Die Eigenschaft der adiabatischen Irreversibilität eines Übergangs ist invariant gegen System-Erweiterung in dem Sinne, daß die Irreversibilität nicht zerstört wird, wenn man ein zweites System mit unter die adiabatische Isolation nimmt und dieses zweite System Kreis-Prozesse durchlaufen läßt.

10. Die metrische Entropie. Es ist nun bemerkenswert, daß die Axiome 1 bis 3 genügen, um einen „Entropie-Erhaltungs-Satz" abzuleiten und damit nach Ziff. 5 eine der empirischen Entropie-Funktionen als *metrisch* auszuzeichnen. Dazu betrachten wir die der adiabatischen Übergangs-Relation $\mathscr{J}([\mathfrak{S}_1, \mathfrak{S}_2], [\mathfrak{S}_1', \mathfrak{S}_2'])$ des Systems $[\mathfrak{Z}_1, \mathfrak{Z}_2]$ zugeordnete „reversible" Übergangs-Relation $\mathscr{J}^*([\mathfrak{S}_1, \mathfrak{S}_2], [\mathfrak{S}_1', \mathfrak{S}_2'])$, die folgendermaßen erklärt ist: Es ist $\mathscr{J}^* = +1$ genau dann, wenn der betreffende Übergang $[\mathfrak{S}_1, \mathfrak{S}_2] \to [\mathfrak{S}_1', \mathfrak{S}_2']$ *adiabatisch-reversibel* ist. In allen anderen Fällen ist $\mathscr{J}^* = -1$. Für diese neue Übergangs-Relation sind auf Grund der Axiome 1, 3a, 3b die Eigenschaften (i), (ii), (iii), (vii) von Ziff. 5 erfüllt, wie man an Hand des dort formulierten Satzes 4 leicht nachweist. Die Eigenschaften (iv), (v), als Aussage über mehr als zwei Systeme, ergeben sich bei Berücksichtigung von Axiom 2. Eigenschaft (vi) erkennt man am besten aus Satz 2, Ziff. 9. Wir haben damit

Satz 4: Die empirischen Entropie-Funktionen aller thermodynamischen Systeme $\mathfrak{Z}_1$, $\mathfrak{Z}_2$,... lassen sich so transformieren: $S_1(\sigma_1)$, $S_2(\sigma_2)$, ..., daß die System-Zusammensetzung eine additive Verknüpfung der entsprechenden Entropie-Funktionen nach sich zieht. Symbolisch: Aus $\mathfrak{Z} = [\mathfrak{Z}_1, \mathfrak{Z}_2, ...]$ folgt $S(z) = S_1(z_1) + S_2(z_2) + \cdots$. Dabei sind die (metrischen) Entropie-Funktionen S_1, S_2, ... bis auf einen gemeinsamen Maßstabs-Faktor und bis auf je eine additive Konstante eindeutig bestimmt.

Die praktische Konsequenz dieses Satzes heben wir noch einmal hervor: Es seien $\mathfrak{S}_1^0$, $\mathfrak{S}_1'$; $\mathfrak{S}_2^0$ drei Entropie-Klassen der beiden Systeme $\mathfrak{Z}_1$, $\mathfrak{Z}_2$, und es sei $[\mathfrak{S}_1^0, \mathfrak{S}_2^0] \rightarrow [\mathfrak{S}_1', \mathfrak{X}]$ ein adiabatisch-reversibler Zustandsübergang des Systems $[\mathfrak{Z}_1, \mathfrak{Z}_2]$, dann besagt Satz 3, daß sich für $\mathfrak{X}$ genau eine Klasse $\mathfrak{S}_2'$ ergibt. Um dieses $\mathfrak{S}_2'$ bei beliebig vorgegebenen $\mathfrak{S}_1^0$, $\mathfrak{S}_1'$; $\mathfrak{S}_2$ zu ermitteln, hat man nur die Möglichkeit, die entsprechenden Experimente wirklich durchzuführen. Dasselbe gilt für alle Systempaare $[\mathfrak{Z}_1, \mathfrak{Z}_i]$, $i = 3, 4, \ldots$. Satz 4 besagt nun, daß für die übrigen Systempaare $[\mathfrak{Z}_k, \mathfrak{Z}_i]$ keine Messungen mehr erforderlich sind, denn bei ihnen läßt sich $\mathfrak{X}$ aus der Gleichung $S_i + S_k = S_i + x(\mathfrak{X})$ berechnen. Ein solches Vorgehen erfordert natürlich die Experimente an den Systemen $[\mathfrak{Z}_1, \mathfrak{Z}_i]$ in der Weise zu führen, wie in Ziff. 5 auseinandergesetzt: Zwei Klassen $\mathfrak{S}_1^0$, $\mathfrak{S}_1'$ des Systems $\mathfrak{Z}_1$ müssen sukzessive mit Klassen-Paaren der anderen Systeme zur Inzidenz gebracht werden, vollkommen analog zur Messung von Strecken mit Hilfe einer Normal-Strecke. Sobald auf diese Weise die Entropie-Funktion S_2 eines Systems $\mathfrak{Z}_2$ bestimmt ist, kann dieses System als ,,Maßstab mit Teilstrichen'' verwendet werden, wodurch die Iteration des Übergangs $\mathfrak{S}_1^0 \rightarrow \mathfrak{S}_1'$ ersetzt werden kann durch monotones Fortschreiten in der S_2-Skala.

Wir benutzen die Gelegenheit, den häufig gebrauchten Begriff des quasistatischen Übergangs einzuführen durch folgende

Definition: Sind $\mathfrak{Z}_1$ und $\mathfrak{Z}_2$ die Teilsysteme eines Gesamtsystems $\mathfrak{Z}$, so werden die adiabatisch-reversiblen Übergänge des Gesamtsystems $\mathfrak{Z}$ als *quasistatische Übergänge* von $\mathfrak{Z}_1$ und $\mathfrak{Z}_2$ bezeichnet.

Man erkennt, daß die Axiome 1 und 3 die Aussage enthalten, daß jeder Übergang eines Systems quasistatisch geführt werden kann. Wenn dies nicht der Fall wäre, ließe sich die Entropie nur für diejenigen Zustands-Teilmengen metrisieren, für welche diese Voraussetzung erfüllt ist. In Praxi wird eines der beiden Systeme $\mathfrak{Z}_1$ und $\mathfrak{Z}_2$ gewöhnlich als ,,Umgebung'' des anderen bezeichnet. Es sei noch erwähnt, daß der Begriff quasistatisch nur kombinatorisch faßbare Eigenschaften benutzt, und nichts mit der Stetigkeit von Kurven im Zustandsraum zu tun hat — eine Auffassung, die in der Literatur häufig anzutreffen ist. Man bemerkt schließlich, daß die angegebene Definition genau den Sachverhalt beschreibt, der in den klassischen Darstellungsweisen der Thermodynamik gemeint ist, wenn von Wärme-Behältern und reversiblen Carnot-Prozessen die Rede ist. Zusammenfassend gilt der

Satz: Die aus den Axiomen 1 bis 3 folgende Struktur der quasistatischen Übergänge eines thermodynamischen Systems sichert die Konstruierbarkeit der metrischen Entropie-Funktion.

Wir weisen darauf hin, daß die Konstruktion der *metrischen* Entropie S von adiabatisch-irreversiblen Übergängen keinen Gebrauch macht. Die Existenz von S ist daher unabhängig von der Wachstums-Eigenschaft unter adiabatischer Isolation.

Analytische Konstruktion der metrischen Entropie. Im Anschluß an Satz 3 kann der Existenznachweis der metrischen Entropie-Funktion auch analytisch geführt werden, und da diese Art der Schlußweise vertrauter sein dürfte, als die oben angewandte, soll sie kurz behandelt werden. Wir betrachten drei Systeme $\mathfrak{B}_1$, $\mathfrak{B}_2$, $\mathfrak{B}_3$ mit den empirischen Entropien $\sigma_1, \sigma_2, \sigma_3$. Die aus diesen zusammengesetzten Systeme haben dann nach Satz 3 wiederum empirische Entropie-Funktionen. Diese seien für

$$[\mathfrak{B}_1, \mathfrak{B}_2]: f(\sigma_1, \sigma_2); \quad [\mathfrak{B}_2, \mathfrak{B}_3]: g(\sigma_2, \sigma_3); \quad [\mathfrak{B}_1, \mathfrak{B}_3]: h(\sigma_1, \sigma_3); \quad [\mathfrak{B}_1, \mathfrak{B}_2, \mathfrak{B}_3]: F(\sigma_1, \sigma_2, \sigma_3).$$

Die entsprechenden Differentiale sind

$$df = f_1(\sigma_1, \sigma_2)\, d\sigma_1 + f_2(\sigma_1, \sigma_2)\, d\sigma_2; \quad dg = \cdots; \quad dh = \cdots;$$
$$dF = F_1(\sigma_1, \sigma_2, \sigma_3)\, d\sigma_1 + F_2(\sigma_1, \sigma_2, \sigma_3)\, d\sigma_2 + F_3(\sigma_1, \sigma_2, \sigma_3)\, d\sigma_3.$$

Betrachtet man *adiabatisch-reversible* Prozesse der zusammengesetzten Systeme, so erhält man zum Beispiel für $[\mathfrak{B}_1, \mathfrak{B}_2]$:

$$df = 0, \quad \text{sowie} \quad dF = 0 \text{ und } d\sigma_3 = 0.$$

Diese Gleichungen liefern

$$\frac{d\sigma_1}{d\sigma_2} = -\frac{f_2}{f_1}, \quad \text{sowie} \quad \frac{d\sigma_1}{d\sigma_2} = -\frac{F_2}{F_1},$$

woraus folgt, daß

$$\frac{F_2(\sigma_1, \sigma_2, \sigma_3)}{F_1(\sigma_1, \sigma_2, \sigma_3)} = \frac{f_2(\sigma_1, \sigma_2)}{f_1(\sigma_1, \sigma_2)} = f^*(\sigma_1, \sigma_2) \quad \text{für } \textit{jedes } \sigma_3.$$

Auf dieselbe Weise erhält man durch Betrachtung von $[\mathfrak{B}_2, \mathfrak{B}_3]$ und $[\mathfrak{B}_1, \mathfrak{B}_3]$ die Beziehungen

$$\frac{F_3(\sigma_1, \sigma_2, \sigma_3)}{F_2(\sigma_1, \sigma_2, \sigma_3)} = \frac{g_3(\sigma_2 . \sigma_3)}{g_2(\sigma_2, \sigma_3)} = g^*(\sigma_2, \sigma_3),$$
$$\frac{F_1(\sigma_1, \sigma_2, \sigma_3)}{F_3(\sigma_1, \sigma_2, \sigma_3)} = \frac{h_1(\sigma_1, \sigma_3)}{h_3(\sigma_1, \sigma_3)} = h^*(\sigma_1, \sigma_3);$$

es gilt somit

$$f^*(\sigma_1, \sigma_2) \cdot g^*(\sigma_2, \sigma_3) \cdot h^*(\sigma_1, \sigma_3) = 1$$

identisch in $\sigma_1, \sigma_2, \sigma_3$. Daraus wiederum schließt man

$$f^* = \frac{\alpha(\sigma_1)}{\beta(\sigma_2)}; \quad g^* = \frac{\beta(\sigma_2)}{\gamma(\sigma_3)}; \quad h^* = \frac{\gamma(\sigma_3)}{\alpha(\sigma_1)},$$

wobei die Funktionen α, β, γ bis auf einen gemeinsamen konstanten Faktor C eindeutig bestimmt sind. Die Eigenschaft der Funktionen $f, g, \ldots$ monoton zunehmend von $\sigma_1, \sigma_2, \sigma_3$ abzuhängen bewirkt, daß alle partiellen Ableitungen $f_1, f_2, \ldots$ positiv sind und α, β, γ somit nur ein Vorzeichen haben können. Es ergibt sich weiter

$$f_1 = h_1 = \frac{C}{\alpha(\sigma_1)}; \quad f_2 = g_2 = \frac{C}{\beta(\sigma_2)}; \quad g_3 = h_3 = \frac{C}{\gamma(\sigma_3)}.$$

Setzt man nun

$$dS_1 = \frac{C}{\alpha(\sigma_1)}\, d\sigma_1; \quad dS_2 = \frac{C}{\beta(\sigma_2)}\, d\sigma_2; \quad dS_3 = \frac{C}{\gamma(\sigma_3)}\, d\sigma_3,$$

bzw. in integrierter Form

$$S_1(\sigma_1) = C \cdot \int_{\sigma_1^0}^{\sigma_1} \frac{d\sigma_1}{\alpha(\sigma_1)} + S_1^0(\sigma_1^0); \quad S_2 = \cdots; \quad S_3 = \cdots;$$

so erhält man neue Entropie-Funktionen $S_k(\sigma_k)$, die außer der Wachstums-Eigenschaft bei adiabatisch-irreversiblen Prozessen wegen

$$df = dS_1 + dS_2, \quad dg = dS_2 + dS_3, \quad dh = dS_1 + dS_3$$

auch eine Erhaltungs-Eigenschaft bei adiabatisch-reversiblen Prozessen besitzen. Man beachte, daß sich die Konstruktion von $S_i(\sigma_i)$ auch durchführen ließe, wenn die monoton wachsende Abhängigkeit der empirischen Entropie eines zusammengesetzten Systems von den Entropien der Teil-Systeme *nicht* gesichert wäre, aber dann würde $S = S_1 + S_2 + S_3$ im allgemeinen nicht mehr *gleichsinnig monoton* von den σ_i abhängen. Bei adiabatisch-irreversiblen Prozessen könnte S dann sowohl zunehmen wie abnehmen.

11. Energetische Isolation. Energie. Die zweite grundlegende thermodynamische bzw. allgemein physikalische System-Einwirkung ist durch die Gesamtheit derjenigen Zustands-Übergänge gegeben, die bei energetischer Isolation des Systems möglich sind. Die zugehörige Übergangs-Relation ist zerfällend oder verzweigungslos; letzteres deshalb, weil *jede energetische Isolation auch eine adiabatische Isolation* ist (aber nicht umgekehrt!). Zwischen den dadurch definierten Klassen-Ketten ist die Relation zerfällend, und hinsichtlich der Energie ist allein dieser Teil der Übergangs-Relation von Interesse.

Formal ist die durch die energetische Isolation gegebene Sachlage genau dieselbe wie bei der Konstruktion der Entropie mittels *adiabatisch-reversibler* Übergänge. Der einzige Unterschied liegt darin, daß es kein Einzelsystem gibt, für das die energetische Isolation eine empirische Variable definiert wie im Fall der adiabatischen Isolation. Aber das ist unwesentlich, denn der in Axiom 3 ausgedrückte Zusammenhang zwischen System-Zusammensetzung und „adiabatisch-reversibler" Isolation ließe die Konstruktion einer metrischen Variablen auch dann zu, wenn die adiabatische Isolation keine empirische Entropie liefern würde. Wir formulieren daher

Axiom 4: Jedes thermodynamische System $\mathfrak{Z}$ besitzt eine System-Einwirkung *energetische Isolation* von der Struktur einer verzweigungslosen Übergangs-Relation. Jede energetische Isolation ist stets auch eine adiabatische Isolation.

Die von der energetischen Isolation erzeugten Klassen-Ketten bezeichnen wir als Energie-Klassen. Zwischen den Energie-Klassen $\mathfrak{E}$ des Systems $\mathfrak{Z} = [\mathfrak{Z}_1, \mathfrak{Z}_2]$ und den Energie-Klassen $\mathfrak{E}_1$, $\mathfrak{E}_2$ von $\mathfrak{Z}_1$, $\mathfrak{Z}_2$ besteht folgender Zusammenhang:

a) Jede Energie-Klasse $\mathfrak{E}$ von $\mathfrak{Z}$ ist die Vereinigungsmenge einer bestimmten Gesamtheit von Klassen-Paaren $[\mathfrak{E}_1, \mathfrak{E}_2]$.

b) In jeder Energie-Klasse $\mathfrak{E}$ von $\mathfrak{Z}$ kommt jede Energie-Klasse von $\mathfrak{Z}_1$ sowohl als von $\mathfrak{Z}_2$ vor und zwar genau einmal.

c) Die energetische Isolation des zusammengesetzten Systems $\mathfrak{Z}$ besitzt die Eigenschaften (i) bis (viii) aus Ziff. 5γ.

Wie in Ziff. 5 gezeigt wurde, haben wir somit

Satz 5: Jedes thermodynamische System $\mathfrak{Z}$ besitzt eine metrische Variable $U(z) = U(\mathfrak{E})$, die Energie von $\mathfrak{Z}$, die gegenüber der System-Zusammensetzung additiv ist: $U([z_1, z_2]) = U_1(z_1) + U_2(z_2) + \text{const}$. Die Energie-Funktionen der einzelnen Systeme sind bis auf einen universellen Maßstabs-Faktor und bis auf je eine additive Konstante bestimmt.

Axiom 4 sowie die Betrachtungen in den Ziff. 8 bis 10 schließen in ihren Formulierungen die Stetigkeit von Entropie und Energie eines Systems ein. Dies ist nicht notwendig; es genügt vielmehr, die Existenz von gewissen Systemen zu fordern, die jede auftretende Entropie- oder Energie-Differenz aufzunehmen imstande sind. Die Formulierung der Axiome würde dadurch natürlich erheblich schwerfälliger.

Die Tatsache, daß jede energetische Isolation auch eine adiabatische Isolation ist, setzt Energie und Entropie in eine bestimmte Beziehung zueinander. Danach kann die Entropie eines Systems bei Übergängen unter energetischer Isolation niemals abnehmen. Diese Aussage hat hier eine weitergehende Bedeutung als im gewohnten physikalischen Sprachgebrauch und zwar infolge der unterschiedlichen Handhabung des Begriffes Isolation. Man ist gewohnt, bei Übergängen unter einer Isolation nur an solche zu denken, die „ohne jede Mitwirkung" der Umgebung des Systems erfolgen können. Es ist unschwer einzusehen, daß eine solche Aussage durch die festverwurzelte Anschauung der zeitlichen Kontinuität aller

Vorgänge zustande kommt, dagegen in einer Theorie, die keine kontinuierliche Zeit-Variable enthält, gar nicht formuliert werden kann. Dieser von seiten der Kombinatorik selbstverständliche Sachverhalt entspricht in der Tat auch der physikalischen Realität, was zum Beispiel daraus erhellt, daß in der gewohnten Darstellung der Thermodynamik die wesentlichen Theoreme erst durch die Hinzunahme von Kreisprozessen zu den dort verwendeten Isolationen zustande kommen[1]. Kreisprozesse aber sind am Einzelsystem gar nicht beschreibbar, und daher muß eine axiomatische Fassung der Theorie sie in den Isolationen enthalten.

In diesem Zusammenhang erweist es sich als zweckmäßig, den Begriff der Umgebung einzuführen durch folgende

Definition: Ist ein System $\mathfrak{Z}_1$ und ein Übergang $z_1 \to z_1'$ in $\mathfrak{Z}_1$ vorgegeben, so heißt jedes System $\mathfrak{Z}_2$ eine (zu dem Übergang $z_1 \to z_1'$ gehörige) *Umgebung* von $\mathfrak{Z}_1$, wenn das zusammengesetzte System $[\mathfrak{Z}_1, \mathfrak{Z}_2]$ unter energetischer Isolation einen Übergang besitzt, bei welchem das Teilsystem $\mathfrak{Z}_1$ den Übergang $z_1 \to z_1'$ erfährt.

Unter energetischer Isolation von $[\mathfrak{Z}_1, \mathfrak{Z}_2]$ gilt also

$$\Delta U_1 + \Delta U_2 = 0, \qquad \Delta S_1 + \Delta S_2 \geqq 0,$$

wenn ΔU_1, usw. die mit dem Übergang verbundene Änderung der Energie U_1 von $\mathfrak{Z}_1$, usw. bezeichnen. Ist $\Delta U_1 = 0$, so auch $\Delta U_2 = 0$, und wenn überdies $\Delta S_2 = 0$, so muß $\Delta S_1 \geqq 0$. Wir haben somit

Satz 6: Bei jedem Übergang eines Systems $\mathfrak{Z}$, zu dem eine Umgebung existiert, deren Entropie-Änderung bei dem betrachteten Übergang Null ist, kann die Entropie von $\mathfrak{Z}$ niemals abnehmen.

Wie oben auseinandergesetzt, ist die energetische Isolation in ihrer experimentellen Realisierung nur schwer zu übersehen. Dagegen gibt es in praxi meist eine weitere System-Einwirkung, die relativ leicht zugänglich ist, nämlich die „räumliche Isolation", welche zu der energetischen in der Beziehung steht, daß jeder Übergang unter räumlicher Isolation auch ein Übergang unter energetischer Isolation ist (nicht dagegen umgekehrt). Die Bildung der Umgebung $\mathfrak{Z}_2$ wird zumeist so vorgenommen, daß $[\mathfrak{Z}_1, \mathfrak{Z}_2]$ räumlich isoliert ist.

12. Die Variabilitätsbereiche von Energie und Entropie. Energie und Entropie sind metrische Variablen und als solche, wie in Ziff. 5 auseinandergesetzt, bis auf einen Maßstabsfaktor und eine additive Konstante bestimmt. Es erhebt sich nunmehr die Frage, ob diese Freiheit noch weiter eingeschränkt werden kann. Wir wollen zeigen, daß dies für das *Vorzeichen des Maßstabsfaktors* zutrifft.

Offensichtlich läßt die Konstruktion einer metrischen Variablen prinzipiell die Freiheit im Vorzeichen des Maßstabsfaktors. Soweit die metrische Struktur einer Variablen X alles ist, was sich über sie aussagen läßt, muß eine Theorie, die diese Variable enthält, invariant gegenüber der Transformation $X \to -X$ formuliert sein. Bei der Entropie ist dies nicht der Fall, da die Anordnung der empirischen Entropie auf der physikalischen Operation der adiabatischen Isolation beruht, und diese Operation eine Richtung auszeichnet. Infolgedessen ist die erwähnte Vorzeichen-Beschränkung des Maßstabsfaktors der Entropie eine Folge der bisherigen Axiome. Für die Energie ist dies aber sicher nicht so, denn

---

[1] Wir möchten darauf hinweisen, daß die Einbeziehung der Zeit-Variablen in die Theorie die logische Situation vom *physikalischen* Standpunkt aus keineswegs vereinfacht. Denn dann stellt sich das Problem, diese Variable zu kennzeichnen, wobei man nicht die Möglichkeit hat, den Begriff des Übergangs auf eine andere Variable zu stützen. Die von uns behauptete physikalische Irrelevanz des Stetigkeits-Begriffes tritt hier in besonders starken Gegensatz zu gewohnten Anschauungen.

Axiom 4 enthält keinerlei Richtungs-Auszeichnung. Wenn also auch der Maß-stabs-Faktor der Energie auf ein Vorzeichen beschränkt sein soll, muß dies auf einem bisher nicht berücksichtigten physikalischen Tatbestand beruhen. Es handelt sich dabei um die Erfahrung, daß die Energie eines physikalischen Systems (mindestens) einseitig beschränkt ist, oder in geläufiger Ausdrucks-weise, *daß es unmöglich ist, einem System beliebig viel Energie zu entziehen.* Diese Formulierung macht schon Gebrauch von der Möglichkeit, die Richtung zur beschränkten Seite der Energie-Klassen-Folge hin „abnehmend" zu nennen. Damit besitzt jedes System einen tiefsten Energie-Wert, und dies wiederum erlaubt, die additive Konstante dadurch festzulegen, daß man diesen tiefsten Energie-Wert gleich Null setzt.

Es kann vorkommen, daß ein System nicht nur einen tiefsten, sondern auch einen höchsten Energie-Wert hat, die Energie des Systems also beidseitig be-schränkt ist. Das ändert an der obigen Festlegung von wachsenden und ab-nehmenden Energie-Werten solange nichts, als es Systeme mit nicht-zweiseitig beschränkter Energie-Skala gibt, die mit dem betrachteten System in energetische Wechselwirkung treten können. Würde man dagegen ausschließlich Systeme mit einem höchsten Wert der Energie betrachten, so wäre man hinsichtlich des Vor-zeichens der Energie in genau derselben Lage wie bei beidseitiger Unbeschränkt-heit. Diese Bemerkung führt unmittelbar zu

Satz 7: Jede Aussage über ein System mit zweiseitig beschränkter Energie, die sich aus Wechselwirkungen dieses Systems mit Systemen der gleichen Art (das heißt auch mit zweiseitig beschränkter Energie) gewinnen läßt, geht wieder in eine richtige Aussage über, wenn die Energie-Skala invertiert wird.

Entropie und Energie sind somit beide bis auf einen positiven Maßstabs-Faktor definiert, aber der Grund für die Richtungs-Auszeichnung ihrer Klassen-Einteilungen ist verschieden. Bei der Entropie ist es die adiabatische Isolation, das heißt die Einwirkung, durch welche die Entropie selbst definiert wird, bei der Energie die mehr „zufällig" erscheinende Eigenschaft physikalischer Systeme, einen tiefsten Energie-Wert zu besitzen. Damit stellt sich die natürliche Frage, ob die Richtung der Entropie-Skala auch mit einer einseitigen Beschränkung in Zusammenhang gebracht werden kann, das heißt, ob die physikalischen Systeme auch einen kleinsten Wert der Entropie haben. Ebenso wie im Fall der Energie kann diese Frage nur durch die Erfahrung beantwortet werden, und diese besagt, daß es *unmöglich ist, einem System (quasistatisch) beliebig viel Entropie zu ent-ziehen.* Der tiefste Wert der Entropie läßt sich dann durch Verfügung über die freie additive Konstante ebenfalls zu Null machen.

Die einseitige Beschränkung der Entropie bewirkt zwar hinsichtlich der Festlegung von wachsenden und abnehmenden Entropie-Werten dasselbe wie die Existenz adiabatisch-irreversibler Prozesse, sofern die Entropie des Systems nicht auch nach oben beschränkt ist, dagegen legt die adiabatische Isolation die Richtung der Entropie ungeachtet derartiger Einschränkungen fest. Daher gibt es keine dem Satz 7 analoge Aussage über die Entropie.

Die beiden Beschränktheits-Aussagen über Energie und Entropie werden schließlich ergänzt durch eine Aussage über ihren Zusammenhang, die eine Ver-allgemeinerung des Nernstschen Wärmesatzes darstellt[1] und besagt, daß die Zustands-Klasse des kleinsten Energie-Wertes in der des kleinsten Entropie-Wertes enthalten ist oder mit ihr zusammenfällt. Wir formulieren somit

---

[1] Vgl. G. Falk: On the Third Law of Thermodynamics. Phys. Rev. eingereicht Nov. 1958.

Axiom 5: Die Entropie eines thermodynamischen Systems besitzt einen kleinsten Wert (der auf Null normiert werden kann). Die Energie eines thermodynamischen Systems ist (mindestens) einseitig beschränkt. Zur unbeschränkten Seite hin heißt die Energie zunehmend. Die Energie-Klasse mit dem kleinsten Energie-Wert des Systems ist enthalten in (oder identisch mit) der Entropie-Klasse des kleinsten Entropie-Wertes. Ist die Energie eines Systems zweiseitig beschränkt, so gilt dasselbe für die Entropie.

Dieses Axiom zusammen mit Satz 7 liefert

Satz 8: Besitzt ein System einen größten Energie-Wert, so haben die Zustände des kleinsten und größten Wertes der Energie denselben Wert der Entropie.

Schließlich sei noch erwähnt: Die einseitige Unbeschränktheit der Energie braucht nicht im mathematischen Sinn gemeint zu sein. Es ist durchaus denkbar, daß jedes physikalische System einen höchsten Energie-Wert besitzt; dennoch kann die Energie solange als einseitig unbeschränkt angesehen werden, als die zu verwirklichenden Energie-Umsetzungen nur einen kleinen Teil der Energie-Skala bedecken. Die Aussagen von Satz 7 und 8 wären dann zwar im Prinzip ausnahmslos gültig, aber gegenstandslos, da die fraglichen Zustände doch nicht erreicht werden können.

II. Die thermodynamischen Koordinaten.

Den Variablen Entropie und Energie eines Systems kommt insofern eine Vorrangstellung vor allen anderen Variablen zu, als jedes thermodynamische, wir können sogar verallgemeinernd sagen, jedes physikalische System diese beiden Variablen besitzt, während sich über andere Variablen keine derartig allgemeine Aussage machen läßt. Wesentlich ist dabei, daß diese beiden Größen völlig gleichberechtigt nebeneinander erscheinen und die Charakterisierung der Entropie keineswegs an andere Variablen, wie die Temperatur, gebunden ist. Die traditionelle Einführung der Entropie *zusammen* mit der Temperatur läßt den Eindruck aufkommen, als sei die Entropie auf Systeme beschränkt, die eine Temperatur besitzen, was keineswegs der Fall ist. Der Grund für diese Art der Einführung der Entropie ist darin zu sehen, daß die Entropie an thermischen Systemen, das heißt an solchen mit Temperatur entdeckt wurde, und daß meßtechnisch Temperatur und Wärmemenge oftmals einfach zu bestimmende Größen sind (was nicht nur ihre historische, sondern auch ihre elementar-didaktische Vorrangstellung erklärt) und daß schließlich die „mechanische Arbeit" als ein vertrauter Begriff gehandhabt wird.

Die bisherigen Axiome sind somit mehr allgemein physikalischen Charakters als kennzeichnend für die im gewöhnlichen Sinne thermodynamischen Systeme, deren Charakterisierung wir uns nunmehr zuwenden wollen. Es handelt sich auch dabei wieder um die Kennzeichnung charakteristischer Variablen.

13. Physikalische Koordinatensysteme. Koordinaten eines Systems sind eine Anzahl von Variablen, die so beschaffen sind, daß sich jeder Zustand eindeutig als Klassen-Durchschnitt dieser Variablen beschreiben läßt, oder in der Sprache der Darstellungen der Variablen, daß jeder Zustand durch ein Werte-n-tupel eindeutig bestimmt ist.

Es ist nun eine triviale Tatsache, daß man nicht nur an Koordinatensystemen von Zustands-Mengen physikalischer Systeme interessiert ist, sondern daß in praxi die Zustands-Mengen sogar mittels Koordinaten definiert werden. Denn jeder quantitative Meßvorgang repräsentiert seiner Natur nach eine Variable, und infolgedessen erfolgt die Charakterisierung der Zustände mittels Variablen. Die Situation scheint im Rückblick auf unseren Ausgang einen Zirkelschluß zu enthalten: Zunächst bezeichneten wir als das wesentliche Problem einer Axiomatik der Thermodynamik, wie jeder physikalischen Theorie, die Charakterisierung von Variablen in einer als gegeben betrachteten Zustands-Menge, und nun stellen wir fest, daß jede Zustandsmenge $\mathfrak{Z}^*$ selbst mittels Variablen definiert wird. Der springende Punkt ist, daß es sich bei der Vorgabe von $\mathfrak{Z}^*$ um *irgendwelche* Variablen handeln kann, während die Theorie eine Charakterisierung *bestimmter* Variablen verlangt. Dabei kann der Fall eintreten, daß Zustände, die zunächst unterschieden werden, im Rahmen der Theorie ihre Unterscheidbarkeit verlieren. Es ist leicht einzusehen, daß dies tatsächlich der realen physikalischen Praxis entspricht; denn es ist irrig anzunehmen, daß die uns so vertrauten physikalischen Variablen etwas sind, was sich ohne Bezugnahme auf eine Theorie oder wenigstens eine theoretische Absicht definieren ließe[1]. Von jedem Variablen-Satz, der als physikalisches Koordinatensystem fungieren soll, ist zu verlangen, daß Energie und Entropie jede der Koordinaten substituieren können. Wir treffen daher folgende

Definition: Eine Auswahl von Variablen $X_1, \ldots, X_n$ eines Systems heißt ein physikalisches Koordinatensystem, wenn bei Konstanthalten von $n-1$ der Variablen und Veränderung einer einzigen X_j die Energie U oder die Entropie S monotone Funktionen von X_j sind.

Diese Abhängigkeit kann auch eine stückweise Monotonie sein, wenn der ganze Variabilitätsbereich von X_j durch Intervalle überdeckt werden kann, innerhalb deren mindestens eine der beiden Größen U oder S monoton von X_j abhängt.

Der Begriff des Konstanthaltens von Variablen bedarf einer zusätzlichen Bemerkung. Ist nämlich $X_1, \ldots, X_n$ ein Koordinatensystem von $\mathfrak{Z}$, so definiert die Gesamtheit aller Zustände, für die $X_1, \ldots, X_{j-1}, X_{j+1}, \ldots, X_n$ fest vorgegebene Werte haben, eine bestimmte „eindimensionale" Untermenge $\mathfrak{X}_j$ von $\mathfrak{Z}$. Konstanthalten aller Koordinaten außer X_j bedeutet nun, daß man nur Übergänge betrachtet, deren Anfangs- und Endzustand zu $\mathfrak{X}_j$ gehören. Nach den „Zwischenzuständen" solcher Übergänge zu fragen — was durch die gewohnte Vorstellung kontinuierlicher Prozeß-Abläufe nur zu leicht geschieht — ist kombinatorisch sinnlos. Man vergleiche die entsprechenden Erläuterungen zum Begriff der Isolation auf S. 142.

Man beachte außerdem, daß obige Definition die Anzahl der Variablen eines Koordinatensystems nicht als kennzeichnende Invariante des Systems festlegt. Denn ein vorgewiesenes physikalisches Objekt kann mit verschiedenen Anzahlen von Variablen der Definition genügen. Als praktisches Beispiel hierfür sei auf

---

[1] Der Ausgang jeder wissenschaftlich-naiven Beobachtung besteht zunächst in der Sammlung aller registrierbaren Merkmale eines realen Objektes und benutzt diese zur Bildung von Variablen. Die Untersuchung der Beziehungen dieser Variablen untereinander deckt dann Zusammenhänge auf, die wiederum als Auswahlprinzip benutzt werden können, indem man sich auf die „interessanten" unter diesen Zusammenhängen beschränkt. Die Entdeckung der fundamentalen Rolle der Energie bei allen mechanischen und thermischen Prozessen ist in diesem Sinn als Gewinnung eines Auswahlprinzips anzusehen. Betrachtet man z.B. ein Gas in einem Gefäß, so bietet unter anderem auch die Gestalt des Gefäßes Möglichkeiten zur Bildung von Variablen, von denen schließlich nur das Volumen als „wesentlich" erkannt wird, wesentlich deshalb, weil die Energie nur vom Volumen, nicht aber von anderen Gestalt-Parametern abhängt. Der ursprüngliche mit „allen" Gestalt-Parametern gebildete Zustandsraum $\mathfrak{Z}^*$ wird dadurch reduziert auf einen solchen von zwei Freiheitsgraden, und dieser ist derjenige, von dem die Theorie handelt.

die Entdeckung neuer Variablen oder auf das Verschwinden von Variablen durch Dissoziation bei genügend hohen Temperaturen und ähnliche Vorgänge hingewiesen.

14. Kontakt-Gleichgewichte und thermodynamische Koordinaten. Wir haben bereits erwähnt, daß der Begriff des thermodynamischen Systems oder vielmehr die Definition seiner Variablen abhängt von der Gesamtheit der jeweils betrachteten und zur Wechselwirkung zugelassenen Systeme. Diese Tatsache wird besonders deutlich bei derjenigen Wechselwirkung, mit deren Hilfe die typisch thermodynamischen Variablen charakterisiert werden, nämlich dem *Kontakt-Gleichgewicht*. In welcher Weise dies geschieht, wird im folgenden auseinandergesetzt.

α) *Orientierende Betrachtungen.* Zunächst ist klar, daß jedes Kontakt-Gleichgewicht, wie Temperatur-, Druck-, oder chemisches Gleichgewicht die in Ziff. 6 behandelten Eigenschaften hat. Damit definiert ein bestimmtes Kontakt-Gleichgewicht in jedem System, welches die fragliche Kontakt-Relation besitzt, eine Klassen-Einteilung derart, daß eine umkehrbar eindeutige Zuordnung zwischen den Klassen der verschiedenen Systeme besteht. Die so erhaltenen Klassen-Einteilungen bilden zwar noch keine physikalischen Variablen (da sie nicht durch einen physikalischen Prozeß angeordnet sind), aber sie bieten einen natürlichen Ausgangspunkt zur Gewinnung von Variablen.

Um die Richtung anzudeuten, in welcher man zum gewünschten Ziel geführt wird, betrachten wir zunächst folgendes Beispiel: In einem aus zwei Exemplaren des gleichen Systems zusammengesetzten System $\mathfrak{Z} = [\mathfrak{Z}_1, \mathfrak{Z}_1]$ sollen sämtliche möglichen Kontakt-Gleichgewichte zwischen den beiden Exemplaren von $\mathfrak{Z}_1$ hergestellt werden; wir nennen sie *innere* Kontakt-Gleichgewichte von $\mathfrak{Z}$. Dadurch wird $\mathfrak{Z}$ auf ein zu $\mathfrak{Z}_1$ äquivalentes System $\mathfrak{Z}_1^*$ reduziert, das heißt, jede Aussage über $\mathfrak{Z}_1$ trifft auch für $\mathfrak{Z}_1^*$ zu, sogar im absoluten Maßstab, wenn die metrischen Variablen von $\mathfrak{Z}_1$ jeweils mit dem Faktor 2 multipliziert werden. Der geschilderte Prozeß ist in gewissem Sinn als Reversion der System-Zusammensetzung anzusehen. Man schließt daraus, daß die Anzahl der möglichen Kontakt-Gleichgewichte eines Systems $\mathfrak{Z}_1$ gleich der Anzahl seiner Freiheitsgrade ist, und man erkennt weiter: Wenn $X_1, \ldots, X_n$ ein geeignetes metrisches Koordinatensystem des Systems $\mathfrak{Z}_1$ von n_1 Freiheitsgraden sind, so bilden die Variablen $X_1 + X_1 = 2X_1, \ldots, X_{n_1} + X_{n_1} = 2X_{n_1}$ ein Koordinatensystem für $\mathfrak{Z}_1^*$. Der nächste verallgemeinernde Schritt besteht darin, anstelle der beiden gleichen Systeme $\mathfrak{Z}_1$ das aus zwei ähnlichen Systemen $\mathfrak{Z}_1$ und $\mathfrak{Z}_2$ zusammengesetzte System $[\mathfrak{Z}_1, \mathfrak{Z}_2]$ zu betrachten. Wir nennen dabei zwei Systeme ähnlich, wenn sie dieselbe Anzahl von Freiheitsgraden haben $(n_1 = n_2)$ und wenn zwischen ihnen auch dieselbe Anzahl (n_1) von verschiedenen Kontakt-Gleichgewichten möglich ist. Bilden dann $X_1^{(1)}, \ldots, X_{n_1}^{(1)}$, und $X_1^{(2)}, \ldots, X_{n_1}^{(2)}$ jeweils ein Koordinatensystem von $\mathfrak{Z}_1$ bzw. $\mathfrak{Z}_2$, und zwar derart, daß $X_j^{(1)}$ und $X_j^{(2)}$ dieselbe metrische Variable bezeichnen, so ist $X_1^{(1)} + X_1^{(2)}, \ldots, X_{n_1}^{(1)} + X_{n_1}^{(2)}$ ein metrisches Koordinatensystem des zusammengesetzten Systems nach Einschalten aller inneren Kontakt-Gleichgewichte, das heißt aller Kontakt-Gleichgewichte zwischen $\mathfrak{Z}_1$ und $\mathfrak{Z}_2$. Die sukzessive Herstellung der Kontakt-Gleichgewichte zwischen $\mathfrak{Z}_1$ und $\mathfrak{Z}_2$ schließlich zeigt, daß es zu jeder Kontakt-Relation eine zugeordnete metrische Variable, $X_j^{(1)}$ bzw. $X_j^{(2)}$ gibt derart, daß nach Herstellung des betreffenden Gleichgewichtes (wodurch die Freiheitsgrade des zusammengesetzten Systems um einen vermindert werden) $X_j^{(1)} + X_j^{(2)}$ zusammen mit den übrigen $X_k^{(1)}, X_l^{(2)} (k, l \neq j)$ als metrisches Koordinatensystem des resultierenden Systems benutzt werden können. Die Dinge sind im konkreten Einzelfall hinreichend bekannt.

Die Betrachtungen zeigen, daß die Kontakt-Gleichgewichte der thermodynamischen Systeme neben ihrer als Kontakt-Relation formulierbaren Eigenschaft noch weitere wesentliche Eigenschaften haben:

1. Ihre Anzahl ist gleich der Anzahl der Freiheitsgrade des Systems.

2. Ein zusammengesetztes System besitzt alle Kontakt-Gleichgewichte seiner Teilsysteme.

3. Jeder Kontakt-Relation ist eindeutig eine metrische Variable zugeordnet.

Wie diese Zuordnung aussieht, soll noch etwas genauer beschrieben werden. Wir wollen die Kontakt-Relationen eines Systems durch Symbole Γ_1, Γ_2, ... bezeichnen, wobei Γ_i ein Repräsentant der Klassen-Einteilung der mit der Nummer i bezeichneten Kontakt-Relation ist; es ist dabei zweckmäßig, die Kontakt-Relationen unabhängig vom Einzelsystem zu betrachten und alle Kontakt-Relationen der betrachteten System-Gesamtheit durchzunumerieren. $\mathfrak{Z}_1$ und $\mathfrak{Z}_2$ mögen beide die Kontakt-Relation Γ_j besitzen. Wenn nun das zu Γ_j gehörige Kontakt-Gleichgewicht zwischen $\mathfrak{Z}_1$ und $\mathfrak{Z}_2$ hergestellt wird, so hat dies zur Folge, daß das zusammengesetzte System $\mathfrak{Z} = [\mathfrak{Z}_1, \mathfrak{Z}_2]$ nur die Zustände einer Teilmenge von $\mathfrak{Z}$ annehmen kann, die wir mit $[\mathfrak{Z}_1, \mathfrak{Z}_2]_{\Gamma_j}$ bezeichnen. Entsprechend reduziert die Herstellung zweier Kontakt-Gleichgewichte Γ_j und Γ_k zwischen $\mathfrak{Z}_1$ und $\mathfrak{Z}_2$ das System $\mathfrak{Z}$ auf $[\mathfrak{Z}_1, \mathfrak{Z}_2]_{\Gamma_j \Gamma_k}$ usw. Alle so durch die Herstellung von inneren Kontakt-Gleichgewichten von $\mathfrak{Z}$ definierten *reduzierten Systeme* $[\mathfrak{Z}_1, \mathfrak{Z}_2]_{\Gamma_j \Gamma_k} \cdots$ sind wieder *thermodynamisch*.

Wenn für $\mathfrak{Z}_1$ und $\mathfrak{Z}_2$ je ein Koordinatensystem $X_1^{(1)}$, $X_2^{(1)}$, ..., $X_{n_1}^{(1)}$, und $X_1^{(2)}$, $X_2^{(2)}$, ..., $X_{n_2}^{(2)}$ gegeben ist, so bildet $X_1^{(1)}$, ..., $X_{n_1}^{(1)}$, $X_1^{(2)}$, ..., $X_{n_2}^{(2)}$ ein Koordinatensystem des zusammengesetzten Systems $[\mathfrak{Z}_1, \mathfrak{Z}_2]$. Wir wollen diese Koordinaten als metrisch annehmen mit der Besonderheit, daß $X_1^{(1)}$ und $X_1^{(2)}$, $X_2^{(1)}$ und $X_2^{(2)}$, ..., $X_n^{(1)}$ und $X_n^{(2)}$ jeweils Variablen derselben Art bezeichnen, das heißt im Sinne eines Erhaltungssatzes einander zugeordnet sind. n ist kleiner oder höchstens gleich der kleineren der beiden Zahlen n_1 und n_2. $\mathfrak{Z}_1$ und $\mathfrak{Z}_2$ sind dann ähnlich, wenn $n = n_1 = n_2$. Die in Eigenschaft 3 gemeinte Zuordnung läßt sich dann folgendermaßen formulieren: Es gibt stets metrische Koordinatensysteme $X_1^{(1)}$, ..., $X_{n_1}^{(1)}$ bzw. $X_1^{(2)}$, ..., $X_{n_2}^{(2)}$ derart, daß $X_j^{(1)} + X_j^{(2)}$, $X_1^{(1)}$, ..., $X_{j-1}^{(1)}$, $X_{j+1}^{(1)}$, ..., $X_{n_1}^{(1)}$, $X_1^{(2)}$, ..., $X_{j-1}^{(2)}$, $X_{j+1}^{(2)}$, ..., $X_{n_2}^{(2)}$ ein Koordinatensystem des Systems $[\mathfrak{Z}_1, \mathfrak{Z}_2]_{\Gamma_j}$ bilden. Und wenn $[\mathfrak{Z}_1, \mathfrak{Z}_2]_{\Gamma_j}$ seinerseits wieder zusammengesetzt wird mit einem System $\mathfrak{Z}_3$, das ebenfalls die Kontakt-Relation Γ_j und somit die metrische Variable $X_j^{(3)}$ besitzt, so bildet $X_j^{(1)} + X_j^{(2)} + X_j^{(3)}$ zusammen mit allen übrigen Variablen (mit Indices $\neq j$) ein Koordinatensystem des reduzierten Systems $[\mathfrak{Z}_1, \mathfrak{Z}_2, \mathfrak{Z}_3]_{\Gamma_j}$.

β) Die Anordnung der Gleichgewichts-Klassen. Kontakt-Variable. Wie eben festgestellt, besitzt jedes thermodynamische System $\mathfrak{Z}$ ein metrisches Koordinatensystem X_1, ..., X_n dessen Variablen in umkehrbar-eindeutiger Weise seinen Kontakt-Relationen, das heißt den Klassen-Einteilungen Γ_1, ..., Γ_n zugeordnet sind. Andererseits liefert nach Ziff. 13 die Ersetzung jeder einzelnen der Variablen X_k durch die Energie U oder durch die Entropie S des Systems wieder ein Koordinatensystem von $\mathfrak{Z}$. Wir betrachten insbesondere den thermodynamisch wichtigen Fall, daß sich jede Koordinate durch U *und* S substituieren läßt. Werden nun alle X_k außer einem X_j konstant[1] gehalten, so können die Zustände des „Rest-Systems" (von *einem* Freiheitsgrad) sowohl durch die metrische Variable X_j als auch durch U und S charakterisiert werden. Dasselbe gilt natürlich für die unter $X_k = $ const möglichen *Übergänge,* von denen jeder durch die Angabe

---

[1] Das Konstanthalten ist hier im oben diskutierten Sinn gemeint.

von $X_j \to X_j'$ oder $U \to U'$ oder $S \to S'$ beschrieben werden kann. Jeder Übergang bestimmt auf diese Weise zwei Differenzen-Quotienten

$$\xi_j = \frac{U' - U}{X_j' - X_j}, \qquad \eta_j = \frac{S' - S}{X_j' - X_j}.$$

Aus diesen Zahlen ξ_j, η_j, die den *Übergängen* $X_j \to X_j'$ zugeordnet sind, ergeben sich durch den Grenzprozeß $(X_j' - X_j) \to 0$ *Zustands*-Funktionen. Wir bezeichnen sie als das zu Γ_j gehörige *Kontakt-Variablen-Paar* $\xi_j(X_1, \ldots, X_n)$, $\eta_j(X_1, \ldots, X_n)$. Zur Rechtfertigung dieser Benennung bleibt zu zeigen, daß die durch $\xi_j = \mathrm{const}$ und $\eta_j = \mathrm{const}$ definierten Klassen-Einteilungen mit der Klassen-Einteilung Γ_j identisch sind.

Da U sowohl als S Funktionen der Koordinaten $X_1, \ldots, X_n$ sind, besagt die Definition von ξ_j bzw. η_j:

$$\left.\begin{aligned} &\xi_j = \frac{\partial U(X_1, \ldots X_n)}{\partial X_j}, \qquad\qquad \eta_j = \frac{\partial S(X_1, \ldots X_n)}{\partial X_j}, \\[2mm] &dU = \xi_1\, dX_1 + \cdots + \xi_n\, dX_n, \quad dS = \eta_1\, dX_1 + \cdots + \eta_n\, dX_n. \end{aligned}\right\} \tag{14.1}$$

Wir betrachten nun zwei Systeme $\mathfrak{Z}_1$ und $\mathfrak{Z}_2$ mit den Koordinatensystemen $X_1^{(1)}, \ldots, X_{n_1}^{(1)}$ und $X_1^{(2)}, \ldots, X_{n_2}^{(2)}$ von derselben Art wie sie in Ziff. 14α benutzt wurden. Dann ist:

$$U_1 = U_1(X_1^{(1)}, \ldots, X_{n_1}^{(1)}), \quad S_1 = S_1(X_1^{(1)}, \ldots, X_{n_1}^{(1)}),$$

$$U_2 = U_2(X_1^{(2)}, \ldots, X_{n_2}^{(2)}), \quad S_2 = S_2(X_1^{(2)}, \ldots, X_{n_2}^{(2)}),$$

$$\xi_j^{(1)} = \frac{\partial U_1}{\partial X_j^{(1)}}, \qquad\qquad \eta_j^{(1)} = \frac{\partial S_1}{\partial X_j^{(1)}},$$

$$\xi_j^{(2)} = \frac{\partial U_2}{\partial X_j^{(2)}}, \qquad\qquad \eta_j^{(2)} = \frac{\partial S_2}{\partial X_j^{(2)}},$$

und da die Energie U und die Entropie S des zusammengesetzten Systems $[\mathfrak{Z}_1, \mathfrak{Z}_2]$ durch $U_1 + U_2$ und $S_1 + S_2$ gegeben sind, folgt nach (14.1)

$$\left.\begin{aligned} dU &= dU_1 + dU_2 = \sum \xi_k^{(1)}\, dX_k^{(1)} + \sum \xi_l^{(2)}\, dX_l^{(2)}, \\ dS &= dS_1 + dS_2 = \sum \eta_k^{(1)}\, dX_k^{(1)} + \sum \eta_l^{(2)}\, dX_l^{(2)}. \end{aligned}\right\} \tag{14.2}$$

Wird nun $[\mathfrak{Z}_1, \mathfrak{Z}_2]$ durch Herstellung des inneren Kontakt-Gleichgewichtes auf $[\mathfrak{Z}_1, \mathfrak{Z}_2]_{\Gamma_j}$ reduziert, so bilden nach Voraussetzung $X_j^{(1)} + X_j^{(2)}$, $X_1^{(1)}, \ldots, X_{j-1}^{(1)}$, $X_{j+1}^{(1)}, \ldots, X_{n_1}^{(1)}, X_1^{(2)}, \ldots, X_{j-1}^{(2)}, X_{j+1}^{(2)}, \ldots, X_{n_1}^{(2)}$ ein Koordinatensystem von $[\mathfrak{Z}_1, \mathfrak{Z}_2]_{\Gamma_j}$. Schreibt man (14.2) in der Form

$$dU = \tfrac{1}{2}(\xi_j^{(1)} - \xi_j^{(2)})\, d(X_j^{(1)} - X_j^{(2)}) + \tfrac{1}{2}(\xi_j^{(1)} + \xi_j^{(2)})\, d(X_j^{(1)} + X_j^{(2)}) + $$
$$+ \sum_{k \neq j} \xi_k^{(1)}\, dX_k^{(1)} + \sum_{l \neq j} \xi_l^{(2)}\, dX_l^{(2)},$$

$$dS = \tfrac{1}{2}(\eta_j^{(1)} - \eta_j^{(2)})\, d(X_j^{(1)} - X_j^{(2)}) + \tfrac{1}{2}(\eta_j^{(1)} + \eta_j^{(2)})\, d(X_j^{(1)} + X_j^{(2)}) + $$
$$+ \sum_{k \neq j} \eta_k^{(1)}\, dX_k^{(1)} + \sum_{l \neq j} \eta_l^{(2)}\, dX_l^{(2)},$$

so erkennt man, daß die durch das Kontakt-Gleichgewicht Γ_j bewirkte Reduktion von $[\mathfrak{Z}_1, \mathfrak{Z}_2]$ auf $[\mathfrak{Z}_1, \mathfrak{Z}_2]_{\Gamma_j}$ den Bedingungen

$$\xi_j^{(1)}(X_1^{(1)}, \ldots, X_{n_1}^{(1)}) = \xi_j^{(2)}(X_1^{(2)}, \ldots, X_{n_2}^{(2)}) \tag{14.3a}$$

bzw.

$$\eta_j^{(1)}(X_1^{(1)}, \ldots, X_{n_1}^{(1)}) = \eta_j^{(2)}(X_1^{(2)}, \ldots, X_{n_2}^{(2)}) \tag{14.3b}$$

äquivalent ist. Diese Gleichungen drücken das Bestehen des Kontakt-Gleichgewichtes Γ_j' zwischen $\mathfrak{B}_1$ und $\mathfrak{B}_2$ aus. Sie erfüllen offensichtlich die Bedingungen 1 bis 3 in Ziff. 14α.

Das Verhältnis der Variablen ξ_j und η_j zueinander ist dadurch charakterisiert, daß die durch $\xi_j = $ const. definierte Klassen-Einteilung eine Verfeinerung der η_j-Einteilung ist oder umgekehrt oder schließlich, daß ξ_j- und η_j-Einteilung identisch sind[1]. Im ersten Fall ist η_j als eindeutige Funktion von ξ_j darstellbar, im zweiten ξ_j als eindeutige Funktion von η_j, im letzten besteht eine umkehrbareindeutige Beziehung zwischen ξ_j und η_j. Nun bestimmen ξ_j und η_j überdies eine *Anordnung* der in Frage stehenden Klassen dadurch, daß in den Gleichungen $\xi_j(X_1, \ldots, X_n) = c$ bzw. $\eta_j(X_1, \ldots, X_n) = c'$ die Konstanten c und c' als Schar-Parameter die reellen Zahlen durchlaufen. Diese Anordnung könnte zwar im Prinzip für ξ_j und η_j verschieden sein, aber dies ist aus physikalischen Gründen undiskutabel, ausgenommen den Fall, in dem die Funktionen $\xi_j(\eta_j)$ oder $\eta_j(\xi_j)$ nur endlich viele Mehrdeutigkeiten aufweisen.

Aus den bisherigen Betrachtungen ist für keine der beiden Variablen ξ_j und η_j eine Bevorzugung vor der anderen herauszulesen. Es ist nicht einmal zu erkennen, ob die durch sie definierte Anordnung der Gleichgewichts-Klassen überhaupt mit einer charakteristischen Eigenschaft physikalischer Prozesse zusammenhängt. Diese Frage läßt sich beantworten, wenn man eine bisher außer Betracht gelassene Eigenschaft von Kontakt-Gleichgewichten berücksichtigt, nämlich die Tatsache, daß sich *jedes innere Kontakt-Gleichgewicht eines Systems unter adiabatischer Isolation einstellen und aufrecht erhalten, aber nur unter Entropie-Entzug wieder rückgängig machen läßt.* Dies tritt als vierte Eigenschaft des thermodynamischen Kontakt-Gleichgewichtes zu den drei oben genannten hinzu. Daß es sich bei diesen Eigenschaften nicht um triviale Ergänzungen der Transitivität des Kontakt-Gleichgewichtes handelt, zeigt das Beispiel des Kontaktes zweier Knudsen-Gase über eine Wandöffnung. Auch dies ist ein transitiver Kontakt (mit gleichen Werten von $p/\sqrt{T}$), aber kein einfaches *thermodynamisches Kontakt-Gleichgewicht*, da die zuletzt genannte Bedingung verletzt ist.

Betrachten wir zum Beispiel einen Ausgleichs-Vorgang, der unter energetischer Isolation zum Kontakt-Gleichgewicht führt. Dies ist ein Übergang zwischen Zuständen mit denselben Werten von $X_k^{(1)}$ und $X_l^{(2)}$ ($k, l \neq j$) und gleichem Wert von U. Die Entropie nimmt bei diesem Übergang zu, und der den Gln. (14.3) genügende Endzustand muß zu einem Maximum der Entropie für die betreffenden Werte von $X_k^{(1)}$, $X_l^{(2)}$ und U gehören; denn andernfalls gäbe es einen Übergang zu einem Zustand mit größerem Wert der Entropie, der vom Kontakt-Gleichgewicht verschieden wäre. Der betrachtete Übergang ist unter energetischer Isolation natürlich irreversibel, aber da nach Axiom 1 und 3 jeder Übergang auch quasistatisch geführt werden kann, ist der wesentliche Teil der obigen Aussage in den Beziehungen zwischen den Variablen und ihren Werten zu sehen. Wir formulieren diese als

Prinzip I: Enthält ein thermodynamisches Koordinatensystem von $[\mathfrak{B}_1, \mathfrak{B}_2]$ ein gleichartiges Variablen-Paar $X_j^{(1)}$, $X_j^{(2)}$, so besitzt das System ein inneres Kontakt-Gleichgewicht, das bei vorgegebenen Werten von $X_k^{(1)}$, $X_l^{(2)}$ ($k, l \neq j$) und U durch den Maximalwert seiner Entropie definiert ist.

---

[1] Diese Möglichkeiten entsprechen den Möglichkeiten, daß U allein oder S allein oder U sowohl als S als Substituenten einer Variablen X_j fungieren können. Der in der Definition von Ziff. 13 erwähnten Erweiterung entspricht die Möglichkeit, daß teilweise die ξ_j-Klassen eine Verfeinerung der η_j-Klassen sind und teilweise umgekehrt die η_j-Klassen eine Verfeinerung der ξ_j-Klassen. Diese Möglichkeit läßt sich durch Aufteilung des Zustandsraumes in geeignete Teilräume auf die obigen Fälle reduzieren.

Oder in analytischer Form: Das Kontakt-Gleichgewicht ist durch das Maximum der Entropie S von $[\mathfrak{Z}_1, \mathfrak{Z}_2]$ unter den Nebenbedingungen

$$dU = 0 \qquad \begin{cases} dX_k^{(1)} = 0 & (k = 1, \ldots, j-1, j+1, \ldots, n_1) \\ dX_l^{(2)} = 0 & (l = 1, \ldots, j-1, j+1, \ldots, n_2) \end{cases} \qquad (14.4)$$

definiert.

Es ist klar, daß sich diese Aussage über die Lage des Kontakt-Gleichgewichtes unmittelbar in eine Aussage über die Richtung möglicher Prozesse verwandeln läßt; denn Prinzip I besagt, daß von den Übergängen zwischen Zuständen mit denselben Werten von U und $X_k^{(1)}, X_l^{(2)} (k, l \neq j)$ und der Nebenbedingung, daß es zu $[\mathfrak{Z}_1, \mathfrak{Z}_2]$ eine Umgebung gibt, deren Entropie im Anfangs- und Endzustand denselben Wert hat, nur diejenigen möglich sind, bei denen die Entropie S von $[\mathfrak{Z}_1, \mathfrak{Z}_2]$ anwächst. Betrachtet man daher Übergänge bei festgehaltenen Koordinaten $X_k^{(1)}$ und $X_l^{(2)} (k, l \neq j)$, so ist, wenn $\xi_j \neq 0$,

$$dX_j^{(1)} = \frac{1}{\xi_j^{(1)}} dU_1, \qquad dX_j^{(2)} = \frac{1}{\xi_j^{(2)}} dU_2$$

und somit

$$dS = \frac{\eta_j^{(1)}}{\xi_j^{(1)}} dU_1 + \frac{\eta_j^{(2)}}{\xi_j^{(2)}} dU_2 = f_j^{(1)} dU_1 + f_j^{(2)} dU_2.$$

Führt man schließlich noch $dU = 0$ ein, so erhält man

$$dS = (f_j^{(1)} - f_j^{(2)}) dU_1 = - (f_j^{(1)} - f_j^{(2)}) dU_2$$

und unter Berücksichtigung des oben Gesagten

Satz 9: Ist $f_j^{(1)} > f_j^{(2)}$, so sind zwischen Zuständen von $[\mathfrak{Z}_1, \mathfrak{Z}_2]$ mit denselben Werten von $U, X_k^{(1)}, X_l^{(2)} (k, l \neq j)$ und der Bedingung der Existenz einer Umgebung mit ungeänderter Entropie nur solche Übergänge möglich, bei denen U_1 zu- und U_2 abnimmt.

Oder: Von Übergängen zwischen Zuständen des Systems $[\mathfrak{Z}_1, \mathfrak{Z}_2]$ mit denselben Werten von $U, X_k^{(1)}, X_l^{(2)} (k, l \neq j)$ und bei Existenz einer Umgebung mit ungeänderter Entropie sind nur diejenigen möglich, die Energie von dem Teilsystem mit kleinerem f_j-Wert zu dem mit höherem f_j-Wert bringen.

Schließlich noch eine Bemerkung über die Fälle $\xi_j^{(1)}, \xi_j^{(2)} \equiv 0$ oder $\eta_j^{(1)}, \eta_j^{(2)} \equiv 0$. Im ersten hängen U_1 bzw. U_2 gar nicht von $X_j^{(1)}$ bzw. $X_j^{(2)}$ ab, und die Nebenbedingung $dU = 0$ ist eine Folge der restlichen Nebenbedingungen in (14.4). Dann müssen nach Ziff. 13 $\eta_j^{(1)}, \eta_j^{(2)}$ natürlich von Null verschieden sein. Somit liefert $dS = \eta_j^{(1)} dX_j^{(1)} + \eta_j^{(2)} dX_j^{(2)} \geqq 0$ den Satz: Ist $\xi_j^{(1)} \equiv \xi_j^{(1)} \equiv 0$, so hat für jeden unter den Nebenbedingungen

$$d(X_j^{(1)} + X_j^{(2)}) = 0, \quad dX_k^{(1)} = 0, \quad dX_l^{(2)} = 0 \quad (k, l \neq j)$$

möglichen Prozeß, für den es eine Umgebung von $[\mathfrak{Z}_1, \mathfrak{Z}_2]$ gibt, deren Entropie keine Änderung erfährt, $dX_j^{(1)}$ das Vorzeichen von $(\eta_j^{(1)} - \eta_j^{(2)})$.

Der Fall $\eta_j^{(1)} \equiv \eta_j^{(2)} \equiv 0$ wirft eine neue Frage auf, da in ihm $dS = 0$ eine Folge der Nebenbedingungen $dX_k^{(1)} = dX_l^{(2)} = 0 \ (k, l \neq j)$ ist und das Prinzip I somit keine Aussage liefert. Man steht dann vor der Alternative, entweder $\eta_j \equiv 0$ auszuschließen oder ein weiteres Prinzip über das Verhalten von U bei konstanter Entropie aufzusuchen. Wir kommen auf diese Frage zurück.

γ) *Die Gibbssche Charakterisierung thermodynamischer Koordinaten mittels Extremalprinzipien.* Der beschriebenen Definition der thermodynamischen Koordinatensysteme läßt sich nach dem Vorgang von GIBBS eine äquivalente Definition unter Benutzung von Extremalprinzipien zur Seite stellen. Man stützt sich dabei auf einen einfachen mathematischen Sachverhalt: Es seien $X_1^{(1)}, \ldots, X_{n_1}^{(1)}$, und $X_1^{(2)}, \ldots, X_{n_2}^{(2)}$ wieder die oben verwendeten Koordinatensysteme der beiden Systeme $\mathfrak{Z}_1$ und $\mathfrak{Z}_2$. Für die Zustands-Änderungen von $\mathfrak{Z}_1$ und $\mathfrak{Z}_2$ bzw. $[\mathfrak{Z}_1, \mathfrak{Z}_2]$ gilt dann:

$$dU = dU_1 + dU_2 = \sum \xi_k^{(1)} dX_k^{(1)} + \sum \xi_l^{(2)} dX_l^{(2)},$$
$$dS = dS_1 + dS_2 = \sum \eta_k^{(1)} dX_k^{(1)} + \sum \eta_l^{(2)} dX_l^{(2)}.$$

Fragt man nun nach den Extrema (bzw. Stellen der Stationarität) von U bzw. S unter den Nebenbedingungen

$$d(X_j^{(1)} + X_j^{(2)}) = 0, \quad dX_k^{(1)} = 0 \quad (k \neq j), \quad dX_l^{(2)} = 0 \quad (l \neq j) \quad (14.5)$$

bei freier Variation von $dX_j^{(1)}$, so findet man unmittelbar

$$\text{für } dU = 0: \quad \xi_j^{(1)} = \xi_j^{(2)}, \quad (14.6\,\text{a})$$

$$\text{für } dS = 0: \quad \eta_j^{(1)} = \eta_j^{(2)}. \quad (14.6\,\text{b})$$

Die Einbeziehung eines dritten Systems zeigt in ebenso einfacher Weise die Transitivität der aus einem solchen Extremalprinzip fließenden Kontakt-Relation, und damit ist evident, daß sich der für die Thermodynamik fundamentale Zusammenhang zwischen Kontakt-Gleichgewichten und zugeordneten metrischen Variablen durch Extremalprinzipien beschreiben läßt.

Man erkennt nun, daß die in (14.5) auftretenden Variablen $X_j^{(1)} + X_j^{(2)}$, $X_k^{(1)} (k \neq j)$, $X_l^{(2)} (l \neq j)$ gerade die thermodynamischen Koordinaten des reduzierten Systems $[\mathfrak{Z}_1, \mathfrak{Z}_2]_{r_j}$ sind. *Gibt man daher die Werte von* $X_j^{(1)} + X_j^{(2)}$, $X_k^{(1)}$, $X_l^{(2)} (k, l \neq j)$ *vor, so ist ein Zustand des reduzierten Systems* $[\mathfrak{Z}_1, \mathfrak{Z}_2]_{r_j}$ *definiert.* Dann müssen aber auch Energie und Entropie von $[\mathfrak{Z}_1, \mathfrak{Z}_2]_{r_j}$ bestimmte Werte haben. Diese Werte sind, wie die obige mathematische Betrachtung zeigt, simultane Extremwerte der Energie U und der Entropie S des *nicht reduzierten* Systems $[\mathfrak{Z}_1, \mathfrak{Z}_2]$ unter den Nebenbedingungen (14.5). Die zum Zustand $X_j^{(1)} + X_j^{(2)}$, $X_k^{(1)}$, $X_l^{(2)}$ von $[\mathfrak{Z}_1, \mathfrak{Z}_2]_{r_j}$ gehörigen Werte von Energie und Entropie können somit durch die simultane Extremal-Eigenschaft von U und S des Systems $[\mathfrak{Z}_1, \mathfrak{Z}_2]$ bis auf eine endliche Mehrdeutigkeit festgelegt werden; denn die Bedingungen (14.5) könnten natürlich mehrere Paare von Extremwerten $\overline{U}$, $\overline{S}$ von U und S liefern. Das Prinzip I kann dann zur Beschränkung dieser Mehrdeutigkeit benutzt werden. Denn ist für das System $[\mathfrak{Z}_1, \mathfrak{Z}_2]_{r_j}$ die Energie ein Substituent der Variablen $X_j^{(1)} + X_j^{(2)}$, so läßt sich der Zustand $X_j^{(1)} + X_j^{(2)}$, $X_k^{(1)}$, $X_l^{(2)}$ auch durch U, $X_k^{(1)}$, $X_l^{(2)}$ beschreiben. Setzt man also für U einen gefundenen Extremwert $\overline{U}$ ein, so muß das Maximum von S unter festgehaltenen $\overline{U}$, $X_k^{(1)}$, $X_l^{(2)}$, was der Nebenbedingung (14.4) entspricht, mit $\overline{S}$ übereinstimmen; anderenfalls ist das Extremwert-Paar $\overline{U}$, $\overline{S}$ zu verwerfen. Man wird somit geneigt sein, die Bedingung zu stellen, daß die simultane Extremal-Forderung für U und S unter den Nebenbedingungen (14.5) zusammen mit der Maximal-Forderung für S unter den Nebenbedingungen (14.4) nur auf eine einzige Lösung führt.

Es bleiben in diesem Zusammenhang noch zwei Fragen zu klären, die Korrelate der beiden Fragen am Schluß von Ziff. 14β darstellen. Die erste betrifft den Fall, daß $dU = d(U_1 + U_2) = 0$ keine unabhängige Bedingung ist, sondern eine Folge der übrigen Bedingungs-Gleichungen in (14.4). Da letztere auch in den Nebenbedingungen (14.5) enthalten sind, ist $dU = 0$ auch eine Folge von (14.5), und die Extremal-Forderung für U und S geht somit in eine Extremal-Forderung für S bei $dU = 0$ über, die durch das Prinzip I wiederum zu einer Maximal-Forderung für S spezifiziert wird. Der zum Zustand $X_j^{(1)} + X_j^{(2)}$, $X_k^{(1)}$, $X_l^{(2)}$

des Systems $[\mathfrak{Z}_1, \mathfrak{Z}_2]_{\Gamma_j}$ gehörige Entropie-Wert ist damit eindeutig festgelegt. Der Energie-Wert bestimmt sich andererseits daraus, daß im betrachteten Fall U nur von $X_k^{(1)}$, $X_l^{(2)}$ $(k, l \neq j)$ abhängt, und somit durch Vorgabe der Werte dieser Koordinaten ebenfalls festgelegt ist.

Die zweite Frage schließlich betrifft den Fall, daß $dS = d(S_1 + S_2) = 0$ eine Folge der Nebenbedingungen $dX_k^{(1)} = dX_l^{(2)} = 0$ $(k, l \neq j)$ ist. Dann liefert die Extremal-Forderung für U und S nur die Aussage, daß U einen Extrem-Wert annimmt, ohne diesen näher zu fixieren. Möglichkeiten auch in diesem Fall zu einer eindeutigen Aussage zu kommen, sind: 1. Die Forderung, daß es zu jeder Werte-Kombination von S, $X_k^{(1)}$, $X_l^{(2)}$, $(k, l \neq j)$ nur einen einzigen Extrem-Wert der Energie gibt. 2. Die Festlegung des fraglichen Wertes von U auf den Minimal-Wert, den U unter den genannten Bedingungen annimmt. 3. Ausschluß des Falles als bei thermodynamischen Systemen nicht vorkommend.

Es ist wichtig, sich klar zu machen, daß diese Möglichkeiten im Prinzip logische Freiheiten der Theorie darstellen, und daß die Hinzunahme jedes dieser Postulate in konsistenter Weise eine Vervollständigung der Theorie liefern würde. Man könnte so etwas verschiedene Theorien aufbauen, die mindestens in allen Aussagen übereinstimmen, in denen diese Postulate nicht benutzt werden.

Forderung 1. kommt jedoch praktisch der Forderung 2. gleich; denn da die Energie ohnehin nach unten beschränkt ist, verlangt 1. die einseitige Unbeschränktheit von U, da es sonst stets ein Maximum *und* ein Minimum der Energie geben würde. Die Forderung 1. verlangt somit ein Minimum von U. Das ist aber im wesentlichen die Aussage von 2. Die Forderung 3. dagegen stellt eine echte Alternative dar. Sie würde zwar die „mechanischen" Systeme ausschließen, das heißt diejenigen Systeme, die nur adiabatisch reversibler Übergänge fähig sind, aber das ist nicht von erheblichem Gewicht, da diese stets als Teilsysteme von thermodynamischen Systemen angesehen werden könnten. Dennoch zeigen die Betrachtungen eine gewisse Rechtfertigung von

Prinzip II: Sind $X_j^{(1)}$, $X_j^{(2)}$ ein Paar gleichartiger thermodynamischer Koordinaten, so hat der Zustand S, $X_k^{(1)}$, $X_l^{(2)}$ $(k, l \neq j)$ des reduzierten Systems $[\mathfrak{Z}_1, \mathfrak{Z}_2]_{\Gamma_j}$ als Energie-Wert das Minimum der Energie U von $[\mathfrak{Z}_1, \mathfrak{Z}_2]$ bei vorgegebenen Werten von S, $X_k^{(1)}$, $X_l^{(2)}$, $(k, l \neq j)$.

Es sei ausdrücklich angemerkt, daß die Prinzipe I und II *nicht* äquivalent, sondern im Rahmen der bisher entwickelten Theorie *unabhängig* voneinander sind. Daß sie in anderen Darstellungen[1] als äquivalent erscheinen, liegt daran, daß in diesen explizite oder implizite weitere Voraussetzungen gemacht werden, die der Aufnahme weiterer Axiome entsprechen. Es handelt sich dabei insbesondere um die Spezialisierung auf thermische Systeme (Ziff. 15), die als zusätzliches *Axiom* betrachtet, die beiden Prinzipe abhängig machen würde.

Die Betrachtungen zu Prinzip II zeigten einen Zusammenhang zwischen der Minimal-Eigenschaft der Energie und ihrer einseitigen Beschränktheit. Damit erhebt sich die Frage nach einer Modifikation von Prinzip II im Falle von Systemen mit zweiseitig beschränkter Energie. Tatsächlich liefert Satz 7 (Ziff. 12) auch unmittelbar folgende

Ergänzung zu Prinzip II: Ist die Energie U des Systems $[\mathfrak{Z}_1, \mathfrak{Z}_2]$ zweiseitig beschränkt, so tritt als Energie-Wert des reduzierten Systems sowohl das Minimum als auch das Maximum von U bei vorgegebenen Werten von S, $X_k^{(1)}$, $X_l^{(2)}$ auf.

Das bedeutet natürlich, daß bei solchen Systemen die Koordinaten $X_k^{(1)}$, $X_l^{(2)}$ $(k, l \neq j)$ zusammen mit S nicht einen, sondern im allgemeinen zwei Zustände von

---

[1] Vgl. z.B. GIBBS: Scientific papers, Vol. I. New York 1906.

$[\mathfrak{Z}_1, \mathfrak{Z}_2]_{\Gamma_j}$ definieren, oder anders ausgedrückt, S ist keine Koordinate von $[\mathfrak{Z}_1, \mathfrak{Z}_2]_{\Gamma_j}$, wohingegen U eine Koordinate ist[1].

$\delta)$ *Axiome.* Blickt man zurück auf obige Diskussionen über die Eigenschaften der Variablen thermodynamischer Koordinatensysteme und vergleicht sie mit der Einführung von Entropie und Energie, so macht man unmittelbar folgende Feststellung: Während Entropie · und Energie mit dem Prozeß der *System-Zusammensetzung* verbunden waren, sind die thermodynamischen Koordinaten mit dem Prozeß der *System-Reduktion* verknüpft. Dies ist für die Formulierung der Axiome von Bedeutung. Wir möchten jedoch ausdrücklich davor warnen, in dieser Unterscheidung einen Wesens-Unterschied der Variablen zu suchen. Wir haben gesehen, daß die Konstruktion *jeder* metrischen Variablen, und damit auch jeder Variablen eines thermodynamischen Koordinatensystems, mit dem Prozeß der System-Zusammensetzung verknüpft ist. Die durch Einstellung innerer Kontakt-Gleichgewichte bewirkte System-Reduktion sucht unter den metrischen Variablen dann die *thermodynamischen Koordinaten* heraus. So kann zum Beispiel die Energie durchaus auch unter diesen durch System-Reduktion ausgewählten Variablen auftreten, und wir werden sehen, daß die *Systeme mit Temperatur* durch diese Eigenschaft ausgezeichnet sind. Zusammenfassend können wir sagen: Die Konstruktion metrischer Variablen basiert auf der System-Zusammensetzung. Die System-Reduktion wählt unter den metrischen Variablen die thermodynamischen Koordinaten aus.

Die Axiome selbst können nach den ausgedehnten Diskussionen ohne Umschweife angegeben werden. Zuvor jedoch folgende

Definition: Besitzt ein System zwei Kontakt-Relationen Γ_j, $\Gamma_{j'}$, die in einem zweiten System dieselbe Einteilung in Kontakt-Klassen erzeugen, so heißen Γ_j und $\Gamma_{j'}$ *gleichartig.*

Metrische Variablen heißen *gleichartig,* wenn sie mittels derselben Wechselwirkung erzeugt werden.

Die Existenz zweier gleichartiger Kontakt-Relationen Γ_j, $\Gamma_{j'}$ in einem System $\mathfrak{Z}$ hat, wie erwähnt, zur Folge, daß $\mathfrak{Z}$ ein *inneres* Kontakt-Gleichgewicht besitzt, dessen Herstellung $\mathfrak{Z}$ auf $\mathfrak{Z}_{\Gamma_j}$ reduziert. Man spricht in diesem Zusammenhang auch von einer *Enthemmung* des Variablen-Paares X_j, $X_{j'}$. Nunmehr formulieren wir

Axiom 6: a) Jedes thermodynamische System besitzt ein thermodynamisches Koordinatensystem $X_1, \dots, X_n$. Sind $X_1^{(1)}, \dots, X_{n_1}^{(1)}$ und $X_1^{(2)}, \dots, X_{n_2}^{(2)}$ thermodynamische Koordinatensysteme zweier Systeme $\mathfrak{Z}_1$ und $\mathfrak{Z}_2$, so bilden die Variablen $X_1^{(1)}, \dots, X_{n_1}^{(1)}, X_1^{(2)}, \dots, X_{n_2}^{(2)}$ ein thermodynamisches Koordinatensystem von $[\mathfrak{Z}_1, \mathfrak{Z}_2]$.

b) Besitzt ein thermodynamisches System $\mathfrak{Z}$ zwei gleichartige Koordinaten X_j, $X_{j'}$, so nimmt für jede Werte-Vorgabe von $X_j + X_{j'}$, $X_k (k \neq j, j')$ die Entropie S von $\mathfrak{Z}$ (als Funktion von X_j) ihr *Maximum* $\overline{S}$ stets für denselben X_j-Wert an, für den die Energie U ihr *Minimum* $\overline{U}$ annimmt.

Die Variablen $X_j + X_{j'}$, $X_k (k \neq j, j')$ bilden ein Koordinatensystem des *reduzierten Systems* $\mathfrak{Z}_{\Gamma_j}$ und $\overline{S} = \overline{S}(X_j + X_{j'}, X_k)$ bzw. $\overline{U} = \overline{U}(X_j + X_{j'}, X_k)$ seine Entropie bzw. Energie.

c) Besitzt ein thermodynamisches Koordinatensystem von $\mathfrak{Z}$ drei gleichartige Variablen X_j, $X_{j'}$, $X_{j''}$, so ist $(\mathfrak{Z}_{\Gamma_j})_{\Gamma_{j''}} = (\mathfrak{Z}_{\Gamma_{j''}})_{\Gamma_{j'}} = \mathfrak{Z}_{\Gamma_{j'}\Gamma_{j''}}$.

---

[1] Beispiele eines derartigen Verhaltens liefern die Kern-Spin-Systeme gewisser Kristalle. Vgl. N.F. Ramsey, Thermodynamics and statistical mechanics at negative absolute temperatures, Phys. Rev. **103**, 20 (1956) und die dort genannte Literatur.

Ergänzung: Ist die Energie U des Systems beidseitig beschränkt, so wird die Bedingung b) abgeändert in: Die Entropie S nimmt (als Funktion von X_j) ihr *Maximum* für diejenigen Werte von X_j an, für welche die Energie U *Extrema* besitzt.

Des Interesses wegen sei erwähnt, daß bei Aufnahme der Forderung 3. (S. 153) anstelle von Prinzip II das Axiom 6 sich unter expliziter Verwendung des Begriffes der Kontakt-Relation formulieren läßt, wobei von den Extremal-Eigenschaften von Entropie und Energie kein Gebrauch gemacht wird:

a) Sind $X_1^{(1)}, \ldots, X_{n_1}^{(1)}$ und $X_1^{(2)}, \ldots, X_{n_2}^{(2)}$ thermodynamische Koordinatensysteme zweier Systeme $\mathfrak{Z}_1$ und $\mathfrak{Z}_2$, so bilden die Variablen $X_1^{(1)}, \ldots, X_{n_1}^{(1)}, X_1^{(2)}, \ldots, X_{n_2}^{(2)}$ ein thermodynamisches Koordinatensystem von $[\mathfrak{Z}_1, \mathfrak{Z}_2]$.

Ein thermodynamisches System besitzt so viele Kontakt-Relationen Γ_j wie thermodynamische Koordinaten, und ein zusammengesetztes System hat alle Kontakt-Relationen seiner Teilsysteme.

b) Für ein System $\mathfrak{Z}$ mit zwei gleichartigen Kontakt-Relationen $\Gamma_j, \Gamma_{j'}$ gilt:

α) das aus $\mathfrak{Z}$ durch Herstellung des zugehörigen inneren Kontakt-Gleichgewichts hervorgehende reduzierte System $\mathfrak{Z}_{\Gamma_j}$ ist wieder thermodynamisch.

β) das thermodynamische Koordinatensystem enthält zwei gleichartige Variablen $X_j, X_{j'}$, derart, daß $X_j + X_{j'}, X_1, \ldots, X_{j-1}, X_{j+1}, \ldots, X_{j'-1}, X_{j'+1}, \ldots, X_n$ ein thermodynamisches Koordinatensystem von $\mathfrak{Z}_{\Gamma_j}$ bilden. Jedes innere Kontakt-Gleichgewicht eines thermodynamischen Systems kann unter adiabatischer Isolation hergestellt und aufrechterhalten werden. Die dabei zum Gleichgewicht führenden Übergänge sind adiabatisch irreversibel.

Diese Formulierung ist allerdings nicht völlig äquivalent mit Axiom 6, da der Begriff des Kontakt-Gleichgewichtes eine Zerlegbarkeit von $\mathfrak{Z}$ in Teilsysteme $\mathfrak{Z} = [\mathfrak{Z}_1, \mathfrak{Z}_2]$ verlangt, die in Axiom 6 nicht gefordert wird.

Schließlich noch eine Bemerkung über die Masse als Variable thermodynamischer Systeme. Wir erwähnten oben (S. 135), daß die Masse eine adiabatisch gehemmte Variable und die adiabatische Übergangsrelation eines Systems mit der Masse als Variable somit verzweigungslos ist. Innerhalb jeder Klassen-Kette läßt sich dann eine metrische Entropie konstruieren. Bei der Zuordnung der Entropie-Klassen verschiedener Ketten bleibt jedoch die prinzipielle Freiheit, *einer* Entropie-Klasse einer Kette eine beliebige Entropie-Klasse einer anderen Kette zuzuordnen. Die Zuordnung aller übrigen Entropie-Klassen der beiden Ketten ist dann festgelegt. Da verschiedene Klassen-Ketten aber verschiedenen Werten der Massen-Variablen m entsprechen, folgt daraus, daß die Abhängigkeit der Entropie S von m willkürlich ist. So überraschend diese Feststellung sein mag, entspricht sie doch genau dem physikalischen Sachverhalt, solange über m nicht mehr bekannt ist als daß es als Variable verwendet werden soll. Weiß man dagegen, daß m eine *thermodynamische Koordinate* des Systems ist, so wird die Willkür in der Zuordnung der Entropie-Klassen verschiedener Ketten weitgehend beseitigt; denn Axiom 6 fordert einen charakteristischen Zusammenhang zwischen der Entropie und jeder metrischen thermodynamischen Koordinate X_j, nämlich den, daß S als Funktion von X_j beim Gleichgewicht ein Maximum annimmt. Die Aussage, daß m eine thermodynamische Koordinate ist, ist somit gleichbedeutend mit der Aussage, daß die Zuordnung der Entropie-Klassen verschiedener Ketten so vorgenommen werden kann, daß Axiom 6 erfüllt ist. Um eine solche Möglichkeit in Praxi festzustellen, wird man sich vorzugsweise auf die zweite Fassung des Axioms 6 (Kleindruck) stützen, da es einfacher ist, die Existenz eines Kontakt-Gleichgewichtes nachzuweisen als die Erfüllbarkeit der Maximalbedingung der Entropie (bzw. der Minimal-Bedingung der Energie) durch geeignete Zuordnung der Entropie-Klassen (bzw. Energie-Klassen). Zusammenfassend kann man sagen: Der Nachweis adiabatischer Ausgleichsprozesse, deren Endzustände alle Eigenschaften des thermodynamischen Kontakt-Gleichgewichtes zeigen und bei denen die Masse einem Erhaltungssatz ($m + m' = $ const) genügt, rechtfertigen die Einführung der Masse als thermodynamische Koordinate. Dieselben Überlegungen gelten natürlich für jede adiabatisch gehemmte Variable, sofern sie als thermodynamische Koordinate fungieren soll.

15. Thermische Systeme. Die bisherigen Axiome legen die Eigenschaften thermodynamischer Systeme insofern fest, als sie ihre Variablen und Gleichgewichte definieren. Andererseits läßt sich jedes thermodynamische Problem als Frage nach gewissen Gleichgewichts-Zuständen und ihrer gegenseitigen Erreichbarkeit unter vorgegebenen Bedingungen fassen. Wir können damit sagen, daß die *Struktur der Thermodynamik mit den bisherigen Axiomen voll gekennzeichnet ist.* Jede weitere Aussage kann in Form von Beschränkungen wohldefinierter Möglichkeiten gefaßt werden, ohne zur Struktur der Theorie, soweit sie die gegenseitigen Beziehungen der Variablen betrifft, beizutragen. Dies ist deshalb bemerkenswert, weil die Temperatur erst jetzt eine wesentliche Rolle spielen wird, nämlich bei der Feststellung der praktischen Wichtigkeit *thermischer* Systeme.

Definition: Ein System, dessen thermodynamisches Koordinatensystem die Energie U als Variable enthält, heißt *thermisch*. Die durch $(\partial S(U, X_2, \ldots, X_n)/\partial U) = 1/T$ definierte Kontakt-Variable T ist die *Temperatur* des Systems.

Für ein solches System ist also

$$dS = \frac{1}{T} dU + \sum_{2}^{n} \eta_k \, dX_k, \qquad \eta_1 = \frac{1}{T}, \qquad X_1 = U \qquad (15.1)$$

und

$$\xi_1 = 1, \qquad \xi_k = 0 \qquad (k = 2, \ldots, n),$$

wenn η_k, ξ_k die in Ziff. 14β eingeführten Kontakt-Variablen-Paare bezeichnen. Wir beweisen sodann

Satz 10: Es ist stets $\dfrac{1}{T} > 0$.

Man betrachtet dazu den Temperatur-Ausgleich zwischen zwei thermischen Systemen $\mathfrak{Z}_1$ und $\mathfrak{Z}_2$, die zweckmäßigerweise als zwei Exemplare desselben Systems gewählt werden. Zunächst ist

$$dS_1 = \frac{1}{T_1} dU_1 + \sum_{2}^{n_1} \eta_k^{(1)} \, dX_k^{(1)}, \qquad dS_2 = \frac{1}{T_2} dU_2 + \sum_{2}^{n_2} \eta_l^{(2)} \, dX_l^{(2)}.$$

Dann gilt für das Gesamtsystem $[\mathfrak{Z}_1, \mathfrak{Z}_2]$

$$\left.\begin{aligned}
dS &= dS_1 + dS_2 = \frac{1}{T_1} dU_1 + \frac{1}{T_2} dU_2 + \sum_{2}^{n_1} \eta_k^{(1)} \, dX_k^{(1)} + \sum_{2}^{n_2} \eta_l^{(2)} \, dX_l^{(2)}, \\
dU &= dU_1 + dU_2 = T_1 \, dS_1 + T_2 \, dS_2 - T_1 \sum_{2}^{n_1} \eta_k^{(1)} \, dX_k^{(1)} - T_2 \sum_{2}^{n_2} \eta_l^{(2)} \, dX_l^{(2)}.
\end{aligned}\right\} \quad (15.2)$$

Nun kann unter der Voraussetzung, daß sich das Gesamtsystem nicht im inneren Gleichgewicht befindet, nach Axiom 6 unter den Nebenbedingungen

a) $dX_k^{(1)} = dX_l^{(2)} = 0, \qquad dU = 0:$ die Entropie S nur zunehmen,

b) $dX_k^{(1)} = dX_l^{(2)} = 0, \qquad dS = 0:$ die Energie U nur abnehmen.

Diese beiden Möglichkeiten liefern mit (15.2) also

$$\text{a)} \quad dS = \left(\frac{1}{T_1} - \frac{1}{T_2}\right) dU_1 = -\left(\frac{T_1 - T_2}{T_1}\right)\frac{1}{T_2} dU_1 > 0,$$

$$\text{b)} \quad dU = (T_1 - T_2)\, dS_1 = \left(\frac{T_1 - T_2}{T_1}\right) dU_1 < 0,$$

woraus unmittelbar $\frac{1}{T_2} > 0$ folgt. Wendet man Satz 10 auf (15.1) an, so erhält man

Satz 11: Für thermische Systeme bilden neben $U, X_2, \ldots, X_n$ auch $S, X_2, \ldots, X_n$ ein thermodynamisches Koordinatensystem.

und schließlich liefert Satz 10, angewendet auf Satz 9

Satz 9a: Sind $\mathfrak{Z}_1$ und $\mathfrak{Z}_2$ thermische Systeme, so können zwischen Zuständen von $[\mathfrak{Z}_1, \mathfrak{Z}_2]$ mit denselben Werten von $U_1 + U_2$, $X_k^{(1)}$, $X_l^{(2)}$ ($k = 2, \ldots, n, l = 2, \ldots, n$) und bei Existenz einer Umgebung von $[\mathfrak{Z}_1, \mathfrak{Z}_2]$ mit ungeänderter Entropie nur solche Übergänge stattfinden, bei denen die Energie von dem Teilsystem mit höherer Temperatur zu dem mit niederer Temperatur übergeht.

Man erkennt in diesem Satz die Clausiussche Formulierung des zweiten Hauptsatzes wieder. Fragt man nach der Thomsonschen Formulierung, so sieht man, daß diese direkt aus Satz 10 folgt; denn $\frac{1}{T} = \frac{\partial S\,(U, X_2, \ldots X_n)}{\partial U} > 0$ besagt ja gerade, daß sich die Entropie bei festen Werten von $X_2, \ldots, X_n$ stets gleichsinnig mit der Energie ändert. Es ist also bei festen $X_2, \ldots, X_n$ unmöglich, einem System nur Energie, dagegen keine Entropie zu entziehen.

Die wichtige Rolle der thermischen Systeme beruht darauf, daß sich jedes thermodynamische System durch Herstellung hinreichend vieler innerer Gleichgewichte auf ein thermisches System reduzieren läßt. Wir wollen diese Tatsache formulieren als

Axiom 7: Jedes thermodynamische System läßt sich durch Herstellung innerer Gleichgewichte auf ein thermisches System reduzieren.

Dieses Axiom stellt keine Aussage über die logische Struktur der Thermodynamik dar, sondern eine solche über die vorkommenden Systeme und besagt somit, daß die Natur keinen Gebrauch macht von allen in der Theorie enthaltenen Möglichkeiten, oder noch anders formuliert: Axiom 7 sondert unter den thermodynamischen Systemen die *realen* aus. Dies gibt dem Axiom gegenüber dem übrigen Teil der Theorie eine etwas schwächere Stellung, denn selbst wenn es sich einmal als unzutreffend erweisen sollte, werden die wesentlichen Aussagen der Theorie davon nicht berührt. Andererseits verleiht das Axiom 7 der Temperatur unter allen Kontakt-Variablen eine offenbare Sonderstellung. Wenn nämlich durch Herstellung innerer Gleichgewichte jedes thermodynamische System auf ein thermisches reduziert werden kann, das heißt auf ein solches, dessen Energie eine thermodynamische Koordinate ist, so muß jedes System zur Energie gleichartige Variablen (d.h. Energien) als thermodynamische Koordinaten enthalten. In den meisten Fällen können diese als Energie-Funktionen geeigneter Teilsysteme angesehen werden, so daß das Gesamtsystem als aus thermischen Systemen zusammengesetzt erscheint.

Schließlich noch ein Wort über die *negativen Temperaturen.* Der Beweis des Satzes 10 zeigt, daß die Positivität der Temperatur aus dem Zusammenspiel der Maximal-Eigenschaft der Entropie S bei konstanter Energie U *und* der Minimal-Eigenschaft von U bei konstantem S gefolgert wurde. Andererseits besagt die Ergänzung in Axiom 6, daß im Fall zweiseitig beschränkter Energie-Funktionen neben dem Minimum auch ein Maximum von U bei konstantem S auftritt. Der obige Beweis liefert dann eine

Ergänzung zu Satz 10: Bei Systemen mit zweiseitig beschränkter Energie nimmt $1/T$ sowohl positive wie negative Werte an.

Man erkennt weiter, daß Satz 9 nunmehr die Richtung des Energie-Überganges zwischen zwei Systemen mit beschränkter Energie so festlegt, daß die Energie unter den genannten Nebenbedingungen nur von dem System mit kleinerem $1/T$-Wert zu dem mit größerem $1/T$-Wert gebracht werden kann, unabhängig davon ob $1/T$ positiv oder negativ ist[1].

Wir haben in Ziff. 5 gesehen, daß es zur Festlegung des Maßstabes einer metrischen Variablen notwendig ist, zwei Standard-Zustände eines Standard-Systems aufzuweisen. Für die Energie wird dieses Problem durch die Mechanik. gelöst: Das Standard-System ist der Massenpunkt, die Standard-Zustände zum Beispiel zwei Zustände bestimmter Geschwindigkeiten. Diese Festlegung hat den Vorteil, daß die Angabe des Systems sowie der Standard-Zustände sehr einfach ist, da sich unter geeigneten Bedingungen fast jedes System als Massenpunkt ansehen läßt. Bei der Entropie ist man nicht in einer derartig glücklichen Lage; man muß also ein spezielles System und zwei Zustände an ihm aufweisen. Die thermischen Systeme erlauben nun eine relativ einfache Lösung dieser Aufgabe, da die Temperatur wegen ihrer Kontakt-Eigenschaft es gestattet, eine charakteristische Temperatur-Klasse eines Systems auf alle anderen thermischen Systeme zu übertragen.

Überdies bieten die thermischen Systeme eine besonders einfache Möglichkeit der Entropie-Bestimmung durch Messung von Temperatur und Energie-Überführungen, oder in gewohnter Terminologie, durch thermische und kalorische Messungen, *sobald an einem einzigen thermischen System die absolute Temperatur als Funktion eines geeigneten Satzes von Variablen festgelegt ist.*

Betrachten wir zunächst das Problem der Herstellung eines absoluten Thermometers. Da T eine aus U und S abgeleitete Größe ist, erfordert die Bestimmung von T die vorherige Festlegung von U und S. Im Hinblick auf die gewohnte Konstruktion der absoluten Temperatur scheint diese Feststellung überraschend[2], aber die ausdrückliche Beschränkung auf quasistatische, d.h. Carnotsche Prozesse bei dieser Konstruktion ist der vorherigen Metrisierung der Entropie (ohne Festlegung der Einheit!) äquivalent. Für die Energie-Änderung eines zusammengesetzten Systems

$$dU = dU_1 + dU_2 = T_1\,dS_1 + T_2\,dS_2 - T_1\sum \eta_k^{(1)}\,dX_k^{(1)} - T_2\sum \eta_l^{(2)}\,dX_l^{(2)}$$

gilt bei Konstant-Halten[3] der Variablen $X_k^{(1)}$, $X_l^{(2)}$

$$dU = T_1\,dS_1 + T_2\,dS_2$$

und bei adiabatisch-reversiblen Prozessen von $\mathfrak{Z}$, das heißt quasistatischen Prozessen von $\mathfrak{Z}_1$ und $\mathfrak{Z}_2$

$$dU = (T_1 - T_2)\,dS_1 = \frac{T_1 - T_2}{T_1}\,dU_1.$$

Messung von dU und dU_1 liefert also $(T_1 - T_2)/T_1$. Setzt man für zwei Standard-Temperatur-Klassen $T_1 - T_2 = 100$, so liefert die Messung von dU und dU_1,

---

[1] Für eine detailliertere Diskussion vgl. N.F. Ramsey: Phys. Rev. **103**, 20 (1956).
[2] Obwohl man es von seiten der statistischen Mechanik nie anders gewohnt war.
[3] Das Konstant-Halten ist hier im Sinne der Erläuterungen von S. 146 gemeint.

wenn die Standard-Temperatur-Klassen die des Siede- und Eis-Punktes von Wasser unter Normal-Bedingungen sind, die gewohnte absolute Temperatur. Damit ist aber auch die *Einheit* der *S*-Skala festgelegt. Im Prinzip ist die Herstellung eines absoluten Thermometers also mit denselben Schwierigkeiten verbunden wie eine direkte Entropiemessung, eine Tatsache, die aus der Tieftemperatur-Physik hinreichend bekannt ist.

Wenn dagegen in einem bestimmten Temperatur-Intervall ein absolutes Thermometer vorliegt, dann ist die Messung der Entropie eines beliebigen thermischen Systems in diesem Intervall nicht mehr an quasistatische Prozesse gebunden, denn $dS = 1/T\,dU$ bei $dX_k = 0$ liefert die Entropieänderung unabhängig davon, ob die Energieänderung dU quasistatisch erfolgt oder nicht.

Als Konsequenz dieser Meßmethode pflegt man der Entropie die Dimension (erg/grad) zu geben. Obwohl Dimensions-Zuordnungen eine gewisse Willkür erlauben und stark dem persönlichen Geschmack unterworfen sind, ist die gewohnte Dimension der Entropie insofern irreleitend, als sie die Meinung unterstützen könnte, daß die Entropie eine aus Temperatur und Energie abgeleitete Größe ist. Unter diesem Gesichtspunkt wäre es daher vorzuziehen, der Temperatur die Dimension (erg/Clausius) zu geben, ebenso wie man dem Druck die Dimension (erg/cm³) und dem chemischen Potential die Dimension (erg/Mol) oder (erg/g) gibt.

16. Zusammenfassung und Überblick. Der erste Schritt in unseren Untersuchungen war die Analyse der Konstruktion metrischer Variablen, die unlösbar mit dem Begriff der System-Zusammensetzung und dem einer „Wechselwirkung mit Erhaltungssatz" verknüpft ist. Die Sachlage wurde an der Energie erläutert, für die der maßgebende Erhaltungssatz nur allzu vertraut ist. Daß auch die Länge oder das Volumen durch Operationen definiert oder wenigstens als metrische Variable nachgewiesen werden, die denselben Formalgesetzen genügen, ist unschwer einzusehen, wenn auch die Bezeichnung „Erhaltungssatz" in diesem Zusammenhang etwas fremdartig klingen mag. Die Masse scheint auf den ersten Blick sich nicht unmittelbar in das Begriffs-Schema einzuordnen, da sie als mechanische Größe durch Messung der auf einen Körper ausgeübten Kraft und der resultierenden Beschleunigung bestimmt wird. Die so definierte Masse ist zwar auch eine Variable, aber zunächst nur eine *empirische*. Daß diese Variable auch *metrisch* ist, folgt erst aus der Erfahrungstatsache, daß zwei Körper gleicher Masse bei einer gewissen Operation des Zusammensetzens die doppelte Masse der Einzelkörper haben; dies ist offenbar der elementare Schritt der Konstruktion einer metrischen Variablen durch Iteration. LaVoisiers Entdeckung der Erhaltung der Masse bei chemischen Umsetzungen besagt, daß diese Massen-Variable auch bei Beschreibung chemischer Umsetzungen als metrische Variable benutzt werden kann. Betrachtet man dagegen Vorgänge, bei denen die Masse keinem Erhaltungssatz genügt, so muß man zu ihrer Beschreibung andere metrische Variablen benutzen, zum Beispiel bei Kern-Umsetzungen die Anzahl der Nukleonen und die Kernladung[1].

Beim Begriff der Zusammensetzung zweier oder mehrerer Systeme handelt es sich zumeist weniger um eine „aktive" physikalische Operation, die mit den Systemen geschieht, als um eine Ausdehnung der jeweils betrachteten Isolation auf beide Systeme zusammen und ihre gleichzeitige Beobachtung. In vielen Fällen trifft die Vorstellung des Nebeneinander-Legens von Systemen und des

---

[1] Die absolute Temperatur ist in unserer Terminologie *keine* metrische Variable, denn ihre Skala kommt nicht zustande durch eine Wechselwirkung, sondern durch eine funktionale Abhängigkeit zwischen den metrischen Größen S und U bei Prozessen $X_k =$ const.

Einschaltens einer Wechselwirkung über eine geeignete „Wand" recht gut das Gemeinte. Die Mischung zweier Gase oder zweier beliebiger Substanzen ist im allgemeinen *keine* System-Zusammensetzung, sondern die Erzeugung eines neuen Systems. In gewohnter physikalischer Ausdrucksweise kann man sagen: Zur Bildung des aus zwei Systemen zusammengesetzten Systems ist es notwendig, daß die Wechselwirkungs-Energie gegenüber der Energie der Einzel-Systeme vernachlässigt werden kann.

Die Rolle der adiabatischen Isolation bei der Konstruktion der Entropie zu erläutern, war der Inhalt des nächsten Kapitels. Die Besonderheit dieser System-Einwirkung besteht darin, daß sie in ihrer einfachsten Form vom Charakter einer vollständigen Übergangs-Relation ist, die unmittelbar eine empirische (nicht-metrische) Variable liefert. Die adiabatisch-reversiblen Übergänge andererseits geben die Möglichkeit, die Klassen gleicher Entropie in metrischer Weise anzuordnen, und es zeigt sich, daß diese Anordnung mit der durch die empirische Entropie gegebenen Anordnung identisch ist.

Die metrischen Variablen sind ihrer Definition nach bis auf lineare Transformationen, das heißt bis auf einen Maßstabs-Faktor und eine additive Konstante festgelegt. Der Maßstabsfaktor kann im allgemeinen positive wie negative Werte haben. (Eine Ausnahme bildet die Entropie, deren positive Richtung durch die adiabatisch-irreversiblen Prozesse in invarianter Weise festgelegt ist.) Daraus ergibt sich das Prinzip: Jede metrische Variable X, die beidseitig unbeschränkt oder beidseitig beschränkt ist, geht so in die Theorie ein, daß diese invariant ist gegenüber der Transformation $X \to -X$ bzw. $X \to -X + \mathrm{const}$, wobei die Konstante so gewählt ist, daß die obere Grenze in die untere Grenze übergeht und umgekehrt. Eine einseitige Beschränktheit der metrischen Variablen dagegen gibt die Möglichkeit einer invarianten Richtungs-Auszeichnung. Energie sowohl als Entropie eines jeden Systems sind nach unten beschränkt. Diese Aussage wird schließlich ergänzt durch ein neues fundamentales Prinzip: *Wenn die Energie eines Systems ihren kleinsten Wert annimmt, so nimmt gleichzeitig auch die Entropie ihren kleinsten Wert an.*

Bislang sind die Betrachtungen allgemein physikalischer Natur und keineswegs charakteristisch für die Thermodynamik bzw. die thermodynamischen Variablen. Letztere werden unter den metrischen Variablen ausgewählt durch eine neue Operation, die System-Reduktion, deren Eigenschaften von der Herstellung des thermischen Gleichgewichtes oder auch der Druck- oder chemischen Gleichgewichte bekannt sind. Wichtig ist, daß bei dieser Auswahl der thermodynamischen unter den metrischen Variablen Entropie und Energie benutzt werden. Äquivalent mit der Auswahl von thermodynamischen (metrischen) Koordinaten ist die Konstruktion von Kontakt- oder „intensiven" Variablen. Als besonders zweckmäßig zur Kennzeichnung der thermodynamischen Koordinaten erweist sich nach dem Vorgang von Gibbs die Verwendung der Minimal- und Maximal-Eigenschaften von Energie und Entropie.

Unter allen Systemen sind die thermischen die wichtigsten. Sie sind dadurch ausgezeichnet, daß sie die Energie als thermodynamische Koordinate besitzen, die zugehörige Kontakt-Variable ist die Temperatur. Jedes thermodynamische System läßt sich durch Herstellung innerer Gleichgewichte auf ein thermisches reduzieren. Dies liefert eine Einschränkung der thermodynamischen Systeme, was auch als Vorzugsstellung der Temperatur unter allen Kontakt-Variablen beschrieben werden kann. Für Systeme mit zweiseitig beschränkter Energie nimmt die reziproke Temperatur sowohl positive wie negative Werte an.

Eine Theorie in axiomatische Form bringen heißt, unter ihren Aussagen und Sätzen eine Auswahl zu treffen derart, daß sich die ganze Theorie aus dieser

Satz-Auswahl, genannt ein Axiomen-System der Theorie, deduktiv ableiten läßt. Jede Auswahl von Sätzen, die dieser Bedingung genügt, ist als Axiomen-System brauchbar, und keine derartige Auswahl ist im Prinzip vor irgendeiner anderen bevorzugt. Damit ist die Auswahl eines Axiomen-Systems einer Theorie letztlich eine Frage der Zweckmäßigkeit oder gar des persönlichen Geschmackes, eine Feststellung, die den Wert axiomatischer Betrachtungen auf den ersten Blick etwas fragwürdig erscheinen läßt. Dieser Eindruck wird gemildert durch die Tatsache, daß es oftmals außerordentlich schwierig ist, die logischen Abhängigkeiten von Aussagen und damit die Grundlagen einer Theorie überhaupt zu durchschauen, so daß man fürs erste mit irgendeinem Axiomen-System zufrieden sein wird. Weiterhin wird man in physikalischen Theorien solche Axiomen-Systeme bevorzugen, deren Axiome einen gewissen Grad physikalischer Evidenz besitzen. Vor allem aber hat, wie wir in der Einleitung sagten, die Axiomatik den Zweck, Erweiterungs-Möglichkeiten einer Theorie einer systematischen Behandlung zugänglich zu machen. Von dieser Möglichkeit haben wir hier Gebrauch gemacht, denn die von uns axiomatisch charakterisierte Thermodynamik stellt eine Erweiterung der klassischen bzw. Carathéodoryschen dar. Dies läßt sich leicht folgendermaßen einsehen: Die klassische wie auch die Carathéodorysche Thermodynamik handelt, wie die Formulierung bereits einzelner Axiome zeigt, nur von thermischen Systemen, das heißt von solchen mit Temperatur, oder von Systemen, die aus thermischen zusammengesetzt sind. Die Entropie ist demgemäß auch nur für solche Systeme erklärt, während unser Aufbau die Konstruktion von Entropie und Energie deutlich von der Charakterisierung der eigentlich thermodynamischen Variablen trennt. Sogar bei letzteren ist es nicht einmal nötig, sich auf thermische oder aus thermischen zusammengesetzte Systeme zu beschränken, und selbst das für die Struktur der Theorie unmaßgebliche Axiom 7 besagt weniger als von den Systemen der klassischen Thermodynamik verlangt wird.

Wenn also auch klassische und Carathéodorysche Thermodynamik äquivalent sind und die Bevorzugung der einen oder der anderen, wenn man so will, eine Frage des Geschmackes ist, gilt dies nicht mehr für eine Stellungnahme zwischen der klassischen bzw. Carathéodoryschen Thermodynamik und der hier gegebenen, da letztere eine umfassendere Theorie darstellt. Es ist kaum zu bezweifeln, daß diese Erweiterung ein natürlicher und eigentlich längst fälliger Schritt ist. Dies wird besonders deutlich durch einen Hinweis auf die statistische Thermodynamik, die sich in unseren Aufbau ebenso einfügt wie die phänomenologische. Es ist ein befriedigender Zug, daß damit das alte Nebeneinander von statistischer und phänomenologischer Thermodynamik auf zwanglose Weise beendet werden kann.

Anhang: Zu CARATHÉODORYS „Untersuchungen über die Grundlagen der Thermodynamik"[1].

Die vielseitige Beachtung sowie die kritischen und kommentierenden Diskussionen, die CARATHÉODORYs „Untersuchungen" in der physikalischen Literatur ausgelöst haben[2], nehmen wir zum Anlaß, den Carathéodoryschen Aufbau der

---

[1] Math. Ann. **67**, 355 (1909).

[2] M. BORN: Phys. Z. **22**, 218, 249, 282 (1921). — T. EHRENFEST-AFANASSJEWA: Z. Physik **33**, 933 (1925). — M. PLANCK: Sitzgsber. preuß. Akad. Wiss., phys.-math. Kl. **1926**, 453. — A. LANDÉ: Handbuch der Physik, Bd. IX, S. 281. Berlin 1926. — S. CHANDRASEKHAR: An introduction to the study of stellar structure. Chicago 1939. — M. BORN: Natural philosophy of cause and chance. Oxford 1949. — H. A. BUCHDAHL: Amer. J. Phys. **17**, 41, 44, 212 (1949); **22**, 182 (1954); **23**, 65 (1955). — Z. Physik **152**, 425 (1958). — P. T. LANDSBERG: Rev. Mod. Phys. **28**, 363 (1956).

Thermodynamik unter Verwendung unserer Terminologie darzustellen, so daß ein Vergleich mit der vorliegenden Axiomatik erleichtert wird.

In der Einleitung seiner Abhandlung betont Carathéodory, daß „die ganze Theorie der Thermodynamik abgeleitet werden kann, ohne die Existenz einer von den gewöhnlichen mechanischen Größen abweichenden physikalischen Größe, der Wärme, vorauszusetzen". Dieses in allen Einzelheiten klar darzulegen, bezeichnet er als das Ziel seiner Untersuchungen. Dabei spielen die Begriffe *quasistatische Arbeit* und *Pfaffsche Form* eine fundamentale Rolle, so daß es gerechtfertigt erscheint, zunächst die formale Seite dieser Begriffe ausführlich zu erläutern (Ziff. A 1, A 2) und dann erst die eigentlich thermodynamischen Überlegungen anzustellen (Ziff. A 3).

A 1. Lineare Prozeß-Belegungen. In der Carathéodoryschen Theorie wird vorausgesetzt, daß die Zustandsmenge eines thermodynamischen Systems ein mehrdimensionales Kontinuum $\mathfrak{Z}_n = \{z_1, \ldots, z_n\}$ bildet, in dem zunächst kein Koordinatensystem vor einem anderen ausgezeichnet ist. Infolgedessen ist die Topologie des Zustandsraumes als ein *Grundbegriff*[1] der Theorie anzusehen. Diese Ausgangsposition bringt mit sich, daß jede Variable[2] von vornherein als stetige bzw. differenzierbare Zustands-Funktion auftritt. Charakteristisch für die Theorie Carathéodorys ist nun die Art und Weise, in der die Entropie-Funktion $S(z)$ und die Temperatur-Funktion $T(z)$ konstruiert werden. Wesentliches Hilfsmittel zu dieser Konstruktion ist die einem System „quasistatisch zugeführte Arbeit" — eine physikalische Größe, welche für jedes orientierte Kurvenstück des Zustandsraumes einen bestimmten Wert besitzt, nicht aber für einzelne Zustände. Eine solche *Zuordnung von Zahlen zu Kurven* wollen wir allgemein *Prozeß-Belegung* nennen und in der Form $\omega = \omega(\mathfrak{C})$ schreiben, um dadurch anzudeuten, daß die Zahlen ω eine Funktion auf der Menge der Kurven $\mathfrak{C}$ repräsentieren. Zur Bezeichnung einer Kurve $\mathfrak{C}$ verwenden wir auch die Schreibweise $\left(z(t), t_a, t_e\right)$, wobei t ein beliebiger Kurven-Parameter ist und t_a, t_e die Parameter-Werte am Anfang bzw. am Ende von $\mathfrak{C}$ bedeuten. Die Prozeß-Belegung der quasistatischen Arbeit besitzt nun einige charakteristische Züge, welche Anlaß geben zu der

Definition: Eine Prozeß-Belegung $\omega(\mathfrak{C})$ soll *linear* heißen, wenn sie folgende Eigenschaften besitzt:

1. Ist einer Kurve $\mathfrak{C}$ die Zahl ω zugeordnet, dann gehört zu der entgegengesetzt orientierten Kurve die Zahl $-\omega$.

2. Setzt sich $\mathfrak{C}$ aus zwei Teilstücken $\mathfrak{C}', \mathfrak{C}''$ so zusammen, daß der Endpunkt von $\mathfrak{C}'$ auf den Anfangspunkt von $\mathfrak{C}''$ fällt, dann gilt für die entsprechenden Zahlen: $\omega = \omega' + \omega''$.

3. ω hängt stetig differenzierbar vom Kurven-Parameter ab.

4. Bei Approximation einer Kurve $\mathfrak{C}$ durch *beliebige* Polygone erhält man auch eine Approximation des ω-Wertes von $\mathfrak{C}$.

Es ist einfach, physikalische Beispiele für lineare Prozeß-Belegungen zu finden; oben wurde bereits die „quasistatisch zugeführte Arbeit" genannt. Die „quasistatisch zugeführte Wärme", die bei Carathéodory, im Gegensatz zur klassischen Thermodynamik, als Grundbegriff vermieden wird, gehört natürlich auch hierher. Weiterhin liefert jede Variable eine lineare Prozeß-Belegung, wenn man jeder

---

[1] Denn hier wird das n-dimensionale Kontinuum nicht durch den Werte-Bereich von n ausgezeichneten Variablen erzeugt, sondern als selbständiger Begriff eingeführt. Es handelt sich also um eine „primäre" Verwendung der Stetigkeit im Sinne unserer Erläuterungen auf S. 120.

[2] Siehe Ziff. 2 und 3.

Kurve $\mathfrak{C}$ die Differenz der Variablen-Werte an den beiden Endpunkten von $\mathfrak{C}$ zuordnet. Es ist nicht ohne Interesse, auf die Rolle der Forderung 4 hinzuweisen. Sie wird zum Beispiel nicht erfüllt von der Prozeß-Belegung „euklidische Länge" L einer Kurve $\mathfrak{C}$, wie das Beispiel in Fig. 3 zeigt. Die approximierenden Polygone haben stets die Länge 2, die approximierte Grundlinie aber die Länge 1. Die Approximation der Bogenlänge $L(\mathfrak{C})$ einer Kurve kann eben nicht durch beliebige Polygone, sondern nur durch *Sehnen*-Polygone erfolgen.

Das adäquate mathematische Beschreibungsmittel für lineare Prozeß-Belegungen sind *Pfaffsche Formen*. Es gilt nämlich der

Satz: Jede lineare Prozeß-Belegung besitzt eine erzeugende Pfaffsche Form, und jede Pfaffsche Form definiert eine lineare Prozeß-Belegung.

Wir betrachten diesen Satz als Motivierung einer mathematischen Einschaltung über Pfaffsche Formen, in deren Verlauf auch der Beweis des Satzes gegeben wird.

A 2. Pfaffsche Formen[1].

In einem n-dimensionalen Punktraum $\mathfrak{Z}_n = \{z_1, \ldots z_n\}$ sind Pfaffsche Formen definiert als linear-homogene Differential-Ausdrücke

$$\delta\omega = Z_1\,dz_1 + \left.\begin{array}{l}\\ +Z_2\,dz_2 + \cdots Z_n\,dz_n\end{array}\right\} \text{(A 2.1)}$$

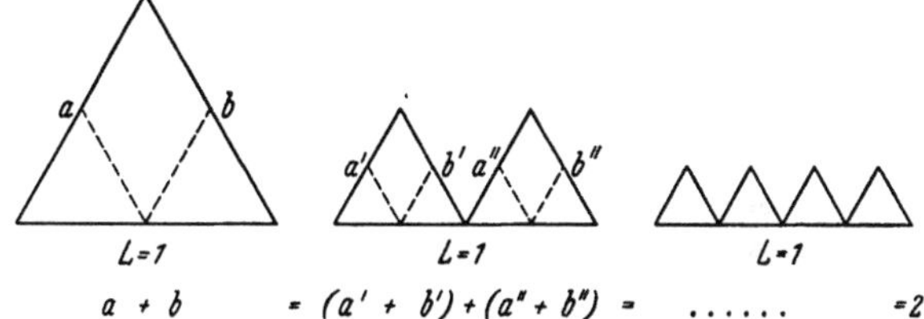

Fig. 3. Die Länge einer Kurve, aufgefaßt als Prozeß-Belegung, erfüllt nicht die Forderungen 1 und 4 der linearen Prozeß-Belegungen. Bei Approximation der Grundlinie des Dreiecks durch die Polygone, die durch fortgesetzte Halbierung aus den beiden anderen Seiten hervorgehen, wird die Länge der Grundlinie offenbar nicht approximiert.

mit *kovarianten Koeffizienten* $Z_i(z)$. Werden also neue Koordinaten z_i' eingeführt, so gilt

$$\delta\omega = \sum Z_i\,dz_i = \sum Z_i'\,dz_i', \quad \text{mit} \quad Z_i' = \sum_j Z_j\frac{\partial z_j}{\partial z_i'}.$$

Jede Pfaffsche Form (A 2.1) definiert demnach ein kovariantes Vektorfeld $\mathbf{Z}(z) = \big(Z_1(z), \ldots, Z_n(z)\big)$, und jedes kovariante Vektorfeld definiert umgekehrt eine Pfaffsche Form mit den Vektor-Komponenten als Koeffizienten. Im Sinn dieser Zuordnung bedeuten Pfaffsche Formen und kovariante Vektorfelder denselben Gegenstand.

Der infinitesimal-skalare Charakter Pfaffscher Formen äußert sich im Endlichen dadurch, daß ihre Linien-Integrale auf orientierten Kurvenstücken $\mathfrak{C} = \big(z(t); t_a, t_e\big)$ koordinaten- und parameter-invariante Zahlen erzeugen:

$$\omega(\mathfrak{C}) = \int_{\mathfrak{C}} \delta\omega = \int_{\mathfrak{C}} \sum Z_i\,dz_i = \int_{t_a}^{t_e} \sum Z_i\,\dot{z}_i(t)\,dt. \tag{A 2.2}$$

Wir betrachten nun ein stetig differenzierbares Vektorfeld $\mathbf{Z}(z)$, das höchstens auf $(n-1)$-dimensionalen Flächen den Wert Null annimmt. Dann legt $\mathbf{Z}(z)$ für jeden Punkt z eindeutig eine (kovariante) Richtung fest, und die *Pfaffsche Gleichung*

$$Z_1\,dz_1 + Z_2\,dz_2 + \cdots Z_n\,dz_n = 0 \tag{A 2.3}$$

charakterisiert für jeden Punkt z das $(n-1)$-dimensionale ebene Büschel der (kontravarianten) Linienelemente $dz = (dz_1, dz_2, \ldots, dz_n)$, die orthogonal zu $\mathbf{Z}(z)$ liegen. Gl. (A 2.3) ist demnach gleichbedeutend mit der Zuordnung einer Hyper-

---

[1] Der mit dem Gegenstand vertraute Leser möge diese Ziffer überschlagen.

ebene $\mathfrak{H}$ zu jedem Punkt z, oder wie wir sagen können, mit einem „Feld" von Hyperebenen $\mathfrak{H}(z)$. Raumkurven $\mathfrak{C}$, die in jedem ihrer Punkte z die Hyperebene $\mathfrak{H}(z)$ tangieren (also Orthogonal-Trajektorien von $\boldsymbol{Z}$) heißen *Integral-Kurven* von (A 2.3).

Im dreidimensionalen euklidischen Raum macht man sich nun leicht klar, daß jedes Vektorfeld $\boldsymbol{V} = (V_1, V_2, V_3)$ zwar Orthogonal-Trajektorien besitzt, aber nur unter besonderen Umständen auch Orthogonal-Flächen. Wenn nämlich eine solche Flächenschar $F(z) = $ const existiert, dann haben $\boldsymbol{V}$ und $\operatorname{grad} F$ dieselbe Richtung, so daß an jedem Punkt z die Darstellung

$$\boldsymbol{V} = N(z)\,\operatorname{grad} F(z)$$

möglich ist, wobei der Faktor $N(z)$ die unterschiedliche Länge von $\boldsymbol{V}$ und $\operatorname{grad} F$ ausgleicht. Daraus folgt aber

$$\boldsymbol{V} \cdot \operatorname{rot} \boldsymbol{V} = (N\operatorname{grad} F) \cdot \operatorname{rot}(N\operatorname{grad} F) = (N\operatorname{grad} F) \cdot (\operatorname{grad} N \times \operatorname{grad} F) = 0,$$

was offensichtlich nicht für jedes Vektorfeld $\boldsymbol{V}$ zutrifft.

Es sei jetzt $\boldsymbol{V} = (V_1, V_2, V_3)$ ein Vektorfeld mit Orthogonal-Flächen $F(z) = $ const im dreidimensionalen euklidischen Raum. Die zugehörige Pfaffsche Gleichung lautet dann

$$V_1\,dz_1 + V_2\,dz_2 + V_3\,dz_3 = 0$$

bzw.

$$N\left(\frac{\partial F}{\partial z_1}\,dz_1 + \frac{\partial F}{\partial z_2}\,dz_2 + \frac{\partial F}{\partial z_3}\,dz_3\right) = 0.$$

Transformation auf neue Koordinaten z_1', z_2', z_3' mit

$$z_1'(z_1, z_2, z_3) = F(z_1, z_2, z_3)$$

ergibt

$$N\,dF = 0.$$

Die linke Seite dieser Gleichung kann durch $N(z)$ dividiert werden, ohne daß sich die zu der Gleichung gehörigen Integral-Kurven ändern. Auf diese Weise wird dem Vektorfeld $\boldsymbol{V} = N \cdot \operatorname{grad} F$ das Vektorfeld $\boldsymbol{V}/N = \operatorname{grad} F$ zugeordnet, das hinsichtlich der Pfaffschen Gleichung mit ihm äquivalent ist: Alle Integrale $\mathfrak{C}$ liegen auf den Flächen $F = $ const, und alle Kurven auf diesen Flächen sind Integrale. Die Funktionen $N(z)$ bzw. $M(z) = 1/N(z)$ heißen *integrierender Nenner* bzw. *Multiplikator* des „integrablen" Vektorfeldes $\boldsymbol{V} = N\operatorname{grad} F$, weil die Pfaffsche Form

$$V_1\,dz_1 + V_2\,dz_2 + V_3\,dz_3$$

bei Division durch N in das vollständige Differential

$$dF = (\partial F/\partial z_1)\,dz_1 + (\partial F/\partial z_2)\,dz_2 + (\partial F/\partial z_3)\,dz_3$$

übergeht. Ganz analog liegen die Verhältnisse im n-dimensionalen, nicht-metrischen Raum $\mathfrak{Z}_n$. Besitzt das kovariante Vektorfeld $\boldsymbol{Z}(z)$ eine Orthogonal-Flächenschar, so läßt sich Gl. (A 2.3) in die Form

$$N\left(\frac{\partial F}{\partial z_1}\,dz_1 + \frac{\partial F}{\partial z_2}\,dz_2 + \cdots + \frac{\partial F}{\partial z_n}\,dz_n\right) = 0$$

oder

$$N\,dF = 0$$

bringen. N repräsentiert eine dem Vektorfeld zugeordnete Größe, die in dem gewählten Koordinatensystem $z_1 = F$, $z_2, \ldots, z_n$ mit der einzigen von Null verschiedenen Komponente von $\boldsymbol{Z}$ übereinstimmt: $Z_1 = N$. Beschreibt man die

Orthogonal-Flächenschar $F = $ const durch einen neuen Parameter $F' = F'(F)$, so geht N als Koeffizient der Pfaffschen Form $N\, dF$ über in

$$N' = N \cdot \frac{\partial F}{\partial F'} = N \cdot f(F). \qquad (A\,2.4)$$

Umgekehrt entspricht einem neuen $N' = N \cdot f(F)$ die Transformation

$$F' = \int\limits_0^F \frac{dF}{f(F)} + F_0'.$$

Es soll nun der in Ziff. A1 angegebene Satz bewiesen werden und zwar zunächst der zweite Teil seiner Behauptung. Wir geben in $\mathfrak{R}_n$ ein beliebiges Koordinatensystem vor und konstruieren die Approximations-Polygone für irgendein Kurvenstück $\mathfrak{C}$ aus Koordinaten-Linien. P_1, P_2, ... sei eine Folge aus derartigen approximierenden Polygonen für die Kurve $\mathfrak{C}$. Zu jedem der Polygone P_i denken wir uns eine Fläche gegeben, die P_i und $\mathfrak{C}$ enthält (Fig. 4). P_i und $\mathfrak{C}$ schließen dann ein Flächenstück F_i dieser Fläche ein (in Fig. 4 schraffiert gezeichnet). Der Folge P_1, P_2, ... ist somit eine gegen Null strebende Folge F_1, F_2, ... zugeordnet. Wählt man nun für die P_i denselben Durchlaufungs-Sinn wie für $\mathfrak{C}$, so gilt für jedes Vektorfeld $\mathbf{Z} = (z_1, \ldots, z_n)$

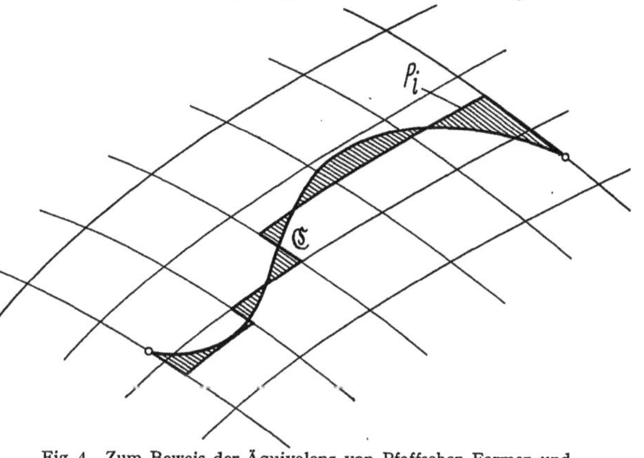

Fig. 4. Zum Beweis der Äquivalenz von Pfaffschen Formen und linearen Prozeß-Belegungen.

$$\int\limits_{\mathfrak{C}} \sum Z_k\, dz_k - \int\limits_{P_i} \sum Z_k\, dz_k = \oint\limits_{(F_i)} \sum Z_k\, dz_k.$$

Nach dem allgemeinen Stokesschen Satz für kovariante Vektorfelder ist aber

$$\oint\limits_{(F_i)} \sum Z_k\, dz_k = \sum_{k,\,l} \int\limits_{F_i} \left(\frac{\partial Z_k}{\partial z_l} - \frac{\partial Z_l}{\partial z_k} \right) dz_l\, dz_k,$$

und da das Integral der rechten Seite für eine nach Null strebende Folge F_i verschwindet, ergibt sich schließlich

$$\int\limits_{\mathfrak{C}} \sum Z_k\, dz_k = \lim_{i \to \infty} \int\limits_{P_i} \sum Z_k\, dz_k$$

unabhängig von der approximierenden Polygon-Folge. Daraus folgt, daß das Linien-Integral

$$\omega(\mathfrak{C}) = \int\limits_{\mathfrak{C}} \sum Z_k\, dz_k$$

jeder Pfaffschen Form eine lineare Prozeß-Belegung ist. Hat man umgekehrt eine lineare Prozeß-Belegung ω, so liefern die partiellen Ableitungen $\partial\omega/\partial z_i$ (in Richtung der benutzten Koordinaten-Linien) die Koeffizienten einer Pfaffschen Form, die durch Integration wieder ω erzeugt.

Jede Hyperflächen-Schar $F(z) = $ const in einem Raum $\mathfrak{Z}_n = \{z\}$ bedeutet in unserer Terminologie[1] eine geordnete Klassen-Einteilung $\{\mathfrak{F}\}$ in $\mathfrak{Z}_n$ und damit eine (nicht-metrische) *Variable*. Jede Funktion $F(z)$, $F'(z) \ldots$, welche die Flächen-Schar $\{\mathfrak{F}\}$ monoton und stetig differenzierbar mit Zahlen belegt, repräsentiert eine zulässige *Darstellung* der Variablen. Ist also in $\mathfrak{Z}_n$ eine lineare Prozeß-Belegung $\omega(\mathfrak{C})$ bzw. eine Pfaffsche Form $\delta\omega = Z_1\, dz_1 + \cdots Z_n\, dz_n$ vorgelegt, *so repräsentiert $\delta\omega$, wenn es integrabel ist, durch seine Orthogonalflächen-Schar eine Variable $\{\mathfrak{F}\}$*, und es muß auf rein rechnerischem Wege möglich sein, aus den Koeffizienten-Funktionen $Z_1(z), \ldots, Z_n(z)$ Darstellungen $F(z)$ dieser Variablen $\{\mathfrak{F}\}$ zu konstruieren. Jede solche Darstellung $F(z)$, $F'(z), \ldots$ legt natürlich auch ein-deutig einen integrierenden Nenner $N(z)$, $N'(z), \ldots$ fest, aber diese Funktionen sind im allgemeinen nicht verschiedene Darstellungen ein und derselben Flächen-Schar, wie aus dem Transformations-Gesetz (A 2.4) hervorgeht. Wenn allerdings eine Darstellung $F(z)$ durch anderweitige Bedingungen bis auf eine additive Konstante eindeutig ausgezeichnet ist (metrische Variable), dann definieren auch die zugehörigen $N(z)$ eindeutig eine Variable $\{\mathfrak{N}\}$, da $N'(z) = N(z)\,f(F) = N(z)\,dF/dF' = N(z)$.

Im Hinblick auf die Thermodynamik ist es erforderlich, den erläuterten Zusammenhang zwischen einer integrablen Pfaffschen Form $\delta\omega$ und der zuge-hörigen Variablen $\{\mathfrak{F}\}$ noch in einer etwas anderen Richtung zu verfolgen. Wir betrachten dazu einen $(n+1)$-dimensionalen Raum $\mathfrak{Z}_{n+1} = \{z_1, \ldots, z_{n+1}\}$, in dem n voneinander unabhängige[2] Variablen $x_1, \ldots, x_n$ ausgezeichnet sind. $\mathfrak{Z}_{n+1}$ läßt sich also durch einen Koordinaten-Raum $\{\xi, x_1, \ldots, x_n\}$ darstellen, wo ξ eine beliebige von $x_1, \ldots, x_n$ unabhängige Variable ist. Jedes Werte-n-tupel $(x_1, \ldots, x_n)$ $= x$ legt dann eine eindimensionale Punkt-Mannigfaltigkeit fest, die wir kurz als *x-Linie* bezeichnen, und der ganze Raum $\mathfrak{Z}_{n+1}$ zerfällt in die n-parametrige Schar dieser Linien. Weiterhin repräsentiert die ξ-Variable eine Schar von Hyperflächen, welche die Schar der x-Linien überall affin-orthogonal durch-setzt, das heißt alle Punkte P von $\mathfrak{Z}_{n+1}$ sind umkehrbar eindeutig gekennzeichnet als Schnittpunkt aus je einer x-Linie und einer ξ-Fläche. Ist dann $\delta\omega = \Xi\, d\xi + X_1\, dx_1 + \cdots X_n\, dx_n$ eine integrable Pfaffsche Form mit der Besonderheit, daß die zugehörige Variable $\{\mathfrak{F}\}$ als ξ-Variable brauchbar ist, so muß offenbar $\Xi \neq 0$ sein, weil nämlich der Vektor $(\Xi, X_1, \ldots, X_n)$ in jedem Punkt des Raumes $\mathfrak{Z}_{n+1}$ proportional ist zum Vektor $\operatorname{grad} F = (\partial F/\partial \xi, \partial F/\partial x_1, \ldots, \partial F/\partial x_n)$ und $\partial F/\partial \xi \neq 0$ vorausgesetzt wurde[3].

Es ist klar, daß die Integral-Kurven einer integrablen Pfaffschen Form es nicht gestatten, von einem Punkt P ausgehend zu jedem Punkt P' einer vollen $(n+1)$-dimensionalen Umgebung von P zu gelangen. Wie wir gesehen haben, bedeutet nämlich die Integrabilität einer Pfaffschen Form die Existenz einer Schar von Hyperflächen derart, daß Integralkurven jeweils vollständig innerhalb einer der Hyperflächen verlaufen. Bemerkenswert ist dieser Tatbestand deshalb, weil auch die Umkehrung richtig ist:

Satz von Carathéodory: Ist eine Pfaffsche Gleichung $d\xi + X_1\, dx_1 + \cdots X_n\, dx_n = 0$ gegeben, wobei die X_i endliche, stetige, differenzierbare Funktionen der x_i sind, und weiß man, daß es in jeder Umgebung eines beliebigen Punktes P des Raumes der (ξ, x_i) Punkte gibt, die man längs Kurven, welche dieser Gleichung genügen,

---

[1] Siehe Ziff. 2 und 3.

[2] Zwei Variablen $f(z)$, $g(z)$ heißen genau dann „unabhängig", wenn ihre Flächen-Scharen $\{\mathfrak{f}\}$, $\{\mathfrak{g}\}$ sich überall durchdringen und sich nicht etwa stückweise oder vollständig überdecken.

[3] Wenn die Flächenschar $\{\mathfrak{F}\}$ alle x-Linien affin-orthogonal durchsetzt, dann heißt das in analytischer Ausdrucksweise, daß jede Darstellung $F(z)$ von $\{\mathfrak{F}\}$ längs der x-Linien nicht-verschwindende Ableitungen besitzt.

nicht erreichen kann, so muß die Pfaffsche Form notwendig einen Multiplikator besitzen, der sie zu einem vollständigen Differential $dF = M \cdot (d\xi + X_1\, dx_1 + \cdots X_n\, dx_n)$ macht, und es gilt dabei $M \neq 0$, $M \neq \infty$.

Der Beweis läßt sich folgendermaßen skizzieren. Man betrachtet allgemeine z-dimensionale Zylinder-Flächen mit den x-Linien als Erzeugenden. Diese Zylinder-Flächen werden in jedem Punkt von dem durch die Pfaffsche Gleichung definierten Hyperflächen-Element in einem Linien-Element geschnitten, da in der Pfaffschen Gleichung der Koeffizient des Differentials $d\xi$ von Null verschieden vorausgesetzt ist. Vorgabe eines Punktes P und einer Zylinder-Fläche bestimmt somit eindeutig eine Integral-Kurve, und umgekehrt legt eine durch P verlaufende Integral-Kurve eine Zylinder-Fläche fest. Betrachtet man nun geschlossene Zylinder-Flächen durch P, dann gibt es zu jeder Pfaffschen Form, die in einer ganzen Umgebung von P nicht integrabel ist, mindestens eine derartige Fläche, deren von P ausgehende Integral-Kurve nicht in sich zurückläuft, sondern die durch P bestimmte x-Linie in einem Punkt $P' \neq P$ schneidet. Stetige Zusammenziehung des Zylinders auf diese x-Linie zeigt, daß das ganze Intervall (P, P') aus untereinander verbindbaren Punkten besteht. Bei Variation von P in einer geeigneten Umgebung überstreichen diese Intervalle ein ganzes Raumstück.

Die hier gegebene Skizze des Begriffes einer Pfaffschen Form enthält alles, was in der Thermodynamik davon Verwendung findet. Die eigentliche Theorie der Pfaffschen Formen, die inhaltlich im wesentlichen mit der Lieschen Theorie der Differential-Gleichungen übereinstimmt, ist hier nicht berührt worden; sie findet indessen in der Thermodynamik auch keine Anwendung.

A 3. CARATHÉODORYS Konstruktion von Entropie und Temperatur. $\alpha)$ *Das Unerreichbarkeits-Axiom.* Für jedes Zustandspaar (z, z') eines Systems $\mathfrak{Z}_n$ ist entscheidbar, ob der Übergang $z \to z'$ unter adiabatischer Isolation realisierbar ist oder nicht, mit anderen Worten, es wird jedem System $\mathfrak{Z}_n$ eine adiabatische Übergangs-Relation[1] $F(z, z') = \pm 1$ zugeordnet und somit auch eine Zerlegung von $\mathfrak{Z}_n$ in maximale adiabatisch-reversible Klassen $\mathfrak{S}$. Es bleibt jedoch offen, ob die Relation F vollständig ist oder wenigstens verzweigungslos. Stattdessen verlangt CARATHÉODORY durch sein *Unerreichbarkeits-Axiom*, daß die adiabatische Übergangs-Relation in jedem thermodynamischen Zustandsraum $\mathfrak{Z}_n$ folgende topologische Eigenschaft besitzt: *Es gibt für keinen Zustand z eine Umgebung $\mathfrak{U}$ derart, daß alle Übergänge $z \to z'$ mit $z' \in \mathfrak{U}$ adiabatisch realisierbar sind.* Daraus folgt sofort

Satz 1: Es gibt keine adiabatisch reversible Zustands-Klasse $\mathfrak{S}$, die einen n-dimensionalen Teilbereich eines thermodynamischen Zustandsraumes $\mathfrak{Z}_n$ vollständig bedeckt.

Der Beweis ist trivial: Wenn es eine Entropie-Klasse $\mathfrak{S}$ gäbe, die ein Raumstück des $\mathfrak{Z}_n$ bedeckt, dann würde sich innerhalb eines solchen Raumbereiches für jeden Zustand z eine Umgebung $\mathfrak{U}$ angeben lassen, die dem Unerreichbarkeits-Axiom widerspricht.

Alle derartigen topologischen Überlegungen zielen offensichtlich auf folgenden Sachverhalt: Wenn $\mathfrak{Z}_n$ ein „vernünftiges" thermodynamisches System ist mit den beliebigen Koordinaten $z_1, \ldots, z_n$, dann zerlegen die adiabatisch-reversiblen Zustands-Veränderungen den Zustandsraum $\mathfrak{Z}_n$ in eine Hyperflächenschar $\sigma(z_1, \ldots, z_n) = $ const. derart, daß jeder adiabatisch-reversible Übergang $z \to z'$

---

[1] Siehe Ziff. 2.

innerhalb einer Fläche $\sigma = \mathrm{const.}$ verläuft, während die adiabatisch-irreversiblen Zustands-Veränderungen stets zwei σ-Flächen miteinander verbinden. Hinsichtlich der letzten Bemerkung beachte man nun, daß jedes Mitglied $\mathfrak{S}$ der Hyperflächenschar den Zustandsraum $\mathfrak{Z}_n$ in zwei Halbräume trennt. Innerhalb der Hyperfläche ($d\sigma = 0$) ist jeder Punkt (Zustand) z mit jedem anderen adiabatisch, genauer sogar adiabatisch-reversibel verbindbar. Es sei jetzt z ein in $\mathfrak{S}$ gegebener Zustand und z' von z aus adiabatisch nicht erreichbar. Der Zustand z' liegt also in einer Hyperfläche $\mathfrak{S}'$, die ganz in einem der beiden durch $\mathfrak{S}$ definierten Halbräume liegt. Da nun z' von z aus adiabatisch nicht erreichbar ist, gilt dasselbe für jedes Zustandspaar von $\mathfrak{S}$ und $\mathfrak{S}'$, und somit gibt es keinen adiabatischen Prozeß, der von $\mathfrak{S}$ nach $\mathfrak{S}'$ oder gar noch über $\mathfrak{S}'$ hinaus führt. Der ganze Teil des Zustandsraumes, der auf der von $\mathfrak{S}$ abgewandten Seite von $\mathfrak{S}'$ liegt, ist demnach von $\mathfrak{S}$ adiabatisch nicht erreichbar. Da es nun in *jeder* Umgebung von z adiabatisch nicht-erreichbare Zustände z' und damit nicht erreichbare Hyperflächen $\mathfrak{S}'$ gibt, bleibt nur folgende Alternative: Entweder gibt es auf *beiden* Seiten von $\mathfrak{S}$ in beliebiger Nähe adiabatisch nicht erreichbare Hyperflächen $\mathfrak{S}'$, dann kann $\mathfrak{S}$ adiabatisch gar nicht verlassen werden, oder es gibt nur auf *einer* Seite von $\mathfrak{S}$ in beliebiger Nähe adiabatisch nicht erreichbare Hyperflächen $\mathfrak{S}'$, so daß $\mathfrak{S}$ adiabatisch nur nach einer Seite hin verlassen werden kann. Ohne weiter in Einzelheiten zu gehen, erkennt man, daß im zweiten Fall die Hyperflächen-Schar $\{\mathfrak{S}\}$ eine aus der adiabatischen Isolation folgende Anordnung besitzt und somit eine Variable ist. Daher liegt es nahe, das Unerreichbarkeits-Axiom ohne große Bedenken in der soeben geschilderten Weise zu ersetzen durch das schärfere Axiom: Es gibt in jedem thermodynamischen Zustandsraum eine empirische Entropie-Funktion[1]. Carathéodory schlägt aber einen anderen Weg ein, dem wir uns nunmehr zuwenden.

β) Einfache Systeme. Es ist klar, daß man nicht daran vorbei kommt, die Existenz der empirischen Entropie entweder axiomatisch direkt zu fordern oder aber auf Grund anderer Axiome zu beweisen. Diese anderen Axiome sind bei Carathéodory in den „Definitionen" genannten Axiomen enthalten, welche den Begriff des *einfachen Systems* und den Begriff der *Deformations-Koordinaten* festlegen. Der wesentliche Punkt ist dabei, daß den einfachen Systemen eine Pfaffsche Form δQ zugeordnet ist, von deren Lösungs-Kurven man weiß, daß sie dem Unerreichbarkeits-Axiom genügen. Daraus folgt dann die eindeutige Existenz und Konstruierbarkeit einer Hyperflächen-Schar (mit $\delta Q = 0$), deren (monotoner) Scharparameter eine empirische Entropie repräsentiert[2]. Wir werden darüber hinaus zeigen, daß der *Existenz-Nachweis* für die empirische Entropie im Rahmen der Carathéodoryschen Theorie sogar *ohne* Benutzung der Pfaffschen Form δQ geführt werden kann. Carathéodory fordert: Ein thermodynamisches System mit den Koordinaten $(\xi, x_1, \ldots, x_n) = (\xi, x)$ sowie der Energie-Funktion $U(\xi, x_1, \ldots, x_n)$ soll ein *einfaches System* heißen und seine Variablen $x_1, \ldots, x_n$ *Deformations-Koordinaten*, wenn vier Eigenschaften a) bis d), von denen wir zunächst nur die ersten beiden formulieren, erfüllt sind.

---

[1] Ein axiomatischer Aufbau der Thermodynamik unter Benutzung dieses schärferen Axioms anstelle von Carathéodorys Unerreichbarkeits-Axiom findet sich in der Dissertation von H. Jung, Die Struktur der Thermodynamik, Aachen 1957. Dort ist auch die Abhängigkeit des Begriffes der Arbeits-Koordinaten vom thermischen Kontakt zum ersten Male erkannt.

Die von H. A. Buchdahl [Z. Physik **152**, 425 (1958)] hervorgehobene Vereinfachung seiner Behandlungsweise gegenüber der Carathéodorys besteht ebenfalls in der Ersetzung des Unerreichbarkeits-Axioms durch das genannte schärfere.

[2] Siehe Ziff. A 2.

a) Die Deformations-Koordinaten $(x_1, \ldots, x_n) = x$ lassen sich *adiabatisch frei bewegen*[1].

b) Wenn man für einen beliebigen Zustand $z^0 = (\xi^0, x^0)$ alle Zustände $z' = (\xi', x')$ aufsucht, die von z^0 aus adiabatisch erreichbar sind und die außerdem in x' übereinstimmen, dann bedecken die zugehörigen ξ'-Werte einen Halbstrahl[2].

Ist also $z^0 = (\xi^0, x^0)$ ein beliebiger Anfangszustand, dann besagt a), daß unter adiabatischer Isolation des Systems mindestens ein Übergang $z^0 \to z' = (\xi', x')$ mit willkürlich vorgegebenem $x' = (x_1', \ldots, x_n')$ erzwungen werden kann.

Zur Veranschaulichung der Eigenschaft b) ist es zweckmäßig, den Zustandsraum $\{\xi, x_1, \ldots, x_n\}$ durch ein Cartesisches Koordinatensystem darzustellen. Jedes Werte-n-tupel $(x_1, \ldots, x_n) = x$ repräsentiert dann eine Gerade, die wir kurz als *Deformations-Linie* x bezeichnen, weil alle ihre Punkte ein und demselben Deformations-Zustand des Systems entsprechen. Weiterhin repräsentiert die ξ-Variable eine Schar von Hyperebenen, welche die n-parametrige Schar der Deformations-Linien orthogonal durchsetzt, so daß alle Zustände des Systems umkehrbar eindeutig gekennzeichnet sind als Schnittpunkte aus je einer Deformations-Linie x und einer Hyperebene ξ. Schließlich denke man sich jeden Übergang $(\xi^0, x^0) \to (\xi', x')$ dargestellt durch ein (beliebig-geformtes) Kurvenstück, das auf der x^0-Linie im Punkt ξ^0 beginnt und auf der x'-Linie im Punkt ξ' endet. In diesem geometrischen Bild besagt b): Die Endpunkte aller adiabatisch möglichen Übergänge, die von einem Punkt z^0 zu einer Deformations-Linie x' verlaufen, bedecken einen Halbstrahl der x'-Linie. Im zweidimensionalen Zustandsraum eines idealen Gases (Fig. 5) lassen sich diese Verhältnisse gut übersehen: Die Endpunkte der adiabatischen Über-

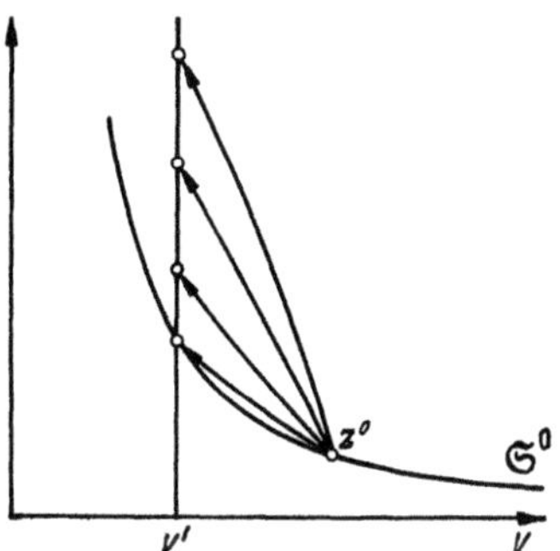

Fig. 5. Zur Veranschaulichung der Eigenschaft b) der einfachen Systeme.

gangs-Gesamtheit liegen auf einer Geraden $V = V' = $ const. und zwar so, daß sie den Halbstrahl bedecken, welcher oberhalb der zu z^0 gehörigen Isentropen $\mathfrak{S}^0$ verläuft. Durch Variation der V'-Linie bei festgehaltenem Anfangszustand z^0 erkennt man weiter, daß die Menge aller adiabatisch erreichbaren Endzustände gegeben ist durch den ganzen Halbraum oberhalb $\mathfrak{S}^0$.

Wir betrachten jetzt an einem beliebigen thermodynamischen System die Gesamtheit der Zustände, die von einem Anfangszustand z^0 aus adiabatisch erreichbar sind und bezeichnen jede solche Gesamtheit als *adiabatischen Bereich* $\mathfrak{A}(z^0)$ mit z^0 als *erzeugenden Zustand*. Diese Definition erlaubt eine äquivalente Formulierung der Eigenschaften a) und b) durch

Satz 2: Jeder adiabatische Bereich $\mathfrak{A}(z^0)$ eines einfachen Systems repräsentiert im Zustandsraum $\mathfrak{Z}_{n+1} = \{\xi, x_1, \ldots, x_n\}$ eine $(n+1)$-dimensionale zusammenhängende Punktmenge, welche den erzeugenden Zustand z^0 und von jeder Deformations-Linie x einen Halbstrahl enthält.

---

[1] Als Beispiel wähle man das System der Fig. 2 mit druck-entkoppelnder und wärmedurchlässiger Zwischenwand. Es ergibt sich dann ein einfaches System mit dreidimensionalem Zustandsraum $\mathfrak{Z}_3 = \{\vartheta, V_1, V_2\}$, für den die Variablen V_1, V_2 Deformations-Koordinaten sind.

[2] Bei Carathéodory ist die Eigenschaft b) im zweiten Teil der Definition einfacher Systeme enthalten, was man erkennt, wenn man sich auf eine geeignete Umgebung des Anfangszustandes z^0 beschränkt. Statt eines Halbstrahles kann auch der Fall eines abgeschlossenen Intervalls einbezogen werden. Dadurch wird an den entscheidenden Überlegungen nichts geändert, wenn man die Betrachtungen auf geeignete $(n+1)$-dimensionale Teilbereiche des Zustandsraumes anwendet. Von dieser Möglichkeit macht Carathéodory wiederholt Gebrauch.

Weiterhin folgt nun aus dem Unerreichbarkeits-Axiom, daß der erzeugende Zustand z^0 eines adiabatischen Bereiches $\mathfrak{A}$ immer dann ein Randpunkt von $\mathfrak{A}$ sein muß, wenn $\mathfrak{A}$ eine zusammenhängende Punktmenge von der Dimension des Zustandsraumes ist, — was gemäß Satz 2 für einfache Systeme stets zutrifft. Schließlich ist leicht einzusehen, daß alle Zustände einer maximalen adiabatisch-reversiblen Zustands-Klasse $\mathfrak{S}$ ein und denselben adiabatischen Bereich $\mathfrak{A}(\mathfrak{S})$ erzeugen. Sind nämlich z', z'' zwei Zustände aus $\mathfrak{S}$, dann lassen sich per definitionem *beide* Übergänge $z' \to z''$, $z'' \to z'$ adiabatisch erzwingen, und somit muß jeder Endzustand z, der durch einen adiabatischen Übergang $z' \to z$ erzeugt wurde, auch von z'' aus adiabatisch erreichbar sein und umgekehrt. Das Beispiel der Fig. 5 kann wiederum zur Veranschaulichung herangezogen werden: Die Isentrope $\mathfrak{S}^0$ erzeugt den über ihr liegenden Halbraum der p-V-Ebene als adiabatischen Bereich $\mathfrak{A}(\mathfrak{S}^0)$. Im übrigen ist es physikalisch nicht selbstverständlich, daß die adiabatisch-reversiblen Zustands-Klassen $\mathfrak{S}$ den Zustandsraum in zwei Halbräume zerlegen; man kann sich nämlich Systeme mit innerer Reibung vorstellen, für die Satz 2 gültig ist, jedes $\mathfrak{S}$ aber nur einen Zustand enthält[1]. Wir haben somit

Satz 3: Jede Entropie-Klasse $\mathfrak{S}$ eines einfachen Systems ist in der Randfläche des zugehörigen adiabatischen Bereiches $\mathfrak{A}(\mathfrak{S})$ enthalten.

Wenn man jetzt die Eigenschaft b) noch dahingehend erweitert, daß sich unter den adiabatischen Übergängen, die von z^0 zu einem Halbstrahl x' verlaufen, auch (mindestens) ein *reversibler* befindet, dann folgt aus Satz 3, daß es nur *genau* ein reversibler sein kann, nämlich der, welcher von z^0 zum Anfangspunkt des Halbstrahls x' führt. Daraus ergibt sich als Verschärfung von Satz 3 die Aussage: Die adiabatischen Bereiche sind Halbräume, die von den Entropie-Klassen $\mathfrak{S}$ begrenzt werden, und die adiabatisch-irreversiblen Prozesse durchlaufen die dadurch definierte Schar $\{\mathfrak{S}\}$ in ein und derselben Richtung. Damit ist die *Existenz* der empirischen Entropie $\sigma(\xi, x)$ bereits gesichert, nicht aber ihre *Konstruierbarkeit*. Um letzteres zu erreichen, benötigt Carathéodory einen neuen Grundbegriff. Er setzt voraus, daß man von jedem Übergang $z \to z'$ entscheiden kann, ob er *quasistatisch-adiabatisch* möglich ist oder nicht, mit anderen Worten, es wird eine weitere Übergangs-Relation eingeführt Wir weisen ausdrücklich darauf hin, daß der Begriff ,,quasistatisch-adiabatisch'' bei Carathéodory ein *neuer Grundbegriff* ist, im Gegensatz zu der von uns benutzten *Definition*[2]. Für einfache Systeme gilt nun weiter

c) Die *Energie U* ist als ξ-*Variable* brauchbar, das heißt die Variablen U, $x_1, \ldots, x_n$ spannen den Zustandsraum $= \{\xi, x_1, \ldots, x_n\}$ auf[3].

d) Bei quasistatisch-adiabatischen Übergängen sind die Deformations-Koordinaten *frei beweglich*, und die zugehörigen Energie-Veränderungen bilden eine lineare Prozeß-Belegung, deren zugehörige Pfaffsche Form δA *nur die Differentiale der Deformations-Koordinaten* enthält: $\delta A = K_1 \, dx_1 + K_2 \, dx_2 + \cdots K_n \, dx_n$.

Ein Übergang $z \to z'$ ist also genau dann ein quasistatisch-adiabatisch möglicher Übergang, wenn sich unter den Integral-Kurven der Pfaffschen Gleichung

$$\delta Q \equiv dU - \delta A = dU - K_1 \, dx_1 - \cdots - K_n \, dx_n = 0 \tag{A 3.1}$$

eine solche befindet, die von z nach z' führt. Dann besitzt aber auch der umgekehrte Übergang $z' \to z$ dieselbe Eigenschaft. Diese Feststellung ist natürlich

---

[1] Siehe Carathéodorys ,,Untersuchungen'', S. 368.

[2] Siehe Ziff. 9.

[3] Carathéodory schränkt diese Forderung dahingehend ein, daß der Zustandsraum $\{\xi, x_1, \ldots, x_n\}$ zumindest in $(n+1)$-dimensionale Teilbereiche zerfällt, innerhalb derer c) erfüllt ist.

trivial als Aussage über Integralkurven, nicht hingegen als Aussage über quasistatisch-adiabatische Prozesse, eine Tatsache, die in der Caratheodoryschen Forderung dadurch zum Ausdruck kommt, daß die Deformations-Koordinaten quasistatisch-adiabatisch frei bewegbar sein sollen. Daß nicht jede Integral-Kurve einer Pfaffschen Gleichung physikalische Bedeutung hat, vergegenwärtigt man sich leicht am Beispiel chemischer Prozesse. Es sei

$$dU = T\,dS - p\,dV - A\,d\lambda$$

das Energie-Differential eines homogenen Systems mit einer Reaktion. Die Lösungs-Gesamtheit der Pfaffschen Gleichung

$$T\,dS = dU + p\,dV + A\,d\lambda = 0$$

enthält zwar alle quasistatisch-adiabatischen Prozesse ($d\lambda = 0$ oder $A = 0$), darüber hinaus aber noch weitere, die sich physikalisch nicht realisieren lassen. Dennoch ist dies für die Thermodynamik solange ohne Bedeutung wie sich alle Punkte einer Entropie-Fläche durch die physikalisch realisierbaren Integralkurven miteinander verbinden lassen. Die Berücksichtigung dieser Besonderheit macht es im Prinzip möglich, auch chemische Prozesse in der Caratheodoryschen Weise zu behandeln. Unter Beachtung von Satz 3 folgt somit

Satz 4: Die (maximalen) Klassen quasistatisch-adiabatisch verbindbarer Zustände bilden eine zu den x-Linien orthogonale Hyperflächen-Schar $\{\mathfrak{F}\}$, die *identisch ist mit der Entropie-Klassen-Schar* $\{\mathfrak{S}\}$. Jede Darstellung $f(z)$ von $\{\mathfrak{F}\}$ repräsentiert somit eine empirische Entropie $\sigma(z)$.

$\gamma)$ *Die thermische Kopplung.* Wir haben im letzten Abschnitt gesehen, daß jedes einfache System $\mathfrak{Z}_{n+1}\{z\} = \{\xi, x_1, \ldots, X_n\}$ eine empirische Entropie σ besitzt, welche ebenso wie die Energie U als ξ-Variable benutzt werden kann. Die Koordinaten-Transformation $(U, x_1, \ldots, x_n) \rightarrow (\sigma, x_1, \ldots, x_n)$ ist also umkehrbar eindeutig, mit anderen Worten $\partial U(\xi, x_1, \ldots, x_n)/\partial \xi \equiv \Theta$ muß für alle Zustände eines einfachen Systems entweder größer oder kleiner als Null sein. Berücksichtigt man Eigenschaft d) sowie Satz 4, so ergibt sich $K_i = \partial U(\sigma, x_1, \ldots, x_n)/\partial x_i$ und somit

$$dU(\sigma, x_1, \ldots, x_n) = \Theta\,d\sigma + K_1\,dx_1 + \cdots K_n\,dx_n. \tag{A3.2}$$

Vergleich mit (A3.1) zeigt, daß Θ ein integrierender Nenner von $dU - \delta A$ ist. Wie bereits erwähnt, kommt es jetzt darauf an, mit Hilfe der thermischen Kopplung unter allen möglichen Variablen-Paaren σ, Θ ein bestimmtes auszuzeichnen.

Die Eigenschaften, welche CARATHÉODORY dem *thermischen Gleichgewicht einfacher Systeme* zuschreibt, lassen sich folgendermaßen formulieren:

$\alpha)$ Zwei einfache Systeme $\mathfrak{Z}_i = \{z^{(i)}\}$, $\mathfrak{Z}_j = \{z^{(j)}\}$ bestimmen durch thermischen Kontakt eindeutig ein *drittes System* $\mathfrak{Z} = \{z\}$, dessen Zustände z von einer Auswahl unter den Zustands-Paaren $[z^{(i)}, z^{(j)}]$ gebildet werden.

$\beta)$ $\mathfrak{Z}_i$, $\mathfrak{Z}_j$ befinden sich genau dann im thermischen Kontakt-Gleichgewicht, wenn zwischen ihren Zustands-Koordinaten eine gewisse Gleichung $f(\sigma^{(i)}, x_1^{(i)}, \ldots, x_{n_i}^{(i)}; \sigma^{(j)}, x_1^{(j)}, \ldots, x_{n_j}^{(j)}) = 0$ erfüllt ist.

$\gamma)$ Die in $\beta)$ enthaltene Alternative $f(z^{(i)}, z^{(j)}) = 0$ bzw. $\neq 0$ ist eine *Kontakt-Relation*[1].

$\delta)$ Die Deformations-Koordinaten von $\mathfrak{Z}_i$, $\mathfrak{Z}_j$ behalten auch am System $\mathfrak{Z}$ die Eigenschaften a), d) der einfachen Systeme, das heißt sie sind adiabatisch und quasistatisch frei beweglich.

---

[1] Siehe Ziff. 6.

ε) Die Energie U und die quasistatisch-adiabatische Arbeit δA von $\mathfrak{Z}$ setzen sich additiv aus den entsprechenden Größen von $\mathfrak{Z}_i$, $\mathfrak{Z}_j$ zusammen.

ζ) $\mathfrak{Z}$ besitzt die Eigenschaften b) und c) der einfachen Systeme[1].

Aus γ) folgt sofort der in Ziff. 6 ausgesprochene Satz über die Einteilung der Zustands-Mengen in Gleichgewichts-Klassen, hier also *Temperatur-Klassen* $\mathfrak{T}$. Jede solche Klassen-Einteilung $\{\mathfrak{T}\}$ eines Zustandsraumes $\mathfrak{Z}_i = \{z^{(i)}\}$ läßt sich natürlich durch eine Zustands-Funktion $\vartheta = \vartheta(z^{(i)})$ darstellen[2]. Überdies ist es möglich, die willkürliche Belegung der $\mathfrak{T}$-Klassen mit den Zahlen ϑ von *einem* System auf alle anderen Systeme so zu übertragen, daß alle im thermischen Gleichgewicht stehenden Klassen ein und dieselbe Zahl ϑ erhalten. Dann ergibt sich als hinreichende und notwendige Bedingung für thermisches Gleichgewicht die Relation

$$\vartheta(z^{(i)}) - \vartheta(z^{(j)}) = 0. \tag{A3.3}$$

Die bisher noch gar nicht benutzte Forderung β) hat offenbar nur die Aufgabe, die Temperatur-Klassen $\vartheta =$ const als Hyperflächen-Schar zu erklären. *Daß deren Anordnung physikalische Bedeutung[3] besitzt, ist durch diese Art der Einführung durchaus nicht gesichert, sondern muß als separate Forderung gestellt werden.* Derselbe Vorwurf trifft übrigens fast alle Formulierungen des „nullten Hauptsatzes".

Das aus zwei einfachen Systemen $\mathfrak{Z}_1$, $\mathfrak{Z}_2$ mit $(n_1 + 1)$ bzw. $(n_2 + 1)$ Freiheitsgraden durch thermischen Kontakt gebildete System $\mathfrak{Z}$ hat gemäß β) und δ) $n_1 + n_2 + 1$ Freiheitsgrade und $n_1 + n_2$ Deformations-Koordinaten. Daß es auch ein einfaches System ist, sichert schließlich ζ).

Wir betrachten nun quasistatisch-adiabatische Prozesse des Systems $\mathfrak{Z}$. Dann besagen δ) und ε), daß diese Prozesse durch die Integral-Kurven der Pfaffschen Gleichung

$$dU^{(1)} - \delta A^{(1)} + dU^{(2)} - \delta A^{(2)} = 0$$

gegeben sind. Diese läßt sich wegen (A3.2) in die Gestalt

$$\Theta^{(1)} d\sigma^{(1)} + \Theta^{(2)} d\sigma^{(2)} = 0$$

bringen. Die zugehörige Pfaffsche Form ist als solche in zwei Variablen trivialerweise integrabel:

$$\Theta^{(1)} d\sigma^{(1)} + \Theta^{(2)} d\sigma^{(2)} = \Theta\, d\sigma. \tag{A3.4}$$

Da $\Theta^{(i)}$ für ein einfaches System nur ein Vorzeichen annehmen kann, ergibt sich eine Einteilung dieser Systeme in zwei Typen: Die adiabatisch-irreversiblen Prozesse längs einer Deformations-Linie sind bei dem einen Typ ($\Theta > 0$) mit Energie-Zufuhr, bei dem anderen ($\Theta < 0$) mit Energie-Abgabe verbunden. Bedenkt man, daß eine Energie-Zufuhr unter adiabatischer Isolation bei konstanten Deformations-Koordinaten einem Reibungs-Vorgang entspricht, so erscheint der Fall $\Theta < 0$ zwar von formaler Natur, aber im Hinblick auf ein Verständnis der Carathéodoryschen Schlußweise nicht unwesentlich.

Gl. (A3.4) zeigt, daß bei Kopplung zweier Systeme $\mathfrak{Z}_1$, $\mathfrak{Z}_2$, deren $\Theta^{(i)}$ verschiedene Vorzeichen haben, adiabatisch reversible Prozesse des Gesamtsystems ($d\sigma = 0$) mit $d\sigma^{(1)}$, $d\sigma^{(2)} > 0$ möglich sind. Irreversible Prozesse an den Einzelsystemen könnten also durch reversible Prozesse des Gesamtsystems rückgängig gemacht werden.

Koppelt man dagegen zwei Systeme $\mathfrak{Z}_1$, $\mathfrak{Z}_2$, deren $\Theta^{(i)}$ gleiches Vorzeichen haben, so ist nach (A3.4) σ eine gleichsinnig monotone Funktion von $\sigma^{(1)}$, $\sigma^{(2)}$.

---

[1] Carathéodory bezeichnet ζ) als *beweisbare* Aussage. Es war den Verfassern nicht möglich, den Beweis zu rekonstruieren.

[2] Siehe Ziff. 3.

[3] Gemeint ist die Ausgleichs-Eigenschaft der Kontakt-Variablen. Siehe auch S. 134.

Diese Monotonie kann wachsend oder fallend sein, je nach dem Vorzeichen von $\Theta^{(i)}/\Theta$. Ist dagegen gesichert, daß $\Theta^{(i)}/\Theta$ stets positiv ist, so kann man auf die monotonwachsende Abhängigkeit der Gesamtentropie σ von den Teilentropien $\sigma^{(1)}, \sigma^{(2)}$ schließen. Hinzunahme eines weiteren einfachen Systems $\mathfrak{Z}_3$ erlaubt dann die direkte Anwendung der Überlegungen von Ziff. 10 (Kleindruck S. 141) und damit die Konstruktion der metrischen Entropien $S^{(i)}(\sigma^{(i)})$ bzw. $S(\sigma^{(1)}, \sigma^{(2)})$. Die eindeutige Konstruierbarkeit der absoluten Temperatur läßt sich dann unmittelbar folgern. Denn da die Entropie S bis auf eine lineare Transformation festliegt, wird unter den integrierenden Nennern $\Theta, \Theta', \ldots$ eindeutig einer und damit eine *Variable* T ausgezeichnet[1]. Aus

$$T^{(1)} dS^{(1)} + T^{(2)} dS^{(2)} = 0 \quad \text{und} \quad dS^{(1)} + dS^{(2)} = 0$$

folgt dann
$$T^{(1)}(S^{(1)}, x_1^{(1)}, \ldots, x_{n_1}^{(1)}) = T^{(2)}(S^{(2)}, x_1^{(2)}, \ldots, x_n^{(2)}). \tag{A3.5}$$

Da gleichzeitig die Bedingung des thermischen Gleichgewichtes (A3.3) gilt und das System $n_1 + n_2 + 1$ Freiheitsgrade hat, müssen $\vartheta^{(i)}$ und $T^{(i)}$ dieselben Hyperflächen-Scharen repräsentieren. T ist somit eine universelle Funktion von ϑ.

Man sieht: Um die empirische Entropie zu metrisieren[2], muß die Entropie eines zusammengesetzten Systems gleichsinnig monoton von den Entropien der Teilsysteme abhängen. Dies ist die Forderung unseres Axiomes 3 c. In CARATHÉODORYs Abhandlung fehlt jedoch eine entsprechende Annahme, und die Theorie ist in einer Weise durchgeführt, die eine solche zusätzliche Forderung scheinbar überflüssig macht. Die Schlußweise ist folgende: Aus (A3.4) oder

$$\frac{\Theta^{(1)}}{\Theta} d\sigma^{(1)} + \frac{\Theta^{(2)}}{\Theta} d\sigma^{(2)} = d\sigma$$

ergibt sich, daß
$$\frac{\Theta^{(1)}}{\Theta} = f(\sigma^{(1)}, \sigma^{(2)}), \qquad \frac{\Theta^{(2)}}{\Theta} = g(\sigma^{(1)}, \sigma^{(2)}).$$

Daraus schließt man unter Einführung der Variablen ϑ in bekannter Weise (CARATHÉODORYs „Untersuchungen" S. 375), daß Θ nur eine Funktion von $\vartheta, \sigma^{(1)}, \sigma^{(2)}$ ist und überdies die Form $\Theta = T(\vartheta) \cdot h(\sigma^{(1)}, \sigma^{(2)})$ hat. Infolgedessen läßt sich schreiben
$$\Theta \, d\sigma = T(\vartheta) \, h(\sigma^{(1)}, \sigma^{(2)}) \, d\sigma(\sigma^{(1)}, \sigma^{(2)})$$
$$= T(\vartheta) \, dS(\sigma^{(1)}, \sigma^{(2)}).$$

Das so konstruierte S braucht aber keineswegs eine *gleichsinnig monotone* Funktion von $\sigma^{(1)}, \sigma^{(2)}$ zu sein. Der Vorteil, daß $\sigma^{(1)}, \sigma^{(2)}$ physikalische Variablen sind, von denen man nach Konstruktion weiß, daß sie unter adiabatischer Isolation nur zunehmen können, verliert dadurch für S jede Bedeutung. Es bleibt dann nichts übrig, als die Forderung, die man für $\sigma(\sigma^{(1)}, \sigma^{(2)})$ zu stellen unterließ, nun für S zu erheben, was sicherlich nicht vorteilhafter ist. CARATHÉODORY versucht statt dessen („Untersuchungen" § 9, S. 377) die Wachstums-Eigenschaft von S unter adiabatischer Isolation mit Argumenten abzuleiten, die in keiner Weise zwingend sind. An dieser Stelle besitzt CARATHÉODORYs Aufbau eine Lücke, die sich nur durch unser Axiom 3 c oder ein gleichwertiges schließen läßt.

A 4. Kritische Bemerkungen. CARATHÉODORYs sowie klassische Thermodynamik sind verschiedene Darstellungen derselben Theorie. Diese Tatsache hat wiederholt die Frage nach der „Äquivalenz" verschiedener Axiome der beiden Darstellungen aufkommen lassen, eine Frage, die zu mancher Verwirrung Anlaß gab.

Betrachtet man eine der klassischen Darstellungsweisen der Thermodynamik, zum Beispiel die PLANCKs, so ist klar, daß es nicht einfach ist, in ihr ein volles

---

[1] Siehe S. 166.
[2] Natürlich unter Erhaltung der Anordnung!

Axiomen-System zu erkennen; denn traditionsgemäß werden nur die beiden Hauptsätze als Axiome herausgestellt, während alle übrigen Voraussetzungen als mehr oder weniger selbstverständlich betrachtet und vielfach nicht einmal erwähnt werden. Ähnliches gilt, wenn auch in geringerem Maße, für CARATHÉODORYs Darstellung; dort wird nur zwei Axiomen — nämlich dem ersten Hauptsatz und dem Unerreichbarkeits-Axiom — auch die Bezeichnung Axiom beigelegt, während die übrigen in Gestalt von „Definitionen" oder beiläufig geforderten Eigenschaften von Koeffizienten Pfaffscher Formen auftreten. In beiden Fällen werden also zwei Axiome hervorgehoben, und daher scheint die Frage naheliegend, ob diese beiden Axiome äquivalent seien in dem Sinn, daß man zum Beispiel den zweiten Hauptsatz durch das Caratheodorysche Unerreichbarkeits-Axiom ersetzen kann und umgekehrt, und dabei wieder Axiomen-Systeme der Thermodynamik resultieren. Ihre Formulierung als Ausschließungs-Prinzipien für gewisse Prozesse suggerierte geradezu, ausschließlich diesen Vergleich in Erwägung zu ziehen. Es ist klar, daß derartige Vergleiche nur sachgemäß sind, wenn bestimmte Axiomen-Systeme der Theorie aufgewiesen werden. In starker Vereinfachung läßt sich das Gemeinte durch folgendes formale Beispiel erläutern: Es liege eine Anzahl von Axiomen $A_1, A_2, \ldots, A_n$ vor, die ein Axiomen-System einer Theorie bilden. In diesem möge man zum Beispiel das Axiom A_1 zerlegen in zwei Axiome B_1 und B_2, so daß diese zusammen mit $A_2, \ldots, A_n$ wieder ein Axiomen-System der Theorie bilden, genauer: B_1 und B_2 seien Folgesätze von $A_1, A_2, \ldots, A_n$, ebenso A_1 Folgesatz von $B_1, B_2, A_2, \ldots, A_n$, nicht aber von $B_1, A_2, \ldots, A_n$ oder $B_2, A_2, \ldots, A_n$. Trivialerweise ist dann auch $A_1, B_2, A_2, \ldots, A_n$ ein Axiomen-System (in dem B_2 sogar weggelassen werden kann), nicht aber $B_1, A_2, \ldots, A_n$. Man kann also B_1 in dem Axiomen-System $B_1, B_2, A_2, \ldots, A_n$ durch A_1 substituieren, nicht aber A_1 durch B_1 in $A_1, A_2, \ldots, A_n$. Hinsichtlich des „Restes" $B_2, A_2, \ldots, A_n$ sind also A_1 und B_1 gleichwertig, nicht aber hinsichtlich $A_2, \ldots, A_n$. Es wäre daher höchst verwirrend, A_1 und B_1 schlechthin „äquivalent" zu nennen. Das Beispiel zeigt nicht nur die sinngemäße Begrenzung des Äquivalenz-Begriffes von Axiomen, sondern überdies die Fragwürdigkeit jeder Argumentation, welche die „Sparsamkeit" in der Aussage eines Axioms zu dessen Gunsten anführt; denn B_1 kann gegenüber A_1 zum Beispiel den Eindruck einer außerordentlich wenig einschneidenden, das heißt allgemeinen Aussage machen, ohne daß damit das Geringste gewonnen wäre.

Ein Rückblick auf Ziff. A3 zeigt, daß jede Kritik des Carathéodoryschen Aufbaues der Thermodynamik sich sowohl gegen das Unerreichbarkeits-Axiom als gegen die fundamentale Rolle der einfachen Systeme richten wird. Dabei ist, um es vorwegzunehmen, das Unerreichbarkeits-Axiom sogar schwererwiegenden Bedenken ausgesetzt als die einfachen Systeme. In diesem Axiom wird vom Stetigkeits-Begriff in einer Weise Gebrauch gemacht, die wir als unphysikalisch ansehen. Außerdem stehen Formulierung und Verwendung des Axioms in einem gewissen Mißverhältnis zueinander. Die Aussage über die Existenz adiabatisch nichterreichbarer Zustände in jeder Umgebung eines Zustandes macht den Eindruck einer außerordentlich tiefliegenden und anschaulich kaum faßbaren topologischen Eigenschaft des Zustands-Raumes. *Verwendet* wird das Axiom aber ausschließlich im Zusammenhang mit den einfachen Systemen, in deren Eigenschaften a) bis d) eine Anzahl von Forderungen enthalten sind, die nichts bewirken als die fragliche Unerreichbarkeits-Aussage auf ihre trivialste Form zu reduzieren[1], nämlich auf eine einparametrige Schar von Hyperflächen, die unter adiabatischer Isolation nur einseitig durchlaufen werden kann, oder in unserer Ausdrucksweise,

---

[1] Man vergleiche hierzu das oben diskutierte formale Modell der Zerlegung des Axioms A_1 in ein sehr allgemeines B_1 und den ergänzenden Rest B_2.

auf die Variable empirische Entropie. Die von CARATHÉODORY als fundamental erkannte Rolle der adiabatischen Isolation findet dadurch eine recht inadäquate axiomatische Charakterisierung; denn sie ist erst durch das Unerreichbarkeits-Axiom *zusammen* mit einigen Forderungen über einfache Systeme gekennzeichnet. Dem Unerreichbarkeits-Axiom allein entspricht dagegen keine physikalische Operation, die seine hervorgehobene Stellung rechtfertigen könnte — wie etwa die energetische Isolation die Stellung des ersten Axioms in CARATHÉODORYs Abhandlung rechtfertigt. Gibt man schließlich zu, daß die Existenz adiabatisch unerreichbarer Zustände in jeder Nachbarschaft eines Zustandes physikalisch oder auch intuitiv keine größere Evidenz besitzt als die simple Behauptung, daß die adiabatische Isolation eine empirische Variable definiert, so fällt es schwer, in der Einführung des Unerreichbarkeits-Axioms mehr zu sehen als ein Zuschneiden des Axiomen-Systems der Thermodynamik auf eine Anwendung des Carathéodoryschen Satzes über Pfaffsche Formen.

Die merkwürdige Rolle des Unerreichbarkeits-Axioms offenbart sich übrigens auch bei der Anwendung der Theorie auf Gase, eine in der Thermodynamik allgemein bevorzugte Klasse von Systemen Die Tatsache, daß ein Gas ein einfaches System mit zwei Freiheitsgraden ist, sichert bereits die Existenz der Adiabaten-Kurven im Zustandsraum, das heißt der empirischen Entropie, ohne vom Unerreichbarkeits-Axiom überhaupt Gebrauch zu machen; denn eine Pfaffsche Form in zwei Variablen ist bekanntlich stets integrierbar. Erst zur Metrisierung dieser Variablen wird das Axiom herangezogen, dann aber in Verbindung mit der thermischen Kopplung einfacher Systeme, die sich hiermit von vergleichbarer Wichtigkeit erweist. Bei dem ganzen Vorgang dient das Unerreichbarkeits-Axiom nur dazu, die empirische Entropie des aus den beiden Gasen im thermischen Gleichgewicht bestehenden Systems zu sichern, und dies wiederum nur zusammen mit anderen Axiomen. Auch hier scheint es somit zweckmäßiger, die Existenz der empirischen Entropie des reduzierten[1] Systems direkt zu fordern. Es ist fast überflüssig hinzuzufügen, daß die Verquickung der Konstruktion von metrischer Entropie und absoluter Temperatur wegen ihres Mangels an Durchsichtigkeit sowohl CARATHÉODORYs als auch den klassischen Darstellungen der Thermodynamik zum Nachteil anzurechnen ist.

Nach dieser in der Sache scharfen Kritik, deren Zweck es ist, unser Abgehen von CARATHÉODORYs axiomatischer Behandlung der Thermodynamik zu begründen, scheinen ein paar Worte des Respektes und der Würdigung angebracht. CARATHÉODORY schrieb seine Abhandlung vor 50 Jahren, zu einer Zeit als die Axiomatik nicht einmal Allgemeingut der Mathematiker war. Er selbst war damals ausschließlich Analytiker, was in der Abhandlung deutlich zu spüren ist. Gerade die Verwendung analytischer und stetigkeits-topologischer Begriffe zur Beschreibung einfacher physikalischer Fakten bildete Anlaß und Ansatzpunkt unserer Kritik. Aber erst die Hervorhebung der kombinatorischen oder „algebraischen" Züge der Theorie erlaubte es, über diese Kritik hinaus zu neuen Ergebnissen vorzustoßen. Es ist bezeichnend, daß CARATHÉODORY in der Mathematik einen ähnlichen Weg gegangen ist, wie seine späteren Arbeiten zur Algebraisierung des Integralbegriffes zeigen. Wir sind daher überzeugt, daß wir der Sache nach völlig im Geist CARATHÉODORYs gehandelt haben, auch wenn dies in gewissem Sinn eine Abkehr von den Methoden seiner Abhandlung bedeutet. Wie hoch deren Wert andererseits einzuschätzen ist, geht daraus hervor, daß sie seit einem halben Jahrhundert die Basis für jede Beschäftigung mit der Axiomatik der Thermodynamik bildet.

---

[1] Ziff. 14.

Prinzipien der statistischen Mechanik.

Von

A. Münster.

Mit 12 Figuren.

1. Einleitung. Als „Statistische Mechanik" bezeichnet man den Teil der theoretischen Physik, der die makroskopischen Eigenschaften der Materie aus ihrer atomistischen Struktur ableitet. Dabei werden in diesem Zusammenhang als „makroskopische Eigenschaften" solche Eigenschaften bezeichnet, die wesentlich durch das Zusammenwirken sehr vieler Teilchen (Atome, Moleküle) bedingt sind (Druck, Temperatur, Dielektrizitätskonstante usw.), im Gegensatz zu Eigenschaften, die für das einzelne Atom oder Molekül definiert werden können (Spektrum, Polarisierbarkeit usw.), wenn sie auch gewöhnlich an Systemen mit großen Teilchenzahlen gemessen werden. Der Gedankengang, welcher der statistischen Mechanik zugrunde liegt, läßt sich kurz in folgender Weise skizzieren. Vom atomistischen Standpunkt betrachtet, ist ein makroskopisches System ein mechanisches System, dessen Eigenschaften durch die Gleichungen der Atom-Mechanik determiniert werden. Ein derartiges mechanisches Problem ist jedoch praktisch unlösbar, nicht nur wegen der im allgemeinen unüberwindlichen mathematischen Schwierigkeiten (etwa 10^{23} Variable!), sondern vor allem im Hinblick auf die Unmöglichkeit, die Anfangs- und Randbedingungen exakt festzulegen. Man kann aber fragen, welche Lösungen der Bewegungsgleichungen mit gewissen „fragmentarischen" Kenntnissen über das betrachtete System (die stets vorausgesetzt werden müssen) vereinbar sind. Auf dieser Grundlage läßt sich ein Wahrscheinlichkeitsaggregat konstruieren, mit dessen Hilfe man gewisse Wahrscheinlichkeitsaussagen über das betrachtete System ableiten kann.

Nach dem Aufbau des Wahrscheinlichkeitsaggregates unterscheidet man in der statistischen Mechanik grundsätzlich zwei Methoden. Wenn die Hamilton-Funktion, bzw. der Hamilton-Operator, nach den Koordinaten der Einzelteilchen separierbar ist, kann man die Einzelteilchen als Elemente eines Wahrscheinlichkeitsaggregates betrachten. Da in der klassischen Mechanik das mechanische Problem eines Moleküls von s Freiheitsgraden in dem s-dimensionalen Phasenraum des Moleküls (μ-Raum nach Ehrenfest [3]) dargestellt wird, bezeichnet man diese Methode als μ-Raum-Statistik. Sie ist durch Anschaulichkeit und Einfachheit ausgezeichnet; ihre Anwendbarkeit ist jedoch auf das ideale Gas und den idealen Kristall, der als System von harmonischen Oszillatoren dargestellt werden kann, beschränkt. Im allgemeinen Falle ist das statistische Problem nur in dem Phasenraum des Gesamtsystems (Gas-Raum, Γ-Raum) darstellbar. Das Wahrscheinlichkeitsaggregat wird dann durch eine Gesamtheit von gedachten Kopien des Originalsystems gebildet, die sich in ihren mechanischen Zuständen in dem durch die vorausgesetzten Kenntnisse über das Originalsystem gegebenen Rahmen unterscheiden können (virtuelle Gesamtheit, statistisches Ensemble). Da die Vorteile der μ-Raum-Statistik im wesentlichen didaktischer Art sind, werden wir der folgenden Darstellung einheitlich die allgemein anwendbare Ensemble-Theorie zugrunde legen und nur gelegentlich den Zusammenhang mit der anderen Methode kurz erläutern.

Die statistische Mechanik ist ursprünglich von Boltzmann und Gibbs auf der Grundlage der Hamiltonschen Mechanik entwickelt worden. Da wir jedoch den Ausgangspunkt als ein Problem der Atom-Mechanik formuliert haben, muß eine streng logische Darstellung des Gebietes notwendig an die Quantenmechanik anknüpfen. Trotzdem werden wir im folgenden aus zwei Gründen die Theorie zuerst auf klassischer Grundlage entwickeln. Einmal ist das Begriffssystem der Quantenstatistik (ähnlich wie das der Quantenmechanik) in möglichst weitgehender Analogie zur klassischen Theorie aufgebaut worden. Die Vorwegnahme der letzteren bringt daher eine wesentliche Vereinfachung der Darstellung und eine Erleichterung des Verständnisses. Zum anderen bleibt für die überwiegende Mehrzahl der Anwendungen von der Quantenstatistik nur eine geringfügige Korrektur an den Resultaten der klassischen Theorie übrig, die sich bereits im Rahmen der letzteren plausibel machen (wenn auch nicht beweisen) läßt. Praktisch benutzt man daher in den meisten Fällen die Methoden der klassischen statistischen Mechanik, deren Kenntnis somit ohnehin unentbehrlich ist.

Die axiomatische Basis der statistischen Mechanik wird naturgemäß zunächst durch die Axiome der Wahrscheinlichkeitsrechnung und der Mechanik gebildet. Auf dieser Grundlage lassen sich allgemeine Sätze über statistische Gesamtheiten ableiten, die für den Aufbau der Theorie von außerordentlicher Bedeutung sind, aber im wesentlichen formalen Charakter besitzen. Bei dem Versuch, die statistische Mechanik zu einer physikalischen Theorie auszugestalten, stößt man auf eine Lücke, die sich trotz vielen Bemühungen bisher auf deduktivem Wege nicht hat schließen lassen. Es spricht alles dafür, daß diese Schwierigkeit prinzipieller Art ist und daher auch durch die Weiterentwicklung der Theorie nicht beseitigt werden wird. Man ist daher gezwungen, die axiomatische Basis durch gewisse Sätze zu erweitern, die als spezifische Axiome der statistischen Mechanik betrachtet werden müssen. Im einzelnen gibt es dafür verschiedene Möglichkeiten, die wir weiter unten erörtern werden.

Der physikalische Gegenstand der statistischen Mechanik ist verhältnismäßig weit gespannt. Die Erfahrung zeigt aber, daß alle makroskopischen Eigenschaften der Materie (in dem oben erläuterten Sinne) wesentlich von der Temperatur abhängen. Die primäre und wichtigste Aufgabe der statistischen Mechanik ist daher die Begründung der phänomenologischen Thermodynamik und ihres Begriffssystems. Unter diesem Gesichtspunkt, und wegen der Unmöglichkeit einer deduktiven Begründung aus der Mechanik wird in neuerer Zeit die statistische Mechanik häufig als „statistische Thermodynamik" bezeichnet. Beide Begriffe unterscheiden sich höchstens durch eine gewisse Akzentverlagerung und werden weitgehend synonym gebraucht.

Die Methode der statistischen Mechanik ist grundsätzlich sowohl auf Gleichgewichtseigenschaften wie auf Nicht-Gleichgewichtszustände anwendbar. Die starke Bindung an die Thermodynamik hat es jedoch mit sich gebracht, daß in der bisherigen Entwicklung die Theorie der Gleichgewichtseigenschaften im Vordergrund gestanden hat. Erst seit etwa zehn Jahren macht sich, im Zusammenhang mit der Entwicklung der Thermodynamik irreversibler Prozesse, ein allgemeineres Interesse an der entsprechenden statistischen Theorie bemerkbar. Im folgenden werden wir das Schwergewicht der Darstellung auf die Theorie der Gleichgewichtseigenschaften legen und die Nicht-Gleichgewichtszustände nur insoweit behandeln, als es zum Verständnis des Gesamtaufbaues der statistischen Mechanik notwendig ist. Für die spezielle Theorie der irreversiblen Prozesse sei auf den Artikel von Meixner in diesem Bande verwiesen.

Von den im vorstehenden angedeuteten allgemeinen Fragen abgesehen, zerfällt die statistische Mechanik in zwei Hauptteile: Die Ableitung der thermodynamischen Prinzipien und weiterer Aussagen von ähnlich allgemeinem Charakter, und die expliziten Berechnungen für konkrete Systeme. Wir beschränken uns in diesem Artikel ausschließlich auf das erste Teilgebiet. Die Rechnungen für konkrete Systeme finden sich in weiteren Bänden dieses Handbuches, welche die thermischen und mechanischen Eigenschaften der Materie zum Gegenstand haben.

Die statistische Begründung der Thermodynamik stößt auf eine eigentümliche Schwierigkeit, insofern die Begriffssysteme der Mechanik und der Thermodynamik weitgehend inkommensurabel sind. Das Problem löst sich dadurch, daß die Anwendung der Statistik die Einführung gewisser Parameter bedingt, die zunächst keine unmittelbare physikalische Bedeutung besitzen. Diese Parameter lassen sich unter gewissen Voraussetzungen, die wir zu untersuchen haben, über die entsprechenden Differentialgleichungen mit den der Mechanik fremden thermodynamischen Größen (Entropie, Temperatur usw.) identifizieren. Man bezeichnet die erwähnten Parameter daher als die statistischen Analoga der thermodynamischen Größen. Bei den Anwendungen der statistischen Mechanik wird der Unterschied gewöhnlich vernachlässigt; man setzt in die Formeln von vornherein an die Stelle der statistischen Parameter die thermodynamischen Zustandsgrößen, was in den meisten Fällen gerechtfertigt ist. Für die Erörterung der grundsätzlichen Fragen liegen die Verhältnisse jedoch anders; wir werden daher streng zwischen beiden Arten von Größen unterscheiden.

A. Klassische statistische Mechanik.

I. Allgemeine Sätze über statistische Gesamtheiten.

2. Phasenraum und virtuelle Gesamtheit. Wir betrachten ein klassisches mechanisches System von n Freiheitsgraden, dessen Zustand oder „Phase"[1] wir durch die Werte von n generalisierten Koordinaten q_i und n generalisierten Impulsen p_i beschreiben. Die Bewegungsgleichungen erscheinen dann in der Hamiltonschen kanonischen Form

$$\frac{dq_i}{dt} = \frac{\partial H}{\partial p_i}, \qquad \frac{dp_i}{dt} = -\frac{\partial H}{\partial q_i}, \tag{2.1}$$

wo H die Hamiltonsche Funktion des Systems bezeichnet. Den euklidischen Raum, der durch die $2n$ rechtwinkligen Cartesischen Koordinaten q_i und p_i aufgespannt wird, bezeichnen wir als den Phasenraum des Systems. Jeder mögliche mechanische Zustand des Systems wird eindeutig durch einen Punkt des Phasenraumes dargestellt. Ist die Lage dieses „Bildpunktes" für einen Zeitpunkt t_0 gegeben, so ist sie damit durch die Bewegungsgleichungen für jeden früheren oder späteren Zeitpunkt t festgelegt. Der Bildpunkt durchläuft also im Laufe der Zeit eine Kurve im Phasenraum, die als Phasenbahn oder Trajektorie bezeichnet wird. Wegen der Eindeutigkeit der Lösungen der Bewegungsgleichungen kann durch jeden Punkt des Phasenraumes nur eine Trajektorie gehen. Die Trajektorien sind daher entweder in sich geschlossen, — dann spricht man von periodischen Systemen —, oder sie besitzen die Struktur eines Knäuels ohne Knoten.

Erste Integrale der Bewegungsgleichungen von der Form

$$F(\boldsymbol{q}, \boldsymbol{p}) = \text{const}, \tag{2.2}$$

---

[1] Dieser Gebrauch des Wortes „Phase" hat weder mit dem Phasenbegriff der Thermodynamik noch mit dem der Wellentheorie etwas zu tun.

die eindeutige stetige Funktionen der q_i und p_i sind, werden uniforme Integrale genannt. Die Existenz eines uniformen Integrals bedingt, daß die Trajektorie vollständig auf der $(2n-1)$-dimensionalen Mannigfaltigkeit $F(q, p) = \text{const}$ verlaufen muß. Es kann also nicht mehr als $2n-1$ unabhängige uniforme Integrale geben, da hierdurch die Trajektorie vollständig festgelegt ist. Allgemein beweisen läßt sich nur für Systeme mit zeitunabhängiger Hamilton-Funktion (konservative Systeme) die Existenz des Energie-Integrals sowie bei Fehlen äußerer Kräfte die dreier Impuls-Integrale und dreier Drehimpuls-Integrale (Erhaltungssätze der klassischen Mechanik).

Gegenstand der statistischen Mechanik sind nicht einzelne Systeme, sondern Gesamtheiten aus vielen (etwa ν) physikalisch gleichartigen Systemen. Dabei besagt der Ausdruck „physikalisch gleichartig", daß für alle Systeme der Gesamtheit die Hamilton-Funktion in gleicher Weise von den generalisierten Koordinaten und Impulsen (und eventuell der Zeit) abhängt. Die Frage nach der physikalischen Interpretation einer solchen virtuellen Gesamtheit stellen wir vorläufig zurück; wir behandeln dieselbe somit zunächst als ein rein formales Konzept. Die virtuelle Gesamtheit wird beschrieben durch eine kontinuierliche Dichtefunktion ϱ der generalisierten Koordinaten und Impulse sowie der Zeit. Es ist zweckmäßig und heute allgemein üblich, diese Funktion auf Eins zu normieren[1]. Wir haben also

$$\varrho = \varrho(q, p, t) \tag{2.3}$$

und

$$\int \varrho(q, p, t)\, d\Omega = 1. \tag{2.4}$$

Der Bruchteil der Phasenpunkte der virtuellen Gesamtheit, der sich zur Zeit t in dem Volumenelement $d\Omega$ des Phasenraumes befindet, ist

$$\frac{d\nu}{\nu} = \varrho(q, p, t)\, d\Omega. \tag{2.5}$$

Gl. (2.5) kann auch interpretiert werden als die Wahrscheinlichkeit, ein beliebig herausgegriffenes System der virtuellen Gesamtheit in dem durch das Volumenelement $d\Omega$ charakterisierten mechanischen Zustand zu finden. Die Funktion $\varrho(q, p, t)$ wird als Phasendichte (density in phase) bezeichnet. Sie ist nach der Definition eine eindeutige, stetige reelle Funktion der Phasenkoordinaten und der Zeit, die nur nicht-negative Werte annehmen kann. Sie muß ferner so beschaffen sein, daß das Integral (2.4) existiert. Mit Hilfe der Phasendichte läßt sich der Phasenmittelwert einer Funktion $F(q, p)$ der generalisierten Koordinaten und Impulse definieren durch die Gleichung

$$\bar{F} = \int F(q, p)\, \varrho(q, p, t)\, d\Omega. \tag{2.6}$$

Ist die Lage der Phasenpunkte einer virtuellen Gesamtheit zur Zeit t_0 durch den Satz q_0, p_0 gegeben, so wird durch die Bewegungsgleichungen zu einer beliebigen Zeit t jedem Punkte des Phasenraumes q_0, p_0 ein Punkt q, p umkehrbar eindeutig zugeordnet. Die Bewegung der virtuellen Gesamtheit im Phasenraum kann daher als eine kontinuierliche Folge von Transformationen, die den Phasenraum in sich selbst überführen, aufgefaßt werden. Dieselben bilden eine endlich-kontinuierliche ein-parametrige Gruppe. Die Gültigkeit der Bewegungsgleichungen (2.1) bedingt jedoch, daß nicht jede kontinuierliche Transformation des Phasenraumes in sich selbst eine mögliche Bewegung der virtuellen Gesamtheit darstellt. Die Abgrenzung der Transformationen, welche mögliche

---

[1] Gibbs [2] versteht unter "density in phase" die nicht normierte Funktion $\nu\varrho$, während er ϱ den Wahrscheinlichkeitskoeffizienten nennt.

Bewegungen der virtuellen Gesamtheit darstellen, bildet den Inhalt eines der grundlegenden Theoreme der statistischen Mechanik, das wir in Ziff. 3 behandeln. Der Einfachheit halber wollen wir von jetzt ab unter Transformationen des Phasenraumes in sich selbst nur solche verstehen, welche dieses Theorem erfüllen, also mögliche Bewegungen der virtuellen Gesamtheit darstellen.

Es kann vorkommen, daß der Phasenraum Γ einen Teil Γ'' besitzt mit der Eigenschaft, daß bei der Bewegung der virtuellen Gesamtheit Γ'' in sich selbst transformiert wird. Man bezeichnet dann Γ'' als invarianten Teil des Phasenraumes Γ. Dieser Begriff spielt eine wichtige Rolle für gewisse Betrachtungen, die mit der Axiomatik der statistischen Mechanik zusammenhängen.

3. Der Liouvillesche Satz. Der grundlegende Satz, der aus der Gesamtheit der kontinuierlichen Transformationen des Phasenraumes in sich selbst diejenigen aussondert, die eine mögliche Bewegung der virtuellen Gesamtheit darstellen, ist bereits 1838 von dem französischen Mathematiker Liouville[1] aufgestellt, aber erst von Boltzmann[2] im Rahmen der statistischen Mechanik angewendet worden. Man kann diesem Satz verschiedene Formulierungen geben; obwohl dieselben mathematisch äquivalent sind, wollen wir sie einzeln betrachten, da sie den Inhalt des Liouvilleschen Satzes unter verschiedenen Gesichtspunkten beleuchten.

Wir fassen die Bewegung der virtuellen Gesamtheit als eine Flüssigkeitsströmung im Phasenraum auf und führen zur Vereinfachung der Notierung den $2n$-dimensionalen Vektor der Phasengeschwindigkeit $\boldsymbol{v}$ mit den Komponenten $\dot{q}_i, \dot{p}_i (i = 1, \ldots, n)$ ein. Da die Systeme der virtuellen Gesamtheit in der Strömung weder entstehen noch verschwinden können, muß für dieselbe zunächst die Kontinuitätsgleichung

$$\operatorname{div} (\varrho\, \boldsymbol{v}) + \frac{\partial \varrho}{\partial t} = 0 \tag{3.1}$$

gelten. Weiter folgt aus den Bewegungsgleichungen

$$\operatorname{div} \boldsymbol{v} = 0. \tag{3.2}$$

Diese Gleichung besagt, daß die aus den Phasenpunkten gebildete „Flüssigkeit" inkompressibel ist. Durch Kombination von (3.1) und (3.2) erhält man

$$\frac{\partial \varrho}{\partial t} = - \sum_{i=1}^{n} \left(\frac{\partial \varrho}{\partial q_i} \dot{q}_i + \frac{\partial \varrho}{\partial p_i} \dot{p}_i \right). \tag{3.3}$$

Mit Benutzung der Bewegungsgleichungen (2.1) und des Symbols der Poisson-Klammern

$$\{J, K\} \equiv \sum_{i=1}^{n} \left(\frac{\partial J}{\partial q_i} \frac{\partial K}{\partial p_i} - \frac{\partial K}{\partial q_i} \frac{\partial J}{\partial p_i} \right) \tag{3.4}$$

kann (3.3) geschrieben werden

$$\frac{\partial \varrho}{\partial t} = - \{\varrho, H\}. \tag{3.5}$$

Der Ausdruck (3.3) bzw. (3.5) für die lokale Änderung der Phasendichte einer virtuellen Gesamtheit stellt die erste Formulierung des Liouvilleschen Satzes dar.

---

[1] J. Liouville: J. de Math. **3**, 348 (1838).
[2] L. Boltzmann: Wien. Ber. **58**, 517 (1868).

Wir betrachten nun die totale Dichteänderung, d.h. die Änderung der Phasendichte in der unmittelbaren Umgebung eines mit der „Flüssigkeit" bewegten Phasenpunktes. Dafür gilt allgemein

$$\frac{d\varrho}{dt} = \frac{\partial\varrho}{\partial t} + \sum_{i=1}^{n}\left(\frac{\partial\varrho}{\partial q_i}\dot{q}_i + \frac{\partial\varrho}{\partial p_i}\dot{p}_i\right). \tag{3.6}$$

Daraus folgt sofort mit (3.3)

$$\frac{d\varrho}{dt} = 0. \tag{3.7}$$

Die Phasendichte in der Umgebung eines bestimmten Phasenpunktes bleibt also während der Strömung konstant. Diese Formulierung des Liouvilleschen Satzes wird nach GIBBS [2] als *Prinzip von der Erhaltung der Phasendichte* (principle of conservation of density in phase) bezeichnet.

GIBBS hat noch eine weitere Fassung des Liouvilleschen Satzes angegeben, die den Begriff der Phasenausdehnung (extension in phase) benutzt. Darunter wird allgemein das zwischen irgendwelchen Grenzen erstreckte Integral

$$\int d\Omega$$

verstanden. Wir betrachten nun eine Phasenausdehnung, die so klein sei, daß darin ϱ als konstant angenommen werden kann. Die Zahl der Systeme der virtuellen Gesamtheit, die in dieses Gebiet des Phasenraumes fallen, ist dann

$$\Delta v = v\,\varrho\int d\Omega. \tag{3.8}$$

Wir denken uns jetzt die Grenzen der Phasenausdehnung durch die darauf liegenden Phasenpunkte fixiert und verfolgen die Bewegung der so bestimmten Phasenausdehnung durch den Phasenraum. Die Zahl der darin befindlichen Phasenpunkte kann sich im Laufe der Zeit nicht verändern. Sie können nämlich einerseits, da sie mechanische Systeme darstellen, weder entstehen noch verschwinden (was wir oben schon benutzt haben). Es können aber im Laufe der Bewegung auch keine Phasenpunkte die Grenzen der Phasenausdehnung überschreiten. Die Bewegung der Grenzen ist nämlich nach der Voraussetzung identisch mit der Bewegung der sie bestimmenden Systeme. Sie verläuft daher auf Trajektorien, und diese können wegen der Eindeutigkeit der mechanischen Bewegung nicht von anderen Trajektorien gekreuzt werden. Es muß daher gelten

$$\frac{d(\Delta v)}{dt} = v\,\frac{d\varrho}{dt}\int d\Omega + v\,\varrho\,\frac{d}{dt}\int d\Omega = 0. \tag{3.9}$$

Daraus folgt in Verbindung mit (3.7)

$$\frac{d}{dt}\int d\Omega = 0. \tag{3.10}$$

Da sich kleine Phasenausdehnungen beliebig kombinieren lassen, gilt (3.10) für Phasenausdehnungen beliebiger Größe. Diese Form des Liouvilleschen Satzes wird als *Prinzip von der Erhaltung der Phasenausdehnung* (principle of conservation of extension in phase) bezeichnet. Es besagt, daß eine Phasenausdehnung mit den oben definierten Grenzen während der Bewegung im Phasenraum stets ihr Volumen beibehält. Der Liouvillesche Satz macht jedoch keine Aussage über die Gestalt dieses Volumens, die sich im Laufe der Zeit ändern wird. Auf diesen letzteren Gesichtspunkt werden wir im Abschnitt A III (S. 226) zurückkommen.

Das Prinzip von der Erhaltung der Phasenausdehnung läßt sich noch auf eine andere Weise formulieren, wenn wir an die in Ziff. 2 besprochene Auffassung der

Bewegung als einer Transformation des Phasenraumes in sich selbst anknüpfen. Bezeichnen wir mit q', p' die Phasenkoordinaten zur Zeit t', mit q'', p'' die zur Zeit $t'' = t' + \tau$, so muß für die in der obigen Weise definierte Phasenausdehnung gelten

$$\int d\Omega' = \int \frac{\partial(q', p')}{\partial(q'', p'')}\, d\Omega''. \tag{3.11}$$

In Verbindung mit Gl. (3.10) folgt daraus

$$\frac{\partial(q', p')}{\partial(q'', p'')} = 1. \tag{3.12}$$

Eine Verschärfung der vorstehenden Aussagen ist möglich, wenn man die Begriffe der Maßtheorie verwendet[1]. Da wir dieselben in Abschnitt A II ohnehin benötigen, wollen wir sie bereits an dieser Stelle verwenden. Es sei M' ein im Lebesgueschen Sinne meßbarer Satz von Phasenpunkten, der durch die Bewegung der virtuellen Gesamtheit in dem Zeitintervall τ in den Satz M'' übergeht. Bezeichnen wir mit $\mathfrak{M}A$ das Lebesguesche Maß eines Satzes A, so ist

$$\mathfrak{M}\, M' = \int\limits_{M'} d\Omega'. \tag{3.13}$$

Aus dem obigen folgt dann unmittelbar

$$\frac{d\,\mathfrak{M}\, M''}{dt} = 0. \tag{3.14}$$

Das Maß meßbarer Punkt-Sätze ist somit eine Invariante der Bewegung der virtuellen Gesamtheit.

Aus dem Liouvilleschen Satz läßt sich eine bemerkenswerte Folgerung ableiten, die wir später (Ziff. 8) benötigen. Es sei M ein im Lebesgueschen Sinne meßbarer Satz von Phasenpunkten (von endlichem Maß) und $F(q, p)$ eine im Lebesgueschen Sinne über Γ integrable Phasenfunktion. In der Zeit τ geht M in den Satz M', das Volumenelement $d\Omega$ in das gleiche Volumenelement $d\Omega'$ über. Wir betrachten nun das Integral

$$\int\limits_{M'} F(q, p)\, d\Omega'. \tag{3.15}$$

Als neue Variable führen wir die Phasenkoordinaten des Punktes ein, der in der Zeit τ in den Punkt q, p übergeht. Dann erhalten wir mit (3.12)

$$\int\limits_{M'} F(q, p)\, d\Omega' = \int\limits_{M} F(q', p')\, d\Omega \tag{3.16}$$

oder

$$\int\limits_{M'} F(q, p)\, d\Omega = \int\limits_{M} F(q', p')\, d\Omega. \tag{3.17}$$

Wenn der Satz M einen invarianten Teil des Phasenraumes darstellt, folgt für beliebiges τ

$$\int\limits_{M} F(q, p)\, d\Omega = \int\limits_{M} F(q', p')\, d\Omega. \tag{3.18}$$

Der Liouvillesche Satz gilt, wie zuerst Boltzmann bemerkt hat, nur für den Phasenraum der generalisierten Koordinaten und Impulse, dagegen nicht für einen beispielsweise aus den generalisierten Koordinaten und Geschwindigkeiten

---

[1] Die wichtigsten Begriffe der Maßtheorie hat J. Lense in Bd. I dieses Handbuchs, S. 83 ff. kurz zusammengestellt.

konstruierten Phasenraum. Dies ist der Hauptgrund dafür, daß man die Hamilton-schen Bewegungsgleichungen als Ausgangspunkt für die Entwicklung der statisti-schen Mechanik wählt.

4. Kanonische Invarianz. Aus der Ableitung des Liouvilleschen Satzes (die nur die Gültigkeit der kanonischen Bewegungsgleichungen voraussetzt) ergibt sich, daß derselbe unabhängig ist von der speziellen Wahl der generalisierten Koordinaten und Impulse. Auf Grund dieses Sachverhaltes läßt sich leicht zeigen, daß Phasendichte und Phasenausdehnung invariant gegen kanonische Trans-formationen sind. Es seien etwa q, p die ursprünglichen, q', p' die daraus durch kanonische Transformation hervorgegangenen Phasenkoordinaten. Die als Funk-tionen dieser Variablen ausgedrückten Phasendichten seien zur Zeit t_1 ϱ_1 und ϱ_1', zur Zeit t_2 ϱ_2 und ϱ_2'. Verstehen wir darunter jetzt die Phasendichten in der Umgebung eines Phasenpunktes, der sich nach den Gleichungen der Mechanik bewegt, so folgt zunächst aus dem Liouvilleschen Satz, Gl. (3.7),

$$\varrho_1 = \varrho_2, \qquad \varrho_1' = \varrho_2'. \tag{4.1}$$

Wir führen nun einen dritten Satz von Variablen q'', p'' ein, der zur Zeit t_1 mit dem Satz q, p, zur Zeit t_2 mit dem Satz q', p' zusammenfällt. Die als Funk-tionen dieser Variablen ausgedrückten Phasendichten seien ϱ_1'' und ϱ_2''. Dann ist

$$\varrho_1 = \varrho_1'', \qquad \varrho_2' = \varrho_2'', \qquad \varrho_1'' = \varrho_2''. \tag{4.2}$$

Daraus folgt in Verbindung mit (4.1)

$$\varrho_1 = \varrho_1', \qquad \varrho_2 = \varrho_2', \tag{4.3}$$

womit die Invarianz der Phasendichte gegen kanonische Transformationen be-wiesen ist.

Die Zahl der Phasenpunkte innerhalb einer kleinen Phasenausdehnung (für welche die Phasendichte als konstant angenommen werden kann) ist bei Be-nutzung der zwei ersten Koordinatensätze

$$\Delta \nu = \nu \varrho \int d\Omega, \tag{4.4}$$

bzw.

$$\Delta \nu' = \nu \varrho' \int d\Omega'. \tag{4.5}$$

Da es sich in beiden Fällen um die gleichen Phasenpunkte handelt, ist $\Delta \nu = \Delta \nu'$ und somit nach (4.3)

$$\int d\Omega = \int d\Omega'. \tag{4.6}$$

Damit ist auch die Invarianz der Phasenausdehnung gegen kanonische Trans-formationen bewiesen. In gleicher Weise wie früher kann dieses Resultat für Phasenausdehnungen beliebiger Größe verallgemeinert werden. Aus der Defini-tion der Phasenausdehnung folgt, daß sie die Dimension der n-ten Potenz einer Wirkung, also $[\mathrm{g\ cm^2\ sec^{-1}}]^n$, besitzt. Ihr Zahlenwert hängt daher von den be-nutzten Einheiten ab.

Bildet man für den Konfigurationsraum den zur Phasenausdehnung analogen Ausdruck

$$\int \cdots \int dq_1 \ldots dq_n = \int \cdots \int \frac{\partial(q_1, \ldots, q_n)}{\partial(q_1', \ldots, q_n')}\, dq_1', \ldots, dq_n', \tag{4.7}$$

so sieht man sofort, daß derselbe nicht die Invarianzeigenschaft der Phasen-ausdehnung besitzt. Man kann aber, wie GIBBS [2] gezeigt hat, durch die folgende Überlegung zu einer Größe gelangen, welche invariant gegen kanonische Trans-formationen ist.

Bezeichnen wir mit E_{kin} die kinetische Energie, so ist

$$p_i = \frac{\partial E_{\mathrm{kin}}}{\partial \dot{q}_i}. \tag{4.8}$$

Es ist somit

$$\frac{\partial p_i'}{\partial p_j} = \frac{\partial}{\partial p_j}\left(\frac{\partial E_{\mathrm{kin}}}{\partial \dot{q}_i'}\right). \tag{4.9}$$

Ferner folgt aus

$$q_i' = q_i'(q_1, \ldots, q_n) \tag{4.10}$$

durch Differentiation

$$\dot{q}_i' = \frac{\partial q_i'}{\partial q_1}\dot{q}_1 + \cdots + \frac{\partial q_i'}{\partial q_n}\dot{q}_n. \tag{4.11}$$

Die $\dot{q}_i'$ sind somit lineare Funktionen der $\dot{q}_i$, wobei die Koeffizienten noch von den q_i abhängen können. Da die kinetische Energie eine homogene quadratische Funktion der generalisierten Geschwindigkeiten ist, müssen nach Gl. (4.8) die $\dot{q}_i$ ihrerseits lineare Funktionen der p_i sein (und umgekehrt). Daraus folgt, daß auch die $\dot{q}_i'$ lineare Funktionen der p_i sind (und umgekehrt). Unter dieser Voraussetzung ist aber

$$\frac{\partial}{\partial p_j}\left(\frac{\partial E_{\mathrm{kin}}}{\partial \dot{q}_i'}\right) = \frac{\partial}{\partial \dot{q}_i'}\left(\frac{\partial E_{\mathrm{kin}}}{\partial p_j}\right) \tag{4.12}$$

oder

$$\frac{\partial p_i'}{\partial p_j} = \frac{\partial \dot{q}_j}{\partial \dot{q}_i'}. \tag{4.13}$$

Da der Wert einer Determinanten durch Vertauschung von Zeilen und Spalten nicht geändert wird, folgt aus (4.13)

$$\frac{\partial(p_1', \ldots, p_n')}{\partial(p_1, \ldots, p_n)} = \frac{\partial(\dot{q}_1, \ldots, \dot{q}_n)}{\partial(\dot{q}_1', \ldots, \dot{q}_n')}. \tag{4.14}$$

Denken wir uns in Gl. (4.11) die gestrichenen und ungestrichenen Koordinaten vertauscht, so ergibt sich unmittelbar

$$\frac{\partial \dot{q}_j}{\partial \dot{q}_i'} = \frac{\partial q_j}{\partial q_i'}. \tag{4.15}$$

Durch Kombination von (4.14) und (4.15) erhalten wir

$$\frac{\partial(p_1', \ldots, p_n')}{\partial(p_1, \ldots, p_n)} = \frac{\partial(\dot{q}_1, \ldots, \dot{q}_n)}{\partial(\dot{q}_1', \ldots, \dot{q}_n')} = \frac{\partial(q_1, \ldots, q_n)}{\partial(q_1', \ldots, q_n')}. \tag{4.16}$$

Es ist daher

$$\left.\begin{aligned}
&\int\cdots\int\left[\frac{\partial(p_1, \ldots, p_n)}{\partial(\dot{q}_1, \ldots, \dot{q}_n)}\right]^{\frac{1}{2}} dq_1 \ldots dq_n \\
&= \int\cdots\int\left[\frac{\partial(p_1, \ldots, p_n)}{\partial(\dot{q}_1, \ldots, \dot{q}_n)}\right]^{\frac{1}{2}}\left[\frac{\partial(p_1', \ldots, p_n')}{\partial(p_1, \ldots, p_n)}\right]^{\frac{1}{2}}\left[\frac{\partial(\dot{q}_1, \ldots, \dot{q}_n)}{\partial(\dot{q}_1', \ldots, \dot{q}_n')}\right]^{\frac{1}{2}} dq_1' \ldots dq_n'.
\end{aligned}\right\} \tag{4.17}$$

Nun gilt aber

$$\left[\frac{\partial(p_1, \ldots, p_n)}{\partial(\dot{q}_1, \ldots, \dot{q}_n)}\right]^{\frac{1}{2}}\left[\frac{\partial(p_1', \ldots, p_n')}{\partial(p_1, \ldots, p_n)}\right]^{\frac{1}{2}}\left[\frac{\partial(\dot{q}_1, \ldots, \dot{q}_n)}{\partial(\dot{q}_1', \ldots, \dot{q}_n')}\right]^{\frac{1}{2}} = \left[\frac{\partial(p_1', \ldots, p_n')}{\partial(\dot{q}_1', \ldots, \dot{q}_n')}\right]^{\frac{1}{2}}. \tag{4.18}$$

Gl. (4.17) geht daher über in

$$\int\cdots\int\left[\frac{\partial(p_1, \ldots, p_n)}{\partial(\dot{q}_1, \ldots, \dot{q}_n)}\right]^{\frac{1}{2}} dq_1 \ldots dq_n = \int\cdots\int\left[\frac{\partial(p_1', \ldots, p_n')}{\partial(\dot{q}_1', \ldots, \dot{q}_n')}\right]^{\frac{1}{2}} dq_1' \ldots dq_n'. \tag{4.19}$$

Wir führen nun die Abkürzung

$$\Delta_{\dot{q}} = \frac{\partial(p_1, \ldots, p_n)}{\partial(\dot{q}_1, \ldots, \dot{q}_n)} \tag{4.20}$$

ein. Man sieht leicht, daß diese Größe die Hessesche Determinante der kinetischen Energie als Funktion der generalisierten Geschwindigkeiten darstellt. Die Gl. (4.19) besagt nun, daß der Ausdruck

$$\Omega_q = \int \cdots \int \Delta_{\dot{q}}^{\frac{1}{2}} \, dq_1 \ldots dq_n \tag{4.21}$$

invariant gegen kanonische Transformationen ist. Diese Größe wird nach GIBBS [2] als Konfigurationsausdehnung (extension in configuration) bezeichnet.

Die Hessesche Determinante $\Delta_{\dot{q}}$ hängt nur von den generalisierten Koordinaten, aber nicht von den generalisierten Geschwindigkeiten oder Impulsen ab. Für ein System von N gleichen, voneinander unabhängigen Massenpunkten der Masse m ist in Cartesischen Koordinaten

$$\frac{\partial p_i}{\partial \dot{q}_j} = m \, \delta_{ij}, \tag{4.22}$$

wo δ_{ij} das Kroneckersche Delta ist. In $\Delta_{\dot{q}}$ verschwinden daher alle Elemente außerhalb der Hauptdiagonalen, und es wird einfach

$$\Delta_{\dot{q}} = m^{3N}. \tag{4.23}$$

Betrachten wir die Konfigurationsausdehnung nur als Funktion des den Massenpunkten zur Verfügung stehenden Volumens V, so ergibt sich

$$\Omega_q^* = m^{\frac{3}{2}N} \int \cdots \int d x_1 \ldots dz_N = (m^{\frac{3}{2}} V)^N. \tag{4.24}$$

Die Hessesche Determinante der kinetischen Energie als Funktion der generalisierten Impulse bezeichnen wir mit Δ_p. Es ist dann

$$\Delta_p = \frac{\partial(\dot{q}_1, \ldots, \dot{q}_n)}{\partial(p_1, \ldots, p_n)} \tag{4.25}$$

und

$$\Delta_p = \Delta_{\dot{q}}^{-1}. \tag{4.26}$$

Auch Δ_p hängt also nur von den generalisierten Koordinaten ab. In ähnlicher Weise wie oben kann man nun zeigen, daß der Ausdruck

$$\Omega_p = \int \cdots \int \Delta_p^{\frac{1}{2}} \, dp_1 \ldots dp_n \tag{4.27}$$

invariant gegen kanonische Transformationen ist. Er wird nach GIBBS [2] als Geschwindigkeitsausdehnung (extension in velocity) bezeichnet. In Analogie zu Gl. (4.24) kann man die Geschwindigkeitsausdehnung als Funktion einer vorgegebenen oberen Grenze der kinetischen Energie E_{kin} betrachten. Es gilt dann, wie wir in Ziff. 25 zeigen werden, die Beziehung

$$\Omega_p^* = \frac{(2\pi E_{\text{kin}})^{\frac{n}{2}}}{\Gamma\left(\frac{n}{2} + 1\right)}. \tag{4.28}$$

Im Gegensatz zu Gl. (4.24) (die nur für den oben betrachteten Spezialfall gilt) ist Gl. (4.28) allgemein gültig.

5. Das statistische Gleichgewicht. Die bisherigen Betrachtungen beziehen sich auf statistische Gesamtheiten, die sich durch die Phasendichte nach Gl. (2.3) und (2.4) beschreiben lassen. Um die Theorie weiter zu entwickeln, ist es notwendig, spezieller Annahmen über die Gesamtheiten einzuführen. Wir betrachten dieselben zunächst wieder rein formal als Aussonderung einer bestimmten Klasse von Gesamtheiten. Die Frage, weshalb grade diese Klasse für den weiteren Aufbau der Theorie verwendet wird, werden wir in Abschnitt A II erörtern.

Wir setzen also von jetzt ab stets voraus, daß die betrachtete Gesamtheit aus konservativen Systemen besteht, daß somit die Hamilton-Funktion nicht explizit von der Zeit abhängt. Da für uniforme Integrale der Bewegungsgleichungen F_i gilt

$$\{F_i, H\} = 0, \tag{5.1}$$

haben wir für konservative Systeme

$$\{H, H\} = 0 \tag{5.2}$$

oder

$$H(\boldsymbol{q}, \boldsymbol{p}) = E, \tag{5.3}$$

wo E die Gesamtenergie des Systems ist. Die zweiten Annahme, die wir machen, besagt, daß die lokale Phasendichte nicht von der Zeit abhängt, daß also

$$\frac{\partial \varrho}{\partial t} = 0 \tag{5.4}$$

ist. Wenn diese Bedingung erfüllt ist, sagt man, daß die Gesamtheit sich im statistischen Gleichgewicht befindet (stationäre Gesamtheit). Aus dem Liouvilleschen Satz, Gl. (3.5), ergibt sich unmittelbar als notwendige und hinreichende Bedingung für das statistische Gleichgewicht

$$\{\varrho, H\} = 0. \tag{5.5}$$

Wir können nun ohne Beschränkung der Allgemeinheit annehmen, daß ϱ sich in der Form

$$\varrho = \varphi(F_1, \ldots, F_r) \tag{5.6}$$

darstellen läßt, wo die F_i zunächst noch nicht näher spezifizierte Funktionen der generalisierten Koordinaten und Impulse sind. Damit wird aus Gl. (5.5)

$$\sum_i \frac{\partial \varrho}{\partial F_i} \cdot \{F_i, H\} = 0. \tag{5.7}$$

Das kann allgemein nur gelten, wenn

$$\{F_i, H\} = 0 \tag{5.8}$$

ist. Für das statistische Gleichgewicht einer virtuellen Gesamtheit aus konservativen Systemen ist somit notwendig und hinreichend, daß die Phasendichte nur von den uniformen Integralen der Bewegungsgleichungen abhängt. Über die Zahl der existierenden uniformen Integrale gibt es keine allgemeine Aussage. Für den Aufbau der Theorie kommen aber explizit offenbar nur solche in Betracht, deren Existenz sich für abgeschlossene konservative Systeme allgemein beweisen läßt, also das Energie-Integral, die drei Impuls-Integrale und die drei Drehimpulsintegrale. Wir beschränken uns in diesem Artikel bei expliziten Formulierungen auf den Fall, daß die Phasendichte nur von der Energie abhängt. Die Gründe für diese Wahl werden wir in Abschnitt A II erörtern. Allgemeinere Fälle sind verschiedentlich in der Literatur (z.B. [2]) diskutiert und in neuerer Zeit ausführlich von Grad[1] behandelt worden. Für die Theorie der Gleichgewichtseigenschaften der Materie (mit der wir es in erster Linie zu tun haben) haben diese Untersuchungen jedoch bisher keine erhebliche Bedeutung erlangt.

---

[1] H. Grad: Comm. Pure Appl. Math. **5**, 455 (1952)

Die wichtigste Eigenschaft einer Gesamtheit im statistischen Gleichgewicht besteht in der Tatsache, daß alle mit einer solchen Gesamtheit gebildeten Phasen-Mittelwerte nach Gl. (2.6) zeitunabhängig sind. Darauf ergibt sich unmittelbar die zentrale Bedeutung derartiger Gesamtheiten für die Theorie der Gleichgewichtseigenschaften der Materie.

II. Axiomatische Grundlagen der statistischen Mechanik.

6. Allgemeine Gesichtspunkte. In Abschnitt A I haben wir aus den Hamiltonschen Bewegungsgleichungen einige Aussagen über statistische Gesamtheiten abgeleitet (wobei wir gelegentlich statistische Begriffe benutzt haben). Diese Ergebnisse sind zunächst rein formaler Natur, denn die Physik hat es mit dem einzelnen System zu tun, das jeweils Gegenstand der experimentellen Untersuchung ist. Die eigentliche Problematik der statistischen Mechanik liegt in der Frage: Wie kann man den Formalismus der Ensemble-Theorie zum Aufbau einer physikalischen Theorie verwenden, die Aussagen über ein Original-System liefert?

. Das Wesen dieser Fragestellung läßt sich am einfachsten aus der historischen Entwicklung der statistischen Theorie verstehen. Dieselbe hat ihren Ursprung bekanntlich in der kinetischen Gastheorie, die nach den ersten Ideen von BERNOULLI (1738) und KRÖNIG (1856) vor allem von CLAUSIUS, MAXWELL und BOLTZMANN entwickelt wurde. Ihre endgültige Formulierung hat sie in der berühmten Boltzmannschen Stoßgleichung (s. Ziff. 14) erfahren, deren Ableitung auf dem sog. Stoßzahlansatz beruht. Die kinetische Gastheorie ist im strengen Sinne die mechanische Theorie eines aus Atomen bestehenden Einzelsystems, welche die vollständige Lösung der Bewegungsgleichungen durch Benutzung wahrscheinlichkeitstheoretischer Begriffe und Überlegungen umgeht. Die Stoßgleichung führt weiter zu dem Boltzmannschen H-Theorem (s. Ziff. 14), nach dem ein System sich in Richtung auf einen Gleichgewichtszustand entwickelt, der durch ein Minimum der Funktion H charakterisiert ist, und in diesem Zustand verharrt. Es wurde bald erkannt, daß diese Aussage nicht richtig sein kann, und daß die Ursache dafür in dem Stoßzahlansatz zu suchen ist in dem Sinne, daß dieser nur als eine statistische Aussage verstanden werden darf. Daraus folgt, daß die Funktion H nicht im Laufe der Zeit monoton auf einen Gleichgewichtswert abfällt, sondern einen viel komplizierteren Verlauf zeigt, der die statistischen Schwankungen des Systems widerspiegelt. Damit ergibt sich die zentrale Frage: Wie läßt sich überhaupt der Gleichgewichtszustand eines Systems in einer atomistischen Theorie beschreiben? In der makroskopischen Thermodynamik wird das Gleichgewicht als der Endzustand definiert, dem ein sich selbst überlassenes, d. h. abgeschlossenes System zustrebt. Die wesentliche Schwierigkeit, auf die ein solches Konzept in der atomistischen Theorie stößt, liegt in der Tatsache, daß hier ein solcher ausgezeichneter Zustand zunächst nicht erkennbar ist. Man kann sich das an einem einfachen Beispiel klarmachen. Wir betrachten zwei, durch ein zunächst geschlossenes Ventil verbundene Gasflaschen, von denen eine leer, die andere mit Wasserstoff gefüllt sei. Öffnen wir das Ventil, so setzt ein Prozeß ein, dessen makroskopischer Endzustand (wenn wir von dem Einfluß der Gravitation absehen) durch einen in beiden Flaschen überall gleichen Wert der Gasdichte charakterisiert wird. In atomistischer Betrachtung wird dieser Zustand immer nur momentan vorhanden sein. Die thermische Bewegung der Moleküle bewirkt ununterbrochen Abweichungen von der gleichmäßigen Dichteverteilung, die als Schwankungen bezeichnet werden. Der Kern des Problems liegt also in der Frage, wie man die Tatsache der Schwankungen bei der

Definition des Gleichgewichts berücksichtigt. Tatsächlich bildet diese Frage, wie wir später (Ziff. 67) noch ausführlicher sehen werden, den Angelpunkt für die statistische Begründung der Thermodynamik.

Da die atomistische Definition des Gleichgewichtes mit dem makroskopischen Konzept konsistent sein muß, darf sie die Schwankungen nicht explizit enthalten. In der Theorie des Einzelsystems gibt es daher offenbar nur zwei Möglichkeiten: Entweder man zeichnet einen mikroskopischen Zustand (etwa den Zustand gleichmäßiger Dichteverteilung) definitionsmäßig als Gleichgewichtszustand aus, oder man bringt die Schwankungen durch Mittelwertbildung zum Verschwinden. Beide Wege sind bereits von Boltzmann beschritten worden. Seine erste Definition[1] besagt

1. Der Gleichgewichtszustand eines Systems ist der unter den Nebenbedingungen konstanter Energie und konstanter Teilchenzahl wahrscheinlichste Zustand.

Dieses Konzept liegt bekanntlich der üblichen Formulierung der μ-Raum-Statistik zugrunde, auf die wir in Ziff. 27 kurz eingehen werden. Es führt, in Übereinstimmung mit den Folgerungen aus der Stoßgleichung, zu der Aussage, daß der Gleichgewichtszustand durch das Minimum der oben erwähnten Funktion H und durch die Maxwell-Boltzmannsche Energieverteilung charakterisiert ist.

Die Definition 1 hat jedoch zwei wesentliche Schwächen. Einmal läßt sich auf dieser Grundlage die Theorie nur für Systeme von unabhängigen Teilchen explizit durchführen. Zum anderen erscheinen hier notwendig die Schwankungen als Abweichungen vom Gleichgewicht, was mit dem makroskopischen Begriff des Gleichgewichtes nur schwer in Einklang zu bringen ist.

Die zweite Boltzmannsche Definition[2] besagt

2. Die Gleichgewichtseigenschaften eines Systems sind die über eine unendlich lange Zeit gebildeten Mittelwerte der betreffenden Eigenschaften.

Man sieht sofort, daß die obigen Einwendungen gegen die Definition 1 hier nicht zutreffen. Dafür treten allerdings gewisse andere Schwierigkeiten auf, die wir weiter unten erörtern. Zunächst ist mit der Definition 2 praktisch wenig anzufangen. Nach dem Grundkonzept der atomistischen Theorie muß jede meßbare Eigenschaft durch eine Phasenfunktion darstellbar sein. Die Berechnung des zeitlichen Mittelwertes würde daher die vollständige Lösung der Bewegungsgleichungen des Systems erfordern. Boltzmann versuchte diese Schwierigkeit zu umgehen durch Einführung der sog. Ergodenhypothese. Dieselbe besagt, daß die Trajektorie eines konservativen Systems durch jeden Punkt der Energiefläche hindurchgeht. Da aus dem Liouvilleschen Satz folgt, daß längs der Trajektorie sich ein Phasenpunkt in gleich großen Volumenelementen gleiche Zeiten aufhält, wird jetzt das Zeitmittel gleich dem Phasenmittel einer statistischen Gesamtheit, die mit konstanter Flächendichte über die Energiefläche verteilt ist. Damit ist in der Tat die Lösung der Bewegungsgleichungen entbehrlich gemacht und das Problem auf die Berechnung der Phasenmittelwerte zurückgeführt, die, wenigstens für Systeme mit separierbarer Hamilton-Funktion, keine prinzipiellen Schwierigkeiten bietet. Die Anwendung der Methode auf Systeme aus unabhängigen Teilchen ergibt wieder das Maxwell-Boltzmannsche Energieverteilungsgesetz.

Die im vorstehenden skizzierte Boltzmannsche Überlegung (die wir im folgenden der Kürze halber als *Ergodentheorie* bezeichnen)[3] macht zum ersten Male (1871) Gebrauch von statistischen Gesamtheiten und bezeichnet daher eigentlich

---

[1] L. Boltzmann: Wien. Ber. **76**, 373 (1877).
[2] L. Boltzmann: Wien. Ber. **58**, 517 (1868).
[3] In der Literatur wird der Ausdruck häufig in engerer Bedeutung gebraucht.

den Beginn der modernen statistischen Mechanik[1]. Allerdings wird die Ensemble-
theorie hier nur als mathematischer Trick eingeführt. Gerade deshalb ist andererseits die Antwort auf die eingangs gestellte Frage hier sehr einfach und unproblematisch: Die Phasenmittelwerte der statistischen Gesamtheit stellen die zeitlichen
Mittelwerte des Einzelsystems in dem oben erläuterten Sinne dar. Der Gedankengang der Ergodentheorie hat etwas außerordentlich Bestechendes. Neben der
einfachen Interpretation des Formalismus der Ensembletheorie scheint er im
Prinzip die Möglichkeit einer lückenlosen deduktiven Begründung der statistischen
Mechanik zu eröffnen. Er enthält zudem eine Reihe von fesselnden mathematischen Problemen. Es erscheint daher verständlich, daß auch heute noch zahlreiche, insbesondere mathematisch orientierte, Theoretiker die Ergodentheorie
als die einzige gesunde Basis der statistischen Mechanik betrachten. Eine genauere Analyse zeigt indessen, daß hier einige grundsätzliche Schwierigkeiten
auftreten, die es zweifelhaft erscheinen lassen, ob die Ergodentheorie eine geeignete Grundlage der statistischen Mechanik darstellt. Da wir in den folgenden
Ziffern die Ergodentheorie ausführlicher behandeln, stellen wir die Erörterung
dieser Fragen zurück und bemerken hier nur, daß sich die Hoffnung auf eine
deduktive Begründung der statistischen Mechanik mit Hilfe der Ergodentheorie
trotz jahrzehntelangen Bemühungen bisher nicht erfüllt hat.

Die eigentliche Entwicklung der Ensemble-Theorie in ihren wesentlichen
Zügen geht bekanntlich auf GIBBS [2] zurück. EINSTEIN[2] hat, was wenig bekannt
ist, gleichzeitig und unabhängig zwei Arbeiten über den gleichen Gegenstand
veröffentlicht. Die Gibbssche Untersuchung besitzt einen stark formalen Charakter, während die physikalischen Gesichtspunkte etwas zurücktreten. Zum Teil
erklärt sich dies jedenfalls (wie GIBBS selbst andeutet) durch die damalige, in
vielen Hinsichten ungeklärte Situation unmittelbar vor dem Auftreten der Quantentheorie. Immerhin ist das zugrunde liegende physikalische Konzept klar zum
Ausdruck gebracht. Im Vorwort heißt es: "The laws of thermodynamics, as
empirically determined, express the approximate and probable behavior of systems
of a great number of particles, or, more precisely, they express the laws of mechanics for such systems as they appear to beings who have not the fineness of
perception to enable them to appreciate quantities of the order of magnitude of
those which relate to single particles, and who cannot repeat their experiments
often enough to obtain any but the most probable results. The laws of statistical
mechanics apply to conservative systems of any number of degrees of freedom,
and are exact. ... The laws of thermodynamics may easily be obtained from the
principles of statistical mechanics, of which they are incomplete expression."
Das heißt in etwas anderer Formulierung: Die Unmöglichkeit, den mechanischen
Zustand eines makroskopischen Systems experimentell exakt zu erfassen, zwingt
dazu, ein solches System durch ein statistisches Ensemble zu beschrieben. Aus
den mit Hilfe eines solchen Ensembles gebildeten Mittelwerten lassen sich die
Gleichungen der Thermodynamik konstruieren. Diese besitzen somit für das
Einzelsystem nur den Charakter von Wahrscheinlichkeitsaussagen. Während
das Ergebnis einer (gedachten) mechanischen Messung in der klassischen Mechanik
das Ergebnis einer zweiten Messung vollständig determiniert, liefert eine thermodynamische Messung nur einen „Erwartungswert" für eine weitere Messung.
Derselbe stellt, strenggenommen, den Mittelwert aus unendlich vielen Messungen
unter gleichen Bedingungen dar; nur für Systeme von sehr vielen Freiheitsgraden

---

[1] Der Ausdruck „statistisch-mechanisch" scheint zuerst von MAXWELL [Cambridge Phil.
Soc. Trans. **12**, 547 (1879)] im Zusammenhang mit der Ensembletheorie gebraucht worden
zu sein.
[2] A. EINSTEIN: Ann. Phys. **9**, 417 (1902); **11**, 170 (1903).

kann er als Voraussage für das Ergebnis einer Einzelmessung interpretiert werden, weil dann die relativen Schwankungen am den Mittelwert asymptotisch verschwinden.

Es ist zunächst eine einfache Konsequenz dieser Gedankengänge, daß der Gleichgewichtszustand eines Systems durch ein im statistischen Gleichgewicht befindliches Ensemble dargestellt wird. Dagegen ist die am Anfang der Entwicklung stehende Frage, wie das Gleichgewicht zustande kommt, im Rahmen der Ensemble-Theorie ein außerordentlich schwieriges Problem. Gibbs hat zu dieser Frage nur qualitative Betrachtungen angestellt, im übrigen seine Theorie jedoch unabhängig davon entwickelt. Einstein, dessen Interpretation der virtuellen Gesamtheit sich im wesentlichen mit der obigen deckt, hat die Existenz des Gleichgewichtszustandes explizit als Erfahrungstatsache eingeführt. Wir werden uns im folgenden zunächst ebenfalls auf diesen Standpunkt stellen und das Problem der Einstellung des Gleichgewichtes gesondert in Abschnitt A III (S. 212) behandeln.

Durch die Ideen von Gibbs und Einstein ist zwar die Ensembletheorie zu einer physikalischen Theorie im eigentlichen Sinne geworden. Abgesehen von dem Problem der Einstellung des Gleichgewichtes, bleiben aber zunächst noch zwei wesentliche Fragen offen:

1. Wie ist für ein durch gegebene makroskopische Parameter charakterisiertes System das entsprechende statistische Ensemble zu konstruieren?

2. Wie ist eine makroskopische Messung zu definieren und die Identifizierung von Phasen-Mittelwerten mit makroskopischen Meßwerten im einzelnen zu verstehen?

Was die Frage 1 betrifft, so führen sowohl Gibbs wie Einstein gewisse Gesamtheiten axiomatisch ein. Diese Betrachtungen von Gibbs über die Gleichgewichtseinstellung machen die Wahl in etwa plausibel, können aber nicht als strenge Rechtfertigung gelten. In der Frage 2, die sowohl für das Problem der Gleichgewichtseinstellung wie im Hinblick auf die Abgrenzung gegen den Grundgedanken der Ergodentheorie wichtig ist, bemerkt Einstein lediglich, daß man meßbare Größen als zeitliche Mittelwerte gewisser Phasenfunktionen anzusehen habe, ohne jedoch das Problem näher zu analysieren.

In den folgenden Jahrzehnten war die statistische Mechanik völlig von der Ergodentheorie beherrscht. P. und T. Ehrenfest [3], Ornstein[1] und Uhlenbeck[2] betrachteten die Ensemble-Theorie in erster Linie als mathematischen Trick zur Berechnung der Zeit-Mittelwerte von Phasenfunktionen. Indessen zeigt die Tatsache, daß selbst Gibbs diesen Gedanken in gewissem Umfange gelten läßt, wie stark damals die verschiedenen Entwicklungen noch ineinander verflochten waren.

Das Verdienst, die grundlegenden Ideen von Gibbs und Einstein erstmalig prägnant und ausführlich dargestellt zu haben, gebührt unstreitig Tolman [12], der zugleich eine klare Antwort auf die obige Frage 1 gegeben hat. Tolman hat den Begriff des „repräsentativen Ensembles" geprägt, welches ein Originalsystem darstellt, das durch makroskopische Parameter, also vom Standpunkt der Mechanik nur ganz fragmentarisch, charakterisiert ist. Er betont, daß die Konstruktion des repräsentativen Ensembles die Einführung eines zusätzlichen Axioms erfordert, also nicht aus den Prinzipien der Mechanik und Statistik deduziert werden kann. Dieses Axiom wird von Tolman die „Hypothese der gleichen a priori-Wahrscheinlichkeiten" genannt. Es besagt, daß gleich große

---

[1] L. S. Ornstein: Diss. Leiden 1908.
[2] G. E. Uhlenbeck: Diss. Leiden 1927.

Gebiete des Phasenraumes gleiche a priori-Wahrscheinlichkeiten besitzen, wenn sie gleich gut unsere fragmentarische Kenntnis des Systems repräsentieren. Mit TOLMANs eigenen Worten: "The essence of the relationship between representative ensemble and system of interest is that the distribution of the members of the ensemble over different states agrees with what is known as to be the actual state of the system of interest but is otherwise uniform in the phase space in accordance with the hypothesis as to equal a priori probabilities." Die obige Frage 2 ist erst in neuerer Zeit von KIRKWOOD[1] ausführlich diskutiert worden.

Fassen wir noch einmal zusammen, so müssen wir zunächst feststellen, daß eine rein deduktive Begründung der statistischen Mechanik bisher nicht gelungen ist. Für die physikalische Interpretation der Ensemble-Theorie haben sich zwei Möglichkeiten ergeben: Die Ergoden-Theorie und die Theorie des repräsentativen Ensembles. In beiden Fällen ist es notwendig, gewisse Axiome zusätzlich einzuführen. In den folgenden Ziffern werden wir beide Theorien ausführlich entwickeln und diskutieren. Das Problem der Einstellung des Gleichgewichtes kann zwar, wenn man die Existenz des Gleichgewichtes als Erfahrungstatsache einführt, methodisch von der eigentlichen Axiomatik getrennt werden. Sachlich bildet es jedoch eine notwendige Ergänzung derselben. Lediglich aus praktischen Gründen behandeln wir diese Fragen erst in Abschnitt A III.

7. Das Ergodenproblem. Wir betrachten ein konservatives System von n Freiheitsgraden und setzen voraus, daß außer dem Energie-Integral keine weiteren uniformen Integrale der Bewegungsgleichungen existieren[2]. Die Trajektorie des Systems verläuft dann vollständig auf der Energiefläche

$$H(\boldsymbol{q}, \boldsymbol{p}) = E. \tag{7.1}$$

Die atomistische Theorie führt notwendig zu der Folgerung, daß jeder meßbaren Größe eine Phasenfunktion $F(\boldsymbol{q}, \boldsymbol{p})$ zugeordnet werden kann. Die Grundannahme der Ergodentheorie besagt, daß das Ergebnis einer makroskopischen Messung an einem im Gleichgewicht befindlichen System durch das längs der Trajektorie genommene Zeitmittel der Funktion $F(\boldsymbol{q}, \boldsymbol{p})$ gegeben ist, wobei das Zeitmittel durch

$$\widehat{F} = \lim_{\tau \to \infty} \frac{1}{\tau} \int_{t_0}^{t_0+\tau} F(\boldsymbol{q}, \boldsymbol{p}) \, dt \tag{7.2}$$

definiert ist. Nehmen wir an, daß $\widehat{F}$ unabhängig von der Wahl des Zeitpunktes t_0 ist, so erfordert die Berechnung des Zeitmittels noch die Kenntnis der Trajektorie, d.h. die Lösung der Bewegungsgleichungen des betrachteten Systems. Die Ergodentheorie postuliert nun, daß die Berechnung des Zeitmittels ersetzt werden kann durch die Berechnung des Phasenmittels einer stationären statistischen Gesamtheit, die mit konstanter Flächendichte über die Energiefläche verteilt ist, während im übrigen Phasenraum die Phasendichte identisch Null ist. Es soll also

$$\widehat{F} = \overline{F} \tag{7.3}$$

sein, wo

$$\overline{F} = \int F(\boldsymbol{q}, \boldsymbol{p}) \, \varrho(\boldsymbol{q}, \boldsymbol{p}) \, d\Omega \tag{7.4}$$

ist und $\varrho(\boldsymbol{q}, \boldsymbol{p})$ die erwähnten speziellen Eigenschaften besitzt. (Die explizite Formulierung, die wir in diesem Zusammenhang nicht benötigen, geben wir in Ziff. 20.) Die als Postulat formulierte Gl. (7.3) stellt das sog. Ergodenproblem dar.

---

[1] J. G. KIRKWOOD: J. Chem. Phys. **14**, 180 (1946).

[2] Diese Voraussetzung dient hier lediglich der Vereinfachung der Darstellung, soll aber zunächst nicht als notwendig für die folgende Überlegung betrachtet werden. Zu der letzteren Frage vgl. Ziff. 9 und 10.

Wir skizzieren zunächst den Boltzmannschen Lösungsversuch[1]. Es seien Γ_1 und Γ_2 zwei Volumenelemente des Phasenraumes derart, daß die „Phasenflüssigkeit" die zur Zeit t_0 Γ_1 füllt, zur Zeit t_0+t das Volumenelement Γ_2 füllt (Fig. 1).

Aus dem Liouvilleschen Satz folgt dann sofort, daß Γ_1 und Γ_2 dasselbe Volumen haben. Es sei nun $ABCD$ eine Trajektorie. Wir denken uns, wie in Ziff. 3, die Grenzen der betrachteten Phasenausdehnung durch die darauf liegenden Phasenpunkte fixiert. Dann muß die Zeit, die ein Phasenpunkt auf der betrachteten Trajektorie von A bis C benötigt, gleich sein der Zeit, die er von B bis D braucht und weiter gleich der Zeit t. Ziehen wir die von B bis C gebrauchte Zeit ab, so folgt, daß die Zeit von A bis B gleich ist der Zeit von C bis D. Das heißt mit anderen Worten, daß der Phasenpunkt längs der Trajektorie gleiche Zeiten in gleich großen Volumenelementen verweilt. Verfolgen wir das System über eine lange Zeit τ, so muß der Phasenpunkt jedesmal, wenn er durch Γ_1 geht, auch Γ_2 passieren. Es ist daher

$$\lim_{\tau \to \infty} \frac{dt(\Gamma_1)}{\tau} = \lim_{\tau \to \infty} \frac{dt(\Gamma_2)}{\tau}. \tag{7.5}$$

Kombinieren wir Γ_1 und Γ_2 zu einem Volumenelement Γ_3, so ist offenbar der Bruchteil von τ, den das System in Γ_3 verweilt, doppelt so groß wie der in Γ_2 verbrachte. Da wir kleine Phasenausdehnungen beliebig kombinieren können (was wir in Ziff. 3 schon benutzt haben) folgt allgemein

$$\lim_{\tau \to \infty} \frac{dt}{\tau} = \frac{d\Omega}{\int d\Omega}, \tag{7.6}$$

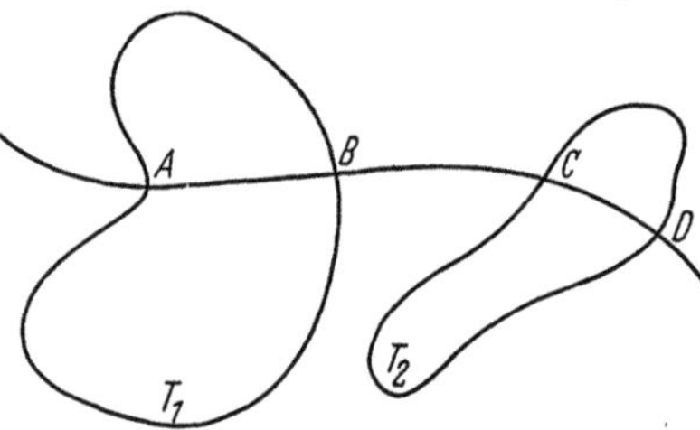
Fig. 1. Zur Ableitung der Gl. (7.6).

wo $\int d\Omega$ die (als endlich vorausgesetzte) Phasenausdehnung bezeichnet, innerhalb deren die Trajektorie verläuft. Damit ist gezeigt, daß allgemein längs der Trajektorie der Bruchteil der Zeit, den das System in einem Volumenelement verbringt, proportional der Größe desselben ist.

Die vollständige Ableitung der Gl. (7.3) erfordert nun noch den Beweis, daß die vorstehende Aussage auf die ganze Energiefläche ausgedehnt werden kann, d.h. einen allgemeinen Satz über den Verlauf der Trajektorie. Die ersten derartigen Sätze waren die Boltzmannsche Ergodenhypothese und die Maxwellsche assumption of continuity of path[2]. Beide sagen das gleiche aus: Die Trajektorie geht durch jeden Punkt der Energiefläche hindurch. Darauf folgt unmittelbar, daß Gl. (7.6) für die ganze Energiefläche gilt und damit weiter die Gültigkeit der Gl. (7.3).

Man kann sich den Inhalt der Ergodenhypothese noch in etwas anderer Weise klarmachen[3]. Aus der Tatsache, daß die hier betrachtete Gesamtheit stationär ist, folgt zunächst

$$\overline{F} = \widehat{\overline{F}}. \tag{7.7}$$

Wegen der Vertauschbarkeit der Mittelwertbildungen (die wir in Ziff. 11 ausführlich beweisen werden) muß

$$\widehat{\overline{F}} = \overline{\widehat{F}} \tag{7.8}$$

sein. Nun ist durch die Bewegungsgleichungen eindeutig das Richtungsfeld im Γ-Raum festgelegt; wir haben

$$dq_1 : dq_2 : \ldots : dp_{n-1} : dp_n = \frac{\partial H}{\partial p_1} : \frac{\partial H}{\partial p_2} : \ldots : -\frac{\partial H}{\partial q_{n-1}} : -\frac{\partial H}{\partial q_n}. \tag{7.9}$$

---

[1] L. Boltzmann: Wien. Ber. **58**, 517 (1868); **63**, 679 (1871).
[2] J. Cl. Maxwell: Cambridge Phil. Soc. Trans. **12**, 547 (1879).
[3] Der folgende Gedankengang rührt im wesentlichen von Maxwell[2] her.

Aus der Ergodenhypothese folgt dann, daß es überhaupt nur eine Trajektorie gibt und daß die Systeme der Gesamtheit sich lediglich durch den Zeitpunkt unterscheiden, in dem sie einen bestimmten Punkt der Trajektorie passieren. Das Zeitmittel ist daher (wenn wir die Unabhängigkeit von t_0 voraussetzen) für alle Systeme der Gesamtheit gleich. Es gilt also

$$\overline{\widehat{F}} = \widehat{F}. \tag{7.10}$$

Daraus folgt in Verbindung mit Gln. (7.7) und (7.8) unmittelbar die Gl. (7.3).

Die Ergodenhypothese in ihrer ursprünglichen Form ist jedoch mathematisch unhaltbar. Ein sehr einfaches und anschauliches Argument dafür ist von TOLMAN [12] angegeben worden. Die Bewegungsgleichungen (2.1) besitzen $2n$ Konstanten der Bewegung. Eine von diesen ist das Energie-Integral und eine fixiert die Zeitskala entlang der Trajektorie. Wählen wir nun auf der Energiefläche einen Punkt P, so sind damit die restlichen $2n-2$ Konstanten festgelegt. Wählen wir andere Werte für diese Konstanten, so entspricht dem etwa ein Punkt Q auf der Energiefläche. Es gibt aber keine Trajektorie, die gleichzeitig durch P und Q geht. Um diese Überlegung zu einem vollständigen Beweis auszugestalten, müßte allerdings noch gezeigt werden, daß die $2n-2$ Konstanten der Bewegung nicht „fast überall" auf der Energiefläche konstant sind.

Der strenge Beweis, daß die Ergodenhypothese falsch ist, ist bereits 1913 von ROSENTHAL[1] und PLANCHEREL[2] mit Hilfe der Maßtheorie geführt worden. Wir betrachten einen Punkt P der Trajektorie und ein kleines Gebiet Γ_1' von endlichem Maß um P. Da die Trajektorie die ganze Energiefläche überdecken soll, kann sie offenbar nicht vollständig in Γ_1' verlaufen. Da ferner für eine nichtperiodische Bewegung kein Punkt der Energiefläche mehrfach durchlaufen wird, muß der in Γ_1' liegende Teil der Trajektorie eine Folge von getrennten Segmenten sein, die einer Folge von endlichen Zeitintervallen entsprechen. Diese Zeitintervalle bilden einen abzählbaren Satz. Das gleiche gilt daher für die Segmente der Trajektorie in Γ_1'. Dieselbe hat somit das Maß Null. Da Γ_1' nach der Voraussetzung ein endliches Maß hat, kann die Trajektorie nicht durch alle Punkte von Γ_1' hindurchgehen. Das gleiche gilt a fortiori für die gesamte Energiefläche.

Schon vor der endgültigen Widerlegung der Ergodentheorie wurde von P. und T. EHRENFEST [3] die sog. *Quasi-Ergodenhypothese* aufgestellt[3]. Danach soll die Trajektorie jedem Punkte der Energiefläche beliebig nahekommen, ohne jedoch wirklich durch jeden Punkt hindurchzugehen. Diese Formulierung vermeidet in der Tat die im vorstehenden angeführten Schwierigkeiten, genügt aber andererseits, um aus Gl. (7.6) die Gl. (7.3) abzuleiten. Der erste ausführliche Beweis dafür wurde von ROSENTHAL[4] gegeben. Derselbe ist allerdings (wegen eines doppelten Grenzüberganges) nicht völlig streng. Wir werden jedoch weiter unten noch auf einem anderen Wege zeigen, daß das Resultat korrekt ist. Durch die Quasi-Ergodenhypothese war somit zwar eine in sich konsistente Formulierung der Theorie erreicht; es handelte sich aber doch zunächst nur um eine Hypothese. Erst 1923 gelang es FERMI[5], zu zeigen, daß eine gewisse Klasse von mechanischen Systemen, die sog. „kanonischen Normalsysteme", sich quasiergodisch verhalten. Den Beweis des Fermischen Satzes bringen wir in Ziff. 10.

---

[1] A. ROSENTHAL: Ann. Phys. **42**, 796 (1913).

[2] M. PLANCHEREL: Ann. Phys. **42**, 1061 (1913).

[3] Es muß jedoch bemerkt werden, daß P. und T. EHRENFEST die Quasi-Ergodenhypothese selbst sehr skeptisch beurteilen.

[4] A. ROSENTHAL: Ann. Phys. **43**, 894 (1914).

[5] E. FERMI: Physik Z. **24**, 261 (1923).

Mit diesem Satz war die Möglichkeit eines quasi-ergodischen Verhaltens nachgewiesen. Obschon Poincaré[1] schon 1892 gezeigt hatte, daß das reduzierte Dreikörperproblem in die Klasse der kanonischen Normalsysteme fällt, blieb nach wie vor die Frage offen, ob die physikalischen Systeme, mit denen es die statistische Mechanik zu tun hat, kanonische Normalsysteme sind oder aus einem anderen Grunde als quasi-ergodisch betrachtet werden können.

Einen wesentlichen Fortschritt erfuhr die Ergodentheorie durch die Untersuchungen von v. Neumann[2] und Birkhoff[3,4], denen es gelang, die Existenz des Zeitmittels und seine Unabhängigkeit von t_0 sowie die Gl. (7.3) direkt, d.h. ohne den Umweg über die Quasi-Ergodenhypothese, unter sehr allgemeinen Voraussetzungen zu beweisen. Durch diese Sätze, die heute gewöhnlich unter der Bezeichnung „Ergodentheorem" zusammengefaßt werden, ist die mathematische Seite des Problems völlig geklärt worden. Dagegen bleibt wieder die Frage offen, ob die gemachte Voraussetzung (die, wie wir sehen werden, das quasi-ergodische Verhalten einschließt) bei den physikalischen Systemen der statistischen Mechanik erfüllt ist. Wir wollen nun die v. Neumann-Birkhoffschen Sätze ausführlich besprechen.

8. Die v. Neumann-Birkhoffschen Sätze. Die v. Neumann-Birkhoffsche Theorie bildet, wie erwähnt, einen ganzen Komplex von Aussagen. Wir werden sie daher der besseren Übersicht halber in einzelne Sätze gliedern. Dies ist um so eher gerechtfertigt, als nicht alle Aussagen die gleichen Voraussetzungen erfordern. Der im folgenden angegebene Beweis stammt von Kolmogoroff [19]. Wir bringen ihn teilweise in der etwas komprimierten Fassung von ter Haar [23].

Theorem 1. Es sei Ω ein invarianter Teil des Phasenraumes von endlichem Volumen und $F(P)$[5] eine Phasenfunktion, die in allen Punkten $P \in \Omega$ definiert und über Ω summierbar ist. Dann existiert

$$\widehat{F}_0 = \lim_{\tau \to \infty} \frac{1}{\tau} \int_0^\tau F(P, t)\, dt \tag{8.1}$$

fast überall in Ω, d.h. für alle Punkte P des Satzes Ω mit Ausnahme höchstens eines Satzes vom Maße Null. Der Grenzwert existiert auch fast überall, wenn $\tau \to -\infty$.

Beweis. Wir definieren

$$\int_n^{n+1} F(P, t)\, dt = x_n = x_n(P), \tag{8.2}$$

$$\int_n^{n+1} |F(P, t)|\, dt = y_n = y_n(P), \tag{8.3}$$

wo n eine beliebige ganze Zahl ist[6]. Zunächst beweisen wir einen Hilfssatz über die y_n, den wir später benötigen.

Lemma 1. Fast überall in Ω ist für $n \to \infty$

$$\frac{1}{n}\, y_n(P) \to 0. \tag{8.4}$$

[1] H. Poincaré: Méthodes Nouvelles de la Mécanique Céleste, Vol. I. Paris 1892.
[2] J. von Neumann: Proc. Nat. Acad. Sci. U.S.A. **18**, 70, 263 (1932).
[3] G. D. Birkhoff: Proc. Nat. Acad. Sci. U.S.A. **17**, 650, 656 (1931).
[4] G. D. Birkhoff u. B. O. Koopman: Proc. Nat. Acad. Sci. U.S.A. **18**, 279 (1932).
[5] Der Einfachheit halber fassen wir im folgenden den Satz aller Phasenkoordinaten in dem Symbol P zusammen.
[6] Die n zerlegen also die Zeit t in endliche Intervalle.

Beweis. Durch die Transformation $t = n + \alpha$ erhalten wir aus (8.3)

$$y_n(P) = \int\limits_0^1 |F(P, n + \alpha)|\, d\alpha = \int\limits_0^1 |F(P_n, \alpha)|\, d\alpha = y_0(P_n). \qquad (8.5)$$

Wir bezeichnen nun mit D_{nn} und D_{n0} zwei Sätze von Punkten P, die zu Ω gehören und den Bedingungen

$$y_n(P) > \varepsilon n \quad \text{bzw.} \quad y_0(P) > \varepsilon n \qquad (8.6)$$

genügen, wo ε eine beliebige feste positive Zahl ist. Man sieht leicht, daß die Strömung der „Phasenflüssigkeit" (d.h. die Transformation des Phasenraumes in sich selbst) in der Zeit n den Satz D_{nn} in den Satz D_{n0} überführt. Nach (8.5) ist nämlich die Ungleichung $y_n(P) > \varepsilon n$ äquivalent der Ungleichung $y_0(P_n) > \varepsilon n$. Die Aussage $P \in D_{nn}$ impliziert somit $P_n \in D_{n0}$ und umgekehrt. Aus dem Liouvilleschen Satz folgt daher

$$\mathfrak{M}\, D_{nn} = \mathfrak{M}\, D_{n0}. \qquad (8.7)$$

Der nächste Schritt besteht in dem Nachweis, daß die Reihe $\sum\limits_{n=1}^{\infty} \mathfrak{M}\, D_{nn}$ konvergiert. Aus (8.7) erhalten wir zunächst

$$\sum\limits_{n=1}^{\infty} \mathfrak{M}\, D_{nn} = \sum\limits_{n=1}^{\infty} \mathfrak{M}\, D_{n0}. \qquad (8.8)$$

Wir bezeichnen nun mit F_m einen Untersatz des Satzes Ω, für den gilt

$$m\,\varepsilon < y_0(P) \leqq (m + 1)\,\varepsilon. \qquad (8.9)$$

Dann wird

$$D_{n0} = \sum\limits_{m=n}^{\infty} F_m, \qquad (8.10)$$

und wir können die Reihe (8.8) schreiben

$$\sum\limits_{n=1}^{\infty} \sum\limits_{m=n}^{\infty} \mathfrak{M}\, F_m = \sum\limits_{m=1}^{\infty} \sum\limits_{n=1}^{m} \mathfrak{M}\, F_m = \sum\limits_{m=1}^{\infty} m\, \mathfrak{M}\, F_m. \qquad (8.11)$$

Für den rechts stehenden Ausdruck haben wir wegen (8.9) die Ungleichung

$$\sum\limits_{m=1}^{\infty} m\, \mathfrak{M}\, F_m = \frac{1}{\varepsilon} \sum\limits_{m=1}^{\infty} m\,\varepsilon\, \mathfrak{M}\, F_m \leqq \frac{1}{\varepsilon} \sum\limits_{m=1}^{\infty} \int\limits_{m\varepsilon < y_0 < (m+1)\,\varepsilon} y_0(P)\, d\Omega \left.\vphantom{\sum\limits_{m=1}^{\infty}}\right\}$$
$$\leqq \frac{1}{\varepsilon} \int\limits_{\Omega} y_0(P)\, d\Omega. \qquad (8.12)$$

Mit Benutzung von (8.5) erhalten wir

$$\frac{1}{\varepsilon} \int\limits_{\Omega} y_0(P)\, d\Omega = \frac{1}{\varepsilon} \int\limits_{\Omega} d\Omega \int\limits_0^1 |F(P, \alpha)|\, d\alpha = \frac{1}{\varepsilon} \int\limits_0^1 d\alpha \int\limits_{\Omega} |F(P, \alpha)|\, d\Omega. \qquad (8.13)$$

Da nach der Voraussetzung Ω ein invarianter Teil des Phasenraumes ist, folgt schließlich mit Gl. (3.18)

$$\frac{1}{\varepsilon} \int\limits_0^1 d\alpha \int\limits_{\Omega} |F(P)|\, d\Omega = \frac{1}{\varepsilon} \int\limits_{\Omega} |F(P)|\, d\Omega. \qquad (8.14)$$

Nach der Voraussetzung ist $F(P)$ eine über Ω summierbare Funktion. Der rechts stehende Ausdruck ist daher eine endliche Zahl, womit die Konvergenz der Reihe (8.8) bewiesen ist. Nach einem Theorem der Maßtheorie folgt daraus, daß jeder Punkt des Satzes Ω, mit Ausnahme höchstens eines Satzes vom Maße Null, zu nicht mehr als einer endlichen Zahl von Sätzen der Folge D_{nn} $(n = 1, 2 \ldots)$ gehört. Das bedeutet, daß für fast alle Punkte $P \in \Omega$ eine Zahl $N = N(P)$ existiert derart, daß für jedes $n > N$ gilt

$$y_n(P) \leqq \varepsilon \, n. \tag{8.15}$$

Da ε beliebig gewählt werden kann, ist damit das Lemma 1 bewiesen.

Wir definieren nun

$$\widehat{F}_{ab} = \frac{1}{b-a} \int\limits_a^b F(P, t) \, dt, \tag{8.16}$$

wo $a < b$ ist. $\widehat{F}_{ab}$ ist somit der über das endliche Zeitintervall $b - a$ gebildete zeitliche Mittelwert von $F(P, t)$. Setzen wir voraus, daß a und b ganze Zahlen sind, so können wir mit Benutzung von (8.2) diesen Mittelwert in der Form

$$\widehat{F}_{ab} = \frac{1}{b-a} \sum_{k=a}^{b-1} x_k \tag{8.17}$$

ausdrücken. Wir beweisen zunächst, daß der Grenzwert $\lim\limits_{n \to \infty} \widehat{F}_{0n}$ fast überall in Ω existiert. Der Beweis beruht auf einem argumentum e contrario; wir zeigen also, daß die gegenteilige Annahme zu einem Widerspruch führt. Den Ausgangspunkt bildet das folgende Lemma.

Lemma 2. Wenn der Grenzwert $\lim\limits_{n \to \infty} \widehat{F}_{0n}(P)$ (für ganzzahlige n) für einen Satz M von positivem Maß nicht existiert, dann muß es für einen beliebigen Punkt $P \in M$ zwei reelle Zahlen α und β $(\alpha < \beta)$ geben derart, daß

$$l(P) = \liminf_{n \to \infty} \widehat{F}_{0n}(P) < \alpha < \beta < L(P) = \limsup_{n \to \infty} \widehat{F}_{0n}(P). \tag{8.18}$$

Es muß ferner einen Unter-Satz M^* des Satzes M, ebenfalls von positivem Maß, geben, mit einem zugehörigen Wertepaar α, β derart, daß die Ungleichungen (8.18) für alle Punkte $P \in M^*$ erfüllt sind.

Da dieser Satz unmittelbar einleuchtend ist, verzichten wir auf die Wiedergabe des Beweises[1]. Es sei nun M_{0n} ein Unter-Satz von M^* derart, daß für alle Punkte $P \in M_{0n}$ gilt

$$\widehat{F}_{0n}(P) > \beta. \tag{8.19}$$

Integrieren wir $\widehat{F}_{0n}(P)$ über den Satz M_{0n}, so erhalten wir mit Gl. (8.5)

$$n \beta \, \mathfrak{M} \, M_{0n} < n \int\limits_{M_{0n}} \widehat{F}_{0n}(P) \, d\Omega = \sum_{k=0}^{n-1} \int\limits_{M_{0n}} x_k(P) \, d\Omega = \sum_{k=0}^{n-1} \int\limits_{M_{kn}} x_0(P_k) \, d\Omega, \tag{8.20}$$

wo M_{kn} der Satz ist, in dem M_{0n} durch die Transformation $P \to P_k$ übergeht. Wir nehmen nun an, daß die Sätze M_{kn} sich nicht überlappen und definieren

$$M_n = \sum M_{kn}. \tag{8.21}$$

Nach dem Liouvilleschen Satz ist

$$\mathfrak{M} \, M_{0n} = \mathfrak{M} \, M_{kn}. \tag{8.22}$$

Aus (8.20) folgt daher

$$\int\limits_{M_n} x_0(P) \, d\Omega > \beta \, \mathfrak{M} \, M_n. \tag{8.23}$$

---

[1] Der Beweis findet sich in [*19*].

Es läßt sich zeigen[1], daß man für jeden Wert von n Summen M_n in solcher Weise finden kann, daß sie M^* erschöpfen. Aus der Ungleichung (8.23) folgt daher

$$\int\limits_{M^*} x_0(P)\, d\Omega > \beta\, \mathfrak{M}\, M^*. \tag{8.24}$$

In analoger Weise kann man die Ungleichung

$$\int\limits_{M^*} x_0(P)\, d\Omega < \alpha\, \mathfrak{M}\, M^* \tag{8.25}$$

beweisen. Unter der Annahme, daß M^* von positivem Maß ist, führen die beiden Ungleichungen (8.24) und (8.25) zu der Folgerung $\alpha > \beta$, im Widerspruch zu der Voraussetzung $\alpha < \beta$. M^* und damit auch M muß daher vom Maße Null sein. Es folgt somit, daß der Grenzwert $\lim\limits_{n \to \infty} \hat{F}_{0n}(P)$ fast überall in Ω existiert.

Als letzten Schritt haben wir jetzt noch die Einschränkung zu beseitigen, daß der Parameter n nur ganzzahlige Werte annehmen kann. Wir betrachten das Integral

$$\frac{1}{\tau}\int\limits_0^\tau F(P, t)\, dt \tag{8.26}$$

und bezeichnen jetzt mit n die größte ganze Zahl, die in τ enthalten ist. Da nach dem Vorhergehenden der Grenzwert

$$\lim\limits_{n \to \infty} \frac{1}{n}\int\limits_0^n F(P, t)\, dt \tag{8.27}$$

fast überall in Ω existiert und das Integral

$$\frac{1}{\tau}\int\limits_0^n F(P, t)\, dt \tag{8.28}$$

sich von dem Integral

$$\frac{1}{n}\int\limits_0^n F(P, t)\, dt \tag{8.29}$$

nur durch einen Faktor unterscheidet, der für $\tau \to \infty$ gegen 1 geht, muß auch der Grenzwert

$$\lim\limits_{\tau \to \infty} \frac{1}{\tau}\int\limits_0^n F(P, t)\, dt \tag{8.30}$$

fast überall in Ω existieren. Es ist nun

$$\left| \frac{1}{\tau}\int\limits_0^\tau F(P, t)\, dt - \frac{1}{\tau}\int\limits_0^n F(P, t)\, dt \right| \leqq \frac{1}{\tau}\int\limits_n^\tau |F(P, t)|\, dt \leqq \frac{1}{n}\int\limits_n^{n+1} |F(P, t)|\, dt. \tag{8.31}$$

Aus Lemma 1 folgt aber

$$\frac{1}{n}\int\limits_n^{n+1} |F(P, t)|\, dt = \frac{1}{n}\, y_n(P) \to 0 \quad \text{für} \quad n \to \infty. \tag{8.32}$$

---

[1] Vgl. [19].

Der Grenzwert

$$\widehat{F}_0 = \lim_{\tau \to \infty} \frac{1}{\tau} \int\limits_0^\tau F(P, t)\, dt$$

existiert somit fast überall in Ω, womit das Theorem 1 bewiesen ist.

Der zweite v. Neumann-Birkhoffsche Satz stellt die Unabhängigkeit des durch Gl. (8.1) definierten Zeitmittels von der Wahl des Anfangspunktes auf der Trajektorie fest. Er ist daher von entscheidender Bedeutung für die praktische Brauchbarkeit dieses Begriffes. Wir geben diesem Satz die Fassung:

Theorem 2. Wenn

$$\widehat{F}_0 = \lim_{\tau \to \infty} \frac{1}{\tau} \int\limits_0^\tau F(P, t)\, dt$$

fast überall in Ω existiert, dann existiert auch

$$\widehat{F}' = \lim_{\tau \to \infty} \frac{1}{\tau} \int\limits_{t_0}^{t_0+\tau} F(P, t)\, dt$$

fast überall in Ω für alle t_0.

Beweis. Wir können ohne Einschränkung der Allgemeinheit $t_0 > 0$ annehmen. Nach der Voraussetzung existiert

$$\lim_{\tau \to \infty} \frac{1}{\tau + t_0} \int\limits_0^{\tau + t_0} F(P, t)\, dt = \widehat{F}_0. \tag{8.33}$$

Nun ist

$$\frac{1}{\tau} \int\limits_0^{\tau+t_0} F(P, t)\, dt - \frac{1}{\tau + t_0} \int\limits_0^{\tau+t_0} F(P, t)\, dt = \frac{t_0}{\tau} \frac{1}{\tau + t_0} \int\limits_0^{\tau+t_0} F(P, t)\, dt, \tag{8.34}$$

und es gilt

$$\lim_{\tau \to \infty} \frac{t_0}{\tau} \frac{1}{\tau + t_0} \int\limits_0^{\tau + t_0} F(P, t)\, dt = 0. \tag{8.35}$$

Es folgt somit

$$\lim_{\tau \to \infty} \frac{1}{\tau} \int\limits_0^{\tau + t_0} F(P, t)\, dt = \widehat{F}_0. \tag{8.36}$$

Ferner ist

$$\frac{1}{\tau} \int\limits_{t_0}^{t_0+\tau} F(P, t)\, dt = \frac{1}{\tau} \int\limits_0^{\tau+t_0} F(P, t)\, dt - \frac{1}{\tau} \int\limits_0^{t_0} F(P, t)\, dt. \tag{8.37}$$

Beim Grenzübergang $\tau \to \infty$ geht der zweite Term der rechten Seite gegen Null, während der erste Term nach Gl. (8.36) gegen $\widehat{F}_0$ strebt. Damit ist das Theorem 2 bewiesen.

Der dritte v. Neumann-Birkhoffsche Satz enthält das eigentliche Ergodentheorem. Dieser Satz erfordert wesentlich engere Voraussetzungen als die beiden vorhergehenden Sätze. Dabei wird der Begriff der „metrischen Transitivität" benutzt, den wir zunächst kurz erläutern wollen. Es sei Ω ein gegen die Gruppe von Transformationen des Phasenraumes in sich selbst invarianter Teil des

Phasenraumes. Die Gruppe heißt metrisch transitiv und Ω metrisch unzerlegbar,
wenn Ω nicht in der Form

$$\Omega = \Omega_1 + \Omega_1 \tag{8.38}$$

dargestellt werden kann, wo Ω_1 und Ω_2 beide von positivem Maß und beide invariant gegen jede Transformation der Gruppe sind. Man kann sich die Bedeutung dieser Definition in folgender Weise noch etwas klarer machen. Da Ω ein
invarianter Satz ist, muß Ω ein Satz von vollständigen Trajektorien sein. Zerlegen wir diesen Satz in zwei Unter-Sätze, von denen jeder wieder aus vollständigen
Trajektorien besteht, so gibt es, wenn Ω metrisch unzerlegbar ist, nur die beiden
folgenden Möglichkeiten: Entweder hat einer der beiden Unter-Sätze das Maß
Null (dann hat der andere das Maß $\mathfrak{M}\Omega$), oder die beiden Unter-Sätze sind nicht
meßbar.

Theorem 3. Wenn die Gruppe der Transformationen des Phasenraumes in
sich selbst metrisch transitiv ist, d.h. wenn der Satz Ω metrisch unzerlegbar ist,
dann ist fast überall in Ω

$$\widehat{F} = \frac{1}{\mathfrak{M}\Omega} \int\limits_{\Omega} F(P)\, d\Omega = \overline{F}. \tag{8.39}$$

Beweis. Wir beweisen zunächst, daß $\widehat{F}$ fast überall in Ω einen konstanten
Wert hat. Wäre dies nämlich nicht der Fall, so müßte es eine reelle Zahl α geben
derart, daß man Ω in zwei Teile Ω_1 und Ω_2 zerlegen könnte, die durch die Bedingungen $\widehat{F} > \alpha$ in Ω_1 und $\widehat{F} \leq \alpha$ in Ω_2 definiert wären. Man kann leicht zeigen[1],
daß dann gelten müßte

$$\mathfrak{M}\,\Omega_1 > 0, \qquad \mathfrak{M}\,\Omega_2 > 0. \tag{8.40}$$

Da nach Theorem 2 $\widehat{F}$ invariant ist gegen alle Transformationen des Phasen
raumes in sich selbst, müssen Ω_1 und Ω_2 Sätze von vollständigen Trajektorien
und damit invariante Teile des Phasenraumes sein. Dies würde aber im Widerspruch stehen zu der Voraussetzung, daß Ω metrisch unzerlegbar ist. Damit ist
die Behauptung bewiesen.

Den konstanten Wert, den $\widehat{F}$ fast überall in Ω besitzt, bezeichnen wir mit a.
Es bleibt dann noch zu beweisen, daß

$$a = \overline{F} \tag{8.41}$$

ist. Dazu definieren wir

$$\widehat{F}_\tau = \frac{1}{\tau} \int\limits_0^\tau F(P, t)\, dt. \tag{8.42}$$

Es ist nun

$$a = \frac{1}{\mathfrak{M}\Omega} \int\limits_{\Omega} a\, d\Omega = \frac{1}{\mathfrak{M}\Omega} \int\limits_{\Omega} (a - \widehat{F}_\tau)\, d\Omega + \frac{1}{\mathfrak{M}\Omega} \int\limits_{\Omega} \widehat{F}_\tau\, d\Omega. \tag{8.43}$$

Ferner haben wir mit Benutzung von (3.18)

$$\left.\begin{aligned}
\frac{1}{\mathfrak{M}\Omega} \int\limits_{\Omega} \widehat{F}_\tau\, d\Omega &= \frac{1}{\tau\,\mathfrak{M}\Omega} \int\limits_0^\tau dt \int\limits_{\Omega} F(P, t)\, d\Omega = \frac{1}{\tau\,\mathfrak{M}\Omega} \int\limits_0^\tau dt \int\limits_{\Omega} F(P)\, d\Omega \\
&= \frac{1}{\mathfrak{M}\Omega} \int\limits_{\Omega} F(P)\, d\Omega = \overline{F}.
\end{aligned}\right\} \tag{8.44}$$

---

[1] Vgl. *[19]*.

Es folgt somit

$$a = \frac{1}{\mathfrak{M}\Omega} \int\limits_{\Omega} (a - \hat{F}_\tau) \, d\Omega + \overline{F}. \tag{8.45}$$

Der erste Term der rechten Seite hängt jedenfalls nicht von τ ab. Es bleibt noch zu zeigen, daß dieser Term gleich Null ist. Es sei $\varepsilon > 0$ eine beliebig kleine Größe und $\Omega_1(\tau)$ der Satz von Punkten $P \in \Omega$, für den gilt

$$|a - \hat{F}_\tau| < \varepsilon. \tag{8.46}$$

Ferner sei

$$\Omega_2(\tau) = \Omega - \Omega_1(\tau). \tag{8.47}$$

Dann ist

$$\left| \int\limits_{\Omega} (a - \hat{F}_\tau) \, d\Omega \right| \leqq \int\limits_{\Omega_1(\tau)} |a - \hat{F}_\tau| \, d\Omega + \int\limits_{\Omega_2(\tau)} |a - \hat{F}_\tau| \, d\Omega \leqq$$
$$\leqq \varepsilon \, \mathfrak{M}\Omega + |a| \, \mathfrak{M}\Omega_2(\tau) + \int\limits_{\Omega_2(\tau)} |\hat{F}_\tau| \, d\Omega. \tag{8.48}$$

Da für $\tau \to \infty$ $\hat{F}_\tau$ fast überall in Ω gegen a strebt, muß für $\tau \to \infty$ gelten $\mathfrak{M}\Omega_2(\tau) \to 0$. Für hinreichend großes τ ist daher $\mathfrak{M}\Omega_2(\tau) < \varepsilon$. Mit Benutzung von (3.17) erhalten wir

$$\int\limits_{\Omega_2(\tau)} |\hat{F}_\tau| \, d\Omega \leqq \frac{1}{\tau} \int\limits_0^\tau dt \int\limits_{\Omega_2(\tau)} |F(P, t)| \, d\Omega = \frac{1}{\tau} \int\limits_0^\tau dt \int\limits_{\Omega_2(\tau, t)} |F(P)| \, d\Omega. \tag{8.49}$$

Hier ist $\Omega_2(\tau, t)$ der Satz, in den der Satz $\Omega_2(\tau)$ durch die Bewegung der „Phasenflüssigkeit" in der Zeit t überführt wird. Nach dem Liouvilleschen Satz ist nun für alle t

$$\mathfrak{M}\Omega_2(\tau, t) = \mathfrak{M}\Omega_2(\tau). \tag{8.50}$$

Für $\tau \to \infty$ gilt daher $\mathfrak{M}\Omega_2(\tau, t) \to 0$, und zwar gleichförmig in bezug auf t. Wegen der absoluten Stetigkeit der Integrale summierbarer Funktionen können wir τ so groß wählen, daß für alle t gilt

$$\int\limits_{\Omega_2(\tau, t)} |F(P)| \, d\Omega < \varepsilon. \tag{8.51}$$

Aus (8.49) folgt, daß dann auch

$$\int\limits_{\Omega_2(\tau)} |\hat{F}_\tau| \, d\Omega < \varepsilon \tag{8.52}$$

ist. Damit wird aus (8.48)

$$\left| \int\limits_{\Omega} (a - \hat{F}_\tau) \, d\Omega \right| \leqq \varepsilon \, \mathfrak{M}\Omega + |a| \, \varepsilon + \varepsilon. \tag{8.53}$$

In Verbindung mit den vorhergehenden Überlegungen ergibt sich daraus, daß die linke Seite dieser Ungleichung beliebig klein wird, wenn wir τ hinreichend groß wählen. Da die linke Seite aber nicht von τ abhängt, muß sie gleich Null sein. Damit ist auch das Theorem 3 bewiesen.

Wir schließen mit einem Satz, den wir früher schon aus dem Liouvilleschen Satz und der Ergodenhypothese abgeleitet haben.

Die v. Neumann-Birkhoffsche Theorie ermöglicht es, demselben eine exakte und hypothesenfreie Formulierung zu geben.

Theorem 4. Unter der Voraussetzung der metrischen Transitivität ist fast überall in Ω

$$\lim_{\tau \to \infty} \frac{\Delta t(\omega)}{\tau} = \frac{\mathfrak{M}\omega}{\mathfrak{M}\Omega}, \tag{8.54}$$

wo $\Delta t(\omega)$ die Zeit ist, die sich der Phasenpunkt in dem Intervall τ in dem Gebiet ω aufhält. Nach den Überlegungen in Ziff. 7 und den vorausgehenden Sätzen ist diese Aussage unmittelbar einleuchtend. Wir verzichten daher auf die Wiedergabe des Beweises.

Die von Neumann-Birkhoffschen Sätze beziehen sich auf den Fall, daß die Trajektorien auf einer Energiefläche verlaufen. HOPF[1] hat ein Ergodentheorem formuliert für den Fall, daß ein Satz von Energieflächen in dem endlichen Intervall zwischen E und $E + \Delta E$ betrachtet wird ($\Delta E \ll E$). Für Einzelheiten muß auf die Literatur verwiesen werden.

9. Das Problem der metrischen Transitivität. Wenn wir von der Klärung der mathematischen Struktur des Ergodenproblems (über deren Bedeutung kein Wort zu verlieren ist) absehen, so führt die von Neumann-Birkhoffsche Theorie zunächst auf die Frage, ob die Voraussetzung der metrischen Transitivität allgemein bei den physikalischen Systemen, mit denen es die statistische Mechanik zu tun hat, erfüllt ist. Diese Frage läßt sich in drei Teilfragen zerlegen:

a) Gibt es überhaupt Flächen, deren kontinuierliche Transformation in sich selbst metrisch transitiv ist?

b) Gibt es Hamiltonsche konservative Systeme, deren Energiefläche metrisch unzerlegbar ist?

c) Kann allgemein für die Systeme der statistischen Mechanik die Energiefläche als metrisch unzerlegbar betrachtet werden?

Die Frage a) läuft auf das Problem hinaus, ob der Begriff der metrischen Transitivität (ähnlich wie die Ergodenhypothese) nicht in sich widerspruchsvoll ist. Sie ist durch eine Untersuchung von OXTOBY und ULAM[2] geklärt worden, die für eine allgemeine Klasse von Flächen (Polyeder von drei und mehr Dimensionen) die Eigenschaft der metrischen Unzerlegbarkeit beweisen konnten. Damit ist zunächst gesichert, daß die Voraussetzung der metrischen Transitivität im Prinzip topologisch erfüllbar und die Diskussion der Fragen b) und c) sinnvoll ist.

Wir wenden uns jetzt der Frage b) zu. Hier stoßen wir sofort auf eine ernste Schwierigkeit, die mit der möglichen Existenz weiterer uniformer Integrale der Bewegungsgleichungen zusammenhängt [19]. Wie schon erwähnt, kann man lediglich die Existenz von je drei Impuls- und Drehimpulsintegralen für abgeschlossene konservative Systeme beweisen. Weitere allgemeine Aussagen lassen sich darüber nicht machen. Wir wollen nun annehmen, daß außer dem Energie-Integral noch k weitere uniforme Integrale existieren. Es ist dann zweckmäßig, dieselben in zwei Gruppen einzuteilen. Wenn das Integral bekannt ist und eine makroskopisch meßbare Größe repräsentiert, kann es (ebenso wie das Energie-Integral) experimentell bestimmt bzw. durch die Versuchsbedingungen vorgegeben werden. Solche Integrale (beispielsweise die Drehimpuls-Integrale) wollen wir kontrollierbare Integrale nennen. Integrale, welche nicht zu diesem Typ gehören, die also keine makroskopisch meßbare Größe repräsentieren oder deren Existenz nicht bekannt ist, bezeichnen wir als freie Integrale.

Wir zeigen nun zunächst, daß die Bedingung der metrischen Transitivität nicht nur hinreichend, sondern auch notwendig für die Gültigkeit der Gl. (8.39) „fast überall in Ω" ist. Wenn der Satz Ω metrisch zerlegbar ist, können wir ihn in zwei invariante Unter-Sätze Ω_1 und Ω_2 aufspalten, von denen jeder positives Maß hat. Es sei nun $F(P)$ eine summierbare Phasenfunktion, die in Ω_1 den

---

[1] E. HOPF: Ergodentheorie. Berlin 1937.
[2] J. C. OXTOBY u. S. M. ULAM: Ann. Math. **42**, 874 (1941).

Wert Null und in Ω_2 den Wert Eins besitzt. Dann können offenbar die Zeitmittel entlang den Trajektorien in Ω_1 nicht den gleichen Wert haben wie die über die Trajektorien in Ω_2. Die ersteren sind Null, die letzteren Eins, während das Phasenmittel irgendeinen intermediären Wert hat. Damit ist die Behauptung bewiesen.

Es sei nun F_i ein freies Integral der Bewegungsgleichungen. Würde F_i auf der Energiefläche einen konstanten Wert besitzen, so wäre es durch das Energie-Integral eindeutig bestimmt und somit kein weiteres unabhängiges uniformes Integral der Bewegungsgleichungen. F_i kann also auf der Energiefläche nicht einen konstanten Wert besitzen, und wegen der Stetigkeit eines uniformen Integrals kann es auch nicht in fast allen Punkten der Energiefläche konstant bleiben. Wenn wir also voraussetzen, daß F_i auf der Energiefläche nicht fast überall konstant ist, können wir eine reelle Zahl α finden derart, daß die Energiefläche in zwei Teile von positivem Maß zerfällt, die durch die Ungleichungen $F_i > \alpha$ und $F_i \leq \alpha$ charakterisiert sind. Da aber F_i ein Integral der Bewegungsgleichungen ist, stellt jeder Teil einen invarianten Satz dar. Die Energiefläche kann daher nicht metrisch unzerlegbar sein. Damit haben wir gezeigt, daß die Voraussetzung der metrischen Transitivität sicher nicht erfüllt ist, wenn ein oder mehrere freie Integrale der Bewegungsgleichungen existieren.

Die Schwierigkeit läßt sich wenigstens teilweise beseitigen, wenn man das Ergodentheorem nur für eine gewisse Klasse von Phasenfunktionen formuliert. Im allgemeinen wird ein bestimmter physikalischer Zustand des Systems durch mehrere, unter Umständen unendlich viele Punkte des Phasenraumes realisiert. Man muß daher voraussetzen, daß jede Phasenfunktion, der eine physikalische Bedeutung zukommt, in allen diesen „äquivalenten" Punkten den gleichen Wert annimmt. Solche Phasenfunktionen bezeichnen wir als *normale Phasenfunktionen*. Nach dem Gesagten bedeutet es keine wesentliche Einschränkung der Theorie, wenn wir das Ergodentheorem nur für normale Phasenfunktionen formulieren. In diesem Falle kann die Forderung der metrischen Transitivität durch eine schwächere Bedingung ersetzt werden.

Wir bezeichnen eine Unterteilung der Energiefläche in zwei invariante Teile von positivem Maß als „normal", wenn alle „äquivalenten" Punkte zu dem gleichen Teil gehören. Eine Fläche, die keine normale Unterteilung zuläßt, nennen wir metrisch unzerlegbar im erweiterten Sinne. Dafür gilt folgendes:

Theorem. Dafür, daß das Zeitmittel einer normalen summierbaren Funktion fast überall in Ω mit dem Phasenmittel zusammenfällt, ist notwendig und hinreichend, daß der Satz Ω metrisch unzerlegbar im erweiterten Sinne ist.

Der Beweis dieses Satzes wird in völliger Analogie zu den Beweisen in Ziff. 8 und in dieser Ziffer geführt. Wir können ihn daher hier übergehen. Man sieht leicht, daß die vorstehende Formulierung des Ergodentheorems auf die gleichen Schwierigkeiten stößt, die wir oben erörtert haben, wenn F_i ein normales Integral ist. Trifft dies jedoch nicht zu, so ist die Frage der metrischen Unzerlegbarkeit im erweiterten Sinne zunächst wieder offen. Es erscheint eine plausible Annahme, daß weitere uniforme Integrale der Bewegungsgleichungen, wenn sie überhaupt existieren, entweder kontrollierbar oder nicht normal sind. Die grundsätzlichen Schwierigkeiten wären damit beseitigt, doch besteht wenig Aussicht auf eine strengere Begründung dieser Hypothese.

Sehen wir von den Schwierigkeiten ab, die sich aus der Existenz weiterer uniformer Integrale ergeben, so bleibt die Frage, ob sich die metrische Transitivität mit Einschluß der Voraussetzung, daß nur das Energie-Integral existiert, allgemeiner beweisen läßt. Dieser Beweis ist schon 1923 von Fermi für eine gewisse Klasse von mechanischen Systemen geführt worden. Wir formulieren und beweisen den Fermischen Satz in der folgenden Ziff. 10. Damit ist die Frage b) ebenfalls positiv zu beantworten. Dagegen ist die Frage c) nach wie vor offen. Wir können daher zusammenfassend feststellen, daß die Möglichkeit der metrischen Transitivität topologisch wie auch für Hamiltonsche Systeme nachgewiesen ist, daß sie aber für die Systeme der statistischen Mechanik letzten Endes doch wieder eine Hypothese darstellt. BIRKHOFF und KOOPMAN[1] haben diesen Sachverhalt prägnant formuliert in dem Satze: "The quasi-ergodic hypothesis has been replaced by its modern version: the hypothesis of metrical transitivity."

10. FERMIs Satz über kanonische Normalsysteme. Ein Hamiltonsches System wird als kanonisches Normalsystem bezeichnet, wenn man durch kanonische Transformation kanonisch konjugierte Variable x_i und y_i einführen kann derart, daß

a) die Hamiltonsche Funktion (und damit die Energie) nicht explizit von der Zeit abhängt,

b) die Hamilton-Funktion in eine Reihe nach Potenzen eines Parameters α entwickelt werden kann gemäß

$$H = H_0 + \alpha H_1 + \alpha^2 H_2 + \cdots \tag{10.1}$$

c) die Hamilton-Funktion und damit alle H_i sowie die rechtwinkligen Lagekoordinaten des Systems in allen x_i mit der gleichen Periode 2π periodisch sind,

d) der Term H_0 in Gl. (10.1) nicht von den x_i abhängt.

Für das ungestörte System ($\alpha = 0$) sind dann die x_i Winkelvariable und die

$$\omega_i = \frac{\partial H_0}{\partial y_i} \tag{10.2}$$

die entsprechenden Kreisfrequenzen. Wir setzen voraus, daß das ungestörte System nicht entartet ist, daß also auf der Fläche $H_0 = $ const eine Gleichung der Form

$$\sum_{i=1}^{n} m_i \omega_i = 0 \tag{10.3}$$

nicht identisch gilt. Dann gilt nach FERMI[2] folgendes

Theorem[3]. Die Energiefläche kanonischer Normalsysteme mit mehr als zwei Freiheitsgraden ($n \geq 3$) ist metrisch unzerlegbar.

Beweis. Wir geben auch hier den Beweis in einer etwas verkürzten Form und beweisen zunächst das folgende Lemma.

Lemma. Für kanonische Normalsysteme mit $n \geq 3$ existiert neben dem Energie-Intervall kein weiteres uniformes Integral der Bewegungsgleichungen.

Beweis. Wir knüpfen an die der vorstehenden Formulierung äquivalente Behauptung an, daß es neben den Energieflächen keine Flächenschar gibt mit der Eigenschaft, daß jede Trajektorie, die von einem Punkt dieser Fläche ausgeht, vollständig auf dieser Fläche verläuft. Es seien

$$\Phi(\boldsymbol{x}, \boldsymbol{y}; \alpha) = 0 \tag{10.4}$$

---

[1] Siehe Zitat S. 194.

[2] E. FERMI: Physikal. Z. **24**, 261 (1923).

[3] Die Fermische Formulierung bezieht sich nur auf das quasi-ergodische Verhalten. Die obige erweiterte Fassung und die entsprechende Ergänzung des Beweises nach TER HAAR [23].

eindeutige analytische und in den x_i periodische Funktionen, die eine Flächen-schar $\mathfrak{S}_\alpha$ im Γ-Raum beschreiben. Wir haben dann zu zeigen, daß die $\mathfrak{S}_\alpha$ mit den Energieflächen zusammenfallen. Entwickeln wir auch Φ nach Potenzen von α, so erhalten wir

$$\Phi = \Phi_0 + \alpha\,\Phi_1 + \alpha^2\,\Phi_2 + \cdots. \tag{10.5}$$

Für die Φ_i gelten die folgenden Sätze:

a) Für gegebene $\mathfrak{S}_\alpha$ kann man Φ_0 stets unabhängig von den x_i wählen.

b) Alle Φ_i können außerhalb $\mathfrak{S}_0$ willkürlich gewählt werden.

Für den Beweis dieser Aussagen müssen wir auf die Originalarbeit verweisen.

Da Φ nach der Voraussetzung ein uniformes Integral der Bewegungsgleichungen ist, muß nach Gl. (5.1) gelten

$$\{H, \Phi\} = 0. \tag{10.6}$$

In Verbindung mit (10.1) und (10.5) folgen daraus die Beziehungen

$$\left.\begin{aligned} \{H_0, \Phi_0\} &= 0, \\ \{H_0, \Phi_1\} + \{H_1, \Phi_0\} &= 0, \\ \cdots\cdots\cdots\cdots\cdots\cdots \end{aligned}\right\} \tag{10.7}$$

Die erste dieser Beziehungen ist naturgemäß trivial, weil weder H_0 noch Φ_0 von den x_i abhängt. Da nach der Voraussetzung H_1 und Φ_1 periodisch in allen x_i sind, können wir für beide Funktionen n-dimensionale Fourier-Entwicklungen ansetzen. Wir haben dann

$$\Phi_1 = \sum_{m_1} \cdots \sum_{m_n} A_m(\mathbf{y})\, e^{i\,\mathbf{m}\cdot\mathbf{x}} \tag{10.8}$$

und

$$H_1 = \sum_{m_1} \cdots \sum_{m_n} B_m(\mathbf{y})\, e^{i\,\mathbf{m}\cdot\mathbf{x}}. \tag{10.9}$$

Hier ist

$$\mathbf{m} = \sum_{i=1}^{n} m_i\,\mathbf{h}_i, \tag{10.10}$$

wo die m_i ganze Zahlen und die $\mathbf{h}_i$ die Einheitsvektoren in Richtung der Koordinatenachsen sind. Mit der Bezeichnung

$$\chi_j = \frac{\partial \Phi_0}{\partial y_j} \tag{10.11}$$

folgt aus den Gln. (10.7) bis (10.9)

$$\sum_{m_1} \cdots \sum_{m_n} e^{i\,\mathbf{m}\cdot\mathbf{x}} \left(A_m \sum_{i=1}^{n} m_i\,\omega_i - B_m \sum_{j=1}^{n} m_j\,\chi_j \right) = 0 \tag{10.12}$$

oder

$$A_m \sum_{i=1}^{n} m_i\,\omega_i = B_m \sum_{j=1}^{n} m_j\,\chi_j. \tag{10.13}$$

Wir setzen voraus, daß $B_m \neq 0$ ist[1]. Dann folgt aus Gl. (10.13), daß in jedem Punkt der Fläche $\mathfrak{S}_0$, in dem $\sum m_i\,\omega_i$ Null ist, auch $\sum m_j\,x_j$ verschwindet.

Der Einfachheit halber setzen wir jetzt $n = 3$. Die folgenden Überlegungen gelten aber auch für $n > 3$, dagegen nicht für $n = 2$. Von den drei Quotienten ω_2/ω_1, ω_3/ω_1, ω_3/ω_2 kann auf der Fläche $\mathfrak{S}_0$ höchstens einer konstant sein, da andernfalls das ungestörte System, im Widerspruch zu der Voraussetzung, entartet wäre. Wir können daher jedenfalls annehmen, daß ω_2/ω_1 und ω_3/ω_1 auf $\mathfrak{S}_0$

---

[1] Der Fall $B_m = 0$ ist von Poincaré (s. Zitat S. 194) diskutiert worden. Die Gültigkeit des folgenden Beweises wird dadurch nicht beeinträchtigt.

veränderlich sind. Dann ist $\mathfrak{S}_0$ dicht mit Punkten belegt, in denen ω_2/ω_1 rationale Werte annimmt. Man kann daher in jedem dieser Punkte zwei ganze Zahlen m_1 und m_2 finden derart, daß

$$m_1\omega_1 + m_2\omega_2 = 0 \tag{10.14}$$

ist. Es folgt dann, daß in diesen Punkten auch

$$m_1\chi_1 + m_2\chi_2 = 0 \tag{10.15}$$

gilt. Es muß daher

$$\frac{\omega_2}{\omega_1} = \frac{\chi_2}{\chi_1} \tag{10.16}$$

sein. Da Gl. (10.16) in einem überall auf $\mathfrak{S}_0$ dichten Punkt-Satz gilt, gilt sie identisch auf $\mathfrak{S}_0$. In analoger Weise leitet man ab, daß auch die Gleichung

$$\frac{\omega_3}{\omega_1} = \frac{\chi_3}{\chi_1} \tag{10.17}$$

auf $\mathfrak{S}_0$ identisch gilt. Aus den Gln. (10.2), (10.11), (10.16) und (10.17) folgt in Verbindung mit der Tatsache, daß H_0 und Φ_0 nicht von den x_i abhängen, daß auf $\mathfrak{S}_0$

$$\Phi_0 = H_0 + c_0 \tag{10.18}$$

gilt, wo c_0 eine Konstante bezeichnet. Da Φ_0 außerhalb $\mathfrak{S}_0$ willkürlich wählbar ist, kann man diese Funktion so wählen, daß Gl. (10.18) im ganzen Phasenraum gilt.

Da wir vorausgesetzt haben, daß Gl. (10.3) nicht identisch auf $\mathfrak{S}_0$ gilt und ferner nach Gl. (10.18)

$$\sum m_i\omega_i = \sum m_i\chi_i \tag{10.19}$$

ist, können wir Gl. (10.13) durch Gl. (10.19) dividieren. Es folgt

$$A_m = B_m \qquad \text{(ausgenommen wenn alle } m_i = 0 \text{ sind).} \tag{10.20}$$

Damit erhalten wir aus Gln. (10.8) und (10.9), zunächst für die Fläche $\mathfrak{S}_0$,

$$\Phi_1(x_1, y_i) = H_1(x_i, y_i) + f_1(y_i). \tag{10.21}$$

Die Funktion Φ_1 kann aber außerhalb $\mathfrak{S}_0$ wieder so gewählt werden, daß Gl. (10.21) im ganzen Phasenraum gültig ist.

Wir nehmen nun an, für einen gewissen Wert von r sei bewiesen worden, daß im ganzen Phasenraum gilt

$$\left.\begin{aligned}
\Phi_r &= H_r + f_r(y_i), \\
\Phi_{r-1} &= H_{r-1} + c_{r-1}, \\
\cdot \quad &\cdot \quad \cdot \quad \cdot \quad \cdot \quad \cdot \\
\Phi_1 &= H_1 + c_1, \\
\Phi_0 &= H_0 + c_0.
\end{aligned}\right\} \tag{10.22}$$

Wir wollen zeigen, daß dann analoge Gleichungen auch für $r+1$ gelten. Setzen wir die Entwicklungen (10.1) und (10.5) in Gl. (10.6) ein, so verschwinden die ersten $r+1$ Terme identisch wegen (10.22), und wir erhalten für den Koeffizienten von α^{r+1}

$$\{H_0, \Phi_{r+1}\} + \{H_1, f_r\} + \{H_{r+1}, H_0\} = \{H_0, \Phi_{r+1} - H_{r+1}\} - \{f_r, H_1\} = 0. \tag{10.23}$$

Behandeln wir diese Gleichung ebenso wie die Gl. (10.7), so finden wir, daß auf $\mathfrak{S}_0$ die partiellen Ableitungen von f_r denen von H_0 proportional sind. Da aber H_0

auf $\mathfrak{S}_0$ konstant ist, muß das gleiche auch für f_r gelten, und weil f_r außerhalb $\mathfrak{S}_0$ beliebig ist, können wir im ganzen Phasenraum setzen

$$\Phi_r = H_r + c_r, \tag{10.24}$$

wo c_r eine Konstante ist. Damit reduziert sich die Gl. (10.23) auf

$$\{H_0, \Phi_{r+1} - H_{r+1}\} = 0. \tag{10.25}$$

Wir setzen nun

$$\Phi_{r+1} - H_{r+1} = \sum_{m_1} \cdots \sum_{m_n} C_m(y)\, e^{i m \cdot x}. \tag{10.26}$$

Dann wird auf $\mathfrak{S}_0$

$$\sum_{m_1} \cdots \sum_{m_n} e^{i m \cdot x}\, C_m(y) \sum_{i=1}^{n} m_i\, \omega_i = 0. \tag{10.27}$$

Nach der Voraussetzung kann aber $\sum m_i \omega_i$ auf $\mathfrak{S}_0$ nur identisch verschwinden, wenn alle $m_i = 0$ sind. Es kann daher nur $C_{0, 0, \dots 0} \neq 0$ sein. Damit erhalten wir für $\mathfrak{S}_0$

$$\Phi_{r+1} = H_{r+1} + f_{r+1}(y). \tag{10.28}$$

Da Φ_{r+1} außerhalb $\mathfrak{S}_0$ willkürlich gewählt werden kann, gilt diese Gleichung für den ganzen Phasenraum. Die Gln. (10.22) gehen für $r = 1$ in die Gln. (10.18) und (10.20) über. Aus den Gln. (10.22), (10.24) und (10.28) folgt daher, daß allgemein

$$\Phi_i = H_i + c_i \tag{10.29}$$

ist, wo die c_i Konstanten sind. Das bedeutet, daß die Gl. (10.4) äquivalent ist der Gleichung

$$H(x, y; \alpha) = \sum_n c_n \alpha^n = c. \tag{10.30}$$

Die $\mathfrak{S}_\alpha$ fallen somit in der Tat mit den Energieflächen zusammen, womit das Lemma bewiesen ist.

Es seien nun σ und σ^* zwei beliebige Gebiete der Energiefläche. Da σ und σ^* beliebig klein gewählt werden können, ist der quasi-ergodische Charakter des Systems bewiesen, wenn wir zeigen, daß es Trajektorien gibt, die von einem Punkt in σ ausgehen und durch σ^* hindurchgehen.

Es sei σ' der Teil der Energiefläche, welcher durch die Trajektorien überdeckt wird, die irgendwo in σ beginnen. Würde σ' die gesamte Energiefläche überdecken, so wäre der quasi-ergodische Charakter des Systems bereits bewiesen. Wir nehmen daher an, daß dies nicht der Fall ist und bezeichnen mit σ'' den nicht überdeckten Teil der Energiefläche. Die Grenzfläche zwischen σ' und σ'' sei $\mathfrak{G}$. Es kann keine Trajektorie geben, die sowohl Punkte von σ' wie auch von σ'' enthält. Um dies zu zeigen, nehmen wir an, daß eine Trajektorie zur Zeit t' den Punkt P' in σ', zur Zeit t'' den Punkt P'' in σ'' passiert. Da die Lösungen der Bewegungsgleichungen analytische Funktionen sind, muß es in σ' ein kleines Gebiet η' um P', in σ'' ein kleines Gebiet η'' um P'' geben derart, daß eine Trajektorie, die zur Zeit t' durch einen beliebigen Punkt Q' in η' geht, zur Zeit t'' durch η'' geht. Da aber η' in σ' liegt, muß es Punkte Q' geben, die auf Trajektorien liegen, die von σ ausgehen. Die Aussage, daß dieselben auch durch η'' gehen, steht daher im Widerspruch zur Definition von σ''.

Es sei nun P ein Punkt von $\mathfrak{G}$, P' und P'' seien Punkte von σ' bzw. σ''. Wir betrachten Trajektorien, die zur Zeit t durch P, P', P'' gehen. Dieselben werden zur Zeit t_1 die Punkte P_1, P_1', P_1'' passieren, zur Zeit t_2 die Punkte P_2, P_2', P_2'' usw. Nach dem oben bewiesenen Satz liegen alle Punkte P_1', P_2', $\dots$ in σ', alle Punkte P_1'', P_2'', $\dots$ in σ''. Wir können P' und P'' beliebig nahe P wählen, so daß auch P_1' und P_1'' beliebig nahe P_1, P_2' und P_2'' beliebig nache P_2 liegen usw.

Dann folgt, daß alle Punkte $P, P_1, P_2 \ldots$ auf $\mathfrak{G}$ liegen und somit eine Trajektorie vollständig auf $\mathfrak{G}$ verläuft. Nach dem oben bewiesenen Lemma gibt es aber außer den Energieflächen keine Flächen, die vollständige Trajektorien enthalten. $\mathfrak{G}$ kann daher nicht existieren oder, anders ausgedrückt, das Gebiet σ' muß die gesamte Energiefläche überdecken. σ^* ist daher notwendig ein Teil von σ', und es muß Trajektorien geben, die von σ ausgehend σ^* passieren. Damit ist der quasi-ergodische Charakter des Systems bewiesen.

Schließlich sieht man leicht, daß aus dem quasi-ergodischen Charakter unmittelbar die metrische Unzerlegbarkeit der Energiefläche folgt. Der quasi-ergodische Charakter bedeutet nämlich, daß jede Trajektorie auf der Energiefläche durch jedes Gebiet $\mathfrak{A}$ von positivem Maß hindurchgeht. Da aber die Trajektorie durch die Gesamtheit der Transformationen $P \to P_t$ entsteht (wo t von $-\infty$ bis $+\infty$ geht), muß die Energiefläche eines quasi-ergodischen Systems metrisch unzerlegbar sein. Damit ist der Fermische Satz vollständig bewiesen.

Bisher ist nur in einem Falle nachgewiesen worden, daß ein mechanisches System die Eigenschaften eines kanonischen Normalsystems besitzt, nämlich für das reduzierte Dreikörperproblem[1]. Es ist jedoch bemerkenswert, daß EIN-STEIN[2] bei seinen Untersuchungen die in diesem Zusammenhang wesentliche Eigenschaft der kanonischen Normalsysteme, nämlich die Gültigkeit des oben bewiesenen Lemmas, explizit voraussetzt.

11. Diskussion der Ergodentheorie. Wir wollen jetzt die Ergodentheorie als Ganzes noch einmal kurz diskutieren. Der wesentliche Gedankengang läßt sich in den folgenden Aussagen zusammenfassen:

1. Ein makroskopisches abgeschlossenes System kann vom Standpunkt der mechanischen Wärmetheorie als ein konservatives System betrachtet werden.

2. Jeder makroskopisch meßbaren Größe ist eine Phasenfunktion zugeordnet.

3. Die Gleichgewichtseigenschaften des Systems sind durch die Zeitmittel der betreffenden Phasenfunktionen nach Gl. (8.1) gegeben.

4. Das Zeitmittel einer Phasenfunktion ist gleich dem Phasenmittel. (Ergodentheorem in engerem Sinne.)

Die Aussage 1 ist sicher nicht streng richtig. Jedes materielle System steht in Wechselwirkung mit elektromagnetischen Feldern[3] und der Gefäßwand. Beide Arten von Wechselwirkung können nicht durch konservative Kräfte beschrieben werden. Man ist also gezwungen, die Beschreibung als konservatives System in irgendeinem Sinne als Näherung aufzufassen. Wesen und Bedeutung dieser Näherung bleiben jedoch im Rahmen der Ergodentheorie völlig unklar.

Die Aussage 2 ist eine selbstverständliche Folgerung der mechanischen Wärmetheorie und erfordert keine weitere Diskussion.

Die Grundlage der Aussage 3 bildet die an sich evidente Tatsache, daß jede makroskopische Messung notwendig einen zeitlichen Mittelwert der betreffenden Phasenfunktion liefert. Die Ergodentheorie behauptet nun, daß dieses „Meßmittel" (das wir in Ziff. 12 streng definieren werden) im Gleichgewicht mit dem durch Gl. (8.1) definierten Zeitmittel identisch ist. Da man naturgemäß nicht über eine unendlich lange Zeit messen kann, muß man zur Rechtfertigung entweder Annahmen über die Relaxationszeiten[4] einführen [23], oder man muß

---

[1] Siehe Zitat S. 194.
[2] Siehe Zitat S. 189.
[3] Auf die Bedeutung dieser Tatsache hat schon GIBBS [2] hingewiesen. Vgl. auch [13].
[4] Die naheliegende Annahme, daß die Meßzeiten von der Größenordnung der POINCARÉ-schen Wiederkehrzeiten t^* (Ziff. 15) sein sollten, kommt wegen deren Größenordnung nicht in Betracht. Die Größenordnung der Relaxationszeiten ist $t^*/N!$.

diese Gleichsetzung statistisch verstehen, was wieder eine Hypothese ist, die im Rahmen der Ergodentheorie nicht begründet werden kann. Auf diese Problematik haben vor allem FOWLER [7] und TOLMAN [22] hingewiesen[1]. JAYNES[2] hat kürzlich gezeigt, daß die Gleichsetzung des Zeitmittels mit dem Meßmittel über eine Zeit von der Größenordnung der Relaxationszeit unzulässig ist; sie läßt sich erst für um viele Größenordnungen längere Meßzeiten rechtfertigen. Die Identifizierung des Meßmittels mit dem Zeitmittel führt zu der Folgerung, daß man Abweichungen von den Gleichgewichtswerten einer meßbaren Eigenschaft des Systems im Gleichgewicht grundsätzlich für unbeobachtbar erklären muß, was offensichtlich absurd ist und den Einbau der Schwankungstheorie fast unmöglich macht[3]. Der Kern der Schwierigkeit liegt darin, daß der Begriff des Gleichgewichts selbst in einer molekularen Theorie nur statistisch verstanden werden kann, und daß eben dieser statistische Charakter, wie besonders TOLMAN betont hat, von der Ergodentheorie verfehlt wird.

Das Ergodentheorem im engeren Sinne haben wir bereits in Ziff. 9 ausführlich erörtert. Wir haben daher lediglich hier nochmals festzustellen, daß das allgemeine Ergodentheorem nach wie vor eine Hypothese ist, weil es nicht gelungen ist, die Voraussetzung der metrischen Transitivität als für physikalische Systeme allgemein gültig zu beweisen. KHINCHIN [19] hat einen sehr bemerkenswerten Versuch unternommen, eine weniger allgemeine Formulierung des Ergodentheorems unabhängig von der Voraussetzung der metrischen Transitivität zu beweisen. Es sei F eine Phasenfunktion eines Systems von N Teilchen mit separierbarer Hamilton-Funktion. F ist dann eine Summenfunktion, d.h. eine Summe von Funktionen, die nur von den Phasenkoordinaten je eines Teilchens abhängen. Bezeichnen wir nun mit Ω_m die Menge der Trajektorien, für die

$$\left| \frac{\hat{F}}{\bar{F}} - 1 \right| > K N^{-\frac{1}{4}} \tag{11.1}$$

ist, so gilt nach KHINCHIN für $N \to \infty$

$$\mathfrak{M}\,\Omega_m = O\,(N^{-\frac{1}{4}}). \tag{11.2}$$

Die wichtigste Eigenschaft dieser Formulierung besteht darin, daß hier das Ergodentheorem ohne die Voraussetzung der metrischen Transitivität als asymptotisches Gesetz für $N \to \infty$ erscheint. Damit entfallen in der Tat verschiedene der vorher erörterten Schwierigkeiten. Es darf aber nicht übersehen werden, daß KHINCHIN (im Gegensatz zu v. NEUMANN-BIRKHOFF) den Beweis nur für Systeme mit separierbarer Hamilton-Funktion durchgeführt hat. Als allgemeine Grundlage der statistischen Mechanik würde daher auch diese Formulierung des Ergodentheorems den Charakter einer Hypothese besitzen.

Aus dem Vorstehenden ergibt sich, daß es bis jetzt jedenfalls nicht gelungen ist, mit Hilfe der Ergodentheorie eine deduktive Begründung der statistischen Mechanik durchzuführen. Wenn man die statistische Mechanik auf der Ergodentheorie aufbauen will, muß man die Aussagen 1, 3 und 4 (letztere eventuell in der Khinchinschen Fassung) als Axiome einführen. Auf die Problematik von 1 und 3 haben wir oben hingewiesen. Es kommt hinzu, daß die Theorie makroskopisch offener Systeme auf dieser Grundlage die sehr künstliche Annahme erfordert, daß das offene System als Teil eines abgeschlossenen Systems betrachtet

---

[1] TER HAAR hat sich in [22] dieser Ansicht angeschlossen, dagegen in [23] den entgegengesetzten Standpunkt vertreten.

[2] E. T. JAYNES: Phys. Rev. (im Druck). Für die Überlassung des Manuskriptes dieser Arbeit bin ich Herrn Dr. JAYNES zu großem Dank verpflichtet.

[3] Dieser Gesichtspunkt ist auch von TER HAAR [22] hervorgehoben worden.

werden kann. Unter diesen Umständen dürfte vom Standpunkt des Physikers die schon von FOWLER [7] und TOLMAN [12] vertretene Ansicht vorzuziehen sein, daß für den Aufbau der statistischen Mechanik eine von der Ergodentheorie völlig unabhängige axiomatische Basis zu wählen ist.

12. Die Theorie des repräsentativen Ensembles. Wir wollen nun die Ideen von GIBBS und TOLMAN, deren Grundgedanken wir in Ziff. 6 skizziert haben, in einer mehr systematischen Weise entwickeln. Den Ausgangspunkt bildet die Überlegung, daß ein nur durch makroskopische Parameter charakterisiertes System vom Standpunkt einer molekularen Theorie nur ganz unvollständig beschrieben ist und daher nur durch eine statistische Gesamtheit repräsentiert werden kann, welche alle mit den Werten der makroskopischen Parameter vereinbaren und dem Originalsystem zugänglichen[1] mechanischen Zustände umfaßt. Wir haben nun zuerst zu untersuchen, wie die mit Hilfe der statistischen Gesamtheit gewonnenen Aussagen physikalisch zu interpretieren sind, d. h. wie sie mit den experimentellen Resultaten zusammenhängen. Die Beantwortung dieser Frage erfordert eine etwas genauere Analyse des Begriffes der Messung einer makroskopischen Eigenschaft. Bezeichnet $F(\boldsymbol{q}, \boldsymbol{p})$ eine Phasenfunktion, die eine meßbare Eigenschaft des Systems repräsentiert, so muß offenbar jede Messung, die man vernünftigerweise als makroskopisch bezeichnen kann, einen zeitlichen Mittelwert dieser Funktion liefern. Dieses „Meßmittel" definieren wir durch

$$\widehat{F}_\tau = \frac{1}{\tau} \int\limits_{t_0}^{t_0+\tau} F(\boldsymbol{q}, \boldsymbol{p})\, dt. \tag{12.1}$$

Dabei ist die Meßzeit τ so zu wählen, daß für eine Änderung derselben um einen Betrag $\varDelta\tau$, der klein ist in der makroskopischen Zeitskala zeitlicher Änderungen des Systems, die Änderung von $\widehat{F}_\tau$ unterhalb der Grenze der Meßgenauigkeit bleibt. Man kann dies auch so ausdrücken, daß τ groß sein muß gegen die Perioden der submikroskopischen Schwankungen, aber klein gegen die Poincarésche Wiederkehrzeit des Systems (vgl. Ziff. 15). Für konkrete Systeme lassen sich diese Festsetzungen noch etwas präzisieren. Für verdünnte Gase muß τ groß gegen die mittlere Stoßzeit sein. Für Flüssigkeiten und flüssige Gemische ergibt sich aus der Theorie der Brownschen Bewegung[2], daß τ groß sein muß gegen ein Zeitintervall, in welchem eine merkliche Korrelation besteht zwischen der auf ein Molekül von allen übrigen Molekülen ausgeübten Kraft am Anfang des Intervalls und der gleichen Kraft am Ende des Intervalls. Die vorstehende Definition des Meßmittels schließt gewisse Annahmen über den Verlauf der Funktion $F(\boldsymbol{q}, \boldsymbol{p})$ ein (Existenz von sog. Plateau-Werten), für die ein allgemeiner Beweis bisher nicht existiert. Sie besitzen daher im Aufbau der Theorie den Charakter eines, allerdings physikalisch einleuchtenden, Axioms.

Wenn eine Messung in dem erläuterten Sinne an sehr vielen physikalisch identischen Systemen unter völlig gleichen Bedingungen durchgeführt wird, müssen die dabei erhaltenen Werte $\widehat{F}_\tau$ nicht untereinander übereinstimmen. Ein eindeutiger Zusammenhang zwischen den Parameterwerten, welche den Zustand des Systems definieren, und weiteren meßbaren makroskopischen Eigenschaften besteht daher nur, wenn die letzteren als Mittelwerte der $\widehat{F}_\tau$ aus zahlreichen gleichartigen Messungen definiert werden. Bilden wir die Gesamtheit der Systeme, an

---

[1] Der Begriff der „Zugänglichkeit" (accessibility) ist von FOWLER [7] eingeführt worden. Näheres darüber in Ziff. 26.

[2] J. G. KIRKWOOD: J. Chem. Phys. **14**, 180 (1946).

denen die Messung ausgeführt wurde, durch eine virtuelle Gesamtheit ab, so ist die makroskopische Eigenschaft durch $\overline{\overline{F}}_\tau$ gegeben.

Wir beschränken uns jetzt auf den Fall des Gleichgewichtes. Nach den vorstehenden Definitionen ist dasselbe empirisch dadurch definiert, daß $\overline{\overline{F}}_\tau$ nicht von der Zeit abhängt; das dieser Situation entsprechende theoretische Modell ist ein Ensemble im statistischen Gleichgewicht. Für diesen Fall kann man leicht zeigen, daß die in der vorstehenden Weise definierten makroskopischen Eigenschaften durch die Phasenmittelwerte der statistischen Gesamtheit gegeben sind. Es ist zunächst

$$\overline{\overline{F}}_\tau = \frac{1}{\tau} \int\limits_{t_0}^{t_0+\tau}\!\!\!\int\int F(\boldsymbol{q},\boldsymbol{p})\, \varrho(\boldsymbol{q}_0,\boldsymbol{p}_0;t_0)\, dt\, d\boldsymbol{q}_0\, d\boldsymbol{p}_0. \tag{12.2}$$

Transformieren wir auf die Integrationsvariablen $\boldsymbol{q},\boldsymbol{p}$, so erhalten wir mit Gl. (3.12)

$$\overline{\overline{F}}_\tau = \frac{1}{\tau} \int\limits_{t_0}^{t_0+\tau}\!\!\!\int\int F(\boldsymbol{q},\boldsymbol{p})\, \varrho(\boldsymbol{q},\boldsymbol{p};t)\, d\boldsymbol{q}\, d\boldsymbol{p}\, dt. \tag{12.3}$$

Da die Phasenmittelwerte einer stationären Gesamtheit zeitunabhängig sind, folgt sofort

$$\overline{\overline{F}}_\tau = \overline{F}. \tag{12.4}$$

Die physikalische Bedeutung der Phasenmittelwerte der Ensemble-Theorie ist damit grundsätzlich geklärt. Es bleibt noch die, praktisch sehr wichtige, Frage, wie man zu einer auf das Einzelsystem anwendbaren Theorie gelangt, wie sie die Begründung der Thermodynamik erfordert. Die Antwort auf diese Frage wird ausschließlich durch die Schwankungen von F, d.h. im wesentlichen durch das Verhalten der Größe

$$\frac{\overline{(F-\overline{F})^2}}{\overline{F}^2} \tag{12.5}$$

bestimmt. Sie läßt sich daher vollständig im Rahmen der statistischen Theorie untersuchen und wir können für die Einzelheiten auf Abschnitt C I (S. 350) verweisen. Das wesentliche Ergebnis besteht in der Aussage, daß die mittleren relativen Schwankungsquadrate (von singulären Stellen abgesehen) asymptotisch verschwinden, wenn die Zahl der Freiheitsgrade gegen Unendlich geht. Die Gleichungen der statistischen Mechanik gehen dann in die Gleichungen der Thermodynamik über. Praktisch ist diese Situation bereits für makroskopische Systeme gegeben, so daß die Aussagen der Ensemble-Theorie in diesem Sinne auf das Einzelsystem anwendbar sind.

Die Anwendung der Ensemble-Theorie auf konkrete Probleme erfordert eine Vorschrift darüber, wie für ein gegebenes Originalsystem, dessen Zustand durch gewisse makroskopische Parameter charakterisiert wird, die repräsentative virtuelle Gesamtheit zu konstruieren ist. Diese Vorschrift hat den Charakter eines Axioms, was nicht überraschen kann. Da nämlich die statistische Mechanik zahlreiche Aussagen ermöglicht, die über die Verknüpfung der makroskopischen Parameter (wie sie die Thermodynamik liefert) hinausgehen, muß für diese Aussagen eine zusätzliche Grundlage (die an die Stelle der fehlenden Kenntnis der mechanischen Zustände tritt) vorhanden sein. Die Notwendigkeit dieser zusätzlichen Annahmen beruht also nicht auf einer Lücke in der klassischen Mechanik, sondern auf der Tatsache, daß wir eine Theorie aufbauen wollen, welche die Anwendung der Mechanik in Fällen erlaubt, in denen die sonst geforderten Kenntnisse über das betrachtete System nur ganz unvollständig vorhanden sind. Der statistische Charakter dieser Theorie macht es möglich, die zusätzlichen Annahmen auf wenige einfache Axiome zu reduzieren, aus denen sich alles weitere deduzieren läßt. Eine Rechtfertigung der Axiome ist nur durch den Vergleich der aus der Theorie abgeleiteten Resultate mit der Erfahrung möglich. Da die statistische Mechanik in keinem Falle zu Widersprüchen mit experimentellen Ergebnissen geführt hat[1], können die Axiome als sichere Grundlage der Theorie betrachtet werden.

Da wir die Thermodynamik aus der statistischen Mechanik entwickeln wollen, können wir zunächst über die mechanische Definition der spezifisch thermodynamischen Größen (Entropie, Temperatur usw.) nichts voraussetzen. Es ist daher notwendig, die makroskopische (vom Standpunkt der Mechanik fragmentarische) Beschreibung des Zustandes in Begriffen zu geben, welche der Mechanik und Thermodynamik gemeinsam sind, um diese Kenntnisse für die Konstruktion der virtuellen Gesamtheit verwerten zu können. Es kommen hier also neben der Zahl der Freiheitsgrade, dem Volumen und gegebenenfalls weiteren äußeren Parametern, nur die Energie und ähnliche Größen in Betracht, die für konservative Systeme uniforme Integrale der Bewegungsgleichungen darstellen. Vom makroskopischen Standpunkt bedeutet dies, daß wir abgeschlossene Systeme betrachten. Für diesen Fall, auf den wir uns zunächst beschränken, nimmt das als Grundlage benötigte Axiom eine besonders einfache Form an. Es besagt, daß die mechanischen Zustände des nicht-konservativen Originalsystems mit hinreichender Genauigkeit dargestellt werden können als Zustände verschiedener in geeigneter Weise ausgewählter konservativer Systeme. Die Bewegung des Originalsystems im Phasenraum wird somit aufgefaßt als ein „Springen" zwischen Trajektorien konservativer Systeme. Diese Vorstellung entstammt eigentlich dem Gedankenkreis der Quantenmechanik, läßt sich aber ohne Schwierigkeit auf die klassische Mechanik übertragen. Durch diese Annahme ist der allgemeine Charakter der zu verwendenden virtuellen Gesamtheit bereits vollständig festgelegt [26]. Da wir uns auf den Fall des Gleichgewichtes beschränken, muß die Gesamtheit stationär sein. Für eine stationäre Gesamtheit aus konservativen Systemen hängt aber nach Ziff. 5 die Phasendichte nur von den uniformen Integralen der Bewegungsgleichungen ab. Das heißt mit anderen Worten, daß die Gesamtheit über alle mit dem makroskopischen Zustand vereinbaren mechanischen Zustände mit konstanter Phasendichte verteilt sein muß. Alle diese Zustände besitzen also die gleiche a priori-Wahrscheinlichkeit. Dies ist die von TOLMAN [12] formulierte „Hypothese der gleichen a priori-Wahrscheinlichkeiten", die wir bereits in Ziff. 6 zitiert haben.

Es bleibt noch die Frage, welche uniformen Integrale bei der Konstruktion der virtuellen Gesamtheit zu berücksichtigen sind. Sie läßt sich naturgemäß nicht allgemein beantworten. Wir werden daher für die Entwicklung der allgemeinen Theorie (mit der wir es in diesem Artikel ausschließlich zu tun haben) lediglich die Existenz des Energie-Integrals voraussetzen. Dies bedeutet keine Einschränkung der Allgemeinheit. Die Impuls- und Drehimpulsintegrale können (von Sonderfällen abgesehen) als nicht existierend vorausgesetzt werden. Im übrigen geht man zweckmäßig aus von der Tatsache, daß jedes derartige Integral die Bewegung auf ein Teilgebiet des Phasenraumes beschränkt. Durch die Existenz solcher Integrale werden also gewisse Teile des Phasenraumes für die Systeme der Gesamtheit unzugänglich. Wir können daher auf die (praktisch undurchführbare) Analyse des mechanischen Problems verzichten und die Frage

---

[1] Das gilt naturgemäß in vollem Umfange erst für die Quantenstatistik.

der Existenz weiterer uniformer Integrale im Rahmen des allgemeineren Problems der „Zugänglichkeit" im Phasenraum behandeln. Allgemein gilt dafür, daß im Falle eines konkreten Problems $\varrho = 0$ gesetzt werden muß für alle Teile des Phasenraumes, welche für das Originalsystem prinzipiell oder praktisch (d.h. in Zeiten von der Größenordnung der Beobachtungszeit) unzugänglich sind. Die Begründung dieser für die Anwendung der statistischen Mechanik außerordentlich wichtigen Festsetzung ist unmittelbar evident, da andernfalls endliche Wahrscheinlichkeiten für nicht realisierbare Zustände und damit unter Umständen völlig andere Mittelwerte resultieren würden. Für die allgemeine Theorie können wir uns mit diesen Feststellungen begnügen. Eine ausführlichere Diskussion der Gesichtspunkte, die bei der Anwendung auf konkrete Systeme von Bedeutung sind, findet sich in Ziff. 26. Das Ergebnis dieser Überlegungen können wir zusammenfassen in dem Satz:

Die makroskopischen Gleichgewichtseigenschaften eines abgeschlossenen Systems sind gegeben durch die Mittelwerte entsprechender Phasenfunktionen über eine im statistischen Gleichgewicht befindliche virtuelle Gesamtheit aus konservativen Systemen, für welche das Energie-Integral vorgegeben ist und die im übrigen auf den praktisch zugänglichen Teil des Phasenraumes beschränkt ist.

Dieses Axiom ermöglicht den Aufbau der Theorie für abgeschlossene Systeme. Die Behandlung offener Systeme erfordert eine weitere Annahme axiomatischer Natur, die wir in Ziff. 67 formulieren werden.

Die weitere Entwicklung der Theorie ist unabhängig davon, ob man die Ergodentheorie oder das oben formulierte Axiom als Grundlage wählt. Dies beruht einfach auf der Tatsache, daß beide Ansätze zum gleichen Formalismus führen. Sie sind auch insofern äquivalent, als sie beide axiomatischer Natur sind. Wir geben jedoch aus den früher erörterten Gründen der in dieser Ziffer entwickelten Axiomatik den Vorzug.

Kürzlich hat Jaynes[1] einen sehr bemerkenswerten Versuch unternommen, die statistische Mechanik deduktiv auf der Grundlage der Informationstheorie[2] aufzubauen. Es erscheint uns verfrüht, diese Ansätze zur Grundlage einer zusammenfassenden Darstellung zu machen. Wir werden jedoch den Begriff der Information im folgenden gelegentlich heranziehen, da er für die Diskussion verschiedener Fragen von außerordentlichem Nutzen ist.

III. Die Einstellung des Gleichgewichtes.

13. Vorbemerkungen. Bei den bisherigen Untersuchungen haben wir im allgemeinen die Existenz des Gleichgewichtes als Erfahrungstatsache vorausgesetzt. Das ist jedoch weder vom logischen noch vom physikalischen Standpunkt sehr befriedigend. Eine Darlegung der Grundlagen der statistischen Mechanik erfordert daher notwendig auch eine Behandlung der Vorgänge, die zum Gleichgewicht führen. Die statistische Theorie dieser „irreversiblen Prozesse" ist in neuester Zeit sehr stark in den Vordergrund des Interesses gerückt; sie wird ausführlich in dem Artikel von Meixner in diesem Bande dargestellt. Wir werden uns daher an dieser Stelle auf die Gesichtspunkte beschränken, die im Zusammenhang mit der Theorie der Gleichgewichtseigenschaften von besonderer Bedeutung sind. Da die ganze Problemstellung sich, wie in Ziff. 6 dargelegt, im Rahmen der kinetischen Gastheorie entwickelt hat, ist es für ein tieferes Verständnis unerläßlich, zunächst diese Gedankengänge zu rekapitulieren. Da die kinetische Gastheorie in Band XII dieses Handbuches ausführlich behandelt wird, geben

---

[1] E. T. Jaynes: Phys. Rev. **106**, 620 (1957).
[2] E. C. Shannon: The Mathematical Theory of Communication. Urbana 1949.

wir hier nur die wesentlichen Gedanken wieder und verweisen für alle Einzelheiten auf den Artikel von GRAD in Bd. XII. Anschließend behandeln wir die Übertragung der Boltzmannschen Ideen in den Rahmen der Ensemble-Theorie und zeigen zum Schluß an einem Beispiel, wie auf diesem Wege die Boltzmannsche Stoßgleichung abgeleitet werden kann.

14. Die Boltzmannsche Stoßgleichung und das H-Theorem. Wir betrachten ein abgeschlossenes System, das aus einem verdünnten einatomigen Gase besteht. Die zwischenmolekulare Wechselwirkung soll durch Zentralkräfte dargestellt werden. Den Zustand des Systems beschreiben wir im μ-Raum der generalisierten Koordinaten und Impulse eines Moleküls in folgender Weise: Wir zerlegen den μ-Raum in kleine aber endliche Volumenelemente $\Delta q_1 \Delta p_1$. Dabei muß das Volumenelement des Konfigurationsraumes Δq_1 so gewählt werden, daß es noch eine große Zahl von Molekülen enthält, daß aber andererseits die makroskopischen Größen (Dichte, Temperatur usw.) innerhalb desselben als konstant betrachtet werden können. Die Zahl der Moleküle im Volumenelement $\Delta q_1 \Delta p_1$ unterliegt infolge der thermischen Bewegung dauernden Schwankungen. Im folgenden verstehen wir unter der Zahl der Moleküle, die sich zur Zeit t im Volumenelement $\Delta q_1 \Delta p_1$ befinden, einen zeitlichen Mittelwert, gebildet über eine kleine, aber endliche Zeit Δt derart, daß Δt groß ist im Vergleich zu der mittleren Zeit, in der ein Molekül Δq_1 durchquert, aber klein im Vergleich zu der Zeit, innerhalb deren sich makroskopische Eigenschaften merklich ändern. Dieses Zeitmittel der Zahl der Moleküle, die sich im Konfigurationsraum an der Stelle q_1 im Volumenelement Δq_1 und im Impulsraum an der Stelle p_1 im Volumenelement Δp_1 befinden, bezeichnen wir mit

$$f(q_1, p_1)\, \Delta q_1\, \Delta p_1. \tag{14.1}$$

Die Funktion $f(q_1, p_1)$ wird als kinetische Verteilungsfunktion bezeichnet. Sie genügt der Gleichung

$$\sum_i f(q_i, p_i)\, \Delta q_i \Delta p_i = N, \tag{14.2}$$

wo N die Zahl der Gasmoleküle ist und die Summierung über alle Volumenelemente des μ-Raumes zu erstrecken ist. In der kinetischen Gastheorie wird $f(q, p)$ als stetige Funktion von q und p betrachtet. Die Gl. (14.2) nimmt dann die Form an

$$\iint f(q, p)\, dq\, dp = N. \tag{14.3}$$

Wenn die kinetische Verteilungsfunktion nicht von q abhängt, ist

$$f(q, p)\, dp = f(p)\, dp \tag{14.4}$$

die Zahl der Moleküle pro Volumeneinheit, deren Impuls (in dem oben erläuterten Sinne) zwischen p und $p + dp$ liegt.

Wir wollen nun die Änderung der Größe $f(q_1, p_1)\, \Delta q_1 \Delta p_1$ in dem endlichen Zeitintervall Δt untersuchen. Diese Änderung beruht auf drei Faktoren.

a) Die Bewegung der Moleküle liefert einen Beitrag

$$-\frac{p_1}{m}\, \frac{\partial f}{\partial q_1}\, \Delta q_1 \Delta p_1 \Delta t. \tag{14.5}$$

b) Bezeichnet X_1 die an der Stelle q_1 auf ein Molekül wirkende äußere Kraft, so ist der Beitrag der äußeren Kräfte zur Änderung von f

$$-X_1 \frac{\partial f}{\partial p_1}\, \Delta q_1 \Delta p_1 \Delta t. \tag{14.6}$$

c) Das schwierigste und gleichzeitig wichtigste Problem ist die Berechnung des Beitrages, der von den molekularen Stößen herrührt. Derselbe kann in der Form

$$-A + B \tag{14.7}$$

geschrieben werden, wobei A die durch Stöße bedingte Abnahme von

$$f(\boldsymbol{q_1}, \boldsymbol{p_1})\, \varDelta \boldsymbol{q_1}\, \varDelta \boldsymbol{p_1}$$

im Zeitintervall $\varDelta t$ bezeichnet, B die entsprechende Zunahme dieser Größe. Da die Rechnung eine Analyse der Stöße erfordert, nehmen wir der Einfachheit halber an, daß die Moleküle elastische Kugeln sind. Das Resultat ist von der Wahl dieses speziellen Modells unabhängig. Bei einem Stoß zwischen zwei Molekülen mit den Impulsen $\boldsymbol{p_1}$ und $\boldsymbol{p_2}$ sind die Impulse nach dem Stoß $\boldsymbol{p_1'}$ und $\boldsymbol{p_2'}$ durch Energie- und Impulssatz bestimmt, wenn die Richtung der Verbindungslinie der Mittelpunkte beider Moleküle, der sog. Zentrilinie, gegeben ist. Bezeichnet $\boldsymbol{\omega}$ den Einheitsvektor in der Richtung der Zentrilinie, so gilt

$$\boldsymbol{\omega} = \frac{\boldsymbol{p_1} - \boldsymbol{p_1'}}{|\boldsymbol{p_1} - \boldsymbol{p_1'}|}. \tag{14.8}$$

Die für die Ableitung grundlegende Annahme besagt nun, daß die Zahl der Stöße im Zeitintervall $\varDelta t$ zwischen Molekülen im Volumenelement $\varDelta \boldsymbol{q_1}\, \varDelta \boldsymbol{p_1}$ und Molekülen im Volumenelement $\varDelta \boldsymbol{q_2}\, \varDelta \boldsymbol{p_2}$, bei denen die Zentrilinie im Raumwinkel $\varDelta \omega$ liegt, gegeben ist durch

$$a_{12 \to 1'2'}\, f(\boldsymbol{q_1}, \boldsymbol{p_1})\, f(\boldsymbol{q_2}, \boldsymbol{p_2})\, \varDelta \boldsymbol{q_1}\, \varDelta \boldsymbol{p_1}\, \varDelta \boldsymbol{q_2}\, \varDelta \boldsymbol{p_2}\, \varDelta \omega\, \varDelta t. \tag{14.9}$$

Die Größe $a_{12 \to 1'2'}$ hängt nur von $|\boldsymbol{p_1} - \boldsymbol{p_2}|$ und der Geometrie des Stoßes ab. Für elastische Kugeln findet man leicht

$$\left. \begin{array}{ll} a_{12 \to 1'2'} = \sigma^2 \dfrac{|\boldsymbol{p_1} - \boldsymbol{p_2}|}{m} \cos \Theta & \text{für} \quad \cos \Theta > 0, \\[2mm] a_{12 \to 1'2'} = 0 & \text{für} \quad \cos \Theta \leqq 0, \end{array} \right\} \tag{14.10}$$

wo σ der Durchmesser der Kugeln und Θ der Winkel zwischen $\boldsymbol{\omega}$ und $\boldsymbol{p_1} - \boldsymbol{p_2}$ ist. Die Annahme, daß Gl. (14.9) zutrifft, wird als „Stoßzahlansatz" oder „Hypothese des molekularen Chaos" bezeichnet. Sie ist komplexer Natur und enthält als wesentliche Elemente die Beschränkung auf Zweierstöße und eine Aussage über die Korrelation im Raum der Molekülpaare, die wir später (Ziff. 19) präzisieren werden.

Um B zu berechnen, machen wir den zu (14.9) analogen Ansatz

$$a_{1'2' \to 12}\, f(\boldsymbol{q_1'}, \boldsymbol{p_1'})\, f(\boldsymbol{q_2'}, \boldsymbol{p_2'})\, \varDelta \boldsymbol{q_1'}\, \varDelta \boldsymbol{p_1'}\, \varDelta \boldsymbol{q_2'}\, \varDelta \boldsymbol{p_2'}\, \varDelta \omega'\, \varDelta t. \tag{14.11}$$

Für das hier betrachtete Molekülmodell ergibt sich sofort aus Gl. (14.8), daß jedem Stoß ein „inverser Stoß" umkehrbar eindeutig zugeordnet ist, bei dem aus den Impulsen $\boldsymbol{p_1'}, \boldsymbol{p_2'}$ die Impulse $\boldsymbol{p_1}, \boldsymbol{p_2}$ entstehen und die Zentrilinie $\boldsymbol{\omega'} = -\boldsymbol{\omega}$ ist. Daraus erhält man

$$a_{12 \to 1'2'} = a_{1'2' \to 12} \equiv a, \tag{14.12}$$

was allgemein eine Folge des Prinzips der mechanischen Umkehrbarkeit ist. Schließlich folgt aus dem Liouvilleschen Satz

$$\varDelta \boldsymbol{q_1'}\, \varDelta \boldsymbol{p_1'}\, \varDelta \boldsymbol{q_2'}\, \varDelta \boldsymbol{p_2'} = \varDelta \boldsymbol{q_1}\, \varDelta \boldsymbol{p_1}\, \varDelta \boldsymbol{q_2}\, \varDelta \boldsymbol{p_2}. \tag{14.13}$$

Um A und B zu erhalten, haben wir nun (14.9) und (14.11) über die Raumwinkel zu integrieren und über die Volumenelemente des μ-Raumes zu summieren.

Diese Summierungen ersetzen wir durch Integrationen. Ferner führen wir an Stelle von a und ω den aus der Theorie der Zweierstöße[1] bekannten Stoßparameter b ein, der für elastische Kugeln durch

$$b = \sigma \cos \vartheta \qquad (14.14)$$

gegeben ist. Wir erhalten dann

$$\left. \begin{aligned} \frac{\partial f}{\partial t} &= - \frac{\boldsymbol{p_1}}{m} \frac{\partial f}{\partial \boldsymbol{q_1}} - \boldsymbol{X_1} \frac{\partial f}{\partial \boldsymbol{p_1}} + \\ &\quad + 2\pi \iint \frac{|\boldsymbol{p_1} - \boldsymbol{p_2}|}{m} \left[f(\boldsymbol{p_1'}) f(\boldsymbol{p_2'}) - f(\boldsymbol{p_1}) f(\boldsymbol{p_2}) \right] d\boldsymbol{p_2}\, b\, db. \end{aligned} \right\} \qquad (14.15)$$

Diese Gleichung wird als Boltzmannsche Stoßgleichung oder Fundamentalgleichung bezeichnet[2]. Sie bildet die Grundlage für die Theorie der Transporterscheinungen in verdünnten Gasen[1].

Die übliche Ableitung der Stoßgleichung, die wir im Vorstehenden skizziert haben, ist in zwei Punkten problematisch. Der eine derselben ist der Stoßzahlansatz, den wir in den folgenden Ziffern diskutieren werden. Auf den zweiten Punkt hat vor allem KIRKWOOD[3] hingewiesen. Er betrifft die Unklarheit in der Definition der Funktion f. Auf der linken Seite der Gl. (14.15) haben wir den ursprünglich auftretenden Differenzenquotienten durch einen Differentialquotienten ersetzt, ohne den Grenzübergang (der mit der ursprünglichen Definition von f nicht zu vereinbaren ist) zu vollziehen. Auf der rechten Seite ist der Stoßzahlansatz nur für die nicht gemittelten f-Funktionen sinnvoll. Wir haben also in der Ableitung eine ziemlich undurchsichtige Mischung von ·gemittelten und nicht gemittelten f-Funktionen. Diese Schwierigkeiten werden wir in Ziff. 19 mit Hilfe der statistischen Mechanik untersuchen.

Wir nehmen jetzt an, daß $\boldsymbol{X} = 0$ und $\partial f/\partial \boldsymbol{q} = 0$ ist und fragen, ob sich allgemein beweisen läßt, daß ein abgeschlossenes System mit beliebig vorgegebener Geschwindigkeitsverteilung durch die Wirkung der molekularen Stöße sich in Richtung auf einen Gleichgewichtszustand verändert. Zur Beantwortung dieser Frage definieren wir eine Funktion

$$H = \sum_i f(\boldsymbol{q_i}, \boldsymbol{p_i}) \ln f(\boldsymbol{q_i}, \boldsymbol{p_i})\, \Delta\boldsymbol{q_i}\, \Delta\boldsymbol{p_i}, \qquad (14.16)$$

die wir unter der Annahme, daß f eine praktisch stetige Funktion im μ-Raum ist, schreiben

$$H = \iint f(\boldsymbol{q}, \boldsymbol{p}) \ln f(\boldsymbol{q}, \boldsymbol{p})\, d\boldsymbol{q}\, d\boldsymbol{p}. \qquad (14.17)$$

Die zeitliche Änderung dieser Funktion ist

$$\frac{dH}{dt} = \iint \frac{\partial f}{\partial t} (\ln f + 1)\, d\boldsymbol{q}\, d\boldsymbol{p}. \qquad (14.18)$$

Mit (14.15) wird daraus [mit $f_1' \equiv f(\boldsymbol{p_1'})$]

$$\frac{dH}{dt} = \iiiint (f_1' f_2' - f_1 f_2)(\ln f_1 + 1)\, 2\pi \frac{|\boldsymbol{p_1} - \boldsymbol{p_2}|}{m}\, b\, db\, d\boldsymbol{q_1}\, d\boldsymbol{p_1}\, d\boldsymbol{p_2}. \qquad (14.19)$$

Durch Vertauschen der Impulse 1 und 2 erhält man

$$\frac{dH}{dt} = \iiiint (f_1' f_2' - f_1 f_2)(\ln f_2 + 1)\, 2\pi \frac{|p_1 - p_2|}{m}\, b\, db\, d\boldsymbol{q_1}\, d\boldsymbol{p_1}\, d\boldsymbol{p_2}. \qquad (14.20)$$

[1] S. CHAPMAN u. T. COWLING: The Mathematical Theory of Non-Uniform Gases. Cambridge 1939.
[2] L. BOLTZMANN: Wien. Ber. **66**, 213 (1872); **72**, 427 (1875).
[3] J. G. KIRKWOOD: J. Chem. Phys. **15**, 72 (1947).

Vertauschen wir gestrichene und ungestrichene Variable und transformieren dann auf die ungestrichenen Variablen zurück, so erhalten wir zwei weitere analoge Gleichungen. Durch Addition aller vier Gleichungen ergibt sich schließlich

$$\frac{dH}{dt} = \frac{1}{4} \iiiint \ln \frac{f(\boldsymbol{p}_1)\,f(\boldsymbol{p}_2)}{f(\boldsymbol{p}_1')\,f(\boldsymbol{p}_2')} \times \\ \times [f(\boldsymbol{p}_1')\,f(\boldsymbol{p}_2') - f(\boldsymbol{p}_1)\,f(\boldsymbol{p}_2)]\,\frac{|\boldsymbol{p}_1 - \boldsymbol{p}_2|}{m}\,2\pi\,b\,db\,d\boldsymbol{q}_1\,d\boldsymbol{p}_1\,d\boldsymbol{p}_2. \qquad (14.21)$$

Der Integrand ist wesentlich negativ oder Null. Es muß daher

$$\frac{dH}{dt} \leqq 0, \qquad (14.22)$$

wobei das Gleichheitszeichen dem erreichten Gleichgewicht entspricht. (10.22) stellt das berühmte Boltzmannsche H-Theorem[1] [1] dar, nach dem die Funktion H in einem abgeschlossenen System bis zum Erreichen des Gleichgewichtes monoton abnimmt. Das Gleichgewicht kann daher auch durch

$$H = \text{Minimum} \qquad (14.23)$$

charakterisiert werden. Aus dieser Formulierung erkennt man sofort die nahe Beziehung der H-Funktion zur Entropie, auf die wir später (Ziff. 16, 17) noch zurückkommen werden. Für das Bestehen des Gleichgewichtes ist notwendig und hinreichend, daß für alle $\boldsymbol{p}_1$ und $\boldsymbol{p}_2$

$$f(\boldsymbol{p}_1)\,f(\boldsymbol{p}_2) = f(\boldsymbol{p}_1')\,f(\boldsymbol{p}_2') \qquad (14.24)$$

ist (Prinzip des detaillierten Gleichgewichtes). Aus (14.23) und (14.24) leitet man leicht die Maxwellsche Geschwindigkeitsverteilung

$$f(\boldsymbol{p}) = \frac{N}{V}\,(2\pi\,m\,kT)^{-\frac{3}{2}}\,e^{-\frac{\boldsymbol{p}^2}{2mkT}} \qquad (14.25)$$

ab.

Die gestellte Frage scheint damit positiv beantwortet zu sein. Tatsächlich liegen die Verhältnisse wesentlich komplizierter. Die historische Diskussion um das H-Theorem, deren wesentliche Gesichtspunkte wir in den folgenden Ziffern skizzieren, hat zu der Erkenntnis geführt, daß die Ansätze der kinetischen Gastheorie nur durch die statistische Mechanik gerechtfertigt und erklärt werden können. Insofern bildet diese Diskussion ein wesentliches Element der Grundlagen der statistischen Mechanik.

15. Umkehreinwand und Wiederkehreinwand. Wir wollen nun zeigen, daß das H-Theorem, wenn es als exaktes Gesetz für das Einzelsystem aufgefaßt wird, im Widerspruch zur Mechanik steht. Zur Vereinfachung der Diskussion nehmen wir an, daß die f-Funktionen nicht als zeitliche Mittelwerte und für endliche Volumenelemente definiert sind, sondern daß sie sich auf die momentane Verteilung spezifizierter Moleküle beziehen. Die wesentlichen Gesichtspunkte werden durch diese Festsetzung nicht berührt. Unter der erwähnten Voraussetzung besteht eine umkehrbar eindeutige Zuordnung zwischen den Werten der f-Funktion und den Punkten des Γ-Raumes. Wir betrachten nun eine Folge von Zeitpunkten $\ldots, t_1, t_2, \ldots, t_n, \ldots$, in denen das System im Γ-Raum die Punkte

$$\ldots, P_1, P_2, \ldots, P_{n-1}, P_n, P_{n+1}, \ldots \qquad (15.1)$$

---

[1] L. Boltzmann: Wien. Ber. **66**, 275 (1872).

passiert. Da jedem Punkte eine f-Funktion entspricht, können wir auch für jeden Punkt P_n die zugehörige H-Funktion berechnen, die wir mit H_n bezeichnen. Dann fordert das H-Theorem

$$\ldots \geqq H_1 \geqq H_2 \ldots \geqq H_{n-1} \geqq H_n \geqq H_{n+1} \geqq \ldots, \tag{15.2}$$

wo das Gleichheitszeichen nur für den Fall des Gleichgewichtes gilt. Wir betrachten nun die Transformation

$$q_k \to q'_k, \qquad p_k \to -\,p'_k. \tag{15.3}$$

Da für konservative Systeme die Hamilton-Funktion eine homogene quadratische Funktion der generalisierten Impulse ist, ist sie invariant gegen die Transformation (15.3). Aus den kanonischen Bewegungsgleichungen folgt dann, daß die Transformation (15.3) einer Umkehrung der Zeitrichtung äquivalent ist. Das vorher betrachtete System durchläuft in transformierten Koordinaten die Punktfolge

$$\ldots P'_{n+1}, P'_n, P'_{n-1}, \ldots, P'_2, P'_1, \ldots. \tag{15.4}$$

Aus der Definition der H-Funktion folgt sofort

$$H_i = H'_i. \tag{15.5}$$

Es ergibt sich somit aus (15.2) und (15.5), daß die Punktfolge (15.4) einer Folge von H-Werten

$$\ldots \leqq H_{n+1} \leqq H_n \leqq H_{n-1} \leqq \ldots \leqq H_2 \leqq H_1 \leqq \ldots \tag{15.6}$$

entspricht. Zu jedem System, das eine monotone Abnahme von H zeigt, läßt sich also ein System konstruieren, das eine monotone Zunahme von H zeigt. Diese Aussage, die eine strenge Folgerung aus den Bewegungsgleichungen darstellt, wird als Loschmidtscher Umkehreinwand[1] bezeichnet. Der Kern desselben liegt in der Tatsache, daß die mechanischen Vorgänge in bezug auf die Zeit umkehrbar sind, während diese Umkehrbarkeit im H-Theorem aufgehoben ist. Die Aufklärung dieser Schwierigkeit bildet eines der wesentlichen Probleme in der Diskussion des H-Theorems.

Wenn ein mechanisches System auf ein endliches Volumen des Phasenraumes beschränkt ist (d.h. weder die q_k noch die p_k unendlich werden können), so gilt das folgende, als Poincaréscher Wiederkehrsatz[2] bezeichnete Theorem:

Ein System möge zu den Zeiten $t_1, t_2, \ldots, t_n, \ldots$ die Punkte $P_1, P_2, \ldots, P_n, \ldots$ im Phasenraum passieren. Δs sei ein beliebiges endliches Längenelement im Phasenraum. Dann kann man stets ein Zeitintervall t^* finden derart, daß das System zu den Zeiten $t_1 + t^*, t_2 + t^*, \ldots, t_n + t^*, \ldots$ die Punkte

$$P''_1, P''_2, \ldots, P''_n, \ldots \tag{15.7}$$

durchläuft, für die

$$|P_i - P''_i| < \Delta s \tag{15.8}$$

gilt.

Der Wiederkehrsatz besagt also, daß unter der obigen Voraussetzung ein System jedem Punkt seiner Phasenbahn nach einer endlichen Zeit wieder beliebig nahe kommt. Für den Beweis des Theorems müssen wir auf die Literatur[3] [15] verweisen. Die Größte t^* wird als Poincarésche Wiederkehrzeit (Poincaré cycle period) bezeichnet. BOLTZMANN[4] hat diese Größe abgeschätzt für ein ideales

[1] J. LOSCHMIDT: Wien. Ber. **73**, 139 (1876); **75**, 67 (1877).
[2] H. POINCARÉ: Acta math. **13**, 67 (1890).
[3] E. ZERMELO: Ann. Phys. **57**, 485 (1896).
[4] L. BOLTZMANN: Ann. Phys. **57**, 773 (1896); **60**, 392 (1897).

Gas unter der Annahme, daß 10^{18} Moleküle in einem Volumen von $1\ cm^3$ eine mittlere Geschwindigkeit von $5 \cdot 10^4\ cm\ sec^{-1}$ beibehalten und dabei die Lagekoordinaten auf $\leq 10^{-7}\ cm$, die Geschwindigkeitskomponenten auf $\leq 10^2\ cm\ sec^{-1}$ reproduzieren. Die dafür erforderliche Zeit beträgt $>10^{10^{19}}$ Jahre. Im übrigen ist über die Größe der Poincaréschen Wiederkehrzeit bei den Systemen der statistischen Mechanik nichts bekannt.

Aus dem Wiederkehrsatz folgt nun wieder unmittelbar, daß das H-Theorem als exaktes Gesetz nicht haltbar ist. Wählen wir nämlich $\varDelta s$ hinreichend klein, o können wir mit beliebiger Annäherung

$$H_i \approx H_i'' \tag{15.9}$$

setzen. Das System durchläuft dann in der Zeitfolge $t_1, t_2, \ldots, t_n, \ldots t_1 + t^*$, $t_2 + t^*, \ldots, t_n + t^*, \ldots$ die H-Werte

$$H_1, H_2, \ldots, H_n, \ldots, H_1'', H_2'', \ldots, H_n'', \ldots \tag{15.10}$$

Aus (15.2) und (15.9) folgt (wenn wir von dem Fall des Gleichgewichtes absehen)

$$H_n < H_1''. \tag{15.11}$$

In dem Zeitintervall zwischen t_n und $t_1 + t^*$ muß daher die H-Funktion zunehmen, im Widerspruch zum H-Theorem. Diese Folgerung aus der Mechanik ist als Wiederkehreinwand von ZERMELO[1] bekannt.

16. Statistische Interpretation des H-Theorems. Die in Ziff. 15 erörterten Schwierigkeiten zwingen dazu, entweder das H-Theorem preiszugeben oder eine genauere Analyse der Ableitung zu versuchen, die eine mit den Prinzipien der Mechanik konsistente Interpretation ermöglicht. Um das letztere Ziel zu erreichen, hat man im wesentlichen drei Wege eingeschlagen: .
a) Untersuchung des stochastischen Problems an einfachen Modellen.
b) Statistische Begründung des H-Theorems und des Stoßzahlansatzes.
c) Behandlung des Problems mit Hilfe der Ensemble-Theorie.
In dieser Ziffer behandeln wir nur die unter a) und b) genannten Methoden.
Die stochastische Analyse einfacher Modelle ist zuerst von P. und T. EHRENFEST[2] [3] eingeführt und in neuerer Zeit von TER HAAR und GREEN[3-5], KAC[6], KLEIN[7,8] u. a. wieder aufgenommen worden. Ihre Bedeutung liegt darin, daß man auf diesem Wege eine exakte Untersuchung der Einzelheiten des Problems durchführen kann. Da die Rechnungen ziemlich umständlich sind, beschränken wir uns auf den einfachsten Fall, das sog. *Ehrenfestsche Urnenmodell*[2] und verzichten auch hier auf eine Wiedergabe der mathematischen Details.
Wir betrachten zwei Urnen A und B, auf die $2N$ von 1 bis $2N$ numerierte Kugeln beliebig verteilt sind. Durch eine geeignete Vorrichtung soll es möglich sein, mit gleichen Wahrscheinlichkeiten $1/2N$ eine der Zahlen 1 bis $2N$ zu ziehen. Dies wird regelmäßig in Zeitabständen τ wiederholt, und jedesmal wird die Kugel, deren Nummer gezogen ist, aus der Urne, in der sie sich befindet, entfernt und in die andere Urne gelegt. Wir charakterisieren den Zustand des Systems durch eine Zahl m, die angibt, daß sich in der Urne A $N + m$ Kugeln befinden.

[1] E. ZERMELO: Ann. Phys. **59**, 793 (1896).
[2] P. u. T. EHRENFEST: Phys. Z. **8**, 311 (1907).
[3] D. TER HAAR u. C. D. GREEN: Proc. Phys. Soc. Lond. A **66**, 153 (1953).
[4] C. D. GREEN u. D. TER HAAR: Physica, Haag **21**, 63 (1954).
[5] D. TER HAAR u. C. D. GREEN: Proc. Cambridge Phil. Soc. **51**, 141 (1955).
[6] M. KAC: Amer. Math. Monthly **54**, 369 (1947).
[7] M. J. KLEIN: Physica, Haag **22**, 569 (1956).
[8] M. J. KLEIN: Phys. Rev. **103**, 17 (1956).

Die Zahl m beschreibt einen Makrozustand; werden zusätzlich die Nummern der $N + m$ Kugeln angegeben, so wird dadurch einer der zu dem Makrozustand gehörenden Mikrozustände charakterisiert. m kann alle ganzzahligen Werte von $-N$ bis $+N$ annehmen. Für das Modell ist die Größe $N + m$ das Analogon der kinetischen Verteilungsfunktion. Wir können also formal schreiben

$$f = N + m \tag{16.1}$$

und entsprechend

$$H = \sum f \ln f = (N + m) \ln (N + m) + (N - m) \ln (N - m). \tag{16.2}$$

Bis auf Terme höherer Ordnung ist dann

$$H \approx \frac{m^2}{N} + \text{const}, \tag{16.3}$$

worin sich die Symmetrie des Problems in bezug auf den Wert $m = 0$ ausdrückt. Wir werden jedoch zunächst die für die Rechnung bequemere Funktion

$$\Delta = |2m| \tag{16.4}$$

betrachten, die einen analogen Verlauf zeigt und ebenfalls der erwähnten Symmetrie Rechnung trägt. Man sieht, daß alle im vorstehenden erwähnten Funktionen in Abhängigkeit von der Zeit $t = s\tau$ ($s = $ Zahl der Ziehungen) Stufenfunktionen

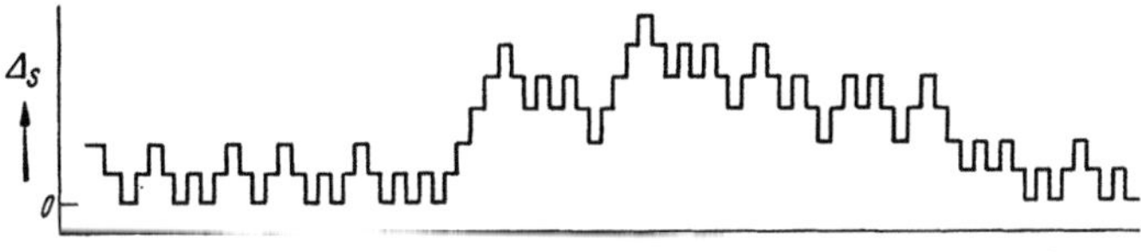

Fig. 2. Zum Ehrenfestschen Urnenmodell.

sind. Das entspricht vollkommen der Definition der kinetischen Funktionen mit Hilfe endlicher Volumenelemente des μ-Raumes. In beiden Fällen werden die stetigen Funktionen erst durch eine nachträgliche „Glättung" erhalten.

Der Verlauf der Funktion $\Delta(s)$ ist von KOHLRAUSCH und SCHRÖDINGER[1] experimentell ermittelt worden. Ein Teil dieser Kurve ist in Fig. 2 dargestellt. Bei jeder Ziehung kann sich $\Delta(s)$ nur um $+2$ oder -2 ändern. Die entsprechenden Übergangswahrscheinlichkeiten sind

$$\left.\begin{aligned} p(\Delta, \Delta - 2) &= \frac{N + \frac{1}{2}\Delta}{2N} = \frac{N + m}{2N}, \\ p(\Delta, \Delta + 2) &= \frac{N - \frac{1}{2}\Delta}{2N} = \frac{N - m}{2N} \end{aligned}\right\} \quad (m \geqq 0). \tag{16.5}$$

Die Wahrscheinlichkeit einer Abnahme ist somit (von dem Fall $m = 0$ abgesehen) stets größer als die einer Zunahme, und zwar um so ausgeprägter, je größer Δ ist. Den vier Folgen

$$\alpha)\quad \Delta - 2 \to \Delta \to \Delta - 2,$$

$$\beta)\quad \Delta - 2 \to \Delta \to \Delta + 2,$$

$$\gamma)\quad \Delta + 2 \to \Delta \to \Delta - 2,$$

$$\delta)\quad \Delta + 2 \to \Delta \to \Delta + 2,$$

---

[1] K. W. F. KOHLRAUSCH u. E. SCHRÖDINGER: Phys. Z. **27**, 306 (1926).

die in Fig. 3 veranschaulicht sind, entsprechen die relativen Wahrscheinlichkeiten.

$$p_\alpha : p_\beta : p_\gamma : p_\delta = \frac{N + \frac{1}{2}\Delta}{N - \frac{1}{2}\Delta} : 1 : 1 : \frac{N - \frac{1}{2}\Delta}{N + \frac{1}{2}\Delta}. \tag{16.6}$$

Man sieht, daß in einer solchen Folge die Wahrscheinlichkeit für das Durchlaufen eines Maximums am größten, die für ein Minimum am kleinsten ist. Die Bevorzugung des Maximums ist wieder um so stärker ausgeprägt, je größer Δ ist.

Betrachten wir, von einem gegebenen Δ-Wert ausgehend, zwei aufeinanderfolgende Ziehungen, so sind für den Endzustand die drei Fälle $\Delta + 4$, Δ, $\Delta - 4$ möglich. Die entsprechenden Übergangswahrscheinlichkeiten sind

$$\left.\begin{aligned}
p(\Delta, \Delta + 4) &= (N - \tfrac{1}{2}\Delta)(N - \tfrac{1}{2}\Delta - 1)/4N^2, \\
p(\Delta, \Delta) &= (N^2 + N - \tfrac{1}{4}\Delta^2)/2N^2, \\
p(\Delta, \Delta - 4) &= (N + \tfrac{1}{2}\Delta)(N + \tfrac{1}{2}\Delta - 1)/4N^2.
\end{aligned}\right\} \tag{16.7}$$

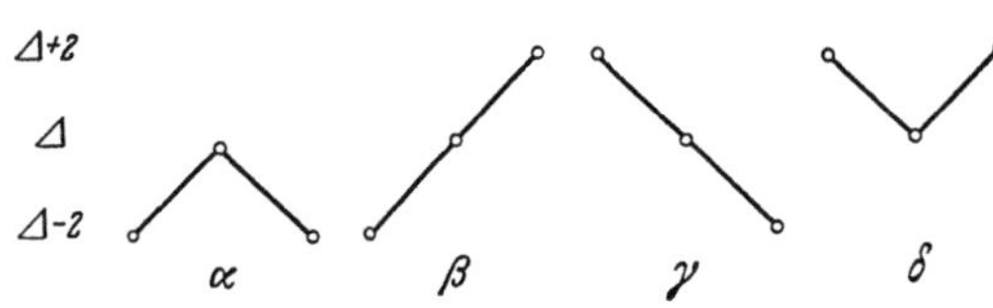

Fig. 3. Zur Diskussion der Übergangswahrscheinlichkeiten des Urnenmodells.

Wenn wir zur Zeit t den Wert Δ vorgeben, so ist der Mittelwert $\overline{\Delta'}$ zur Zeit $t + 2\tau$

$$\left.\begin{aligned}
\overline{\Delta'} = (\Delta + 4)\, p(\Delta, \Delta + 4) + \\
+ \Delta\, p(\Delta, \Delta) + \\
+ (\Delta - 4)\, p(\Delta, \Delta - 4)
\end{aligned}\right\} \tag{16.8}$$

oder mit Gl. (16.7)

$$\overline{\Delta'} = \Delta - \Delta(2N - 1)/N^2. \tag{16.9}$$

Definieren wir die mittlere Änderungsgeschwindigkeit von Δ durch

$$\overline{\left(\frac{d\Delta}{dt}\right)} = \frac{\overline{\Delta'} - \Delta}{2\tau}, \tag{16.10}$$

so wird

$$\overline{\left(\frac{d\Delta}{dt}\right)} = -\frac{\Delta(2N - 1)}{2\tau N^2}. \tag{16.11}$$

Da der Quotient auf der rechten Seite wesentlich positiv ist, haben wir für jedes Zeitintervall 2τ im Mittel stets eine Abnahme von Δ. Erst wenn wir die Mittelwertbildung auf der linken Seite außer acht lassen und die Gleichung integrieren, wird daraus eine monotone (exponentielle) Abnahme bis zum Gleichgewichtswert $\Delta = 0$. Diese Aussage stellt das Analogon des H-Theorems dar. Man sieht, daß dieselbe im Widerspruch zu den Gln. (16.5) steht und daher nicht richtig sein kann. Die vorstehenden Überlegungen zeigen einmal, wie eine solche Aussage zustande kommen kann und weiter, was ihr tatsächlicher Inhalt ist.

Die Analyse läßt sich noch weiterführen, wenn man von der stochastischen Grundgleichung des Problems ausgeht. Wir bezeichnen mit $P(n|m, s)$ die Wahrscheinlichkeit, daß nach s Ziehungen in der Urne A $N + m$ Kugeln sind, wenn sich anfangs $N + n$ Kugeln darin befanden. Die stochastische Gleichung[1] lautet dann

$$P(n|m, s) = \frac{N + m + 1}{2N}\, P(n|m + 1, s - 1) + \frac{N - m + 1}{2N}\, P(n|m - 1, s - 1). \tag{16.12}$$

---

[1] Über stochastische Prozesse vgl. den Artikel von Ramakrishnan in diesem Bande.

Diese Gleichung drückt die Wahrscheinlichkeit zur Zeit $s\tau$ durch die Wahrscheinlichkeiten zur Zeit $(s-1)\tau$ aus. Die Lösung der Gl. (16.12) für die Anfangsbedingung $P(n\,|\,m, 0) = \delta_{nm}$ ist von KAC[1] gegeben worden. KAC hat gezeigt, daß für jeden Anfangszustand sich schließlich eine stationäre Verteilung einstellt, die durch

$$P_0(m) = \alpha \, \frac{(2N)!}{(N+m)!\,(N-m)!} \equiv \alpha \, G_m \qquad (16.13)$$

gegeben ist. Hier bezeichnet $P_0(m)$ die stationäre Wahrscheinlichkeit, daß die Urne A $N+m$ Kugeln enthält, und $\alpha = 2^{-2N}$ ist ein Normierungsfaktor. Man bemerkt, daß die zu $P_0(m)$ proportionale Größe G_m einfach die Zahl der Mikrozustände ist, die zu dem durch m definierten Makrozustand gehören. Der Ausdruck stationär bedeutet in diesem Zusammenhang, daß die Verteilung (16.13) durch die stochastische Gleichung nicht geändert wird. Es ist nämlich

$$P_0(m) = \frac{N+m+1}{2N}\, P_0(m+1) + \frac{N-m+1}{2N}\, P_0(m-1). \qquad (16.14)$$

Dagegen ist der Wert von m nicht fixiert und zeigt weiter dauernde Schwankungen. Der Sinn dieser Aussage ist der folgende: Wenn wir etwa von einem großen n-Wert ausgehen, so haben wir zunächst, wie sich schon aus (16.5) ergibt, eine überwiegende Tendenz zur Abnahme von m, welche die Wahrscheinlichkeit für das Auftreten einzelner m-Werte bestimmt. Wenn der stationäre Zustand erreicht ist, ist der Einfluß des Ausgangszustandes verschwunden, und die Wahrscheinlichkeit für das Auftreten bestimmter m-Werte hängt nach Gl. (16.13) lediglich von der Zahl der zugehörigen Mikrozustände ab. Daraus folgt bereits, daß große Abweichungen von dem Wert $m=0$ im stationären Zustand sehr selten sind. Dies wird durch die Berechnung der Wiederkehrzeit[2] bestätigt, für die sich nach KAC

$$\Theta^* \sim \tau/G_m \qquad (16.15)$$

ergibt.

Wir wollen nun diese Modell-Analyse des H-Theorems in Beziehung setzen zu der ersten Boltzmannschen Definition des Gleichgewichtes (Ziff. 6). Danach ist der Gleichgewichtszustand der wahrscheinlichste Zustand, für den also (unter entsprechenden Nebenbedingungen) G_m ein Maximum bzw. H ein Minimum wird. Für das Urnenmodell ist dieser Zustand durch $m=0$ gegeben. Nun ist aber $P(n\,|\,m, s) = \delta_{m0}$ keine stationäre Lösung der stochastischen Gleichung (16.12). Man muß also annehmen, daß das Gleichgewicht immer nur momentan vorhanden ist und daß dauernd die als Schwankungen bezeichneten Abweichungen vom Gleichgewicht auftreten. Da in diesem Konzept die thermodynamischen Funktionen für das durch einen bestimmten m-Wert charakterisierte Gleichgewicht definiert werden, muß man konsequenterweise auch Schwankungen der thermodynamischen Funktionen zulassen. Nach BOLTZMANN hat man also allgemein für die Entropie unseres Modells zu setzen

$$S = \ln G_m = \ln \frac{(2N)!}{(N+m)!\,(N-m)!}, \qquad (16.16)$$

---

[1] M. KAC: Amer. Math. Monthly **54**, 369 (1947), abgedruckt in Selected Papers on Noise and Stochastic Processes (Herausgeber N. WAX), New York 1954.

[2] Die hier gemeinte Wiederkehrzeit (recurrence time) ist die mittlere Zeit zwischen dem Auftreten zweier identischer Makrozustände. Sie ist für mechanische Systeme von der Poincaréschen Wiederkehrzeit t^* (Ziff. 15) zu unterscheiden, die sich auf die Wiederkehr eines Phasenpunktes im Γ-Raum bezieht. Vgl. CHANDRASEKHAR [15].

oder für große N mit Benutzung der Stirlingschen Formel

$$S = -(N+m)\ln(N+m) - (N-m)\ln(N-m). \tag{16.17}$$

Der Vergleich mit (16.2) ergibt sofort

$$S = -H. \tag{16.18}$$

Man sieht daraus, daß die Behauptung, H nehme mit der Zeit monoton bis zu einem Gleichgewichtswert ab, auch mit der Boltzmannschen Definition des Gleichgewichtes unvereinbar ist.

Die Analyse des H-Theorems führt ohne weiteres zu der Frage, worin nun eigentlich das Wesen eines irreversiblen Prozesses besteht. Smoluchowski[1] und ihm folgend Chandrasekhar [15] definieren einen Prozeß als irreversibel, wenn der Ausgangszustand durch eine mittlere Wiederkehrzeit charakterisiert wird, die lang im Vergleich zur Beobachtungszeit ist. Die damit erreichte Abgrenzung gegenüber den „gewöhnlichen" Schwankungen ist ziemlich scharf, da für große N die Funktion G_m an der Stelle $m=0$ ein äußerst steiles Maximum besitzt. Trotzdem ist diese Unterscheidung logisch etwas unbefriedigend, weil eben auch im stationären Zustand beliebig große Werte $|m| \leq N$ als Schwankungen grundsätzlich auftreten können, wenn sie auch praktisch nicht beobachtet werden. Die Diskussion der stochastischen Gleichung (16.12) zeigt nun, daß der entscheidende Punkt für den irreversiblen Charakter eines Prozesses die Anfangsbedingung $P(n|m, 0) = \delta_{nm}$ ist. Das heißt mit anderen Worten: Wenn wir maximale Information besitzen, daß zur Zeit $t=0$ das Modell sich im Zustand n befindet, dann tragen die Wahrscheinlichkeitsaussagen über die Entwicklung des Systems den oben ausführlich erläuterten Charakter eines irreversiblen Prozesses. Haben wir zur Zeit $t=0$ keine Information, so müssen alle Mikrozustände als gleich wahrscheinlich betrachtet werden. Wir befinden uns im stationären Zustand, und die Wahrscheinlichkeit eines bestimmten m-Wertes wird durch Gl. (16.13) bestimmt. Man sieht an diesem einfachen Beispiel, daß der Unterschied zwischen den Begriffen „Irreversibler Prozeß" und „Schwankung" nicht in der physikalischen Struktur des Systems, sondern in der Information zur Zeit $t=0$ seine Wurzel hat. Dieser Begriff ist der Mechanik fremd, bildet aber ein wesentliches Element der Wahrscheinlichkeitstheorie. Es ist daher verständlich, daß der scheinbar der Mechanik widersprechende Begriff des „Irreversiblen Prozesses" lediglich eine Folge der Anwendung stochastischer Methoden ist. Erst die spezielle Betrachtung der Wiederkehrzeiten ermöglicht dann die physikalisch anschaulichere Definition von Smoluchowski.

Aus der vorstehenden Deutung ergibt sich, daß auch die Aussage des H-Theorems von der Information her verstanden werden muß und daß dies der eigentliche Sinn der statistischen Interpretation ist. Der Entwicklung in Richtung auf den stationären Zustand entspricht offenbar eine Verminderung der Information. Wir werden diesen Gedanken in Ziff. 17 weiter verfolgen.

Die stochastische Analyse des Urnenmodells, die wir im vorstehenden skizziert haben, ist in neuerer Zeit mit Erfolg auf Modelle von mehr physikalischem Charakter ausgedehnt worden. Dabei tritt allerdings die Schwierigkeit auf, daß sich für diese Modelle bisher nicht hat beweisen lassen, daß es sich um Markoffsche Prozesse[2] handelt (was für das Urnenmodell trivial ist). Für Einzel-

---

[1] M. von Smoluchowski: Phys. Z. **13**, 1069 (1912).

[2] Ein stochastischer Prozeß wird als Markoffscher Prozeß bezeichnet, wenn die bedingte Wahrscheinlichkeit, daß eine Variable y zur Zeit t_n zwischen y_n und $y_n + dy_n$ liegt, unter der Voraussetzung, daß den Zeiten $t_1, t_2, \ldots, t_{n-1}$ die Werte $y_1, y_2, \ldots, y_{n-1}$ zugeordnet sind, außer von y_n und t_n nur von y_{n-1} und t_{n-1} abhängt. Er stellt nach Chandrasekhar [15] „ein schrittweises Entfalten der Übergangswahrscheinlichkeiten" dar.

heiten müssen wir auf die Originalarbeiten [22], [23][1] verweisen. Die einschlägigen stochastischen Probleme sind ausführlich in der zusammenfassenden Darstellung von CHANDRASEKHAR [15] sowie in dem Artikel von RAMAKRISHNAN in diesem Bande behandelt.

Wir kommen nun zu der oben (S. 218) unter b) angeführten Methode. Dieselbe versucht (da die detaillierte Analyse für ein Gas praktisch undurchführbar ist) durch allgemeinere statistische Überlegungen den Widerspruch zwischen H-Theorem und Mechanik aufzulösen. Dieser Weg ist bereits von BOLTZMANN eingeschlagen worden. P. und T. EHRENFEST [3] haben die Boltzmannschen Überlegungen systematisch dargestellt und kritisch analysiert. Man kann im wesentlichen drei Gedankengänge unterscheiden, die wir kurz skizzieren.

Die Problematik des älteren H-Theorems beruht rein formal auf dem Stoßzahlansatz. An einfachen Modellen kann man dies leicht explizit zeigen und auch nachweisen, daß (vom Falle des Gleichgewichtes abgesehen) der Stoßzahlansatz nicht allgemein gültig sein kann [3], [23]. BOLTZMANN versuchte nun, das Problem der Interpretation des H-Theorems zunächst völlig unabhängig vom Stoßzahlansatz zu diskutieren.

Der erste Gedankengang[2] enthält die folgenden Schritte:

a) Die auf Grund der Zellenteilung des μ-Raumes definierte Maxwell-Boltzmannsche Verteilung nimmt im Γ-Raum den weitaus überwiegenden Teil der Energieschale zwischen E und $E + \Delta E$ ein.

Diese Aussage folgt unmittelbar aus der bekannten Boltzmannschen Ableitung mit Hilfe der Komplexionen im μ-Raum.

b) Auf der Energiefläche $H(\boldsymbol{q}, \boldsymbol{p}) = E$ ist der Inhalt der zur Maxwell-Boltzmannschen Verteilung gehörenden Gebiete weitaus größer als der aller übrigen Gebiete.

P. und T. EHRENFEST [3] und neuerdings TER HAAR [22], [23] haben darauf hingewiesen, daß b) nicht streng aus a) folgt und daher nicht als bewiesen angesehen werden kann. Dieses Bedenken dürfte gegenstandslos sein, nachdem KHINCHIN [19] das Maxwell-Boltzmannsche Verteilungsgesetz streng für die Energiefläche abgeleitet hat.

c) Unter der Voraussetzung, daß Gl. (7.6) gültig ist, folgt dann, daß auch im zeitlichen Ablauf die Maxwell-Boltzmannsche Verteilung weitaus überwiegt. Dadurch wird verständlich, daß man, ausgehend von einer abweichenden Verteilung, fast immer eine Annäherung an die Maxwell-Boltzmannsche Verteilung beobachtet.

Die Schwäche dieses Gedankenganges liegt darin, daß er die Gültigkeit der Gl. (7.6) und damit des Ergoden-Theorems voraussetzt. Er kann daher nach den Ausführungen in Ziff. 9 nicht zur Begründung des H-Theorems dienen.

Der zweite Boltzmannsche Gedankengang[3, 4] umfaßt eine Diskussion der Treppenkurve der H-Werte. Dabei werden im wesentlichen die Eigenschaften herausgearbeitet, die wir oben für das Urnenmodell erörtert haben. Sie führen daher zu einer analogen Interpretation des H-Theorems. Für ein Gas besitzen diese Überlegungen jedoch hypothetischen Charakter. Sie zeigen lediglich, daß das H-Theorem bei korrekter Interpretation nicht im Widerspruch zur Mechanik stehen muß. Für die Einzelheiten sei auf EHRENFEST [3] und TOLMAN [12] verwiesen.

---

[1] Siehe die auf S. 218 zitierten Arbeiten von TER HAAR und GREEN.
[2] L. BOLTZMANN: Wien. Ber. **76**, 373 (1877).
[3] L. BOLTZMANN: Nature, Lond. **51**, 413, 581 (1894); **52**, 221 (1895).
[4] L. BOLTZMANN: Math. Ann. **50**, 325 (1898).

Gleichzeitig mit den zuletzt erwähnten Überlegungen hat Boltzmann[1-3] sich mit der Frage auseinandergesetzt, wie im Rahmen der neuen Interpretation des H-Theorems der Stoßzahlansatz zu verstehen ist. Seine Andeutungen lassen sich nach P. und T. Ehrenfest [3] zusammenfassen in den Sätzen:

a) Der Stoßzahlansatz gibt für jedes Zeitelement Δt die wahrscheinlichsten Werte der Stoßzahlen an. Das H-Theorem liefert für jedes Δt den wahrscheinlichsten Wert der H-Änderung.

b) Die wirkliche Zahl der Stöße und die wirkliche H-Änderung schwanken um diese wahrscheinlichsten Werte nach Maßgabe der zwar viel kleineren, aber doch von Null verschiedenen Wahrscheinlichkeiten anderer Werte.

Wenn man von der etwas unscharfen Terminologie absieht, stimmen diese Aussagen überein mit den Ergebnissen, die man für einfache Modelle (soweit es sich um das H-Theorem handelt, bereits für das Urnen-Modell) rechnerisch ableiten kann. Für Gase besitzen sie jedoch den Charakter von, wenn auch plausiblen, Hypothesen. Die Aussage a) ist von Jeans[4] wesentlich schärfer gefaßt worden; einen eigentlichen Beweis hat auch er nicht gebracht. In neuerer Zeit hat Tolman [12] die obigen Gedanken in etwas anderer Form entwickelt. Wir erläutern seine Methode an dem Beispiel des Ansatzes (14.9) für elastische Kugeln. Dieser Ausdruck wird formal als für verdünnte Gase gültig angenommen. Die Größe $a_{1\,2\to1'\,2'}$ wird jedoch interpretiert als die Wahrscheinlichkeit, daß der Schwerpunkt des Moleküls 2 sich in dem Volumen $\sigma^2 \dfrac{|\boldsymbol{p}_1 - \boldsymbol{p}_2|}{m} \cos \Theta$ befindet. Diese Wahrscheinlichkeit hängt ab von der räumlichen Verteilung der Moleküle. Tolman nimmt an, daß die wahrscheinlichste Verteilung die gleichmäßige Verteilung ist. Die Wahrscheinlichkeit, innerhalb eines Volumens V ein Molekül im Volumenelement $\Delta\boldsymbol{q}_1$ zu finden, ist dann

$$W = \frac{\Delta\boldsymbol{q}_1}{V}. \tag{16.19}$$

Daraus folgt unmittelbar die Gl. (14.10). Tolman schließt nun, in Übereinstimmung mit der obigen Aussage a), daß der Stoßzahlansatz die wahrscheinlichste Stoßzahl gibt. Die Aussage des H-Theorems soll „im Mittel", aber nicht in jedem Einzelfall gültig sein.

Auch diese Ableitung erscheint wenig befriedigend und kann kaum als strenger Beweis gelten. Zunächst läßt sich mit Hilfe der molekularen Verteilungsfunktionen [26] leicht zeigen, daß die Anwendung der Gl. (16.19) auf das hier in Frage stehende Problem nicht korrekt ist. Davon abgesehen bleiben sowohl die Bedeutung des Ausdruckes „wahrscheinlichste Stoßzahl" wie die statistische Natur der Größe dH/dt ziemlich vage.

Schließlich ist noch ein Gesichtspunkt von Bedeutung, auf den schon P. und T. Ehrenfest [3] hingewiesen haben und der besonders ausführlich von Tolman [12] erörtert worden ist. Jede tiefer gehende Diskussion des H-Theorems muß, ebenso wie die Modellrechnungen, implizit oder explizit die Annahme einführen, daß man es mit Markoffschen Prozessen zu tun hat. Das ist nicht nur, für sich genommen, eine Hypothese, sondern es ist auch unklar, wie sich dieselbe zur Gültigkeit der Bewegungsgleichungen, die ja die Trajektorie bestimmen, wenn ein Punkt bekannt ist, verhält.

---

[1] Siehe Fußnote 3, S. 223.
[2] Siehe Fußnote 4, S. 223.
[3] L. Boltzmann: Ann. Phys. **57**, 773 (1896); **60**, 392 (1897).
[4] J. H. Jeans: Dynamical Theory of Gases. Cambridge 1904.

Zusammenfassend können wir feststellen, daß durch die eingangs unter a) und b) genannten Methoden die Widersprüche zwischen der ursprünglichen Formulierung des H-Theorems und der Mechanik beseitigt worden sind. Die Überlegungen enthalten aber, soweit sie sich auf Gase beziehen, zahlreiche Hypothesen; eine einwandfreie Ableitung des H-Theorems existiert bisher nicht.

17. Tolman's Verallgemeinerung des H-Theorems. Das Boltzmannsche H-Theorem kann nach seiner Ableitung von vornherein nur Gültigkeit für verdünnte Gase beanspruchen. Wir wollen nun im Anschluß an Tolman [12] das Verhalten nicht-stationärer Gesamtheiten untersuchen unter dem speziellen Gesichtspunkt, eine dem H-Theorem analoge allgemeine Gesetzmäßigkeit aufzufinden. Wir können erwarten, daß wir auf diesem Wege auch wesentliche Aufschlüsse über das Boltzmannsche H-Theorem erhalten werden.

Wir betrachten eine Gesamtheit aus konservativen Systemen, die das gleiche Energieintegral besitzen. Alle Trajektorien verlaufen somit auf der Hyperfläche $H(\boldsymbol{q}, \boldsymbol{p}) = E$. Im Gegensatz zu früheren Betrachtungen postulieren wir aber jetzt nicht mehr, daß die Phasendichte auf der Energiefläche konstant ist, sondern wir betrachten $\varrho(\boldsymbol{q}, \boldsymbol{p})$ als eine — zunächst beliebige — Funktion der Phasenkoordinaten. Aus der Bedingung (5.8) folgt unmittelbar, daß die betrachtete Gesamtheit im allgemeinen nicht stationär ist.

Bisher haben wir die Verteilung der Gesamtheit im Phasenraum durch die Phasendichte $\varrho(\boldsymbol{q}, \boldsymbol{p})$ beschrieben, die sich auf den im Sinne der klassischen Mechanik scharf charakterisierten Zustand der Systeme bezieht. Wir nennen diese Größe von jetzt ab die „*feine Dichte*" (fine-grained density). Für das Folgende ist es notwendig, explizit zu berücksichtigen, daß Messungen (auch bei weitgehender Idealisierung) nur eine wesentlich ungenauere Charakterisierung erlauben. Aus solchen Messungen läßt sich also nicht $\varrho(\boldsymbol{q}, \boldsymbol{p})$, sondern nur der jeweilige Mittelwert von ϱ über ein kleines, aber endliches Volumenelement des Phasenraumes $\Delta\Omega$ ableiten. Diese Größe, die durch

$$P = \frac{\int \varrho(\boldsymbol{q}, \boldsymbol{p})\, d\Omega}{\Delta\Omega} \tag{17.1}$$

definiert ist, bezeichnen wir als „*grobe Dichte*" (coarse-grained density). Die Notwendigkeit, zwischen feiner und grober Dichte zu unterscheiden (die erstmalig von P. und T. Ehrenfest [3] klar ausgesprochen wurde), wird aus der Art der Problemstellung verständlich. In der Gleichgewichtsstatistik wird die Phasendichte aus gewissen Axiomen abgeleitet; als Beobachtungsgrößen treten dabei nur Parameter von der Art der thermodynamischen Zustandsgrößen auf. Bei dem hier interessierenden Problem muß die nicht nur vom Energie-Integral abhängende Verteilung im Phasenraum als im Prinzip aus den Beobachtungen konstruierbar angesehen werden, und das Konzept der Information spielt, wie wir schon in Ziff. 16 gesehen haben, eine wichtige Rolle für die Betrachtungen.

Aus der Definition von P folgt, daß

$$\frac{\Delta v}{v} = P \Delta \Omega \tag{17.2}$$

die Wahrscheinlichkeit ist, ein beliebig herausgegriffenes System der Gesamtheit in dem kleinen, aber endlichen Volumenelement $\Delta\Omega$ zu finden. Denken wir uns den gesamten Phasenraum in Zellen $\Delta\Omega_m$ unterteilt, so ist

$$\sum_m P_m \Delta\Omega_m = 1, \tag{17.3}$$

wo P_m die grobe Dichte in der m-ten Zelle ist. Ähnlich wie in der kinetischen Gastheorie (Ziff. 14) ist es auch hier zweckmäßig, P_m als stetige Funktion der Phasenkoordinaten zu betrachten. Die Gl. (17.3) kann dann geschrieben werden

$$\int P(\boldsymbol{q}, \boldsymbol{p})\, d\Omega = 1. \tag{17.4}$$

Wir definieren nun zwei Funktionen

$$\sigma = \int \varrho \ln \varrho \, d\Omega \tag{17.5}$$

und

$$\overline{H} = \sum_m P_m \ln P_m \, \Delta\Omega_m, \tag{17.6}$$

bzw.

$$\overline{H} = \int P \ln P \, d\Omega. \tag{17.7}$$

Formal sind σ und $\overline{H}$ die Phasenmittelwerte von $\ln \varrho$ bzw. $\ln P$. Da, wie wir zeigen werden, σ und P für die gleichmäßige Verteilung ein Minimum besitzen und andererseits, wie wir in Ziff. 5 gezeigt haben, die gleichmäßige Verteilung der feinen Dichte stationär ist, ist die Bedeutung dieser Funktionen für das hier interessierende Problem unmittelbar evident. Das Verhalten dieser Funktionen bildet im wesentlichen den Gegenstand der folgenden Untersuchung.

Obwohl beiden Funktionen σ und $\overline{H}$ die erwähnte Extremaleigenschaft zukommt, verhalten sie sich in bezug auf ihre zeitliche Änderung grundsätzlich verschieden. Betrachten wir die Funktion σ zu den Zeiten t' und t'' und fassen die Bewegung wie früher als eine Transformation des Phasenraumes in sich selbst auf, so haben wir

$$\sigma(t') = \int \varrho' \ln \varrho' \, d\Omega' = \int \varrho'' \ln \varrho'' \, J \, d\Omega'', \tag{17.8}$$

wo J die Jacobische Determinante bezeichnet. Mit Benutzung des Liouvilleschen Satzes Gl. (3.12) folgt daraus[1]

$$\sigma(t') = \sigma(t''). \tag{17.9}$$

Für die Funktion $\overline{H}$ läßt sich ein analoges Resultat nicht ableiten, weil für P der Liouvillesche Satz nicht gilt und nicht allgemein $P = \varrho$ ist. Das Ergebnis (17.9) erscheint zunächst merkwürdig und legt die Frage nahe, ob sich überhaupt irreversible Prozesse mit Hilfe der Ensemble-Theorie beschreiben lassen. Der Schlüssel zu der Schwierigkeit liegt in der schon früher (Ziff. 3) gemachten Bemerkung, daß aus dem Liouvilleschen Satz nur die Erhaltung des Volumens einer Phasenausdehnung folgt, während er keine Aussage über die Gestalt dieses Volumens macht. Gibbs [2] hat für diesen Sachverhalt ein anschauliches Beispiel gegeben. Wenn man zu Wasser einen unlöslichen Farbstoff (etwa chinesische Tusche) hinzufügt, so sind beide im ersten Augenblick scharf getrennt. Durch hinreichend langes Rühren erreicht man schließlich einen von der Art des Rührens unabhängigen Endzustand, in welchem bei grober Betrachtung der Farbstoff gleichmäßig über das ganze Gefäß verteilt erscheint. Erst bei mikroskopischer Betrachtung stellt man fest, daß die Verteilung nicht gleichmäßig ist; die Farbstoff-Partikelchen sind durch Zwischenräume, die nur Wasser enthalten, voneinander getrennt. Das von dem Farbstoff eingenommene Volumen hat sich beim Rühren nicht verändert, wohl aber die Gestalt dieses Volumens. Die feine Dichte ist nach wie vor ungleichmäßig, aber die grobe Dichte hat sich in Richtung auf eine gleichmäßige Verteilung entwickelt. Man wird also nur von einer Funktion der groben Dichte erwarten können, daß sie die für einen

---

[1] S. H. Burbury: Phil. Mag. **6**, 251 (1903).

irreversiblen Prozeß charakteristischen Eigenschaften zeigt. Die Funktion P gibt uns somit die Möglichkeit, die Kluft zwischen der grundsätzlich reversiblen Mechanik und den makroskopischen irreversiblen Prozessen zu überbrücken.

Die mathematische Grundlage der nun folgenden Überlegung bildet die Tatsache, daß die Funktion

$$y = x\,\mathrm{e}^x - \mathrm{e}^x + 1 \qquad (17.10)$$

für $x \neq 0$ positiv ist, während sie für $x = 0$ Null wird. Die letztere Aussage ist unmittelbar evident, die erste folgt, wenn wir bemerken, daß dy/dx das gleiche Vorzeichen wie x hat.

Wir nehmen an, daß durch eine Beobachtung der Zustand des Originalsystems zur Zeit t' innerhalb der oben erwähnten Genauigkeitsgrenzen bestimmt worden ist. Wenn wir nun das repräsentative Ensemble nach dem Prinzip der gleichen a priori-Wahrscheinlichkeiten konstruieren, so muß $\varrho = \text{const}$ sein in allen Zellen $\Delta \Omega_i$, welche mit der durch die Messung erreichten (unvollständigen) Kenntnis des Systems vereinbar sind, dagegen Null im übrigen Teil des Phasenraumes. Das Ensemble zur Zeit t_i repräsentiert also die maximale durch die Beobachtung gewonnene Information. Es ist dann

$$\varrho' = P' \qquad (17.11)$$

und

$$\bar{H}' = \int \varrho' \ln \varrho' \, d\Omega. \qquad (17.12)$$

Verfolgen wir die Bewegung dieser Gesamtheit im Laufe der Zeit, so muß nach dem Liouvilleschen Satz die mit jeder Zelle zur Zeit t' verknüpfte Phasenausdehnung erhalten bleiben. Ihre Gestalt wird sich jedoch ändern, was nur möglich ist, wenn Zellen teilweise überdeckt werden, für die vorher $\varrho' = P' = 0$ war. Zu einer späteren Zeit t'' muß daher

$$\varrho'' \neq P'' \qquad (17.13)$$

sein. Es ist dann

$$\bar{H}'' = \int P'' \ln P'' \, d\Omega. \qquad (17.14)$$

Wir schreiben nun (17.13) in der Form

$$\varrho'' = P'' \,\mathrm{e}^{\Delta \eta}, \qquad (17.15)$$

wo $\Delta \eta$ eine Funktion der Phasenkoordinaten ist. Transformieren wir nun Gl. (17.12) mit Benutzung des Liouvilleschen Satzes auf zweimal gestrichene Variable, subtrahieren die Gl. (17.14) und addieren die aus (2.4) und (17.4) folgende Gleichungen

$$\int (P'' - \varrho'') \, d\Omega = 0, \qquad (17.16)$$

so erhalten wir

$$\bar{H}' - H'' = \int P'' (\Delta \eta \,\mathrm{e}^{\Delta \eta} - \mathrm{e}^{\Delta \eta} + 1) \, d\Omega \qquad (17.17)$$

oder mit Benutzung des oben bewiesenen Satzes

$$\bar{H}' - \bar{H}'' > 0. \qquad (17.18)$$

Sofern also Teile des Phasenraumes, für die zur Zeit t' $\varrho' = P' = 0$ war, dem System zugänglich sind, muß für jede Zeit $t'' > t'$ der Wert der Funktion $\bar{H}$ kleiner sein als zur Zeit t'. Diese Aussage wird von TOLMAN [*12*] als generalisiertes *H*-Theorem bezeichnet. Die Funktion $\bar{H}$ kann nicht beliebig, sondern nur bis zu einem Minimum abnehmen. Bilden wir nämlich

$$\delta \bar{H} = \int (\ln P + 1) \, \delta P \, d\Omega = 0 \qquad (17.19)$$

mit der Nebenbedingung
$$\delta E = 0, \tag{17.20}$$
so haben wir als Lösung
$$P = \text{const}, \tag{17.21}$$

also eine gleichmäßige grobe Dichte auf der zugänglichen Energiefläche. Wenn wir berücksichtigen, daß die Größe $\Delta\eta$ ein Maß unserer Information, bezogen auf den Zeitpunkt t', darstellt, so erscheint die Folgerung plausibel, daß $\overline{H}$ monoton abfällt, bis der durch Gl. (17.21) definierte stationäre Zustand erreicht ist. Auch das Gibbssche Modell legt einen solchen Schluß nahe. Ein eigentlicher Beweis existiert indessen für diese weitergehende Aussage nicht. Setzen wir sie zunächst einmal voraus, so ergibt sich eine bemerkenswerte Folgerung. Wir werden später (Ziff. 26) zeigen, daß für ein stationäres Ensemble die Größe

$$S = -k \int P \ln P \, d\Omega \tag{17.22}$$

(wo k die Boltzmannsche Konstante bezeichnet) die Eigenschaften der thermodynamischen Entropie besitzt. Wenn wir auf Grund dieser Formulierung den Entropiebegriff für nicht-stationäre Gesamtheiten verallgemeinern, haben wir

$$\overline{H} = -S/k. \tag{17.23}$$

Die obige (hypothetische) Aussage bedeutet dann, daß bei einem irreversiblen Prozeß die (verallgemeinerte) Entropie monoton bis zum Gleichgewichtswert zunimmt, während gleichzeitig die Information abnimmt. Wir sehen somit auch hier wieder, daß das Konzept der Information eine zentrale Bedeutung für das Verständnis der irreversiblen Prozesse und auch der Entropie besitzt[1-3].

Der Begriff des Gleichgewichtes ist hier wesentlich verschieden von dem in Ziff. 16 erörterten Boltzmannschen Konzept. Das durch Gl. (17.21) charakterisierte Gleichgewicht umfaßt tatsächlich alle Zustände, die das System (für die gegebenen Werte der markoskopischen Parameter) überhaupt einnehmen kann[4]. Die Schwankungen erscheinen als eine Gleichgewichtseigenschaft (nicht als Abweichungen vom Gleichgewicht), und der irreversible Prozeß ist lediglich durch den Gesichtspunkt der Information herausgehoben. Insofern haben wir eine vollkommene Analogie zur stationären Lösung der stochastischen Gleichung des Urnenmodells (Ziff. 16). Während die H-Treppenkurve des Einzelmodells, wie wir in Ziff. 16 gesehen haben, sicher nicht monoton abfällt, ist ein solcher Verlauf für die Funktion $\overline{H}$ ohne weiteres denkbar (wenn auch nicht bewiesen), da $\overline{H}$ der Mittelwert von $\ln P$ ist.

Wir wollen nun noch zeigen, daß aus der erweiterten Fassung des generalisierten H-Theorems das Boltzmannsche H-Theorem in seiner ursprünglichen Fassung, aber mit der durch die Ensemble-Theorie gegebenen statistischen Interpretation folgt. Dazu definieren wir den Satz der molekularen Verteilungsfunktionen[5]

$$f^{(N)} \equiv P, \ldots, f^{(N)} = \iint P \, d\mathbf{q}^{(N-n)} \, d\mathbf{p}^{(N-n)}, \ldots, f^{(1)} = \iint P \, d\mathbf{q}^{(N-1)} \, d\mathbf{p}^{(N-1)}. \tag{17.24}$$

---

[1] L. Szilard: Z. Physik **53**, 840 (1929).
[2] L. Brillouin: J. Appl. Phys. **22**, 334 (1951); **24**, 1152 (1953).
[3] E. T. Jaynes: Phys. Rev. **106**, 620 (1957).
[4] Aus diesem Grunde halten P. u. T. Ehrenfest [3] die Gibbssche Funktion $\overline{H}$ für ungeeignet zur Beschreibung irreversibler Prozesse.
[5] Die durch Gl. (17.24) definierten molekularen Verteilungsfunktionen sind Wahrscheinlichkeitsdichten für spezifizierte Moleküle. Vgl. dazu Ziff. 18.

Die Funktion $f^{(1)}$ unterscheidet sich von der in Ziff. 14 benutzten kinetischen Verteilungsfunktion dadurch, daß sie mit Hilfe einer statistischen Gesamtheit definiert ist und die Mittelung über die Zeit nicht enthält. Die Verwendung der groben Dichte entspricht der Zellenteilung des μ-Raumes. Für ein hinreichend verdünntes Gas ist

$$f^{(N)} = \prod_{i=1}^{N} f^{(1)}. \tag{17.25}$$

Setzen wir dies in Gl. (17.7) ein, so folgt

$$\overline{H} = \sum_{j=1}^{N} \iint f^{(1)}(\boldsymbol{q}_j, \boldsymbol{p}_j) \ln f^{(1)}(\boldsymbol{q}_j, \boldsymbol{p}_j) \, d\boldsymbol{q}_j \, d\boldsymbol{p}_j \times \int \ldots \int \prod_{\substack{i=1 \\ i \neq j}}^{N} [f^{(1)}(\boldsymbol{q}_i, \boldsymbol{p}_i) \, d\boldsymbol{q}_i \, d\boldsymbol{p}_i]. \tag{17.26}$$

Wegen der Normierung der molekularen Verteilungsfunktionen ist das multiple Integral gleich Eins. Definieren wir nun in Analogie zu Gl. (14.17)

$$H_B = \iint f^{(1)}(\boldsymbol{q}_1, \boldsymbol{p}_1) \ln f^{(1)}(\boldsymbol{q}_1, \boldsymbol{p}_1) \, d\boldsymbol{q}_1 \, d\boldsymbol{p}_1 \tag{17.27}$$

und setzen voraus, daß alle Moleküle gleich sind, so erhalten wir

$$\overline{H} = N H_B. \tag{17.28}$$

Aus der erweiterten Fassung des generalisierten H-Theorems folgt dann

$$\frac{dH_B}{dt} \leqq 0, \tag{17.29}$$

wobei das Gleichheitszeichen für den stationären Zustand gilt. Damit wäre die statistische Interpretation des H-Theorems dahin präzisiert, daß die ursprüngliche Fassung (10.22) eine Aussage darstellt, die sich auf eine durch die grobe Dichte charakterisierte statistische Gesamtheit bezieht.

Obwohl die im Vorstehenden skizzierte Theorie einleuchtend ist und ein geschlossenes Bild der statistischen Theorie irreversibler Vorgänge und der Gleichgewichtseigenschaften gibt, enthält sie einige Punkte, die erhebliche Schwierigkeiten machen und noch der Klärung bedürfen. Zunächst ist, wie schon erwähnt, die erweiterte Fassung des generalisierten H-Theorems eine plausible, aber bisher unbewiesene Hypothese[1]. Zwei weitere Einwände sind von Kirkwood [16], [25] gemacht worden. Wir nehmen der Einfachheit halber an, daß die Gesamtheit zur Zeit t' sich in der Zelle $\Delta\Omega_1$ des Phasenraumes befindet. Da die Gesamtheit nicht stationär ist, muß sie sich zur Zeit $t''' < t'$ über viele Zellen des Phasenraumes erstreckt haben. Daraus würde folgen $\overline{H}''' < \overline{H}'$, im Widerspruch zu (17.18). Zu diesem Einwand bemerkt Kirkwood [16] selbst, daß t' als der Zeitpunkt der maximalen Information angesehen werden muß, was mit der oben entwickelten Auffassung übereinstimmt. Der zweite Einwand besagt, daß man genau das Volumen, das die Gesamtheit zur Zeit $t'' > t'$ einnimmt, als eine der Zellen wählen könne, in die der Phasenraum unterteilt wird. Dann würde $\overline{H}' < \overline{H}''$ sein, abermals im Widerspruch zu (17.18), und die $\overline{H}$-Kurve würde zur Zeit t'' eine Spitze zeigen. Diese Schwierigkeit wird teilweise beseitigt durch die obige Festsetzung über die Konstruktion des Ensembles zur Zeit t'. Danach muß für jede Wahl der Zelleneinteilung das Volumen, in dem $P \neq 0$ ist, zur Zeit $t'' > t'$ größer als zur Zeit t' (im Grenzfall gleich diesem Volumen) sein. Eine solche Vorschrift bedeutet keine Einschränkung der Allgemeinheit, sondern liegt in der Natur der Sache begründet. Wir werden in Ziff. 19 sehen, daß die durch Zeitmittelung

---

[1] Diese Ansicht wird auch von ter Haar [23] und Becker [24] vertreten.

gewonnene grobe Dichte ebenfalls spezielle Vorschriften erfordert. Was bisher nicht bewiesen wurde, ist der monotone Abfall von $\overline{H}$, so daß intermediäre Maxima nicht auszuschließen sind. Es erscheint denkbar, daß für diese Frage die Wahl der Zellenteilung eine Rolle spielt, so daß vielleicht noch eine weitere Vorschrift erforderlich ist.

Eine ernste Schwierigkeit liegt auch in der Frage, ob die durch (17.21) definierte Verteilung stationär ist. In Ziff. 5 haben wir das Kriterium für das statistische Gleichgewicht mit Hilfe des Liouvilleschen Satzes für die feine Dichte abgeleitet. Da im Falle der Gl. (17.21) ϱ notwendig eine Funktion der Phasenkoordinaten ist, also nicht nur von den uniformen Integralen der Bewegungsgleichungen abhängt, ist die Gesamtheit in bezug auf die feine Dichte sicher nicht stationär. Das schließt nicht notwendig aus, daß die durch (17.21) definierte Verteilung stationär ist. Da aber für P der Liouvillesche Satz nicht gilt, fehlt uns hier vorläufig ein Kriterium zur Untersuchung dieser Frage.

Das Problem des H-Theorems ist somit auch in der Ensemble-Theorie erst teilweise gelöst. Grundsätzlich erscheinen hier die Aussichten für eine einheitliche Darstellung des Problems wesentlich günstiger als bei den in Ziff. 16 erörterten statistischen Überlegungen. Ob indessen die angedeuteten Lücken sich schließen lassen, muß zur Zeit als offene Frage betrachtet werden. Bemerkenswert ist in diesem Zusammenhang eine neuere Untersuchung von Born und Hooton[1]. Diese Autoren konnten zeigen, daß eine Gesamtheit aus mehrfach periodischen Systemen, die zur Zeit $t=0$ im Raume der Winkel- und Wirkungsvariablen durch Gauß-Verteilungen darstellbar ist, sich so bewegt, daß schließlich scharfe Werte für die Konstanten der Bewegung und gleichmäßige Verteilung für alle damit verträglichen q und p erreicht werden.

18. Kirkwood's Theorie der Transporterscheinungen: Grundlagen. Kirkwood[2] hat eine statistische Theorie der Transporterscheinungen auf der Grundlage der Ensemble-Theorie entwickelt, deren Grundkonzept eine gewisse Analogie zu den Überlegungen in Ziff. 17 zeigt. Der gemeinsame Ausgangspunkt besteht in der Erkenntnis, daß man durch einen geeigneten Mittelungsprozeß „geglättete" Phasenfunktionen einführen muß, wenn man makroskopische irreversible Prozesse mit Hilfe der Ensemble-Theorie beschreiben will. In Ziff. 16 haben wir in diesem Sinne die grobe Dichte benutzt. An Stelle dieser Mittelung im Phasenraum führt Kirkwood eine zeitliche Mittelung der Funktionen über ein noch zu definierendes Zeitintervall τ ein. Wir haben also etwa im einfachsten Falle eine an jeder Stelle des Phasenraumes zeitlich geglättete Phasendichte. Das Ziel der Kirkwoodschen Untersuchungen ist von vornherein die Behandlung konkreter physikalischer Probleme. Erst über die Ableitung der Boltzmannschen Stoßgleichung (Ziff. 19) ergibt sich wieder ein Ausblick auf das H-Theorem. Der komplizierte Formalismus macht es jedoch schwierig, ein geschlossenes Bild über das Zustandekommen des irreversiblen Charakters zu gewinnen. Wir werden auf diese Frage noch einmal zurückkommen.

Wir beginnen mit einer Analyse des Begriffes der Messung einer makroskopischen Eigenschaft, die eine Verallgemeinerung der Betrachtung in Ziff. 12 im Hinblick auf die gegenwärtige Problemstellung darstellt. Es seien q_0, p_0 die Phasenkoordinaten zur Zeit t_0, L_s ein Operator, der dieselben in die Phasenkoordinaten zur Zeit $t_0 + s$, die wir mit q, p bezeichnen, transformiert. Wir haben also

$$q = L_s q_0, \quad p = L_s p_0. \tag{18.1}$$

---

[1] M. Born u. D. J. Hooton: Z. Physik **142**, 201 (1955).
[2] J. G. Kirkwood: J. Chem. Phys. **14**, 180 (1946); **15**, 72 (1947).

Für die mathematische Theorie des Operators L_s, deren Einzelheiten wir im folgenden nicht benötigen, müssen wir auf die Originalarbeit [wo dieser Operator mit $\exp(-is\,L_0)$ bezeichnet ist] und die dort zitierte Literatur verweisen. Wir definieren ferner zwei Operatoren α_ε und α_τ, deren Anwendung auf eine Phasenfunktion die durch Gl. (2.6) und (12.1) definierten Mittelwerte liefert. Wir haben also

$$\alpha_\varepsilon F = \iint F(\boldsymbol{q}_0, \boldsymbol{p}_0)\, \varrho\,(\boldsymbol{q}_0, \boldsymbol{p}_0; t_0)\, d\boldsymbol{q}_0\, d\boldsymbol{p}_0 \tag{18.2}$$

und

$$\alpha_\tau F = \frac{1}{\tau} \int\limits_0^\tau F(L_s\,\boldsymbol{q}_0, L_s\,\boldsymbol{p}_0)\, ds. \tag{18.3}$$

Eine meßbare makroskopische Eigenschaft identifizieren wir, wie in Ziff. 12, mit dem Phasenmittel des Zeitmittels. Wir schreiben also

$$(F)_{\text{mess}} = \alpha_\varepsilon \alpha_\tau F. \tag{18.4}$$

Dazu muß bemerkt werden, daß der Begriff „meßbare makroskopische Eigenschaft" sich nicht von vornherein auf eine Einzelmessung, sondern auf eine Vielzahl von Messungen unter gleichen Bedingungen bezieht. Die Frage, inwieweit die Größe $(F)_{\text{mess}}$ als Eigenschaft des Einzelsystems betrachtet werden kann, muß, wie schon in Ziff. 12 erwähnt, mit Hilfe der Schwankungstheorie untersucht werden.

Die auf der rechten Seite der Gl. (18.4) stehende Größe lautet explizit

$$\alpha_\varepsilon \alpha_\tau F = \frac{1}{\tau} \int\limits_0^\tau\!\!\iint F(L_s\,\boldsymbol{q}_0, L_s\,\boldsymbol{p}_0)\, \varrho\,(\boldsymbol{q}_0, \boldsymbol{p}_0; t_0)\, ds\, d\boldsymbol{q}_0\, d\boldsymbol{p}_0. \tag{18.5}$$

Vertauschen wir die beiden Mittelwertbildungen, so haben wir

$$\alpha_\tau \alpha_\varepsilon F = \iint F(\boldsymbol{q}_0, \boldsymbol{p}_0)\, \hat{\varrho}_\tau(\boldsymbol{q}_0, \boldsymbol{p}_0; t_0)\, d\boldsymbol{q}_0\, d\boldsymbol{p}_0 \tag{18.6}$$

mit

$$\hat{\varrho}_\tau(\boldsymbol{q}_0, \boldsymbol{p}_0; t_0) = \frac{1}{\tau} \int \varrho\,(\boldsymbol{q}_0, \boldsymbol{p}_0; t_0 + s)\, ds. \tag{18.7}$$

Diese Funktion ist die eingangs erwähnte zeitlich geglättete Phasendichte. Aus Gl. (18.5) erhalten wir durch Transformation auf die Variablen $\boldsymbol{q}, \boldsymbol{p}$ mit Benutzung von Gl. (3.12) und (18.1)

$$\alpha_\varepsilon \alpha_\tau F = \iint F(\boldsymbol{q}, \boldsymbol{p})\, \hat{\varrho}_\tau(\boldsymbol{q}, \boldsymbol{p}; t_0)\, d\boldsymbol{q}\, d\boldsymbol{p}. \tag{18.8}$$

Die rechte Seite dieser Gleichung ist identisch mit der von Gl. (18.6). Es ist somit

$$(F)_{\text{mess}} = \alpha_\varepsilon \alpha_\tau F = \alpha_\tau \alpha_\varepsilon F \tag{18.9}$$

und

$$\alpha_\varepsilon \alpha_\tau - \alpha_\tau \alpha_\varepsilon = 0. \tag{18.10}$$

Die Operatoren α_ε und α_τ sind somit vertauschbar. Für stationäre Gesamtheiten folgt aus Gl. (18.9) unmittelbar das Resultat der Ziff. 12

$$(F)_{\text{mess}} = \alpha_\varepsilon F. \tag{18.11}$$

Diese Gleichung gilt indessen nicht allgemein für stationäre Transportvorgänge. Es ist nämlich möglich, daß $\hat{\varrho}_\tau$ zeitunabhängig ist, obwohl dies für ϱ nicht zutrifft.

Die allgemeine Gültigkeit der Vertauschungsrelation (18.10) ist an die Voraussetzung geknüpft, daß die Mittelwerte über alle Freiheitsgrade des Systems gebildet werden. Wird nur über einen Teil der Koordinaten gemittelt, so kann Gl. (18.10) ungültig werden. Diese Tatsache spielt für gewisse irreversible Vorgänge eine bedeutsame Rolle.

An Stelle der Operatoren α_ε und α_τ führen wir nun die folgende, teilweise etwas vereinfachte Bezeichnung ein:

$$\left. \begin{aligned} \alpha_\tau F &= \hat{F}, & \alpha_\varepsilon F &= \overline{F}, \\ \alpha_\varepsilon \alpha_\tau F &= \overline{\hat{F}}, & \alpha_\tau \alpha_\varepsilon F &= \hat{\overline{F}}. \end{aligned} \right\} \tag{18.12}$$

Wir machen also die Mittelung über die Zeit τ nicht mehr durch einen Index kenntlich. Ferner beschränken wir uns auf Einkomponentensysteme.

Aus der Phasendichte leiten wir, wie schon in Ziff. 17, durch sukzessive Integration molekulare Verteilungsfunktionen niederer Ordnung ab. Dieselben sind hier, im Gegensatz zu Ziff. 17, aus der feinen Dichte abgeleitet, was wir in der Bezeichnung nicht besonders zum Ausdruck bringen. Wir haben also

$$\left. \begin{aligned} f^{(N)}(\boldsymbol{q}^{(N)}, \boldsymbol{p}^{(N)}; t) &\equiv \varrho(\boldsymbol{q}, \boldsymbol{p}; t), \\ \cdots \cdots \cdots \cdots \cdots \\ f^{(n)}(\boldsymbol{q}^{(n)}, \boldsymbol{p}^{(n)}; t) &= \iint f^{(N)}(\boldsymbol{q}^{(N)}, \boldsymbol{p}^{(N)}; t)\, d\boldsymbol{q}^{(N-n)}\, d\boldsymbol{p}^{(N-n)}, \\ \cdots \cdots \cdots \cdots \cdots \\ f^{(1)}(\boldsymbol{q}_1, \boldsymbol{p}_1; t) &= \iint f^{(N)}(\boldsymbol{q}^{(N)}, \boldsymbol{p}^{(N)}; t)\, d\boldsymbol{q}^{(N-1)}\, d\boldsymbol{p}^{(N-1)}. \end{aligned} \right\} \tag{18.13}[1]$$

Die entsprechenden zeitlich geglätteten Funktionen sind

$$\hat{f}^{(n)}(\boldsymbol{q}^{(n)}, \boldsymbol{p}^{(n)}; t) = \frac{1}{\tau} \int_0^\tau f^{(n)}(\boldsymbol{q}^{(n)}, \boldsymbol{p}^{(n)}; t+s)\, ds. \tag{18.14}$$

Haben wir nun eine Funktion $\varphi(\boldsymbol{q}^{(n)}, \boldsymbol{p}^{(n)})$, die also nur von den generalisierten Koordinaten und Impulsen eines Satzes von n Molekülen abhängt, so erhalten wir aus den Gln. (18.8), (18.10), (18.13) und (18.14)

$$\hat{\overline{\varphi}} = \iint \varphi(\boldsymbol{q}^{(n)}, \boldsymbol{p}^{(n)})\, \hat{f}^{(n)}(\boldsymbol{q}^{(n)}, \boldsymbol{p}^{(n)}; t)\, d\boldsymbol{q}^{(n)}\, d\boldsymbol{p}^{(n)}. \tag{18.15}$$

Zwischen den Verteilungsfunktionen $f^{(N)}$ und $f^{(n)}$ besteht der Zusammenhang

$$\left. \begin{aligned} f^{(N)}(\boldsymbol{q}^{(n)}, \boldsymbol{p}^{(n)}, \boldsymbol{q}^{(N-n)}, \boldsymbol{p}^{(N-n)}; t) &= f^{(n/N)}(\boldsymbol{q}^{(n)}, \boldsymbol{p}^{(n)}, \boldsymbol{q}^{(N-n)}, \boldsymbol{p}^{(N-n)}; t) \times \\ &\quad \times f^{(n)}(\boldsymbol{q}^{(n)}, \boldsymbol{p}^{(n)}; t). \end{aligned} \right\} \tag{18.16}$$

Hier ist $f^{(n/N)}$ die Wahrscheinlichkeitsdichte im Phasenraum des Satzes von $N-n$ Molekülen, wenn der Satz n in seinem Unter-Raum fixiert ist [2].

Es ist zu beachten, daß alle aufgeführten molekularen Verteilungsfunktionen aus der „speziellen" Phasendichte abgeleitet sind, welche Phasen, die durch einfache Vertauschung gleicher Moleküle auseinander hervorgehen, als verschieden betrachtet [3]. Die obigen molekularen Verteilungsfunktionen sind daher Wahrscheinlichkeitsdichten für spezifizierte Moleküle. Man kann die $f^{(n)}$ naturgemäß auch so definieren, daß der Satz n nicht spezifiziert ist. In der Theorie der Gleichgewichtseigenschaften ist dies allgemein üblich [26]. Wir setzen von jetzt ab

---

[1] n bezeichnet hier und in Ziff. 19 nicht die Zahl der Freiheitsgrade.

[2] Zu diesem Typ von molekularen Verteilungsfunktionen gehört im Konfigurationsraum der Paare die radiale Verteilungsfunktion fluider Phasen.

[3] Näheres darüber in Ziff. 25.

voraus, daß die Moleküle nur translatorische Freiheitsgrade besitzen. Die molekulare Verteilungsfunktion $\hat{f}^{(1)}(\boldsymbol{q}_1, \boldsymbol{p}_1; t)$ stellt die Wahrscheinlichkeitsdichte eines bestimmten Moleküls in seinem (6-dimensionalen) Phasenraum dar. Die mittlere molekulare Konzentration an der Stelle $\boldsymbol{q}_1$ zur Zeit t ist daher

$$c(\boldsymbol{q}_1, t) = N \int \hat{f}^{(1)}(\boldsymbol{q}_1, \boldsymbol{p}_1; t)\, d\boldsymbol{p}_1. \tag{18.17}$$

Die Dichte des Massenstromes an der Stelle $\boldsymbol{q}_1$ ist zur Zeit t

$$\boldsymbol{j} = N \int \boldsymbol{p}_1 \hat{f}^{(1)}(\boldsymbol{q}_1, \boldsymbol{p}_1; t)\, d\boldsymbol{p}_1. \tag{18.18}$$

Diese Größe ist vor allem für die Untersuchung von Diffusion und Konvektion von Bedeutung.

Da die Boltzmannsche Stoßgleichung (in unserer jetzigen Sprache) eine Gleichung für die Funktion $\hat{f}^{(1)}$ ist, können wir die weitere Untersuchung auf diese Funktion beschränken. Wir leiten zunächst eine streng gültige Integro-Differentialgleichung für $\hat{f}_1$ ab, die sich beträchtlich von der Stoßgleichung unterscheidet. In Ziff. 19 werden wir dann untersuchen, durch welche Annahmen bzw. Näherungen man zu der Boltzmannschen Gleichung gelangt.

Wir gehen aus von der Liouvilleschen Gleichung (3.3), die wir für die folgende Ableitung schreiben

$$\frac{\partial f^{(N)}}{\partial t} + \sum_{i=1}^{N} \left[\frac{\boldsymbol{p}_i}{m} \frac{\partial}{\partial \boldsymbol{q}_i} f^{(N)} + (\boldsymbol{X}_i + \boldsymbol{F}_i) \frac{\partial}{\partial \boldsymbol{p}_i} f^{(N)} \right] = 0. \tag{18.19}$$

Hier ist $\boldsymbol{X}_i$ (wie in Ziff. 14) die auf das Molekül i wirkende äußere Kraft und $\boldsymbol{F}_i$ die auf das Molekül i von den übrigen Molekülen ausgeübte Kraft. Wir setzen voraus, daß an der Begrenzung des dem System zugänglichen Phasenraumes (bzw. Unter-Raumes) die Flächenintegrale von $f^{(N)}$ und allen Strömen nach Gl. (18.18) verschwinden. Integrieren wir nun die Gl. (18.19) über die Phasenkoordinaten des Satzes $N-n$ und bilden das zeitliche Mittel nach Gl. (18.3), so erhalten wir mit Benutzung der Formel

$$\int \varphi\, d\boldsymbol{f} = \int \frac{\partial \varphi}{\partial \boldsymbol{q}}\, d\boldsymbol{q} \tag{18.20}$$

(wo $d\boldsymbol{f}$ der Vektor des Oberflächenelementes, $d\boldsymbol{q}$ das Volumenelement des betrachteten Raumes ist) die Gleichung

$$\frac{\partial \hat{f}^{(n)}}{\partial t} + \frac{\boldsymbol{p}^{(n)}}{m} \frac{\partial}{\partial \boldsymbol{q}^{(n)}} \hat{f}^{(n)} + \boldsymbol{X}^{(n)} \frac{\partial}{\partial \boldsymbol{p}^{(n)}} \hat{f}^{(n)} = \frac{\partial}{\partial \boldsymbol{p}^{(n)}} \Phi^{(n)} \tag{18.21}$$

mit

$$\left. \begin{aligned} \Phi^{(n)} = -\frac{1}{\tau} \int_0^\tau \!\!\! \int\!\!\int \boldsymbol{F}^{(n)}(\boldsymbol{q}^{(n)}, \boldsymbol{q}^{(N-n)}) \times \\ \times f^{(N)}(\boldsymbol{q}^{(n)}, \boldsymbol{p}^{(n)}, \boldsymbol{q}^{(N-n)}, \boldsymbol{p}^{(N-n)}; t+s)\, d\boldsymbol{q}^{(N-n)}\, d\boldsymbol{p}^{(N-n)}\, ds. \end{aligned} \right\} \tag{18.22}$$

Hier ist $\boldsymbol{X}^{(n)}$ die gesamte äußere und $\boldsymbol{F}^{(n)}$ die gesamte zwischenmolekulare Kraft, beide als Vektoren im n-dimensionalen Konfigurationsraum des Satzes n betrachtet. $\boldsymbol{p}^{(n)}/m$ ist entsprechend ein Vektor im n-dimensionalen Impulsraum. Für $n = 1$ wird aus (18.21) und (18.22)

$$\frac{\partial \hat{f}^{(1)}}{\partial t} + \frac{\boldsymbol{p}_1}{m} \frac{\partial}{\partial \boldsymbol{q}_1} \hat{f}^{(1)} + \boldsymbol{X}_1 \frac{\partial}{\partial \boldsymbol{p}_1} \hat{f}^{(1)} = \frac{\partial}{\partial \boldsymbol{p}_1} \Phi^{(1)} \tag{18.23}$$

mit

$$\Phi^{(1)} = -\frac{1}{\tau} \int\limits_0^\tau \!\!\int\!\!\int F_1(q_1, q^{(N-1)}) \times \\ \times f^{(N)}(q_1, p_1, q^{(N-1)}, p^{(N-1)}; t+s)\, dq^{(N-1)}\, dp^{(N-1)}\, ds. \Bigg\} \quad (18.24)$$

Wir setzen nun voraus, daß das Potential der zwischenmolekularen Wechselwirkung $U^{(N)}$ sich darstellen läßt in der Form

$$U^{(N)} = \sum_{i<j} u_{ij}, \\ u_{ij} = u(r_{ij}), \quad r_{ij} = |q_j - q_i|. \Bigg\} \quad (18.25)$$

Diese Annahme (Zentralkräfte zwischen starren Molekülen) dient der Vereinfachung der Rechnung und stellt für einfachere Moleküle mit einigermaßen kugelsymmetrischem Wechselwirkungspotential eine sehr gute Näherung dar. Wir können dann schreiben

$$\Phi^{(1)} = N\,\Phi_{12} \qquad (18.26)$$

mit

$$\Phi_{12} = -\frac{1}{\tau} \int\limits_0^\tau \!\!\int\!\!\int F(r_{12})\, f^{(N)}(q_1, p_1, q_2, p_2, q^{(N-2)}, p^{(N-2)}; t+s) \times \\ \times d q^{(N-2)}\, d p^{(N-2)}\, d q_2\, d p_2\, ds. \Bigg\} \quad (18.27)$$

Aus dem Liouvilleschen Satz erhalten wir

$$f^{(N)}(q_1, p_1, q_2, p_2, q^{(N-2)}, p^{(N-2)}; t+s) \\ = f^{(N)}(q_{01}, p_{01}, q_{02}, p_{02}, q_0^{(N-2)}, p_0^{(N-2)}; t). \Bigg\} \quad (18.28)$$

Die Anwendung von Gl. (18.16) liefert

$$f^{(N)}(q_{01}, p_{01}, q_{02}, p_{02}, q_0^{(N-2)}, p_0^{(N-2)}, t) \\ = f^{(2/N)}(q_{01}, p_{01}, q_{02}, p_{02}, q_0^{(N-2)}, p_0^{(N-2)}; t)\, f^{(2)}(q_{01}, p_{01}, q_{02}, p_{02}; t). \Bigg\} \quad (18.29)$$

Schließlich definieren wir noch die Korrelationsfunktion im Phasenraum der Paare

$$(1 + \vartheta_{12}) = \frac{f_{12}^{(2)}(q_{01}, \ldots, p_{02}; t)}{f_1^{(1)}(q_{01}, p_{01}; t)\, f_2^{(1)}(q_{02}, p_{02}; t)} . \qquad (18.30)$$

Wir integrieren nun auf der rechten Seite der Gl. (18.27) über den gesamten Phasenraum des Systems, indem wir für die Phasenkoordinaten des Moleküls 1 δ-Funktionen einführen. Transformieren wir das Integral auf die Variablen $q_{01}, \ldots, p_0^{(N-2)}$, so erhalten wir mit Benutzung der Gln. (3.12), (18.28) und (18.29):

$$\Phi_{12} = \int \cdots \int K_{12}(q_{01}, \ldots, p_{02}; t)\, f_1^{(1)}(q_{01}, p_{01}; t) \times \\ \times f_2^{(1)}(q_{02}, p_{02}; t)\, d q_{01}\, d p_{01}\, d q_{02}\, d p_{02}, \Bigg\} \quad (18.31)$$

wo

$$K_{12} = -\frac{1}{\tau} \int\limits_0^\tau \!\!\int\!\!\int F_{12}(t+s)\, \delta(q_{01} + \varDelta q_1 - q_1)\, \delta(p_{01} + \varDelta p_1 - p_1)\, (1 + \vartheta_{12}) \times \\ \times f^{(2/N)}(q_{01}, \ldots, p_{02}, q_0^{(N-2)}, p_0^{(N-2)}; t)\, d q_0^{(N-2)}\, d q_0^{(N-2)} \Bigg\} \quad (18.32)$$

und

$$F_{12}(t+s) = F_{12}(r_{12}) = F_{12}(q_{01}, \ldots, p_0^{(N-2)}; s), \\ \varDelta q_1 = \int\limits_0^s \frac{p_1}{m}\, d s', \quad \varDelta p_1 = \int\limits_0^s F_1\, d s' + X_1 s \Bigg\} \quad (18.33)$$

ist. Die Gln. (18.23), (18.26), (18.31) bis (18.33), die unter der Voraussetzung (18.25) eine strenge Folgerung aus der Liouvilleschen Gleichung sind, stellen den Ausgangspunkt für die Ableitung der Boltzmannschen Stoßgleichung dar.

19. KIRKWOOD's Ableitung der Boltzmannschen Stoßgleichung. Das Problem der statistischen Ableitung der Stoßgleichung ist durch die Überlegungen in Ziff. 18 auf die Berechnung des Integrals (18.31) reduziert worden. Diese Berechnung ist ebenfalls von KIRKWOOD[1] durchgeführt worden, wobei verschiedene Annahmen eingeführt werden müssen, um das gewünschte Resultat zu erhalten.

Wir machen zunächst folgende Annahmen:

a) Es treten nur zwischenmolekulare Kräfte kurzer Reichweite auf. Es kann also eine endliche Größe r_0 definiert werden derart, daß für $r_{ij} > r_0$ $F_{ij} \approx 0$ gesetzt werden darf.

b) Es treten praktisch nur Zweierstöße auf. Es ist somit

$$\frac{\partial \Delta \boldsymbol{p}_1}{\partial s} = \boldsymbol{F}_{12}. \tag{19.1}$$

Die Gültigkeit dieser Annahme hängt davon ab, daß das Verhältnis der Wahrscheinlichkeit, in einem Gebiet der Lineardimension r_0 des dreidimensionalen Konfigurationsraumes drei oder mehr Moleküle zu finden, zu der Wahrscheinlichkeit, in diesem Gebiet zwei Moleküle zu finden, praktisch Null ist. Man kann annehmen, daß dies für geringe Abweichungen vom Gleichgewicht und hinreichende Verdünnung erfüllt ist, obwohl ein strenger Beweis fehlt.

c) $f_1^{(1)}$ und $f_2^{(1)}$ hängen nur makroskopisch von $\boldsymbol{q}_1$ und $\boldsymbol{q}_2$ ab. In dem Integranden auf der rechten Seite der Gl. (18.32) kann daher $\delta(\boldsymbol{q}_{01} + \Delta\boldsymbol{q}_1 - \boldsymbol{q}_1)$ durch $\delta(\boldsymbol{q}_{01} - \boldsymbol{q}_1)$ ersetzt werden.

Anstelle der Variablen $\boldsymbol{q}_{01}$ und $\boldsymbol{q}_{02}$ führen wir die Variablen $\boldsymbol{q}_{01}$ und

$$\boldsymbol{r}_{12} = \boldsymbol{q}_{02} - \boldsymbol{q}_{01} \tag{19.2}$$

ein. Da die Jacobische Determinante dieser Transformation Eins ist, erhalten wir aus den Gln. (18.31) bis (18.33) mit den vorstehenden Annahmen

$$\frac{\partial}{\partial \boldsymbol{p}_1} \Phi_{12} = \frac{1}{\tau} \iiint \psi_{12} f_1^{(1)}(\boldsymbol{q}_1, \boldsymbol{p}_{01}; t) f_2^{(1)}(\boldsymbol{q}_1, \boldsymbol{p}_{02}; t)\, d\boldsymbol{r}_{12}\, d\boldsymbol{p}_{01}\, d\boldsymbol{p}_{02} \tag{19.3}$$

mit

$$\left.\begin{aligned}
\psi_{12} &= (1 + \vartheta_{12}) \int\limits_0^\tau \frac{\partial \Delta \boldsymbol{p}_1}{\partial s}\, \frac{\partial}{\partial \boldsymbol{p}_1}\, \delta(\boldsymbol{p}_{01} + \Delta\boldsymbol{p}_1 - \boldsymbol{p}_1)\, ds \\
&= (1 + \vartheta_{12})\left[\delta(\boldsymbol{p}_{01} + \Delta\boldsymbol{p}_1(\tau) - \boldsymbol{p}_1) - \delta(\boldsymbol{p}_{01} - \boldsymbol{p}_1)\right].
\end{aligned}\right\} \tag{19.4}$$

Wegen Gl. (19.1) hängt in dem Integranden der Gl. (18.32) nur $f^{(2/N)}$ von dem Koordinatensatz $\boldsymbol{q}^{(N-2)}, \boldsymbol{p}^{(N-2)}$ ab; diese Integration ergibt Eins, da $f^{(2/N)}$ normiert ist.

An dieser Stelle führen wir eine weitere Annahme ein:

d) Das Integral $\int \vartheta_{12}\, d\boldsymbol{r}_{12}$ ist beschränkt, und sein Beitrag zur rechten Seite der Gl. (19.3) kann durch geeignete Wahl von τ beliebig klein gemacht werden. Diese Annahme, daß die Korrelation vernachlässigbar ist, erscheint unter den bei b) angeführten Voraussetzungen plausibel, ist aber ebenfalls nicht bewiesen.

---

[1] J. G. KIRKWOOD: J. Chem. Phys. **15**, 72 (1947).

Die Gl. (19.3) kann nun geschrieben werden

$$\frac{\partial}{\partial \boldsymbol{p_1}} \Phi_{12} = \frac{1}{\tau} \iiint [\delta(\boldsymbol{p_1'} - \boldsymbol{p_1}) - \delta(\boldsymbol{p_{01}} - \boldsymbol{p_1})] \times$$
$$\times f_1^{(1)}(\boldsymbol{q_1}, \boldsymbol{p_{01}}; t)\, f_2^{(1)}(\boldsymbol{q_1}, \boldsymbol{p_{02}}; t)\, d\boldsymbol{r_{12}}\, d\boldsymbol{p_{01}}\, d\boldsymbol{p_{02}} + O\left(\frac{\tau_k}{\tau}\right). \tag{19.5}$$

Hier ist τ_k ein endliches Zeitintervall, dessen Größe durch das unter d) erwähnte Korrelationsintegral bestimmt wird, und

$$\boldsymbol{p_1'} = \boldsymbol{p_1} + \varDelta \boldsymbol{p_1}(\tau). \tag{19.6}$$

Der erste Term auf der rechten Seite der Gl. (19.5) ist, wie wir zeigen werden, für hinreichend großes τ unabhängig von τ, so daß in der Tat der Beitrag des zweiten Terms beliebig klein gemacht werden kann.

Die weitere Rechnung erfordert eine Diskussion der Lösungen der Bewegungsgleichung für den Zweierstoß, Gl. (19.1). Wir geben hier nur die für das Weitere wesentlichen Resultate und verweisen für Einzelheiten auf die Literatur[1]. Es sei $m_{12} = m/2$ die reduzierte Masse und

$$\boldsymbol{p_{12}} = \tfrac{1}{2}(\boldsymbol{p_2} - \boldsymbol{p_1}) \tag{19.7}$$

der relative Impuls. Diese Größe bleibt konstant gleich $\boldsymbol{p_{12}^0}$, bis das Molekül 2 in eine um das Molekül 1 beschriebene Kugel vom Radius r_0 eindringt. Wenn dies nach den Anfangsbedingungen möglich ist, findet innerhalb einer Zeit τ_s, die von den Stoßparametern abhängt, ein vollständiger Stoß statt. Dabei nimmt der relative Impuls um $\varDelta p^*$ zu und bleibt dann konstant $\boldsymbol{p_{12}'}$. Wir haben also

$$\varDelta \boldsymbol{p^*} = \boldsymbol{p_{12}'} - \boldsymbol{p_{12}^0}, \qquad p_{12}' = p_{12}^0 \tag{19.8}$$

und nach dem Impulssatz

$$\varDelta \boldsymbol{p_2} = -\varDelta \boldsymbol{p_1} = \varDelta \boldsymbol{p^*}. \tag{19.9}$$

Wir beschreiben die relative Konfiguration der beiden Moleküle 1 und 2 durch Zylinderkoordinaten z, b, ε, deren Ursprung in dem Molekül 1 liegt. Die z-Achse ist antiparallel zu $\boldsymbol{p_{12}^0}$. Die radiale Koordinate b ist der durch die Gleichung

$$b\, p_{12}^0 = |\boldsymbol{r_{12}} \times \boldsymbol{p_{12}^0}| \tag{19.10}$$

definierte Stoßparameter und ε sein Azimut. Wenn die Bedingung

$$(r_0^2 - b^2)^{\frac{1}{2}} \leqq z \leqq (p_{12}^0/m_{12})\,\tau + (r_0^2 - b^2)^{\frac{1}{2}}, \qquad 0 \leqq b \leqq r_0 \tag{19.11}$$

nicht erfüllt ist, dringt das Molekül 2 im Zeitintervall τ nicht in die Kugel vom Radius r_0 ein, $\varDelta \boldsymbol{p_1}$ bleibt Null und der Integrand der Gl. (19.5) verschwindet. Wenn die Ausgangskonfiguration der Bedingung (19.11) genügt und in einem Volumenelement

$$(p_{12}^0/m_{12})\,(\tau - \tau_s)\, b\, db\, d\varepsilon \tag{19.12}$$

liegt, findet im Zeitintervall τ ein vollständiger Stoß mit den durch Gl. (19.8) und (19.9) gegebenen Impulsänderungen statt. Wenn die Bedingung

$$(p_{12}^0/m_{12})\,(\tau - \tau_s) + (r_0^2 - b^2)^{\frac{1}{2}} < z \leqq (p_{12}^0/m_{12})\,\tau + (r_0^2 - b^2)^{\frac{1}{2}} \tag{19.13}$$

erfüllt ist und die Ausgangskonfiguration in einem Volumenelement

$$(p_{12}^0/m_{12})\,\tau_s\, b\, db\, d\varepsilon \tag{19.14}$$

---

[1] S. Chapman and T. Cowling: The Mathematical Theory of Non-Uniform Gases. Cambridge 1939.

liegt, finden im Zeitintervall τ unvollständige Stöße statt. Ist die Bedingung

$$- (r_0^2 - b^2)^{\frac{1}{2}} \leq z < + (r_0^2 - b^2)^{\frac{1}{2}} \tag{19.15}$$

erfüllt und liegt die Ausgangskonfiguration in einem Volumenelement

$$2 (r_0^2 - b^2)^{\frac{1}{2}} \, b \, db \, d\varepsilon \tag{19.16}$$

so können im Zeitintervall τ unvollständige Stöße vollständig ablaufen oder mehrfach periodische Bahnen mit beschränktem Δp_{12} durchlaufen werden. Wenn wir singuläres Verhalten ausschließen, bleibt der Integrand für die Gebiete (19.14) und (19.16) endlich.

Mit den Ausdrücken (19.12), (19.14), (19.16) erhalten wir aus Gl. (19.5)

$$\frac{\partial}{\partial p_1} \Phi_{12} = \int \cdots \int \frac{p_{12}^0}{m_{12}} \left[\delta (p_1' - p_1) - \delta (p_{01} - p_1) f_1^{(1)}(q_1, p_{01}; t) \times \right. \\ \left. \times f_2^{(1)}(q_1, p_{02}; t) \, dp_{01} \, dp_{02} \, b \, db \, d\varepsilon + O\left(\frac{\overline{\tau}_s}{\tau}\right). \right\} \tag{19.17}$$

Hier ist $\overline{\tau}_s$ eine endliche Zeit von der Größenordnung einer mittleren Stoßzeit oder von τ_k, wenn dies größer ist und

$$p_1' = p_{01} - \Delta p^*. \tag{19.18}$$

Man bemerkt, daß der erste Term der rechten Seite der Gl. (19.17) unabhängig von τ ist. Durch geeignete Wahl von τ kann daher der zweite Term beliebig klein gegen den ersten gemacht werden. Es muß jetzt eine Annahme über die Zeit τ eingeführt werden.

e) Die Zeit τ ist groß gegen $\overline{\tau}_s$ bzw. τ_k zu wählen, so daß auf den rechten Seiten der Gln. (19.5) und (19.17) die zweiten Terme vernachlässigt werden können. Andererseits darf τ nicht beliebig groß werden; wenn τ die Größenordnung von t^*, der Wiederkehrzeit des betrachteten Zustandes erreicht, wird die Bahn auch im entgegengesetzten Sinne durchlaufen werden, so daß die beiden Beiträge zum Integral sich herausheben.

Um die Integration über die δ-Funktion $\delta (p_1 - p_1)$ auszuführen, ist es notwendig, das Integral auf die Variablen

$$p_i' = p_{01} - \Delta p^*, \qquad p_2' = p_{02} + \Delta p^* \tag{19.19}$$

zu transformieren. Die Jacobische Determinante dieser Transformation ist Eins. Die Ausführung der Integration ergibt dann, wenn wir die restlichen Impulsvariablen p_2' und p_{02} mit dem gemeinsamen Symbol p_2 bezeichnen,

$$\frac{\partial}{\partial p_1} \Phi_{12} = \int\int\int_0^{2\pi} \frac{p_{12}}{m_{12}} \left[f_1^{(1)}(q_1, p_1 + \Delta p^*; t) \, f_2^{(1)}(q_1, p_2 - \Delta p^*; t) - \right. \\ \left. - f_1^{(1)}(q_1, p_1; t) \, f_2^{(1)}(q_1, p_2; t) \right] dp_2 \, b \, db \, d\varepsilon. \right\} \tag{19.20}$$

Dabei haben wir die Terme $O(\overline{\tau}_s|\tau)$ vernachlässigt. Setzen wir den Ausdruck (19.20) mit Benutzung von (18.26) in Gl. (18.23) ein und mitteln wieder über die Zeit τ, so erhalten wir [da durch die Mittelwertbildung die linke Seite von (18.23) nicht berührt wird]

$$\frac{\partial \hat{f}_1^{(1)}}{\partial t} = - \frac{p_1}{m} \frac{\partial \hat{f}_1^{(1)}}{\partial q_1} - X_1 \frac{\partial \hat{f}_1^{(1)}}{\partial p_1} + 2\pi N \int\int \frac{|p_1 - p_2|}{m} \left[\widehat{f_1^{(1)*} f_2^{(1)*}} - \widehat{f_1^{(1)} f_2^{(1)}} \right] dp_2 \, b \, db \tag{19.21}$$

mit

$$f_1^{(1)*} = f_1^{(1)}(q_1, p_1 + \Delta p^*; t), \qquad f_2^{(1)*} = f_2^{(1)}(q_1, p_2 - \Delta p^*; t). \tag{19.22}$$

Die Gl. (19.21) unterscheidet sich von der Boltzmannschen Stoßgleichung noch durch die Mittelwertbildung über die Produkte im Integranden der rechten Seite. Da wir geringe Abweichungen vom Gleichgewicht voraussetzen (Annahme b), können wir schreiben

$$f_1^{(1)} = {}^0f_1^{(1)} + \varphi_1^{(1)}, \qquad f_2^{(1)} = {}^0f_2^{(1)} + \varphi_2^{(1)}, \tag{19.23}$$

wo ${}^0f_1^{(1)}$ und ${}^0f_2^{(1)}$ die (zeitunabhängigen) Gleichgewichtsverteilungen sind. Daraus folgt

$$\widehat{f_1^{(1)}\, f_2^{(1)}} = {}^0f_1^{(1)}\, {}^0f_2^{(1)} + {}^0f_1^{(1)}\, \hat\varphi_2^{(1)} + {}^0f_2^{(1)}\, \hat\varphi_1^{(1)} + O\,(\widehat{\varphi_1^{(1)}\, \varphi_2^{(1)}}). \tag{19.24}$$

Bis zu Termen erster Ordnung gilt somit

$$\widehat{f_1^{(1)}\, f_2^{(1)}} = \hat f_1^{(1)}\, \hat f_2^{(1)}. \tag{19.25}$$

Damit erhalten wir schließlich in der gleichen Näherung

$$\left.\begin{aligned}
\frac{\partial \hat f_1^{(1)}}{\partial t} = {}&- \frac{\boldsymbol{p}_1}{m}\, \frac{\partial \hat f_1^{(1)}}{\partial \boldsymbol{q}_1} - \boldsymbol{X}_1\, \frac{\partial \hat f_1^{(1)}}{\partial \boldsymbol{p}_1} + \\
&+ 2\pi N \iint \frac{|\boldsymbol{p}_1 - \boldsymbol{p}_2|}{m}\, [\hat f_1^{(1)*}\, \hat f_2^{(1)*} - \hat f_1^{(1)}\, \hat f_2^{(1)}]\, d\boldsymbol{p}_2\, b\, db.
\end{aligned}\right\} \tag{19.26}$$

Diese Gleichung stimmt mit der Boltzmannschen Stoßgleichung, Gl. (14.15), überein. Der Faktor N auf der rechten Seite ist durch die Verwendung „spezieller" molekularer Verteilungsfunktionen (s. Ziff. 18) in der vorstehenden Ableitung bedingt. Das Ergebnis läßt sich zusammenfassen in der Aussage, daß unter den Annahmen a) bis e) die Boltzmannsche Stoßgleichung aus den Prinzipien der statistischen Mechanik als eine Näherung für die zeitlich geglätteten molekularen Verteilungsfunktionen folgt, die bis zu Termen erster Ordnung in der Abweichung vom Gleichgewicht gültig ist. Obwohl damit gegenüber den älteren Ableitungen ein außerordentlicher Fortschritt erreicht ist, kann nicht übersehen werden, daß auch die Kirkwoodsche Ableitung noch in mancher Hinsicht unbefriedigend ist[1]. Problematisch erscheinen vor allem die folgenden Punkte:

α) Die Annahmen b) bis d) werden nicht bewiesen bzw. die Voraussetzungen ihrer Gültigkeit nicht geklärt.

β) Die Bedeutung der zweiten Mittelung über die Zeit τ, die zu Gl. (19.21) führt, bleibt unklar.

γ) Der Übergang von Gl. (19.21) zu Gl. (19.26) setzt voraus, daß auch für die nicht geglättete Funktion $f^{(1)}$ die Gleichgewichtsverteilung ${}^0f^{(1)}$ definiert werden kann. Nach den Überlegungen in Ziff. 16 und 17 muß man annehmen, daß diese Funktion nicht existiert.

Sehen wir von diesen Schwierigkeiten ab, so folgt aus Gl. (19.26) nach einer der in Ziff. 14 durchgeführten analogen Rechnung das H-Theorem, wobei H durch die Gleichung

$$H = \iint \hat f^{(1)}(\boldsymbol{q}_1, \boldsymbol{p}_1)\, \ln \hat f(\boldsymbol{q}_1, \boldsymbol{p}_1)\, d\boldsymbol{q}_1\, d\boldsymbol{p}_1 \tag{19.27}$$

definiert ist. Die Gültigkeit des H-Theorems ist indessen dadurch eingeschränkt, daß Gl. (19.26) nur als bis zu Termen erster Ordnung korrekt angenommen wird. Das Problem der Begründung des H-Theorems wird daher auch durch die Kirkwoodsche Theorie nicht gelöst.

Es bleibt noch die Frage, wie in der Kirkwoodschen Theorie der irreversible Charakter zustande kommt. Kirkwood und ihm folgend de Boer [20] nehmen

---

[1] Für eine Diskussion dieser Frage habe ich Herrn Prof. Ono zu danken.

an, daß entscheidend hierfür die Annahme e) mit dem Ausschluß langer Poincaré-Zyklen ist. Diese Auffassung ist zweifellos insofern gerechtfertigt, als nur unter einer derartigen Voraussetzung gesichert wird, daß die rechte Seite der Gl. (18.23) außerhalb des Gleichgewichtes von Null verschieden ist. Dagegen scheint nicht geklärt zu sein, ob die erwähnte Annahme für sich bereits genügt, um den irreversiblen Charakter zu sichern, und welche Rolle dabei möglicherweise noch die übrigen Annahmen (insbesondere d) spielen.

In neuerer Zeit haben sich noch verschiedene Autoren[1-9] mit der strengen Ableitung der Boltzmannschen Stoßgleichung beschäftigt. Wir können darauf hier nicht eingehen und müssen auf die Originalarbeiten sowie auf den Artikel von H. S. GREEN in Band X dieses Handbuchs verweisen.

IV. Die mikrokanonische Gesamtheit.

20. Definition und allgemeine Beziehungen. Die Betrachtungen in Ziff. 12 zeigen, daß den natürlichen Ausgangspunkt der statistischen Mechanik die Untersuchung einer virtuellen Gesamtheit bildet, deren Phasendichte im Energie-Intervall zwischen E und $E + \Delta E$ einen konstanten Wert besitzt, während sie im restlichen Teil des Phasenraumes identisch Null ist. Eine in dieser Weise verteilte Gesamtheit wird nach GIBBS [2] als *mikrokanonische Gesamtheit* bezeichnet. Sie entspricht einem Originalsystem, für welches die Energie, das Volumen (und gegebenenfalls weitere äußere Parameter) sowie die Zahl der Moleküle (Zahl der Freiheitsgrade) vorgegeben sind. Der fundamentalen Bedeutung der mikrokanonischen Gesamtheit im Rahmen der statistischen Theorie entspricht die Tatsache, daß die hier auftretenden makroskopischen Zustandsgrößen die gleichen sind, welche als unabhängige Variable der Entropie, der thermodynamischen Fundamentalgröße zugeordnet sind.

Die obige Definition, welche die mikrokanonische Gesamtheit über die „Energieschale" zwischen den Energieflächen E und $E + \Delta E$ erstreckt, trägt der Tatsache Rechnung, daß ein makroskopisch abgeschlossenes System nicht im strengen Sinne ein konservatives System darstellt. Naturgemäß ist ΔE in jedem Falle gegenüber E als äußerst kleine Größe anzusehen. In der klassischen Statistik ist es (im Gegensatz zur Quantenstatistik, s. Ziff. 43) möglich und zweckmäßig, $\Delta E \rightarrow 0$ anzunehmen, da eine prinzipielle Grenze für die Genauigkeit der Energiemessung hier nicht existiert und die praktisch vorhandene Ungenauigkeit auch bei den übrigen makroskopischen Parametern nicht in die Theorie eingebaut wird. Dieser Grenzübergang führt zu der mit konstanter Flächendichte auf der Energiefläche verteilten Gesamtheit, die wir in Teil A II und A III schon häufig betrachtet haben. Mathematisch bedingt der Grenzübergang eine gewisse Komplikation, welche die älteren Darstellungen der Theorie [2], [9] ziemlich schwerfällig macht. Man kann diese Schwierigkeit jedoch in einfacher Weise vermeiden, wenn man von vornherein die Phasendichte mit Hilfe der Diracschen δ-Funktion formuliert[10] [26].

[1] J. YVON: La théorie statistique des fluides et l'équation d'état (Actualités scientifiques et industrielles Nr. 203), Paris 1935.

[2] M. BORN u. H. S. GREEN: Proc. Roy. Soc. Lond., Ser. A **188**, 10 (1946); **190**, 455 (1947); **191**, 168 (1947); **192**, 166 (1947).

[3] M. BORN u. H. S. GREEN: A General Kinetic Theory of Liquids. Cambridge 1949.

[4] H. S. GREEN: The Molecular Theory of Fluids. Amsterdam 1952.

[5] N. BOGOLJUBOW: J. Phys. USSR. **10**, 265 (1946).

[6] H. KÜMMEL: Z. Physik **143**, 219 (1955).

[7] M. S. GREEN: J. Chem. Phys. **25**, 836 (1956).

[8] P. ST. PÜTTER: Ann. Phys. **19**, 247 (1956).

[9] R. BROUT: Physica, Haag **22**, 509 (1956).

[10] A. MÜNSTER: Z. Physik **137**, 386 (1954).

Wir bezeichnen das Phasenvolumen als Funktion einer oberen Grenze der Energie mit Ω^*. Es ist also gegeben durch die Phasenausdehnung

$$\Omega^* = \int\limits^{E} d\Omega, \tag{20.1}$$

wo das Integral über alle Phasen mit einer Energie $\leqq E$ zu erstrecken ist. Ferner definieren wir eine Funktion

$$\Phi = \ln \frac{d\Omega^*}{dE}, \tag{20.2}$$

die als Gibbssche Energiefunktion bezeichnet wird. Eine analoge Größe kann aus der Geschwindigkeitsausdehnung als Funktion einer oberen Grenze der kinetischen Energie (Ziff. 4) abgeleitet werden durch die Gleichung

$$\Phi_p = \ln \frac{d\Omega_p^*}{dE_{\mathrm{kin}}}. \tag{20.3}$$

Die Phasendichte der mikrokanonischen Gesamtheit definieren wir durch die Gleichung

$$\varrho = C \cdot \delta(E^* - E), \tag{20.4}$$

wo E^* den Wert von E für die mikrokanonische Gesamtheit und C den Normierungsfaktor bezeichnet. Für diesen ergibt sich aus Gln. (2.4) und (20.2) in Verbindung mit der Definition der δ-Funktion

$$C = \mathrm{e}^{-\Phi}, \tag{20.5}$$

wo Φ jetzt und im weiteren der Wert dieser Größe für $E = E^*$ ist. Entsprechend verstehen wir unter Ω^* von jetzt ab das bis zur oberen Grenze E^* erstreckte Integral über $d\Omega$, also das von der mikrokanonischen Energiefläche umschlossene Phasenvolumen.

Der mikrokanonische Mittelwert einer Phasenfunktion F ist nach Gln. (2.6), (20.4) und (20.5) gegeben durch

$$\overline{F} = \mathrm{e}^{-\Phi} \int F\, \delta(E^* - E)\, d\Omega. \tag{20.6}$$

Aus dieser allgemeinen Gleichung lassen sich verschiedene weitere Formulierungen ableiten, die unmittelbar für spezielle Anwendungen benutzt werden können. Führen wir den $2n$-dimensionalen Phasenvektor $\boldsymbol{R}$ mit den Komponenten $q_1, \ldots, q_n, p_1, \ldots, p_n$ ein, so kann die Gleichung der Energiefläche geschrieben werden

$$E(\boldsymbol{R}) = E^*(\boldsymbol{R}^*). \tag{20.7}$$

Mit der Definition

$$|\operatorname{grad} E| = \left[\left(\frac{\partial E}{\partial q_1} \right)^2 + \cdots + \left(\frac{\partial E}{\partial p_n} \right)^2 \right]^{\frac{1}{2}} \tag{20.8}$$

haben wir dann

$$\delta(E^* - E) = \sum \frac{\delta(\boldsymbol{R} - \boldsymbol{R}^*)}{|\operatorname{grad} E|}, \tag{20.9}$$

wo $\delta(\boldsymbol{R} - \boldsymbol{R}^*)$ die $2n$-dimensionale δ-Funktion ist und die Summierung über alle (als einfach vorausgesetzten) Nullstellen der Funktion $(E^* - E)(\boldsymbol{R})$ zu erstrecken ist. Da diese Nullstellen ein Kontinuum bilden, muß die Summierung durch ein Integral über die Fläche (20.7) ersetzt werden. Führen wir dies Integral in Gl. (20.6) ein, so wird

$$\overline{F} = \mathrm{e}^{-\Phi} \iint \frac{F}{|\operatorname{grad} E|}\, \delta(\boldsymbol{R} - \boldsymbol{R}^*)\, d\Omega\, df, \tag{20.10}$$

wo df das $(2n-1)$-dimensionale Flächenelement bezeichnet. Daraus folgt[1]

$$\overline{F} = e^{-\Phi} \int \frac{F}{|\operatorname{grad} E|}\, df. \tag{20.11}$$

Wir nehmen nun an, daß eine Funktion u nur von den generalisierten Koordinaten abhängt. Im besonderen kann u also eine Funktion der potentiellen Energie oder wegen

$$E^* - U = E_{\text{kin}} \tag{20.12}$$

auch der kinetischen Energie sein. Für diesen Fall läßt sich aus (20.6) eine weitere nützliche Mittelwertsformel ableiten. Auf Grund der Definitionen (4.21) und (4.27) kann Gl. (20.6) geschrieben werden

$$\bar{u} = e^{-\Phi} \iint u\, \delta(E^* - E)\, d\Omega_p\, d\Omega_q. \tag{20.13}$$

Nun ist für konstantes U nach (20.12) $dE_{\text{kin}} = dE$. Mit Benutzung von (20.3) wird daher aus (20.13)

$$\bar{u} = e^{-\Phi} \iint u\, e^{\Phi_p}\, \delta(E^* - E)\, dE\, d\Omega_q \tag{20.14}$$

oder

$$\bar{u} = e^{-\Phi} \int_{\Omega_q = 0}^{U = E^*} u\, e^{\Phi_p}\, d\Omega_q. \tag{20.15}$$

In Ziff. 25 werden wir diese Beziehung in ihrer expliziten Form, d.h. ohne Benutzung der Funktionen Φ und Φ_p ableiten.

21. Äquipartitionstheorem und Virialsatz. Wir wollen jetzt Gl. (20.11) benutzen, um zwei wichtige Sätze abzuleiten. Es sei r_i eine beliebige Phasenkoordinate und

$$F_i = r_i \frac{\partial E}{\partial r_i}. \tag{21.1}$$

Definieren wir

$$F = \sum_{i=1}^{2n} F_i = \boldsymbol{R} \cdot \operatorname{grad} E, \tag{21.2}$$

so erhalten wir aus Gl. (20.11) für den Mittelwert

$$\overline{F} = e^{-\Phi} \int \boldsymbol{R}\, d\boldsymbol{f}, \tag{21.3}$$

wo $d\boldsymbol{f}$ der Vektor des Flächenelementes ist. Mit Benutzung des Gaußschen Satzes wird daraus

$$\overline{F} = 2n\, e^{-\Phi}\, \Omega^*. \tag{21.4}$$

Daraus folgt

$$\overline{q_i \frac{\partial E}{\partial q_i}} = \overline{p_j \frac{\partial E}{\partial p_j}} = e^{-\Phi}\, \Omega^*. \tag{21.5}$$

Diese Gleichung wird als *verallgemeinertes Äquipartitionstheorem* bezeichnet[2] [*12*]. Sie besagt, daß die Mittelwerte der Größen (21.1) für alle Phasenkoordinaten untereinander gleich sind und daß sie gleich sind einer Größe, welche nur von der Gesamtenergie eines Systems der mikrokanonischen Gesamtheit abhängt.

Wir nehmen jetzt an, daß die potentielle Energie eine quadratische Funktion von m generalisierten Koordinaten ist. Dann können wir die Hamilton-Funktion schreiben

$$H(\boldsymbol{q}, \boldsymbol{p}) = \frac{1}{2} \sum_{i=1}^{n} p_j \frac{\partial H}{\partial p_i} + \frac{1}{2} \sum_{j=1}^{m} q_i \frac{\partial H}{\partial q_i}. \tag{21.6}$$

---

[1] Man beachte, daß $d\boldsymbol{R} \equiv d\Omega$ ist.
[2] R. Tolman: Phys. Rev. **11**, 261 (1918).

Der Vergleich mit (21.5) zeigt, daß im Mittel jeder Term den gleichen Beitrag

$$\bar{\varepsilon}_i = \tfrac{1}{2}\,e^{-\Phi}\,\Omega^* \tag{21.7}$$

zur Gesamtenergie liefert. Diese Aussage ist das Äquipartitionstheorem der klassischen Statistik. Wie wir in den folgenden Ziffern zeigen werden, ist die Größe $e^{-\Phi}\Omega^*$ ein statistisches Analogon der Temperatur, die mit der thermodynamischen Temperatur T durch die Gleichung

$$e^{-\Phi}\Omega^* = kT \tag{21.8}$$

zusammenhängt, wo

$$k = (1,38044 \pm 0,00007) \cdot 10^{-16}\ \text{erg grad}^{-1} \tag{21.9}$$

die Boltzmannsche Konstante ist. Damit kann (21.7) in der geläufigeren Form

$$\bar{\varepsilon}_i = \tfrac{1}{2}\,kT \tag{21.10}$$

geschrieben werden.

Das Äquipartitionstheorem ist unter den angegebenen Voraussetzungen eine strenge und allgemeine Folgerung aus der klassischen statistischen Mechanik. Die schon GIBBS [2] bekannte Tatsache, daß diese Aussage im Widerspruch zur experimentellen Erfahrung steht, zeigt, daß eine molekulare Theorie auch bei der Beschreibung makroskopischer Eigenschaften von der Quantenmechanik ausgehen muß.

In formaler Analogie zur kinetischen Energie

$$E_{\text{kin}} = \frac{1}{2}\sum_{i=1}^{n} p_i\,\frac{\partial E}{\partial p_i} \tag{21.11}$$

kann man eine Größe

$$[V] = \frac{1}{2}\sum_{i=1}^{n} q_i\,\frac{\partial E}{\partial q_i} \tag{21.12}$$

definieren, die man nach CLAUSIUS[1] das Virial des Systems nennt. Die physikalische Bedeutung des Virials ergibt sich aus der Tatsache, daß $-\partial E/\partial q_i$ die generalisierte Kraftkomponente in Richtung der i-ten Koordinate darstellt. Aus Gl. (21.5), (21.11) und (21.12) folgt nun unmittelbar

$$\bar{E}_{\text{ki}} = \overline{[V]}. \tag{21.13}$$

Im Mittel ist die kinetische Energie eines Systems gleich seinem Virial. Dieses von CLAUSIUS[1] aufgefundene Theorem wird als Virialsatz bezeichnet. Die große Bedeutung desselben liegt in der Tatsache, daß er eine direkte Berechnung des Druckes ermöglicht, ohne daß man den Umweg über die thermodynamischen Funktionen zu gehen hat.

Als einfaches Anwendungsbeispiel behandeln wir das einatomige ideale Gas. Für diesen Fall lautet der Virialsatz in Cartesischen Koordinaten

$$\tfrac{1}{2}\sum_{i=1}^{N} m_i\,\overline{(\dot{x}_i^2 + \dot{y}_i^2 + \dot{z}_i^2)} = -\tfrac{1}{2}\sum_{i=1}^{N} \overline{(X_i x_i + Y_i y_i + Z_i z_i)}. \tag{21.14}$$

Es sind hier lediglich die von der Wand des Behälters ausgeübten Kräfte zu berücksichtigen. Diese sind nur für solche Werte der Koordinaten von Null verschieden, bei denen sich die Moleküle an der Wand, d.h. an der Oberfläche des Volumens befinden. Bezeichnen wir mit d das Flächenelement der Oberfläche,

---

[1] R. CLAUSIUS: Pogg. Ann. **141**, 124 (1870).

mit α, β, γ die Richtungscosinus der nach außen gerichteten Normalen, mit P den Druck, so sind die Komponenten der von diesem Flächenelement ausgeübten, über alle Moleküle summierten mittleren Kraft

$$d\overline{X} = -P\alpha\,df, \quad d\overline{Y} = -P\beta\,df, \quad d\overline{Z} = -P\gamma\,df. \tag{21.15}$$

Die zugehörigen Koordinaten sind die Werte von x, y, z für das Flächenelement. Mit Benutzung des Radiusvektors $\mathbf{r}$ ergibt sich dann

$$\overline{[V]} = \tfrac{1}{2} P \int \mathbf{r}\,df \tag{21.16}$$

oder, mit Benutzung des Gaußschen Satzes

$$\overline{[V]} = \tfrac{3}{2} P V. \tag{21.17}$$

Da die linke Seite der Gl. (21.14) durch das Äquipartitionstheorem gegeben ist erhalten wir schließlich

$$PV = e^{-\Phi}\Omega^*, \tag{21.18}$$

die Zustandsgleichung des idealen Gases. Durch Vergleich mit der empirischen Form der Zustandsgleichung folgt daraus die Gl. (21.8).

22. Die empirische Temperatur. Wir betrachten ein abgeschlossenes System, das aus zwei Teilsystemen besteht, die miteinander Energie austauschen können. Die Teilsysteme bezeichnen wir durch die Indices 1 und 2, während die auf das Gesamtsystem bezogenen Größen ohne Index geschrieben werden. Der mikrokanonische Mittelwert der Funktion $e^{-\Phi_{p1}}\Omega^*_{p1}$ ist nach Gl. (20.6)

$$\overline{e^{-\Phi_{p1}}\Omega^*_{p1}} = e^{-\Phi} \int e^{-\Phi_{p1}}\Omega^*_{p1}\,\delta(E^* - E)\,d\Omega. \tag{22.1}$$

Nun gilt

$$d\Omega = d\Omega_1\,d\Omega_2 = d\Omega_{p1}\,d\Omega_{q1}\,d\Omega_2. \tag{22.2}$$

Es wird daher

$$\overline{e^{-\Phi_{p1}}\Omega^*_{p1}} = e^{-\Phi} \iiint e^{-\Phi_{p1}}\Omega^*_{p1}\,\delta(E^* - E)\,d\Omega_{p1}\,d\Omega_{q1}\,d\Omega_2. \tag{22.3}$$

Nach der Voraussetzung ist

$$E_{\mathrm{kin}1} + U_1 + E_2 = E^*. \tag{22.4}$$

Für konstante U_1 und E_2 ist daher $dE_{\mathrm{kin}1} = dE$. Mit Benutzung von (20.3) folgt dann aus (22.3)

$$\overline{e^{-\Phi_{p1}}\Omega^*_{p1}} = e^{-\Phi}\Omega^*. \tag{22.5}$$

In analoger Weise leitet man ab

$$\overline{e^{-\Phi_{p2}}\Omega^*_{p2}} = e^{-\Phi}\Omega^*. \tag{22.6}$$

Schließlich folgt unmittelbar aus Gl. (20.15)

$$\overline{e^{-\Phi_{p}}\Omega_{p}} = e^{-\Phi}\Omega^*. \tag{22.7}$$

Der Mittelwert der Funktion $e^{-\Phi_{p}}\Omega^*_{p}$ besitzt somit im Gleichgewicht für zwei in thermischer Berührung stehende Teilsysteme und das daraus gebildete Gesamtsystem den gleichen Wert. Dieser Mittelwert bzw. die Größe $e^{-\Phi}\Omega^*$ ist daher ein statistisches Analogon der empirischen Temperatur. Der Nachweis, daß $e^{-\Phi}\Omega^*$ auch die Eigenschaften der thermodynamischen Temperatur besitzt, erfordert die Bestimmung des statistischen Analogons der Entropie. Diese Frage bildet den Gegenstand der folgenden Ziffern.

23. Der Satz von der adiabatischen Invarianz des Phasenvolumens. Die Energie eines Systems muß grundsätzlich als Funktion der äußeren Parameter x_j betrachtet werden, von denen der wichtigste (und häufig einzige) das Volumen V ist. Einer Änderung des äußeren Parameters x_j widersetzt sich das System mit einer generalisierten Kraft

$$X_j = - \left(\frac{\partial E}{\partial x_j} \right)_{q,p}, \qquad (23.1)$$

Dieselbe hängt von der Phase des Systems ab; makroskopische Bedeutung hat daher nur der mit Hilfe einer statistischen Gesamtheit gebildete Mittelwert dieser Größe. Im allgemeinen wird eine stationäre Gesamtheit nach einer Änderung der äußeren Parameter nicht mehr im statistischen Gleichgewicht sein. Um diese Schwierigkeit zu umgehen, nehmen wir an, daß die Änderung der Parameter unendlich langsam verläuft, wodurch die Störung des Gleichgewichtes (auch bei einer endlichen Parameteränderung) beliebig klein wird [2]. Man spricht dann von einer *adiabatischen Parameteränderung*[1]. Wir setzen ferner in dieser Ziffer voraus, daß die Änderung der Energie eindeutig durch die Änderung der äußeren Parameter bestimmt ist. Es werden somit alle Prozesse ausgeschlossen, welche die Energie ohne Änderung der äußeren Parameter ändern[2].

Der mikrokanonische Mittelwert der durch (23.1) definierten generalisierten Kraft ist nach Gl. (20.11)

$$\overline{X}_j = - \mathrm{e}^{-\Phi} \int \frac{\left(\frac{\partial E}{\partial x_j} \right)_{q,p}}{|\operatorname{grad} E|} \, df. \qquad (23.2)$$

Für eine Änderung der äußeren Parameter muß daher eine (mittlere) Arbeit

$$\overline{\varDelta A} = - \sum_j \overline{X}_j \varDelta x_j \qquad (23.3)$$

an dem System geleistet werden. Die Energie des Systems wird dadurch im Mittel um einen Betrag

$$\overline{\varDelta E} = \overline{\varDelta A} = \mathrm{e}^{-\Phi} \frac{\sum_j \left(\frac{\partial E}{\partial x_j} \right)_{q,p} \varDelta x_j}{|\operatorname{grad} E|} \, df \qquad (23.4)$$

geändert.

Da das Phasenvolumen eine Funktion der Energie ist, bewirkt eine Änderung der äußeren Parameter nach (23.4) auch eine Änderung des Phasenvolumens, indem das System auf eine benachbarte Energiefläche gebracht wird. Die Parameteränderung modifiziert aber auch die Form der Hamilton-Funktion und damit die Gestalt der Energiefläche. Wir können daher schreiben

$$\overline{\varDelta \Omega^*} = \left(\frac{\partial \Omega^*}{\partial E} \right)_x \overline{\varDelta E} + \sum_j \left(\frac{\partial \Omega^*}{\partial x_j} \right)_E \varDelta x_j, \qquad (23.5)$$

wobei zu beachten ist, daß hier nur die äußeren Parameter unabhängige Variable sind. Die durch den zweiten Term gegebene Änderung des Phasenvolumens ist gleich dem Volumen zwischen den beiden Energieflächen[3]

$$E(\boldsymbol{q}, \boldsymbol{p}, x) = E \qquad (23.6)$$

---

[1] In der Ausdrucksweise der Thermodynamik handelt es sich um einen quasistatischen Prozeß.

[2] Thermodynamisch heißt ein derartiger Prozeß adiabatisch. Wir haben es also mit einem quasistatisch-adiabatischen Prozeß zu tun.

[3] Da das Symbol E^* nur im Hinblick auf die Benutzung der δ-Funktion eingeführt wurde, schreiben wir im Interesse einer übersichtlichen Bezeichnung stattdessen einfach E, wenn dadurch kein Mißverständnis entstehen kann.

und

$$E'(\boldsymbol{q}, \boldsymbol{p}, x + \Delta x) = E'(\boldsymbol{q}, \boldsymbol{p}, x) + \sum_i \frac{\partial E'}{\partial x_j} \Delta x_j = E, \qquad (23.7)$$

wo $E(\boldsymbol{q}, \boldsymbol{p}, x)$ und $E'(\boldsymbol{q}, \boldsymbol{p}, x)$ verschiedene Funktionen der Phasenkoordinaten sind. Es ist daher, wenn Δs den (mit Vorzeichen versehenen) Abstand der beiden Flächen bezeichnet,

$$\sum_j \left(\frac{\partial \Omega^*}{\partial x_j} \right)_E \Delta x_j = \int \Delta s \, df. \qquad (23.8)$$

Andererseits ist

$$E' - E = \Delta s \, |\operatorname{grad} E'| \qquad (23.9)$$

und nach (23.7)

$$E' - E = - \sum_j \frac{\partial E'}{\partial x_j} \Delta x_j. \qquad (23.10)$$

Da E' und E sich nur wenig unterscheiden, sind die Ableitungen beider Größen bis auf Glieder höherer Ordnung gleich. Es folgt daher aus (23.8) bis (23.10)

$$\sum_j \left(\frac{\partial \Omega^*}{\partial x_j} \right)_E \Delta x_j = - \int \frac{\sum_j \frac{\partial E}{\partial x_j} \Delta x_j}{|\operatorname{grad} E|} df. \qquad (23.11)$$

Durch Einsetzen von (23.4) und (23.11) in Gl. (23.5) und Berücksichtigung von (20.2) ergibt sich schließlich

$$\overline{\Delta \Omega^*} = 0. \qquad (23.12)$$

Die beiden in Gl. (23.5) dargestellten Effekte kompensieren sich somit. Das Phasenvolumen ist invariant gegen adiabatische Parameteränderungen.

In der Ausdrucksweise der Thermodynamik wird ein Prozeß von der hier betrachteten Art als quasistatisch-adiabatisch bezeichnet. Die einzige thermodynamische Größe, die bei einem solchen Prozeß notwendig konstant bleibt, ist die Entropie. Wir können daher schließen, daß eine ein-eindeutige Funktion des Phasenvolumens ein statistisches Analogon der Entropie darstellt. In Ziff. 24 werden wir dies vollständig und für beliebige Zustandsänderungen beweisen.

24. Entropie und absolute Temperatur. Aus der molekularen Theorie der Wärme folgt unmittelbar, daß die Energie eines Systems auch bei konstanten Werten der äußeren Parameter geändert werden kann. Während für die in Ziff. 23 behandelten quasistatisch-adiabatischen Prozesse gilt

$$dE = dA \qquad (24.1)$$

ist allgemein

$$dE = d'Q + d'A, \qquad (24.2)$$

wo das Symbol d' bedeutet, daß die betreffenden infinitesimalen Größen keine exakten Differentiale sind. Als zugeführte Wärme wird die Differenz $dE - d'A$ definiert. (I. Hauptsatz der Thermodynamik.)

Es bleibt jetzt noch zu zeigen, daß für ein durch die mikrokanonische Gesamtheit dargestelltes Originalsystem auch der II. Hauptsatz der Thermodynamik in dem früher (Ziff. 12) definierten Sinne der statistischen Theorie gültig ist. Wir haben also für die mikrokanonische Gesamtheit Größen zu definieren, welche die Eigenschaften der Entropie und der absoluten Temperatur besitzen. Die fraglichen Größen müssen daher einer Differentialgleichung der Form

$$dS = \frac{1}{T} dE + \sum_j \frac{X_j}{T} dx_j \qquad (24.3)$$

genügen. Für zwei in thermischer Berührung stehende Teilsysteme, wie wir sie in Ziff. 22 betrachtet haben, muß gelten

$$dS = dS_1 + dS_2. \tag{24.4}$$

Schließlich muß S homogen vom ersten Grade in E, V und N sein. Diese Forderung geht allerdings über den II. Hauptsatz hinaus. Wir werden daher später (Ziff. 25) darauf zurückkommen. Auch die Gl. (24.4) können wir erst später diskutieren. Für die Teilsysteme gelten nämlich nicht mehr die gleichen Randbedingungen wie für das durch die mikrokanonische Gesamtheit dargestellte Gesamtsystem, da die Energie eines Teilsystems nicht vorgegeben ist. Die statistische Definition der Entropie als Funktion der Energie, mit der wir es hier zu tun haben, ist daher für ein Teilsystem zunächst nicht eindeutig durchführbar. Tatsächlich beruht Gl. (24.4) auf der Voraussetzung, daß die thermodynamischen Funktionen unabhängig von den Randbedingungen definierbar sind. Da wir diese Frage in allgemeiner Form erst in Teil C I (S. 350) behandeln, reduziert sich unsere Aufgabe auf die statistische Begründung der Gl. (24.3).

Da wir jetzt (im Gegensatz zu Ziff. 23) auch die Energie als unabhängige Variable zu betrachten haben, gilt allgemein

$$d\Omega^* = \frac{\partial \Omega^*}{\partial E}\, dE + \sum_j \frac{\partial \Omega^*}{\partial x_j}\, dx_j. \tag{24.5}$$

Mit Benutzung von (20.2) läßt sich diese Gleichung schreiben

$$d\ln\Omega^* = \frac{1}{e^{-\Phi}\Omega^*}\, dE + \sum_j \frac{\partial \ln\Omega^*}{\partial x_j}\, dx_j. \tag{24.6}$$

Da nach Gl. (23.2) und (23.11)

$$\sum_j \frac{\partial \ln\Omega^*}{\partial x_j} = \frac{1}{e^{-\Phi}\Omega^*} \sum_j \overline{X}_j \tag{24.7}$$

ist, folgt

$$d\ln\Omega^* = \frac{1}{e^{-\Phi}\Omega^*}\, dE + \sum_j \frac{\overline{X}_j}{e^{-\Phi}\Omega^*}\, dx_j. \tag{24.8}$$

Der Vergleich von Gl. (24.3) und (24.8) zeigt in Verbindung mit den Ergebnissen der Ziff. 22, daß die Größe $e^{-\Phi}\Omega^*$ ein statistisches Analogon der absoluten Temperatur ist und $\ln\Omega^*$ ein statistisches Analogon der Entropie (II. Gibbssches Entropie-Analogon). Der formelmäßige Zusammenhang zwischen der Entropie und dem Logarithmus des Phasenvolumens hängt von der Wahl der Temperaturskala ab. Für die übliche Skala ergibt sich aus Gl. (21.8) und (24.3)

$$dS = k\, d\ln\Omega^*. \tag{24.9}$$

Wir wollen dieses grundlegende Ergebnis noch auf einem von den Überlegungen der Ziff. 23 unabhängigen Wege ableiten[1] [2], [26]. Nach Gl. (2.4), (20.4) und (20.5) ist

$$e^{\Phi} = \int \delta(E^* - E)\, d\Omega. \tag{24.10}$$

Durch Differentiation nach einem äußeren Parameter folgt daraus

$$\frac{\partial e^{\Phi}}{\partial x_j} = \int X_j\, \delta'(E^* - E)\, d\Omega \tag{24.11}$$

---

[1] A. Münster: Z. Physik **137**, 386 (1954).

oder

$$\frac{\partial e^{\Phi}}{\partial x_j} = \int \overline{X}_j \, e^{\Phi(E)} \, \delta'(E^* - E) \, dE. \qquad (24.12)^{[1]}$$

Das ergibt

$$\frac{\partial e^{\Phi}}{\partial x_j} = \frac{\partial}{\partial E}(\overline{X}_j \, e^{\Phi}) \qquad (24.13)$$

oder mit (20.2)

$$\frac{\partial^2 \Omega^*}{\partial E \, \partial x_j} = \frac{\partial}{\partial E}(\overline{X}_j \, e^{\Phi}). \qquad (24.14)$$

Diese Gleichung hat die Lösung

$$\frac{\partial \Omega^*}{\partial x_j} = \overline{X}_j \, e^{\Phi} + F_j, \qquad (24.15)$$

wo F_j eine Funktion der äußeren Parameter ist. Setzen wir diesen Ausdruck in Gl. (24.5) ein, so erhalten wir

$$d\Omega^* = e^{\Phi}\left(dE + \sum_j X_j \, dx_j\right) + \sum_j F_j \, dx_j. \qquad (24.16)$$

Um die Funktionen F_j zu bestimmen, betrachten wir eine Zustandsänderung, bei der die äußeren Parameter willkürlich variiert werden und die Energie stets den niedrigsten Wert annimmt, der mit den jeweiligen Werten der äußeren Parameter vereinbar ist[2]. Während dieses Vorganges ist

$$d\Omega^* = 0. \qquad (24.17)$$

Ferner ist die kinetische Energie dauernd null und daher die Gesamtenergie nur eine Funktion der äußeren Parameter. Es gilt daher für die betrachtete Änderung

$$dE = \sum_j \frac{\partial E}{\partial x_j} \, dx_j. \qquad (24.18)$$

Schließlich sind unter diesen Umständen die generalisierten Kräfte für alle Systeme der Gesamtheit gleich. Es gilt also

$$X_j = \overline{X}_j. \qquad (24.19)$$

Wir haben somit

$$dE + \sum_j \overline{X}_j \, dx_j = 0. \qquad (24.20)$$

Aus Gl. (24.16), (24.17) und (24.20) folgt, zunächst für die betrachtete Zustandsänderung,

$$\sum_j F_j \, dx_j = 0. \qquad (24.21)$$

Da aber die Änderungen der Parameter willkürlich sind und die F_j nicht von der Energie abhängen, muß allgemein gelten

$$F_j = 0. \qquad (24.22)$$

Gl. (24.15) geht daher in Gl. (24.7) über, und wir erhalten wieder die Gl. (24.8) und (24.9)

---

[1] Durch die Schreibweise $\Phi(E)$ deuten wir an, daß Φ an dieser Stelle, im Sinne der allgemeinen Definition (20.2), als Funktion der Integrationsvariablen E aufzufassen ist.

[2] Der folgende Beweis beruht auf der Betrachtung einer Zustandsänderung am absoluten Nullpunkt. Diese ist hier jedoch nicht im Sinne einer physikalischen Theorie, sondern rein formal als Untersuchung der Randbedingungen zu verstehen.

25. Explizite Formulierung der Funktionen Ω^* und Φ. Auf Grund der Ergebnisse von Ziff. 24 ist es im Prinzip möglich, die Entropie eines beliebigen Systems als Funktion von E, V und N zu berechnen. Zur Durchführung einer solchen Rechnung benötigt man jedoch eine explizite Darstellung der Funktionen Ω^* und Φ, welche dieselben auf die (als bekannt angenommene) Hamilton-Funktion und die Werte der äußeren Parameter zurückführt. Diese Ausdrücke lassen sich in einfacher Weise mit Hilfe der Fourier-Zerlegung der δ-Funktion ableiten[1] [26].

Wir gehen aus von Gl. (20.6), die wir für eine Funktion u der generalisierten Koordinaten (oder der potentiellen oder der kinetischen Energie) schreiben

$$\bar{u} = e^{-\Phi} \iint u\, \delta(E^* - U - E_{\mathrm{kin}})\, d\Omega_p\, d\Omega_q. \qquad (25.1)$$

Für die kinetische Energie können wir allgemein setzen

$$E_{\mathrm{kin}} = \tfrac{1}{2} \sum_{i=1}^{n} a_i\, \dot{q}_i^2 = \tfrac{1}{2} \sum_{i=1}^{n} a_i^{-1} p_i^2. \qquad (25.2)$$

Gehen wir damit in Gl. (25.1) ein, so erhalten wir durch Fourier-Zerlegung der δ-Funktion

$$\bar{u} = e^{-\Phi} \int d\Omega_q\, u\, \frac{1}{2\pi} \int_{-\infty}^{+\infty} dk\, e^{ik(E^*-U)} \int_{-\infty}^{+\infty} e^{-ik\sum_i \frac{p_i^2}{2a_i}}\, d\Omega_p. \qquad (25.3)$$

Nach Ausführung des letzten Integrals wird daraus

$$\bar{u} = (2\pi)^{\frac{n}{2}} e^{-\Phi} \int d\Omega_q\, u\, \frac{1}{2\pi} \int_{-\infty}^{+\infty} (ik)^{-\frac{n}{2}}\, e^{ik(E^*-U)}\, dk. \qquad (25.4)$$

Der Integrand des zweiten Integrals hat im Ursprung eine Singularität. Der Integrationsweg ist unterhalb derselben vorbeizuführen und kann über die positive imaginäre Achse geschlossen werden. Wir setzen voraus $E^* - U > 0$ und schreiben

$$\frac{n}{2} = l + 1, \quad ik = z \qquad (25.5)$$

wo l eine positive ganze Zahl ist. Dann wird aus (24.4)

$$\bar{u} = (2\pi)^{\frac{n}{2}} e^{-\Phi} \int d\Omega_q\, u\, \frac{1}{2\pi i} \oint \frac{e^{z(E^*-U)}}{z^{l+1}}\, dz. \qquad (25.6)$$

Der Integrationsweg umschließt den Pol $(l+1)$-ter Ordnung des Integranden bei $z = 0$. Mit Hilfe des Residuensatzes erhalten wir aus (25.6)

$$\bar{u} = \frac{(2\pi)^{\frac{n}{2}}}{\Gamma\left(\frac{n}{2}\right)}\, e^{-\Phi} \int_{\Omega_q=0}^{U=E^*} u\, (E^* - U)^{\frac{n}{2}-1}\, d\Omega_q. \qquad (25.7)$$

Diese Gleichung stellt die explizite Formulierung der früher abgeleiteten Mittelwertsformel Gl. (20.15) dar. Durch Vergleich beider Ausdrücke ergibt sich

$$e^{\Phi_p} = \frac{(2\pi)^{\frac{n}{2}}}{\Gamma\left(\frac{n}{2}\right)}\, E_{\mathrm{kin}}^{\frac{n}{2}-1}. \qquad (25.8)$$

Setzen wir (25.8) ind (20.3) ein und integrieren, so erhalten wir Gl. (4.28).

---

[1] A. Münster: Z. Physik **137**, 386 (1954).

Für $u=1$ folgt aus (25.7)

$$e^{\Phi} = \frac{(2\pi)^{\frac{n}{2}}}{\Gamma\left(\frac{n}{2}\right)} \int\limits_{\Omega_q=0}^{U=E^*} (E^* - U)^{\frac{n}{2}-1}\, d\Omega_q. \tag{25.9}$$

Diese Beziehung läßt sich mit (20.2) kombinieren und liefert nach Integration

$$\Omega^* = \frac{(2\pi)^{\frac{n}{2}}}{\Gamma\left(\frac{n}{2}+1\right)} \int\limits_{\Omega_q=0}^{U=E^*} (E^* - U)^{\frac{n}{2}}\, d\Omega_q. \tag{25.10}$$

Die aus Gl. (24.9) und (25.10) berechnete Entropie ist indessen nicht homogen vom ersten Grade in E, V, N, was sich an dem weiter unten zu besprechenden Beispiel des einatomigen idealen Gases leicht explizit zeigen läßt. Sie gibt daher insbesondere keine Erklärung für das sog. *Gibbssche Paradoxon*, daß gleiche Gase (im Gegensatz zu ungleichen Gasen) sich bei konstantem Druck ohne Entropieänderung vermischen. Dieser Widerspruch zur Erfahrung wird dadurch verursacht, daß an Gl. (25.10) noch zwei Korrekturen anzubringen sind, deren strenge Begründung erst durch die Quantenstatistik erfolgen kann, obschon sie sich im Rahmen der klassischen Theorie verständlich machen lassen.

Bei dem bisher verwendeten Phasenbegriff erscheinen zwei Zustände des Systems, die durch einfache Vertauschung gleicher Teilchen auseinander hervorgehen, als verschiedene Phasen. Nach GIBBS [2] wird diese Definition der Phase als *„spezielle Phase"* (specific phase) bezeichnet. GIBBS [2] hat auch bereits die Möglichkeit diskutiert, daß die Gesamtheit aller speziellen Phasen, die durch Vertauschung gleicher Teilchen aus einander hervorgehen, als eine einzige Phase aufzufassen ist. Er bezeichnet die so definierte Phase als *„generelle Phase"* (generic phase) und vermutet, daß diese Definition dem Wesen der statistischen Mechanik besser entspricht. Aus der Quantenstatistik folgt nun, daß, soweit die klassische Statistik als Näherung anwendbar ist, dabei unter „Phase" die generelle Phase zu verstehen ist, was letzten Endes auf die Nichtunterscheidbarkeit gleicher Teilchen zurückgeht. Die Folgerung, die daraus für die Entropieformel zu ziehen ist, ergibt sich nun sehr einfach. Da (für ein Einkomponentensystem) $N!$ Phasen durch Vertauschung der Teilchen aus einander hervorgehen, haben wir lediglich Gl. (25.10) durch $N!$ zu dividieren, um daraus den im obigen Sinne korrekten Ausdruck für die Entropie zu erhalten. Um vollständige Übereinstimmung mit der aus der Quantenstatistik abgeleiteten „halbklassischen Näherung" zu erhalten, ist es außerdem noch notwendig, das Phasenvolumen in Einheiten h^n (wo h das Plancksche Wirkungsquantum ist) zu messen. Die Einführung einer solchen Maßeinheit (nicht ihr Zahlenwert) läßt sich im Rahmen der klassischen Theorie plausibel machen durch die Forderung, daß in Gl. (24.9) unter dem Logarithmus eine dimensionslose Größe stehen sollte. Der Einfachheit halber definieren wir von jetzt ab die Größe Ω^* so, daß sie die Division durch $N!\,h^n$ einschließt.

Bei der Betrachtung der Gl. (25.10) fällt auf, daß in die Berechnung der Entropie eines Systems von vorgegebener Energie auch Zustände mit einer Energie $<E^*$ eingehen. Da im Sinne der Quantenstatistik eine Zelle der Größe h^n einen mechanischen Zustand des Systems repräsentiert, bedeutet dies, daß bei der Berechnung der Entropie Zustände mitgezählt werden, die das System in Wirklichkeit nicht einnehmen kann. Dieser scheinbare Widerspruch findet seine Aufklärung durch die geometrische Tatsache, daß bei einem Gebilde von sehr

hoher Dimensionszahl der Logarithmus des Gesamtvolumens von der gleichen Größenordnung ist, wie der Logarithmus des Volumens einer Schale an der Oberfläche. Aus Gl. (25.9) und (25.10) findet man in der Tat, daß asymptotisch

$$\ln \Omega^* = \Phi \qquad (n \to \infty) \tag{25.11}$$

ist. Wir können daher auch schreiben

$$dS = k \, d\Phi. \tag{25.12}$$

Der Logarithmus der Ableitung des Phasenvolumens nach der Energie der umschließenden mikrokanonischen Energieschale stellt somit ebenfalls ein statistisches Analogon der Entropie dar (III. Gibbssches Entropie-Analogon). Vom physikalischen Standpunkt ist, wie schon Gibbs [2] bemerkt hat, die Analogie zwischen S und Φ als die primäre anzusehen. Tatsächlich kann diese Analogie auch unabhängig von den Gln. (25.9) und (25.10) direkt nachgewiesen werden. Die Größe Ω^* wäre dann aus der zuerst definierten Größe Φ durch Integration abzuleiten. Der umgekehrte Weg ist jedoch mathematisch einfacher und daher auch hier benutzt worden.

Wir können jetzt die Entropieformel in ihrer endgültigen expliziten Gestalt anschreiben. Dabei beschränken wir uns der Einfachheit halber auf den Fall eines Einkomponentensystems aus einatomigen Molekülen ohne innere Freiheitsgrade (Massenpunkte). Es ergibt sich dann mit Benutzung von Gl. (4.23)

$$S = k \ln \Omega^* = k \, \Phi = k \ln \left[\frac{(2\pi m)^{\frac{3}{2} N}}{h^{3N} N! \, (\frac{3}{2} N)!} \int \cdots \int (E^* - U)^{\frac{3}{2} N} \, d\boldsymbol{q}_1 \ldots d\boldsymbol{q}_N \right]. \tag{25.13}$$

Diese Gleichung läßt sich noch auf eine etwas übersichtlichere Form bringen, wenn wir die Stirlingsche Formel anwenden und die mittlere Energie eines Moleküls $\bar{\varepsilon} = E^*/N$ einführen. Dann ergibt sich

$$S = k \, N \ln \left(\frac{4 \pi m \bar{\varepsilon} \, e}{3 h^2} \right)^{\frac{3}{2}} + k \ln \Lambda, \tag{25.14}$$

wo

$$\Lambda = \frac{1}{N!} \int \cdots \int \left(1 - \frac{U}{E^*} \right)^{\frac{3}{2} N} d\boldsymbol{q}_1 \ldots d\boldsymbol{q}_N \tag{25.15}$$

ist. Die vorstehenden Gleichungen sind für den betrachteten Fall der korrekte Ausdruck der halbklassischen Näherung. Die Verallgemeinerung für mehratomige Moleküle macht grundsätzlich keine Schwierigkeiten, wenn die hinzukommenden Freiheitsgrade ebenfalls klassisch behandelt werden können.

Für das einatomige ideale Gas ist $U = 0$ und somit

$$\ln \Lambda = \ln \frac{V^N}{N!} = N \ln (v \, e), \tag{25.16}$$

wo $v = V/N$ das Volumen pro Molekül bezeichnet. In Verbindung mit Gl. (25.14) folgt daraus für die Entropie

$$S = k \, N \ln \left[\left(\frac{4 \pi m \bar{\varepsilon}}{3 h^2} \right)^{\frac{3}{2}} v \, e^{\frac{5}{2}} \right]. \tag{25.17}$$

Diese Gleichung stellt die Entropie als thermodynamisches Potential, d.h. als Funktion der unabhängigen Variablen E, V, N dar. Sie zeigt, daß nach Anbringen der oben besprochenen Korrektur S in der Tat eine homogene Funktion ersten Grades ist. Führen wir in (25.17) das Äquipartitionstheorem Gl. (21.10) ein, so erhalten wir

$$S = k \, N \left[\ln \frac{V}{N} + \frac{5}{2} + \ln \left(\frac{2 \pi m k T}{h^2} \right)^{\frac{3}{2}} \right]. \tag{25.18}$$

In dieser Form, als Funktion von T, V, N ist die Entropie kein thermodynamisches Potential sondern die Ableitung der freien Energie nach HELMHOLTZ.

Die Auswertung der Gl. (25.13) für allgemeinere Fälle stellt ein schwieriges Problem dar. Es ist jedoch STREETER und MAYER[1] gelungen, auf dieser Grundlage die exakte Theorie realer Gase zu entwickeln.

26. Das thermodynamische Gleichgewicht. Zugänglichkeit und Hemmungen. Die Bedingung des thermodynamischen Gleichgewichtes lautet bekanntlich in ihrer allgemeinsten Form

$$(\delta S)_{E,V,N} \lneqq 0. \tag{26.1}$$

Hier bezeichnet das Symbol δ eine virtuelle Verrückung in dem aus der Mechanik bekannten Sinne. Für homogene Systeme (auf die wir uns vorläufig beschränken) können solche Verrückungen im Hinblick auf die angeschriebenen Nebenbedingungen nur in Verschiebungen der inneren Parameter (Reaktionslaufzahlen, Ordnungsgrad usw.) bestehen. Die für das Folgende wesentlichen Eigenschaften der virtuellen Verrückung sind:

a) Sie muß möglich sein, d.h. den allgemeinen Bedingungen des Systems entsprechen.

b) Sie ist keine Funktion der Zeit.

c) Die Entropie muß auch für den variierten Zustand definiert sein.

Wir fragen nun, ob die in den vorhergehenden Ziffern untersuchten statistischen Analoga der Entropie auch die Begründung der Gleichgewichtsbedingung (26.1) ermöglichen. Diese Frage ist nicht trivial, weil wir bisher die Existenz des Gleichgewichtes vorausgesetzt und das System durch ein stationäres Ensemble abgebildet haben. Es ist daher jetzt notwendig, abgeänderte Verteilungen zu untersuchen, welche den virtuellen Verrückungen des Originalsystems entsprechen, die aber, im Hinblick auf die Nebenbedingungen, ebenfalls auf die mikrokanonische Energiefläche beschränkt sind. Die Begründung der Bedingung (26.1) besteht dann in dem Nachweis, daß eine der Entropie analoge statistische Funktion für die stationäre (d.h. hier die mikrokanonische) Verteilung einen Extremwert annimmt. Letzten Endes handelt es sich also darum, die Bedingung für das statistische Gleichgewicht nicht durch die stationäre Lösung der Liouvilleschen Gleichung (Ziff. 5), sondern durch die Extremaleigenschaften einer geeigneten Maßfunktion auszudrücken.

Man sieht nun leicht, daß die in Ziff. 23 und 24 definierten statistischen Analoga der Entropie für diesen Zweck nicht verwendbar sind. Dieselben enthalten nämlich nicht mehr explizit die Verteilung; sie sind vielmehr rein geometrische Größen des Phasenraumes, die erst im Zusammenhang mit der mikrokanonischen Verteilung ihre physikalische Bedeutung erhalten. Wir werden daher zunächst eine neue statistische Formel für die Entropie aufstellen und zeigen, daß dieselbe im Falle der mikrokanonischen Verteilung in die bisherigen Definitionen übergeht. Später (Ziff. 67) werden wir sehen, daß die Gültigkeit dieser Formel nicht darauf beschränkt ist, sondern daß sie den allgemeinsten statistischen Ausdruck für die Entropie überhaupt darstellt.

Die neue Formel lautet

$$S = -k \int \varrho \ln \varrho \, d\Omega. \tag{26.2}$$

Die rechts stehende Funktion haben wir bereits im Zusammenhang mit der Diskussion des H-Theorems (Ziff. 17) eingeführt. Wir haben dort gesehen, daß diese

---

[1] S. F. STREETER u. J. E. MAYER: J. Chem. Phys. **7**, 1025 (1939). — Vgl. auch Bd. XII dieses Handbuches.

Funktion wegen des Liouvilleschen Satzes die Einstellung des Gleichgewichtes nicht beschreiben kann. Um diese Schwierigkeit zu umgehen, haben wir eine analoge Funktion unter Verwendung der groben Dichte P konstruiert und mit deren Hilfe das verallgemeinerte H-Theorem formuliert. Es liegt daher die Frage nahe, ob nicht auch in Gl. (26.2) konsequenterweise P an Stelle von ϱ einzusetzen ist. Formal würde dadurch in den Ableitungen nichts geändert, da für die mikrokanonische Gesamtheit auch P durch Gl. (20.4) zu definieren wäre. Die wesentliche Schwierigkeit liegt, wie schon in Ziff. 17 erwähnt, darin, daß für P kein unabhängiges Kriterium des statistischen Gleichgewichtes bekannt ist. Andererseits handelt es sich hier nicht um die zeitliche Entwicklung des Systems, sondern um die Frage, ob die stationäre Verteilung auf der Energiefläche sich als Extremaleigenschaft darstellen läßt. Die in Ziff. 17 erwähnte Schwierigkeit ist daher hier bedeutungslos. Schließlich muß auch im Sinne der Betrachtungsweise von Ziff. 17 $P = \varrho$ gesetzt werden, wenn das Gleichgewicht nicht als Endzustand eines irreversiblen Prozesses, sondern als der von vornherein gegebene Zustand maximaler Information betrachtet wird.

Wir definieren nun eine neue Größe

$$\eta = \ln \varrho, \tag{26.3}$$

die man nach Gibbs [2] den Wahrscheinlichkeitsindex nennt. Damit wird aus (26.2)

$$S = -k \int \eta \, e^\eta \, d\Omega = -k \bar{\eta}. \tag{26.4}$$

Die Entropie ist somit (bis auf den von der Temperaturskala abhängigen Proportionalitätsfaktor) gleich dem negativen mittleren Wahrscheinlichkeitsindex der betreffenden Verteilung. Setzt man in Gl. (26.2) die Phasendichte der mikrokanonischen Gesamtheit ein, so ergibt sich

$$S = k \int \Phi \, e^{-\Phi} \, \delta(E^* - E) \, d\Omega = k \Phi \tag{26.5}$$

in Übereinstimmung mit Gl. (25.12).

Wir betrachten nun eine Verteilung auf der Energiefläche, die durch

$$\varrho = e^{-\Phi + \Delta\eta} \, \delta(E^* - E) \tag{26.6}$$

gegeben ist, wo $\Delta\eta$ eine willkürliche Funktion der Phasenkoordinaten bezeichnet. Dabei muß wegen (2.4) gelten

$$\int e^{-\Phi} \, \delta(E^* - E) \, d\Omega = \int e^{-\Phi + \Delta\eta} \, \delta(E^* - E) \, d\Omega = 1. \tag{26.7}$$

Wir wollen jetzt zeigen, daß für jeden derartigen Nachbarzustand die nach Gl. (26.2) berechnete Entropie kleiner ist als die des ursprünglichen Gleichgewichtszustandes, die durch Gl. (26.5) gegeben ist. Es ist also die Ungleichung

$$\int (-\Phi + \Delta\eta) \, e^{-\Phi + \Delta\eta} \, \delta(E^* - E) \, d\Omega > \int -\Phi \, e^{-\Phi} \, \delta(E^* - E) \, d\Omega \tag{26.8}$$

zu beweisen. Da für die Integration Φ eine Konstante ist, kann (26.8) geschrieben werden

$$\int (-\Phi + \Delta\eta) \, e^{\Delta\eta} \, \delta(E^* - E) \, d\Omega > \int -\Phi \, \delta(E^* - E) \, d\Omega. \tag{26.9}$$

In analoger Weise kann in (26.7) der Faktor $e^{-\Phi}$ gekürzt werden. Multiplizieren wir dann diese Gleichung mit dem konstanten Faktor $(-\Phi + 1)$, so folgt

$$\int (-\Phi + 1) \, e^{\Delta\eta} \, \delta(E^* - E) \, d\Omega = \int (-\Phi + 1) \, \delta(E^* - E) \, d\Omega. \tag{26.10}$$

Wird schließlich (26.10) von (26.9) subtrahiert, so ergibt sich

$$\int (\Delta\eta \, e^{\Delta\eta} - \Delta\eta + 1)\, \delta(E^* - E)\, d\Omega > 0. \tag{26.11}$$

Da der Klammerausdruck, wie in Ziff. 17 gezeigt (abgesehen von dem trivialen Fall $\Delta\eta = 0$) notwendig positiv ist, ist auch das Integral positiv, womit die Behauptung bewiesen ist. Die stationäre Verteilung einer Gesamtheit auf der Energiefläche, d.h. die mikrokanonische Verteilung, ist also dadurch charakterisiert, daß die Funktion (26.2) den durch (26.5) gegebenen maximalen Wert annimmt. Die durch Gl. (26.2) definierte Entropie ist primär eine Eigenschaft der Gesamtheit (was in der Quantenstatistik noch schärfer hervortritt), im Gegensatz zur Energie, die für jedes System als Integral der Bewegungsgleichungen definiert ist. Zu einer Eigenschaft eines einzelnen Systems wird die Entropie erst dadurch, daß ein System durch eine repräsentative Gesamtheit dargestellt, d.h. statistisch beschrieben wird. Vom Standpunkt der Mechanik gesehen, ist daher das Konzept der Entropie an die Einführung der statistischen Methode geknüpft. Auch in diesem Sinne ist sie ein unmittelbarer Ausdruck der unvollständigen Information.

Bisher haben wir angenommen, daß das Gleichgewicht als makroskopischer Zustand eindeutig definiert ist. Die experimentelle Erfahrung zeigt indessen, daß dies keineswegs der Fall ist. Das System Wasserstoff-Sauerstoff-Wasserdampf verhält sich unterschiedlich, je nachdem ob wir es mit oder ohne Katalysator untersuchen. In beiden Fällen spricht man von thermodynamischem Gleichgewicht. Ein anderes Beispiel bieten zweiatomige Gase, bei denen sich der Freiheitsgrad der Schwingung verhältnismäßig langsam ins Gleichgewicht setzt mit den Freiheitsgraden der Rotation und Translation. Die Berechnung der Molwärme aus Messungen der Schallgeschwindigkeit liefert daher ein anderes Ergebnis als die Anwendung einer Strömungsmethode. Viel häufiger und im allgemeinen auch komplizierter sind derartige Erscheinungen in kondensierten Phasen. Man muß daher schließen, daß der Begriff ,,Gleichgewicht'' sich physikalisch sinnvoll nur im Hinblick auf bestimmte Vorgänge definieren läßt. Der Begriff eines ,,absoluten Gleichgewichtes'' oder ,,Gleichgewichtes im Hinblick auf alle denkbaren Vorgänge'' hat keine physikalische Bedeutung.

Um den Sachverhalt zunächst phänomenologisch zu erfassen, gehen wir davon aus, daß wir einen beliebigen Zustand des betrachteten Systems vollständig beschreiben können, wenn wir außer den thermodynamischen Zustandsvariablen die ,,inneren Parameter'' ξ_i benutzen. Rein formal können wir annehmen, daß wir über einen vollständigen Satz der ξ_i verfügen derart, daß jede denkbare Zustandsänderung sich als Änderung eines oder mehrerer dieser Parameter darstellen läßt. Dieselben ändern sich im Laufe der Zeit in Richtung auf die Werte, die sie (bei gegebenen Werten der thermodynamischen Zustandsvariablen) im Gleichgewicht annehmen. Wir bezeichnen diese Gleichgewichtswerte mit ξ_{0i} und schreiben

$$\xi_i - \xi_{0i} = C \cdot e^{-\frac{t}{\tau_i}}, \tag{26.12}$$

wo τ_i die betreffende Relaxationszeit ist. Auf dieser Grundlage können wir die inneren Parameter in drei Klassen einteilen, nämlich, wenn t_e die Größenordnung der Dauer einer Messung ist,

$$\left.\begin{array}{l} 1.\ \tau_i \gg t_e, \\[4pt] 2.\ \tau_i \approx t_e, \\[4pt] 3.\ \tau_i \ll t_e. \end{array}\right\} \tag{26.13}$$

Die Anwendung der Thermodynamik setzt grundsätzlich voraus, daß unter den gewählten Bedingungen keine Parameter der Klasse 2 auftreten. Für die Parameter der Klasse 3 kann definitionsgemäß momentane Gleichgewichtseinstellung vorausgesetzt werden; über die Parameter der Klasse 3 braucht außer ihrer allgemeinen Charakteristik nichts bekannt zu sein, nicht einmal ihre Zahl. Es gibt also gelegentlich für das gleiche System verschiedene Möglichkeiten der thermodynamischen Behandlung, von denen jede durch ein Gleichungssystem

$$\xi_k = \text{const}, \qquad \left(\frac{\partial S}{\partial \xi_l}\right)_{E,\,V,\,N} = 0 \qquad\qquad (26.14)$$

(wo der Index k die Parameter der Klasse 1, der Index l die der Klasse 3 bezeichnet) definiert ist. Jedes derselben gilt für Zustandsänderungen, die auf der betreffenden Entropiefläche verlaufen.

Für die Anwendung der statistischen Mechanik (soweit sie auf der Verwendung von stationären Gesamtheiten beruht) gilt zunächst der gleiche Gesichtspunkt, daß unter den gewählten Bedingungen keine Parameter der Klasse 2 auftreten dürfen. Von den Parametern der Klasse 3 kann ein Teil bereits durch die Wahl des Modells, welches der statistischen Rechnung zugrunde gelegt wird, eliminiert werden. Wählt man etwa für die Moleküle eines zweiatomigen Gases als Modell starre Rotatoren, so hat der Γ-Raum $10N$ Dimensionen. Die Freiheitsgrade, welche sich auf die Struktur der Atome beziehen (Elektronen Kernspin, Kernreaktionen) sind ebenso wie Schwingungen und Dissoziationen bereits a priori ausgeschlossen. In einem gewissen Temperaturgebiet ist dieses Modell angemessen, um die Ergebnisse der Messungen der Schallgeschwindigkeit darzustellen. Soweit die Parameter der Klasse 1 nicht durch das Modell eliminiert werden, bewirken sie, daß Teile des gewählten Phasenraumes für das System praktisch unzugänglich sind. In diesen Teilen muß die Phasendichte identisch Null gesetzt werden. Die statistische Betrachtung bezieht sich also stets nur auf den praktisch zugänglichen Teil des Phasenraumes. Auch die statistische Gleichgewichtsbedingung ist in diesem Sinne zu verstehen; die in (26.6) auftretende Größe $\Delta\eta$, welche die Abweichung von der gleichmäßigen Verteilung beschreibt, kann daher ebenfalls nur in dem zugänglichen Teil des Phasenraumes von Null verschieden sein. Damit ist auch der möglichen Existenz weiterer uniformer Integrale der Bewegungsgleichungen Rechnung getragen.

Wenn die Unzugänglichkeit im Phasenraum nicht prinzipieller Natur ist (was nur in der Quantenstatistik vorkommt), sagt man, daß das System einer *Hemmung* unterliegt. Eine solche Hemmung kann in verschiedener Weise aufgehoben werden: durch Änderung der Meßmethode, durch Einbringen eines Katalysators oder durch einen äußeren Eingriff. In jedem Falle gilt der Satz: Durch das Aufheben einer Hemmung wird die Entropie entweder vermehrt oder sie bleibt ungeändert. Dieser Satz folgt unmittelbar aus den Definitionen der Entropie und der Hemmung unter Berücksichtigung der Tatsache, daß für die Berechnung der Entropie die generelle Phase (vgl. S. 249) zu verwenden ist.

Die Freiheit in der Wahl des Modells hat zur Folge, daß auch der Begriff der absoluten Entropie keinen physikalischen Sinn hat. An diesem, von der Thermodynamik her bekannten, Sachverhalt wird auch durch die statistische Theorie nichts geändert. Es ist zwar richtig, daß die statistischen Formeln für die Entropie keine unbestimmte Integrationskonstante mehr enthalten. Man kann aber durch Übergang zu anderen Modellen (z.B. vom Massenpunkt zum Atom mit Elektronenhülle und Kernspin) beliebig neue Terme einführen. Physikalische Bedeutung haben dieselben nur insoweit, als es Vorgänge gibt, bei denen die betreffenden Terme ihren Wert ändern. Beispielsweise haben sich die

Terme für den Kernspin bei gewöhnlichen chemischen Gleichgewichten grundsätzlich heraus, so daß man sie hier ebensogut weglassen könnte. Andererseits werden aber die thermodynamischen Formeln durch die statistische Theorie in dem Sinne ergänzt, als sie eine Vorausberechnung gewisser Größen (chemische Konstanten, Dampfdruckkonstanten) ermöglicht, die sonst nur empirisch bestimmt werden können. Diese Rechnungen lassen sich allerdings erst mit Hilfe der Quantenstatistik durchführen.

27. Mikrokanonische Gesamtheit und μ-Raum-Statistik. Die bisherigen Entwicklungen dieses Abschnitts gelten ganz allgemein. Wir wollen jetzt noch kurz den Spezialfall betrachten, daß die Hamilton-Funktion nach den Koordinaten der Einzelteilchen separierbar ist, sich also in der Form

$$H(\boldsymbol{q}, \boldsymbol{p}) = \sum_{i=1}^{N} H_i(\boldsymbol{q}_i, \boldsymbol{p}_i) \tag{27.1}$$

darstellen läßt. Physikalisch bedeutet diese Voraussetzung, daß die Wechselwirkung zwischen den Teilchen (deren Existenz vorausgesetzt werden muß!) in der Hamilton-Funktion als vernachlässigbare Störung behandelt wird.

Wenn die Voraussetzung (27.1) erfüllt ist, läßt sich, wie schon in Ziff. 16 erwähnt, nach BOLTZMANN[1] die statistische Theorie im μ-Raum entwickeln. Diese Methode hat den Vorzug formaler und begrifflicher Einfachheit; sie wird daher im Unterricht vielfach bevorzugt. Wir können hier auf die Einzelheiten nicht eingehen (vgl. die am Schluß des Artikels zitierten Lehrbücher) und wollen lediglich die Beziehung der μ-Raum-Statistik zur Ensemble-Theorie kurz erörtern. Dazu übertragen wir die Überlegungen, die sonst im μ-Raum durchgeführt werden, in den Γ-Raum.

Ein Mikrozustand, der einer Verteilung der unterscheidbar gedachten Teilchen auf die Zellen des μ-Raumes entspricht, wird im Γ-Raum durch ein endliches Volumenelement der mikrokanonischen „Energieschale" zwischen E und $E + \Delta E$ dargestellt. Die Wahrscheinlichkeit, das Originalsystem in einem bestimmten Mikrozustand zu finden, ist

$$W_{\text{mikro}} = \varrho \, \Delta\Omega. \tag{27.2}$$

Da innerhalb der mikrokanonischen Energieschale $\varrho = \text{const}$ ist, formuliert diese Gleichung das Grundpostulat der μ-Raum-Statistik, daß alle Mikrozustände die gleiche a priori-Wahrscheinlichkeit besitzen. Von einem gegebenen Mikrozustand kann man zu weiteren Mikrozuständen gelangen, indem man Teilchen mit verschiedenen Werten der Koordinaten und Impulse miteinander vertauscht. Die Gesamtheit aller durch derartige Operationen auseinander hervorgehenden Mikrozustände definiert einen Makrozustand. Das entsprechende Zustandsgebiet im Γ-Raum wird nach EHRENFEST [3] als Γ-Stern bezeichnet. Die Wahrscheinlichkeit eines Makrozustandes ist

$$W_{\text{makro}} = \varrho \, \frac{N!}{\Pi N_i!} \, \Delta\Omega, \tag{27.3}$$

wo N_i die Zahl der Moleküle in der i-ten Zelle des μ-Raumes bezeichnet. Das Volumen der gesamten Energieschale ist

$$\Omega = e^{\Phi} \, \Delta E, \tag{27.4}$$

das Volumen eines Γ-Sternes

$$\Omega_D = \frac{N!}{\Pi N_i!} \, \Delta\Omega, \tag{27.5}$$

---

[1] L. BOLTZMANN: Wien. Ber. **76**, 373 (1877).

und die Phasendichte
$$\varrho = \Omega^{-1} = e^{-\Phi} (\Delta E)^{-1}. \tag{27.6}$$

Die Gl. (27.3) kann daher geschrieben werden

$$W_{\mathrm{makro}} = \frac{\Omega_D}{\Omega} \tag{27.7}$$

mit
$$\sum \Omega_D = \Omega. \tag{27.8}$$

Der charakteristische Zug der μ-Raum-Statistik, der sie grundsätzlich von der Ensemble-Theorie unterscheidet, besteht nun darin, daß das Gleichgewicht mit einem bestimmten Makrozustand, der Ω_D zu einem Maximum macht, identifiziert wird. Die Gleichgewichtseigenschaften ergeben sich dann aus dem Ansatz

$$\delta \Omega_D = 0 \tag{27.9}$$

mit den Nebenbedingungen

$$\sum \delta N_i = 0, \qquad \sum \varepsilon_i \, \delta N_i = 0, \tag{27.10}$$

wo ε_i die Energie eines Moleküls in der i-ten Zelle des μ-Raumes ist. Daraus folgt mit Benutzung der Stirlingschen Formel unmittelbar das Maxwell-Boltzmannsche Energieverteilungsgesetz

$$\frac{N_i}{N} = \frac{e^{-\frac{\varepsilon_i}{kT}}}{\sum e^{-\frac{\varepsilon_i}{kT}}}, \tag{27.11}$$

das wir in Ziff. 28 mit Hilfe der Ensembletheorie ableiten werden. Für die Entropie erhält man

$$S = k \ln \Omega_{D\,\mathrm{max}}. \tag{27.12}$$

Diese Gleichung ist als *Boltzmannsches Entropiegesetz* bekannt.

Im Rahmen der Ensemble-Theorie läßt sich Gl. (27.12), wie die ganze Methode der μ-Raum-Statistik, als eine Näherung auffassen, deren Ergebnisse für die Gleichgewichtseigenschaften makroskopischer Systeme praktisch übereinstimmen mit denen der exakten Theorie. Für die mikrokanonische Gesamtheit ist nämlich, wenn man auf den Grenzübergang verzichtet,

$$S = k (\Phi + \ln \Delta E) = k \ln \Omega, \tag{27.13}$$

wo ΔE lediglich eine thermodynamisch bedeutungslose additive Konstante darstellt. Andererseits ist für sehr großes N

$$\Omega_{D\,\mathrm{max}}/\Omega \approx 1, \tag{27.14}$$

so daß die Gln. (27.12) und (27.13) praktisch identisch sind.

Die Methoden und Ergebnisse der μ-Raum-Statistik lassen sich daher ohne Schwierigkeit in die Ensemble-Theorie einbauen. Diese Tatsache darf aber nicht darüber hinwegtäuschen, daß zwischen den beiden Methoden ein tiefgehender Unterschied der Grundkonzepte besteht, den wir in Ziff. 6, 16, 17 ausführlich erörtert haben. Er drückt sich darin aus, daß in der Boltzmannschen Auffassung die Gl. (27.12) nicht eine Näherung, sondern exakt gültig ist. Für die Theorie der Schwankungen und der irreversiblen Vorgänge gewinnt dieser Unterschied entscheidende Bedeutung.

28. Das Maxwell-Boltzmannsche Energieverteilungsgesetz. Die in Ziff. 27 angedeutete Ableitung des Maxwell-Boltzmannschen Energieverteilungsgesetzes ist mathematisch recht anfechtbar und daher verschiedentlich (z.B. [5], [6]) scharf kritisiert worden. Sie macht auch nicht durchsichtig, daß Gl. (27.11) ein asymptotisches Grenzgesetz für $N \to \infty$ ist. Der erste einwandfreie Beweis ist von DARWIN und FOWLER[1] im Rahmen der Quantenstatistik geführt worden (vgl. Ziff. 49) [6], [7]. Er läßt sich ohne weiteres auf den klassischen Fall übertragen, wenn man die von BOLTZMANN eingeführte Zelleneinteilung des μ-Raumes beibehält, was allerdings nicht ganz unproblematisch erscheint. Die mathematisch strenge Ableitung mit Hilfe der mikrokanonischen Gesamtheit ist erst in neuerer Zeit von KHINCHIN [19] durchgeführt worden. Wir skizzieren im folgenden den Gedankengang der Ableitung, ohne auf die mathematischen Details einzugehen, für die auf KHINCHIN's Buch [19] verwiesen werden muß.

Wir betrachten ein abgeschlossenes System, das aus zwei Teilsystemen 1 und 2 in thermischer Berührung besteht. Es sei $f_1(E_1)$ eine Funktion der Energie des Teilsystems 1. Für den mikrokanonischen Mittelwert dieser Funktion gilt dann nach (20.6) und (22.2)

$$\bar{f}_1 = \mathrm{e}^{-\Phi} \iint f_1\, \delta(E^* - E)\, d\Omega_2\, d\Omega_1 \tag{28.1}$$

oder, da für konstantes E_1 $dE_2 = dE$ ist,

$$\bar{f} = \mathrm{e}^{-\Phi} \iint f_1\, \mathrm{e}^{\Phi_2}\, \delta(E^* - E)\, dE\, d\Omega_1. \tag{28.2}$$

Für $f_1 = 1$ folgt daraus

$$\mathrm{e}^{\Phi} = \int \mathrm{e}^{\Phi_1(E_1)}\, \mathrm{e}^{\Phi_2(E^* - E_1)}\, dE_1. \tag{28.3}$$

Die rechte Seite stellt ein Faltungsintegral dar, das wir symbolisch als Faltungsprodukt schreiben

$$\mathrm{e}^{\Phi} = \mathrm{e}^{\Phi_1} * \mathrm{e}^{\Phi_2}. \tag{28.4}$$

Diese Darstellung läßt sich für Zerlegung in beliebig viele Teilsysteme mit thermischer Kopplung verallgemeinern. Unter der Voraussetzung (27.1) können wir letzten Endes die Teilchen selbst als Teilsysteme in diesem Sinne ansehen. Es gilt daher, wenn für das i-te Teilchen die Ableitung seines Phasenvolumens nach der Energie mit $g_i(\varepsilon)$ bezeichnet wird

$$\mathrm{e}^{\Phi} = \overset{*}{\prod_i} g_i. \tag{28.5}$$

Wir zerlegen jetzt das System in zwei Teilsysteme derart, daß das Teilsystem 1 aus einem einzelnen Teilchen, das Teilsystem 2 aus den restlichen $N-1$ Teilchen besteht. Die Wahrscheinlichkeitsdichte der Energie ε ist dann nach (28.3)

$$w(\varepsilon) = g(\varepsilon)\, \frac{\mathrm{e}^{\Phi_2(E^* - \varepsilon)}}{\mathrm{e}^{\Phi(E^*)}}. \tag{28.6}$$

Für ein ideales Gas aus Massenpunkten kann man die expliziten Ausdrücke nach Ziff. 25 einsetzen und erhält dann

$$g(\varepsilon) = 4\pi m\, V\, (2m\,\varepsilon)^{\frac{1}{2}} \tag{28.7}$$

und

$$w(\varepsilon) = g(\varepsilon)\, \frac{\Gamma(\tfrac{3}{2} N)}{\Gamma(\tfrac{3}{2}N - \tfrac{3}{2})} \cdot \frac{(E^* - \varepsilon)^{\frac{3}{2}N - \frac{5}{2}}}{(2\pi m)^{\frac{3}{2}}\, V\, E^{* \frac{3}{2}N - 1}}. \tag{28.8}$$

Diese Gleichung, die zuerst von SCHLÜTER[2] abgeleitet wurde, stellt (für den angegebenen Spezialfall) das exakte Energieverteilungsgesetz der klassischen

[1] C. S. DARWIN u. R. H. FOWLER: Phil. Mag. **44**, 450, 823 (1922); **45**, 1 (1923).
[2] A. SCHLÜTER: Z. Naturforsch. **3**a, 350 (1948).

Statistik dar, das für Systeme von beliebig vielen Freiheitsgraden gültig ist. Die Ableitung des Maxwell-Boltzmannschen Gesetzes läuft nun hinaus auf das Problem, für die rechte Seite der Gl. (28.6) einen asymptotischen Ausdruck für $N \to \infty$ zu finden.

Der Grundgedanke der Khinchinschen Lösung besteht darin, das Problem durch eine geeignete Transformation auf eine Anwendung des zentralen Limes-Satzes der Wahrscheinlichkeitstheorie zu reduzieren. Wenn dies gelungen ist, kann man die Existenzbeweise einfach übernehmen und hat lediglich noch zu zeigen, daß die statistischen Funktionen die in der Wahrscheinlichkeitstheorie gemachten Voraussetzungen erfüllen. Da die Anwendung des Limes-Satzes Wahrscheinlichkeitsdichten von unabhängigen „Zufallsvariablen" (random variables) voraussetzt, ist sie nicht unmittelbar bei Gl. (28.6) möglich. Wir können aber das Problem leicht auf eine dazu geeignete Form bringen durch Anwendung des Faltungssatzes der Laplace-Transformation. Es sei $e^{\Phi^{(m)}}$ die Ableitung des Phasenvolumens nach der Energie für ein Teilsystem von m Teilchen (von KHINCHIN Strukturfunktion genannt). Die Laplace-Transformierte der Strukturfunktion sei

$$Q^{(m)} = \int\limits_0^\infty e^{-\alpha E^{(m}} \, e^{\Phi^{(m)}} \, dE^{(m)} \tag{28.9}$$

(erzeugende Funktion nach KHINCHIN).

Aus bekannten Eigenschaften der Laplace-Transformation folgen über $Q^{(m)}$ unmittelbar die folgenden Sätze

a) $Q^{(m)}(\alpha)$ ist eine positive und monoton abnehmende Funktion von α.

b) $Q^{(m)}(\alpha) \to \infty$ für $\alpha \to 0$.

c) Für $\alpha > 0$ besitzt $Q^{(m)}(\alpha)$ Ableitungen aller Ordnungen.

Für den Spezialfall $m = 1$ schreiben wir

$$Q^{(1)} \equiv f(\alpha) = \int\limits_0^\infty e^{-\alpha \varepsilon} g(\varepsilon) \, d\varepsilon. \tag{28.10}$$

Aus den Gln. (28.5) und (28.10) folgt dann in Verbindung mit dem Faltungssatz

$$Q^{(N)}(\alpha) = \prod_{i=1}^N f_i(\alpha), \tag{28.11}$$

und aus (28.3)

$$Q^{(N)}(\alpha) = f(\alpha) \, Q^{(N-1)}(\alpha). \tag{28.12}$$

Wir definieren nun noch eine neue Klasse von Funktionen durch die Gleichung

$$W^{(m)}(E^{(m)}) = \begin{cases} Q^{(m)-1} e^{-\alpha E^{(m)}} e^{\Phi^{(m)}} & \text{für } E^{(m)} \geqq 0, \\ 0 & \text{für } E^{(m)} \leqq 0. \end{cases} \tag{28.13}$$

Da für $\alpha > 0$

$$W^{(m)}(E^{(m)}) \geqq 0, \qquad \int W^{(m)}(E^{(m)}) \, dE^{(m)} = 1, \tag{28.14}$$

stellt $W^{(m)}$ für jedes $\alpha > 0$ eine Wahrscheinlichkeitsdichte dar. Aus Gl. (28.3) leitet man ab

$$W^{(N)}(E) = \int W_2(E - E_1) \, W_1(E_1) \, dE_1 \tag{28.15}$$

und daraus durch Induktion mit Benutzung von (28.10)

$$W^{(N)}(E) = \int \prod_{i=1}^{N-1} [W_i^{(1)}(\varepsilon) \, d\varepsilon] \, W_n^{(1)}(\varepsilon). \tag{28.16}$$

Die $W^{(m)}$, die nach KHINCHIN als konjugierte Verteilungsgesetze bezeichnet werden, sind somit Wahrscheinlichkeitsdichten für die Summen unabhängiger Zufallsveränderlicher; sie lassen sich aus den Strukturfunktionen unabhängiger Teilsysteme definieren. Da

$$e^{\Phi^{(m)}} = Q^{(m)} e^{\alpha E^{(m)}} W^{(m)} (E^{(m)}) \tag{28.17}$$

ist, lassen sich für ein vorher definiertes α aus $Q^{(N)}$ und $W^{(N)}$ die Funktion e^{Φ} und damit alle mikrokanonischen Mittelwerte berechnen. Ein asymptotisches Gesetz für $W^{(N)}$ ist daher auch ein asymptotisches Gesetz für e^{Φ}.

Der Mittelwert von $E^{(m)}$ ist

$$\overline{E^{(m)}} = \int E^{(m)} W^{(m)} dE^{(m)} = - \frac{d \ln Q^{(m)}}{d\alpha}, \tag{28.18}$$

das mittlere Schwankungsquadrat (Dispersion)

$$\overline{E^{(m)\,2}} - \overline{E^{(m)}}^{\,2} = \int (E^{(m)} - \overline{E^{(m)}})^2 W^{(m)} dE^{(m)} = \frac{d^2 \ln Q^{(m)}}{d\alpha^2}. \tag{28.19}$$

Für (28.18) gilt der wichtige Satz:

Für jedes gegebene $\overline{E^{(m)}} > 0$ hat die Gleichung

$$\frac{Q^{(m)\prime}(\alpha)}{Q(\alpha)} = \overline{E^{(m)}} \tag{28.20}$$

eine und nur eine positive Wurzel, oder

Für jede positive Zahl $\overline{E^{(m)}}$ gibt es eine und nur eine konjugierte Funktion $W^{(m)}$, für die $\overline{E^{(m)}}$ der durch (28.18) definierte Mittelwert ist.

Auf Grund des vorstehenden Satzes können wir die bisher offen gelassene Konstante α eindeutig fixieren durch die Festsetzung, daß α die positive Wurzel der Gleichung

$$-\frac{d \ln Q}{d\alpha} = E^* \tag{28.21}$$

ist.

Schließlich machen wir noch die folgenden (praktisch immer erfüllten) Voraussetzungen über die Strukturfunktionen $g(\varepsilon)$:

a) $g(\varepsilon)$ ist analytisch.

b) Es gibt eine Zahl r derart, daß $g(\varepsilon) = O(\varepsilon^r)$ wird für $\varepsilon \to \infty$.

Damit haben wir das Problem auf eine Form gebracht, die unmittelbar die Anwendung des zentralen Limes-Satzes der Wahrscheinlichkeitstheorie ermöglicht.

Der Limes-Satz besagt in der für unsere Zwecke geeigneten, von KHINCHIN [*19*] angegebenen Formulierung:

Es sei $x_1, x_2, \ldots$ eine Folge von unabhängigen Zufallsvariablen mit den Wahrscheinlichkeitsdichten $u_1(x), u_2(x), \ldots$ und den charakteristischen Funktionen $g_1(t), g_2(t), \ldots$, wo

$$g_i(t) = \int e^{itx} u_i(x) dx \qquad (i = 1, 2, \ldots) \tag{28.22}$$

ist. Ferner sei

$$\left. \begin{array}{l} \int x\, u_i(x)\, dx = a_i, \\[4pt] \int (x - a_i)^2\, u_i(x)\, dx = b_i, \\[4pt] \text{usw. bis zur 5. Ordnung} \end{array} \right\} \quad (i = 1, 2, \ldots). \tag{28.23}$$

Es sei

$$A_n = \sum_{i=1}^n a_n, \qquad B_n = \sum_{i=1}^n b_n, \tag{28.24}$$

und $U_n(x)$ bezeichne die Wahrscheinlichkeitsdichte für die Summe der ersten n Zufallsvariablen. Unter gewissen Voraussetzungen gilt dann für $n \to \infty$

$$U_n(x) = \frac{1}{(2\pi B_n)^{\frac{1}{2}}} e^{-\frac{(x-A_n)^2}{2 B_n}} + \begin{cases} O\left(\dfrac{1+|x-A_n|}{n^{\frac{3}{2}}}\right) & \text{für} \quad |x-A_n| < 2\ln^2 n, \\ O\left(\dfrac{1}{n}\right) & \text{für} \quad \text{alle } x. \end{cases} \tag{28.25}$$

Dabei sind A_n und B_n $O(n)$.

Man kann nun leicht zeigen, daß die Voraussetzungen dieses Satzes bei unserem Problem erfüllt sind und (28.25) unmittelbar den asymptotischen Ausdruck ür $W^{(N)}$ liefert, wenn wir setzen

$$A_n = -\frac{d\ln Q}{d\alpha} \equiv A, \qquad B_n = \frac{d^2\ln Q}{d\alpha} \equiv B. \tag{28.26}$$

Mit (28.17) erhalten wir dann als asymptotischen Ausdruck für $e^{\Phi(E)}$ die Gleichung

$$e^{\Phi(E)} = Q(\alpha)\, e^{\alpha E} \left[\frac{1}{(2\pi\, d^2\ln Q(\alpha)/d\alpha^2)^{\frac{1}{2}}} e^{-\frac{\left[E+\left(\frac{d\ln Q}{d\alpha}\right)\right]^2}{2\, d^2\ln Q/d\alpha^2}} + \right.$$
$$\left. + \begin{cases} O\left(\dfrac{1+|E-A_N|}{N^{\frac{3}{2}}}\right) & \text{für} \quad |E-A_N| < 2\ln^2 N \\ O\left(\dfrac{1}{N}\right) & \text{für} \quad \text{alle } E \end{cases} \right] \tag{28.27}$$

Bezeichnen wir mit β die positive Wurzel der Gl. (28.21) und mit Φ den Wert von $\Phi(E)$ auf der betrachteten mikrokanonischen Energiefläche, so erhalten wir

$$e^{\Phi} = Q(\beta)\, e^{\beta E^*} \left[\frac{1}{(2\pi B)^{\frac{1}{2}}} + O(N^{-\frac{3}{2}}) \right]. \tag{28.28}$$

Diese Gleichung ist die Grundformel, mit deren Hilfe wir sofort das Maxwell-Boltzmannsche Gesetz ableiten können.

Wir betrachten wieder zwei Teilsysteme und definieren zunächst

$$\bar{\varepsilon}_i = -\left(\frac{d\ln f_i(\alpha)}{d\alpha}\right)_{\alpha=\beta}, \qquad b_i = \left(\frac{d^2\ln f_i(\alpha)}{d\alpha^2}\right)_{\alpha=\beta}. \tag{28.29}$$

Ferner haben wir

$$Q_1 = \prod_{i=1}^{N_1} f_i(\beta), \qquad Q_2 = \prod_{i=N_1+1}^{N} f_i(\beta), \tag{28.30}$$

$$E^* = -\frac{d\ln Q}{d\beta} = \sum_{i=1}^{N} \bar{\varepsilon}_i, \tag{28.31}$$

$$\bar{E}_1 = -\frac{d\ln Q_1}{d\beta} = \sum_{i=1}^{N_1} \bar{\varepsilon}_i, \qquad \bar{E}_2 = -\frac{d\ln Q_2}{d\beta} = \sum_{i=N_1+1}^{N} \bar{\varepsilon}_i, \tag{28.32}$$

$$\bar{E}_1 + \bar{E}_2 = E^* \tag{28.33}$$

und entsprechend

$$B = \frac{d^2\ln Q}{d\beta^2} = \sum_{i=1}^{N} b_i, \tag{28.34}$$

$$B_1 = \frac{d^2\ln Q_1}{d\beta^2} = \sum_{i=1}^{N_1} b_i, \qquad B_2 = \frac{d^2\ln Q_2}{d\beta^2} = \sum_{i=N_1+1}^{N} b_i, \tag{28.35}$$

$$B_1 + B_2 = B. \tag{28.36}$$

Die Größen E^* und B sind somit von der Ordnung N.

Wir nehmen nun wieder an, daß das Teilsystem 1 aus einem Molekül, das Teilsystem 2 aus den restlichen $N-1$ Molekülen besteht. Für $N \to \infty$ wird dann

$$N_2 \approx N, \quad \bar{E}_2 \approx E^*, \quad B_2 \approx B. \tag{28.37}$$

Wir gehen nun aus von Gl. (28.6) und berechnen die dort auftretenden Größen mit Hilfe der asymptotischen Formeln (28.27) bzw. (28.28). Wir erhalten dann

$$e^{\Phi(E^*)} = \frac{Q\,e^{\beta E^*}}{(2\pi B)^{\frac{1}{2}}}\,[1 + o(1)] \tag{28.38}$$

$$e^{\Phi_2(E^*-\varepsilon)} = Q_2\,e^{\beta(E^*-\varepsilon)}\left[\frac{1}{(2\pi B_2)^{\frac{1}{2}}}\,e^{-\frac{(\varepsilon-\bar\varepsilon)^2}{2 B_2}} + O\left(\frac{1}{N}\right)\right]. \tag{28.39}$$

Setzen wir voraus

$$\varepsilon - \bar\varepsilon = O(N^{\frac{1}{4}}) \tag{28.40}$$

so wird aus (28.39)

$$e^{\Phi_2(E^*-\varepsilon)} = \frac{Q_2\,e^{\beta(E^*-\varepsilon)}}{(2\pi B_2)^{\frac{1}{2}}}\,[1 + o(1)]. \tag{28.41}$$

Aus (28.6), (28.38) und (28.41) folgt wegen (28.30) und (28.37)

$$w(\varepsilon) = \frac{g(\varepsilon)\,e^{-\beta\varepsilon}}{f(\beta)}\,[1 + o(1)] \quad (N \to \infty,\ \varepsilon - \bar\varepsilon = O(N^{\frac{1}{4}})). \tag{28.42}$$

Dies ist die exakte Form des Maxwell-Boltzmannschen Energieverteilungsgesetzes, die damit völlig streng bewiesen ist. Von der Formulierung als asymptotisches Grenzgesetz abgesehen, ist die Voraussetzung (28.40) bemerkenswert. Sie zeigt, daß die übliche Formulierung des Maxwell-Boltzmannschen Gesetzes nicht für beliebig große Abweichungen vom Mittelwert $\bar\varepsilon$ gilt. Für die Integration entstehen daraus keine Schwierigkeiten. Ist $\varphi(\boldsymbol{q}_1, \boldsymbol{p}_1)$ eine Funktion der Phasenkoordinaten eines Moleküls und setzen wir voraus

$$\varphi(\boldsymbol{q}_1, \boldsymbol{p}_1) = O(\varepsilon^k) \quad \text{für} \quad \varepsilon \to \infty, \tag{28.43}$$

wo k eine feste Zahl ist, so gilt, wie man ebenfalls streng beweisen kann [19], allgemein die Formel

$$\bar\varphi = \iint \varphi(\boldsymbol{q}_1, \boldsymbol{p}_1)\,\frac{e^{-\beta\varepsilon}}{f(\beta)}\,d\boldsymbol{q}_1\,d\boldsymbol{p}_1 + O\left(\frac{1}{N}\right). \tag{28.44}$$

Im Besonderen ist

$$\bar\varepsilon = \int \varepsilon\,\frac{g(\varepsilon)\,e^{-\beta\varepsilon}}{f(\beta)}\,d\varepsilon + O\left(\frac{1}{N}\right). \tag{28.45}$$

Wir haben jetzt noch den Parameter β physikalisch zu interpretieren. Man kann dazu von der Definitionsgleichung (28.21) ausgehen. Wir wollen hier einen anderen Weg einschlagen, um den Zusammenhang mit den früheren Betrachtungen herzustellen. Dazu entwickeln wir $\ln \Omega^*(E^* - \varepsilon)$ an der Stelle $\varepsilon = 0$ in eine Taylorsche Reihe[1]. Das ergibt

$$\ln \Omega^*(E^* - \varepsilon) = \ln \Omega^*(E^*) - \gamma\,\varepsilon + \frac{1}{2}\,\frac{d\gamma}{dE}\,\varepsilon^2 - \cdots \tag{28.46}$$

wo

$$\gamma = (e^{-\Phi}\,\Omega^*)^{-1} = (kT)^{-1} \tag{28.47}$$

ist. Daraus folgt

$$e^{\Phi(E^*-\varepsilon)} = e^{\Phi(E^*)}\,e^{-\gamma\varepsilon}\,[1 + o(1)] \quad (N \to \infty). \tag{28.48}[2]$$

Der Vergleich mit Gl. (28.41) zeigt, daß

$$\gamma = \beta = (kT)^{-1} \tag{28.49}$$

ist.

---

[1] Von hier bis zum Schluß dieser Ziffer beziehen sich die Funktionen Ω^* und Φ durchweg auf das Restsystem aus $N-1$ Teilchen. Der Index 2 ist der Einfachheit halber weggelassen.

[2] Dieser Formel ist schon von BOLTZMANN [Wien. Ber. **63**, 679 (1871)] gefunden worden.

V. Die kanonische Gesamtheit.

29. Definition und allgemeine Beziehungen. Die Methode der mikrokanonischen Gesamtheit ist, unbeschadet ihrer grundsätzlichen Bedeutung für die Theorie, für explizite Rechnungen ziemlich unbequem. Wenn man (wie es meistens der Fall ist) nur an den asymptotischen Resultaten interessiert ist, kann man sich von der Nebenbedingung $E = $ const freimachen durch Einführung einer (reellen oder komplexen) erzeugenden Funktion. Diese Methode, die wir bereits in Ziff. 28 benutzt haben, läuft darauf hinaus, daß man eine über den ganzen Phasenraum verteilte Gesamtheit benutzt, die asymptotisch die gleichen Mittelwerte liefert wie die mikrokanonische Gesamtheit. Wir wollen diese Gesamtheit jetzt systematisch untersuchen.

Wichtiger als die vorstehende formale Überlegung ist die Tatsache, daß die Theorie der mikrokanonischen Gesamtheit gewisse Randbedingungen voraussetzt, die in Wirklichkeit oft nicht erfüllt sind. Das Konzept des abgeschlossenen Systems wird durch jede Temperaturmessung und erst recht durch einen Thermostaten illusorisch gemacht. Der Formalismus der Thermodynamik ist von solchen Randbedingungen unabhängig. Die bisherigen statistischen Überlegungen lassen eine solche Verallgemeinerung nur unter sehr einschränkenden Voraussetzungen zu. Wenn wir unser System als kleines Teilsystem eines abgeschlossenen Gesamtsystems betrachten können, so lassen sich bei vernachlässigbarer Kopplung unmittelbar die Überlegungen von Ziff. 28 anwenden. Wenn man sich von diesen Voraussetzungen freimachen will, ist es notwendig, an dieser Stelle ein neues Axiom einzuführen. Da wir die allgemeine Formulierung und physikalische Begründung desselben erst in Teil C behandeln, gehen wir an dieser Stelle auf die verschiedenen Versuche, die Phasendichte der fraglichen Gesamtheit aus Postulaten abzuleiten[1,2] [*17*], nicht ein. Wir setzen daher vorläufig axiomatisch für die Phasendichte einer stationären Gesamtheit, die ein in bezug auf die Energie offenes System repräsentiert,

$$\varrho = e^{\frac{\psi - E}{\Theta}} \tag{29.1}$$

mit

$$e^{-\frac{\psi}{\Theta}} = \int\limits_{\text{Alle Phasen}} e^{-\frac{E}{\Theta}}\, d\Omega. \tag{29.2}$$

Eine in dieser Weise verteilte Gesamtheit bezeichnet man nach Gibbs [*2*] als *kanonische Gesamtheit*. Die rechte Seite der Gl. (29.2) wird häufig *Gibbssches Phasenintegral* genannt. Die Größe Θ heißt der *Verteilungsmodul*; sie ist, wie wir später (Ziff. 52, 54) ausführlich beweisen werden, ein Analogon der Temperatur. Es gilt

$$\Theta = kT. \tag{29.3}$$

Die Systeme der kanonischen Gesamtheit sind konservative Systeme, die sich unabhängig voneinander im Phasenraum bewegen. Für jedes derselben existiert das Energie-Integral; es bewegt sich daher auf einer Fläche konstanter Energie. In diesem Sinne kann man sagen, daß die kanonische Gesamtheit aus einer Vielzahl von mikrokanonischen Gesamtheiten aufgebaut ist. Es ist daher auch möglich, die für die mikrokanonische Gesamtheit definierten Größen über die kanonische Gesamtheit zu mitteln. Das durch die kanonische Gesamtheit abgebildete Originalsystem kann dagegen auch nicht näherungsweise als konservativ betrachtet werden. Für dieses System existiert kein Energie-Integral, und seine Hamilton-

---

[1] R. Eisenschitz: J. Chem. Phys. **18**, 858 (1950).
[2] A. Münster: Proc. Cambridge Phil. Soc. **46**, 319 (1950).

Funktion hat keine anschauliche Bedeutung. Es ist jedoch üblich, als Hamilton-Funktion des Systems die Energie als Funktion der generalisierten Koordinaten und Impulse zu bezeichnen, obwohl dieselbe keine Konstante der Bewegung ist. Es handelt sich dabei streng genommen, um die Hamilton-Funktion der Systeme der kanonischen Gesamtheit, die für jedes derselben eine Konstante der Bewegung ist.

Wir berechnen mit Hilfe der Gl. (29.1) für ein System von n Freiheitsgraden den Mittelwert der kinetischen Energie. Dafür gilt

$$\overline{E}_{\mathrm{kin}} = \int E_{\mathrm{kin}}\, e^{\frac{\psi - E}{\Theta}}\, d\Omega. \tag{29.4}$$

Mit Benutzung von Gl. (25.2) erhalten wir daraus

$$\overline{E}_{\mathrm{kin}} = \frac{n}{2}\,\Theta. \tag{29.5}$$

Die mittlere kinetische Energie pro Freiheitsgrad ist daher

$$\overline{\varepsilon} = \tfrac{1}{2}\Theta = \tfrac{1}{2}kT. \tag{29.6}$$

Wir erhalten somit wieder das Äquipartitionstheorem.

30. Thermodynamische Analoga. Wir betrachten nun, wie schon in Ziff. 23, die Energie als Funktion der äußeren Parameter x_j. Differenzieren wir die Gl. (29.2), so ergibt sich

$$e^{-\frac{\psi}{\Theta}}\left(-\frac{1}{\Theta}\,d\psi + \frac{\psi}{\Theta^2}\,d\Theta\right) = \frac{1}{\Theta^2}\,d\Theta\int E\,e^{-\frac{E}{\Theta}}\,d\Omega - \sum_j \frac{1}{\Theta}\,dx_j\int \frac{\partial E}{\partial x_j}\,e^{-\frac{E}{\Theta}}\,d\Omega. \tag{30.1}$$

Führen wir die durch Gl. (23.1) definierten generalisierten Kräfte und die kanonischen Mittelwerte gemäß Gl. (29.4) ein, so folgt

$$d\psi = \frac{\psi}{\Theta}\,d\Theta - \frac{\overline{E}}{\Theta}\,d\Theta - \sum_j \overline{X}_j\,dx_j. \tag{30.2}$$

Nach der Definition des Wahrscheinlichkeitsindex, Gl. (26.3), ist für die kanonische Gesamtheit

$$\eta = \frac{\psi - E}{\Theta} \tag{30.3}$$

und

$$\overline{\eta} = \frac{\psi - \overline{E}}{\Theta}. \tag{30.4}$$

Damit kann Gl. (30.2) geschrieben werden

$$d\psi = \overline{\eta}\,d\Theta - \sum_j \overline{X}_j\,d x_j. \tag{30.5}$$

Die vorstehenden Beziehungen vergleichen wir mit den thermodynamischen Formeln

$$-S = \frac{F - E}{T} \tag{30.6}$$

und

$$dF = -S\,dT - \sum_j X_j\,dx_j. \tag{30.7}$$

Unter den gemachten Voraussetzungen muß (bei Abwesenheit äußerer Felder) die mittlere Energie eines Systems der kanonischen Gesamtheit offenbar mit der thermodynamischen inneren Energie des Originalsystems identifiziert werden. Da Θ ein Analogon der Temperatur ist, zeigen die obigen Formeln, daß $-\bar{\eta}$ ein Analogon der Entropie ist (I. Gibbssches Entropie-Analogon). Die Größe ψ ist ein Analogon der freien Energie nach Helmholtz. Das mittlere Verhalten einer kanonischen Gesamtheit folgt somit den Gesetzen der Thermodynamik.

Die in Ziff. 25 besprochenen quantenstatistischen Korrekturen müssen naturgemäß auch in den Formeln für die kanonische Gesamtheit angebracht werden. An die Stelle des Gibbsschen Phasenintegrals tritt dann der Ausdruck

$$Q = \frac{1}{h^n \prod\limits_{i=1}^{m} N_i!} \int e^{-\frac{E}{kT}} d\Omega, \tag{30.8}$$

der als *Verteilungsfunktion* (partition function) des Gesamtsystems bezeichnet wird. Die freie Energie nach Helmholtz ist dann

$$F = -kT \ln Q. \tag{30.9}$$

Diese Formeln sind häufig der zweckmäßigste Ausgangspunkt für die Berechnung der thermodynamischen Eigenschaften konkreter Systeme. In Ziff. 60 werden wir die „halbklassische Näherung" der Gl. (30.8) aus der strengen Quantenstatistik ableiten.

Da die Integration über die Impulse stets ausführbar ist, kann die Verteilungsfunktion auch geschrieben werden (wenn wir uns auf Massenpunkte beschränken)

$$Q = \lambda^{-3N} \frac{Q_\tau}{N!} \tag{30.10}$$

mit

$$\lambda = \frac{h}{(2\pi m kT)^{\frac{1}{2}}}. \tag{30.11}$$

Das eigentliche Problem liegt somit in der Berechnung der Größe

$$Q_\tau = \int e^{-\frac{U}{kT}} d\mathbf{q}, \tag{30.12}$$

die als Konfigurationsintegral oder Verteilungsfunktion der potentiellen Energie bezeichnet wird. Der mit einem geeigneten Normierungsfaktor versehene Integrand stellt die Wahrscheinlichkeitsdichte einer speziellen (d.h. durch spezifizierte Moleküle definierten) Konfiguration des Systems dar. Man erhält

$$\varrho_{\text{spez}}^{(N)} = \frac{1}{\lambda^{3N} N! Q} e^{-\frac{U}{kT}}. \tag{30.13}$$

Diese Gleichung bildet den Ausgangspunkt für die Entwicklung der sog. molekularen Verteilungsfunktionen, die in dem Artikel von Münster in Bd. XIII dieses Handbuches behandelt werden.

31. Das thermodynamische Gleichgewicht. Die in Ziff. 30 gegebene Definition der Entropie ist, wie man leicht feststellt, wieder ein Spezialfall der allgemeinen Gl. (26.2). Nach der in Ziff. 26 benutzten Methode kann man zeigen, daß auch in diesem Falle die statistisch definierte Entropie für die stationäre (d.h. hier die kanonische) Verteilung einen Extremwert annimmt, wenn wir die Nebenbedingung

$$\int E \varrho \, d\Omega = \int E \varrho \, e^{\Delta\eta} d\Omega = \bar{E} \tag{31.1}$$

berücksichtigen.

Da jedoch, wie wir gesehen haben, die kanonische Verteilung unmittelbar mit der freien Energie nach HELMHOLTZ verknüpft ist, wollen wir hier die entsprechende Form der Gleichgewichtsbedingung

$$(\delta F)_{T,V,N} \gtreqless 0 \tag{31.2}$$

statistisch begründen [2]. Dazu müssen wir zeigen, daß die Größe $\bar\eta + \overline{E}/\Theta$ für die kanonische Verteilung einen kleineren Wert besitzt als für jede andere Verteilung mit dem Wahrscheinlichkeitsindex $(\psi - E)/\Theta + \Delta\eta$, wo $\Delta\eta$ wieder eine willkürliche Funktion der Phasenkoordinaten ist. Die zu beweisende Behauptung lautet also

$$\frac{\psi}{\Theta} < \int \left(\frac{\psi}{\Theta} + \Delta\eta\right) e^{\frac{\psi - E}{\Theta} + \Delta\eta}\, d\Omega, \tag{31.3}$$

wo

$$\int e^{\frac{\psi - E}{\Theta} + \Delta\eta}\, d\Omega = \int e^{\frac{\psi - E}{\Theta}}\, d\Omega = 1 \tag{31.4}$$

ist. Da ψ/Θ eine Konstante ist, kann (31.3) wegen der Bedingung (31.4) geschrieben werden

$$\int \Delta\eta\, e^{\frac{\psi - E}{\Theta} + \Delta\eta}\, d\Omega > 0. \tag{31.5}$$

Dieser Ausdruck läßt sich durch nochmalige Anwendung von (31.4) auf die Form bringen

$$\int (\Delta\eta\, e^{\Delta\eta} + 1 - e^{\Delta\eta})\, e^{\frac{\psi - E}{\Theta}}\, d\Omega > 0. \tag{31.6}$$

Da der Klammerausdruck (abgesehen von dem Fall $\Delta\eta = 0$) notwendig positiv ist (vgl. Ziff. 17), ist damit die Behauptung bewiesen.

32. Der Zusammenhang zwischen kanonischer und mikrokanonischer Gesamtheit. Der formale Zusammenhang zwischen kanonischer und mikrokanonischer Verteilung läßt sich am einfachsten übersehen, wenn wir die Gl. (29.2) schreiben

$$e^{-\frac{\psi}{\Theta}} = \int e^{-\frac{E}{\Theta} + \Phi}\, dE. \tag{32.1}$$

$e^{-\frac{\Theta}{\psi}}$ stellt somit die Laplace-Transformierte von e^{Φ} dar. An die Stelle der scharfen Energie E^* tritt dadurch eine Verteilung über alle Energiewerte, deren Parameter das Temperatur-Analogon Θ ist. In den thermodynamischen Gleichungen entspricht diesem Übergang die Einführung der Temperatur als unabhängige Variable an Stelle der Energie. Im Hinblick auf die Begründung einer einheitlichen Thermodynamik wollen wir untersuchen, wie die früher mit Hilfe der mikrokanonischen Gesamtheit abgeleiteten Analoga der thermodynamischen Größen mit denen zusammenhängen, die sich mit Hilfe der kanonischen Gesamtheit ergeben. Als Beispiel wählen wir die Analoga der Temperatur [2]. Es sei u eine beliebige Funktion der Energie, gegebenenfalls auch des Verteilungsmoduls und der äußeren Parameter. Dann ist

$$\bar u = \int_0^\infty u\, e^{\frac{\psi - E}{\Theta} + \Phi}\, dE. \tag{32.2}$$

Andererseits haben wir

$$\int_0^\infty \left(\frac{du}{dE} - \frac{u}{\Theta} + u\frac{d\Phi}{dE}\right) e^{\frac{\psi - E}{\Theta} + \Phi}\, dE = \left| u\, e^{\frac{\psi - E}{\Theta} + \Phi} \right|_0^\infty \tag{32.3}$$

Es folgt somit

$$\overline{\frac{du}{dE}} - \frac{\bar{u}}{\Theta} + \overline{u\,\frac{d\Phi}{dE}} = \left| u\,\mathrm{e}^{\frac{\psi-E}{\Theta}+\Phi}\right|_0^\infty. \tag{32.4}$$

Es sei nun $u = \mathrm{e}^{-\Phi}\,\Omega^*$. Die Größe auf der rechten Seite von (32.4) ist dann, bis auf einen konstanten Faktor, gleich $\mathrm{e}^{-\frac{E}{\Theta}}\Omega^*$. Schreiben wir nun $\Theta' = 2\Theta$, so gilt

$$\mathrm{e}^{-\frac{E}{\Theta}}\Omega^* = \mathrm{e}^{-\frac{E}{\Theta}}\int_0^E \mathrm{e}^{\Phi}\,dE \leqq \mathrm{e}^{-\frac{E}{\Theta'}}\int_0^E \mathrm{e}^{-\frac{E}{\Theta'}+\Phi}\,dE \leqq \mathrm{e}^{-\frac{E}{\Theta'}}\,\mathrm{e}^{-\frac{\psi'}{\Theta'}}, \tag{32.5}$$

wo ψ' der zu dem Verteilungsmodul Θ' gehörende Wert von ψ ist. Da für $E \to \infty$ die rechte Seite verschwindet, muß dies auch für die links stehende Größe gelten. Die auf der rechten Seite von (32.4) stehende Größe verschwindet somit an der oberen Grenze. Sie verschwindet aber auch an der unteren Grenze, denn es ist

$$\mathrm{e}^{-\frac{E}{\Theta}}\Omega^* \leqq \int_0^E \mathrm{e}^{-\frac{E}{\Theta}+\Phi}\,dE. \tag{32.6}$$

Es wird daher

$$1 - \overline{\mathrm{e}^{-\Phi}\Omega^*\frac{d\Phi}{dE}} - \overline{\mathrm{e}^{-\Phi}\frac{\Omega^*}{\Theta}} + \overline{\mathrm{e}^{-\Phi}\Omega^*\frac{d\Phi}{dE}} = 0 \tag{32.7}$$

oder

$$\overline{\mathrm{e}^{-\Phi}\Omega^*} = \Theta. \tag{32.8}$$

Das Temperatur-Analogon der knanonischen Gesamtheit Θ ist somit gleich dem kanonischen Mittelwert des Temperatur-Analogons der mikrokanonischen Gesamtheit $\mathrm{e}^{-\Phi}\Omega^*$. Für die Energie liegen die Verhältnisse ähnlich. Der scharfen Energie E^* der mikrokanonischen Gesamtheit entspricht eine mittlere Energie $\bar{E}$ in den mit Hilfe der kanonischen Gesamtheit gewonnenen thermodynamischen Formeln. Es ist daher evident, daß eine einheitliche, auf das Einzelsystem anwendbare Thermodynamik nur möglich ist, wenn die Schwankungen um die Mittelwerte vernachlässigt werden können. Wir wollen hier nur kurz die Schwankungen der Energie betrachten und verweisen für die allgemeine Behandlung des Problems auf Abschnitt C I (S. 350).

Für den kanonischen Mittelwert der Energie gilt definitionsgemäß

$$\bar{E} = \int E\,\mathrm{e}^{\frac{\psi-E}{\Theta}}\,d\Omega \tag{32.9}$$

und nach Gl. (30.4) und (30.5)

$$\bar{E} = \psi - \Theta\,\frac{\partial\psi}{\partial\Theta}. \tag{32.10}$$

Durch Differentiation von (32.9) nach Θ erhält man mit Benutzung von (32.10) für das mittlere Schwankungsquadrat

$$\overline{(E - \bar{E})^2} = \Theta^2\,\frac{\partial\bar{E}}{\partial\Theta}. \tag{32.11}$$

Für das relative mittlere Schwankungsquadrat ergibt sich mit Benutzung des Äquipartitionstheorems

$$\frac{\overline{(E - \bar{E})^2}}{\bar{E}_{\mathrm{kin}}^2} = \frac{2}{n}\,\frac{\Theta}{\bar{E}_{\mathrm{kin}}}\,\frac{\partial\bar{E}}{\partial\Theta}. \tag{32.12}$$

Die Gleichung zeigt, daß das relative mittlere Schwankungsquadrat der Energie mit wachsender Größe des Systems wie $1/n$ gegen Null geht[1]. Eine kanonische

---

[1] Dabei ist vorausgesetzt, daß $\partial\bar{E}/\partial\Theta$ endlich ist, was im allgemeinen zutrifft. Den Fall, daß diese Voraussetzung nicht erfüllt ist, behandeln wir in Abschnitt C II, S. 371.

Gesamtheit von makroskopischen Systemen verhält sich daher gegenüber thermodynamischen Messungen wie eine Gesamtheit von Systemen der gleichen Energie $\bar{E}$, d.h. wie eine mikrokanonische Gesamtheit. Beide Beschreibungen werden dann äquivalent, womit die Grundlage einer einheitlichen Thermodynamik gegeben ist. Die Gesetze der Thermodynamik stellen somit im strengen Sinne Grenzgesetze für unendlich große Systeme dar; in Teil C werden wir diese Frage ausführlich behandeln. Praktisch, d.h. im Hinblick auf die Meßgenauigkeit ist die erwähnte Voraussetzung bereits bei den makroskopischen Systemen, mit denen es die Thermodynamik zu tun hat, erfüllt. Dagegen muß bei Systemen von submikroskopischen Dimensionen sorgfältig geprüft werden, inwieweit die Anwendung thermodynamischer Begriffe noch einen Sinn hat.

B. Quantenstatistik.

I. Allgemeine Sätze über quantenstatistische Gesamtheiten.

33. Quantenmechanik und Quantenstatistik. Im Teil A haben wir gesehen, daß die klassische statistische Mechanik eine Begründung der allgemeinen Gleichungen der Thermodynamik ermöglicht. In dem dadurch gegebenen Rahmen treten aber verschiedene schwerwiegende Widersprüche zur experimentellen Erfahrung auf, von denen wir einige schon früher erwähnt haben. Die wichtigsten derselben betreffen

a) die Erklärung des Gibbsschen Paradoxons (Ziff. 25);

b) die Temperaturabhängigkeit der spezifischen Wärme von Gasen, insbesondere von Gasen aus zweiatomigen Molekülen;

c) die Temperaturabhängigkeit der spezifischen Wärme von Kristallen;

d) die Massenabhängigkeit des zweiten Virialkoeffizienten;

e) die Berechnung der Dampfdruckkonstanten und chemischen Konstanten.

Bei den unter b) und c) genannten Problemen handelt es sich um direkte Folgerungen aus dem Äquipartitionstheorem (Ziff. 21). Wegen der zentralen Stellung dieses Theorems muß hier ein Widerspruch zur experimentellen Erfahrung notwendig die ganze Theorie in Frage stellen. Dieser schon lange bekannte Sachverhalt ist zweifel os in erster Linie verantwortlich für die innere Unsicherheit, welche die Entwicklung der klassischen statistischen Mechanik charakterisiert. GIBBS [2] hat bekanntlich aus diesem Grunde jede Verwendung von Molekülmodellen (die für explizite Berechnungen unentbehrlich ist) abgelehnt und sich auf allgemeine Untersuchungen beschränkt. Lord KELVIN[1] ist durch diese Schwierigkeiten veranlaßt worden, ernstlich an den Grundlagen der Theorie zu zweifeln. P. und T. EHRENFEST [3] kommen zu dem Schluß, daß die in der klassischen Theorie verwendeten Gesamtheiten (im Hinblick auf die Begründung durch die Ergodenhypothese ergodische Gesamtheiten genannt) aufzugeben sind.

Die Diskussion dieser Schwierigkeiten für das spezielle Problem der Hohlraumstrahlung hat bekanntlich mit der Entdeckung des Wirkungsquantums durch PLANCK[2] die Entwicklung ausgelöst, die schließlich zur Beseitigung aller oben erwähnten Widersprüche zwischen Theorie und Experiment geführt hat. Wir können diese sich über mehr als dreißig Jahre erstreckende Entwicklung hier nicht im einzelnen verfolgen. Ihre Ergebnisse bilden in der Form, in der sie

---

[1] Lord KELVIN: Vorlesungen über Molekularmechanik, deutsch von B. WEINSTEIN, S. 419f. Leipzig 1909.
[2] Vgl. die von M. PLANCK selbst gegebene Darstellung dieser Entdeckung. Naturwiss. **31**, 153 (1943).

sich heute darstellen, den Inhalt dieses Abschnittes. Wir wollen jedoch in dieser Ziffer noch einige Bemerkungen machen, die sich auf die allgemeine Charakterisierung der Quantenstatistik und ihr Verhältnis zur klassischen statistischen Mechanik beziehen.

Das vom Standpunkt der älteren Betrachtungen vielleicht überraschendste Ergebnis der erwähnten Entwicklung liegt in der Tatsache, daß der Unterschied zwischen Quantenstatistik und klassischer statistischer Mechanik sich auf die einfache Formel bringen läßt: An die Stelle der Hamiltonschen Mechanik tritt als Grundlage die Quantenmechanik. Damit ist zunächst gesagt, daß das Verhältnis zwischen Mechanik und statistischer Mechanik in der Quantentheorie grundsätzlich das gleiche ist wie in der klassischen Theorie. Darüber hinaus kann der formale Apparat der klassischen Statistik, insbesondere die Definition der statistischen Gesamtheiten und die Ableitungen der thermodynamischen Analoga, im wesentlichen einfach übernommen werden. Lediglich dort, wo die spezielle Form der Hamiltonschen Mechanik in Erscheinung tritt, bedingt die Einführung der Quantenmechanik gewisse Modifikationen. Unter Voraussetzungen, die bei den meisten Anwendungen erfüllt sind (näheres Ziff. 60), bleibt von diesen Modifikationen nur eine geringfügige Korrektur an der klassischen statistischen Mechanik übrig, die wir bereits in Ziff. 25 und 30 eingeführt haben. Die Arbeiten von Boltzmann und Gibbs besitzen daher auch für die Quantenstatistik grundlegende Bedeutung[1]. Wir können uns also in vielem jetzt kürzer fassen und unser Augenmerk in erster Linie auf die charakteristischen Züge der Quantenstatistik richten.

Die Quantenmechanik ist selbst eine statistische Theorie; ihre Aussagen beziehen sich auf virtuelle Gesamtheiten. Es liegt daher die Frage nahe, inwiefern hier noch Raum ist für eine eigene statistische Mechanik. Die Antwort darauf ergibt sich aus der Tatsache, daß eine quantenmechanische Gesamtheit grundsätzlich verschieden ist von einer klassischen statistischen Gesamtheit. Die letztere beschreibt ein Originalsystem, dessen mechanischer Zustand unvollständig bekannt ist, aber im Prinzip vollständig bekannt sein könnte (vgl. Ziff. 12). Die Systeme der Gesamtheit befinden sich daher in vollständig definierten mechanischen Zuständen; es wird angenommen, daß das Originalsystem sich (zum wenigstens näherungsweise) jeweils in einem dieser Zustände befindet; man weiß nur nicht, in welchem. Eine derartige Auffassung ist für die quantenmechanische Gesamtheit nicht durchführbar [vgl. z.B.[2]]. Die Ψ-Funktion der Quantenmechanik repräsentiert die maximale Kenntnis, die für das Originalsystem erreicht werden kann. Sie ist daher identisch mit der Beschreibung des quantenmechanischen Zustandes des Originalsystems. Bei verschiedenen Ψ-Funktionen befindet sich das System in verschiedenen Zuständen, bei gleicher Ψ-Funktion befindet es sich im gleichen Zustand. Die Aussage der Ψ-Funktion bezieht sich zwar auf ein Wahrscheinlichkeitsaggregat, die quantenmechanische Gesamtheit. Es ist aber nicht möglich, den Systemen dieser Gesamtheit einen über die Aussage der Ψ-Funktion hinaus definierten mechanischen Zustand zuzuschreiben, den wir lediglich nicht kennen. Jede derartige Annahme führt, wie man an einfachen Beispielen leicht zeigen kann[2], auf Widersprüche mit anderen Behauptungen der Quantenmechanik. Für alle Systeme der quantenmechanischen Gesamtheit muß daher die gleiche Ψ-Funktion und damit der gleiche quantenmechanische Zustand angenommen werden. Verglichen mit der klassischen Gesamtheit, führt die quantenmechanische Gesamtheit nur ein

---

[1] Im Hinblick auf die Gibbssche Statistik ist dies wohl zuerst von M. Delbrück und G. Molière [*29*] ausführlich begründet worden.

[2] E. Schrödinger: Naturwiss. **23**, 807, 823, 844 (1935).

Schattendasein. Sie ist nicht mehr und nicht weniger als ein „Katalog der Erwartung"[1]. Mit diesem Sachverhalt hängt ein zweiter Unterschied gegenüber der klassischen statistischen Mechanik zusammen. Die Ψ-Funktion, welche die quantenmechanische Gesamtheit beschreibt, ist keine Wahrscheinlichkeitsdichte, sondern eine Wahrscheinlichkeitsamplitude. Die Wahrscheinlichkeitsdichte, etwa für den Koordinatensatz q ist dagegen

$$W(\boldsymbol{q}, t) = \Psi^* \Psi. \tag{33.1}$$

Daraus ergibt sich die Möglichkeit einer Interferenz der Wahrscheinlichkeiten, die der klassischen Theorie völlig fremd ist.

Bisher haben wir die Ψ-Funktion und damit maximale Kenntnis des Originalsystems vorausgesetzt. Man spricht dann nach VON NEUMANN[2] von einer *reinen Gesamtheit*. Dieselbe kann dadurch charakterisiert werden, daß sie sich nicht durch Vermischen zweier anderer, von ihr verschiedener Gesamtheiten erzeugen läßt. Wenn aber die maximale Kenntnis des Originalsystems nicht verfügbar ist, wird dasselbe durch eine Gesamtheit dargestellt, deren Systeme verschiedene Ψ-Funktionen und damit verschiedene quantenmechanische Zustände besitzen. Eine solche *gemischte Gesamtheit*[2] zeigt eine viel weitgehendere Analogie zur klassischen Gesamtheit. Diese gemischten Gesamtheiten sind nun der eigentliche Gegenstand der Quantenstatistik; daraus erklärt sich der schon erwähnte Parallelismus mit der klassischen statistischen Mechanik. Die Quantenmechanik ermöglicht es jedoch, den reinen und den gemischten Fall durch den gleichen Formalismus darzustellen, so daß hier Mechanik und statistische Mechanik viel enger verknüpft erscheinen als es in der klassischen Theorie der Fall ist. Dabei ergibt sich noch ein wesentlicher Unterschied gegenüber der klassischen Theorie, der wegen seiner grundsätzlichen Bedeutung erwähnt werden muß. In der klassischen Mechanik folgt aus der maximalen Kenntnis des Gesamtsystems ohne weiteres auch maximale Kenntnis für die Teilsysteme. Dieser Satz gilt nicht mehr in der Quantenmechanik. Es kann daher vorkommen, daß das Gesamtsystem durch eine reine Gesamtheit, ein Teilsystem durch eine gemischte Gesamtheit dargestellt wird. In zugespitzten Fällen führt dieser Sachverhalt zu Antinomien, die als Paradoxon von EINSTEIN-PODOLSKY-ROSEN[3] bekannt sind. Eine Diskussion dieser Frage geht über den Rahmen dieses Artikels hinaus; wir begnügen uns daher mit dem Hinweis auf einige Literaturstellen[4,5]. Wir erwähnen diese Schwierigkeiten hier, weil sie zeigen, daß die Quantenmechanik, ungeachtet aller Erfolge, nicht eine formal abgeschlossene Theorie ist wie die Hamiltonsche Mechanik. Diese Tatsache macht sich in gewissem Ausmaß auch in der Quantenstatistik noch bemerkbar.

Wir machen schließlich noch eine Bemerkung über die Rolle des Modells in der Quantenmechanik, die für das Verständnis der Quantenstatistik von Bedeutung ist. Bekanntlich sind viele Aussagen der Quantenmechanik an einem Modell im Sinne der klassischen Mechanik völlig unanschaulich. Trotzdem ist ein solches Modell auch in der Quantenmechanik unentbehrlich. Durch das Modell wird nämlich festgelegt, welche Größen für das betrachtete System als prinzipiell meßbar gelten sollen. Diese Wahl des Modells findet ihren Ausdruck in der expliziten Formulierung der Schrödinger-Gleichung. Auch die Ergebnisse der Quantenstatistik sind daher notwendig von der Wahl des Modells abhängig und

---

[1] Siehe Fußnote 2, S. 268.
[2] J. v. NEUMANN: Göttinger Nachr. **1927**, 245, 273.
[3] A. EINSTEIN, B. PODOLSKY u. N. ROSEN: Phys. Rev. **47**, 777 (1935).
[4] E. SCHRÖDINGER: Naturwiss. **23**, 807, 823, 844 (1935).
[5] N. BOHR: Phys. Rev. **48**, 696 (1935).

besitzen in diesem Sinne ebensowenig einen „absoluten" Charakter wie die der klassischen Statistik.

Im folgenden beschränken wir uns auf die Formulierung der Quantenstatistik, welche auf der nicht-relativistischen Quantenmechanik des Partikelbildes beruht. Wir setzen also Gültigkeit der zeitabhängigen Schrödinger-Gleichung

$$\mathsf{H}\,\Psi + \frac{h}{2\pi i}\,\frac{\partial \Psi}{\partial t} = 0 \tag{33.2}$$

voraus und werden den Formalismus der zweiten Quantelung nicht in Betracht ziehen.

34. Die Dichtematrix. Wir betrachten zunächst eine reine Gesamtheit, deren Systeme sich alle im gleichen quantenmechanischen Zustand befinden. Die Gesamtheit ist dann vollständig durch die Ψ-Funktion charakterisiert. Unter ziemlich allgemeinen Voraussetzungen läßt sich die Ψ-Funktion nach den Funktionen eines vollständigen normierten Orthogonalsystems ψ_n entwickeln. Wir haben dann

$$\Psi(\boldsymbol{q}, t) = \sum_n a_n(t)\,\psi_n(\boldsymbol{q}) \tag{34.1}$$

mit

$$a_n(t) = \int \psi_n^*(\boldsymbol{q})\,\Psi(\boldsymbol{q}, t)\,d\boldsymbol{q}, \tag{34.2}$$

wo ψ_n^* die zu ψ_n konjugiert komplexe Funktion ist. Die $a_n(t)$ stellen ebenfalls Wahrscheinlichkeitsamplituden dar. Es ist (wenn wir den Index n jetzt als Argument schreiben)

$$W(n, t) = a^*(n, t)\,a(n, t) \tag{34.3}$$

die Wahrscheinlichkeit, das System in dem durch den Index n charakterisierten Zustand zu finden. Sind die ψ_n Eigenfunktionen des Operators F, so gibt (34.3) die Wahrscheinlichkeit, den Eigenwert F_n dieses Operators zu messen[1]. Es ist, wie unmittelbar aus der Normierung der Ψ-Funktion folgt,

$$\sum_n a^*(n, t)\,a(n, t) = 1. \tag{34.4}$$

Es sei nun F ein hermitischer Operator. Die Elemente der Matrix, welche diesen Operator im System der ψ_n darstellt, seien

$$F_{mn} = \int \psi^*(m, \boldsymbol{q})\,\mathsf{F}\psi(n, \boldsymbol{q})\,d\boldsymbol{q}. \tag{34.5}$$

Diese Matrix ist ebenfalls hermitisch; es gilt also

$$F_{mn} = F_{nm}^*. \tag{34.6}$$

Der zur Wahrscheinlichkeitsamplitude $a(m, t)$ gehörige transformierte Operator $\mathsf{F}^{(a)}$ ist dann definiert durch die Gleichung

$$\mathsf{F}^{(a)}\,a(m, t) = \sum_n F_{mn}\,a(n, t). \tag{34.7}$$

In dieser „allgemeinen Sprache" ist der Erwartungswert der Größe F

$$\overline{F} = \sum_m a^*(m, t)\,\mathsf{F}^{(a)}\,a(m, t). \tag{34.8}$$

---

[1] Dabei ist vorausgesetzt, daß der Eigenwert F_n nicht entartet ist. Die Verallgemeinerung für den Fall der Entartung liegt auf der Hand.

Für die Wahrscheinlichkeitsamplituden $a(m, t)$ gilt die transformierte Schrödinger-Gleichung

$$\mathsf{H}^{(a)} a(m, t) + \frac{h}{2\pi i} \frac{\partial a(m, t)}{\partial t} = 0, \tag{34.9}$$

wo $\mathsf{H}^{(a)}$ der gemäß Gl. (34.7) definierte transformierte Hamilton-Operator ist.

Die Gl. (34.1) stellt eine Zerlegung des Vektors $\Psi(\boldsymbol{q}, t)$ nach den Grundvektoren $\psi(n, \boldsymbol{q})$ im Hilbert-Raum dar. Wir wollen untersuchen, wie sich die Komponenten transformieren, wenn wir von der Basis ψ_n zu einer Basis φ_l übergehen, der die Wahrscheinlichkeitsamplituden $b(l, \boldsymbol{q})$ entsprechen mögen. Die Grundvektoren sind verknüpft durch die Gleichungen

$$\varphi_l = \sum_n \psi_n U_{nl}, \qquad \psi_n = \sum_l \varphi_l U_{ln}^{-1}. \tag{34.10}$$

Dabei sind die Elemente der Transformationsmatrizen

$$U_{ml} = \int \psi_m^* \varphi_l \, d\boldsymbol{q}, \qquad U_{kn}^{-1} = U_{nk}^* = \int \varphi_k^* \psi_n \, d\boldsymbol{q}, \tag{34.11}$$

und es ist

$$\sum_n U_{nl}^* U_{nk} = \delta_{lk}, \qquad \sum_l U_{nl}^* U_{ml} = \delta_{nm}. \tag{34.12}$$

Bezeichnen wir nun den Vektor mit den Komponenten $a(n, t)$ mit Ψ_a, den Vektor mit den Komponenten $b(l, t)$ mit Ψ_b, so lauten die Transformationsgleichungen

$$\begin{aligned} \Psi_b &= \mathsf{U}^{-1} \Psi_a = \mathsf{U}^\dagger \Psi_a. \\ \Psi_a &= \mathsf{U} \Psi_b = (\mathsf{U}^{-1})^\dagger \Psi_b, \end{aligned} \right\} \tag{34.13}$$

wo U die durch (34.11) und (34.12) definierte unitäre Matrix und $\mathsf{U}^\dagger$ ihre Adjungierte ist. Für die Matrix (34.5) lautet die Transformation von der Basis ψ_n auf die Basis φ_l

$$\begin{aligned} \mathsf{F}_b &= \mathsf{U}^{-1} \mathsf{F}_a \mathsf{U}, \\ \mathsf{F}_a &= \mathsf{U} \mathsf{F}_b \mathsf{U}^{-1}. \end{aligned} \right\} \tag{34.14}$$

Wir betrachten schließlich noch die zeitliche Änderung des Erwartungswertes einer Größe F, deren Operator $\mathsf{F}(\boldsymbol{q}, \boldsymbol{p})$ nicht explizit von der Zeit abhängt. Mit Hilfe der zeitabhängigen Schrödinger-Gleichung (33.2) findet man

$$\frac{d\bar{F}}{dt} = \frac{2\pi}{ih} \int \Psi^* [\mathsf{F}, \mathsf{H}] \Psi \, d\boldsymbol{q} = \frac{2\pi}{ih} \overline{[\mathsf{F}, \mathsf{H}]}, \tag{34.15}$$

wo

$$[\mathsf{F}, \mathsf{H}] = \mathsf{F}\mathsf{H} - \mathsf{H}\mathsf{F} \tag{34.16}$$

der Kommutator der Operatoren F und H ist. Ist F der Hamilton-Operator, so folgt aus (34.15)

$$\frac{d\bar{H}}{dt} = 0. \tag{34.17}$$

Für ein System, dessen Hamilton-Operator nicht explizit von der Zeit abhängt (konservatives System) ist somit der Erwartungswert der Energie zeitlich konstant (Energiesatz der Quantenmechanik). Entsprechend definiert man allgemein als eine Konstante der Bewegung eine meßbare Größe, deren Operator mit dem Hamilton-Operator vertauschbar ist.

Wir gehen nun zur gemischten Gesamtheit über, deren ν Systeme sich nicht mehr im gleichen quantenmechanischen Zustand befinden. Dieselbe läßt sich nach VON NEUMANN[1] durch eine Matrix ρ beschreiben, die das Analogon der

---

[1] J. v. NEUMANN: Göttinger Nachr. **1927**, 245, 273.

klassischen Phasendichte darstellt und daher als *Dichtematrix* bezeichnet wird. In der allgemeinen Sprache der Transformationstheorie sind die Elemente der Dichtematrix definiert durch die Gleichungen

$$\varrho_{nm} = \frac{1}{\nu} \sum_{i=1}^{\nu} a_i^*(m,t)\, a_i(n,t) = \overline{a_m^* a_n}. \tag{34.18}$$

Die Dichtematrix ist hermitisch; sie kann endlich, aber auch unendlich sein. Aus der Definition folgt in Verbindung mit (34.4) die Normierung

$$\sum_n \overline{a_n^* a_n} = \operatorname{spur} \rho = 1. \tag{34.19}$$

In dieser Form hängt die Dichtematrix von der Wahl des Orthogonalsystems ψ_n ab. Es ist

$$W(n) = \overline{a_n^* a_n} \tag{34.20}$$

die Wahrscheinlichkeit, ein willkürlich aus der Gesamtheit herausgegriffenes System in dem durch die Eigenfunktion $\psi(n, \boldsymbol{q})$ charakterisierten Eigenzustand zu finden. Ist $\psi(n, \boldsymbol{q})$ eine Eigenfunktion des Operators F, so gibt (34.20) die Wahrscheinlichkeit, an einem willkürlich herausgegriffenen System den Eigenwert F_n zu messen. Für den quantenstatistischen Erwartungswert $\overline{F}$ ergibt sich aus (34.7), (34.8) und (34.18)

$$\overline{F} = \sum_m \sum_n F_{mn} \overline{a_m^* a_n} \tag{34.21}$$

oder

$$\overline{F} = \operatorname{spur} \mathsf{F}\rho = \operatorname{spur} \rho\,\mathsf{F}. \tag{34.22}$$

Die Gln. (34.19) und (34.22) zeigen, daß der Integration der Phasendichte über den Phasenraum in der klassischen Statistik die Bildung der Spur der analogen Matrix in der Quantenstatistik entspricht.

Der Übergang von der Basis ψ_n mit der Wahrscheinlichkeitsamplitude $a(n,t)$ zu einer Basis φ_l mit der Wahrscheinlichkeitsamplitude $b(l,t)$ wird durch die Transformationsgleichungen

$$\rho_b = \mathsf{U}^{-1} \rho_a \mathsf{U} \tag{34.23}$$

und

$$\mathsf{F}_b \rho_b = \mathsf{U}^{-1} \mathsf{F}_a \rho_a \mathsf{U} \tag{34.24}$$

dargestellt, wo U die durch (34.11) und (34.12) definierte unitäre Matrix ist. Da die Spur einer Matrix invariant gegen unitäre Transformationen ist, sind die Gln. (34.19) und (34.22) unabhängig von der Wahl des Orthogonalsystems. Dieser Satz entspricht der Invarianz der Phasendichte gegen kanonische Transformationen in der klassischen Theorie.

Eine hermitische Matrix läßt sich stets durch eine unitäre Transformation auf Diagonalform bringen. Da nach Gl. (34.20) die Elemente der Hauptdiagonalen der Dichtematrix stets positiv sind und ihre Summe notwendig gleich Eins ist, müssen für die Eigenwerte der Dichtematrix, die wir mit ϱ_d bezeichnen, die beiden Relationen gelten

$$\varrho_d \geqq 0, \quad \sum_d \varrho_d = 1. \tag{34.25}$$

Zu einer oft nützlichen Darstellung der Dichtematrix gelangt man, wenn man als Basis die n-dimensionalen Diracschen δ-Funktionen (Koordinaten-Eigenfunktionen) wählt. Wir haben dann

$$\varphi_l = \delta(\boldsymbol{q} - \boldsymbol{q}_l), \quad \varphi_k^* = \delta(\boldsymbol{q} - \boldsymbol{q}_k). \tag{34.26}$$

Mit (34.11) ergibt sich für die Elemente der Transformationsmatrix

$$\left.\begin{array}{l} U_{ml} = \int \psi_m^*(\boldsymbol{q})\, \delta(\boldsymbol{q} - \boldsymbol{q}_l)\, d\boldsymbol{q} = \psi_m^*(\boldsymbol{q}_l), \\[4pt] U_{nk}^* = \int \psi_n(\boldsymbol{q})\, \delta(\boldsymbol{q} - \boldsymbol{q}_k)\, d\boldsymbol{q} = \psi_n(\boldsymbol{q}_k). \end{array}\right\} \tag{34.27}$$

Die Größen $\psi_m^*(\boldsymbol{q}_l)$ und $\psi_n(\boldsymbol{q}_k)$ sind die Werte der Eigenfunktionen ψ_n für die speziellen Koordinatensätze $\boldsymbol{q}_l$ und $\boldsymbol{q}_k$. Bezeichnen wir die Elemente der transformierten Matrix mit $\varrho(\boldsymbol{q}, \boldsymbol{q}')$ (wo jetzt $\boldsymbol{q}$ für $\boldsymbol{q}_l$, $\boldsymbol{q}'$ für $\boldsymbol{q}_k$ steht), so folgt aus (34.23) und (34.27)

$$\varrho(\boldsymbol{q}, \boldsymbol{q}') = \sum_m \sum_n \varrho_{nm}\psi_n(\boldsymbol{q})\, \psi_m^*(\boldsymbol{q}'). \tag{34.28}$$

Mit (34.1) und (34.18) erhalten wir schließlich

$$\varrho(\boldsymbol{q}, \boldsymbol{q}') = \overline{\Psi^*(\boldsymbol{q}, t)\, \Psi(\boldsymbol{q}', t)}. \tag{34.29}$$

In dieser Darstellung ist die Dichtematrix tatsächlich der Kern eines Integraloperators. Die Normierungsrelation (34.19) lautet daher jetzt

$$\int \varrho(\boldsymbol{q}, \boldsymbol{q})\, d\boldsymbol{q} = 1. \tag{34.30}$$

Die Bedeutung der Darstellung (34.29) liegt darin, daß sie, wie man an Hand der Gl. (34.28) leicht verifiziert, unabhängig von der Wahl des Orthogonalsystems ist.

Die Dichtematrix kann als Darstellung eines Dichteoperators ρ aufgefaßt werden. Wir haben dann

$$\varrho_{nm} = \int \psi_n^*(\boldsymbol{q})\, \rho\, \psi_m(\boldsymbol{q})\, d\boldsymbol{q} \tag{34.31}$$

und

$$\varrho(\boldsymbol{q}, \boldsymbol{q}') = \sum_n \psi_n^*(\boldsymbol{q}')\, \rho\, \psi_n(\boldsymbol{q}). \tag{34.32}$$

Der Ausdruck (34.32) ist wieder von der Wahl des Orthogonalsystems unabhängig.

Wenn sich die Gesamtheit in einem reinen Zustand befindet, werden die Elemente der Dichtematrix

$$\varrho_{nm} = a_m^*\, a_n. \tag{34.33}$$

Die Gln. (34.19), (34.22) bis (34.24) gelten naturgemäß auch für diesen Fall. Wir haben aber jetzt noch den speziellen Zusammenhang

$$\sum_k \varrho_{nk}\varrho_{km} = \sum_k a_k^*\, a_n\, a_m^*\, a_k = a_m^*\, a_n = \varrho_{nm} \tag{34.34}$$

oder

$$\rho\,\rho = \rho. \tag{34.35}$$

Denken wir uns jetzt die Dichtematrix auf Diagonalform gebracht, so ergibt sich unmittelbar, daß Gl. (34.35) nur erfüllt sein kann, wenn alle Eigenwerte entweder Null oder Eins sind. Da aber auch für die transformierte Matrix Gl. (34.19) gilt, folgt, daß ein Eigenwert gleich Eins und alle übrigen gleich Null sind. Im Hinblick auf Gl. (34.20) bedeutet dies, daß alle Systeme der Gesamtheit sich in dem gleichen, durch den nicht verschwindenden Eigenwert der Dichtematrix definierten, quantenmechanischen Zustand befinden. Die Gl. (34.35) stellt daher die notwendige und hinreichende Bedingung für den reinen Zustand dar.

35. Der Liouvillesche Satz der Quantenstatistik. Die zeitliche Änderung der Dichtematrix ergibt sich unmittelbar durch Anwendung der transformierten Schrödinger-Gleichung (34.9) auf (34.18). Man erhält dann

$$\frac{\partial \varrho_{nm}}{\partial t} = -\frac{2\pi i}{h} \sum_k \left(H_{nk}\varrho_{km} - \varrho_{nk}H_{km}\right), \tag{35.1}$$

wo die Elemente der Matrix H durch Gl. (34.5) aus dem Hamilton-Operator definiert sind. Mit Benutzung der Definition (34.16) kann Gl. (35.1) übersichtlicher geschrieben werden

$$\frac{\partial \rho}{\partial t} = -\frac{2\pi}{ih}\left[\rho, \mathsf{H}\right].$$ (35.2)

Diese Gleichung stellt das quantenstatistische Analogon des Liouvilleschen Satzes in der Form der Gl. (3.5) dar.

Es sei nun F ein Operator, der auch explizit von der Zeit abhängt. Dann gilt die Heisenbergsche Bewegungsgleichung

$$\frac{d\mathsf{F}}{dt} = \frac{2\pi}{ih}\left[\mathsf{F}, \mathsf{H}\right] + \frac{\partial \mathsf{F}}{\partial t}.$$ (35.3)

Setzen wir für F den Dichteoperator, so erhalten wir durch Kombination mit Gl. (35.2)

$$\frac{d\rho}{dt} = 0.$$ (35.4)

Diese Gleichung ist das quantenstatistische Analogon des Prinzips von der Erhaltung der Phasendichte.

36. Das statistische Gleichgewicht. Wir beschränken unsere Betrachtung jetzt auf den Fall, daß die durch Gl. (34.22) definierten Mittelwerte notwendig zeitunabhängig sind. Dazu muß die Beziehung

$$\frac{\partial \rho}{\partial t} = 0$$ (36.1)

erfüllt sein. Wenn dies der Fall ist, sagt man wieder, daß die Gesamtheit sich im statistischen Gleichgewicht befindet. Als notwendige und hinreichende Bedingung dafür ergibt sich aus Gl. (35.2)

$$\left[\rho, \mathsf{H}\right] = 0.$$ (36.2)

Allgemein ist also eine Gesamtheit im statistischen Gleichgewicht, wenn der Dichteoperator mit dem Hamilton-Operator vertauschbar ist. Wir greifen nun wieder den speziellen Fall einer Gesamtheit aus konservativen Systemen heraus. Um zu expliziten Aussagen zu gelangen, ist es notwendig, die Dichtematrix als Funktion einer anderen Matrix F (oder mehrerer Matrizen) auszudrücken. Wir schreiben

$$\rho = \varphi(\mathsf{F})$$ (36.3)

als symbolische Darstellung der Potenzreihe

$$\rho = a_0 + a_1 \mathsf{F} + a_2 \mathsf{F}^2 + a_3 \mathsf{F}^3 + \cdots,$$ (36.4)

die für Matrizen erklärt ist. Für eine Gesamtheit aus konservativen Systemen wird damit die hinreichende und notwendige Bedingung des statistischen Gleichgewichtes, daß die Dichtematrix nur von den Konstanten der Bewegung abhängt. Der wichtigste Fall ist wieder

$$\rho = \varphi(\mathsf{H}).$$ (36.5)

Wählen wir hier als Basis die Energie-Eigenfunktionen, so ist

$$H_{nm} = \int \psi_n^* \mathsf{H} \psi_m \, d\boldsymbol{q} = E_m \delta_{nm}.$$ (36.6)

Durch Einsetzen in Gl. (36.4) folgt daraus

$$\varrho_{nm} = a_0 \delta_{nm} + a_1 E_m \delta_{nm} + a_2 E_m^2 \delta_{nm} + a_3 E_m^3 \delta_{nm} + \cdots$$ (36.7)

oder in der Schreibweise der Gl. (36.3)

$$\varrho_{nm} = \varphi(E_m)\,\delta_{nm}.\qquad(36.8)$$

In diesem Falle reduziert sich somit die Dichtematrix auf die Hauptdiagonale.

37. Makroskopische Variable. Bisher haben wir angenommen, daß die Genauigkeit von Messungen lediglich durch die Heisenbergschen Unbestimmtheitsrelationen begrenzt ist. Wegen der Kleinheit des Wirkungsquantums spielt diese Beschränkung keine unmittelbare Rolle für die makroskopischen Messungen, mit denen wir es in der statistischen Mechanik zu tun haben. Wir haben jedoch bereits in der klassischen Theorie gesehen, daß es für gewisse Überlegungen notwendig ist, die begrenzte Genauigkeit makroskopischer Messungen explizit zu berücksichtigen. In der Quantenstatistik ist dieser Sachverhalt naturgemäß noch stärker ausgeprägt, weil hier die Messung in die Theorie eingebaut ist. Aus diesem Grunde ist auch die Definition makroskopischer Variablen in der Quantenstatistik weniger einfach als in der klassischen Theorie. Da wir im folgenden diesen Begriff, der zuerst durch VON NEUMANN[1] eingeführt wurde, verschiedentlich benötigen, wollen wir ihn bereits an dieser Stelle etwas ausführlicher erörtern. Wir folgen dabei im wesentlichen einer neueren Untersuchung von VAN KAMPEN[2].

Wir definieren zunächst den Begriff einer langsam mit der Zeit veränderlichen Größe. Es sei F eine meßbare Größe, ΔF ihre experimentelle Unsicherheit und Δt die zu ihrer Messung erforderliche Zeit. Dann ist jedenfalls

$$\Delta F \geqq \dot{F}\,\Delta t.\qquad(37.1)$$

F soll eine langsam veränderliche Größe heißen, wenn die Bedingung

$$\Delta F \gg \dot{F}\,\Delta t\qquad(37.2)$$

erfüllt ist. Nach der Unbestimmtheitsrelation ist

$$\delta E\,\delta t \approx \frac{h}{2\pi}\qquad(37.3)^3$$

und somit

$$\Delta E\,\Delta t \gtrsim \frac{h}{2\pi}\qquad(37.4)$$

Nun ist

$$\delta E\,\delta F \approx [\mathsf{H}, \mathsf{F}] \approx \frac{h}{2\pi}\,\dot{\mathsf{F}}.\qquad(37.5)$$

Mit (37.4) und (37.5) läßt sich die Bedingung (37.2) schreiben

$$\Delta E\,\Delta F \gg \delta E\,\delta F.\qquad(37.6)$$

Für zwei meßbare Größen F und G gilt entsprechend

$$\Delta F\,\Delta G \gg \delta F\,\delta G.\qquad(37.7)$$

Wir betrachten nun ein vollkommen abgeschlossenes System, das durch einen zeitunabhängigen Hamilton-Operator H beschrieben wird, dessen Eigenwerte E_n und dessen Eigenfunktionen ψ_n seien. Wir können dann die Gl. (34.1) schreiben

$$\Psi(\boldsymbol{q}, t) = \sum_n a_n \psi_n\, \mathrm{e}^{-\frac{2\pi i E_n}{h} t},\qquad(37.8)$$

---

[1] J. v. NEUMANN: Z. Physik **57**, 30 (1929).

[2] G. N. VAN KAMPEN: Physica, Haag **20**, 603 (1954).

[3] Es ist zu beachten, daß diese Relation nicht aus dem mathematischen Formalismus der Quantenmechanik folgt, sondern lediglich durch gewisse Überlegungen plausibel gemacht wird. Diese Sonderstellung der Zeit hängt mit den in Ziff. 33 erwähnten Dingen zusammen.

wo die a_n jetzt Konstanten sind, die der Normierungsrelation

$$\sum_n |a_n|^2 = 1 \tag{37.9}$$

genügen. Wir setzen voraus, daß die ψ_n entweder symmetrisch oder antisymmetrisch sind. Die Eigenwerte ordnen wir nach dem Schema[1]

$$0 < E_1 < E_2 < E_3 < \cdots . \tag{37.10}$$

Da für ein System von sehr vielen Freiheitsgraden das Energiespektrum sehr dicht ist (Quasi-Kontinuum), können wir die Eigenwerte zu Gruppen $[\Delta E]_K$ zusammenfassen derart, daß innerhalb jeder Gruppe eine große Zahl von Eigenwerten liegt, der überdeckte Wertebereich aber von der Größenordnung der experimentellen Unsicherheit ist. Eine solche Gruppe von Eigenwerten nennen wir, in Anlehnung an die klassische Terminologie, eine *Energieschale*. Es soll also möglich sein, experimentell zwischen verschiedenen Energieschalen, aber nicht zwischen den Eigenwerten innerhalb einer Energieschale zu unterscheiden. Einer makroskopischen Energiemessung entspricht somit eine Diagonalmatrix, deren Elemente für alle ψ_n einer Schale $[\Delta E]_K$ gleich sind einem mittleren Werte E_K. Dieser Operator ist mit der gleichen Genauigkeit definiert, mit der die Messungen ausgeführt werden können; wir bezeichnen ihn mit $\{H\}$. Es sei F ein im Sinne von (37.7) langsam veränderlicher Operator. Stellen wir ihn im System der ψ_n

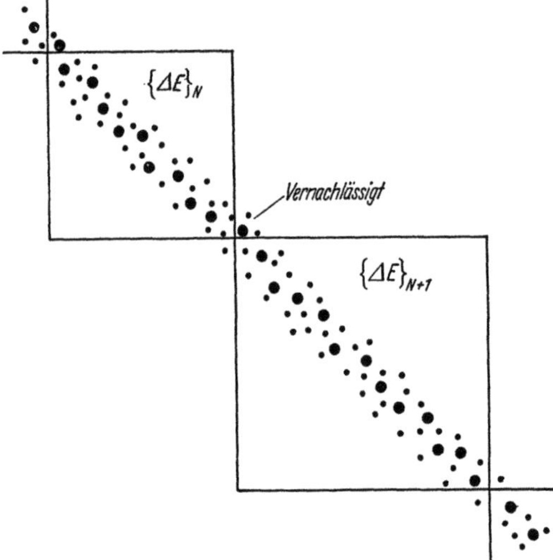

Fig. 4. Zur Definition eines makroskopischen Operators.

dar, so sind die Matrixelemente F_{nm} nur in einem schmalen Streifen längs der Hauptdiagonalen merklich von Null verschieden. Aus Gl. (35.3) folgt nämlich für einen nicht explizit von der Zeit abhängigen Operator F

$$-\frac{ih}{2\pi}\dot{F}_{nm} = (HF - FH)_{nm} = (E_n - E_m) F_{nm}. \tag{37.11}$$

Aus (37.5) ergibt sich die Abschätzung der Größenordnung

$$(E_n - E_m) F_{nm} \approx \delta E \, \delta F. \tag{37.12}$$

Die Größe $|E_n - E_m|$ stellt nun ein natürliches Maß für die Entfernung eines Matrixelementes F_{nm} von der Hauptdiagonalen dar. Wenn dieser Abstand von der Größenordnung der Dicke der Energieschale, also

$$|E_n - E_m| \approx \Delta E \tag{37.13}$$

ist, so folgt aus (37.12), daß F_{nm} klein ist gegen die experimentelle Unsicherheit ΔF und vernachlässigt werden kann. Die noch verbleibenden Matrixelemente bilden einen Streifen längs der Hauptdiagonalen, der schmal ist im Vergleich zu der Dicke der Energieschalen (Fig. 4).

---

[1] Eine zufällige Entartung können wir uns durch eine kleine Änderung der Gefäßdimensionen entfernt denken.

Die Zahl der Elemente, welche benachbarte Energieschalen verknüpfen, ist daher relativ klein und soll ebenfalls vernachlässigt werden. Es resultiert dann ein Operator, welcher die direkte Summe von Untermatrizen $\mathsf{F}^{(K)}$ darstellt, von denen jede nur in der entsprechenden Energieschale $[\Delta E]_K$ wirkt. In dieser Näherung ist F mit dem makroskopischen Energie-Operator $\{\mathsf{H}\}$ vertauschbar. Es besteht daher keine Beschränkung in der gleichzeitigen Meßbarkeit von E und F.

Wir betrachten nun im besonderen die Energieschale $[\Delta E]_K$. Die zugehörigen Eigenvektoren bezeichnen wir mit $\psi_r^{(K)}$. Es sei χ_λ ein othonormales Funktionensystem, das mit $\psi_r^{(K)}$ durch die unitäre Transformation

$$\chi_\lambda = \sum_r U_{r\lambda}^{(K)} \psi_r^{(K)} \tag{37.14}$$

zusammenhängt und so gewählt ist, daß $F^{(K)}$ in dieser Darstellung diagonal wird. Es soll also gelten

$$F_{\lambda\varkappa}^{(K)} = \int \chi_\lambda^* F^{(K)} \chi_\varkappa \, d\boldsymbol{q} = F_\lambda \delta_{\lambda\varkappa}. \tag{37.15}$$

Die Eigenwerte F_λ fassen wir wieder, entsprechend der experimentellen Unsicherheit ΔF, in Gruppen $[\Delta F]_\Lambda$ zusammen. Jeder solchen „F-Zelle" wird ein in der Größenordnung ΔF willkürlicher Eigenwert F_Λ zugeschrieben. Den so definierten makroskopischen Operator bezeichnen wir mit $\{\mathsf{F}\}$.

Die F-Zellen sind lineare Unterräume des Hilbert-Raumes. Alle Vektoren in jeder F-Zelle sind Eigenvektoren der makroskopischen Operatoren $\{\mathsf{H}\}$ und $\{\mathsf{F}\}$ mit den Eigenwerten E_K und F_Λ.

Es sei nun G ein zweiter langsam veränderlicher Operator. Wir vernachlässigen von vornherein wieder alle Matrixelemente

$$G_{r,s}^{(K,L)} = \int \psi_r^{*(K)} \mathsf{G} \, \psi_s^{(L)} d\boldsymbol{q} \quad (K \neq L) \tag{37.16}$$

die verschiedene Energieschalen (K und L) miteinander verknüpfen. In jeder Schale wird durch die Transformation (37.14) die Untermatrix $G_{r,s}^{(K)}$ in eine Matrix $G_{\lambda\varkappa}^{(K)}$ überführt, die im allgemeinen nicht diagonal ist. Die Nicht-Diagonalelemente lassen sich mit Hilfe der Vertauschungsrelation für F und G abschätzen. Man hat

$$(FG - GF)_{\lambda\varkappa} = (F_\lambda - F_\varkappa) \, G_{\lambda\varkappa}. \tag{37.17}$$

Die linke Seite dieser Gleichung ist von der Größenordnung $\delta F \, \delta G$. Für ein Matrixelement, das in der Energieschale $[\Delta E]_K$ zwei verschiedene F-Zellen Λ und M verknüpft, ergibt sich daher

$$\frac{G_{\Lambda\varrho,M\sigma}^{(K)}}{\Delta G} \approx \frac{\delta F \, \delta G}{\Delta F \, \Delta G} \quad (\Lambda \neq M). \tag{37.18}$$

Auch diese Matrixelemente können daher vernachlässigt werden, wenn (was wir voraussetzen) die experimentellen Unsicherheiten groß gegen die quantenmechanischen Unbestimmtheiten sind. Die noch verbleibenden Elemente von G bilden eine Reihe von Untermatrizen $\mathsf{G}^{(K,\Lambda)}$, von denen jede als Operator in der entsprechenden Zelle $[\Delta F]_K$ wirkt.

In jeder F-Zelle kann $\mathsf{G}^{(K,\Lambda)}$ durch die unitäre Transformation

$$\xi_\gamma = \sum_\varrho U_{\varrho\gamma}^{(\Lambda)} \chi_\varrho^{(\Lambda)} \tag{37.19}$$

auf Diagonalform gebracht werden, so daß wir haben

$$G_{\gamma\beta}^{(K,\Lambda)} = \int \xi_\gamma^* \mathsf{G}^{(K,\Lambda)} \xi_\beta \, d\boldsymbol{q} = G_\gamma \delta_{\gamma\beta}. \tag{37.20}$$

Die G_ν fassen wir, entsprechend der experimentellen Unsicherheit ΔG, in Gruppen $[\Delta G]_\Gamma$ zusammen. Die ξ_ν jeder Zelle spannen einen Unter-Raum des Hilbert-Raumes auf, für den die makroskopischen Operatoren $\{H\}$, $\{F\}$, $\{G\}$ gleichzeitig mit der Genauigkeit, mit welcher die entsprechenden Größen gemessen werden können, definiert sind.

In der beschriebenen Weise kann man für weitere langsam veränderliche Variable fortfahren. Dabei muß für jedes Paar von nicht vertauschbaren Operatoren (37.7) gelten. Der Einfachheit halber nehmen wir an, daß die Zellen durch einen vollständigen Satz[1] von unabhängigen langsam veränderlichen Operatoren definiert worden sind. Die auf diesem Wege definierten Zellen bilden das quantenstatistische Analogon der Zellen des Γ-Raumes in der klassischen Statistik. Wir bezeichnen mit Ω die Zahl der Zustände (d.h. der linear unabhängigen Eigenfunktionen des Hamilton-Operators in einer Energieschale), mit M die Zahl der Zellen in der Energieschale und mit ω_i die Zahl der Zustände in der i-ten Zelle. Es ist also

$$\sum_{i=1}^{M} \omega_i = \Omega \,. \tag{37.21}$$

Durch die unitäre Transformation

$$\psi_n = \sum_{i=1}^{M} \sum_{j=1}^{\omega_i} U_{ij,\,n}\,\varphi_{ij}, \qquad \varphi_{ij} = \sum_n U_{n,\,ij}^{-1}\,\psi_n \tag{37.22}$$

kann für jede Zelle ein orthonormales Funktionensystem definiert werden. Die Gln. (37.22) sind naturgemäß nur streng gültig, wenn über alle ψ_n bzw. φ_{ij} im Hilbert-Raum summiert wird. Die Beschränkung auf langsam veränderliche Operatoren rechtfertigt jedoch, im Rahmen der bisherigen Näherung zu setzen

$$U_{ij,\,n} = 0, \tag{37.23}$$

wenn ψ_n nicht zu $[\Delta E]_K$ gehört.

II. Axiomatische Grundlagen der Quantenstatistik.

38. Problemstellung. In Abschnitt I haben wir die quantenstatistische Ensembletheorie lediglich unter formalen Gesichtspunkten betrachtet. Wenn man sie zu einer physikalischen Theorie ausgestalten will, stößt man auf Probleme, die denen der klassischen statistischen Mechanik völlig analog sind. Es handelt sich vor allem um die Fragen

1. Wie ist eine makroskopische Gleichgewichtseigenschaft zu definieren?

2. Wie ist für ein gegebenes System die Dichtematrix zu konstruieren?

Es ist in der Quantenstatistik ebensowenig wie in der klassischen Theorie gelungen, diese Fragen auf deduktivem Wege schlüssig zu beantworten. Man benötigt daher auch in der Quantenstatistik eine eigene axiomatische Basis. Die dazu führenden Überlegungen entsprechen weitgehend denen der klassischen Theorie. Wir unterscheiden sie wieder als Ergodentheorie und Hypothese der gleichen a priori-Wahrscheinlichkeiten.

Die Ergodentheorie beantwortet die erste der obigen Fragen mit der Aussage, daß die makroskopischen Gleichgewichtseigenschaften durch die (über eine unendliche Zeit erstreckten) Zeitmittel der betreffenden Observablen gegeben sind. Die direkte Berechnung dieser Zeitmittel (deren exakte Definition wir in Ziff. 39

---

[1] Damit ist gemeint, daß alle langsam veränderlichen Größen als Funktionen des Satzes ausgedrückt werden können.

ge·ben) .würde die Lösung der Schrödinger-Gleichung des Systems erfordern, die praktisch undurchführbar ist. In der ursprünglichen Formulierung der klassischen Theorie wird die Lösung der Bewegungsgleichungen durch eine allgemeine Aussage über diese Lösungen, die Ergodenhypothese, ersetzt. Die analoge Behauptung der Quantenstatistik [14], [28] besagt, daß für ein abgeschlossenes System jeder Zustand innerhalb der Energieschale $[\varDelta E]_K$ von jedem anderen Zustand in der gleichen Energieschale (direkt oder über andere Zustände) erreicht werden kann. Die für das Zustandekommen der Übergänge notwendige Störung des Hamilton-Operators kann in einem abgeschlossenen System durch das elektromagnetische Feld geliefert werden oder auch durch Wechselwirkungen innerhalb des Systems.

Bezeichnen wir nun die Wahrscheinlichkeit, das System im Zustand i zu finden, mit p_i und mit w_{ij} die Übergangswahrscheinlichkeit vom Zustand i zum Zustand j, so gilt (vgl. Ziff. 42)

$$\frac{dp_i}{dt} = \sum_j w_{ij}(p_j - p_i) \tag{38.1}$$

mit

$$w_{ij} = w_{ji}. \tag{38.2}$$

Die stationäre Lösung dieser Gleichung ist

$$p_i = \Omega^{-1}. \tag{38.3}$$

Im Gleichgewicht verweilt das System somit im Mittel gleiche Zeiten in jedem der Ω Zustände. Die Berechnung des Zeitmittels kann daher durch die Mittelung über eine quantenstatistische mikrokanonische Gesamtheit ersetzt werden.

Im Gegensatz zur klassischen Theorie kann man nicht beweisen, daß das obige Analogon der Ergodenhypothese falsch ist; es stellt aber eine reine Hypothese dar und kann insofern eher zur Quasi-Ergodenhypothese in Parallele gesetzt werden. Die Situation ist jedoch dadurch komplizierter, daß auch die Ableitung der Gl. (38.1) gewisse Annahmen einschließt. Aus beiden Gründen erscheint es daher auch für die Quantenstatistik befriedigender, den Weg der klassischen von Neumann-Birkhoffschen Theorie einzuschlagen und zu versuchen, direkt die Gleichsetzung des Zeitmittels mit dem mikrokanonischen Ensemble-Mittel zu rechtfertigen.

39. Das Ergodentheorem der Quantenstatistik. Der dem klassischen von Neumann-Birkhoffschen Ergodentheorem entsprechende quantenstatistische Satz wurde zuerst durch VON NEUMANN[1] abgeleitet. Der Beweis wurde später von PAULI und FIERZ[2] vereinfacht. Neuerdings haben FIERZ[3] sowie FARQUHAR und LANDSBERG[4] gezeigt, daß gewisse Voraussetzungen, die VON NEUMANN als wesentlich angenommen hatte, entbehrlich sind. Wir schließen uns hier an die Arbeit von FARQUHAR und LANDSBERG[4] an, wobei wir für einige mathematische Details auf die Originalarbeit verweisen müssen.

Das von Neumannsche Ergodentheorem bezieht sich auf die in Ziff. 37 eingeführten makroskopischen Variablen, d.h. auf ein System, dessen Zustände als Zellen innerhalb der Energieschale $[\varDelta E]_K$ definiert sind. Formal steht es daher dem Hopfschen Ergodentheorem der klassischen Theorie am nächsten. Das

J. v. NEUMANN: Z. Physik **57**, 30 (1929).
[2] W. PAULI u. M. FIERZ: Z. Physik **106**, 572 (1937).
[3] M. FIERZ: Helv. phys. Acta **28**, 705 (1955).
[4] I.E. FARQUHAR u. P.T. LANDSBERG: Proc. Roy. Soc. Lond., Ser. A **239**, 134 (1957).

Problem des Beweises liegt in der Tatsache, daß die Zellenteilung der Energie-schale weitgehend willkürlich ist. Es wird nun angenommen, daß jedem makro-skopischen Beobachter eine unitäre Transformation (37.22) eineindeutig zuge-ordnet ist. Der von Neumannsche Satz besagt dann, daß die Wahrscheinlichkeit, einen makroskopischen Beobachter zu finden, der eine Verletzung des Ergoden-theorems konstatiert, extrem klein ist unter folgenden Voraussetzungen:

a) $\omega_i \gg M$ für alle i.

b) Alle makroskopischen Beobachter sind gleich wahrscheinlich.

c) Die Ω Eigenwerte des (mikroskopischen) Hamilton-Operators sind alle verschieden.

d) Die $\Omega(\Omega-1)$ nicht verschwindenden Differenzen, die aus diesen Eigen-werten gebildet werden können, sind ebenfalls sämtlich verschieden.

Tatsächlich sind für das makroskopische Ergodentheorem die Voraussetzun-gen c) und d) entbehrlich, während a) (was schon von Neumann vermutet hatte) durch eine schwächere Bedingung ersetzt werden kann. Um das Problem mathe-matisch zu formulieren, betrachten wir ein abgeschlossenes System, von dem durch eine makroskopische Energiemessung bekannt ist, daß es sich in der Energie-schale $[\Delta E]_K$ befindet. Wir können dann die Ψ-Funktion entwickeln in der Form

$$\Psi(\boldsymbol{q}, t) = \sum_{n=1}^{\Omega} a_n \psi_n \, \mathrm{e}^{-\frac{2\pi i E_n}{h} t} = \sum_{i=1}^{M} \sum_{j=1}^{\omega_i} b_{ij} \varphi_{ij}, \tag{39.1}$$

wo

$$\sum_{n=1}^{\Omega} |a_n|^2 = \sum_{i=1}^{M} \sum_{j=1}^{\omega_i} |b_{ij}|^2 = 1 \tag{39.2}$$

ist. Die dem makroskopischen Beobachter zugeordnete unitäre Transformation (die wir schon in Ziff. 37 eingeführt haben) lautet

$$\psi_n = \sum_{i=1}^{M} \sum_{j=1}^{\omega_i} U_{ij,n} \varphi_{ij}, \qquad \varphi_{ij} = \sum_{n=1}^{\Omega} U_{n,ij}^{-1} \psi_n, \tag{39.3}$$

wo für die Transformationsmatrizen die Gln. (34.10) und (34.11) gelten. Aus (39.1) und (39.3) folgt

$$b_{ij} = \sum_{n=1}^{\Omega} a_n U_{ij,n} \, \mathrm{e}^{-\frac{2\pi i E_n}{h} t}. \tag{39.4}$$

Damit wird die Wahrscheinlichkeit, das System in der i-ten Zelle zu finden

$$P_i = \sum_{j=1}^{\omega_i} |b_{ij}|^2 = \sum_{n,\,m=1}^{\Omega} a_n^* a_m \, \mathrm{e}^{-\frac{2\pi i (E_m - E_n)}{h} t} \sum_{j=1}^{\omega_i} U_{ij,n}^* U_{ij,m}. \tag{39.5}$$

Das Zeitmittel von P_i ist definiert durch die Gleichung

$$\overline{P_i} = \lim_{\tau \to \infty} \frac{1}{\tau} \int_0^{\tau} P_i \, dt. \tag{39.6}$$

Es sei nun $\{F\}$ ein von P_i unabhängiger makroskopischer Operator, der in der φ_{ij}-Darstellung diagonal ist mit dem Eigenwert F_i in der i-ten Zelle. Dann ist das Zeitmittel der Observablen F

$$\widehat{F} = \sum_{i=1}^{M} F_i \, \widehat{P_i}. \tag{39.7}$$

Der über die mikrokanonische Gesamtheit gebildete Mittelwert ist

$$\overline{F} = \sum_{i=1}^{M} F_i \frac{\omega_i}{\Omega}. \tag{39.8}$$

Zum Beweise des Ergodentheorems genügt es daher, zu zeigen, daß die Wahrscheinlichkeit, einen Beobachter zu finden, welcher eine Verletzung der Gleichung

$$P_i = \frac{\omega_i}{\Omega} \tag{39.9}$$

konstatiert, äußerst klein ist. Es ist dazu notwendig, gewisse Mittelwerte über die Gesamtheit der makroskopischen Beobachter bzw. der unitären Transformationen U zu bilden. Diese Mittelwerte bezeichnen wir durch das Symbol $\langle\ \rangle$.

Die $U_{ij,\,n}$ sind Komponenten von unitären Einheitsvektoren im Hilbert-Raum. Die Größen $y_i \equiv \sum_{j=1}^{\omega_i} |U_{ij,\,n}|^2$ sind die Längenquadrate der Projektionen dieser Vektoren auf den Unter-Raum der ω_i Zustände der betrachteten Zelle. Die obige Annahme b) bedeutet nun, daß die Wahrscheinlichkeit, Zelleneinteilungen zu finden, für welche die entsprechenden unitären Einheitsvektoren U in einem gegebenen Raumwinkel des Hilbert-Raumes liegen, proportional ist der Fläche der Einheitskugel im Hilbert-Raume, welche durch den Raumwinkel begrenzt wird. Damit äquivalent ist die Behauptung, daß die Wahrscheinlichkeitsverteilung der Transformationen U invariant ist gegen alle unitären Transformationen des durch die Energieschale definierten Ω-dimensionalen Funktionenraumes.

Nach diesen vorbereitenden Bemerkungen formulieren wir das folgende

Theorem. Wenn

$$\omega_i \gg 1 \quad \text{(alle } i\text{)} \tag{39.10}$$

und Voraussetzung b) erfüllt ist, so ist

$$Z_i \equiv \frac{\langle(\hat{P}_i - \omega_i/\Omega)^2\rangle}{(\omega_i/\Omega)^2} \ll 1. \tag{39.11}$$

Beweis. Aus Gl. (39.5) und (39.6) erhalten wir

$$\hat{P}_i = \sum_{n=1}^{\Omega} |a_i|^2 \sum_{j=1}^{\omega_i} |U_{ij,\,n}|^2 + \sum_{n \neq m} a_n^* a_m \sum_{j=1}^{\omega_i} U_{ij,\,n}^* U_{ij,\,m}. \tag{39.12}$$

Der zweite Term der rechten Seite verschwindet, wenn keine Entartung vorliegt. Es sei p die Zahl der voneinander verschiedenen Eigenwerte E_k und g_k der zugehörige Entartungsgrad. Die Zahl der Summanden in der Summe über n und m ist dann

$$l = \sum_{k=1}^{p} g_k(g_k - 1). \tag{39.13}$$

Die Elemente von U sind nach Gl. (34.11) skalare Produkte, für die wir symbolisch schreiben

$$(\psi_n \cdot \varphi_{ij}) \equiv \int \psi_n^* \varphi_{ij}\, d\mathbf{q}. \tag{39.14}$$

Wir betrachten nun den Mittelwert

$$\langle U_{ij,\,n}^* U_{ij,\,m}\rangle = \langle(\psi_n \cdot \varphi_{ij})(\varphi_{ij} \cdot \psi_m)\rangle. \tag{39.15}$$

Wenn die ursprüngliche Basis φ_{ij} einer unitären Transformation unterworfen und dann der Mittelwert der Gl. (39.15) über alle neuen Basen φ'_{ij} gebildet wird, so bleibt nach der Voraussetzung der Mittelwert ungeändert. Es ist also, wenn V eine unitäre Matrix bezeichnet,

$$\left.\begin{aligned}\langle(\psi_n \cdot \varphi_{ij})(\varphi_{ij} \cdot \psi_m)\rangle &= \langle(\psi_n \cdot \mathsf{V}^{-1}\varphi_{ij})(\mathsf{V}^{-1}\varphi_{ij} \cdot \psi_m)\rangle \\ &= \langle(\mathsf{V}\psi_n \cdot \varphi_{ij})(\varphi_{ij} \cdot \mathsf{V}\psi_m)\rangle.\end{aligned}\right\} \tag{39.16}$$

Ist im besonderen V eine Diagonalmatrix mit $V_{ii}=-1$ und $V_{kk}=1$ für alle $k \neq i$, so wird

$$\langle(\psi_n \cdot \varphi_{ij})(\varphi_{ij} \cdot \psi_m)\rangle = -\langle(\psi_n \cdot \varphi_{ij})(\varphi_{ij} \cdot \psi_m)\rangle. \tag{39.17}$$

Es muß daher

$$\langle U_{ij,n}^* \, U_{ij,m}\rangle = 0 \tag{39.18}$$

sein. Aus Gln. (39.12) und (39.18) folgt dann

$$\langle \hat{P}_i\rangle = \sum_{j=1}^{\omega_i}\langle|U_{ij,n}|^2\rangle \sum_{n=1}^{\Omega}|a_n|^2. \tag{39.19}$$

Die Berechnung der Mittelwerte beruht auf dem schon von von Neumann[1] sowie Pauli und Fierz[2] abgeleiteten Resultat, daß unter der Voraussetzung b) die Wahrscheinlichkeitsdichte für y gegeben ist durch

$$W(y) = C \cdot y^{\omega_i-1}(1-y)^{\Omega-\omega_i-1}, \tag{39.20}$$

wo C ein Normierungsfaktor ist. Daraus erhält man zunächst

$$\langle|U_{ij,n}|^2\rangle = \frac{\omega_i}{\Omega} \tag{39.21}$$

und damit unter Berücksichtigung von (39.2) aus (39.19)

$$\langle \hat{P}_i\rangle = \frac{\omega_i}{\Omega}. \tag{39.22}$$

Damit ist zunächst gezeigt, daß die Behauptung (39.9) unter den angegebenen Voraussetzungen jedenfalls im Mittel richtig ist. Der Beweis der weitergehenden Behauptung (39.11) erfordert ziemlich umständliche Rechnungen, für deren Einzelheiten wir auf die Arbeit von Farquhar und Landsberg[3] verweisen müssen. Wir geben hier lediglich das Endresultat wieder. Schreiben wir zur Abkürzung

$$A \equiv \sum_{n=1}^{\Omega}|a_n|^4, \tag{39.23}$$

$$B \equiv \sum_{n \neq m}^{\Omega}|a_n^* \, a_m|^2, \tag{39.24}$$

und bezeichnen mit q eine reelle Zahl, für die $-1 \leq q \leq +1$ gilt, so erhält man

$$Z_i = \frac{(\Omega-\omega_i)\,[A\Omega - 1 + B\Omega(1+q)]}{\omega_i(\Omega^2-1)}. \tag{39.25}$$

Für $\Omega \geq 3$ folgt daraus

$$Z_i < \frac{2(\Omega-\omega_i)}{\omega_i\Omega}. \tag{39.26}$$

Unter der Voraussetzung (39.10) folgt daraus unmittelbar die Behauptung (39.11).

---

[1] J. v. Neumann: Z. Physik 57, 30 (1929).
[2] W. Pauli u. M. Fierz: Z. Physik 106, 572 (1937).
[3] I. E. Farquhar u. P. T. Landsberg: Proc. Roy. Soc. Lond., Ser. A 239, 134 (1957).

Für die Diskussion des Ergodentheorems der Quantenstatistik können wir weitgehend auf die Ausführungen in Ziff. 11 verweisen, die sich entsprechend auf das hier vorliegende Problem übertragen lassen. Wir heben daher nur einige Punkte noch hervor. Daß die Indentifizierung des Meßmittels mit dem Zeitmittel auch in der Quantenstatistik unzulässig ist, hat neuerdings JAYNES[1] betont. Die entscheidende Problematik im Beweis des eigentlichen Ergodentheorems liegt in der Voraussetzung b), die, wie kürzlich FIERZ[2] betont hat, physikalisch kaum zu rechtfertigen ist. Er vermutet, daß mit Hilfe einer Gesamtheit von zufälligen Störungen (die wieder auf eine Gesamtheit von unitären Transformationen abgebildet werden kann), dem physikalischen Gesichtspunkt besser Rechnung getragen wird. Dieser Gedanke ist von EKSTEIN[3] ausgeführt worden, wobei auch die Beschränkung auf makroskopische Observable aufgegeben wird. Indessen bleibt auch hier der kritische Punkt die Tatsache, daß die Mittelung über die Gesamtheit der Störungen wieder nach der oben angedeuteten von Neumannschen Methode durchgeführt wird. Die Grundlage der Ableitung bildet somit auch in diesem Falle eine, wenn auch vielleicht plausible, Hypothese.

FARQUHAR und LANDSBERG[4] haben die Frage nach einem mikroskopischen Kriterium für die Gültigkeit des Ergodentheorems, welches der Bedingung der metrischen Transitivität in der klassischen Theorie entsprechen würde, aufgeworfen. Ein solches Kriterium ist von KLEIN[5] formuliert worden, der in seiner Analyse ebenfalls explizit ein Störungsglied in den Hamilton-Operator einführt. Es wird angenommen, daß das System mit seiner „Umgebung" (das „Gesamtsystem") abgeschlossen ist und daß im Störungsglied alle Matrixelemente, welche verschiedene Energieschalen verknüpfen, verschwinden. Das Ergodentheorem kann nun formuliert werden: Das Zeitmittel des Erwartungswertes eines beliebigen Operators F ist gleich dem mikrokanonischen Mittelwert $\bar{F}$ dann, und nur dann, wenn jeder zeitunabhängige Operator des Gesamtsystems R so beschaffen ist, daß seine Projektion auf den Funktionenraum des Systems R_s ein konstantes Vielfaches des Einheitsoperators ist. Für den Beweis und die Diskussion dieser Aussage müssen wir auf die Originalarbeit verweisen. TER HAAR [23] hat darauf hingewiesen, daß KLEINs Analyse nicht die Tatsache berücksichtigt, daß der Beobachter es mit makroskopischen Variablen zu tun hat. Es erscheint in der Tat wenig befriedigend, daß zwar eine Energieschale endlicher Ausdehnung eingeführt, im übrigen aber ausschließlich mit mikroskopischen Operatoren gerechnet wird. Auch FARQUHAR und LANDSBERG[4] haben die Notwendigkeit betont, bei der Ableitung des Ergodentheorems „grobe" Wahrscheinlichkeiten zu benutzen. Auffallend ist auch, daß hier, im Gegensatz zu den sonstigen klassischen und quantenstatistischen Formulierungen, eine absolute strenge Gültigkeit des Ergodentheorems behauptet wird[6].

40. Die Hypothese der gleichen a priori-Wahrscheinlichkeiten und der willkürlichen Phasenverteilung. Die Aussagen der Quantenmechanik beziehen sich

[1] E. T. JAYNES: Phys. Rev. (im Druck). Für die Überlassung des Manuskriptes dieser Arbeit bin ich Herrn Dr. JAYNES zu großem Dank verpflichtet.

[2] M. FIERZ: Helv. phys. Acta **28**, 705 (1955).

[3] H. EKSTEIN: Phys. Rev. **107**, 333 (1957).

[4] I. E. FARQUHAR u. P. T. LANDSBERG: Proc. Roy. Soc. Lond., Ser. A **239**, 134 (1957).

[5] M. J. KLEIN: Phys. Rev. **87**, 111 (1952).

[6] *Zusatz bei der Korrektur:* Kürzlich haben BOCCHIERI und LEINGER (noch nicht veröffentlicht) gezeigt, daß bereits die bloße Mittelung über die Beobachter gemäß Voraussetzung b) das Ergodentheorem liefert und somit die zeitliche Entwicklung des Systems (d. h. die Gültigkeit der Schrödinger-Gleichung) in diesem „Beweis" überhaupt nicht eingeht. Für die Überlassung des Manuskriptes habe ich den Autoren herzlich zu danken. Ferner sei auf neue Untersuchungen von LUDWIG [Z. Naturforsch. **12a**, 662 (1957); Z. Physik **150**, 346 (1958)] hingewiesen.

jeweils auf scharfe Zeitpunkte. Wenn wir bei der Begründung der Quantenstatistik von dem Konzept der Ergodentheorie keinen Gebrauch machen wollen, ist es daher notwendig, in anderer Form die Tatsache zu berücksichtigen, daß eine makroskopische Messung stets einen Mittelwert des betreffenden Erwartungswertes über die Meßzeit τ liefert. Dies kann in folgender Weise geschehen (vgl. [1]). Wir definieren zunächst den zeitlichen Mittelwert der in Ziff. 39 eingeführten Wahrscheinlichkeit, das System in der Zelle i zu finden, durch die Gleichung

$$\widehat{P}_{i\tau} = \frac{1}{\tau} \int\limits_0^\tau P_i \, dt. \tag{40.1}$$

Den hier der Deutlichkeit halber angeschriebenen Index τ lassen wir in Zukunft fort und verstehen alle zeitlichen Mittelwerte im Sinne der vorstehenden Definition. Ist das System in einem reinen Zustand mit der Ψ-Funktion

$$\Psi(\boldsymbol{q}, t) = \sum_{ij} b_{ij} \varphi_{ij}, \tag{40.2}$$

so ist der Erwartungswert eines makroskopischen Operators $\{\mathsf{F}\}$ für einen bestimmten Zeitpunkt

$$\left.\begin{aligned} \overline{F} &= \sum_{ij,\,i'j'} b_{ij}^* \int \varphi_{ij}^* \{\mathsf{F}\} \varphi_{i'j'} \, d\boldsymbol{q} \, b_{i'j'} \\ &= \sum_i F_i \sum_j |b_{ij}|^2 = \sum_i F_i P_i. \end{aligned}\right\} \tag{40.3}$$

Für eine makroskopische Messung ist dann der Erwartungswert

$$\widehat{\overline{F}} = \sum_i F_i \sum_j \widehat{|b_{ij}|^2} = \sum_i F_i \widehat{P}_i. \tag{40.4}$$

Die $\widehat{P}_i$ können wir als Diagonalelemente eines „groben" Dichte-Operators $\widehat{\mathsf{P}}$ in der φ_{ij}-Darstellung auffassen. Wir können dann leicht zu einer gemischten Gesamtheit übergehen, indem wir definieren

$$\widehat{P}_i = \frac{1}{\nu} \sum_{l=1}^\nu \widehat{P}_l(i) = \overline{\sum_j \widehat{|b_{ij}|^2}}. \tag{40.5}$$

Wir haben dann für das Ensemble-Mittel des Zeitmittels (die Bezeichnung für den quantenmechanischen Mittelwert lassen wir weg)

$$\widehat{\overline{F}} = \operatorname{spur} \{\mathsf{F}\} \, \widehat{\mathsf{P}} = \operatorname{spur} \widehat{\mathsf{P}} \{\mathsf{F}\}. \tag{40.6}$$

Da $\{\mathsf{F}\}$ als langsam veränderlicher Operator definiert ist, erhalten wir

$$\widehat{\overline{F}} = \operatorname{spur} \widehat{\{\mathsf{F}\}\,\mathsf{P}} = \frac{1}{\tau} \int\limits_0^\tau [\operatorname{spur} \{\mathsf{F}\}\,\mathsf{P}] \, dt. \tag{40.7}$$

Setzen wir voraus, daß die statistische Gesamtheit in bezug auf den groben Dichteoperator im statistischen Gleichgewicht ist, so folgt

$$\widehat{\overline{F}} = \overline{F}. \tag{40.8}$$

Eine makroskopische Gleichgewichtseigenschaft ist durch eine zeitunabhängige Größe $\widehat{\overline{F}}$ definiert. Die der Gl. (12.4) analoge Gl. (40.8) besagt, daß diese Gleichgewichtseigenschaften durch die Ensemble-Mittelwerte einer im Hinblick auf den

---

[1] N. G. van Kampen: Physica, Haag **20**, 603 (1954).

groben Dichteoperator stationären Gesamtheit gegeben sind. Die Frage, inwieweit damit das Verhalten eines konkreten Systems beschrieben werden kann, wird auch im Rahmen der Quantenstatistik durch die Schwankungstheorie beantwortet. Für makroskopische Systeme können die Mittelwerte der Ensemble-Theorie praktisch als Eigenschaften des Einzelsystems betrachtet werden, wie es für die Begründung der Thermodynamik notwendig ist.

Die erste der in Ziff. 38 formulierten Fragen ist damit beantwortet. Es bleibt noch die Frage, wie für ein gegebenes System die Dichtematrix zu konstruieren ist. Wir beschränken uns dabei an dieser Stelle auf den Fall des Gleichgewichtes. Im Hinblick auf die Begründung der Thermodynamik haben wir daher zunächst nur den Fall zu betrachten, daß das durch gewisse äußere Parameter und die Zahl der Freiheitsgrade charakterisierte System durch eine makroskopische Energiemessung auf die Energieschale $[\varDelta E]_K$ festgelegt ist. Im Prinzip lassen sich dann die Diagonalelemente der groben Dichtematrix aus weiteren Messungen bestimmen. Tatsächlich liegt das Problem jedoch anders: Man benötigt einen allgemeinen Ausdruck für die feine Dichtematrix, um aus den dynamischen Eigenschaften gegebener Systeme ihre thermodynamischen Funktionen abzuleiten. Es handelt sich also darum, die unvollständige, durch den groben Dichteoperator beschriebene Kenntnis des Systems durch eine statistische Gesamtheit darzustellen, deren feiner Dichteoperator in Übereinstimmung mit der vorhandenen Kenntnis des Systems steht. An dieser Stelle muß, wie in der klassischen Theorie, das zusätzliche Axiom der statistischen Mechanik eingeführt werden. Für die Quantenstatistik lautet es in der allgemeinen Formulierung [12]: Alle mit der gegebenen Kenntnis des Systems in gleicher Weise vereinbaren Quantenzustände des Systems besitzen gleiche a priori-Wahrscheinlichkeit und willkürliche (random) a priori-Phasen.

Wir betrachten zunächst den Fall, daß aus Messungen eines vollständigen Satzes von makroskopischen Observablen die Elemente der groben Dichtematrix P_i bekannt sind. Nach dem obigen Axiom müssen dann alle zur Zelle i gehörigen Diagonalelemente der feinen Dichtematrix ϱ in der φ_{ij}-Darstellung untereinander gleich sein. Es muß also gelten

$$\overline{|b_{ij}|^2} = \varrho_{jj} = \varrho_{kk} = \overline{|b_{ik}|^2} \quad (j, k = 1 \text{ bis } \omega_i). \tag{40.9}$$

Für die in der Zelle liegenden Nicht-Diagonalelemente gilt allgemein

$$\varrho_{jk} = \frac{1}{\nu} \sum b_{ij} b_{ik} = \frac{1}{\nu} \sum r_j r_k \, e^{i(\alpha_j - \alpha_k)} \quad (j \neq k), \tag{40.10}$$

wo r_j der Modul und α_j die Phase von b_{ij} ist. Die Annahme willkürlicher a priori-Phasen besagt dann, daß die Phasen so über die Systeme der Gesamtheit verteilt sind, daß bei der Mittelung die Nichtdiagonalelemente der Dichtematrix verschwinden. Wir haben also für die Elemente der feinen Dichtematrix

$$\varrho_{jk} = (P_i/\omega_i)\, \delta_{jk}. \tag{40.11}$$

Im Falle des Gleichgewichtes wird in der Quantenstatistik der ein zeitabhängiges Störungsglied enthaltende Hamilton-Operator des Originalsystems ersetzt durch den ungestörten Hamilton-Operator einer Gesamtheit aus konservativen Systemen, die sich im statistischen Gleichgewicht befindet. Wenn durch eine makroskopische Energiemessung festgestellt ist, daß die Energie des Systems zwischen E_K und $E_K + \varDelta E$ liegt, so ist auch die repräsentative Gesamtheit auf die Energieschale $[\varDelta E]_K$ beschränkt. Aus der Hypothese der gleichen a priori-Wahrscheinlichkeiten und der willkürlichen a priori-Phasen folgt dann, daß für die Basis der Energie-Eigenfunktionen die innerhalb $[\Delta E]_K$ liegenden Diagonalelemente der Dichtematrix untereinander gleich sind und alle übrigen verschwinden, während alle Nicht-Diagonalelemente ebenfalls verschwinden. Die Dichtematrix der quantenstatistischen mikrokanonischen Gesamtheit ist daher

$$\varrho_{nm} = \Omega^{-1}\,\delta_{nm} \quad \text{für} \quad E_K \leqq E_n \leqq E_K + \Delta E, \quad \left.\begin{array}{c} \\ \\ \end{array}\right\} \qquad (40.12.)$$
$$\varrho_{nm} = 0 \quad \text{in allen übrigen Fällen}.$$

Man bemerkt, daß hier die Hypothese der willkürlichen a priori-Phasen eine Folgerung ist, die sich notwendig aus der Darstellung durch eine stationäre Gesamtheit konservativer Systeme ergibt. In diesem Falle muß nach Ziff. 36 die Dichtematrix, wenn sie (wie hier angenommen) nur von dem Hamilton-Operator abhängt, für die Basis der Energie-Eigenfunktionen diagonal werden. Dagegen ist ein analoger Schluß für die a priori-Wahrscheinlichkeiten nicht zulässig, wenn man der Energieschale, wie es in der Quantenstatistik wenigstens grundsätzlich notwendig ist, eine endliche Dicke zuschreibt. Man kann lediglich folgern, daß alle zu einem (entarteten) Eigenwert E_n gehörenden Quantenzustände die gleiche a priori-Wahrscheinlichkeit besitzen. TOLMAN [12] hat darauf hingewiesen, daß man die Dichtematrix der mikrokanonischen Gesamtheit auch in der Form

$$\varrho_{nm} = \varrho_0\, e^{-\frac{(E_n - E_K)^2}{\sigma^2}}\,\delta_{nm} \qquad (40.13)$$

ansetzen kann (wo ϱ_0 und σ Konstanten sind) und daß man damit für hinreichend kleines σ praktisch die gleichen Ergebnisse erhält wie mit Gl. (40.12)[1]. Dieser Sachverhalt hängt mit der Willkür in der Wahl der Größe ΔE zusammen. Wenn auch schon im Hinblick auf die Unbestimmtheitsrelation

$$\delta E\,\delta t \geqq \frac{h}{2\pi} \qquad (40.14)$$

grundsätzlich daran festgehalten werden muß, daß ΔE endlich ist, so darf die Willkür in der Wahl dieser Größe doch in den makroskopischen Gleichungen nicht mehr in Erscheinung treten. Tatsächlich liegt, wie wir später (Ziff. 47) zeigen werden, der Einfluß der Unschärfe von ΔE auf die Definition der thermodynamischen Funktionen weit unterhalb der experimentellen Nachweisgrenze. Praktisch kann man daher so rechnen, als gehörten alle Systeme der mikrokanonischen Gesamtheit zum gleichen Eigenwert E_K, dessen Entartungsgrad dann Ω ist (vgl. Ziff. 48). Für das hier betrachtete Problem hat daher die Hypothese der gleichen a priori-Wahrscheinlichkeiten, insoweit sie über die Folgerung aus der Verwendung einer stationären Gesamtheit von konservativen Systemen hinausgeht, keine physikalische Bedeutung.

Wir haben die Hypothese der gleichen a priori-Wahrscheinlichkeiten und willkürlichen a priori-Phasen als Axiom eingeführt. Obwohl wir damit auf eine Deduktion verzichtet haben, kann man durch verschiedene Überlegungen zeigen, daß die erwähnte Hypothese plausibel ist. Was zunächst die Annahme willkürlicher a priori-Phasen betrifft, so kann man sie eigentlich nicht als spezifisches Axiom der Quantenstatistik betrachten, da sie schon in der Quantenmechanik unentbehrlich ist [12], [23]. Um dies zu zeigen, nehmen wir an, daß wir durch eine exakte Messung der Größe F an einem quantenmechanischen System festgestellt haben, daß zur Zeit $t = 0$ dasselbe den Eigenwert F_k des Operators F

---

[1] Es ist erwähnenswert, daß GIBBS [2] die Theorie der mikrokanonischen Gesamtheit mit einer Phasendichte von der Form (40.13) entwickelt und erst am Schluß den Grenzübergang $\sigma \to 0$ ausführt.

hat. Entwickeln wir die Ψ-Funktion nach Eigenfunktionen von F, so ist für $t = 0$ die Wahrscheinlichkeit des Zustandes k

$$a_k^* a_k = 1, \tag{40.15}$$

während alle übrigen Terme verschwinden. Die Phase α_k kann dabei willkürlich gewählt werden; es gibt keine Möglichkeit, dieselbe experimentell zu bestimmen. Die Entwicklung des Systems ist dann durch die transformierte Schrödinger-Gleichung (34.9) bestimmt, die jetzt lautet

$$\frac{\partial a_n}{\partial t} = -\frac{2\pi i}{h} H_{nk} a_k = -\frac{2\pi i}{h} H_{nk} r_k e^{i\alpha_k}. \tag{40.16}$$

Die Lösung dieser Gleichung kann geschrieben werden

$$a_n(t) = \left[e^{-\frac{2\pi i H}{h} t} \right]_{nk} a_k(0). \tag{40.17}$$

Man sieht daraus, daß auch die Wahrscheinlichkeitsvoraussage für jeden Zustand n zu einer späteren Zeit t durch die Wahl von α_k nicht beeinflußt wird. Man kann daher den Anfangszustand als ein „Gemisch" willkürlich verteilter Phasen auffassen.

Wir betrachten nun die sog. „uniforme Gesamtheit", deren Dichtematrix durch die Gleichung

$$\varrho_{nm} = \varrho_0 \delta_{nm} \tag{40.18}$$

gegeben ist. Alle Diagonalelemente sind somit untereinander gleich, alle Nicht-Diagonalelemente Null. Da eine Normierung hier unzweckmäßig ist, kann ϱ_0 aufgefaßt werden als die relative Wahrscheinlichkeit, ein aus der Gesamtheit herausgegriffenes System in einem Zustand n zu finden. Die uniforme Gesamtheit, die schon von Neumann[1] und später Dirac[2] untersucht hat, kann in unserem jetzigen Zusammenhang betrachtet werden als eine Gesamtheit, die bei Fehlen jeder Kenntnis über das Originalsystem nur auf Grund der Hypothese der gleichen a priori-Wahrscheinlichkeiten und der willkürlichen a priori-Phasen konstruiert worden ist. Wir wollen nun zeigen, daß die hierdurch bestimmten Eigenschaften der uniformen Gesamtheit sich weder mit der Zeit noch mit einer Transformation der Basis verändern. Die erste Behauptung folgt unmittelbar aus Gl. (35.1), die nach Einsetzen von (40.18) lautet

$$\left. \begin{aligned} \frac{\partial \varrho_{nm}}{\partial t} &= -\frac{2\pi i}{h} \varrho_0 \sum_k (H_{nk} \delta_{km} - \delta_{nk} H_{km}) \\ &= -\frac{2\pi i}{h} \varrho_0 (H_{nm} - H_{nm}) = 0. \end{aligned} \right\} \tag{40.19}$$

Die uniforme Gesamtheit befindet sich somit im statistischen Gleichgewicht. Um die zweite Behauptung zu beweisen, schreiben wir die Dichtematrix (40.18)

$$\rho = \varrho_0 E, \tag{40.20}$$

wo E die Einheitsmatrix ist, und führen eine unitäre Transformation U durch. Dann erhalten wir

$$\rho' = U^\dagger \rho U = U^\dagger \varrho_0 E U = \varrho_0 U^\dagger E U = \varrho_0 U^\dagger U = \varrho_0 E. \tag{40.21}$$

---

[1] J. v. Neumann: Göttinger Nachr. 1927, 245.
[2] P. A. M. Dirac: Proc. Cambridge Phil. Soc. 25, 62 (1929).

Für die Ableitung beider Resultate ist die Annahme willkürlicher a priori-Phasen wesentlich. Gleiche a priori-Wahrscheinlichkeiten würden in der gewählten Darstellung auch durch eine Dichtematrix der Form

$$\varrho_{nm} = \varrho_0 \delta_{nm} + B_{nm}(1 - \delta_{nm}) \tag{40.22}$$

erreicht. Da aber im allgemeinen die Operatoren H und B nicht vertauschbar sind, ist eine solche Gesamtheit nicht stationär. Ebensowenig ist die Matrix (40.22) invariant gegen unitäre Transformationen. Das obige Ergebnis zeigt, daß die Quantenmechanik kein Argument gegen die Hypothese der gleichen a priori-Wahrscheinlichkeiten und der willkürlichen a priori-Phasen liefert und somit diese Grundlage der Quantenstatistik mit der mechanischen Basis konsistent ist

In der klassischen Theorie können Ergodentheorie und Hypothese der gleichen a priori-Wahrscheinlichkeiten formal als äquivalente axiomatische Formulierungen betrachtet werden. Diese Verhältnisse sind in der Quantenstatistik etwas weniger übersichtlich, weil hier das Ergodentheorem keine Aussage über die bei der Konstruktion der Dichtematrix anzunehmende Verteilung der Phasen ermöglicht. Die formale Äquivalenz läßt sich jedoch retten, wenn man die Hypothese der willkürlichen a priori-Phasen als zur Quantenmechanik gehörig betrachtet. Diese Aussage bezieht sich jedoch nur auf die Gleichgewichts-Statistik abgeschlossener Systeme. Die Behandlung offener Systeme erfordert eine Verbreiterung der axiomatischen Basis; dafür ist das Ergodentheorem ohne Bedeutung.

III. Die Einstellung des Gleichgewichtes.

41. Das quantenmechanische H-Theorem. Für ein volles Verständnis der statistischen Theorie der Gleichgewichtseigenschaften ist es auch in der Quantenstatistik notwendig, die Frage der Einstellung des Gleichgewichtes zu erörtern. Der Kern des Problems ist der gleiche wie in der klassischen Theorie. Auch in der Quantenmechanik gilt das Prinzip der Umkehrbarkeit mechanischer Vorgänge. Betrachten wir zwei konservative Systeme $'$ und $''$ mit dem gleichen reellen Hamilton-Operator und bezeichnen mit $W(\boldsymbol{q}, t)$ die Wahrscheinlichkeit, zur Zeit t den Koordinatensatz $\boldsymbol{q}$ zu finden, so ist

$$\left.\begin{aligned} W''(\boldsymbol{q}, t) &= W'(\boldsymbol{q}, -t), \\ W''(\boldsymbol{p}, t) &= W'(-\boldsymbol{p}, -t), \\ \bar{F}''(\boldsymbol{q}, \boldsymbol{p}, t) &= \bar{F}'(\boldsymbol{q}, -\boldsymbol{p}, -t). \end{aligned}\right\} \tag{41.1}$$

Diese Aussagen lassen sich leicht mit Hilfe der zeitabhängigen Schrödinger-Gleichung beweisen [*12*]. Es entsteht also auch hier die Frage, wie die irreversible Einstellung des makroskopischen Gleichgewichtes im Hinblick auf die Gln. (41.1) zu verstehen ist. In den letzten Jahren ist die Entwicklung auf diesem Gebiet wieder sehr in Fluß gekommen. Wir werden daher, wie in der klassischen Theorie, nur die älteren Untersuchungen etwas ausführlicher darstellen und uns für die neueste Entwicklung mit einigen Andeutungen und Literaturhinweisen begnügen.

Um die Analogien und Unterschiede gegenüber der klassischen Theorie zu verdeutlichen, erscheint es zweckmäßig, mit einer Formulierung des H-Theorems zu beginnen, die sich auf ein ideales Gas und (im Sinne der Quantenmechanik) auf ein einzelnes System bezieht. Dieses Theorem ist zuerst von Pauli[1] aufgestellt worden, dessen Ableitung wir im wesentlichen hier folgen. Wir betrachten

---

[1] W. Pauli: Sommerfeld-Festschrift. Leipzig 1928.

also ein System aus gleichen Teilchen, dessen Hamilton-Operator die Form

$$\mathsf{H} = \mathsf{H}_0 + \mathsf{V} \tag{41.2}$$

mit

$$\mathsf{H}_0 = \sum_{i=1}^{N} \mathsf{H}_i \tag{41.3}$$

hat, wo H_i der Hamilton-Operator des i-ten Teilchens ist. Der Wechselwirkungs-Operator V soll in erster Näherung vernachlässigbar sein, so daß wir den Zustand des Systems durch die Besetzungszahlen N_j der Eigenwerte ε_j des ungestörten Operators H_i beschreiben können. Die Berücksichtigung von V ist dagegen wesentlich für die Analyse des Mechanismus, welcher die Einstellung des Gleichgewichtes ermöglicht. Diese Annahmen entsprechen vollkommen denen der klassischen Theorie.

Ein (ungestörter) Eigenwert des Gesamtsystems kann geschrieben werden

$$E_n = \sum_j N_j \varepsilon_j, \qquad \sum_j N_j = N. \tag{41.4}$$

Zu jedem ε_j sollen g_j Eigenfunktionen $\psi_j(\boldsymbol{q}_i)$ gehören, wobei wir zwischen Entartung und Fastentartung nicht unterscheiden wollen. Bei der Konstruktion der Eigenfunktion des Gesamtsystems muß berücksichtigt werden, daß diese nur symmetrisch (Bose-Einstein-Statistik) oder antisymmetrisch (Fermi-Dirac-Statistik) sein darf. Wir haben daher, wenn P_q die Folge der Permutationsoperatoren bezeichnet, welche die $\boldsymbol{q}_i$ vertauschen und $|P|$ die Zahl der zu dem jeweiligen Permutationsoperator gehörigen Vertauschungen (Transpositionen)

$$\Psi_n = C_{\mathrm{DE}} \cdot \sum_{P_q} \mathsf{P}_q \psi'_n \qquad \text{(Bose-Einstein)} \tag{41.5}$$

und

$$\Psi_n = C_{\mathrm{FD}} \sum_{P_q} (-1)^{|P_q|} \mathsf{P}_q \psi'_n \qquad \text{(Fermi-Dirac)} \tag{41.6}$$

mit

$$\psi'_n = \prod_i \psi_j(\boldsymbol{q}_i), \tag{41.7}$$

wo C_{BE} und C_{FD} die entsprechenden Normierungsfaktoren sind. Für diese ergibt sich [12]

$$|C_{\mathrm{BE}}|^2 = [N! \prod n_\varkappa!]^{-1} \tag{41.8}$$

und

$$|C_{\mathrm{FD}}|^2 = N!^{-1}, \tag{41.9}$$

wo $n_\varkappa$ die Zahl der Faktoren $\psi_\varkappa$ auf der rechten Seite der Gl. (41.7) ist. In Analogie zur klassischen μ-Raum-Statistik (Ziff. 27) können wir auch hier Makrozustände einführen. Die Zahlen der zugehörigen Eigenfunktionen sind

$$\Omega_D = \prod_j \frac{(N_j + g_j - 1)!}{N_j!(g_j - 1)!} \qquad \text{(Bose-Einstein)} \tag{41.10}$$

bzw.

$$\Omega_D = \prod_j \frac{g_j!}{N_j!(g_j - N_j)!} \qquad \text{(Fermi-Dirac)}. \tag{41.11}$$

Als Analogon der klassischen H-Funktion können wir dann unmittelbar definieren

$$H = -\ln \Omega_D. \tag{41.12}$$

Wir erhalten nun, wenn wir noch den Fall der Maxwell-Boltzmann-Statistik in unsere Formeln einschließen

$$H = \sum_j \left\{ \gamma g_j \ln \frac{g_j}{N_j} - (N_j + \gamma g_j) \ln \left(\frac{g_j}{N_j} + \gamma \right) \right\} \tag{41.13}$$

mit

$$\gamma = \left\{ \begin{array}{ll} 0 & (\text{Maxwell-Boltzmann}) \\ 1 & (\text{Bose-Einstein}) \\ -1 & (\text{Fermi-Dirac}). \end{array} \right\} \tag{41.14}$$

Wir bezeichnen nun mit $Z_{jk \to j'k'}$ die Zahl der Übergänge (Stöße) pro Zeiteinheit, bei denen die Gruppen g_j und g_k jede ein Teilchen verlieren und die Gruppen $g_{j'}$ und $g_{k'}$ je ein Teilchen gewinnen. Solche Übergänge sind nur möglich, wenn die Energiebedingung

$$\varepsilon_j + \varepsilon_k = \varepsilon_{j'} + \varepsilon_{k'} \tag{41.15}$$

erfüllt ist. Für die Größe $Z_{jk \to j'k'}$ machen wir den Ansatz

$$Z_{jk \to j'k'} = A_{jk \to j'k'} N_j N_k (g_{j'} + \gamma N_{j'}) (g_{k'} + \gamma N_{k'}) \tag{41.16}$$

mit

$$A_{jk \to j'k'} = A_{j'k' \to jk}, \tag{41.17}$$

den wir zunächst ohne weitere Begründung als Analogon des klassischen Stoßzahlansatzes einführen. Wir berechnen nun zunächst die zeitliche Änderung der Größe N_j. Mit Benutzung von (41.16) und (41.17) erhalten wir dafür

$$\left. \begin{array}{l} \dfrac{dN_j}{dt} = \sum_{k, j'k'} A_{jk \to j'k'} [N_{j'} N_{k'} (g_j + \gamma N_j) (g_k + \gamma N_k) - \\[2mm] \qquad\qquad - N_j N_k (g_{j'} + \gamma N_{j'}) (g_{k'} + \gamma N_{k'})]. \end{array} \right\} \tag{42.18}$$

Dabei ist die Summierung über alle k und über alle möglichen Paare j', k' zu erstrecken, welche für gegebene j und k der Bedingung (41.15) genügen. Für die zeitliche Änderung der Funktion H erhalten wir aus Gl. (41.13)

$$\frac{dH}{dt} = \sum_j \frac{dN_j}{dt} \ln \frac{N_j}{g_j + \gamma N_j}. \tag{41.19}$$

Setzen wir Gl. (41.18) hier ein, so folgt

$$\left. \begin{array}{l} \dfrac{dH}{dt} = \sum_j \sum_{k, j'k'} A_{jk \to j'k'} [N_{j'} N_{k'} (g_j + \gamma N_j) (g_k + \gamma N_k) - \\[2mm] \qquad\qquad - N_j N_k (g_{j'} + \gamma N_{j'}) (g_{k'} + \gamma N_{k'})] \ln \dfrac{N_j}{g_j + \gamma N_j}. \end{array} \right\} \tag{41.20}$$

Führen wir an Stelle der getrennten Summierung über j und k eine Summierung über alle Paare j, k ein, so wird

$$\left. \begin{array}{l} \dfrac{dH}{dt} = \sum_{jk, j'k'} A_{jk \to j'k'} [N_{j'} N_{k'} (g_j + \gamma N_j) (g_k + \gamma N_k) - \\[2mm] \qquad\qquad - N_j N_k (g_{j'} + \gamma N_{j'}) (g_{k'} + \gamma N_{k'})] \ln \dfrac{N_j N_k}{(g_j + \gamma N_j)(g_k + \gamma N_k)}. \end{array} \right\} \tag{41.21}$$

Bilden wir nun das arithmetische Mittel aus (41.21) und den Ausdruck, der daraus durch Vertauschen der Indices j, k und j', k' entsteht, so erhalten wir schließlich

$$\frac{dH}{dt} = \frac{1}{2} \sum_{jk, j'k'} A_{jk \to j'k'} \left[N_{j'} N_{k'} (g_j + \gamma N_j)(g_k + \gamma N_k) - \right.$$
$$\left. - N_j N_k (g_{j'} + \gamma N_{j'})(g_{k'} + \gamma N_{k'}) \right] \ln \frac{N_j N_k (g_{j'} + \gamma N_{j'})(g_{k'} + \gamma N_{k'})}{N_{j'} N_{k'} (g_j + \gamma N_j)(g_k + \gamma N_k)} . \tag{41.22}$$

Auf der rechten Seite dieser Gleichung können weder $A_{jk \to j'k'}$ und die N_j (wegen ihrer physikalischen Bedeutung) noch (im Falle der Fermi-Dirac-Statistik) die $(g_j + \gamma N_j)$ (wegen des Pauli-Prinzips) negativ sein. Der Ausdruck auf der rechten Seite hat daher die Form $\sum A(x - y) \ln \frac{x}{y}$, wobei keine der Größen A, x, y negativ werden kann. Daraus folgt unmittelbar

$$\frac{dH}{dt} \leqq 0, \tag{41.23}$$

wobei das Gleichheitszeichen dem erreichten Gleichgewicht entspricht. Für diesen letzteren Fall ergibt sich als Lösung der Gl. (41.22) mit der Nebenbedingung (41.15) die Verteilungsformel

$$N_j = \frac{g_j}{e^{\alpha + \beta \varepsilon_j} + \gamma}, \tag{41.24}$$

die wir später (Ziff. 50 und 57) nach anderen Methoden ableiten werden. Für $\gamma = 0$ geht sie in das Maxwell-Boltzmannsche Verteilungsgesetz Gl. (27.11) über. Die Beziehung (41.23), das quantenmechanische H-Theorem, führt zunächst auf eine analoge Problematik wie das klassische Boltzmannsche H-Theorem. Es ist daher notwendig, die dem Stoßzahlansatz entsprechende Gl. (41.16) durch eine genauere Analyse zu rechtfertigen. Wir skizzieren hier die Ableitung und verweisen für eine ausführliche Diskussion des Problems auf TOLMAN [12]. Die grundlegende Annahme besagt, daß der Operator V die Form

$$\mathsf{V} = \sum_{l < m} \mathsf{V}_{lm} \tag{41.25}$$

hat, wo l und m Indices der Teilchen sind und die Summierung über alle Paare von Teilchen zu erstrecken ist. Die Gl. (41.25) entspricht der Annahme von Zweierstößen in der klassischen Theorie. Wir benötigen nun das Matrixelement $V_{a'a}$, das dem Übergang von einem Zustand a mit $n_\varkappa$ Eigenfunktionen $\psi_\varkappa$, n_λ Eigenfunktionen ψ_λ, $n_{\varkappa'}$ Eigenfunktionen $\psi_{\varkappa'}$, und $n_{\chi'}$ Eigenfunktionen $\psi_{\chi'}$ zu einem Zustand a' mit $n_\varkappa - 1$ Faktoren $\psi_\varkappa$, $n_\lambda - 1$ Faktoren ψ_λ, $n_{\varkappa'} + 1$ Faktoren $\psi_{\varkappa'}$ und $n_{\chi'} + 1$ Faktoren $\psi_{\chi'}$ entspricht. Mit Gl. (41.5) bis (41.7) und (41.25) erhalten wir dafür den allgemeinen Ausdruck

$$V_{a'a} = C'^* C \int \sum_{P_q} (\pm 1)^{|P_q|} P_q \prod_i \psi_j^*(\boldsymbol{q}_i) \sum_{l < m} \mathsf{V}_{lm} \sum_{P_q} (\pm 1)^{|P_q|} P_q \prod_i \psi_j(\boldsymbol{q}_i) \, d\boldsymbol{q}. \tag{41.26}$$

Für den angenommenen Übergang können wir dann im Falle der Fermi-Dirac-Statistik explizit schreiben

$$V_{a'a} = C_{FD}'^* C_{FD} \int \sum_{P_q} (-1)^{|P_q|} P_q \psi_{\varkappa'}^*(\boldsymbol{q}_1) \psi_\lambda^*(\boldsymbol{q}_2) \psi_\mu^*(\boldsymbol{q}_3) \dots \psi_\sigma^*(\boldsymbol{q}_N) \times$$
$$\times \sum_{l < m} \mathsf{V}_{lm} \sum_{P_q} (-1)^{|P_q|} P_q \psi_\varkappa(\boldsymbol{q}_1) \psi_\lambda(\boldsymbol{q}_2) \psi_\mu(\boldsymbol{q}_3) \dots \psi_\sigma(\boldsymbol{q}_N) \, d\boldsymbol{q}. \tag{41.27}$$

Für die Bose-Einstein-Statistik lautet der entsprechende Ausdruck

$$
\left.
\begin{aligned}
V_{a'a} = C_{\mathrm{BE}}'^{*} C_{\mathrm{BE}} \int \sum_{P_q} \mathsf{P}_q \, \psi_{\varkappa'}^{*\,n_{\varkappa'}+1}\, \psi_{\lambda'}^{*\,n_{\lambda'}+1}\, \psi_{\varkappa}^{*\,n_{\varkappa}-1}\, \psi_{\lambda}^{*\,n_{\lambda}-1}\, \psi_{\mu}^{*\,n_{\mu}} \dots \psi_{\sigma}^{*\,n_{\sigma}} \times \\
\times \sum_{l<m} \mathsf{V}_{lm} \sum_{P_q} \mathsf{P}_q \, \psi_{\varkappa'}^{n_{\varkappa'}}\, \psi_{\lambda'}^{n_{\lambda'}}\, \psi_{\varkappa}^{n_{\varkappa}}\, \psi_{\lambda}^{n_{\lambda}}\, \psi_{\mu}^{n_{\mu}} \dots \psi_{\sigma}^{n_{\sigma}}\, d\boldsymbol{q}.
\end{aligned}
\right\} \quad (41.28)
$$

Mit Berücksichtigung der Gln. (41.8) und (41.9) für die Normierungsfaktoren erhält man daraus

$$
V_{a'a} = |I_1 \pm I_2|^2\, n_{\varkappa} n_{\lambda} (n_{\varkappa'} \pm 1)(n_{\lambda'} \pm 1) \tag{41.29}
$$

mit

$$
I_1 = \iint \psi_{\varkappa'}^{*}(\boldsymbol{q}_1)\, \psi_{\lambda'}^{*}(\boldsymbol{q}_2)\, \mathsf{V}_{lm} \psi_{\varkappa}(\boldsymbol{q}_1)\, \psi_{\lambda}(\boldsymbol{q}_2)\, d\boldsymbol{q}_1\, d\boldsymbol{q}_2 \tag{41.30}
$$

und

$$
I_2 = \iint \psi_{\varkappa'}^{*}(\boldsymbol{q}_2)\, \psi_{\lambda'}^{*}(\boldsymbol{q}_1)\, \mathsf{V}_{lm} \psi_{\varkappa}(\boldsymbol{q}_1)\, \psi_{\lambda}(\boldsymbol{q}_2)\, d\boldsymbol{q}_1\, d\boldsymbol{q}_2. \tag{41.31}
$$

Dabei bezieht sich in Gl. (41.29) das positive Vorzeichen auf die Bose-Einstein-Statistik, das negative auf die Fermi-Dirac-Statistik.

Wir fragen nun nach der Wahrscheinlichkeit, ein System, das sich zur Zeit $t=0$ in einem der durch die Besetzungszahlen (der Eigenfunktionen!) $n_{\varkappa}$, n_{λ}, $n_{\varkappa'}$, $n_{\lambda'}$ gekennzeichneten Zustände befindet, zur Zeit t in einem der durch $n_{\varkappa'}+1$, $n_{\lambda'}+1$, $n_{\varkappa}-1$, $n_{\lambda}-1$ gekennzeichneten Zustände anzutreffen. Die Zustände zur Zeit $t=0$ bezeichnen wir mit dem Index i (bzw. i' für die konjugiert komplexe Größe), die Zustände zur Zeit t mit dem Index f, die zugehörigen Wahrscheinlichkeitsamplituden mit $a_i(t)$ und $a_f(t)$. Die Zustände i sollen, wie vorher durch die Besetzungszahlen N_j, N_k, $N_{j'}$, $N_{k'}$, die Zustände f durch N_j-1, N_k-1, $N_{j'}+1$, $N_{k'}+1$ spezifiziert sein. Wir nehmen an, daß eine Messung zur Zeit $t=0$ uns die Anfangsbedingungen

$$
\sum_i W_i(0) = \sum_i a_i^{*}(0)\, a_i(0) = 1, \tag{41.32}
$$

$$
a_f(0) = 0 \quad (\text{alle } f) \tag{41.33}
$$

liefert. Die Diracsche Methode der Variation der Konstanten ergibt dann für die Wahrscheinlichkeitsamplituden zur Zeit t

$$
a_f(t) = \sum_i V_{fi} \frac{e^{\frac{2\pi i}{h}(E_f - E_i)t} - 1}{E_i - E_f}\, a_i(0), \tag{41.34}
$$

wo E_i und E_f die zu den Zuständen i und f gehörenden Eigenwerte des ungestörten Hamilton-Operators und V_{fi} die Elemente der Störungsmatrix sind. Durch Multiplikation mit der konjugiert komplexen Größe

$$
a_f^{*}(t) = \sum_{i'} V_{fi'}^{*} \frac{e^{-\frac{2\pi i}{h}(E_f - E_{i'})t} - 1}{E_{i'} - E_f}\, a_{i'}^{*}(0) \tag{41.35}
$$

und Summierung über alle in Betracht kommenden f erhalten wir für die gesuchte Wahrscheinlichkeit

$$
\left.
\begin{aligned}
\sum_f W_f(t) &= \sum_f a_f^{*}(t)\, a_f(t) \\
&= \sum_f \sum_{i'} \sum_i V_{fi'}^{*} V_{fi} \frac{\left(e^{-\frac{2\pi i}{h}(E_f - E_{i'})} - 1\right)\left(e^{\frac{2\pi i}{h}(E_f - E_i)} - 1\right)}{(E_{i'} - E_f)(E_i - E_f)}\, a_{i'}^{*}(0)\, a_i(0).
\end{aligned}
\right\} \quad (41.36)
$$

Eine exakte Berechnung dieses Ausdruckes würde die Kenntnis der Größen $a_{i'}^*(0)$ und $a_i(0)$ erfordern, zu deren Bestimmung Gl. (41.32) nicht ausreicht. Es ist daher notwendig, an dieser Stelle von der Quantenstatistik Gebrauch zu machen, obwohl dies im Zusammenhang des hier untersuchten Problems etwas inkonsequent erscheint [22]. Wir konstruieren also die Dichtematrix nach dem Prinzip der gleichen a priori-Wahrscheinlichkeiten und der willkürlichen a priori-Phasen. Bezeichnen wir mit ω_i die Gesamtzahl der Anfangszustände i, so haben wir

$$\varrho_{ii'} = \overline{a_{i'}^*(0)\, a_i(0)} = \begin{cases} \dfrac{1}{\omega_i}\, \delta_{ii'} & \text{für die } \omega_i \text{ Anfangszustände,} \\[2mm] 0 & \text{für alle übrigen Zustände.} \end{cases} \tag{41.37}$$

Damit reduziert sich Gl. (41.36) auf

$$P_f(t) \equiv \sum_f \overline{W_f(t)} = \frac{4}{\omega_i} \sum_f \sum_i |V_{fi}|^2 \, \frac{\sin^2 \dfrac{\pi}{h}(E_f - E_i)\, t}{(E_f - E_i)^2}. \tag{41.38}$$

In diese Gleichung führen wir nun für V_{fi} Gl. (41.29) ein und setzen

$$E_f - E_i = \varepsilon_j + \varepsilon_k - \varepsilon_{j'} + \varepsilon_{k'}. \tag{41.39}$$

Ferner setzen wir

$$n_\varkappa = N_j/g_j, \qquad n_\lambda = N_k/g_k, \qquad n_{\varkappa'} = N_{j'}/g_{j'}, \qquad n_{\lambda'} = N_{k'}/g_{k'}. \tag{41.40}$$

Wir ersetzen also $n_\varkappa, n_\lambda, n_{\varkappa'}, n_{\lambda'}$ durch die Mittelwerte der betreffenden Gruppe von Zuständen. Diese Größen können dann vor die Summierung über $\varkappa, \lambda, \varkappa', \lambda'$ gezogen werden, und die Summierung über die Anfangszustände i hebt sich gegen den Faktor $1/\omega_i$. Wir haben dann

$$P_f(t) = 4|I_1 \pm I_2|^2 \frac{N_j}{g_j} \frac{N_k}{g_k}\left(1 \pm \frac{N_{j'}}{g_{j'}}\right)\left(1 \pm \frac{N_{k'}}{g_{k'}}\right) \sum_{\varkappa,\,\lambda,\,\varkappa',\,\lambda'} \frac{\sin^2 \dfrac{\pi}{h}(\varepsilon_j + \varepsilon_k - \varepsilon_{j'} - \varepsilon_{k'})\, t}{(\varepsilon_j + \varepsilon_k - \varepsilon_{j'} - \varepsilon_{k'})^2}. \tag{41.41}$$

Ersetzen wir die Summe durch ein Integral, so wird

$$\sum_{\varkappa,\,\lambda,\,\varkappa',\,\lambda'} \frac{\sin^2 \dfrac{\pi}{h}(\varepsilon_j + \varepsilon_k - \varepsilon_{j'} - \varepsilon_{k'})\, t}{(\varepsilon_j + \varepsilon_k - \varepsilon_{j'} - \varepsilon_{k'})^2} = \frac{g_j\, g_k\, g_{j'}\, g_{k'}}{\varDelta \varepsilon}\, \frac{\pi}{h}\, t, \tag{41.42}$$

wo $\varDelta \varepsilon$ das einer Gruppe von Zuständen entsprechende Energieintervall ist, und wir erhalten

$$P_f(t) = \frac{4\pi^2}{h}\, \frac{|I_1 \pm I_2|^2}{\varDelta \varepsilon}\, N_j N_k\, (g_{j'} \pm N_{j'})\, (g_{k'} \pm N_{j'})\, t. \tag{41.43}$$

Diese wichtige Beziehung zeigt, daß in erster Näherung, für nicht zu lange und nicht zu kurze Zeiten, die Wahrscheinlichkeit, das System in dem spezifizierten Endzustand zu finden, proportional der Zeit wächst. Aus (41.43) folgt unmittelbar die Gl. (41.16), und (41.17) ergibt sich aus dem hermitischen Charakter des Operators V.

Die vorstehende Ableitung ist insofern befriedigend, als sie klar den statistischen Charakter des H-Theorems hervortreten läßt. Insofern kann man sie in Parallele setzen zu den in Ziff. 16 angedeuteten Überlegungen. Andererseits kann nicht übersehen werden, daß die eigentlich entscheidende Frage, wie das H-Theorem mit dem Prinzip der mechanischen Umkehrbarkeit zu vereinbaren ist, doch weitgehend ungeklärt bleibt. Zunächst enthält die Ableitung eine Anzahl von Näherungen, deren Tragweite nicht leicht zu übersehen ist. Immerhin dürfte die Mehrzahl derselben nur periphere Bedeutung haben. Den Kern des Problems berührt jedoch die Art, wie von der Diracschen Störungstheorie Gebrauch gemacht wird[1]. Nach ihrer Ableitung können die Gln. (41.34) bis (41.36) nur für ein *kleines* Zeitintervall $t = 0$ bis $t = \tau$ gültig sein. Der Zeitablauf bis zum Erreichen des Gleichgewichtes muß also aus Intervallen τ zusammengesetzt werden, und am Beginn jedes Intervalls hat man von neuem die Hypothese der willkürlichen a priori-Phasen einzuführen. In dem oben skizzierten Gedankengang ist dieses Verfahren für das Zustandekommen des irreversiblen Charakters wesentlich. Es bleibt aber unklar, welches seine physikalische Bedeutung und Rechtfertigung ist und ob es wirklich notwendig ist, um zu einem irreversiblen Prozeß zu gelangen. In Ziff. 44 werden wir auf diese Fragen nochmals zurückkommen. Die hier gegebene Formulierung des H-Theorems führt zwangsläufig wieder auf die erste Boltzmannsche Definition des Gleichgewichtes (Ziff. 6). Da unter diesem Gesichtspunkt die Verhältnisse in der Quantenstatistik sich nicht wesentlich von denen der klassischen Theorie unterscheiden, können wir für die Diskussion des Problems auf die Ausführungen in Ziff. 16 und 17 verweisen.

42. „Master-equation" und H-Theorem. Wir wollen nun das Problem des H-Theorems vom Standpunkt der quantenstatistischen Ensemble-Theorie betrachten. Der erste Schritt dazu besteht, ähnlich wie in der klassischen Theorie, notwendig in der Einführung einer groben Dichtematrix. Wir haben dieselbe bereits in Ziff. 40 benutzt. Es ist aber üblich, sie in diesem Zusammenhang etwas anders zu definieren. Zum Unterschied von der früheren Größe werden wir jetzt die Matrixelemente doppelt indicieren. Wir definieren also für die Darstellung der Energie-Eigenfunktionen

$$P_{kl} = \frac{P_i}{\omega_i} = \delta_{kl} \sum_j \varrho_{jj}/\omega_i. \qquad (42.1)^2$$

Die Matrix P (die wir nicht besonders kennzeichnen) soll bis zum Schluß des Abschnitts B III im Sinne dieser Definition verstanden werden. Aus der Definition folgt sofort

$$\text{spur } \mathsf{P} = 1. \qquad (42.2)$$

Ferner definieren wir zwei Funktionen

$$\sigma = \text{spur } (\rho \ln \rho) \qquad (42.3)$$

und

$$\overline{H} = \text{spur } (\mathsf{P} \ln \mathsf{P}). \qquad (42.4)$$

Wir betrachten zunächst die Funktion σ und zeigen, daß dieselbe sich im Laufe der Zeit nicht ändert. Der Beweis beruht auf der Tatsache, daß die feinen Dichtematrizen für zwei Zeitpunkte t' und t'', die wir mit ρ' und ρ'' bezeichnen, durch eine unitäre Transformation zusammenhängen. Es ist also

$$\rho'' = \mathsf{U}^\dagger \, \rho' \, \mathsf{U} \qquad (42.5)$$

mit

$$\mathsf{U}^\dagger \mathsf{U} = \mathsf{U} \mathsf{U}^\dagger = 1. \qquad (42.6)$$

---

¹ N. G. van Kampen: Physica, Haag **20**, 603 (1954).

² Man bemerkt, daß wir die durch Gl. (42.1) definierte *grobe* Dichtematrix in Ziff. 40 als *feine* Dichtematrix bezeichnet haben. Dies erklärt sich dadurch, daß wir von verschiedenen Überlegungen her zu dieser Größe gelangt sind. In Ziff. 40 handelt es sich um ein Axiom, das die Konstruktion der feinen Dichtematrix auf Grund einer makroskopischen Beobachtung ermöglicht. Hier ist dagegen die feine Dichtematrix im Sinne der abstrakten Ensemble-Theorie des Teiles B I gemeint. Die Gleichsetzung der feinen mit der durch (42.1) definierten groben Dichtematrix für die Zeit der Beobachtung t' bildet den Ausgangspunkt für die Ableitung des verallgemeinerten H-Theorems. Vgl. Ziff. 44.

Der durch Gl. (34.22) definierte quantenstatistische Erwartungswert ist nun invariant gegen unitäre Transformationen. Wegen

$$\text{spur}\,(\mathsf{A}\,\mathsf{B}) = \text{spur}\,(\mathsf{B}\,\mathsf{A}) \tag{42.7}$$

ist nämlich

$$\begin{aligned}
\text{spur}\,(\rho''\,\mathsf{F}'') &= \text{spur}\,(\mathsf{U}^\dagger\,\rho'\,\mathsf{U}\,\mathsf{U}^\dagger\,\mathsf{F}'\,\mathsf{U}) = \text{spur}\,(\mathsf{U}^\dagger\,\rho'\,\mathsf{F}'\,\mathsf{U}) \\
&= \text{spur}\,(\mathsf{F}'\,\mathsf{U}\,\mathsf{U}^\dagger\,\rho') = \text{spur}\,(\rho'\,\mathsf{F}').
\end{aligned} \right\} \tag{42.8}$$

Setzen wir für F jetzt $\ln\rho$, so folgt mit Gl. (42.3)

$$\sigma' = \sigma'' \tag{42.9}$$

in Analogie zu der klassischen Gl. (17.9). Die Funktion σ kann daher nicht zur Formulierung des H-Theorems benutzt werden; es ist vielmehr notwendig, die mit Hilfe der groben Dichtematrix definierte Funktion $\overline{H}$ zugrunde zu legen.

Nach diesen vorbereitenden Bemerkungen betrachten wir zunächst eine Ableitung des verallgemeinerten H-Theorems, die ebenfalls von Pauli[1] herrührt und in nahem Zusammenhang mit der in Ziff. 41 benutzten Methode steht, andererseits aber auf eine Gleichung führt, die in den neuesten Untersuchungen eine besondere Rolle spielt. Die Grundlage der Ableitung bildet wieder die Anwendung der Diracschen Störungstheorie, die zum Konzept der zeitproportionalen Übergänge führt. Wir formulieren zunächst diese Überlegung in einer für unseren jetzigen Zweck geeigneten Form. Den Hamilton-Operator nehmen wir wieder in der Form

$$\mathsf{H} = \mathsf{H}_0 + \mathsf{V} \tag{42.10}$$

an, wo aber jetzt der Störungsoperator eine Wechselwirkung mit der Umgebung beschreibt, welche Übergänge zwischen den stationären Zuständen des Gesamtsystems, d.h. den Eigenfunktionen des Operators H_0 ermöglicht. Die Methode der Variation der Konstanten wird nun in der gleichen Weise wie in Ziff. 41 angewendet, wobei lediglich die dort gegebene Spezifikation der Anfangs- und Endzustände jetzt wegfällt. Die Wahrscheinlichkeit, das System zur Zeit t in einem der Zustände j zu finden, ist somit für die Anfangsbedingungen (41.32) und (41.33) wieder

$$P_j(t) = \frac{4}{\omega_i}\sum_n\sum_k |V_{kn}|^2\,\frac{\sin^2\dfrac{\pi}{h}(E_k - E_n)\,t}{(E_k - E_n)^2}, \tag{42.11}[2]$$

wobei der Index i die Gruppe der Anfangszustände mit den Eigenwerten E_k, der Index j die Gruppe der Endzustände mit den Eigenwerten E_n bezeichnet. Die Summierungen ersetzen wir näherungsweise durch Integrationen, indem wir für die Zahlen der Zustände in den Energie-Intervallen dE_k und dE_n schreiben

$$dn_k = \sigma_k\,dE_k, \qquad dn_n = \sigma_n\,dE_n. \tag{42.12}$$

Für die Integrationen betrachten wir $|V_{kn}|^2$, σ_k und σ_n näherungsweise als Konstanten. Wir haben dann

$$P_j(t) = \frac{4}{\omega_i}\,|V_{kn}|^2\,\sigma_k\,\sigma_n\int_E^{E+\varDelta E}\int_E^{E+\varDelta E}\frac{\sin^2\dfrac{\pi}{h}(E_k - E_n)\,t}{(E_k - E_n)^2}\,dE_k\,dE_n. \tag{42.13}$$

---

[1] W. Pauli: Sommerfeld-Festschrift. Leipzig 1928.
[2] Der Einfachheit halber sind die Eigenwerte als nicht entartet angenommen. Die Verallgemeinerung macht keine Schwierigkeiten. Vgl. [12].

Für eine hinreichend lange Zeit können wir annehmen, daß die Differenz $(E_k - E_n)$ alle Werte von $-\infty$ bis $+\infty$ annimmt. Wir integrieren daher zunächst für konstantes E_n zwischen diesen Grenzen über E_k und dann über E_n zwischen den angeschriebenen Grenzen. Damit erhalten wir schließlich

$$P_j(t) = \frac{1}{\omega_i} \frac{4\pi^2}{h} |V_{kn}|^2 \sigma_k \sigma_n \, \Delta E \, t \tag{42.14}$$

oder in abgekürzter Schreibweise

$$P_j = \frac{T_{ij}}{\omega_i} t. \tag{42.15}$$

Die Wahrscheinlichkeit, das System in einem der Zustände der Gruppe j zu finden, wächst somit wieder proportional der Zeit. Für die Anfangsbedingung, daß sich das System zur Zeit $t=0$ in einem der Zustände der Gruppe j befindet, erhält man in gleicher Weise

$$P_i = \frac{T_{ij}}{\omega_j} t. \tag{42.16}$$

Wenn die Beobachtung zur Zeit $t=0$ ergibt, daß sich das System mit den Wahrscheinlichkeiten $P_i, P_j, \ldots$ in einer der Gruppen $i, j, \ldots$ befindet, so tritt an Stelle der Gl. (41.37) für die Dichtematrix die Form

$$\varrho_{kl} = (P_i/\omega_i) \, \delta_{kl}, \tag{42.17}$$

die wir schon in Ziff. 40 [Gl. (40.11)] eingeführt haben. Bezeichnen wir mit Z_{ij} die mittlere Zahl der Übergänge pro Zeiteinheit von der Gruppe i zur Gruppe j, so ergibt eine der obigen analoge Rechnung

$$Z_{ij} = \frac{T_{ij} P_i}{\omega_i} \tag{42.18}$$

oder, mit

$$A_{ij} = \frac{T_{ij}}{\omega_i \omega_j}, \tag{42.19}$$

$$Z_{ij} = A_{ij} \omega_j P_i. \tag{42.20}$$

Aus dem hermitischen Charakter des Störungsoperators folgt schließlich

$$T_{ij} = T_{ji} \tag{42.21}$$

und

$$A_{ij} = A_{ji}. \tag{42.22}$$

Es ist nun

$$\frac{dP_i}{dt} = \sum_{j \neq i} (Z_{ji} - Z_{ij}), \tag{42.23}$$

und wir erhalten

$$\frac{dP_i}{dt} = \sum_j (A_{ji} \omega_i P_j - A_{ij} \omega_j P_i) \tag{42.24}$$

oder in etwas kompakterer Form

$$\frac{dP_i}{dt} = \sum_j (W_{ji} P_j - W_{ij} P_i), \tag{42.25}$$

wobei jetzt die Symmetriebeziehung

$$W_{ij} \omega_i = W_{ji} \omega_j \tag{42.26}$$

lautet. Setzen wir

$$p_k = P_i/\omega_i, \qquad w_{kn} = W_{ij}/\omega_i \qquad (42.27)$$

und behandeln k und n als durchlaufende Indices, so wird aus Gl. (42.25)

$$\frac{dp_k}{dt} = \sum_n (w_{nk}\,p_n - w_{kn}\,p_k) \qquad (42.28)$$

mit

$$w_{kn} = w_{nk}, \qquad (42.29)$$

was wir bereits in Ziff. 38 benutzt haben.

Die Differentialgleichung erster Ordnung (42.24) [bzw. (42.25) oder (42.28)], die in der neueren Literatur als „*master equation*“ bezeichnet wird, spielt für die quantenstatistische Theorie der Transportvorgänge eine ähnliche zentrale Rolle wie die Boltzmannsche Stoßgleichung in der klassischen Gastheorie. Die Frage ihrer Begründung und ihrer Beziehung zur Schrödinger-Gleichung (die das Prinzip der mechanischen Umkehrbarkeit involviert) ist daher in den letzten Jahren Gegenstand zahlreicher Untersuchungen gewesen. Wir werden in Ziff. 44 kurz darauf zurückkommen.

Man sieht nun leicht, daß das verallgemeinerte H-Theorem eine unmittelbare Folgerung aus der master equation ist. Aus der Definition (42.4) folgt nämlich mit Berücksichtigung von (42.1) und $\sum P_i = 1$

$$\frac{d\bar{H}}{dt} = \sum_i (\ln P_i - \ln \omega_i)\,\frac{dP_i}{dt} \qquad (42.30)$$

und mit Gl. (42.24)

$$\frac{d\bar{H}}{dt} = \sum_{i,j} (A_{ji}\,\omega_i\,P_j - A_{ij}\,\omega_j\,P_i)\,\ln\frac{P_i}{\omega_i}. \qquad (42.31)$$

Bilden wir das arithmetische Mittel aus diesem Ausdruck und dem daraus durch Vertauschen der Indices erhaltenen, so folgt schließlich mit Berücksichtigung von (42.22)

$$\frac{d\bar{H}}{dt} = \frac{1}{2} \sum_{i,j} A_{ij}(\omega_i\,P_j - \omega_j\,P_i)\,\ln\frac{\omega_j P_i}{\omega_i P_j}. \qquad (42.32)$$

Da alle auf der rechten Seite vorkommenden Größen nach ihrer physikalischen Bedeutung positiv sein müssen, folgt in gleicher Weise wie in Ziff. 41

$$\frac{d\bar{H}}{dt} \leqq 0. \qquad (42.33)$$

Das Gleichheitszeichen gilt wieder für den Fall des Gleichgewichtes, der durch die Gleichungen

$$\frac{P_j}{\omega_j} = \frac{P_i}{\omega_i} \quad \text{für } A_{ij} \neq 0 \qquad (42.34)$$

charakterisiert ist. Mit Benutzung von (42.20) und (42.22) folgt daraus

$$Z_{ij} = Z_{ji}. \qquad (42.35)$$

Diese Gleichung besagt, daß im Gleichgewicht die mittleren Zahlen der Übergänge pro Zeiteinheit von der Gruppe i zur Gruppe j und in umgekehrter Richtung einander gleich sind. Sie ist ein weiteres Beispiel für das schon in Ziff. 10 erwähnte „Prinzip des detaillierten Gleichgewichtes“, welches besagt, daß im Gleichgewicht die mittlere Zahl der Prozesse pro Zeiteinheit, die eine Situation A vernichten und eine Situation B herbeiführen, gleich ist der Zahl der Prozesse, die eine Situation B vernichten und A herbeiführen. Dieses Prinzip, dessen Grundgedanke auf Kirchhoff[1] zurückgeht und das zuerst wohl von Richardson[2] formuliert wurde, hat sich für die Behandlung zahlreicher Probleme der Physik und physikalischen Chemie[3] als außerordentlich fruchtbar erwiesen. Es darf jedoch nicht übersehen werden, daß dasselbe schon im Gleichgewicht nicht allgemein gültig ist[4-6]. Für stationäre Nicht-Gleichgewichtszustände muß es grundsätzlich als unzutreffend betrachtet werden[7].

Die durch Gl. (42.33) gegebene Formulierung des H-Theorems nimmt eine intermediäre Stellung ein zwischen dem „gaskinetischen" H-Theorem der Ziff. 41 und dem verallgemeinerten H-Theorem der Quantenstatistik, das wir in Ziff. 44 behandeln. Formal beruht die Ableitung auf den gleichen Prinzipien wie die von (41.23). Die Diskussion am Schluß von Ziff. 41 läßt sich daher unmittelbar auf den gegenwärtigen Fall übertragen. Die H-Funktion ist aber, im Gegensatz zu der früheren Formulierung, jetzt durch Gl. (42.4) und damit für eine quantenstatistische Gesamtheit definiert. (42.33) führt daher notwendig auf die Gibbssche Definition des Gleichgewichtes (vgl. Ziff. 17). Die Beziehung beider Formulierungen zueinander ist leicht zu übersehen, wenn wir Gl. (42.4) in der Form

$$\bar{H} = \sum_i P_i \ln P_i - \sum_i P_i \ln \omega_i \qquad (42.36)$$

schreiben. Da die Größe Ω_D der Ziff. 41 gleich ist der Zahl der Zustände in einer Gruppe ω_i, können wir die Definition (41.12) hier in der Form

$$H_{\text{Syst}} = -\ln \omega_i \qquad (42.37)$$

einführen. Dann wird aus (42.36)

$$\bar{H} = \bar{H}_{\text{Syst}} + \sum_i P_i \ln P_i, \qquad (42.38)$$

wo $\bar{H}_{\text{Syst}}$ der Mittelwert der für das Einzelsystem definierten H-Funktion über die Gesamtheit ist. Aus (42.38) folgt, daß

$$\bar{H} = H_{\text{Syst}} \qquad (42.39)$$

wird, wenn die Messung ergibt, daß sich das Originalsystem in der Gruppe i befindet und daher $P_i = 1$ und $P_{j \neq i} = 0$ ist. Weiter ergibt sich

$$\frac{d\bar{H}}{dt} = \frac{d\bar{H}_{\text{Syst}}}{dt} + \sum_i \frac{dP_i}{dt} \ln P_i. \qquad (42.40)$$

Diese Gleichung zeigt, daß die Abnahme von $\bar{H}$ auf zwei Effekten beruht: Der Abnahme von H_{Syst} für die Systeme der Gesamtheit und der Tendenz zur Abnahme von $\sum_i P_i \ln P_i$ durch eine gleichmäßigere Verteilung über die verschiedenen Gruppen von Zuständen. Gl. (42.40) läßt den statistischen Charakter des H-Theorems für das Einzelsystem schärfer hervortreten als die in Ziff. 41 gegebene Ableitung, bei der die Statistik gewissermaßen nur durch eine Hintertür eingeführt wurde.

[1] G. Kirchhoff: Pogg. Ann. 109, 148 (1860).
[2] O.W. Richardson: Phil. Mag. 27, 476 (1914).
[3] Zusammenstellung der Literatur bei ter Haar [23].
[4] H.A. Lorentz: Wien. Ber. 95, 115 (1887).
[5] J. Hamilton u. H.W. Peng: Proc. Roy. Ir. Acad. A 49, 197 (1944).
[6] H.G. van Kampen: Physica 20, 603 (1954).
[7] M.J. Klein: Phys. Rev. 97, 1446 (1955).

Die Aussage (42.33) geht wesentlich über das in Ziff. 44 zu behandelnde verallgemeinerte H-Theorem hinaus, indem sie einen monotonen Abfall der Funktion $\bar{H}$ bis zum Gleichgewicht behauptet. Diese größere Allgemeinheit ist indessen nur scheinbar. Selbst wenn wir von der Problematik der hier gegebenen Ableitung absehen, so zeigen die neueren Untersuchungen (vgl. Ziff. 45), daß die master equation nur als Näherung bei Berücksichtigung der niedrigsten Ordnung der Störung gültig ist. Die Einführung des Konzepts der Übergangswahrscheinlichkeiten bedingt überdies, daß die Begründung der mikrokanonischen Gesamtheit aus dem H-Theorem hier nur mit Hilfe der Ergodenhypothese erfolgen kann, wie wir dies bereits in Ziff. 38 angedeutet haben. Schließlich bleibt unklar, ob und inwieweit spezifisch quantenmechanische Gesichtspunkte in das H-Theorem eingreifen. Für die Diskussion der in dieser Ziffer erörterten Fragen sei auch auf eine neuere Arbeit von Landsberg[1] verwiesen.

43. Das Kleinsche Lemma. Für die Ableitung des verallgemeinerten H-Theorems benötigen wir ein Lemma, das zuerst von Klein[2] bewiesen wurde.

Wir betrachten zwei Dichtematrizen ρ' und ρ'', die das System zu den Zeiten t und t'' darstellen. Ihre Elemente seien ϱ'_{kl} und ϱ''_{kl} und es sei

$$\varrho'_{kl} = \varrho'_{kk}\,\delta_{kl}, \tag{43.1}$$

während ρ'' nicht notwendig eine Diagonalmatrix ist. Da die Diagonalelemente ϱ'_{kk} und ϱ''_{kk} die Wahrscheinlichkeiten sind, das System im Zustand k zu finden, können sie nicht negativ sein. Mit Rücksicht auf die Eigenschaft von (17.10) kann daher auch der Ausdruck

$$Q_{kn} = \varrho'_{kk}(\ln \varrho'_{kk} - \ln \varrho''_{nn} - 1) + \varrho''_{nn} \tag{43.2}$$

nicht negativ sein. Die Matrixelemente ϱ''_{kl} sind definitionsgemäß durch die Wahrscheinlichkeitsamplituden zur Zeit t'' bestimmt, die ihrerseits aus den Wahrscheinlichkeitsamplituden zur Zeit t' durch Integration der zeitabhängigen Schrödinger-Gleichung erhalten werden. Die durch Gl. (40.17) gegebene Lösung läßt sich bekanntlich als unitäre Transformation schreiben, wenn man den Operator

$$\mathsf{U} = \mathrm{e}^{-\frac{2\pi i \mathsf{H}}{h}t} \tag{43.3}$$

einführt. Wir haben dann

$$a_n(t'') = \sum_l U_{nl}\, a(t') \tag{43.4}$$

und für die Elemente der Dichtematrix

$$\varrho''_{nn} = \sum_k |U_{nk}|^2 \varrho'_{kk} + \sum_{l \neq k} U^*_{nl} U_{nk}\, \varrho'_{kl}. \tag{43.5}$$

Wegen Gl. (43.1) reduziert sich (43.5) auf

$$\varrho''_{nn} = \sum_k |U_{nk}|^2 \varrho'_{kk}. \tag{43.6}$$

Multiplizieren wir nun Gl. (43.2) mit der wesentlich positiven Größe $|U_{nk}|^2$ und summieren über alle n und k, so erhalten wir

$$\left.\begin{aligned}
\sum_{k,n} |U_{nk}|^2 Q_{kn} = \sum_{k,n} |U_{nk}|^2 \varrho'_{kk} \ln \varrho'_{kk} - \sum_{k,n} |U_{nk}|^2 \varrho'_{kk} \ln \varrho''_{nn} - \\
- \sum_{k,n} |U_{nk}|^2 \varrho'_{kk} + \sum_{k,n} |U_{nk}|^2 \varrho''_{nn} \geqq 0.
\end{aligned}\right\} \tag{43.7}$$

---

[1] P. T. Landsberg: Phys. Rev. **96**, 1420 (1954).
[2] O. Klein: Z. Physik **72**, 767 (1931).

Da für unitäre Matrizen
$$\sum_n U_{nk}^* U_{nl} = \delta_{kl} \qquad (43.8)$$

ist, folgt mit Benutzung von Gl. (43.6)

$$\sum_k \varrho'_{kk} \ln \varrho'_{kk} - \sum_n \varrho''_{nn} \ln \varrho''_{nn} - \sum_k \varrho'_{kk} + \sum_n \varrho''_{nn} \geqq 0 \qquad (43.9)$$

oder, wegen der für ϱ' und ϱ'' gültigen Normierungsrelation (34.19),

$$\sum_k \varrho'_{kk} \ln \varrho'_{kk} - \sum_n \varrho''_{nn} \ln \varrho''_{nn} \geqq 0. \qquad (43.10)$$

Diese Relation wird als *Kleinsches Lemma* bezeichnet. Sie erscheint auf den ersten Blick überraschend mit Rücksicht auf die klassische Beziehung

$$\int \varrho' \ln \varrho' \, d\Omega - \int \varrho'' \ln \varrho'' \, d\Omega = 0. \qquad (43.11)$$

Das quantenstatistische Analogon dieser Beziehung ist jedoch die aus (42.9) folgende Gleichung

$$\sum_k [\varrho' \ln \varrho']_{kk} - \sum_n [\varrho'' \ln \varrho'']_{nn} = 0, \qquad (43.12)$$

die nur dann mit (43.10) identisch wird, wenn ρ'' eine Diagonalmatrix ist. Die in (43.10) auftretende Größe $\sum_k \varrho_{kk} \ln \varrho_{kk}$ ist dagegen im allgemeinen nicht die Spur eines quantenmechanischen Operators. Sie hat daher kein Gegenstück in der klassischen Theorie, und (43.10) stellt einen spezifisch quantenmechanischen Effekt dar. Da der Wert der Summe $\sum_k \varrho_{kk} \ln \varrho_{kk}$ um so kleiner ist, je mehr Zustände besetzt sind, bezeichnet Tolman [12] diesen Effekt als die ,,quantenmechanische Ausbreitung der feinen Wahrscheinlichkeit". Dieselbe tritt, wie man sofort aus Gl. (43.4) ableitet, bereits bei einem einzelnen System auf und hängt mit der Störung des Systems durch die Beobachtung zusammen. Sie hat daher ihren Ursprung in dem statistischen Charakter der Quantenmechanik und ist zu unterscheiden von den Effekten, die mit der Einführung der statistischen Mechanik in der Form der Quantenstatistik zusammenhängen. Dieser Unterschied zwischen den beiden Effekten ergibt sich besonders deutlich daraus, daß (43.10) offenbar mit der Unbestimmtheitsrelation zusammenhängt, während makroskopische Messungen (mit denen es die Quantenstatistik zu tun hat), wie wir in Ziff. 37 ausführlich erörtert haben, dadurch charakterisiert sind, daß die Unbestimmtheitsrelation keine Rolle gegenüber der experimentellen Unsicherheit spielt und daher alle makroskopischen Operatoren vertauschbar sind.

44. Das verallgemeinerte H-Theorem. Wir nehmen nun an, daß eine Beobachtung zur Zeit t' uns die Elemente der groben Dichtematrix P' geliefert hat. Nach der Hypothese der gleichen a priori-Wahrscheinlichkeiten und der willkürlichen a priori-Phasen ist dann die feine Dichtematrix zur Zeit t'

$$\rho' = \mathsf{P}'. \qquad (44.1)$$

Es wird somit
$$\overline{H}' = \mathrm{spur}\,(\mathsf{P}' \ln \mathsf{P}') = \mathrm{spur}\,(\rho' \ln \rho'). \qquad (44.2)$$

Wir bilden nun
$$\overline{H}' - \overline{H}'' = \mathrm{spur}\,(\mathsf{P}' \ln \mathsf{P}') - \mathrm{spur}\,(\mathsf{P}'' \ln \mathsf{P}''). \qquad (44.3)$$

Da P stets eine Diagonalmatrix ist, erhalten wir mit (44.2)

$$\left. \begin{aligned} \overline{H}' - \overline{H}'' &= \mathrm{spur}\,(\rho' \ln \rho') - \mathrm{spur}\,(\rho'' \ln \mathsf{P}'') \\ &= \sum_k \varrho'_{kk} \ln \varrho'_{kk} - \sum_n \varrho''_{nn} \ln P''_{nn}. \end{aligned} \right\} \qquad (44.4)$$

Führen wir hier das Kleinsche Lemma (43.10) ein und addieren rechts noch die aus den Normierungsrelationen folgende Beziehung

$$\text{spur } \mathsf{P}'' - \text{spur } \rho'' = 0, \tag{44.5}$$

so wird

$$\bar{H}' - \bar{H}'' \geqq \sum_n \left(\varrho''_{nn} \ln \varrho''_{nn} - \varrho''_{nn} \ln P''_{nn} - \varrho''_{nn} + P''_{nn} \right). \tag{44.6}$$

Für den rechts stehenden Ausdruck ergibt sich aus (17.10) mittels der Substitution $x = \ln \alpha - \ln \beta$

$$\alpha \ln \alpha - \alpha \ln \beta - \alpha + \beta \begin{cases} > 0 \ \text{ für } \alpha \neq \beta, \\ = 0 \ \text{ für } \alpha = \beta. \end{cases} \tag{44.7}$$

Wir erhalten somit

$$\bar{H}' - \bar{H}'' \geqq 0. \tag{44.8}$$

Diese Aussage stellt das verallgemeinerte H-Theorem der Quantenstatistik dar. Sie besagt, daß der aus einer Beobachtung zur Zeit t' berechnete Wert von $\bar{H}$ größer ist als der aus einer Beobachtung zu einer späteren Zeit t'' berechnete, wenn sich nicht das System von vornherein im Gleichgewicht befindet. Nur in diesem Falle ist nämlich $\rho'' = \mathsf{P}''$ und damit in (44.6) bis (44.8) das Gleichheitszeichen gültig. Die weitergehende Behauptung, daß $\bar{H}$ bis zum Gleichgewicht (im Sinne der Gibbsschen Definition) monoton abfällt, ist zwar plausibel, hat sich aber bisher ebensowenig wie in der klassischen Statistik, beweisen lassen. Da die qualitativen Überlegungen im Prinzip den früheren (Ziff. 17) ähnlich sind, verzichten wir auf eine Wiedergabe und verweisen für Einzelheiten auf Tolman [12].

Die Ableitung von (44.8) zeigt, daß in der Quantenstatistik das Problem der Einstellung des Gleichgewichtes komplizierter ist als in der klassischen Theorie, da zwei Effekte dazu betragen: Die quantenmechanische Ausbreitung der feinen Wahrscheinlichkeit und der zunehmende Unterschied zwischen ρ und P. Dieser Sachverhalt ist in den Diskussionen nicht immer berücksichtigt worden.

Von weiteren Untersuchungen zum quantenstatistischen H-Theorem seien noch die Überlegungen von von Neumann[1] sowie Pauli und Fierz[2] erwähnt, die neuerdings von Farquhar und Landsberg[3] wieder aufgegriffen worden sind. Sie knüpfen letzten Endes an die Boltzmannschen Gedankengänge (Ziff. 16) an, die das H-Theorem mit dem Ergodentheorem verknüpfen. Einen anderen Weg haben Born und Green[4,5] eingeschlagen. Sie definieren

$$\bar{H} = \sum_k \varrho_{kk} \ln \varrho_{kk} \tag{44.9}$$

und zeigen dann mit Hilfe der Diracschen Störungstheorie, daß diese Funktion monoton bis zum Gleichgewicht abnimmt. Wenn man die Definition (44.9) voraussetzt, ist diese Aussage, wie Pauli[6] bemerkt hat, einfach eine Folgerung aus dem Kleinschen Lemma. Die Gl. (44.9) wird aus Gl. (42.4) erhalten, wenn man aller ω_i gleich Eins setzt. Sie kann auch aus (42.3) abgeleitet werden durch die Annahme, daß ρ zu allen Zeiten diagonal ist.

[1] J. v. Neumann: Z. Physik **57**, 30 (1929).
[2] W. Pauli u. M. Fierz: Z. Physik **106**, 572 (1937).
[3] I. E. Farquhar u. P. T. Landsberg: Proc. Roy. Soc. Lond., Ser. A **239**, 134 (1957).
[4] M. Born: Ann. Physik **3**, 107 (1948).
[5] M. Born u. H. S. Green: Proc. Roy. Soc. Lond., Ser. A **192**, 166 (1948).
[6] W. Pauli: Nuovo Cim. **6**, Suppl., 166 (1949).

Die Born-Greensche Definition der H-Funktion verwischt daher, wie TER HAAR [23] betont hat, den Unterschied zwischen ρ und P. Sie erscheint ziemlich willkürlich und dürfte sich kaum in befriedigender Weise begründen lassen.

In der neuesten Zeit beschäftigen sich, wie schon erwähnt, zahlreiche Arbeiten mit dem Problem der quantenstatistischen Theorie irreversibler Prozesse, wobei allerdings das H-Theorem selbst etwas in den Hintergrund getreten ist. Eine Darstellung dieser Entwicklung würde nach Umfang und Gegenstand den Rahmen dieses Artikels weit überschreiten. Wir begnügen uns daher mit einigen Hinweisen auf Arbeiten, die im Zusammenhang mit den hier erörterten Problemen von besonderem Interesse sind.

Auf der Grundlage der Analyse des Begriffes der makroskopischen Variablen (Ziff. 37) hat VAN KAMPEN[1] eine Ableitung der master equation (42.25) durchgeführt, die auch dadurch bemerkenswert ist, daß die durch Gl. (40.1) definierten Zeitmittel der P_i benutzt werden. Hier wird also eine in doppeltem Sinne grobe Dichte eingeführt, was in der klassischen Theorie, wie es scheint, bisher noch nicht versucht worden ist. Von einer Ableitung im mathematischen Sinne kann allerdings kaum gesprochen werden, da die Hypothese der willkürlichen Phasenverteilung für alle Zeiten benutzt und lediglich durch qualitative Betrachtungen plausibel gemacht wird. KÜMMEL[2] hat diese Überlegungen noch weiter ausgearbeitet. Im Gegensatz dazu ist es VAN HOVE[3] gelungen, die master equation streng nach den Methoden der Quantenmechanik für gewissen Klassen von Anfangszuständen abzuleiten. Die Grundlage der Ableitung bildet eine Aussage über den Störungsoperator V für den Grenzfall eines Systems aus unendlich vielen Teilchen, die sich für gewisse Fälle explizit verifizieren läßt. Dieser Gesichtspunkt ist von besonderem Interesse, weil er die schon früher[4] geäußerte Vermutung bestätigt, daß die aus dem auch in der Quantenmechanik gültigen Wiederkehrsatz[5, 6] für das Verständnis derirre versiblen Prozesse entstehende Schwierigkeit allein durch den erwähnten Grenzübergang beseitigt werden kann. VAN HOVE zeigt, daß die speziellen Anfangszustände ersetzt werden können durch die Hypothese willkürlicher a priori-Phasen, die aber nur für die Zeit $t = 0$ gefordert wird und die master equation zu einer statistischen Gleichung macht. In einer kürzlich erschienenen Arbeit hat VAN HOVE[7] seine Analyse auf Störungen beliebiger Ordnung ausgedehnt und eine „generalisierte master equation" abgeleitet, aus der (42.25) als Spezialfall für verschwindende Störung erhalten werden kann.

Der Zusammenhang zwischen dem irreversiblen Charakter eines Prozesses und dem Verlust an Information (vgl. Ziff. 16, 17) ist für die Quantenstatistik in neuester Zeit von FANO[8] und besonders eingehend von JAYNES[9] erörtert worden. Für Einzelheiten müssen wir auf die Originalarbeiten verweisen.

IV. Die mikrokanonische Gesamtheit der Quantenstatistik.

45. Definition. Der Virialsatz. Wir betrachten ein im makroskopischen Sinne abgeschlossenes System, das sich im thermodynamischen Gleichgewicht befindet und für welches die Energie, die Teilchenzahlen und die äußeren Parameter

---

[1] N. G. VAN KAMPEN: Physica, Haag **20**, 603 (1954).
[2] H. KÜMMEL: Z. Naturforsch. **11**a, 15 (1956).
[3] L. VAN HOVE: Physica, Haag **21**, 517 (1955).
[4] E. W. MONTROLL u. M. S. GREEN: Ann. Rev. Phys. Chem. **5**, 449 (1954).
[5] S. ONO: Mem. Fac. Engng. Kyushu **11**, 125 (1949).
[6] P. BOCCHIERI u. A. LOINGER: Phys. Rev. **107**, 337 (1957).
[7] L. VAN HOVE: Physica, Haag **23**, 441 (1957).
[8] U. FANO: Rev. Mod. Phys. **29**, 74 (1957).
[9] E. T. JAYNES: Phys. Rev. **108**, 171 (1957).

vorgegeben sind. Die Energiebedingung können wir präzise so formulieren, daß alle Matrixelemente des Störungsoperators V verschwinden sollen, bei denen einer oder beide Zustände außerhalb des Intervalls E bis $E + \Delta E$ liegen. Ein solches System wird nach Ziff. 40 durch eine mikrokanonische Gesamtheit dargestellt, deren Elemente für die Basis der Energie-Eigenfunktionen dargestellt werden durch

$$\left.\begin{aligned} \varrho_{nm} &= \Omega^{-1}\delta_{nm} \quad \text{für} \quad E \leqq E_n \leqq E + \Delta E, \\ \varrho_{nm} &= 0 \quad \text{in allen übrigen Fällen.} \end{aligned}\right\} \tag{45.1}$$

Hier ist Ω^{-1} die Zahl der linear unabhängigen Eigenfunktionen des Hamilton-Operators für das ungestörte System. Für den mikrokanonischen Mittelwert einer durch den Operator F dargestellten Observablen ergibt sich aus Gl. (34.22) und (45.1)

$$\overline{F} = \Omega^{-1}\,\text{spur}\,\mathsf{F}. \tag{45.2}$$

Wir wollen diese Gleichung benutzen, um den Virialsatz auf der Grundlage der Quantenstatistik abzuleiten[1]. Dazu gehen wir aus von der Schrödinger-Gleichung des ungestörten Systems die wir explizit schreiben

$$-\sum_{i=1}^{n}\frac{h^2}{8\pi^2 m}\frac{\partial^2\psi}{\partial q_i^2} + (U - E)\psi = 0, \tag{45.3}$$

wo U die potentielle Energie des Gesamtsystems und n die Zahl der Freiheitsgrade ist. Der Einfachheit halber setzen wir voraus, daß alle q_i kanonische quantenmechanische Koordinaten[2] sind. Lassen wir auf Gl. (45.3) den Operator $q_j\psi^*\dfrac{\partial}{\partial q_j}$ wirken, so folgt

$$-\sum_{i}\frac{h^2}{8\pi^2 m}q_j\psi^*\frac{\partial^3\psi}{\partial q_i^2\,\partial q_j} + q_j\psi^*\frac{\partial U}{\partial q_j}\psi + q_j\psi^*(U - E)\frac{\partial\psi}{\partial q_j} = 0. \tag{45.4}$$

Andererseits ergibt sich durch Multiplikation der zu (45.3) konjugiert komplexen Gleichung mit $q_j\dfrac{\partial\psi}{\partial q_j}$

$$q_j\psi^*(U - E)\frac{\partial\psi}{\partial q_j} = \sum_{i}\frac{h^2}{8\pi^2 m}q_j\frac{\partial\psi}{\partial q_j}\frac{\partial^2\psi^*}{\partial q_i^2}. \tag{45.5}$$

Setzen wir dies in (45.4) ein und summieren über j, so erhalten wir

$$-\sum_{i}\frac{h^2}{8\pi^2 m}\sum_{j}q_j\left(\psi^*\frac{\partial^3\psi}{\partial q_i^2\,\partial q_j} - \frac{\partial\psi}{\partial q_j}\frac{\partial^2\psi^*}{\partial q_i^2}\right) + \psi^*\left(\sum_{j}q_j\frac{\partial U}{\partial q_j}\right)\psi = 0. \tag{45.6}$$

Nun ist

$$\psi^{*2}\frac{\partial}{\partial q_i}\frac{\sum\limits_{j}q_j\dfrac{\partial\psi}{\partial q_j}}{\psi^*} = \psi^*\frac{\partial\psi}{\partial q_i} + \sum_{j}\left(\psi^* q_j\frac{\partial^2\psi}{\partial q_i\,\partial q_j} - \frac{\partial\psi}{\partial q_j}\frac{\partial\psi^*}{\partial q_i}\right). \tag{45.7}$$

Daraus folgt durch Differentiation

$$\sum_{j}q_j\left(\psi^*\frac{\partial^3\psi}{\partial^2 q_i\,\partial q_j} - \frac{\partial\psi}{\partial q_j}\frac{\partial^2\psi^*}{\partial q_i^2}\right) = -2\psi^*\frac{\partial^2\psi}{\partial q_i^2} + \frac{\partial}{\partial q_i}\left(\psi^{*2}\frac{\partial}{\partial q_i}\frac{\sum\limits_{j}q_j\dfrac{\partial\psi}{\partial q_j}}{\psi^*}\right). \tag{45.8}$$

---

[1] J. Slater: J. Chem. Phys. **1**, 687 (1933).
[2] Quantenmechanische Koordinaten und Impulse werden als kanonisch bezeichnet, wenn sie über den ganzen Wertebereich von $-\infty$ bis $+\infty$ physikalisch definiert sind [*12*].

Bei der Integration von $q_i = -\infty$ bis $q_i = +\infty$ verschwindet der zweite Term der rechten Seite, da ψ^{*2} an den Grenzen Null wird. Setzen wir (45.8) in (45.6) ein und integrieren über den Konfigurationsraum, so ergibt sich

$$-2 \sum_i \frac{h^2}{8\pi^2 m} \int \psi^* \frac{\partial^2}{\partial q_i^2} \psi \, d\mathbf{q} = \int \psi^* \left(\sum_j q_i \frac{\partial U}{\partial q_j} \right) \psi \, d\mathbf{q}. \tag{45.9}$$

Wir definieren nun den Virial-Operator durch

$$[\mathsf{V}] = \frac{1}{2} \sum_i q_i \frac{\partial U}{\partial q_i}. \tag{45.10}$$

Da

$$\mathsf{E}_{\mathrm{kin}} = -\frac{h^2}{4\pi^2} \sum_i \frac{1}{2m} \frac{\partial^2}{\partial q_i^2} \tag{45.11}$$

der Operator der kinetischen Energie ist, folgt aus (45.9)

$$\mathsf{E}_{\mathrm{kin}} = [\mathsf{V}] \tag{45.12}$$

und daraus durch Kombination mit (45.2)

$$\overline{E}_{\mathrm{kin}} = [\overline{V}] \tag{45.13}$$

in Übereinstimmung mit Gl. (21.14). Der Virialsatz gilt somit auch in der Quantenstatistik streng und allgemein, im Gegensatz zum Äquipartitionstheorem, das hier nur ein asymptotisches Grenzgesetz darstellt.

46. Adiabatische Änderung der Parameter. Die Ableitung der thermodynamischen Analoga für die mikrokanonische Gesamtheit entspricht im wesentlichen den Gedankengängen der klassischen Theorie. Wir können uns daher jetzt kürzer fassen. Die wichtigste Voraussetzung, die eingeführt werden muß, besteht in der Annahme, daß Ω als stetige differenzierbare Funktion der Energie aufgefaßt werden kann. Da das Energieintervall von E bis $E + \Delta E$ physikalisch durch eine makroskopische Energiemessung definiert ist, bedeutet die erwähnte Annahme zunächst einfach, daß wir von der mikroskopischen zur makroskopischen Beschreibung zurückkehren. Darüber hinaus müssen die Energie-Eigenwerte so dicht liegen, daß sie für die makroskopische Beschreibung als Kontinuum aufgefaßt werden können. Diese Annahme kann im allgemeinen für makroskopische Systeme als erfüllt angesehen werden. Lediglich bei den allertiefsten Temperaturen, bei denen nur die untersten Niveaus besetzt sind, ist eine besondere Diskussion erforderlich.

Die Energie muß natugemäß auch hier als Funktion der äußeren Parameter x_j betrachtet werden. Daraus ergibt sich unmittelbar als Definition einer mittleren generalisierten Kraft

$$\overline{X}_j = -\Omega^{-1} \sum_{m=1}^{\Omega} \frac{\partial E_m}{\partial x_j}. \tag{46.1}$$

Die durch eine Änderung der äußeren Parameter bedingte mittlere Änderung der Energie ist dann

$$\overline{\Delta E} = \overline{\Delta A} = \Omega^{-1} \sum_j \sum_m \frac{\partial E_m}{\partial x_j} \Delta x_j. \tag{46.2}$$

Wir betrachten nun wieder eine adiabatische Parameteränderung, bei der nur die äußeren Parameter unabhängige Variable sind und die Änderung der Energie durch Gl. (46.2) bestimmt ist. Dann gilt

$$\overline{\Delta \Omega} = \left(\frac{\partial \Omega}{\partial E} \right)_x \overline{\Delta E} + \sum_j \left(\frac{\partial \Omega}{\partial x_j} \right)_E \Delta x_j. \tag{46.3}$$

Der zweite Term der rechten Seite gibt die Änderung der Zahl der Zustände der Energie E durch Änderung der äußeren Parameter bei konstantem E. Diese ist zunächst gleich der Summe über alle Zustände, deren Energie durch die Parameteränderung beeinflußt wird. Dafür kann aber die mittlere Energieänderung multipliziert mit der negativen Ableitung von Ω nach E gesetzt werden. Die beiden Terme auf der rechten Seite der Gl. (46.3) sind somit entgegengesetzt gleich, und es folgt

$$\overline{\Delta\Omega} = 0. \tag{46.4}$$

Die Größe Ω ist somit invariant gegen adiabatische Parameteränderungen. Sie besitzt daher für quasistatisch-adiabatische Prozesse die Invarianzeigenschaft der Entropie.

47. Thermodynamische Analoga. Für beliebige Zustandsänderungen gilt

$$d\ln\Omega = \frac{\partial\ln\Omega}{\partial E}\,dE + \sum_j \frac{\partial\ln\Omega}{\partial x_j}\,dx_j, \tag{47.1}$$

wo jetzt E und die x_j unabhängige Variable sind. Nun ist

$$\sum_j \frac{\partial\ln\Omega}{\partial x_j}\,dx_j = \frac{\partial\ln\Omega}{\partial E}\sum_j \overline{X}_j\,dx_j. \tag{47.2}$$

Die Größe $\ln\Omega$ ist daher, wie der Vergleich mit (47.1) zeigt, ein statistisches Analogon der Entropie, wenn $\partial\ln\Omega/\partial E$ mit der reziproken absoluten Temperatur identifiziert werden kann[1]. Die Anpassung an die übliche Temperaturskala ergibt dann

$$S = k\ln\Omega. \tag{47.3}$$

In diese Entropiedefinition geht implizit die eigentlich nur qualitativ bestimmte Größe ΔE ein. Man muß daher fragen, welchen Einfluß diese Unbestimmtheit auf die thermodynamische Größe S hat. Für die Untersuchung dieser Frage formulieren wir zunächst den in Ziff. 46 postulierten Übergang von der arithmetischen zur geometrischen Verteilung durch die Gleichung

$$\Omega = \Omega(E)\,\Delta E, \tag{47.4}$$

wo E jetzt [im Gegensatz zu Gl. (47.1)] als in der „mikroskopischen Skala" gemessen zu denken ist. Es muß dann gelten

$$\ln\int_0^E \Omega(E)\,dE - \ln\frac{E}{\Delta E} \leqq \ln\left[\Omega(E)\,\Delta E\right] \leqq \ln\int_0^E \Omega(E)\,dE. \tag{47.5}$$

Wenn auf der linken Seite der zweite Term gegen den ersten vernachlässigt werden kann, gilt das Gleichheitszeichen, und wir können als Entropiedefinition schreiben

$$S = k\ln\int_0^E \Omega(E)\,dE. \tag{47.6}$$

Die unbestimmte Größe ΔE kommt in diesem Augenblick nicht mehr vor. Die physikalische Bedeutung der beiden Entropiedefinitionen liegt darin, daß die Entropie in Gl. (47.3) proportional dem Logarithmus der Zahl der Quantenzustände im Energieintervall von E bis $E + \Delta E$ gesetzt wird, die also das System tatsächlich einnehmen kann, während sie in Gl. (47.6) proportional dem Logarithmus der Zahl aller Quantenzustände gesetzt wird, die mit einer Energie $\leqq E$

---

[1] Für separierbare Systeme wird dies in Ziff. 49 bewiesen.

erreichbar sind. Die Gültigkeit der Voraussetzung hängt einmal von der Größenordnung von $\ln \Omega$ ab, zum anderen von der Größenordnung von $\ln (E/\Delta E)$, die praktisch durch die Genauigkeit der makroskopischen Energiemessung bestimmt wird. Beispielsweise ist für ein Mol Helium bei 273° K und 1 Atm $\ln \Omega \approx 10^{25}$. Würde die Energie auf 100 Stellen genau gemessen, so hätte man $\ln (E/\Delta E) \approx 10^2$. Selbst in diesem Falle liegt somit der Unterschied der aus den beiden Entropiedefinitionen berechneten Zahlenwerte weit unterhalb der experimentellen Nachweisgrenze. Vom praktischen Standpunkt ist somit die Unbestimmtheit der Größe ΔE jedenfalls bedeutungslos für die Definition der Entropie.

Die Frage, wieweit man im Prinzip ΔE verkleinern kann, führt zunächst [wegen (40.14)] zu der Folgerung, daß dies nur auf Kosten der Beobachtungszeit möglich ist. Es ergibt sich aber darüber hinaus, daß die Unschärfe der Energie die notwendige Voraussetzung für einen endlichen Wert der Entropie ist. Ein ideal abgeschlossenes System, das sich in einem reinen Eigenzustand der Energie befindet, hat nach Gl. (47.3) die Entropie Null. Jede Messung an dem System hebt jedoch die Isolierung auf, sie bewirkt eine Störung und damit unkontrollierbare Übergänge in benachbarte Zustände von annähernd der gleichen Energie. Wir werden somit auch hier wieder auf die Vorstellung geführt, daß die Entropie ein Maß für den Verlust an Information darstellt.

Der aus der klassischen Theorie bekannte Zusammenhang zwischen H-Funktion und thermodynamischer Entropie führt in der Quantenstatistik bei Anwendung auf Gl. (42.4) zu der Formel

$$S = - k \operatorname{spur} (\rho \ln \rho), \tag{47.7}$$

wobei wir für den Fall des Gleichgewichtes die feine Dichtematrix mit der groben Dichtematrix identifiziert haben (vgl. Ziff. 40). Die Gl. (47.7), die bereits von Neumann[1] angegeben hat, stellt die allgemeinste quantenstatistische Definition der Entropie dar. Man sieht, daß Einsetzen der Dichtematrix der mikrokanonischen Gesamtheit (45.1) dann sofort auf die von uns unabhängig eingeführte Definition (47.3) führt.

48. Berechnung von Ω für separierbare Systeme. Für explizite Berechnungen ist die mikrokanonische Gesamtheit der Quantenstatistik bisher nur bei Systemen, deren Hamilton-Operator sich nach den Koordinaten der Einzelteilchen separieren läßt, verwendet worden. Der wichtigste Unterschied gegenüber der klassischen Theorie besteht, abgesehen von der Tatsache, daß wir jetzt diskrete Zustände haben, darin, daß die Symmetriebedingungen der Quantenmechanik, die wir schon in Ziff. 41 erwähnt haben, berücksichtigt werden müssen. Bei den Anwendungen liegen jedoch häufig Verhältnisse vor, für welche die Symmetriebedingungen praktisch zu vernachlässigen sind. Es ist daher zweckmäßig, diesen Fall gesondert zu behandeln.

Wir erläutern die Verhältnisse zunächst an dem einfachen Beispiel eines Systems aus zwei Teilchen. Die Eigenwerte des Hamilton-Operators seien

$$E_{m,n} = \varepsilon_m + \varepsilon_n, \tag{48.1}$$

die zugehörigen Eigenfunktionen

$$\psi_{m,n} = \psi_m(q_1)\,\psi_n(q_2)\,, \qquad \psi_{n,m} = \psi_n(q_1)\,\psi_m(q_2)\,. \tag{48.2}$$

Wir haben somit Austauschentartung. Die Eigenfunktionen genügen indessen nicht der Forderung nach Invarianz meßbarer Größen gegen Teilchenvertauschung.

---

[1] J. v. Neumann: Göttinger Nachr. **1927**, 245, 273.

Dabei ist zu beachten, daß die hier gemeinte „Nichtunterscheidbarkeit" keine „intrinsic property" der Teilchen ist. A priori sind gleiche Teilchen definitionsgemäß nicht unterscheidbar. Die obige Invarianzforderung ist vielmehr eine Folgerung aus der Unbestimmtheitsrelation, wie die klassische Unterscheidbar keit der Teilchen aus der Gültigkeit der klassischen Bewegungsgleichungen folgt In unserem Falle lautet die Invarianzforderung

$$\psi_{m,n}^{*}\,\psi_{m,n} = \psi_{n,m}^{*}\,\psi_{n,m}\,. \tag{48.3}$$

Die allgemeine Lösung der Schrödinger-Gleichung ist

$$\psi_{m,n} = a\,\psi_{m}(\boldsymbol{q_1})\,\psi_{n}(\boldsymbol{q_2}) + b\,\psi_{m}(\boldsymbol{q_2})\,\psi_{n}(\boldsymbol{q_1})\,. \tag{48.4}$$

Man erkennt hier sofort, daß nur die beiden Wertepaare $a=1/\sqrt{2}$, $b=1/\sqrt{2}$ und $a=1/\sqrt{2}$, $b=-1/\sqrt{2}$ die Invarianzforderung erfüllen. Die möglichen Eigenfunktionen sind daher die symmetrische Lösung

$$\psi_{s} = \frac{1}{\sqrt{2}}\left[\psi_{m}(\boldsymbol{q_1})\,\psi_{n}(\boldsymbol{q_2}) + \psi_{m}(\boldsymbol{q_2})\,\psi_{n}(\boldsymbol{q_1})\right] \tag{48.5}$$

und die antisymmetrische Lösung

$$\psi_{a} = \frac{1}{\sqrt{2}}\left[\psi_{m}(\boldsymbol{q_1})\,\psi_{n}(\boldsymbol{q_2}) - \psi_{m}(\boldsymbol{q_2})\,\psi_{n}(\boldsymbol{q_1})\right]\,. \tag{48.6}$$

Aus der zeitabhängigen Schrödinger-Gleichung folgt nun, da der Hamilton-Operator symmetrisch in den $\boldsymbol{q_i}$ ist: Ein System aus gleichen Teilchen, das sich in einem durch eine symmetrische Eigenfunktion charakterisierten Zustand befindet, kann niemals und unter keinen Umständen in einen durch eine antisymmetrische Eigenfunktion charakterisierten Zustand übergehen und umgekehrt. Der Symmetriecharakter der Eigenfunktionen ist somit eine Eigenschaft der das System aufbauenden Teilchen selbst. Zu jedem Eigenwert des Hamilton-Operators gehört daher (wenn die Eigenwerte der Einzelteilchen nicht entartet sind) nur eine Eigenfunktion. Die Austauschentartung ist aufgehoben. Vom Standpunkt der statistischen Mechanik gesehen, haben wir hier den Fall einer prinzipiellen Unzugänglichkeit, der in der klassischen Theorie (vgl. Ziff. 26) nicht vorkommt. Die Quantenstatistik zerfällt daher in die Statistik mit symmetrischen Eigenfunktionen (Bose-Einstein-Statistik) und die Statistik mit antisymmetrischen Eigenfunktionen (Fermi-Dirac-Statistik).

Wir betrachten nun den Spezialfall, daß die Eigenfunktionen ψ_m und ψ_n sich praktisch nicht überlappen. Es soll also ψ_m groß sein für solche Werte der Koordinaten, für die ψ_n praktisch Null ist, und umgekehrt. Das heißt als Formel geschrieben

$$\psi_{m}^{*}(\boldsymbol{q_1})\,\psi_{n}(\boldsymbol{q_1}) \approx \psi_{n}^{*}(\boldsymbol{q_1})\,\psi_{m}(\boldsymbol{q_1}) \approx \psi_{m}^{*}(\boldsymbol{q_2})\,\psi_{n}(\boldsymbol{q_2}) \approx \psi_{n}^{*}(\boldsymbol{q_2})\,\psi_{m}(\boldsymbol{q_2}) \approx 0. \tag{48.7}$$

Wir berechnen nun für die beiden Eigenfunktionen (48.5) und (48.6) die Wahrscheinlichkeitsdichten mit Berücksichtigung von (48.7). Es ergibt sich dann

$$\psi_{s}^{*}\,\psi_{s} = \psi_{a}^{*}\,\psi_{a} = \tfrac{1}{2}\left[|\psi_{m}(\boldsymbol{q_1})|^{2}\,|\psi_{n}(\boldsymbol{q_2})|^{2} + |\psi_{m}(\boldsymbol{q_2})|^{2}\,|\psi_{n}(\boldsymbol{q_1})|^{2}\right]\,. \tag{48.8}$$

Der Unterschied im Verhalten von Teilchen mit symmetrischen Eigenfunktionen und solchen mit antisymmetrischen Eigenfunktionen verschwindet somit, wenn die Eigenfunktionen der Einzelteilchen sich nicht überlappen. Wir können noch einen Schritt weitergehen und die Wahrscheinlichkeit berechnen, ein Teilchen im Volumenelement $d\boldsymbol{q}^{(m)}$ zu finden, wo ψ_m groß ist, und das andere Teilchen im Volumenelement $d\boldsymbol{q}^{(n)}$, wo ψ_n groß ist. Dann ergibt sich mit (48.8)

$$\begin{aligned}
|\psi_s|^2 (d\boldsymbol{q}_1^{(m)} d\boldsymbol{q}_2^{(n)} + d\boldsymbol{q}_2^{(m)} d\boldsymbol{q}_1^{(n)}) &= |\psi_a|^2 (d\boldsymbol{q}_1^{(m)} d\boldsymbol{q}_2^{(n)} + d\boldsymbol{q}_2^{(m)} d\boldsymbol{q}_1^{(n)}) \\
&= \tfrac{1}{2} |\psi_m(\boldsymbol{q}_1)|^2 |\psi_n(\boldsymbol{q}_2)|^2 d\boldsymbol{q}_1^{(m)} d\boldsymbol{q}_2^{(n)} + \tfrac{1}{2} |\psi_m(\boldsymbol{q}_2)|^2 |\psi_n(\boldsymbol{q}_1)|^2 d\boldsymbol{q}_2^{(m)} d\boldsymbol{q}_1^{(n)}.
\end{aligned} \right\} \quad (48.9)$$

Den gleichen Ausdruck erhält man auch, wenn man bei der Berechnung von den Eigenfunktionen (48.2) ausgeht. Das heißt mit anderen Worten: Wenn die Voraussetzung (48.7) erfüllt ist, können wir näherungsweise einem bestimmten Teilchen die Energie ε_m und die Eigenfunktion ψ_m, dem anderen die Energie ε_n und die Eigenfunktion ψ_n zuschreiben. Bleiben diese Verhältnisse über eine im Vergleich zur Dauer einer Messung lange Zeit erhalten, so können die Teilchen näherungsweise als individuell unterscheidbar im Sinne der klassischen Theorie betrachtet werden. Physikalisch erklärt sich dies dadurch, daß, wenn (48.7) gilt, praktisch keine Stöße zwischen den Teilchen auftreten und damit die Ursache der Nichtunterscheidbarkeit fortfällt. Beispiele für das beschriebene Verhalten sind Gasmoleküle in verschiedenen Gefäßen oder die Atome eines Kristalls auf ihren Gitterplätzen. Auch die (nach Abseparation der Schwerpunktsbewegung verbleibenden) inneren Freiheitsgrade der Gasmoleküle gehören hierher. In derartigen Fällen verwendet man Eigenfunktionen von der Art der Gl. (48.2), welche (48.7) exakt erfüllen. Da aber jetzt die Symmetrisierung der Eigenfunktionen entfällt, muß die Austauschentartung berücksichtigt werden. Man spricht dann von lokalisierten Teilchen oder quantisierter Maxwell-Boltzmann-Statistik.

Die vorstehenden Überlegungen lassen sich ohne Schwierigkeit für ein System aus N Teilchen verallgemeinern. Die Eigenwerte des Hamilton-Operators haben jetzt die Form

$$E_l = N_m \varepsilon_m + N_n \varepsilon_n + \cdots + N_r \varepsilon_r, \tag{48.10}$$

wo N_m die Zahl der Teilchen mit dem Eigenwert ε_m ist und

$$N_m + N_n + \cdots + N_r = N \tag{48.11}$$

gilt. Eine (48.2) entsprechende Eigenfunktion hat die Gestalt

$$\psi_l = \psi_m(\boldsymbol{q}_1) \ldots \psi_m(\boldsymbol{q}_{N_m}) \psi_n(\boldsymbol{q}_{N_m+1}) \ldots \psi_n(\boldsymbol{q}_{N_m+N_n}) \ldots \psi_r(\boldsymbol{q}_N). \tag{48.12}$$

Für lokalisierte Teilchen ist daher die Zahl der zu dem Eigenwert (48.10) gehörenden Eigenfunktionen, wenn wir neben der Austauschentartung noch die Entartung der ε_i berücksichtigen,

$$\Omega_D = \frac{N!}{\prod_i N_i!} \prod_i g_i^{N_i} \quad (\text{Maxwell-Boltzmann}). \tag{48.13}$$

Für nicht lokalisierte Teilchen sind die möglichen Eigenfunktionen durch Gl. (41.5) und (41.6) gegeben. Daß nur diese Eigenfunktionen möglich sind, folgt aus der Tatsache, daß der Permutationsoperator nur die Eigenwerte $+1$ und -1 hat. Durch einfache kombinatorische Überlegung findet man, daß die Zahl der zum Eigenwert (48.10) gehörenden symmetrischen Eigenfunktionen durch

$$\Omega_D = \prod_i \frac{(N_i + g_i - 1)!}{N_i!(g_i - 1)!} \quad (\text{Bose-Einstein}) \tag{48.14}$$

gegeben ist, die Zahl der antisymmetrischen Eigenfunktionen durch

$$\Omega_D = \prod_i \frac{g_i!}{N_i!(g_i - N_i)!} \qquad (N_i \leqq g_i) \qquad \text{(Fermi-Dirac)}. \tag{48.15}$$

Es ist nun

$$\Omega = \sum \Omega_D. \tag{48.16}$$

Die Mittelwertformel (45.2) kann daher geschrieben werden

$$\overline{F} = \Omega^{-1} \sum F_D \Omega_D, \tag{48.17}$$

wenn wir voraussetzen, daß die Diagonalelemente der Matrix F für alle Eigenfunktionen des Satzes Ω_D jeweils gleich sind. Die Anwendung der mikrokanonischen Gesamtheit auf separierbare Systeme führt somit auf das mathematische Problem, Summen der obigen Fakultätenausdrücke zu berechnen.

49. Die Darwin-Fowlersche Methode für lokalisierte Teilchen. Das am Schluß von Ziff. 48 erwähnte mathematische Problem läßt sich mit Hilfe erzeugender Funktionen lösen. Bei Verwendung reeller Funktionen ist, wie in Ziff. 28, eine Reduktion auf den zentralen Limessatz der Wahrscheinlichkeitstheorie möglich [30]. Wir führen hier die Rechnung nach einer von Darwin und Fowler[1] stammenden Methode durch, die komplexe erzeugende Funktionen benutzt und deren Prinzip auf zahlreiche Probleme der statistischen Mechanik anwendbar ist. Dabei beschränken wir uns auf die Wiedergabe des wesentlichen Gedankenganges und verweisen für Details, insbesondere die mathematischen Existenzbeweise, auf die Fowlersche Monographie [7].

Als einfaches Beispiel betrachten wir etwa ein System aus N lokalisierten linearen harmonischen Oszillatoren. Die Eigenwerte ε_i sind hier nicht entartet. Wir führen die Rechnung zunächst unter der (nicht streng zutreffenden) Annahme durch, daß das System eine (mikroskopisch) scharfe Energie E besitzt. Am Schluß werden wir zeigen, daß die durch Berücksichtigung des endlichen Intervalls von E bis $E + \Delta E$ entstehende Korrektur zu vernachlässigen ist. Die zu berechnenden Größen sind

$$\Omega = \sum \Omega_D = \sum \frac{N!}{\prod_i N_i!}, \tag{49.1}$$

$$\Omega \overline{N_j} = \sum N_j \frac{N!}{\prod_i N_i!}, \tag{49.2}$$

$$\Omega N \overline{\varepsilon} = \sum \left(\sum_i N_i \varepsilon_i \right) \frac{N!}{\prod_i N_i!}. \tag{49.3}$$

Dabei gelten die Nebenbedingungen

$$\sum_i N_i = N, \tag{49.4}$$

$$\sum N_i \varepsilon_i = E. \tag{49.5}$$

Der Einfachheit halber nehmen wir an, daß die ε_i ganzzahlige Vielfache einer passend gewählten Energieeinheit ohne gemeinsamen Faktor sind. Wir definieren nun die Funktion einer komplexen Variablen x durch

$$f(x) = \sum_i x^{\varepsilon_i}. \tag{49.6}$$

---

[1] C. S. Darwin u. R. H. Fowler: Phil. Mag. **44**, 450, 823 (1922); **45**, 1 (1923).

Mit Hilfe des Multinomialtheorems sieht man sofort, daß

$$[f(x)]^N = \left[\sum_i x^{\varepsilon_i}\right]^N \tag{49.7}$$

die erzeugende Funktion für Ω und Ω der Koeffizient von x^E in der Entwicklung von $[f(x)]^N$ nach Potenzen von x ist.

Die Gl. (49.2) formen wir nun um in

$$\Omega\,\overline{N_j} = N \sum \frac{(N-1)!}{\prod\limits_i N_i!} \tag{49.8}$$

mit den Nebenbedingungen

$$\sum_i N_i = N - 1, \tag{49.9}$$

$$\sum_i N_i \varepsilon_i = E - \varepsilon_i. \tag{49.10}$$

Dann folgt sofort, daß $\Omega\overline{N_j}$ der Koeffizient von $x^{E-\varepsilon_j}$ in der Entwicklung von $N[f(x)]^{N-1}$ nach Potenzen von x ist. Äquivalent damit ist die Aussage, daß $\Omega\overline{N_j}$ der Koeffizient von x^E in der Entwicklung von $N x^{\varepsilon_j}[f(x)]^{N-1}$ ist.

Wir schreiben nun die Gl. (49.7) in der Form

$$[f(x)]^N = \sum \frac{N!}{\prod\limits_i N_i!}\, x^{\sum_i N_i \varepsilon_i}. \tag{49.11}$$

Daraus folgt

$$x\frac{\partial}{\partial x}[f(x)]^N = \sum \left(\sum_i N_i \varepsilon_i\right) \frac{N!}{\prod\limits_i N_i!}\, x^{\sum_i N_i \varepsilon_i}. \tag{49.12}$$

Der Vergleich mit (49.3) zeigt, daß $\Omega N\overline{\varepsilon}$ der Koeffizient von x^E in der Entwicklung von $x\partial[f(x)]^N/\partial x$ nach Potenzen von x ist. Damit sind alle erzeugenden Funktionen bestimmt.

Die Anwendung des Residuensatzes liefert nun sofort

$$\Omega = \frac{1}{2\pi i} \oint \frac{[f(x)]^N}{x^{E+1}}\, dx, \tag{49.13}$$

$$\Omega\,\overline{N_j} = \frac{1}{2\pi i} \oint \frac{N x^{\varepsilon_j}[f(x)]^{N-1}}{x^{E+1}}\, dx, \tag{49.14}$$

$$\Omega N\overline{\varepsilon} = \frac{1}{2\pi i} \oint \frac{x\dfrac{\partial}{\partial x}[f(x)]^N}{x^{E+1}}\, dx. \tag{49.15}$$

Die Berechnung dieser Integrale erfolgt nach der Sattelpunktmethode. Wir betrachten zuerst das Integral (49.13). Der Integrand hat auf der positiven reellen Achse Singularitäten an den Stellen $x=0$ und $x=1$. Dazwischen liegt (wie man beweisen kann) ein und nur ein Minimum, etwa bei $x=\vartheta$. ϑ ist also die einzige positive reelle Wurzel der Gleichung

$$\frac{\partial}{\partial x}\left\{\frac{[f(x)]^N}{x^E}\right\} = 0 \tag{49.16}$$

oder

$$-E\,\frac{[f(x)]^N}{x^{E+1}} + N\,\frac{[f(x)]^{N-1}f'(x)}{x^E} = 0. \tag{49.17}$$

Wir erhalten somit als Bestimmungsgleichung für ϑ

$$N\,\vartheta\,\frac{\partial \ln f(\vartheta)}{\partial \vartheta} = E.\tag{49.18}$$

Der Wert des Parameters ϑ ist eindeutig durch die Energie E bestimmt. Wir wählen jetzt als Integrationsweg einen Kreis vom Radius $|x|=\vartheta$ und setzen

$$x = \vartheta\,e^{i\alpha},\tag{49.19}$$

$$\frac{dx}{x} = i\,d\alpha.\tag{49.20}$$

Ferner definieren wir

$$\varphi(x) = \frac{[f(x)]^{N/E}}{x}\,.\tag{49.21}$$

In dem nach Polarkoordinaten transformierten Integral wird dann der Integrand einfach $[\varphi(x)]^E$. Den Logarithmus dieses Integranden entwickeln wir an der Stelle $x=\vartheta$ in eine Taylorsche Reihe. Da für kleine α

$$x - \vartheta = i\,\vartheta\,\alpha\tag{49.22}$$

ist, erhalten wir

$$[\varphi(x)]^E = [\varphi(\vartheta)]^E \exp\left[-\frac{1}{2}E\,\frac{\varphi''(\vartheta)}{\varphi(\vartheta)}\,\vartheta^2\,\alpha^2 + i\,E\,A\,\alpha^3 - O(E\,\alpha^4)\right],\quad(49.23)$$

wo der kubische Term nicht mehr explizit angegeben ist. Der Integrand fällt von ϑ aus nach beiden Seiten exponentiell ab, und zwar um so steiler, je größer E ist. Der Punkt ϑ ist somit ein Sattelpunkt. Das Integral (49.13) nimmt nun die Form an

$$\Omega = \frac{1}{2\pi}\int\limits_{-\pi}^{+\pi}[\varphi(\vartheta,\alpha)]^E\,d\alpha.\tag{49.24}$$

Für hinreichend großes E können die Integrationsgrenzen durch $-\infty$ und $+\infty$ ersetzt werden. Wir erhalten dann mit (49.23)

$$\Omega = \frac{1}{2\pi}[\varphi(\vartheta)]^E \int\limits_{-\infty}^{+\infty}[1 + i\,A\,E\,\alpha^3 - O(E\,\alpha^4)]\,e^{-\frac{1}{2}E\,\frac{\varphi''(\vartheta)}{\varphi(\vartheta)}\,\vartheta^2\,\alpha^2}\,d\alpha.\tag{49.25}$$

Man sieht leicht, daß bei der Integration alle Terme mit ungraden Potenzen von α verschwinden. Die Ausführung des Integrals liefert

$$\Omega = \frac{[\varphi(\vartheta)]^E}{[2\pi E\,\vartheta^2\,\varphi''(\vartheta)/\varphi(\vartheta)]^{\frac{1}{2}}}\,[1 - O(E^{-1})].\tag{49.26}$$

Für hinreichend großes E können der zweite und alle höheren Terme in der Kammer vernachlässigt werden. Der vor der Klammer stehende Ausdruck ist somit der exakte asymptotische Wert des Integrals (49.13) für $E \to \infty$. Wir haben zwar bei der obigen Ableitung nur die unmittelbare Umgebung des Sattelpunktes in Betracht gezogen. Man kann aber beweisen, daß das obige Ergebnis erhalten bleibt, wenn das Verhalten des Integranden auf dem ganzen Integrationsweg berücksichtigt wird.

Wir gehen jetzt zur Untersuchung des Integrals (49.14) über. Setzen wir zur Abkürzung

$$g(x) = \frac{1}{\ln x}\,\frac{\partial \ln f(x)}{\partial \varepsilon_j}\,,\tag{49.27}$$

so können wir schreiben

$$\Omega\,\overline{N}_j = \frac{N}{2\pi i} \oint \frac{g(x)\,[f(x)]^N}{x^E}\,\frac{dx}{x}. \tag{49.28}$$

Die Entwicklung von $g(x)$ enthält nur Potenzen von x, die niedriger als x^E oder x^N sind. Auch in (49.28) besitzt daher der Integrand auf der positiven reellen Achse ein Minimum zwischen $x=0$ und $x=1$, das jetzt definiert ist als einzige positive reelle Wurzel der Gleichung

$$\frac{\partial}{\partial x}\left\{\frac{g(x)\,[f(x)]^N}{x^E}\right\} = 0. \tag{49.29}$$

Daraus folgt, wenn wir die Stelle des Minimums mit ϑ^* bezeichnen

$$N\,\vartheta^*\,\frac{\partial \ln f(\vartheta^*)}{\partial \vartheta^*} + \vartheta^*\,\frac{\partial \ln g(\vartheta^*)}{\partial \vartheta^*} = E. \tag{49.30}$$

Da der zweite Term der linken Seite nicht von N oder E abhängt, wird der erste mit wachsendem N gegen den zweiten beliebig groß. Es gilt daher, wie der Vergleich mit (49.18) zeigt, für $N \to \infty$ asymptotisch

$$\vartheta^* = \vartheta. \tag{49.31}$$

Der Sattelpunkt liegt dann an der gleichen Stelle wie bei dem Integral (49.13). Von (49.28) ausgehend, führen wir jetzt die gleiche Rechnung wie vorher durch, wobei wir $g(x)$ einfach als Faktor unter dem Integralzeichen stehen lassen. Es ergibt sich dann

$$\Omega\,\overline{N}_j = \frac{N\,[\varphi(\vartheta^*)]^E}{[2\pi E\vartheta^{*2}\varphi''(\vartheta^*)/\varphi(\vartheta^*)]^{\frac{1}{2}}}\left[g(\vartheta^*) - \frac{g''(\vartheta^*)}{4E\,\varphi''(\vartheta^*)/\varphi(\vartheta^*)} - O(E^{-1})\right]. \tag{49.32}$$

Kombinieren wir diese Gleichung mit (49.26) und gehen zur Grenze $E \to \infty$ über, so folgt mit Benutzung von (49.31) und (49.27)

$$\frac{\overline{N}_j}{N} = \frac{\vartheta^{\varepsilon_j}}{f(\vartheta)}. \tag{49.33}$$

Es bleibt jetzt noch das Integral (49.15), das wir schreiben können

$$\Omega\,N\,\overline{\varepsilon} = \frac{1}{2\pi i} \oint \frac{N x\,\dfrac{\partial \ln f(x)}{\partial x}\,[f(x)]^N}{x^E}\,\frac{dx}{x}. \tag{49.34}$$

Die formale Analogie dieses Problems mit dem vorhergehenden liegt offen zutage. Wir haben lediglich den Faktor $g(x)$ jetzt anders zu definieren und können dann in genau der gleichen Weise vorgehen. Wir schreiben daher direkt das asymptotische Resultat für $E \to \infty$ an. Es ergibt sich in Verbindung mit Gl. (49.26)

$$\overline{\varepsilon} = \vartheta\,\frac{\partial \ln f(\vartheta)}{\partial \vartheta}. \tag{49.35}$$

Damit ist das Problem formal gelöst. Die Verallgemeinerung für den Fall, daß die Eigenwerte der Einzelteilchen entartet sind, ergibt sich unmittelbar, wenn die Funktion $f(x)$ durch die Gleichung

$$f(x) = \sum g_i\,x^{\varepsilon_i} \tag{49.36}$$

definiert wird.

Die physikalische Bedeutung des Parameters ϑ ergibt sich, wenn man die vorstehende Rechnung (in Analogie zu der Betrachtung in Ziff. 22) für zwei Teilsysteme 1 und 2 in thermischer Berührung durchführt. Die Nebenbedingungen sind dann

$$\sum N_{1i} = N_1, \qquad \sum N_{2i} = N_2 \tag{49.37}$$

und

$$\sum N_{1i}\varepsilon_{1i} + \sum N_{2i}\varepsilon_{2i} = E. \tag{49.38}$$

Man findet, daß im Gleichgewicht der Parameter ϑ für das Gesamtsystem und beide Teilsysteme den gleichen Wert besitzt. ϑ besitzt daher die *Eigenschaft der empirischen Temperatur*. Nach Gl. (49.26) und (49.21) ist nun asymptotisch

$$\ln \Omega = \ln \frac{[f(\vartheta)]^N}{\vartheta^E} \tag{49.39}$$

und somit

$$\frac{\partial \ln \Omega}{\partial E} = \ln 1/\vartheta. \tag{49.40}$$

Wir sind daher berechtigt, die links stehende Größe (bis auf einen Faktor) mit der reziproken absoluten Temperatur zu identifizieren, wovon wir in Ziff. 47 schon Gebrauch gemacht haben. Es ergibt sich dann

$$\vartheta = e^{-\frac{1}{kT}}. \tag{49.41}$$

Damit wird aus Gl. (49.33) (wenn wir die Entartung einbeziehen)

$$\frac{\overline{N_j}}{N} = \frac{g_j e^{-\frac{\varepsilon_j}{kT}}}{\sum_i' g_i e^{-\frac{\varepsilon_i}{kT}}} = \frac{g_j e^{-\frac{\varepsilon_j}{kT}}}{f(T)}. \tag{49.42}$$

Für ein quantenmechanisches System aus lokalisierten Teilchen gilt somit das Maxwell-Boltzmannsche Energieverteilungsgesetz. Die Größe $f(T)$ wird als *Verteilungsfunktion (partition function) des Einzelmoleküls* bezeichnet.

Wir haben jetzt noch zu zeigen, daß der durch die Beschränkung auf den scharfen Energiewert E bedingte Fehler vernachlässigbar ist. Dazu deuten wir die exakte Rechnung an. Die Energiebedingung lautet jetzt

$$E \leqq \sum_i N_i \varepsilon_i \leqq E + \Delta E. \tag{49.43}$$

Die Größe Ω ist gleich der Summe der Koeffizienten aller Potenzen von x von x^E bis $x^{E+\Delta E}$ einschließlich. Wir haben also

$$\Omega = \frac{1}{2\pi i} \oint \frac{h(x)[f(x)]^N}{x^{E+1}} dx \tag{49.44}$$

mit

$$h(x) = 1 + \frac{1}{x} + \cdots + \frac{1}{x^{\Delta E}}. \tag{49.45}$$

Das Problem ist daher formal identisch mit dem der Gl. (49.28), und wir erhalten als asymptotischen Wert des Integrals

$$\Omega = \frac{[\varphi(\vartheta)]^E}{[2\pi E \vartheta^2 \varphi''(\vartheta)|\varphi(\vartheta)]^{\frac{1}{2}}} h(\vartheta). \tag{49.46}$$

Daraus folgt

$$\ln \Omega = N \ln f(\vartheta) - E \ln \vartheta - \tfrac{1}{2} \ln E + \ln h(\vartheta) + c, \tag{49.47}$$

wo c eine Größe ist, die nicht von N oder E abhängt. Aus der Voraussetzung $E \gg \Delta E$ folgt nun, daß die Summe in Gl. (49.45) weniger als N Terme haben muß, deren größter nach Gl. (49.41) $e^{\Delta E/kT}$ ist. Es ist daher

$$\ln h(\vartheta) < \ln N + \ln \frac{\Delta E}{kT}. \qquad (49.48)$$

Mit wachsendem N und E wird somit $\ln h(\vartheta)$ beliebig klein gegen die beiden ersten Terme auf der rechten Seite von (49.47), und wir erhalten wieder die Gl. (49.39). Bei der Bildung der Mittelwerte (49.28) und (49.34) hebt sich der „Extrafaktor" $h(\vartheta)$ einfach heraus.

50. Die Darwin-Fowlersche Methode für nicht lokalisierte Teilchen. Wir betrachten ein System aus N nicht lokalisierten Teilchen ohne Spin, dessen Gesamtenergie E sei. Die Eigenwerte der Einzelteilchen $\varepsilon_0, \varepsilon_1, \ldots, \varepsilon_i$ seien wieder ganze Zahlen ohne gemeinsamen Faktor und nicht entartet. Im Falle der Bose-Einstein-Statistik[1,2] ist dann stets $\Omega_D = 1$ und Ω gleich der Zahl der Sätze N_i, welche den Nebenbedingungen

$$\sum_i N_i = N \qquad (50.1)$$

und

$$\sum_i N_i \, \varepsilon_i = E \qquad (50.2)$$

genügen. Man sieht leicht, daß Ω der Koeffizient von $z^N x^E$ in der Entwicklung der Funktion

$$f(z, x) = \prod_i (1 + z\, x^{\varepsilon_i} + z^2\, x^{2\,\varepsilon_i} + z^3\, x^{3\,\varepsilon_i} + \cdots) = \prod_i (1 - z\, x^{\varepsilon_i})^{-1} \qquad (50.3)$$

ist, wo x und z komplexe Variable bezeichnen. Die erzeugende Funktion für $\Omega \overline{N}_j$ ist

$$z\, x^{\varepsilon_j}(1 - z\, x^{\varepsilon_j})^{-1} \prod_i (1 - z\, x^{\varepsilon_i})^{-1}. \qquad (50.4)$$

Für die Fermi-Diracsche Statistik[3,4] gilt ebenfalls, daß alle $\Omega_D = 1$ sind. Die N_i sind aber hier auf die Werte 0 und 1 beschränkt.

Dieser Unterschied gegenüber der Bose-Einstein-Statistik ist jedoch nur sekundärer Natur und lediglich eine Folgerung aus dem verschiedenen Symmetriecharakter der Eigenfunktionen. Gentile[5] hat eine Statistik angegeben, bei welcher die maximale Besetzungszahl eines Eigenwertes gleich einer beliebigen positiven ganzen Zahl N^* gesetzt wird. Formal sind darin Bose-Einstein- und Fermi-Dirac-Statistik als Sonderfälle für $N^* = 1$ und $N^* = N$ enthalten. Alle übrigen Fälle werden als „intermediäre Statistiken" bezeichnet. Diese haben jedoch, wie Sommerfeld[6] und Schubert[7] bemerkt haben, keinerlei physikalische Bedeutung, weil eben das wesentliche nicht die maximale Besetzungszahl, sondern der Symmetriecharakter der Eigenfunktionen ist, der nur symmetrisch oder antisymmetrisch sein kann.

Die Größe Ω ist jetzt gleich der Zahl der Sätze der N_i, welche den Nebenbedingungen (50.1) und (50.2) und weiter der zusätzlichen Bedingung genügen, daß nur die Werte 0 und 1 auftreten. Daraus ergibt sich als erzeugende Funktion für Ω

$$f(z, x) = \prod_i (1 + z\, x^{\varepsilon_i}) \qquad (50.5)$$

[1] S.N. Bose: Z. Physik **26**, 178 (1924).
[2] A. Einstein: Sitzgsber. preuß. Akad. Wiss. **1924**, 261.
[3] E. Fermi: Z. Physik **36**, 902 (1926).
[4] P.A.M. Dirac: Proc. Roy. Soc. Lond., Ser. A **112**, 661 (1926).
[5] G. Gentile: Nuovo Cim. **17**, 493 (1940).
[6] A. Sommerfeld: Ber. dtsch. chem. Ges. **75**, 1988 (1942).
[7] G. Schubert: Z. Naturforsch. **1**, 113 (1946).

und als erzeugende Funktion für $\Omega \overline{N}_j$

$$z\, x^{\varepsilon_j}(1 + z\, x^{\varepsilon_j}) \prod_i (1 + z\, x^{\varepsilon_i}). \tag{50.6}$$

Nachdem wir die Grundformeln festgestellt haben, läßt sich die weitere Rechnung für beide Formen der Statistik gemeinsam durchführen. Dabei soll stets das untere Vorzeichen für die Bose-Einstein-, das obere für die Fermi-Dirac-Statistik gelten. Wir schreiben also die erzeugende Funktion für Ω

$$\prod_i (1 \pm z\, x^{\varepsilon_i})^{\pm 1}, \tag{50.7}$$

die für $\Omega \overline{N}_j$

$$z\, x^{\varepsilon_j}(1 \pm z\, x^{\varepsilon_j})^{-1} \prod_i (1 \pm z\, x^{\varepsilon_i})^{\pm 1}. \tag{50.8}$$

Der Residuensatz ergibt nun

$$\Omega = \left(\frac{1}{2\pi i}\right)^2 \oint\oint \frac{\prod\limits_i (1 \pm z\, x^{\varepsilon_i})^{\pm 1}}{z^N x^E} \frac{d\,x}{x} \frac{d\,z}{z}, \tag{50.9}$$

$$\Omega \overline{N}_j = \left(\frac{1}{2\pi i}\right)^2 \oint\oint \frac{z\, x^{\varepsilon_j}(1 \pm z\, x^{\varepsilon_j})^{-1} \prod\limits_i (1 \pm z\, x^{r_i})^{\pm 1}}{z^N x^E} \frac{d\,x}{x} \frac{d\,z}{z}. \tag{50.10}$$

Die Integrationswege sind geschlossene Kurven um die Pole des Integranden bei $x=0$ und $z=0$, die keine weiteren Singularitäten einschließen dürfen. Die Auswertung der Integrale erfolgt wieder nach der Sattelpunktmethode. Setzen wir

$$z^{-N} x^{-E} \prod_i (1 \pm z\, x^{\varepsilon_i})^{\pm 1} = e^{Y(z,\,x)} \tag{50.11}$$

und bezeichnen mit Z und ϑ die Koordinaten der Sattelpunkte, so ergibt sich

$$\Omega = \frac{e^{Y(Z,\,\vartheta)}}{2\pi (AB - C^2)^{\frac{1}{2}}} [1 + \cdots], \tag{50.12}$$

wo die nicht ausgeschriebenen höheren Glieder für $N \to \infty$, $E \to \infty$ asymptotisch verschwinden. Dabei ist

$$A = Z^2 \left(\frac{\partial^2 Y}{\partial z^2}\right)_{\substack{z=Z\\x=\vartheta}}, \qquad B = \vartheta^2 \left(\frac{\partial^2 Y}{\partial x^2}\right)_{\substack{z=Z\\x=\vartheta}}, \qquad C = Z\vartheta \left(\frac{\partial^2 Y}{\partial x\, \partial z}\right)_{\substack{z=Z\\x=\vartheta}} \tag{50.13}$$

und

$$AB > C^2. \tag{50.14}$$

Die Parameter Z und ϑ sind die positiven reellen Wurzeln der Gleichungen

$$Z\frac{\partial}{\partial Z} \ln \prod_i (1 \pm Z\, \vartheta^{\varepsilon_i})^{\pm 1} = \sum_i \frac{Z\vartheta^{\varepsilon_i}}{1 \pm Z\vartheta^{\varepsilon_i}} = N, \tag{50.15}$$

$$\vartheta\frac{\partial}{\partial Z} \ln \prod_i (1 \pm Z\, \vartheta^{\varepsilon_i})^{\pm 1} = \sum_i \frac{\varepsilon_i Z\vartheta^{\varepsilon_i}}{1 \pm Z\vartheta^{\varepsilon_i}} = E. \tag{50.16}$$

Aus (50.10) erhält man

$$\overline{N}_j = \frac{Z\vartheta^{\varepsilon_j}}{1 \pm Z\vartheta^{\varepsilon_j}}. \tag{50.17}$$

Wenn die Eigenwerte der Einzelteilchen entartet sind, wird die erzeugende Funktion für Ω

$$f(z, x) = \prod_i (1 \pm z\, x^{\varepsilon_i})^{\pm g_i}, \tag{50.18}$$

und es ergibt sich als allgemeines Verteilungsgesetz für nicht lokalisierte Teilchen

$$\overline{N}_j = \frac{g_j Z \vartheta^{\varepsilon_j}}{1 \pm Z \vartheta^{\varepsilon_j}}. \tag{50.19}$$

Auch hier ist, wie man nach der in Ziff. 49 erwähnten Methode zeigt

$$\vartheta = e^{-\frac{1}{kT}}. \tag{50.20}$$

Ferner gilt

$$Z = e^{\frac{\mu}{kT}}, \tag{50.21}$$

wo μ das chemische Potential pro Molekül ist, während Z häufig als *absolute Aktivität*[1] bezeichnet wird. Daß die durch Gl. (50.21) definierte Größe μ die Eigenschaften des chemischen Potentials besitzt, werden wir später beweisen (Ziff. 64).

Mit Benutzung von (50.20) und (50.21) können wir das Verteilungsgesetz schreiben

$$\overline{N}_j = \frac{g_j}{e^{\frac{\mu - \varepsilon_j}{kT}} \pm 1}. \tag{50.22}$$

Da wir das ideale Bose-Einstein-Gas in anderem Zusammenhang (Ziff. 78) ausführlich behandeln werden, beschränken wir uns hier auf eine kurze Diskussion des Verteilungsgesetzes der Fermi-Dirac-Statistik. Der Einfachheit halber nehmen wir an, daß sich das Gas in einem kubischen Behälter der Kantenlänge L befindet. Die Eigenwerte der Einzelteilchen sind dann

$$\varepsilon_k = \tfrac{1}{3}\,\varepsilon_1 (k_x^2 + k_y^2 + k_z^2) \tag{50.23}$$

mit

$$\varepsilon_1 = \frac{3h^2}{8mL^2}. \tag{50.24}$$

Das Eigenwertspektrum läßt sich mit für die meisten Zwecke ausreichender Genauigkeit durch eine stetige differenzierbare Dichtefunktion $g(\varepsilon)$ approximieren, für die sich

$$g(\varepsilon) = 4\pi \frac{mV}{h^3} (2m\,\varepsilon)^{\frac{1}{2}} \tag{50.25}$$

ergibt, was bis auf den Faktor $1/h^3$ mit der klassischen Gleichung (28.7) übereinstimmt. Wir können daher in dieser Näherung das Verhalten des Gases in dem in *Zellen der Größe* h^3 unterteilten μ-Raum veranschaulichen. Die Zahl der μ-Raum-Zellen mit Energien $\varepsilon \leqq \varepsilon^*$ (wo ε^* eine beliebig vorgegebene Energie bezeichnet) ist nach (50.25)

$$\int_0^{\varepsilon^*} g(\varepsilon)\, d\varepsilon = \frac{4\pi}{3} V \left(\frac{2m\,\varepsilon^*}{h^2}\right)^{\frac{3}{2}}. \tag{50.26}$$

---

[1] In der Literatur wird diese Größe gewöhnlich mit λ bezeichnet. Häufig ist es zweckmäßiger, die normierte Größe $Z = \frac{f(T)}{V} e^{\frac{\mu}{kT}}$ zu verwenden, die als *Fugazität* bezeichnet wird. Es gilt dann $Z \to \varrho$ für $\varrho \to 0$, wo ϱ die molekulare Dichte ist.

Während sich ein klassisches Gas mit sinkender Temperatur beliebig in Richtung abnehmender Impulskoordinaten konzentriert, ist dies für ein Fermi-Dirac-System nur bis zu einer gewissen Grenze möglich,weil jede Zelle maximal mit einem Teilchen (im Falle von Elektronen wegen der zwei möglichen Spinorientierungen mit zwei Teilchen) besetzt werden kann. Ein derartiges Gas muß daher eine durch die Statistik bedingte Nullpunktsenergie besitzen, die sich leicht aus Gl. (50.26) berechnen läßt. Bezeichnen wir mit ε_F die maximale Energie der besetzten Zellen („Fermi-Energie"), so lautet das Verteilungsgesetz für $T=0$

$$\left. \begin{aligned} \overline{N}_j &= g_j \quad \text{für} \quad \varepsilon_j \leqq \varepsilon_F, \\ \overline{N}_j &= 0 \quad \text{für} \quad \varepsilon_j > \varepsilon_F. \end{aligned} \right\} \tag{50.27}$$

Die Auftragung von $\overline{N}_j/g_j$ gegen ε_j ergibt das in Fig. 5 dargestellte Bild.

Mit steigender Temperatur wird die Stufe mehr und mehr abgeflacht. Die Grenzenergie ε_F ergibt sich, wenn die rechte Seite von (50.26), (die wir hier für die Anwendung auf Elektronen noch mit dem Faktor 2 multipliziert denken) gleich der Teilchenzahl N und $\varepsilon^* = \varepsilon_F$ gesetzt wird. Wir erhalten dann

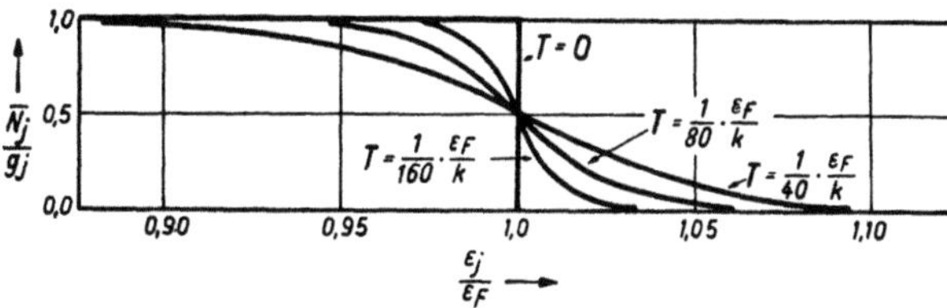

Fig. 5. Fermi-Verteilung.

$$\varepsilon_F = \frac{h^2}{8m}\left(\frac{3}{\pi}\frac{N}{V}\right)^{\frac{2}{3}}. \tag{50.28}$$

Für Elektronen kann Gl. (50.25) geschrieben werden

$$g_{\text{el}}(\varepsilon) = \frac{3}{2} N \frac{\varepsilon^{\frac{1}{2}}}{\varepsilon_F^{\frac{3}{2}}}. \tag{50.29}$$

Für die Nullpunktsenergie des Elektronengases ergibt sich

$$^0E = \int\limits_0^{\varepsilon_F} \varepsilon\, g_{\text{el}}(\varepsilon)\, d\varepsilon = \frac{3h^2}{40m} N \left(\frac{3}{\pi}\frac{N}{V}\right)^{\frac{2}{3}} \tag{50.30}$$

oder

$$^0E = \tfrac{3}{5} N \varepsilon_F = 359{,}7 \cdot 10^3\, V_m^{-\frac{2}{3}}\,\text{cal}, \tag{50.31}$$

wo V_m das Molvolumen bezeichnet.

Der Unterschied zwischen Bose-Einstein- und Fermi-Dirac-Statistik muß bedeutungslos werden, wenn die Zahl der μ-Raum-Zellen, die mit der mittleren thermischen Energie eines Teilchens bei der betreffenden Temperatur erreichbar sind, sehr groß ist gegen die Zahl der Teilchen. Wir können erwarten, daß dann die in Ziff. 25 definierte „halbklassische Näherung" gilt. Die Bedingung dafür ergibt sich, da

$$\overline{\varepsilon} = \tfrac{3}{2} kT \tag{50.32}$$

ist, unmittelbar aus (50.26) in der Form

$$2\pi V \left(\frac{2m}{h^2}\right)^{\frac{3}{2}} \int\limits_0^{\frac{3}{2}kT} \varepsilon^{\frac{1}{2}}\, d\varepsilon \gg N \tag{50.33}$$

oder

$$\frac{N}{V}\left(\frac{h^2}{2\pi m kT}\right)^{\frac{3}{2}} \ll 1. \tag{50.34}$$

Aus Gl. (50.19) sieht man nun, daß das Verteilungsgesetz der Quantenstatistik in das Maxwell-Boltzmannsche Verteilungsgesetz übergeht, wenn

$$Z \ll \vartheta^{-\varepsilon_j} \tag{50.35}$$

ist. Dann wird nämlich

$$\overline{N_j} = Z\, g_j\, \vartheta^{\varepsilon_j}. \tag{50.36}$$

Nun ist aber

$$\sum_i \overline{N_i} = \sum_i Z\, g_i\, \vartheta^{\varepsilon_i} = Z\, f(\vartheta) = N, \tag{50.37}$$

wo $f(\vartheta)$ die durch Gl. (49.36) (mit $x = \vartheta$) definierte Verteilungsfunktion des Einzelmoleküls ist. Durch Elimination von Z folgt aus Gl. (50.36) und (50.37)

$$\frac{\overline{N_j}}{N} = \frac{g_j\, e^{-\frac{\varepsilon_j}{kT}}}{\sum_i g_i\, e^{-\frac{\varepsilon_i}{kT}}} \tag{50.38}$$

das Maxwell-Boltzmannsche Verteilungsgesetz. Da mit beliebig wachsender Temperatur $\vartheta^{-\varepsilon_j}$ gegen Eins geht, kann die Bedingung (50.35) auch geschrieben werden

$$Z \ll 1. \tag{50.39}$$

Um zu zeigen, daß diese Bedingung mit (50.34) identisch ist, berechnen wir die Verteilungsfunktion des Einzelmoleküls

$$f(T) = \sum_{k_x=1}^{\infty} \sum_{k_y=1}^{\infty} \sum_{k_z=1}^{\infty} e^{-\frac{\varepsilon_1(k_x^2 + k_y^2 + k_z^2)}{3kT}} \tag{50.40}$$

mit Hilfe der Approximation des kontinuierlichen Eigenwertspektrums (50.25), die auch die Grundlage für die Ableitung der Bedingung (50.34) bildet. Es ergibt sich

$$f(T) = 2\pi \left(\frac{2m}{h^2}\right)^{\frac{3}{2}} V \int_0^{\infty} \varepsilon^{\frac{1}{2}}\, e^{-\frac{\varepsilon}{kT}}\, d\varepsilon = \left(\frac{2\pi m\, kT}{h^2}\right)^{\frac{3}{2}} V. \tag{50.41}$$

Durch Einsetzen in Gl. (50.37) erhält man

$$Z = \frac{N}{V} \left(\frac{h^2}{2\pi m\, kT}\right)^{\frac{3}{2}}. \tag{50.42}$$

Die Bedingungen (50.34) und (50.39) sind somit in der Tat identisch. Die durch die Quantenstatistik bedingten Abweichungen von der Zustandsgleichung des idealen Gases werden häufig als *Gasentartung* bezeichnet. Die Bedingung (50.34) bzw. (50.39) heißt danach das *Entartungskriterium*. Da für Wasserstoff unter Normalbedingungen bei 9° K $Z = 0{,}005$ ist, finden die in dieser Ziffer entwickelten Formeln nur für die Statistik des Elektronen- und Photonengases Anwendung.

V. Die kanonische Gesamtheit der Quantenstatistik.

51. Definition. Die kanonische Gesamtheit wird in der Quantenstatistik unter den gleichen Gesichtspunkten eingeführt wie in der klassischen Theorie. Sie erscheint einerseits als mathematischer Kunstgriff, um durch Einführung einer erzeugenden Funktion asymptotische Resultate für die mikrokanonische Gesamtheit zu erhalten. In diesem Sinne haben wir sie tatsächlich bereits in Ziff. 49 und 50 benutzt. Andererseits stellt sie die angemessene physikalische Beschreibung eines Systems dar, das mit seiner Umgebung im thermischen Gleichgewicht steht, also ungehindert Energie mit ihr austauschen kann. Obwohl verschiedentlich[1] [*12*]

---

[1] E.C. Kemble: Phys. Rev. **56**, 1013, 1146 (1939).

versucht worden ist, die Dichtematrix der kanonischen Gesamtheit durch quantenmechanische Betrachtungen zu begründen, geben wir einer axiomatischen Einführung den Vorzug. Wir setzen also

$$\rho = e^{\frac{\psi - H}{\Theta}} \qquad (51.1)$$

oder explizit

$$\rho = e^{\frac{\psi}{\Theta}}\left(1 - \frac{H}{\Theta} + \frac{1}{2!}\frac{H^2}{\Theta^2} - \frac{1}{3!}\frac{H^3}{\Theta^3} + \cdots\right), \qquad (51.2)$$

wo die Größe Θ, wie früher, als Verteilungsmodul bezeichnet wird. Aus Ziff. 36 ergibt sich, daß die kanonische Gesamtheit stationär ist. Die Normierungsrelation (34.19) liefert für ψ/Θ die Gleichung

$$e^{-\frac{\psi}{\Theta}} = \text{spur } e^{-\frac{H}{\Theta}} . \qquad (51.3)$$

Für die Basis der Energie-Eigenfunktionen erhält man die spezielle Form

$$\varrho_{nm} = e^{\frac{\psi - E_m}{\Theta}}\, \delta_{nm} . \qquad (51.4)$$

Die Wahrscheinlichkeit, ein willkürlich herausgegriffenes System in einem durch den Index n charakterisierten Eigenzustand der Energie zu finden, ist dann

$$W(n) = \varrho_{nn} = e^{\frac{\psi - E_n}{\Theta}} . \qquad (51.5)$$

Ferner gilt

$$\sum_n \varrho_{nn} = \sum_n e^{\frac{\psi - E_n}{\Theta}} = 1 \qquad (51.6)$$

oder

$$e^{-\frac{\psi}{\Theta}} = \sum_n e^{-\frac{E_n}{\Theta}} . \qquad (51.7)$$

In den vorstehenden Formeln beziehen sich die Indices auf die Energie-Eigenfunktionen. Für manche Betrachtungen ist es zweckmäßig, den Entartungsgrad Ω_l einzuführen. Wir haben dann für die Wahrscheinlichkeit, einen Energie-Eigenwert E_l zu finden:

$$W(E_l) = \Omega_l e^{\frac{\psi - E_l}{\Theta}} , \qquad (51.8)$$

während Gl. (51.7) jetzt lautet

$$e^{-\frac{\psi}{\Theta}} = \sum_l \Omega_l e^{-\frac{E_l}{\Theta}} , \qquad (51.9)$$

wo über die Gruppen entarteter (oder fast entarteter) Zustände zu summieren ist. Diese Schreibweise zeigt deutlich den Aufbau der kanonischen Gesamtheit aus einer Vielzahl von mikrokanonischen Gesamtheiten, den wir schon in Ziff. 29 erörtert haben.

Für den kanonischen Mittelwert einer Größe F haben wir als Spezialfall von (34.22)

$$\bar{F} = \sum_n F_{nn}\, e^{\frac{\psi - E_n}{\Theta}} = \sum_l F_l \Omega_l e^{\frac{\psi - E_l}{\Theta}} , \qquad (51.10)$$

wo F_{nn} das Matrixelement des Operators F für die Basis der Energie-Eigenfunktionen ist und F_l den über die Ω_l zum Energie-Eigenwert E_l gehörigen Eigenfunktionen gebildeten Mittelwert der Größe F bezeichnet, der durch Gl. (45.2) definiert ist. Für die mittlere Energie eines Systems der kanonischen Gesamtheit ergibt sich daraus die Formel

$$\bar{E} = \sum_l E_l \Omega_l \, e^{\frac{\psi - E_l}{\Theta}} \,. \tag{51.11}$$

Betrachten wir die Energie-Eigenwerte wieder als stetige differenzierbare Funktionen der äußeren Parameter x_j, so erhalten wir durch Differentiation der Gl. (51.9)

$$e^{-\frac{\psi}{\Theta}}\left(-\frac{1}{\Theta}\,d\psi + \frac{\psi}{\Theta^2}\,d\Theta\right) = \frac{1}{\Theta^2}\,d\Theta \sum_l E_l \Omega_l \, e^{-\frac{E_l}{\Theta}} - \frac{1}{\Theta}\sum_j dx_j \sum_l \frac{\partial E_l}{\partial x_j}\Omega_l \, e^{-\frac{E_l}{\Theta}} \tag{51.12}$$

oder mit Benutzung von (46.1) und (51.10)

$$d\psi = \frac{\psi}{\Theta}\,d\Theta - \frac{\bar{E}}{\Theta}\,d\Theta - \sum_j \bar{X}_j\,dx_j\,. \tag{51.13}$$

Setzen wir

$$\eta = \frac{\psi - \bar{E}}{\Theta}\,, \tag{51.14}$$

so wird

$$d\psi = \bar{\eta}\,d\Theta - \sum_j \bar{X}_j\,dx_j\,. \tag{51.15}$$

Im folgenden setzen wir, wenn nichts besonderes bemerkt ist, als Basis stets die Energie-Eigenfunktionen voraus und werden dann die Matrix-Elemente nur einfach indizieren.

52. Die empirische Temperatur. Wir wollen jetzt die physikalische Bedeutung des Verteilungsmoduls Θ untersuchen. Dazu betrachten wir zwei Systeme 1 und 2, die zunächst voneinander getrennt seien und durch kanonische Gesamtheiten mit den Dichtematrizen

$$\varrho_n^1 = e^{\frac{\psi_1 - E_n^1}{\Theta}} \tag{52.1}$$

und

$$\varrho_m^2 = e^{\frac{\psi_2 - E_m^2}{\Theta}} \tag{52.2}$$

dargestellt werden. Beide Systeme können wir in Gedanken als ein einziges System auffassen, für dessen Dichtematrix gilt

$$\varrho_{n,m}^{1,2} = e^{\frac{\psi_1 + \psi_2 - (E_n^1 + E_m^2)}{\Theta}}\,. \tag{52.3}$$

Die durch diesen Ausdruck dargestellte kanonische Gesamtheit kommt dadurch zustande, daß jedes System der Gesamtheit 1 mit jedem System der Gesamtheit 2 kombiniert wird. Werden nun die beiden Originalsysteme durch Herstellen einer wärmeleitenden Verbindung vereinigt, so tritt im Hamilton-Operator ein zusätzliches Störungsglied H_{12} auf. Die Dichtematrix ist dann für die Eigenfunktionen des ungestörten Hamilton-Operators nicht mehr diagonal. Der Definition der empirischen Temperatur liegt aber die Annahme zugrunde, daß diese Störung völlig vernachlässigt werden kann[1]. Führen wir diese Annahme hier ein, so folgt, daß auch nach Herstellen der wärmeleitenden Verbindung das Gesamtsystem

---

[1] Sie wird dadurch gerechtfertigt, daß die Thermodynamik nur für den Grenzfall unendlich großer Systeme gilt. Näheres s. Teil C I.

durch die Dichtematrix (52.3) dargestellt wird. Wenn aber für die getrennten Systeme einzeln gilt

$$\varrho_n^1 = e^{\frac{\psi_1 - E_n^1}{\Theta_1}} \tag{52.4}$$

und

$$\varrho_m^2 = e^{\frac{\psi_2 - E_m^2}{\Theta_2}} , \tag{52.5}$$

so haben wir für das in Gedanken konstruierte Gesamtsystem

$$\varrho_{n,m}^{1,2} = e^{\frac{\psi_1 - E_n^1}{\Theta_1} + \frac{\psi_2 - E_m^2}{\Theta_2}} . \tag{52.6}$$

Wird nun wieder eine wärmeleitende Verbindung hergestellt, so muß das Gesamtsystem im Gleichgewicht durch eine kanonische Gesamtheit dargestellt werden, d. h. durch eine Dichtematrix von der Form der Gl. (52.3). Diese fällt aber dann und nur dann mit (52.6) zusammen, wenn $\Theta_1 = \Theta_2$ ist. Es gilt daher der Satz: Die gemeinsame Dichtematrix zweier Systeme wird durch Herstellen einer wärmeleitenden Verbindung dann und nur dann nicht geändert, wenn der Verteilungsmodul Θ für beide Systeme den gleichen Wert hat. Damit ist Θ als Analogon der empirischen Temperatur identifiziert. Der Zusammenhang mit der Temperatur des Gasthermometers, der durch Berechnung der Zustandsgleichung des idealen Gases erhalten werden kann, ist, wie in der klassischen Theorie, durch die Gleichung

$$\Theta = kT \tag{52.7}$$

gegeben.

53. Adiabatische Änderung der Parameter. Wir führen jetzt eine der Ziff. 46 entsprechende Überlegung für die kanonische Gesamtheit durch und betrachten eine adiabatische Änderung der äußeren Parameter x_j. Ganz allgemein ergibt sich zunächst für eine Variation der kanonischen Verteilung aus Gl. (51.14)

$$\delta \bar{\eta} = \frac{\delta \psi}{\Theta} - \frac{\delta \bar{E}}{\Theta} - \frac{\psi - \bar{E}}{\Theta^2} \delta \Theta . \tag{53.1}$$

Andererseits ist, wenn wir (51.13) entsprechend umschreiben,

$$\frac{\delta \psi}{\Theta} + \frac{1}{\Theta} \sum_j \bar{X}_j \, \delta x_j - \frac{\psi - \bar{E}}{\Theta^2} \delta \Theta = 0 . \tag{53.2}$$

Aus diesen beiden Gleichungen folgt

$$- \delta \bar{\eta} = \frac{\delta E}{\Theta} + \frac{1}{\Theta} \sum_j \bar{X}_j \, \delta x_j . \tag{53.3}$$

Da die kanonische Gesamtheit aus mikrokanonischen Gesamtheiten aufgebaut ist, wird nach Gl. (46.4) bei dem betrachteten Prozeß die Zahl der Eigenfunktionen nicht geändert. Es finden somit lediglich eine Verschiebung der Eigenwerte statt. Setzen wir nun voraus, daß sich wieder eine kanonische Verteilung einstellt[1] und berücksichtigen, daß definitionsgemäß die Energie nur durch die Variation der äußeren Parameter geändert wird, so haben wir (bis auf Glieder höherer Ordnung)

$$\delta \bar{E} = \sum_n \varrho_n \left(\sum_j \frac{\partial E_n}{\partial x_j} \delta x_j \right) = - \sum_j \bar{X}_j \, \delta x_j . \tag{53.4}$$

---

[1] Daß diese Annahme für $n \to \infty$ korrekt ist, wurde auf der Grundlage der klassischen Statistik von L. Rosenfeld [Proc. Kon. Nederl. Acad. Wetensch. **45**, 970 (1942)] gezeigt.

Durch Einsetzen in (53.3) erhalten wir dann

$$\delta\bar{\eta} = 0. \tag{53.5}$$

Für die kanonische Gesamtheit besitzt somit die Größe $\bar{\eta}$ die Invarianzeigenschaft der Entropie gegen quasistatisch-adiabatische Prozesse.

54. Thermodynamische Analoga. Die an einem System der kanonischen Gesamtheit im n-ten Eigenzustand geleistete Arbeit ist

$$dA_n = dE_n = - \sum_j X_j\, dx_j. \tag{54.1}$$

Wir betrachten nun Prozesse, bei denen die an dem System geleistete Arbeit nicht notwendig gleich der Änderung der inneren Energie ist. Die mittlere Arbeit ist dann

$$d'\bar{A} = \sum_n \varrho_n\, dA_n = - \sum_j \bar{X}_j\, dx_j, \tag{54.2}$$

wo das Symbol d' andeutet, daß es sich nicht um ein exaktes Differential handelt. Die dem System zugeführte Wärme definieren wir durch

$$d'Q^* = d\bar{E} - d'\bar{A} \tag{54.3}$$

(I. Hauptsatz der Thermodynamik). Nun ist (51.11)

$$d\bar{E} = \sum_n (\varrho_n\, dE_n + E_n\, d\varrho_n). \tag{54.4}$$

Ferner folgt aus (54.1) und (54.2)

$$d'\bar{A} = \sum_n \varrho_n\, dE_n. \tag{54.5}$$

Setzen wir (54.4) und (54.5) in (54.3) ein und benutzen die aus (51.6) folgende Beziehung

$$\sum_n d\varrho_n = 0, \tag{54.6}$$

so erhalten wir

$$d'Q^* = \sum_n (E_n - \psi)\, d\varrho_n. \tag{54.7}$$

Wir wollen jetzt zeigen, daß $d'Q/\Theta = - d\bar{\eta}$ und ein vollständiges Differential ist. Aus Gl. (51.4) bekommen wir

$$E_n - \psi = - \Theta \ln \varrho_n. \tag{54.8}$$

Das ergibt mit (54.7)

$$\frac{d'Q^*}{\Theta} = - \sum_n \ln \varrho_n\, d\varrho_n. \tag{54.9}$$

Benutzen wir jetzt Gl. (51.14) und (54.6), so folgt

$$\frac{d'Q^*}{\Theta} = - d\left(\sum_n \varrho_n \ln \varrho_n\right) = - d\bar{\eta}, \tag{54.10}$$

womit die Behauptung bewiesen ist. Da Θ die Eigenschaft der empirischen Temperatur besitzt und ein integrierender Nenner für $d'Q^*$ ist, muß Θ ein Analogon der absoluten Temperatur und $-\bar{\eta}$ ein Analogon der Entropie sein. (II. Hauptsatz der Thermodynamik.)

Aus Gl. (54.10) sieht man, daß die Entropiedefinition

$$S = - k\bar{\eta} \tag{54.11}$$

wieder ein Spezialfall der allgemeinen Gleichung

$$S = - k \operatorname{spur} (\rho \ln \rho) \tag{54.12}$$

ist. Wir wollen nun mit Hilfe dieser Beziehung noch die Frage der Additivität der Entropie kurz erörtern[1]. Wir betrachten wieder zwei Teilsysteme 1 und 2. Der Satz ϱ_n bezeichne ein vollständiges normiertes Orthogonalsystem im Konfigurationsraum des Systems 1, der Satz χ_m ein analoges System im Konfigurationsraum des Systems 2. Dann bildet der Satz $\psi_{nm} = \varrho_n \chi_m$ ein vollständiges normiertes Orthogonalsystem im Konfigurationsraum des Gesamtsystems $1+2$. Der Zustand des Gesamtsystems werde durch eine Dichtematrix $\rho_{1,2}$ beschrieben, deren Elemente wir für die Basis der ψ_{nm} mit $\varrho_{1,2}$ $(nm, n' m')$ bezeichnen. Die Dichtematrix für das Teilsystem 1 ist dann definiert durch

$$\varrho_1(n, n') = \sum_m \varrho_{1,2}(n\,m, n'\,m), \tag{54.13}$$

die für das Teilsystem 2 durch

$$\varrho_2(m, m') = \sum_n \varrho_{1,2}(n\,m, n\,m'). \tag{54.14}$$

Die Orthogonalsysteme φ_n und χ_m seien nun so gewählt, daß sie ρ_1 bzw. ρ_2 auf Diagonalform bringen. Im allgemeinen Fall ist dann $\rho_{1,2}$ für die Basis ψ_{nm} keine Diagonalmatrix. Die Ausdrücke für die Entropien sind dann

$$S_1 = - k \sum_n \varrho_1(n, n) \ln \varrho_1(n, n), \tag{54.15}$$

$$S_2 = - k \sum_m \varrho_2(m, m) \ln \varrho_2(m, m), \tag{54.16}$$

$$S_{1,2} = - k \operatorname{spur} (\rho_{1,2} \ln \rho_{1,2}). \tag{54.17}$$

Wir betrachten jetzt den Fall, der durch

$$\rho_{1,2} = \rho_1 \times \rho_2 \tag{54.18}[2]$$

beschrieben wird. Es folgt dann unmittelbar, daß für die Basis ψ_{nm} auch $\rho_{1,2}$ eine Diagonalmatrix ist. Wir haben also

$$\varrho_{12}(n, m, n\,m) = \varrho_1(n, n)\,\varrho_2(m, m) \tag{54.19}$$

und erhalten mit (54.17)

$$S_{1,2} = - k \sum_n \sum_m \varrho_1(n, n)\,\varrho_2(m, m) \ln [\varrho_1(n, n)\,\varrho_2(m, m)] \tag{54.20}$$

oder wegen (34.19)

$$S_{1,2} = - k \left[\sum_n \varrho_1(n, n) \ln \varrho_1(n, n) + \sum_m \varrho_2(m, m) \ln \varrho_2(m, m) \right]. \tag{54.21}$$

Der Vergleich mit (54.15) und (54.16) ergibt dann

$$S_{1,2} = S_1 + S_2. \tag{54.22}$$

Für dieses Ergebnis ist die Bedingung (54.18) nicht nur hinreichend, sondern, wie man beweisen kann[1], auch notwendig. Sie besagt, daß die Zustände der Systeme 1 und 2 unabhängig voneinander sind in dem Sinne, daß $\rho_{1,2}$ keine

---

[1] E.C. KEMBLE: Phys. Rev. **56**, 1146 (1939).
[2] $A \times B$ bezeichnet das direkte Produkt zweier Matrizen.

Aussage darüber enthält, daß das Ergebnis einer Messung an 1 die Erwartung für das Ergebnis einer Messung an 2 beeinflussen könnte. Dies trifft nicht nur dann zu, wenn die Teilsysteme 1 und 2 völlig unabhängig sind, sondern auch, wie sich aus Ziff. 52 ergibt, wenn sie miteinander im thermischen Gleichgewicht stehen. Man kann daher sagen: Die Entropie eines Gesamtsystems ist gleich der Summe der Entropien der Teilsysteme, wenn diese völlig voneinander unabhängig oder miteinander im thermischen Gleichgewicht sind.

Wenn wir den Fall einschließen, daß (54.18) nicht erfüllt ist, so gilt allgemein, wie wir hier nicht beweisen wollen[1],

$$S_{1,2} \leqq S_1 + S_2. \tag{54.23}$$

Diese Beziehung ist auf der Grundlage der klassischen Statistik bereits von GIBBS [2] abgeleitet worden.

Nachdem die statistischen Analoga der Entropie und absoluten Temperatur identifiziert worden sind, folgt aus der in Ziff. 51 abgeleiteten Gleichung

$$d\psi = \bar{\eta}\, d\Theta - \sum_j \bar{X}_j\, dx_j \tag{54.24}$$

unmittelbar, daß die Größe ψ ein statistisches Analogon der freien Energie nach HELMHOLTZ darstellt.

55. Das thermodynamische Gleichgewicht. Die mit Hilfe der freien Energie nach HELMHOLTZ formulierte thermodynamische Gleichgewichtsbedingung

$$(\delta F)_{T,V,N} \geqq 0 \tag{55.1}$$

läßt sich ohne Schwierigkeit auch mit Hilfe der Quantenstatistik begründen. Dazu müssen wir zeigen, daß die Größe $\bar{\eta} + \bar{E}/\Theta$ für die kanonische Gesamtheit ein Minimum wird gegenüber variierten Gesamtheiten, die sich nicht im statistischen Gleichgewicht befinden. Eine solche variierte Gesamtheit kann etwa beschrieben werden durch die Dichtematrix

$$\rho' = e^{\frac{\psi - \mathsf{H}}{\Theta}} \cdot e^{\mathsf{z}}, \tag{55.2}$$

wo z eine mit H nicht vertauschbare, aber sonst beliebige quantenmechanische Matrix ist. Dabei gilt

$$\mathrm{spur}\, \rho' = 1. \tag{55.3}$$

Für die Durchführung des Beweises[2] ist es zweckmäßig, neue Operatoren n' und n einzuführen durch die Gleichungen

$$\rho' = e^{\mathsf{n}'}, \quad \rho = e^{\mathsf{n}}. \tag{55.4}$$

Wir haben dann

$$e^{\mathsf{n}'} = e^{\mathsf{n}} \cdot e^{\mathsf{z}}. \tag{55.5}$$

Es ist nun

$$\bar{\eta}' + \frac{\bar{E}'}{\Theta} = \mathrm{spur}\,(\rho' \ln \rho') + \Theta^{-1}\, \mathrm{spur}\, \rho'\mathsf{H}. \tag{55.6}$$

Wir erhalten daher mit (55.3)

$$\left(\bar{\eta}' + \frac{\bar{E}'}{\Theta}\right) - \left(\bar{\eta} + \frac{\bar{E}}{\Theta}\right) = \mathrm{spur}\,(\mathsf{n}'e^{\mathsf{n}'} - \mathsf{n}\,e^{\mathsf{n}}) + \Theta^{-1}\,\mathrm{spur}\,(\mathsf{H}\,e^{\mathsf{n}'} - \mathsf{H}\,e^{\mathsf{n}}). \tag{55.7}$$

---

[1] E.C. KEMBLE: Phys. Rev. **56**, 1146 (1939).
[2] Die in [12] und [26] gegebenen Beweise sind nicht korrekt.

Der erste Term der rechten Seite kann geschrieben werden

$$\text{spur}\,(n'\,e^{n'} - n\,e^{n}) = \text{spur}\,[(n' - n)\,e^{n'}] + \text{spur}\,[n\,(\rho' - \rho)]$$
$$= \text{spur}\,[(n' - n)\,e^{n'}] + \text{spur}\left[\frac{\psi - H}{\Theta}\,(\rho' - \rho)\right]. \tag{55.8}$$

Mit Berücksichtigung von (34.19) und (55.3) folgt dann

$$\left(\bar{\eta}' + \frac{\bar{E}'}{\Theta}\right) - \left(\bar{\eta} + \frac{\bar{E}}{\Theta}\right) = \text{spur}\,[(n' - n)\,e^{n'}]. \tag{55.9}$$

Für jeden hermitischen Operator A gilt nun [29]

$$[e^{A}]_{ii} \geqq e^{A_{ii}}, \tag{55.10}$$

wobei sich das Gleichheitszeichen auf den Fall bezieht, daß A eine Diagonalmatrix ist. Wir wählen eine Basis, für welche der Operator n' diagonal wird. Dann erhalten wir mit (34.19) und (55.3)

$$\text{spur}\,[(n' - n)\,e^{n'}] = \text{spur}\,[(n' - n)\,e^{n'} - e^{n'} + e^{n}]$$
$$= \sum_{i} [(\eta_i' - \eta_{ii})\,e^{\eta_i'} - e^{\eta_i'} + (e^{n})_{ii}]$$
$$\geqq \sum_{i} [(\eta_i' - \eta_{ii})\,e^{\eta_i'} - e^{\eta_i'} + e^{\eta_{ii}}]$$
$$= \sum_{i} e^{\eta_{ii}}\,[(\eta_i' - \eta_{ii} - 1)\,e^{\eta_i' - \eta_{ii}} + 1], \tag{55.11}$$

wobei wir die Diagonalelemente von n' mit η_i', die von n mit η_{ii} bezeichnet haben. Da in der gewählten Darstellung auch z diagonal ist, erhalten wir aus (55.5) und (55.10)

$$\eta_i' - \eta_{ii} \geqq z_i. \tag{55.12}$$

Damit wird schließlich

$$\left(\bar{\eta}' + \frac{\bar{E}'}{\Theta}\right) - \left(\bar{\eta} + \frac{\bar{E}}{\Theta}\right) \geqq \sum_{i} e^{\eta_{ii}}\,[z_i\,e^{z_i} - e^{z_i} + 1]. \tag{55.13}$$

Der rechts in eckigen Klammern stehende Ausdruck hat wieder die Form der Funktion (17.10). Es folgt daher

$$\left(\bar{\eta}' + \frac{\bar{E}'}{\Theta}\right) - \left(\bar{\eta} + \frac{\bar{E}}{\Theta}\right) \geqq 0, \tag{55.14}$$

womit die Behauptung bewiesen ist, da das Gleichheitszeichen nur für den Fall $z_i = 0$ (alle i) gilt.

56. Verteilungsfunktion und Slater-Summe. Aus Ziff. 51 und 54 sieht man, daß die thermodynamischen Eigenschaften eines Systems durch die Funktion

$$Q = \text{spur}\,(e^{-\beta H}) \tag{56.1}$$

mit

$$\beta = (kT)^{-1} \tag{56.2}$$

bestimmt werden. Diese Funktion wird, wie die analoge in der klassischen Theorie, die (kanonische) Verteilungsfunktion des Gesamtsystems genannt. Sie hängt mit der freien Energie nach HELMHOLTZ zusammen durch die Gleichung

$$F = - kT \ln Q. \tag{56.3}$$

Daraus folgt für die Entropie

$$S = k\left(\ln Q + T\frac{\partial \ln Q}{\partial T}\right), \qquad (56.4)$$

für die innere Energie

$$E = -kT^2\frac{\partial \ln Q}{\partial T} \qquad (56.5)$$

und für den Druck

$$P = kT\frac{\partial \ln Q}{\partial V}. \qquad (56.6)$$

Mit Benutzung der Verteilungsfunktion kann der Dichteoperator der kanonischen Gesamtheit geschrieben werden

$$\rho = Q^{-1}e^{-\beta\mathsf{H}}. \qquad (56.7)$$

Die Elemente der Dichtematrix (34.31) sind dann

$$\varrho_{nm} = Q^{-1}\int \varphi_n^* \, e^{-\beta\mathsf{H}}\, \varphi_m\, d\boldsymbol{q}, \qquad (56.8)$$

wo der Satz φ_n ein beliebiges vollständiges normiertes Orthogonalsystem bezeichnet. Für die Basis der Energie-Eigenfunktionen wird

$$\varrho_{nm} = Q^{-1}e^{-\beta E_m}\,\delta_{nm}. \qquad (56.9)$$

Die entsprechenden Formeln für die Verteilungsfunktion sind

$$Q = \sum_n \int \varphi_n^* \, e^{-\beta\mathsf{H}}\, \varphi_n\, d\boldsymbol{q} \qquad (56.10)$$

und (für die Basis der Energie-Eigenfunktionen)

$$Q = \sum_n e^{-\frac{E_n}{kT}} = \sum_l \Omega_l \, e^{-\frac{E_l}{kT}}. \qquad (56.11)$$

Für die spezielle Darstellung (34.29) bzw. (34.32) ergibt sich mit (56.7)

$$\varrho(\boldsymbol{q}, \boldsymbol{q}') = Q^{-1}\sum_n \varphi_n^*(\boldsymbol{q}')\, e^{-\beta\mathsf{H}}\, \varphi_n(\boldsymbol{q}). \qquad (56.12)$$

Das Diagonalelement

$$\varrho(\boldsymbol{q}, \boldsymbol{q}) = Q^{-1}\sum_n \varphi_n^* \, e^{-\beta\mathsf{H}}\, \varphi_n \qquad (56.13)$$

stellt die Wahrscheinlichkeitsdichte der Konfiguration q dar. Definiert man die (normierte) Slater-Summe für Massenpunkte durch

$$S(\boldsymbol{q}) = N!\,\lambda^{3N}\sum_n \varphi_n^*(\boldsymbol{q})\, e^{-\beta\mathsf{H}}\, \varphi_n(\boldsymbol{q}), \qquad (56.14)$$

wo

$$\lambda = \frac{h}{(2\pi m\, kT)^{\frac{1}{2}}} \qquad (56.15)$$

ist, so kann (56.13) geschrieben werden

$$\varrho(\boldsymbol{q}, \boldsymbol{q}) = \frac{1}{\lambda^{3N} N! \, Q}\, S(\boldsymbol{q}), \qquad (56.16)$$

in formaler Analogie zu dem klassischen Ausdruck (30.13). Die Slater-Summe ist somit das quantenstatistische Analogon der klassischen Größe $e^{-\beta U}$. Auf Grund dieser Analogie kann das quantenstatistische Konfigurationsintegral definiert werden durch

$$Q_{\tau} = \int S(\boldsymbol{q})\, d\boldsymbol{q}. \qquad (56.17)$$

Damit erhält man für die Verteilungsfunktion den Ausdruck

$$Q = \lambda^{-3N} \frac{Q_\tau}{N!} \tag{56.18}$$

in formaler Übereinstimmung mit der klassischen Gl. (30.10).

VI. Die Berechnung der kanonischen Verteilungsfunktion.

57. Separierbare Systeme. Für Systeme, deren Hamilton-Operator nach den Koordinaten der Einzelteilchen separierbar ist, läßt sich die Verteilungsfunktion ohne Schwierigkeit nach der Darwin-Fowlerschen Methode berechnen. Wir haben zunächst

$$Q = \sum_l \Omega_l \, e^{-\frac{E_l}{kT}} \tag{57.1}$$

mit

$$E_l = \sum_k N_{lk}\, \varepsilon_k, \tag{57.2}$$

wobei wir die ε_k als nicht entartet voraussetzen. Dabei gilt die Nebenbedingung

$$\sum_k N_{lk} = N. \tag{57.3}$$

Mit der Abkürzung

$$e^{-\frac{\varepsilon_k}{kT}} = y_k \tag{57.4}$$

wird die halbklassische Näherung

$$Q = \frac{1}{N!} \sum_l \frac{N!}{\prod_k N_{lh}!} \prod_k y_k^{N_{lk}} = \frac{1}{N!} \left(\sum_k y_k\right)^N. \tag{57.5}$$

Für die Bose-Einstein- und Fermi-Dirac-Statistik ist $\Omega_l = 1$ und somit

$$Q = \sum_l \prod_k y_k^{N_{lk}}. \tag{57.6}$$

Dabei gilt die Nebenbedingung (57.3) und für die Fermi-Dirac-Statistik die zusätzliche Bedingung, daß die N_{lk} nur die Werte 0 und 1 annehmen können. Wir definieren nun eine erzeugende Funktion

$$\Xi = \sum_{N=0}^{\infty} z^N Q^{(N)}, \tag{57.7}$$

wo z eine komplexe Variable und $Q^{(N)}$ die Verteilungsfunktion eines Systems von N Teilchen bezeichnet. Setzen wir hier zunächst (57.6) ein, so erhalten wir (nach Vertauschen von Summierung und Produktbildung)

$$\Xi = \prod_k \sum_l (z\, y_k)^{N_{lk}}. \tag{57.8}$$

Im Falle der Bose-Einstein-Statistik ist dieser Ausdruck ein Produkt aus geometrischen Reihen. Wir haben also

$$\Xi_{\mathrm{BE}} = \prod_k (1 - z\, y_k)^{-1}. \tag{57.9}$$

Für die Fermi-Dirac-Statistik gilt einfach

$$\Xi_{\mathrm{FD}} = \prod_k (1 + z\, y_k). \tag{57.10}$$

Die erzeugende Funktion für die Verteilungsfunktion der halbklassischen Näherung ist

$$\Xi_{MB} = e^{z \sum_k y_k},$$ (57.11)

wie man durch Entwicklung der e-Funktion sofort verifiziert. Da

$$\lim_{\gamma \to 0} (1 + \gamma z y_k)^{\frac{1}{\gamma}} = e^{z y_k}$$ (57.12)

ist, kann die erzeugende Funktion für Q allgemein geschrieben werden

$$\Xi = \prod_k (1 + \gamma z y_k)^{\frac{1}{\gamma}}$$ (57.13)

mit

$$\gamma = \begin{cases} 0 & \text{(Maxwell-Boltzmann)} \\ -1 & \text{(Bose-Einstein)} \\ +1 & \text{(Fermi-Dirac)} . \end{cases}$$ (57.14)

Der Residuensatz liefert dann

$$Q^{(N)} = \frac{1}{2\pi i} \oint \frac{\Xi}{z^{N+1}} \, dz.$$ (57.15)

Das Integral läßt sich wieder nach der Sattelpunktmethode auswerten. Man erhält asymptotisch den Ausdruck

$$\ln Q = \frac{1}{\gamma} \sum_k \ln (1 + \gamma Z y_k) - N \ln Z,$$ (57.16)

wo Z die Koordinate des Sattelpunktes ist, die wieder der Gl. (50.21) genügt. Aus der kanonischen Mittelwertsformel (51.10) und Gl. (57.5) folgt

$$\overline{N}_k = y_k \frac{\partial \ln Q}{\partial y_k}.$$ (57.17)

Wir erhalten somit als allgemeine Verteilungsformel für separierbare Systeme

$$\overline{N}_k = \frac{Z \, e^{-\frac{\varepsilon_k}{kT}}}{1 + \gamma Z \, e^{-\frac{\varepsilon_k}{kT}}}.$$ (57.18)

Für $\gamma = \pm 1$ erhält man daraus die Verteilungsformel der Bose-Einstein- und Fermi-Dirac-Statistik (50.17). Für $\gamma = 0$ kommt man auf Gl. (50.37) (mit $g_i = 1$) und damit zum Maxwell-Boltzmannschen Verteilungsgesetz. Für die freie Energie nach Helmholtz erhält man aus Gl. (57.16)

$$F = - kT \left[\sum_k \ln \left(1 + \gamma Z \, e^{-\frac{\varepsilon_k}{kT}}\right)^{\frac{1}{\gamma}} - N \ln Z \right].$$ (57.19)

Wir machen zu der vorstehenden Ableitung noch zwei Bemerkungen. Zunächst darf nicht übersehen werden, daß die Sattelpunktmethode nur asymptotische Resultate liefert. Gl. (57.18) stellt daher nicht das exakte Verteilungsgesetz

der kanonischen Gesamtheit für endliche Systeme dar. Letzteres wurde von
SAKAI[1] sowie ANSBACHER und EHRENBERG[2] abgeleitet und lautet

$$\overline{N}_j^{(N)} = \left[\frac{Q^{(N+1)}}{Q^{(N)}} \frac{\overline{N}_j^{(N+1)}}{\overline{N}_j^{(N)}} e^{\frac{\varepsilon_j}{kT}} \pm 1 \right]^{-1}. \tag{57.20}$$

Der direkte Übergang von dieser Gleichung zu der asymptotischen Formel ist
von FRASER[3], LANDSBERG[4,5] sowie ANSBACHER und LANDSBERG[6] diskutiert
worden.

Die zweite Bemerkung betrifft die Anwendung der Sattelpunktmethode.
Dieselbe führt im Falle des Bose-Einstein-Gases bei sehr tiefen Temperaturen
auf schwierige Konvergenzprobleme, auf die zuerst SCHUBERT[7] hingewiesen hat.
Das Problem ist von einer Reihe von Autoren[4,8,9] untersucht worden, ohne daß
eine völlige Klärung erreicht worden wäre. Erst in neuester Zeit ist von FORD
und BERLIN[10] ein Weg angegeben worden, der diese Schwierigkeit vermeidet.

58. Übersicht über die allgemeinen Methoden. Für nicht separierbare Systeme
ist die Berechnung der kanonischen Verteilungsfunktion in der Quantenstatistik
noch wesentlich schwieriger als in der klassischen Theorie. Dies beruht einerseits
darauf, daß das Eigenwertspektrum des Hamilton-Operators nicht bekannt ist,
andererseits auf der Notwendigkeit, die quantenmechanischen Symmetriebedingungen zu berücksichtigen. Das Problem ist, zumal in den letzten Jahren,
von zahlreichen Autoren bearbeitet worden. Wir können es hier nur insoweit
erörtern, als es mit den in diesem Artikel behandelten allgemeinen Fragen zusammenhängt und geben zunächst eine kurze Übersicht über die Methoden.

Man kann im wesentlichen drei Typen von Entwicklungen unterscheiden:

$\alpha)$ cluster-Entwicklung (Virial-Entwicklung),

$\beta)$ Hochtemperatur-Entwicklung,

$\gamma)$ Tieftemperatur-Entwicklung.

$\alpha)$ *Die cluster-Entwicklung* ist eine formal der ebenso bezeichneten klassischen
Entwicklung nachgebildete Entwicklung nach Potenzen der Fugazität Z oder
der Dichte ϱ an der Stelle $Z = \varrho = 0$. Sie ist für die Quantenstatistik zuerst von
KAHN und UHLENBECK[11] auf der Grundlage der Gl. (56.18) formuliert worden.
Da die Koeffizienten (quantenstatistische cluster-Integrale) die Slater-Summen
für 2, 3, ... n, ... Teilchen enthalten, ist damit das Problem auf die sukzessive
Lösung des 2-, 3-, ..., n-, ...-Körper-Problems reduziert. Bisher ist eine strenge
Lösung nur für das 2-Körper-Problem gefunden worden, welche die exakt quanten-
statistische Berechnung des II. Virialkoeffizienten ermöglicht[12,13].

In neuester Zeit haben HUANG, YANG und LUTTINGER[14,15] eine Methode entwickelt, die für ein Bose-Einstein-Gas aus harten Kugeln die näherungsweise

[1] T. SAKAI: Proc. Phys.-Math. Soc. Japan **22**, 193 (1940).
[2] F. ANSBACHER u. W. EHRENBERG: Phil. Mag. **40**, 626 (1949).
[3] A.R. FRASER: Phil. Mag. **42**, 165 (1951).
[4] P.T. LANDSBERG: Proc. Cambridge Phil. Soc. **50**, 65 (1954).
[5] P.T. LANDSBERG: Phys. Rev. **93**, 1170 (1954).
[6] F. ANSBACHER u. P.T. LANDSBERG: Phys. Rev. **96**, 1707 (1954).
[7] G. SCHUBERT: Z. Naturforsch. **1**, 113 (1946).
[8] G. LEIBFRIED: Z. Naturforsch. **2a**, 305 (1947).
[9] R.B. DINGLE: Proc. Cambridge Phil. Soc. **45**, 275 (1949).
[10] J. FORD u. T.H. BERLIN: J. Chem. Phys. **27**, 931 (1957).
[11] B. KAHN u. G.E. UHLENBECK: Physica, Haag **5**, 399 (1938).
[12] G.E. UHLENBECK u. E. BETH: Physica, Haag **3**, 729 (1936).
[13] E. BETH u. G.E. UHLENBECK: Physica, Haag **4**, 915 (1937).
[14] K. HUANG u. C.N. YANG: Phys. Rev. **105**. 767 (1957).
[15] K. HUANG, C.N. YANG u. J.M. LUTTINGER: Phys. Rev. **105**, 776 (1957).

Berechnung aller Virialkoeffizienten auf der Grundlage der Lösung des 2-Körper-Problems erlaubt.

Montroll und Ward[1] haben die cluster-Entwicklung in einer Form dargestellt, welche die Lösung des Eigenwertproblems überhaupt entbehrlich macht. Sie benutzen die weiter unten zu besprechende Methode der Blochschen Gleichung. Diese wird mit Hilfe einer Greenschen Funktion gelöst, die einen Feynmanschen Propagator[2] darstellt. Die Berechnung der auf diesem Wege definierten „cluster-Integrale" stößt, wie in der klassischen Theorie, auf große Schwierigkeiten, ist aber für gewisse Typen (die den klassischen „Ring-Integralen" entsprechen) durchführbar. Da die cluster-Entwicklung auf Gase beschränkt ist und jenseits des Kondensationspunktes divergiert, gehen wir im folgenden nicht weiter darauf ein und verweisen auf den Artikel von J. E. Mayer in Bd. XII dieses Handbuches.

β) *Hochtemperatur-Entwicklungen* heißen Entwicklungen nach Potenzen von h, deren erster Term der entsprechende klassische Ausdruck bzw. die halbklassische Näherung ist. Ihre Bedeutung liegt einmal in der strengen Begründung der halbklassischen Näherung (die wir bisher lediglich plausibel gemacht haben) und weiter darin, daß sie eine verhältnismäßig einfache Berechnung der quantenstatistischen Korrekturen für mäßig tiefe Temperaturen ermöglichen. In diesem Sinne können sie auch bei der Berechnung des II. Virialkoeffizienten verwendet werden, wodurch dann die Lösung des 2-Körper-Problems entbehrlich gemacht wird. Sie divergieren bei sehr tiefen Temperaturen. Hochtemperatur-Entwicklungen sind zuerst von Wigner[3] für die Konfigurations-Wahrscheinlichkeitsdichte (56.16) und von Kirkwood[4] für die Verteilungsfunktion angegeben worden. Letzterer hat (im Gegensatz zu Wigner[3]) auch die Symmetriebedingungen berücksichtigt. In neuerer Zeit haben Goldberger und Adams[5] die Hochtemperatur-Entwicklung nach einer anderen Methode, aber ohne Berücksichtigung der Symmetriebedingungen, abgeleitet.

γ) *Den Tieftemperatur-Entwicklungen* ist gemeinsam, daß als erster Term die Verteilungsfunktion eines Systems ohne Wechselwirkung (d.h. eines idealen Gases) auftritt. Im übrigen erscheinen sie in sehr verschiedenen Formen. Man kann jedoch ihre Äquivalenz nachweisen und zeigen, daß sie sich als Entwicklung nach Potenzen eines Kopplungsparameters darstellen lassen. Sie sind vor allem für Probleme der Tieftemperatur-Physik, bei denen die Quanteneffekte dominierend werden, wichtig.

Die ersten Terme einer Tieftemperatur-Entwicklung sind zuerst von Green[5] angegeben worden. Die vollständige Entwicklung haben kurz darauf Goldberger und Adams[6] abgeleitet. Green[7] hat dann nach einer anderen Methode eine äquivalente Entwicklung erhalten. Diese Reihen berücksichtigen nicht die Symmetriebedingungen und sind außerdem in ihrer Struktur so kompliziert, daß sie für explizite Berechnungen kaum in Betracht kommen. Eine sehr einfache Ableitung der Tieftemperatur-Entwicklung, die auch eine Berücksichtigung der

---

[1] E. W. Montroll u. J. C. Ward: Phys. Fluids **1**, 55 (1958). Den Herren Prof. Montroll und Dr. Ward bin ich für die Überlassung des Manuskriptes dieser Arbeit zu großem Dank verpflichtet.

[2] R. P. Feynmann: Phys. Rev. **76**, 749 (1949).

[3] E. Wigner: Phys. Rev. **40**, 749 (1932).

[4] J. G. Kirkwood: Phys. Rev. **44**, 31 (1933).

[5] H. S. Green: J. Chem. Phys. **19**, 955 (1951).

[6] M. Goldberger u. E. N. Adams: J. Chem. Phys. **20**, 240 (1952).

[7] H. S. Green: J. Chem. Phys. **20**, 1274 (1952).

Symmetriebedingungen ermöglicht, ist von SIEGERT[1] angegeben worden. Da jedoch die Methode auf einem Iterationsverfahren beruht, wird kein Ausdruck für den allgemeinen Term erhalten. Eine verhältnismäßig einfache Formulierung welche die Symmetriebedingungen berücksichtigt, ist dann von CHESTER[2] gefunden worden. Die Frage der Konvergenz, die von CHESTER diskutiert worden ist, scheint allerdings nicht völlig geklärt zu sein.

Die für die Ableitung der Entwicklungen benutzten mathematischen Methoden sind im Einzelnen sehr verschiedenartig. Die Mehrzahl derselben beruht jedoch letzten Endes auf der Tatsache, daß man das Problem der quantenstatistischen Verteilungsfunktion (bzw. der Dichtematrix oder Slater-Summe) auf eine Differentialgleichung reduzieren kann, die formal mit der zeitabhängigen Schrödinger-Gleichung identisch ist. Im folgenden behandeln wir zunächst diesen grundlegenden Zusammenhang und geben dann einige typische Beispiele.

59. Die Blochsche Gleichung. In Ziff. 56 haben wir gezeigt, daß die Verteilungsfunktion durch

$$Q = \mathrm{spur}\,(\mathrm{e}^{-\beta \mathsf{H}}) = \sum_n \int \varphi_n^* \, \mathrm{e}^{-\beta \mathsf{H}} \, \varphi_n \, d\boldsymbol{q} \tag{59.1}$$

und die Dichtematrix durch

$$\varrho(\boldsymbol{q}, \boldsymbol{q}') = Q^{-1} \sum_n \varphi_n^*(\boldsymbol{q}') \, \mathrm{e}^{-\beta \mathsf{H}} \, \varphi_n(\boldsymbol{q}) \tag{59.2}$$

gegeben ist, wo der Satz φ_n ein vollständiges normiertes Orthogonalsystem bezeichnet. Setzen wir

$$u(\beta) = \mathrm{e}^{-\beta \mathsf{H}} \, u(0) \tag{59.3}$$

mit

$$u(0) \equiv \varphi, \tag{59.4}$$

so nehmen die Gln. (59.1) und (59.2) die einfachere Form an

$$Q = \sum_n \int u_n^*(0) \, u_n(\beta) \, d\boldsymbol{q} \tag{59.5}$$

$$\varrho(\boldsymbol{q}, \boldsymbol{q}') = Q^{-1} \sum_n u_n^*(0, \boldsymbol{q}') \, u_n(\beta, \boldsymbol{q}). \tag{59.6}$$

Für die Funktion u erhält man durch Differentiation aus Gl. (59.3) die Differentialgleichung

$$\mathsf{H}\, u + \frac{\partial u}{\partial \beta} = 0. \tag{59.7}$$

Diese für die Quantenstatistik außerordentlich wichtige Beziehung ist zuerst von BLOCH[3] abgeleitet worden und wird heute als *Blochsche Gleichung* bezeichnet. Sie ist formal mit der zeitabhängigen Schrödinger-Gleichung identisch, in der jetzt lediglich die Größe β an die Stelle von $2\pi i t/h$ tritt. Die Bedeutung der Blochschen Gleichung liegt einmal darin, daß sie für die Berechnung der Verteilungsfunktion die Bestimmung der Eigenwerte des Hamilton-Operators entbehrlich macht und das Problem auf eine Summierung bzw. Integration reduziert. Zum anderen ermöglicht sie durch ihre formale Identität mit der Schrödinger-Gleichung, daß mathematische Methoden der Quantenmechanik auch für die Quantenstatistik nutzbar gemacht werden.

Für die in Ziff. 60 zu besprechende Hochtemperatur-Entwicklung ist es zweckmäßig, als Basis die Eigenfunktionen des Impuls-Operators zu wählen.

---

[1] A. J. F. SIEGERT: J. Chem. Phys. **20**, 572 (1952).
[2] G. V. CHESTER: Phys. Rev. **93**, 606 (1954).
[3] F. BLOCH: Z. Physik **74**, 295 (1932).

In diesem Falle werden die vorstehenden Formeln dadurch etwas modifiziert, daß wir ein kontinuierliches Eigenwertspektrum haben. Außerdem erfordert die Berücksichtigung der Symmetriebedingungen einige Sorgfalt. Wir wollen daher die Ableitung für diesen Spezialfall etwas genauer durchführen[1]. Die Eigenfunktionen des Impuls-Operators sind, wenn wir die Teilchen als Massenpunkte betrachten,

$$\chi(\boldsymbol{p}, \boldsymbol{q}) = \frac{1}{h^{\frac{3}{2}N}} e^{\frac{2\pi i}{h} \boldsymbol{p}\boldsymbol{q}}. \tag{59.8}$$

Bezeichnen wir mit $\psi_n(\boldsymbol{q})$ die Eigenfunktionen des Hamilton-Operators, so lauten die Transformationsgleichungen

$$\psi_n(\boldsymbol{q}) = \int \varphi_n(\boldsymbol{p}) \chi(\boldsymbol{p}, \boldsymbol{q}) d\boldsymbol{p}, \tag{59.9}$$

$$\varphi_n(\boldsymbol{p}) = \int \psi_n(\boldsymbol{q}) \chi^*(\boldsymbol{p}, \boldsymbol{q}) d\boldsymbol{q}. \tag{59.10}$$

Für ein System aus nicht lokalisierten gleichen Teilchen ist nun

$$\psi(\boldsymbol{q}) = (\pm 1)^{|P_q|} \mathsf{P}_q \psi(\boldsymbol{q}), \tag{59.11}$$

wo P_q die Folge der Permutationsoperatoren, welche die $\boldsymbol{q}_i$ vertauschen, und $|P_q|$ die Zahl der jeweils zugehörigen Transpositionen bezeichnet. Das positive Vorzeichen bezieht sich auf die Bose-Einstein-, das negative auf die Fermi-Dirac-Statistik. Aus (59.11) folgt

$$\psi(\boldsymbol{q}) = \frac{1}{N!} \sum_{P_q} (\pm 1)^{|P_q|} \mathsf{P}_q \psi(\boldsymbol{q}). \tag{59.12}$$

Einsetzen von (59.9) auf der rechten Seite ergibt

$$\psi(\boldsymbol{q}) = \frac{1}{\sqrt{N!}} \int \varphi(\boldsymbol{p}) X(\boldsymbol{p}, \boldsymbol{q}) d\boldsymbol{p}, \tag{59.13}$$

wo

$$X(\boldsymbol{p}, \boldsymbol{q}) = \frac{1}{h^{\frac{3}{2}N} \sqrt{N!}} \sum_{P_q} (\pm 1)^{|P_q|} \mathsf{P}_q e^{\frac{2\pi i}{h} \boldsymbol{p}\boldsymbol{q}} = \frac{1}{h^{\frac{3}{2}N} \sqrt{N!}} \sum_{P_p} (\pm 1)^{|P_q|} \mathsf{P}_p e^{\frac{2\pi i}{h} \boldsymbol{p}\boldsymbol{q}} \tag{59.14}$$

ist. Hier bezeichnet P_p die Folge der Permutationsoperatoren, welche die p_i vertauschen. Aus (59.14) folgt nun

$$\mathsf{P}_p X(\boldsymbol{p}, \boldsymbol{q}) = (\pm 1)^{|P_p|} X(\boldsymbol{p}, \boldsymbol{q}). \tag{59.15}$$

Es ist daher

$$\mathsf{P}_p [\varphi(\boldsymbol{p}) X(\boldsymbol{p}, \boldsymbol{q})] = (\pm 1)^{|P_p|} [\mathsf{P}_p \varphi(\boldsymbol{p})] X(\boldsymbol{p}, \boldsymbol{q}). \tag{59.16}$$

Ferner gilt für eine beliebige Funktion der Impulse $F(\boldsymbol{p})$, wenn die Reihenfolge der Integration über die einzelnen Impulse vertauschbar ist,

$$\int F(\boldsymbol{p}) d\boldsymbol{p} = \frac{1}{N!} \int \sum_{P_p} \mathsf{P}_p F(\boldsymbol{p}) d\boldsymbol{p}. \tag{59.17}$$

Aus Gl. (59.13), (59.16) und (59.17) folgt

$$\psi(\boldsymbol{q}) = \frac{1}{N!} \int \Phi(\boldsymbol{p}) X(\boldsymbol{p}, \boldsymbol{q}) d\boldsymbol{p} \tag{59.18}$$

mit

$$\Phi(\boldsymbol{p}) = \frac{1}{\sqrt{N!}} \sum_{P_p} (\pm 1)^{|P_p|} \mathsf{P}_p \varphi(\boldsymbol{p}). \tag{59.19}$$

---

[1] J.G. Kirkwood: Phys. Rev. **44**, 31 (1933).

Setzen wir hier (59.10) ein, so wird mit Benutzung von (59.14)

$$\Phi(\boldsymbol{p}) = \int \psi(\boldsymbol{q})\, X^*(\boldsymbol{p},\boldsymbol{q})\, d\boldsymbol{q}. \tag{59.20}$$

Die Gln. (59.18) und (59.20) entsprechen den früheren Gln. (59.9) und (59.10). Die Funktion $\Phi(\boldsymbol{p})$ besitzt in den p_i die gleichen Symmetrieeigenschaften wie $\psi(\boldsymbol{q})$ in den $\boldsymbol{q}_i$. Durch Einsetzen von (59.18) in die Gleichung

$$Q = \sum_n \int \psi_n^*(\boldsymbol{q})\, \mathrm{e}^{-\beta H}\, \psi(\boldsymbol{q})\, d\boldsymbol{q} \tag{59.21}$$

ergibt sich

$$Q = \frac{1}{N!}\sum_n \iint \psi_n^*(\boldsymbol{q})\, \Phi(\boldsymbol{p})\, \mathrm{e}^{-\beta H}\, X(\boldsymbol{p},\boldsymbol{q})\, d\boldsymbol{q}\, d\boldsymbol{p}, \tag{59.22}$$

wo die Summierung jetzt nur über die symmetrischen (bzw. antisymmetrischen) Eigenfunktionen zu erstrecken ist. Mit Benutzung von (59.20) wird daraus

$$Q = \frac{1}{N!}\iint \sum_n \int \psi_n^*(\boldsymbol{q})\, \psi_n(\boldsymbol{q}')\, X^*(\boldsymbol{p},\boldsymbol{q})\, \mathrm{e}^{-\beta H}\, X(\boldsymbol{p},\boldsymbol{q})\, d\boldsymbol{q}\, d\boldsymbol{p}\, d\boldsymbol{q}'. \tag{59.23}$$

Da die ψ_n jeweils ein vollständiges Funktionensystem im Unter-Raum der symmetrischen (bzw. antisymmetrischen) Funktionen bilden, gilt

$$\sum_n \psi_n^*(\boldsymbol{q})\, \psi_n(\boldsymbol{q}') = \delta(\boldsymbol{q}-\boldsymbol{q}'), \tag{59.24}$$

und wir erhalten

$$Q = \frac{1}{h^{3N}(N!)^2}\sum_{P'}\sum_P (\pm 1)^{|P|+|P'|}\iint \mathrm{e}^{-\frac{2\pi i}{h}\,\mathsf{P}'\boldsymbol{p}\boldsymbol{q}}\,\mathrm{e}^{-\beta H}\,\mathrm{e}^{\frac{2\pi i}{h}\,\mathsf{P}\boldsymbol{p}\boldsymbol{q}}\, d\boldsymbol{q}\, d\boldsymbol{p}. \tag{59.25}{}^{[1]}$$

Die rechte Seite dieser Gleichung ist die mit Hilfe der Impuls-Eigenfunktionen unter Berücksichtigung der Symmetriebedingungen dargestellte Spur des Operators $\mathrm{e}^{-\beta H}$. Sie kann als quantenstatistisches Analogon des klassischen Phasenintegrals Gl. (29.2) aufgefaßt werden.

Wir setzen nun

$$v(\mathsf{P}) = \mathrm{e}^{-\beta H}\chi(\mathsf{P}) \tag{59.26}$$

mit

$$\chi(\mathsf{P}) = \mathrm{e}^{\frac{2\pi i}{h}\,\mathsf{P}\boldsymbol{p}\boldsymbol{q}}. \tag{59.27}$$

Die Differentiation von (59.25) nach β ergibt dann wieder die Blochsche Gleichung

$$\mathsf{H}\,v + \frac{\partial v}{\partial \beta} = 0 \tag{59.28}$$

mit der Randbedingung

$$\lim_{\beta=0} v(\mathsf{P}) = \chi(\mathsf{P}). \tag{59.29}$$

60. Entwicklung für hohe Temperaturen. KIRKWOOD[2] hat eine Lösung der Blochschen Gleichung (59.28) angegeben, welche v als Potenzreihe in h darstellt und eine strenge Begründung der halbklassischen Näherung ermöglicht. Wir gehen aus von dem Ansatz

$$v = w\,\chi(\mathsf{P})\,\mathrm{e}^{-\beta H(\boldsymbol{q},\boldsymbol{p})}, \tag{60.1}$$

wo

$$H(\boldsymbol{q},\boldsymbol{p}) = E_{\mathrm{kin}} + U(\boldsymbol{q}) \tag{60.2}$$

---

[1] Die Indices an den Permutationsoperatoren lassen wir jetzt weg, da es gleichgültig ist, ob wir dieselben auf die p_i oder q_i wirken lassen.

[2] J. G. KIRKWOOD: Phys. Rev. **44**, 31 (1933).

die klassische Hamilton-Funktion ist. Die Blochsche Gleichung (59.28) liefert dann

$$\mathsf{H}\left[w\,\chi(\mathsf{P})\,e^{-\beta H}\right] - w\,\chi(\mathsf{P})\,H\,e^{-\beta H} + \chi(\mathsf{P})\,e^{-\beta H}\frac{\partial w}{\partial \beta} = 0. \tag{60.3}$$

Schreiben wir nun den Hamilton-Operator

$$\mathsf{H} = -\frac{h^2}{8\pi^2 m}\sum_{i=1}^{N}\Delta_i + U = \mathsf{H}_0 + U \tag{60.4}$$

(wo Δ_i der Laplace-Operator für die Koordinaten eines Teilchens ist) und benutzen (60.2), so wird aus (60.3)

$$\left.\begin{aligned}
&\mathsf{H}_0\left[w\,\chi(\mathsf{P})\,e^{-\beta H}\right] + U\,w\,\chi(\mathsf{P})\,e^{-\beta H} - U\,w\,\chi(\mathsf{P})\,e^{-\beta H} - \\
&\qquad - w\,\chi(\mathsf{P})\,E_{\mathrm{kin}}\,e^{-\beta H} + \chi(\mathsf{P})\,e^{-\beta H}\frac{\partial w}{\partial \beta} = 0.
\end{aligned}\right\} \tag{60.5}$$

Da der Operator H_0 nur auf die Koordinaten wirkt und E_{kin} nur von den Impulsen abhängt, hebt sich der Faktor $e^{-\beta E_{\mathrm{kin}}}$ heraus und wir erhalten

$$\mathsf{H}_0\left[w\,\chi(\mathsf{P})\,e^{-\beta U}\right] - w\,\chi(\mathsf{P})\,E_{\mathrm{kin}}\,e^{-\beta U} + \chi(\mathsf{P})\,e^{-\beta U}\frac{\partial w}{\partial \beta} = 0. \tag{60.6}$$

Für das erste Glied der linken Seite ergibt sich mit Benutzung des Nabla-Operators

$$\left.\begin{aligned}
&\mathsf{H}_0\left[w\,\chi(\mathsf{P})\,e^{-\beta U}\right] \\
&= \frac{h^2}{8\pi^2 m}\left[\chi(\mathsf{P})\sum_{i=1}^{N}\Delta_i\,e^{-\beta U}\,w + 2\sum_{i=1}^{N}\nabla_i\,\chi(\mathsf{P})\,\nabla_i\,e^{-\beta U}\,w + e^{-\beta U}\,w\sum_{i=1}^{N}\Delta_i\,\chi(\mathsf{P})\right].
\end{aligned}\right\} \tag{60.7}$$

Nun ist wegen (59.27)

$$-\frac{i\,h}{2\pi}\sum_{i=1}^{N}\nabla_i\,\chi(\mathsf{P})\cdot\nabla_i = \chi(\mathsf{P})\left(\mathsf{P}\sum_{i=1}^{N}\boldsymbol{p}_i\cdot\nabla_i\right). \tag{60.8}$$

Ferner gilt

$$\mathsf{H}_0\,\chi(\mathsf{P}) = -\frac{h^2}{8\pi^2 m}\sum_{i=1}^{N}\Delta_i\,\chi(\mathsf{P}) = E_{\mathrm{kin}}\,\chi(\mathsf{P}), \tag{60.9}$$

wobei unter der Voraussetzung

$$\lambda \ll L \tag{60.10}$$

(L = Kantenlänge des Behälters) E_{kin} mit der ebenso bezeichneten klassischen Funktion der Impulse näherungsweise identifiziert werden kann. Damit wird

$$\left.\begin{aligned}
\mathsf{H}_0\left[w\,\chi(\mathsf{P})\,e^{-\beta U}\right] &= -\frac{h^2}{8\pi^2 m}\chi(\mathsf{P})\sum_{i=1}^{N}\Delta_i\,e^{-\beta U}\,w - \\
&\quad - \frac{i\,h}{2\pi m}\chi(\mathsf{P})\left[\mathsf{P}\sum_{i=1}^{N}\boldsymbol{p}_i\nabla_i\,e^{-\beta U}\,w\right] + e^{-\beta U}\,w\,E_{\mathrm{kin}}\,\chi(\mathsf{P}).
\end{aligned}\right\} \tag{60.11}$$

Durch Einsetzen dieses Ausdruckes in Gl. (60.6) folgt schließlich für w die Differentialgleichung

$$\frac{\partial w}{\partial \beta} - e^{\beta U}\left[\frac{i\,h}{2\pi m}\left(\mathsf{P}\sum_{i=1}^{N}\boldsymbol{p}_i\cdot\nabla_i\,e^{-\beta U}\,w\right) + \frac{h^2}{8\pi^2 m}\sum_{i=1}^{N}\Delta_i\,e^{-\beta U}\,w\right] = 0 \tag{60.12}$$

mit der Randbedingung

$$\lim_{\beta=0} w = 1. \tag{60.13}$$

Wegen der Randbedingung läßt sich (60.12) umformen in die Integralgleichung

$$w = 1 + \frac{ih}{2\pi m} \int_0^\beta e^{Ut} \left(\mathsf{P} \sum_{i=1}^N \boldsymbol{p}_i \cdot \boldsymbol{V}_i\, e^{-Ut} w \right) dt + \frac{h^2}{8\pi^2 m} \int_0^\beta e^{Ut} \sum_{i=1}^N \Delta_i\, (e^{-Ut} w)\, dt. \quad (60.14)$$

Diese Gleichung läßt sich durch Iteration lösen mit Hilfe des Ansatzes

$$w = \sum_{k=0}^\infty \left(\frac{h}{2\pi} \right)^k w_k. \quad (60.15)$$

Es ergibt sich dann

$$w_0 = 1, \quad (60.16)$$

$$w_1 = -\frac{i\beta^2}{2m} \left(\mathsf{P} \sum_{i=1}^N \boldsymbol{p}_i \cdot \boldsymbol{V}_i\, U \right), \quad (60.17)$$

$$\left. \begin{aligned} w_2 = -\frac{1}{2m} \Bigg\{ &\frac{\beta^2}{2} \sum_{i=1}^N \Delta_i\, U - \\ &-\frac{\beta^3}{3} \left[\sum_{i=1}^N (\boldsymbol{V}_i\, U)^2 + \frac{1}{m} \left(\mathsf{P} \sum_{i=1}^N \boldsymbol{p}_i \cdot \boldsymbol{V}_i \right)^2 U \right] + \frac{\beta^4}{4m} \left(\mathsf{P} \sum_{i=1}^N \boldsymbol{p}_i \cdot \boldsymbol{V}_i\, U \right)^2 \Bigg\}, \end{aligned} \right\} \quad (60.18)$$

usw. Der allgemeine Term läßt sich mit Hilfe der Rekursionsformel

$$w_k = \frac{1}{m} \int_0^\beta e^{Ut} \left[\frac{1}{2} \sum_{i=1}^N \Delta_i\, (e^{-Ut} w_{k-2}) + i \left(\mathsf{P} \sum_{i=1}^N \boldsymbol{p}_i \cdot \boldsymbol{V}_i\, e^{-Ut} w_{k-1} \right) \right] \quad (60.19)$$

berechnen. Schreiben wir die Gl. (60.1)

$$v(\mathsf{P}) = \chi(\mathsf{P})\, e^{-\beta H} \sum_{k=0}^\infty \left(\frac{h}{2\pi} \right)^k w_k(\mathsf{P}), \quad (60.20)$$

so erhalten wir durch Einsetzen in (59.25)

$$Q = \frac{1}{h^{3N}(N!)^2} \iint e^{-\beta H} \sum_{P'} \sum_P (\pm 1)^{|P'|+|P|}\, e^{\frac{2\pi i}{h}(P-P')\boldsymbol{p}\boldsymbol{q}} \sum_{k=0}^\infty \left(\frac{h}{2\pi} \right)^k w_k(\mathsf{P})\, d\boldsymbol{q}\, d\boldsymbol{p}. \quad (60.21)$$

Die Summierung über P und P' führen wir nun in folgender Weise durch: Zunächst fassen wir alle Terme zusammen, in denen P und P' dieselbe Permutation bezeichnen. Die Zahl dieser Terme ist $N!$. Da hier $\mathsf{P} = \mathsf{P}'$ ist, wird der Exponentialfaktor gleich eins. Bei der Integration über die Impulse liefern alle Terme das gleiche Resultat, da $H(\boldsymbol{q}, \boldsymbol{p})$ in den Impulsen symmetrisch ist und es physikalisch bedeutungslos sein muß, welches p_i mit einem gegebenen q_i verknüpft ist. Für die Funktionen $w_k(\mathsf{P})$ kann daher bei den fraglichen Termen einfach ihr Wert für die identische Permutation, den wir mit w_k bezeichnen, eingesetzt werden.

An zweiter Stelle fassen wir alle Terme zusammen, bei denen sich P und P' durch die Permutation zweier Teilchen unterscheiden. In diesem Falle bleiben in dem Exponenten des Exponentialfaktors die Glieder mit den Koordinaten und Impulsen dieser Teilchen stehen. Für ein bestimmtes Teilchenpaar sind dann den Operatoren P und P' folgende Möglichkeiten zugeordnet:

$$
\begin{array}{cc@{\qquad}cc}
\mathsf{P} & \mathsf{P}' & \mathsf{P} & \mathsf{P}' \\[4pt]
\left.\begin{array}{c} \boldsymbol{p}_j\, \boldsymbol{q}_j \\ \boldsymbol{p}_l\, \boldsymbol{q}_l \end{array}\right. & \left.\begin{array}{c} \boldsymbol{p}_j\, \boldsymbol{q}_l \\ \boldsymbol{p}_l\, \boldsymbol{q}_j \end{array}\right\} \text{(I)} &
\left.\begin{array}{c} \boldsymbol{p}_j\, \boldsymbol{q}_l \\ \boldsymbol{p}_l\, \boldsymbol{q}_j \end{array}\right. & \left.\begin{array}{c} \boldsymbol{p}_j\, \boldsymbol{q}_j \\ \boldsymbol{p}_l\, \boldsymbol{q}_l. \end{array}\right\} \text{(II)}
\end{array}
$$

Den beiden Fällen (I) und (II) entsprechen verschiedene Operatoren $\mathbf{P}$ und damit verschiedene Funktionen $w_k(\mathbf{P})$, die wir mit $w_k(jl)$ und $w_k(lj)$ bezeichnen. Die spezielle Wahl des Paares ist ohne Bedeutung; es muß daher einfach der für ein bestimmtes Paar angeschriebene Ausdruck über alle möglichen Paare summiert werden. Man kann in der beschriebenen Weise weiter fortfahren. Wir verzichten jedoch hier auf die Ausrechnung der höheren Glieder. Es ergibt sich dann

$$Q = \frac{1}{h^{3N} N!} \iint e^{-\beta H} \left\{ \sum_{k=0}^{\infty} \left(\frac{h}{2\pi}\right)^k w_k \pm \right.$$
$$\pm \frac{1}{2N!} \sum_{j \neq l} \left[e^{\frac{2\pi i}{h}(\mathbf{p}_j - \mathbf{p}_l)(\mathbf{q}_j - \mathbf{q}_l)} \sum_{k=0}^{\infty} \left(\frac{h}{2\pi}\right)^k w_k(jl) + \right.$$
$$\left. \left. + e^{-\frac{2\pi i}{h}(\mathbf{p}_j - \mathbf{p}_l)(\mathbf{q}_j - \mathbf{q}_l)} \sum_{k=0}^{\infty} \left(\frac{h}{2\pi}\right)^k w_k(lj) \right] + \cdots \right\} d\mathbf{q}\, d\mathbf{p}. \tag{60.22}$$

Wir setzen jetzt hier die Werte für w_k nach Gl. (60.16) bis (60.18) ein und berücksichtigen dabei für die identischen Permutationen die drei ersten Terme, für den zweiten Teil des obigen Ausdruckes dagegen nur die beiden ersten Terme. Schreiben wir

$$\mathbf{r}_{jl} = \mathbf{q}_j - \mathbf{q}_l \tag{60.23}$$

und integrieren über die Impulse, so folgt

$$Q = \frac{1}{\lambda^{3N} N!} \int e^{-\beta U} \left\{ \left[1 - \frac{h^2 \beta^2}{48\pi^2 m} \sum_{i=0}^{N} \left[\Delta_i U - \frac{\beta}{2}(\nabla_i U)^2 \right] + \cdots \right] \pm \right.$$
$$\left. \pm \frac{1}{N!} \sum_{j \neq l} e^{-\frac{4\pi^2 m r_{jl}^2}{\beta h^2}} \left[1 + \frac{\beta}{2} r_{jl}(\nabla_j U) - \nabla_l U) + \cdots \right] + \cdots \right\} d\mathbf{q}. \tag{60.24}$$

Für $\lambda \to 0$ wird daraus

$$Q = \frac{1}{h^{3N} N!} \iint e^{-\frac{H(\mathbf{q},\mathbf{p})}{kT}} d\mathbf{q}\, d\mathbf{p} + O(\lambda^2). \tag{60.25}$$

Die Gl. (60.24) stellt die gesuchte Hochtemperatur-Entwicklung dar. Gl. (60.25) zeigt, daß dieselbe für $\lambda \to 0$ asymptotisch in die halbklassische Näherung übergeht, die damit in voller Allgemeinheit aus der Quantenstatistik begründet ist. Man sieht aus Gl. (60.25), daß an der Grenze jeder linear unabhängigen Eigenfunktion ein Volumen h^{3N} in dem klassischen Phasenraum entspricht. Die Berücksichtigung der Nichtunterscheidbarkeit reduziert sich auf eine Multiplikation des Phasenintegrals mit $1/N!$. Die Gültigkeit dieser Näherung reicht jedoch erheblich weiter als die der vollständigen halbklassischen Näherung, da nach Gl. (60.24) der den Unterschied zwischen Bose-Einstein- und Fermi-Dirac-Statistik enthaltende letzte Term von viel höherer (nämlich exponentieller) Ordnung verschwindet als die Quantenkorrektur des zweiten Terms. Die Vernachlässigung der Symmetriebedingungen[1,2] ist daher für die Hochtemperatur-Entwicklung (wenn man den Faktor $1/N!$ hinzufügt) in gewissem Umfang gerechtfertigt.

Eine experimentelle Prüfung der Entwicklung (60.24) läßt sich an Hand der Messungen des zweiten Virialkoeffizienten leichter Gase (H_2, D_2, He) bei mäßig tiefen Temperaturen durchführen. Als untere Temperaturgrenze der Anwendbarkeit kann man nach de Boer [20] etwa annehmen für Helium 40° K, für

[1] E. Wigner: Phys. Rev. **40**, 749 (1932).
[2] M. Goldberger u. E.N. Adams: J. Chem. Phys. **20**, 240 (1952).

Wasserstoff $75°$ K und für Deuterium $45°$ K. Es ergibt sich eine sehr gute Übereinstimmung mit den experimentellen Ergebnissen. Insbeondere wird die Folgerung bestätigt, daß (bei gleichen zwischenmolekularen Kräften) der zweite Virialkoeffizient von der Masse abhängt. Für Einzelheiten sei auf Bd. XII verwiesen.

61. Entwicklung für tiefe Temperaturen. Die älteren Tieftemperatur-Entwicklungen[1] berücksichtigen, wie in Ziff. 58 erwähnt, nicht die Symmetriebedingungen. Dieses Verfahren ist verschiedentlich[2,3] kritisiert worden. Explizite Rechnungen (z.B. [4]) zeigen, daß bei sehr tiefen Temperaturen der Typ der Statistik von entscheidender Bedeutung wird, was auch nach den Ergebnissen für separierbare Systeme zu erwarten ist. Wir geben daher ein Beispiel einer Tieftemperatur-Entwicklung, welche die Symmetriebedingungen berücksichtigt, aber auch die quantisierte Maxwell-Boltzmann-Statistik einschließt[3].

Wir schreiben zunächst den Hamilton-Operator in der Form

$$\mathsf{H} = \mathsf{H}_0 + g\,\mathsf{U}, \tag{61.1}$$

wo H_0 wieder der Operator der kinetischen Energie und g ein Kopplungsparameter ist. Wir haben dann

$$e^{-\beta(\mathsf{H}_0 + g\mathsf{U})} = e^{-\beta\mathsf{H}} = \sum_{\nu=0}^{\infty} \frac{(-1)^\nu}{\nu!}\, \beta^\nu \mathsf{H}^\nu. \tag{61.2}$$

Entwickeln wir an der Stelle $g = 0$ in eine Taylorsche Reihe, so folgt

$$e^{-\beta\mathsf{H}} = \sum_{n=0}^{\infty} \frac{g^n}{n!} \sum_{\nu=0}^{\infty} \frac{(-1)^\nu}{\nu!}\, \beta^\nu (D^n \mathsf{H}^\nu)_{g=0}, \tag{61.3}$$

wo zur Abkürzung $D = \partial/\partial g$ geschrieben ist. Bilden wir die Spur dieses Operators, so erhalten wir

$$Q = \operatorname{spur} e^{-\beta\mathsf{H}} = \sum_{n=0}^{\infty} \frac{g^n}{n!} \sum_{\nu=0}^{\infty} \frac{(-1)^\nu}{\nu!}\, \beta^\nu \operatorname{spur} (D^n \mathsf{H}^\nu)_{g=0}. \tag{61.4}$$

Es ist nun

$$D\,\mathsf{H} = \mathsf{U}, \qquad D^n \mathsf{H} = 0 \ \text{ für } \ n > 1, \qquad (\mathsf{H})_{g=0} = \mathsf{H}_0. \tag{61.5}$$

Der Ausdruck $(D^n \mathsf{H}^\nu)_{g=0}$ ergibt sich dann als eine Summe von Termen der Form

$$\mathsf{H}_0^{\nu_1'}\,\mathsf{U}\,\mathsf{H}_0^{\nu_2'}\,\mathsf{U}\,\mathsf{H}_0^{\nu_3'}\dots\mathsf{H}_0^{\nu_n'}\,\mathsf{U}\,\mathsf{H}_0^{\nu_{n+1}'}, \tag{61.6}$$

da H_0 und U nicht vertauschbar sind. Dabei sind die ν_k' ganze Zahlen, die alle Werte von 0 bis $\nu - n$ annehmen können mit der Nebenbedingung

$$\sum_{k=1}^{n+1} \nu_k' = \nu - n. \tag{61.7}$$

Da wir nur an der Spur des Operators interessiert sind, können wir (61.6) auch schreiben

$$\mathsf{H}_0^{\nu_1}\,\mathsf{U}\,\mathsf{H}_0^{\nu_2}\,\mathsf{U}\,\mathsf{H}_0^{\nu_3}\dots\mathsf{H}_0^{\nu_n}\,\mathsf{U} \tag{61.8}$$

[1] H.S. Green: J. Chem. Phys. **19**, 955 (1951). — M.L. Goldberger u. E.N. Adams: J. Chem. Phys. **20**, 240 (1952). — H.S. Green: J. Chem. Phys. **20**, 1274 (1952).
[2] A.J.F. Siegert: Privatmitteilung.
[3] G.V. Chester: Phys. Rev. **93**, 606 (1954).
[4] J.H. de Boer u. E.G.D. Cohen: Physica, Haag **17**, 993 (1951).

mit der Nebenbedingung

$$\sum_{k=1}^{n} \nu_k = \nu - n.$$ (61.9)

Wir erhalten somit

$$(D^n H^\nu)_{g=0} = (n-1)!\,\nu \sum_{\nu_k} H_0^{\nu_1} U H_0^{\nu_2} U \ldots H_0^{\nu_n} U \quad (n \geq 1),$$ (61.10)

wo die Summierung über alle Sätze der ν_k zu erstrecken ist, welche der Nebenbedingung (61.9) genügen. Der Term für $n = 1$ in Gl. (61.4) ist einfach spur $e^{-\beta H_0}$.

Um die Spur von (61.10) zu berechnen, benutzen wir wieder die Eigenfunktionen der kinetischen Energie, also die Lösungen der Wellengleichung

$$H_0 \chi_l = E_l \chi_l.$$ (61.11)[1]

Je nachdem, ob wir einfache Produktfunktionen verwenden oder, wie in Ziff. 60 beschrieben, die Symmetriebedingungen einführen, erhalten wir die Formeln für die quantisierte Maxwell-Boltzmann-, die Bose-Einstein- und die Fermi-Dirac-Statistik. Da die weitere Rechnung alle drei Fälle einschließt, schreiben wir für die Eigenfunktionen einfach χ_l. Wir erhalten dann aus Gl. (69.10)

$$\text{spur}\,(D^n H^\nu)_{g=0} = (n-1)!\,\nu \sum_{l_j} \sum_{\nu_k} E_{l_1}^{\nu_1} E_{l_2}^{\nu_2} \ldots E_{l_n}^{\nu_n} U_{l_1 l_2} \ldots U_{l_n l_1}.$$ (61.12)

Hier ist die erste Summierung über alle Sätze der l_j zu erstrecken, und die Matrixelemente $U_{l_i l_j}$ sind gegeben durch

$$U_{l_i l_j} = \int \chi_{l_i}^* U \chi_{l_j}\, d\boldsymbol{q}.$$ (61.13)

Man bemerkt, daß der Typ der Statistik an zwei Stellen von Bedeutung ist: Bei der Bildung der Summe $\sum_{l_j}$ und bei der Berechnung der Matrixelemente $U_{l_i l_j}$. Die Summierung über die Sätze ν_k ist mathematisch äquivalent dem Problem des (endlichen) idealen Bose-Einstein-Gases (Ziff. 57). Wir können daher eine in diesem Zusammenhang von Dingle[2] entwickelte Methode benutzen. Schreiben wir

$$\text{spur}\,(D^n H^\nu)_{g=0} = (n-1)!\,\nu \sum_{l_j} U_{l_1 l_2} \ldots U_{l_n l_1} \Phi(l_j),$$ (61.14)

so ist $\Phi(l_j)$ der Koeffizient von $z^{\nu-n}$ in der Entwicklung von $\prod_{i=1}^{n} (1 - z E_{l_i})^{-1}$. Wir zerlegen nun die erzeugende Funktion in Partialbrüche gemäß

$$\prod_{i=1}^{n} (1 - z E_{l_i})^{-1} = \sum_{j=1}^{n} A_j (1 - z E_{l_j})^{-1},$$ (61.15)

wobei wir zunächst voraussetzen, daß alle E_{l_j} untereinander verschieden sind. Dann ist

$$A_j = \prod_{i \neq j=1}^{n} E_{l_j}^{n-1} (E_{l_j} - E_{l_i})^{-1}.$$ (61.16)

Aus (61.15) und (61.16) folgt

$$\prod_{i=1}^{n} (1 - z E_{l_i})^{-1} = \sum_{j=1}^{n} \left[\prod_{i \neq j}^{n} \frac{E_{l_j}^{n-1}}{(E_{l_j} - E_{l_i})} \right] (1 - z E_{l_j})^{-1}.$$ (61.17)

---

[1] Es sind die Lösungen gemeint, die freien Teilchen entsprechen (free particle wave functions).

[2] R. B. Dingle: Proc. Cambridge Phil. Soc. 45, 275 (1949).

Da der Koeffizient von $z^{\nu-n}$ in $(1 - z E_{l_j})^{-1}$ einfach $E_{l_j}^{\nu-n}$ ist, erhalten wir

$$\Phi(l_j) = \sum_{j=1}^{n} E_{l_j}^{\nu-1} \prod_{i \neq j}^{n} (E_{l_j} - E_{l_i})^{-1}. \tag{61.18}$$

Damit wird schließlich

$$\text{spur}\,(D^n H^\nu)_{g=0} = (n-1)!\,\nu \sum_{l_j} U_{l_1 l_2} \ldots U_{l_n l_1} \sum_{j=1}^{n} E_{l_j}^{\nu-1} \prod_{i \neq j}^{n} (E_{l_j} - E_{l_i})^{-1}. \tag{61.19}$$

Die Entwicklung der Verteilungsfunktion kann jetzt geschrieben werden

$$Q = Q_0 + \sum_{n=1}^{\infty} g^n Q_n \tag{61.20}$$

mit

$$Q_0 = \sum_{l} e^{-\beta E_l} \tag{61.21}$$

und

$$Q_n = -\frac{\beta}{n} \sum_{l_j} U_{l_1 l_2} \ldots U_{l_n l_1} \sum_{j} \frac{e^{-\beta E_{l_j}}}{\prod_{i \neq j}^{n} (E_{l_j} - E_{l_i})}. \tag{61.22}$$

Die Untersuchung des Grenzüberganges $E_{l_i} \to E_{l_j}$ zeigt, daß in diesem Falle auch der Zähler verschwindet und ein endlicher Grenzwert erhalten wird. Die Einschränkung $E_{l_i} \neq E_{l_j}$ kann daher fallen gelassen werden.

Das Problem der Tieftemperatur-Entwicklung ist damit formal gelöst in einer Weise, die sowohl die Berücksichtigung der Symmetriebedingungen einschließt wie das Anschreiben des allgemeinen Terms [Gl. (61.22)] ermöglicht und im Prinzip auch einer expliziten Berechnung zugänglich ist. Wesentlich schwieriger ist die Frage nach der Konvergenz der Entwicklung zu beantworten, die ebenfalls von CHESTER[1] untersucht worden ist. Wir können hier nur kurz die Ergebnisse anführen und müssen für Einzelheiten auf die Originalarbeit verweisen. Die Existenz der einzelnen Terme der Entwicklung hängt im wesentlichen ab von dem Charakter der Potentialfunktion U. Wenn dieselbe keine Singularitäten besitzt, so existieren die Integrale (61.13) und damit ziemlich sicher auch die einzelnen Terme. Dieser Fall ist indessen physikalisch von geringerem Interesse, da er für die Wechselwirkung der Moleküle nicht zutrifft. Wird U in einem endlichen Gebiete des Konfigurationsraumes positiv unendlich (Moleküle mit inkompressiblem Kern), so müssen die Matrixelemente $U_{l_i l_j}$ mit Hilfe der Eigenfunktionen starrer Kugeln definiert werden, um die Existenz der einzelnen Terme zu sichern. In diesem Falle muß auch der Term Q_0 modifiziert werden. Wird schließlich U nur in isolierten Punkten positiv unendlich, so divergieren die einzelnen Terme, und die Reihe ist, wenn überhaupt, nur brauchbar, indem man zuerst summiert und dann den Grenzübergang zu den isolierten Punkten vollzieht. Ein wirklich brauchbares Kriterium für die Konvergenz der Reihe als Ganzes ist bisher nicht gefunden worden.

Eine experimentelle Prüfung im strengen Sinne (wie für die Hochtemperatur-Entwicklung) ist bisher für die Tieftemperatur-Entwicklung nicht durchgeführt worden. Die hier beschriebene Entwicklung ist von CHESTER[2] als Grundlage für eine Theorie des flüssigen Heliums benutzt worden. Für diese Frage sei auf den Artikel von MÜNSTER in Bd. XIII dieses Handbuches verwiesen.

---

[1] G. V. CHESTER: Phys. Rev. **93**, 606 (1954).
[2] G. V. CHESTER: Phys. Rev. **100**, 455 (1955).

62. Feynmans Lösung der Blochschen Gleichung. Zum Schluß dieses Abschnittes behandeln wir noch kurz eine von Feynman[1] angegebene Lösung der Blochschen Gleichung, die ein eindrucksvolles Beispiel für die Anwendung mathematischer Methoden der Quantenmechanik in der Quantenstatistik darstellt. Sie ist dadurch von besonderem Interesse, daß sie die Grundlage der Feynmanschen Theorie des Heliums[1] (vgl. den Artikel von Münster in Bd. XIII dieses Handbuches) wie auch der in Ziff. 58 erwöhnten cluster-Entwicklung von Montroll und Ward[2] darstellt.

Zum besseren Verständnis skizzieren wir zunächst die quantenmechanische Theorie[3,4]. Zur Notierung bemerken wir, daß im folgenden q_1, q_2, q_3 (abweichend von der sonst in diesem Artikel gebrauchten Bezeichnungsweise) die vollständigen Koordinatensätze zu den Zeiten t_1, t_2, t_3 bezeichnen. Eine Funktion $f(q_1, t_1)$ chreiben wir abgekürzt $f(1)$ usw.

Die zeitabhängige Schrödinger-Gleichung

$$\mathsf{H}\,\Psi + \frac{h}{2\pi i}\,\frac{\partial \Psi}{\partial t} = 0 \tag{62.1}$$

hat als lineare Differentialgleichung die Lösung

$$\Psi(q_2, t_2) = \int K(q_2, t_2; q_1, t_1)\,\Psi(q_1, t_1)\,d q_1, \tag{62.2}$$

wo $K(q_2, t_2; q_1, t_1)$ die Greensche Funktion von (62.1) ist. Für konservative Systeme können wir schreiben

$$\Psi(q, t_1) = \sum_n a_n \psi_n(q), \tag{62.3}$$

wo die ψ_n die Eigenfunktionen des Hamilton-Operators sind. Dann wird

$$\Psi(q, t_2) = \sum_n e^{-\frac{2\pi i E_n}{h}(t_2 - t_1)} a_n \psi_n(q). \tag{62.4}$$

Wegen

$$a_n = \int \psi_n^*(q_1)\,\Psi(q_1, t_1)\,d q_1 \tag{62.5}$$

findet man durch Vergleich mit (62.2)

$$K(2, 1) = \sum_n \psi_n(q_2)\,\psi_n^*(q_1)\,e^{-\frac{2\pi i E_n}{h}(t_2 - t_1)}. \tag{62.6}$$

Es ist zweckmäßig, zu definieren

$$K(2, 1) = 0 \quad \text{für} \quad t_2 < t_1. \tag{62.7}$$

Allgemein kann $K(2, 1)$ definiert werden als Lösung der Differentialgleichung

$$\left[\frac{h}{2\pi i}\,\frac{\partial}{\partial t_2} + \mathsf{H}(2)\right] K(2, 1) = \frac{h}{2\pi i}\,\delta(q_2 - q_1)\,\delta(t_2 - t_1), \tag{62.8}$$

welche für $t_2 < t_1$ verschwindet. Dabei bedeutet das Argument 2 bei dem Hamilton-Operator, daß derselbe nur auf die Variablen des Satzes 2 wirkt.

Wir betrachten nun ein einfaches Störungsproblem. Ein Teilchen soll sich in einem schwachen Potential $U(q, t)$ bewegen derart, daß U nur für $t_1 < t < t_2$ von Null verschieden ist. Wir entwickeln $K(2, 1)$ nach Potenzen von U in der Form

$$K(2, 1) = K_0(2, 1) + K_{\mathrm{I}}(2, 1) + \cdots. \tag{62.9}$$

[1] R.P. Feynman: Phys. Rev. **91**, 1291 (1953).
[2] E.W. Montroll u. J.C. Ward: Phys. Fluids **1**, 55 (1958).
[3] R.P. Feynman: Rev. Mod. Phys. **20**, 367 (1948).
[4] R.P. Feynman: Phys. Rev. **76**, 749 (1949).

Der erste Term $K_0(2, 1)$ entspricht einem freien Teilchen. Um die Störung erster Ordnung zu berechnen, nehmen wir zunächst an, daß U nur in dem infinitesimalen Intervall Δt_3 zwischen t_3 und $t_3 + \Delta t_3 (t_1 < t_3 < t_2)$ von Null verschieden ist. Da das Teilchen von t_1 bis t_3 frei ist, haben wir

$$\Psi(3) = \int K_0(3, 1)\, \Psi(1)\, d\boldsymbol{q}_1. \tag{62.10}$$

Für das Intervall Δt_3 können wir die Lösung von (62.1) schreiben

$$\Psi(\boldsymbol{q}, t_3 + \Delta t_3) = e^{-\frac{2\pi i \mathsf{H}}{h} \Delta t_3} \Psi(\boldsymbol{q}, t_3) = \left(1 - \frac{2\pi i}{h} \mathsf{H}_0 \Delta t_3 - \frac{2\pi i}{h} \mathsf{U}\, \Delta t_3\right) \Psi(\boldsymbol{q}, t_3), \tag{62.11}$$

wo H_0 wieder der Hamilton-Operator des freien Teilchens ist. Das Störpotential bewirkt somit, daß die Wahrscheinlichkeitsamplitude $\psi(\boldsymbol{q}, t_3 + \Delta t_3)$ einen Zusatzterm

$$\Delta \Psi = -\frac{2\pi i}{h}\, U(\boldsymbol{q}_3, t_3)\, \Psi(\boldsymbol{q}_3, t_3)\, \Delta t_3 \tag{62.12}$$

erhält, der als die von dem Potential gestreute Amplitude bezeichnet wird. Da nach der Zeit $t_3 + \Delta t_3$ das Teilchen wieder frei ist, gilt

$$\Psi(\boldsymbol{q}_2, t_2) = \int K_0(\boldsymbol{q}_2, t_2; \boldsymbol{q}_3, t_3 + \Delta t_3)\, \Psi(\boldsymbol{q}_3, t_3 + \Delta t_3)\, d\boldsymbol{q}_3. \tag{62.13}$$

Durch Einsetzen von (62.10) in (62.12) und von (62.12) in (62.13) findet man für die durch das Störpotential verursachte Änderung der Wellenfunktion in 2

$$\Delta \Psi(2) = -\frac{2\pi i}{h} \iint K_0(2, 3)\, U(3)\, K_0(3, 1)\, \Psi(1)\, d\boldsymbol{q}_1 d\boldsymbol{q}_3 \Delta t_3. \tag{62.14}$$

Wenn das Störpotential über ein endliches Zeitintervall wirkt, kann die gesamte Änderung der Wellenfunktion als Summe der von den einzelnen Intervallen Δt_3 herrührenden Effekte betrachtet werden. Man hat dann auch über t_3 zu integrieren und erhält mit Gl. (62.2)

$$K_{\mathrm{I}}(2, 1) = -\frac{2\pi i}{h} \iint K_0(2, 3)\, U(3)\, K_0(3, 1)\, d\boldsymbol{q}_3 dt_3. \tag{62.15}$$

In ähnlicher Weise können die höheren Terme in (62.9), die Zwei- und Mehrfachstreuungen entsprechen, berechnet werden.

Die Greensche Funktion $K(2, 1)$, die auch als *Propagator* bezeichnet wird, hat eine anschauliche Bedeutung. Nach den Regeln der Quantenmechanik ist die Wahrscheinlichkeitsamplitude, daß das System, dessen Zustand $\Psi(\boldsymbol{q}, t)$ sei, bei einer Messung in dem Eigenzustand $\chi(\boldsymbol{q}, t)$ gefunden wird

$$\int \chi^*(\boldsymbol{q}, t)\, \Psi(\boldsymbol{q}, t)\, d\boldsymbol{q}. \tag{62.16}$$

Beziehen wir diese Gleichung auf die Zeit $t = t_2$ und benutzen Gl. (62.2), so erhalten wir

$$\iint \chi^*(2)\, K(2, 1)\, \Psi(1)\, d\boldsymbol{q}_1 d\boldsymbol{q}_2 \tag{62.17}$$

als Wahrscheinlichkeitsamplitude, daß ein System, welches zur Zeit t_1 im Zustand $\Psi(1)$ war, zur Zeit t_2 im Zustand $\chi(2)$ gefunden wird. $K(2, 1)$ kann daher interpretiert werden als die Wahrscheinlichkeitsamplitude, daß ein System, das sich zur Zeit t_1 in der Konfiguration q_1 befindet, zur Zeit t_2 die Konfiguration q_2 erreicht. Diese Amplitude ist, wie FEYNMAN[1] gezeigt hat, gleich der Summe

---

[1] R.P. FEYNMAN: Phys. Rev. **91**, 1291 (1953).

von $e^{\frac{2\pi i}{h} S}$ über alle Trajektorien, welche die Konfigurationen q_1 und q_2 verbinden, wo

$$S = \int_{t_1}^{t_2} L\, dt \qquad (62.18)$$

die klassische Wirkungsfunktion ist. Es ist also

$$K(2, 1) = \frac{1}{C} \int e^{\frac{2\pi i}{h} S} \mathscr{D} q(t) \qquad (62.19)$$

$$\text{alle}$$
$$\text{Trajektorien}$$

wo das Integral über alle Trajektorien [symbolisch durch $\int \ldots \mathscr{D} q(t)$ bezeichnet] so zu bilden ist, daß $q(t_1) = q_1$ und $q(t_2) = q_2$ ist und $1/C$ einen Normierungsfaktor bezeichnet. Derselbe ist dadurch festgelegt, daß für freie Teilchen [wie man unmittelbar aus (62.6) ableitet]

$$K_0(2, 1) = \left[\frac{ih}{m}(t_2 - t_1)\right]^{-\frac{2}{3}N} e^{\frac{1}{2} \frac{2\pi i m}{h} \frac{(q_2 - q_1)^2}{(t_2 - t_1)}} \qquad (62.20)$$

gelten muß.

Die im vorstehenden skizzierte Theorie läßt sich nun unmittelbar für die Lösung der Blochschen Gleichung verwenden, wenn wir beachten, daß hier β an Stelle von $2\pi it/h$ tritt. Wir definieren einen Propagator

$$K(q_2, \beta_2; q_1, \beta_1) \equiv K(2, 1) \qquad (62.21)$$

mit der Eigenschaft

$$\Psi(q_2, \beta_2) = \int K(q_2, \beta_2; q_1, \beta_1)\, \Psi(q_1, \beta_1)\, d q_1. \qquad (62.22)$$

$K(2, 1)$ ist die Greensche Funktion der Blochschen Gleichung und genügt der Differentialgleichung

$$\left[\frac{\partial}{\partial \beta_2} + \mathsf{H}(2)\right] K(2, 1) = \delta(q_2 - q_1)\, \delta(\beta_2 - \beta_1), \qquad (62.23)$$

wobei wir definieren

$$K(2, 1) = 0 \quad \text{für} \quad \beta_2 < \beta_1. \qquad (62.24)$$

Es ist nun

$$K(2, 1) = \sum_n \psi_n^*(q_2)\, \psi_n(q_1)\, e^{-E_n(\beta_2 - \beta_1)} \quad (\beta_2 > \beta_1) \qquad (62.25)$$

und

$$K(2, 1) \to \delta(q_2 - q_1) \quad \text{für} \quad \beta_2 \to \beta_1. \qquad (62.26)$$

Aus Gl. (59.1) und (62.25) folgt

$$Q = \int K(q, \beta; q, 0)\, d q. \qquad (62.27)$$

Die Bestimmung des Propagators reduziert somit die Berechnung der Verteilungsfunktion auf ein multiples Integral im Konfigurationsraum. Der Propagator kann hier allgemein interpretiert werden als die Wahrscheinlichkeitsamplitude, daß ein System, welches sich bei der Temperatur β_1 in der Konfiguration q_1 befindet, bei langsamem Abkühlen auf die Temperatur β_2 die Konfiguration q_2 erreicht. Der in (62.27) auftretende Term $K(q, \beta; q, 0)$ stellt im besonderen die Wahrscheinlichkeitsamplitude dar, daß das System, wenn es bei unendlich hoher Temperatur mit der Konfiguration q startet, nach langsamem Abkühlen auf die Temperatur β wieder in die gleiche Konfiguration zurückkehrt. Die Verteilungsfunktion ist dann eine Mittelung über alle möglichen Anfangskonfigurationen für einen solchen Prozeß.

Wir wenden nun die Störungstheorie an und setzen

$$\mathsf{H}(2) = \mathsf{H}_0(2) + \mathsf{U}(2) \tag{62.28}$$

und

$$K(2, 1) = K_0(2, 1) + K_{\mathrm{I}}(2, 1). \tag{62.29}$$

Dann ist

$$\left[\frac{\partial}{\partial \beta} + \mathsf{H}_0(2)\right] K_0(2, 1) = \delta(\boldsymbol{q}_2 - \boldsymbol{q}_1)\, \delta(\beta_2 - \beta_1). \tag{62.30}$$

Aus (62.28) bis (62.30) folgt mit (62.24)

$$\left[\frac{\partial}{\partial \beta} + \mathsf{H}_0(2)\right] K_{\mathrm{I}}(2, 1) = -\,\mathsf{U}(2)\, K(2, 1). \tag{62.31}$$

Die Lösung dieser Gleichung ist, wie man durch Einsetzen verifiziert und auch auf einem dem zu (62.15) führenden analogen Wege direkt ableiten kann

$$K_{\mathrm{I}}(2, 1) = -\int_{\beta_1}^{\beta_2}\!\!\!\int K_0(2, 3)\, U(3)\, K(3, 1)\, d\boldsymbol{q}_3\, d\beta_3. \tag{62.32}$$

Mit Rücksicht auf (62.29) kann diese Gleichung geschrieben werden

$$K(2, 1) = K_0(2, 1) - \int_{\beta_1}^{\beta_2}\!\!\!\int K_0(2, 3)\, U(3)\, K(3, 1)\, d\boldsymbol{q}_3\, d\beta_3. \tag{62.33}$$

Die Integralgleichung läßt sich durch Iteration lösen.

Der Propagator für freie Teilchen ist jetzt

$$K_0(2, 1) = \left[\frac{h^2}{2\pi m}(\beta_2 - \beta_1)\right]^{-\frac{3}{2}N} e^{-\frac{2\pi^2 m}{h^2}\frac{(\boldsymbol{q}_2 - \boldsymbol{q}_1)^2}{(\beta_2 - \beta_1)}}. \tag{62.34}$$

Einsetzen dieses Ausdruckes in Gl. (62.27) führt auf die Formeln der quantisierten Maxwell-Boltzmann-Statistik. Für die Bose-Einstein- und Fermi-Dirac-Statistik müssen an dieser Stelle die Symmetriebedingungen eingeführt werden. Für die Einzelheiten dieses Verfahrens sei auf die Arbeit von MONTROLL und WARD[1] verwiesen.

VII. Die große kanonische Gesamtheit.

63. Definition. Der Formalismus der kanonischen Gesamtheit führt auf Probleme, die den bei der mikrokanonischen Gesamtheit auftretenden weitgehend analog sind. Die oft unbequeme Nebenbedingung konstanter Teilchenzahl legt die Benutzung einer (reellen oder komplexen) erzeugenden Funktion nahe. Vom physikalischen Standpunkt erscheint die Beschränkung auf Systeme, welche mit ihrer Umgebung zwar Energie (Wärme), aber keine Materie austauschen können, häufig nicht gerechtfertigt. Es liegt daher nahe, den formalen Apparat der statistischen Mechanik nochmals zu erweitern und Gesamtheiten einzuführen, die aus kanonischen Gesamtheiten mit Teilchenzahlen von Null bis unendlich aufgebaut sind. Dieser Gedanke ist im Rahmen der klassischen statistischen Mechanik bereits von GIBBS [2] durchgeführt worden, von dem auch die Bezeichnung „große kanonische Gesamtheit" (grand canonical ensemble) herrührt. Die systematische Übertragung der Gibbsschen Ansätze in die Quantenstatistik ist von DELBRÜCK und MOLIÈRE [29] sowie TOLMAN [12] durchgeführt worden.

---

[1] Siehe Zitat S. 340.

Der Wahrscheinlichkeitsansatz für die große kanonische Gesamtheit ist von verschiedenen Postulaten ausgehend begründet worden[1-3]. Aus den in Ziff. 29 und 51 erwähnten Gründen ziehen wir auch hier eine axiomatische Einführung vor. Wir setzen also für die Wahrscheinlichkeit, daß ein beliebig herausgegriffenes System der Gesamtheit die Teilchenzahlen $N_1, \ldots, N_m$ (wo m die Zahl der Komponenten ist) besitzt und sich in einem zu dem Eigenwert E_n gehörenden Quantenzustand befindet

$$W = e^{\frac{\Omega + \mu_1 N_1 + \cdots + \mu_m N_m - E_n}{\Theta}} . \tag{63.1}[4]$$

Für den Parameter Ω ergibt sich daraus unmittelbar

$$e^{-\frac{\Omega}{\Theta}} = \sum_{N_1=0}^{\infty} \cdots \sum_{N_m=0}^{\infty} e^{\frac{\mu_1 N_1 + \cdots + \mu_m N_m}{\Theta}} \left[\sum_n e^{-\frac{E_n}{\Theta}} \right] \tag{63.2}[5]$$

oder in kompakter Form

$$e^{-\frac{\Omega}{\Theta}} = \sum_N e^{\frac{\mu N}{\Theta}} Q^{(N)} . \tag{63.3}$$

Die große kanonische Gesamtheit, die durch die vorstehenden Gleichungen definiert ist, befindet sich im statistischen Gleichgewicht, da sie, wie die Formeln zeigen, aus kanonischen Gesamtheiten aufgebaut ist, die man ihrerseits wieder aus mikrokanonischen Gesamtheiten zusammensetzen kann. Die Definition der großen kanonischen Gesamtheit überschreitet jedoch offensichtlich den Rahmen der statistischen Mechanik im eigentlichen Sinne, insofern man darunter eine Statistik über die mechanischen Phasen bzw. quantenmechanischen Zustände versteht. Man sieht dies unmittelbar daran, daß (im klassischen Fall) W nicht mehr eine Phasendichte darstellt. Tatsächlich haben wir diesen Schritt bereits mit der Einführung der kanonischen Gesamtheit getan, und es ist mehr zufällig, daß diese Verteilung noch im Phasenraum darstellbar ist. Da nämlich die in der statistischen Mechanik benutzten virtuellen Gesamtheiten definitionsgemäß aus konservativen Systemen bestehen, führt die mechanische Theorie nur zu Aussagen über die mikrokanonische Gesamtheit. Jede verallgemeinerte, aus sich in irgendeiner Hinsicht unterscheidenden mikrokanonischen Gesamtheiten aufgebaute Gesamtheit erfordert zu ihrer Begründung zusätzliche Überlegungen bzw. Postulate. Wir werden in Ziff. 67 auf diese Frage zurückkommen. In der Quantenstatistik sind die Verhältnisse weniger übersichtlich, weil hier die kanonische Gesamtheit mit Hilfe des Dichteoperators definiert wird. Es ist versucht worden, auch die große kanonische Gesamtheit der Quantenstatistik mit Hilfe eines Dichteoperators unter Verwendung der aus der Theorie der zweiten Quantelung bekannten Jordan-Klein- und Jordan-Wigner-Matrizen zu definieren [22], [23]. Da aber hier die Eigenwerte (im Gegensatz zum Hamilton-Operator) von vornherein bekannt sind, ist eine solche Formulierung praktisch jedenfalls bedeutungslos. Ob man grundsätzlich für die verallgemeinerten Gesamtheiten an dem Konzept des Dichteoperators festhalten soll, erscheint nach den obigen Überlegungen zum mindesten fraglich. Diese Frage ist jedoch bisher noch nicht untersucht worden. Wir werden im folgenden (wie bereits in den

---

[1] R.C. Tolman: Phys. Rev. **57**, 1160 (1940).
[2] A. Münster: Proc. Cambridge Phil. Soc. **46**, 319 (1950).
[3] R. Becker: Z. phys. Chem. **196**, 181 (1950).
[4] Wir benutzen hier im Anschluß an Gibbs [2] das Symbol Ω. Eine Verwechslung mit dem Phasenvolumen ist durch den Zusammenhang ausgeschlossen.
[5] Die Summierung in der eckigen Klammer erstreckt sich über die linear unabhängigen Eigenfunktionen.

obigen Definitionsgleichungen) für die kanonische Verteilungsfunktion die Basis der Energie-Eigenfunktionen voraussetzen und damit auf den Operator-Formalismus verzichten.

Der über eine große kanonische Gesamtheit gebildete Mittelwert ist definiert durch die Gleichung

$$\bar{u} = \sum_{N_1} \cdots \sum_{N_m} \left[\sum_n u_n \, e^{\frac{\Omega + \mu_1 N_1 + \cdots + \mu_m N_m - E_n}{\Theta}} \right]. \tag{63.4}$$

Im besonderen gilt für die mittlere Energie eines durch die große kanonische Gesamtheit dargestellten Systems

$$\bar{E} = \sum_{N_1} \cdots \sum_{N_m} \left[\sum_n E_n \, e^{\frac{\Omega + \mu_1 N_1 + \cdots + \mu_m N_m - E_n}{\Theta}} \right] \tag{63.5}$$

und für die mittlere Molekülzahl der Komponente i

$$\bar{N}_i = \sum_{N_1} \cdots \sum_{N_m} \left[\sum_n N_i \, e^{\frac{\Omega + \mu_1 N_1 + \cdots + \mu_m N_m - E_n}{\Theta}} \right]. \tag{63.6}$$

64. Thermodynamische Analoga. Wir betrachten wieder die Energie eines Quantenzustandes als differenzierbare Funktion der äußeren Parameter x_j. Wenn Molekülzahlen und Quantenzustand vorgegeben sind, ist dann die generalisierte Kraft

$$X_j = - \frac{\partial E_n}{\partial x_j}. \tag{64.1}$$

Differenzieren wir nun Gl. (63.2), so erhalten wir

$$d\Omega = \frac{\Omega + \mu_1 \bar{N}_1 + \cdots + \mu_m \bar{N}_m - \bar{E}}{\Theta} \, d\Theta + \bar{N}_1 \, d\mu_1 - \cdots - \bar{N}_m \, d\mu_m - \sum \bar{X}_j \, dx_j. \tag{64.2}$$

Wir definieren jetzt eine Größe[1]

$$\bar{H} = \frac{\Omega + \mu_1 \bar{N}_1 + \cdots \mu_m \bar{N}_m - \bar{E}}{\Theta} \tag{64.3}$$

und können dann schreiben

$$d\Omega = \bar{H} d\Theta - \bar{N}_1 \, d\mu_1 - \cdots - \bar{N}_m \, d\mu_m - \sum_j \bar{X}_i \, dx_i. \tag{64.4}$$

Thermodynamisch gilt, wenn wir als einzigen äußeren Parameter das Volumen annehmen,

$$G = E - TS + PV, \quad F = E - TS \tag{64.5}$$

und ferner

$$G = N_1 \mu_1 + \cdots + N_m \mu_m. \tag{64.6}$$

Daraus folgt

$$PV = G - F = N_1 \mu_1 + \cdots + N_m \mu_m - F \tag{64.7}$$

und wegen

$$dF = - S \, dT - P \, dV + \mu_1 \, dN_1 + \cdots + \mu_m \, dN_m \tag{64.8}$$

durch Differentiation von (64.7)

$$d(PV) = S \, dT + P \, dV + N_1 \, d\mu_1 + \cdots + N_m \, d\mu_m. \tag{64.9}$$

Die Funktion

$$PV = PV(T, V, \mu) \tag{64.10}$$

---

[1] Diese Größe ist nicht mit der ebenso bezeichneten Größe des generalisierten H-Theorems (Ziff. 44) zu verwechseln.

ist somit ein thermodynamisches Potential und zwar für die unabhängigen Variablen T, V und μ_i. Beschränken wir auch in Gl. (64.4) die äußeren Parameter auf das Volumen, so erhalten wir durch Vergleich mit (64.9) die folgenden Beziehungen zwischen statistischen und thermodynamischen Größen:

$$\Omega \to -PV, \quad \overline{H} \to -S/k, \quad \Theta \to kT$$
$$\overline{N_i} \to N_i, \quad \mu_i \to \mu_i, \quad \overline{X}_j \to \overline{X}. \tag{64.11}[1]$$

Die Deutungen von $\overline{N}_i$ und $\overline{X}_j$ sind unmittelbar evident. Für den Beweis, daß $-\overline{H}$ die Eigenschaften der Entropie und Θ die Eigenschaften der empirischen Temperatur besitzt, können wir auf Ziff. 52 bis 54 verweisen, da hier jetzt keine wesentlichen neuen Gesichtspunkte auftreten. Es bleibt daher nur noch die Identifizierung der statistischen Parameter μ_i mit den chemischen Potentialen der Komponenten zu rechtfertigen. Dazu betrachten wir zwei Systeme ′ und ″, die sich im thermodynamischen Gleichgewicht befinden und deren Gleichgewichtseigenschaften durch zwei große kanonische Gesamtheiten dargestellt werden. Die Systeme sollen zunächst völlig voneinander getrennt sein, aber die gleichen Werte der Parameter μ_i und Θ besitzen. Die entsprechenden Wahrscheinlichkeiten sind

$$W' = e^{\frac{\Omega' + \mu_1 N_1' + \cdots + \mu_m N_m' - E_n'}{\Theta}} \tag{64.12}$$

und

$$W'' = e^{\frac{\Omega'' + \mu_1 N_1'' + \cdots + \mu_m N_m'' - E_n''}{\Theta}}. \tag{64.13}$$

Die beiden Systeme lassen sich in Gedanken zu einem Gesamtsystem ′″ zusammenfassen; die Systeme der entsprechenden virtuellen Gesamtheit werden gebildet, indem jedes System der ersten mit jedem System der zweiten Gesamtheit kombiniert wird[2]. Die zugehörige Wahrscheinlichkeit ist dann (wegen der Unabhängigkeit der beiden Systeme)

$$W''' = W' \cdot W'' \tag{64.14}$$

oder

$$W''' = e^{\frac{\Omega''' + \mu_1 N_1''' + \cdots + \mu_m N_m''' - E_n'''}{\Theta}}. \tag{64.15}$$

mit

$$\Omega''' = \Omega' + \Omega'', \quad N_i''' = N_i' + N_i'', \quad E_n''' = E_n' + E_n''. \tag{64.16}$$

Das Gesamtsystem wird somit ebenfalls durch eine große kanonische Gesamtheit dargestellt. Das thermodynamische Gleichgewicht des Gesamtsystems beruht jedoch lediglich auf dem der Teilsysteme und der vorausgesetzten Unabhängigkeit derselben. Insofern hat die Gl. (64.15) nur formale Bedeutung.

Wir denken uns jetzt die beiden Einzelsysteme durch eine wärmeleitende Wand verbunden. Die damit verbundene Wechselwirkungsenergie lassen wir wie früher (Ziff. 52) gegen Null gehen. Auch dann gelten für das Gesamtsystem wieder die Gl. (64.15) und (64.16). Sie haben aber jetzt in bezug auf die Verteilung der Energie (d. h. für die kleinen kanonischen Gesamtheiten) eine physikalische Bedeutung, und wir können aus ihrer Gültigkeit auf thermodynamisches Gleichgewicht des Gesamtsystems schließen. Die notwendige und hinreichende Bedingung dafür ist

$$\Theta' = \Theta'' = \Theta'''. \tag{64.17}$$

---

[1] Wenn mehrere äußere Parameter zu berücksichtigen sind, gilt $\Omega \to -\sum_j X_j x_j$.

[2] Die neue Gesamtheit ist in bezug auf die Systemzahl eine Gesamtheit höherer Ordnung.

Wir wollen jetzt annehmen, daß die Wand zwischen den Einzelsystemen beweglich ist, so daß dieselben Kräfte aufeinander ausüben können[1]. Das mechanische Gleichgewicht des Gesamtsystems erfordert, daß diese Kräfte im Mittel verschwinden. Es muß daher

$$\overline{X}'_j = \overline{X}''_j = \overline{X}'''_j \quad \text{(für alle } j\text{)} \tag{64.18}$$

sein.

Die Wand zwischen den beiden Einzelsystemen können wir, insofern sie den Austausch von Materie verhindert, als unendlich hohen Potentialwall betrachten. Machen wir nun die Wand für alle Teilchen durchlässig, so daß auch Übergang von Teilchen zwischen den Einzelsystemen möglich ist, so wird damit wieder die Energiefunktion (bzw. der Hamilton-Operator) modifiziert. Läßt man aber die Systeme immer größer werden, so wird auch dieser „Wandeffekt" beliebig klein gegen die Energie E'_n und E''_n. Für das Gesamtsystem kommen wir dann wieder auf die Gln. (64.15) und (64.16). Diese haben jetzt in vollem Umfang ihre physikalische Bedeutung. Das Gesamtsystem wird durch eine große kanonische Gesamtheit mit den Parametern der Einzelsysteme dargestellt. Die Wahrscheinlichkeiten der Molekülzahlen für die Einzelsysteme werden durch die Ermöglichung des Molekülaustausches nicht geändert. Im Mittel findet also kein Transport von Materie statt; die Einzelsysteme und das Gesamtsystem befinden sich auch in dieser Beziehung im Gleichgewicht. Die notwendige und hinreichende Bedingung dafür ist, wie sich analog zu den früheren Überlegungen ergibt,

$$\mu'_i = \mu''_i = \mu'''_i. \tag{64.19}$$

Damit haben wir gezeigt, daß die statistischen Parameter μ_i in der Tat die wesentliche Eigenschaft der chemischen Potentiale besitzen und die Beziehungen (64.11) somit gerechtfertigt sind. Gleichzeitig haben wir die allgemeinen Bedingungen für das thermodynamische Gleichgewicht heterogener Systeme abgeleitet.

In dem zuletzt betrachteten Fall wird die Wand zwischen den Einzelsystemen zu einer reinen Fiktion, welche lediglich dazu dient, die Volumina zu definieren und damit die Einzelsysteme abzugrenzen. Die Überlegung läßt sich aber auch auf den Fall anwenden, daß nur gewisse Molekülarten die Wand passieren und damit zwischen den Systemen ausgetauscht werden können. Es ergibt sich dann, daß die Gln. (64.15) und (64.16) nur im Hinblick auf die Verteilung dieser Moleküle physikalische Bedeutung haben. Die Gleichgewichtsbedingungen (64.19) sind daher auf die betreffenden Molekülarten beschränkt. In diesem Falle muß die Wand als semipermeable Membran wirklich vorhanden sein. Sie muß auch, da sie die Volumina definiert, starr sein und kann daher von dem System ausgeübte Kräfte aufnehmen. Die Gleichgewichtsbedingungen (64.18) müssen (und werden auch im allgemeinen) daher jetzt nicht mehr erfüllt sein. Das einfachste Beispiel für diesen Sachverhalt liefert die Erscheinung des osmotischen Druckes.

Wenn wir die statistischen Parameter gemäß (64.11) durch die entsprechenden thermodynamischen Größen ersetzen, können wir die Gl. (63.2) schreiben

$$P V = kT \ln \varXi \tag{64.20}$$

mit

$$\varXi = \sum_{N_1} \cdots \sum_{N_m} e^{\frac{\mu_1 N_1 + \cdots + \mu_m N_m}{\Theta}} \left[\sum_n e^{-\frac{E_n}{kT}} \right]. \tag{64.21}$$

---

[1] Die Einzelsysteme werden in diesem Falle nicht mehr durch eine große kanonische Gesamtheit dargestellt. Wir haben hier analoge Verhältnisse wie bei der Identifizierung der Temperatur mit Hilfe der mikrokanonischen Gesamtheit. Vgl. Ziff. 22.

Die Gl. (64.20) entspricht völlig der Gl. (56.3). Man nennt daher nach einem Vorschlag von Fowler[1] Ξ die *große Verteilungsfunktion (grand partition function)*. Dieselbe läßt sich in etwas anderer Weise schreiben, wenn wir die schon in Ziff. 50 und 57 benutzten absoluten Aktivitäten einführen. Wegen

$$Z_i = e^{\frac{\mu_i}{kT}} \tag{64.22}$$

haben wir dann

$$\Xi = \sum_{N_1} \ldots \sum_{N_m} Z_1^{N_1} \ldots Z_m^{N_m} Q^{(N)} . \tag{64.23}$$

Mit Benutzung des Darwin-Fowlerschen Temperatur-Analogons (Ziff. 49)

$$\vartheta = e^{-\frac{1}{kT}} \tag{64.24}$$

ergeben sich dann für die Mittelwerte (63.5) und (63.6) die völlig symmetrischen Formeln

$$\overline{E} = \vartheta \, \frac{\partial \ln \Xi}{\partial \vartheta} \tag{64.25}$$

und

$$\overline{N}_i = Z_i \, \frac{\partial \ln \Xi}{\partial Z_i} . \tag{64.26}$$

Der Übergang zur halbklassischen Näherung vollzieht sich hier in gleicher Weise wie bei der kleinen kanonischen Gesamtheit.

Die Methode der großen kanonischen Gesamtheit nimmt heute wegen ihrer Allgemeinheit und Anpassungsfähigkeit eine zentrale Stellung in der statistischen Mechanik ein. Merkwürdigerweise geriet sie nach den grundlegenden Untersuchungen von Gibbs zunächst für längere Zeit fast völlig in Vergessenheit. Die wichtigste Anwendung vor dem Auftreten der Quantenmechanik ist die von Zernike[2] entwickelte Theorie der Lichtstreuung von Vielkomponenten-Systemen. In der neueren Literatur findet sie sich zwar vereinzelt [*11*], [*12*], [*29*]; ihre große Bedeutung ist aber wohl erst durch Fowler[3] wieder in vollem Umfange erkannt und gewürdigt worden. Er nennt sie "the most general and powerful of all the methods yet devised in statistical mechanics".

65. Beziehungen zur kanonischen Gesamtheit. Die Definitionsgleichung der Entropie für die große kanonische Gesamtheit läßt sich, wie man aus (63.1), (63.3) und (64.11) erkennt, auf die gleiche Form bringen wie früher bei der mikrokanonischen und kanonischen Gesamtheit. Wir haben wieder

$$S = -k \sum_{N_1} \ldots \sum_{N_m} \sum_n W \ln W , \tag{65.1}$$

wobei W die Verallgemeinerung der klassischen Phasendichte bzw. des Diagonalelements der Dichtematrix darstellt. Naturgemäß liegt die Frage nahe, ob die Gln. (47.7), (54.12) und (65.1) über die formale Übereinstimmung hinaus bei expliziter Berechnung zum gleichen Resultat führen. In unmittelbarem Zusammenhang damit steht eine weitere Frage. Offenbar bildet die Gl. (64.21) die statistische Grundlage für den durch Gl. (64.7) gegebenen thermodynamischen Zusammenhang zwischen dem Potential PV und der freien Energie nach Helmholtz. Da PV als thermodynamische Funktion für die Molekülzahlen $\overline{N}_1, \ldots, \overline{N}_m$ definiert ist, sollte man Gültigkeit von Gl. (64.7) erwarten, wenn die freie Energie

<hr>

[1] R. H. Fowler: Proc. Cambridge Phil. Soc. **34**, 382 (1938).
[2] F. Zernike: Diss. Amsterdam 1915. (Arch. Néerl., Ser. III A, Bd. I B.)
[3] R. H. Fowler: Proc. Cambridge Phil. Soc. **34**, 382 (1938).

nach HELMHOLTZ als $-kT \ln Q^{(N)}$ definiert wird. Man kann jedoch leicht zeigen, daß dies nicht zutrifft. Auch die Größe $-kT \overline{\ln Q^{(N)}}$ erfüllt nicht exakt die Gl. (64.7) und ist außerdem von der zuerst erwähnten Größe verschieden. Es ist daher zunächst unklar, wie man im Rahmen der großen kanonischen Gesamtheit überhaupt die freie Energie nach HELMHOLTZ zu definieren hat. Darüber hinaus scheint eine grundsätzliche Schwierigkeit in der statistischen Begründung der Thermodynamik aufzutreten.

Die Lösung der Schwierigkeit liegt in der Tatsache, daß die Gleichungen der Thermodynamik Grenzgesetze für unendlich große Systeme sind. Wenn man mit den statistischen Funktionen zur Grenze übergeht, erhält man für die mit Hilfe der verschiedenen Gesamtheiten definierten Größen die von der Thermodynamik geforderten Zusammenhänge. Die beim Grenzübergang verschwindenden Terme sind, wie man zeigen kann, durch die statistischen Schwankungen bedingt. Wir wollen daher die ganze Frage in allgemeiner Form in Zusammenhang mit der Schwankungstheorie in Teil C untersuchen und hier lediglich kurz die Schwankungen der Molekülzahlen betrachten, welche die große kanonische Gesamtheit von der kanonischen Gesamtheit unterscheiden.

Wir schreiben zunächst die Gl. (63.2)

$$\sum_{N_1} \cdots \sum_{N_m} e^{\frac{\Omega + \mu_1 N_1 + \cdots + \mu_m N_m - \psi}{\Theta}} = 1 \tag{65.2}$$

mit

$$e^{-\frac{\psi}{\Theta}} = \sum_n e^{-\frac{E_n}{\Theta}}. \tag{65.3}$$

Differentiation von (65.2) nach μ_i ergibt

$$\sum_{N_1} \cdots \sum_{N_m} \left(\frac{\partial \Omega}{\partial \mu_i} + N_i \right) e^{\frac{\Omega + \mu_1 N_1 + \cdots + \mu_m N_m - \psi}{\Theta}} = 0 \tag{65.4}$$

oder mit (63.6)

$$\overline{N}_i = -\frac{\partial \Omega}{\partial \mu_i}. \tag{65.5}$$

Durch nochmalige Differentiation von (65.4) nach μ_i erhält man mit Benutzung der Gl. (65.5)

$$\sum_{N_1} \cdots \sum_{N_m} \left[\frac{\partial^2 \Omega}{\partial \mu_i^2} + \frac{(N_i - \overline{N}_i)^2}{\Theta} \right] e^{\frac{\Omega + \mu_1 N_1 + \cdots + \mu_m N_m - \psi}{\Theta}} = 0 \tag{65.6}$$

und daraus

$$\overline{(N_i - \overline{N}_i)^2} = -\Theta \frac{\partial^2 \Omega}{\partial \mu_i^2} = \Theta \frac{\partial \overline{N}_i}{\partial \mu_i}. \tag{65.7}$$

Das mittlere relative Schwankungsquadrat der molekularen Dichte $\varrho_i = N_i/V$ wird dann

$$\frac{\overline{(\varrho_i - \overline{\varrho}_i)^2}}{\overline{\varrho}_i^2} = \frac{1}{\overline{N}_i} \frac{\Theta}{\overline{\varrho}_i} \frac{\partial \overline{\varrho}_i}{\partial \mu_i}. \tag{65.8}$$

Diese wichtige Beziehung zeigt, daß mit unbegrenzt wachsender Größe des Systems ($\overline{N}_i \to \infty$) die mittleren relativen Schwankungsquadrate der molekularen Dichten asymptotisch wie $(\overline{N}_i)^{-1}$ verschwinden[1]. Die große kanonische Gesamt-

---

[1] Dabei wird vorausgesetzt, daß $\partial \varrho_i / \partial \mu_i$ endlich ist, was im allgemeinen zutrifft. Den Fall, daß diese Voraussetzung nicht erfüllt ist, behandeln wir in Ziff. 76.

heit verhält sich dann praktisch wie eine kleine kanonische Gesamtheit mit den Molekülzahlen $\overline{N}_i$, und die nach beiden Methoden erhaltenen Ergebnisse werden identisch. Sie sind somit unabhängig von den speziellen Randbedingungen, wie es die Thermodynamik verlangt.

C. Allgemeine Begründung der Thermodynamik. Schwankungen. Phasenumwandlungen.

I. Begründung der Thermodynamik. Theorie der Schwankungen.

66. Generalisierte Zustandsgrößen. Für jede der bisher untersuchten statistischen Gesamtheiten (d.h. für die mikrokanonische, die kanonische und die große kanonische Gesamtheit) läßt sich, wie wir gesehen haben, eine Größe definieren, welche die Eigenschaften der thermodynamisch definierten Entropie besitzt. Diese Größen sind jedoch (obwohl sie sich durch eine gemeinsame Formel darstellen lassen) nicht untereinander gleich. Ähnliche Schwierigkeiten ergeben sich, wenn man versucht, mit Hilfe der großen kanonischen Gesamtheit die freie Energie nach HELMHOLTZ zu definieren (Ziff. 65). Diese Verhältnisse sind, wie schon erwähnt, durch die statistischen Schwankungen bedingt, welche bei den einzelnen Gesamtheiten in verschiedener Form auftreten. Wir haben an Beispielen gezeigt, daß unter gewissen Voraussetzungen die mittleren relativen Schwankungsquadrate asymptotisch verschwinden, wenn wir zur Grenze unendlich großer Systeme übergehen. Es ist daher zu erwarten, daß dann die mit Hilfe der verschiedenen Gesamtheiten definierten Größen identisch werden und damit eine von den Randbedingungen unabhängige Thermodynamik entsteht. Wir wollen dieses Problem jetzt in einer strengeren und allgemeineren Form behandeln und dabei gleichzeitig das zur Ergänzung der Sätze in Ziff. 12 bzw. Ziff. 40 notwendige Postulat formulieren, welches die Grundlage aller über die mikrokanonische Gesamtheit hinausgehenden Ansätze der statistischen Mechanik bildet[1] [26].

Die Methode, die wir benutzen, besteht im wesentlichen darin, daß wir mit Hilfe einer Transformationstheorie für die statistische Mechanik eine allgemeine Sprache definieren und diese zur Begründung der Thermodynamik verwenden. Naturgemäß muß dazu auch die Thermodynamik in einer allgemeinen Sprache formuliert werden. Wir beschreiben zunächst kurz diese Formulierung[2], die wir als Grundlage für das Folgende benötigen. Dabei muß eine von der früheren abweichende Notierung benutzt werden. Wir betrachten ein homogenes System aus m Komponenten. Alle thermodynamischen Eigenschaften desselben lassen sich ableiten aus einer Gleichung

$$E = E(S, X_2, \ldots, X_r) \qquad (r \geq m + 2) \tag{66.1}$$

die man nach GIBBS[3] die Fundamentalgleichung des Systems nennt[4]. E und S haben die gleiche Bedeutung wie früher; die X_i sind, wenn keine äußeren Felder vorhanden sind und Grenzflächenerscheinungen vernachlässigt werden können[5], das negative Volumen und die Molekülzahlen. Der Einfachheit halber setzen wir das System als isotrop voraus. Dann ist E homogen vom ersten Grade in S

---

[1] A. MÜNSTER: Z. Physik **136**, 179 (1953).
[2] H. A. C. McKAY: J. Chem. Phys. **3**, 715 (1935).
[3] J. W. GIBBS: On the Equilibrium of Heterogeneous Substances, Collected Works, Vol. I. New York 1948.
[4] GIBBS gebraucht diesen Ausdruck auch für die übrigen thermodynamischen Potentiale.
[5] Die letztere Voraussetzung machen wir durchweg im folgenden.

und den X_i. Diese Größen werden daher als *generalisierte Koordinaten* oder *extensive Parameter* bezeichnet. Die Ableitungen

$$T = \left(\frac{\partial E}{\partial S}\right)_{X_i}, \qquad P_i' = \left(\frac{\partial E}{\partial X_i}\right)_{S, X_j} \left.\begin{array}{l} \\ \\ \end{array}\right\} \qquad (66.2)$$
$$(i = 2, 3, \ldots, r) \qquad (j = 2, \ldots, i-1, i+1, \ldots, r)$$

bezeichnen wir als *generalisierte Kräfte* oder *intensive Parameter*. Wir haben dann

$$dE = T\,dS + \sum_{i=2}^{r} P_i'\,dX_i. \qquad (66.3)$$

Man kann nun einen oder mehrere intensive Parameter als unabhängige Variable einführen und erhält dann durch eine Folge von Legendre-Transformationen die thermodynamischen Potentiale

$$\Psi_k = E - TS - \sum_{i=2}^{k} P_i' X_i. \qquad (66.4)$$

Die hinreichende Bedingung für die Existenz dieser Transformation ist, daß die Jacobische Determinante nicht verschwindet, also

$$|E_{ij}| \equiv \frac{\partial(T, P_2', \ldots, P_k')}{\partial(S, X_2, \ldots, X_k)} \neq 0. \qquad (66.5)$$

Die Theorie der mikrokanonischen Gesamtheit führt, wie wir gesehen haben, auf eine zu (66.1) inverse Funktion, nämlich die Entropie. Für die statistische Mechanik ist es daher zweckmäßig, die Fundamentalgleichung in der Form

$$S = S(X_1, X_2, \ldots, X_r) \qquad (66.6)$$

zu schreiben. Dabei ist es zweckmäßig, unter X_1 die Energie zu verstehen. Das Volumen ist hier mit positivem Vorzeichen zu nehmen. (66.6) führt auf einen zweiten Satz von intensiven Parametern

$$P_i = \left(\frac{\partial S}{\partial X_i}\right)_{X_j}, \qquad (i = 1, 2, \ldots, r) \quad (j = 1, \ldots, i-1, i+1, \ldots, r) \qquad (66.7)$$

welche im einfachsten Falle den thermodynamischen Größen $1/T$, P/T, $-\mu_i/T$ entsprechen[1]. Durch Legendre-Transformation erhält man aus (66.6) die generalisierten Massieu-Planckschen Funktionen

$$\Phi_k = S - \sum_{i=1}^{k} P_i X_i. \qquad (66.8)$$

Die bekanntesten Beispiele dafür sind die Funktionen $-F/T$ und $-G/T$[2]. Die Bedingung für die Existenz dieser Transformation lautet

$$|S_{ij}| \equiv \frac{\partial(P_1, \ldots, P_k)}{\partial(X_1, \ldots, X_k)} \neq 0. \qquad (66.9)$$

---

[1] Die durch Gl. (66.7) verknüpften Größen X_i und P_i bezeichnen wir als konjugierte Parameter.

[2] Diese beiden Funktionen sind sechs Jahre vor den Gibbsschen Untersuchungen von Massieu [C. R. Acad. Sci., Paris **69**, 858 (1869)] eingeführt worden. Die letztere ist vor allem von Planck (vgl. Thermodynamik, 10. Aufl., Berlin 1954) viel verwendet worden. Guggenheim (Thermodynamics, 2nd ed. Amsterdam 1950) bezeichnet daher $-F/T$ als Massieusche Funktion, $-G/T$ als Plancksche Funktion. Vgl. auch vorstehenden Beitrag von Guggenheim, S. 1.

Für die Entropie gilt nach (66.6) und (66.7)

$$dS = \sum_{i=1}^{r} P_i \, dX_i. \tag{66.10}$$

Differenziert man Gl. (66.8) so folgt durch Kombination mit (66.10)

$$d\Phi_k = -\sum_{i=1}^{k} X_i \, dP_i + \sum_{j=k+1}^{r} P_j \, dX_j. \tag{66.11}$$

Daraus ergeben sich unmittelbar die generalisierten Maxwellschen Relationen

$$\frac{\partial^2 \Phi_k}{\partial P_i \, \partial X_j} = -\frac{\partial X_i}{\partial X_j} = \frac{\partial P_j}{\partial P_i}, \tag{66.12}$$

$$\frac{\partial^2 \Phi_k}{\partial P_i \, \partial P_m} = -\frac{\partial X_i}{\partial P_m} = -\frac{\partial X_m}{\partial P_i}, \tag{66.13}$$

$$\frac{\partial^2 \Phi_k}{\partial X_j \, \partial X_n} = \frac{\partial P_j}{\partial X_n} = \frac{\partial P_n}{\partial X_j}. \tag{66.14}$$

Aus (66.8) folgt

$$\Phi_k = \Phi_l - \sum_{i=l+1}^{k} P_i X_i \qquad (k > l). \tag{66.15}$$

Nun ist nach (66.11)

$$\frac{\partial \Phi_k}{\partial P_i} = -X_i. \tag{66.16}$$

Es wird somit

$$\Phi_k = \Phi_l + \sum_{i=l+1}^{k} P_i \frac{\partial \Phi_k}{\partial P_i}. \tag{66.17}$$

Dies ist die allgemeine Form der bekannten Gibbs-Helmholtzschen Gleichung.

67. Transformationstheorie der Verteilungsfunktionen. Die früher (Ziff. 47, 56, 64) abgeleiteten Gleichungen

$$S = k \ln \Omega, \tag{67.1}$$

$$-F/T = k \ln Q, \tag{67.2}$$

$$PV/T = k \ln \Xi \tag{67.3}$$

zeigen, daß jede der bisher untersuchten Gesamtheiten in der gleichen Weise mit einer Massieu-Planckschen Funktion verknüpft ist. Es liegt daher nahe, nun in analoger Weise zu den übrigen Massieu-Planckschen Funktionen weitere statistische Gesamtheiten zu konstruieren. Dieser Gedanke ist zuerst von Guggenheim[1] ausgeführt worden (vgl. auch [25]). Wir wollen ihn hier in einer systematischeren Form unter dem Gesichtspunkt einer Transformationstheorie entwickeln[2] [26]. Dabei beschränken wir uns auf den Fall der halbklassischen Näherung; die wichtigsten Ergebnisse sind jedoch auch für die strenge Quantenstatistik gültig.

Den Ausgangspunkt bildet auch hier die Theorie der mikrokanonischen Gesamtheit. Es ist jedoch zweckmäßig, im Hinblick auf das Folgende einige der früheren Formulierungen etwas abzuändern. Zunächst führen wir in die Grundgleichungen explizit alle extensiven Parameter (also nicht nur die Energie) ein.

---

[1] E. A. Guggenheim: J. Chem. Phys. **7**, 103 (1939).
[2] A. Münster: Z. Physik **136**, 179 (1953).

Zweitens betrachten wir alle extensiven Parameter (auch die Molekülzahlen) als stetig veränderliche Variable. Wir können dann definieren

$$\Omega^{**} = \int\limits_0^{X_r} dX_r \ldots \int\limits_0^{X_2} dX_2 \cdot \frac{1}{h^n \, \Pi \, N_i!} \int\limits_0^E d\Omega \qquad (67.4)$$

und

$$\Phi = \ln \frac{\partial^r \Omega^{**}}{\partial E \, \partial X_2 \ldots \partial X_r}. \qquad (67.5)$$

Die Größe Φ der Gl. (67.5) ist (bis auf den quantenstatistischen Korrekturfaktor) mit der durch Gl. (20.2) definierten Größe identisch. Für die Phasendichte haben wir jetzt zu schreiben

$$\varrho = C \cdot \delta(E^* - E)\, \delta(X_2^* - X) \ldots \delta(X_r^* - X). \qquad (67.6)$$

Die X_i^* sind hier vorgegebene Werte der äußeren Parameter. Die Normierungsbedingung (2.4) ergibt dann, wie früher,

$$C = e^{-\Phi}, \qquad (67.7)$$

wo Φ jetzt und im weiteren der Wert dieser Größe für $X_i = X_i^*$ (alle i) ist. Der mikrokanonische Mittelwert einer Phasenfunktion F ist dann

$$\overline{F} = e^{-\Phi} \int \delta(X_r^* - X_r)\, dX_r \ldots \int \delta(X_2^* - X_2)\, dX_2 \int F\, \delta(E^* - E)\, d\Omega, \qquad (67.8)$$

was lediglich eine formale Umschreibung der Gl. (20.6) darstellt. Damit ist der Zusammenhang mit der früheren Schreibweise festgelegt, und wir können die weiteren Ergebnisse von Abschnitt A IV einfach übernehmen. Eine besondere Bemerkung erfordern lediglich noch die Gln. (28.3) und (28.4). Die neuen Formulierungen ermöglichen es, den allgemeinen Fall zu betrachten, daß die beiden Teilsysteme eines abgeschlossenen Gesamtsystems nicht nur in bezug auf die Energie, sondern in bezug auf jeden anderen extensiven Parameter oder auch in bezug auf k extensive Parameter gleichzeitig im Austausch stehen. Dabei muß $k \leq r - 1$ sein, da ein extensiver Parameter zur Definition der Teilsysteme benötigt wird. An die Stelle der Gl. (28.3) tritt dann ein k-faches Faltungsintegral. Wir können daher schreiben

$$e^{\Phi} = e^{\Phi'} \binom{*}{k} e^{\Phi''} \qquad (67.9)$$

wo das Symbol $\binom{*}{k}$ ein k-faches Faltungsprodukt bezeichnet.

Der Einfachheit halber bezeichnen wir im folgenden mit S den Quotienten aus der Entropie und der Boltzmann-Konstanten. Wir haben dann

$$S = \Phi \qquad (67.10)$$

und die intensiven Parameter sind definiert durch

$$P_i = \overline{p}_i = \frac{\partial \Phi}{\partial X_i}. \qquad (67.11)$$

An Stelle der Gl. (24.8) schreiben wir jetzt

$$d\Phi = \sum_{i=1}^{r} \overline{p}_i\, dX_i = \sum_{i=1}^{r} P_i\, dX_i. \qquad (67.12)$$

Die $\overline{p}_i$ sind, wie wir für Temperatur (Ziff. 22) und äußere Kräfte (Ziff. 23) explizit gezeigt haben, Mittelwerte gewisser Phasenfunktionen. Nur diese Mittelwerte besitzen die Eigenschaften der thermodynamischen intensiven Parameter. Man kann zwar diesen Mittelwerten Schwankungsgrößen zuordnen, z.B. $\overline{(p_i-\bar{p}_i)^2}$. Dieselben beziehen sich aber auf die betreffenden Phasenfunktionen und nicht auf die als Mittelwerte definierten intensiven Parameter. Die mikrokanonische Gesamtheit ist somit dadurch ausgezeichnet, daß sie allen thermodynamischen Parametern scharfe Werte zuteilt und die Statistik sich (im klassischen Fall) nur auf den Phasenraum der generalisierten Koordinaten und Impulse bezieht. Wir haben lediglich „Schwankungen innerer Parameter".

Das Originalsystem, welches durch die mikrokanonische Gesamtheit abgebildet wird, ist experimentell ausschließlich durch Messungen extensiver Parameter bestimmt. Versucht man, einen mit Hilfe der Theorie der mikrokanonischen Gesamtheit berechneten intensiven Parameter, etwa die Temperatur, zu messen, so zeigt sich, daß dies nur möglich ist, indem man einen Energieaustausch des Systems mit dem Thermometer zuläßt. Das System ist also in bezug auf die Energie nicht mehr abgeschlossen, und die Energie wird damit notwendig unbestimmt. Wenn daher ein System experimentell durch eine Temperaturmessung charakterisiert ist, kann es nicht mehr durch eine mikrokanonische Gesamtheit abgebildet werden. Ähnlich liegen die Verhältnisse für die übrigen Paare von konjugierten Parametern. Will man etwa den Druck messen, so muß man, damit sich das Manometer einstellen kann, das Volumen veränderlich machen. Für die Messung des chemischen Potentials (bzw. der Aktivität oder Fugazität) muß das System mit einer zweiten Phase, welche die betreffende Komponente enthält, ins Gleichgewicht gebracht werden, so daß ein Austausch von Molekülen dieser Komponente möglich ist (Messung des Dampfdruckes, des osmotischen Druckes, des Potentials einer galvanischen Zelle). In all diesen Fällen werden die jeweiligen extensiven Parameter unbestimmt.

Der im vorstehenden beschriebene Sachverhalt bildet die Grundlage für die Formulierung des Postulates, welches die notwendige Ergänzung der in Ziff. 12 und 40 entwickelten Grundlagen bildet. Wir geben ihm die Fassung:

Durch die Messung eines intensiven Parameters P_i wird der dazu konjugierte extensive Parameter X_i notwendig unbestimmt. Das System wird dann nicht mehr durch die Funktion Φ, sondern durch eine generalisierte Verteilungsfunktion beschrieben, welche P_i als Parameter enthält und aus e^Φ durch Laplace-Transformation erhalten wird. Die generalisierte Verteilungsfunktion für k intensive Parameter lautet somit

$$\Xi_k \equiv e^{\Phi_k} = \int\limits_0^\infty \cdots \int\limits_0^\infty e^{-\sum\limits_{i=1}^k P_i X_i}\, e^\Phi \prod_{i=1}^k dX_i. \tag{67.13}$$

Der vorstehende Satz zeigt eine gewisse Analogie zu der quantenmechanischen Komplementarität zwischen kanonisch konjugierten Variablen. Der wesentliche Unterschied liegt darin, daß er nicht umkehrbar ist. Das bedeutet, daß die Messung eines extensiven Parameters den dazu konjugierten intensiven Parameter nicht unbestimmt macht und die Parameter P_1 bis P_k somit in jedem Falle scharfe Größen sind. Dieser Sachverhalt drückt sich in dem unterschiedlichen mathematischen Formalismus aus. Während die Fourier-Transformation von der q- in die p-Sprache sich völlig symmetrisch umkehren läßt, lautet die Umkehrung von (67.13)

$$e^\Phi = (2\pi i)^{-k} \int\limits_{c-i\infty}^{c+i\infty} \cdots \int e^{\sum\limits_{i=1}^k P_i X_i}\, e^{\Phi_k} \prod_{i=1}^k dP_i, \tag{67.14}$$

wo die P_i jetzt als komplexe Variable aufzufassen sind und die Integration über eine zur imaginären Achse parallele Gerade innerhalb des Regularitätsstreifens zu erstrecken ist. Da aber nur der Realteil von P_i physikalische Bedeutung besitzt, stellt (67.14) eine formale Operation dar, für welche die physikalische Größe $\Re(P_i)$ den konstanten Wert c hat. Dagegen werden im Falle der Gl. (67.13) alle nicht zu dem Satz P_1 bis P_k gehörenden intensiven Parameter (wir bezeichnen sie mit P_j) ebenfalls unbestimmt, da sie als Ableitungen von Φ definiert sind und diese Größe mit den Parametern X_1 bis X_k unscharf wird. Die Parameter P_1 bis P_k werden, im Gegensatz dazu, durch Gl. (67.13) neu definiert. In Ziff. 32 haben wir diese Verhältnisse am Beispiel der Temperatur-Analoga der mikrokanonischen und kanonischen Gesamtheit bereits ausführlich besprochen. Die Untersuchung der Frage, ob die obigen Laplace-Transformationen existieren, erfordert naturgemäß Aussagen über die Funktionen e^{Φ_k}. Wir werden im folgenden darauf nur insoweit eingehen, als sie von physikalischem Interesse ist, und verweisen im übrigen auf die mathematische Literatur[1].

Aus Gl. (67.13) folgt für die Wahrscheinlichkeitsdichte, an dem System Parameterwerte zwischen X_1 und $X_1 + dX_1, \ldots, X_k$ und $X_k + dX_k$ zu finden,

$$W_k = e^{-\Phi_k - \sum\limits_{i=1}^{k} P_i X_i} e^{\Phi}. \qquad (67.15)$$

Der physikalische Sinn der Gln. (67.13) und (67.15) besteht, wie nach dem früheren ohne weiteres klar ist, darin, daß jetzt das System in bezug auf die k Größen X_i offen gegenüber einem unendlich großen Reservoir ist. Jedes Φ_k gehorcht, wie man durch Differentiation von (67.13) (bei konstantem X_1 bis X_k) leicht ableitet, einer Differentialgleichung

$$d\Phi_k = -\sum\limits_{i=1}^{r} \overline{X_i}\, dP_i + \sum\limits_{j=k+1}^{k} \overline{P_j}\, dX_j, \qquad (67.16)$$

wo die Mittelwerte durch

$$\bar{u} = \int \cdots \int u\, e^{-\Phi_k - \sum\limits_{i=1}^{k} P_i X_i} e^{\Phi} \prod\limits_{i=1}^{k} dX_i \qquad (67.17)$$

definiert sind. Für ein Gesamtsystem, das aus zwei Teilsystemen mit vernachlässigbarer Wechselwirkung besteht, setzt Φ_k sich additiv aus den entsprechenden Funktionen der Teilsysteme Φ_k' und Φ_k'' zusammen. Nach dem Faltungssatz ist nämlich

$$e^{\Phi_k'} \cdot e^{\Phi_k''} = \int \cdots \int e^{-\sum\limits_{i=1}^{k} P_i X_i} e^{\Phi'} \binom{*}{k} e^{\Phi''} \prod\limits_{i=1}^{k} dX_i \qquad (67.18)$$

und daraus folgt mit (67.9) und (67.13)

$$\Phi_k = \Phi_k' + \Phi_k''. \qquad (67.19)$$

Φ_k besitzt daher, wenn die in Gl. (67.16) auftretenden Mittelwerte mit den entsprechenden thermodynamischen Zustandsgrößen identifiziert werden, die Eigenschaften einer generalisierten Massieu-Planckschen Funktion für die unabhängigen Variablen P_i, X_i. In Tabelle 1 sind die sich aus (67.13) ergebenden Spezialfälle (ohne Berücksichtigung äußerer Felder) in der üblichen Schreibweise zusammengestellt.

Von den in dieser Tabelle neu aufgeführten Gesamtheiten haben die meisten bisher keine selbständige Bedeutung erlangt. Lediglich die Gesamtheit für die unabhängigen Variablen $1/T$, P/T, N_i, welche der Funktion $-G/T$ entspricht,

---

[1] G. Doetsch: Handbuch der Laplace-Transformation. Basel 1950—1956.

Tabelle 1. *Generalisierte Verteilungsfunktionen und Massieu-Plancksche Funktionen.*

Verteilungsfunktion	Massieu-Plancksche Funktion	Differentialgleichung	Unabhängige Variable
e^{Φ}	S	$dS = \dfrac{dE}{T} + \dfrac{P}{T}\,dV - \sum_i \dfrac{\mu_i}{T}\,dN_i$	E, V, N_i
$\displaystyle\int e^{-\frac{E}{kT}}\,e^{\Phi}\,dE$	$S - \dfrac{E}{T} = -\dfrac{F}{T}$	$d\left(-\dfrac{F}{T}\right) = -E\,d\left(\dfrac{1}{T}\right) + \dfrac{P}{T}\,dV - \sum_i \dfrac{\mu_i}{T}\,dN_i$	$\dfrac{1}{T}, V, N_i$
$\displaystyle\int e^{-\frac{PV}{kT}}\,e^{\Phi}\,dV$	$S - \dfrac{PV}{T}$	$d\left(S - \dfrac{PV}{T}\right) = \dfrac{dE}{T} - V\,d\left(\dfrac{P}{T}\right) - \sum_i \dfrac{\mu_i}{T}\,dN_i$	$E, \dfrac{P}{T}, N_i$
$\displaystyle\int \cdots \int e^{\frac{\Sigma \mu_i N_i}{kT}}\,e^{\Phi}\prod_i dN_i$	$S + \dfrac{\Sigma \mu_i N_i}{T}$	$d\left(S + \dfrac{\Sigma \mu_i N_i}{T}\right) = \dfrac{dE}{T} + \dfrac{P}{T}\,dV + \sum_i N_i\,d\left(\dfrac{\mu_i}{T}\right)$	$E, V, -\dfrac{\mu_i}{T}$
$\displaystyle\iint e^{\frac{-E-PV}{kT}}\,e^{\Phi}\,dE\,dV$	$S - \dfrac{E}{T} - \dfrac{PV}{T} = -\dfrac{G}{T}$	$d\left(-\dfrac{G}{T}\right) = -E\,d\left(\dfrac{1}{T}\right) - V\,d\left(\dfrac{P}{T}\right) - \sum_i \dfrac{\mu_i}{T}\,dN_i$	$\dfrac{1}{T}, \dfrac{P}{T}, N_i$
$\displaystyle\int \cdots \int e^{\frac{-E+\Sigma \mu_i N_i}{kT}}\,e^{\Phi}\,dE\prod_i dN_i$	$S - \dfrac{E}{T} + \dfrac{\Sigma \mu_i N_i}{T} = \dfrac{PV}{T}$	$d\left(\dfrac{PV}{T}\right) = -E\,d\left(\dfrac{1}{T}\right) + \dfrac{P}{T}\,dV + \sum_i N_i\,d\left(\dfrac{\mu_i}{T}\right)$	$\dfrac{1}{T}, V, -\dfrac{\mu_i}{T}$
$\displaystyle\int \cdots \int e^{\frac{-PV+\Sigma \mu_i N_i}{kT}}\,e^{\Phi}\,dV\prod_i dN_i$	$S - \dfrac{PV}{T} + \dfrac{\Sigma \mu_i N_i}{T} = \dfrac{E}{T}$	$d\left(\dfrac{E}{T}\right) = \dfrac{dE}{T} - V\,d\left(\dfrac{P}{T}\right) + \sum_i N_i\,d\left(\dfrac{\mu_i}{T}\right)$	$E, \dfrac{P}{T}, -\dfrac{\mu_i}{T}$

Entnommen aus: A. Münster: Z. Physik **136**, 184 (1953).

ist gelegentlich benutzt worden[1,2]. Byers Brown[3] hat versucht, die Gesamtheiten, welche an Stelle von V als unabhängige Variable P/T haben (pressure ensembles), auf einem anderen Wege einzuführen. Diese Methode führt aber, wie man leicht zeigen kann[4], zu falschen Schwankungsformeln und muß daher als verfehlt betrachtet werden. Für $k=r$ divergiert das Laplace-Integral, was man leicht am Beispiel eines idealen Gases verifiziert[5]. Die daraus sich ergebende Folgerung, daß $\varXi_r$ nicht existiert, ist tatsächlich unmittelbar evident. Wenn nämlich das System in bezug auf sämtliche X_i offen ist, besitzt es keine Abgrenzung mehr gegen das Reservoir und ist daher nicht mehr definierbar. Man kann natürlich durch ein "cut off" die Konvergenz erzwingen [25]; es erscheint aber fraglich, ob damit viel gewonnen ist. Eine geistreiche Anwendung im Sinne eines mathematischen Kunstgriffes ist kürzlich von Longuet-Higgins[6] gemacht worden. Eine allgemeine Untersuchung dieses Problems hat Sack[7] durchgeführt.

Um auch für die Gl. (67.14) ein konkretes Beispiel zu geben, betrachten wir die Berechnung der kanonischen Verteilungsfunktion aus der großen Verteilungsfunktion eines Einkomponentensystems. Die Gl. (67.14) lautet dann

$$Q = \frac{1}{2\pi i} \int\limits_{c-i\infty}^{c+i\infty} e^{-\frac{N\mu}{kT}} \varXi\, d\left(-\frac{\mu}{kT}\right) \qquad (67.20)$$

wo $-\mu/kT$ jetzt als komplexe Variable aufzufassen ist. Wegen

$$Q = \frac{f(T)}{V} \frac{Q_\tau}{N!} \qquad (67.21)$$

wird daraus

$$\frac{Q_\tau}{N!} = \frac{V}{f(T)} \frac{1}{2\pi i} \int\limits_{c-i\infty}^{c+i\infty} e^{-\frac{N\mu}{kT}} \varXi\, d\left(-\frac{\mu}{kT}\right). \qquad (67.22)$$

Gehen wir nun mit Hilfe der konformen Abbildung

$$z = \frac{f(T)}{V} e^{\frac{\mu}{kT}} \qquad (67.23)$$

auf die z-Ebene über, so wird der Integrationsweg ein Kreis vom Radius $\frac{f(T)}{V} e^{-c}$ um den Ursprung, der im Sinne des Uhrzeigers zu durchlaufen ist. Aus Gl. (67.22) wird dann

$$\frac{Q_\tau}{N!} = \frac{1}{2\pi i} \oint \frac{\varXi}{z^{N+1}}\, dz, \qquad (67.24)$$

wo jetzt (wegen der Vorzeichenumkehrung) der Integrationsweg entgegen dem Sinne des Uhrzeigers zu durchlaufen ist. Der Integrand ist auf dem Integrationswege regulär und hat im Ursprung einen Pol $(N+1)$-ter Ordnung; der Wert des Integrals ist daher durch den Residuensatz gegeben. Für reale Gase (vgl. den Artikel von J.E. Mayer in Bd. XII dieses Handbuches) ist

$$\varXi = \exp\left(N \sum_l v\, b_l\, z^l\right), \qquad (67.25)$$

[1] L. van Hove: Physica, Haag **16**, 137 (1950).
[2] M. B. Lewis u. A. J. F. Siegert: Phys. Rev. **101**, 1227 (1956).
[3] W. Byers Brown: Mol. Phys. **1**, 68 (1958).
[4] A. Münster: Mol. Phys. **2**, 1 (1959).
[5] I. Prigogine: Physica, Haag **16**, 133 (1950).
[6] H. C. Longuet-Higgins: Mol. Phys. **1**, 83 (1958).
[7] R. A. Sack: Mol. Phys. **2**, 8 (1959). Für die Überlassung des Manuskriptes dieser Arbeit habe ich Herrn Dr. Sack herzlich zu danken.

wo die b_l die sog. cluster-Integrale sind und $v = V/N$ ist. Damit wird

$$\frac{Q_\tau}{N!} = \frac{1}{2\pi i} \oint \exp\left[N \sum v\, b_l\, z^l - (N+1)\ln z\right] dz. \tag{67.26}$$

Mit Hilfe der Sattelpunktmethode ergibt sich daraus als asymptotischer Wert für $N \to \infty$

$$\frac{Q_\tau}{N!} = \frac{\exp\left(N \sum v\, b_l\, Z^l\right)}{Z^N \left(2\pi N \sum l^2\, v\, b_l\, Z^l\right)^{\frac{1}{2}}}. \tag{67.27}$$

Die Sattelpunktskoordinate Z ist die durch Gl. (67.23) definierte (reelle) Fugazität, die jetzt als Funktion von T und v durch die Gleichung

$$\sum l\, v\, b_l\, Z^l = 1 \tag{67.28}$$

bestimmt ist.

68. Die Thermodynamik als asymptotischer Grenzfall. Die durch Gl. (67.16) definierten Massieu-Planckschen Funktionen entsprechen, wie schon mehrfach betont, verschiedenen physikalischen Situationen. Zu jeder dieser Situationen gehört zunächst sozusagen eine eigene Thermodynamik. Die Größe

$$S = - \int W_k' \ln W_k' \prod_{k+1}^{r} \delta(X_j^* - X_j)\, d\Omega^{**} \tag{68.1}$$

(mit $W_k' = W_k\, e^{-\Phi}$) besitzt in jedem Falle die Eigenschaften der Entropie. Die Ableitung von (66.12) bis (66.14) zeigt, daß auch die generalisierten Maxwellschen Relationen gelten. Die aus der Massieu-Planckschen Funktion abgeleiteten Größen sind aber jetzt ebenfalls nur für die betreffenden Randbedingungen definiert. Dies beruht darauf, daß in der statistischen Theorie die Massieu-Planckschen Funktionen nicht durch die Legendre-Transformation, sondern durch die Laplace-Transformation zusammenhängen. Das Problem der Begründung einer einheitlichen Thermodynamik läuft also auf die Untersuchung der Frage hinaus, ob und unter welchen Bedingungen diese Transformationen identisch werden.

Um diese Untersuchung durchzuführen[1], schreiben wir Gl. (67.13) in der zu (66.15) analogen Form

$$\Xi_k \equiv e^{\Phi_k} = \int \cdots \int e^{-\sum\limits_{i=l+1}^{k} P_i X_i + \Phi_l} \prod_{i=l+1}^{k} dX_i \qquad (l < k). \tag{68.2}$$

Die Funktion Φ_l entwickeln wir an der Stelle $\overline{X}_{l+1}, \ldots, \overline{X}_k$ in eine Taylorsche Reihe. Mit den Bezeichnungen

$$\delta X_i = X_i - \overline{X}_i \tag{68.3}$$

und

$$\Phi_l^0 = \Phi_l(\overline{X}_{l+1}, \ldots, \overline{X}_k) \tag{68.4}$$

haben wir dann

$$\Phi_l = \Phi_l^0 + \sum_{i=l+1}^{k} \left(\frac{\partial \Phi_l}{\partial X_i}\right)_0 \delta X_i + \frac{1}{2} \sum_i \sum_j \left(\frac{\partial^2 \Phi_l}{\partial X_i\, \partial X_j}\right)_0 \delta X_i\, \delta X_j + \cdots, \tag{68.5}$$

wo der Index 0 die Ableitung an der Stelle $\overline{X}_{l+1}, \ldots, \overline{X}_k$ bezeichnet. Im folgenden setzen wir die Existenz dieser Entwicklung und ihre gleichmäßige Konvergenz

---

[1] A. Münster: Z. Physik **136**, 179 (1953).

in dem ganzen in Betracht kommenden Bereich der Variablen voraus. Wir brechen mit dem quadratischen Gliede ab und erhalten dann aus (68.2)

$$\left. \begin{aligned} e^{\Phi_k} = e^{-\sum\limits_{i=l+1}^{k} P_i \bar{X}_i + \Phi_l^0} \times \\ \times \int \cdots \int e^{\sum\limits_{i=l+1}^{k} \left[\left(\frac{\partial \Phi_l}{\partial X_i} \right)_0 - P_i \right] \delta X_i + \frac{1}{2} \sum\limits_i \sum\limits_j \left(\frac{\partial^2 \Phi_l}{\partial X_i \partial X_j} \right)_0 \delta X_i \delta X_j} \prod\limits_{i=l+1}^{k} dX_i. \end{aligned} \right\} \quad (68.6)$$

Wir führen jetzt neue Variable

$$\xi_i = \frac{\delta X_i}{\bar{X}_i} \qquad (68.7)$$

ein und schreiben zur Abkürzung

$$B_{ij}^l = - \frac{1}{2} \bar{X}_i \bar{X}_j \left(\frac{\partial^2 \Phi_l}{\partial X_i \partial X_j} \right)_0 . \qquad (68.8)$$

Zwischen der zu (68.2) gehörenden Wahrscheinlichkeitsdichte $W_{k,l}$ und der entsprechenden Wahrscheinlichkeitsdichte im ξ-Raum $W_{k,l}(\xi)$ besteht die Beziehung

$$W_{k,l}(\xi) = \prod\limits_{i=l+1}^{k} \bar{X}_i W_{k,l}. \qquad (68.9)$$

Es ergibt sich somit

$$W_{k,l}(\xi) = C \exp \left\{ \sum\limits_{i=l+1}^{k} \left[\left(\frac{\partial \Phi_l}{\partial X_i} \right)_0 - P_i \right] \bar{X}_i \xi_i - \sum\limits_i \sum\limits_j B_{ij}^l \xi_i \xi_j \right\}, \qquad (68.10)$$

wo

$$C = \prod\limits_{i=l+1}^{k} \bar{X}_i \exp \left[- \Phi_k - \sum P_i \bar{X}_i + \Phi_l^0 \right] \qquad (68.11)$$

ist[1]. Die in (68.10) auftretende quadratische Form $\tilde{\xi} \, \mathsf{B}^l \, \xi$ verwandeln wir in eine Summe von Quadraten durch die kongruente Transformation

$$\xi = \mathsf{A} \, \eta, \qquad (68.12)$$

$$\tilde{\mathsf{A}} \mathsf{B}^l \mathsf{A} = [\alpha_{ij} \, \delta_{ij}] . \qquad (68.13)$$

Die Tatsache, daß diese Transformation nicht eindeutig ist, spielt hier keine Rolle, da die hier wesentliche Eigenschaft, die Zahl der positiven, negativen und verschwindenden Koeffizienten α_{ij} nach dem Sylvesterschen Trägheitssatz in jedem Falle erhalten bleibt. Für die α_{ij} gilt

$$\alpha_{l+1} = \Delta_{l+1}, \qquad \alpha_{l+2} = \Delta_{l+2}/\Delta_{l+1}, \ldots, \alpha_k = \Delta_k/\Delta_{k-1}, \qquad (68.14)$$

wo $\Delta_k \equiv |B_l|$ ist und die Δ_i die Hauptminoren dieser Determinante bezeichnen. Es ist ferner

$$\Pi \, d\xi_i = \frac{\partial (\xi_{l+1}, \ldots, \xi_k)}{\partial (\eta_{l+1}, \ldots, \eta_k)} \Pi \, d\eta_i = |A| \, \Pi \, d\eta_i. \qquad (68.15)$$

Die Transformation des ersten Terms im Exponenten von (68.10) ergibt

$$\sum\limits_i \left[\left(\frac{\partial \Phi_l}{\partial X_i} \right)_0 - P_i \right] \bar{X}_i \xi_i = 2 \sum\limits_j A_j \eta_j \qquad (68.16)$$

mit

$$A_j = \frac{1}{2} \sum\limits_i \left[\left(\frac{\partial \Phi_l}{\partial X_i} \right)_0 - P_i \right] \bar{X}_i A_{ij}. \qquad (68.17)$$

---

[1] Wenn C durch Gl. (68.11) definiert ist, erfüllt (68.10) naturgemäß nicht exakt die Normierungsbedingung (68.23). Die Integrationsgrenzen sind jetzt -1 und $+\infty$.

Um die Größe A_j zu untersuchen, gehen wir aus von der aus der Theorie der Laplace-Transformation bekannten Formel

$$P_{l+1} e^{\Phi_{l+1}} - e^{\Phi_l(0)} = \int e^{-P_{l+1}X_{l+1}+\Phi_l} \frac{\partial \Phi_l}{\partial X_{l+1}} dX_{l+1}. \tag{68.18}$$

Mit $e^{\Phi_l(0)} = 0$ kann diese Gleichung geschrieben werden

$$P_{l+1} = \overline{\frac{\partial \Phi_l}{\partial X_{l+1}}}, \tag{68.19}$$

wo sich die Mittelwertbildung auf die Verteilung mit $l+1$ intensiven Parametern bezieht. Wir entwickeln nun $\partial \Phi_l/\partial X_{l+1}$ an der Stelle $X_{l+1} = \overline{X}_{l+1}$. Das ergibt

$$\frac{\partial \Phi_l}{\partial X_{l+1}} = \left(\frac{\partial \Phi_l}{\partial X_{l+1}}\right)_0 + \left(\frac{\partial^2 \Phi}{\partial X_{l+1}^2}\right)_0 \delta X_{l+1} + \frac{1}{2}\left(\frac{\partial^3 \Phi_l}{\partial X_{l+1}^3}\right)_0 (\delta X_{l+1})^2 + \cdots. \tag{68.20}$$

Mit

$$\Phi_l/\overline{X}_{l+1} \equiv \varphi_l \tag{68.21}$$

und Benutzung von (68.7) kann (68.20) geschrieben werden

$$\frac{\partial \Phi_l}{\partial X_{l+1}} = \left(\frac{\partial \Phi_l}{\partial X_{l+1}}\right)_0 + \left(\frac{\partial^2 \varphi_l}{\partial \xi_{l+1}^2}\right)_0 \xi_{l+1} + \frac{1}{2}\left(\frac{\partial^3 \varphi_l}{\partial \xi_{l+1}^3}\right)_0 \xi_{l+1}^2 + \cdots. \tag{68.22}$$

Durch Einsetzen in Gl. (68.18) folgt dann mit (67.15) und der Normierungsbedingung

$$\int \cdots \int W_{k,l}(\xi) \prod d\xi_i = 1 \tag{68.23}$$

die Beziehung

$$P_{l+1} = \left(\frac{\partial \Phi_l}{\partial X_{l+1}}\right)_0 + \frac{1}{2}\left(\frac{\partial^3 \varphi_l}{\partial \xi_{l+1}^3}\right)_0 \overline{\xi_{l+1}^2} + \cdots. \tag{68.24}$$

Da $\overline{\xi_{l+1}^2}$ das mittlere relative Schwankungsquadrat des extensiven Parameters X_{l+1} ist, haben wir damit das Problem auf die Schwankungstheorie (Ziff. 69) zurückgeführt. Wir benötigen hier jedoch nur die Gleichung

$$\overline{\xi_{l+1}^2} = - \frac{1}{\overline{X}_{l+1}^2} \frac{\partial \overline{X}_{l+1}}{\partial P_{l+1}}. \tag{68.25}$$

Wir setzen nun voraus, daß die in Gl. (68.10) auftretende quadratische Form positiv definit und somit $|B^l|$ mit allen Hauptminoren positiv ist. Dann können wir (68.13) in der speziellen Gestalt

$$\tilde{A} B^l A = E \tag{68.26}$$

benutzen, wo E die Einheitsmatrix ist. Aus (68.26) folgt

$$|A| = |B^l|^{-\frac{1}{2}}. \tag{68.27}$$

Damit wird aus (68.6), wenn $-a_{l+1}, \ldots, -a_k$ die transformierten Integrationsgrenzen sind,

$$e^{\Phi_k} = e^{-\sum\limits_{i=l+1}^{k} P_i \overline{X}_i + \Phi_l^0} \frac{\prod \overline{X}_i}{|B^l|^{\frac{1}{2}}} \int\limits_{-a_{l+1}}^{\infty} \cdots \int\limits_{-a_k}^{\infty} e^{\sum\limits_{j}(2A_j \eta_j - \eta_j^2)} \prod_j d\eta_j. \tag{68.28}$$

Das Integral können wir für unsere Zwecke durch seinen maximalen Wert $\pi^{\frac{1}{2}(k-l)}$ ersetzen. Dann folgt

$$\Phi_k = \Phi_l^0 + \sum_{i=l+1}^{k} P_i \overline{X}_i - \frac{1}{2}\ln|B^l| + \sum_{i=l+1}^{k} \ln X_i + \sum A_j^2 + \frac{1}{2}(k-l)\ln \pi. \tag{68.29}$$

Aus Gl. (68.25) sieht man, daß die A_j beschränkt sind, wenn dies für $(\overline{X}_j)^{-1}\partial\overline{X}_j/\partial P_j$ gilt. Daß letzteres unter den gemachten Voraussetzungen zutrifft, werden wir in Ziff. 69 zeigen. Nach Division durch einen mittleren extensiven Parameter[1] erhalten wir daher für die spezifischen Massieu-Planckschen Funktionen

$$\varphi_k = \varphi_l^0 - \sum_{i=l+1}^{k} P_i \bar{x}_i + O(\overline{X}_{l+1}^{-1} \sum \ln \overline{X}_i),\qquad(68.30)$$

wo

$$\bar{x}_i = \overline{X}_i/\overline{X}_{l+1}\qquad(68.31)$$

ist.

Die Gl. (68.30) zeigt, daß die Laplace-Transformation der statistischen Theorie für $\overline{X}_i \to \infty$ in die Legendre-Transformation der Thermodynamik übergeht, wenn

a) die Entwicklungen (68.5) existieren und in dem gesamten Variablenbereich gleichmäßig konvergieren,

b) $|B^l/\overline{X}_{l+1}|$ und alle Hauptminoren > 0 sind[2],

c) die Grenzwerte der spezifischen Massieu-Planckschen Funktionen existieren.

Die Thermodynamik stellt somit den Grenzfall dar, dem sich die statistische Theorie nähert, wenn das betrachtete System beliebig groß wird. Die physikalische Voraussetzung für die Anwendbarkeit der Thermodynamik besteht, wie Gl. (68.24) unmittelbar zeigt, darin, daß die statistischen Schwankungen um die Mittelwerte vernachlässigt werden können.

Die früher mit Hilfe spezieller Verteilungen erhaltenen Ergebnisse (empirische Temperatur, thermodynamisches Gleichgewicht usw.) können nun einfach übernommen werden, so daß sich eine nochmalige Ableitung an dieser Stelle erübrigt.

Im folgenden behandeln wir als notwendige Ergänzung der vorstehenden Ableitung zunächst die Schwankungstheorie. Die Bedingungen b) und c) sind eng mit dem Problem der Phasenumwandlungen verknüpft. Wir werden sie daher in Abschnitt C II ausführlich erörtern.

69. Schwankungen extensiver Parameter. Die statistische Begründung der Thermodynamik erfordert, wie wir gesehen haben, eine Untersuchung der Schwankungserscheinungen. Dieselben führen darüber hinaus zu einer Reihe von direkt beobachtbaren Effekten (Brownsche Bewegung, Lichtstreuung von Gasen und Flüssigkeiten, Schroteffekt usw.). Schließlich sind sie auch im Zusammenhang mit den allgemeinen Fragen, die wir in Abschnitt C II behandeln, von entscheidender Bedeutung. Wir wollen daher die Formulierungen der Ziff. 67 benutzen, um die Theorie der Schwankungen in einer ganz allgemeinen Form zu entwickeln[3].

Wir betrachten zunächst die Schwankungen der extensiven Parameter. Für die mikrokanonische Gesamtheit besitzen alle extensiven Parameter feste Werte, so daß hier (d.h. physikalisch in einem abgeschlossenen System) diese Schwankungen nicht auftreten[4]. Gehen wir aber zu den durch Gl. (67.15) definierten Gesamtheiten über, so ist das Originalsystem in bezug auf wenigstens einen extensiven Parameter offen. Der Wert desselben ist nur statistisch bestimmt; wir können aus Gl. (67.15) seinen Mittelwert und die Schwankungen um diesen Mittel-

---

[1] Der Parameter $\overline{X}_{l+1}$ kann etwa das Volumen oder die Molekülzahl einer Komponente sein. Er kann auch durch die Summe aller oder einiger extensiver Parameter (z.B. der Molekülzahlen) ersetzt werden.

[2] Die Division der Determinanten-Elemente durch $\overline{X}_{l+1}$ (vgl. Fußnote 1) erfolgt im Hinblick auf Ziff. 76.

[3] A. MÜNSTER: Z. Physik **136**, 179 (1953).

[4] Dagegen sind allen mit Hilfe der mikrokanonischen Gesamtheit gebildeten Mittelwerten naturgemäß auch Schwankungen zugeordnet.

wert berechnen. Dabei ist zu beachten, daß in den thermodynamischen Gleichungen nur die Mittelwerte dieser Größen auftreten. Es hat daher, obwohl es sich physikalisch um die gleichen Größen (z.B. Molekülzahlen) handelt, keinen Sinn (wie es gelegentlich geschieht) von thermodynamischen Schwankungen zu sprechen.

Eine allgemeine Formel, welche ein beliebiges Korrelationsmoment mit den Parametern der Verteilung verknüpft, läßt sich nicht angeben. Man kann die Berechnung nur mit Hilfe von Iterationsverfahren durchführen. Ein solches ergibt sich unmittelbar aus der Theorie der Laplace-Transformation. Es ist nämlich, wenn wir uns der Einfachheit halber auf den Fall einer Variablen beschränken,

$$(-1)^n \frac{\partial n}{\partial P_{l+1}^n} e^{\Phi_{l+1}} = \int e^{-P_{l+1} X_{l+1} + \Phi_l} X_{l+1}^n \, dX_{l+1}. \tag{69.1}$$

Daraus folgt für $n=1$

$$\frac{\partial \Phi_{l+1}}{\partial P_{l+1}} = -\overline{X}_{l+1}. \tag{69.2}$$

Für $n=2$ bekommt man mit Benutzung von (69.2)

$$\frac{\partial^2 \Phi_{l+1}}{\partial P_{l+1}^2} = \overline{(X_{l+1} - \overline{X}_{l+1})^2} \tag{69.3}$$

oder mit (67.16) und (68.3)

$$\overline{(\delta X_{l+1})^2} = -\frac{\partial \overline{X}_{l+1}}{\partial P_{l+1}}. \tag{69.4}$$

Man sieht leicht, daß die früher abgeleiteten Gl. (32.11) und (65.7) Spezialfälle von (69.4) sind. Die Weiterführung des Verfahrens und die Verallgemeinerung für mehrere Variable liegen auf der Hand.

Eine allgemeine Rekursionsformel läßt sich in folgender Weise ableiten. Aus Gl. (62.15) folgt

$$\begin{aligned} \frac{\partial}{\partial P_j} \int \cdots \int \prod_i (\delta X_i)^{r_i} e^{-\sum_i P_i X_i + \Phi_l} \prod_i dX_i \\ = \int \cdots \int \left\{ -X_j \prod_i (\delta X_i)^{r_i} - \sum_i \left[r_i (\delta X_i)^{r_i-1} \prod_m (\delta X_m)^{r_m} \frac{\partial \overline{X}_i}{\partial P_j} \right] \right\} e^{-\sum_i P_i X_i + \Phi_l} \prod_i dX_i \\ (m \neq i) \quad (\textstyle\sum r_i = n). \end{aligned} \tag{69.5}$$

Schreiben wir für ein Korrelationsmoment n-ter Ordnung zur Abkürzung

$$\langle \delta X_i \rangle_n = \overline{\prod_i (\delta X_i)^{r_i}}, \qquad (\textstyle\sum r_i = n), \tag{69.6}$$

so wird aus (69.5)

$$\frac{\partial}{\partial P_j} \left[\langle \delta X_i \rangle_n e^{\Phi_k} \right] = \left[-\overline{X_j \prod (\delta X_i)^{r_i}} - \sum_i \overline{r_i (\delta X_i)^{r_i-1} \prod_m (\delta X_m)^{r_m}} \frac{\partial \overline{X}_i}{\partial P_j} \right] e^{\Phi_k}. \tag{69.7}$$

Mit Benutzung von (67.16) folgt

$$\frac{\partial}{\partial P_j} \langle \delta X_i \rangle_n - \langle \delta X_i \rangle_n \overline{X}_j = -\overline{X_j \prod_i (\delta X_i)^{r_i}} - \sum_i \overline{r_i (\delta X_i)^{r_i-1} \prod_m (\delta X_m)^{r_m}} \frac{\partial \overline{X}_i}{\partial P_j}. \tag{69.8}$$

Nun ist

$$\overline{X_j \prod_i (\delta X_i)^{r_i}} - \overline{X}_j \langle \delta X_i \rangle_n = \langle \delta X_i \rangle_{n+1}, \tag{69.9}$$

wo $\langle \delta X_i \rangle_{n+1}$ sich von $\langle \delta X_i \rangle_n$ dadurch unterscheidet, daß r_j durch $r_j + 1$ ersetzt wird. Damit ergibt sich aus (69.8) die Rekursionsformel

$$\langle \delta X_i \rangle_{n+1} = - \frac{\partial}{\partial P_j} \langle \delta X_i \rangle_n - \sum_i r_i \langle \delta X_i \rangle_{n-1} \frac{\partial \overline{X}_i}{\partial P_j}, \tag{69.10}$$

wo unter $\langle \delta X_i \rangle_{n-1}$ alle Korrelationsmomente zu verstehen sind, die aus $\langle \delta X_i \rangle_n$ dadurch hervorgehen, daß jeweils ein r_i durch $r_i - 1$ ersetzt wird. Gl. (69.10) ist allgemein gültig für $n \geqq 1$. Sie führt auf Gl. (69.4), wenn man beachtet, daß $\langle \delta X_i \rangle_1 = 0$ und $\langle \delta X_i \rangle_0 = 1$ ist.

Als Beispiel für die Anwendung der Gl. (69.10) betrachten wir die Schwankungen der Molekülzahl der Komponente i in einem System, das durch eine große kanonische Gesamtheit dargestellt wird. Für diesen Fall lautet die Gl. (69.10)

$$\overline{(N_i - \overline{N}_i)^{n+1}} = \Theta \frac{\partial}{\partial \mu_i} \overline{(N_i - \overline{N}_i)^n} + n \, \overline{(N_i - \overline{N}_i)^{u-1}} \, \Theta \, \frac{\partial \overline{N}_i}{\partial \mu_i}. \tag{69.11}$$

Daraus ergibt sich, wenn wir den Operator

$$\Delta_i = \Theta \frac{\partial}{\partial \mu_i} \tag{69.12}$$

einführen,

$$\overline{(N_i - \overline{N}_i)^2} = \Delta_i \overline{N}_i, \tag{69.13}$$

$$\overline{(N_i - \overline{N}_i)^3} = \Delta_i^2 \overline{N}_i, \tag{69.14}$$

$$\overline{(N_i - \overline{N}_i)^4} = \Delta_i^3 \overline{N}_i + 3 \, (\Delta_i \overline{N}_i)^2 \tag{69.15}$$

usw.

Unter den Schwankungsgrößen sind weitaus am wichtigsten die zweiten Momente. Wir wollen diese daher noch etwas eingehender betrachten. Aus Gl. (69.10) folgt zunächst in Verbindung mit (68.7)

$$\overline{\xi_i \xi_j} = - \frac{1}{2} \, \frac{2}{\overline{X}_i \overline{X}_j} \, \frac{\partial \overline{X}_i}{\partial P_i}. \tag{69.16}$$

Die rechte Seite dieser Gleichung (ohne den Faktor $\frac{1}{2}$) fassen wir als Element einer Matrix $\mathbf{Q}$ auf. Bezeichnen wir die zu $\mathbf{Q}$ reziproke Matrix mit $\mathbf{B}^*$, so ist

$$B_{ij}^* = - \frac{1}{2} \overline{X}_i \overline{X}_j \frac{\partial P_j}{\partial \overline{X}_i}. \tag{69.17}$$

Zwischen $\mathbf{Q}$ und $\mathbf{B}^*$ besteht die Beziehung

$$\mathbf{Q} = \frac{\hat{\mathbf{B}}^*}{|B^*|}, \tag{69.18}$$

wo $\hat{\mathbf{B}}^*$ aus $\mathbf{B}^*$ dadurchentsteht, daß zuerst zum Element B_{ij}^* in $|B^*|$ der Kofaktor gebildet und die dann resultierende Matrix transponiert wird. Damit wird aus (69.16)

$$\overline{\xi_i \xi_j} = \frac{1}{2} \frac{|B^*|_{ij}}{|B^*|}, \tag{69.19}$$

wo $|B^*|_{ij}$ der Kofaktor des Matrixelementes B_{ij}^* ist. Für diese Transformation muß vorausgesetzt werden, daß die Matrix $\mathbf{B}^*$ nicht singulär und somit

$$|B^*| \neq 0 \tag{69.20}$$

ist.

Für manche Überlegungen ist es zweckmäßig, die zweiten Momente direkt mittels der Wahrscheinlichkeitsdichte (67.15) unter Benutzung der Entwicklung (68.5) zu berechnen. Dieses Verfahren geht im Prinzip auf Smoluchowski[1] und Einstein[2] zurück. Wir brechen auch hier die Reihe mit dem quadratischen Gliede ab und setzen ferner die zweiten Momente als so klein voraus, daß im Hinblick auf Gl. (68.24) die Gl. (68.10) mit hinreichender Näherung geschrieben werden kann

$$W_{k,l}(\xi) = C' \exp\left(- \sum_i \sum_j B_{ij}^l \xi_i \xi_j\right). \tag{69.21}$$

Für den Normierungsfaktor C' ergibt sich dann aus (68.23) mit Benutzung der Transformation (68.26)

$$C' = \left[\frac{|B^l|}{\pi^{(k-l)}}\right]^{\frac{1}{2}}. \tag{69.22}$$

Damit wird

$$\overline{\xi_j \xi_i} = - \left[\frac{|B^l|}{\pi^{(k-l)}}\right]^{\frac{1}{2}} \frac{\partial}{\partial B_{ij}^l} \int \cdots \int \exp\left(- \sum_i \sum_j B_{ij}^l \xi_i \xi_j\right) \prod_{i=l+1}^{k} d\xi_i \tag{69.23}$$

oder

$$\overline{\xi_j \xi_i} = \frac{1}{2} \frac{|B^l|_{ij}}{|B^l|}. \tag{69.24}$$

Aus Gl. (69.16) folgt zunächst, daß die zweiten Momente für $\overline{X}_j \to \infty$ wie $(\overline{X}_j)^{-1}$ verschwinden, wenn $\partial^2 \Phi_k/\partial P_i\, \partial P_j$ positiv endlich ist. Analoge Aussagen ergeben sich, wie man leicht verifiziert, für die höheren Momente. Damit folgt aus Gl. (68.24)

$$\lim_{\text{alle }\overline{X}_i \to \infty} [\mathbf{B}^* - \mathbf{B}^l] = \mathbf{O}. \tag{69.25}$$

Wir wollen nun [abweichend von Gl. (69.21)] mit $W_{k,l}(\xi)$ den *exakten* Ausdruck für die Wahrscheinlichkeitsdichte im ξ-Raum bezeichnen. Dann folgt aus Gl. (69.19), (69.24) und (69.25)

$$\lim_{\text{alle }\overline{X}_i \to \infty} \left\{\int \cdots \int \xi_i \xi_j W_{k,l}(\xi) \prod d\xi_i - \left[\frac{|B^l|}{\pi^{(k-l)}}\right]^{\frac{1}{2}} \int \cdots \int \xi_i \xi_j\, e^{-\sum_i \sum_j B_{ij}^l \xi_i \xi_j} \prod_i d\xi_i\right\} = 0. \tag{69.26}$$

Die Wahrscheinlichkeitsdichte $W_{k,l}(\xi)$ nähert sich also asymptotisch der Normalverteilung (69.21). Daraus folgt, daß die Bedingung b (am Schluß der Ziff. 68) notwendig und hinreichend dafür ist, daß für $\overline{X}_j \to \infty$ die zweiten Momente wie $(\overline{X}_j)^{-1}$ verschwinden. Unter dieser Voraussetzung ist $(\overline{X}_j)^{-1} \partial \overline{X}_j/\partial P_j$ und damit auch die durch Gl. (68.17) definierte Größe A_j beschränkt, was wir bereits in Ziff. 68 benutzt haben.

70. Schwankungen intensiver Parameter. Die Theorie der Schwankungen intensiver Parameter hat nicht die weitreichende allgemeine Bedeutung wie die der Schwankungen extensiver Parameter. Sie ist aber für sich von Interesse und auch dadurch bemerkenswert, daß sich in der Literatur viele einander widersprechende Auffassungen darüber finden. Wir wollen daher etwas näher darauf eingehen. Zunächst fassen wir noch einmal die hier wesentlichen Punkte aus den vorhergehenden Untersuchungen zusammen. In der mikrokanonischen Gesamtheit gibt es Schwankungen von Phasenfunktionen, aber keine Schwankungen intensiver Parameter, da diese als Mittelwerte von Phasenfunktionen bzw. Ableitungen von Φ (oder $\ln \Omega^*$) definiert sind[3]. In einer Gesamtheit mit

---

[1] M. v. Smoluchowski: Ann. Physik **25**, 205 (1908).
[2] A. Einstein: Ann. Physik **33**, 1275 (1910).
[3] Eine andere Ansicht findet sich bei ter Haar [22].

k unscharfen extensiven Parametern X_i besitzen die dazu konjugierten intensiven Parameter P_i definitionsgemäß scharfe Werte als Parameter der Verteilung. Nur für die intensiven Parameter P_j, welche den jeweils fixierten Parametern X_j konjugiert sind, lassen sich Schwankungen definieren. Es gilt also der Satz: Von zwei konjugierten Parametern kann stets nur einer Schwankungen zeigen. Für ein System, das durch eine kanonische Gesamtheit dargestellt wird und somit durch eine Temperaturmessung charakterisiert ist, sind daher Schwankungen der Temperatur nicht definierbar. Dem widerspricht auch nicht die Tatsache, daß ein hinreichend empfindliches ideales Thermometer zweifellos Schwankungen anzeigen würde. Jedes „Thermometer" zeigt nämlich primär Änderungen der Energie und damit gegebenenfalls auch Schwankungen der Energie an. Die Messung der Temperatur kann daher, streng genommen, nur so definiert werden, daß eine Statistik der Energiewerte aufgenommen und daraus der Parameter der Verteilung errechnet wird. Lediglich bei offenen Systemen mit fixierter Energie können Temperaturschwankungen im eigentlichen Sinne auftreten. Derartige Systeme sind aber, wie schon GUGGENHEIM[1] bemerkt hat, physikalisch kaum realisierbar. Die in der Literatur[2,3] verschiedentlich diskutierten Temperaturschwankungen beruhen auf der Annahme, daß man in Gl. (69.21) für die ξ_i beliebige thermodynamische Variable einsetzen könne. Diese Annahme ist jedoch ganz willkürlich und berücksichtigt nicht die statistische Definition der thermodynamischen Größen.

Um die Schwankungen der intensiven Parameter zu berechnen, gehen wir aus von Gl. (67.13), die wir schreiben

$$\int \cdots \int e^{-\Phi_k - \sum\limits_{i=1}^{k} P_i X_i + \Phi} \prod\limits_{i=1}^{k} dX_i = 1 . \tag{70.1}$$

Daraus folgt durch Differentiation nach dem extensiven Parameter $X_j (j > k)$

$$\int \cdots \int e^{-\Phi_k - \sum\limits_{i=1}^{k} P_i X_i + \Phi} \left(-\frac{\partial \Phi_k}{\partial X_j} + \frac{\partial \Phi}{\partial X_j} \right) \prod\limits_{i=1}^{k} dX_i = 0 \tag{70.2}$$

oder mit (67.11)

$$\frac{\partial \Phi_k}{\partial X_j} = \overline{P_i} \tag{70.3}$$

in Übereinstimmung mit (67.16). Nachmalige Differentiation von (70.2) nach X_j ergibt

$$\int \cdots \int e^{-\Phi_k - \sum\limits_{i=1}^{k} P_i X_i + \Phi} \left[\left(-\frac{\partial \Phi_k}{\partial X_j} + \frac{\partial \Phi}{\partial X_j} \right)^2 + \left(-\frac{\partial^2 \Phi_k}{\partial X_j^2} + \frac{\partial^2 \Phi}{\partial X_j^2} \right) \right] \prod\limits_{i=1}^{k} dX_i = 0 \tag{70.4}$$

oder mit (70.3)

$$\overline{(P_j - \overline{P_j})^2} = \frac{\partial \overline{P_j}}{\partial X_j} - \frac{\partial \overline{P_j}}{\partial X_j} . \tag{70.5}$$

Differenzieren wir (70.2) nach einem zweiten extensiven Parameter aus der Reihe X_{k+1} bis X_r, den wir mit $X_{j'}$ bezeichnen, so folgt

$$\overline{(P_j - \overline{P_j}) (P_{j'} - \overline{P_{j'}})} = \frac{\partial \overline{P_j}}{\partial X_{j'}} - \frac{\overline{\partial P_j}}{\partial X_{j'}} = \frac{\partial \overline{P_{j'}}}{\partial X_j} - \frac{\overline{\partial P_{j'}}}{\partial X_j} . \tag{70.6}$$

---

[1] E. A. GUGGENHEIM: J. Chem. Phys. **7**, 103 (1939).
[2] L. LANDAU u. E. LIFSHITZ: Statistical Physics. Oxford 1938.
[3] M. v. LAUE: Phys. Z. **18**, 542 (1917).

Schließlich ergibt die Differentiation der Gl. (70.2) nach einem intensiven Parameter P_i $(i = 1, \ldots, k)$

$$\int \cdots \int e^{-\Phi_k - \sum\limits_{i=1}^{k} P_i X_i + \Phi} \left[\left(-\frac{\partial \Phi_k}{\partial X_j} + \frac{\partial \Phi}{\partial X_j} \right) \left(-\frac{\partial \Phi_k}{\partial P_i} - X_i \right) - \frac{\partial^2 \Phi_k}{\partial X_j \partial P_i} \right] \prod_{i=1}^{k} dX_i = 0 \quad (70.7)$$

oder

$$(P_j - \overline{P}_j)(X_i - \overline{X}_i) = -\frac{\partial^2 \Phi_k}{\partial X_j \partial P_i} = -\frac{\partial \overline{P}_j}{\partial P_i} = -\frac{\partial \overline{X}_i}{\partial X_j}. \quad (70.8)$$

Von den auf der rechten Seite der Gl. (70.5) stehenden Größen hat die erste eine einfache thermodynamische Bedeutung: Sie stellt die Änderung des mittleren intensiven Parameters bei Änderung des konjugierten äußeren Parameters dar, also beispielsweise für eine kanonische Gesamtheit die Änderung des durch den Verteilungsmodul Θ dividierten mittleren Druckes mit dem Volumen bei konstanten Θ. Dagegen läßt sich der zweite Term nicht in der Sprache der Thermodynamik ausdrücken. Er ist nur rechnerisch, unter konkreten Annahmen über die Natur des Systems, zugänglich. Es ist daher nicht möglich, über die Schwankungen der intensiven Parameter allgemeinere Aussagen zu machen. Man kann sie, im Gegensatz zu den Schwankungen extensiver Parameter, weder mit den thermodynamischen Zustandsgrößen verknüpfen noch ihr asymptotisches Verhalten für unendlich große Systeme in allgemeiner Form diskutieren. Die Gl. (70.6) zeigt, daß die Schwankungen der verschiedenen intensiven Parameter im allgemeinen nicht unabhängig voneinander sind. Im übrigen liegen die Verhältnisse jedoch ähnlich wie bei Gl. (70.5). Für die Schwankungsgröße der Gl. (70.8) treten diese Schwierigkeiten nicht auf. Die Gleichung zeigt, daß zwischen den Schwankungen eines extensiven und eines intensiven Parameters notwendig Korrelation besteht. Man kann daher sagen, daß das "Öffnen" des Systems in bezug auf gewisse extensive Parameter zwangsläufig Schwankungen der nicht fixierten intensiven Parameter nach sich zieht.

Wir wollen nun das konkrete Beispiel der *Schwankungen des Druckes* in einer kanonischen Gesamtheit etwas ausführlicher betrachten und dabei auch kurz auf die in der Literatur vertretenen Auffassungen eingehen. Wir bedienen uns dabei der Notierung der Abschnitte A IV und A V, bezeichnen aber sofort mit V das Volumen und mit

$$P = -\left(\frac{\partial E}{\partial V} \right)_{p, q} \quad (70.9)$$

die entsprechende generalisierte Kraft. Mikrokanonische Mittelwerte kennzeichnen wir durch *einen*, kanonische durch *zwei* Querstriche. Durch zweimalige Differentiation der Gleichung

$$\int e^{\frac{\psi - E}{\Theta}} \, d\Omega = 1 \quad (70.10)$$

erhält man, wie schon Gibbs [2] gezeigt hat,

$$\int e^{\frac{\psi - E}{\Theta}} \left[\left(\frac{\partial^2 \psi}{\partial V^2} \right)_{\Theta} - \left(\frac{\partial^2 E}{\partial V^2} \right)_{p, q} + \frac{1}{\Theta} \left(\frac{\partial \psi}{\partial V} - \frac{\partial E}{\partial V} \right)^2 \right] d\Omega = 0 \quad (70.11)$$

oder wegen (70.10) und

$$\left(\frac{\partial \psi}{\partial V} \right)_{\Theta} = \overline{\overline{\left(\frac{\partial E}{\partial V} \right)_{p, q}}} = -\overline{\overline{P}}, \quad (70.12)$$

$$\overline{\overline{(P - \overline{\overline{P}})^2}} = -\Theta \left(\frac{\partial^2 \psi}{\partial V^2} - \overline{\overline{\frac{\partial^2 E}{\partial V^2}}} \right) = \Theta \left(\frac{\partial \overline{\overline{P}}}{\partial V} - \overline{\overline{\frac{\partial P}{\partial V}}} \right). \quad (70.13)$$

Man sieht, daß im Prinzip die Verhältnisse hier ganz ähnlich liegen wie bei Gl. (70.5). Auch in (70.13) hat der erste Term der rechten Seite eine einfache thermodynamische Bedeutung, während der zweite nur über explizite Rechnung zugänglich ist. Eine solche Rechnung ist wohl zuerst von FOWLER [7] durchgeführt worden unter den beiden Annahmen, daß es sich um ein ideales Gas handelt und das Potential an der Wand des Behälters die Form

$$u_w = C \cdot r^{-6} \tag{70.14}$$

hat, wo r den Abstand von der Wand und C eine Konstante bezeichnet. Es ergibt sich dann

$$\frac{\overline{\overline{(P - \overline{\overline{P}})^2}}}{\overline{\overline{P}}{}^2} \sim N^{-\frac{2}{3}}. \tag{70.15}$$

Das mittlere relative Schwankungsquadrat verschwindet somit wieder asymptotisch für unendlich große Systeme, aber von niedrigerer Ordnung als bei extensiven Parametern. WERGELAND[1,2] hat eine analoge Rechnung für einen kugelförmigen Behälter vom Radius R mit dem Wandpotential

$$u_w = \begin{cases} C \cdot (r - R) & \text{für } r > R \\ 0 & \text{für } r < R \end{cases} \tag{70.16}$$

durchgeführt, das für $C \rightarrow \infty$ unendlich steil wird (vollkommen starre Wand). Er erhält

$$\overline{\overline{\left(\frac{\partial^2 E}{\partial V^2}\right)}}_{p,q} = \frac{N\Theta}{V} \left[\frac{C}{\Theta} \cdot \left(\frac{V}{36\pi}\right)^{\frac{1}{3}} + \cdots \right]. \tag{70.17}$$

Nach Gl. (70.13) muß daher $\overline{\overline{(P - \overline{\overline{P}})^2}}$ mit C gegen Unendlich gehen. Dieses spezielle Ergebnis wie überhaupt die starke Abhängigkeit der durch Gl. (70.13) definierten Schwankungsgröße vom Wandpotential erscheinen zunächst etwas befremdend. Die Erklärung liegt in der Tatsache, daß P durch Gl. (70.9) als rein mechanische Größe definiert ist, die weder im Sinne der kinetischen Gastheorie noch der statistischen Mechanik einen Druck darstellt[3] und als solcher auch nicht meßbar ist. Die Größe $(P - \overline{\overline{P}})^2$ ist daher nicht die Schwankung eines inneren Parameters und kann nur in einem verallgemeinerten Sinne als Druckschwankung bezeichnet werden. Es besteht kein Grund für die Annahme, daß diese Größe das gleiche asymptotische Verhalten zeigt wie die mittleren relativen Schwankungsquadrate extensiver Parameter. Schon GIBBS [2] hat darauf hingewiesen, daß $\overline{\partial^2 P/\partial V^2}$ sehr große Werte annehmen kann. Ein physikalisch vernünftiges Resultat kann aber nur für einen einigermaßen realistischen Potentialansatz erwartet werden, und das trifft bei Gl. (70.16) für $C \rightarrow \infty$ nicht mehr zu.

GIBBS [2] hat auch bereits bemerkt, daß die Größe $(P - \overline{\overline{P}})^2$ sich aufspalten läßt nach

$$\overline{\overline{(P - \overline{\overline{P}})^2}} = \overline{\overline{(P - \overline{P})^2}} + \overline{\overline{(\overline{P} - \overline{\overline{P}})^2}}. \tag{70.18}$$

Das „anomale" Verhalten ist offenbar durch den ersten Term der rechten Seite bedingt. Der zweite Term stellt die eigentliche Druckschwankung, d.h. die Schwankung des für die mikrokanonische Gesamtheit definierten Druckes im

---

[1] H. WERGELAND: Kon. Nor. Vid. Selsk. **28**, Nr. 21 (1955).
[2] Für eine Korrespondenz über diese Arbeit habe ich Herrn Prof. WERGELAND herzlich zu danken.
[3] Sie hat naturgemäß die Dimension eines Druckes im Sinne der Mechanik.

Rahmen der kanonischen Gesamtheit dar, für die ein den Schwankungen extensiver Parameter ähnliches Verhalten vermutet werden kann [2]. Diese Vermutung läßt sich für ein ideales Gas aus Massenpunkten leicht explizit bestätigen [22]. Wegen $E = E_{\text{kin}}$ erhalten wir aus Gl. (29.5) und (32.12)

$$\frac{\overline{\overline{(E - \overline{\overline{E}})^2}}}{\overline{\overline{E}}} = \frac{2}{3N}. \tag{70.19}$$

Der Virialsatz (Ziff. 21) ergibt

$$\tfrac{3}{2} \overline{P} V = E, \tag{70.20}$$

was man auch aus Gl. (24.8) und (25.10) ableiten kann. Aus (70.19) und (70.20) folgt

$$\frac{\overline{\overline{(\overline{P} - \overline{\overline{P}})^2}}}{\overline{\overline{P^2}}} = \frac{2}{3N}, \tag{70.21}$$

womit die obige Vermutung für den betrachteten Fall bestätigt ist.

WERGELAND[1] betrachtet die obigen aus Gl. (70.13) abgeleiteten Resultate als physikalisch absurd und schlägt eine Modifikation vor, welche die Ableitung bei konstantem p, q durch eine Ableitung für konstante Wirkungsintegrale (Phasenintegrale) ersetzt. Letztere kann auf Grund der Invarianz des Phasenvolumens bzw. der Größe $\overline{\eta}$ gegen adiabatische Parameteränderungen (Ziff. 23, 46, 53) mit der thermodynamischen Ableitung bei konstanter Entropie identifiziert werden. Damit erhält Gl. (70.13) die Form

$$\overline{\overline{(P - \overline{\overline{P}})^2}} = \Theta \left[\left(\frac{\partial \overline{\overline{P}}}{\partial V} \right)_\Theta - \left(\frac{\partial \overline{\overline{P}}}{\partial V} \right)_{\overline{\eta}} \right]. \tag{70.22}$$

Für das ideale Gas führt dieser Ausdruck in der Tat wieder auf Gl. (70.21). Trotzdem erscheint das Verfahren (auch abgesehen von dem Problem der Existenz der Wirkungsintegrale für nicht periodische Systeme) weitgehend willkürlich. Eine deduktive Ableitung der Schwankungsgleichung aus Gl. (70.10) führt notwendig entweder auf (70.13) oder auf Gl. (70.5), die wir weiter unten diskutieren.

LANDAU und LIFSHITZ [11] geben für die Druckschwankung die Formel

$$\overline{\overline{(\overline{P} - \overline{\overline{P}})^2}} = - \Theta \left(\frac{\partial \overline{\overline{P}}}{\partial V} \right)_{\overline{\eta}}, \tag{70.23}$$

die bei Anwendung auf das ideale Gas einen von (70.21) abweichenden Zahlenfaktor liefert. Man übersieht leicht, daß keine der in Ziff. 67 behandelten Gesamtheiten auf diese Formel führen kann. Tatsächlich beruht sie auf der schon erwähnten unkorrekten Verallgemeinerung der Gl. (69.21). Das gleiche Verfahren führt zu der Aussage, daß zwischen Schwankungen des Druckes und des Volumens Korrelation besteht, was dem am Anfang dieser Ziffer formulierten allgemeinen Satz widerspricht.

Wir haben jetzt noch die Gl. (70.5) zu diskutieren, die wir ebenfalls auf die kanonische Gesamtheit anwenden. Berechnet man für das ideale Gas die rechte Seite mit Hilfe des Virialsatzes, so findet man, daß dieselbe identisch verschwindet. Dieses zunächst überraschende Ergebnise erklärt sich unmittelbar durch Gl. (21.18).

---

[1] H. WERGELAND: Kon. Nor. Vid. Selsk. **28**, Nr. 21 (1955).

Es ist aber nicht völlig exakt, weil wir bei der Ableitung von (70.5) für die mikrokanonische Gesamtheit das Φ-System (vgl. Ziff. 25) zugrunde gelegt haben. Der in diesem System durch die Gleichung

$$\overline{P} = \frac{\partial \Phi / \partial V}{\partial \Phi / \partial E} \tag{70.24}$$

definierte Druck erfüllt jedoch nicht exakt den Virialsatz. Für das ideale Gas ist nämlich, wie man leicht aus Gl. (25.9) ableitet,

$$\overline{P} = \frac{2}{3} \frac{E}{V} \left(1 - \frac{2}{3N}\right)^{-1}. \tag{70.25}$$

Damit wird dann

$$\frac{\overline{\left(P \cdot \dfrac{\partial \Phi}{\partial E} - \dfrac{\overline{\overline{P}}}{\Theta}\right)^2}}{(\overline{P}/\Theta)^2} = \frac{2}{3 N^2} + \text{Terme höherer Ordnung}. \tag{70.26}$$

71. Schwankungen innerer Parameter. Als innere Parameter bezeichnen wir in diesem Zusammenhang alle statistischen Größen, die (bzw. deren Mittelwerte) nicht Analoga thermodynamischer Zustandsgrößen sind. Aus der großen Zahl der hierher gehörenden Schwankungsprobleme greifen wir zwei Beispiele heraus: Die Schwankungen des lokalen Brechungsindex in einem Volumenelement ΔV eines Vielkomponentensystems und die Schwankungen der Besetzungszahlen der Eigenwerte der Einzelteilchen eines idealen Gases.

α) *Schwankungen des Brechungsindex.* Das erste Problem ist für die Theorie der Lichtstreuung von Bedeutung. Es gilt nämlich für die sog. Trübung (turbidity)

$$\tau = \frac{32 \pi^3 n^2}{3 \lambda^4} \Delta V \overline{(\delta n)^2}, \tag{71.1}$$

wo n der Brechungsindex und λ die Lichtwellenlänge im Vakuum ist. Die Berechnung von $\overline{(\delta n)^2}$ läßt sich mit Hilfe der großen kanonischen Gesamtheit nach den Formeln der Ziff. 69 durchführen. Wir betrachten das Volumenelement ΔV als System, dessen Temperatur und Volumen vorgegeben sind, und schreiben

$$\overline{\delta N_i \delta N_j} = \Theta \frac{\partial \overline{N_i}}{\partial \mu_j} \tag{71.2}$$

oder

$$\overline{\delta N_i \delta N_j} = \Theta \frac{|\Delta|_{ij}}{|\Delta|} \tag{71.3}$$

mit

$$|\Delta| = \frac{\partial (\mu_1, \ldots, \mu_m)}{\partial (\overline{N}_1, \ldots, \overline{N}_m)}. \tag{71.4}$$

Wir setzen nun

$$\delta n = \sum_{i=1}^{m} \frac{\partial n}{\partial N_i} \delta N_i, \tag{71.5}$$

wo die Ableitungen an der Stelle $N_i = \overline{N}_i$ zu nehmen sind. Daraus folgt

$$\overline{(\delta n)^2} = \sum_i \sum_j \frac{\partial n}{\partial N_i} \frac{\partial n}{\partial N_j} \overline{\delta N_i \delta N_j} \tag{71.6}$$

oder mit Gl. (71.3)

$$\overline{(\delta n)^2} = \frac{\Theta}{|\Delta|} \sum_i \sum_j \frac{\partial n}{\partial N_i} \frac{\partial n}{\partial N_j} |\Delta|_{ij}. \tag{71.7}$$

Die Doppelsumme entsteht mit negativem Vorzeichen durch zweifache Laplace-Entwicklung der Determinante

$$|D| = \begin{vmatrix} 0 & \dfrac{\partial n}{\partial N_1} & \cdots & \dfrac{\partial n}{\partial N_m} \\[2mm] \dfrac{\partial n}{\partial N_1} & \dfrac{\partial \mu_1}{\partial \overline{N}_1} & \cdots & \dfrac{\partial \mu_1}{\partial \overline{N}_m} \\[2mm] \cdots & \cdots & \cdots & \cdots \\[2mm] \dfrac{\partial n}{\partial N_m} & \dfrac{\partial \mu_m}{\partial \overline{N}_1} & \cdots & \dfrac{\partial \mu_m}{\partial \overline{N}_m} \end{vmatrix}. \tag{71.8}$$

Es wird somit

$$\overline{(\delta n)^2} = - \Theta \frac{|D|}{|\Delta|}. \tag{71.9}$$

Durch Einsetzen in (71.1) erhält man

$$\tau = - \frac{32\pi^3 n^2}{3\lambda^4} \Delta V \Theta \frac{|D|}{|\Delta|}, \tag{71.10}$$

die zuerst von Zernike[1] abgeleitete allgemeine Streuformel für Vielkomponentensysteme.

Die Ableitung der Gl. (71.1) setzt voraus, daß die Lineardimension von ΔV klein ist gegen die Wellenlänge λ. Damit wird das am Schluß von Ziff. 32 angedeutete Problem aktuell, of die Identifizierung der statistischen Parameter mit thermodynamischen Größen noch einen Sinn hat. Diese Frage ist für den obigen Fall von Tolman [12] und Münster[2] [26] untersucht worden. Man findet, daß in Gl. (71.10) μ mit dem chemischen Potential identifiziert werden kann, solange nicht die Größe $\partial^2 F/\partial N^2$ sehr klein wird. In der Umgebung kritischer Phasen ist diese Voraussetzung nicht mehr erfüllt. Da hier aber auch die Gl. (71.1) nicht mehr gültig ist, hat diese Einschränkung keine Bedeutung.

$\beta)$ *Schwankungen der Besetzungszahlen in einem idealen Gas.* Auch das zweite der oben erwähnten Probleme läßt sich mit Hilfe der großen kanonischen Gesamtheit behandeln. Wir betrachten zunächst den Fall des idealen Bose-Einstein-Gases. Die große Verteilungsfunktion kann hier geschrieben werden (vgl. Ziff. 57)

$$\Xi_{\text{BE}} = \sum_{N_k=0}^{\infty} \cdots \sum e^{\sum\limits_k N_k \frac{\mu - \varepsilon_k}{\Theta}} = \prod_k \left(1 - e^{\frac{\mu - \varepsilon_k}{\Theta}}\right)^{-1}. \tag{71.11}$$

Daraus folgt

$$\overline{N}_k = - \Theta \frac{\partial \ln \Xi}{\partial \varepsilon_k} = \frac{1}{e^{-\frac{\mu - \varepsilon_k}{\Theta}} - 1} \tag{71.12}$$

und weiter

$$\overline{(N_k - \overline{N}_k)^2} = \Theta^2 \frac{\partial^2 \ln \Xi}{\partial \varepsilon_k^2} = \overline{N}_k + \overline{N}_k^2. \tag{71.13}$$

Wir erhalten somit

$$\frac{\overline{(N_k - N_k)^2}}{\overline{N}_k^2} = \frac{1}{\overline{N}_k} + 1 \qquad \text{(Bose-Einstein)}. \tag{71.14}$$

Für das ideale Fermi-Dirac-Gas haben wir

$$\Xi_{\text{FD}} = \sum_{N_k=0}^{1} \cdots \sum e^{\sum\limits_k N_k \left(\frac{\mu - \varepsilon_k}{\Theta}\right)} = \prod_k \left(1 + e^{\frac{\mu - \varepsilon_k}{\Theta}}\right). \tag{71.15}$$

---

[1] F. Zernike: Diss. Amsterdam 1915 (Arch. néerl., Ser. III A, Bd. IB).
[2] A. Münster: Z. Physik **136**, 179 (1953).

In gleicher Weise wie oben erhalten wir durch Differentiation von $\ln \Xi$ nach ε_k/Θ

$$\overline{N}_k = \frac{1}{e^{-\frac{\mu-\varepsilon_k}{\Theta}}+1} \qquad (71.16)$$

und

$$\overline{(N_k-\overline{N}_k)^2} = N_k - \overline{N}_k^2 . \qquad (71.17)$$

Es wird daher

$$\frac{\overline{(N_k-\overline{N}_k)^2}}{\overline{N}_k^2} = \frac{1}{\overline{N}_k}-1 \quad (0 < \overline{N}_k \leqq 1) \quad \text{(Fermi-Dirac)}. \qquad (71.18)$$

Für die quantisierte Maxwell-Boltzmann-Statistik ergibt sich

$$\frac{\overline{(N_k-\overline{N}_k)^2}}{\overline{N}_k} = \frac{1}{\overline{N}_k} \quad \text{(Maxwell-Boltzmann)}. \qquad (71.19)$$

Man bemerkt, daß die Schwankungen innerer Parameter, im Gegensatz zu den früher betrachteten Fällen, auch bei beliebig wachsender Größe des Systems endlich bleiben können. In dem ersten Beispiel beruht dies einfach darauf, daß das betrachtete Volumen ΔV als endlich vorausgesetzt wird. In den Fällen der Gln. (71.14) und (71.18) handelt es sich dagegen um einen quantenstatistischen Effekt, der bei Ausschluß der Symmetriebedingungen nicht auftritt [Gl. (71.19)].

II. Phasenumwandlungen. Stabilitätsbedingungen.

72. Thermodynamische Beschreibung der Umwandlungen. Bisher haben wir stets ausdrücklich oder stillschweigend vorausgesetzt, daß die betrachteten Systeme im thermodynamischen Sinne homogen sind. Diese Voraussetzung ist nicht für alle Werte der Zustandsvariablen erfüllbar. Wir haben daher noch die Frage zu behandeln, wie sich die unter dem Begriff der Umwandlungen zusammengefaßten Erscheinungen (Phasenumwandlungen, Ordnungs-Unordnungs-Umwandlungen in Kristallen, λ-Punkt des Heliums usw.) im Rahmen der statistischen Mechanik darstellen. Im Interesse einer zusammenhängenden Darstellung rekapitulieren wir zunächst einige Gesichtspunkte der Thermodynamik.

Wir haben mehrfach erwähnt, daß der Formalismus der Thermodynamik unabhängig von den Randbedingungen ist und die verschiedenen thermodynamischen Potentiale oder Massieu-Planckschen Funktionen gleichwertig für die Beschreibung eines Systems sind, so daß ihre Wahl durch den Gesichtspunkt der Zweckmäßigkeit bestimmt wird. Dies gilt jedoch ohne Einschränkung nur für homogene Systeme. Wenn es sich dagegen um Phasenumwandlungen und heterogene Systeme handelt, ergeben sich, je nach der Wahl der Funktion, verschiedene Gesichtspunkte, die naturgemäß untereinander konsistent sind, aber den gleichen Sachverhalt gewissermaßen von verschiedenen Seiten beleuchten. Wir wollen dies an zwei Beispielen veranschaulichen. Die Gleichgewichtsbedingungen für heterogene Systeme werden bekanntlich von Gibbs[1] mit Hilfe der inneren Energie abgeleitet, indem für jede Phase die Fundamentalgleichung angeschrieben und auf dieses Gleichungssystem die allgemeine Gleichgewichtsbedingung angewandt wird. Es ist aber nicht möglich, etwa die innere Energie eines Mols Argon als stetige Funktion von Entropie und Volumen über die Umwandlungspunkte hinweg darzustellen. Beispielsweise bricht im Kondensationspunkt die Funktion ab und ist dann für einen gewissen Bereich der Variablen

---

[1] J. W. Gibbs: On the Equilibrium of Heterogeneous Substances, Collected Works, Vol. I. New Haven 1948.

(welcher der Verdampfungsentropie und dem Verdampfungsvolumen entspricht) nicht definierbar. Man kann zwar durch Einbeziehen metastabiler und instabiler Zustände formal über diese Lücke interpolieren. Trotz seinem praktischen Nutzen hat dies Verfahren jedoch, wie wir sehen werden, keinerlei physikalische Bedeutung. Es ist zwar richtig, daß beim Verfolgen der Kondensation in einem geschlossenen System innere Energie und Entropie bei konstantem Volumen sich stetig ändern. Daß es sich dabei aber nicht um die Fortsetzung der für das homogene Gebiet gültigen Funktion handeln kann, sieht man am einfachsten daraus, daß hier, im heterogenen Gebiet, keine Zustandsgleichung im üblichen Sinne existiert, dagegen aber eine zusätzliche Beziehung zwischen den Ableitungen der inneren Energie, den intensiven Parametern T und P. Wählt man als unabhängige Variable zwei intensive Parameter, etwa T und P, so ist das zugehörige thermodynamische Potential, in diesem Falle die freie Energie nach GIBBS G, über das ganze Zustandsgebiet stetig. In den Punkten, die einer Phasenumwandlung entsprechen, hat diese Funktion Singularitäten[1] derart, daß die ersten Ableitungen unstetig, die zweiten Ableitungen unendlich sind. Stellen wir G_m als Funktion von T und P dar, so sind (in der Projektion auf die T, P-Ebene) die Existenzgebiete zweier Phasen, die wir mit ′ und ″ bezeichnen, durch eine Kurve getrennt, die wir die Koexistenzkurve nennen. Für ein Fortschreiten längs dieser Kurve gilt wegen der Stetigkeit von G_m

$$dG_m' = dG_m'' \tag{72.1}$$

und somit

$$- S_m' dT + V_m' dP = - S_m'' dT + V_m'' dP \tag{72.2}$$

oder

$$\frac{dP}{dT} = \frac{S_m' - S_m''}{V_m' - V_m''} = \frac{H_m'' - H_m'}{T(V_m'' - V_m')} \tag{72.3}$$

(Clausius-Clapeyronsche Gleichung). Analoge Beziehungen erhält man für die Variablenpaare T, μ und P, μ. Man sieht, daß in dieser Darstellungsweise das vom homogenen System aus gesehene Einsetzen der Phasenumwandlung im Vordergrund steht, während das heterogene Gebiet der üblichen Zustandsdiagramme zu einer Linie degeneriert. Die Umwandlungen erscheinen als singuläre Stellen im gesamten Zustandsgebiet.

Auf dieser Grundlage hat EHRENFEST[2] allgemein Umwandlungen n-ter Ordnung definiert durch die Festsetzung, daß im Umwandlungspunkt die freie Energie nach GIBBS und ihre Ableitungen bis zu den $(n-1)$-ten stetig, die n-ten Ableitungen unstetig und die $(n+1)$-ten Ableitungen unendlich sind. Der Fall $n = 1$ umfaßt also die gewöhnlichen Phasenumwandlungen, die daher auch als *Umwandlung erster Ordnung* bezeichnet werden. Man sieht nun leicht, daß für die Umwandlungen höherer Ordnung zu Gl. (72.3) analoge Beziehungen gelten müssen. Der Einfachheit halber beschränken wir uns auf den wichtigsten Fall $n = 2$. Die Voraussetzung muß zunächst dahin präzisiert werden, daß die ersten Ableitungen von G_m über ein endliches Stück der Gleichgewichtskurve (nicht nur in einem isolierten Punkt) stetig sind. Dann haben wir

$$d\left(\frac{\partial G_m}{\partial T}\right) = - dS_m = \frac{\partial^2 G_m}{\partial T^2} dT + \frac{\partial^2 G_m}{\partial P \partial T} dP \tag{72.4}$$

und

$$d\left(\frac{\partial G_m}{\partial P}\right) = d V_m = \frac{\partial^2 G}{\partial T \partial P} dT + \frac{\partial^2 G_m}{\partial P^2} dP. \tag{72.5}$$

---

[1] Diese ziemlich eingebürgerte Ausdrucksweise ist im Reellen nicht ganz scharf. Wir können sie unbedenklich verwenden, da wir später (Ziff. 75) die ins Komplexe fortgesetzten Funktionen behandeln werden.

[2] P. EHRENFEST: Commun. Kamerlingh Onnes Lab. Univ. Leiden Suppl. **75**b (1933).

Daraus folgt

$$\frac{dT_{\mathrm{II}}}{dP_{\mathrm{II}}} = \frac{V_m\,T(\gamma''-\gamma')}{C_p''-C_p'} \tag{72.6}$$

und

$$\frac{dT_{\mathrm{II}}}{dP_{\mathrm{II}}} = \frac{\varkappa''-\varkappa'}{\gamma''-\gamma'} \tag{72.7}$$

wo T_{II} und P_{II} Temperatur und Druck der Umwandlung zweiter Ordnung bezeichnen, C_p die Molwärme bei konstantem Druck, γ den kubischen Ausdehnungskoeffizienten und $\varkappa$ die isotherme Kompressibilität. Gl. (72.6) und (72.7) sind die *Ehrenfestschen Gleichungen für Umwandlungen zweiter Ordnung*.

Die Ehrenfestschen Gleichungen bedeuten zunächst die Anwendung der Thermodynamik auf einen vorher definitionsmäßig festgelegten Sachverhalt. Insofern sind sie zweifellos korrekt. Weniger einfach ist die Frage zu beantworten, ob dieser Sachverhalt in der Natur tatsächlich vorkommt. Es sind zwar experimentell zahlreiche Umwandlungen höherer Ordnung festgestellt worden; insbesondere gehören dazu die sog. kooperativen Erscheinungen in Kristallen und der λ-Punkt des Heliums (vgl. den Artikel von MÜNSTER in Bd. XIII dieses Handbuches). Die Festlegung der Ordnung einer solchen Umwandlung aus experimentellen Daten stößt jedoch auf außerordentliche Schwierigkeiten. Eine weitere

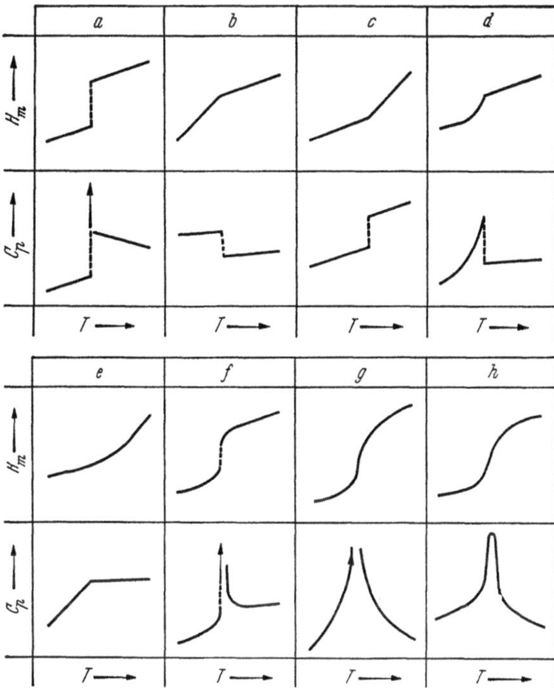

Fig. 6a — h. Verschiedene Typen von Umwandlungen. a Umwandlung erster Ordnung; b Umwandlung zweiter Ordnung; c Umwandlung zweiter Ordnung; d Umwandlung zweiter Ordnung (λ-Punkt); e Umwandlung dritter Ordnung; f Anomale Umwandlung erster Ordnung; g Onsagersche Umwandlung; h Diffuse Umwandlung.

Frage ist, ob das Ehrenfestsche Schema alle denkbaren Umwandlungstypen umfaßt. Diese Frage ist schon im Rahmen rein thermodynamischer Betrachtungen zu verneinen. Dieser Gesichtspunkt ist vor allem von TISZA [*31*] ausführlich erörtert worden. Wir können jedoch hier darauf nicht näher eingehen. Aus der statistischen Theorie sind zwei Umwandlungstypen bekannt, die sich nicht in das Ehrenfestsche Schema einordnen lassen: Die sog. anomale Umwandlung erster Ordnung der Mayerschen Kondensationstheorie[1], die auch bei der Einstein-Kondensation vorliegt (vgl. Ziff. 78) und die von ONSAGER[2] aus der exakten Theorie des zweidimensionalen Ising-Modells abgeleitete Umwandlung.

Alle im vorstehenden erwähnten Umwandlungen sind auf Grund der allgemeinen Voraussetzungen der Thermodynamik definiert. Sie beziehen sich also,

---

[1] J. E. MAYER u. S. F. HARRISON: J. Chem. Phys. **6**, 87 (1938).
[2] L. ONSAGER: Phys. Rev. **65**, 117 (1944).

streng genommen, auf unendlich große Systeme und setzen vollständige Einstellung definierter innerer Gleichgewichte voraus. Die Bedeutung der letzteren Voraussetzung, die vor allem in kristallinen Phasen oft schwer zu erfüllen ist, ist von TEMPERLEY [33] eingehend diskutiert worden. Wenn die erwähnten Voraussetzungen nicht mit genügender Annäherung erfüllt sind, kann es zu dem Bilde einer diffusen Umwandlung kommen, bei der von einem Umwandlungspunkt eigentlich nicht mehr gesprochen werden kann. In Fig. 6 sind die wichtigsten der erwähnten Umwandlungstypen schematisch dargestellt.

In nahem Zusammenhang mit den Phasenumwandlungen steht das Problem der thermodynamischen Stabilität. Wir beschränken uns auch hier auf einige Bemerkungen, die für das Verständnis der statistischen Theorie nützlich sind. Die allgemeine Stabilitätsbedingung lautet bekanntlich in der von GIBBS[1] angegebenen Form:

Wenn man in dem Ausdruck

$$E - TS + PV - \sum \mu_i N_i \qquad (72.8)$$

den Größen T, P, μ_i solche Werte zuteilen kann, daß dieser Ausdruck für die betrachtete Phase Null, für jede andere (aus den gleichen Komponenten gebildete) Phase positiv ist, so ist die betrachtete Phase *absolut stabil*. Ist der Ausdruck (72.8) für gewisse andere Phasen ebenfalls Null, so ist die betrachtete im Hinblick auf diese *im indifferenten Gleichgewicht*. Sie können mit der ursprünglichen koexistieren, und T, P, μ_i sind dann die gemeinsamen Werte von Temperatur, Druck und chemischen Potentialen.

Es kann vorkommen, daß die Stabilitätsbedingung erfüllt ist in bezug auf Phasen, die sich in ihren Eigenschaften nur um infinitesimale Beträge von der ursprünglichen unterscheiden (banachbarte Phasen), dagegen nicht für solche, die durch diskontinuierliche Änderungen aus der ursprünglichen hervorgehen. Man spricht dann von *metastabilen* Phasen. Bekannte Beispiele dafür sind der übersättigte Dampf und die unterkühlte Flüssigkeit. Die hier eingeführte Unterscheidung zwischen absolut stabil und metastabil hat jedoch aus den in Ziff. 26 erörterten Gründen nur einen relativen Charakter; im Aufbau der statistischen Theorie spielt sie keine wesentliche Rolle. Wesentlich anders liegen die Verhältnisse für die Bedingungen der Stabilität in bezug auf kontinuierliche Änderungen. Da eine Phase, welche diesen Bedingungen nicht genügt, sich im labilen Gleichgewicht (im strengen Sinne) befindet, ist sie definitionsgemäß nicht über eine endliche Zeit existenzfähig. Die entsprechenden Stabilitätsbedingungen stellen daher allgemein gültige Aussagen über die Zustandsgrößen dar, die für jede wirklich vorkommende Phase erfüllt sind. Dieselben beziehen sich auf die höheren Ableitungen der thermodynamischen Potentiale (bzw. Massieu-Planckschen Funktionen), während die Gleichgewichtsbedingungen sich auf die ersten Ableitungen beziehen. Man spricht daher auch wohl von Gleichgewichtsbedingungen höherer Ordnung. Von jetzt ab gebrauchen wir den Ausdruck „Stabilitätsbedingungen" nur noch in diesem Sinne. Wählen wir als thermodynamisches Potential die innere Energie und bezeichnen mit $\Delta^2 E$ das Differential zweiter Ordnung, so lautet die Stabilitätsbedingung

$$\Delta^2 E \, (S, V, N_i)_{N_1} > 0, \qquad (72.9)$$

wo N_1 konstant gehalten wird, um Änderungen auszuschließen, welche lediglich die Menge der Phase verändern. An Stelle von N_1 kann auch etwa V oder $\sum N_i$

---

[1] J. W. GIBBS: On the Equilibrium of Heterogeneous Substances, Collected works, Vol. I. Hew Haven 1948.

(alle i) treten. Für eine stabile Phase muß somit die aus den zweiten Ableitungen der inneren Energie gebildete quadratische Form positiv definit sein. Damit äquivalent ist die Aussage, daß die aus den zweiten Ableitungen gebildete Determinante mit allen Hauptminoren positiv sein muß.

Die Stabilitätsbedingungen lassen sich auch mit Hilfe der übrigen thermodynamischen Potentiale formulieren. Wir übertragen sie gleich in die allgemeine Sprache der Ziff. 66 und haben dann die zuerst von SCHOTTKY, ULICH und WAGNER[1] angegebene allgemeine Formulierung

$$[\varDelta^2 \varPsi_k(X_{k+1}, \ldots, X_{r-1})]_{P_1', \ldots, P_k', X_r} - [\varDelta^2 \varPsi_k(P_1', \ldots, P_k')]_{X_{k+1}, \ldots, X_r} > 0. \qquad (72.10)$$

Bemerkenswert ist hier das Verschwinden der Glieder, welche Produkte aus extensiven und intensiven Parametern enthalten. Wir haben somit eine Aufspaltung in zwei quadratische Formen, je eine für extensive und für intensive Parameter.

Vom Standpunkt der statistischen Theorie ist es zweckmäßiger, die Stabilitätsbetrachtungen mit Hilfe der Entropie durchzuführen und somit auch hier die Fundamentalgleichung in der Form der Gl. (66.6) anzuwenden. An die Stelle von (72.9) tritt dann die Bedingung

$$- \varDelta^2 S (X_1, \ldots, X_{r-1}) > 0. \qquad (72.11)$$

Das besagt, daß die Determinante

$$\left| \frac{- \partial^2 S}{\partial X_i \, \partial X_j} \right| \qquad (72.12)$$

mit allen Hauptminoren positiv sein muß. Die Verallgemeinerung dieser Aussage für beliebige Massieu-Plancksche Funktionen werden wir auf statistischem Wege ableiten (Ziff. 77).

Die Grenze des Stabilitätsbereiches ist dadurch gegeben. daß $\varDelta^2 E$ oder $\varDelta^2 S$ bzw. die entsprechenden Determinanten aus den zweiten Ableitungen Null werden. Nach den obigen Ausführungen sollte diese Grenze im allgemeinen nicht mit dem Einsetzen einer gewöhnlichen Phasenumwandlung (z.B. Kondensation) zusammenfallen, da wir im letzteren Falle nur eine Grenze der Stabilität gegen diskontinuierliche Änderungen haben. Diese Auffassung ist in der Thermodynamik außerordentlich fruchtbar gewesen. Sie führt auf die vor allem von der holländischen Schule[2] viel benutzte Interpolation der thermodynamischen Funktionen über das heterogene Gebiet hinweg mit Benutzung von metastabilen und instabilen Zuständen, für welche die van der Waalssche Gleichung das bekannteste Beispiel bietet. Für die statistische Mechanik ist diese Begriffsbildung indessen unbrauchbar, weil hier die metastabilen Zustände unter einem ganz anderen Gesichtspunkt (Ziff. 26) betrachtet werden. Wenn aber metastabile Zustände ausgeschlossen werden, so muß man auch vom Standpunkt der Thermodynamik die Koexistenzkurve als Grenze des Stabilitätsbereiches betrachten. Diese Auffassung stützt sich unmittelbar auf die experimentelle Erfahrung, denn im Umwandlungspunkt (etwa im Kondensationspunkt) wird $(\partial P/\partial V)_{T,N} = 0$, im Widerspruch zu der aus der freien Energie nach HELMHOLTZ abgeleiteten Stabilitätsbedingung (vgl. Tabelle 2 auf S. 393). Wir werden sehen, daß die statistische Mechanik ebenfalls auf diese Darstellungsweise führt.

73. Aufgaben und Möglichkeiten der statistischen Theorie. Das Problem der Phasenumwandlungen kann in der statistischen Mechanik unter zwei völlig

---

[1] W. SCHOTTKY, H. ULICH u. C. WAGNER: Thermodynamik. Berlin 1929.
[2] J. D. VAN DER WAALS u. PH. KOHNSTAMM: Lehrbuch der Thermostatik. Leipzig 1927.

verschiedenen Gesichtspunkten betrachtet werden. Die historisch ältere Betrachtungsweise, die wohl zuerst von Mie[1] angewandt wurde, setzt die Koexistenz von Phasen mit gegebenen Eigenschaften voraus und beschränkt sich darauf, die Gleichgewichtsverteilung der Moleküle auf die beiden Phasen zu untersuchen. Diese Methode hat den Vorteil, daß sie in geeigneten Fällen auf verhältnismäßig einfache Weise explizite Resultate liefert. Insbesondere ermöglicht sie die Ableitung der Dampfdruckformel für Kristalle und damit die Berechnung der auch im Zusammenhang mit dem Nernstschen Wärmesatz wichtigen Dampfdruckkonstanten. Für Einzelheiten sei auf die Literatur (etwa [13] oder [26]) verwiesen. Die Methode beantwortet aber nicht die Frage, wie überhaupt das Auftreten von Umwandlungen zu erklären ist, oder in einer prägnanten Formulierung[2]: „Wie können die Gasmoleküle wissen, wann sie zu kondensieren haben?"

Diese Problemstellung ist erstmalig in der berühmten Mayerschen Kondensationstheorie [14], [26] formuliert worden, die auch den ersten Versuch zu einer Lösung darstellt. Seitdem ist eine umfangreiche Literatur über dieses Gebiet entstanden, ohne daß eine abschließende Klärung erreicht worden wäre. Die Phasenumwandlungen sind auch heute noch eines der aktuellsten — und schwierigsten — Probleme der statistischen Mechanik. Immerhin sind in den letzten zwanzig Jahren erhebliche Fortschritte erzielt worden, die den Versuch einer zusammenfassenden Darstellung sinnvoll erscheinen lassen. Es ist naturgemäß im Rahmen dieses Artikels nicht möglich, auf die zahlreichen Kontroversen im einzelnen einzugehen. Wir müssen uns daher insoweit mit einigen Hinweisen begnügen und beschränken uns darauf, ein einigermaßen repräsentatives Bild von dem gegenwärtigen Stand der Theorie zu geben, das in seinen wesentlichen Zügen als gesichert gelten kann.

Von einer statistischen Theorie der Phasenumwandlungen wird man vor allem die Beantwortung der folgenden Fragen erwarten:

a) Wie ist das Zustandekommen einer Umwandlung physikalisch, d.h. in diesem Falle in den Begriffen der statistischen Mechanik, zu verstehen?

b) Welche mathematischen Eigenschaften müssen die Analoga der thermodynamischen Funktionen besitzen, wenn sie eine Phasenumwandlung beschreiben sollen? Kann man allgemeine Aussagen über Möglichkeit und Voraussetzungen solcher Eigenschaften machen?

c) Welche Form haben die genannten Funktionen für konkrete Systeme, welche Phasenumwandlungen zeigen?

In diesem Artikel haben wir nur die Fragen a) und b) ausführlich zu erörtern. Lediglich zur Erläuterung der allgemeinen Theorie behandeln wir ein konkretes Beispiel, die Einstein-Kondensation. Im übrigen muß für spezielle Systeme auf die Artikel von Mayer (Bd. XII) und Münster (Bd. XIII) verwiesen werden. Die Beantwortung der Fragen a) und b) ist naturgemäß nicht unabhängig voneinander; es handelt sich dabei eigentlich nur um zwei verschiedene Aspekte der gleichen Frage. Die grundlegenden Kontroversen über die Theorie der Phasenumwandlungen beziehen sich formal auf die Frage b); ihnen entsprechen aber auch verschiedene Antworten auf die Frage a). Sieht man von hier unwichtigen Details ab, so bezieht sich die wesentliche Meinungsverschiedenheit auf das Problem, ob der Umwandlungspunkt durch eine Singularität der thermodynamischen Funktionen im Sinne der Funktionentheorie charakterisiert ist oder die Unstetigkeit in der Steigung etwa der Funktion $P(T, \mu)$ in anderer Weise erklärt werden muß. Ein allgemeiner strenger Beweis ist bisher für keine der beiden Auffassungen

---

[1] G. Mie: Ann. Physik **11**, 657 (1903).
[2] M. Born u. K. Fuchs: Proc. Roy. Soc. Lond., Ser. A **166**, 391 (1938).

geführt worden, was der Natur der Sache nach kaum überraschen kann. Die letztere, die u. a. von GREEN[1], RODRIGUEZ[2], GEILIKMAN[3], KATSURA[4] vertreten wird, hat als wichtigste Stütze die exakte Theorie des sog. Husimi-Temperley-Modells, die von TEMPERLEY[5] und KATSURA[4] entwickelt wurde. Diese Theorie unterliegt jedoch gewissen Einwendungen, auf die hier nicht eingegangen werden kann, welche jedoch ihre Beweiskraft zweifelhaft erscheinen lassen. Zugunsten der ersten Auffassung spricht zunächst, daß in den beiden einzigen Fällen, in denen bisher die Theorie der Umwandlungen mathematisch vollständig durchgeführt werden konnte, dem zweidimensionalen Ising-Modell und der Einstein-Kondensation, die Umwandlungspunkte durch Singularitäten im Sinne der Funktionentheorie charakterisiert sind. Allerdings handelt es sich in beiden Fällen nicht um Phasenumwandlungen im eigentlichen Sinne, sondern (vom Standpunkt des Ehrenfestschen Schemas) um Zwischentypen zwischen den Umwandlungen erster und zweiter Ordnung. Für das dreidimensionale Gittermodell einer festen Lösung kann man zeigen, daß die Entmischung durch eine Singularität bestimmt wird[6]. Auch für ein tieferes physikalisches Verständnis der Umwandlungen hat bisher nur diese Auffassung brauchbare Ansatzpunkte geliefert. Wir werden sie daher, unbeschadet der noch vorhandenen Lücken und Unsicherheiten, im folgenden zugrundelegen.

Die allgemeine Theorie, auf die wir uns hier beschränken, kann naturgemäß keine Aussagen darüber machen, ob eine Umwandlung stattfindet. Sie hat lediglich zu zeigen, daß es überhaupt möglich ist, im Rahmen der statistischen Mechanik das Auftreten von Umwandlungen zu verstehen und weiter die Voraussetzungen dafür zu klären, soweit dies ohne speziellere Annahmen über die Natur des Systems möglich ist. Gewisse Annahmen müssen allerdings, schon im Hinblick auf die Konvergenzprobleme, auch in der allgemeinen Theorie gemacht werden. Die Zulässigkeit und hinreichende Allgemeinheit dieser Annahmen muß dann unter physikalischen Gesichtspunkten beurteilt werden.

Der erste Schritt der Theorie besteht in dem Nachweis, daß die Bedingung c) am Schluß von Ziff. 68 erfüllt ist, d.h., daß die Grenzwerte der spezifischen Massieu-Planckschen Funktionen existieren. Darüber hinaus werden wir zeigen, daß sie stetige, monotone Funktionen der betrachteten Variablen sind. Im zweiten Schritt machen wir Gebrauch von der schon in Ziff. 68 eingeführten Fortsetzung der Verteilungsfunktionen ins Komplexe und beweisen wieder die Existenz der so verallgemeinerten Grenzfunktionen. Es läßt sich dann zeigen, daß dieselben unter gewissen Voraussetzungen Singularitäten auf der positiven reellen Achse besitzen. Die physikalische Interpretation dieser Voraussetzungen bildet den dritten Schritt. Er führt zur Verknüpfung der Phasenumwandlungen mit der Theorie der Schwankungen und den thermodynamischen Stabilitätsbedingungen.

74. Existenz des Grenzwertes der kanonischen Verteilungsfunktion. Der für alle weiteren Untersuchungen grundlegende Beweis, daß der aus der kanonischen Verteilungsfunktion berechnete Grenzwert der freien Energie nach HELMHOLTZ pro Teilchen existiert, ist zuerst von VAN HOVE[7] geführt worden. Da die Verteilungsfunktion der kinetischen Energie in diesem Zusammenhang trivial ist, können wir uns auf die Untersuchung des Konfigurationsintegrals Gl. (30.12) beschränken.

---

[1] H. S. GREEN: Proc. Roy. Soc. Lond., Ser. A **189**, 103 (1947).

[2] A. E. RODRIGUEZ: Proc. Roy. Soc. Lond., Ser. A **196**, 73 (1949).

[3] B. T. GEILIKMAN: Statisticheskaya Teoriya Fazovykh Pevrasheneny. Moskva 1954.

[4] S. KATSURA: Progr. Theor. Phys. **13**, 571 (1955).

[5] H. N. V. TEMPERLEY: Proc. Phys. Soc. Lond. A **67**, 233 (1954).

[6] A. MÜNSTER u. K. SAGEL: Z. phys. Chem., N. F. **7**, 267 (1956).

[7] L. VAN HOVE: Physica, Haag **15**, 951 (1949).

Wir betrachten ein System von N gleichen Teilchen der Masse m, die in ein Gebiet D des dreidimensionalen Raumes eingeschlossen sind. Dann ist

$$Q_\tau(N, D) = \int_D e^{-\frac{U^{(N)}}{kT}} d\boldsymbol{q}. \tag{74.1}$$

Wir setzen

$$\frac{Q_\tau(N, D)}{N!} = e^{N f(N, D)}. \tag{74.2}$$

Die hierdurch definierte Größe $f(N, D)$ hängt mit der freien Energie nach Helmholtz pro Teilchen zusammen durch die Gleichung

$$f(N, D) = -\frac{F/N}{kT} + 3 \ln \lambda, \tag{74.3}$$

wobei die Moleküle als Massenpunkte angenommen sind. Wir machen nun die folgenden Voraussetzungen:

a) Für jedes N ist

$$\left.\begin{aligned} U^{(N)} &= \sum_{1 \leq i < j \leq N} \sum U^{(2)}(\boldsymbol{q}_i, \boldsymbol{q}_j) + \sum_{1 \leq i < j < k \leq N} \sum \sum U^{(3)}(\boldsymbol{q}_i, \boldsymbol{q}_j, \boldsymbol{q}_k) + \cdots \\ &+ \sum_{1 \leq i_1 < i_2 \ldots i_\lambda \leq N} \sum \cdots \sum U^{(\lambda)}(\boldsymbol{q}_{i_1}, \boldsymbol{q}_{i_2}, \ldots, \boldsymbol{q}_{i_\lambda}), \end{aligned}\right\} \tag{74.4}$$

wo λ eine von N unabhängige ganze Zahl ist. Die Funktionen $U^{(l)}(\boldsymbol{q}_1, \ldots, \boldsymbol{q}_l)$ sind unabhängig von N, symmetrisch und stetig in $\boldsymbol{q}_l, \ldots, \boldsymbol{q}_1$ und invariant gegen eine Translation

$$U^{(l)}(\boldsymbol{q}_1 + \boldsymbol{q}', \ldots, \boldsymbol{q}_l + \boldsymbol{q}') = U^{(l)}(\boldsymbol{q}_1, \ldots, \boldsymbol{q}_l). \tag{74.5}$$

Die Funktionen $e^{-U^{(l)}/kT}$ sind beschränkt.

b) Es existiert ein von Null verschiedener Abstand σ derart, daß $U^{(2)}(\boldsymbol{q}_1, \boldsymbol{q}_2) = +\infty$ wird, wenn $|\boldsymbol{q}_1 - \boldsymbol{q}_2| \leq \sigma$ ist. Die Summe $\sum\limits_{i \geq 2} U^{(2)}(\boldsymbol{q}_1, \boldsymbol{q}_i)$ wird beliebig klein, wenn alle Abstände $|\boldsymbol{q}_1 - \boldsymbol{q}_i|$ $(i \geq 2)$ einen gewissen Wert überschreiten.

c) Es existiert ein Abstand r_1 derart, daß jede der Funktionen $U^{(l)}(\boldsymbol{q}_1, \ldots, \boldsymbol{q}_l)$ $(l = 3, \ldots, \lambda)$ Null wird, wenn der Koordinaten-Satz $\boldsymbol{q}_1, \ldots, \boldsymbol{q}_l$ sich so in zwei Unter-Sätze zerlegen läßt, daß jeder Punkt des ersten Unter-Satzes einen Abstand $> r_1$ von jedem Punkt des zweiten Unter-Satzes hat.

d) Das Gebiet D wächst beliebig mit N derart, daß der Quotient $V(D)/N$ [wo $V(D)$ das Volumen des Gebietes ist] gegen einen endlichen Wert v (das Volumen pro Molekül) strebt. $v_{\min}$ sei der kleinste mit b) vereinbare Wert von v.

e) Für jedes kubische Gitter im dreidimensionalen Raum gilt $\lim\limits_{N \to \infty} N_s/N = 0$, wo N_s die Zahl der Würfel ist, die Randpunkte von D enthalten.

Es gilt nun folgendes

Theorem 1. Wenn die Voraussetzungen a) bis e) erfüllt sind, so konvergiert $f(N, D)$ für $N \to \infty$ und $v > v_{\min}$ gegen einen endlichen Grenzwert $f(v)$, der (außer von der Temperatur und den zwischenmolekularen Kräften) nur von v abhängt. $f(v)$ ist eine stetige nicht abnehmende Funktion von v. Ihre Ableitung nach v existiert und ist proportional dem Druck; sie ist eine nicht zunehmende Funktion von v.

Beweis. Der Einfachheit halber ersetzen wir die Voraussetzung b) in ihrem zweiten Teil durch die engere Voraussetzung $U^{(2)}(\boldsymbol{q}_1, \boldsymbol{q}_2) = 0$ für $|\boldsymbol{q}_1 - \boldsymbol{q}_2| \geq r_2$. Den größeren der beiden Abstände r_1 und r_2 bezeichnen wir mit r_3.

Wir konstruieren nun im Raume ein kubisches Gitter R_a mit der Gitter-konstanten $a > r_3$. Es sei Γ_1' das Gebiet, das von den Zellen im Inneren von D gebildet wird, Γ_2' das Gebiet, das von Γ_1' und den Zellen, die Randpunkte von D enthalten, gebildet wird. Die Volumina dieser Gebiete genügen für jedes N den Relationen

$$V(\Gamma_1') \leqq V(D) \leqq V(\Gamma_2'), \quad V(\Gamma_2') = V(\Gamma_1') + N_s a^3. \tag{74.6}$$

Da a sich nicht mit N ändert, folgt aus den Voraussetzungen d) und e)

$$\lim_{N\to\infty} V(\Gamma_1')/N = \lim_{N\to\infty} V(\Gamma_2')/N = v. \tag{74.7}$$

Ferner ist

$$f(N, \Gamma_1') \leqq f(N, D) \leqq f(N, \Gamma_2'). \tag{74.8}$$

Wir bezeichnen mit Γ ein Gebiet, das von der Gesamtheit der Zellen $\gamma_1, \ldots, \gamma_\nu$ in R_a gebildet wird und betrachten die Größe $f(N, \Gamma)$. Für eine gegebene Konfiguration $\boldsymbol{q}_1, \ldots, \boldsymbol{q}_N$ in Γ streichen wir in der Summe (74.4) die Terme, welche sich auf Gruppen von Teilchen beziehen, die sich über mehr als eine Zelle erstrecken. Die dann entstehende Funktion bezeichnen wir mit $\tilde{U}^{(N)}(\boldsymbol{q}_1, \ldots, \boldsymbol{q}_N)$ und setzen

$$e^{N\tilde{f}(N,\Gamma)} = \frac{1}{N!} \int_\Gamma e^{-\frac{\tilde{U}^{(N)}}{kT}} d\boldsymbol{q}. \tag{74.9}$$

Da dieses Integral keine Wechselwirkung zwischen verschiedenen Zellen enthält, kann es in Produkte von Integralen, die jeweils über die einzelnen Zellen $\gamma_1, \ldots, \gamma_\nu$ erstreckt werden, zerlegt werden. Wir erhalten dann

$$e^{N\tilde{f}(N,\Gamma)} = \sum_{n_\varkappa} \prod_{\varkappa=1}^{\nu} \frac{1}{n_\varkappa!} \int_{\gamma_\varkappa} e^{-\frac{U^{(n_\varkappa)}}{kT}} d\boldsymbol{q}^{(n_\varkappa)}. \tag{74.10}$$

Hier ist $n_\varkappa$ die Zahl der Teilchen in der Zelle $\gamma_\varkappa$, und die Summe ist über alle Sätze der $n_\varkappa$ zu erstrecken, die der Nebenbedingung

$$\sum_{\varkappa=1}^{\nu} n_\varkappa = N \tag{74.11}$$

genügen. Die Integrale in (74.10) können wegen (74.5) über beliebige Zellen γ erstreckt werden. Mit (74.1) und (74.2) erhalten wir dann

$$e^{N\tilde{f}(N,\Gamma)} = \sum_{n_\varkappa} e^{\sum_{\varkappa=1}^{\nu} n_\varkappa f(n_\varkappa, \gamma)}. \tag{74.12}$$

Die Größe $\tilde{f}(N, \Gamma)$ unterscheidet sich von $f(N, \Gamma)$ nur um einen Ausdruck, der [bei fixiertem $V(\Gamma)/N$] beliebig klein wird, wenn a hinreichend groß wird. Es gilt nämlich

$$|f(N, \Gamma) - \tilde{f}(N, \Gamma)| \leqq K \frac{V(\Gamma)}{N a} + \varepsilon\left(1 - \frac{r_3}{a}\right), \tag{74.13}$$

wo K eine von N und a unabhängige positive endliche Größe und $\varepsilon(x)$ eine Funktion ist, die der Relation $\lim_{x=1} \varepsilon(x) = 0$ genügt und nur von den zwischenmolekularen Kräften abhängt. Für den Beweis der Ungleichung (74.13) muß auf die Originalarbeit verwiesen werden.

Um $\tilde{f}(N, \Gamma)$ zu berechnen, formen wir zunächst die Gl. (74.12) um. In jedem Term der Summe bezeichnen wir mit α_k die Zahl der $n_\varkappa$, die den Wert k haben ($k = 0, 1, \ldots, k_{\max} \leq 3\,a^3/4\pi\sigma^3$). Dann ergibt sich

$$e^{N\tilde{f}(N,\Gamma)} = \sum_{\alpha_k} \frac{\nu!}{\prod\limits_{k \geq 0} \alpha_k!}\, e^{\sum\limits_{k \geq 1} \alpha_k k f(k,\gamma)} , \tag{74.14}$$

wo jetzt die Summierung über alle Sätze der α_k zu erstrecken ist, die den Nebenbedingungen

$$\sum_{k \geq 0} \alpha_k = \nu, \qquad \sum_{k \geq 1} k\alpha_k = N \tag{74.15}$$

genügen. Bezeichnen wir mit $P_{\max}$ den größten der in Gl. (74.14) auftretenden Faktoren $e^{\sum \alpha_k k f(k,\gamma)}$, so gilt

$$\frac{1}{N} \ln P_{\max} \leq \tilde{f}(N, \Gamma) \leq \frac{1}{N} \ln P_{\max} + \ln \left(\sum_{\alpha_k} \frac{\nu!}{\prod \alpha_k!} \right)^{1/N} . \tag{74.16}$$

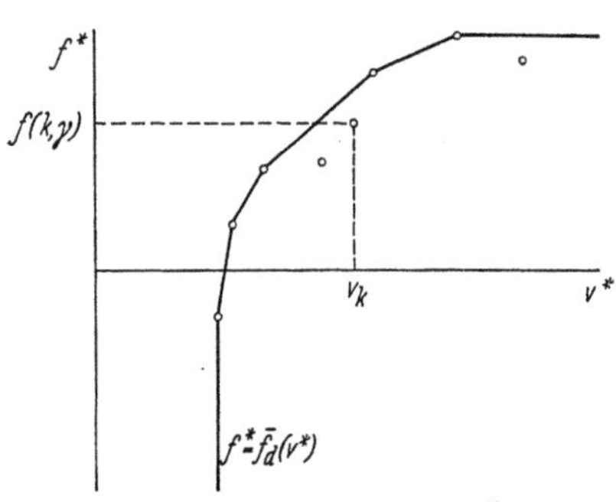

Fig. 7. Darstellung der Funktion $\tilde{f}_d(v^*)$.

Der Residuensatz ergibt

$$\sum_{\alpha_k} \frac{\nu!}{\prod \alpha_k!} = \frac{1}{2\pi i} \oint \frac{dz}{(1 - z)^\nu\, z^{N+1}} , \tag{74.17}$$

wo der Integrationsweg eine geschlossene Kurve um den Ursprung ist, welche den Punkt $z = 1$ nicht einschließt. Wählen wir als Integrationsweg einen Kreis um $z = 0$ vom Radius $e^{-\nu/N}$, so folgt

$$\left[\sum_{\alpha_\nu} \frac{\nu!}{\prod a_k!} \right]^{1/N} \leq e^{\nu/N} (1 - e^{-\nu/N})^{-\nu/N}. \tag{74.18}$$

Mit (74.16) erhalten wir dann

$$0 \leq \tilde{f}(N, \Gamma) - \frac{1}{N} \ln P_{\max} \leq \varepsilon_1 \left(\frac{\nu}{N} \right), \tag{74.19}$$

wo

$$\varepsilon_1(x) = x - x \ln (1 - e^{-x}) \tag{74.20}$$

ist. Ferner haben wir

$$\frac{\nu}{N} = \frac{V(\Gamma)}{N a^3} . \tag{74.21}$$

Wir führen nun neue Variable

$$\beta_k = \frac{k\alpha_k}{N} \qquad (k = 1, 2, \ldots, k_{\max}) \tag{74.22}$$

ein. Dann wird

$$\frac{1}{N} \sum_{k \geq 1} \alpha_k k f(k, \gamma) = \sum_{k \geq 1} \beta_k f(k, \gamma), \tag{74.23}$$

und die Nebenbedingungen (74.15) lauten

$$\sum_{k \geq 1} \beta_k = 1, \qquad \sum_{k \geq 1} \beta_k v_k \leq \frac{V(\Gamma)}{N}, \qquad v_k = \frac{a^3}{k} . \tag{74.24}$$

In einem Koordinatensystem f^*, v^* tragen wir die $k_{\max}$-Punkte mit den Koordinaten $v^* = v_k$, $f^* = f(k, \gamma)$ auf und konstruieren das gegen die v^*-Achse konkave[1] Polygon, das sich auf die größtmögliche Zahl dieser Punkte stützt

---

[1] In der Originalarbeit wird einfach die Bezeichnung konvex gebraucht, was zu Mißverständnissen Anlaß geben kann. Die obige Bezeichnung ist eindeutig.

und an den Enden jeweils in eine Parallele zu der Halbachse $v^* \geqq 0$ bzw. $f^* \leqq 0$ ausläuft (Fig. 7).

Die durch dieses Polygon definierte Funktion bezeichnen wir mit $\tilde{f}_d(v^*)$. Dann ist

$$\tilde{f}_d\left(\frac{V(\Gamma)-a^3}{N}\right) \leqq \frac{1}{N} \ln P_{\max} \leqq \tilde{f}_d\left(\frac{V(\Gamma)}{N}\right). \tag{74.25}$$

Mit Benutzung von (74.8), (74.13), (74.19) erhalten wir schließlich

$$\left.\begin{aligned} \tilde{f}_d\left(\frac{V(\Gamma_1)-a^3}{N}\right) - K\frac{V(\Gamma_1)}{Na} - \varepsilon\left(1-\frac{r_3}{a}\right) \leqq f(N,D) \leqq \\ \leqq \tilde{f}_d\left(\frac{V(\Gamma_2)}{N}\right) + K\frac{V(\Gamma_2)}{Na} + \varepsilon\left(1-\frac{r_3}{a}\right) + \varepsilon_1\left(\frac{V(\Gamma_2)}{Na^3}\right). \end{aligned}\right\} \tag{74.26}$$

Wir setzen nun voraus $v > v_{\min}$ und lassen $N \to \infty$ gehen, Wegen (74.7) konvergieren dann die äußeren Glieder der Ungleichung (74.26) gegen die endlichen Grenzwerte

$$\left.\begin{aligned} \tilde{f}_d(v) - K\frac{v}{a} - \varepsilon\left(1-\frac{r^3}{a}\right), \\ \tilde{f}_d(v) + K\frac{v}{a} + \varepsilon\left(1-\frac{r_3}{a}\right) + \varepsilon_1\left(\frac{v}{a^3}\right). \end{aligned}\right\} \tag{74.27}$$

Die Differenz derselben wird beliebig klein, wenn a hinreichend groß wird. Daraus folgt, daß auch $f(N_1, D_1) - (N_2, D_2)$ beliebig klein wird, wenn N_1 und N_2 hinreichend groß sind. Nach dem Cauchyschen Konvergenzkriterium existiert daher der endliche Grenzwert

$$\lim_{N \to \infty} f(N,D) = f(v). \tag{74.28}$$

Derselbe genügt der Ungleichung

$$-K\frac{v}{a} - \varepsilon\left(1-\frac{r_3}{a}\right) \leqq f(v) - \tilde{f}_d(v) \leqq K\frac{v}{a} + \varepsilon\left(1-\frac{r_3}{a}\right) + \varepsilon_1\left(\frac{v}{a^3}\right). \tag{74.29}$$

Daraus folgt

$$f(v) = \lim_{a \to \infty} \tilde{f}_d(v), \tag{74.30}$$

womit gezeigt ist, das $f(v)$ nicht von der Form des Gebietes D abhängt.

Die Funktion $\tilde{f}_d(v)$ ist für $v > v_{\min}$ nach ihrer Konstruktion stetig und nicht abnehmend; sie besitzt eine nicht zunehmende Ableitung. Diese Eigenschaften bleiben bei den Grenzübergang $a \to \infty$ erhalten und kommen daher nach (74.30) auch der Funktion $f(v)$ zu. Daß die Ableitung proportional dem Druck ist, folgt unmittelbar aus Ziff. 30 bzw. 54.

Damit ist das Theorem unter der an Stelle von b) eingeführten engeren Voraussetzung vollständig bewiesen. Für die Erweiterung des Beweises auf den Fall der Voraussetzung b) müssen wir auf die Originalarbeit verweisen. Mit zwei zusätzlichen Voraussetzungen (welche sich auf die Änderung der Oberfläche und des Oberfläche-Volumen-Verhältnisses von D beim Grenzübergang beziehen) kann man zeigen[1], daß $f(N, D)$ in jedem geschlossenen Intervall $v_1 \leqq v \leqq v_2$ gleichmäßig gegen den Grenzwert $f(v)$ konvergiert, wo

$$v_{\min} < v_1 < v_2 < \infty \tag{74.31}$$

ist.

---

[1] M. B. Lewis u. A. J. F. Siegert: Phys. Rev. **101**, 1227 (1956).

Analoge Existenz-Theoreme sind für die große Verteilungsfunktion von Yang und Lee[1], für die Verteilungsfunktion der „Druck-Gesamtheit" (mit den Parametern $1/T$, P/T, N) von Lewis und Siegert[2] bewiesen worden. Voraussetzungen und Beweismethode sind ähnlich wie in dem hier behandelten Fall. Wir verzichten daher auf die Wiedergabe und verweisen für Einzelheiten auf die Originalarbeiten. Da wir die Ergebnisse aber im folgenden benötigen, wollen wir sie in einem weiteren Satz zusammenfassend formulieren. Wir benutzen dazu die Bezeichnungsweise von Ziff. 67. $\mathcal{Z}_{l+1}$ sei eine Laplace-Transformierte der kanonischen Verteilungsfunktion, also die Verteilungsfunktion einer Gesamtheit mit zwei intensiven Parametern. Wir setzen

$$Z = \mathrm{e}^{-P_{l+1}} \tag{74.32}$$

und betrachten $\mathcal{Z}$ als Funktion der (reellen) Variablen Z. Dann gilt das folgende

Theorem 2. Für alle positiven reellen Werte von Z konvergiert $X_{l+2}^{-1} \ln \mathcal{Z}_{l+1}$ für $X_{l+2} \to \infty$ gegen einen endlichen Grenzwert, der eine stetige monoton ansteigende Funktion von Z ist.

Die letzte Behauptung folgt unmittelbar aus Gl. (69.2).

Alle angeführten Existenzbeweise machen die Voraussetzung eines unendlich steilen Abstoßungspotentials der zwischenmolekularen Kräfte („inkompressibler Kern"). Diese Annahme ist ziemlich unrealistisch und physikalisch wenig befriedigend. Ihr eigentlicher Inhalt ist aber nicht eine physikalische Aussage über das System, sondern die Festlegung einer im Hinblick auf das Verhalten im Endlichen sinnvollen Randbedingung[3]. Im folgenden werden wir daher die Voraussetzung auf diesen Kern reduzieren und von der früheren Formulierung keinen Gebrauch mehr machen.

75. Theorie der Phasenumwandlungen I. Wir behandeln das Problem der Phasenumwandlungen zunächst unter rein formalen Gesichtspunkten, indem wir Existenz und analytisches Verhalten der Grenzfunktionen in der komplexen Ebene untersuchen. Da die allgemeine Theorie bisher fast ausschließlich für Gesamtheiten mit zwei intensiven Parametern formuliert worden ist, werden wir uns im wesentlichen auf diesen Fall beschränken und nur gelegentlich Ergebnisse, die für spezielle Probleme mit Hilfe der kanonischen Gesamtheit berechnet worden sind, heranziehen. Wir haben also die Funktion einer komplexen Variablen z

$$X_{l+2}^{-1} \ln \mathcal{Z}_{l+1}(z) = \frac{\Phi_{l+1}(z)}{X_{l+2}} \tag{75.1}$$

und insbesondere ihr Verhalten for $X_{l+2} \to \infty$ zu untersuchen, wo

$$\mathrm{Re}\, z \equiv Z = \mathrm{e}^{-P_{l+1}} \tag{75.2}$$

ist. Yang and Lee[1] haben in einer grundlegenden Arbeit gezeigt, daß das analytische Verhalten der Funktion (75.1) auf der positiven reellen Achse durch die Verteilung der Nullstellen von $\mathcal{Z}_{l+1}$ in der komplexen Ebene bestimmt wird. Wir formulieren das von ihnen aufgestellte

Theorem. Wenn ein Gebiet $\mathfrak{G}$ der komplexen Ebene, welches ein Stück der positiven reellen Achse enthält, frei ist von Nullstellen der Funktion $\mathcal{Z}_{l+1}$, so

---

[1] C. N. Yang u. T. D. Lee: Phys. Rev. **87**, 404 (1952).
[2] Siehe Fußnote 1, S. 381.
[3] Eine analoge Bedeutung haben die in der statistischen Mechanik benutzten Annahmen über das Verhalten der betrachteten Systeme für $T \to \infty$.

existieren innerhalb dieses Gebietes die Grenzwerte

$$\lim_{X_{l+2}\to\infty} X_{l+2}^{-1}\ln\Xi_{l+1},\qquad \lim_{X_{l+2}\to\infty}\left(\frac{\partial}{\partial\ln z}\right)X_{l+2}^{-1}\ln\Xi_{l+1},$$
$$\lim_{X_{l+2}\to\infty}\left(\frac{\partial}{\partial\ln z}\right)^2 X_{l+2}^{-1}\ln\Xi_{l+1},\ \ldots \tag{75.3}$$

und sind reguläre Funktionen von z. Die Operationen $(\partial/\partial\ln z)$ und $\lim\limits_{X_{l+2}\to\infty}$ sind in $\mathfrak{G}$ vertauschbar.

Beweis. Da der ursprüngliche Beweis von YANG und LEE unnötig starke Voraussetzungen macht, geben wir hier einen von MÜNSTER[1] stammenden Beweis wieder. Zunächst beweisen wir zwei Hilfssätze. Wir betrachten die Reihe

$$\sum_{n=0}^{\infty} b_n(X_{l+2})\, z^n = S_{X_{l+2}}(z), \tag{75.4}$$

wo

$$|b_n(X_{l+2})| \leq A\,\sigma^{-n} \tag{75.5}$$

ist. Wir setzen voraus, daß für alle reellen z zwischen $-\sigma$ und $+\sigma\ \lim\limits_{X_{l+2}\to\infty} S_{X_{l+2}}(z)$ existiert. Dann gilt

Lemma 1. Der Grenzwert

$$\lim_{X_{l+2}\to\infty} b_n(X_{l+2}) = b_n \tag{75.6}$$

existiert für jedes n.

Beweis. Aus der Voraussetzung folgt, daß b_0 existiert und gleich $\lim\limits_{X_{l+2}\to\infty} S_{X_{l+2}}(0)$ ist. Um die Existenz von b_1 zu beweisen, nehmen wir eine reelle Zahl ε zwischen 0 und $\sigma/2$. Nach der Voraussetzung existiert dann ein Parameterwert $X_{l+2}^{(0)}$ derart, daß für alle Werte $X_{l+2}^{(1)}$ und $X_{l+2}^{(2)} > X_{l+2}^{(0)}$ gilt

$$|S_{X_{l+2}^{(1)}}(\varepsilon) - S_{X_{l+2}^{(2)}}(\varepsilon)| < \varepsilon^2, \tag{75.7}$$

$$|b_0(X_{l+2}^{(1)}) - b_0(X_{l+2}^{(2)})| < \varepsilon^2. \tag{75.8}$$

Nun ist

$$\left|\sum_{n=2}^{\infty} b_n(X_{l+2}^{(1)})\,\varepsilon^n\right| \leq \frac{A\,\varepsilon^2}{1-\varepsilon\sigma^{-1}} \leq 2A\,\varepsilon^2. \tag{75.9}$$

Die gleiche Beziehung gilt, wenn $X_{l+2}^{(1)}$ durch $X_{l+2}^{(2)}$ ersetzt wird. Da

$$\varepsilon\, b_1(X_{l+2}^{(1)}) = S_{X_{l+2}^{(1)}}(\varepsilon) - b_0(X_{l+2}^{(1)}) - \sum_{n=2}^{\infty} b_n(X_{l+2}^{(1)})\,\varepsilon^n \tag{75.10}$$

ist, folgt mit (75.7) bis (75.9)

$$\varepsilon\,|b_1(X_{l+2}^{(1)}) - b_1(X_{l+2}^{(2)})| < (2+4A)\,\varepsilon^2|. \tag{75.11}$$

Nach dem Cauchyschen Kriterium existiert daher $\lim\limits_{X_{l+2}\to\infty} b_1(X_{l+2})$. Die gleiche Methode ist auf alle übrigen $b_l(X_{l+2})$ anwendbar. Damit ist das Lemma 1 bewiesen.

Lemma 2. Für alle $|z| < \sigma$ existiert

$$\lim_{X_{l+2}\to\infty} S_{X_{l+2}}(z) = \sum_{l=0}^{\infty} b_l z^l. \tag{75.12}$$

Die Reihe auf der rechten Seite konvergiert für alle $|z| < \sigma$.

---

[1] A. MÜNSTER: Z. Physik **136**, 179 (1953).

Beweis. Da nach der Voraussetzung die Reihe $\sum b_n(X_{l+2})\, z^n$ gleichmäßig in z für $|z| < \sigma$ konvergiert, folgt das Lemma unmittelbar aus einem bekannten Satz über doppelte Grenzübergänge.

Wir gehen nun zum Beweis des obigen Theorems über. Für endliche Werte von X_{l+2} ist jedenfalls $\mathit{\Xi}_{l+1}$ eine ganze Funktion. Der Produktsatz von WEIERSTRASS liefert daher

$$\mathit{\Xi}_{l+1} = \prod_{\nu=1}^{\infty} \left\{ \left(1 - \frac{z}{z_\nu}\right) \exp\left[\frac{z}{z_\nu} + \frac{1}{2}\left(\frac{z}{z_\nu}\right)^2 + \cdots + \frac{1}{k_\nu}\left(\frac{z}{z_\nu}\right)^{k_\nu}\right]\right\}^{\alpha_\nu}. \tag{75.13}$$

Hier bezeichnen die z_ν die Nullstellen von $\mathit{\Xi}_{l+1}$ und α_ν ihre Ordnung. Die k_ν sind eine Folge von positiven ganzen Zahlen, die so gewählt sind, daß die Reihe

$$\sum_{\nu=1}^{\infty} \alpha_\nu \left(\frac{z}{z_\nu}\right)^{k_\nu+1} \tag{75.14}$$

für jedes z absolut konvergiert.

Wir entwickeln nun Φ_{l+1}/X_{l+2} an einer auf der positiven reellen Achse gelegenen Stelle $z = a$ nach Potenzen von $y = z - a$. Das ergibt

$$\Phi_{l+1}/X_{l+1} = \sum_{n=0}^{\infty} b(X_{l+2})\, y^n \tag{75.15}$$

mit

$$b_0(X_{l+2}) = X_{l+2}^{-1} \sum_{\nu=1}^{\infty} \alpha_\nu \ln \frac{y_\nu}{y_\nu+a} + \sum_{\nu=1}^{\infty} \frac{\alpha_\nu}{k_\nu}\left(\frac{a_\nu}{y_\nu+a}\right)^{k_\nu} \tag{75.16}$$

und

$$b_n(X_{l+2}) = - X_{l+2}^{-1} \left[n^{-1} \sum_{\nu=1}^{\infty} \alpha_\nu \left(\frac{1}{y_\nu}\right)^n + \sum_{\nu=n}^{\infty} \frac{(k_\nu-1)!}{n!} \frac{\alpha_\nu a^{k_\nu-n}}{(y_\nu+a)^{k_\nu}} \right]. \tag{75.17}$$

Wir benötigen nun Aussagen über die Nullstellen von $\mathit{\Xi}_{l+1}$. Da nach (69.2) Φ_{l+1} eine monoton ansteigende Funktion von Z ist, können auf der positiven reellen Achse keine Nullstellen von $\mathit{\Xi}_{l+1}$ liegen. Über die Zahl M der in der komplexen Ebene liegenden Nullstellen beweisen wir das folgende

Lemma 3. Wenn die Größe

$$\frac{\partial \ln \mathit{\Xi}_{l+1}}{\partial \ln y} = \overline{X}_{l+1}(y) \tag{75.18}$$

für $|y| \to \infty$ gleichmäßig in arg y gegen einen positiven reellen endlichen Grenzwert $\overline{X}^*_{l+1}$ konvergiert, so ist die Zahl M der Nullstellen der Funktion $\mathit{\Xi}_{l+1}$ gegeben durch

$$M = \overline{X}^*_{l+1}. \tag{75.19}$$

Beweis. Die Zahl der Nullstellen ist allgemein

$$M = \frac{1}{2\pi i} \oint^{(\mathfrak{C})} \frac{y\, \mathit{\Xi}'_{l+1}(y)}{\mathit{\Xi}_{l+1}(y)} \frac{dy}{y}, \tag{75.20}$$

wo $\mathfrak{C}$ ein geschlossener, alle Nullstellen umschließender Weg ist, auf dem $\mathit{\Xi}_{l+1}$ regulär und von Null verschieden und innerhalb dessen $\mathit{\Xi}_{l+1}$ regulär ist. Mit Benutzung von (75.18) können wir (75.20) schreiben

$$M = \frac{1}{2\pi i} \oint^{(\mathfrak{C})} \frac{\overline{X}_{l+1}(y)}{y}\, dy. \tag{75.21}$$

Aus der Voraussetzung folgt, daß alle Nullstellen im Endlichen liegen. Da sie sich (wegen der Regularität von Ξ_{l+1}) dort nirgends häufen können, muß auch ihre Zahl endlich sein. Lassen wir nun $\mathfrak{C}$ im Unendlichen verlaufen (d. h. integrieren wir über den unendlich fernen Kreis), so ist die Voraussetzung von (75.20) sicher erfüllt, und es folgt aus (75.21) die Gl. (75.19), womit das Lemma bewiesen ist.

Wir betrachten jetzt einen Kreis C mit $z = a$ als Mittelpunkt innerhalb des von Nullstellen freien Gebietes $\mathfrak{G}$. Der Radius des Kreises sei σ. Dann ist

$$|y_\nu| \geqq \sigma, \tag{75.22}$$

und wir erhalten mit Benutzung von Lemma 3 aus (75.17)

$$\left| b_n(X_{l+2}) \right| \leqq \frac{\overline{X}_{l+1}^*}{X_{l+2}} \left[n^{-1}\sigma^{-1} + \frac{(k_M - 1)!}{n!} \left(\frac{a}{\sigma + a} \right)^{k_M} \right] \qquad (n \geqq 1). \tag{75.23}$$

Da $\overline{X}_{l+1}^*/X_{l+2}$ beschränkt ist, können wir Gl. (75.23) schreiben

$$\left| b_n(X_{l+2}) \right| < A\,\sigma^{-1}, \tag{75.24}$$

wo A und σ positive Konstanten sind. Nach Lemma 1 folgt daraus, daß innerhalb von C der Grenzwert $\lim\limits_{X_{l+2} \to \infty} b_n(X_{l+2})$ für jedes n existiert. Nach Lemma 2 folgt dann weiter, daß innerhalb von C der Grenzwert $\lim\limits_{X_{l+2} \to \infty} (X_{l+2}^{-1} \ln \Xi_{l+1})$ existiert und eine reguläre Funktion von z ist. Durch analytische Fortsetzung läßt sich diese Aussage auf das ganze Gebiet $\mathfrak{G}$ erweitern. Damit ist das Theorem bewiesen.

Für endliche Werte von X_{l+2} umfaßt das Gebiet $\mathfrak{G}$ die ganze positive reelle Achse, da in diesem Falle Φ_{l+1}/X_{l+2} unter allen Umständen dort regulär ist. Die Nullstellen können sich dann also nicht beliebig der positiven reellen Achse nähern. Dagegen kann dies eintreten für $X_{l+2} \to \infty$. Nehmen wir an, dies sei im Punkte Z^* der Fall. Dann läßt sich, wie der obige Beweis zeigt, Z^* auf keine Weise in den Regularitätsbereich der Funktion $\lim\limits_{X_{l+2} \to \infty} \Phi_{l+1}/X_{l+2}$ einbeziehen und ist daher definitionsgemäß ein singulärer Punkt dieser Funktion. Es liegt in der Natur der Sache, daß sich im Rahmen einer derart allgemeinen Betrachtung keine Aussagen darüber machen lassen, unter welchen Voraussetzungen das beschriebene Verhalten eintritt und welches die analytische Natur der Singularität ist. Immerhin kann man einige damit zusammenhängende Fragen unter Heranziehung gewisser Erfahrungstatsachen erörtern.

Nach Lemma 2 ist Z^* die erste auf der positiven reellen Achse liegende Singularität der Funktion $\lim \Phi_{l+1}/X_{l+2}$ und damit, solange wir die b_l als positiv voraussetzen dürfen, nach einem bekannten Satz der Konvergenzradius der Potenzreihe $\Sigma b_l z^l$. Die vorstehende Einschränkung braucht uns nicht weiter zu beschäftigen, da sie jederzeit durch analytische Fortsetzung umgangen werden kann. Nur der Einfachheit halber behalten wir die obige Formulierung bei. IKEDA[1] und KATSURA[2] haben bezweifelt, daß der Punkt, in dem sich die Nullstellen der positiven reellen Achse beliebig nähern, und die erste Singularität von $\lim \Phi_{l+1}/X_{l+2}$ auf der positiven reellen Achse zusammenfallen. Nach dem obigen Beweis scheint uns ein solcher Zweifel kaum begründbar zu sein.

Aus physikalischen Gründen muß $\Sigma b_l Z^l$ in Z^* noch konvergieren. Setzen wir dies voraus, so ist die linksseitige Stetigkeit von $\lim \Phi_{l+1}/X_{l+2}$ durch den

---

[1] K. IKEDA: Proc. Int. Conf. Theor. Phys., p. 544, Tokyo 1954.
[2] S. KATSURA: Progr. Theor. Phys. **13**, 571 (1955).

Abelschen Grenzwertsatz gesichert. Man kann danach annehmen, daß die Singularität ein Verzweigungspunkt vom regulären Typ ist. Tatsächlich verhält es sich so in den beiden einzigen Fällen, die sich explizit behandeln lassen (jedoch keine Umwandlungen erster Ordnung sind); ein allgemeiner Beweis existiert nicht. Bei Umwandlungen erster Ordnung muß weiter die Ableitung $\lim \dfrac{\partial \Phi_{l+1}/X_{l+2}}{\partial \ln Z}$ in Z^* unstetig sein. Die Möglichkeit eines solchen Verhaltens läßt sich an dem Modell des zweidimensionalen „Gitter-Gases" (das wieder auf dem Ising-Modell

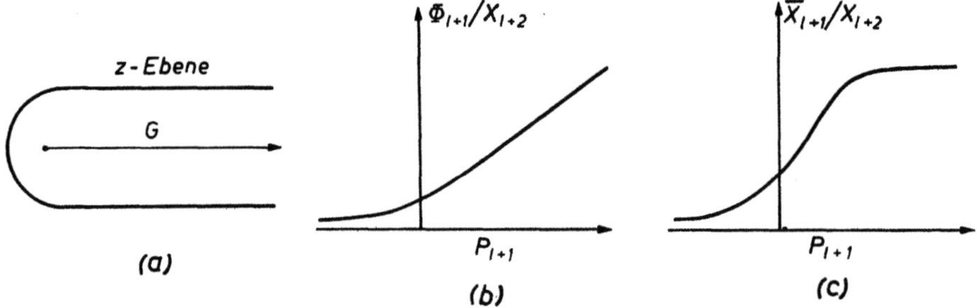

Fig. 8a—c. Verhalten der Funktion Φ_{l+1} für ein endliches System. Es treten keine Phasenumwandlungen auf.

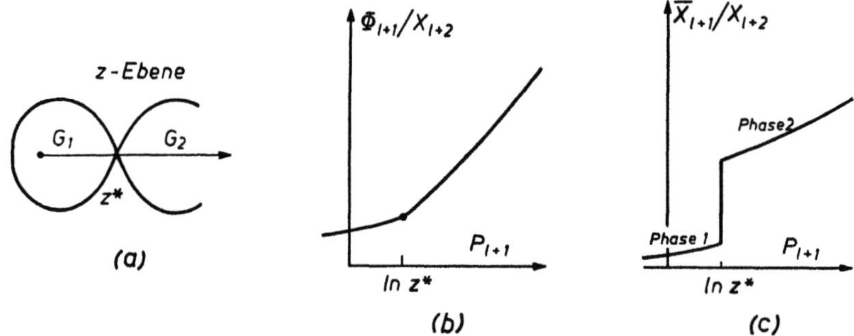

Fig. 9a—c. Verhalten von Φ_{l+1} für ein unendlich großes System in der Umgebung eines Umwandlungspunktes.

beruht) demonstrieren[1], ohne daß die analytische Natur der Singularität bekannt ist. Setzen wir ein solches Verhalten voraus, so folgt daraus, wegen der Stetigkeit von Φ_{l+1}/X_{l+2} auf der positiven reellen Achse, sofort

$$\frac{dP_{l+1}}{dP_l} = -\frac{(X_l'' - X_l')/X_{l+2}}{(\overline{X}_{l+1} - \overline{X}_{l+1})/X_{l+2}}, \tag{75.25}$$

die Clausius-Clapeyronsche Gleichung in generalisierten Zustandsgrößen. Das beschriebene Verhalten ist in Fig. 8 und 9 dargestellt.

Verstehen wir unter Ξ_{l+1} die große kanonische Verteilungsfunktion, so sind bei Anwendung auf das reale Gas die Koeffizienten b_l die bekannten cluster-Integrale der Mayerschen Kondensationstheorie. Der Kondensationspunkt ist dann die auf der positiven reellen Achse liegende Singularität, die den Konvergenzkreis von $\Sigma b_l Z^l$ bestimmt. Eine cluster-Entwicklung im Sinne Mayers (die den Grenzübergang impliziert) kann daher grundsätzlich nicht zu einer Darstellung der flüssigen Phase führen, weil eine analytische Fortsetzung über Z^* hinaus nicht möglich ist. Das schließt natürlich nicht aus, daß für einfache Modelle,

---

[1] T. D. Lee u. C. N. Yang: Phys. Rev. **87**, 410 (1952).

die in bezug auf den Umwandlungspunkt symmetrisch sind (Gitter-Gas, Gitter-modell einer binären Lösung) sich cluster-Reihen für beide Phasen explizit angeben lassen. Die in der Literatur vielfach anzutreffende Behauptung, daß eine Berücksichtigung der Volumenabhängigkeit der cluster-Integrale die Darstellung der flüssigen Phase ermöglichen würde, ist im Rahmen der hier entwickelten Theorie sicher unrichtig. Sie hat überhaupt nur einen Sinn, wenn man auf den Grenzübergang verzichtet und an eine Theorie für endliche Systeme unter expliziter Berücksichtigung der Grenzflächen denkt. Inwieweit etwas derartiges durchführbar ist, mag dahingestellt bleiben.

76. Theorie der Phasenumwandlungen II. Von den in Ziff. 73 formulierten Fragen wird durch die in Ziff. 75 entwickelte Theorie nur b) in gewissem Umfang beantwortet. Dagegen trägt sie kaum etwas bei zur Beantwortung der Frage a). Wir wollen daher jetzt das Problem unter einem mehr physikalischen Gesichtspunkt betrachten[1,2]. Wir beschränken uns auch hier auf Einkomponentensysteme und Gesamtheiten mit zwei intensiven Parametern, für die $k-l=1$ ist. Die in Ziff. 68 und 69 abgeleitete Bedingung lautet dann

$$B_{11}^l/\overline{X}_{l+1} > 0 \quad \text{bzw.} \quad B_{11}^*/\overline{X}_{l+1} > 0. \tag{76.1}$$

Wenn diese Bedingung erfüllt ist, verschwinden für $\overline{X}_{l+1} \to \infty$ die zweiten relativen Korrelationsmomente wie $(\overline{X}_{l+1})^{-1}$. Aus den Schwankungsgleichungen (69.19) und (69.22) ergibt sich unmittelbar, daß keine der beiden Größen B_{11}^l und B_{11}^* negativ werden kann. Für B_{11}^l folgt dies auch aus Gl. (68.28), da für negatives B_{11}^l das Integral divergieren würde. Wir haben daher lediglich den Grenzfall

$$B_{11}^l/\overline{X}_{l+1} = 0 \tag{76.2}$$

zu diskutieren. Wir nehmen zunächst an, daß $\overline{X}_{l+1}$ endlich ist. Dann bedeutet (76.2) einfach, daß die aus der mit dem quadratischen Glied abbrechenden Reihenentwicklung (68.5) abgeleitete Schwankungsgleichung (69.24) ungültig wird und $\overline{\xi_{l+1}^2}$ nicht mehr proportional $\overline{X}_{l+1}^{-1}$ ist. Die Größe B_{11}^* der exakten Schwankungsgleichung (69.19) ist unter diesen Umständen eine Funktion von $\overline{X}_{l+1}$ bzw. X_{l+2}. Sie kann aber nicht Null werden, weil das mittlere relative Schwankungsquadrat aus physikalischen Gründen endlich bleiben muß. Wir lassen nun $\overline{X}_{l+1}$ bzw. X_{l+2} gegen Unendlich gehen. Auch in diesem Falle ist die aus der Reihenentwicklung abgeleitete Gl. (69.24) nicht mehr gültig. Wegen (69.25) wird aber jetzt auch $B_{11}^*/\overline{X}_{l+1}$ Null. Da die Bedingung (69.20) nicht mehr erfüllt ist, wird Gl. (69.19) unbestimmt. Aus (69.16) sieht man, daß in diesem Falle $\overline{\xi_{l+1}^2}$ jedenfalls für $\overline{X}_{l+1} \to \infty$ nicht mehr wie $\overline{X}_{l+1}^{-1}$ verschwindet. Die Parameterwerte, für welche dies eintritt, sind durch Gl. (76.2) bestimmt. Sie definieren eine Kurve, deren Gleichung wir symbolisch schreiben

$$f(P_l, P_{l+1}) = 0. \tag{76.3}$$

Man sieht nun leicht, daß die in Ziff. 75 erörterten Singularitäten von $\lim \Phi_{l+1}/X_{l+2}$ auf der durch (76.3) definierten Kurve liegen. Wie wir gesehen haben, ist diese Funktion in einem von Nullstellen der Funktion Ξ_{l+1} freien Gebiet regulär, weil hier

$$|b_n(X_{l+2})| < A\,\sigma^{-1} \tag{76.4}$$

---

[1] A. Münster: Z. Naturforsch. **6a**, 139 (1951); **7a**, 613 (1952).
[2] A. Münster: Z. Physik **136**, 179 (1953).

ist, wo A und σ positive Konstanten sind. Nun ist aber längs der positiven reellen Achse

$$|b_2(X_{l+2})| = \frac{1}{2}\, e^{2\,P_{l+1}} \left[\frac{\partial^2(\Phi_{l+1}/X_{l+2})}{\partial P_{l+1}^2} + \frac{\partial(\Phi_{l+1}/X_{l+2})}{\partial P_{l+1}} \right]. \tag{76.5}$$

Da der zweite Term der Klammer aus physikalischen Gründen stets beschränkt ist, ist (76.4) notwendig und hinreichend dafür, daß $\overline{\xi_{l+1}^2}$ wie $\overline{X_{l+1}^{-1}}$ verschwindet, und umgekehrt[1]. Die Funktion lim Φ_{l+1}/X_{l+2} muß daher auf der Kurve (76.3) singulär werden.

Es zeigt sich somit, daß ein Umwandlungspunkt durch ein „anomales" Verhalten der Schwankungen extensiver Parameter charakterisiert ist in dem Sinne, daß diese Schwankungen hier außerordentlich groß werden. *Vom Standpunkt der statistischen Mechanik besteht also das Wesen einer Umwandlung darin, daß für gewisse Werte der Parameter die Verteilung „instabil" wird und enorme Schwankungen auftreten.* Man kann leicht noch auf einem anderen Wege sehen, daß ein solches Verhalten nur für singuläre Werte der Parameter möglich ist und daß umgekehrt Umwandlungen im Sinne der Thermodynamik nur für den Grenzfall unendlich großer Systeme definierbar sind. Der Einfachheit halber benutzen wir die große kanonische Gesamtheit und gehen von Gl. (65.8) aus. Es sei bei konstantem Θ zwischen μ' und μ''

$$\frac{\overline{(\varrho - \bar\varrho)^2}}{\bar\varrho^2} > \overline{N}^{-\frac{1}{2}}. \tag{76.6}$$

Setzen wir dies in (65.8) ein und integrieren zwischen den angegebenen Grenzen, so folgt

$$\tfrac{1}{2}\Theta\,(\overline{N}''^{-\frac{1}{2}} - \overline{N}'^{-\frac{1}{2}}) > \mu' - \mu'' \tag{76.7}$$

und daraus für $\overline{N} \to \infty$

$$\mu' = \mu'', \tag{76.8}$$

womit die Behauptung bewiesen ist. Eine analoge Betrachtung kann man für die Schwankungen der Energiedichte durchführen. Schreiben wir $\varepsilon = E/V$, so gilt bei Berücksichtigung des Äquipartitionstheorems

$$\frac{\overline{(\varepsilon - \bar\varepsilon)^2}}{\bar\varepsilon_{\mathrm{kin}}^2} = \frac{2}{\bar n}\,\frac{\Theta}{\bar\varepsilon_{\mathrm{kin}}}\,\frac{\partial\varepsilon}{\partial\Theta}, \tag{76.9}$$

wo $\bar n$ die mittlere Zahl der Freiheitsgrade ist. Wir nehmen jetzt an, daß zwischen Θ' und Θ''

$$\frac{\overline{(\varepsilon - \bar\varepsilon)^2}}{\bar\varepsilon_{\mathrm{kin}}^2} > \bar n^{-\frac{1}{2}} \tag{76.10}$$

ist. Dann folgt aus (76.9) durch Integration zwischen diesen Grenzen (da $\bar n$ eine Funktion von Θ ist)

$$\bar\varepsilon' - \bar\varepsilon'' > \int\limits_{\Theta''}^{\Theta'} \frac{\bar n^{\frac{3}{2}}}{V}\,d\Theta. \tag{76.11}$$

Für endliche Θ ist die linke Seite dieser Ungleichung stets endlich. Sie kann daher für $\bar n \to \infty$ nur erfüllt sein, wenn

$$\Theta' = \Theta'' \tag{76.12}$$

ist.

Genauere Aussagen über das Verhalten der Schwankungsgrößen im Umwandlungspunkt lassen sich ohne speziellere Annahmen aus der Theorie nicht

---

[1] Dabei ist vorausgesetzt, daß P_{l+1} endlich ist.

deduzieren. Man kann aber auch hier noch etwas weiter kommen, wenn man die empirischen Tatsachen zu Hilfe nimmt. Wir betrachten etwa ein Teilvolumen V, das in Gedanken in einem sehr viel größeren Volumen (dem Reservoir) eines kondensierenden Gases abgegrenzt ist. Von dem Einfluß der Schwerkraft sehen wir ab. V tauscht mit dem Reservoir Energie und Materie aus; es wird daher im Gleichgewicht durch eine große kanonische Gesamtheit beschrieben. Solange das System homogen ist, gilt für die Dichteschwankungen in V die Gl. (68.5). Das mittlere relative Schwankungsquadrat verschwindet wie $\overline{N}^{-1}$. Im Kondensationspunkt enthält V Gas und Flüssigkeitströpfchen in einem dauernd wechselnden Mengenverhältnis. Auch wenn wir V beliebig groß werden lassen, ändert sich daran nichts. Man muß daher annehmen, daß bei einer Umwandlung erster Ordnung das mittlere relative Schwankungsquadrat der molekularen Dichten für $\overline{N} \to \infty$ einem endlichen Grenzwert zustrebt. Wenn die Dichtedifferenz der beiden Phasen gegen Null geht, d.h. im kritischen Punkt, muß auch der Grenzwert der Schwankungsgröße gegen Null gehen, er wird aber im kritischen Punkt von niedrigerer Ordnung als $\overline{N}^{-1}$ verschwinden. Von J. E. MAYER u. Mitarb. wurde bekanntlich angenommen, daß zwischen dem kritischen Punkt und dem Erscheinen des Meniscus ein Gebiet liegt, in welchem die Isothermen horizontal ohne Unstetigkeit in der Steigung verlaufen. Dies würde eine Koexistenz der (durch die Ausdehnung des horizontalen Stückes der Isothermen definierten) Folge von unendlich benachbarten Phasen bedeuten (sog. *anomale Umwandlung erster Ordnung*). Auch in diesem Falle sollte die Schwankungsgröße noch mit $\overline{N} \to \infty$ verschwinden, aber von noch niedrigerer Ordnung als im kritischen Punkt.

Die obige Aussage über Umwandlungen erster Ordnung hat sich bisher nicht streng beweisen lassen, weil die vollständige explizite Behandlung einer solchen Umwandlung noch in keinem Falle gelungen ist. Auf die beiden anderen Beispiele werden wir in Ziff. 78 zurückkommen.

Der Zusammenhang zwischen Phasenumwandlungen und Schwankungen führt noch zu weiteren bemerkenswerten Gesichtspunkten. Man findet zunächst, daß ein klassisches System, dessen Hamilton-Funktion nach den Koordinaten der Einzelteilchen separierbar ist, unter keinen Umständen eine Phasenumwandlung zeigen kann. Für das ideale Maxwell-Boltzmann-Gas gilt nämlich

$$\frac{\overline{(N - \overline{N})^2}}{\overline{N}^2} = \frac{1}{\overline{N}}. \tag{76.13}$$

Phasenumwandlungen sind also jedenfalls kooperative Erscheinungen. Indessen hat diese Feststellung nur den Charakter einer notwendigen, aber nicht den einer hinreichenden Bedingung. Die Berücksichtigung der ersten Virialkoeffizienten in der Zustandsgleichung eines realen Gases ergibt noch keine Phasenumwandlung. In Ziff. 79 werden wir zeigen, daß ein eindimensionales System überhaupt keine Phasenumwandlungen zeigen kann. Um die Verhältnisse deutlicher zu machen, führen wir die molekularen Verteilungsfunktionen ein, für deren ausführliche Behandlung auf den Artikel von MÜNSTER in Band XIII dieses Handbuches verwiesen sei. Es sei

$$\varrho^{(n)} = \overline{\varrho}^n \, g(Z, \boldsymbol{q}^{(n)}) \tag{76.14}$$

die Wahrscheinlichkeitsdichte, bei der Fugazität $Z n$ nicht spezifizierte Moleküle bei dem Koordinatensatz $\boldsymbol{q}^{(n)}$ anzutreffen. Dabei gilt die Normierung

$$\lim_{V \to \infty} \left[\frac{1}{V^n} \int \cdots \int g^{(n)}(Z, \boldsymbol{q}^{(n)}) \, d\boldsymbol{q}^{(n)} \right] = 1. \tag{76.15}$$

Nun gilt für das mittlere relative Schwankungsquadrat der Molekülzahlen

$$\frac{\overline{(N-\overline{N})^2}}{\overline{N}^2} = \frac{1}{V^2} \iint [g^2(\boldsymbol{q}_1, \boldsymbol{q}_2) - g^{(1)}(\boldsymbol{q}_1)\, g^{(1)}(\boldsymbol{q}_2)]\, d\boldsymbol{q}_1\, d\boldsymbol{q}_2 + \frac{1}{\overline{N}}. \qquad (76.16)$$

Die molekularen Verteilungsfunktionen für Einzelmoleküle und Paare lassen sich aber nicht für sich in geschlossener Form berechnen, sondern sie werden durch alle höheren molekularen Verteilungsfunktionen bis zur Ordnung N bestimmt, was sich im einzelnen verschieden darstellen läßt. Um eine Berechnung zu ermöglichen, wurde von Kirkwood[1] das sog. Superpositionsprinzip eingeführt. Wir setzen

$$g^{(n)}(Z, \boldsymbol{q}^{(n)}) = e^{-\frac{W^{(n)}}{kT}} \qquad (76.17)$$

und

$$W^{(n)} = \sum_{\binom{\nu}{n}} w^{(n)}(Z, \boldsymbol{q}^{\binom{\nu}{n}}). \qquad (76.18)$$

Hier bezeichnet $\boldsymbol{q}^{\binom{\nu}{n}}$ einen aus den Koordinaten von ν Molekülen bestehenden Unter-Satz des Koordinatensatzes $\boldsymbol{q}^{(n)}$, und die Summierung ist über alle möglichen Unter-Sätze zu erstrecken. Für fluide Phasen ist $w^{(1)} = W^{(1)} = 0$ und somit $w^{(2)} = W^{(2)}$. Das Superpositionsprinzip besagt nun, daß $w^{(n)} = 0$ ist für $n \geq 3$. Das heißt mit anderen Worten, daß alle Korrelationen höherer Ordnung durch Paar-Korrelationen dargestellt werden. Für Gittermodelle kondensierter Phasen ist dem Superpositionsprinzip äquivalent das Bethesche (quasi-chemische) Näherungsverfahren[2]. Man kann nun zeigen[3], daß für eindimensionale Systeme (bei denen keine Phasenumwandlungen auftreten) das Superpositionsprinzip exakt gültig ist. Andererseits ergibt die Anwendung des Superpositionsprinzips auf zwei- und dreidimensionale Systeme (z.B. [4],[5]) im wesentlichen immer das gleiche Bild: An der Stelle, an der eine Phasenumwandlung zu erwarten ist, erhält man Isothermen von dem aus der van der Waalsschen Zustandsgleichung bekannten Typ. Es zeigt sich somit, daß Phasenumwandlungen in einem ganz extremen Sinne kooperative Erscheinungen sind und daß jede Reduktion dieses kooperativen Charakters (die im homogenen Gebiet eine sehr gute Näherung sein kann) das Bild mehr oder weniger verfälschen muß. Das eindruckvollste Beispiel dafür ist die Berechnung der spezifischen Wärme des zweidimensionalen Ising-Modells, für die sich im Umwandlungspunkt nach der Betheschen Methode eine endliche Unstetigkeit (Umwandlung zweiter Ordnung), nach der exakten Theorie von Onsager[6] eine logarithmische Unendlichkeitsstelle ergibt. Aus diesem Sachverhalt erklären sich ohne weiteres die außerordentlichen mathematischen Schwierigkeiten, auf welche die explizite Behandlung konkreter Phasenumwandlungen führt. Physikalisch bedeutet dieser kooperative Charakter, daß der Prozeß der Umwandlungen sich gewissermaßen selbst katalysiert, daß für gewisse Werte der Parameter die lokalen Schwankungen sich nicht, wie sonst, kompensieren, sondern verstärken, bis die vorhandene Phase instabil wird und eine neue Phase auftritt. In Ziff. 77 werden wir zeigen, daß diese Interpretation nicht nur ein anschauliches Bild gibt, sondern im strengen Sinne die Deutung des Stabilitätsbegriffes der Thermodynamik darstellt.

[1] J. G. Kirkwood: J. Chem. Phys. 3, 300 (1935).
[2] T. Murakami u. S. Ono: Mem. Fac. Engrg. Kyushi 12, 309, 319 (1951).
[3] Z. W. Salsburg, R. W. Zwanzig u. J. G. Kirkwood: J. Chem. Phys. 21, 1098 (1953).
[4] A. E. Rodriguez: Proc. Roy. Soc. Lond., Ser. A 196, 73 (1949).
[5] E. A. Guggenheim u. M. L. McGlashan: Proc. Roy. Soc. Lond., Ser. A 206, 335 (1951).
[6] L. Onsager: Phys. Rev. 65, 117 (1944).

Zuvor wollen wir noch kurz auf die Frage eingehen, welche Rolle die zwischenmolekularen Kräfte für das Zustandekommen einer Phasenumwandlung (oder Umwandlung höherer Ordnung) spielen. In der van der Waalsschen Theorie erscheinen die zwischenmolekularen Kräfte als das (gegenüber dem idealen Gas) wesentliche neue Element, welches die Kondensation erklärt. Das ist zwar richtig, kann aber leicht mißverstanden werden. Jede Umwandlung ist, wie wir gesehen haben, wesentlich ein statistisches Phänomen und hat daher primär überhaupt nichts mit zwischenmolekularen Kräften zu tun. Die Richtigkeit dieser paradox klingenden Behauptung ergibt sich unmittelbar aus der Kondensation des idealen Bose-Einstein-Gases, die wir in Ziff. 78 behandeln. Die zwischenmolekularen Kräfte sind lediglich eine (im Rahmen der klassischen Statistik allerdings die einzige) Möglichkeit, Korrelationen beliebig hoher Ordnung und damit den kooperativen Charakter herzustellen. Wenn sie überhaupt eine Rolle spielen, wird man fragen, ob zwischen dem Charakter dieser Kräfte und dem Typ der Umwandlung ein Zusammenhang besteht. Diese Frage ist noch wenig geklärt; möglicherweise läßt sie sich überhaupt nicht eindeutig beantworten. TEMPERLEY [33] hält es für möglich, daß Kräfte kurzer Reichweite und Kräfte großer Reichweite zu grundsätzlich verschiedenen Umwandlungstypen führen, wobei die Frage der Grenzziehung zwischen den beiden Wechselwirkungstypen offen gelassen wird. Die experimentellen Ergebnisse scheinen indessen mehr gegen eine solche Zuordnung zu sprechen. Umwandlungen erster Ordnung treten bei Systemen mit den verschiedenartigsten Wechselwirkungen auf. Der λ-Punkt des Heliums scheint nach neueren Untersuchungen vom gleichen Typ zu sein wie die Überstruktur-Umwandlung des β-Messings[1]. Der Nachweis der „kritischen Opaleszenz" in festen binären Legierungen[2] zeigt, daß die Natur des kritischen Punktes der Entmischung hier die gleiche sein muß wie in Gemischen organischer Flüssigkeiten. Dabei ist sicherlich der „Mechanismus der Korrelation", d.h. die Wechselwirkung, in beiden Fällen völlig verschieden, während in dem vorhergehenden Beispiel bei Helium die Korrelation wesentlich auf den quantenmechanischen Symmetriebedingungen beruhen dürfte. Dieser Sachverhalt ist von erheblicher Bedeutung, weil eine vollständige theoretische Behandlung von Umwandlungen nur für die einfachsten Modelle in Betracht kommt. Man kann hoffen, auf diesem Wege zu einem Verständnis der wesentlichen Züge zu gelangen, auch wenn die Rechnung sich für kompliziertere Systeme nicht durchführen läßt.

77. Die thermodynamischen Stabilitätsbedingungen. Wenn wir die statistischen Parameter mit den thermodynamischen Zustandsvariablen identifizieren, so wird die zweite der am Schluß von Ziff. 68 formulierten Bedingungen identisch mit den Stabilitätsbedingungen der Thermodynamik für extensive Parameter, wobei lediglich die thermodynamischen Potentiale durch die Massieu-Planckschen Funktionen ersetzt sind. Für den Falle $l=0$ kommt man unmittelbar auf die allgemeine Bedingung (72.11), der die Entropie genügen muß. Im Falle der transformierten Funktionen ergibt sich ohne Rechnung aus den Transformationsgleichungen, daß in der aus den zweiten Ableitungen gebildeten quadratischen Form keine aus Ableitungen nach extensiven und intensivenParametern gemischten Terme auftreten, sondern ein Zerfall in zwei quadratische Formen stattfindet, wie es aus der Thermodynamik [vgl. (72.10)] bekannt ist. Die Stabilitätsbedingungen für die Ableitungen nach extensiven Parametern sind dann durch die Bedingung b) in Ziff. 68 gegeben. Wir haben daher jetzt noch die

---

[1] L. ONSAGER: Privatmitteilung.
[2] A. MÜNSTER u. K. SAGEL: Mol. Phys. **1**, 23 (1958).

Stabilitätsbedingungen für die Ableitungen nach intensiven Parametern zu begründen[1]. Dazu bilden wir die Umkehrung der Transformation (69.18) und schreiben

$$\mathbf{B}^* = \frac{\hat{\mathbf{Q}}}{|Q|}, \qquad (77.1)$$

wo $\hat{\mathbf{Q}}$ aus $\mathbf{Q}$ (Ziff. 69) dadurch entsteht, daß zuerst zum Element Q_{ij} in $|Q|$ der Kofaktor gebildet und die dann resultierende Matrix transponiert wird. Betrachten wir nun etwa das Matrixelement B_{kk}^*, so folgt aus (77.1) in Verbindung mit Bedingung b) in Ziff. 68 und Gl. (69.25), daß $B_{kk}^* > 0$ und somit $|Q|$ und der erste Hauptminor gleiches Vorzeichen besitzen müssen. Wir konstruieren nun eine Matrix $^{k-1}\mathbf{B}^*$, welche dem ersten Hauptminor von $|B^*|$ entspricht und schreiben für diese eine der Gl. (77.1) analoge Beziehung an. Im Nenner der rechten Seite erscheint dann der erste Hauptminor von $|Q|$ und im Zähler die nach der obigen Vorschrift aus $^{k-1}\mathbf{Q}$ gebildete Matrix. Wir haben also

$$^{k-1}\mathbf{B}^* = \frac{^{k-1}\hat{\mathbf{Q}}}{^{k-1}|Q|} \qquad (77.2)$$

und können daraus auf analoge Weise schließen, daß auch der erste und zweite Hauptminor von $|Q|$ gleiches Vorzeichen haben müssen. Fährt man so fort, so ergibt sich schließlich in Übereinstimmung mit Gl. (69.16), daß das Matrixelement $Q_{11} > 0$ ist. Wir erhalten somit das Resultat, daß

$$\left| \frac{\partial^2 \Phi_k}{\partial P_i \, \partial P_j} \right| \quad \text{und alle Hauptminoren} > 0 \qquad (77.3)$$

sein muß. Die durch eine beliebige Massieu-Plancksche Funktion ausgedrückten Stabilitätsbedingungen fordern somit, daß die quadratischen Formen

$$\Delta^2 \Phi_k (P_1, \ldots, P_k) \qquad (77.4)$$

und

$$- \Delta^2 \Phi_k (X_{k+1}, \ldots, X_{r+1}) \qquad (77.5)$$

beide positiv definit sein müssen. Dies ist die zu (72.10) analoge allgemeine Formulierung der Stabilitätsbedingungen; die Übereinstimmung beider ist leicht festzustellen. In Tabelle 2 sind als Beispiel einige Spezialfälle für ein binäres System in der üblichen thermodynamischen Schreibweise zusammengestellt.

Die statistische Deutung der Stabilitätsbedingungen ergibt sich aus der allgemeinen Schwankungstheorie der Ziff. 69 und der in Ziff. 75 und 76 entwickelten Theorie der Phasenumwandlungen. Vom Standpunkt der statistischen Theorie beziehen sich die Stabilitätsbedingungen zunächst auf das asymptotische Verschwinden der mittleren relativen Schwankungsquadrate der extensiven Parameter für unendlich große Systeme. Andererseits stellt sich die Bildung neuer Phasen statistisch als Auftreten größerer Schwankungen dar. Die mittleren relativen Schwankungsquadrate streben endlichen Grenzwerten zu, wenn die neue Phase durch diskontinuierliche Änderung aus der ursprünglichen hervorgeht; sie verschwinden von kleinerer Ordnung als $\overline{X}_i^{-1}$, wenn es sich um benachbarte Phasen handelt. Damit ist unmittelbar die Verbindung zu dem thermodynamischen Konzept der Stabilität gegeben. Wir können sie formulieren in dem Satz; Eine thermodynamisch stabile Phase ist statistisch dadurch charakterisiert, daß die mittleren relativen Schwankungsquadrate der extensiven

---

[1] A. Münster: Z. Physik **136**, 179 (1953).

Parameter für unendlich große Systeme von der Ordnung eines reziproken extensiven Parameters bzw. seines Mittelwertes verschwinden.

Die vorstehenden Betrachtungen zeigen, daß die Thermodynamik sich nur unter der Voraussetzung, daß die Stabilitätsbedingungen erfüllt sind, statistisch begründen läßt[1]. Aus Ziff. 68 ergibt sich nämlich, daß für instabile Phasen die Integrale der Verteilungsfunktionen divergieren, während man aus Ziff. 69 sieht, daß die Schwankungen dann imaginär werden. Eine korrekte statistische Berechnung der thermodynamischen Funktionen kann daher niemals instabile Zustände liefern[2]. Die Einführung derselben in die Thermodynamik bedeutet eine rein formale Fiktion ohne physikalische Grundlage.

78. Die Einstein-Kondensation als Modell einer Umwandlung. Die vollständige mathematische Analyse einer Umwandlung ist, wie schon erwähnt, bisher nur in zwei Fällen gelungen: Bei der Umwandlung des zweidimensionalen Ising-Modells und bei

[1] Wir sehen ab von den Grenzfällen der Phasenumwandlungen.

[2] Dieser Satz ist zuerst von VAN HOVE [Physica, Haag **15**, 951 (1949)] für die kanonische Gesamtheit in voller Klarheit ausgesprochen und bewiesen worden.

Tabelle 2. *Formulierung der Stabilitätsbedingungen für ein binäres System mit Hilfe einiger Massieu-Planckscher Funktionen.*

Massieu-Plancksche Funktion	Ableitungen nach extensiven Parametern	Ableitungen nach intensiven Parametern
S	$-\begin{vmatrix} \dfrac{\partial^2 S}{\partial E^2} & \dfrac{\partial^2 S}{\partial E\,\partial V} & \dfrac{\partial^2 S}{\partial E\,\partial N_1} \\[4pt] \dfrac{\partial^2 S}{\partial V\,\partial E} & \dfrac{\partial^2 S}{\partial V^2} & \dfrac{\partial^2 S}{\partial V\,\partial N_1} \\[4pt] \dfrac{\partial^2 S}{\partial N_1\,\partial E} & \dfrac{\partial^2 S}{\partial N_1\,\partial V} & \dfrac{\partial^2 S}{\partial N_1^2} \end{vmatrix} > 0,$ $\quad -\begin{vmatrix} \dfrac{\partial^2 S}{\partial E^2} & \dfrac{\partial^2 S}{\partial E\,\partial V} \\[4pt] \dfrac{\partial^2 S}{\partial V\,\partial E} & \dfrac{\partial^2 S}{\partial V^2} \end{vmatrix} > 0,$ $\quad -\dfrac{\partial^2 S}{\partial E^2} > 0,\ -\dfrac{\partial^2 S}{\partial V^2} > 0,$ $\quad -\dfrac{\partial^2 S}{\partial N_1^2} > 0$	—
$S - \dfrac{E}{T} = -\dfrac{F}{T}$	$\begin{vmatrix} \dfrac{\partial^2 (F/T)}{\partial V^2} & \dfrac{\partial^2 (F/T)}{\partial V\,\partial N_1} \\[4pt] \dfrac{\partial^2 (F/T)}{\partial N_1\,\partial V} & \dfrac{\partial^2 (F/T)}{\partial N_1^2} \end{vmatrix} > 0,$ $\quad \dfrac{\partial^2 (F/T)}{\partial V^2} > 0$ $\quad \dfrac{\partial^2 (F/T)}{\partial N_1^2} > 0$	$-\dfrac{\partial^2 (F/T)}{\partial (1/T)^2} > 0$
$S - \dfrac{E}{T} - \dfrac{PV}{T} = -\dfrac{G}{T}$	$\dfrac{\partial^2 (G/T)}{\partial N_1^2} > 0$	$-\begin{vmatrix} \dfrac{\partial^2 (G/T)}{\partial (1/T)^2} & \dfrac{\partial^2 (G/T)}{\partial (1/T)\,\partial (P/T)} \\[4pt] \dfrac{\partial^2 (G/T)}{\partial (1/T)\,\partial (P/T)} & \dfrac{\partial^2 (G/T)}{\partial (P/T)^2} \end{vmatrix} > 0,$ $\quad -\dfrac{\partial^2 (G/T)}{\partial (1/T)^2} > 0$ $\quad -\dfrac{\partial^2 (G/T)}{\partial (P/T)^2} > 0$

Entnommen aus: A. MÜNSTER: Z. Physik **136**, 200 (1953).

der Umwandlung des idealen Bose-Einstein-Gases, die gewöhnlich als Einstein-Kondensation bezeichnet wird. Wir wollen hier das letztere Problem, das mathematisch wesentlich einfacher ist, etwas ausführlicher betrachten. Für die Theorie des Ising-Modells sei auf den Artikel von Münster in Bd. XIII dieses Handbuches verwiesen.

Daß im idealen Bose-Einstein-Gas eine Umwandlung auftritt, ist zuerst von Einstein[1] bemerkt worden. London[2] hat vermutet, daß die Einstein-Kondensation in naher Beziehung zu dem λ-Punkt des Heliums steht. Wir werden indessen hier diesen physikalischen Gesichtspunkt (für dessen Erörterung wir auf den Artikel von Münster in Bd. XIII dieses Handbuches verweisen) völlig außer acht lassen und die Einstein-Kondensation lediglich als ein Modell (und in gewissem Sinne als eine Prüfung) für die in den vorhergehenden Ziffern entwickelte allgemeine Theorie betrachten. Aus diesem Grunde werden wir auch auf eine Reihe von methodischen Fragen nicht eingehen und uns auf die Behandlung des Problems mit Hilfe der großen kanonischen Gesamtheit[3-6] beschränken.

Wir betrachten ein System von N untereinander gleichen Massenpunkten der Masse m mit vernachlässigbarer Wechselwirkung, welches der Bose-Einstein-Statistik gehorcht. Die große Verteilungsfunktion des Systems lautet nach Gl. (57.9)

$$\Xi = \prod_k \left(1 - e^{\frac{\mu - \varepsilon_k}{kT}}\right)^{-1}. \tag{78.1}$$

Daraus folgt mit (64.20) und (64.26)

$$PV = -kT \sum_k \ln \left(1 - e^{\frac{\mu - \varepsilon_k}{kT}}\right) \tag{78.2}$$

und

$$\overline{N} = \sum_k \left(e^{-\frac{\mu - \varepsilon_k}{kT}} - 1\right)^{-1}. \tag{78.3}$$

Wir nehmen an, daß sich das System in einem kubischen Volumen der Kantenlänge L befindet und schreiben die Eigenwerte der Teilchen

$$\varepsilon_k = \tfrac{1}{3} \varepsilon_1 (k_x^2 + k_y^2 + k_z^2) \tag{78.4}$$

mit

$$\varepsilon_1 = \frac{3h^2}{8mL^2}. \tag{78.5}$$

Wir approximieren zunächst, wie in Ziff. 50, das Eigenwertspektrum durch die stetige Funktion

$$g(\varepsilon) = 4\pi \frac{mV}{h^3} (2m\varepsilon)^{\frac{1}{2}}. \tag{78.6}$$

Gehen wir damit in Gl. (78.2) ein, so erhalten wir

$$PV = -2\pi \left(\frac{2m}{h^2}\right)^{\frac{3}{2}} kT V \int_0^\infty \varepsilon^{\frac{1}{2}} \ln \left(1 - e^{\frac{\mu - \varepsilon}{kT}}\right) d\varepsilon. \tag{78.7}$$

[1] A. Einstein: Sitzgsber. preuß. Akad. Wiss. **3**, 18 (1925).
[2] F. London: Phys. Rev. **54**, 947 (1938).
[3] R. Becker: Z. Physik **128**, 120 (1950).
[4] D. ter Haar: Proc. Roy. Soc. Lond., Ser. A **212**, 552 (1952).
[5] P. T. Landsberg: Proc. Cambridge Phil. Soc. **50**, 65 (1954).
[6] A. Münster: Z. Physik **144**, 197 (1956).

Die Integration läßt sich ohne weiteres mit Hilfe der Reihenentwicklung des Logarithmus ausführen und ergibt

$$PV = \frac{kTV}{\lambda^3}\,g(Z),\tag{78.8}$$

wo λ durch Gl. (56.15) definiert und

$$g(Z) = \sum_{l=1}^{\infty} l^{-\frac{5}{2}}\,e^{\frac{l\mu}{kT}} = \sum_{l=1}^{\infty} l^{-\frac{5}{2}}\,(\lambda^3 Z)^l\tag{78.9}$$

ist. Z bezeichnet wieder die durch Gl. (67.23) definierte (reelle) Fugazität. Definieren wir nun cluster-Integrale durch

$$b_l = l^{-\frac{5}{2}}\,\lambda^{3(l-1)} = l^{-\frac{5}{2}}\left(\frac{h^2}{2\pi m\,kT}\right)^{\frac{3}{2}(l-1)},\tag{78.10}$$

so wird aus (78.8)

$$PV = kTV\sum b_l Z^l\tag{78.11}$$

in formaler Übereinstimmung mit der für reale Gase geltenden Gl. (67.25). Tatsächlich kann man die Gl. (78.11) auch aus der allgemeinen quantenstatistischen Theorie der realen Gase ableiten, indem man in die Definitionsgleichung der cluster-Integrale die Slater-Summen des idealen Bose-Einstein-Gases einführt[1]. Aus (78.8) folgt

$$\bar{N} = \frac{V}{\lambda^3}\,Z\,g'(Z),\tag{78.12}$$

wo

$$Z\,g'(Z) = \sum_{l=1}^{\infty} l^{-\frac{3}{2}}\,(\lambda^3 Z)^l\tag{78.13}$$

ist. Entsprechend erhält man aus (78.11)

$$\bar{N} = V\sum l\,b_l Z^l.\tag{78.14}$$

Für das Auftreten eines Umwandlungspunktes ist nach Ziff. 75 das Konvergenzverhalten der cluster-Entwicklung entscheidend. Um dasselbe zu untersuchen, setzen wir (in Analogie zur Theorie der realen Gase)

$$b_l = f(l,\,T)\,b_0^l,\tag{78.15}$$

wo

$$f(l,\,T) = l^{-\frac{5}{2}}\,\lambda^{-3}\tag{78.16}$$

und

$$b_0 = \lambda^3\tag{78.17}$$

ist. Für den Konvergenzradius ergibt sich dann aus dem Theorem von Cauchy-Hadamard

$$b_0 Z^* = 1.\tag{78.18}$$

Die Reihe (78.11) hat daher (bei endlichen Temperaturen) einen endlichen Konvergenzradius. Da alle Koeffizienten positiv reell sind, liegt notwendig eine der den Konvergenzradius bestimmenden Singularitäten auf der positiven reellen Achse. Es findet daher bei der Fugazität

$$Z^* = \lambda^{-3} = \left(\frac{2\pi m\,kT}{h^2}\right)^{\frac{3}{2}}\tag{78.19}$$

---

[1] B. Kahn: Diss. Utrecht 1938.

eine Umwandlung statt[1]. Über die Natur der Singularität und damit über die Art der Umwandlung läßt sich nur allgemein aussagen, daß an der Stelle $Z^* P$ endlich und stetig, $\bar{\varrho}$ endlich sein muß.

Die vorstehende Ableitung entspricht völlig der Mayerschen Kondensationstheorie für reale Gase. Die vom Volumen unabhängigen cluster-Integrale b_l [Gl. (78.10)] sind mit den durch Gl. (75.6) definierten Grenzwerten identisch. Die Gl. (78.11) ist eine unmittelbare Anwendung von Lemma 2 in Ziff. 75. Die Vorteile und Grenzen der cluster-Methode sind klar zu erkennen. Die Vorteile liegen in der sehr allgemeinen Anwendbarkeit. Die Schwächen sind einmal die Tatsache, daß die Bedeutung der Singularität der cluster-Reihe $\sum b_l Z^l$ nicht streng bewiesen werden kann, andererseits die Unmöglichkeit einer Aussage über den Charakter der Singularität und damit der Umwandlung. Die erste Frage haben wir bereits in Ziff. 75 und 76 ausführlich behandelt. Die zweite läßt sich naturgemäß nur für konkrete Systeme diskutieren.

Wir wollen nun die Theorie der Einstein-Kondensation in einer ganz anderen Form entwickeln und dabei zeigen, daß die Folgerungen der cluster-Theorie korrekt sind. Gleichzeitig werden wir die hier gelassenen Lücken ausfüllen und damit die allgemeine Theorie der Ziff. 75 und 76 erläutern und bestätigen.

Nach Gl. (78.3) ist der mittlere Bruchteil der Teilchen, der sich im Grundzustand befindet, gegeben durch

$$\frac{\overline{N_1}}{\overline{N}} = \frac{1}{\overline{N}} \frac{1}{e^{-\frac{\mu - \varepsilon_1}{kT}} - 1}. \tag{78.20}$$

Wegen (78.5) folgt daraus

$$\lim_{V \to \infty} \frac{\overline{N_1}}{\overline{N}} = \frac{1}{\overline{N}} \frac{1}{e^{-\frac{\mu}{kT}} - 1}. \tag{78.21}$$

Es ist zweckmäßig, die Form des auf der rechten Seite dieser Gleichung stehenden Ausdruckes auch für endliche Systeme beizubehalten. Dazu setzen wir

$$\mu' = \mu - \varepsilon_1, \quad \nu = \mu'/kT, \tag{78.22}$$

wo

$$\lim_{V \to \infty} \mu' = \mu \tag{78.23}$$

ist. Der Ersatz von μ durch μ' in den allgemeinen Formeln bedeutet dann eine Verschiebung des Nullpunktes der Energieskala, die wir in der Notierung nicht besonders hervorheben. Wesentlich ist, daß beim Grenzübergang $V \to \infty$ für $T, \mu = \mathrm{const}$ die Größe ν nicht konstant bleibt.

Wir können nun für die mittlere molekulare Dichte schreiben

$$\bar{\varrho} = V^{-1} \left[(e^{-\nu} - 1)^{-1} + \sum_{k=2}^{M-1} \left(e^{-\nu + \frac{\varepsilon_k}{kT}} - 1\right)^{-1} + \sum_{k=M}^{\infty} \left(e^{-\nu + \frac{\varepsilon_k}{kT}} - 1\right)^{-1} \right], \tag{78.24}$$

wo $M \geqq 2$ eine im übrigen beliebige positive endliche ganze Zahl ist. Wenn M hinreichend groß gewählt wird, können wir im letzten Term der rechten Seite das diskrete Spektrum mit beliebiger Genauigkeit durch ein Kontinuum approxi-

---

[1] Die Reihe (78.9) konvergiert in dem auf dem Konvergenzkreis liegenden Punkte $\lambda^3 Z = 1$. Man hat dann die Riemannsche ζ-Funktion $\zeta(s)$ für $s = \frac{5}{2}$.

mieren. Wir bekommen dann mit Benutzung von (78.5)

$$\bar{\varrho} = V^{-1}\left[(e^{-\nu}-1)^{-1} + \sum_{k=2}^{M}\left(e^{-\nu+\frac{\varepsilon_k}{kT}}-1\right)^{-1} + \right.$$
$$\left. + 2\pi\left(\frac{2m}{h^2}\right)^{\frac{3}{2}}V\int_{\varepsilon_M}^{\infty}\varepsilon^{\frac{1}{2}}\left(e^{-\nu+\frac{\varepsilon}{kT}}-1\right)^{-1}d\varepsilon\right]. \tag{78.25}$$

Multiplizieren wir beide Seiten dieser Gleichung mit $(e^{-\nu}-1)$ und gehen zur Grenze über, so verschwinden auf der rechten Seite die beiden ersten Terme, und es wird $\varepsilon_M \to 0$ für $V \to \infty$. Wir erhalten somit

$$\lim_{V\to\infty}(e^{-\nu}-1)\,\bar{\varrho} = (e^{-\nu}-1)\,\lambda^{-3}Z\,g'(Z). \tag{78.26}$$

Für $\nu < 0$ bzw. $Z < Z^*$ geht diese Gleichung in Gl. (78.12) über, womit die Korrektheit der früheren Ableitung bestätigt und das Lemma 2 der Ziff. 75 für diesen Fall noch einmal explizit bewiesen ist. Man bemerkt, daß die Benutzung des kontinuierlichen Eigenwertspektrums (78.6) bereits den Grenzübergang zum unendlich großen System und damit die Einführung der Grenzwerte der cluster-Integrale impliziert.

Wir gehen nun zur Untersuchung der Singularität über. Aus Gl. (78.24) folgt zunächst, daß für endliche Systeme $\nu < 0$ sein muß. In dem damit zugelassenen Wertebereich ist $\bar{\varrho}$ bzw. PV eine durchweg reguläre Funktion von μ, wie es die allgemeine Theorie fordert. Nur für $V \to \infty$ kann ν den Wert Null annehmen; aus (78.9), (78.19) und (79.23) ergibt sich, daß dann auch $\mu/kT = 0$ ist und dieser Wert der Singularität der cluster-Entwicklung bzw. der Funktion $g(Z)$ entspricht. Damit ist für diesen Fall explizit gezeigt, daß die erste Singularität von $\lim\limits_{V\to\infty}\bar{\varrho}$ auf der positiven reellen Achse mit der Singularität der cluster-Entwicklung $\sum b_l Z_l$ zusammenfällt. Die Funktion

$$g(\nu) \equiv g(\tfrac{5}{2},\nu) = \sum_{l=1}^{\infty}l^{-\frac{5}{2}}e^{l\nu} \tag{78.27}$$

die das analytische Verhalten der cluster-Entwicklung bestimmt, ist für $\nu < 0$ regulär und hat an der Stelle $\nu = 0$ einen Verzweigungspunkt. Das gleiche gilt für die Ableitungen, von denen

$$g'(\nu) \equiv g(\tfrac{3}{2},\nu) = \sum_{l=1}^{\infty}l^{-\frac{3}{2}}e^{l\nu} \tag{78.28}$$

ebenso wie $g(\nu)$, an der Stelle $\nu = 0$ endlich ist, während

$$g''(\nu) \equiv g(\tfrac{1}{2},\nu) = \sum_{l=1}^{\infty}l^{-\frac{1}{2}}e^{l\nu} \tag{78.29}$$

und alle höheren Ableitungen hier unendlich werden. In der Umgebung der singulären Stelle gelten, wie Robinson[1] gezeigt hat, die Entwicklungen

$$g(\nu) \equiv g(\tfrac{5}{2},\nu) = 2{,}36(-\nu)^{\frac{3}{2}} + 1.34 + 2{,}61\nu - 0{,}730\nu^2 - \cdots, \tag{78.30}$$

$$g'(\nu) \equiv g(\tfrac{3}{2},\nu) = -3{,}54(-\nu)^{\frac{1}{2}} + 2{,}61 - 1{,}46\nu - 0{,}104\nu^2 - \cdots, \tag{78.31}$$

$$g''(\nu) \equiv g(\tfrac{1}{2},\nu) = 1{,}77(-\nu)^{-\frac{1}{2}} - 1{,}46 - 0{,}208\nu + \cdots. \tag{78.32}$$

---

[1] J. E. Robinson: Phys. Rev. **83**, 678 (1951).

Es sei nun $v = V/\overline{N}$ das mittlere Volumen pro Molekül. Wir definieren

$$v_c = \frac{\lambda^3}{g'(0)} = \frac{\lambda^3}{2{,}61} \tag{78.33}$$

und zeigen zunächst, daß an der Stelle $v = 0$ notwendig $v \leqq v_c$ ist. Dazu schreiben wir die Gl. (78.25) in einer etwas vereinfachten Form. Wir setzen

$$\sum_{k=2}^{M-1} \left(e^{-\nu + \frac{\varepsilon_k}{kT}} - 1 \right)^{-1} = f(V) \tag{78.34}$$

und haben dann

$$\lim_{V \to \infty} V^{-1} (e^{-\nu} - 1)\, f(V) = 0. \tag{78.35}$$

Ferner schreiben wir

$$2\pi \left(\frac{2m}{h^2} \right)^{\frac{3}{2}} \int_{\varepsilon_M}^{\infty} \varepsilon^{\frac{1}{2}} \left(e^{-\nu + \frac{\varepsilon}{kT}} - 1 \right)^{-1} d\varepsilon = \lambda^{-3} g'(\nu) + \varphi(V), \tag{78.36}$$

wo

$$\lim_{V \to \infty} \varphi(V) = 0 \tag{78.37}$$

ist. Da wir jetzt nur den Fall $0 \leqq -\nu \ll 1$ betrachten, können wir $e^{-\nu}$ entwickeln und mit dem linearen Gliede abbrechen. Dann wird

$$-\nu \overline{N} = 1 - \nu \lambda^{-3} v\, g'(\nu) - \nu [f(V) + \varphi(V)]. \tag{78.38}$$

Führen wir hier die Entwicklung (78.31) ein, so erhalten wir mit (78.33)

$$\nu \overline{N} = -1 + \frac{\nu \overline{N} v}{\lambda^3} \left[\frac{\lambda^3}{v_c} - 3{,}54(-\nu)^{\frac{1}{2}} \right] + \nu f(V) + \nu \varphi(V). \tag{78.39}$$

Man sieht nun leicht, daß für $\nu = 0$ (d.h. $\mu/kT = 0$, $V \to \infty$) nicht $v > v_c$ sein kann. Schreiben wir nämlich mit der Annahme $v > v_c$ Gl. (78.39) in der Form

$$\overline{\varrho} \left(1 - \frac{v}{v_c} \right) = -(Vv)^{-1} - 3{,}54 \lambda^{-3} (-\nu)^{\frac{1}{2}} + V^{-1} f(V) + V^{-1} \varphi(V), \tag{78.40}$$

so steht auf der linken Seite eine negative endliche Größe. Setzen wir nun $\mu/kT = 0$, so verschwinden beim Grenzübergang $V \to \infty$ auf der rechten Seite der zweiten und der letzte Term, während die beiden anderen positiven Grenzwerten (mit Einschluß von Null) zustreben. Die Gleichung ist daher nicht mehr erfüllt und die Behauptung bewiesen. Es folgt, daß bei gegebener Temperatur v_c das mittlere Volumen pro Molekül ist, für welches (bei dem unendlich großen System) $\nu = 0$ bzw. $Z = Z^*$ wird. Die für den Charakter der Umwandlung entscheidende Frage ist nun, wie ν gegen Null geht, wenn wir für $\mu/kT = 0$ zur Grenze $V \to \infty$ übergehen. Führen wir diesen Grenzübergang in Gl. (78.39) aus, so verschwindet auf der rechten Seite der letzte Term, während der vorletzte einem negativen endlichen Grenzwert zustrebt und $v \to v_c$ geht. Es folgt daher

$$-\nu = O(\overline{N}^{-\frac{2}{3}}). \tag{78.41}$$

Wir betrachten jetzt die Ableitungen des thermodynamischen Potentials PV nach μ an der Stelle $\mu = 0$. Aus Gl. (78.39) haben wir

$$\frac{1}{V} \frac{\partial(PV)}{\partial \mu} = \overline{\varrho} = \lambda^{-3} g'(0) - 3{,}54 \lambda^{-3} (-\nu)^{\frac{1}{2}} - (Vv)^{-1} + V^{-1} [f(V) + \varphi(V)]. \tag{78.42}$$

Mit Benutzung von (78.33), (78.34) und (78.41) folgt daraus

$$\bar{\varrho} = v_c^{-1} - O(\bar{N}^{-\frac{1}{3}}) \qquad (\mu/kT = 0,\ V \to \infty). \tag{78.43}$$

Es wird somit

$$\lim_{V \to \infty} \bar{\varrho} = v_c^{-1} \qquad (\mu/kT = 0) \tag{78.44}$$

in Übereinstimmung mit dem früheren Resultat. Durch Differentiation der Gl. (78.42) erhält man in analoger Weise

$$\frac{\partial \bar{\varrho}}{\partial \mu} = O(\bar{N}^{\frac{1}{3}}) \qquad (\mu/kT = 0,\ V \to \infty) \tag{78.45}$$

$$\frac{\partial^2 \bar{\varrho}}{\partial \mu^2} = O(\bar{N}) \qquad (\mu/kT = 0,\ V \to \infty), \tag{78.46}$$

und allgemein

$$\frac{1}{V}\frac{\partial^n (PV)}{\partial \mu^n} = O(\bar{N}^{\frac{1}{3}(2n-3)}) \qquad (\mu/kT = 0,\ V \to \infty,\ n \geqq 2). \tag{78.47}$$

Bisher haben wir lediglich die Isothermen der Einstein-Kondensation betrachtet. Aus Gl. (78.19) bzw. (78.33), sieht man, daß die Umwandlung sich auch erreichen läßt, indem man bei vorgegebener Fugazität bzw. vorgegebenem Volumen die Temperatur erniedrigt. Man kann also (für das unendlich große System) eine Umwandlungstemperatur T_c definieren durch

$$T_c = \frac{h^2}{2\pi m k} Z^{\frac{2}{3}} \tag{78.48}$$

oder

$$T_c = \frac{h^2}{2\pi m k} (2{,}61\,v)^{-\frac{2}{3}}. \tag{78.49}$$

In Verbindung mit Gl. (78.33) ergibt sich daraus die bemerkenswerte Beziehung (für $T \geqq T_c$, $v \geqq v_c$)

$$\left[\frac{T}{T_c(v)}\right]^{\frac{3}{2}} = \frac{v}{v_c(T)} = 2{,}61 \left(\frac{2\pi m k T}{h^2}\right)^{\frac{3}{2}}. \tag{78.50}$$

Für einen durch T und v gegebenen Zustand des Gases ist also das Verhältnis der Temperatur zur Umwandlungstemperatur bei dem betreffenden Volumen, zur Potenz $\frac{3}{2}$ erhoben, gleich dem Verhältnis des Volumens zum Umwandlungsvolumen bei der betreffenden Temperatur.

Die Gl. (78.48) hat zur Folge, daß im Rahmen der großen kanonischen Gesamtheit Temperaturen $T < T_c$ nicht definierbar sind, da Z für T_c jeweils einen maximalen Wert erreicht, der nicht überschritten werden kann. Wir werden auf diesen Punkt noch einmal zurückkommen. Da für die unabhängigen Variablen T und v keine derartige Beschränkung existiert, kann das Gebiet $T < T_c$ mit Hilfe der kleinen kanonischen Gesamtheit untersucht werden. Wir wollen dies hier nicht explizit durchführen, werden aber einige der auf diese Weise erhaltenen Ergebnisse für $T < T_c$ zur Abrundung des Bildes in die folgenden Formeln mit einschließen. Die hier gegebenen Ableitungen sind streng nur für $T = T_c$ gültig.

Für $T \leqq T_c$ gilt eine einfache Beziehung zwischen der Temperatur und dem Bruchteil der Teilchen, der sich im Grundzustand befindet[1]. Wir schreiben zunächst die Gl. (78.38) mit Benutzung von (78.20) in der Form

$$\bar{N} = \bar{N}_1 + V \lambda^{-3} g'(v) + f(V) + \varphi(V). \tag{78.51}$$

---

[1] F. London: Phys. Rev. **54**, 947 (1938).

Mit (78.33) wird daraus

$$1 = \frac{\overline{N}_1}{\overline{N}} + \frac{v}{v_c}\frac{g'(v)}{g'(0)} + \overline{N}^{-1}[f(V) + \varphi(V)]. \tag{78.52}$$

Gehen wir an der Stelle $\mu/kT = 0$ zur Grenze $V \to \infty$ über, so verschwinden die beiden letzten Terme[1], und es wird zunächst

$$\lim_{V \to \infty} \frac{\overline{N}_1}{\overline{N}} = 1 - \frac{v}{v_c}. \qquad (v \leqq v_c). \tag{78.53}$$

Mit Gl. (78.50) folgt dann

$$\lim_{V \to \infty} \frac{\overline{N}_1}{\overline{N}} = 1 - \left(\frac{T}{T_c}\right)^{\frac{3}{2}} \qquad (T \leqq T_c). \tag{78.54}$$

Wir betrachten jetzt die mittlere Energie des Bose-Einstein-Gases und ihre Abhängigkeit von der Temperatur. Aus der Theorie der großen Verteilungsfunktion folgt allgemein

$$\overline{E} = kT^2\left[\frac{\partial(PV/kT)}{\partial T}\right]_{V,\mu/T}. \tag{78.55}$$

Für $T > T_c$ erhält man daraus mit Benutzung von Gl. (78.8)

$$\overline{E} = \tfrac{3}{2}kT\,V\,\lambda^{-3}g(v). \tag{78.56}$$

Zu dem gleichen Ausdruck gelangt man auch durch Anwendung des Virialsatzes (Ziff. 45). Für die Ableitungen der mittleren Energie nach der Temperatur bei konstantem μ/T ergibt sich

$$\left(\frac{\partial \overline{E}}{\partial T}\right)_{V,\mu/T} = \frac{15}{4}\,kV\,\lambda^{-3}g(v) \tag{78.57}$$

und

$$\left(\frac{\partial^2 \overline{E}}{\partial T^2}\right)_{V,\mu/T} = \frac{45}{8}\,kT^{-1}\,V\,\lambda^{-3}g(v). \tag{78.58}$$

Aus Gl. (78.56) folgt, daß die mittlere Energie bei konstantem μ/T mit $T^{\frac{5}{2}}$ abnimmt. Es ist bemerkenswert, daß dies auch für das klassische ideale Gas gilt. Dagegen ist die Temperaturabhängigkeit der mittleren Energie verschieden, wenn an Stelle der Größe μ/T das Volumen pro Molekül konstant gehalten wird. Wir wollen die ziemlich umständliche Umrechnung auf die unabhängige Variable v hier nicht durchführen und notieren lediglich das Ergebnis. Man erhält

$$\overline{E} = \tfrac{3}{2}NkT(1 - 0{,}177\,\lambda^3 v^{-1} - 0{,}003\,\lambda^6 v^{-2} - \cdots), \tag{78.59}$$

$$\left(\frac{\partial \overline{E}}{\partial T}\right)_{N,V} = \frac{3}{2}Nk(1 + 0{,}088\,\lambda^3 v^{-1} + 0{,}007\,\lambda^6 v^{-2} + \cdots), \tag{78.60}$$

$$\left(\frac{\partial^2 \overline{E}}{\partial T^2}\right)_{N,V} = -\frac{3}{2}NkT^{-1}(0{,}133\,\lambda^3 v^{-1} + 0{,}02\,\lambda^6 v^{-2} + \cdots). \tag{78.61}$$

Für $T \leqq T_c$ muß wieder eine genauere Untersuchung durchgeführt werden, bei der wir von Gl. (78.2) auszugehen haben. Mit Gl. (78.55) folgt daraus zunächst

$$\overline{E} = \sum_k \varepsilon_k\left(e^{-v+\frac{\varepsilon_k}{kT}} - 1\right)^{-1}. \tag{78.62}$$

---

[1] Das folgt aus Gl. (78.41). Die allgemeinere Formulierung bei der Diskussion der Gl. (78.40) war notwendig, weil Gl. (78.41) hier noch nicht zur Verfügung stand.

Diese Gleichung schreiben wir in Analogie zu (78.24)

$$E/\overline{V} = V^{-1}\left[\varepsilon_1(e^{-\nu}-1)^{-1} + \sum_{k=1}^{M-1}\varepsilon_k\left(e^{-\nu+\frac{\varepsilon_k}{kT}}-1\right)^{-1} + \right.$$
$$\left. + 2\pi\left(\frac{2m}{h^2}\right)^{\frac{3}{2}}V\int_{\varepsilon_M}^{\infty}\varepsilon^{\frac{3}{2}}\left(e^{-\nu+\frac{\varepsilon}{kT}}-1\right)^{-1}d\varepsilon\right]. \qquad (78.63)$$

Wir setzen nun

$$2\pi\left(\frac{2m}{h^2}\right)^{\frac{3}{2}}V\int_{\varepsilon_M}^{\infty}\varepsilon^{\frac{3}{2}}\left(e^{-\nu+\frac{\varepsilon}{kT}}-1\right)^{-1}d\varepsilon = \frac{3}{2}\frac{kTV}{\lambda^3}g(\nu) + \varphi^*(V), \qquad (78.64)$$

wo

$$\lim_{V\to\infty}\varphi^*(V) = 0 \qquad (78.65)$$

ist. Dann wird aus (78.63)

$$\overline{E}/V = V^{-1}\left[\varepsilon_n(e^{-\nu}-1)^{-1} + \sum_{k=2}^{M-1}\varepsilon_k\left(e^{-\nu+\frac{\varepsilon_k}{kT}}-1\right)^{-1} + \tfrac{3}{2}kTV\lambda^{-3}g(\nu) + \varphi^*(V)\right]. \qquad (78.66)$$

Gehen wir zur Grenze über, so folgt für $T>T_c$ (da dann $\nu<0$ ist) wieder Gl. (78.56). Für $T=T_c$ wird beim Grenzübergang nach Gl. (78.48) $\nu=0$, und es folgt

$$\lim_{V\to\infty}\overline{E}/V = \frac{3}{2}kT_c\lambda^{-3}g(0) = \frac{3}{2}\cdot 1{,}34\,kT_c\left(\frac{2\pi mkT_c}{h^2}\right)^{\frac{3}{2}}, \qquad (78.67)$$

wo T_c durch Gl. (78.48) oder Gl. (78.49) definiert ist. Für $T\leqq T_c$ erhält man mit Hilfe der kleinen kanonischen Gesamtheit

$$\lim_{N,V\to\infty}\overline{E}/V = \frac{3}{2}\cdot 1{,}34\,kT\left(\frac{2\pi mkT}{h^2}\right)^{\frac{3}{2}}. \qquad (78.68)$$

Dieser Ausdruck schließt sich, wie es sein muß, stetig an (78.67) an. Mit dem Einsetzen der Einstein-Kondensation hängt somit die Energiedichte nur noch von der absoluten Temperatur ab. Sie bleibt daher bei einer isothermen Kompression konstant.

Das Verhalten der Ableitungen von $\overline{E}$ nach der Temperatur hängt wieder davon ab, welche Zustandsvariablen konstant gehalten werden. Da für alle Umwandlungstemperaturen $\mu/kT=0$ ist, folgt, daß man durch Abkühlen bei konstantem μ/T überhaupt keine Umwandlung erreicht. Die Funktion $\overline{E}=\overline{E}(T)_{\mu/T}$ ist also stets regulär und liefert keine Aussagen über die Umwandlungstemperatur der Einstein-Kondensation. Um die Änderung der Energie mit der Temperatur bei konstantem Volumen zu untersuchen, schreiben wir zunächst die Gl. (78.68) mit Benutzung von (78.49)

$$\overline{\varepsilon} = \frac{3}{2}\frac{1{,}34}{2{,}61}\,kT\left(\frac{T}{T_c}\right)^{\frac{3}{2}}. \qquad (78.69)$$

Da für gegebenes V und $N\,T_c$ eine Konstante ist, folgt daraus

$$\left(\frac{\partial\overline{\varepsilon}}{\partial T}\right)_{N,V} = 1{,}93\,k\left(\frac{T}{T_c}\right)^{\frac{3}{2}} \qquad (T\leqq T_c) \qquad (78.70)$$

und

$$\left(\frac{\partial^2\overline{\varepsilon}}{\partial T^2}\right)_{N,V} = 2{,}89\,kT^{-1}\left(\frac{T}{T_c}\right)^{\frac{3}{2}} \qquad (T\leqq T_c). \qquad (78.71)$$

Der Vergleich dieser Beziehungen mit den Gln. (78.59) bis (78.61) zeigt, daß bei konstantem Volumen und konstanter Teilchenzahl für $T = T_c$ $\bar\varepsilon$ und $\partial\bar\varepsilon/\partial T$ stetig sind, während $\partial^2\bar\varepsilon/\partial T^2$ eine endliche Unstetigkeit besitzt.

Es bleibt noch die Frage zu erörtern, welchem Umwandlungstyp die Einstein-Kondensation entspricht[1]. Wir betrachten dazu zunächst noch einmal die Isothermen. Die P-Z-Isothermen sind durch die Gl. (78.11) gegeben. Sie brechen jeweils im Punkte

$$P = 1{,}34\,kT\,\lambda^{-3} \tag{78.72}$$

ab, in dem die Einstein-Kondensation einsetzt. Die Umwandlungspunkte liegen auf einer Kurve, als deren Gleichung sich aus (78.19) und (78.72)

$$P = 1{,}34\,\frac{h^2}{2\pi m}\,Z^{\frac{5}{3}} \tag{78.73}$$

ergibt. Die P-v-Isothermen lassen sich nach den aus der Theorie der realen Gase (vgl. den Artikel von J. E. MAYER in Bd. XII dieses Handbuches) bekannten Methoden berechnen. In Fig. 10 ist eine solche Isotherme, sowie eine Isotherme des klassischen idealen Gases für die gleiche Temperatur dargestellt.

Man sieht, daß die Kurve des idealen Bose-Einstein-Gases in der Tat eine auffallende Ähnlichkeit mit der Isotherme eines realen Gases in der Umgebung des Kondensationspunktes hat. Zwei wesentliche Unterschiede sind jedoch hervorzuheben. Einmal brechen die Isothermen des idealen Bose-Einstein-Gases mit dem horizontalen Stück ab; es fehlt der der flüssigen Phase entsprechende Zweig. Zum anderen erfolgt die Annäherung an den horizontalen Verlauf stetig, während sie bei realen Gasen jedenfalls in einiger Entfernung vom kritischen Punkt, unstetig ist. Als Gleichung der Kurve, auf der im P-v-Diagramm die Umwandlungspunkte liegen, ergibt sich aus den Gln. (78.33) und (78.72)

$$P = 1{,}34\,\frac{h^2}{2\pi m}\,(2{,}61\,v)^{-\frac{5}{3}}. \tag{78.74}$$

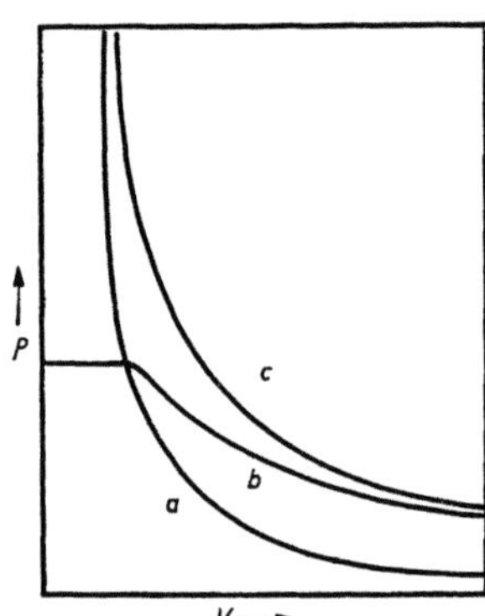

Fig. 10. Kurve a: Grenzkurve der Einstein-Kondensation; Kurve b: Isotherme des idealen Bose-Einstein-Gases; Kurve c: Isotherme des klassischen idealen Gases für die gleiche Temperatur wie Kurve b.

Diese Kurve ist ebenfalls in Fig. 10 dargestellt.

In Ziff. 76 haben wir gezeigt, daß Umwandlungen eine spezielle Art von Schwankungserscheinungen darstellen. Wir wollen dieselben daher jetzt für den Fall der Einstein-Kondensation untersuchen. Für das mittlere relative Schwankungsquadrat der molekularen Dichte ergibt sich aus Gln. (65.8) und (78.45)

$$\overline{\frac{(\varrho - \bar\varrho)^2}{\bar\varrho^2}} = O(\bar N^{-\frac{2}{3}}) \qquad (\mu/kT = 0,\ V \to \infty). \tag{78.75}$$

Ferner ist nach Gl. (78.45)

$$\lim_{V \to \infty} \frac{\partial\mu}{\partial\bar\varrho} = 0. \tag{78.76}$$

---

[1] Im Hinblick auf das oben erörterte Verhalten der Molekülwärme bei konstantem Volumen wird die Einstein-Kondensation in der Literatur häufig als Umwandlung dritter Ordnung bezeichnet. Diese Terminologie erscheint wenig zweckmäßig, weil das Ehrenfestsche Schema (für Einkomponentensysteme) thermodynamische Potentiale mit zwei intensiven Parametern voraussetzt, also etwa die freie Energie nach GIBBS. Die Molekülwärme bei konstantem Druck wird aber im Umwandlungspunkt der Einstein-Kondensation unendlich. [L. GOLDSTEIN: J. Chem. Phys. **14**, 276 (1946).]

An der Stelle $\mu/kT = 0$ bzw. $Z = Z^*$ wird somit die Grenze des Stabilitätsbereiches erreicht. Das relative mittlere Schwankungsquadrat der molekularen Dichte verschwindet zwar für $\overline{N} \to \infty$, aber von kleinerer Ordnung als $\overline{N}^{-1}$. Nach Ziff. 76 bedeutet dies, daß es sich um eine Phasenumwandlung handelt, bei welcher zu der ursprünglichen Phase benachbarte Phasen (d.h. Phasen, die sich von der ursprünglichen beliebig wenig unterscheiden) gebildet werden. Daraus folgt unmittelbar, daß zwischen den koexistierenden Phasen keine endliche Grenzflächenspannung bestehen kann. Wir können daher das dem horizontalen Stück der P-v-Isothermen entsprechende Zustandsgebiet als „quasi-heterogenes Gebiet" bezeichnen. Es ist erwähnenswert, daß die Möglichkeit eines Gleichgewichtes zwischen benachbarten Phasen bereits von GIBBS[1] bemerkt wurde, der sie jedoch als unwahrscheinlich betrachtet und nicht näher untersucht hat. Schließlich muß hier nochmals erwähnt werden, daß die P-v-Isothermen mit horizontaler Tangente in den Umwandlungspunkt einmünden und entsprechend $\partial\overline{\varrho}/\partial\mu$ sich stetig dem Wert Unendlich nähert. Eine derartige Umwandlung ist von MAYER und STREETER[2] als „anomale Umwandlung erster Ordnung" bezeichnet worden. Ihr Auftreten ist zuerst von MAYER und HARRISON[3] bei realen Gasen für ein Gebiet zwischen der kritischen Temperatur T_k und einer Temperatur T_m, bei welcher der Meniscus erscheint, vermutet worden. Dieses Problem ist jedoch bisher weder theoretisch noch experimentell völlig geklärt worden, während sich die Einstein-Kondensation streng als anomale Umwandlung erster Ordnung identifizieren läßt.

Wir betrachten nun die Schwankungen der Energiedichte. Für die Quantenstatistik muß die Gl. (76.9) naturgemäß in der Form

$$\frac{\overline{(E/V - \overline{E}/V)^2}}{(\overline{E}/V)^2} = -\frac{kT^2}{V(\overline{E}/V)^2}\left(\frac{\partial(\overline{E}/V)}{\partial T}\right)_{\frac{\mu}{T}} \tag{78.77}$$

geschrieben werden. Aus Gl. (78.66) erhält man

$$\left.\begin{aligned}\left(\frac{\partial(\overline{E}/V)}{\partial T}\right)_{\frac{\mu}{T}} &= \frac{\varepsilon_1^2}{VkT^2(e^{-\nu}-1)^2} + \sum_{k=2}^{M-1}\frac{\varepsilon_k^2}{VkT^2\left(e^{-\nu+\frac{\varepsilon_k}{kT}}-1\right)^2} + \\ &\quad + \frac{15}{4}k\lambda^{-3}g(\nu) + \frac{1}{V}\frac{\partial\varphi^*(V)}{\partial T}.\end{aligned}\right\} \tag{78.78}$$

Setzen wir dies in Gl. (78.77) ein und gehen an der Stelle $\mu/kT = 0$ zur Grenze über, so folgt

$$\frac{\overline{(E/V - \overline{E}/V)^2}}{(\overline{E}/V)^2} = O(\overline{N}^{-1}). \tag{78.79}$$

Das mittlere relative Schwankungsquadrat der Energiedichte verschwindet somit im Umwandlungspunkt wie $\overline{N}^{-1}$, d.h. von der gleichen Ordnung wie im homogenen Gebiet. Dieses Ergebnis scheint auf den ersten Blick der allgemeinen Theorie der Ziff. 76 zu widersprechen, bietet aber tatsächlich eine Bestätigung derselben. Aus Gl. (78.68) folgt nämlich, daß alle koexistierenden Phasen die gleiche Energiedichte besitzen. Unter diesen Umständen kann daher, wenn die

[1] J. W. GIBBS: On the Equilibrium of Heterogeneous Substances, Collected Works, Vol. I. New Haven 1948.
[2] J. E. MAYER u. F. S. STREETER: J. Chem. Phys. 7, 1019 (1939).
[3] J. E. MAYER u. S. F. HARRISON: J. Chem. Phys. 6, 87 (1938).

Verknüpfung von Phasenumwandlungen und Schwankungserscheinungen, welche die Grundlage der allgemeinen Theorie bildet, richtig ist, hier im Umwandlungspunkt keine zusätzliche Schwankung der Energiedichte auftreten. Es handelt sich dabei um eine spezielle Eigentümlichkeit der Einstein-Kondensation. Durch Integration der Gl. (78.70) zeigt man leicht, daß alle koexistierenden Phasen auch die gleiche Entropiedichte besitzen.

Schließlich ist noch eine spezielle Eigenschaft der Einstein-Kondensation die Tatsache, daß die Grenze des quasi-heterogenen Gebietes mit der Grenze des Zustandsfeldes zusammenfällt. Da das quasi-heterogene Gebiet in einer Darstellung mit zwei intensiven Parametern zu einer Linie (der Koexistenzkurve) entartet (vgl. Ziff. 72), ergibt sich unmittelbar, daß die P-Z-Isothermen bei der Fugazität Z^* enden und die diese Endpunkte verbindende Koexistenzkurve (78.73) das Zustandsfeld begrenzt. Das Verhalten im quasi-heterogenen Gebiet kann also, wie erwähnt, in dieser Darstellung nicht erfaßt werden.

Es ist von vornherein zu erwarten, daß die anomale Umwandlung erster Ordnung in naher Beziehung zu der Umwandlung am kritischen Punkt steht. Letztere ist für das Gittermodell einer zweidimensionalen binären Lösung (kritischer Punkt der Entmischung) in gewissem Umfang der strengen expliziten Behandlung zugänglich, da dieses Problem mathematisch dem zweidimensionalen Ising-Modell äquivalent ist. Aus der Onsagerschen Theorie des Ising-Modells läßt sich eine Formel für das mittlere relative Schwankungsquadrat der Energiedichte am kritischen Punkt ableiten [26]. Man findet, daß dasselbe für $N \to \infty$ wie $N^{-1} \ln N$ verschwindet. Für die Schwankungen der molekularen Dichten hat sich die Rechnung bisher nicht durchführen lassen. Man kann hier aber wohl ein analoges Verhalten vermuten. Sieht man diese Ergebnisse als repräsentativ an, so könnte man sagen, daß die mittleren relativen Schwankungsquadrate bei der anomalen Umwandlung erster Ordnung wie $N^{-1} \cdot N^x (0 < x < 1)$ verschwinden, im kritischen Punkt dagegen wie $N^{-1} \cdot \ln N$ und im homogenen Gebiet wie $N^{-1} \cdot N^0$. Danach würde es plausibel erscheinen, wenn unterhalb des kritischen Punktes die anomale Umwandlung erster Ordnung als eine Art Zwischenstufe aufträte. Indessen muß eine solche Verallgemeinerung vorläufig als rein hypothetisch betrachtet werden, obschon nach neuesten Messungen[1] die Möglichkeit eines derartigen Verhaltens im Falle der Kondensation nicht auszuschließen ist.

79. Das eindimensionale System. Unter einem eindimensionalen System versteht man allgemein ein System von Teilchen, die zu einer linearen Kette aneinandergereiht sind und nur einen Freiheitsgrad der Schwerpunktstranslation besitzen. Solche Systeme lassen sich naturgemäß physikalisch nicht realisieren; sie sind aber theoretisch aus zwei Gründen von großem Interesse. Zunächst kann man eine Reihe von mathematischen Problemen der statistischen Mechanik für den eindimensionalen Fall exakt lösen, während man für den physikalisch interessierenden dreidimensionalen Fall auf Näherungsmethoden angewiesen ist. Man kann daher erwarten, daß das Studium der eindimensionalen Systeme sowohl über die Natur der Probleme wie über das Wesen und die Brauchbarkeit der Näherungsmethoden wertvolle Aufschlüsse liefert. Sodann besitzen die eindimensionalen Systeme die Eigenschaft, daß ihre statistischen bzw. thermodynamischen Funktionen durchweg regulär sind; sie können daher weder Phasenumwandlungen noch Umwandlungen höherer Ordnung zeigen. Diese Tatsache ist für das Verständnis der Umwandlungen von großer Bedeutung. Die statistische

---

[1] R. H. Wentorf: J. Chem. Phys. **24**, 607 (1956).

Mechanik der eindimensionalen Systeme ist daher von zahlreichen Autoren[1-17] unter verschiedenen Gesichtspunkten studiert worden. Wir geben hier als Beispiel eine sehr allgemeine Methode wieder, die von GÜRSEY[9] entwickelt worden ist.

Das betrachtete System bestehe aus N gleichen Teilchen der Masse m, die sich entlang einer Geraden der Länge L bewegen können. Die Koordinate des i-ten Teilchens bezeichnen wir mit q_i, seinen Impuls mit p_i. Dann lautet die halbklassische Verteilungsfunktion

$$Q = \frac{1}{h^N N!} \int\limits_0^L dq_1 \ldots \int\limits_0^L dq_N \ldots \int\limits_{-\infty}^{+\infty} \ldots \int\limits_{-\infty}^{+\infty} e^{-\frac{H(\boldsymbol{q},\boldsymbol{p})}{kT}} dp_1 \ldots dp_n. \tag{79.1}$$

Für die Hamilton-Funktion setzen wir

$$H(\boldsymbol{q}, \boldsymbol{p}) = \sum_{i=1}^N \frac{p_i^2}{2m} + U(q_1, \ldots, q_N) + U', \tag{79.2}$$

wo U das Potential der zwischenmolekularen Wechselwirkung und U' das Potential der von der „Wand" (d.h. der Begrenzung des Systems) auf die Teilchen ausgeübten Kraft ist. Nach Ausführung der Integration über die Impulse wird aus (79.1)

$$Q = \left(\frac{2\pi m kT}{h^2}\right)^{\frac{1}{2}N} \frac{Q_\tau}{N!}, \tag{79.3}$$

wo

$$Q_\tau = \int\limits_0^L \ldots \int\limits_0^L e^{-\frac{U+U'}{kT}} dq_1 \ldots dq_N \tag{79.4}$$

wieder das Konfigurationsintegral ist. Wir nehmen nun an, daß die potentielle Energie der zwischenmolekularen Wechselwirkung U sich darstellen läßt als Summe der Wechselwirkungsenergien zwischen je zwei nächsten Nachbarn in der Kette. Physikalisch bedeutet dies eine Beschränkung auf zwischenmolekulare Kräfte kurzer Reichweite (van der Waalssche Kräfte). Diese Annahme stellt naturgemäß nicht den allgemeinsten Fall dar; sie ist aber wesentlich für die hier benutzte mathematische Methode. Wir numerieren die Teilchen so, daß

$$0 < q_1 < q_2 < \cdots < q_N < L \tag{79.5}$$

ist. Dann besagt die obige Annahme

$$U = \sum_{i=1}^{N-1} u(q_{i+1} - q_i). \tag{79.6}$$

[1] K. F. HERZFELD u. M. GOEPPERT-MAYER: J. Chem. Phys. **2**, 38 (1934).
[2] E. ISING: Z. Physik **31**, 253 (1935).
[3] L. TONKS: Phys. Rev. **50**, 955 (1936).
[4] T. NAGAMIYA: Proc. Phys.-Math. Soc. Japan **22**, 705 (1940).
[5] H. TAKAHASI: Proc. Phys.-Math. Soc. Japan **24**, 60 (1942).
[6] H. N. V. TEMPERLEY: Proc. Cambridge Phil. Soc. **40**, 239 (1944).
[7] G. S. RUSHBROOKE u. H. D. URSELL: Proc. Cambridge Phil. Soc. **44**, 263 (1948).
[8] H. HARTMANN: Z. Naturforsch. **3a**, 617 (1948).
[9] F. GÜRSEY: Proc. Cambridge Phil. Soc. **46**, 182 (1950).
[10] L. VAN HOVE: Physica, Haag **16**, 137 (1950).
[11] R. KIKUCHI: J. Chem. Phys. **19**, 1230 (1951).
[12] Z. W. SALSBURG, R. W. ZWANZIG u. J. G. KIRKWOOD: J. Chem. Phys. **21**, 1098 (1954).
[13] A. RAMAKRISHNAN: Phil. Mag. **45**, 401 (1954).
[14] I. PRIGOGINE u. S. LAFLEUR: Bull. Acad. Roy. Belg., Cl. Sci. **40**, 484, 497 (1954).
[15] R. KIKUCHI: J. Chem. Phys. **23**, 2327 (1955).
[16] H. C. LONGUET-HIGGINS: Mol. Phys. **1**, 83 (1958).
[17] L. SAROLÉA u. A. MÜNSTER: Mol. Phys. (im Druck).

Den Einfluß der „Wand" stellen wir in der Weise dar, daß wir zwei weitere, den übrigen gleiche Teilchen in den Punkten $q=0$ und $q=L$ fixiert denken. Dann wird die gesamte potentielle Energie

$$U + U' = u(q_1) + \sum_{i=1}^{N-1} u(q_{i+1} - q_i) + u(L - q_N). \tag{79.7}$$

Über das zwischenmolekulare Wechselwirkungspotential $u(q)$[1] setzen wir noch voraus, daß

$$\lim_{q \to 0} u(q) = \infty \tag{79.8}$$

und

$$\lim_{q \to \infty} u(q) = 0 \tag{79.9}$$

ist in solcher Weise, daß alle auftretenden Integrale konvergieren. Diese Bedingungen werden von den üblicherweise verwendeten Wechselwirkungspotentialen erfüllt und können ohne weiteres als physikalisch vernünftig betrachtet werden. Gl. (79.8) besagt, daß die Moleküle undurchdringlich sind; daraus folgt, daß die einmal gewählte Reihenfolge der Moleküle (79.5) erhalten bleibt und sich nicht etwa im Laufe der Zeit ändert. Es sind also nur solche Konfigurationen des Systems möglich, bei denen die Reihenfolge der Moleküle erhalten bleibt.

Ist $f(q_1, \ldots, q_N)$ eine in den unabhängigen Variablen symmetrische Funktion, so gilt, wie man durch Induktion beweist,

$$\int_0^L \ldots \int_0^L f(q_1, \ldots, q_N) \, dq_1 \ldots dq_N = N! \int_0^L dq_N \int_0^{q_N} dq_{N-1} \ldots \int_0^{q_2} f(q_1, \ldots, q_N) \, dq_1. \tag{79.10}$$

Aus Gln. (79.4), (79.7) und (79.10) folgt

$$Q_\tau = N! \int_0^L e^{-\frac{u(L-q_N)}{kT}} dq_N \int_0^{q_N} e^{-\frac{u(q_N - q_{N-1})}{kT}} dq_{N-1} \ldots \int_0^{q_2} e^{-\frac{u(q_2 - q_1)}{kT}} e^{-\frac{u(q_1)}{kT}} dq_1. \tag{79.11}$$

Das Integral der rechten Seite ist ein $N+1$-faches Faltungsprodukt. Wir erhalten daher durch Laplace-Transformation

$$\int_0^\infty e^{-sL} Q_\tau \, dL = N! \, [\varphi(s)]^{N+1} \tag{79.12}$$

mit

$$\varphi(s) = \int_0^\infty e^{-sq} e^{-\frac{u(q)}{kT}} dq. \tag{79.13}$$

Die Anwendung der Umkehrungsformel der Laplace-Transformation auf Gl. (79.12) ergibt

$$Q_\tau = \frac{N!}{2\pi i} \int_{c-i\infty}^{c+i\infty} e^{Ls} [\varphi(s)]^{N+1} ds. \tag{79.14}$$

Die Berechnung des Integrals erfolgt durch Verwandlung in ein Kurvenintegral und Anwendung des Residuensatzes. Wir setzen $s = R e^{i\alpha}$ (mit $-\pi \leq \alpha \leq +\pi$). Dann muß für die Umwandlung in ein Kurvenintegral vorausgesetzt werden, daß $|e^{Ls} [\varphi(s)]^{N+1}| < C R^{-k}$ ist, wo C eine Konstante bezeichnet und $k \geq 1$ sowie $L > 0$ gilt.

---

[1] q bezeichnet hier den Abstand zweier benachbarter Moleküle.

Als einfaches Beispiel betrachten wir ein System aus harten Kugeln vom Durchmesser σ. Es ist dann $u(q) = \infty$ für $q < \sigma$ und $u(q) = 0$ für $q > \sigma$. Damit wird aus (79.13)

$$\varphi(s) = s^{-1} e^{-\sigma s}. \qquad (79.15)$$

Die Voraussetzung für die Umformung ist daher erfüllt, wenn $L > (N+1)\sigma$ ist, und wir bekommen

$$Q_\tau = \frac{N!}{2\pi i} \oint e^{Ls} \left[e^{-\sigma s} s^{-1} \right]^{N+1} ds. \qquad (79.16)$$

Unter der Voraussetzung $L > (N+1)\sigma$ ergibt die Anwendung des Residuensatzes

$$Q_\tau = [L - (N+1)\sigma]^N. \qquad (79.17)$$

Setzen wir diesen Ausdruck in Gl. (79.3) ein, so erhalten wir für die Verteilungsfunktion

$$Q = \left(\frac{2\pi m kT}{h^2} \right)^{\frac{1}{2}N} \frac{[L - (N+1)\sigma]^N}{N!}. \qquad (79.18)$$

Daraus folgt für die freie Energie nach Helmholtz

$$F = -NkT \left\{ \ln[L - (N+1)\sigma] - \ln N + 1 + \ln \frac{(2\pi m kT)^{\frac{1}{2}}}{h} \right\}. \qquad (79.19)$$

Der Druck ist

$$P = -\frac{\partial F}{\partial L} = \frac{NkT}{L - (N+1)\sigma}. \qquad (79.20)$$

Führen wir die Länge pro Molekül ein durch die Gleichung

$$l = \frac{L}{N+1} \qquad (79.21)$$

und vernachlässigen 1 gegen N, so erhalten wir die thermische Zustandsgleichung in der einfachen Form

$$P(l - \sigma) = kT. \qquad (79.22)$$

Dieses Ergebnis wurde bereits von Tonks[1] erhalten. Man sieht aus den vorstehenden Gleichungen unmittelbar, daß das eindimensionale System aus harten Kugeln keine Umwandlungen zeigen kann.

Wir wollen die Rechnung jetzt für ein beliebiges Wechselwirkungspotential durchführen, von dem wir (außer den früher eingeführten Annahmen) lediglich voraussetzen, daß es die Bedingungen für die Umformung des Integrals (79.14) in ein Kurvenintegral erfüllt. Wir nehmen also an, daß Gl. (79.14) geschrieben werden kann

$$Q_\tau = \frac{N!}{2\pi i} \oint e^{Ls} [\varphi(s)]^{N+1} ds. \qquad (79.23)$$

Um $\lim_{N \to \infty} \ln [Q_\tau/N!]^{1/(N+1)}$ (für konstantes l) zu berechnen, bedienen wir uns einer Methode, die von Kahn und Uhlenbeck[2] für die Theorie der Kondensation (vgl. den Artikel von J. E. Mayer in Band XII dieses Handbuches) entwickelt wurde. Es sei

$$S(z) = \sum_{N=0}^{\infty} \frac{Q_\tau^{(N)}}{N!} z^{N+1} \qquad (l = \text{const}). \qquad (79.24)$$

---

[1] L. Tonks: Phys. Rev. **50**, 955 (1936).
[2] B. Kahn u. G. E. Uhlenbeck: Physica, Haag **5**, 399 (1938).

Nach dem Theorem von Cauchy-Hadamard gilt für den Konvergenzradius R dieser Reihe

$$\frac{1}{R} \lim_{N\to\infty} \left(\frac{Q_\tau^{(N)}}{N!}\right)^{\frac{1}{N+1}}. \tag{79.25}$$

Wir haben also den Abstand der ersten Singularität der Funktion $S(z)$ vom Ursprung zu bestimmen. Dazu setzen wir zunächst (79.23) in (79.24) ein und erhalten

$$S(z) = \frac{1}{2\pi i} \sum_{N=0}^{\infty} \oint [e^{ls}\varphi(s)z]^{N+1}\, ds. \tag{79.26}$$

Wir wählen nun den Integrationsweg so, daß stets

$$|z\, e^{ls}\varphi(s)| \leqq a < 1 \tag{79.27}$$

ist. Dann gilt, wenn r die Länge des Integrationsweges bezeichnet

$$\oint [e^{ls}\varphi(s)z]^{N+1}\, ds \leqq r\, a^{N+1}. \tag{79.28}$$

Unter der Voraussetzung (79.27) ist die Reihe (79.26) daher gleichmäßig konvergent, und wir können die Reihenfolge von Summierung und Integration vertauschen. Damit erhalten wir

$$S(z) = \frac{1}{2\pi i} \oint \sum [e^{ls}\varphi(s)z]^{N+1}\, ds = \frac{1}{2\pi i} z \oint (\varphi^{-1}e^{-ls} - z)^{-1}\, ds. \tag{79.29}$$

Innerhalb des durch (79.27) definierten Integrationsweges liegt ein Pol des Integranden an der Stelle $s = s_0$, welche durch die Gleichung

$$z - \frac{e^{-ls_0}}{\varphi(s_0)} = 0 \tag{79.30}$$

bestimmt wird. Das Residuum des Integranden an dieser Stelle ist $-e^{ls_0}/[l\varphi(s_0) + \varphi'(s_0)]$. Die Anwendung des Residuensatzes auf (79.29) ergibt daher

$$S(z) = -\left[l + \frac{\varphi'(s_0)}{\varphi(s_0)}\right]^{-1} \quad \text{mit} \quad z = e^{-ls_0}/\varphi(s_0). \tag{79.31}$$

Da $S(z)$ durch Gl. (79.24) als eine Potenzreihe mit positiven reellen Koeffizienten definiert ist, muß die dem Ursprung nächste Singularität, welche den Konvergenzkreis der Reihe bestimmt, auf der positiven reellen Achse liegen. Der Ausdruck (79.31) kann als analytische Fortsetzung der Reihe (79.24) betrachtet werden. Er zeigt, daß die den Konvergenzkreis bestimmende Singularität ein Pol an der Stelle

$$z = R = \frac{e^{-l\xi}}{\varphi(\xi)} \tag{79.32}$$

ist, wo ξ durch die Gleichung

$$l + \frac{1}{\varphi(\xi, T)} \frac{\partial\varphi(\xi, T)}{\partial\xi} = 0 \tag{79.33}$$

bestimmt wird. Wenn wir dieses Ergebnis mit Gl. (79.25) kombinieren, so folgt zunächst

$$\lim_{N\to\infty} \frac{Q_\tau}{N!} = e^{L\xi}\, [\varphi(\xi, T)]^{N+1} \tag{79.34}$$

und weiter mit (79.3) als asymptotischer Ausdruck für die freie Energie nach Helmholtz

$$F = -N\, kT \ln\varphi(\xi, T) - kT\, L\, \xi - N\, kT \ln \frac{(2\pi m\, kT)^{\frac{1}{2}}}{h}. \tag{79.35}$$

Der Parameter ξ ist aber nichts anderes als der durch kT dividierte Druck. Aus (79.35) folgt nämlich

$$- \frac{\partial F}{\partial L} = P = kT\,\xi. \tag{79.36}$$

In Verbindung mit (79.33) und der Gleichung

$$\varphi(\xi, T) = \int_0^\infty e^{-\xi q}\, e^{-\frac{u(q)}{kT}}\, dq \tag{79.37}$$

stellt Gl. (79.36) die thermische Zustandsgleichung dar. Man kann (79.36) aber auch als thermodynamische Interpretation des Parameters ξ auffassen und den letzteren damit aus den thermodynamischen Funktionen eliminieren. Dann ergibt sich aus (79.35) für die freie Energie nach GIBBS $G(T, P) = F + PL$

$$G = - N\,kT \left[\ln \varphi(T, P) + \ln \frac{(2\pi m\,kT)^{\frac{1}{2}}}{h} \right]. \tag{79.38}$$

Aus den Gln. (79.35) und (79.38) sieht man unmittelbar, daß die thermodynamischen Potentiale hier durchweg reguläre Funktionen der Zustandsgrößen sind und daher Umwandlungen nicht stattfinden können. Wir wollen darüber hinaus noch auf der Grundlage der allgemeinen Theorie (Ziff. 76 und 77) zeigen, daß das durch Gl. (79.38) beschriebene System den Stabilitätsbedingungen genügt und daß speziell keine Phasenumwandlungen möglich sind. Dazu müssen wir nach (72.10) bzw. (77.4) beweisen, daß $\partial^2 (G/N)/\partial P^2 < 0$ ist und weiter, daß diese Größe für alle endlichen Werte der Zustandsgrößen endlich bleibt. Wir gehen aus von Gl. (79.37). Da $\varphi(\xi, T)$ eine positive Funktion ist, gilt das gleiche auch für

$$- \frac{\varphi'(\xi)}{\varphi(\xi)} = \frac{\displaystyle\int_0^\infty x\,e^{-\xi x} f(x)\, dx}{\displaystyle\int_0^\infty e^{-\xi x} f(x)\, dx}. \tag{79.39}$$

Dabei haben wir die Integrationsvariable jetzt mit x bezeichnet und zur Abkürzung $f(x) = e^{-\frac{u}{kT}}$ gesetzt. Aus Gl. (79.33) und (79.39) folgt

$$\left.\begin{aligned} \varphi^2 \frac{\partial l}{\partial \xi} &= \int_0^\infty x\,e^{-\xi x} f(x)\, dx \int_0^\infty y\,e^{-\xi y} f(y)\, dy - \\ &\quad - \int_0^\infty e^{-\xi x} f(x)\, dx \int_0^\infty y^2\,e^{-\xi y} f(y)\, dy. \end{aligned}\right\} \tag{79.40}$$

Vertauschen wir auf der rechten Seite x und y und addieren die beiden Ausdrücke, so erhalten wir

$$2\varphi^2 \frac{\partial l}{\partial \xi} = - \int_0^\infty\!\!\int_0^\infty e^{-\xi(x+y)} f(x)\, f(y)\, (x - y)^2\, dx\, dy. \tag{79.41}$$

Daraus ergibt sich, daß $(\partial l/\partial \xi)_T < 0$ sein muß und weiter, im Hinblick auf die Voraussetzungen über das zwischenmolekulare Potential, daß die genannte Größe endlich bleibt. Mit (79.36) folgt dann

$$\left(\frac{\partial^2 (G/N)}{\partial P^2} \right)_T = \left(\frac{\partial l}{\partial P} \right)_T < 0 \quad \text{und endlich,} \tag{79.42}$$

womit die Behauptungen bewiesen sind.

Die weitere Diskussion, auf deren Einzelheiten wir hier nicht eingehen können, führt zu dem Ergebnis, daß die Eigenschaften des Systems sich asymptotisch für $T \to \infty$ denen eines idealen Gases, für $T \to 0$ denen eines linearen Kristalls nähern. Die Isothermen eines linearen Systems sind von Gürsey unter der Annahme eines „kasten"-förmigen Anziehungspotentials der Breite τ und der Tiefe u_0, wie es in Fig. 11 dargestellt ist, berechnet worden.

Einige derselben sind in den reduzierten Zustandsgrößen

$$T' = \frac{kT}{|u_0|}, \qquad P' = \frac{\tau P}{|u_0|}, \qquad l' = \frac{l}{\tau} \tag{79.43}$$

in Fig. 12 dargestellt.

Man sieht, daß bei sehr tiefen Temperaturen die Neigung derselben sich in einem äußerst kleinen Bereich sehr stark ändert. Man kann dies wohl als die erste Andeutung einer Phasenumwandlung auffassen, die im zwei- und dreidimensionalen Fall zu echten Singularitäten in den thermodynamischen Funktionen führt.

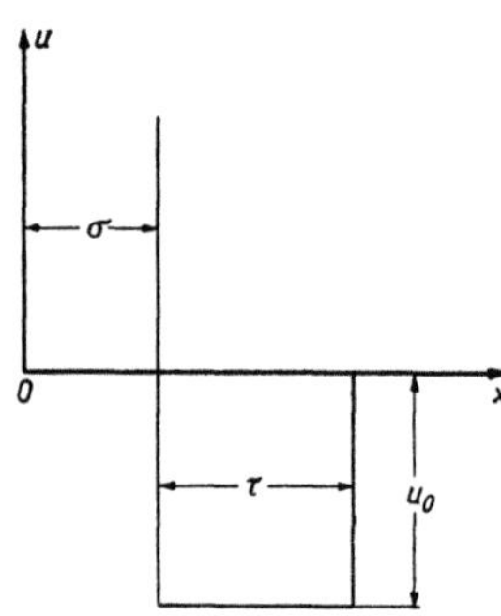

Fig. 11. Kastenpotential (square well potential).

Fig. 12. Isothermen eines eindimensionalen Systems.

Daß ein linearer Kristall für $N \to \infty$ nur am absoluten Nullpunkt existieren kann, läßt sich auch ganz unabhängig von dem hier erörterten Problem der Umwandlungen nachweisen[1]. Geht man von dem Born-von Kármánschen Modell eines linearen Kristalls[2] mit Hookeschen Kräften zwischen den Gitterbausteinen aus, so findet man, daß für $N \to \infty$ und jedes $T > 0$ die kristalline Struktur verschwindet und die Atomverteilung einer fluiden Phase entspricht. Man kann danach vermuten, daß das in Ziff. 76 erwähnte Superpositionsprinzip für Kristalle eine sehr schlechte Näherung darstellt.

Verzeichnis der Formelsymbole.

Thermodynamische Größen.

Es sind nur solche Bezeichnungen aufgeführt, die von allgemeinerer Bedeutung sind.

A	An dem System geleistete Arbeit
E	Innere Energie
F	Freie Energie nach Helmholtz
G	Freie Energie nach Gibbs, freie Enthalpie
H	Enthalpie, Wärmeinhalt
P	Druck
Q	Dem System zugeführte Wärme
R	Gaskonstante

---

[1] L. Saroléa u. A. Münster: Mol. Phys. (im Druck)
[2] M. Born u. Th. von Kármán: Phys. Z. **13**, 297 (1912); **14**, 15 (1913).

S	Entropie
T	Absolute Temperatur
V	Volumen
Z	Fugazität, absolute Aktivität
μ_i	Chemisches Potential

Statistische Größen.

E	Energie
$H(\boldsymbol{q}, \boldsymbol{p})$	Hamilton-Funktion
N	Teilchenzahl
N_L	Loschmidtsche Zahl
Q	Verteilungsfunktion (Zustandssumme) des Gesamtsystems
Q_τ	Verteilungsfunktion der potentiellen Energie, Konfigurationsintegral
S	Slater-Summe
U	Potentielle Energie
$[V]$	Virial
W	Wahrscheinlichkeit
$f(T)$	Verteilungsfunktion (Zustandssumme) des Einzelmoleküls
h	Plancksches Wirkungsquantum
k	Boltzmannsche Konstante
m	Masse eines Teilchens
n	Zahl der Freiheitsgrade des Systems
n	Molekülzahl einer Untergruppe von Molekülen
p_i	Generalisierter Impuls des i-ten Freiheitsgrades
$\boldsymbol{p}_i$	Satz der Impulskoordinaten des i-ten Moleküls, Ortsvektor im Impulsraum des i-ten Moleküls
$\boldsymbol{p}$	Vollständiger Satz der Impulskoordinaten des Systems, Ortsvektor im Impulsraum des Systems
q_i	Generalisierte Koordinate des i-ten Freiheitsgrades
$\boldsymbol{q}_i$	Satz der Ortskoordinaten des i-ten Moleküls, Ortsvektor im Konfigurationsraum des i-ten Mileküls
$\boldsymbol{q}$	Vollständiger Satz der Ortskoordinaten des Systems, Ortsvektor im Konfigurationsraum des Systems
$\boldsymbol{q}^{(n)}$	Satz der Ortskoordinaten von n Molekülen, Ortsvektor im Konfigurationsraum einer Gruppe von n Molekülen
$\Theta = kT$	Verteilungsmodul der kanonischen und der großen kanonischen Gesamtheit
Ξ	Große Verteilungsfunktion
Φ	Gibbssche Energiefunktion
Ω	Zahl der Eigenfunktionen zwischen E und $E + \Delta E$
Ω^*	Volumen des Phasenraumes bis zur maximalen Energie E

$\beta = \dfrac{1}{kT}$

ε	Energie eines Teilchens

$\varepsilon = E/V$ Energiedichte

$\vartheta = e^{-\frac{1}{kT}}$ Darwin-Fowlersches Temperaturanalogon

$\lambda = \dfrac{h}{\sqrt{2\pi m\, kT}}$

ϱ	Phasendichte

$\varrho = N/V$ Molekulare Dichte

ψ	Parameter der kanonischen Gesamtheit
ψ_n	Eigenfunktion des Hamilton-Operators

Ein rechter oberer Index in Klammern, z.B. $q^{(n)}$, bedeutet im allgemeinen, daß die betreffende Größe sich in dem jeweils erläuterten Sinne auf eine Gruppe von n Molekülen bezieht.

Mathematische Symbole.

a	Skalare Größe
$\boldsymbol{a}$	Vektor
A	Operator
A, $[A_{ij}]$	Matrix mit den Elementen A_{ij}
$\boldsymbol{\alpha}$	Matrix mit einer Reihe oder Spalte (Vektor)
$|A|$	Determinante mit den Elementen A_{ij}

$|A|_{ij}$ Kofaktor des Matrixelementes A_{ij}
$\sum$ Summierung
$\prod$ Produkt
$\boldsymbol{a \cdot b}$ Skalares Produkt
$\boldsymbol{a \times b}$ Vektorielles Produkt
$\mathsf{A \times B}$ Direktes Produkt
$[\mathsf{A, B}]$ Kommutator
$\int F \, d\boldsymbol{q}_i$ Integral von F über den Konfigurationsraum des Moleküls i
$\int F \, d\boldsymbol{q}$ Integral von F über den Konfigurationsraum des Systems
$\int F \, d\boldsymbol{q}^{(n)}$ Integral von F über den Konfigurationsraum von n Molekülen
$\int F \, d\Omega$ Integral von F über den Phasenraum des Systems
$$\frac{\partial(x_1, x_2, \ldots)}{\partial(y_1, y_2, \ldots)}, \quad \text{Jacobische Determinante}$$

Literatur.

(Monographien und zusammenfassende Darstellungen.)

[1] Boltzmann, L.: Vorlesungen über Gastheorie. Leipzig 1896—1898.
[2] Gibbs, J. W.: Elementary Principles in Statistical Mechanics. New Haven 1902. (Collected works, Vol. II. New Haven 1948.)
[3] Ehrenfest, P. u. T.: Enzyklopädie der mathematischen Wissenschaften, Bd. IV, Teil 32. Leipzig 1911.
[4] Hertz, P.: Repertorium der Physik von Weber-Gans, Bd. I/2. Leipzig 1916.
[5] Herzfeld, K. F.: Müller-Pouillet's Lehrbuch der Physik, Bd. III/2. Braunschweig 1925.
[6] Fowler, R. H.: Statistische Mechanik (Deutsch von O. Halpern u. H. Smereker). Leipzig 1931.
[7] Fowler, R. H.: Statistical Mechanics, 2nd ed. Cambridge 1936.
[8] Smekal, A.: Handbuch der Physik von Geiger-Scheel, Bd. IX. Berlin 1926.
[9] Fürth, R.: Handbuch der Physik von Geiger-Scheel, Bd. IV. Berlin 1929.
[10] Haas, A., and G. F. Donnan: Commentary on the Scientific Writings of J. W. Gibbs. New York 1936.
[11] Landau, L., and E. Lifshitz: Statistical Physics. Oxford 1938.
[11] Tolman, R. C.: The Principles of Statistical Mechanics. Oxford 1938.
[13] Fowler, R. A., and E. A. Guggenheim: Statistical Thermodynamics. Cambridge 1939.
[14] Mayer, J. E., and M. Goeppert-Mayer: Statistical Mechanics. New York 1940.
[15] Chandrasekhar, S.: Rev. Mod. Phys. 15, 1 (1943).
[16] Kirkwood, J. G.: Selected Topics in Statistical Mechanics. (Vorlesungen an der Princeton University 1947, als Manuskript vervielfältigt.)
[17] Schrödinger, E.: Statistical Thermodynamics. Cambridge 1948.
[18] Rushbrooke, G. S.: Introduction to Statistical Mechanics. Oxford 1949.
[19] Khinchin, A. J.: Mathematical Foundations of Statistical Mechanics. New York 1949.
[20] Boer, J. de: Rep. Progr. Phys. 12, 305 (1949).
[21] Kar, K. C.: Statistical Mechanics. Calcutta 1952.
[22] Haar, D. ter: Elements of Statistical Mechanics. New York 1954.
[23] Haar, D. ter: Rev. Med. Phys. 27, 289 (1955).
[24] Becker, R.: Theorie der Wärme. Berlin 1955.
[25] Hill, T. L.: Statistical Mechanics. New York 1956.
[26] Münster, A.: Statistische Thermodynamik. Berlin 1956.
[27] Frenkel, J. I.: Statistische Physik (Deutsch von H. Jancke). Berlin 1957.

Spezielle Literatur zu Teil B.

[28] Jordan, P.: Statistische Mechanik auf quantentheoretischer Grundlage. Braunschweig 1933.
[29] Delbrück, M., u. M. Moliére: Abh. preuß. Akad. Wiss. 1936, Nr. 1.
[30] Chintschin, A. J. (Khinchin): Mathematische Grundlagen der Quantenstatistik. Berlin 1956.

Spezielle Literatur zu Teil C.

[31] Phase Transformations in Solids. Herausgeg. von R. Smoluchowski, J. E. Mayer, W. A. Weyl. New York 1951.
[32] Comptes Rendus de la 2e Réunion „Changements de Phases", Paris 1952.
[33] Temperley, H. N. V.: Changes of State. London 1956.

Thermodynamik der irreversiblen Prozesse.

Von

J. MEIXNER und H. G. REIK.

Mit 4 Figuren.

A. Einleitung.

1. Vorbemerkungen. Die Thermodynamik irreversibler Prozesse ist eine makroskopische Theorie der Zustände und Zustandsänderungen in der kontinuierlich ausgebreiteten Materie. Sie ist eine Erweiterung der Theorie thermodynamisch-mechanischer Gleichgewichtszustände, d. h. der klassischen Thermodynamik und kann auf hydrodynamische und aerodynamische Probleme, auf Erscheinungen der Elastizitätslehre und der Theorie der kontinuierlichen Medien in elektromagnetischen Feldern, sowie auf Vorgänge in Bifluiden, d. h. in Gasplasmen, Supraleitern und Superfluiden angewandt werden. Wenn auch die Theorie dieser Erscheinungen großenteils vor der Entwicklung der Thermodynamik irreversibler Prozesse bestand, so kann man doch sagen, daß alle diese Erscheinungen methodisch und inhaltlich von der Thermodynamik irreversibler Prozesse in einheitlicher Weise umfaßt werden.

Die Voraussetzungen, die die Thermodynamik irreversibler Prozesse bei der Behandlung der genannten Erscheinungen macht, sind allgemein und von großer Tragweite: Man nimmt nur an, daß sich die Massenelemente eines Mediums als thermodynamische Systeme behandeln lassen, die mit ihrer Umgebung in bezug auf gewisse physikalische Größen in Austausch stehen. Das Erfülltsein oder Nichterfülltsein dieser Voraussetzungen bestimmt den Gültigkeitsbereich der Theorie.

Die Voraussetzung, daß die Massenelemente thermodynamische Systeme sind, besagt im einzelnen, daß die Werte der extensiven thermodynamischen Variablen, nämlich der inneren Energie, des Volumens (beziehungsweise der Dehnungen) und der Massen der Bestandteile die intensiven Variablen, d.h. die Temperatur, den Druck, bzw. die Spannungen und die chemischen Potentiale über die Relationen der Thermodynamik bestimmen. Damit existiert natürlich auch für jedes Massenelement ein thermodynamisches Potential; für die genannten extensiven Variablen ist es die Entropie.

Die Formulierung der bisher für die einzelnen Massenelemente ausgesprochenen Voraussetzungen für das kontinuierliche Medium besagt dann, daß der thermodynamische Zustand eines Mediums durch Felder beschrieben werden kann. Da diese auch von der Zeit abhängen dürfen, sind auch die Zustandsänderungen erfaßt.

Den extensiven Variablen der Massenelemente entsprechen die Dichtefelder. Diese umfassen das Feld der Energiedichte $\varrho u(\boldsymbol{r}, t)$, das Dichtefeld $\varrho(\boldsymbol{r}, t)$ (bzw. die Dehnungsfelder) und die Felder der Partialdichten $\varrho_i(\boldsymbol{r}, t)$. Die Verteilung der intensiven Parameter im Medium wird durch das Temperaturfeld, das Druckfeld (die Felder der Spannungen) und die Felder der chemischen Potentiale dargestellt. Die genannten Felder werden im folgenden als Zustandsfelder bezeichnet.

Befindet sich das Medium in elektromagnetischen Feldern, so sind diese zur Beschreibung des Zustandes mit hinzuzunehmen. Die Verteilung des thermodynamischen Potentials wird durch ein Potentialfeld beschrieben.

Wegen der Sonderstellung, die die Entropie unter den Potentialen der Thermodynamik einnimmt, und wegen der hervorragenden Rolle der Entropieproduktion in der Thermodynamik der irreversiblen Prozesse, sind die Dichtefelder und das Potentialfeld $\varrho s (\mathbf{r}, t)$ für die Formulierung der Theorie besonders wichtig. Zu diesen Feldern tritt das Feld der Strömungsgeschwindigkeit $\mathbf{v}(\mathbf{r}, t)$, das den mechanischen Zustand des Kontinuums kennzeichnet.

Die Aufgabe der Thermodynamik irreversibler Prozesse besteht in der Aufstellung von Gleichungen für die räumlich-zeitliche Veränderung dieser Felder. Die zusammenfassende Beschreibung vieler wechselwirkender thermodynamischer Systeme durch Variablenfelder führt somit auf eine Aufgabenstellung, die von den klassischen Feldtheorien, z.B. der Elektrodynamik, wohlbekannt ist. Man gewinnt die Gesetze für diese Felder aus einigen allgemeinen Prinzipien, nämlich den Erhaltungssätzen, dem zweiten Hauptsatz der Thermodynamik und einem Satz der Schwankungstheorie. Man geht dazu in folgender Weise vor: Es werden zunächst die Erhaltungs- bzw. Wachstumssätze der Physik, differentiell für nicht abgeschlossene Systeme, die miteinander im Austausch stehen, formuliert. Man erhält dann Bilanzgleichungen die ihrer Herkunft nach für alle Materialien gültig sind. Sie führen die zeitlichen Veränderungen der Dichtefelder und des Potentialfeldes auf äußere Einflüsse, auf Wechselwirkungsprozesse zwischen den Massenelementen (z.B. Energieaustausch) und auf Homogenprozesse (chemische Reaktionen) zurück. Zur mathematischen Formulierung der Bilanzgleichungen benötigt man deshalb neue Felder, die die Prozesse beschreiben. So werden Diffusionsprozesse, d.h. Materieaustausch, durch die Vektorfelder der Diffusionsstromdichten, Energieaustausch durch das Vektorfeld der Energiestromdichte, die Impulsübertragung durch das Tensorfeld der Impulsstromdichte dargestellt. Die chemischen Reaktionen setzt man durch Skalarfelder chemischer Produktionsdichten in Rechnung.

Die Gesamtheit der Felder, die den ablaufenden Prozessen zugeordnet sind, bezeichnen wir als *Prozeßfelder.* Man sieht, daß die *Bilanzgleichungen* der Thermodynamik irreversibler Prozesse noch kein vollständiges Gleichungssystem bilden. Sie sind also durch weitere Beziehungen, die *phänomenologischen Gleichungen,* zu ergänzen. In sie gehen die Materialeigenschaften von speziellen Medien ein. Sie stellen den zweiten wichtigen Typ von Grundgleichungen der Thermodynamik irreversibler Prozesse dar.

Auch die phänomenologischen Gleichungen lassen sich in systematischer Weise aufstellen. Zu ihrer Herleitung macht man expliziten Gebrauch von der Voraussetzung, daß das kontinuierliche Medium aus thermodynamischen Systemen zusammengesetzt ist. Daraus zieht man den Schluß, daß alle Prozesse unter Entropieproduktion ablaufen müssen. Diese Forderung führt insbesondere für kleine Gleichgewichtsabweichungen zu linearen phänomenologischen Gleichungen für die Prozeßfelder. Ihre Koeffizienten genügen gewissen Symmetriebeziehungen, den *Onsager-Casimirschen Reziprozitätsbeziehungen,* die aus der Schwankungstheorie folgen. Zusammen mit den Bilanzgleichungen stellen die phänomenologischen Gleichungen ein vollständiges System von Grundgleichungen dar.

In diesem Artikel werden wir im wesentlichen die Formulierung und Anwendung der phänomenologischen Seite der Thermodynamik irreversibler Prozesse behandeln. Auf die Theorie der diskontinuierlichen Systeme gehen wir jedoch nicht ein, da sie ein einfacher Spezialfall der entwickelten Theorie kontinuierlicher Medien ist. Wir verweisen vielmehr auf die ausführlichen Darstellungen von PRIGOGINE, DE GROOT und HAASE. Ferner werden wir einige Gebiete der statistischen Theorie irreversibler Prozesse insbesondere die Begründung der Onsager-Casimirschen Reziprozitätsbeziehungen behandeln. Wir verzichten jedoch auf die quantenmechanische Theorie von Systemen, die sich in Wechselwirkung mit einer klassisch beschreibbaren Umgebung befinden. Auch die Versuche, die Irreversibilität von Abläufen aus „first principles" zu verstehen, haben wir nur referierend berücksichtigt. Die Entwicklung auf diesem Gebiet ist zu sehr im Fluß.

2. Historisches. Die phänomenologische Beschreibung der irreversiblen Prozesse geht bis zu NEWTON und der Aufstellung des Reibungsgesetzes für viskose Flüssigkeiten zurück. Spezielle phänomenologische Ansätze für einzelne irreversible Prozesse finden sich in FOURIERs Wärmeleitungsgesetz, im Fickschen Diffusionsgesetz und im Ohmschen Gesetz. Obwohl mit ihrer Hilfe schon frühzeitig irreversible Prozesse erfolgreich behandelt werden konnten — man denke an die technische Bedeutung der Wärmeleitungstheorie — sind einheitliche Gesichtspunkte nicht vorhanden. Diese deuten sich erst mit dem Auftauchen der Begriffe Entropiestromdichte und lokale Entropieproduktion an. Letztere wurde in der älteren Literatur (im Anschluß an CLAUSIUS) als die durch die absolute Temperatur dividierte nicht kompensierte Wärme bezeichnet. Mit BERTRAND[1] (1887) beginnt der Gedanke Gestalt zu gewinnen, daß Entropie eine im Medium räumlich verteilte Quantität ist, die strömen kann und nur positive Quellen besitzt. BERTRAND bemerkt ausdrücklich, daß diese Auffassung nicht thermostatisch begründet werden kann[2]. Für verschiedene Vorgänge gelang es mit dieser Vorstellung Ausdrücke für die lokale Entropieproduktion zu gewinnen. Mit DUHEM[3] findet diese Entwicklung einen gewissen Abschluß. Um dieselbe Zeit entsteht die erste Arbeit von JAUMANN[4], die umfangreiche und allgemeine Untersuchungen über geschlossene Systeme physikalischer und chemischer Differentialgesetze enthält. Sie rückt die differentielle Form der Erhaltungssätze in den Vordergrund und macht von den Begriffen der Entropieströmung und der lokalen Entropieproduktion klaren Gebrauch. Daneben läuft eine andere Entwicklung, die heute als Pseudothermostatik bezeichnet wird. Der Name rührt daher, daß diese Theorie Aussagen über gekoppelte irreversible Prozesse scheinbar aus der Thermostatik gewinnt. Man postuliert nämlich in ihr als neues Prinzip, daß die Entropieproduktion sich aus den für die einzelnen Prozesse charakteristischen Anteile additiv zusammensetzen läßt, so daß die Kopplung

---

[1] J.L.F. BERTRAND: Thermodynamique. Paris 1887.

[2] Kap. XII, S. 266: Quand tous les points du corps que l'on étudie n'ont pas conservé, pendant la transformation des températures égales pour tous, que deviennent le rapport dQ/T de la quantité de chaleur fournie au corps à la température T, et l'intégrale de ce rapport? Faut-il décomposer le corps en éléments infiniment petits et réunir les intégrales relative à chacun d'eux? La démonstration des théorèmes n'y autorise pas.

Lorsque le corps, en changeant de volume n'exerce pas sur les corps environnants une action égale à celle qu'il pourrait vaincre; lorsque, en le comprimant, on exerce sur lui une action superieure à la résistance dont il est capable, les molécules, dans le cas d'un gaz particulièrement prennent des vitesses finies inégales en grandeur, différentes en direction, et, pendant la durée de cette perturbation le mot pression, appliqué aux corps cesse d'avoir un sene defini.

[3] P. DUHEM: Energétique. Paris 1911.

[4] G. JAUMANN: Sitzgsber. Akad. Wiss. Wien, Math.-naturw. Kl., Abt. II A **120**, 385 (1911). Siehe auch die spätere Arbeit von JAUMANN: Denkschr. Akad. Wiss. Wien, Math.-naturw. Kl. **95**, 461 (1918) und die Arbeit von E. LOHR: Denkschr. Akad. Wiss. Wien, Math.-naturw. Kl. **93**, 339 (1916); **99**, 11, 59 (1924); Festschr. Techn. Hochsch. Brünn, S. 176 1924.

zwischen irreversiblen Prozessen keine zusätzliche (positive oder negative) Entropieproduktion bedingt. Man sieht also die durch die Kopplung hervorgerufenen Effekte als reversibel an. Ein klassisches Beispiel hierzu ist die zweite Thomsonsche Beziehung[1] in der Theorie der thermoelektrischen Effekte, deren Herleitung schon von Boltzmann[2] kritisiert worden ist. In größerem Umfang wurden pseudothermostatische Überlegungen von Eastman und Wagner in der Theorie von Wärmeleitungs- und Diffusionsvorgängen durchgeführt[3,4].

Thomson und insbesondere Wagner weisen selbst darauf hin, daß ihre Ergebnisse nicht auf rein thermostatischen Schlüssen aufgebaut sind und daher einer sorgfältigen experimentellen Bestätigung bedürfen. — Wenn auch die pseudothermostatischen Betrachtungen in den erwähnten Beispielen zu richtigen Ergebnissen geführt haben, so weiß man heute doch, daß dies eine mehr oder minder zufällige Ursache hat. Das der Theorie zugrunde liegende Prinzip ist nämlich so unbestimmt formuliert, daß es nicht eindeutig und nicht mit eindeutigen Ergebnissen angewandt werden kann. Seine Anwendung setzt ein gewisses Gefühl für das richtige Ergebnis voraus. Bei komplizierteren gekoppelten Prozessen versagt es überhaupt. — Man weiß heute, daß alle Aussagen der Pseudothermostatik, soweit sie richtig sind, ihre Begründung aus der Thermodynamik der irreversiblen Prozesse, insbesondere aus der Anwendung der Onsagerschen Symmetriebeziehungen in den phänomenologischen Gleichungen erhalten.

Die heutige Form der Thermodynamik irreversibler Prozesse hat sich im Anschluß an die Arbeiten von Onsager[5] entwickelt. Die Bedeutung dieser Arbeiten beruht im wesentlichen auf zwei Feststellungen: Die Entropieproduktion wird nicht mehr als sekundäre Größe betrachtet, die positiv ist, wenn ein irreversibler Prozeß abläuft. Die Entropieproduktion bestimmt vielmehr, wie schnell die irreversiblen Prozesse ablaufen. Diesen Wandel in der Auffassung hat Richard Becker besonders hervorgehoben[6]. Die zweite Feststellung ist eine Symmetrieaussage für die Gesetze irreversibler Abläufe. Die Onsagerschen Symmetriebeziehungen wurden von Casimir[7] ergänzt. Er erkannte die Notwendigkeit der Unterscheidung zwischen geraden und ungeraden Variabeln, die die Abweichungen vom Gleichgewicht beschreiben.

Erst ein Jahrzehnt nach den grundlegenden Arbeiten von Onsager begann sich die Thermodynamik der irreversiblen Prozesse systematisch zu entwickeln. Den Anstoß hierzu hat Eckart[8] gegeben. Aufbauend auf seinen und Onsagers Arbeiten hat die phänomenologische Seite der Theorie mit Arbeiten von Meixner (1939 bis 1954), Prigogine und Mitarbeitern (1947 bis 1954), de Groot und Mitarbeitern (1945 bis 1954) einen gewissen Abschluß erreicht. Von besonderen Entwicklungen möchten wir hier folgende erwähnen: Die Aufstellung von Extremalprinzipien durch Prigogine und Mazur sowie Wergeland, die allgemeine thermodynamische Theorie der Relaxationserscheinungen von Meixner, die relativistische Formulierung der Thermodynamik irreversibler Prozesse durch Eckart, und unter Hinzunahme von Diffusion und elektromagnetischen Erscheinungen durch de Groot und Kluitenberg, die Anwendung auf elektrochemische Erscheinungen durch Haase und Holtan.

[1] W. Thomson: Proc. Roy. Soc. Edinburgh 3, 225 (1854).,
[2] L. Boltzmann: Sitzgsber. Akad. Wiss. Wien, Math.-naturw. Kl. Abt. II 96, 1258 (1887).
[3] E. D. Eastman: Amer. Chem. Soc. 48, 1482 (1926); 49, 794 (1927); 50, 283, 292 (1928).
[4] C. Wagner: Ann. Phys. [5] 3, 629 (1929); 6, 370 (1930).
[5] L. Onsager: Phys. Rev. 37, 405 (1931); 38, 2265 (1931).
[6] R. Becker: Theorie der Wärme, S. 289. Berlin 1955.
[7] H. B. G. Casimir: Rev. Mod. Phys. 17, 343 (1945).
[8] C. Eckart: Phys. Rev. 58, 267, 269, 919, 924 (1940).

Die Theorie der chemischen Reaktionen lief lange parallel zu und unabhängig von der Entwicklung der Thermodynamik irreversibler Prozesse. Die Berührung zwischen diesen beiden Gebieten beschränkt sich im wesentlichen auf das Prinzip vom detaillierten Gleichgewicht. Sein Zusammenhang mit den Symmetrie-relationen der Thermodynamik irreversibler Prozesse wurde schon von ONSAGER diskutiert und später für den Fall von Gasreaktionen von MEIXNER untersucht.

Die Begründung der Thermodynamik der irreversiblen Prozesse auf statistisch-mechanischer und kinetischer Basis ist heute in vollem Fluß. Es handelt sich dabei einerseits um die Vertiefung der Beweise von ONSAGER und CASIMIR, in jüngster Zeit aber bevorzugt um die Herleitung der Thermodynamik irreversibler Prozesse aus tieferliegenden Prinzipien und damit um das Verständnis der Irreversibilität überhaupt. Spezielle statistische und kinetische Theorie haben als Modellbeispiele schon früher die Entwicklung beeinflußt. Dies gilt in besonderem Maße für die kinetische Gastheorie. So findet man bei BOLTZMANN[1] (1895) neben dem Begriff der Entropieproduktion deutlich die Idee einer Entropieströmung, während ENSKOG[2] (1917) die „regelmäßigen" und „unregelmäßigen" Veränderungen der Entropie gaskinetisch formuliert und die bekannte phänomenologische Gestalt der Entropiebilanz herleitet.

Zusammenfassende Darstellungen der Thermodynamik irreversibler Prozesse oder spezieller Aspekte dieser Theorie stammen von PRIGOGINE (1947)[3], TOLMAN und FINE (1948)[4], DE GROOT (1951)[5], DENBIGH (1951)[6], HAASE (1952)[7], MONTROLL und GREEN (1954)[8], MEIXNER (1954)[9], COX (1955)[10], PRIGOGINE (1955)[11], DAVIES (1956)[12], KOFINK (1957)[13] und EISENSCHITZ (1958)[14].

B. Die Thermodynamik der irreversiblen Prozesse in kontinuierlichen Medien mit inneren Umwandlungen.

3. Der thermodynamisch-mechanische Zustand der kontinuierlichen Materie. Im folgenden werden fluide kontinuierliche Medien betrachtet, die aus mehreren Bestandteilen zusammengesetzt sind. Die Bezeichnung kontinuierliches Medium bringt zum Ausdruck, daß der von der Materie eingenommene Raum in folgender Weise in Volumenelemente eingeteilt werden kann.

Die Materie in solchen Volumenelementen soll alle Eigenschaften eines thermodynamischen Systems besitzen; diese Annahme hat zur Folge, daß der thermodynamische Zustand durch die innere Energie U, das Volumen V und die Massen m_i der Bestandteile vollständig gekennzeichnet ist. Dies ist der Fall, wenn die Volumenelemente einerseits so groß sind, daß die Schwankungen in ihnen noch keine Rolle spielen, andererseits so klein, daß sie praktisch homogen

---

[1] L. BOLTZMANN: Vorlesungen über Gastheorie, S. 124f. Leipzig 1895.

[2] D. ENSKOG: Kinetische Theorie der Vorgänge in mäßig verdünnten Gasen. Diss. Uppsala 1917. S. 108f.

[3] I. PRIGOGINE: Etude Thermodynamique des Phénomènes irréversibles. Liège 1947.

[4] R.C. TOLMAN u. P.C. FINE: Rev. Mod. Phys. **20**, 81 (1948).

[5] S.R. DE GROOT: Thermodynamics of irreversibles Processes. Amsterdam 1951.

[6] K.G. DENBIGH: Thermodynamics of the Steady State. London 1951.

[7] R. HAASE: Ergebn. exakt. Naturw. **22**, 56 (1952).

[8] E.W. MONTROLL u. M.S. GREEN: Amer. Rev. Phys. Chem. **5**, 449 (1954).

[9] J. MEIXNER: Thermodynamik irreversibler Prozesse. Aachen 1954.

[10] R.T. COX: Statistical Mechanics of Irreversible Change. Baltimore 1955.

[11] I. PRIGOGINE: Introduction to Thermodynamics of Irreversibles Process. Springfield 1955.

[12] R.O. DAVIES: Rep. Progr. Phys. **19**, 326 (1956).

[13] W. KOFINK: Die Physik in Einzelberichten, H. 3, 25, 1957.

[14] R. EISENSCHITZ: Statistical Theory of Irreversible Processes. London: Oxford University Press 1958.

von Materie erfüllt sind und der Zustand überall im Volumenelement praktisch derselbe ist. In dieser Annahme liegt die wesentlichste Voraussetzung der Thermodynamik irreversibler Prozesse und ihre Gültigkeit bedarf in jedem Falle einer Prüfung (s. dazu Ziff. 7β)[1].

Zu bestimmten Werten von U, V, m_i gehören wohldefinierte Werte der Temperatur T, des Drucks p und der (auf die Masseneinheit bezogenen) chemischen Potentiale μ_i der Bestandteile und damit ein wohldefinierter Wert des thermodynamischen Potentials. Dieses ist bei den gewählten Variablen die Entropie S. Die Änderung von S ist mit den Änderungen von U, V, m_i durch die Gibbssche Fundamentalgleichung

$$dS = \frac{1}{T}\,dU + \frac{p}{T}\,dV - \sum_{i=1}^{n}\frac{\mu_i}{T}\,dm_i \tag{3.1}$$

verknüpft. Die Potentialeigenschaft der Entropie kommt in den Relationen

$$\left(\frac{\partial S}{\partial U}\right)_{V,m_i} = \frac{1}{T}, \qquad \left(\frac{\partial S}{\partial V}\right)_{U,m_i} = \frac{p}{T}, \qquad \left(\frac{\partial S}{\partial m_i}\right)_{U,V,m_j} = -\frac{\mu_i}{T} \tag{3.2}$$

zum Ausdruck.

Besonders betont sei, daß (3.1) nicht nur dann gilt, wenn eine Zustandsänderung eines und desselben Volumen- oder Massenelements betrachtet wird. Es könnte sich sogar um räumlich getrennte Elemente handeln, deren Entropien, innere Energien, Volumina und Massen sich um dS, dU, dV, dm_i unterscheiden, wenn nur T, p, μ_i in beiden Fällen gleiche (oder höchstens um differentielle Beträge verschiedene) Werte besitzen. Damit erledigt sich die häufig ausgesprochene Bemerkung, daß im Rahmen der Thermodynamik irreversibler Prozesse die Gibbssche Gleichung nur in besonderen Bezugssystemen [nämlich nur für Massenelemente, die sich mit der baryzentrischen Geschwindigkeit bewegen (s. unten)], gültig sei.

Die vorausgesetzten Eigenschaften genügend kleiner Raumbereiche erlauben in bekannter Weise die Dichte ϱ, die Partialdichten ϱ_i der Bestandteile, die spezifische innere Energie u (sowie die Energiedichte ϱu) und die spezifische Entropie s (sowie die Entropiedichte ϱs) zu definieren. Die Zusammensetzung des Mediums kann auch durch die Massenkonzentrationen oder Massenbrüche der Komponenten,

$$\gamma_i = \frac{\varrho_i}{\varrho} \qquad (i = 1, 2, \ldots, n-1) \tag{3.3}$$

beschrieben werden. Unter Benutzung der Dichtegrößen oder der spezifischen Größen läßt sich die Gibbssche Fundamentalgleichung ohne weiteres auf die Massen- oder Volumeneinheit anwenden. Man erhält so für die Entropieänderung der Masseneinheit

$$ds = \frac{1}{T}\,du + \frac{p}{T}\,d\frac{1}{\varrho} - \sum_{i=1}^{n-1}\frac{\mu_i - \mu_n}{T}\,d\gamma_i \tag{3.4}$$

und für die der Volumeneinheit

$$d(\varrho s) = \frac{1}{T}\,d(\varrho u) - \sum_{i=1}^{n}\frac{\mu_i}{T}\,d\varrho_i. \tag{3.5}$$

---

[1] Vgl. hierzu E. Mach: Prinzipien der Wärmelehre, Leipzig 1900, S. 71 ff., insbesondere S. 77: „Überall, wo wir ein Kontinuum vorzufinden glauben, heißt das nur, daß wir an den kleinsten wahrnehmbaren Teilen des betreffenden Systems noch analoge Beobachtungen anstellen und ein analoges Verhalten bemerken können, wie an größeren. Wieweit dies sich fortsetzt wird nur die Erfahrung entscheiden können.“

Die letzte Gleichung folgt aus (3.4) mittels der thermodynamischen Beziehung

$$u - TS + pV - \sum_{i=1}^{n} \mu_i m_i = 0. \qquad (3.6)$$

Nach Einführung der Dichten und der spezifischen Größen braucht man nicht mehr von der Größe der homogenen Raumbereiche zu sprechen. Man kann vielmehr den lokalen thermodynamischen Zustand durch die unabhängigen Größen ϱ_i, u und das zugehörige Potential s beschreiben.

Der thermodynamische Zustand eines kontinuierlichen Mediums mit orts- und zeitabhängigem Zustand wird damit durch die Felder $\varrho_i(r, t)$, $u(r, t)$, und das zugehörige Potentialfeld $s(r, t)$ beschrieben.

Da die Medien im allgemeinen bewegt sind, ist neben ihrem thermodynamischen Zustand auch ihr Bewegungszustand zu kennzeichnen. Zunächst bietet sich an, daß man die Geschwindigkeiten v_i der Bestandteile (das sind die mittleren Geschwindigkeiten der Atome oder Moleküle usw. der betreffenden Art in hinreichend kleinen homogenen Volumenelementen) als Funktionen von Ort und Zeit angibt. Bei Diffusionsvorgängen sind diese Geschwindigkeiten nicht alle einander gleich. Die mittlere Massengeschwindigkeit (Schwerpunktsgeschwindigkeit oder baryzentrische Geschwindigkeit) berechnet sich zu

$$v = \sum_{i=1}^{n} \gamma_i v_i. \qquad (3.7)$$

Sie spielt eine bevorzugte Rolle, da ihre Änderung durch die mechanische Bewegungsgleichung (die Impulsbilanz) beschrieben wird. Sie, oder besser das Geschwindigkeitsfeld $v(r, t)$, kennzeichnet also den mechanischen Zustand des kontinuierlichen Mediums. Die Geschwindigkeiten v_i bestimmen in der in (3.7) gegebenen Kombination den mechanischen Zustand, beschreiben aber überdies Bewegungen der einzelnen Bestandteile gegeneinander. Diese Relativbewegungen werden durch die Diffusionsstromdichten

$$J_i = \varrho_i(v_i - v) \qquad (3.8)$$

ausgedrückt. Sie kennzeichnen keine Zustände, sondern beschreiben irreversible Prozesse. Man hat sie daher (ebensowenig wie die einzelnen v_i) unter die den mechanischen Zustand charakterisierenden Größen aufzunehmen.

Eine (formale) Beschreibung des Bewegungszustandes durch ein anderes Geschwindigkeitsfeld kann für manche speziellen Probleme zweckmäßig sein; z.B. wird man bei der Behandlung der Diffusion in Gasen die mittlere Molekulargeschwindigkeit, in Kristallen die Gittergeschwindigkeit auszeichnen. Die „Bewegung" ist aber dann nicht mehr rein mechanischer Natur und die auf diese Bewegung bezogene Diffusion nicht mehr rein thermischer Natur. Darauf hat zuerst PRIGOGINE[1] hingewiesen.

4. Die Erhaltungssätze und ihre Bilanzgleichungen. $\alpha)$ *Allgemeine Gestalt einer Bilanz.* Die Erhaltungssätze für Masse, Impuls und Energie eines Gesamtsystems lassen sich als Bilanzgleichungen formulieren. Um sie aufzustellen, betrachten wir zunächst die zeitliche Veränderung physikalischer Größen in einem von Materie erfüllten ortsfesten Volumen.

Die Änderung einer extensiven Größe $\mathscr{E} = \int \varrho\,\varepsilon\,dV$ kommt durch Einströmen dieser Größe durch die feste Oberfläche und durch Produktion im Inneren des

---

[1] I. PRIGOGINE: Etude Thermodynamique des Phénomènes irréversibles, S. 89. Paris u. Liège 1947.

betrachteten Volumens zustande. Bezeichnet $\boldsymbol{\Phi}_\varepsilon$ die Dichte des ε-Stromes, σ_ε die Dichte der ε-Erzeugung, so erhält man für die zeitliche Änderung von $\mathscr{E}$ die Bilanzgleichung

$$\frac{d}{dt}\,\mathscr{E} = \frac{d}{dt}\int \varrho\,\varepsilon\,dV = -\oint \boldsymbol{\Phi}_\varepsilon \cdot d\boldsymbol{F} + \int \sigma_\varepsilon\,dV\,. \tag{4.1}$$

Nach Anwendung des Gaußschen Satzes erhält man wegen der willkürlichen Größe und Gestalt des ortsfesten Volumens die differentielle Formulierung der Bilanzgleichung

$$\frac{\partial}{\partial t}\,(\varrho\,\varepsilon) = -\operatorname{div}\boldsymbol{\Phi}_\varepsilon + \sigma_\varepsilon\,. \tag{4.2}$$

(4.1) und (4.2) sind *lokale Formulierungen einer Bilanz*. Die Erhaltungssätze für Masse, Impuls, Gesamtenergie und die Kontinuitätsgleichungen für die Massen der Bestandteile, der inneren Energie und der Entropie erhält man in Gestalt von Bilanzen, wenn man in (4.1) bzw. (4.2) für ε die spezifische Masse, die Komponenten des spezifischen Impulses usw. wählt. Man hat dann mit den entsprechenden Strom- und Produktionsdichten in (4.1) und (4.2) einzugehen. Teilt man die Produktionsdichten in einen inneren und äußeren Anteil auf (der äußere wird durch eingeprägte Kraftfelder bedingt), so besitzen Erhaltungssätze die innere Produktionsdichte Null.

Ist ε insbesondere die spezifische Masse (die Masse pro Masseneinheit, d. h. gleich 1), so ist die zugehörige Stromdichte $\boldsymbol{\Phi}_1 = \varrho\,\boldsymbol{v}$, während $\sigma_1 = 0$ ist. Somit folgt aus (4.2)

$$\frac{\partial\varrho}{\partial t} = -\operatorname{div}\varrho\,\boldsymbol{v}\,. \tag{4.3}$$

Man kann die Bilanz auch auf ein beliebiges vorgegebenes Geschwindigkeitsfeld $\boldsymbol{\varphi}$ beziehen, mit dem sich das Integrationsvolumen, also auch seine Oberfläche, mitbewegt. Natürlich ist dann anstelle von $\boldsymbol{\Phi}_\varepsilon$ die auf dieses Geschwindigkeitsfeld bezogene Stromdichte $\boldsymbol{\Phi}_\varepsilon - \varrho\,\varepsilon\,\boldsymbol{\varphi}$ in (4.1), (4.2) einzusetzen. Wird speziell $\boldsymbol{\varphi} = \boldsymbol{v}$ gewählt, so erhält man die substantielle Formulierung:

$$\frac{d}{dt}\int \varrho\,\varepsilon\,dV^* = -\int (\boldsymbol{\Phi}_\varepsilon - \varrho\,\varepsilon\,\boldsymbol{v}) \cdot d\boldsymbol{F}^* + \int \sigma_\varepsilon\,dV^*\,. \tag{4.4}$$

Man nennt sie so, weil das Integrationsvolumen stets konstante Masse enthält. Der Stern bei $d\boldsymbol{F}^*$ und dV^* soll andeuten, daß sich Flächen- und Volumenelemente mit dem Feld $\boldsymbol{v}\,(\boldsymbol{r},\,t)$ mitbewegen. Die Aufteilung in Massenelemente $\varrho\,dV^*$, welche bei dieser Bewegung zeitlich konstant gewählt werden können, läßt erkennen daß die Zeitdifferentiation in (4.4) nur auf ε wirkt. Man kann daher für die linke Seite von (4.4) schreiben

$$\frac{d}{dt}\int \varrho\,\varepsilon\,dV^* = \int \varrho\,\frac{D\varepsilon}{Dt}\,dV^*\,. \tag{4.5}$$

Das Symbol D/Dt deutet an, daß es sich um die substantielle Differentiation handelt. Der Gaußsche Satz führt jetzt auf die differentielle Formulierung

$$\varrho\,\frac{D\varepsilon}{Dt} = -\operatorname{div}(\boldsymbol{\Phi}_\varepsilon - \varrho\,\varepsilon\,\boldsymbol{v}) + \sigma_\varepsilon\,. \tag{4.6}$$

(4.5) und (4.6) sind *substantielle Formulierungen einer Bilanz*. Eliminiert man σ_ε aus (4.6) und (4.2) unter Berücksichtigung von (4.3), so erhält man als Zusammenhang zwischen lokaler und substantieller Änderung von ε

$$\frac{D\varepsilon}{Dt} = \frac{\partial\varepsilon}{\partial t} + \boldsymbol{v}\cdot\operatorname{grad}\varepsilon\,. \tag{4.7}$$

Diese Gleichung kann als Operator-Gleichung geschrieben werden

$$\frac{D}{Dt} = \frac{\partial}{\partial t} + \boldsymbol{v} \cdot \mathrm{grad} \tag{4.8}$$

und läßt sich dann unmittelbar auch auf vektorielle oder tensorielle Größen ε anwenden. Die bei den vorhergehenden Formeln für diese Fälle erforderlichen Modifikationen sind ersichtlich.

Die substantielle Differentiation (4.8), auf die Dichte ϱ angewandt, gibt unter Berücksichtigung von (4.3) die Beziehung

$$\frac{D\varrho}{Dt} = -\varrho \, \mathrm{div}\, \boldsymbol{v}. \tag{4.9}$$

$\beta)$ *Die Bilanzen für die Partialdichten.* Wir setzen nun voraus, daß die Komponente i nicht notwendig erhalten bleibt, sondern daß sie eine (positive oder negative) Produktion infolge chemischer Reaktionen erfahren kann. Ihre Größe pro Volumen- und Zeiteinheit wird als chemische Produktionsdichte Γ_i bezeichnet. Sie hängt vom lokalen Zustand ab. Für alle indifferenten (d. h. nicht reagierenden) Komponenten ist natürlich $\Gamma_i = 0$.

Die Bilanzgleichung für die Komponente i ergibt sich dann nach (4.6) oder (4.2) mit $\varepsilon = \varrho_i/\varrho = \gamma_i$, $\boldsymbol{\Phi}_i = \varrho_i \boldsymbol{v}_i$, $\sigma_i = \Gamma_i$ zu

$$\varrho \frac{D\gamma_i}{Dt} = -\mathrm{div}\, \boldsymbol{J}_i + \Gamma_i \tag{4.10}$$

oder

$$\frac{\partial \varrho_i}{\partial t} = -\mathrm{div}\, \varrho_i \boldsymbol{v}_i + \Gamma_i = -\mathrm{div}\, \varrho_i \boldsymbol{v} - \mathrm{div}\, \boldsymbol{J}_i + \Gamma_i. \tag{4.11}$$

Die $\boldsymbol{J}_i$ sind die in (3.8) eingeführten Diffusionsstromdichten. (4.10) ist eine besonders anschauliche Formulierung. Man erkennt an ihr, daß die substantielle Änderung der Konzentration verschwindet, falls keine Diffusion und keine Reaktion stattfindet. Wegen der Massenerhaltung gilt stets

$$\sum_{i=1}^{n} \Gamma_i = 0, \quad \sum_{i=1}^{n} \boldsymbol{J}_i = 0. \tag{4.12}$$

Durch Summierung aller Gleichungen (4.11) entsteht daher (4.3).

$\gamma)$ *Die Bewegungsgleichung (Impulsbilanz).* Zu ihrer Herleitung knüpft man wieder an (4.4) und und wählt für ε den Impuls pro Masseneinheit $\varepsilon = \boldsymbol{v}$. Dann ist $\boldsymbol{\Phi}_{\boldsymbol{v}}$ die Impulsstromdichte durch die mitbewegte Oberfläche, d. h. die Kraft pro Flächeneinheit, die durch das umgebende Medium ausgeübt wird. Sie läßt sich durch Tangential- und Normalspannungen beschreiben, die man durch den Drucktensor P darstellt. Sind auch noch eingeprägte äußere Kräfte vorhanden, die zur Masse proportional sind, so sei mit $\boldsymbol{F}_i$ die auf die Masseneinheit der i-ten Komponente wirkende Kraft bezeichnet. Ihr Beitrag zur Impulsbilanz $\varrho_i \boldsymbol{F}_i$ wird in die äußere Impulsproduktion aufgenommen. Dann entsteht aus (4.4) unter Berücksichtigung von (4.5) wie früher

$$\varrho \frac{D\boldsymbol{v}}{Dt} = -\mathrm{Div}\, \mathsf{P} + \sum_{i=1}^{n} \varrho_i \boldsymbol{F}_i. \tag{4.13}$$

Div bedeutet die Tensordivergenz. Unter Berücksichtigung des konvektiven Impulstransports $\varrho \boldsymbol{v}\boldsymbol{v}$ ($\boldsymbol{v}\boldsymbol{v}$ ist das dyadische Produkt des Vektors $\boldsymbol{v}$ mit sich selbst) erhält man die (4.2) entsprechende Formulierung für die lokale Änderung der Impulsdichte

$$\frac{\partial}{\partial t} \varrho \boldsymbol{v} = -\mathrm{Div}\, (\mathsf{P} + \varrho \boldsymbol{v}\boldsymbol{v}) + \sum_{i=1}^{n} \varrho_i \boldsymbol{F}_i. \tag{4.14}$$

δ) Die Bilanz für die Gesamtenergie. Die Aufstellung der Bilanzgleichung für die innere Energie erfordert eine Vorüberlegung. Für die Summe aller Energien gilt ein Erhaltungssatz, d.h. eine Bilanzgleichung mit der inneren Produktionsdichte Null. Beim Vorliegen äußerer Kräfte berücksichtigt man deren Beitrag in der äußeren Energieproduktionsdichte, deren Größe pro Volumen und Zeiteinheit gleich ihrer Leistungsdichte

$$\sum_{i=1}^{n} \varrho_i \boldsymbol{F}_i \cdot \boldsymbol{v}_i = \left(\sum_{i=1}^{n} \varrho_i \boldsymbol{F}_i\right) \cdot \boldsymbol{v} + \sum_{i=1}^{n} \boldsymbol{F}_i \cdot \boldsymbol{J}_i \tag{4.15}$$

ist. Potentielle Energien in äußeren Feldern sind damit weder in der Energiedichte noch im Energiestrom enthalten. Für ε hat man die Summe aus spezifischer kinetischer und innerer Energie $\varepsilon = \frac{1}{2} v^2 + u$ zu setzen. $\boldsymbol{\Phi}_e$ enthält den konvektiven Anteil $\varepsilon \cdot \boldsymbol{v}$ und den Arbeitsfluß $\mathsf{P} \cdot \boldsymbol{v}$ infolge der Wechselwirkung mit der Umgebung. Weitere Anteile zu $\boldsymbol{\Phi}_e$ werden in der Energiestromdichte $\boldsymbol{J}_u$ zusammengefaßt. Man sieht ohne weiteres, daß solche Anteile vorhanden sein können, indem man ein ruhendes Medium betrachtet. Dann ist $\boldsymbol{J}_u$ der einzige Anteil und stellt die Wärmeleitung im Temperaturfeld dar.

(4.6) und (4.2) geben dann die äquivalenten Formulierungen der Energiebilanz für die spezifische Energie und für die Energiedichte

$$\varrho \frac{D}{Dt}\left(\frac{1}{2} v^2 + u\right) = - \operatorname{div}\left[\mathsf{P} \cdot \boldsymbol{v} + \boldsymbol{J}_u\right] + \sum_{i=1}^{n} \varrho_i \boldsymbol{F}_i \cdot \boldsymbol{v} + \sum_{i=1}^{n} \boldsymbol{F}_i \cdot \boldsymbol{J}_i \tag{4.16}$$

$$\frac{\partial}{\partial t} \varrho \left(\frac{1}{2} v^2 + u\right) = - \operatorname{div}\left[\varrho\left(\frac{1}{2} v^2 + u\right)\boldsymbol{v} + \mathsf{P} \cdot \boldsymbol{v} + \boldsymbol{J}_u\right] + \sum_{i=1}^{n} \varrho_i \boldsymbol{F}_i \cdot \boldsymbol{v} + \sum_{i=1}^{n} \boldsymbol{F}_i \cdot \boldsymbol{J}_i. \tag{4.17}$$

ε) Die Bilanz der inneren Energie. Multipliziert man die Bewegungsgleichung (4.13) skalar mit $\boldsymbol{v}$, so ergibt sich

$$\varrho \frac{D}{Dt} \frac{v^2}{2} = - \boldsymbol{v} \cdot \operatorname{Div} \mathsf{P} + \sum \varrho_i \boldsymbol{F}_i \cdot \boldsymbol{v}. \tag{4.18}$$

Auch diese Beziehung kann in eine Bilanz, nämlich für die kinetische (konvektive) Energie umgeformt werden. Wir verzichten jedoch auf ihre Formulierung und geben sofort die durch Subtraktion der Gln. (4.16) und (4.18) entstehende Bilanz der inneren Energie an. Sie lautet in substantieller und lokaler Fassung

$$\varrho \frac{Du}{Dt} = - \operatorname{div} \boldsymbol{J}_u + \sum_{i=1}^{n} \boldsymbol{F}_i \cdot \boldsymbol{J}_i - p \operatorname{div} \boldsymbol{v} - (\mathsf{P} - p\,\mathsf{I}) : \operatorname{Grad} \boldsymbol{v}, \tag{4.19}$$

$$\frac{\partial}{\partial t}(\varrho u) = - \operatorname{div}(\varrho u \boldsymbol{v} + \boldsymbol{J}_u) + \sum_{i=1}^{n} \boldsymbol{F}_i \cdot \boldsymbol{J}_i - p \operatorname{div} \boldsymbol{v} - (\mathsf{P} - p\,\mathsf{I}) : \operatorname{Grad} \boldsymbol{v}. \tag{4.20}$$

Mit I ist der Einheitstensor bezeichnet.

Die Produktion der inneren Energie wird in einleuchtender Weise durch denjenigen Anteil der Arbeit der äußeren Kräfte bewirkt, der nicht zur Beschleunigung dient, sowie durch Umwandlung konvektiver Energie in innere Energie durch Kompression und Reibung.

5. Die Entropiebilanz. Ein mit der baryzentrischen Geschwindigkeit mitbewegter Beobachter stellt an seinem Ort Änderungen $\dfrac{Du}{Dt}, \dfrac{D}{Dt}\dfrac{1}{\varrho}, \dfrac{D\gamma_i}{Dt}$ der spezifischen Größen $u, \dfrac{1}{\varrho}, \gamma_i$ fest. Diese sind nach der Gibbsschen Fundamentalgleichung (3.4) von einer Änderung der spezifischen Entropie

$$\frac{Ds}{Dt} = \frac{1}{T} \frac{Du}{Dt} + \frac{p}{T} \frac{D}{Dt} \frac{1}{\varrho} - \sum_{i=1}^{n} \frac{\mu_i}{T} \frac{D\gamma_i}{Dt} \tag{5.1}$$

begleitet. Eliminiert man $\dfrac{Du}{Dt}$, $\dfrac{D}{Dt}\dfrac{1}{\varrho}$, $\dfrac{D\gamma_i}{Dt}$ vermittels (4.19), (4.9), (4.10), so wird die substantielle Änderung der spezifischen Entropie durch die die irreversiblen Vorgänge beschreibenden Größen $\boldsymbol{J}_u$, $\boldsymbol{J}_i$, Γ_i, $\mathsf{P}-p\,\mathsf{I}$ ausgedrückt. Man erhält zunächst

$$\left. \begin{aligned} \varrho\,\frac{Ds}{Dt} &= -\frac{1}{T}\operatorname{div}\boldsymbol{J}_u + \sum_{i=1}^{n}\boldsymbol{J}_i\cdot\frac{\boldsymbol{F}_i}{T} - (\mathsf{P}-p\,\mathsf{I}):\frac{1}{T}\operatorname{Grad}\boldsymbol{v} - \\ &\quad - \sum_{i=1}^{n}\frac{\mu_i}{T}\,[\Gamma_i - \operatorname{div}\boldsymbol{J}_i] \end{aligned} \right\} \qquad (5.2)$$

und nach Umformung

$$\left. \begin{aligned} \varrho\,\frac{Ds}{Dt} &= -\operatorname{div}\left[\frac{1}{T}\Big(\boldsymbol{J}_u - \sum_{i=1}^{n}\mu_i\,\boldsymbol{J}_i\Big)\right] + \boldsymbol{J}_u\cdot\operatorname{grad}\frac{1}{T} + \\ &\quad + \sum_{i=1}^{n}\boldsymbol{J}_i\cdot\Big(\frac{\boldsymbol{F}_i}{T} - \operatorname{grad}\frac{\mu_i}{T}\Big) - (\mathsf{P}-p\,\mathsf{I}):\frac{1}{T}\operatorname{Grad}\boldsymbol{v} - \\ &\quad - \sum_{i=1}^{n}\Gamma_i\frac{\mu_i}{T}\,. \end{aligned} \right\} \qquad (5.3)$$

Sowohl (5.2) als auch (5.3) haben die Gestalt einer Bilanzgleichung für die Entropie, wie ein Vergleich mit (4.6) zeigt. In (5.2) wäre die ganze rechte Seite als Entropieproduktion zu deuten und die Entropiestromdichte gleich $\boldsymbol{\Phi}_s = \varrho s \boldsymbol{v}$ zu setzen. Es gäbe dann nur einen Entropietransport durch Konvektion allein. Hingegen kann (5.3) in der Form

$$\varrho\,\frac{Ds}{Dt} = -\operatorname{div}\boldsymbol{J}_s + \sigma_s \qquad (5.4)$$

mit

$$\boldsymbol{J}_s = \frac{1}{T}\Big(\boldsymbol{J}_u - \sum_{i=1}^{n}\mu_i\,\boldsymbol{J}_i\Big) \qquad (5.5)$$

und

$$\sigma_s = \boldsymbol{J}_u\cdot\operatorname{grad}\frac{1}{T} + \sum_{i=1}^{n}\boldsymbol{J}_i\cdot\Big(\frac{\boldsymbol{F}_i}{T} - \operatorname{grad}\frac{\mu_i}{T}\Big) - (\mathsf{P}-p\,\mathsf{I}):\frac{1}{T}\operatorname{Grad}\boldsymbol{v} - \sum_{i=1}^{n}\Gamma_i\frac{\mu_i}{T} \qquad (5.6)$$

als Bilanzgleichung für die Entropie mit der totalen Entropiestromdichte unter Einschluß des konvektiven Entropietransports

$$\boldsymbol{\Phi}_s = \boldsymbol{J}_s + \varrho s \boldsymbol{v} \qquad (5.7)$$

und mit der Entropieproduktionsdichte σ_s aufgefaßt werden.

Es sind formal auch noch andere Schreibweisen von (5.2) als Bilanzgleichung denkbar, die wieder andere Definitionen für die Entropiestromdichte und die Entropieproduktionsdichte liefern.

Eine sinnvolle Definition muß jedoch die Eigenschaft haben, daß die Entropieproduktionsdichte in jedem Raumpunkt positiv oder allenfalls gleich Null ist. Dies gilt sicher nicht für die rechte Seite von (5.2). Man sieht dies, indem man $\boldsymbol{J}_i = 0$, $\Gamma_i = 0$, $\mathsf{P} = p\,\mathsf{I}$ setzt. Das Feld der Energiestromdichte kann dann stets so gewählt werden, daß wenigstens in gewissen Raumbereichen $\operatorname{div}\boldsymbol{J}_u > 0$ ist; damit würde dort die Entropieproduktion negativ sein. Man sieht hieran und an analogen Beispielen, daß die gleiche Schwierigkeit immer auftritt, solange in der Entropieproduktion Divergenzen zugelassen werden. Die einzige Möglichkeit, überhaupt die Forderung positiver lokaler Entropieerzeugung stellen zu können, bietet offenbar die Interpretation (5.6). Damit ist dann auch zwangsläufig die Definition (5.5) beziehungsweise (5.7) der Entropiestromdichte anzunehmen. Daß sich diese Forderung wirklich realisieren läßt, wird in der nächsten Ziffer gezeigt werden.

In ganz natürlicher Weise ergibt sich die Bilanzgleichung (5.4) mit der Entropiestromdichte (5.5) und der Entropieproduktionsdichte (5.6) aus den kinetischen Theorien. Bezüglich der kinetischen Gastheorie verweisen wir auf Prigogine[1] und Reik[2].

Für ein abgeschlossenes System führt natürlich jede Form der Entropiebilanz auf die Aussage des zweiten Hauptsatzes. Besonders deutlich sieht man dies indessen, wenn man von der Bilanzgleichung (5.4) ausgeht. Für ein abgeschlossenes System gibt nämlich das Raumintegral über div $\boldsymbol{J}_s$ keinen Beitrag, da für ein solches an der Oberfläche des Systems $\boldsymbol{J}_s = 0$ ist. Die Entropieänderung ist also hier gleich $\int \sigma_s \, dV$. Man sieht unmittelbar, daß dieser Ausdruck nicht negativ ist, falls σ_s selbst nicht negativ ist.

Der substantiellen Formulierung (5.4) der Entropiebilanz läßt sich wieder eine lokale Formulierung

$$\frac{\partial}{\partial t} (\varrho \, s) = - \operatorname{div} \boldsymbol{\Phi}_s + \sigma_s \tag{5.8}$$

mit denselben Ausdrücken für $\boldsymbol{\Phi}_s$, und σ_s wie in (5.7) mit (5.5) und in (5.6) angeben. Man kann sie auch direkt herleiten, wenn man die Gibbsschen Fundamentalgleichung in der Form (3.5) auf die lokalen Änderungen anwendet

$$\frac{\partial}{\partial t} (\varrho \, s) = \frac{1}{T} \frac{\partial}{\partial t} (\varrho \, u) - \sum_{i=1}^{n} \frac{\mu_i}{T} \frac{\partial \varrho_i}{\partial t} \tag{5.9}$$

und die hierin vorkommenden Differentialquotienten nach der Zeit mit Hilfe der lokalen Bilanzgleichungen (4.20), (4.11) ausdrückt.

Die Entropieproduktionsdichte (5.6) hat eine charakteristische Gestalt; sie ist eine Summe von Produkten. Die Faktoren $\boldsymbol{J}_u, \boldsymbol{J}_i, p\,\mathsf{I} - \mathsf{P}, \varGamma_i$ werden als „Flüsse", die anderen Faktoren werden als „Kräfte" bezeichnet. Die Kräfte sind nicht Kräfte im mechanischen Sinne, sondern thermodynamische Kräfte, die die Abweichungen vom thermodynamischen Gleichgewichtszustand kennzeichnen[3].

Es ist zu beachten, daß die hier gewählten Flüsse wegen (4.12) nicht unabhängig sind. Unabhängige Diffusionsstromdichten erhält man, wenn man etwa $\boldsymbol{J}_n$ mit Hilfe von (4.12) eliminiert. Dann lautet der zweite Term in (5.6)

$$\sum_{i=1}^{n-1} \boldsymbol{J}_i \left(\frac{\boldsymbol{F}_i - \boldsymbol{F}_n}{T} - \operatorname{grad} \frac{\mu_i - \mu_n}{T} \right).$$

Beim letzten Term von (5.6) fallen ohnehin alle Summanden heraus, die Umwandlungsgeschwindigkeiten indifferenter Komponenten beschreiben. Die verbleibenden $\varGamma_i$ lassen sich durch r unter ihnen ausdrücken, wenn r die Zahl der unabhängigen Reaktionen (inneren Umwandlungen) ist. Sie ist wegen (4.12) um mindestens eins kleiner als die Zahl der reagierenden Bestandteile.

---

[1] I. Prigogine: Physica, Haag **15**, 272 (1949).
[2] H. G. Reik: Z. Physik **148**, 156, 333 (1957).
[3] Gelegentlich werden auch die mit T multiplizierten thermodynamischen Kräfte als Kräfte bezeichnet.

6. Die linearen phänomenologischen Ansätze und die Onsager-Casimirschen Reziprozitätsbeziehungen. Der Übersichtlichkeit halber soll zunächst die Entropieproduktion, nach Reduktion auf unabhängige Flüsse und Kräfte, in der abgekürzten Form

$$\sigma_s = \sum_{i=1}^{m} Y_i X_i \tag{6.1}$$

geschrieben werden. Die Y_i bezeichnen skalare Flüsse und kartesische Komponenten der vektoriellen oder tensoriellen Flüsse. Sie umfassen also die früheren Flüsse Γ_i sowie die Komponenten von $p\mathbf{I} - \mathbf{P}$, $\mathbf{J}_u$ und $\mathbf{J}_i$, soweit diese unabhängig sind. Die X_i sind die skalaren Kräfte und die kartesischen Komponenten der vektoriellen und tensoriellen Kräfte, wie sie sich nach Elimination der abhängigen Flüsse aus (5.6) ergeben.

Die bekannten Gesetze der Wärmeleitung, der inneren Reibung, der Diffusion und der chemischen Reaktionen sind Beispiele für funktionale Beziehungen zwischen Flüssen und zugeordneten Kräften. Es handelt sich nun darum, diese Beziehungen auf den vorliegenden umfassenderen Fall zu verallgemeinern. Zunächst ist klar, daß im thermodynamischen Gleichgewicht alle Kräfte X_i verschwinden. Mit ihnen verschwinden aber auch die Ausgleichsvorgänge, also die Flüsse Y_i. Dies legt nahe anzunehmen, daß auch im vorliegenden Fall funktionale Zusammenhänge nur zwischen Flüssen und Kräften bestehen. Es ist aber, um nichts an Allgemeinheit des Ansatzes zu verlieren, notwendig zuzulassen, daß ein Fluß Y_i nicht nur von der zugeordneten Kraft X_i, sondern möglicherweise von allen Kräften $X_1, X_2, \ldots, X_m$ abhängt. Damit wird

$$Y_i = Y_i(X_1, X_2, \ldots X_m) \tag{6.2}$$

mit

$$0 = Y_i(0, 0, \ldots 0). \tag{6.3}$$

Bei kleinen Abweichungen vom thermodynamischen Gleichgewicht, also bei genügend kleinen Werten der X_k ist es berechtigt, die Flüsse Y_i nach den Kräften X_k zu entwickeln und nach den linearen Gliedern abzubrechen. Da das konstante Glied der Entwicklung wegen (6.3) verschwindet, ergibt sich

$$Y_i = \sum_{k=1}^{m} L_{ik} X_k \quad (i = 1, 2, , \ldots m). \tag{6.4}$$

Dies sind die sog. linearen phänomenologischen Gleichungen; die Entwicklungskoeffizienten nennt man phänomenologische Koeffizienten.

Es ist eine besondere Aufgabe, den Gültigkeitsbereich der linearen Gl. (6.4) abzustecken und die Berechtigung der allgemeineren Ansätze (6.2) außerhalb des linearen Bereiches zu prüfen. Dies ist natürlich nicht im Bereich einer makroskopischen Kontinuumstheorie möglich; nur kinetische Theorien, im Einzelfall auch experimentelle Erfahrungen, können über diese Fragen Aufschluß geben.

Den linearen Ansätzen sind *zwei wichtige Einschränkungen aufzuerlegen*. Zunächst folgt durch Einsetzen von (6.4) in (6.1)

$$\sigma_s = \sum_{i=1}^{m} L_{ik} X_i X_k. \tag{6.5}$$

Die X_i sind unabhängige Kräfte, also willkürlich vorgebbar. Damit stets $\sigma_s \geqq 0$ ist, muß man daher verlangen, daß die quadratische Form in den X_i in (6.5) positiv definit oder wenigstens nicht-negativ definit ist. Eine hinreichende Bedingung dafür ist, daß alle Hauptminoren der symmetrisierten Determinante

$|L_{ik} + L_{ki}|$ positiv oder wenigstens nicht-negativ sind. Eine selbstverständliche Folgerung ist

$$L_{ii} \geqq 0, \qquad (i = 1, 2, \ldots m);\tag{6.6}$$

von den übrigen Aussagen sind meist nur jene über zweireihige Hauptminoren von praktischer Bedeutung

$$L_{ii} L_{kk} \geqq \tfrac{1}{4}(L_{ik} + L_{ki})^2.\tag{6.7}$$

Bei unabhängigen Flüssen und Kräften ist das Gleichheitszeichen auszuschließen, es sei denn, daß der Ablauf irreversibler Vorgänge teilweise oder ganz gehemmt ist. Im letzteren Fall sind alle $L_{ik} = 0$. Die Entropieproduktion verschwindet nur im gehemmten oder im ungehemmten Gleichgewicht; im ersten Fall sind alle $Y_i = 0$, ohne daß alle X_i zu verschwinden brauchen. Im zweiten Fall sind auch alle $X_i = 0$.

Eine weitere Einschränkung für die Koeffizienten der linearen Ansätze (6.4) geben die Onsager-Casimirschen[1] Reziprozitätsbeziehungen (im folgenden stets mit O.C.R.B. abgekürzt)

$$L_{ik} = \varepsilon_i \varepsilon_k L_{ki}.\tag{6.8}$$

Hierin ist $\varepsilon_i = 1$ wenn die Kraft X_i in bezug auf Zeitumkehr gerade, $\varepsilon_i = -1$, wenn sie ungerade ist. Man nennt Kräfte der ersten Art auch Kräfte vom α-Typ, solche der zweiten Art Kräfte vom β-Typ. Beispielsweise in (5.6) sind alle Kräfte gerade mit Ausnahme von Grad v; die Geschwindigkeit kehrt ja mit der Zeitumkehr ihr Vorzeichen um.

Soweit die X_i eigentliche thermodynamische Variable sind, nennt man sie entsprechend auch α- und β-Variable.

Sind speziell alle ε_i vom selben Vorzeichen, so nennt man die Beziehungen (6.8) Onsagersche Reziprozitätsbeziehungen. Im allgemeinen Fall werden die Koeffizienten L_{ik} mit $\varepsilon_i \varepsilon_k = +1$ auch Onsagersche, jene mit $\varepsilon_i \varepsilon_k = -1$ Casimirsche Koeffizienten genannt.

In der quadratischen Form der Entropieproduktionsdichte $\sigma_s = \sum\limits_{i=1}^{m} \sum\limits_{k=1}^{m} L_{ik} X_i X_k$ fallen gemischte Glieder mit Casimirschen Koeffizienten wegen (6.8) stets heraus.

Der Fall linear abhängiger Flüsse (z.B. der J_i mit $\Sigma J_i = 0$) und Kräfte kann stets auf den Fall linear unabhängiger Flüsse und Kräfte zurückgeführt werden; dann sind die O.C.R.B. in der oben gegebenen Form unmittelbar anwendbar. Dies ist jedoch gelegentlich umständlich. Es läßt sich indessen zeigen (s. Ziff. 8), daß die O.C.R.B. auch ohne diese Zurückführung gelten, wenn nur die Kräfte entweder vom α-Typ oder vom β-Typ, aber nicht von gemischter Natur sind. Für Kräfte einheitlichen Typs (Onsager-Fall) ist dies von Hooyman und de Groot (s. Ziff. 8) gezeigt worden.

Eine Modifikation der O.C.R.B. (6.8) tritt bei irreversiblen Vorgängen im äußeren Magnetfeld (magnetische Induktion B) oder in rotierenden Systemen (Winkelgeschwindigkeit ω) oder in Systemen mit ungeraden Parametern p_i (Ziff. 25) ein. Sie lauten dann

$$L_{ik}(B, \omega, p_i) = \varepsilon_i \varepsilon_k L_{ki}(-B, -\omega, -p_i).\tag{6.9}$$

7. Die linearen phänomenologischen Ansätze für fluide Medien mit inneren Umwandlungen. $\alpha)$ *Phänomenologische Ansätze.* Nach den Vorbereitungen in Ziff. 6 können nun aus der Entropieproduktionsdichte (5.6) die linearen phänomenologischen Ansätze für fluide Medien mit inneren Umwandlungen abgelesen werden.

---

[1] L. Onsager: Phys. Rev. **37**, 405 (1931); **38**, 2265 (1931). — H.B.G. Casimir: Rev. Mod. Phys. **17**, 343 (1945).

Es ist jedoch zu beachten, daß die von Anfang an vorausgesetzte Isotropie der fluiden Medien den linearen Ansätzen noch weitere Einschränkungen auferlegt: Es lassen sich nur Größen gleichen Transformationsverhaltens gegenüber Drehungen im Raum miteinander linear verknüpfen (P. Curiesches Gesetz). Da die in (5.6) auftretenden Flüsse und Kräfte skalar, vektoriell oder tensoriell von zweiter Stufe sind, so folgt: Die vektoriellen Flüsse $\boldsymbol{J_i}$ und $\boldsymbol{J_u}$ können nur von den vektoriellen Kräften linear abhängen; die skalaren Flüsse Γ_i können nur von den skalaren Kräften und von der skalaren Invarianten (Spur) der tensoriellen Kräfte linear abhängen, die tensoriellen Flüsse können nur von den tensoriellen Kräften und von skalaren Kräften, die mit dem Einheitstensor multipliziert sind, linear abhängen. Dies führt zu folgenden Ansätzen (mit etwas abgeänderter Bedeutung der L_{ik} gegenüber Ziff. 6) für die vektoriellen Flüsse

$$\boldsymbol{J_i} = \sum_{k=1}^{n} L_{ik}\left(\frac{\boldsymbol{F_k}}{T} - \operatorname{grad}\frac{\mu_k}{T}\right) + L_{iu}\operatorname{grad}\frac{1}{T} \quad (i = 1, 2, \ldots n), \quad (7.1)$$

$$\boldsymbol{J_u} = \sum_{k=1}^{n} L_{uk}\left(\frac{\boldsymbol{F_k}}{T} - \operatorname{grad}\frac{\mu_k}{T}\right) + L_{uu}\operatorname{grad}\frac{1}{T} \quad (7.2)$$

mit den O.C.R.B. (wir schließen hier und im folgenden äußere Magnetfelder und andere ungerade Parameter aus)

$$L_{ik} = L_{ki} \quad (i, k = 1, 2, \ldots n), \quad L_{iu} = L_{ui} \quad (i = 1, 2, \ldots n) \quad (7.3)$$

und den aus der Abhängigkeit der Diffusionstromdichten folgenden Eigenschaften

$$\sum_{i=1}^{n} L_{ik} = 0 \quad (k = 1, 2, \ldots n) \quad \sum_{i=1}^{n} L_{iu} = 0. \quad (7.4)$$

Die aus den Koeffizienten L_{ik}, L_{uk}, L_{uu}, aufgebaute Matrix hat eine nicht-negativ definite quadratische Form. Die Gln. (7.1) und (7.2) geben eine vollständige Beschreibung der Thermodiffusion und des Diffusionsthermoeffekts bei praktisch-reibungsfreien Vorgängen (s. Ziff. 14).

Der Tensor der Reibungsspannungen $p\,\delta_{\alpha\beta} - P_{\alpha\beta}$ kann von den Tensoren $\partial v_\alpha/\partial x_\beta$, $\partial v_\beta/\partial x_\alpha$, $\delta_{\alpha\beta}\operatorname{div}\boldsymbol{v}$, $\delta_{\alpha\beta}\,\mu_i$ linear abhängen. Da er wegen Schwerpunkts- und Drehimpulssatz symmetrisch ist ($P_{\alpha\beta} = P_{\alpha\beta}$), so ergibt sich folgender linearer Ansatz

$$p\,\delta_{\alpha\beta} - P_{\alpha\beta} = \eta\left(\frac{\partial v_\alpha}{\partial x_\beta} + \frac{\partial v_\beta}{\partial x_\alpha}\right) + \left(\zeta - \frac{2}{3}\eta\right)\delta_{\alpha\beta}\operatorname{div}\boldsymbol{v} + \delta_{\alpha\beta}\sum_{i=1}^{n}\Lambda_i\left(-\frac{\mu_i}{T}\right). \quad (7.5)$$

Analog sind die chemischen Produktionen lineare Funktionen der μ_i/T und der Spur des Tensors $\dfrac{1}{T}\dfrac{\partial v_\alpha}{\partial x_\beta}$, also

$$\Gamma_i = \Lambda_i'\frac{1}{T}\operatorname{div}\boldsymbol{v} + \sum_{k=1}^{n}\Lambda_{ik}\left(-\frac{\mu_k}{T}\right) \quad (i = 1, 2, \ldots n). \quad (7.6)$$

Hier sind die Onsager-Casimirschen Reziprozitätsbeziehungen anzuwenden, da die Kräfte $\partial v_\alpha/\partial x_\beta$ vom β-Typ, die μ_i/T solche vom α-Typ sind. Daher folgt

$$\Lambda_i' = -\Lambda_i \quad (i = 1, 2, \ldots n); \quad \Lambda_{ik} = \Lambda_{ki} \quad (i, k = 1, 2, \ldots n). \quad (7.7)$$

Wegen $\sigma > 0$ außerhalb des Gleichgewichtes beim Fehlen von Hemmung sind η und ζ positiv.

Wegen der linearen Abhängigkeit der Γ_i' nach (4.12) gilt allgemein

$$\sum_{i=1}^{n} \Lambda_i' = 0, \quad \sum_{i=1}^{n} \Lambda_{ik} = 0 \quad (k = 1, 2, \ldots n). \tag{7.8}$$

Weitere Einschränkungen für diese Koeffizienten ergeben sich bei Kenntnis der Zahl und Art der möglichen chemischen Reaktionen, da dann die Γ_i neben (4.12) noch weiteren, stöchiometrischen Bedingungen (7.10), (7.11) unterliegen.

Beim Fehlen chemischer Reaktionen ist (7.5) der bekannte Ansatz für den Reibungstensor in einem fluiden Medium mit der *Schubviskosität* η und der *Volumenviskosität* ζ (auch *Druckviskosität* oder *zweite Viskosität* genannt). Da die thermodynamischen Kräfte, welche durch das Geschwindigkeitsfeld hervorgerufen werden, durch den Tensor $\frac{1}{T} \frac{\partial v_\alpha}{\partial x_\beta}$ gegeben sind, so sind nicht η und ζ selbst, sondern $T\eta$ und $T\zeta$ phänomenologische Koeffizienten. Die Koeffizienten Λ_i, welche nur in chemisch reagierenden Medien auftreten, können als *chemische Viskositäten* bezeichnet werden. Ein Beitrag der chemischen Reaktionen zum Drucktensor existiert tatsächlich, wie man sich an Hand von speziellen Beispielen klar machen kann[1]. Insbesondere bei langsamen Vorgängen erweist sich dieser Beitrag als proportional zu div $\boldsymbol{v}$ und kann mit dem Koeffizienten ζ vereinigt werden. Auf diese Weise ist es möglich, wenigstens einen Teil der bei langsamen Vorgängen gemessenen Volumenviskosität durch chemische Reaktionen oder allgemeinere innere Umwandlungen zu erklären (s. Ziff. 22).

Setzt man die phänomenologischen Gesetze in den Ausdruck (5.6) für die Entropieproduktion ein, so erhält man

$$\left.\begin{aligned}
\sigma = {} & \sum_{i,k=1}^{n} L_{ik} \left(\frac{\boldsymbol{F}_i}{T} - \operatorname{grad} \frac{\mu_i}{T} \right) \cdot \left(\frac{\boldsymbol{F}_k}{T} - \operatorname{grad} \frac{\mu_k}{T} \right) + \\
& + 2 \sum_{i=1}^{n} L_{iu} \left(\frac{\boldsymbol{F}_i}{T} - \operatorname{grad} \frac{\mu_i}{T} \right) \cdot \operatorname{grad} \frac{1}{T} + \frac{\eta}{2T} \sum_{\alpha,\beta=1}^{3} \left(\frac{\partial v_\alpha}{\partial x_\beta} + \frac{\partial v_\beta}{\partial x_\alpha} \right)^2 + \\
& + \frac{1}{T} \left(\zeta - \frac{2}{3} \eta \right) (\operatorname{div} \boldsymbol{v})^2 + \sum_{i,k=1}^{n} \Lambda_{ik} \frac{\mu_i}{T} \frac{\mu_k}{T}.
\end{aligned}\right\} \tag{7.9}$$

Die Terme von (7.9) die sich auf die innere Reibung beziehen, werden als *Rayleighsche Dissipationsfunktion* bezeichnet. (7.9) stellt also eine Verallgemeinerung dieser Funktion dar, in der auch der Entropieproduktion durch Diffusion, Wärmeleitung und chemische Reaktionen Rechnung getragen ist. Wie schon in Ziff. 6 bemerkt, fällt die Kopplung zwischen chemischen Reaktionen und Reibungsdrucktensor, die durch Casimirsche Koeffizienten vermittelt wird, aus dem Ausdruck für die verallgemeinerte Dissipationsfunktion heraus.

Die linearen phänomenologischen Ansätze (7.1), (7.2), (7.5), (7.6) sind symmetrisch in den Bestandteilen. Insbesondere bei solchen Problemen, bei denen die Diffusion keine wesentliche Rolle spielt, ist es zweckmäßig, die Gesamtheit der Produktionen Γ_i auf unabhängige unter ihnen zu reduzieren. Wenn r unabhängige Reaktionen stattfinden, dann sind r Bestandteile unabhängig variierbar. Ihre Dichten können durch geeignete Numerierung als $\varrho_1, \varrho_2, \ldots \varrho_r$ gewählt werden, ihre Produktionen sind dann $\Gamma_1, \Gamma_2 \ldots \Gamma_r$. Alle übrigen Produktionen sind von diesen linear abhängig. Sie lassen sich daher in folgender Weise ausdrücken:

$$\Gamma_i - \sum_{k=1}^{r} \nu_{ik} \Gamma_k = 0 \quad (i = r + 1, r + 2, \ldots n). \tag{7.10}$$

---

[1] J. Meixner: Z. Physik **131**, 456 (1951). — Acustica **2**, 101 (1952).

Insbesondere sind alle $\nu_{ik}=0$ $(k=1, 2, \ldots r)$ für einen Bestandteil i $(i=r+1, \ldots n)$ der an keiner Reaktion teilnimmt. Die Gln. (7.10) können auch als Reaktionsgleichungen gelesen werden, wenn man dazu die Γ_i als Symbole für die Masseneinheit des betreffenden Bestandteils ansieht.

Setzt man die Γ_i $(i=r+1, \ldots n)$ aus (7.10) in die erste Beziehung (4.12) ein, so ergibt sich wegen der Unabhängigkeit der $\Gamma_1, \ldots \Gamma_r$

$$\sum_{i=r+1}^{n} \nu_{ik} + 1 = 0 \qquad (k=1, 2, \ldots r). \tag{7.11}$$

Dies ist die bekannte Aussage, daß die Summe der stöchiometrischen Koeffizienten einer Reaktionsgleichung verschwindet.

Der Beitrag der chemischen Reaktionen zur Entropieproduktionsdichte (5.6) schreibt sich nun

$$\left.\begin{aligned} -\frac{1}{T}\sum_{i=1}^{n}\Gamma_i\mu_i &= -\frac{1}{T}\sum_{k=1}^{r}\left(\mu_k + \sum_{i=r+1}^{n}\nu_{ik}\mu_i\right) \\ &\equiv \frac{1}{T}\sum_{k=1}^{r}\Gamma_k A_k. \end{aligned}\right\} \tag{7.12}$$

Die Größen A_k die durch (7.12) definiert sind, werden *Affinitäten* genannt. Sie sind also die den unabhängigen $\Gamma_1, \Gamma_2, \ldots, \Gamma_r$ zugeordneten Kräfte.

Damit reduzieren sich die linearen phänomenologischen Ansätze (7.5) und (7.6) auf

$$\left.\begin{aligned} p\,\delta_{\alpha\beta} - P_{\alpha\beta} &= \eta\left(\frac{\partial v_\alpha}{\partial x_\beta} - \frac{\partial v_\beta}{\partial x_\alpha}\right) + \left(\zeta - \frac{2}{3}\eta\right)\delta_{\alpha\beta}\operatorname{div}\boldsymbol{v} + \\ &\quad + \delta_{\alpha\beta}\sum_{k=1}^{r}\frac{\lambda_k}{T}A_k \qquad (\alpha, \beta = 1, 2, 3), \end{aligned}\right\} \tag{7.13}$$

$$\Gamma_k = -\frac{\lambda_k}{T}\operatorname{div}\boldsymbol{v} + \sum_{i=1}^{r}\frac{\lambda_{ki}}{T}A_i. \tag{7.14}$$

Die Koeffizinenten λ_k nennen wir die den Affinitäten zugeordneten chemischen Viskositäten. Sie lassen sich aus den den chemischen Potentialen zugeordneten chemischen Viskositäten Λ_k berechnen.

Ist keine der chemischen Reaktionen gehemmt, so hat die symmetrische Matrix λ_{ik} eine positiv-definite quadratische Form, während für die Matrix Λ_{ik} in (7.6) nur festgestellt werden konnte, daß ihre quadratische Form nicht negativ ist.

In der quadratischen Form der Entropieproduktionsdichte (7.9) tritt nunmehr an Stelle des letzten Gliedes der Ausdruck

$$\sum_{i=1}^{r}\sum_{k=1}^{r}\lambda_{ik}\frac{A_i}{T}\frac{A_k}{T}.$$

β) Beziehungen zur Kinetik. In den vorhergehenden Ziffern haben wir den vollständigen Satz von Grundgleichungen der Thermodynamik irreversibler Prozesse — die Bilanzgleichungen und die phänomenologischen Gleichungen — abgeleitet. Von den in Ziff. 1 und 3 genannten thermodynamischen Voraussetzungen wurde dabei an zwei Stellen Gebrauch gemacht, nämlich bei der Aufstellung der Entropiebilanz und bei der Ableitung der phänomenologischen Gleichungen aus dem Ausdruck für die Entropieproduktion. Man kann die Frage stellen, wie sich diese Voraussetzungen mit Hilfe von kinetischen Theorien begründen lassen.

Diese Frage ist für den Fall der kinetischen Gastheorie von Meixner, Prigogine und Reik behandelt worden. Wir geben im folgenden einen Überblick über die von diesen Autoren benutzten Argumente und ihre Ergebnisse.

Allen Arbeiten ist gemeinsam, daß sie sich auf die Behandlung bestimmter Typen von kinetischen Verteilungsfunktionen, nämlich auf die Chapman-Enskog-schen Lösungen der Boltzmannschen Fundamentalgleichung beschränken. Diese Verteilungen haben die Eigenschaft, daß sie von der Variablen c (der Molekulargeschwindigkeit) und von Parametern abhängen, die thermodynamische Variable bzw. deren Ableitung nach Ortskoordinaten sind. Die Orts- und Zeitabhängigkeit dieser Verteilungsfunktionen wird also durch die Zustandsfelder, z.B. durch $\varrho(r, t)$, $T(r, t)$, $v(r, t)$ vollständig bestimmt. Insbesondere läßt sich die Verteilungsfunktion durch eine Reihe

$$f = f^{(0)} + f^{(1)} + f^{(2)} + \cdots = f^{(0)} (1 + \Phi^{(1)} + \Phi^{(2)} + \cdots) \tag{7.15}$$

darstellen, wobei $f^{(0)}$ die lokale Gleichgewichtsverteilung

$$\ln f^{(0)} = - \frac{m(c - v)^2}{2kT} - \ln Z \tag{7.16}$$

mit der Zustandssumme Z ist. Sie enthält bekanntlich als Parameter nur die thermodynamischen Variablen, die Funktion $f^{(1)}$ enthält daneben auch $\partial T/\partial x_\alpha$, $\partial v_\beta/\partial x_\alpha$ ($\alpha, \beta = 1, 2, 3$) als Parameter, $f^{(2)}$ hängt von Variablen und Ableitungen bis einschließlich der zweiten Ordnung ab.

Meixner[1] geht mit der Entwicklung (7.15) in die kinetische Definition der Entropiedichte

$$\varrho\, s = - k \int f \ln f\, d(c) \tag{7.17}$$

ein und erhält

$$\varrho\, s = \varrho\, (s_0 + s_1) = - k \int f^{(0)} \ln f^{(0)}\, d(c) - \frac{k}{2} \int f^{(0)} (\Phi^{(1)})^2\, d(c), \tag{7.18}$$

wobei das erste Glied der rechten Seite mit der thermodynamisch definierten Entropiedichte identisch ist. Das zweite Glied hängt quadratisch von den Ableitungen grad T, Grad v ab. Nach Meixner sind die Massenelemente dann thermodynamische Systeme, wenn der Zusatzterm s_1 klein gegenüber der thermodynamischen Entropie s_0 bleibt. Fragt man nach Bedingungen hierfür, so erhält man Ungleichungen für die Beträge der Geschwindigkeits- und Temperaturgradienten. Diese Bedingungen für die Steilheitseigenschaften des Temperatur- und Geschwindigkeitsfeldes hat Meixner anschaulich formuliert: Die Massenelemente eines einatomigen gasförmigen Mediums sind dann thermodynamische Systeme, wenn die Temperatur- bzw. die Geschwindigkeitsänderung über eine mittlere freie Weglänge klein gegenüber der absoluten Temperatur bzw. der Schallgeschwindigkeit bleibt. Damit sind noch ziemlich steile Temperatur- und Geschwindigkeitsfelder zugelassen. So findet Meixner für Gase unter Normalbedingungen, daß Temperaturgradienten bis 10^5 Grad/cm noch mit den Methoden der Thermodynamik irreversibler Prozesse behandelbar sind. Selbst unter diesen extremen Bedingungen sind die Massenelemente thermodynamische Systeme.

Zur Aufstellung der Entropiebilanz mit Hilfe der Gibbsschen Fundamentalgleichung sind die Meixnerschen Bedingungen notwendig. Es ist aber damit noch nicht bewiesen, daß sie auch hinreichend sind, da es bei der Aufstellung der *Entropiebilanz* nicht auf die relative Größe der Anteile zur Entropiedichte sondern auf die relative Größe der Anteile zur *Entropieänderung* ankommt.

---

[1] J. Meixner: Ann. Physik (5) **39**, 333 (1941). — Z. phys. Chem., Abt. B **53**, 235 (1943).

Diese Schwierigkeit wird in der Behandlung von PRIGOGINE[1] teilweise umgangen. PRIGOGINE betrachtet nämlich den Ausdruck, den die kinetische Gastheorie für die Entropieproduktion ableitet und vergleicht ihn mit der thermodynamisch gewonnenen Beziehung. Dazu muß er die Struktur von $f^{(1)}$ in bezug auf die Parameterabhängigkeit kennen und mit $f^{(1)}$ in die Ausdrücke für den Wärmestrom und den Reibungsdrucktensor einerseits und in den Ausdruck für die Entropieproduktion andererseits eingehen. Für das von ihm betrachtete Modell findet er, daß der thermodynamische und der kinetische Ausdruck für die Entropieproduktion dann übereinstimmen, wenn die wirklich vorliegende Verteilung f durch $f^{(0)} + f^{(1)}$ genügend genau approximiert wird. Damit findet man für die Steilheitseigenschaften der Felder $T(\mathbf{r}, t)$, $\mathbf{v}(\mathbf{r}, t)$ wieder die Meixnerschen Bedingungen. Aus der Prigogineschen Behandlung kann man entnehmen, daß die Meixnerschen Steilheitskriterien nicht nur notwendig, sondern auch hinreichend sind.

Ein drittes Argument wird von REIK benutzt. Er geht von der Bemerkung aus, daß man die Dichtefelder und die Prozeßfelder der Thermodynamik irreversibler Prozesse durch geeignete Mittelbildung über die Verteilungsfunktion erhält. So bestehen für Energiedichte bzw. Energiestromdichte eines Einkomponentensystems die Beziehungen

$$\varrho\, u = \frac{m}{2} \int (\mathbf{c} - \mathbf{v})^2 f\, d(\mathbf{c}) , \tag{7.19}$$

$$\mathbf{J}_u = \frac{m}{2} \int (\mathbf{c} - \mathbf{v})^2 (\mathbf{c} - \mathbf{v})\, f\, d(\mathbf{c}) . \tag{7.20}$$

Die Gleichungen für die zeitliche Veränderung der Dichtefelder erhält man durch entsprechende Mittelwertbildung über die Boltzmannsche Fundamentalgleichung. Sie entsprechen den Bilanzgleichungen der Thermodynamik irreversibler Prozesse. Die Definitionsgleichung für die Stromdichten, z. B. für $\mathbf{J}_u$ sind die Analoga zu den phänomenologischen Gleichungen der Thermodynamik irreversibler Prozesse. Der volle Satz dieser Gleichungen wird als Mittelwertsgleichungen der kinetischen Theorie bezeichnet.

Nach REIK[2] ist die Behandlung eines gasförmigen Mediums mit Hilfe der Thermodynamik irreversibler Prozesse vom Standpunkt der kinetischen Theorie solange zulässig, als der volle Satz von Grundgleichungen — d. h. der Bilanzgleichung und der phänomenologischen Gleichungen — strukturgleich ist mit den diesen Grundgleichungen entsprechenden Mittelwertsgleichungen der kinetischen Theorie. Den Bilanzgleichungen für Masse, Impuls und Energie entsprechen in der kinetischen Theorie die Mittelwertsgleichung für die Stoßinvarianten. Diese Mittelwertsgleichungen stimmen für alle Verteilungstypen (auch für nicht-Enskogsche) mit der Bilanzgleichung der Thermodynamik irreversibler Prozesse überein.

Der Entropiebilanz entspricht in der kinetischen Theorie die Mittelwertgleichung für den Logarithmus der Verteilungsfunktion. Sie nimmt deshalb für verschiedene Näherungen der Verteilungsfunktion verschiedene Formen an. Dasselbe trifft für die den phänomenologischen Gleichungen entsprechenden Mittelwertsgleichungen zu. Die Charakterisierung der Thermodynamik irreversibler Prozesse vom Standpunkt der kinetischen Theorie wird damit — was den Satz von Bilanzgleichungen anlangt — auf die Charakterisierung der Entropiebilanz reduziert. Daneben sind natürlich auch die phänomenologischen Gleichungen und ihre kinetischen Analoga zu untersuchen. Soll für eine gewisse

---

[1] I. PRIGOGINE: Physica, Haag **15**, 272 (1949).
[2] H. G. REIK: Z. Physik **148**, 156, 333 (1957).

Näherung der Verteilungsfunktion die Gibbssche Fundamentalgleichung bestehen, so muß $\varrho\,s$ eine Linearkombination der Stoßinvarianten sein. Diese Forderung ist nur für die Gleichgewichtsverteilung (7.16) erfüllt. Die Charakterisierung der thermodynamischen Näherung im Rahmen der Gastheorie besteht also in der Ermittlung derjenigen Verteilungstypen, für die die Entropiebilanz eine Bilanz für den Mittelwert von $\ln f^{(0)}$ wird. Selbstverständlich ist dies für $f = f^{(0)}$ erfüllt. Dieser Fall ist trivial, er entspricht dem lokalen thermodynamischen Gleichgewicht und den Bewegungen in ideal fluiden Gasen. Gibt es nicht-triviale Verteilungen, für die die Entropiebilanzgleichung die erwähnte Eigenschaft besitzt, so verschwindet in ihr die Entropieproduktion durch solche Stöße, bei denen die beteiligten Moleküle das betrachtete Massenelement nicht verlassen. Da nach (7.16) die Faktoren der Stoßinvarianten in $\ln f^{(0)}$ Funktionen des Ortes sind, findet Entropieproduktion statt, wenn das Massenelement Moleküle mit seiner Umgebung austauscht. Der damit verbundene Austausch an innerer Energie und Impuls kann aber durch die Größen $\boldsymbol{J}_u$, P beschrieben werden. Man wird infolgedessen erwarten, daß für diese nicht trivialen Verteilungstypen die Entropiebilanzgleichung der kinetischen Gastheorie in die der Thermodynamik irreversibler Prozesse übergeht, insbesondere, daß sich die Entropieproduktion als Bilinearform aus Strömen und zugeordneten Kräften schreiben läßt. Diese Erwartung wird durch die Rechnung bestätigt.

Berechnet man durch Mittelbildung über die einzelnen Gleichungen des Enskogschen Approximationsverfahrens die zu den Verteilungstypen $f^{(0)} + f^{(1)}$, $f^{(0)} + f^{(1)} + f^{(2)}$ gehörenden Entropiebilanzgleichungen, so findet man wieder Übereinstimmung zwischen Thermodynamik und Kinetik für den Fall, daß die Verteilung durch $f^{(0)} + f^{(1)}$ hinreichend genau approximiert wird. Man beachte, daß bei diesem Verfahren die Entropiebilanzgleichungen der Thermodynamik und der Kinetik nicht verglichen werden. Vielmehr wird die thermodynamische Entropiebilanzgleichung aus der kinetischen durch geeignete Spezialisierung gewonnen. Weiter ist zu bemerken, daß man nur von den Enskogschen Gleichungen Gebrauch zu machen hat. Ihre Lösungen müssen (im Gegensatz zu den Verfahren von Meixner und Prigogine) nicht explizit bekannt sein. Das Reiksche Verfahren ist deshalb auch auf kompliziertere Gase mit inneren Freiheitsgasen anwendbar. Aus der kinetischen Theorie ergibt sich des weiteren, daß die Definitionsgleichungen der Stromdichten für die betrachteten Näherungen der Verteilungsfunktion mit den phänomenologischen Gleichungen der Thermodynamik irreversibler Prozesse strukturgleich werden.

Die Untersuchung höherer Näherungen der Lösung der Boltzmann-Gleichung bildet eine naturgemäße Erweiterung der Thermodynamik irreversibler Prozesse. Die Aufstellung der phänomenologischen Gleichungen für solche Näherungen wurde zuerst von Lennard-Jones[1] und Burnett[2] durchgeführt. Ein vereinfachtes Verfahren hierzu haben Chapman und Cowling[3] angegeben. Später haben Kohler[4] und Reik[5] die phänomenologischen Gleichungen in höherer Näherung mit der Bilanz für s_1 in Beziehung gesetzt.

8. Lineare Transformationen von Flüssen und Kräften. α) *Erhaltung der O.C.R.B. bei unabhängigen Flüssen und Kräften.* In vielen Fällen ist es zweckmäßig, die Flüsse Y_i und die Kräfte X_i durch gewisse Linearkombinationen

---

[1] I. E. Lennard-Jones: Phil. Trans. Roy. Soc. Lond. A **223**, 1 (1923).

[2] D. Burnett: Proc. Lond. Math. Soc. **40**, 382 (1935).

[3] S. Chapman u. T. G. Cowling: The Mathematical Theory of Non-Uniform Gases. Cambridge 1939.

[4] M. Kohler Z. Physik **127**, 201, 215 (1949).

[5] H. G. Reik: Z. Naturforsch. **12**a, 663 (1957).

derselben zu ersetzen. Beispiele finden sich in den Ziff. 9 und 18. Es entsteht dann die Frage ob die O.C.R.B. auch in den neuen Flüssen und Kräften gelten, wenn sie in den ursprünglichen Flüssen und Kräften richtig waren. Diese Frage ist zu bejahen, wenn nur Kräfte bzw. Flüsse gleichen Typs miteinander linear kombiniert werden. Andere Transformationen, welche Flüsse verschiedenen Typs miteinander mischen, sind ohnehin nicht von Bedeutung.

Zum Beweis dieser Transformationseigenschaft der O.C.R.B. gehen wir von dem Ausdruck für die Entropieproduktion

$$\sigma = \sum_{i=1}^{N} Y_i X_i + \sum_{i=1}^{n} y_i x_i \tag{8.1}$$

aus, in dem nun mit X_i die Kräfte vom α-Typ, mit x_i die Kräfte vom β-Typ bezeichnet sind. Die zugeordneten Flüsse sind entsprechend Y_i und y_i genannt. Sowohl die Y_i, y_i als auch die X_i, x_i seien als linear unabhängig vorausgesetzt. Nun gehen wir zu neuen Kräften X'_i, über, die mit den ursprünglichen über die linearen Beziehungen

$$X_i = \sum_{k=1}^{N} C_{ik} X'_k \quad (i=1,2,\ldots,N), \qquad x_i = \sum_{k=1}^{n} c_{ik} x'_k \quad (i=1,2,\ldots,n) \tag{8.2}$$

mit det $C_{ik} \neq 0$, det $c_{ik} \neq 0$ zusammenhängen. Sie verknüpfen nur Kräfte gleichen Typs. Die Summen gehen stets über alle Indices, die in einem Produkt zweimal vorkommen. Der Bereich der Summation 1 bis N, 1 bis n wird im folgenden nicht mehr besonders angegeben. Einsetzen von (8.2) in (8.1) führt dann auf

$$\sigma = \sum (\sum Y_i C_{ik}) X'_k + \sum (\sum y_i c_{ik}) x'_k. \tag{8.3}$$

Die den neuen Kräften zugeordneten Flüsse Y'_k, y'_k sind damit festgelegt; denn die Kräfte sind ja als Koeffizienten der Flüsse in der Entropieproduktionsdichte definiert. Somit muß gelten

$$Y'_k = \sum C_{ik} Y_i; \quad (k=1,2,\ldots,N), \qquad y'_k = \sum c_{ik} y_i \quad (k=1,2,\ldots,n). \tag{8.4}$$

Zwischen den ursprünglichen Flüssen und Kräften gelten die linearen phänomenologischen Ansätze

$$\left.\begin{aligned}
Y_i &= \sum L_{ik}^{(1)} X_k + \sum L_{ik}^{(2)} x_k \quad (i=1,2,\ldots,N), \\
y_i &= \sum L_{ik}^{(3)} X_k + \sum L_{ik}^{(4)} x_k \quad (i=1,2,\ldots,n).
\end{aligned}\right\} \tag{8.5}$$

Setzt man diese in (8.4) ein und eliminiert die X_i und x_i mittels (8.2), so gehen die phänomenologischen Ansätze (8.5) für den neuen Satz von Flüssen und Kräften über in

$$\left.\begin{aligned}
Y'_k &= \sum_{m=1}^{N} l_{km}^{(1)} X'_m + \sum_{m=1}^{n} l_{km}^{(2)} x'_m, \\
y'_k &= \sum_{m=1}^{N} l_{km}^{(3)} X'_m + \sum_{m=1}^{n} l_{km}^{(4)} x'_m.
\end{aligned}\right\} \tag{8.6}$$

Die neuen phänomenologischen Koeffizienten sind durch die ursprünglichen und durch die Elemente der Transformations-Matrix bestimmt. Es gilt

$$\left.\begin{aligned}
l_{km}^{(1)} &= \sum_{i=1}^{N} \sum_{l=1}^{N} C_{ik} L_{il}^{(1)} C_{lm}, \qquad & l_{km}^{(2)} &= \sum_{i=1}^{N} \sum_{l=1}^{n} C_{ik} L_{il}^{(2)} c_{lm}, \\
l_{km}^{(3)} &= \sum_{i=1}^{n} \sum_{l=1}^{N} c_{ik} L_{il}^{(3)} C_{lm}, \qquad & l_{km}^{(4)} &= \sum_{i=1}^{n} \sum_{l=1}^{n} c_{ik} L_{il}^{(4)} c_{lm}.
\end{aligned}\right\} \tag{8.7}$$

Hieran erkennt man, daß aus dem Bestehen der O.C.R.B.

$$L_{ik}^{(1)} = L_{ki}^{(1)}, \qquad L_{ik}^{(2)} = - L_{ki}^{(3)}, \quad L_{ik}^{(4)} = L_{ki}^{(4)} \tag{8.8}$$

für die ursprünglichen Flüsse und Kräfte wieder die O.C.R.B.

$$l_{ik}^{(1)} = l_{ki}^{(1)}, \qquad l_{ik}^{(2)} = - l_{ki}^{(3)}, \quad l_{ik}^{(4)} = l_{ki}^{(4)} \tag{8.9}$$

für die neuen Flüsse und Kräfte folgen.

Diese Aussage bleibt auch bei Vorhandensein eines Magnetfeldes $\boldsymbol{B}$ erhalten, wenn nur die Transformationsmatrizen C_{ik}, c_{ik} bei Ersetzung von $\boldsymbol{B}$ durch $-\boldsymbol{B}$ ungeändert bleiben. Die Elemente der Transformationsmatrix müssen also gerade Funktionen von $\boldsymbol{B}$ sein.

β) Linear abhängige Flüsse und Kräfte. In den Anwendungen benutzt man häufig aus Zweckmäßigkeitsgründen linear abhängige Flüsse oder Kräfte. Ein Beispiel sind die Diffusionsstromdichten (3.8) mit der Bedingung (4.12). Wir zeigen nun, daß man auch bei linear abhängigen Flüssen oder Kräften die O.C.R.B. anwenden darf.

Wir gehen wieder von der Entropieproduktionsdichte (8.1) aus und nehmen zunächst an, daß zwischen den Flüssen Y_i genau eine lineare Beziehung besteht. Sie kann ohne Beschränkung der Allgemeinheit in der Form

$$Y_1 = \sum_{i=2}^{N} a_i Y_i \tag{8.10}$$

geschrieben werden. Diesen Fall führen wir auf den vorhergehenden zurück, indem wir Y_1 aus (8.1) mit Hilfe von (8.10) eliminieren. Dann ergibt sich

$$\sigma = \sum_{i=2}^{N} Y_i (X_i + a_i X_1) + \sum_{i=1}^{n} y_i x_i. \tag{8.11}$$

Nach (6.1) sind also den unabhängigen Flüssen Y_i ($i = 2, \dots, n$) die Kräfte $X_i + a_i X_1$ zugeordnet. Von der Entropieproduktionsdichte (8.11) ausgehend, erhält man also die linearen phänomenologischen Gleichungen

$$Y_i = \sum_{k=2}^{N} L_{ik}^{(1)} (X_k + a_k X_1) + \sum_{k=1}^{n} L_{ik}^{(2)} x_k \qquad (i = 2, \dots, N), \tag{8.12}$$

$$y_i = \sum_{k=2}^{N} L_{ik}^{(3)} (X_k + a_k X_1) + \sum_{k=1}^{n} L_{ik}^{(4)} x_k \qquad (i = 1, 2, \dots, n). \tag{8.13}$$

Für die phänomenologischen Koeffizienten $L_{ik}^{(j)}$ ($j = 1, 2, 3, 4$) bestehen die O.C.R.B.

Durch Einsetzen von (8.12) in (8.10) und durch Umordnen von (8.12) und (8.13) folgen die phänomenologischen Gleichungen für die ursprünglichen abhängigen Flüsse

$$\left. \begin{aligned}
Y_1 &= \sum_{i,k=2}^{N} a_i a_k L_{ik}^{(1)} X_1 + \sum_{i,k=2}^{N} a_i L_{ik}^{(1)} X_k + \sum_{i=2}^{N} \sum_{k=1}^{n} a_i L_{ik}^{(2)} x_k, \\
Y_i &= \sum_{k=2}^{N} a_k L_{ik}^{(3)} X_1 + \sum_{k=2}^{N} L_{ik}^{(1)} X_k + \sum_{k=1}^{n} L_{ik}^{(2)} x_k, \qquad (i = 2, 3, \dots, N), \\
y_i &= \sum_{k=2}^{N} a_k L_{ik}^{(3)} X_1 + \sum_{k=2}^{N} L_{ik}^{(3)} X_k + \sum_{k=1}^{n} L_{ik}^{(4)} x_k, \qquad (i = 1, 2, \dots, n).
\end{aligned} \right\} \tag{8.14}$$

Aus den O.C.R.B. für die Koeffizienten in (8.12) und (8.13) ergibt sich unmittelbar, daß auch für die phänomenologischen Gleichungen (8.14) die O.C.R.B.

gelten. Ebenso sieht man, daß die Koeffizientenmatrix in (8.14) eine nicht-negativ definite quadratische Form hat. Sie ist allerdings nicht positiv definit; sie verschwindet nämlich für $X_k + a_k X_1 = 0$ $(k = 2, \ldots, N)$ und $x_k = 0$ $(k = 1, \ldots, n)$.

Dieser Beweis kann ohne weiteres auf den Fall ausgedehnt werden, daß sowohl zwischen den Y_i $(i = 1, 2, \ldots, N)$ als auch zwischen den x_i $(i = 1, 2, \ldots, n)$ mehrere unabhängige lineare Beziehungen bestehen.

Wir betrachten nun den Fall linear abhängiger Kräfte. Er soll am einfachsten Beispiel auseinandergesetzt werden. Sei

$$\sigma = Y_1 X_1 + Y_2 X_2 \tag{8.15}$$

und $X_1 = X_2$. Die phänomenologischen Gleichungen sind

$$Y_1 = L_1 X_1, \qquad Y_2 = L_2 X_2 \tag{8.16}$$

mit der Bedingung $L_1 + L_2 > 0$. Für die Gln. (8.16) gelten zwar in der angeschriebenen Form die O.C.R.B.; es ist aber nicht notwendig $L_1 > 0$, $L_2 > 0$. Diese Gleichungen lassen sich aber wegen der linearen Abhängigkeit der Kräfte, $X_1 = X_2$, stets so schreiben, daß die O.C.R.B. erfüllt bleiben, nämlich als

$$Y_1 = (L_1 + l) X_1 - l X_2, \qquad Y_2 = - l X_1 + (L_2 + l) X_2 \tag{8.17}$$

mit beliebigem l und man kann l sogar so wählen, daß die zum Koeffizientenschema in (8.17) gehörende quadratische Form positiv definit ist.

Allgemein gilt folgendes *Theorem:* Die O.C.R.B. können sowohl bei abhängigen Flüssen als auch bei abhängigen Kräften in derselben Weise wie bei unabhängigen Flüssen und Kräften angesetzt werden; die Matrix der Koeffizienten kann ohne Beschränkung der Allgemeinheit so gewählt werden, daß ihre quadratische Form nicht-negativ definit ist.

Auf den allgemeinen Beweis soll hier nicht eingegangen werden. Er ist für den Fall, daß alle Kräfte von gleichem Typ sind, von HOOYMAN und DE GROOT[1] gegeben worden.

9. Beispiele zur Transformation von Flüssen und Kräften. *α) Transformation des Flusses der inneren Energie.* Die Energiestromdichte $\boldsymbol{J}_u$ spielt in den Anwendungen eine verhältnismäßig untergeordnete Rolle. Man geht daher häufig zu anderen Flüssen $\boldsymbol{J}'$ über, welche Linearkombinationen von $\boldsymbol{J}_u$ und den Diffusionsstromdichten $\boldsymbol{J}_i$ sind:

$$\boldsymbol{J}' = C_{uu} \boldsymbol{J}_u + \sum_{i=1}^{n} C_{iu} \boldsymbol{J}_i. \tag{9.1}$$

Die Diffusionsstromdichten $\boldsymbol{J}_i' = \boldsymbol{J}_i$ behält man bei. Dazu gehen wir von den phänomenologischen Gleichungen (7.1) und (7.2)

$$\left.\begin{array}{l} \boldsymbol{J}_i = \sum\limits_{k=1}^{n} L_{ik} \boldsymbol{X}_k + L_{iu} \boldsymbol{X}_u, \\[2mm] \boldsymbol{J}_u = \sum\limits_{k=1}^{n} L_{uk} \boldsymbol{X}_k + L_{uu} \boldsymbol{X}_u \end{array}\right\} \tag{9.2}$$

mit den Kräften

$$\boldsymbol{X}_k = \frac{\boldsymbol{F}_k}{T} - \operatorname{grad} \frac{\mu_k}{T}, \qquad \boldsymbol{X}_u = \operatorname{grad} \frac{1}{T} \tag{9.3}$$

aus. Wegen der Invarianz der Bilinearform (8.1) der Entropieproduktion sind die den neuen Flüssen $\boldsymbol{J}'$ und $\boldsymbol{J}_k'$ zugeordneten Kräfte

$$\boldsymbol{X}_k' = \boldsymbol{X}_k + \left(c - \frac{C_{ku}}{C_{uu}}\right) \boldsymbol{X}_u, \qquad \boldsymbol{X}' = \frac{1}{C_{uu}} \boldsymbol{X}_u, \tag{9.4}$$

---

[1] G. J. HOOYMAN u. S. R. DE GROOT: Physica, Haag **21**, 73 (1955).

wobei c wegen der linearen Abhängigkeit der Diffusionsstromdichten willkürlich ist. Da c aus den Ergebnissen herausfällt, so wollen wir ohne Beschränkung der Allgemeinheit $c=0$ setzen. Die phänomenologischen Gleichungen lauten dann

$$\left.\begin{aligned} \boldsymbol{J}_i' = \boldsymbol{J}_i &= \sum_{k=1}^{n} l_{ik}\boldsymbol{X}_k' + l_i \boldsymbol{X}', \\ \boldsymbol{J}' &= \sum_{k=1}^{n} l_k \boldsymbol{X}_k' + l\,\boldsymbol{X}' \end{aligned}\right\} \tag{9.5}$$

mit den O.C.R.B. $l_{ik}=l_{ki}$. Die Koeffizientenmatrix in (9.5) hat wieder eine nicht-negativ definite quadratische Form. Man findet für die Koeffizienten l_{ik}, l_i, l die Ausdrücke

$$\left.\begin{aligned} l_{ik} &= L_{ik}, \qquad l_i = \sum_{k=1}^{n} L_{ik} C_{ku} + L_{iu} C_{uu}, \\ l &= \sum_{i=1}^{n}\sum_{k=1}^{n} C_{ku} C_{iu} L_{ik} + 2 C_{uu} \sum_{i=1}^{n} C_{iu} L_{iu} + C_{uu}^2 L_{uu}. \end{aligned}\right\} \tag{9.6}$$

Wegen der linearen Abhängigkeit (4.12) der $\boldsymbol{J}_i$ gilt wieder

$$\sum_{i=1}^{n} l_{ik}=0, \qquad \sum_{i=1}^{n} l_i = 0. \tag{9.7}$$

β) *Wichtige Spezialfälle.* Wählt man in (9.1) $C_{iu}=-\dfrac{\mu_i}{T}$; $C_{uu}=\dfrac{1}{T}$, so ist $\boldsymbol{J}'$ die Entropiestromdichte

$$\boldsymbol{J}' \equiv \boldsymbol{J}_s = \frac{1}{T}\left(\boldsymbol{J}_u - \sum_{i=1}^{n} \mu_i \boldsymbol{J}_i\right). \tag{9.8}$$

Die den Stromdichten $\boldsymbol{J}_s$, $\boldsymbol{J}_i$ zugeordneten Kräfte sind

$$\boldsymbol{X}' = T\,\mathrm{grad}\,\frac{1}{T}, \qquad \boldsymbol{X}_i' = \frac{1}{T}\,(\boldsymbol{F}_i - \mathrm{grad}\,\mu_i). \tag{9.9}$$

Von besonderer Bedeutung ist der reduzierte Wärmestrom

$$\boldsymbol{J}' = \boldsymbol{J}_{\mathrm{red}} = \boldsymbol{J}_u - \sum_{i=1}^{n} h_i \boldsymbol{J}_i. \tag{9.10}$$

Man erhält ihn aus (9.1) mit der Wahl $C_{uu}=1$, $C_{iu}=-h_i$; die h_i sind partiellen spezifischen Enthalpien

$$h_i = \mu_i - T\left(\frac{\partial \mu_i}{\partial T}\right)_{p,\gamma_j} = -T^2\left(\frac{\partial}{\partial T}\,\frac{\mu_i}{T}\right)_{p,\gamma_j}. \tag{9.11}$$

Den Stromdichten $\boldsymbol{J}_{\mathrm{red}}$, $\boldsymbol{J}_i$ sind die Kräfte

$$\boldsymbol{X}' = \mathrm{grad}\,\frac{1}{T}, \qquad \boldsymbol{X}_i' = \frac{1}{T}\,(\boldsymbol{F}_i - \mathrm{grad}_T\,\mu_i) \tag{9.12}$$

zugeordnet. Unter grad_T ist der Gradient bei konstant gehaltener Temperatur zu verstehen, wenn T, p und γ_i die unabhängigen Variablen sind.

γ) *Überführungsenergien.* Die Transformation (9.1) kann auch so gewählt werden, daß die Stromdichte $\boldsymbol{J}'$ und die Diffusionsstromdichten $\boldsymbol{J}_i$ entkoppelt sind. Man setzt dazu $C_{uu}=1$; $C_{iu}=-Q_i$, d.h.

$$\boldsymbol{J}' = \boldsymbol{J}_u - \sum_{i=1}^{n} Q_i \boldsymbol{J}_i \tag{9.13}$$

und wählt die Q_k so, daß die Gleichungen

$$-\sum_{k=1}^{n} L_{ik}\, Q_k + L_{iu} = 0 \qquad (9.14)$$

erfüllt sind. Wegen (7.4) lassen sich die Gln. (9.14) nicht eindeutig nach den Q_k auflösen; es bleibt in allen ein gemeinsamer Summand unbestimmt. Man nennt die Größen Q_k spezifische Überführungsenergien. Die phänomenologischen Gleichungen (9.5) lauten dann

$$\left.\begin{aligned}
\boldsymbol{J}_i' \equiv \boldsymbol{J}_i &= \sum_{k=1}^{n} L_{ik}\, \boldsymbol{X}_k', \\
\boldsymbol{J}' &= \left(L_{uu} - 2\sum_{i=1}^{n} L_{iu}\, Q_i + \sum_{i=1}^{n}\sum_{k=1}^{n} Q_i\, Q_k\, L_{ik}\right)\boldsymbol{X}' \\
&= \left(L_{uu} - \sum_{i=1}^{n}\sum_{k=1}^{n} Q_i\, Q_k\, L_{ik}\right)\boldsymbol{X}'
\end{aligned}\right\} \qquad (9.15)$$

mit

$$\left.\begin{aligned}
\boldsymbol{X}' &= \boldsymbol{X}_u, \\
\boldsymbol{X}_i' &= \boldsymbol{X}_i + Q_i\, \boldsymbol{X}_u.
\end{aligned}\right\} \qquad (9.16)$$

Man kann die Transformation von $\boldsymbol{J}_u$ auf $\boldsymbol{J}'$ auch in zwei Schritten durchführen, indem man erst auf den reduzierten Wärmestrom und dann auf $\boldsymbol{J}'$ transformiert. So ergibt sich

$$\boldsymbol{J}' = \boldsymbol{J}_u - \sum_{i=1}^{n} \boldsymbol{J}_i\, h_i - \sum_{i=1}^{n} q_i\, \boldsymbol{J}_i. \qquad (9.17)$$

Wird in den Formeln (9.1), (9.4) und (9.6) $C_{uu} = 1$, $C_{iu} = -h_i$ gewählt, sind also (9.5) die phänomenologischen Gleichungen für die $\boldsymbol{J}_i$ zusammen mit $\boldsymbol{J}_{\text{red}}$, dann sind die q_i definiert durch

$$\sum_{k=1}^{n} l_{ik}\, q_k + l_i = 0, \qquad (9.18)$$

und es wird

$$\boldsymbol{J}_i = \sum_{k=1}^{n} l_{ik}\, \boldsymbol{X}_k'. \qquad (9.19)$$

Die q_i nennt man gelegentlich spezifische Überführungswärmen. Vergleich von (9.13) und (9.17) gibt bei geeigneter Wahl eines ohnehin unbestimmt bleibenden gemeinsamen Summanden $Q_i = h_i + q_i$ $(i = 1, 2, \ldots n)$.

δ) *Invarianz der Grundgleichungen bei einer ε-Substitution.* Der Nullpunkt der Energiezählung ist bisher unbestimmt geblieben. Daher bleibt zu untersuchen, wie sich die Grundgleichungen der Thermodynamik irreversibler Prozesse ändern, wenn die Nullpunktsenergien der unabhängigen Bestandteile (etwa der Elemente) um willkürliche Beiträge geändert werden. Im Anschluß an ECKART[1] wird dieser Vorgang als *ε-Substitution* bezeichnet.

Die Kontinuitätsgleichungen (4.10) für die Partialdichten und die Bewegungsgleichung (4.13) bleiben bei einer ε-Substitution ungeändert, da sie keine energetischen Größen enthalten. Dasselbe gilt für die phänomenologischen Gleichungen (7.14) und (7.15). Die ε-Substitution macht sich daher nur in der Energie- und der Entropiebilanz, sowie in den phänomenologischen Gleichungen (7.1) und

---

[1] C. ECKART: Phys. Rev. **58**, 267, 269 (1940).

(7.2) bemerkbar. Sei ε_i die Verschiebung der Nullpunktsenergie des Bestandteils i pro Masseneinheit. Die ε_i der unabhängigen Bestandteile können willkürlich gewählt werden. Die anderen ε_i sind dann durch (7.10) bestimmt. Insbesondere gilt

$$\sum_{i=1}^{n} \varepsilon_i \, \Gamma_i = 0, \tag{9.20}$$

d.h., die gesamte Produktion von Nullpunktsenergie durch chemische Umwandlungen verschwindet.

Die spezifische Energie u^* in der neuen Zählung ist dann durch

$$u^* = u + \sum_{i=1}^{n} \gamma_i \, \varepsilon_i \tag{9.21}$$

gegeben.

In der Bilanz der inneren Energie (4.19) ist zu schreiben

$$\left. \begin{aligned} \varrho \, \frac{D u}{D t} + \operatorname{div} \boldsymbol{J}_u &= \varrho \, \frac{D u^*}{D t} - \varrho \sum_{i=1}^{n} \varepsilon_i \, \frac{D \gamma_i}{D t} + \operatorname{div} \boldsymbol{J}_u \\ &= \varrho \, \frac{D u^*}{D t} + \operatorname{div} \Big(\boldsymbol{J}_u + \sum_{i=1}^{n} \varepsilon_i \, \boldsymbol{J}_i \Big), \end{aligned} \right\} \tag{9.22}$$

wobei (4.10) und (9.20) benutzt sind. Wegen (9.22) gilt also die Energiegleichung (4.19) ungeändert, wenn man u^* statt u schreibt und die Stromdichte $\boldsymbol{J}_u$ durch $\boldsymbol{J}^* = \boldsymbol{J}_u + \sum_{i=1}^{n} \varepsilon_i \, \boldsymbol{J}_i$ ersetzt. Es liegt also eine Transformation von der in Abschnitt α behandelten Art vor.

Da sich auch die spezifischen Enthalpien h_i bei der ε-Substitution um ε_i zu h_i^* ändern, so bleibt der reduzierte Wärmestrom

$$\boldsymbol{J}_{\text{red}} = \boldsymbol{J}_u - \sum_{i=1}^{n} h_i \, \boldsymbol{J}_i = \boldsymbol{J}_u^* - \sum h_i^* \, \boldsymbol{J}_i \tag{9.23}$$

gegenüber der ε-Substitution ungeändert. Dasselbe trifft für den Entropiestrom zu. Auch die μ_i ändern sich nur um konstante Werte ε_i. Damit bleibt $\operatorname{grad}_T \mu_i$ ungeändert; die phänomenologischen Koeffizienten in (9.6) und (9.10) werden daher von der ε-Substitution nicht betroffen.

Die eben genannten Eigenschaften des reduzierten Wärmestroms haben mehrere Autoren[1] veranlaßt, den Satz der Stromdichten $\boldsymbol{J}_{\text{red}}, \boldsymbol{J}_i$ vor den äquivalenten Sätzen $\boldsymbol{J}_u, \boldsymbol{J}_i$; $\boldsymbol{J}_s, \boldsymbol{J}_i$ etc. zu bevorzugen. Diese Wahl bringt noch weitere formale Vorteile mit sich: Die mit dem reduziertem Wärmestrom definierten Wärmeleitfähigkeiten sind sowohl im Zustand der homogenen Durchmischung als auch im stationären Zustand der Thermodiffusion stets positiv; auch die Bedingung des mechanischen Gleichgewichts läßt sich in dem $\boldsymbol{J}_{\text{red}}, \boldsymbol{J}_i$ zugeordneten Satz von Kräften besonders einfach formulieren (s. Ziff. 13). Wir möchten aber betonen, daß die Wahl $\boldsymbol{J}_{\text{red}}, \boldsymbol{J}_i$ auch in vielen Anwendungen recht zweckmäßig ist. Durch die meisten Randwertprobleme der Thermodynamik irreversibler Prozesse wird die Verwendung der Felder $\varrho(\boldsymbol{r}, t)$, $\gamma_i(\boldsymbol{r}, t)$, $\boldsymbol{v}(\boldsymbol{r}, t)$, $T(\boldsymbol{r}, t)$ zur Beschreibung des kontinuierlichen Mediums nahegelegt. Schreibt man die Energiegleichung in eine Gleichung für das Temperaturfeld um (vgl. Ziff. 11β), so treten in dieser $\boldsymbol{J}_u$ und $\boldsymbol{J}_i$ in den Gliedern erster Ordnung gerade in der Linearkombination auf, die den reduzierten Wärmestrom definiert.

---

[1] P. Mazur u. I. Prigogine: J. Phys. Radium 12, 616 (1951). — H. A. Tolhoek u. S. R. De Groot: Physica, Haag 18, 780 (1952). — R. Haase: Z. Naturforsch. 8a, 729 (1953).

Auch die kinetische Behandlung der Mehrkomponentengase führt bei der Chapman-Enskogschen Lösung der Boltzmannschen Fundamentalgleichung zunächst auf den reduzierten Wärmestrom, wenn man die Parameter der gestörten Verteilungen durch die Ableitungen der Felder $\varrho(\boldsymbol{r}, t)$, $\gamma_i(\boldsymbol{r}, t)$, $\boldsymbol{v}(\boldsymbol{r}, t)$, $T(\boldsymbol{r}, t)$ bestimmt. Aus allem diesem ergibt sich jedoch, daß die ausgezeichnete Bedeutung des reduzierten Wärmestroms auf den Anwendungen beruht; im Rahmen der allgemeinen Theorie spielt er natürlich keine ausgezeichnete Rolle.

ε) Hauptachsentransformation der inneren Variablen. Für ein ruhendes fluides Medium $(\boldsymbol{v}=0)$ ohne Transporterscheinungen $(\boldsymbol{J}_u=0;\ \boldsymbol{J}_i=0)$ und ohne äußere Kräfte $(\boldsymbol{F}_i=0)$ vereinfachen sich Grundgleichungen und phänomenologische Gleichungen erheblich. Alle Größen sind dann ortsunabhängig. Die Gln. (4.3), (4.10), (4.14), (4.20) und (7.1) (7.2), (7.13) und (7.14) reduzieren sich dann auf das Gleichungssystem

$$\frac{\partial \gamma_i}{\partial t} = + \sum_{k=1}^{r} \frac{\lambda_{ik}}{T\varrho}\, A_k \qquad (i=1, 2, \ldots r), \qquad \lambda_{ik} = \lambda_{ki} \tag{9.24}$$

für die r unabhängigen inneren Umwandlungen und auf die Energiegleichung in der Form

$$u = u(s, \varrho_1, \varrho_2, \ldots \varrho_r) = \text{const.} \tag{9.25}$$

Die übrigen Partialdichten ϱ_i bestimmen sich mit Hilfe von (7.10). Die Gibbssche Fundamentalgleichung (3.4) lautet nach Reduktion auf die unabhängigen Konzentrationen wegen (7.10)

$$ds = \frac{1}{T}\, du + \frac{p}{T}\, d\frac{1}{\varrho} + \sum_{i=1}^{r} \frac{A_i}{T}\, d\gamma_i. \tag{9.26}$$

Unter den gegebenen Bedingungen vereinfacht sich dieser Ausdruck zu

$$ds = \sum_{i=1}^{r} \frac{A_i}{T}\, d\gamma_i.$$

Im chemischen Gleichgewicht hat s in bezug auf die Variablen $\gamma_1 \ldots \gamma_r$ ein Maximum; d.h., es ist $\partial s/\partial \gamma_i = 0$ $(i=1, 2, \ldots r)$. Die Bedingung des chemischen Gleichgewichts hat also $A_i = 0$ zur Folge; nach (9.24) ist dann $\partial\gamma_i/\partial t = 0$. Das Verschwinden der Affinitäten ist daher bei ungehemmten Reaktionen notwendig und hinreichend für das chemische Gleichgewicht. Die Gleichgewichtskonzentrationen nennen wir γ_{i0}.

In der Umgebung des Gleichgewichts kann man die Entropie bei konstanter Energie und konstantem Volumen entwickeln:

$$s = s_0 - \tfrac{1}{2} \sum_{i=1}^{r} \sum_{k=1}^{r} S_{ik}(\gamma_i - \gamma_{i0})(\gamma_k - \gamma_{k0}) \tag{9.27}$$

unter Vernachlässigung von Gliedern höherer Ordnung.

Die Matrix S_{ik} sei ohne Beschränkung der Allgemeinheit symmetrisch angenommen. Sie hat nach dem zweiten Hauptsatz eine positiv definite quadratische Form. Mit ihrer Hilfe lassen sich nach (9.25) und (9.26) die Affinitäten durch die Konzentrationen ausdrücken

$$A_i = - \sum_{k=1}^{r} T S_{ik}(\gamma_k - \gamma_{k0}) \qquad (i=1, 2, \ldots r). \tag{9.28}$$

Statt der γ_i kann man nun irgendwelche r unabhängige Linearkombinationen

$$\gamma_i' = \sum_{k=1}^{r} C_{ik}\gamma_k \qquad (i=1, 2, \ldots r), \qquad \det C_{ik} \neq 0 \tag{9.29}$$

zur Kennzeichnung des inneren Zustandes verwenden. Die den neuen inneren Variablen γ_i' zugeordneten Affinitäten A_i' sind wieder aus der Gibbsschen Fundamentalgleichung bestimmt:

$$\sum_{i=1}^{r} A_i\, d\gamma_i = \sum_{i=1}^{r} A_i'\, d\gamma_i', \qquad A_i' = T\left(\frac{\partial s}{\partial \gamma_i'}\right)_{u,\,\varrho}.$$

Dann folgt

$$A_i = \sum_{k=1}^{r} C_{ki} A_k' \qquad (i = 1, 2, \ldots n). \tag{9.30}$$

Die neuen Kräfte A_i' zu den neuen Flüssen $\partial\gamma_i/\partial t$ sind also jetzt aus der Gleichgewichtsthermodynamik bestimmt, d.h. aus der Invarianz der Bilinearform $\sum_{i=1}^{r} A_i\, d\gamma_i$. Man sieht jedoch, daß dies äquivalent der früheren Bestimmung aus der Invarianz der Bilinearform der Entropieproduktionsdichte $\frac{1}{T}\sum_{i=1}^{r} A_i\, \Gamma_i$ ist, da die Γ_i gerade proportional zu den $\partial\gamma_i/\partial t$ sind.

Eine analoge Äquivalenz kann man auch für allgemeinere irreversible Prozesse mit Transporterscheinungen herleiten[1].

Bezeichnet man mit λ^{ik} die zu λ_{ik} reziproke Matrix, so läßt sich die Entropieproduktion $\sum_{i=1}^{r} \frac{A_i}{T}\, \frac{\partial\gamma_i}{\partial t}$ durch die quadratische Form

$$\varrho \sum_{i=1}^{r} \sum_{k=1}^{r} \lambda^{ik}\, \frac{\partial\gamma_i}{\partial t}\, \frac{\partial\gamma_k}{\partial t}$$

darstellen. Nach Sätzen über die simultane Hauptachsentransformation zweier quadratischer Formen, von denen wenigstens eine positiv definit ist, kann man nun die Transformation (9.29) so wählen, daß C_{ik} in die mit $1/T$ multiplizierte Einheitsmatrix übergeht, während λ^{ik} in eine Diagonalmatrix transformiert wird. Diese hat nur positive Elemente, da mit λ_{ik} auch λ^{ik} eine positiv definite quadratische Form hat. (9.28) und (9.24) gehen dann über in

$$A_i' = -\,(\gamma_i' - \gamma'_{\,0}), \qquad \frac{\partial\gamma_i'}{\partial t} = \frac{1}{\tau_i}\, A_i'. \tag{9.31}$$

Die Integration dieser Differentialgleichungen zeigt, daß sich jede Variable γ'_i ihrem Gleichgewichtswert γ_{i0}' nach einer Exponentialfunktion mit der Zeitkonstanten τ_i nähert. Daraus folgt, daß die Konzentrationen $\gamma_i(t)$ sich auf ihre Gleichgewichtswerte nach einer Summe von Exponentialfunktionen mit Exponenten $-t/\tau_i$ einstellen, sofern sie dem Gleichgewicht bereits genügend nahe sind.

Die Hauptachsentransformation der inneren Variablen ist besonders nützlich für die Herleitung allgemeiner Theoreme in der Theorie der Relaxationserscheinungen (Ziff. 23). Ihr Nachteil ist, daß die Transformationskoeffizienten C_{ik} im allgemeinen von der Temperatur abhängen. Dieser Nachteil ist jedoch bei den Relaxationserscheinungen nicht wesentlich, da man dort in der Regel nur periodische Vorgänge mit kleinen Amplituden, also praktisch konstanter Temperatur betrachtet.

Die oben durchgeführte Hauptachsentransformation war nur möglich, weil die inneren Variablen als Konzentrationen und damit als Variable vom α-Typ vorausgesetzt waren. Sind unter den inneren Variablen zusätzlich solche vom β-Typ, so lassen sich nur die beiden Matrizen der Onsagerschen Koeffizienten zusammen mit der Matrix S_{ik} in (9.28) auf Hauptachsen bringen.

---

[1] G. J. Hooyman, S. R. de Groot u. P. Mazur: Physica, Haag **21**, 362 (1955).

10. Elastische Medien mit inneren Umwandlungen. Die Behandlung der irreversiblen Prozesse in elastischen Medien mit inneren Umwandlungen erfolgt analog der der fluiden Medien in den Ziff. 4, 5 und 7. Die Kontinuitätsgleichung für die Dichte (4.9), die Bilanzen für die Partialdichten (4.10) und die Bewegungsgleichung (4.13) können unmittelbar übernommen werden. Auch die Bilanz der inneren Energie bleibt ungeändert, wenn man die beiden, sich ohnehin weghebenden, Glieder $\pm p\,\mathrm{div}\,\boldsymbol{v}$ in (4.19) wegläßt.

Beschränkt man sich auf kleine Verschiebungen des elastischen Mediums (lineare Elastizitätstheorie), so lautet die Gibbssche Fundamentalgleichung

$$ds = \frac{1}{T}\,du - \frac{1}{T\varrho}\sum_{\alpha=1}^{3}\sum_{\beta=1}^{3}\sigma_{\alpha\beta}\,d\varepsilon_{\alpha\beta} - \sum_{i=1}^{n}\frac{\mu_i}{T}\,d\gamma_i . \tag{10.1}$$

Die $\varepsilon_{\alpha\beta}$ sind die Komponenten des symmetrischen Dehnungstensors. Er hat die Eigenschaft

$$\frac{D\varepsilon_{\alpha\beta}}{Dt} = \frac{1}{2}\left(\frac{\partial v_\alpha}{\partial x_\beta} + \frac{\partial v_\beta}{\partial x_\alpha}\right) \qquad (\alpha,\beta = 1,2,3) . \tag{10.2}$$

Die $\sigma_{\alpha\beta}$ sind die Komponenten des symmetrischen Spannungstensors, wie sie sich nach der Gleichgewichtsthermodynamik bei gegebenen u, $\varepsilon_{\alpha\beta}$, γ_i berechnen.

In der Entropiebilanz hat man an Stelle von $p\,\mathrm{div}\,\boldsymbol{v}$ zu schreiben

$$-\sum_{\alpha=1}^{3}\sum_{\beta=1}^{3}\sigma_{\alpha\beta}\frac{D\varepsilon_{\alpha\beta}}{Dt} = -\sum_{\alpha=1}^{3}\sum_{\beta=1}^{3}\sigma_{\alpha\beta}\frac{\partial v_\alpha}{\partial x_\beta} . \tag{10.3}$$

Der Ausdruck $\mathsf{P}-p\,\mathsf{I}$, oder in Komponenten $P_{\alpha\beta}-p\,\delta_{\alpha\beta}$, in der Entropieproduktionsdichte (5.6) ist somit durch $P_{\alpha\beta}+\sigma_{\alpha\beta}$ zu ersetzen.

Bei anisotropen Medien ist das Curiesche Gesetz (S. 427) nicht mehr anwendbar. Man hat daher zunächst die skalaren Flüsse und alle Komponenten der vektoriellen und tensoriellen Flüsse als lineare Funktionen aller skalaren Kräfte und aller Komponenten der vektoriellen und tensoriellen Kräfte anzusetzen. Eine mehr oder minder große Reduktion der phänomenologischen Koeffizienten ergibt sich aus der Forderung, daß die phänomenologischen Gleichungen gegenüber den Decktransformationen des Kristallsystems kovariant sein müssen.

Die Gleichungen der Thermodynamik irreversibler Prozesse für elastische Medien sind wegen der größeren Zahl von Variablen schwerfälliger als für fluide Medien. Man muß überdies bei Kristallen von der baryzentrischen Geschwindigkeit auf die Gittergeschwindigkeit und auf die Diffusion gegen das Gitter umrechnen, weil diese der Anschauung und den experimentellen Gegebenheiten besser gerecht werden. Andererseits treten Vereinfachungen durch die Beschränkung auf kleine Dehnungen und Dehnungsgeschwindigkeiten ein. Dann darf man D/Dt durch $\partial/\partial t$ ersetzen, man kann die Entropie um einen Bezugszustand nach der inneren Energie, den Dehnungen und den inneren Variablen entwickeln und nach den quadratischen Gliedern abbrechen[1] und gewinnt so lineare Gleichungen.

Die so formulierte Theorie enthält eine Reihe wichtiger Spezialfälle. Bei Hemmung aller irreversiblen Vorgänge, d.h. der inneren Umwandlungen, der Diffusion, der inneren Reibung und Wärmeleitung — dann sind alle phänomenologischen Koeffizienten gleich Null — erhält man die Grundgleichungen der Thermoelastizität. In diesem Fall findet keine Energiedissipation statt. Laufen jedoch irgendwelche der genannten irreversiblen Prozesse ab, so hat man die

---

[1] J. Meixner: Z. Naturforsch. **9a**, 654 (1954).

Theorie der Thermoelastizität mit innerer Dämpfung. Der Beitrag der Wärmeleitung zur inneren Dämpfung wurde für einige spezielle Probleme von Päsler[1] untersucht. Der Einfluß innerer Umwandlungen wurde am Beispiel des α-Eisens (raumzentriertes kubisches Gitter mit geringer Konzentration von Fremdatomen auf Zwischengitterplätzen verschiedener Art) von Polder[2] behandelt. Später hat Manz[3] die innere Dämpfung in kubischen Kristallen mit den Methoden der Thermodynamik irreversibler Prozesse systematisch diskutiert. Dabei wurden auch Wärmeleitung und Diffusion berücksichtigt. Eine Klassifikation der Relaxationsvariablen in allgemeinen Kristallen haben Falk und Meixner[4] durchgeführt. Eine Anwendung auf die mechanische Relaxation von Eis hat Bass[5] gegeben.

Häufig bezeichnet man die innere Dämpfung als innere Reibung des elastischen Mediums. Diese Redeweise ist indessen mißverständlich. Die innere Reibung im Sinne der Hydrodynamik wird ja durch eine besondere phänomenologische Gleichung für den Tensor $P_{\alpha\beta} + \sigma_{\alpha\beta}$ (Reibungsdrucktensor) mit Viskositätstensoren als Koeffizienten beschrieben.

Wir bemerken hierzu, daß man für genügend langsame Vorgänge durch Elimination der inneren Variablen Gleichungen erhält, die formal den Gleichungen ohne innere Umwandlungen und Diffusion äquivalent sind. Die inneren Variablen äußern sich dann in einem zusätzlichen Anteil zum Viskositätstensor[6]. Bei langsamen Vorgängen kann also ein Teil der inneren Reibung auf nicht explizit berücksichtigte innere Umwandlungen zurückgeführt werden. Es ist natürlich denkbar, daß sich die gesamte Viskosität als Folge von nicht explizit betrachteten inneren Variablen erklären läßt. Wir müssen jedoch diese Frage offen lassen. Siehe dazu auch Ziff. 22.

C. Anwendung der Thermodynamik irreversibler Prozesse in fluiden Medien.

11. Zusammenstellung der Gleichungen. In den folgenden Ziffern müssen wir fortwährend auf die im Vorangehenden abgeleiteten Gleichungen der Thermodynamik irreversibler Prozesse zurückgreifen. Wir stellen deshalb die allgemeinen Gleichungen nochmals zusammen, ergänzen sie durch die Gleichung für das Temperaturfeld und spezialisieren sie auf den wichtigen Fall von Zweikomponentensystemen. Anschließend geben wir ein einfaches Verfahren für thermodynamische Umrechnungen.

α) *Fluide Medien mit beliebig vielen Komponenten.* Die Bilanzgleichungen lauten:

$$\frac{D\varrho}{Dt} + \varrho \operatorname{div} \boldsymbol{v} = 0, \tag{11.1}$$

$$\frac{D\varrho_i}{Dt} + \varrho_i \operatorname{div} \boldsymbol{v} + \operatorname{div} \boldsymbol{J}_i = \Gamma_i \quad (i = 1, 2, \ldots n), \tag{11.2}$$

$$\varrho \frac{D\gamma_i}{Dt} + \operatorname{div} \boldsymbol{J}_i = \Gamma_i \quad (i = 1, 2, \ldots n), \tag{11.3}$$

$$\varrho \frac{D\boldsymbol{v}}{Dt} + \operatorname{Div} \mathsf{P} = \sum_{i=1}^{n} \varrho_i \boldsymbol{F}_i, \tag{11.4}$$

<hr>

[1] M. Päsler: Z. Physik **72**, 357 (1944); **124**, 105 (1947).
[2] D. Polder: Philips Res. Rep. **1**, 5 (1945).
[3] B. Manz: Zur Theorie der irreversiblen Prozesse in kubischen Kristallen mit einer Anwendung auf die Schallabsorption. Diss. Aachen 1956.
[4] G. Falk u. J. Meixner: Z. Naturforsch. **11**a, 782 (1956).
[5] R. Bass: Z. Physik **153**, 16 (1958).
[6] J. Meixner: Proc. Roy. Soc. Lond., Ser. A **226**, 57 (1954).

$$\varrho \frac{Du}{Dt} + \operatorname{div} \boldsymbol{J}_u = \sum_{i=1}^{n} \boldsymbol{F}_i \cdot \boldsymbol{J}_i - p \operatorname{div} \boldsymbol{v} - (\mathsf{P} - p\,\mathsf{I}) : \operatorname{Grad} \boldsymbol{v}, \tag{11.5}$$

$$\varrho \frac{Ds}{Dt} + \operatorname{div} \boldsymbol{J}_s = \sigma_s, \tag{11.6}$$

$$\sum_{i=1}^{n} \Gamma_i = 0, \quad \sum_{i=1}^{n} \boldsymbol{J}_i = 0, \quad \boldsymbol{J}_s = \frac{1}{T}\Big(\boldsymbol{J}_u - \sum_{i=1}^{n} \boldsymbol{J}_i \mu_i\Big), \quad \boldsymbol{J}_{\mathrm{red}} = \boldsymbol{J}_u - \sum_{i=1}^{n} h_i \boldsymbol{J}_i. \tag{11.7}$$

Die bilineare Form der Entropieproduktionsdichte ist

$$\left. \begin{aligned} \sigma_s = {}&\boldsymbol{J}_{\mathrm{red}} \cdot \operatorname{grad} \frac{1}{T} + \sum_{i=1}^{n} \boldsymbol{J}_i \cdot \frac{1}{T}\,(\boldsymbol{F}_i - \operatorname{grad}_T \mu_i) - \\ &- \frac{1}{T}\,(\mathsf{P} - p\,\mathsf{I}) : \operatorname{Grad} \boldsymbol{v} - \sum_{i=1}^{n} \Gamma_i \frac{\mu_i}{T}. \end{aligned} \right\} \tag{11.8}$$

Daraus folgen die linearen phänomenologischen Ansätze

$$\boldsymbol{J}_i = \sum_{k=1}^{n} \frac{l_{ik}}{T}\,(\boldsymbol{F}_k - \operatorname{grad}_T \mu_k) + l_i \operatorname{grad} \frac{1}{T}, \tag{11.9}$$

$$\boldsymbol{J}_{\mathrm{red}} = \sum_{k=1}^{n} \frac{l_k}{T}\,(\boldsymbol{F}_k - \operatorname{grad}_T \mu_k) + l \operatorname{grad} \frac{1}{T}, \tag{11.10}$$

$$p\,\delta_{\alpha\beta} - P_{\alpha\beta} = \eta\Big(\frac{\partial v_\alpha}{\partial x_\beta} + \frac{\partial v_\beta}{\partial x_\alpha}\Big) + \Big(\zeta - \frac{2}{3}\,\eta\Big)\delta_{\alpha\beta} \operatorname{div} \boldsymbol{v} + \delta_{\alpha\beta} \sum_{i=1}^{n} \Lambda_i\Big(-\frac{\mu_i}{T}\Big), \tag{11.11}$$

$$\Gamma_i = -\frac{\Lambda_i}{T} \operatorname{div} \boldsymbol{v} + \sum_{k=1}^{n} \Lambda_{ik}\Big(-\frac{\mu_k}{T}\Big) \quad (i = 1, 2, \dots n). \tag{11.12}$$

Sind r unabhängige chemische Reaktionen möglich, so lassen sich, bei geeigneter Numerierung der Komponenten, $\Gamma_{r+1}, \dots \Gamma_n$ durch $\Gamma_1 \dots \Gamma_r$ ausdrücken:

$$\Gamma_i = \sum_{k=1}^{r} \nu_{ik} \Gamma_k \quad (i = r + 1, \dots n). \tag{11.13}$$

Mit Hilfe der Affinitäten

$$-A_k = \mu_k + \sum_{i=r+1}^{n} \nu_{ik}\mu_i \quad (k = 1, 2, \dots r) \tag{11.14}$$

lassen sich (11.8), (11.11) und (11.12) auch in folgender Form schreiben

$$\left. \begin{aligned} \sigma_s = {}&\boldsymbol{J}_{\mathrm{red}} \cdot \operatorname{grad} \frac{1}{T} + \sum_{i=1}^{n} \boldsymbol{J}_i \cdot \frac{1}{T}\,(\boldsymbol{F}_i - \operatorname{grad}_T \mu_i) - \\ &- \frac{1}{T}\,(\mathsf{P} - p\,\mathsf{I}) : \operatorname{Grad} \boldsymbol{v} + \sum_{i=1}^{r} \Gamma_i \frac{A_i}{T}, \end{aligned} \right\} \tag{11.15}$$

$$p\,\delta_{\alpha\beta} - P_{\alpha\beta} = \eta\Big(\frac{\partial v_\alpha}{\partial x_\beta} + \frac{\partial v_\beta}{\partial x_\alpha}\Big) + \Big(\zeta - \frac{2}{3}\,\eta\Big)\delta_{\alpha\beta} \operatorname{div} \boldsymbol{v} + \delta_{\alpha\beta} \sum_{i=1}^{r} \lambda_i \frac{A_i}{T}, \tag{11.16}$$

$$\Gamma_i = -\frac{\lambda_i}{T} \operatorname{div} \boldsymbol{v} + \sum_{k=1}^{r} \lambda_{ik} \frac{A_k}{T} \quad (i = 1, 2, \dots r). \tag{11.17}$$

Als quadratische Form der Entropieproduktionsdichte ergibt sich

$$\sigma_s = \sum_{i=1}^{n} \sum_{k=1}^{n} \frac{l_{ik}}{T^2} (\boldsymbol{F}_i - \mathrm{grad}_T\,\mu_i)(\boldsymbol{F}_k - \mathrm{grad}_T\,\mu_k) + 2\sum_{i=1}^{n} \frac{l_i}{T}(\boldsymbol{F}_i - \mathrm{grad}_T\,\mu_i)\cdot\mathrm{grad}\frac{1}{T} +$$
$$+ l\left(\mathrm{grad}\frac{1}{T}\right)^2 + \frac{\eta}{2T}\sum_{\alpha=1}^{3}\sum_{\beta=1}^{3}\left(\frac{\partial v_\alpha}{\partial x_\beta} + \frac{\partial v_\beta}{\partial x_\alpha}\right)^2 + \frac{1}{T}\left(\zeta - \frac{2}{3}\eta\right)(\mathrm{div}\,\boldsymbol{v})^2$$
$$+ \sum_{i=1}^{r}\sum_{k=1}^{r}\frac{\lambda_{ik}}{T}A_i A_k. \tag{11.18}$$

Die phänomenologischen Koeffizienten haben bei Fehlen eines Magnetfeldes die Eigenschaften

$$l_{ik} = l_{ki}, \quad \lambda_{ik} = \lambda_{ki}, \quad \sum_{i=1}^{n} l_{ik} = 0, \quad \sum_{i=1}^{n} l_i = 0; \quad l > 0, \quad \eta > 0, \quad \zeta > 0. \tag{11.19}$$

Die Matrix λ_{ik} hat, wenn keine Reaktion gehemmt ist, eine positiv definite quadratische Form. Die Matrix der Koeffizienten in (11.9) und (11.10) von $n+1$ Reihen und Spalten hat eine nicht-negativ definite quadratische Form und ist vom Rang n; die Matrix l_{ik} ist vom Rang $n-1$ (sonst wäre wenigstens eine Diffusion oder die Wärmeleitung gehemmt).

β) Die Temperaturgleichung. Im Hinblick auf spätere Anwendungen geben wir noch die Gleichung für das Temperaturfeld. Ersetzt man die spezifische innere Energie in (11.11) durch die spezifische Enthalpie

$$h = u + \frac{p}{\varrho}, \tag{11.20}$$

so folgt mit Hilfe von (11.1)

$$\varrho\frac{Dh}{Dt} - \frac{Dp}{Dt} + \mathrm{div}\,\boldsymbol{J}_u = \sum_{i=1}^{n}\boldsymbol{F}_i\cdot\boldsymbol{J}_i - (\mathsf{P} - p\mathsf{I}):\mathrm{Grad}\,\boldsymbol{v}. \tag{11.21}$$

Faßt man h als Funktion von T, p, γ_i auf und bezeichnet mit c_p die spezifische Wärme bei konstantem Druck und konstanten Konzentrationen und mit h_i die partiellen spezifischen Enthalpien der Komponenten [s. Gl. (9.11)], so läßt sich (11.21) umschreiben zu

$$\left.\begin{aligned}
&\varrho\,c_p\frac{DT}{Dt} + \left[\varrho\left(\frac{\partial h}{\partial p}\right)_{T,\gamma_i} - 1\right]\frac{Dp}{Dt} + \mathrm{div}\,\boldsymbol{J}_{\mathrm{red}} + \sum_{i=1}^{n} h_i\,\Gamma_i \\
&\qquad = \sum_{i=1}^{n}\boldsymbol{J}_i(\boldsymbol{F}_i - \mathrm{grad}\,h_i) - (\mathsf{P} - p\,\mathsf{I}):\mathrm{Grad}\,\boldsymbol{v}.
\end{aligned}\right\} \tag{11.22}$$

Durch Übergang zu lokaler Beschreibung ergibt sich

$$\left.\begin{aligned}
&\varrho\,c_p\frac{\partial T}{\partial t} + \left[\varrho\left(\frac{\partial h}{\partial p}\right)_{T,\gamma_i} - 1\right]\left(\frac{\partial p}{\partial t} + \boldsymbol{v}\cdot\mathrm{grad}\,p\right) + \mathrm{div}\,\boldsymbol{J}_{\mathrm{red}} + \sum_{i=1}^{n} h_i\,\Gamma_i \\
&\qquad = -\varrho\,c_p\,\boldsymbol{v}\cdot\mathrm{grad}\,T + \sum_{i=1}^{n}\boldsymbol{J}_i\cdot(\boldsymbol{F}_i - \mathrm{grad}\,h_i) - (\mathsf{P} - p\,\mathsf{I}):\mathrm{Grad}\,\boldsymbol{v}.
\end{aligned}\right\} \tag{11.23}$$

Für spätere Verwendung (s. Ziff. 17) sei noch eine zu (11.23) äquivalente Formulierung angegeben, bei welcher der Energiestrom, ergänzt durch den konvektiven Enthalpiestrom, eine Vorzugsstellung einnimmt

$$\left.\begin{aligned}
&\varrho\,c_p\frac{\partial T}{\partial t} + \left[\varrho\left(\frac{\partial h}{\partial p}\right)_{T,\gamma_i} - 1\right]\frac{\partial p}{\partial t} - \boldsymbol{v}\cdot\mathrm{grad}\,p + \mathrm{div}\,(\boldsymbol{J}_u + \varrho\,h\cdot\boldsymbol{v}) \\
&\qquad = \sum_{i=1}^{n}[\boldsymbol{F}_i\cdot\boldsymbol{J}_i + h_i\,\mathrm{div}\,(\boldsymbol{J}_i + \varrho_i\boldsymbol{v})] + (p\,\mathsf{I} - \mathsf{P}):\mathrm{Grad}\,\boldsymbol{v} - \sum_{i=1}^{n} h_i\,\Gamma_i.
\end{aligned}\right\} \tag{11.24}$$

Man kann auch schreiben

$$\tilde{\boldsymbol{J}}_u \equiv \boldsymbol{J}_u + \varrho\, h\, \boldsymbol{v} = \boldsymbol{J}_{\mathrm{red}} + \sum_{i=1}^{n} \varrho_i\, h_i\, \boldsymbol{v}_i, \tag{11.25}$$

eine Darstellung, die besonders dann zweckmäßig ist, wenn alle v_i bis auf eines verschwinden. Dieser Fall liegt gerade in der Metalltheorie, Ziff. 17, vor.

$\gamma)$ *Fluide Medien mit zwei Komponenten.* Wir ändern die Bezeichnungen geringfügig ab, indem wir γ für γ_1, A für A_1, $\boldsymbol{J}$ für $\boldsymbol{J}_1$ und Γ für Γ_1 schreiben. Wir eliminieren $\boldsymbol{J}_2$ und können die chemischen Potentiale μ_1 und μ_2, da nur deren Differenz auftritt, nach (11.14) durch die Affinität

$$A = \mu_2 - \mu_1 \tag{11.14a}$$

ausdrücken. Damit nehmen die Gleichungen des vorhergehenden Abschnitts, soweit sie im folgenden benötigt werden, die spezielle Form an:

$$\frac{D\varrho}{Dt} + \varrho \operatorname{div} \boldsymbol{v} = 0, \tag{11.1a}$$

$$\varrho\, \frac{D\gamma}{Dt} + \operatorname{div} \boldsymbol{J} = \Gamma, \tag{11.3a}$$

$$\varrho\, \frac{D\boldsymbol{v}}{Dt} + \operatorname{Div} \mathsf{P} = \varrho_1 \boldsymbol{F}_1 + \varrho_2 \boldsymbol{F}_2, \tag{11.4a}$$

$$\varrho\, \frac{Du}{Dt} + \operatorname{div} \boldsymbol{J}_u = (\boldsymbol{F}_1 - \boldsymbol{F}_2) \cdot \boldsymbol{J} - p \operatorname{div} \boldsymbol{v} - (\mathsf{P} - p\,\mathsf{I}) : \operatorname{Grad} \boldsymbol{v}, \tag{11.5a}$$

$$\varrho\, \frac{Ds}{Dt} + \operatorname{div} \boldsymbol{J}_s = \sigma_s, \tag{11.6a}$$

$$\boldsymbol{J}_s = \frac{1}{T}\,(\boldsymbol{J}_u + A\,\boldsymbol{J}), \qquad \boldsymbol{J}_{\mathrm{red}} = \boldsymbol{J}_u + (h_2 - h_1)\,\boldsymbol{J}, \tag{11.7a}$$

$$\boldsymbol{J} = \frac{l_{11}}{T}\,(\boldsymbol{F}_1 - \boldsymbol{F}_2 + \operatorname{grad}_T A) + l_1 \operatorname{grad} \frac{1}{T}, \tag{11.9a}$$

$$\boldsymbol{J}_{\mathrm{red}} = \frac{l_1}{T}\,(\boldsymbol{F}_1 - \boldsymbol{F}_2 + \operatorname{grad}_T A) + l \operatorname{grad} \frac{1}{T}, \tag{11.10a}$$

$$p\,\delta_{\alpha\beta} - P_{\alpha\beta} = \eta\left(\frac{\partial v_\alpha}{\partial x_\beta} + \frac{\partial v_\beta}{\partial x_\alpha}\right) + \left(\zeta - \frac{2}{3}\,\eta\right)\delta_{\alpha\beta}\operatorname{div} \boldsymbol{v} + \delta_{\alpha\beta}\,\frac{\lambda_1}{T}\,A, \tag{11.16a}$$

$$\Gamma = -\frac{\lambda_1}{T}\operatorname{div} \boldsymbol{v} + \frac{\lambda_{11}}{T}\,A. \tag{11.17a}$$

Die quadratische Form der Entropieproduktion lautet

$$\left.\begin{aligned}
\sigma_s &= \frac{l_{11}}{T^2}(\boldsymbol{F}_1 - \boldsymbol{F}_2 + \operatorname{grad}_T A)^2 + \frac{2 l_1}{T}(\boldsymbol{F}_1 - \boldsymbol{F}_2 + \operatorname{grad}_T A)\cdot\operatorname{grad}\frac{1}{T} + l\left(\operatorname{grad}\frac{1}{T}\right)^2 + \\
&\quad + \frac{\eta}{2T}\sum_{\alpha=1}^{3}\sum_{\beta=1}^{3}\left(\frac{\partial v_\alpha}{\partial x_\beta} + \frac{\partial v_\beta}{\partial x}\right)^2 + \frac{1}{T}\left(\zeta - \frac{2}{3}\,\eta\right)(\operatorname{div}\boldsymbol{v})^2 + \frac{\lambda_{11}}{T}\,A^2.
\end{aligned}\right\} \tag{11.18a}$$

Für die phänomenologischen Koeffizienten bestehen folgende Ungleichungen

$$l_{11} > 0, \quad l > 0, \quad l_{11}\,l - l_1^2 > 0, \quad \eta > 0, \quad \zeta > 0, \quad \lambda_{11} > 0. \tag{11.19a}$$

Wir vermerken ferner die Beziehung

$$\left(\frac{\partial h}{\partial \gamma}\right)_{T,p} = h_1 - h_2 = T^2\left(\frac{\partial}{\partial T}\left(\frac{A}{T}\right)\right)_{P,\gamma}. \tag{11.26}$$

δ) *Thermodynamische Umrechnungen.* Thermodynamische Umrechnungen werden im folgenden meist nicht im einzelnen durchgeführt. Wir wollen jedoch an einem Beispiel eine einfache Methode angeben, mit der man Umrechnungsformeln prüfen oder herleiten kann. Wir beschränken uns auf drei unabhängige Variable T, v, γ. Für diese ist die spezifische freie Energie f ein thermodynamisches Potential mit dem vollständigen Differential

$$df = -s\,dT - p\,dv - A\,d\gamma. \tag{11.27}$$

In einem Bezugszustand T^+, v^+, γ^+ mögen die abhängigen Variablen die Werte s^+, p^+, $A^+ = 0$ haben. Dann läßt sich die freie Energie in der Umgebung des Bezugszustandes nach Potenzen und Produkten von $T - T^+$, $v - v^+$, $\gamma - \gamma^+$ entwickeln. Mit Hilfe von (11.27) erhält man aus dieser Entwicklung für $s - s^+$, $v - v^+$, A bei Berücksichtigung von linearen Gliedern allein

$$-(s - s^+) = f_{11}(T - T^+) + f_{12}(v - v^+) + f_{13}(\gamma - \gamma^+), \tag{11.28}$$

$$-(p - p^+) = f_{21}(T - T^+) + f_{22}(v - v^+) + f_{23}(\gamma - \gamma^+), \tag{11.29}$$

$$-A = f_{31}(T - T^+) + f_{32}(v - v^+) + f_{33}(\gamma - \gamma^+), \tag{11.30}$$

mit Koeffizienten $f_{ik} = f_{ki}$, die nur von T^+, v^+, γ^+ abhängen. Sie berechnen sich als zweite Differentialquotienten der spezifischen freien Energie nach den entsprechenden Variablen.

Alle ersten Differentialquotienten irgendeiner der sechs Größen s, p, A, T, v, γ nach irgendeiner anderen von ihnen bei Konstanthaltung von irgend zwei weiteren lassen sich durch die Koeffizienten f_{ik} ausdrücken. Zum Beispiel ist die spezifische Wärme $c_{p\gamma}$ bei konstantem Druck und bei konstanter Konzentration durch

$$\frac{1}{T}c_{p\gamma} = \left(\frac{\partial s}{\partial T}\right)_{p,\gamma} = -f_{11} + \frac{f_{12}^2}{f_{22}} \tag{11.31}$$

gegeben. Zur Herleitung dieser Beziehung setzt man in den Gln. (11.28), (11.29) $p = p^+$, $\gamma = \gamma^+$, drückt $v - v^+$ mit (11.29) durch $T - T^+$ aus und setzt in (11.28) ein. Die spezifische Wärme $c_{v\gamma}$ bei konstantem Volumen und konstanter Konzentration wird $-Tf_{11}$. Nun läßt sich auch leicht die Differenz der spezifischen Wärmen

$$c_{p\gamma} - c_{v\gamma} = T\frac{f_{12}^2}{f_{22}} \tag{11.32}$$

durch andere thermodynamische Größen ausdrücken. Für f_{12}/f_{22} findet man $-(\partial v/\partial T)_{p\gamma}$, wenn man in (11.29) $p = p^+$, $\gamma = \gamma^+$ setzt. Andererseits gilt nach (11.29) $f_{12} = -(\partial p/\partial T)_{v\gamma}$. Damit entsteht aus (11.31) die bekannte Relation

$$c_{p\gamma} - c_{v\gamma} = T\left(\frac{\partial v}{\partial T}\right)_{p,\gamma}\left(\frac{\partial p}{\partial T}\right)_{v,\gamma}. \tag{11.33}$$

Als weiteres Beispiel einer Relation, die sich auf diese Weise schnell beweisen läßt, geben wir

$$\frac{c_{p\gamma}}{c_{v\gamma}} = \frac{\chi_{T\gamma}}{\chi_{s\gamma}}. \tag{11.34}$$

$\chi_{T\gamma}$ und $\chi_{s\gamma}$ sind die isotherme und die adiabatische Kompressibilität bei konstanter Konzentration.

Die Gln. (11.28) bis (11.30) reichen grundsätzlich für alle Umrechnungen bei drei unabhängigen Variablen aus. Es kann jedoch gelegentlich bequemer sein, die (11.28) bis (11.30) entsprechenden Gleichungen für andere Potentiale und die

ihnen zugeordneten unabhängigen Variablen zu verwenden. Solche Gleichungs-
sätze findet man, indem man in (11.28) bis (11.30) eine oder mehrere der Sub-
stitutionen

$$1. \quad T - T^+ \to s - s^+, \qquad s - s^+ \to -(T - T^+),$$

$$2. \quad v - v^+ \to p - p^+, \qquad p - p^+ \to -(v - v^+),$$

$$3. \quad \gamma - \gamma^+ \to A, \qquad A \to -(\gamma - \gamma^+)$$

vornimmt. Die Koeffizientenmatrix f_{ik} geht dann in eine neue, aber wiederum
symmetrische Koeffizientenmatrix über.

12. Reversible Bewegungen des Kontinuums. Wir behandeln zunächst die
Frage, ob es Bewegungen des Kontinuums gibt, die reversibel verlaufen. Sie
kann auf zwei Weisen formuliert werden, die sich als äquivalent herausstellen
werden:

1. Unter welchen Bedingungen sind die Gleichungen in Ziff. 11 invariant
gegen Zeitumkehr?

2. Unter welchen Bedingungen ist die Entropieproduktionsdichte identisch
gleich Null?

Invarianz gegen Zeitumkehr bedeutet Invarianz gegen die Ersetzungen

$$t \to -t, \quad \boldsymbol{v} \to -\boldsymbol{v}, \quad \Gamma_i \to -\Gamma_i, \quad \boldsymbol{J}_i \to -\boldsymbol{J}_i, \quad \boldsymbol{J}_{\mathrm{red}} \to -\boldsymbol{J}_{\mathrm{red}}.$$

Wir nehmen an, daß die äußeren Kräfte invariant gegen Zeitumkehr sind und
daß kein Magnetfeld vorhanden ist. Dann bleiben die Bilanzgleichungen (11.1)
bis (11.5) bei dieser Ersetzung ungeändert. In (11.9) und (11.10) ändern die linken
Seiten ihr Vorzeichen, während die rechten Seiten sich nicht ändern. Daraus
folgt $\boldsymbol{J}_i = 0$, $\boldsymbol{J}_{\mathrm{red}} = 0$. Analog folgt aus (11.17) $A_i = 0$ $(i = 1, 2, \ldots, r)$ und damit
aus (11.16) $p\,\delta_{\alpha\beta} = P_{\alpha\beta}$. Daher ergibt sich aus (11.15) $\sigma = 0$ als Bedingung der
Umkehrbarkeit.

Aus dem folgenden wird hervorgehen, daß auch die Umkehrung gilt: Ist
$\sigma = 0$, so sind die Gleichungen in Ziff. 11 invariant gegen Zeitumkehr.

Die Bedingungen für reversible Bewegung können auf verschiedene Weisen
realisiert sein. Einen Grenzfall stellen die idealen-fluiden Medien dar. In ihnen
sind alle irreversiblen Vorgänge per definitionem vollständig gehemmt. Damit
verschwinden alle phänomenologischen Koeffizienten mit Ausnahme der Casimirschen, also der λ_i. In ideal-fluiden Medien gibt es daher weder Wärmeleitung
noch Diffusion; jedoch sind innere Reibung und innere Umwandlungen möglich,
wenn nicht alle $\lambda_i = 0$ sind. Diese Prozesse haben aber keinen dissipativen Charakter, da sich ihre Beiträge zur Entropieproduktion kompensieren. Für diese
speziellen Medien entwickeln sich die Felder $\varrho(\boldsymbol{r}, t)$, $\gamma_i(\boldsymbol{r}, t)$, $\boldsymbol{v}(\boldsymbol{r}, t)$, $u(\boldsymbol{r}, t)$ aus
beliebigen Anfangsbedingungen bei $t = 0$ nach Gleichungen, die gegen Zeitumkehr
invariant sind, also in reversibler Weise.

Das Verhalten fluider Medien kann bei vielen Problemen durch Nullsetzen der
phänomenologischen Koeffizienten gut approximiert werden, nämlich immer
dann, wenn die mit der Entropieproduktion verbundenen thermischen Effekte
keinen wesentlichen Einfluß auf den Vorgang haben. In solchen Fällen hat man
angenäherte Reversibilität. Die Hydrodynamik idealer Flüssigkeiten ist ein
klassisches Beispiel für diesen Grenzfall.

Ein anderer Grenzfall betrifft fluide Medien, bei denen keiner der irreversiblen
Prozesse gehemmt ist. Vorgänge ohne Entropieproduktion gibt es in solchen
Medien nur für eine ausgezeichnete Klasse von Anfangsbedingungen, d.h. nur
eine echte und relativ kleine Unterklasse aller möglichen Bewegungen ist reversibel. Sie soll im folgenden näher untersucht werden.

Verschwindende Entropieproduktion hat nach (11.18) wegen des positiv definiten Charakters der quadratischen Form der Matrix λ_{ik}, wegen des nicht-negativ definiten Charakters der quadratischen Form der Matrix l_{ik} vom Rang $n-1$ und wegen $l>0$, $\eta>0$, $\zeta\geqq 0$ zur Folge, daß im ganzen Medium

$$\operatorname{grad}\frac{1}{T}=0,\qquad \boldsymbol{F}_i-\operatorname{grad}_T\mu_i=\text{unabhängig von }i\quad (i=1,2,\ldots,n),\qquad (12.1)$$

$$A_i=0,\qquad \zeta\operatorname{div}\boldsymbol{v}=0,\qquad (12.2)$$

$$\frac{\partial v_\alpha}{\partial x_\beta}+\frac{\partial v_\beta}{\partial x_\alpha}=0\quad(\alpha\neq\beta),\qquad \frac{\partial v_1}{\partial x_1}=\frac{\partial v_2}{\partial x_2}=\frac{\partial v_3}{\partial x_3}\qquad (12.3)$$

gilt. Aus (11.9), (11.10), (11.6) folgt somit

$$\boldsymbol{J}_{\text{red}}=0,\qquad \boldsymbol{J}_i=0\quad(i=1,2,\ldots,n),\qquad p\,\delta_{\alpha\beta}=P_{\alpha\beta}\quad(\alpha,\beta=1,2,3).\qquad (12.4)$$

Damit ist auch bewiesen, daß die Bedingung $\sigma=0$ die Invarianz der Bewegungsgleichung gegen Zeitumkehr liefert.

In Systemen, in welchen ungehemmte chemische Reaktionen möglich sind, können die spezifischen Kräfte $\boldsymbol{F}_i$ nicht unabhängig sein. Solche Systeme mit unabhängigen $\boldsymbol{F}_i$ könnten weder reversible Bewegungen ausführen noch im thermodynamischen Gleichgewicht sein, da die zweite Beziehung (12.1) und die erste Beziehung (12.2) einander widersprechen. Berücksichtigt man nämlich neben diesen Beziehungen noch (11.14), so folgt

$$\boldsymbol{F}_k=-\sum_{i=r+1}^{n} \nu_{ik}\boldsymbol{F}_i\quad(k=1,2,\ldots r).\qquad (12.5)$$

Diese Gleichung ist insbesondere dann erfüllt, wenn alle $\boldsymbol{F}_i$ einander gleich sind [vgl. (11.13)], wie im Fall der Schwerkraft; sie gilt bei Elektrolyten wenigstens im Gleichgewicht und allgemeiner im raumladungsfreien Zustand wegen der Elektroneutralitätsbedingung (16.2).

Leiten sich die spezifischen Kräfte aus skalaren Potentialen her,

$$\boldsymbol{F}_i=-\operatorname{grad}\Phi_i\quad(i=1,2,\ldots,n),\qquad (12.6)$$

so schreibt sich (12.5) in der Form

$$\operatorname{grad}\left(\Phi_k+\sum_{i=r+1}^{n}\nu_{ik}\Phi_i\right)=0\quad(k=1,2,\ldots,r),\qquad (12.7)$$

und die zweite Beziehung (12.1) geht über in

$$\operatorname{grad}\left(\Phi_i+\mu_i\right)=\text{unabhängig von }i.\qquad (12.8)$$

Die Integration der partiellen Differentialgleichungen (12.3) gibt

$$\boldsymbol{v}(\boldsymbol{r},t)=\boldsymbol{a}+\boldsymbol{b}\times\boldsymbol{r}+c\,\boldsymbol{r}+2\boldsymbol{r}(\boldsymbol{d}\cdot\boldsymbol{r})-\boldsymbol{d}\,r^2\qquad (12.9)$$

mit zeitabhängigen Vektoren $\boldsymbol{a}(t)$, $\boldsymbol{b}(t)$, $\boldsymbol{d}(t)$ und einer skalaren Zeitfunktion $c(t)$. Haben die Kräfte skalare Potentiale, so bestehen zwischen den Koeffizienten in (12.9) Beziehungen. Zu ihrer Herleitung benutzen wir die Gibbs-Duhemsche Relation

$$s\,dT-\frac{1}{\varrho}\,dp+\sum_{i=1}^{n}\gamma_i\,d\mu_i=0.\qquad (12.10)$$

Aus ihr folgt nach (12.1) wegen $\operatorname{grad}T=0$

$$\operatorname{grad}p=\sum_{i=1}^{n}\varrho_i\operatorname{grad}\mu_i.\qquad (12.11)$$

Damit läßt sich die Bewegungsgleichung (11.4) zunächst in

$$\frac{D\boldsymbol{v}}{Dt} = -\sum_{i=1}^{n} \gamma_i \operatorname{grad}(\Phi_i + \mu_i) \tag{12.12}$$

umformen. Wegen (12.8) gilt weiter

$$\frac{D\boldsymbol{v}}{Dt} = -\operatorname{grad}(\Phi_i + \mu_i) \qquad (i = 1, 2, \ldots, n). \tag{12.13}$$

Somit ist

$$\operatorname{rot}\frac{D\boldsymbol{v}}{Dt} = 0. \tag{12.14}$$

Einsetzen von (12.9) gibt daher

$$\dot{\boldsymbol{b}} + 2c\,\boldsymbol{b} = 0, \quad \boldsymbol{d} = 0. \tag{12.15}$$

Eine weitere Einschränkung der Bewegung folgt aus $A_i = 0$ [s. (12.2)] und (11.1) und (11.3) im Verein mit $Ds/Dt = 0$ (wegen $\boldsymbol{J}_s = 0$, $\sigma = 0$). Die an den Reaktionen nicht beteiligten Bestandteile haben substantiell unveränderliche Konzentrationen. Es ist daher bei Zugrundelegung der unabhängigen Variablen s, ϱ, γ_1, γ_2, ..., γ_r

$$\frac{\partial A_i}{\partial \varrho}\frac{D\varrho}{Dt} + \sum_{k=1}^{r}\frac{\partial A_i}{\partial \gamma_k}\frac{D\gamma_k}{Dt} = 0 \qquad (i = 1, 2, \ldots, r). \tag{12.16}$$

Einsetzen von (11.1) und (11.3) gibt

$$\left[\frac{\partial A_i}{\partial \varrho} - \sum_{k=1}^{r}\frac{\partial A_i}{\partial \gamma_k}\frac{\lambda_k}{T\varrho^2}\right]\operatorname{div}\boldsymbol{v} = 0 \qquad (i = 1, 2, \ldots, r). \tag{12.17}$$

Wäre $\operatorname{div}\boldsymbol{v} \neq 0$, so würde gelten

$$\frac{\partial A_i}{\partial \varrho} - \sum_{k=1}^{r}\frac{\partial A_i}{\partial \gamma_k}\frac{\lambda_k}{T\varrho^2} = 0 \qquad (i = 1, 2, \ldots, r). \tag{12.18}$$

Diese Gleichungen können nur für Medien mit sehr speziellen thermodynamischen Eigenschaften erfüllt sein. Insbesondere für den Fall verschwindender Koeffizienten λ_k würde (12.18) bedeuten, daß sich das Reaktionsgleichgewicht bei konstant gehaltener Entropie nicht verschiebt, wenn man das Volumen verändert.

Bei Systemen, in welchen wenigstens *eine* Reaktion möglich ist, gilt also für die reversiblen Bewegungen im allgemeinen $\operatorname{div}\boldsymbol{v} = 0$ (auch wenn $\zeta = 0$ ist). Dies bedeutet für (12.5) unter Berücksichtigung von (12.15)

$$\boldsymbol{b} = \text{const}, \quad c = 0. \tag{12.19}$$

Dann sind alle ϱ_i und u und damit auch die Temperatur substantiell konstant und wegen $\operatorname{grad} T = 0$ folgt auch $\partial T/\partial t = 0$. Ferner sind alle μ_i substantiell konstant. (12.13) lautet nun

$$\dot{\boldsymbol{a}} + \boldsymbol{b}\times\boldsymbol{a} + \boldsymbol{b}(\boldsymbol{b}\cdot\boldsymbol{r}) - \boldsymbol{b}^2\boldsymbol{r} + \operatorname{grad}(\Phi_i + \mu_i) = 0 \qquad (i = 1, 2, \ldots, n) \tag{12.20}$$

oder nach den Koordinaten integriert

$$(\dot{\boldsymbol{a}} + \boldsymbol{b}\times\boldsymbol{a})\cdot\boldsymbol{r} + \tfrac{1}{2}(\boldsymbol{b}\cdot\boldsymbol{r})^2 - \tfrac{1}{2}\boldsymbol{b}^2\boldsymbol{r}^2 + \Phi_i + \mu_i = \psi_i(t) \qquad (i = 1, 2, \ldots, n) \tag{12.21}$$

mit „Integrationskonstanten" $\psi_i(t)$. Anwendung des Operators D/Dt gibt schließlich

$$\ddot{\boldsymbol{a}}\cdot\boldsymbol{r} + \boldsymbol{a}\cdot\dot{\boldsymbol{a}} + (\boldsymbol{a} + \boldsymbol{b}\times\boldsymbol{r})\cdot\operatorname{grad}\Phi_i = \dot{\psi}_i(t). \tag{12.22}$$

Diese Gleichung kann als partielle Differentialgleichung für die Potentiale Φ_i aufgefaßt werden. Wir verzichten auf die Diskussion des allgemeinen Falles und begnügen uns damit, zwei einfache Fälle näher zu betrachten, nämlich den kräftefreien Fall, $\Phi_i = 0$ und das homogene Feld, etwa das Schwerefeld mit grad $\Phi_i = -\boldsymbol{g}$. Im kräftefreien Fall ergibt sich dann aus (12.22) $\ddot{\boldsymbol{a}} = 0$. Kräftefreie reversible Bewegungen sind also Überlagerungen einer gleichförmig beschleunigten Bewegung $\dot{\boldsymbol{a}}$ und einer gleichförmigen Rotation um eine feste Achse. Das Medium verhält sich dabei wie ein starrer Körper. Ein mitbewegter Beobachter stellt vollkommenes mechanisches und thermodynamisches Gleichgewicht unter dem Einfluß der Trägheitskräfte fest. Im Schwerefeld hingegen ist $\ddot{\boldsymbol{a}} + \boldsymbol{g} \times \boldsymbol{b} = 0$. Hierzu gehören Bewegungen mit gleichförmiger Beschleunigung und gleichförmiger Rotation um die Richtung von $\boldsymbol{g}$, aber auch die Bewegungen komplizierterer Natur.

Die obige Diskussion erfaßt die möglichen reversiblen Bewegungen in Medien, welche entweder eine nicht-verschwindende Volumenviscosität haben oder in denen chemische Reaktionen (dazu sind auch Anregungsprozesse zu rechnen) ablaufen. Allgemeinere reversible Bewegungen (mit div $\boldsymbol{v} \neq 0$) kann man wegen (12.2) nur in nicht-reagierenden Medien mit verschwindender Volumenviscosität erhalten, also in Mischungen von Edelgasen unter geringem Druck. Für diese allgemeinen Lösungen mit div $\boldsymbol{v} \neq 0$ ist

$$\boldsymbol{a} = 0, \quad \boldsymbol{b} = 0, \quad c = \frac{1}{t}$$

ein charakteristisches Beispiel. (Wegen einer allgemeinen Diskussion vgl. Meixner[1].)

Die Gesamtheit der reversiblen Bewegungen in einatomigen Gasen hat bereits Boltzmann[2] angegeben. Er erhielt sie jedoch durch eine andere Fragestellung, nämlich bei der Untersuchung derjenigen Felder $\varrho_i(\boldsymbol{r}, t)$, $v(\boldsymbol{r}, t)$, $T(\boldsymbol{r}, t)$, für die die lokale Maxwell-Verteilung Lösung der Boltzmannschen Fundamentalgleichung ist.

13. Das mechanische Gleichgewicht; schleichende Bewegungen. Verschwindet in einem fluiden Medium die Entropieproduktionsdichte, so sind höchstens Bewegungen vom Typ (12.9) möglich. Stellt man die zusätzliche Bedingung $D\boldsymbol{v}/Dt = 0$, daß die Trägheitskräfte verschwinden, so reduziert sich der Ansatz (12.9) auf $\boldsymbol{v} = \text{const}$, d.h. unabhängig von Ort und Zeit. Aus (12.13) folgt dann

$$\boldsymbol{F}_i = \operatorname{grad} \mu_i, \tag{13.1}$$

und es ist nach (12.2) und (11.14)

$$\mu_k = - \sum_{i=r+1}^{n} v_{ik} \mu_i \quad (k = 1, 2, \ldots, r). \tag{13.2}$$

Daneben besteht natürlich die Beziehung (12.5) zwischen den spezifischen Kräften $\boldsymbol{F}_i$.

Aus den Gln. (13.1) und (13.2) bestimmt sich die räumliche Verteilung der Bestandteile im äußeren Kraftfeld. Ein bekanntes Beispiel hierfür ist das isotherme atmosphärische Gleichgewicht.

Man bezeichnet den Zustand mit $D\boldsymbol{v}/Dt = 0$ und $\sigma = 0$ als mechanisch-thermodynamisches Gleichgewicht.

---

[1] J. Meixner: Ann. Physik (5), **41**, 409 (1942).
[2] L. Boltzmann: Wien. Ber. **74**, 503 (1876).

Ist die Bedingung $D\boldsymbol{v}/Dt = 0$ allein erfüllt, d.h., verschwinden die Trägheitskräfte, ohne daß gleichzeitig die Entropieproduktion zu verschwinden braucht, so spricht man in der Literatur von mechanischem Gleichgewicht. Es sondert jedoch aus den Bewegungen des fluiden Mediums eine verhältnismäßig unwichtige Unterklasse von Bewegungen aus.

Viel wichtiger ist eine andere Klasse von Bewegungen, die man als schleichende Bewegungen bezeichnet. Sie sind dadurch definiert, daß man die Trägheitskräfte vernachlässigen kann $(D\boldsymbol{v}/Dt \approx 0)$ und daß die lokale Beschleunigung und die örtlichen Ableitungen der Geschwindigkeit hinreichend klein sind. Von der experimentell schwer erfaßbaren chemischen Viscosität wollen wir im folgenden absehen, d.h., wir setzen $\lambda_i = 0$. Soweit bei den Experimenten die Haftbedingung an festen Wänden keine Rolle spielt, darf man $P_{\alpha\beta} = p\,\delta_{\alpha\beta}$ setzen. Dann folgt aus der Bewegungsgleichung (4.13) $\operatorname{grad} p \approx \sum\limits_{i=1}^{n} \varrho_i \boldsymbol{F}_i$. Falls äußere Kräfte fehlen oder ihr Einfluß wie bei der Schwerkraft bei den üblichen Ausdehnungen der Systeme vernachlässigt werden kann, so ist das System bei schleichender Bewegung annähernd isobar.

Die Bedingung für schleichende Bewegungen kann in eine Beziehung zwischen den Diffusionskräften für den Satz $\boldsymbol{J}_{\mathrm{red}}, \boldsymbol{J}_i$

$$\boldsymbol{X}_i' = \frac{1}{T}\left(\boldsymbol{F}_i - \operatorname{grad}_T \mu_i\right) \tag{13.3}$$

umgeformt werden. Wegen der Voraussetzungen $\lambda_i = 0$ und $P_{\alpha\beta} = p\,\delta_{\alpha\beta}$ gilt dann nach (11.9) und (11.11)

$$\varrho\,\frac{D\boldsymbol{v}}{Dt} = \sum_{i=1}^{n} \varrho_i \boldsymbol{F}_i - \operatorname{grad} p = T \sum_{i=1}^{n} \varrho_i \boldsymbol{X}_i' + \sum \varrho_i \operatorname{grad}_T \mu_i - \operatorname{grad} p.$$

Wegen (12.11) heben sich die beiden letzten Glieder weg und es entsteht

$$\varrho\,\frac{D\boldsymbol{v}}{Dt} = T \sum_{i=1}^{n} \varrho_i \boldsymbol{X}_i'. \tag{13.4}$$

Für schleichende Bewegungen ist daher

$$\sum \varrho_i \boldsymbol{X}_i' \approx 0.$$

Diese Beziehung gilt, wie aus ihrer Herleitung hervorgeht, auch für nichtisotherme Systeme[1]. Bei isothermen Systemen gilt sie sogar für einen beliebigen Satz von Diffusionskräften.

Die Auszeichnung der Klasse der schleichenden Bewegungen beruht darauf, daß man bei vielen Problemen, und zwar meist durch den Einfluß von Randbedingungen der Undurchlässigkeit für Materie, zwei Stadien des Ablaufs unterscheiden kann. Im ersten Stadium klingt die makroskopische mechanische Bewegung (Strömungsbewegung), die durch anfänglich vorhandene größere Druckunterschiede zustande kam, durch die innere Reibung ab. Im zweiten Stadium bleiben nur die relativ langsamen Vorgänge der Wärmeleitung und Diffusion übrig, während Geschwindigkeit und Beschleunigung relativ klein sind. In diesem Stadium hat man schleichende Bewegung. In welchem Stadium die chemischen Reaktionen wesentlich sind, hängt natürlich von der Größe der Reaktionsgeschwindigkeiten ab.

---

[1] I. PRIGOGINE: Etude thermodynamique des phénomènes irréversibles. Paris u. Lüttich 1947.

14. Thermodiffusion und Diffusionsthermoeffekt. Die phänomenologischen Gleichungen für Energiestrom und Diffusionsströme bilden die Grundlage für die phänomenologische Behandlung der Thermodiffusion und des Diffusionsthermoeffektes. Sie werden im folgenden für ein nicht reagierendes Zweikomponentensystem diskutiert. Wir schließen uns in den Bezeichnungen an Ziff. 11γ an, ersetzen aber die chemischen Potentiale nicht durch die Affinität, da diese bei nicht reagierenden Mischungen nur den Charakter einer Abkürzung für die Differenz $\mu_2 - \mu_1$ hat. Nach (11.9a) und (11.10a) ist also

$$\boldsymbol{J} = \frac{l_{11}}{T} \left(\boldsymbol{F}_1 - \boldsymbol{F}_2 - \operatorname{grad}_T (\mu_1 - \mu_2) \right) + l_1 \operatorname{grad} \frac{1}{T}, \tag{14.1}$$

$$\boldsymbol{J}_{\text{red}} = \frac{l_1}{T} \left(\boldsymbol{F}_1 - \boldsymbol{F}_2 - \operatorname{grad}_T (\mu_1 - \mu_2) \right) + l \operatorname{grad} \frac{1}{T}. \tag{14.2}$$

Wählt man als unabhängige Variable T, P, γ, führt man die partiellen spezifischen Volumina

$$v_i = \left(\frac{\partial \mu_i}{\partial p} \right)_{T, \gamma} \qquad (i = 1, 2) \tag{14.3}$$

ein und macht von der Gibbs-Duhemschen Relation

$$\gamma (d\mu_1)_{T, p} + (1 - \gamma) (d\mu_2)_{T, p} = 0 \tag{14.4}$$

Gebrauch, so lassen sich (14.1) und (14.2) umformen zu

$$\boldsymbol{J} = \frac{l_{11}}{T} \left(\boldsymbol{F}_1 - \boldsymbol{F}_2 - (v_1 - v_2) \operatorname{grad} p - \frac{1}{1 - \gamma} \left(\frac{\partial \mu_1}{\partial \gamma} \right)_{T, p} \operatorname{grad} \gamma \right) + l_1 \operatorname{grad} \frac{1}{T}, \tag{14.5}$$

$$\boldsymbol{J}_{\text{red}} = \frac{l_1}{T} \left(\boldsymbol{F}_1 - \boldsymbol{F}_2 - (v_1 - v_2) \operatorname{grad} p - \frac{1}{1 - \gamma} \left(\frac{\partial \mu_1}{\partial \gamma} \right)_{T, p} \operatorname{grad} \gamma \right) + l \operatorname{grad} \frac{1}{T}. \tag{14.6}$$

Diese Gleichungen machen deutlich, wie Diffusion und Wärmeleitung von äußeren Kräften, vom Druck-, Konzentrations- und Temperaturgradienten abhängen. Die acht Koeffizienten in (14.5) und (14.6) lassen sich auf nur drei unabhängige phänomenologische Koeffizienten l_{11}, l_1, l zurückführen. Dies ist eine wichtige Aussage der Thermodynamik irreversibler Prozesse.

Es ist üblich, diese phänomenologischen Koeffizienten durch drei Größen λ, D_{12} und α auszudrücken:

$$\left. \begin{aligned} \varrho\, T\, (1 - \gamma)\, D_{12} &= l_{11} \left(\frac{\partial \mu_1}{\partial \gamma} \right)_{T, p}, \\ \varrho\, T \gamma\, (1 - \gamma)\, \alpha\, D_{12} &= l_1, \\ T^2 \lambda &= l. \end{aligned} \right\} \tag{14.7}$$

Durch Betrachtung spezieller Fälle ergibt sich aus (14.5) und (14.6) die Bedeutung von λ als der Wärmeleitfähigkeit bei homogener Durchmischung, von D_{12} als des Diffusionskoeffizienten bei räumlich konstanter Temperatur. α ist der sog. Thermodiffusionsfaktor. Bei den üblichen Experimenten zur Thermodiffusion und zur Diffusionsthermik ist $\boldsymbol{F}_1 = \boldsymbol{F}_2$ und die Druckdiffusion vernachlässigbar. Unter diesen Voraussetzungen schreiben sich (14.5) und (14.6) in den neuen Bezeichnungen

$$\boldsymbol{J} = - \varrho\, D_{12} \operatorname{grad} \gamma - \frac{\varrho D_{12}}{T} \gamma\, (1 - \gamma)\, \alpha \operatorname{grad} T, \tag{14.8}$$

$$\boldsymbol{J}_{\text{red}} = - \varrho\, \gamma \left(\frac{\partial \mu_1}{\partial \gamma} \right)_{T, p} \alpha\, D_{12} \operatorname{grad} \gamma - \lambda \operatorname{grad} T. \tag{14.9}$$

In dieser Gestalt der phänomenologischen Gleichungen ist die Symmetrie des Koeffizientenschemas verlorengegangen, obwohl natürlich die O.C.R.B. berücksichtigt sind. An diesem Beispiel wird besonders deutlich, daß die Anwendbarkeit der O.C.R.B. die Schreibweise der phänomenologischen Gleichungen in Flüssen und thermodynamisch konjugierten Kräften voraussetzt.

Die Bedingung der positiven Entropieproduktion bedeutet, daß die quadratische Form der Koeffizientenmatrix in (14.5) und (14.6) positiv definit ist, also $l_{11} > 0$, $l > 0$, $l_{11} l > l_1^2$. Mit (14.7) entsteht hieraus

$$\lambda > 0, \quad D_{12} > 0, \quad \alpha^2 < \frac{T}{\varrho\gamma(1-\gamma)} \frac{1}{\left(\dfrac{\partial \mu_1}{\partial \ln \gamma}\right)_{T,p}} \frac{\lambda}{D_{12}}. \tag{14.10}$$

Die letzte Ungleichung findet sich zuerst explicit bei HAASE[1]; die spezielle Beziehung (14.13) ist von MEIXNER[2] angegeben worden. Während also die Bedingung positiver Entropieproduktion für λ und D_{12} Vorzeichenaussagen liefert, läßt sie für α beide Vorzeichen zu, gibt aber eine Schranke für den Betrag von α.

Für ideale Gasmischungen ist

$$\left(\frac{\partial h}{\partial p}\right)_{T,\gamma} = 0, \quad \mu_i = \mu_i(T, p, 0) + \frac{RT}{M_i} \ln x_i \tag{14.11}$$

mit den durch

$$x_i \sum_{k=1}^{n} \frac{\gamma_k}{M_k} = \frac{\gamma_i}{M_i}, \quad \sum_{i=1}^{n} x_i = 1 \tag{14.12}$$

definierten Molenbrüchen x_i. Damit geht (14.10) über in

$$\alpha^2 < \frac{1}{\gamma'(1-\gamma)} \frac{\lambda}{\varrho D_{12} R} \left[\gamma(M_2 - M_1) + M_1\right]. \tag{14.13}$$

Setzt man für λ und D die Ausdrücke ein, welche die elementare kinetische Gastheorie liefert[3],

$$\lambda = (3 + f_1 + f_2) \frac{R\eta}{2\left[\gamma(M_2 - M_1) + M_1\right]}, \quad D_{12} = \frac{\eta}{\varrho}, \tag{14.14}$$

so folgt weiter

$$\alpha^2 < \frac{3}{2\gamma(1-\gamma)} + \frac{f_1 + f_2}{2\gamma(1-\gamma)}. \tag{14.15}$$

f_1 und f_2 bezeichnen die Anzahl der zur spezifischen Wärme beitragenden Rotations- und Schwingungsfreiheitsgrade. Das Minimum der rechten Seite von (14.15) als Funktion von γ liegt bei $6 + 2(f_1 + f_2)$ während die gemessenen Werte des Thermodiffusionsfaktors α kaum über 0,4 hinausgehen; damit ist α in der Regel wesentlich kleiner als die in (14.15) angegebenen Schranke.

Andererseits ergibt sich mit den in verdünnten Lösungen geltenden Größenordnungen

$$\lambda \approx 10^{-3} \frac{\text{cal}}{\text{cm sec Grad}}, \quad D_{12} \approx 10^{-5} \frac{\text{cm}^2}{\text{sec}}, \quad \varrho \approx 1 \frac{\text{g}}{\text{cm}^3},$$

$$T \approx 300° \text{K}, \quad \frac{\partial \mu}{\partial \gamma} \approx 5 \text{ cal/g}, \quad \gamma_1 \approx \frac{1}{10}$$

die Ungleichung

$$\alpha^2 < 6 \cdot 10^4. \tag{14.16}$$

---

[1] R. HAASE: Z. Elektrochem. **54**, 450 (1950).
[2] J. MEIXNER: Ann. Physik [5] **39**, 333 (1941).
[3] Vgl. A. EUCKEN: Lehrbuch der chemischen Physik, II, 1. Leipzig 1948.

Gemessene Werte von $|\alpha|$ liegen zwischen 1 und 10[1], so daß auch bei verdünnten Lösungen α^2 erheblich kleiner als die Schranke in (14.16) ist.

Ist das fluide Medium anfänglich in Ruhe und wird es durch begrenzende Wände auch weiterhin „in Ruhe" gehalten, ist also die Konvektion durch Diffusion vernachlässigbar, so lassen sich für hinreichend kleine Temperatur- und Konzentrationsgradienten die Gleichungen in Ziff. 11γ sehr vereinfachen. Dann ist $v \approx 0$, grad $p \approx 0$, und es bleiben nur die Gln. (11.3 a) mit $\Gamma = 0$, (14.8) und (14.9). Mit den hier angebrachten Vernachlässigungen geben sie

$$\frac{\partial \gamma}{\partial t} = \operatorname{div}\left(D_{12}\operatorname{grad}\gamma + \frac{\gamma(1-\gamma)}{T}\alpha D_{12}\operatorname{grad} T\right), \qquad (14.17)$$

$$\varrho\, c_p \frac{\partial T}{\partial t} + \left(\varrho\,\frac{\partial h}{\partial p} - 1\right)\frac{\partial p}{\partial t} = \operatorname{div}\left[\varrho\,\gamma\left(\frac{\partial \mu_1}{\partial \gamma}\right)_{T,p}\alpha D_{12}\operatorname{grad}\gamma + \lambda\operatorname{grad} T\right]. \qquad (14.18)$$

Man kann auch noch das Glied mit $\partial p/\partial t$ vernachlässigen und wegen der vorausgesetzten kleinen Konzentrations- und Temperaturgradienten alle Koeffizienten als zeitlich und örtlich konstant annehmen.

Nach Gl. (14.18) gibt ein Konzentrationsgradient auch thermische Effekte (Diffusionsthermoeffekt), nach (14.17) gibt auch ein Temperaturgradient Anlaß zu Konzentrationsänderungen (Thermodiffusionseffekt). Beide Effekte sind durch denselben Faktor α bestimmt; dies ist die Folge der O.C.R.B.

Die Gl. (14.18) geht für grad $\gamma = 0$ in die gewöhnliche Wärmeleitungsgleichung über. Wie schon erwähnt, ist λ die Wärmeleitfähigkeit für homogene Durchmischung. In einem Zustand, in welchem keine Diffusion erfolgt, also $J = 0$ ist, sind Konzentrations- und Temperaturgradient nach (14.8) proportional. Eliminiert man für diesen Fall grad γ aus (14.8) und (14.18), so ergibt sich wieder eine Wärmeleitungsgleichung, aber mit der veränderten Wärmeleitfähigkeit

$$\lambda' = \lambda - \frac{\varrho\gamma^2(1-\gamma)}{T}\left(\frac{\partial \mu_1}{\partial \gamma}\right)_{T,p}\alpha^2 D_{12}.$$

Sie ist die Wärmeleitfähigkeit für stationäre Diffusion. Auch sie ist nach (14.10) positiv und unterscheidet sich nach dem oben Gesagten von λ in der Regel um nicht mehr als 1,5% bei idealen Gasmischungen.

Wir betrachten nun eine Reihe von Sonderfällen.

α) *Thermodiffusion ohne Konvektion.* Das fluide Medium sei zu Beginn ($t = 0$) homogen durchmischt, und es sei ein Temperaturgradient in x-Richtung hergestellt. Seien die Temperaturen T_0 bei $x = 0$ und T bei $x = l$ zeitlich konstant vorgeschrieben und seien die dort befindlichen Wände für beide Bestandteile undurchlässig, also $J = 0$ für $x = 0$ und $x = l$. Man kann damit rechnen, daß dann grad γ stets kleiner bleibt als der Wert, welcher zum stationären Zustand mit dem jeweiligen Temperaturgradienten gehört, also $|\operatorname{grad}\gamma| < \frac{1}{T}\gamma(1-\gamma) \times \alpha|\operatorname{grad} T|$. Wegen der Abschätzung des Thermodiffusionsfaktors darf man daher in (14.18) das Glied mit grad γ vernachlässigen. Die Berechnung der Konzentration γ als Funktion von x, t läuft also darauf hinaus, daß man erst die gewöhnliche Wärmeleitungsgleichung mit den gegebenen Anfangs- und Randbedingungen integriert und mit der so gewonnenen Funktion $T(x, t)$ Gl. (14.17) integriert. Wegen Einzelheiten sei auf die Literatur verwiesen[2].

---

[1] E.L. Dougherty jr., u. H.G. Drickamer: J. Chem. Phys. **23**, 295 (1955). Vgl. auch die dort zitierten Arbeiten.

[2] W.H. Furry, R. Clark Jones u. L. Onsager: Phys. Rev. **55**, 1083 (1939). L. Waldmann: Z. Naturforsch. **4a**, 105 (1949).

β) *Der Diffusionsthermoeffekt.* Hier geht man von einem fluiden Medium konstanter Temperatur, aber mit Konzentrationsunterschieden in x-Richtung aus. In diesem Falle wird man annehmen dürfen, daß in (14.18) das Glied mit grad T klein gegen jenes mit grad γ ist. Dasselbe Argument wie in α) erlaubt dann in (14.17) das Glied mit grad T zu vernachlässigen. Diese Gleichung ist dann die gewöhnliche Diffusionsgleichung und kann unmittelbar integriert werden. Mit der so gewonnenen Funktion $\gamma(x, t)$ wird dann (14.18) gelöst.

γ) *Thermodiffusion mit Konvektion; Trennrohr.* Die Gleichungen der Thermodynamik irreversibler Prozesse in fluiden Medien lassen sich auf die Thermodiffusion im Trennrohr[1] oder auf den stationären Diffusionsthermoeffekt anwenden. Auch bei diesen Beispielen handelt es sich um schleichende Bewegungen. Insbesondere für das Trennrohr sind jedoch zwei Punkte hervorzuheben.

1. Die Ausbildung einer stationären Geschwindigkeitsverteilung über den Querschnitt ist nur möglich, wenn man die innere Reibung nicht vernachlässigt; die Geschwindigkeit ergibt sich nämlich als umgekehrt proportional zur Viskosität. Dies hängt mit dem Haften der Strömung an der Wand zusammen.

2. Während bei der „ruhenden" Diffusion und Wärmeleitung in α) und β) die Gleichungen aus Ziff. 11 für kleine grad T und grad γ linearisiert werden konnten, ist dies in der Theorie des Trennrohrs nicht mehr zulässig. Seine Wirksamkeit beruht gerade auf der Mitnahme des Gliedes zweiter Ordnung $D\gamma/Dt = \boldsymbol{v} \cdot$ grad γ in Gl. (11.3a). Für eine hinreichend genaue Theorie ist es ferner notwendig die Veränderlichkeit des Faktors vor grad T in (14.8) zu berücksichtigen. Damit bleibt in (11.3) ein weiteres nicht-lineares Glied proportional zu grad $\gamma \cdot$ grad T übrig. Wegen der Durchführung der Rechnungen sei auf die angegebene Literatur verwiesen.

15. Extremaleigenschaften des stationären Zustandes. Man nennt einen Zustand, bei dem alle lokalen Zeitableitungen verschwinden, einen stationären Zustand. Unter gewissen Voraussetzungen ist er gegenüber benachbarten nichtstationären Zuständen hinsichtlich der Entropieproduktion ausgezeichnet. Um diese Eigenschaft herzuleiten und zu formulieren, betrachten wir ein fluides Medium, das in einer raumfesten Hülle eingeschlossen ist. Die Bewegung sei so weit abgeklungen, daß folgende Annahmen erfüllt sind.

1. Die baryzentrische Geschwindigkeit ist so klein, daß die substantiellen Differentialquotienten nach der Zeit durch die lokalen ersetzt werden können.

2. Die Beiträge der Schubviscosität, der Volumenviscosität und der chemischen Viskosität zum Drucktensor sind vernachlässigbar, d.h. es ist $P_{\alpha\beta} = p\,\delta_{\alpha\beta}$.

3. Die zeitliche Änderung des Druckes ist vernachlässigbar, $\partial p/\partial t = 0$.

4. Die äußeren Kräfte hängen nicht von der Zeit ab.

Wir betrachten eine Funktion P, die sich formal aus der Entropieproduktionsdichte (5.6) ergibt, wenn man die Kräfte durch ihre Zeitableitungen ersetzt. Der zweite Term ist wegen der zweiten Annahme zu streichen:

$$P = \boldsymbol{J_u} \cdot \operatorname{grad} \frac{\partial}{\partial t}\left(\frac{1}{T}\right) + \sum_{i=1}^{n} \boldsymbol{J_i} \frac{\partial}{\partial t}\left[\frac{\boldsymbol{F_i}}{T} - \operatorname{grad} \frac{\mu_i}{T}\right] - \sum_{i=1}^{n} \Gamma_i \frac{\partial}{\partial t}\left(\frac{\mu_i}{T}\right). \quad (15.1)$$

Diese Funktion besitzt unter noch festzulegenden Oberflächenbedingungen eine Extremaleigenschaft. Um sie herzuleiten, integrieren wir (15.1) über den Bereich des fluiden Mediums und beseitigen gleichzeitig durch Anwendung des Gaußschen

---

[1] P. DEBYE: Ann. Physik **36**, 284 (1939). — R. CLARK JONES u. W.H. FURRY: Rev. Mod. Phys. **18**, 151 (1946).

Satzes die Gradienten. Dann folgt

$$\int P\,dV = P_O + P_V \tag{15.2}$$

mit dem Oberflächenintegral

$$P_O = \oint \left[\boldsymbol{J}_u \frac{\partial}{\partial t}\left(\frac{1}{T}\right) - \sum_{i=1}^{n} \boldsymbol{J}_i \frac{\partial}{\partial t}\left(\frac{\mu_i}{T}\right) \right] d\boldsymbol{f} \tag{15.3}$$

und dem Volumenintegral

$$P_V = \int \left[-\frac{\partial}{\partial t}\left(\frac{1}{T}\right)\operatorname{div}\boldsymbol{J}_u + \sum_{i=1}^{n} \boldsymbol{J}_i \cdot \boldsymbol{F}_i \frac{\partial}{\partial t}\left(\frac{1}{T}\right) + \sum_{i=1}^{n} \frac{\partial}{\partial t}\left(\frac{\mu_i}{T}\right)\operatorname{div}\boldsymbol{J}_i - \left. - \sum_{i=1}^{n} \Gamma_i \frac{\partial}{\partial t}\left(\frac{\mu_i}{T}\right) \right] dV. \tag{15.4}$$

Ersetzt man nun div $\boldsymbol{J}_u$ nach (11.5) und div $\boldsymbol{J}_i$ nach (11.2), so folgt, wenn noch div $\boldsymbol{v}$ nach (11.1) ausgedrückt wird,

$$P_V = \int \left[\varrho \frac{\partial}{\partial t}\left(\frac{1}{T}\right)\left(\frac{\partial u}{\partial t} + p\,\frac{\partial}{\partial t}\,\frac{1}{\varrho}\right) - \varrho \sum_{i=1}^{n} \frac{\partial}{\partial t}\,\frac{\mu_i}{T}\left(\frac{\partial \varrho_i}{\partial t} - \frac{\varrho_i}{\varrho}\,\frac{\partial \varrho}{\partial t}\right) \right] dV. \tag{15.5}$$

Nach (11.20) ist

$$h = u + \frac{p}{\varrho}, \qquad dh = c_p\,dT + \left(\frac{\partial h}{\partial p}\right)_{T,\gamma_i} dp + \sum_{i=1}^{n} h_i\,d\gamma_i. \tag{15.6}$$

Wegen der dritten Annahme wird also zunächst

$$P_V = \int \left[\varrho \frac{\partial}{\partial t}\left(\frac{1}{T}\right)\left(c_p \frac{\partial T}{\partial t} + \sum_{i=1}^{n} h_i \frac{\partial \gamma_i}{\partial t}\right) - \varrho \sum_{i=1}^{n} \frac{\partial}{\partial T}\left(\frac{\mu_i}{T}\right)\frac{\partial T}{\partial t}\,\frac{\partial \gamma_i}{\partial t} - \right. - \frac{1}{T} \sum_{i=1}^{n}\sum_{j=1}^{n} \frac{\partial \mu_i}{\partial \varrho_j}\,\frac{\partial \varrho_j}{\partial t}\left(\frac{\partial \varrho_i}{\partial t} - \frac{\varrho_i}{\varrho}\,\frac{\partial \varrho}{\partial t}\right) \right] dV.$$

Der letzte Term im Integral verschwindet wegen der Gibbs-Duhemschen Beziehung, der zweite und der dritte Term heben sich wegen (9.11) auf und so bleibt, nach geringer Umformung,

$$P_V = -\int \left[\frac{\varrho\,c_p}{T^2}\left(\frac{\partial T}{\partial t}\right)^2 + \sum_{i,j=1}^{n} \frac{\partial \mu_i}{\partial \varrho_j}\,\frac{\partial \varrho_i}{\partial t}\,\frac{\partial \varrho_j}{\partial t} \right] dV. \tag{15.7}$$

Da c_p positiv ist und die Matrix $\partial \mu_i / \partial \varrho_j$ bei thermodynamischer Stabilität gegen Entmischung eine positiv definite quadratische Form hat, so gilt stets

$$P_V \leqq 0. \tag{15.8}$$

Das Gleichheitszeichen besteht dann und nur dann, wenn die Temperatur T und die Partialdichten ϱ_i nicht von der Zeit abhängen, wenn also der Zustand stationär ist.

Damit ist folgendes Resultat gewonnen: Verschwindet die Normalkomponente von

$$\boldsymbol{J}_u \frac{\partial}{\partial t}\,\frac{1}{T} - \sum_{i=1}^{n} \boldsymbol{J}_i \frac{\partial}{\partial t}\,\frac{\mu_i}{T} \tag{15.9}$$

auf der ganzen raumfesten Oberfläche, die das fluide Medium begrenzt, so ist unter den Annahmen 1 bis 4 das Raumintegral der Funktion P stets negativ, solange Zustandsänderungen ablaufen. Es verschwindet, wenn der stationäre

Zustand erreicht ist. Die Funktion P hat also ein Maximum im stationären Zustand immer wenn die Normalkomponente des Ausdrucks (15.9) verschwindet.

Dies ist beispielsweise dann erfüllt, wenn über die ganze Oberfläche zeitunabhängige Werte der Temperatur und der chemischen Potentiale vorgeschrieben sind, oder auch wenn zeitunabhängige Werte der Temperatur allein vorgeschrieben sind, aber die Oberfläche für alle Bestandteile undurchlässig ist, oder wenn auf einem Teil der Oberfläche der eine Satz von Bedingungen, auf dem restlichen Teil der andere Satz von Bedingungen vorgeschrieben ist usw.

Die Bedeutung dieses Ergebnisses von GLANSDORFF und PRIGOGINE[1] — siehe dazu auch GLANSDORFF und PASSELECQ[2] — liegt darin, daß es für beliebige Abweichungen vom thermodynamischen Gleichgewicht gilt, sofern nur die Annahmen 1 bis 4 erfüllt sind. Es ist jedoch zu bemerken, daß die Bildungsvorschrift für P nicht invariant gegenüber linearen Transformationen von Flüssen und Kräften ist; sie gilt nur für die Darstellung (5.6) der Entropieproduktion.

Nun sollen noch zwei weitere Annahmen eingeführt werden:

5. Die phänomenologischen Koeffizienten Λ_i in (11.12) sind entweder gleich Null oder vernachlässigbar.

6. Die Gradienten der Temperaturen und der chemischen Potentiale sowie die Affinitäten sind überall so klein, daß die linearen phänomenologischen Gln. (7.1), (7.2) und (7.15) gelten. Dies schließt nicht aus, daß weit entfernte Raumpunkte sehr verschiedene Temperaturen und chemische Potentiale haben, wenn nur das System genügend groß ist.

Unter den zusätzlichen Annahmen 5 und 6 besteht zwischen der Funktion P und der Entropieproduktion ein einfacher Zusammenhang. In der formalen Schreibweise der Ziff. 8 lautet er

$$\sigma = \sum Y_i X_i, \quad P = \sum Y_i \frac{\partial X_i}{\partial t}, \quad Y_i = \sum L_{ik} X_k \qquad (15.10)$$

mit $L_{ik} = L_{ki}$. Man überzeugt sich nun leicht, daß

$$P = \frac{1}{2} \frac{\partial \sigma}{\partial t} \qquad (15.11)$$

ist, wenn man von den zeitlichen Differentialquotienten der phänomenologischen Koeffizienten L_{ik} absieht. Diese würden ja nur zu zusätzlichen Termen Anlaß geben, welche wegen Annahme 6 von dritter Ordnung klein sind.

Damit folgt der *Satz*: Verschwindet die Normalkomponente des Ausdrucks (15.9) auf der ganzen Oberfläche, welche das fluide Medium begrenzt und sind die Annahmen 1 bis 6 erfüllt, so nimmt die gesamte Entropieproduktion im fluiden Medium mit der Zeit ab, so lange der stationäre Zustand nicht erreicht ist. Im stationären Zustand hat also die gesamte Entropieproduktion einen minimalen Wert gegenüber den Nachbarzuständen zu selben Oberflächenbedingung.

Insbesondere folgt hieraus die Stabilität des stationären Zustandes; das fluide Medium kann sich aus ihm nicht entfernen, da dazu die Entropieproduktion zunehmen müßte.

Dieser Satz, der auf PRIGOGINE[3] zurückgeht und allgemein von WERGELAND[4] und in besonders übersichtlicher Weise von MAZUR[5] bewiesen wurde, hat vor allem

[1] P. GLANSDORFF u. I. PRIGOGINE: Physica, Haag **20**, 773 (1954).

[2] P. GLANSDORFF u. J. PASSELECQ: Bull. Acad. Roy. Belg. Cl. Sci. **43**, 188 (1957).

[3] I. PRIGOGINE: Symposium on Thermodynamique. Brussels, Union of Physics, Unesco 1948, 87.

[4] H. WERGELAND: Kgl. norske Vidensk. Selsk. **24**, 110 (1951).

[5] P. MAZUR: Bull. Acad. Roy. Belg. Cl. Sci. **38**, 182 (1952).

Anwendung in der Theorie der Stoßwellen[1], der Gasentladungen[2] und auf die Entwicklung lebender Organismen gefunden[3,4]. Den Schlußfolgerungen gegenüber ist in diesem Fall eine gewisse Zurückhaltung angebracht, da gerade bei Organismen die Voraussetzung der Nähe des chemischen Gleichgewichts nicht erfüllt ist (s. a. Haase[5]).

Von Denbigh[6] und Nielsen[7] sind Beispiele angegeben worden, für welche der Satz von der minimalen Entropieproduktion nicht gilt, wenn man die Annahme 6 fallen läßt. Daß die Annahmen 1 bis 3 für diesen Satz wesentlich sind, erkennt man jedoch schon an folgendem Beispiel. Man betrachte ein Gas in einem geschlossenen Gefäß, das nur aus einer Komponente besteht, und in welchem als irreversible Prozesse daher nur Wärmeleitung und innere Reibung möglich sind. Man stelle nun einen Anfangszustand her, welcher isotherm ist und bei dem v überall den Wert Null hat, ohne daß gleichzeitig das System bei Fehlen äußerer Kräfte isobar ist. Dann ist in diesem Augenblick die Entropieproduktion gleich Null. Wegen der Druckunterschiede setzt dann eine Bewegung ein, es bilden sich Temperaturgradienten aus, und es erfolgt eine positive Entropieproduktion. Die gesamte Entropieproduktion nimmt also in diesem Beispiel wenigstens im Anfangsstadium mit der Zeit zu.

Die wesentliche Ursache hierfür liegt in der Bewegungsgleichung, also in den Trägheitseffekten. Ein Beispiel aus einem ganz anderen Bereich läßt die Verhältnisse deutlicher werden. Man betrachte dazu die einfachen linearen elektrischen Netzwerke in Fig. 1.

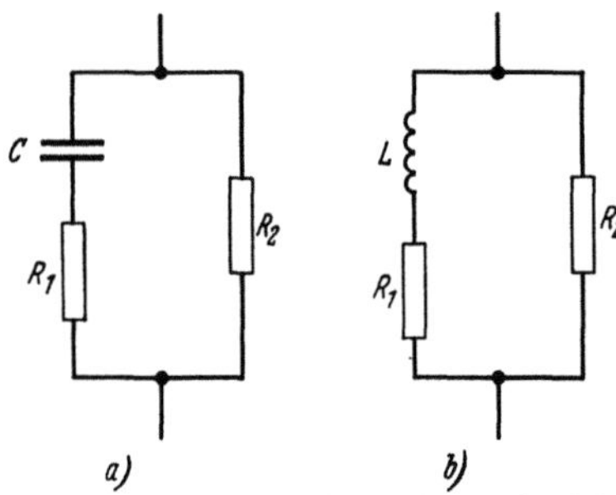

Fig. 1 a und b. Beispiele für Netzwerke, bei denen die Entropieproduktion im stationären Zustand ein Minimum bzw. ein Maximum besitzt.

Die Rechnung zeigt, daß im ersten Netzwerk die Joulesche Wärmeleistung (= absolute Temperatur × Entropieproduktion) mit der Zeit abnimmt, während sie im zweiten Netzwerk zunimmt, falls die Spannung konstant gehalten wird und beliebige Anfangswerte der Ladung auf dem Kondensator bzw. des Stroms in der Induktivität vorgeschrieben werden. Dieser Sachverhalt trifft auch dann zu, wenn das erste Netzwerk beliebig aus Kondensatoren und Widerständen, das zweite Netzwerk beliebig aus Induktivitäten und Widerständen zusammengesetzt ist. Voraussetzung für die elementare Behandlung ist, daß die Joulesche Wärme abgeleitet wird und die Widerstände isotherm bleiben. — Für Netzwerke, die alle drei Arten von Elementen enthalten, gilt kein allgemeiner Satz über die zeitliche Veränderung der Entropieproduktion.

16. Elektrochemische Effekte. Systeme mit geladenen Bestandteilen (positiven und negativen Ionen, Elektronen) werden elektrochemische Systeme genannt. Zu ihnen gehören die Elektrolyte, das Plasma und die Metalle. Bei der Behandlung dieser Systeme mit Hilfe der Thermodynamik der irreversiblen Prozesse spielt die Kraftfelddiffusion eine wichtige Rolle.

Eine vereinfachte Behandlung ist für solche Effekte möglich, bei denen sich das elektrische Feld nur langsam ändert und die Eigenschaft der Elektroneutrali-

[1] R. Haase: Z. Naturforsch. 6a, 522 (1951).
[2] H. Maecker u. T. Peters: Z. Physik 144, 612 (1956).
[3] I. Prigogine u. J.M. Wiame: Experientia 2, 451 (1946).
[4] R. Haase: Z. Elektrochem. 55, 566 (1951).
[5] R. Haase: Z. Naturforsch. 8a, 729 (1953).
[6] K.S. Denbigh: Trans. Faraday Soc. 48, 389 (1952).
[7] A.E. Nielsen: Bull. Acad. Roy. Belg., Cl. Sci. 40, 539 (1954).

tät im Makroskopischen besteht. Dann sind die Gleichungen aus Ziff. 11 anwendbar. Insbesondere läßt sich das elektrische Feld E aus einem Gradienten herleiten

$$E = - \operatorname{grad} \varphi. \tag{16.1}$$

Für die spezifischen Ladungen e_i (Ladung des Bestandteils i pro Masseneinheit dieses Bestandteils) gilt

$$\sum_{i=1}^{n} \varrho_i e_i = 0, \tag{16.2}$$

und die Kräfte F_i in (11.4), (11.5), (11.8) bis (11.10) sind, falls von anderen als den elektrischen Kräften (also z.B. von der Schwerkraft) abgesehen wird,

$$F_i = - e_i \operatorname{grad} \varphi. \tag{16.3}$$

Damit lassen sich die linearen phänomenologischen Ansätze (11.9) und (11.10) so schreiben

$$J_i = - \sum_{k=1}^{n} \frac{l_{ik}}{T} \operatorname{grad}_T \tilde{\mu}_k + l_i \operatorname{grad} \frac{1}{T}, \tag{16.4}$$

$$J_{\mathrm{red}} = - \sum_{k=1}^{n} \frac{l_k}{T} \operatorname{grad}_T \tilde{\mu}_k + l \operatorname{grad} \frac{1}{T}. \tag{16.5}$$

$\tilde{\mu}_k$ ist das sog. elektrochemische Potential des Bestandteils k; es ist definiert durch

$$\tilde{\mu}_k = \mu_k + e_k \varphi. \tag{16.6}$$

Bei den im folgenden behandelten Beispielen besteht mechanisches Gleichgewicht, $\frac{Dv}{Dt} = 0$, und es kann von der inneren Reibung abgesehen werden. Dann ist nach (11.4) und (11.11)

$$0 = \operatorname{grad} p + \sum_{i=1}^{n} \varrho_i F_i \tag{16.7}$$

und wegen (16.3) und der Elektroneutralität (16.2) folgt

$$\operatorname{grad} p = 0. \tag{16.8}$$

Für die Anwendungen, insbesondere für die Behandlung der reinen Kraftfelddiffusion und der stationären Zustände, ist es meist zweckmäßiger, chemisches und elektrisches Potential nicht zusammenzufassen. Die Gln. (16.4) und (16.5) lauten dann unter Berücksichtigung von (16.6) und unter Einführung der Überführungswärmen nach (9.18)

$$J_i = \sum_{k=1}^{n} \frac{l_{ik}}{T} \left(- e_k \operatorname{grad} \varphi - \operatorname{grad}_{T,p} \mu_k + T q_k \operatorname{grad} \frac{1}{T} \right), \tag{16.8}$$

$$J_{\mathrm{red}} = \sum_{i,k=1}^{n} \frac{l_{ik} q_k}{T} \left(- e_k \operatorname{grad} \varphi - \operatorname{grad}_{T,p} \mu_k \right) + l \operatorname{grad} \frac{1}{T}. \tag{16.9}$$

α) *Definition der Überführungszahlen.* Nimmt man einen homogen durchmischten und isothermen Elektrolyten an, so kann eine Ionendiffusion nur durch das Kraftfeld E hervorgerufen werden. Sie ist durch

$$J_i = - \sum_{k=1}^{n} \frac{l_{ik}}{T} e_k \operatorname{grad} \varphi \tag{16.10}$$

gegeben. Die Ionenart i gibt den Beitrag

$$e_i \boldsymbol{J}_i = - \sum_{k=1}^{n} \frac{l_{ik}}{T} e_i e_k \operatorname{grad} \varphi \qquad (16.11)$$

zum Gesamtstrom $\boldsymbol{I}$

$$\boldsymbol{I} = \sum_{i=1}^{n} e_i \boldsymbol{J}_i = - \sum_{i,k=1}^{n} \frac{l_{ik}}{T} e_i e_k \operatorname{grad} \varphi. \qquad (16.12)$$

Der relative Anteil der Ionenart i am Gesamtstrom unter den vorgenannten Bedingungen $T = \text{konstant}$, $\mu_k = \text{konstant}$ wird als *Überführungszahl* t_i bezeichnet. Er ist unabhängig von der Feldstärke, da

$$t_i \boldsymbol{I} = e_i \boldsymbol{J}_i = \frac{e_i \sum\limits_{k=1}^{n} l_{ik} e_k}{\sum\limits_{j,k=1}^{n} e_j l_{jk} e_k} \boldsymbol{I}, \qquad (16.13)$$

und es gilt für ihn per definitionem

$$\sum_{i=1}^{n} t_i = 1. \qquad (16.14)$$

β) Das Thermodiffusionspotential. Befindet sich der Elektrolyt in einer Zelle bei offenem äußeren Stromkreis, so wird bei beliebigen Anfangsbedingungen hinsichtlich Konzentrationen und Temperatur zunächst die Elektroneutralität hergestellt. Damit verschwindet auch der Gesamtstrom, $\boldsymbol{I} = \sum\limits_{i=1}^{n} e_i \boldsymbol{J}_i = 0$. Setzt man in (16.12) die Gln. (16.8) ein, so folgt wenn man noch die Definition (16.13) der Überführungszahlen heranzieht und nach $\operatorname{grad} \varphi$ auflöst,

$$\operatorname{grad} \varphi = - \sum_{k=1}^{n} \frac{t_k}{e_k} \operatorname{grad}_{T,p} \mu_k - \sum_{k=1}^{n} \frac{1}{T} \frac{t_k q_k}{e_k} \operatorname{grad} T. \qquad (16.15)$$

Hierbei ist von den O.C.R.B. $l_{ik} = l_{ki}$ Gebrauch gemacht.

Nach der Elektrodynamik ist im stromlosen Zustand die Summe aus elektrischer Feldstärke und eingeprägter elektrischer Feldstärke gleich Null. Der Ausdruck auf der rechten Seite von (16.15) ist daher eine eingeprägte elektrische Feldstärke

$$\boldsymbol{E}^e = - \sum_{k=1}^{n} \frac{t_k}{e_k} \operatorname{grad}_{T,p} \mu_k - \sum_{k=1}^{n} \frac{1}{T} \frac{t_k q_k}{e_k} \operatorname{grad} T. \qquad (16.16)$$

Sie ist durch die Gradienten der Temperatur und der Konzentration dargestellt.

In den Anwendungen werden anstelle der spezifischen Ladung und der spezifischen chemischen Potentiale und Überführungswärmen meist die auf das Mol bezogenen Größen e_k', μ_k', q_k' benutzt. Diese hängen mit den spezifischen Größen über die Beziehungen

$$\left. \begin{aligned} e_k' &= \mathsf{F} \cdot z_k = M_k e_k, \\ \mu_k' &= M_k \mu_k, \\ q_k' &= M_k q_k \end{aligned} \right\} \qquad (16.17)$$

zusammen. F ist die Faradaysche Konstante, z_k die elektrochemische Wertigkeit, M_k das Ionengewicht.

Erweitert man die einzelnen Summanden der rechten Seite von (16.15) mit den zugehörigen Ionengewichten M_k, beachtet (16.17) und multipliziert schließlich die ganze Gleichung mit der Faradayschen Konstanten F, so erhält man für (16.15) den äquivalenten Ausdruck

$$F \operatorname{grad} \varphi = - \sum_{k=1}^{n} \frac{t_k}{z_k} \operatorname{grad}_{T,p} \mu_k' - \sum_{k=1}^{n} \frac{1}{T} \frac{t_k q_k'}{z_k} \operatorname{grad} T. \tag{16.18}$$

Gl. (16.15) bzw. (16.18) gibt die Potentialverteilung $\varphi(r)$ im Zustand verschwindender Stromdichte bei vorgegebenen Konzentrations- und Temperaturfeldern.

Der erste Term der rechten Seite von (16.15) bzw. (16.18) ist nur von den Gradienten der Konzentrationen abhängig. Er ist das Diffusionspotential in einem Medium konstanter Temperatur. Die Theorie dieses Anteils zum Potential ist zuerst von PLANCK[1] und HENDERSON[2] gegeben worden. Der zweite Term hängt nur vom Temperaturgradienten ab und wird deshalb als Thermopotential bezeichnet.

PRIGOGINE[3], HAASE[4] und DE GROOT[5] haben die Thermodynamik der irreversiblen Prozesse auf diese Erscheinung angewandt. Die verschiedene Wahl von Flüssen und Kräften in den Arbeiten dieser Autoren hat zu verschiedenen Ausdrücken für das Thermodiffusionspotential geführt; ihre Äquivalenz wurde von HOLTAN[6] nachgewiesen.

Von HANSSEN[7] ist eine Modifikation der Gl. (16.15) für die Potentialverteilung im stromlosen Zustand angegeben worden. Sie kann als Spezialfall des folgenden allgemeineren Falles aufgefaßt werden: Läßt man in (16.3) als spezifische Kräfte nicht nur elektrische zu, gilt also

$$\boldsymbol{F}_i = - e_i \operatorname{grad} \varphi + \boldsymbol{\Psi}_i(r), \tag{16.19}$$

so hängt die Potentialverteilung im stromlosen Zustand nicht nur von den Gradienten der Konzentration und der Temperatur, sondern auch von den Zusatzkräften $\psi_i(r)$ ab. Diese modifizieren aber die Bedingung des mechanischen Gleichgewichtes. Somit wird die Potentialverteilung auch druckabhängig.

Den von HANSSEN betrachteten Spezialfall erhält man, wenn die Zusatzkräfte durch ein auf magnetische Momente wirkendes inhomogenes Magnetfeld $\boldsymbol{H}$

$$\boldsymbol{\Psi}_i = \operatorname{grad}(\boldsymbol{M}_i \cdot \boldsymbol{H}) \tag{16.20}$$

($\boldsymbol{M}_i =$ spezifisches magnetisches Moment) hervorgerufen werden. Die dabei auftretenden Effekte sind von HANSSEN in zwei Arbeiten experimentell studiert worden.

γ) *Der Soret-Effekt.* Das Verschwinden nicht nur der elektrischen Stromdichte, sondern auch aller einzelnen Diffusionsstromdichten $\boldsymbol{J}_i$ ist die Bedingung für den eingestellten Soret-Effekt. Da nur $n-1$ Diffusionsstromdichten unabhängig sind, so ergeben sich aus (16.8) die Bedingungen

$$\left.\begin{aligned}(e_k - e_n) \operatorname{grad} \varphi + \operatorname{grad}_{T,p}(\mu_k - \mu_n) + \frac{1}{T}(q_k - q_n) \operatorname{grad} T = 0 \\ (k = 1, 2, \ldots n-1).\end{aligned}\right\} \tag{16.21}$$

---

[1] M. PLANCK: Ann. Physik (3) **40**, 561 (1890). — Berl. Ber. **1927**, 285; **1929**, 9.

[2] P. HENDERSON: Z. phys. Chem. **59**, 118 (1907).

[3] I. PRIGOGINE: Etude thermodynamique des phénomènes irréversibles, S. 31 ff. Lüttich 1947.

[4] R. HAASE: Ergebn. exakt. Naturw. **26**, 56 (1952).

[5] S. R. DE GROOT: Thermodynamics of irreversible processes. Amsterdam 1951.

[6] H. HOLTAN jr.: Electric potentials in thermocouples and thermocells, S. 33. Diss. Utrecht 1953.

[7] K. J. HANSSEN: Z. Naturforsch. **9a**, 323, 919 (1954).

Zusammen mit der Elektroneutralität (16.2) hat man also n Gleichungen, mit deren Hilfe sich die Gradienten der Konzentrationen und von φ durch grad T ausdrücken lassen.

Der wichtigste Fall $n=3$ ist ausführlich von DE GROOT[1] behandelt. Er entspricht einem Lösungsmittel, in welchem ein Salz gelöst ist, das in zwei Ionenarten zerfällt. Der Fall $n=2$ mit nicht elektrisch neutralen Komponenten ist in diesem Zusammenhang trivial; denn die Elektroneutralität fordert $\gamma_1 e_1 + \gamma_2 e_2 = 0$, andererseits ist $\gamma_1 + \gamma_2 = 1$. Falls $e_1 \neq 0$, $e_2 \neq 0$, sind also γ_1 und γ_2 festgelegt; es wird daher grad $\gamma_1 = $ grad $\gamma_2 = 0$.

δ) *Die elektromotorische Kraft elektrolytischer Zellen.* Zum Abschluß behandeln wir die Gleichungen für die gesamte EMK einer elektrolytischen Zelle im stromlosen Zustand. Die Zelle bestehe aus einem elektrolytisch leitenden Medium,

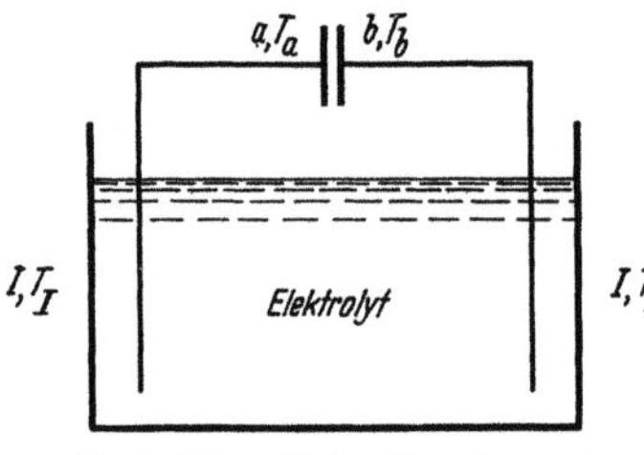

Fig. 2. Schematische Darstellung einer nicht-isothermen elektrolytischen Zelle.

das durch zwei gleiche metallische Elektroden abgeschlossen ist (Fig. 2). Die gesamte elektromotorische Kraft (die eingeprägte Spannung) setzt sich aus zwei homogenen und zwei heterogenen Anteilen zusammen. Ein Homogenanteil ist die Thermospannung $\Delta \varphi_t$ im metallischen Teil. Sie tritt auf, wenn die beiden Elektroden I und II verschiedene Temperaturen besitzen. Der zweite Homogenanteil ist die durch Konzentrations- und Temperaturunterschiede im Elektrolyten hervorgerufene Thermodiffusionsspannung $\Delta \varphi_{td}$. Hierzu addieren sich die beiden heterogenen Anteile $\Delta \varphi_h$, die Potentialsprünge an den Elektroden I und II. Man erhält also für die gesamte elektromotorische Kraft

$$V_{k,a}^2 = \Delta \varphi_t + \Delta \varphi_{td} + \Delta \varphi_h. \tag{16.22}$$

Zur Berechnung der homogenen Anteile setzen wir voraus, daß die Temperaturunterschiede $T - T_m$ im ganzen System klein sind gegenüber der mittleren Temperatur T_m und daß die Überführungszahlen und -wärmen wenig von der Temperatur abhängen. Dann liefert (16.18) für die Spannung zwischen der Elektrode I und a den Ausdruck

$$F \Delta \varphi_{\mathrm{I},a} = \frac{q'_{el}}{T_m} (T_\mathrm{I} - T_a) ;$$

in gleicher Weise ergibt sich die Spannung zwischen b und II

$$F \Delta \varphi_{b,\mathrm{II}} = \frac{q'_{el}}{T} (T_a - T_\mathrm{II}) .$$

Die gesamte Thermospannung erhält man durch Addition:

$$F \Delta \varphi_t = \frac{q'_{el}}{T_m} (T_\mathrm{I} - T_\mathrm{II}) . \tag{16.23}$$

Im Elektrolyten stellt sich nach (16.18) zwischen den Elektroden II und I die Spannung

$$F \Delta \varphi_{td} = - \sum_{k=1}^{n} \frac{t_k}{z_k} \Delta_{T,p} \mu'_k \Big|_\mathrm{I}^\mathrm{II} - \sum_{k=1}^{n} \frac{t_k q'_k}{T_m z_k} (T_\mathrm{II} - T_\mathrm{I}) \tag{16.24}$$

ein. Dabei ist $\Delta_{T,p} \mu_k|_\mathrm{I}^\mathrm{II}$ der durch die Konzentrationsunterschiede an den Elektroden II und I verursachte Unterschied im chemischen Potential des Bestandteils k.

---

[1] Siehe Fußnote 5, S. 461.

Zur Berechnung der heterogenen Anteile setzen wir voraus, daß an der Elektrodenreaktion nur der Bestandteil 1 teilnimmt und daß die potentialbestimmende Heterogenreaktion

$$1 = 1^{z_1^+} + z_1\,\mathrm{el}$$

ungehemmt ist (1 bezeichnet das Metall, el das Elektron). Unter dieser Voraussetzung stellt sich an beiden Elektroden elektrochemisches Gleichgewicht ein. Damit erhält man für die Potentialsprünge bei I und II die Ausdrücke

$$z_1 F\,\Delta\varphi_{h,\,\mathrm{I},\,\mathrm{II}} = (-\mu'_{z_1^+} - z_1\mu'_{\mathrm{el}} + \mu'_1)_{\mathrm{I},\,\mathrm{II}}\,.$$

Durch Addition erhält man den heterogenen Anteil

$$F\cdot\Delta\varphi_h = \frac{1}{z_1}\,\Delta\mu'_{z_1^+}\Big|_{\mathrm{I}}^{\mathrm{II}} + \Delta\mu'_{\mathrm{el}}\Big|_{\mathrm{I}}^{\mathrm{II}} - \frac{1}{z_1}\,\Delta\mu'_1\Big|_{\mathrm{I}}^{\mathrm{II}}, \tag{16.25}$$

wobei die Beziehungen

$$\Delta\mu'_{\mathrm{el}}\big|_{\mathrm{I}}^{\mathrm{II}} = -s'_{\mathrm{el}}(T_{\mathrm{II}} - T_{\mathrm{I}}), \qquad \Delta\mu'_1\big|_{\mathrm{I}}^{\mathrm{II}} = -s'(T_{\mathrm{II}} - T_{\mathrm{I}}) \tag{16.25 a}$$

bestehen. In $\Delta\mu_{z_1^+}$ hat man eine zusätzliche Abhängigkeit von der Zusammensetzung mitzuberücksichtigen. Damit wird

$$\Delta\mu'_{z_1^+}\big|_{\mathrm{I}}^{\mathrm{II}} = -s'_{z_1^+}(T_{\mathrm{II}} - T_{\mathrm{I}}) + \Delta_{T,\,p}\,\mu_{z_1^+}\big|_{\mathrm{I}}^{\mathrm{II}} \tag{16.25 b}$$

und

$$F\Delta\varphi_h = \frac{1}{z_1}\,\Delta_{T,\,p}\,\mu'_{z_1^+}\Big|_{\mathrm{I}}^{\mathrm{II}} - \frac{1}{z_1}\,(s'_{z_1^+} + z_1 s'_e - s'_1)(T_{\mathrm{II}} - T_{\mathrm{I}})\,. \tag{16.26}$$

Mit der Abkürzung

$$\Delta s' = s'_{z_1^+} + z_1 s'_{\mathrm{el}} - s'_1$$

ergibt sich durch Addition von (16.23), (16.24), (16.26) der Ausdruck für die gesamte elektromotorische Kraft der betrachteten elektrischen Zelle

$$\left.\begin{aligned} F V^e_{b,\,a} ={}& -\sum_{k=1}^{n} \frac{t_k - \delta_{1k}}{z_k}\,\Delta_{T,\,p}\,\mu_k\Big|_{\mathrm{I}}^{\mathrm{II}}\\ & -\left[\sum_{k=1}^{n}\frac{t_k q'_k}{T_m z_k} + \frac{q'_{\mathrm{el}}}{T_m} + \frac{\Delta s'}{z_1}\right](T_{\mathrm{II}} - T_{\mathrm{I}})\,. \end{aligned}\right\} \tag{16.27}$$

(16.27) ist die allgemeine Gleichung für die Spannung einer Thermodiffusionszelle, δ_{lk} ist das Kronecker-Symbol. Für reine Thermozellen nimmt sie die Form

$$F V^c_{b,\,a} = -\left[\sum_{k=1}^{n}\frac{t_k q'_k}{T_m z_k} + \frac{q'_{\mathrm{el}}}{T_m} + \frac{\Delta s'}{z_1}\right](T_{\mathrm{II}} - T_{\mathrm{I}}) \tag{16.28}$$

an, in der die Spezialfälle rein metallischer Thermozellen und rein elektrolytischer Thermozellen enthalten sind.

Diskussion der Gl. (16.28). Der Auswertung von Gl. (16.28) steht im Wege, daß die Überführungswärmen q'_k der elektrolytischen Ionen und der Elektronen sowie die Reaktionsentropie $\Delta s'$ im allgemeinen nicht bekannt sind. Insbesondere enthält $\Delta s'$ die Molentropien der Elektronen und der Ionen 1. Diese Schwierigkeit kann teilweise dadurch umgangen werden, daß man in Gl. (16.28) den Term

$$\left[\sum_{k=1}^{n}\frac{t_k}{z_k}\,s'_k + s'_{\mathrm{el}}\right](T_{\mathrm{II}} - T_{\mathrm{I}})$$

addiert und subtrahiert. Damit erhält man

$$
\begin{aligned}
F V_{b,a}^{e} &= -\left[\sum_{k=1}^{n} \frac{t_k}{z_k}\left(\frac{q_k'}{T_m} + s_k'\right) + \left(\frac{q_{el}'}{T_m} + s_{el}'\right) + \frac{\Delta s'}{z_1} - \sum_{k=1}^{n} \frac{t_k}{z_k} s_k' - s_{el}'\right](T_{II} - T_{I}) \\
&= -\left[\sum_{k=1}^{n} \frac{t_k}{z_k} s_k^{*\prime} + s_{el}^{*\prime} + \frac{\Delta s'}{z_1} - \sum_{k=1}^{n} \frac{t_k}{z_k} s_k' - s_{el}'\right](T_{II} - T_{I}),
\end{aligned}
\tag{16.29}
$$

wenn man durch $s^* = \dfrac{q'}{T} + s'$ die Überführungsentropien s^* einführt. Aus den drei letzten Termen fallen die Ionen- und Elektronenentropien heraus; sie enthalten nunmehr die Molentropie der metallischen Elektroden und die Salzentropien in der Elektrolytlösung. Dies sieht man leicht am Beispiel der Thermozelle

$$
\mathrm{Pb}\,|_{T_I}\,\mathrm{PbCl_2}\,|_{T_{II}}\,\mathrm{Pb}.
$$

(16.29) gibt für diesen Spezialfall, da $\mathrm{PbCl_2}$ ein reiner Anionenleiter ist,

$$
F V_{b,a}^{e} = \left[s_{Cl}^{*\prime} - s_{el}^{*\prime} - \frac{1}{2} s_{PbCl_2}' + \frac{1}{2} s_{Pb}'\right][T_{II} - T_{I}].
$$

Mit Gl. (16.29) hat Holtan[1] eine Reihe von Experimenten an Thermozellen diskutiert. Er ging dabei so vor, daß er die drei letzten Terme in Gl. (16.29) rechnerisch ermittelte und die Differenz

$$
F V_{b,a}^{e} + \left[\frac{\Delta s_1'}{z_1} - \sum_{k=1}^{n} \frac{t_k}{z_k} s_k' - s_{el}'\right](T_{II} - T_{I})
$$

zwischen den gemessenen Spannungen und dem thermostatischen Anteil bestimmte. Dabei ergab sich, daß der Term

$$
\left[\sum_{k=1}^{n} \frac{t_k}{z_k} s_k^{*\prime} + s_{el}^{*\prime}\right](T_{II} - T_{I})
$$

in (16.29) in vielen Fällen zu vernachlässigen war. Da die drei letzten Terme in Gl. (16.29) sich jedoch nicht allein auf die reversible Elektrodenspannung beziehen (man beachte das Auftreten der Überführungszahl), kann daraus nicht der Schluß gezogen werden, daß die irreversiblen Anteile zur Gesamtspannung zu vernachlässigen sind.

Haase[2] hat die allgemeinen Formeln nochmals zusammengestellt und — im Hinblick auf die Diskussion von Messungen an speziellen Elektrolytsystemen, insbesondere Thermozellen — die chemischen Potentiale der Komponenten durch Molenbrüche und Aktivitätskoeffizienten ausgedrückt.

17. Thermoelektrische, galvanomagnetische und thermomagnetische Effekte. α) *Die allgemeinen Gleichungen.* Die Theorie der elektrochemischen Effekte läßt sich unmittelbar auf die Elektrizitäts- und Wärmeleitung in Metallen übertragen. Gegenüber den flüssigen elektrolytischen Lösungen treten leichte Verallgemeinerungen ein. An die Stelle des hydrostatischen Drucks tritt der Spannungstensor und in den phänomenologischen Gleichungen ist die Anisotropie der Metalle zu berücksichtigen. Im folgenden sollen auch äußere Magnetfelder zugelassen sein, so daß die O.C.R.B. in ihrer allgemeineren Form (6.9) zugrundezulegen sind.

---

[1] H. Holtan jr.: Electric Potentials in Thermocouples and Thermocells. Thesis, Utrecht 1953.

[2] R. Haase: Z. phys. Chem., N.F. **11**, 379 (1957); **13**, 21 (1958); **14**, 292 (1958).

Wie in der vorhergehenden Ziffer betrachten wir „ruhende Leiter" im mechanischen Gleichgewicht und im elektrisch neutralen Zustand. Damit bleiben von den in Ziff. 11 angegebenen Gleichungen als wesentlich nur die Energiegleichung (11.5), bzw. an ihrer Stelle die Temperaturgleichung (11.24) in der spezialisierten Gestalt

$$\varrho\, c_p\, \frac{\partial T}{\partial t} + \operatorname{div}\left(\boldsymbol{J}_{\mathrm{red}} + \sum_{i=1}^{n} \varrho_i\, h_i\, \boldsymbol{v}_i\right) = \sum_{i=1}^{n} [\boldsymbol{J}_i \cdot \boldsymbol{F}_i + h_i\, \operatorname{div} \varrho_i\, \boldsymbol{v}_i] \qquad (17.1)$$

und die phänomenologischen Gleichungen

$$\boldsymbol{J}_i = \sum_{k=1}^{n} \frac{\mathsf{l}_{ik}}{T}\,(\boldsymbol{F}_k - \operatorname{grad}_T \mu_k) + \mathsf{l}_i\, \operatorname{grad} \frac{1}{T}\,, \qquad (17.2)$$

$$\boldsymbol{J}_{\mathrm{red}} = \sum_{k=1}^{n} \frac{\mathsf{l}_k'}{T}\,(\boldsymbol{F}_k - \operatorname{grad}_T \mu_k) + \mathsf{l}\, \operatorname{grad} \frac{1}{T}\,, \qquad (17.3)$$

wobei die spezifischen Kräfte durch das elektrische Feld

$$\boldsymbol{F}_k = e_k\, \boldsymbol{E} \qquad (17.4)$$

bestimmt sind. Mit $e_1, e_2, \ldots$ sind wieder die spezifischen Ladungen bezeichnet.

Wegen der vorausgesetzten Anisotropie des Materials ist jeder phänomenologische Koeffizient $\mathsf{l}_{ik}, \mathsf{l}_i, \mathsf{l}_k', \mathsf{l}$ selbst ein Tensor mit neun Komponenten. Die O.C.R.B. machen über diese Tensoren die Aussagen

$$\mathsf{l}_{ik}(\boldsymbol{B}) = \tilde{\mathsf{l}}_{ki}(-\boldsymbol{B}), \qquad \mathsf{l}_i'(\boldsymbol{B}) = \tilde{\mathsf{l}}_i(-\boldsymbol{B}), \qquad \mathsf{l}(\boldsymbol{B}) = \tilde{\mathsf{l}}(-\boldsymbol{B})\,. \qquad (17.5)$$

Hierin ist $\boldsymbol{B}$ die magnetische Induktion des äußeren Magnetfeldes und die Tilde bedeutet die Bildung des adjungierten, d.h. an der Hauptdiagonale gespiegelten Tensors.

$\beta)$ *Spezialisierung der Temperaturgleichung.* Die thermoelektrischen, galvanomagnetischen und thermomagnetischen Effekte in Metallen kommen durch Transport von Elektronen im elektrischen, magnetischen und thermischen Feld zustande. Bei diesen Effekten spielt die Diffusion der anderen Bestandteile gegeneinander eine untergeordnete Rolle. Wir können daher ihre Geschwindigkeiten bei ruhendem Leiter alle gleich Null setzen. Bezeichnet man die Komponente der Leitungselektronen mit dem Index 1, so gilt also

$$\boldsymbol{v}_2 = \boldsymbol{v}_3 = \boldsymbol{v}_4 = \cdots = \boldsymbol{v}_n = 0\,, \qquad \varrho\, \boldsymbol{v} = \varrho_1\, \boldsymbol{v}_1 \qquad (17.6)$$

und, wegen (17.6) und (3.8),

$$\boldsymbol{J}_1 = (\varrho - \varrho_1)\, \boldsymbol{v}\,, \qquad \boldsymbol{J}_i = -\varrho_i\, \boldsymbol{v} \quad (i = 2, 3, \ldots n)\,. \qquad (17.7)$$

Wegen der vorausgesetzten Elektroneutralität ist

$$\sum_{i=1}^{n} \varrho_i\, e_i = 0 \qquad (17.8)$$

und damit

$$\operatorname{div} \boldsymbol{I} = 0\,. \qquad (17.9)$$

Mit (17.6) ergibt sich schließlich für die elektrische Stromdichte

$$\boldsymbol{I} = \sum_{i=1}^{n} e_i\, \varrho_i\, \boldsymbol{v}_i = e_1\, \varrho_1\, \boldsymbol{v}_1 = e_1\, \varrho\, \boldsymbol{v} \qquad (17.10)$$

und weiter mit (17.7)

$$I = \frac{e_1 \varrho}{\varrho - \varrho_1} \boldsymbol{J}_1 = - \frac{e_1 \varrho}{\varrho_i} \boldsymbol{J}_i .$$ (17.11)

Wegen (17.6) bis (17.11) vereinfacht sich dann (17.1) zu

$$\varrho \, c_p \frac{\partial T}{\partial t} + \operatorname{div} \boldsymbol{J}' = \boldsymbol{E} \cdot \boldsymbol{I}$$ (17.12)

mit

$$\boldsymbol{J}' = \boldsymbol{J}_{\mathrm{red}} + \frac{h_1}{e_1} \boldsymbol{I} = \boldsymbol{J}_u + \frac{h}{e_1} \boldsymbol{I} .$$ (17.13)

Eine weitere Verkürzung dieser Gleichung ergibt sich, wenn man in den Energiestrom den Strom der potentiellen Energie $\varphi \boldsymbol{I}$, oder den Poyntingschen Vektor $\boldsymbol{E} \times \boldsymbol{H}$ aufnimmt, die beide, langsam veränderliche Felder vorausgesetzt, die gleiche Divergenz $- \boldsymbol{E} \cdot \boldsymbol{I}$ haben. Dann entsteht

$$\varrho \, c_p \frac{\partial T}{\partial t} + \operatorname{div} \boldsymbol{J}'' = 0$$ (17.14)

mit

$$\boldsymbol{J}'' = \boldsymbol{J}_{\mathrm{red}} + \left(\varphi + \frac{h_1}{e_1} \right) \boldsymbol{I} = \boldsymbol{J}_u + \left(\varphi + \frac{h}{e_1} \right) \boldsymbol{I} .$$ (17.15)

γ) *Reduktion der phänomenologischen Ansätze.* Wenn alle Gitterbestandteile, wie oben vorausgesetzt, gleiche Geschwindigkeit haben sollen, dann liegt eine Hemmung gegen Diffusion vor. Sie bedingt zusätzliche Beziehungen zwischen den phänomenologischen Koeffizienten. Verlangt man also, gemäß (17.7), daß identisch $\boldsymbol{J}_2/\varrho_2 = \boldsymbol{J}_3/\varrho_3 = \cdots = \boldsymbol{J}_n/\varrho_n$, berücksichtigt (4.12) und die O.C.R.B. (17.5), so lassen sich alle phänomenologischen Koeffizienten auf l_{11}, l_1, l_1' und l zurückführen, und es ergibt sich

$$\left. \begin{aligned}
\boldsymbol{J}_1 &= \mathsf{l}_{11} \left[\boldsymbol{X}_1 + \frac{1}{\varrho_1 - \varrho} \sum_{k=2}^{n} \varrho_k \boldsymbol{X}_k \right] + \mathsf{l}_1 \operatorname{grad} \frac{1}{T} , \\
\boldsymbol{J}_i &= \frac{\mathsf{l}_{11} \varrho_i}{\varrho_1 - \varrho_2} \left[\boldsymbol{X}_1 + \frac{1}{\varrho_1 - \varrho} \sum_{k=2}^{n} \varrho_k \boldsymbol{X}_k \right] + \frac{\mathsf{l}_1 \varrho_i}{\varrho_1 - \varrho} \operatorname{grad} \frac{1}{T} , \quad (i = 2, 3, \ldots n) \\
\boldsymbol{J}_{\mathrm{red}} &= \mathsf{l}_1' \left[\boldsymbol{X}_1 + \frac{1}{\varrho_1 - \varrho} \sum_{k=2}^{n} \varrho_k \boldsymbol{X}_k \right] + \mathsf{l} \operatorname{grad} \frac{1}{T} .
\end{aligned} \right\}$$ (17.16)

Dabei ist zur Abkürzung

$$\boldsymbol{X}_k = \boldsymbol{F}_k - \operatorname{grad}_T \mu_k \quad (k = 1, 2, \ldots n)$$ (17.17)

gesetzt. Beachtet man noch die Bedingung (13.5) des mechanischen Gleichgewichts

$$\sum_{k=1}^{n} \varrho_k \boldsymbol{X}_k = 0 ,$$ (17.18)

schreibt von $\boldsymbol{J}_1$ mit Hilfe von (17.11) auf die elektrische Stromdichte $\boldsymbol{I}$ um, und setzt zur Abkürzung

$$\mathsf{S}_1 = \frac{e_1^2 \varrho^2}{(\varrho - \varrho_1)^2} \frac{\mathsf{l}_{11}}{T} , \quad \mathsf{S}_2 = \frac{e_1 \varrho \, \mathsf{l}_1}{(\varrho - \varrho_1) \, T} , \quad \mathsf{S}_3 = \frac{e_1 \varrho \, \mathsf{l}_1'}{(\varrho - \varrho_1) \, T} , \quad \mathsf{S}_4 = \frac{1}{T} , \quad (17.19)$$

so folgt

$$\boldsymbol{I} = \mathsf{S}_1 \left(\boldsymbol{E} - \operatorname{grad}_T \frac{\mu_1}{e_1} \right) + \mathsf{S}_2 \, T \operatorname{grad} \frac{1}{T} ,$$ (17.20)

$$\boldsymbol{J}_{\mathrm{red}} = \mathsf{S}_3 \left(\boldsymbol{E} - \operatorname{grad}_T \frac{\mu_1}{e_1} \right) + \mathsf{S}_4 \, T \operatorname{grad} \frac{1}{T} .$$ (17.21)

Die in (17.19) definierten Tensoren S_1 bis S_4 haben wegen (17.5) die Eigenschaften

$$\mathsf{S}_1(\boldsymbol{B}) = \tilde{\mathsf{S}}_1(-\boldsymbol{B}), \qquad \mathsf{S}_2(\boldsymbol{B}) = \tilde{\mathsf{S}}_3(-\boldsymbol{B}), \qquad \mathsf{S}_4(\boldsymbol{B}) = \tilde{\mathsf{S}}_4(-\boldsymbol{B}). \qquad (17.22)$$

In den Gln. (17.12), (17.13), (17.20), (17.21), (17.22) ist die Theorie der thermoelektrischen, galvanomagnetischen und thermomagnetischen Effekte enthalten. Sie haben, von linearen Transformationen der Kräfte und Stromdichten abgesehen, die Form, die der Behandlung dieser Effekte üblicherweise zugrundegelegt wird (vgl. die am Ende dieser Ziffer angegebene Literatur). Diese Gleichungen folgen auch direkt aus der Elektronentheorie der Metalle (vgl. MEIXNER[1]).

δ) *Lineare Transformation von Kräften und Stromdichten.* Da in den Temperaturgleichungen (17.12) und (17.14) die Energiestromdichten $\boldsymbol{J}'$ bzw. $\boldsymbol{J}''$ auftreten, geben wir die phänomenologischen Gleichungen auch für diese Sätze von Flüssen und zugeordneten Kräften an. Dazu braucht man zuerst die Entropieproduktionsdichte, für welche man durch Umrechnung auf $\boldsymbol{I}$ und $\boldsymbol{J}_{\text{red}}$ die Gleichung

$$\sigma = \boldsymbol{I} \cdot \frac{1}{T}\left(\boldsymbol{E} - \operatorname{grad}_T \frac{\mu_1}{e_1}\right) + \boldsymbol{J}_{\text{red}} \cdot \operatorname{grad} \frac{1}{T} \qquad (17.23)$$

findet. Sie lautet mit $\boldsymbol{J}'$ bzw. $\boldsymbol{J}''$ an Stelle von $\boldsymbol{J}_{\text{red}}$

$$\sigma = \boldsymbol{I} \frac{1}{T}\left(\boldsymbol{E} - T \operatorname{grad} \frac{\mu_1}{e_1 T}\right) + \boldsymbol{J}' \cdot \operatorname{grad} \frac{1}{T}, \qquad (17.24)$$

bzw.

$$\sigma = -\boldsymbol{I} \operatorname{grad} \frac{e_1 \varphi + \mu_1}{e_1 T} + \boldsymbol{J}'' \operatorname{grad} \frac{1}{T}. \qquad (17.25)$$

Dementsprechend lauten die phänomenologischen Gleichungen

$$\boldsymbol{I} = \mathsf{S}_1'\left(\boldsymbol{E} - T \operatorname{grad} \frac{\mu_1}{e_1 T}\right) + \mathsf{S}_2' \, T \operatorname{grad} \frac{1}{T}, \qquad (17.26)$$

$$\boldsymbol{J}'' = \mathsf{S}_3'\left(\boldsymbol{E} - T \operatorname{grad} \frac{\mu_1}{e_1 T}\right) + \mathsf{S}_4' \, T \operatorname{grad} \frac{1}{T}, \qquad (17.27)$$

oder auch

$$\boldsymbol{I} = -\mathsf{S}_1'' \, T \operatorname{grad} \frac{e_1 \varphi + \mu_1}{e_1 T} + \mathsf{S}_2'' \, T \operatorname{grad} \frac{1}{T}, \qquad (17.28)$$

$$\boldsymbol{J}'' = -\mathsf{S}_3'' \, T \operatorname{grad} \frac{e_1 \varphi + \mu_1}{e_1 T} + \mathsf{S}_4'' \, T \operatorname{grad} \frac{1}{T}. \qquad (17.29)$$

Für die Tensoren S_i' und S_i'' ($i = 1, 2, 3, 4$), welche sich übrigens leicht durch die S_i aus (17.20) und (17.21) ausdrücken lassen, gelten ebenfalls die Beziehungen (17.22).

Eine weitere, gelegentlich nützliche Schreibweise der phänomenologischen Gleichungen benützt die elektrische Stromdichte und die gesamte Entropiestromdichte einschließlich des konvektiven Anteils [vgl. (5.7)]

$$\boldsymbol{J}_s' = \frac{1}{T}\left(\boldsymbol{J}_{\text{red}} + \sum_{i=1}^{n} (h_i - \mu_i)\,\boldsymbol{J}_i\right) = \frac{1}{T}\,\boldsymbol{J}_{\text{red}} - \frac{1}{c}\left(\frac{\partial \mu_1}{\partial T}\right)_{p,\gamma_i} \boldsymbol{I} = \frac{1}{T}\left(\boldsymbol{J}' - \frac{\mu_1}{e_1}\,\boldsymbol{I}\right). \qquad (17.30)$$

Damit entsteht

$$\left.\begin{aligned}
\boldsymbol{I} &= \mathsf{S}_1'''\left(\boldsymbol{E} - \operatorname{grad} \frac{\mu_1}{e_1}\right) + \mathsf{S}_3''' \, T \operatorname{grad} \frac{1}{T}, \\[2mm]
\boldsymbol{J}_s' &= \mathsf{S}_3'''\left(\boldsymbol{E} - \operatorname{grad} \frac{\mu_1}{e_1}\right) + \mathsf{S}_4''' \, T \operatorname{grad} \frac{1}{T}.
\end{aligned}\right\} \qquad (17.31)$$

---

[1] J. MEIXNER: Ann. Physik [5] **35**, 701 (1939); [5] **40**, 165 (1941).

ε) *Die thermischen Effekte.* Die thermischen Effekte werden durch (17.12) oder (17.14) beschrieben. Im Hinblick auf die experimentellen Verhältnisse drücken wir den Fluß $\boldsymbol{J}'$ bzw. $\boldsymbol{J}''$ durch die elektrische Stromdichte $\boldsymbol{I}$ und den Temperaturgradienten aus. Tatsächlich ist es einfacher von (17.12) und (17.26), (17.27) auszugehen statt von (17.14) und (17.28), (17.29). Die phänomenologischen Koeffizienten S_i' und S_i'' hängen nämlich vom elektrischen Potential φ ab. Diese Abhängigkeit kann jedoch in (17.26) und (17.27) in erster Näherung vernachlässigt werden. In (17.28) und (17.29) ist sie hingegen sehr wohl zu berücksichtigen (man vergleiche dazu etwa die Behandlung bei Domenicali[1]).

Zuvor führen wir im Anschluß an die üblichen Bezeichnungen in der Thermoelektrizität folgende Größen ein

$$\rho = (S_1')^{-1}, \qquad \varepsilon = \frac{\mu_1}{e_1 T} I - \frac{1}{T}\, \rho \cdot S_2', \left.\begin{aligned}\\ \\ \end{aligned}\right\}$$
$$\pi = \frac{\mu_1}{e_1} I - \tilde{\rho}_1 \cdot \tilde{S}_3', \qquad T\lambda = S_4' - S_3' \cdot \rho \cdot S_2'. \qquad (17.32)$$

Damit gehen die phänomenologischen Gleichungen (17.26) und (17.27) über in

$$\begin{aligned} \boldsymbol{E} &= \rho\, \boldsymbol{I} + \operatorname{grad}\frac{\mu_1}{e_1} - \varepsilon \cdot \operatorname{grad} T, \\ \boldsymbol{J}' &= \left(\frac{\mu_1}{e_1} - \tilde{\pi}\right)\cdot \boldsymbol{I} - \lambda \cdot \operatorname{grad} T. \end{aligned} \left.\begin{aligned}\\ \\ \\ \end{aligned}\right\} \qquad (17.33)$$

Einsetzen in die Temperaturgleichung (17.12) gibt

$$\varrho\, c_p \frac{\partial T}{\partial t} = \tilde{\boldsymbol{I}}\rho\, \boldsymbol{I} + \tilde{\boldsymbol{I}}\mu \operatorname{grad} T + \operatorname{div}_T \tilde{\pi} \cdot \boldsymbol{I} + \operatorname{div}(\lambda \operatorname{grad} T) \qquad (17.34)$$

mit

$$\mu = \frac{\partial \pi}{\partial T} - \varepsilon. \qquad (17.35)$$

Hierin sind alle bekannten thermischen Effekte enthalten. Der erste Term auf der rechten Seite von (17.34) gibt die Joulesche Wärmeleistung in der Volumeneinheit bei der Stromdichte $\boldsymbol{I}$, wobei ρ als Tensor des spezifischen Widerstands zu deuten ist. Der letzte Term trägt ersichtlich der Wärmeleitung Rechnung; bei verschwindender Stromdichte $\boldsymbol{I}=0$ entsteht ja die bekannte Wärmeleitungsgleichung. λ ist deshalb der Wärmeleitfähigkeitstensor. Der zweite Term gibt die Thomson-Wärme; sie ist proportional zur Stromdichte und zum Temperaturgradienten. μ ist der Thomson-Tensor, im isotropen Fall der Thomson-Koeffizient. Der dritte Term schließlich enthält drei Effekte, die jedoch durch denselben Tensor π (Peltier-Tensor, im isotropen Fall Peltier-Koeffizient) beschrieben werden. Man schreibt dazu diesen Term

$$\operatorname{div}_T \tilde{\pi} \cdot \boldsymbol{I} = \sum_{\alpha,\,\beta=1}^{3} \left[\left(\frac{\partial \pi_{\beta\alpha}}{\partial x_\alpha}\right)_{T,B} I_\beta + \left(\frac{\partial \pi_{\beta\alpha}}{\partial B_\gamma}\right)_{T,M} \frac{\partial B_\gamma}{\partial x_\alpha} I_\beta + \pi_{\beta\alpha}\frac{\partial I_\beta}{\partial x_\alpha}\right]. \qquad (17.36)$$

Der erste Anteil tritt dort auf, wo sich das Material ändert, ist also besonders stark in Übergangsschichten ausgeprägt; dort spricht man dann von Peltier-Wärme. Der zweite Anteil gibt Wärmeeffekte an Stellen veränderlichen Magnetfeldes; er kann ebenfalls als Peltier-Wärme bezeichnet werden, wenn man das gleiche, aber verschieden magnetisierte Material als verschiedene Materialien ansieht. Der letzte Anteil ist die Bridgman-Wärme. Sie tritt nur im anisotropen Material auf, wobei ein isotropes Material im Magnetfeld ebenfalls anisotrop genannt wird.

---

[1] C.A. Domenicali: Rev. Mod. Phys. **26**, 237 (1954).

Die Onsagerschen Reziprozitätsbeziehungen (17.22) lauten, auf die neuen Größen umgerechnet

$$\varrho_{\alpha\beta}(\boldsymbol{B}) = \varrho_{\beta\alpha}(-\boldsymbol{B}), \quad T\,\varepsilon_{\alpha\beta}(\boldsymbol{B}) = \pi_{\alpha\beta}(-\boldsymbol{B}), \quad \lambda_{\alpha\beta}(\boldsymbol{B}) = \lambda_{\beta\alpha}(-\boldsymbol{B}). \qquad (17.37)$$

Damit folgt aus (17.35)

$$\mu(\boldsymbol{B}) \equiv T\,\frac{\partial\varepsilon(\boldsymbol{B})}{\partial T}. \qquad (17.38)$$

ζ) *Diskussion der phänomenologischen Gleichungen.* Für die weitere Diskussion werden die phänomenologischen Gleichungen zweckmäßig in der Gestalt (17.33) zugrundegelegt. Die Koeffizienten sind Funktionen der Temperatur, des Magnetfeldes und des Materials. Ihre Druckabhängigkeit kann unberücksichtigt bleiben, da ohnehin vorausgesetzt ist, daß das betrachtete System isobar ist.

Ist $\boldsymbol{B}=0$ und sind die Materialien des Systems isotrop, so vereinfachen sich die Gleichungen zu

$$\boldsymbol{E} = \varrho\,\boldsymbol{I} + \frac{1}{e_1}\,\operatorname{grad}\mu_1 - \varepsilon\,\operatorname{grad} T, \qquad (17.39)$$

$$\boldsymbol{J}' = \frac{\mu_1}{e_1}\,\boldsymbol{I} - \pi\,\boldsymbol{I} - \lambda\,\operatorname{grad} T. \qquad (17.40)$$

Die Koeffizienten sind nun skalar. Ihre Bedeutung ist leicht erkennbar. ϱ ist der spezifische Widerstand, λ die Wärmeleitfähigkeit. Der Peltier-Koeffizient π gibt im wesentlichen den Wärmetransport, der mit der Einheit der Stromdichte im isothermen Material verknüpft ist. Gl. (17.39) stellt das Ohmsche Gesetz dar mit der eingeprägten elektrischen Feldstärke

$$\boldsymbol{E}^e = -\frac{1}{e_1}\,\operatorname{grad}\mu_1 + \varepsilon\,\operatorname{grad} T. \qquad (17.41)$$

Die eingeprägte elektromotorische Kraft in einem geschlossenen Stromkreis ergibt sich hieraus zu

$$\oint \boldsymbol{E}^e \cdot d\boldsymbol{r} = \oint \varepsilon\,\operatorname{grad} T \cdot d\boldsymbol{r} = \oint \varepsilon\,dT.$$

Besteht der Stromkreis aus zwei Leitern A und B und sind die beiden Kontaktstellen auf den Temperaturen T_1 und T_2, so läßt sich das letzte Integral auch in folgender Gestalt schreiben

$$\oint \boldsymbol{E}^e \cdot d\boldsymbol{r} = \int\limits_{T_\mathrm{I}}^{T_\mathrm{II}} \varepsilon_A(T)\,dT + \int\limits_{T_\mathrm{II}}^{T_\mathrm{I}} \varepsilon_B(T)\,dT = \int\limits_{T_\mathrm{I}}^{T_\mathrm{II}} [\varepsilon_A(T) - \varepsilon_B(T)]\,dT.$$

Hierin drückt sich zunächst das Gesetz der thermoelektrischen Spannungsreihe aus, aber auch die Tatsache, daß die eingeprägte thermoelektrische Kraft nur von den Temperaturen T_1 und T_2 der beiden Kontaktstellen abhängt. Es kommt also nicht auf den Temperaturverlauf im homogenen Material an. Diese Tatsache wird häufig dahingehend formuliert, daß der Sitz der thermoelektrischen Kraft nur in den Kontaktstellen der Leiter ist.

In anisotropen Materialien sind die Verhältnisse komplizierter. Es treten auch Quereffekte verschiedener Art auf. In einem stromdurchflossenen Einkristallstab, der nach der x_1-Richtung orientiert ist, können sich transversale Feldstärkenkomponenten und transversale Temperaturgefälle ausbilden. Sie haben ihren Ursprung teils in der Anisotropie des Widerstandstensors, teils in der Anisotropie der absoluten Thermokraft und hängen im einzelnen noch von den Randbedingungen an der Oberfläche des Einkristalls ab. Typische Grenzfälle sind die isotherme und die adiabatisch isolierte Oberfläche. Die Anwesenheit eines Magnetfeldes bringt keine wesentliche zusätzliche Komplikation außer der, daß die Anisotropieverhältnisse noch von der Orientierung des Magnetfeldes abhängen.

Die zahlreichen Effekte welche im isotropen oder gar im anisotropen Material ohne und mit Magnetfeld auftreten, sind von Meissner[1] ausführlich diskutiert worden. Weiter führende Behandlungen finden sich bei Kohler[2] und bei Meixner[3]. Ferner sei auf die Arbeiten von Callen[4] (Systematik der Effekte im isotropen Körper ohne Magnetfeld), Mazur und de Groot[5], Mazur und Prigogine[6], Jan[7], Poppelbaum[8] (magnetogalvanische und magnetothermische Transversaleffekte), Fieschi[9] und auf die schon erwähnte neuere zusammenfassende Darstellung von Domenicali[10] hingewiesen.

D. Eigenschaften von Medien mit inneren Variablen.

18. Mehrkomponentensysteme mit sehr schnellen Reaktionen. Die in Ziff. 11 zusammengestellten Grundgleichungen der Thermodynamik irreversibler Prozesse lassen sich auf eine geringere Zahl reduzieren, wenn einige Reaktionen so schnell ablaufen, daß die entsprechenden Affinitäten an jeder Stelle praktisch verschwinden. Diese Reduktion führen wir an einem einfachen Beispiel durch. Wir nehmen an, daß *alle* Komponenten des fluiden Mediums durch chemische Reaktionen ineinander umwandelbar sind, daß alle Reaktionen sehr schnell verlaufen, und daß die spezifischen Kräfte auf die Komponenten einander gleich sind. Damit sind geladene Komponenten im elektrischen Feld ausgeschlossen. Dann sind (an jeder Stelle und zu jeder Zeit) alle chemischen Potentiale μ_i einander stets gleich und alle Affinitäten A_k verschwinden. Ein solches System verhält sich dann wie ein Einkomponentensystem.

Zum Beweis sind die Grundgleichungen in Ziff. 11 entsprechend zu spezialisieren. Die Bilanzen der Gesamtdichte (11.1), des Impulses (11.4) und der inneren Energie (11.5) sind bereits identisch mit den Bilanzen eines Einkomponentensystems. Gl. (11.11) gibt den Tensor der Reibungsdrucke in derselben Gestalt wie für ein Einkomponentensystem. Die Bilanzen der Massenbrüche (11.3) zusammen mit (11.17) lauten

$$\varrho\,\frac{D\gamma_i}{Dt} + \operatorname{div} \boldsymbol{J}_i = -\frac{\lambda_i}{T}\operatorname{div}\boldsymbol{v} - \sum_{k=1}^{r} \lambda_{ik}\frac{A_k}{T} \qquad (i = 1, 2, \dots r). \qquad (18.1)$$

Sie geben nun keine zusätzliche Aussage sondern präzisieren nur den Grenzwert des letzten Terms in (18.1) für $A_k \to 0$ und Reaktionsgeschwindigkeiten $\to \infty$. Es bleiben also nur noch die phänomenologischen Gleichungen für Diffusions- und Energiestromdichte zu behandeln. Da alle μ_k/T und daher auch ihre Gradienten im chemischen Gleichgewicht einander gleich sind, benützt man am besten den Satz von Energie- und Diffusionsstromdichten (7.1) und (7.2). Auf Grund der Koeffizientenbeziehungen (7.4) folgt dann

$$\boldsymbol{J}_i = L_{iu}\operatorname{grad}\frac{1}{T}, \qquad \boldsymbol{J}_u = L_{uu}\operatorname{grad}\frac{1}{T}. \qquad (18.2)$$

[1] W. Meissner: Handbuch der Experimentalphysik, XI, 2. Leipzig 1935.
[2] M. Kohler: Ann. Physik 27, 201 (1936); 40, 196 (1941).
[3] J. Meixner: Ann. Physik 35, 701 (1939); 40, 165 (1941).
[4] H. B. Callen: Phys. Rev. 85, 16 (1952).
[5] P. Mazur u. S. R. de Groot: Physica, Haag 19, 961 (1953).
[6] P. Mazur u. I. Prigogine: J. Phys. Radium 12, 616 (1951)
[7] J. P. Jan: Helv. phys. Acta 25, 677 (1952); 26, 281 (1953).
[8] W. Poppelbaum: Comparaison directe de l'effet Hall et de l'effet Corbino. Thèse Lausanne 1954.
[9] R. Fieschi: Nuovo Cim. Suppl. (X), 1, 1 (1955).
[10] C. A. Domenicali: Rev. Mod. Phys. 26, 237 (1954).

Die Diffusionsstromdichten sind hier von geringem Interesse, da sie wieder nur in
(18.1) eingehen. Damit ist also gezeigt, daß man für $A_k = 0$ tatsächlich aus
(11.1), (11.4) bis (11.6) (mit $\lambda_i = 0$) und der zweiten Gl. (18.2) die Grundgleichungen
der Thermodynamik irreversibler Prozesse für ein Einkomponentensystem erhält.
Insbesondere zeigt sich, daß L_{uu}/T^2 die Wärmeleitfähigkeit dieses Einkomponentensystems ist. Sie läßt sich durch die phänomenologischen Koeffizienten
der realen, im Umwandlungsgleichgewicht vorhandenen Komponenten ausdrücken.
Es gelten hierfür die Umrechnungsformeln (9.6) mit $C_{uu} = 1$, $C_{iu} = -h_i$ (s.
Ziff. 9β). Somit wird

$$L_{uu} = l + 2 \sum_{i=1}^{n} l_i h_i + \sum_{i,k=1}^{n} l_{ik} h_i h_k. \tag{18.3}$$

Um diese Beziehung zu erläutern, wollen wir die Wärmeleitfähigkeit, welche
dieselbe Mischung für den Fall völlig gehemmter Reaktion hat, berechnen.
Nach (11.3) ist für diesen Fall

$$\varrho \frac{D\gamma_i}{Dt} + \operatorname{div} \boldsymbol{J}_i = 0 \tag{18.4}$$

und nach (11.23) gilt bei gleichem und zeitlich konstantem Druck

$$\varrho \, c_p \frac{\partial T}{\partial t} + \operatorname{div} \boldsymbol{J}_{\mathrm{red}} = - \sum_{i=1}^{n} \boldsymbol{J}_i \cdot \operatorname{grad} h_i. \tag{18.5}$$

Für kleine Konzentrations- und Temperaturgradienten ist die rechte Seite von
zweiter Ordnung klein. Sie kann deshalb vernachlässigt werden. Gl. (18.5) ist
also eine Wärmeleitungsgleichung mit der „Wärmestromdichte" $\boldsymbol{J}_{\mathrm{red}}$. Bei örtlich konstanten Konzentrationen ist die phänomenologische Gleichung (11.10)
ein Fourierscher Wärmeleitungsansatz mit der Wärmeleitfähigkeit l/T^2.

Die Beziehung (18.3) erlaubt demnach die Wärmeleitfähigkeit einer Mischung,
zwischen deren Komponenten sehr schnelle Reaktionen ablaufen, zu berechnen,
wenn man die Transporteigenschaften der Mischung, d.h. die Koeffizienten
l, l_i, l_{ik} bei völlig gehemmter Reaktion kennt. Zwei Beispiele hierzu werden in
Ziff. 19 behandelt.

19. Die Euckensche und die Nernstsche Gleichung. α) *Die Wärmeleitfähigkeit
der idealen Gase.* Die kinetische Gastheorie liefert für einatomige ideale Gase
einen Zusammenhang zwischen der Wärmeleitfähigkeit λ, der Schubviskosität η
und der spezifischen Wärme c_p. Er lautet

$$\lambda = f \cdot c_p \eta. \tag{19.1}$$

f liegt unabhängig vom angenommenen Wechselwirkungspotential zwischen den
Molekülen sehr nahe an $\frac{3}{2}$. Eucken[1] hat in qualitativer Übereinstimmung mit
der Erfahrung vorgeschlagen, die Wärmeleitfähigkeit mehratomiger Moleküle
durch drei Terme darzustellen, die den Transport von Translations-, Rotations-
und Schwingungsenergie der Moleküle beschreiben sollen:

$$\lambda = (\tfrac{3}{2} c_{p,t} + c_r + c_s) \eta. \tag{19.2}$$

$c_{p,t}$ ist die spezifische Wärme der Translation, c_r die der Rotation und c_s die der
Schwingung. Chapman und Cowling[2] sowie Schaefer[3] haben angeregt, statt
dessen

$$\lambda = \tfrac{3}{2} c_{p,t} \eta + \varrho \, D_r c_r + \varrho \, D_s c_s \tag{19.3}$$

[1] A. Eucken: Phys. Z. **14**, 1, 324 (1913).
[2] S. Chapman u. T.G. Cowling: The Mathematical Theory of Non-Uniform Gases,
S. 237. Cambridge 1939.
[3] Kl. Schaefer: Z. phys. Chem., Abt. B **53**, 149 (1943).

anzusetzen. D_r und D_s haben den Charakter von Diffusionskoeffizienten. Eine einheitlichere Darstellung von (19.3) ist

$$\lambda = 1{,}2\varrho\, D_t\, c_{p,t} + \varrho\, D_r\, c_r + \varrho\, D_s\, c_s. \tag{19.4}$$

Hierbei ist D_t der theoretische Diffusionskoeffizient für ein Gasmodell mit starren glatten Kugeln ($D_t = \tfrac{5}{4}\,\eta/\varrho$). Den Ansätzen (19.2) bis (19.4) liegt die Vorstellung zugrunde, daß der Transport von Rotations- und Schwingungsenergie durch reine Diffusion erfolgt, während beim Transport der Translationsenergie auch die Persistenz der Geschwindigkeiten [in (19.4) mit dem Faktor 1,2] wirksam wird.

Die Beziehung (19.4) läßt sich unter geeigneten Annahmen über den Diffusionsvorgang aus der allgemeinen Gl. (18.3) gewinnen. Gleichzeitig wird den Koeffizienten D_r und D_s eine wohldefinierte physikalische Bedeutung im Rahmen des gewählten Modells gegeben.

Man faßt dazu das im gewöhnlichen Sprachgebrauch indifferent genannte ideale Gas als eine Mischung von chemisch reagierenden Komponenten auf. Diese sind die Moleküle in den verschiedenen Rotations- und Schwingungszuständen. Die chemischen Reaktionen sind dann Übergänge zwischen den verschiedenen Niveaus.

Es sollen nur Moleküle mit einem Schwingungsfreiheitsgrad betrachtet werden. Bei nicht zu hohen Temperaturen genügt es ferner, nur Moleküle im Grundzustand und im ersten angeregten Zustand der Schwingung anzunehmen. Die Massenkonzentration der Moleküle im Rotationszustand i und im Grundzustand der Schwingung sei γ_i, im ersten angeregten Schwingungszustand γ_i^*. Diesen Konzentrationen sind in üblicher Weise die spezifischen chemischen Potentiale μ_i und μ_i^*, sowie die spezifischen Enthalpien h_i und h_i^* zugeordnet. Für sie gilt

$$h_i = h(T,p) + u_i, \qquad h_i^* = h(T,p) + v + u_i, \tag{19.5}$$

wenn $h(T,p)$ die spezifische Enthalpie des Grundzustandes, u_i die spezifische Anregungsenergie der Rotation im Zustand i und v die Anregungsenergie des Schwingungszustandes pro Masseneinheit ist. Sinngemäß sind die phänomenologischen Gleichungen (11.9) umzuschreiben in

$$\boldsymbol{J}_i = -\sum_k \frac{l_{ik}}{T}\,\mathrm{grad}_T\,\mu_k - \sum_{k^*} \frac{l_{ik}^*}{T}\,\mathrm{grad}_T\,\mu_k^* + l_i\,\mathrm{grad}\,\frac{1}{T}, \tag{19.6}$$

$$\boldsymbol{J}_i^* = -\sum_k \frac{l_{ki}^*}{T}\,\mathrm{grad}_T\,\mu_k - \sum_{k^*} \frac{l_{ik}^{**}}{T}\,\mathrm{grad}_T\,\mu_k^* + l_i^*\,\mathrm{grad}\,\frac{1}{T}. \tag{19.7}$$

Die Koeffizienten genügen den O.C.R.B. und den Summenrelationen (11.19). Setzt man $L_{uu}/T^2 = \lambda_A$ d.h. gleich der Wärmeleitfähigkeit bei verschwindenden Affinitäten, $l/T^2 = \lambda$ d.h. gleich der Wärmeleitfähigkeit bei gehemmtem Gleichgewicht und örtlich konstanten Konzentrationen, so geht (18.3) in den neuen Bezeichnungen über in

$$T^2\,\lambda_A = T^2\,\lambda + 2\sum_k (l_k + l_k^*)\,u_k + \sum_{i,k} (l_{ik} + 2l_{ik}^* + l_{ik}^{**})\,u_i u_k + \sum_{i,k} l_{ik}^{**}\,v^2. \tag{19.8}$$

Hierin sind (19.5) und (11.19) benutzt. Der zweite Term in (19.8) gibt die einander gleichen Beiträge von Diffusionsthermik und Thermodiffusion, der dritte den Beitrag der Diffusion von Rotationsenergie und der letzte den Beitrag der Diffusion der Schwingungsenergie zur Wärmeleitfähigkeit,

Nun soll aus der allgemeinen Formel (19.8) die spezielle Gl. (19.4) begründet werden. Auf Grund der bekannten Formeln für die Rotations- und Schwingungswärme

$$c_r = \frac{M}{kT^2}\left[\sum_i \frac{\gamma_i}{\gamma}\,u_i^2 - \left(\sum_i \frac{\gamma_i}{\gamma}\,u_i\right)^2\right], \qquad c_s = \frac{M}{kT^2}\,\gamma\gamma^*v^2\,,$$
$$\gamma = \sum_i \gamma_i\,, \qquad \gamma^* = \sum_i \gamma_i^* \tag{19.9}$$

läßt sich eine gewisse Verwandtschaft zwischen (19.8) und (19.4) erkennen.

Eine volle Übereinstimmung ergibt sich natürlich nur unter zusätzlichen Voraussetzungen über die phänomenologischen Koeffizienten. Bevor wir diese benennen, bemerken wir, daß sich die phänomenologischen Koeffizienten l_{ik}, l_{ik}^*, l_{ik}^{**} in erster Näherung durch die binären Diffusionskoeffizienten für jeweils zwei allein vorhandene Komponenten ausdrücken lassen[1]. Damit ist zwar die Zahl der Koeffizienten noch nicht reduziert; sie hängen aber nur noch von den Eigenschaften eines Paares von Komponenten ab. Die Zahl der unabhängigen Koeffizienten wird nun durch folgende Voraussetzung erheblich eingeschränkt. Alle binären Diffusionskoeffizienten für Moleküle ohne Schwingungsanregung seien gleich D, für Moleküle mit Schwingungsanregung gleich D^{**} und für Diffusion einer Komponente ohne Schwingungsanregung gegen eine solche mit Schwingungsanregung gleich D^*. Dann schreibt sich (19.8) bis auf den Thermodiffusions- und Diffusionsthermikanteil genau wie (19.4), wenn man

$$D_s = D^*, \qquad D_r = \gamma\left[\frac{\gamma}{D} + \frac{\gamma^*}{D^*}\right]^{-1} + \gamma^*\left[\frac{\gamma}{D^*} + \frac{\gamma^*}{D^{**}}\right] \tag{19.10}$$

setzt. Da im allgemeinen $\gamma^* \ll 1$ ist, so gilt in erster Näherung $D_r = D$, d.h. die Diffusion der Rotationsenergie wird durch die Selbstdiffusion der Moleküle im Grundzustand der Schwingung bewirkt. Dies gibt eine Rechtfertigung für die Schaefersche Deutung des Koeffizienten D_r.

Wegen Einzelheiten der Rechnung und einer Abschätzung des Thermodiffusionseinflusses sei auf MEIXNER[2] verwiesen. Umfangreiche experimentelle Daten zur Euckenschen Gleichung, die im Sinne der obigen Überlegungen ausgewertet werden könnten, finden sich bei BENNETT und VINES[3].

β) Die Wärmeleitfähigkeit reagierender fluider Mischungen mit zwei Komponenten. Die Beziehung (18.3) nimmt eine besonders einfache und übersichtliche Gestalt für reagierende fluide Mischungen mit zwei Komponenten an (Bezeichnungen nach Ziff. 11γ). Die phänomenologischen Koeffizienten l, l_1, l_{11} lassen sich nach (14.7) durch die Wärmeleitfähigkeit λ, den Diffusionskoeffizienten D und den Thermodiffusionsfaktor α ausdrücken. Es ergibt sich

$$\lambda_A = \lambda + \frac{\alpha}{T}\,\gamma(1-\gamma)(h_1 - h_2)\,\varrho D + \frac{1}{T}\left(\frac{\partial\gamma}{\partial\mu_1}\right)_{T,p}\gamma(h_1 - h_2)^2\,\varrho D. \tag{19.11}$$

Diese Beziehung kann noch in verschiedener Weise umgeschrieben werden. Wir benutzen hierzu (11.26), (11.17a) und (14.4). Dann folgt

$$\frac{\partial\mu_1}{\partial\gamma} = (1-\gamma)\left(\frac{\partial A}{\partial\gamma}\right)_{T,p}. \tag{19.12}$$

[1] C.F. CURTISS u. J.O. HIRSCHFELDER: J. Chem. Phys. 17, 550 (1949).
[2] J. MEIXNER: Z. Naturforsch. 8a, 69 (1953).
[3] L.A. BENNETT u. R.G. VINES: J. Chem. Phys. 23, 1587 (1955).

Ferner definieren wir die spezifischen Wärmen $c_{p\gamma}$ bei konstantem Druck und konstanter Konzentration und c_{pA} bei konstantem Druck und konstanter Affinität $A = 0$ durch

$$c_{p\gamma} = T\left(\frac{\partial s}{\partial T}\right)_{p,\gamma} = \left(\frac{\partial h}{\partial T}\right)_{p,\gamma}, \qquad c_{pA} = T\left(\frac{\partial s}{\partial T}\right)_{p,A} = \left(\frac{\partial h}{\partial T}\right)_{p,A}. \tag{19.13}$$

Ihre Differenz nennt man auch innere spezifische Wärme bei konstantem Druck. Für diese gilt

$$c_{pi} = c_{pA} - c_{p\gamma} = -\frac{1}{T}\left(\frac{\partial \gamma}{\partial A}\right)_{T,p}\left(\frac{\partial h}{\partial \gamma}\right)^2. \tag{19.14}$$

Damit nimmt (19.11) die Gestalt an

$$\lambda_A = \lambda + \frac{2}{T}\gamma(1-\gamma)\alpha\varrho D + c_{pi}\varrho D. \tag{19.15}$$

Eine andere Schreibweise ist

$$\lambda_A = \lambda + 2\delta\sqrt{\lambda c_{pi}\varrho D} + c_{pi}\varrho D. \tag{19.16}$$

In ihr ist die Ungleichung (14.10) für den Thermodiffusionsfaktor in der Form

$$\frac{1}{T}\gamma(1-\gamma)\left(\frac{\partial h}{\partial \gamma}\right)_{T,p}\alpha\varrho D = \delta\sqrt{\lambda c_{pi}\varrho D}, \qquad |\delta| < 1 \tag{19.17}$$

herangezogen. Je nach dem Vorzeichen des Thermodiffusionsfaktors α, d.h. von δ ist der zweite Term in (19.16) positiv oder negativ. Wegen $c_{pi} > 0$ ist aber stets $\lambda_A > \lambda$.

Besonders geeignet für die Diskussion von (19.16) sind natürlich chemische Gasreaktionen; denn für Gase gibt die kinetische Gastheorie brauchbare Werte von λ und D. Berechnet man c_{pi} speziell für dissoziierende Gase, so ergibt sich für den Term $c_{pi}\varrho D$ in (19.16) ein Ausdruck, den Nernst[1] in einer klassischen Arbeit auf elementare Weise hergeleitet hat.

Für dissoziierendes Stickstofftetroxyd, $N_2O_4 \rightleftharpoons 2NO_2$, bei Atmosphärendruck und mit einer Temperatur für welche $\gamma = 0,5$ ist, findet man $c_{pi}\varrho D \approx 10\lambda$. Man erhält daher aus (19.16) $\lambda_A \approx (11 + 6,4\delta)\lambda$. Wenn man auch bei reagierenden Gasmischungen annimmt, daß nicht nur $|\delta| < 1$, sondern wie bei allen indifferenten Gasmischungen $< \frac{1}{8}$ oder sogar $\ll \frac{1}{8}$ ist, so ist der Beitrag von Thermodiffusion und Diffusionsthermik $< 10\%$ oder wahrscheinlicher noch von der Größenordnung 1% des Wertes von λ_A. Relativ am größten ist der Einfluß dieser Effekte, wenn $c_{pi}\varrho D \approx \lambda$. Dann wird $\lambda_A = 2(1+\delta)\lambda$ mit einem Beitrag dieser beiden Effekte $< 17\%$. Jedenfalls kann man schließen, daß diese beiden Effekte dann für die Wärmeleitung zu berücksichtigen oder wenigstens abzuschätzen sind, wenn man Wärmeleitfähigkeiten auf wenigstens 2% genau berechnen oder deuten will.

Bei hohen und bei tiefen Temperaturen ist c_{pi} klein. Die chemische Reaktion wirkt sich daher auf die Wärmeleitfähigkeit λ_A nur in einem beschränkten Temperaturbereich aus, in dem im Gleichgewicht vergleichbare Mengen beider Bestandteile vorhanden sind. Der Bereich ist um so schmaler, je größer die Reaktionsenthalpie ist.

Die thermodynamische Theorie der Wärmeleitfähigkeit reagierender Mischungen ist fast gleichzeitig von Franck[2], Meixner[3] und Prigogine und Buess[4] gegeben worden. Franck[5] sowie Franck und Spalthoff[6] haben die

---

[1] W. Nernst: Boltzmann-Festschrift, S. 904, Leipzig 1904.
[2] E. U. Franck: Z. phys. Chem. **201**, 16 (1952).
[3] J. Meixner: Z. Naturforsch. **7**a, 553 (1952).
[4] I. Prigogine u. R. Buess: Bull. Acad. Roy. Belg., Cl. Sci. **38**, 701 (1952).
[5] E. U. Franck: Chem.-Ing.-Tech. **25**, 238 (1953). — Z. Elektrochem. **55**, 636 (1951).
[6] E. U. Franck u. W. Spalthoff: Naturwiss. **40**, 580 (1953).

Beziehung (19.11) am Beispiel des dissoziierenden Stickstofftetroxyds, Fluors, Jods und Fluorwasserstoffs mit Erfolg nachgeprüft. Eine interessante Anwendung auf die Wärmeleitfähigkeit realer Gase haben WAELBROECK, LAFLEUR und PRIGOGINE[1] gegeben, indem sie die Clusterbildung, welche insbesondere in der Nähe des kritischen Punktes merklich ist, als chemische Reaktion behandeln. Für Argon bei 90° K wird die Zunahme von λ_A von 0 bis zu 1 atm auf 6% abgeschätzt.

20. Die Wärmeleitfähigkeit bei verzögerter Gleichgewichtseinstellung. In Ziff. 19 und 20 wurden fluide Mischungen mit chemisch reagierenden Komponenten betrachtet, bei denen die Gleichgewichtseinstellung unendlich schnell erfolgt. Diese Voraussetzung ist natürlich nie genau erfüllt. Insbesondere bei Gasreaktionen ist die charakteristische Zeit für die Gleichgewichtseinstellung wenigstens von der Größe der Stoßzeit, da ja die chemische Umwandlung nur bei Stößen erfolgt. Es gibt zwei interessante Effekte, bei denen die verzögerte Gleichgewichtseinstellung eine Rolle spielt; das sind die Wärmeleitung in kleinen Gefäßen, bzw. bei geringen Drucken, und die Erscheinung der Volumenviskosität. Bevor wir diese Effekte behandeln, soll ein geeignetes Maß für die Dauer der Gleichgewichtseinstellung, die Relaxationszeit, angegeben werden. Dabei beschränken wir uns auf zwei reagierende Komponenten.

α) *Die Relaxationszeit.* In einem homogenen, in Ruhe befindlichen Medium ist die Änderung der Konzentration nach (11.3a) und (11.17a) durch

$$\varrho \frac{\partial \gamma}{\partial t} = + \lambda_{11} \frac{A}{T} \tag{20.1}$$

gegeben. In (20.1) sind ohnehin kleine Abweichungen vom chemischen Gleichgewicht vorausgesetzt. Aus diesem Grunde ist A klein und man darf entwickeln

$$A(T, p, \gamma) = \left(\frac{\partial A}{\partial \gamma}\right)_{T,p} (\gamma - \bar{\gamma}). \tag{20.2}$$

$\bar{\gamma}$ ist die Gleichgewichtskonzentration bei gegebenem T, p. $(\partial A/\partial \gamma)_{T,p}$ bezieht sich ebenfalls auf den Gleichgewichtszustand. Einsetzen von (20.2) in (20.1) gibt eine Differentialgleichung für γ, die bei festgehaltenen T, p linear ist und deren Koeffizienten in guter Näherung konstant sind. Sie läßt sich in der Gestalt

$$\tau \frac{\partial \gamma}{\partial t} = - (\gamma - \bar{\gamma}) \tag{20.3}$$

schreiben mit

$$\tau = \tau_{T,p} = - \frac{T\varrho}{\lambda_{11}} \left(\frac{\partial \gamma}{\partial A}\right)_{T,p}. \tag{20.4}$$

τ ist ersichtlich die Zeit, in der die Gleichgewichtsabweichung von γ auf $1/e$ zurückgeht. Man nennt sie Relaxationszeit bei konstanter Temperatur und konstantem Druck. Sie ist durch den phänomenologischen Koeffizienten und im übrigen durch thermodynamische Daten bestimmt.

Man kann die Gleichgewichtseinstellung auch unter anderen Bedingungen, z.B. bei konstanter innerer Energie und Dichte verfolgen. Dann hat man in (20.2) überall T, P durch u, ϱ zu ersetzen und $\bar{\gamma}$ ist die Gleichgewichtskonzentration zu gegebenen Werten von u und ϱ. Für das zeitliche Verhalten von γ findet man wieder (20.3). Doch ist an Stelle von τ die Relaxationszeit

$$\tau' = - \frac{T\varrho}{\lambda_{11}} \left(\frac{\partial \gamma}{\partial A}\right)_{u,\varrho} \tag{20.5}$$

[1] F. WAELBROECK, S. LAFLEUR u. I. PRIGOGINE: Physica, Haag **21**, 667 (1955).

zu setzen. τ' ist die Relaxationszeit bei konstanter innerer Energie und Dichte. Das Verhältnis τ'/τ enthält den phänomenologischen Koeffizienten λ_{11} nicht mehr. τ' läßt sich daher aus τ unter Benutzung rein thermodynamischer Daten berechnen. Das Verhältnis τ'/τ ist in der Regel von der Größenordnung 1, so daß man wahlweise τ oder τ' zur qualitativen Charakterisierung der Reaktionsgeschwindigkeit wählen kann.

β) *Stationäre Wärmeleitung bei verzögerter Gleichgewichtseinstellung.* Die stationäre Wärmeleitung in einer Mischung mit zwei reagierenden Komponenten wird nach (11.3a), (11.5a) und (20.4) wegen $v = 0$ und $\operatorname{grad} p = 0$ durch die Gleichungen

$$\operatorname{div} \boldsymbol{J}_u = 0, \quad \operatorname{div} \boldsymbol{J} = -\frac{\varrho}{\tau} \left(\frac{\partial \gamma}{\partial A}\right)_{T,p} A \tag{20.6}$$

beschrieben. Berücksichtigt man die Beziehung

$$\operatorname{grad} A = \left(\frac{\partial A}{\partial T}\right)_{p\gamma} \operatorname{grad} T + \operatorname{grad}_T A, \tag{20.7}$$

so lauten die phänomenologischen Gleichungen (11.9a) und (11.10a)

$$\boldsymbol{J} = \frac{l_{11}}{T} \operatorname{grad} A - \left(\frac{l_{11}}{T}\left(\frac{\partial A}{\partial T}\right)_{p,\gamma} + \frac{l_1}{T^2}\right) \operatorname{grad} T, \tag{20.8}$$

$$\boldsymbol{J}_{\mathrm{red}} = \frac{l_1}{T} \operatorname{grad} A - \left(\frac{l_1}{T}\left(\frac{\partial A}{\partial T}\right)_{p,\gamma} + \frac{l}{T^2}\right) \operatorname{grad} T. \tag{20.9}$$

Man erhält daher aus (20.6) die folgenden Differentialgleichungen für die Felder $T(\boldsymbol{r})$ und $A(\boldsymbol{r})$:

$$\operatorname{div}\left[-\lambda_A \operatorname{grad} T + \frac{1}{T}\left(l_1 + \frac{\partial h}{\partial \gamma} l_{11}\right) \operatorname{grad} A\right] = 0, \tag{20.10}$$

$$\operatorname{div}\left[\frac{l_{11}}{T} \operatorname{grad} A - \left(\frac{l_{11}}{T}\left(\frac{\partial A}{\partial T}\right)_{p,\gamma} + \frac{l_1}{T^2}\right) \operatorname{grad} T\right] + \frac{\varrho}{\tau}\left(\frac{\partial \gamma}{\partial A}\right)_{T,p} A = 0. \tag{20.11}$$

Diese Gleichungen sind lineare partielle Differentialgleichungen für die Felder A und T, wenn man die Koeffizienten konstant annimmt. Diese Voraussetzung ist bei kleinen Temperaturgradienten, wie sie bei Wärmeleitungsexperimenten vorliegen, und damit kleinen Werten von A und $\operatorname{grad} A$ erfüllt. In den Koeffizienten dürfen dann die Gleichgewichtswerte, d.h. $A = 0$, eingesetzt werden. Dann ist nach (11.26) $T(\partial A/\partial T)_{p,\gamma} = (\partial h/\partial \gamma)_{T,p}$.

Es soll nun speziell die eindimensionale stationäre Wärmeleitung zwischen zwei Platten bei $x = 0$ und $x = l$ mit den Temperaturen T_0 und T_1 betrachtet werden. Dann sind (20.10) und (20.11) zwei gewöhnliche lineare Differentialgleichungen für die Funktionen $T(x)$ und $A(x)$. Ihre Integration ist elementar. Das allgemeine Integral enthält vier Integrationskonstanten, die aus den Randbedingungen bei $x = 0$ und $x = l$ zu bestimmen sind. Zwei der Randbedingungen sind $T = T_0$ für $x = 0$ und $T = T_1$ für $x = l$, wenn man keine Temperatursprünge an den Platten annimmt. Zwei weitere Randbedingungen folgen daraus, daß an der Plattenoberfläche keine Anreicherung einer Komponente erfolgen darf. Dies drückt sich durch

$$\boldsymbol{n} \cdot \boldsymbol{J} = \varepsilon A \tag{20.12}$$

aus; $\boldsymbol{n}$ ist ein auf der Plattenoberfläche senkrechter Einheitsvektor und ε kennzeichnet die Geschwindigkeit der Oberflächenreaktion. Wegen (20.8) geht (20.12) über in

$$\boldsymbol{n} \cdot (\operatorname{grad} A - r \operatorname{grad} T) = q A \tag{20.13}$$

oder

$$\frac{\partial A}{\partial x} - r\frac{\partial T}{\partial x} = qA \quad (x=0),$$
$$\frac{\partial A}{\partial x} - r\frac{\partial T}{dx} = -qA \quad (x=l) \tag{20.14}$$

mit

$$T r = \frac{l_1}{l_{11}} + \left(\frac{\partial h}{\partial \gamma}\right)_{T,p}, \quad q = \frac{T\varepsilon}{l_{11}}. \tag{20.15}$$

Während r eine Eigenschaft der reagierenden Mischung allein ist, hängt q von ε und damit auch von den Eigenschaften der Plattenoberflächen ab. Die effektive Wärmeleitfähigkeit des Gemisches ist dem Meßvorgang entsprechend durch

$$\lambda^* = -\frac{l}{T_1 - T_0}\,|J_u| \tag{20.16}$$

definiert. Für sie folgt nach elementarer Rechnung

$$\frac{\lambda_A}{\lambda^*} - 1 = \left(\frac{\lambda_A}{\lambda'} - 1\right)\left[1 + \frac{\lambda_A}{\lambda'}\cdot\frac{q}{\sigma}\right]^{-2}\frac{2}{\sigma l}\,\mathrm{Tan}\,\frac{\sigma l}{2}. \tag{20.17}$$

Hierin ist $\lambda' = \lambda(1 - \delta^2)$ die Wärmeleitfähigkeit ohne chemische Reaktion bei stationärer Thermodiffusion. σ ist durch

$$\sigma^2 = \frac{1}{\tau D}\frac{\lambda_A}{\lambda'} \tag{20.18}$$

gegeben.

Die effektive Wärmeleitfähigkeit λ^* hängt also von der Reaktionsgeschwindigkeit im Gas nur über die Größe σl, von der Oberflächenreaktion über die Größe q/σ ab. Der Größe σl kann für den Fall reagierender Gase eine anschauliche Deutung gegeben werden. Ist Λ die mittlere freie Weglänge, τ_s die Stoßzeit und τ_l die mittlere Zeit, welche ein Molekül braucht, um im Zickzackweg die Entfernung l zurückzulegen, so gilt nach der elementaren kinetischen Gastheorie

$$3\tau_s D = \Lambda^2, \quad \tau_l = \tau_s\frac{l^2}{\Lambda^2}. \tag{20.19}$$

In τ_l ist die Persistenz der Geschwindigkeiten nicht berücksichtigt; sie ändert nichts Wesentliches an den folgenden Aussagen. Mit (20.19) folgt aus (20.18)

$$(\sigma l)^2 = 3\frac{\lambda_A}{\lambda'}\frac{\tau_l}{\tau}. \tag{20.20}$$

Läßt man die Wandreaktion unberücksichtigt, setzt also $q=0$, so folgen die in Ziff. 19 behandelten Grenzfälle unmittelbar für $\tau=0$ oder $\tau=\infty$. Dann ist $\sigma l = \infty$ oder $\sigma l = 0$ und nach (20.17) $\lambda^* = \lambda_A$ oder $\lambda^* = \lambda'$. Für die Dissoziation von $N_2O_4 \rightleftharpoons 2\,NO_2$ bei Atmosphärendruck und $\gamma = 0,5$ ist $\lambda_A = 11\,\lambda'$. Mit $l = 1$ cm und $\Lambda \approx 10^{-4}$ cm wird $\tau_l \approx 10^8\,\tau_s$ und $\lambda^*/\lambda_A = \left[1 + 0,003\,\sqrt{\frac{\tau_l}{\tau_s}}\right]^{-1}$. Merkliche Unterschiede zwischen λ_A und λ^* ergeben sich daher erst für $\tau > 10^6\,\tau_s$.

21. Die Volumenviskosität. Die verzögerte Gleichgewichtseinstellung macht sich auch im Tensor der Reibungsdrucke durch das Auftreten einer zusätzlichen Volumenviskosität bemerkbar. Ihr Zustandekommen soll der Einfachheit halber ebenfalls am Beispiel einer Mischung von zwei reagierenden Komponenten erläutert werden. Diffusion und Wärmeleitung sollen dabei unberücksichtigt bleiben. Für die Reduktion der Gleichungen in Ziff. 11 auf die eines Einkomponentensystems ist es nun wesentlich, daß man die unabhängigen thermodynamischen Variablen des Einkomponentensystems von vornherein festlegt. Zweckmäßig wählt man die spezifische innere Energie u und die Dichte ϱ, weil dann (11.1a) und (11.5a) ungeändert für das ersatzweise betrachtete Einkomponentensystem gelten. Der thermodynamische Druck $p(u, \varrho, A)$ des Zweikomponentensystems darf jedoch nicht mit dem thermodynamischen Druck $p(u, \varrho)$ des Einkomponentensystems identifiziert werden, außer wenn $A = 0$ ist. Man hat daher zu setzen

$$p(u, \varrho, A) = p_1(u, \varrho) + \left(\frac{\partial p}{\partial A}\right)_{u,\varrho} A. \tag{21.1}$$

Der Tensor der Reibungsdrucke des Einkomponentensystems ist daher nicht mehr $p\,\delta_{\alpha\beta} - P_{\alpha\beta}$, sondern $p_1\,\delta_{\alpha\beta} - P_{\alpha\beta}$. Aus (11.16) und (21.1) ergibt sich für ihn

$$\left. \begin{aligned} p_1\delta_{\alpha\beta} - P_{\alpha\beta} = \eta\left(\frac{\partial v_\alpha}{\partial x_\beta} + \frac{\partial v_\beta}{\partial x_\alpha}\right) + \left(\zeta - \frac{2}{3}\eta\right)\delta_{\alpha\beta}\,\mathrm{div}\,\boldsymbol{v} + \\ + \left[\left(\frac{\partial p}{\partial A}\right)_{u,\varrho} - \frac{\lambda_1}{T}\right]\delta_{\alpha\beta}\,A. \end{aligned} \right\} \tag{21.2}$$

Seine endgültige Form gewinnen wir, indem wir die Affinität durch Daten des Einkomponentensystems ausdrücken. Aus (11.3a) und (11.17a) folgt mit (20.5) wegen der Vernachlässigung der Diffusion

$$\varrho\,\frac{D\gamma}{Dt} = -\frac{\lambda_1}{T}\,\mathrm{div}\,\boldsymbol{v} - \frac{\varrho}{\tau'}\left(\frac{\partial\gamma}{\partial A}\right)_{u,\varrho} A. \tag{21.3}$$

Eine Vereinfachung dieser Gleichung ist möglich, wenn die Kompressionen und Dilatationen hinreichend langsam sind, d.h., wenn $\mathrm{div}\,\boldsymbol{v}$ hinreichend klein wird. Da nämlich Wärmeleitung und Diffusion als gehemmt angenommen sind, so können Störungen des chemischen Gleichgewichts nur über Dichteänderungen erfolgen. Andererseits gehen Gleichgewichtsabweichungen wegen der großen Reaktionsgeschwindigkeit schnell wieder zurück. A bleibt deswegen klein und wir können γ auf der linken Seite von (21.3) durch die Gleichgewichtskonzentration $\bar{\gamma}(u, \varrho)$ ersetzen. Man denke sich dazu alle Gleichgewichtsabweichungen nach Potenzen von τ' entwickelt und behalte nur die ersten Entwicklungsglieder bei.

Nun wird $D\bar{\gamma}/Dt$ durch Du/Dt und $D\varrho/Dt$ ausgedrückt. Es gilt

$$\frac{D\bar{\gamma}}{Dt} = \left(\frac{\partial\bar{\gamma}}{\partial u}\right)_S \frac{Du}{Dt} + \left(\frac{\partial\bar{\gamma}}{\partial\varrho}\right)_u \frac{D\varrho}{Dt} = -\left(\frac{\partial\bar{\gamma}}{\partial u}\right)_\varrho \frac{p_1}{\varrho}\,\mathrm{div}\,\boldsymbol{v} - \left(\frac{\partial\bar{\gamma}}{\partial\varrho}\right)_u \varrho\,\mathrm{div}\,\boldsymbol{v}. \tag{21.4}$$

Bei der Umformung sind (11.1a) und (11.5a) benutzt worden. Glieder mit $p_1\delta_{\alpha\beta} - P_{\alpha\beta}$ sind hierbei vernachlässig worden; sie würden Terme liefern, die quadratisch in den Ortsableitungen der Geschwindigkeitskomponenten sind. Aus (21.3) und (21.4) gewinnt man eine Beziehung zwischen der Affinität und $\mathrm{div}\,\boldsymbol{v}$. Sie lautet nach einfachen thermodynamischen Umrechnungen

$$A = -\frac{\tau'}{\varrho}\left(\frac{\partial A}{\partial\gamma}\right)_{u,\varrho}\left[\frac{\lambda_1}{T} - \left(\frac{\partial p}{\partial A}\right)_{u,\varrho}\right]\mathrm{div}\,\boldsymbol{v}. \tag{21.5}$$

Damit nimmt (21.2) die Gestalt

$$p_1\delta_{\alpha\beta} - P_{\alpha\beta} = \eta\left(\frac{\partial v_\alpha}{\partial x_\beta} + \frac{\partial v_\beta}{\partial x_\alpha}\right) + \left(\zeta + \zeta_r - \frac{2}{3}\eta\right)\delta_{\alpha\beta}\,\mathrm{div}\,\boldsymbol{v} \tag{21.6}$$

an, wobei

$$\zeta_r = -\frac{\tau'}{\varrho}\left(\frac{\partial A}{\partial\gamma}\right)_{\varrho,u}\left[\left(\frac{\partial p}{\partial A}\right)_{u,\varrho} - \frac{\lambda_1}{T}\right]^2 \geqq 0. \tag{21.7}$$

Das Ergebnis (21.6) lautet in Worten: Ein fluides Medium aus zwei reagierenden Komponenten kann bei Dichteänderungen, die gemessen an τ hinreichend langsam sind, $\dfrac{1}{\varrho}\left|\dfrac{\partial \varrho}{\partial t}\right| \ll \dfrac{1}{\tau}$, wie ein Einkomponentensystem mit einer um ζ_r vergrößerten Volumenviskosität behandelt werden.

Dieses Ergebnis ist insbesondere aus historischen Gründen, aber auch wegen seiner hydrodynamischen Folgen bemerkenswert. Es ist zwar, solange die Theorie der reibenden Flüssigkeiten besteht, bekannt, daß der phänomenologische Ansatz für den Tensor der Reibungsdrucke zwei Viskositätskonstanten zuläßt, genau so wie man für ein isotropes elastisches Medium zwei elastische Konstanten hat. Während aber die Existenz von zwei elastischen Konstanten experimentell leicht bestätigt werden kann, führt die Annahme einer von Null verschiedenen Volumenviskosität nicht zu sehr augenfälligen Effekten.

Die Deutung der Volumenviskosität als phänomenologischer Ersatz für die explizite Betrachtung von chemischen Reaktionen und inneren Umwandlungen ist zum erstenmal wohl von MANDELSTAM und LEONTOVITCH[1] ausgesprochen und von FRENKEL und OBRASTZOV[2] weiter verfolgt worden. Unabhängig von diesen Autoren hat TISZA[3] für reagierende Mischungen mit zwei Komponenten, insbesondere für Gase mit thermischer Relaxation, die Volumenviskosität als Funktion der Frequenz angegeben und MEIXNER[4] hat sie für Gase mit beliebig vielen inneren Umwandlungen bei hohen Frequenzen berechnet.

Es erhebt sich natürlich die Frage, ob die gesamte Volumenviskosität $\zeta + \zeta_r$ als Folge von inneren Umwandlungen verstanden werden kann, ob also auch ζ wieder auf innere Umwandlungen zurückgeführt werden kann, die noch wesentlich schneller ablaufen, als die in Ziff. 11 berücksichtigten. Diese Frage ist offen, aber wohl auch nicht sehr wichtig. Eine andere Frage ist indessen, ob eine chemische Viskosität, ausgedrückt durch die phänomenologischen Koeffizienten in (11.11) und (11.12), existiert. Obwohl es schwer sein dürfte, diese Koeffizienten experimentell zu erfassen, kann man gute theoretische Gründe für ihre Existenz angeben und auch sie auf vernachlässigte innere Umwandlungen zurückzuführen. MEIXNER[5] hat zu diesem Zweck eine Mischung von drei Komponenten mit zwei chemischen Reaktionen, von denen eine sehr schnell gegen die andere ist, betrachtet. Sie soll weder Volumenviskosität noch chemische Viskosität besitzen. Dann lassen sich die Gleichungen der Ziff. 11 für Vorgänge, die langsam im Vergleich zur schnellen Reaktion erfolgen, auf die Grundgleichungen für ein Zweikomponentensystem mit einer von Null verschiedenen Volumenviskosität und mit im allgemeinen nicht verschwindender chemischer Viskosität reduzieren.

22. Schallabsorption in reagierenden Mischungen und dynamische Zustandsgleichungen. Die Schallabsorption bedeutet einen Verlust der Schallwelle an mechanischer Energie. Er ist durch die dissipativen Vorgänge in dem die Schallwelle tragenden Medium bestimmt. Man kann daher die Schallabsorption mit den Methoden der Thermodynamik irreversibler Prozesse als Funktion der Frequenz, der thermodynamischen Daten und der phänomenologischen Koeffizienten berechnen. Die Hinzunahme kinetischer Überlegungen ist nur für extreme Versuchsbedingungen notwendig.

Für ein Einkomponentensystem hat bereits KIRCHHOFF[6] die entsprechenden Rechnungen durchgeführt. Die Schallabsorption durch innere Umwandlungen

[1] L.I. MANDELSTAM u. M.A. LEONTOVITCH: J. Exp. Theor. Phys. USSR. 7, 438 (1937).
[2] J. FRENKEL u. J. OBRASTZOV: J. Phys. USSR. 2, 131 (1940).
[3] L. TISZA: Phys. Rev. 61, 531 (1942).
[4] J. MEIXNER: Ann. Physik [5] 43, 470 (1943). — Acustica 2, 101 (1952).
[5] J. MEIXNER: Z. Physik 131, 456 (1951).
[6] G. KIRCHHOFF: Pogg. Ann. 134, 177 (1868).

allein — verzögerte Einstellung der inneren Freiheitsgrade oder eines Reaktions-
gleichgewichtes — ist insbesondere von Einstein[1], Kneser[2] und Rutgers[3]
behandelt worden. Herzfeld und Rice[4] untersuchen die Gesamtabsorption
durch Transporterscheinungen und innere Umwandlungen für ein Zweikomponentensystem bei niedrigen Frequenzen. Damköhler[5] betrachtet Mischungen
mit beliebig vielen reagierenden Komponenten und Reaktionen, behandelt
allerdings den Beitrag der Transporterscheinungen etwas summarisch. Eine
systematische Theorie mittels der Thermodynamik irreversibler Prozesse hat
Meixner[6] entwickelt. Diese Theorie zeigt insbesondere, daß die Beiträge der
Transporterscheinungen und der inneren Umwandlungen zur Schallabsorption
nicht streng additiv sind und erlaubt, die Abweichungen von der Additivität
abzuschätzen.

α) *Schallabsorption durch Transporterscheinungen.* Die klassische Kirchhoff-
sche Theorie gibt für den Absorptionskoeffizienten der Amplitude pro Längen-
einheit der ebenen Schallwelle

$$\frac{k^2}{\omega^2} = \frac{1}{2}\left(\frac{4}{3}\eta + \zeta\right)\frac{1}{c_0^3\varrho} + \frac{\lambda(\varkappa - 1)}{2\varrho\,c_p\,c_0^3}. \tag{22.1}$$

c_0 ist die Schallgeschwindigkeit für sehr kleine Kreisfrequenzen ω; c_p und c_v
sind die spezifischen Wärmen bei konstantem Druck und Volumen und $\varkappa = c_p/c_v$.
Gl. (22.1) gilt für beliebige fluide Einkomponentensysteme mit nicht zu großer
Zähigkeit. Die innere Reibung gibt bei Gasen in der Regel, bei Flüssigkeiten
wegen $\varkappa \approx 1$ stets den Hauptbeitrag zur Absorption. Beide Anteile haben die
gleiche Frequenzabhängigkeit. Mit der Absorption ist natürlich auch eine Dispersion der Schallgeschwindigkeit verbunden; diese wird jedoch bei Gasen erst
für Wellenlängen von der Größenordnung der mittleren freien Weglänge merklich.

β) *Schallabsorption durch innere Umwandlung.* Wir wenden uns nun der
Schallabsorption in einem Zweikomponentensystem zu. Transporterscheinungen
und chemische Viscosität werden vernachlässigt. Unter der Voraussetzung
kleiner Amplituden sind v, grad ϱ, grad p, $\partial\gamma/\partial t$, A von erster Ordnung kleine
Größen. Unter konsequenter Vernachlässigung von Größen mindestens zweiter
Ordnung folgt dann durch Elimination von v aus (11.1a) und (11.4a) mit $F_i = 0$

$$\frac{\partial^2\varrho}{\partial t^2} = \text{div grad } p. \tag{22.2}$$

Die Entropieproduktion in (11.6a) ist als Produkt zweier kleiner Größen erster
Ordnung von zweiter Ordnung klein. Deshalb gilt in konsequenter Näherung

$$\frac{\partial s}{\partial t} = 0; \tag{22.3}$$

d.h. die Zustandsänderungen erfolgen isentrop. Schließlich liefert (21.3)

$$\tau\frac{\partial\gamma}{\partial t} = -\left(\frac{\partial\gamma}{\partial A}\right)_{s\varrho} A, \tag{22.4}$$

[1] A. Einstein: Sitzgsber. preuß. Akad. Wiss., Berlin, 38, 1920.
[2] H. O. Kneser: Ann. Physik [5] **11**, 761 (1931).
[3] A. J. Rutgers: Ann. Physik [5] **16**, 350 (1933).
[4] K. F. Herzfeld u. F. O. Rice: Phys. Rev. **31**, 691 (1928).
[5] G. Damköhler: Z. Elektrochem. **48**, 212 (1940).
[6] Siehe Fußnote 4, S. 479.

wenn man berücksichtigt, daß

$$\left(\frac{\partial \gamma}{\partial A}\right)_{u,\varrho} = \left(\frac{\partial \gamma}{\partial A}\right)_{S,\varrho}$$

für $A = 0$ gilt.

γ) *Dynamische Zustandsgleichung.* Wir betrachten zunächst nur die zeitlichen Änderungen der thermodynamischen Größen. Sie mögen von gewissen Gleichgewichtswerten γ_0, s_0, p_0, ϱ_0 nur wenig abweichen. Wegen (22.3) und wegen

$$A = \left(\frac{\partial A}{\partial \varrho}\right)_{S,\gamma} (\varrho - \varrho_0) + \left(\frac{\partial A}{\partial \gamma}\right)_{S,\varrho} (\gamma - \gamma_0) \tag{22.5}$$

folgt dann aus (22.4)

$$\gamma - \gamma_0 + \tau \frac{\partial \gamma}{\partial t} = \left(\frac{\partial \gamma}{\partial \varrho}\right)_{S,A} (\varrho - \varrho_0). \tag{22.6}$$

Hieraus und mit p an Stelle von A in (22.5) ergibt sich dann weiter durch Elimination von γ nach elementarer thermodynamischer Umrechnung

$$p - p_0 + \tau \frac{\partial p}{\partial t} = \left(\frac{\partial p}{\partial \varrho}\right)_{S,A} \left[\varrho - \varrho_0 + \tau_1 \frac{\partial \varrho}{\partial t}\right] \tag{22.7}$$

mit

$$\tau_1 = \left(\frac{\partial \gamma}{\partial A}\right)_{S,p} \left(\frac{\partial A}{\partial \gamma}\right)_{S,\varrho} \tau. \tag{22.8}$$

Man bezeichnet Gleichungen der Art (22.7) als dynamische Zustandsgleichungen. Gl. (22.7) ist speziell die dynamische *thermische* Zustandsgleichung bei konstanter Entropie. Bei sehr langsamen Zustandsgleichungen geht sie nämlich in die statische adiabatische Zustandsgleichung für kleine Abweichungen vom Bezugszustand p_0, ϱ_0, $A = 0$ über. Bei zeitabhängigen Vorgängen ersetzt sie die explizite Betrachtung der inneren Umwandlung. Die physikalische Bedeutung der beiden Relaxationszeiten τ und τ_1 wird an (22.7) besonders anschaulich: Sie sind die Relaxationszeiten für die Einstellung des Drucks bzw. der Dichte auf das Gleichgewicht bei konstant gehaltener Dichte bzw. bei konstant gehaltenem Druck. In beiden Fällen läßt sich (22.7) leicht integrieren und gibt exponentielle Einstellung auf p_0 bzw. ϱ_0 mit einer Zeitkonstanten τ bzw. τ_1.

Als nächstes betrachten wir zeitlich harmonische Vorgänge mit der Zeitabhängigkeit $e^{i\omega t}$. Für diese folgt aus (22.7)

$$\frac{p - p_0}{\varrho - \varrho_0} = \left(\frac{\partial p}{\partial \varrho}\right)_{S,A} \frac{1 + i\omega\tau_1}{1 + i\omega\tau}. \tag{22.9}$$

Die linke Seite sei mit $[\varrho \chi_S(\omega)]^{-1}$ bezeichnet. Dann hat $\chi_S(\omega)$ die Bedeutung einer komplexen und frequenzabhängigen isentropen Kompressibilität.

In den Grenzfällen $\omega = 0$ und $\omega \to \infty$ nimmt $\chi_S(\omega)$ die reellen Werte

$$\chi_S(0) = \frac{1}{\varrho}\left(\frac{\partial \varrho}{\partial p}\right)_{S,A}, \quad \chi_S(\infty) = \frac{1}{\varrho}\left(\frac{\partial \varrho}{\partial p}\right)_{S,A} \frac{\tau}{\tau_1} = \frac{1}{\varrho}\left(\frac{\partial \varrho}{\partial p}\right)_{S,\gamma} \tag{22.10}$$

an.

δ) *Die komplexe Schallgeschwindigkeit.* Ebene Schallwellen gewinnt man nun aus (22.2) und (22.7) mit dem Ansatz $p - p_0$ und $\varrho - \varrho_0$ proportional zu $e^{i(\omega t - kx)}$. Damit folgt aus (22.6) und (22.9)

$$\frac{k^2}{\omega^2} = \frac{\varrho - \varrho_0}{p - p_0} = \varrho_0 \chi_S(\omega). \tag{22.11}$$

Zerlegt man k in Real- und Imaginärteil, $k = k_1 + ik_2$, so ergibt sich die reelle Schallgeschwindigkeit aus

$$\frac{1}{v_s(\omega)} = \frac{k_1}{\omega} = \mathrm{Re}\,[\varrho_0 \chi_S(\omega)]^{\frac{1}{2}} \tag{22.12}$$

und der Absorptionskoeffizient der Amplitude zu

$$k_2 = -\,\omega\, \mathrm{Im}\, [\varrho_0 \chi_S(\omega)]^{\frac{1}{2}}.\tag{22.13}$$

Hierin ist also

$$\chi_S(\omega) = \chi_S(0)\,\frac{1+i\,\omega\,\tau}{1+i\,\omega\,\tau_1} = \chi_S(\infty) + [\chi_S(0) - \chi_S(\infty)]\,\frac{1}{1+i\,\omega\,\tau_1}.\tag{22.14}$$

Es sei noch bemerkt, daß man das Zweikomponentensystem wie ein Einkomponentensystem mit einer komplexen Volumenviscosität

$$\zeta = \tau_1\,\varrho\,\frac{v_s^2(\infty) - v_s^2(0)}{1+i\,\omega\,\tau_1}$$

behandeln kann. Man erkennt hieran besonders deutlich, bis zu welchen Grenzen die Annahme einer konstanten reellen Volumenviscosität zulässig ist, nämlich für harmonische Vorgänge mit $\omega\tau_1 \ll 1$.

Der Absorptionskoeffizient hat die größten Werte für $\omega\tau_1 \approx 1$, während er für $\omega = 0$ und $\omega \to \infty$ verschwindet. Der Bereich der maximalen Absorption ist experimentell in vielen Fällen erreichbar. Bei Schwingungsrelaxation in Gasen werden τ_1-Werte von etwa 10^{-3} sec bis 10^{-6} sec beobachtet. Wegen experimenteller Ergebnisse und Meßanordnungen sei auf eine zusammenfassende Darstellung von Kneser[1] verwiesen.

Die Berechnung der Schallabsorption unter gleichzeitiger Berücksichtigung von innerer Umwandlung und von Transporterscheinungen führt auf kompliziertere Ergebnisse. In erster Näherung kann man den Absorptionskoeffizienten in zwei Anteile zerlegen, von denen der eine durch Transportphänomene, der andere durch innere Umwandlungen zustandekommt. Dabei treten Fehler bis zu 10% und mehr auf, so daß eine Abschätzung der Abweichungen von der Additivität nicht überflüssig ist (Meixner[2]).

Zwischen Absorptionskoeffizient und Entropieproduktion besteht ein enger Zusammenhang. Die pro Volumen- und Zeiteinheit absorbierte Energie $2\,k_2\varrho_0\overline{v^2}c_s$ ist gleich der negativen Divergenz der zeitlich gemittelten Energiestromdichte $\varrho_0\overline{v^2}c_s$. Sie ist andererseits gleich der mittleren Energiedissipation $T\bar\sigma$. Hieraus folgt

$$k_2 = \frac{T\bar\sigma}{2\,\varrho_0\,\overline{v^2}\,c_s}.$$

$\frac{1}{2}\varrho v^2$ ist die mittlere kinetische Energiedichte. Diese Beziehung erlaubt eine vereinfachte Behandlung des Gesamtproblems der Schallabsorption, wenn die ebene Welle für den Spezialfall der inneren Umwandlung ohne Transporterscheinungen bereits berechnet ist. Setzt man sie nämlich in die vollständige Formel (11.8) für die Entropieproduktion ein, so wird diese bereits in den in den Transportkoeffizienten linearen Gliedern (s. Meixner[2]) richtig.

23. Frequenzabhängigkeit thermodynamischer Größen. In der vorhergehenden Ziffer haben wir gezeigt, daß das Verhalten des fluiden Mediums bei zeitlich harmonischer Beanspruchung durch eine Kompressibilität beschrieben werden kann, die nicht nur von den thermodynamischen Daten des Bezugszustandes, sondern auch von der Frequenz abhängt. In der Frequenzabhängigkeit äußert sich der Verzicht auf die explizite Behandlung der inneren Variablen.

Die Abhängigkeit thermodynamischer Koeffizienten, (z.B. der Dielektrizitätskonstanten, magnetischen Permeabilität, des Elastizitäts- und Schubmoduls)

---

[1] H.O. Kneser: Ergebn. exakt. Naturw. **22**, 121 (1949).

[2] J. Meixner: Ann. Physik [5] **43**, 470 (1943).

von der Frequenz ist eine häufig beobachtete Erscheinung. Ihre Theorie wird im folgenden entwickelt.

α) *Systeme mit geraden inneren Variablen.* Wir betrachten dazu ein homogenes System in der Nähe eines Bezugszustandes, in dem die inneren Variablen im Gleichgewicht sind. Die Abweichungen der extensiven Variablen (Entropie, Volumen usw.) von ihren Werten im Bezugszustand seien mit $x_1, x_2, \ldots$ bezeichnet und zu dem Spaltenvektor x zusammengefaßt. Die Abweichungen der thermodynamisch konjugierten Variablen von ihren Werten im Bezugszustand seien $y_1, y_2, \ldots$; ihre Gesamtheit stellen wir durch den Spaltenvektor y dar. Entsprechend bezeichnen wir mit α_1, α_2 die Differenzen der inneren Variablen α_i gegen ihre Werte im Bezugszustand und die ihnen zugeordneten Affinitäten mit $A_1, A_2, \ldots$ Diese verschwinden ohnehin im Bezugszustand, da dieser als thermodynamischer Gleichgewichtszustand vorausgesetzt ist. Wir fassen auch diese, den inneren Zustand charakterisierenden Größen zu den Spaltenvektoren α und A zusammen. Die Zahl ihrer Komponenten ist natürlich unabhängig von der der Vektoren x, y. Die transponierten oder Zeilenvektoren seien durch eine Tilde, z.B. $\tilde{x}$, bezeichnet. Betrachtet man einen Zeilenvektor als einzeilige Matrix, einen Spaltenvektor als einspaltige Matrix, so kann man nach den Regeln der Matrizenrechnung das skalare Produkt $\sum_i x_i y_i$ einfach durch $\tilde{x} \cdot y$ ausdrücken. Andererseits ist das dyadische Produkt $x\,y$ eine Matrix mit den Elementen $x_i y_k$.

Die Gibbssche Fundamentalgleichung lautet in dieser Schreibweise

$$d u = \tilde{y} \cdot dx - \tilde{A} \cdot d\alpha. \tag{23.1}$$

Ist u^+ die spezifische Energie im Bezugszustand, so läßt sich die spezifische Energie u in der Umgebung des Bezugszustandes näherungsweise durch

$$u - u^+ + \tfrac{1}{2}\tilde{x}\,\mathsf{a}\,x + \tilde{x}\,\mathsf{b}\,\alpha + \tfrac{1}{2}\tilde{\alpha}\,\mathsf{c}\,\alpha \tag{23.2}$$

darstellen. Hierin sind Glieder, die wenigstens von dritter Ordnung in x_i und α_i klein sind, vernachlässigt; $\mathsf{a}, \mathsf{b}, \mathsf{c}$ sind Matrizen, von denen a und c ohne Beschränkung der Allgemeinheit als symmetrisch angenommen werden können. Bezeichnen wir die transponierten Matrizen ebenfalls mit einer Tilde, so wird also $\tilde{\mathsf{a}} = \mathsf{a}$, $\tilde{\mathsf{c}} = \mathsf{c}$ vorausgesetzt.

In der Umgebung des Bezugszustandes gelten nach (23.1) und (23.2) die Zustandsgleichungen

$$y = \mathsf{a}x + \mathsf{b}\alpha, \quad -A = \mathsf{b}x + \mathsf{c}\alpha, \tag{23.3}$$

wobei von Gliedern, die mindestens von zweiter Ordnung klein sind, abgesehen ist.

Wir nehmen zunächst an, daß alle inneren Variablen gerade sind. Dann liest man aus der Entropieproduktion (11.18) die phänomenologischen Gleichungen

$$\dot{\alpha} = \mathsf{V}A \tag{23.4}$$

ab. V ist wegen der O.C.R.B. eine symmetrische Matrix.

Bei harmonischen Zustandsänderungen mit der Frequenz ω ist $\dot{\alpha} = i\omega\alpha$. Eliminiert man nun α und A aus (23.4) und (23.3), so folgt

$$y = \{\mathsf{a} - \mathsf{b}\,(\mathsf{c} + i\,\omega\,\mathsf{V}^{-1})^{-1}\,\tilde{\mathsf{b}}\}\,x \equiv \alpha(\omega) \cdot x. \tag{23.5}$$

Andererseits ist im thermodynamischen Gleichgewicht

$$y = \{\mathsf{a} - \mathsf{b}\,\mathsf{c}^{-1}\,\tilde{\mathsf{b}}\}\,x \equiv \alpha(0) \cdot x. \tag{23.6}$$

Die Elemente der Matrix $\alpha(0) = a - b\,c^{-1}\,\tilde{b}$ sind thermodynamische Koeffizienten für den Fall des thermodynamischen Gleichgewichts. Die Analogie zwischen (23.6) und (23.5) berechtigt dazu, die Elemente der Matrix $\alpha(\omega)$ als thermodynamische Koeffizienten zur Frequenz ω zu bezeichnen. Man sieht unmittelbar, daß dieses Koeffizientenschema wegen $a = \tilde{a}$, $c = \tilde{c}$, $V = \tilde{V}$ symmetrisch ist. Es ist nicht unwichtig zu bemerken, daß die Matrix $\alpha(\omega)$ für $\omega = 0$ (d.h. im ungehemmten thermodynamischen Gleichgewicht) und $\omega \to \infty$ (dies ist gleichbedeutend mit dem vollständig gehemmten Gleichgewicht) aus thermodynamischen Gründen symmetrisch ist. Ihre Symmetrie für andere Frequenzen ist dahingegen eine Folge der O.C.R.B. für die Matrix V.

Die Struktur der Matrix $\alpha(\omega)$ wird noch deutlicher, wenn wir uns die inneren Variablen so gewählt denken, daß c eine Einheitsmatrix und V eine Diagonalmatrix mit den positiven Diagonalelementen $1/\tau_1$, $1/\tau_2$ ist (s. Ziff. 9ε). Dann wird

$$\alpha_{ik}(\omega) = a_{ik} - \sum_j \frac{1}{1 + i\,\omega\,\tau_j}\, b_{ij}\, b_{kj}. \tag{23.7}$$

β) *Systeme mit geraden und ungeraden inneren Variablen.* Eine Komplikation tritt ein, wenn neben geraden Variablen auch noch ungerade innere Variable β_i mit thermodynamisch konjugierten Affinitäten B_i eine Rolle spielen. Sie seien in den Spaltenvektoren β, $\boldsymbol{B}$ zusammengefaßt. Die innere Energie läßt sich dann in der Umgebung des Gleichgewichtszustandes näherungsweise darstellen als

$$u = u^+ + \tfrac{1}{2}\tilde{x}\,a\,x + \tilde{x}\,b\,\alpha + \tfrac{1}{2}\tilde{\alpha}\,c\,\alpha + \tfrac{1}{2}\tilde{\beta}\,e\,\beta. \tag{23.8}$$

Gemischte Terme in x und β sowie α und β können nicht auftreten, da u selbst gerade gegen Zeitumkehr ist. Aus der Gibbsschen Fundamentalgleichung folgen nun die Zustandsgleichungen

$$y = a\,x + b\,\alpha, \quad -A = b\,x + c\,\alpha, \quad -B = e\,\beta, \tag{23.9}$$

während man aus der Entropieproduktion die phänomenologischen Gleichungen (jetzt zweckmäßigerweise nach den Affinitäten aufgelöst)

$$A = R\,\dot{\alpha} + S\,\dot{\beta}, \quad B = -\tilde{S}\,\dot{\alpha} + T\,\dot{\beta} \tag{23.10}$$

abliest. Die O.C.R.B. haben zur Folge, daß die Matrizen R und T symmetrisch sind, während die Matrix $\tilde{S}$ wie bereits durch die Bezeichnung zum Ausdruck gebracht, zu S transponiert ist.

Setzt man wieder harmonische Vorgänge der Frequenz ω voraus und eliminiert α, β, A, B aus (23.9) und (23.10), so folgt

$$y = \alpha(\omega)\, x \tag{23.11}$$

mit
$$\alpha(\omega) = a - b\,[c + i\,\omega\,R - \omega^2\,S\,(e + i\,\omega\,T)^{-1}\,\tilde{S}]^{-1}\,\tilde{b}. \tag{23.12}$$

Auch jetzt ist die Matrix $\alpha(\omega)$ der thermodynamischen Koeffizienten zur Frequenz ω eine symmetrische Matrix. Ihre Struktur ist jedoch nicht mehr so einfach wie die in (23.7) angegebene. Der Grund hierfür ist, daß man zwar durch geeignete Wahl der inneren Variablen c und e zu Einheitsmatrizen, R und T zu Diagonalmatrizen machen kann, daß aber dann keine Freiheit mehr besteht, um auch S in eine einfache Gestalt zu bringen.

γ) *Allgemeine Bemerkungen über lineare dissipative Systeme.* Ausgehend von der expliziten Betrachtung von inneren Variablen α, β haben wir festgestellt, daß die thermodynamischen Koeffizienten $\alpha(\omega)$ sich in der Gestalt (23.12) darstellen lassen. Man kann schwächere, aber doch sehr nützliche Aussagen über die Existenz und die Eigenschaften der Matrix $\alpha(\omega)$ machen, ohne von inneren Variablen und ihrem Verhalten zu sprechen. Man geht hierzu von einigen allgemeinen und physikalisch einleuchtenden Postulaten aus. Diese sind die Linearität des Systems (Superpositionsprinzip), die Unveränderlichkeit des Materials und die Eigenschaft der nicht-negativen Entropieproduktion (Dissipativität)[1].

Diese Postulate, sind bei den bisher betrachteten Systemen erfüllt: Die Linearität ist dadurch sichergestellt, daß wir nur kleine Abweichungen vom Gleichgewichtszustand betrachten; die Unveränderlichkeit des Materials kommt in der Zeitunabhängigkeit der Matrizen a, b, c, e, R, S, T zum Ausdruck; die Dissipativität ist dadurch gewährleistet, daß die Matrizen R und T positiv definite quadratische Formen haben.

Man kann auf Grund dieser allgemeinen Postulate und gewisser physikalisch plausibler Einschränkungen für die Klasse der zugelassenen Funktionen $x(t)$, $y(t)$ sowie der zugelassenen linearen Transformationen $x(t) \to y(t)$ zeigen[2], daß für Vorgänge mit $x(t) = 0$ und damit $y(t) = 0$ für $t \leqq 0$ (Einschaltvorgänge) die Beziehung besteht

$$\int_0^\infty e^{-pt}\,y(t)\,dt = p\,Z(p)\int_0^\infty e^{-pt}\,x(t)\,dt. \tag{23.13}$$

Die Matrix $Z(p)$ der komplexen Variablen p wird als Impedanzmatrix des Systems bezeichnet. Ihre quadratische Form ist reell für reelle positive p und ist regulär analytisch mit positivem Realteil für $\mathrm{Re}\,p > 0$. Für harmonische Vorgänge gilt (23.11) mit $\alpha(\omega) = i\omega\,Z(i\omega)$. Diese Impedanzmatrix charakterisiert also in vollem Umfang das Verhalten linearer dissipativer Systeme.

$\delta)$ *Die Energiedissipation als Potential.* Während die innere Energie in den Variablen x, α, β eine Potentialeigenschaft für die thermodynamischen Beziehungen (23.9) hat, gilt dies nicht für die Entropieproduktion

$$\sigma = \frac{1}{T}\left(\tilde{A}\cdot\dot{\alpha} + \tilde{B}\cdot\dot{\beta}\right) = \frac{1}{T}\left(\tilde{\dot{\alpha}}\,R\,\dot{\alpha} + \tilde{\dot{\beta}}\,T\,\dot{\beta}\right) \tag{23.14}$$

in den Variablen $\dot{\alpha}$, $\dot{\beta}$; dies geht schon daraus hervor, daß der Ausdruck für σ die Matrix S nicht enthält. σ kann diese Matrix deshalb bei irgendwelchen Differentiationsprozessen nicht produzieren. Es gibt jedoch eine bemerkenswerte andere Schreibweise der Gln. (23.9) und (23.10), in der auch die Entropieproduktion den Charakter eines Potentials besitzt. Um sie zu gewinnen, führt man erst gerade innere Variable χ als Zeitintegrale der ungeraden Variablen β ein, $\dot{\chi} = \beta$. Es ist allerdings zu bemerken, daß man damit nicht auf den früher betrachteten Fall von geraden inneren Variablen zurückkommt; denn nun wird die innere Energie von den Zeitableitungen innerer Variablen abhängig:

$$u = u^+ + \tfrac{1}{2}\tilde{x}\,a\,x + \tilde{x}\,b\,\alpha + \tfrac{1}{2}\tilde{\alpha}\,c\,\alpha + \tfrac{1}{2}\tilde{\dot{\chi}}\,e\,\dot{\chi}. \tag{23.15}$$

Nach (23.9) ist $-B = e\,\dot{\chi}$; daher läßt sich $\dot{\beta}$ nach (23.10) durch $\dot{\chi}$ und $\dot{\alpha}$ ausdrücken. Die Entropieproduktion (23.14) kann daher in folgender Weise dargestellt werden:

$$2D = T\sigma = \tilde{\dot{\alpha}}\,r\,\dot{\alpha} + 2\,\tilde{\dot{\alpha}}\,s\,\dot{\chi} + \tilde{\dot{\chi}}\,t\,\dot{\chi}. \tag{23.16}$$

Hierin ist

$$\left.\begin{aligned} r &= \tilde{r} = R + S\,T^{-1}\tilde{S},\\ s &= -\,S\,T^{-1}e,\\ t &= \tilde{t} = e\,T^{-1}e. \end{aligned}\right\} \tag{23.17}$$

---

[1] J. Meixner: Z. Physik **139**, 30 (1954).
[2] H. König u. J. Meixner: Math. Nachr. **19**, 265 (1958). — J. Meixner u. H. König: Rheologica Acta **1**, 190 (1958).

Man nennt $\frac{1}{2} T\sigma = D$ auch die Dissipationsfunktion. Den durch die Gln. (23.9) und (23.10) ausgedrückten Sachverhalt kann man nun durch die Potentiale u in (23.15) und D in (23.16) ausdrücken. Man faßt, um zu einer kompakten Schreibweise zu kommen, die drei Spaltenvektoren $\boldsymbol{x}$, $\boldsymbol{\alpha}$, $\boldsymbol{X}$ zu einem neuen Spaltenvektor $\boldsymbol{z}$ zusammen. Ferner bildet man einen analogen Spaltenvektor mit $\boldsymbol{y}$, 0, 0 an Stelle von $\boldsymbol{x}$, $\boldsymbol{\alpha}$, $\boldsymbol{X}$. Er sei der Einfachheit halber ebenfalls mit $\boldsymbol{y}$ bezeichnet. Dann gilt

$$\frac{\partial u}{\partial z_i} + \frac{d}{dt}\frac{\partial u}{\partial \dot{z}_i} + \frac{\partial D}{\partial \dot{z}_i} = y_i. \tag{23.18}$$

In Analogie zum Formalismus der Lagrangeschen Gleichungen der Mechanik würde man beim zweiten Term ein negatives Zeichen erwarten. Dies kann man leicht erreichen, wenn man nicht u als Potential wählt, sondern konsequenterweise $u - \dot{\chi}\,\mathsf{e}\,\dot{\chi}$.

Die Gln. (23.18) lauten in unserem Falle explizite

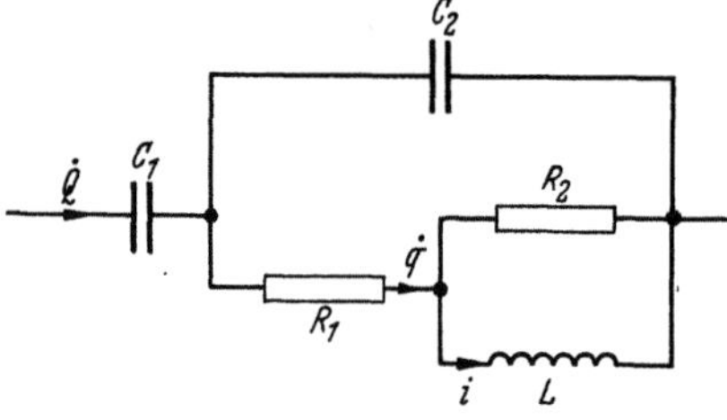

Fig. 3. Elektrisches Netzwerk mit Kapazitäten C_1 und C_2, Widerständen R_1 und R_2 und einer Induktivität L als Modell für ein System mit einer geraden und einer ungeraden inneren Variablen.

$$\left.\begin{aligned} \boldsymbol{y} &= \mathsf{a}\,\boldsymbol{x} + \mathsf{b}\,\boldsymbol{\alpha}, \\ 0 &= \mathsf{b}\,\boldsymbol{x} + (\mathsf{c}\,\boldsymbol{\alpha} + \mathsf{r}\,\dot{\boldsymbol{\alpha}}) + \mathsf{s}\,\dot{\chi}, \\ 0 &= \mathsf{s}\,\dot{\boldsymbol{\alpha}} + \mathsf{t}\,\dot{\chi} + \mathsf{e}\,\ddot{\chi}. \end{aligned}\right\} \tag{23.19}$$

Ihre Übereinstimmung mit (23.9) und (23.10) ist leicht festzustellen.

ε) *Verallgemeinerte Affinitäten.* Der Formalismus der Thermodynamik irreversibler Prozesse scheint mit der Einführung der inneren Variablen χ an Stelle der β nicht mehr anwendbar zu sein. Man kann ihn jedoch in einfacher Weise aufrechterhalten, wenn man nach dem Vorgang von Machlup und Onsager[1] verallgemeinerte Affinitäten einführt:

$$-A_\alpha = \frac{\partial u}{\partial \alpha} + \frac{d}{dt}\frac{\partial u}{\partial \dot{\alpha}} = \mathsf{b}\,\boldsymbol{x} + \mathsf{c}\,\boldsymbol{\alpha}, \tag{23.20}$$

$$-A_\chi = \frac{\partial u}{\partial \chi} + \frac{d}{dt}\frac{\partial u}{\partial \dot{\chi}} = \mathsf{e}\,\ddot{\chi}. \tag{23.21}$$

Dann lautet die Entropieproduktion

$$\sigma = \frac{1}{T}(\tilde{A}_\alpha \cdot \dot{\alpha} + \tilde{A}_\chi\,\dot{\chi}),$$

und als phänomenologische Gleichungen gewinnt man durch formale Anwendung der üblichen Vorschrift

$$A_\alpha = \mathsf{r}\,\dot{\alpha} + \mathsf{s}\,\dot{\chi}, \qquad A_\chi = \tilde{\mathsf{s}}\,\dot{\alpha} + \mathsf{t}\,\dot{\chi}.$$

Ihre Idendität mit den Gln. (23.9) und (23.10) prüft man leicht nach.

ζ) *Analogie mit elektrischen Netzwerken.* Die Analogie der Gln. (23.18) mit den Gleichungen für ein elektrisches Netzwerk ist offensichtlich. Sie ist nicht verwunderlich, da auch Netzwerke als thermodynamische Systeme — und zwar solche besonders durchsichtiger Struktur — aufgefaßt werden können. Wir geben als Beispiel ein Netzwerk (Fig. 3), das ein thermodynamisches System mit einer äußeren und je einer geraden und einer ungeraden inneren Variablen darstellt.

<hr>

[1] S. Machlup u. L. Onsager: Phys. Rev. **91**, 1512 (1954).

Die angelegte Spannung sei U, der Eingangsstrom $\dot{Q}$, der Strom durch den Widerstand R_1 sei $\dot{q}$ und jener durch die Induktivität L sei i. Dann ist Q äußere extensive Variable, q ist eine gerade, i eine ungerade innere Variable. Es gelten die zu (23.19) bis (23.21) völlig analogen Gleichungen

$$
\left.
\begin{aligned}
U &= \left(\frac{1}{C_1} + \frac{1}{C_2}\right) Q - \frac{1}{C_2} q, \\
-A &= -\frac{1}{C_2} Q + \frac{1}{C_2} q, \\
-B &= L i, \\
B &= -L\dot{q} + \frac{L^2}{R_2} i, \qquad A = R_1\dot{q} + L i.
\end{aligned}
\right\}
\tag{23.22}
$$

Es sei noch bemerkt, daß die nach MACHLUP und ONSAGER definierten Affinitäten gleich dem obigen A und gleich Li, also gerade die Spannungen an den dissipationsfreien Elementen C_2 und L bedeuten.

24. Transportphänomene als Relaxationserscheinungen. Die irreversiblen Prozesse kann man in zwei Klassen einteilen. Die Prozesse der ersten Art sind durch den *lokalen Zustand* bestimmt; sie können also auch im homogenen Zustand eines Systems ablaufen. Man bezeichnet Prozesse dieser Art und Effekte die auf ihnen beruhen, als *Relaxationserscheinungen*. Da der lokale Zustand durch skalare thermodynamische Variable beschrieben wird, erscheinen in den phänomenologischen Gleichungen die zeitlichen Änderungen dieser thermodynamischen Variablen als Flüsse.

Die irreversiblen Prozesse der zweiten Art sind die *Transportphänomene*. Man kann sie durch Stromdichten beschreiben, die durch die *Gradienten der Zustandsfelder* bestimmt werden. Die phänomenologischen Gleichungen für die Transportphänomene verknüpfen deshalb vektorielle oder tensorielle Größen. Sie haben nicht die Eigenschaft, daß die Flüsse Zeitableitungen von Zustandsvariablen sind.

In gewisser Hinsicht ist die Einteilung der irreversiblen Prozesse in Relaxationserscheinungen und Transportphänomene formal. Dies geht daraus hervor, daß man auch die Transportphänomene als Relaxationserscheinungen behandeln kann. Man braucht dazu nur den Begriff der inneren Variablen dadurch zu erweitern, daß man die Werte der Zustandsfelder in verschiedenen Raumpunkten des Systems, d.h. in hinreichend kleinen Volumenelementen, als innere Variable behandelt. Wird also der homogene Zustand etwa durch die Variablen T, p, γ_1, γ_2, γ_3, $\ldots$, γ_n beschrieben, so wird das System im inhomogenen Zustand durch die Gesamtheit der Werte $T(\mathbf{r})$, $p(\mathbf{r})$, $\gamma_1(\mathbf{r})$, $\gamma_2(\mathbf{r})$, $\ldots \gamma_n(\mathbf{r})$ charakterisiert. Hierbei läuft $\mathbf{r}$ über das ganze Systemvolumen.

Die Transportphänomene werden nun nicht mehr durch räumliche Stromdichten, sondern durch Übergangswahrscheinlichkeiten (z.B. für innere Energie und Materie) zwischen verschiedenen Volumenelementen des Systems dargestellt. Diese Beschreibung der Transporterscheinungen hat mehrere Vorteile:

1. Sie gibt eine vertiefte Einsicht in die Transportphänomene, weil man die besonderen Voraussetzungen für die Übergangswahrscheinlichkeiten erkennen kann, welche mit den phänomenologischen Ansätzen für vektorielle und tensorielle Phänomene [z.B. (11.9), (11.10), (11.16)] gemacht worden sind.

2. Sie erlaubt, die O.C.R.B., welche ursprünglich nur für phänomenologische Gleichungen mit skalaren Variablen bewiesen waren, auch für die Beschreibung der Transportphänomene durch Stromdichten nutzbar zu machen. Die Folgerungen sind nicht trivial[1].

---

[1] H.B.G. CASIMIR: Rev. Mod. Phys. **17**, 343 (1945).

3. Sie steht in methodischer Analogie zur thermodynamischen Behandlung der kinetischen Fundamentalgleichungen (s. Ziff. 25).

Wir entwickeln diese Form der Beschreibung ausführlich an dem besonders durchsichtigen Beispiel der Wärmeleitung im festen Körper bei äußerem Magnetfeld. Casimir hat an diesem Beispiel zum erstenmal die O.C.R.B. auf Transportphänomene konsequent angewandt. Wir schließen uns teilweise an eine spätere Darstellung von Mazur und de Groot[1] an.

α) *Die allgemeine Wärmeleitungsgleichung.* Für das Folgende setzen wir voraus, daß das System ruht, daß außer der Wärmeleitung kein irreversibler Prozeß abläuft und daß die thermische Ausdehnung zu vernachlässigen ist. Das System sei adiabatisch isoliert; das bedeutet, daß die Energie U des Systems konstant ist, also

$$\frac{du}{dt} = \int \varrho\,(\boldsymbol{r})\,\frac{\partial u(\boldsymbol{r},t)}{\partial t}\,dV = 0. \tag{24.1}$$

Die zeitliche Änderung der Entropie S ist dann

$$\frac{dS}{dt} = \int \varrho\,(\boldsymbol{r})\,\frac{\partial s(\boldsymbol{r},t)}{\partial t}\,dV = \int \frac{1}{T(\boldsymbol{r},t)}\,\varrho\,(\boldsymbol{r})\,\frac{\partial u(\boldsymbol{r},t)}{\partial t}\,dV. \tag{24.2}$$

Weicht die Temperatur im ganzen System wenig von einem Wert T_0 ab, und setzen wir $T(\boldsymbol{r}) = T_0 + \Delta T(\boldsymbol{r})$, wobei $|\Delta T(\boldsymbol{r})| \ll T_0$, so folgt aus (24.2) und (24.1) bis auf Glieder, die von höherer als zweiter Ordnung klein sind,

$$\frac{dS}{dt} = \int \left(-\frac{\Delta T(\boldsymbol{r},t)}{T_0^2}\right)\varrho\,(\boldsymbol{r})\,\frac{\partial u(\boldsymbol{r},t)}{\partial t}\,dV. \tag{24.3}$$

Dieser Ausdruck stellt die Entropieproduktion im gesamten, adiabatisch isolierten System dar. Er hat die übliche Struktur einer Summe von Flüssen $\varrho\,(\boldsymbol{r})\,\dfrac{\partial u(\boldsymbol{r},t)}{\partial t}\,dV$ multipliziert mit thermodynamischen Kräften $-\dfrac{\Delta T(\boldsymbol{r},t)}{T_0^2}$, die hier natürlich als Integral erscheint. Flüsse und Kräfte sind skalar; insbesondere sind die Flüsse die Zeitableitungen der thermodynamischen Variablen $\varrho\,(r)\,u\,(r)$. Diese sind ihrerseits zu den thermodynamischen Kräften konjugiert.

Die Flüsse sind wegen (24.1) nicht unabhängig. Daher lautet die Gleichgewichtsbedingung für das Temperaturfeld nicht $\Delta T(\boldsymbol{r}) = 0$ sondern $\Delta T(\boldsymbol{r}) =$ konst. Als phänomenologische Gleichungen haben wir

$$\varrho\,(\boldsymbol{r})\,\frac{\partial u(\boldsymbol{r},t)}{\partial t} = -\int K\big(\boldsymbol{r},\boldsymbol{B}\,(\boldsymbol{r});\boldsymbol{r}',\boldsymbol{B}\,(\boldsymbol{r}')\big)\,\Delta\,T(\boldsymbol{r}',t)\,dV' \tag{24.4}$$

anzusetzen. Der Kern K ist in der Kontinuumsformulierung das Analogon zur Matrix der phänomenologischen Koeffizienten; in ihm ist der konstante Faktor $1/T_0^2$ aufgenommen. Er bestimmt den Energieübergang vom Punkte $\boldsymbol{r}'$ zum Punkte $\boldsymbol{r}$; man muß daher zulassen, daß er von der magnetischen Induktion $\boldsymbol{B}$ des äußeren Magnetfeldes in $\boldsymbol{r}$ und in $\boldsymbol{r}'$ abhängt.

Die O.C.R.B. lauten nach (6.9):

$$K(\boldsymbol{r},\boldsymbol{B};\boldsymbol{r}',\boldsymbol{B}') = K(\boldsymbol{r}',-\boldsymbol{B}';\boldsymbol{r},-\boldsymbol{B}), \tag{24.5}$$

wobei $\boldsymbol{B}\,(\boldsymbol{r})$ durch $\boldsymbol{B}$, $\boldsymbol{B}\,(\boldsymbol{r}')$ durch $\boldsymbol{B}'$ abgekürzt ist. Wegen der linearen Abhängigkeit der Flüsse nach (24.1) gelten zu (7.3) analoge Relationen, nämlich

$$\int K(\boldsymbol{r},\boldsymbol{B};\boldsymbol{r}',\boldsymbol{B}')\,dV' = 0. \tag{24.6}$$

Man erhält diese Gleichung aus (24.4), wenn man $\Delta T(r') =$ konst. $\neq 0$ setzt.

---

[1] P. Mazur u. S. R. de Groot: Physica, Haag **19**, 961 (1953).

$\beta)$ *Die Fouriersche Wärmeleitungsgleichung als Spezialfall.* Die Wärmeleitungsgleichung (24.4) geht in eine partielle Differentialgleichung zweiter Ordnung in den Raumkoordinaten x_α über, wenn man den Kern K durch den formalen Ansatz

$$K(\boldsymbol{r}, \boldsymbol{B}; \boldsymbol{r}', \boldsymbol{B}') = -a(\boldsymbol{r}, \boldsymbol{B})\, \delta(\boldsymbol{r}' - \boldsymbol{r}) - \sum_{\alpha=1}^{3} b_\alpha(\boldsymbol{r}, \boldsymbol{B})\, \frac{\partial}{\partial x_\alpha'}\, \delta(\boldsymbol{r}' - \boldsymbol{r}) - $$
$$\left. - \sum_{\alpha,\beta=1}^{3} c_{\alpha\beta}(\boldsymbol{r}, \boldsymbol{B})\, \frac{\partial^2}{\partial x_\alpha' \partial x_\beta'}\, \delta(\boldsymbol{r}' - \boldsymbol{r}) \right\} \qquad (24.7)$$

darstellt. $\delta(\boldsymbol{r}' - \boldsymbol{r})$ ist eine räumliche δ-Funktion. Die Koeffizienten $a(\boldsymbol{r}, \boldsymbol{B})$, $b_\alpha(\boldsymbol{r}, \boldsymbol{B})$, $c_{\alpha\beta}(\boldsymbol{r}, \boldsymbol{B})$ sind nicht frei wählbar. Wegen (24.6) ist $a(\boldsymbol{r}, \boldsymbol{B}) = 0$. Um die O.C.R.B. (24.5) auf den Ansatz (24.7) anzuwenden, multiplizieren wir (24.5) mit einer willkürlichen Funktion $f(\boldsymbol{r}')$, integrieren nach dV', setzen (24.7) ein und entfernen die δ-Funktionen durch Ausführung der Integration nach dV'. Aus der sich ergebenden Identität für $f(\boldsymbol{r})$ erhält man durch Koeffizientenvergleich

$$c_{\alpha\beta}(\boldsymbol{r}, \boldsymbol{B}) + c_{\beta\alpha}(\boldsymbol{r}, \boldsymbol{B}) = c_{\alpha\beta}(\boldsymbol{r}, -\boldsymbol{B}) + c_{\beta\alpha}(\boldsymbol{r}, -\boldsymbol{B}), \qquad (24.8)$$

$$b_\alpha(\boldsymbol{r}, \boldsymbol{B}) + b_\alpha(\boldsymbol{r}, -\boldsymbol{B}) + \sum_{\beta=1}^{3} \left(\frac{\partial c_{\alpha\beta}(\boldsymbol{r}, \boldsymbol{B})}{\partial x_\beta} + \frac{\partial c_{\beta\alpha}(\boldsymbol{r}, \boldsymbol{B})}{\partial x_\beta} \right) = 0, \qquad (24.9)$$

$$\sum_\alpha \frac{\partial b_\alpha(\boldsymbol{r}, B)}{\partial x_\alpha} + \sum_\alpha \frac{\partial^2 c_{\alpha\beta}(\boldsymbol{r}, \boldsymbol{B})}{\partial x_\alpha \partial x_\beta} = 0. \qquad (24.10)$$

Aus (24.10) folgt

$$b_\alpha(\boldsymbol{r}, \boldsymbol{B}) = -\sum_{\beta=1}^{3} \left(\frac{\partial c_{\alpha\beta}(\boldsymbol{r}, \boldsymbol{B})}{\partial x_\beta} + \frac{\partial \Psi_{\alpha\beta}(\boldsymbol{r}, \boldsymbol{B})}{\partial x_\beta} \right) \qquad (24.11)$$

mit einem antisymmetrischen Tensor $\Psi_{\alpha\beta} = -\Psi_{\beta\alpha}$. Setzt man (24.11) in (24.9) ein, so ergibt sich

$$\sum_\beta \frac{\partial}{\partial x_\beta} \left[\Psi_{\alpha\beta}(\boldsymbol{r}, \boldsymbol{B}) - \Psi_{\beta\alpha}(\boldsymbol{r}, -\boldsymbol{B}) + c_{\alpha\beta}(\boldsymbol{r}, \boldsymbol{B}) - c_{\beta\alpha}(\boldsymbol{r}, -\boldsymbol{B}) \right] = 0. \qquad (24.12)$$

Die Wärmeleitungsgleichung (24.4) gibt nach Einsetzen von (24.7), Ausführung der Integrationen und Berücksichtigung von (24.8), (24.11) und (24.12)

$$\varrho(\boldsymbol{r})\, \frac{\partial u(\boldsymbol{r},t)}{\partial t} = \sum_{\alpha,\beta=1}^{3} \frac{\partial}{\partial x_\beta} \left(\lambda_{\alpha\beta}(\boldsymbol{r}, \boldsymbol{B})\, \frac{\partial \Delta T(\boldsymbol{r},t)}{\partial x_\alpha} \right), \qquad (24.13)$$

also eine Fouriersche Wärmeleitungsgleichung mit dem Tensor

$$\lambda_{\alpha\beta} \equiv c_{\alpha\beta} + \Psi_{\alpha\beta}.$$

Letzterer besitzt auf Grund von (24.8) und (24.12) die Eigenschaften

$$\lambda_{\alpha\beta}(\boldsymbol{r}, \boldsymbol{B}) + \lambda_{\beta\alpha}(\boldsymbol{r}, \boldsymbol{B}) = \lambda_{\alpha\beta}(\boldsymbol{r}, -\boldsymbol{B}) + \lambda_{\beta\alpha}(\boldsymbol{r}, -\boldsymbol{B}) \qquad (24.14)$$

und

$$\sum_{\beta=1}^{3} \frac{\partial \lambda_{\alpha\beta}(\boldsymbol{r}, \boldsymbol{B})}{\partial x_\beta} = \sum_{\beta=1}^{3} \frac{\partial \lambda_{\beta\alpha}(\boldsymbol{r}, -\boldsymbol{B})}{\partial x_\beta}. \qquad (24.15)$$

Der Tensor $\lambda_{\alpha\beta}(\boldsymbol{r}, \boldsymbol{B})$ ist nicht notwendig der Wärmeleitungstensor des Fourierschen Ansatzes. Wir nehmen an, der Wärmeleitungstensor unterscheide sich von $\lambda_{\alpha\beta}(\boldsymbol{r}, \boldsymbol{B})$ um einen Tensor $\nu_{\alpha\beta}(\boldsymbol{r}, \boldsymbol{B})$. Dann lautet der Fouriersche Ansatz

$$J_{u\alpha} = -\sum_{\beta=1}^{3} \left(\lambda_{\alpha\beta}(\boldsymbol{r}, \boldsymbol{B}) + \nu_{\alpha\beta}(\boldsymbol{r}, \boldsymbol{B}) \right) \frac{\partial \Delta T(\boldsymbol{r}, t)}{\partial x_\beta}. \qquad (24.16)$$

Setzt man ihn in die Bilanzgleichung der inneren Energie $\varrho \dfrac{\partial u}{\partial t} = - \operatorname{div} J_u$, ein, so erhält man (24.13) genau dann, wenn

$$v_{\alpha\beta}(r, \boldsymbol{B}) + v_{\beta\alpha}(r, \boldsymbol{B}) = 0 \qquad (\alpha, \beta = 1, 2, 3), \tag{24.17}$$

$$\sum_{\beta=1}^{3} \frac{\partial v_{\alpha\beta}}{\partial x_\beta} = 0 \qquad (\alpha = 1, 2, 3). \tag{24.18}$$

Der Wärmeleitungstensor ist also aus der Wärmeleitungsgleichung (24.13) nur bis auf einen antisymmetrischen Tensor bestimmt, der der Bedingung (24.18) unterliegt.

Man überzeugt sich leicht, daß wegen (24.17) und (24.18) nicht nur (24.13), sondern auch die Beziehungen (24.14) und (24.15) mit dem Wärmeleitfähigkeitstensor $\lambda_{\alpha\beta} + v_{\alpha\beta}$ an Stelle von $\lambda_{\alpha\beta}$ gelten. Aus diesem Grunde bleiben die Gln.(24.13) bis (24.16) richtig, wenn wir unter $\lambda_{\alpha\beta}$ den Wärmeleitfähigkeitstensor verstehen und $v_{\alpha\beta} = 0$ setzen.

Es sei sehr bemerkenswert, daß wir für den Wärmeleitfähigkeitstensor durch Anwendung der O.C.R.B. keineswegs die in (6.9) angenommene Beziehung $\lambda_{\alpha\beta}(\boldsymbol{B}) = \lambda_{\beta\alpha}(-\boldsymbol{B})$ erhalten. Es ist aber noch merkwürdiger, daß die Beziehungen (24.14) für $\boldsymbol{B} = 0$ und (24.15) für Systeme im homogenen Zustand inhaltsleer sind. Tatsächlich erhält man weitergehende Folgerungen, wenn man (24.15) nicht als Gleichung für ein bestimmtes System auffaßt, sondern wenn man den Wärmeleitfähigkeitstensor als Funktion des Materials, der Temperatur und des Magnetfeldes ansieht und (24.15) als Identität für beliebige Materialverteilungen und Felder $T(r)$, $\boldsymbol{B}(r)$ auffaßt. Dann folgt aus (24.15), daß $\lambda_{\alpha\beta}(r, \boldsymbol{B}) - \lambda_{\beta\alpha}(r, -\boldsymbol{B})$ vom Material, von der Temperatur und vom Magnetfeld unabhängig, also ein universeller Tensor ist. Im isotropen Körper ohne Magnetfeld ist $\lambda_{\alpha\beta}$ diagonal, daher verschwindet dieser universelle Tensor.

$\gamma)$ *Verallgemeinerungen.* Physikalisch vernünftiger ist es, den Kern $K(r, \boldsymbol{B};$ $r', \boldsymbol{B}')$ nicht durch uneigentliche Funktionen darzustellen, sondern als eine Funktion von r und $r' - r$ mit einem sehr schmalen und steilen Maximum für $r' \approx r$ anzunehmen. Die Breite dieses Maximums kennzeichnet die Abstände, über die ein unmittelbarer Energieaustausch erfolgt. Im atomistischen Bild ist die Größenordnung dieser Abstände gleich der der Entfernung, über welche im Elementarprozeß Energie transportiert wird, also von der Größenordnung der mittleren freien Weglänge in Gasen oder Gitterkonstanten in festen Körpern. Man kann dann den Kern K in seiner Abhängigkeit von $r' - r$ nach Hermiteschen Orthogonalfunktionen, die mit einer Gauß-Funktion dieser Breite multipliziert sind, entwickeln. In erster Näherung kommt man dann mit denselben Gliedern aus, die dem Ansatz (24.7) entsprechen, wenn die Temperaturänderung über die Breite hinreichend klein ist[1].

Die Wärmeleitungsgleichung (24.4) umfaßt neben der Fourierschen Wärmeleitung auch noch Mechanismen, bei denen im Elementarprozeß die Energie über beliebige Entfernungen übertragen wird. Ein solcher Fall liegt beim Wärmeaustausch durch Strahlung im durchsichigen Medium vor.

Die Behandlung der Wärmeleitungsgleichung in diesem Abschnitt ist ein Beispiel für viele ähnliche Probleme. Solche sind andere makroskopische Transportprozesse, aber auch Transportprozesse im Phasenraum, die durch kinetische

---

[1] Diese Nahewirkungsauffassung hat bereits Stokes zur Begründung einer Wärmeleitungsgleichung zweiter Ordnung in den Raumkoordinaten benutzt. G. G. Stokes: Cambridge and Dublin Math. J. 6, 215 (1851).

Fundamentalgleichungen beschrieben werden. Bei den makroskopischen Transportprozessen hat man es im Elementarprozeß mit Übertragung von Materie, Energie, Impuls, Ladung über sehr kleine Entfernungen zu tun. Man kann daher stets Ansätze der Art (24.7) machen. Dasselbe gilt bei der Behandlung der Brownschen Bewegung hinsichtlich der Koordinaten- und Impulsdifferenzen im Elementarprozeß und bei den kinetischen Fundamentalgleichungen in der Regel in bezug auf die Koordinaten (s. dazu auch Ziff. 25).

Die Herleitung der Folgerungen aus den O.C.R.B. ist allerdings, wie auch im Falle der Wärmeleitung etwas schwerfällig. DE GROOT und VAN KAMPEN[1] haben indessen gezeigt, daß man diese Folgerungen für Transporterscheinungen auch direkt an den Bilanzgleichungen und den phänomenologischen Gleichungen ziehen kann, ohne daß man erst den Umweg über die allgemeinen Integralkerne K gehen muß. Die Begründung hierfür läßt sich entweder aus den O.C.R.B., oder durch unmittelbare Herleitung aus der Schwankungstheorie nach DE GROOT und MAZUR[2] gewinnen.

25. Kinetische Fundamentalgleichungen. Unter kinetischen Fundamentalgleichungen verstehen wir Gleichungen, die die zeitliche Entwicklung von Verteilungsfunktionen im μ-Raum beschreiben. Beispiele sind die Boltzmannsche Gleichung der kinetischen Gastheorie, die Blochsche Gleichung der Elektronentheorie, die Fokker-Plancksche und die Kramerssche Gleichung der Brownschen Bewegung.

Wir zeigen im folgenden, daß die kinetischen Fundamentalgleichungen phänomenologische Ansätze im Sinne der Thermodynamik der irreversiblen Prozesse sind und besprechen die Eigenschaften der phänomenologischen Koeffizienten. Bei den genannten Beispielen treten sie im allgemeinen als Integralkern in Erscheinung; vgl. auch Ziff. 24.

α) *Fundamentalgleichungen als phänomenologische Gleichungen.* Wir betrachten Systeme im μ-Raum mit einer Hamilton-Funktion $H(q_\alpha, p_\alpha; x_i)$. Sie soll gerade gegenüber der Ersetzung $p_\alpha \to -p_\alpha$ sein. Die x_i seien äußere Parameter, z.B. Volumen oder elektrische Feldstärke. Eine Gesamtheit von N solchen Systemen wird durch eine Verteilungsfunktion $f(q_\alpha, p_\alpha, t)$ im Phasenraum des einzelnen Systems beschrieben. Damit ergibt sich die innere Energie U und die Entropie S der Gesamtheit zu

$$U = N \int H f\, d\tau, \quad S = -N k \int f \log f\, d\tau \qquad (25.1)$$

mit Integration über den ganzen Phasenraum. Die Verteilungsfunktion ist damit zu 1 normiert, d.h.

$$\int f\, d\tau = 1. \qquad (25.2)$$

Die Gleichgewichtsverteilung $f^{(0)}(q_\alpha, p_\alpha; x_i, T)$ zur Temperatur T ist durch

$$\log f^{(0)}(q_\alpha, p_\alpha; x_i, T) = -\frac{H(q_\alpha, p_\alpha; x_i)}{kT} - \log Z(x_i, T) \qquad (25.3)$$

mit der Zustandssumme Z gegeben.

Die kinetischen Fundamentalgleichungen haben die Struktur

$$\frac{df}{dt} \equiv \frac{\partial f}{\partial t} + \sum_\alpha \left(\frac{\partial f}{\partial q_\alpha} \frac{\partial H}{\partial p_\alpha} - \frac{\partial f}{\partial p_\alpha} \frac{\partial H}{\partial q_\alpha} \right) = \left(\frac{\partial f}{\partial t} \right)_{\text{Stöße}}. \qquad (25.4)$$

---

[1] S. R. DE GROOT u. VAN KAMPEN: Physica, Haag **21**, 39 (1954).
[2] S. R. DE GROOT u. P. MAZUR: Physica, Haag **19**, 961 (1953).

Besteht keine Wechselwirkung zwischen den Systemen der Gesamtheit, so reduziert sich (25.4) auf die Liouvillesche Gleichung $df/dt = 0$. Eine Wechselwirkung äußert sich in einer zusätzlichen Änderung der Verteilungsfunktion, die in (25.4) mit $(\partial f/\partial t)_{\text{Stöße}}$ bezeichnet ist, auch wenn die Wechselwirkung nicht unmittelbar zwischen den Systemen, sondern indirekt, z.B. über ein Temperaturbad, erfolgt.

Ansätze für $(\partial f/\partial t)_{\text{Stöße}}$ kann man aus der Thermodynamik irreversibler Prozesse herleiten. Den Anschluß an den üblichen Formalismus erhalten wir, wenn wir eine Zustandsänderung betrachten, bei der sich U um dU, S um dS, x_i um dx_i und f um δf ändern. Aus (25.1) folgt dann

$$dU - T\,dS - \sum y_i\,dx_i = N\int \mu\,\delta f\,d\tau, \tag{25.5}$$

wobei

$$y_i = \frac{\partial U}{\partial x_i} = N\int \frac{\partial H(q_\alpha, p_\alpha; x_i)}{\partial x_i}\,f\,d\tau, \tag{25.6}$$

$$\mu(q_\alpha, p_\alpha; x_i, T) = H(q_\alpha, p_\alpha; x_i) - kT\log f(q_\alpha, p_\alpha) \tag{25.7}$$

$$= kT\,[\log f(q_\alpha, p_\alpha) - \log f^{(0)}(q_\alpha, p_\alpha; x_i, T)] - kT\log Z. \tag{25.8}$$

Die Beziehung (25.5) ist eine Gibbssche Fundamentalgleichung. Die y_i sind also zu x_i konjugierte Variable. Jedem Punkt des Phasenraumes q_α, p_α ist eine innere Variable $f(q_\alpha, p_\alpha)$ zugeordnet. Den Koeffizienten $\mu(q_\alpha, p_\alpha; x_i, T)$ nennen wir das chemische Potential im Punkte q_α, p_α. Im thermodynamischen Gleichgewicht ist das chemische Potential über den ganzen Phasenraum konstant. Wir setzen nun reversiblen Arbeitsaustausch mit der Umgebung voraus, d.h. $dU = \sum_i y_i\,dx_i$.

Dann ergibt sich die Entropieproduktion aus (25.5) und (25.7) wegen (25.2) zu

$$\left(\frac{dS}{dt}\right)_{\text{irr}} = N\,k\int \frac{\partial f}{\partial t}\,(\log f^{(0)} - \log f)\,d\tau. \tag{25.9}$$

Sie hat die übliche Gestalt einer Summe aus Produkten von Flüssen $\partial f/\partial t$ und thermodynamischen Kräften $N\,k\,(\log f^{(0)} - \log f)$. Beschränken wir uns auf kleine Abweichungen vom Gleichgewicht, so ist

$$\log f^{(0)} - \log f \approx \frac{f^{(0)} - f}{f^{(0)}}. \tag{25.10}$$

Man liest somit aus (25.9) und (25.10) lineare phänomenologische Gleichungen der Form

$$\frac{df(q_\alpha, p_\alpha, t)}{dt} = \int K(q_\alpha, p_\alpha; q'_\alpha, p'_\alpha; x_i, T)\,\frac{f^{(0)}(q_\alpha, p_\alpha, x_i, T) - f(q_\alpha, p_\alpha, t)}{f^{(0)}(q_\alpha, p_\alpha, x_i, T)}\,d\tau' \tag{25.11}$$

ab. Der Integralkern K gibt die Übergangseigenschaften zwischen den Punkten q_α, p_α und q'_α, p'_α des Phasenraumes an. Seine Abhängigkeit von x_i, T werden wir im folgenden nicht mehr explizite anschreiben.

Die phänomenologischen Gleichungen (25.11) haben tatsächlich die Struktur der Gl. (25.4); man kann nämlich mit Hilfe von (25.3) zeigen, daß die Inhomogenität von (25.11) eine scheinbare ist.

β) *Allgemeine Eigenschaften der Integralkerne.* Der Integralkern in (25.11) besitzt einige allgemeine Eigenschaften. Zunächst genügt K den O.C.R.B. Sie sind hier gegenüber (24.5) zu modifizieren, da die p_α ungerade Parameter sind. Den Fall eines Magnetfeldes lassen wir jetzt außer acht. Dann ist also

$$K(q_\alpha, p_\alpha; q'_\alpha, p'_\alpha) = K(q'_\alpha, -p'_\alpha; q_\alpha, -p_\alpha). \tag{25.12}$$

Eine weitere allgemeine Eigenschaft ergibt sich, wenn man (25.11) über den Phasenraum integriert und (25.2) beachtet. Es entsteht dann eine Identität, die für beliebige normierte Verteilungsfunktionen $f(q'_\alpha, p'_\alpha, t)$ gilt. Daraus schließt man auf

$$\int K(q_\alpha, p_\alpha; q'_\alpha, p'_\alpha)\, d\tau' = C\, f^{(0)}(q_\alpha, p_\alpha) \tag{25.13}$$

mit einer beliebigen Konstanten C. Definiert man nun einen neuen Kern $\widetilde{K} = K - C\, f^{(0)}(H)\, f^{(0)}(H')$, so gilt für ihn ebenfalls (25.11), während (25.13) in eine entsprechende Gleichung mit verschwindender rechter Seite übergeht. Bezeichnet man $\widetilde{K}$ wieder mit K, so bleibt also (25.11) bestehen und statt (25.13) gilt ohne Beschränkung der Allgemeinheit

$$\int K(q_\alpha, p_\alpha; q'_\alpha, p'_\alpha)\, d\tau' = 0. \tag{25.14}$$

Wegen (25.14) gilt dann auch

$$\int K(q_\alpha, p_\alpha; q'_\alpha, p'_\alpha)\, d\tau = 0. \tag{25.15}$$

Wir zerlegen nun den Kern K in einen symmetrischen und in einen antisymmetrischen Anteil bezüglich Vertauschung von q_α, p_α mit q'_α, p'_α. Der antisymmetrische Anteil gibt keinen Beitrag zur Entropieproduktion, entspricht also einem „reversiblen Anteil" der Bewegung. Da die reversible Bewegung einer Gesamtheit von mechanischen Systemen durch eine Liouvillesche Gleichung $df/dt = 0$ beschrieben wird, so liegt es nahe, anzunehmen, daß der antisymmetrische Anteil von K entweder verschwindet oder daß er einem Zusatz zur Hamilton-Funktion entspricht, der auch von vornherein in diese aufgenommen werden kann. Wir wollen diese Annahme machen, da sie sich für alle bekannten Fundamentalgleichungen begründen läßt. Man vergleiche hierzu die Diskussionen von BERGMANN und LEBOWITZ[1], MEIXNER[2], aber auch das Gegenbeispiel von LEBOWITZ und BERGMANN[3].

Unter der Voraussetzung, daß der Kern symmetrisch ist, folgt mit Hilfe von (25.12) weiter, daß

$$K(q, p; q', p') = K(q, -p; q', -p'). \tag{25.16}$$

γ) Spezielle Eigenschaften der Integralkerne. Eingangs sind verschiedene Fundamentalgleichungen der Struktur (25.4) genannt worden. Sie unterscheiden sich nur in der expliziten Gestalt der rechten Seite, also durch den phänomenologischen Ansatz, d.h. durch den Kern $K(q_\alpha, p_\alpha; q'_\alpha, p'_\alpha)$. Der Kern der Boltzmannschen oder der Blochschen Gleichung muß nahe Stöße im Ortsraum darstellen. Die Abhängigkeit von q_α und q'_α wird sich also durch einen Ansatz $\delta(q' - q) \times K_1(q_\alpha; p_\alpha p'_\alpha)$ ausdrücken lassen, wobei K_1 bei homogenen Systemen von q unabhängig ist. Der Kern der Kramersschen Gleichung, oder jeder Gleichung, die sich auf Brownsche Bewegung oder auf Systeme im Γ-Raum bezieht, muß kleinen Stößen sowohl im Orts- als auch im Geschwindigkeitsraum entsprechen. Wir wollen die Kramerssche Gleichung näher behandeln und machen daher einen Ansatz, der nur nahe Stöße im Impuls- und im Ortsraum zuläßt. Er ist analog zu (24.7), und lautet unter Berücksichtigung von (25.4):

$$K(q_\alpha, p_\alpha; q'_\alpha, p'_\alpha) = \delta(q'_\alpha - q_\alpha) \times \left. \begin{array}{l} \\ \times \left[\sum_\alpha b_\alpha(q_\alpha, p_\alpha) \dfrac{\partial}{\partial p'_\alpha} \delta(p'_\alpha - p_\alpha) + \sum_{\alpha,\beta} c_{\alpha\beta}(q_\alpha, p_\alpha) \dfrac{\partial^2}{\partial p'_\alpha \partial p'_\beta} \delta(p'_\alpha - p_\alpha) \right]. \end{array} \right\} \tag{25.17}$$

[1] P.G. BERGMANN u. J.L. LEBOWITZ: Phys. Rev. **99**, 578 (1955).
[2] J. MEIXNER: Z. Physik **149**, 624 (1957).
[3] J.L. LEBOWITZ u. P.G. BERGMANN: Ann. of Physics **1**, 1 (1957).

Anwendung der O.C.R.B. (25.12) und Einsetzen in (25.11) gibt die Kramerssche Gleichung[1]

$$\frac{df}{dt} = \sum_{\alpha,\beta} \frac{\partial}{\partial p_\alpha} L_{\alpha\beta}(q_\alpha, p_\alpha) \left[\frac{\partial H}{\partial p_\beta} f + kT \frac{\partial f}{\partial p_\beta} \right] \qquad (25.18)$$

mit

$$L_{\alpha\beta}(q_\alpha, p_\alpha) = [2kT f^{(0)}]^{-1} [c_{\alpha\beta}(q_\alpha, p_\alpha) + c_{\beta\alpha}(q_\alpha, -p_\alpha)]. \qquad (25.19)$$

Hieraus folgt die Symmetrieeigenschaft $L_{\alpha\beta}(q_\alpha, p_\alpha) = L_{\beta\alpha}(q_\alpha, -p_\alpha)$. Gilt (25.16), so ist die Matrix $L_{\alpha\beta}$ symmetrisch und gerade in den p_α. Überdies folgert man aus der positiven Entropieproduktion, daß diese Matrix eine positiv definite quadratische Form hat.

Man kann von (26.9) leicht zu einer Beschreibung der Verteilung durch diskrete innere Variable übergehen. Dazu entwickelt man $f(q_\alpha, p_\alpha, t)$ nach einem Orthogonalsystem von Funktionen $\Phi_n(q_\alpha, p_\alpha)$, $\Psi_n(q_\alpha, p_\alpha)$ mit geeigneter Gewichtsfunktion. Die Φ_n seien gerade, die Ψ_n ungerade bei der Ersetzung $p_\alpha \to -p_\alpha$. Die Koeffizienten der Φ_α sind dann gerade, die der Ψ_n ungerade innere Variable. Man kann ihnen in einfacher Weise Affinitäten zuordnen und den in Ziff. 23 dargestellten Formalismus für innere Variable und Affinitäten für Parameter x_i und konjugierte Variable y_i in vollem Umfang bestätigen.

Einzelheiten finden sich bei Meixner[2] und für den einfacheren Fall der Smoluchowski-Gleichung bereits bei Prigogine und Mazur[3].

E. Relativistische Thermodynamik der irreversiblen Prozesse.

26. Vorbemerkungen. Irreversible Prozesse laufen meist so langsam ab, daß eine relativistische Behandlung kaum angebracht ist. Trotzdem gibt es einige gute theoretische Gründe, die eine relativistische Formulierung der Theorie irreversibler Prozesse wünschenswert erscheinen lassen. Der wichtigste unter diesen ist, daß die Gesetze für das Verhalten der kontinuierlichen Materie im elektromagnetischen Feld wegen der Lorentz-Kovarianz der Maxwellschen Gleichungen nur in einer relativistischen Formulierung durchsichtig werden. Aber auch für die kontinuierliche Materie ohne elektromagnetische Felder ist der Versuch einer relativistischen Fassung der Gesetze aus Gründen der Systematik sehr wohl angebracht. Praktisch bedeutsam kann die relativistische Theorie irreversibler Prozesse für die Theorie des Plasmas bei schnellen Strömungsvorgängen sein.

Aufbauend auf einer Arbeit von Eckart[4], welche Materialien, die nur aus einer Komponente bestehen, behandelt, hat Kluitenberg[5] in seiner gehaltvollen und ausführlichen Dissertation die relativistische Behandlung für kontinuierliche Medien mit einer beliebigen Zahl von chemischen Komponenten in elektromagnetischen Feldern gegeben. Ihre Ergebnisse sollen im folgenden kurz auseinandergesetzt werden. Auf die Durchführung der häufig langwierigen Umrechnungen wird verzichtet.

Eine charakteristische Abweichung von der nicht-relativistischen Theorie rührt daher, daß der Massenerhaltungssatz (wegen der Trägheit der Energie) nunmehr kein unabhängiger Erhaltungssatz ist, sondern mit dem Energieerhaltungssatz zusammenfließt. An seine Stelle treten die Erhaltungssätze für die

[1] H. A. Kramers: Physica, Haag **7**, 284 (1940).
[2] Siehe Fußnote 2, S. 493.
[3] I. Prigogine u. P. Mazur: Physica, Haag **19**, 241 (1953).
[4] C. Eckart: Phys. Rev. **58**, 919 (1940).
[5] G. A. Kluitenberg: Relativistic Thermodynamics of Irreversible Processes, Diss. Leiden 1954, und S. R. de Groot: Physica, Haag **19**, 689, 1079 (1953); **20**, 199 (1954).

Anzahlen der letzten Bestandteile des Kontinuums, also etwa der Elektronen und der verschiedenen Atomkerne, falls Zerstrahlung und Kernumwandlungen ausgeschlossen bleiben. Diese Erhaltungssätze formuliert man zweckmäßig für die Ruhmassendichten an Stelle der Teilchenzahlen. Die Ruhmasse eines Atoms oder Ions ist *definiert* als Summe der Ruhmassen des Kerns und der Hüllenelektronen. Damit sind dann die Ruhmassendichten ϱ^j ($j = 1, 2, \ldots, n$) der einzelnen Komponenten der kontinuierlichen Mischung als Ruhmasse pro Volumeneinheit erklärt[1]. Wegen der Lorentz-Kontraktion sind die ϱ^j nicht invariant.

Die gesamte Ruhmassendichte ist

$$\varrho = \sum_{j=1}^{n} \varrho^j. \tag{26.1}$$

Der Materiefluß der Komponente j ist in bekannter Weise durch den Vierervektor

$$m_\alpha^j = \varrho^j v_\alpha^j \quad (\alpha = 1, 2, 3), \qquad m_4^j = i c \varrho^j, \quad (j = 1, \ldots, n) \tag{26.2}$$

gegeben, wenn $\boldsymbol{v}^j$ (mit den Komponenten v_1^j, v_2^j, v_3^j) die Geschwindigkeit der Komponente j bedeutet.

Der gesamte Massenfluß ist durch den Vierervektor $m_\alpha = \sum_j m_\alpha^j$ gegeben. Definiert man in üblicher Weise die baryzentrische Geschwindigkeit $\boldsymbol{v}$ durch

$$\varrho \, \boldsymbol{v} = \sum_{j=1}^{n} \varrho^j \boldsymbol{v}^j, \tag{26.3}$$

so wird

$$m_\alpha = \varrho \, v_\alpha \quad (\alpha = 1, 2, 3), \qquad m_4 = i c \, \varrho. \tag{26.4}$$

Für die auf die baryzentrische Bewegung bezogenen Materieflüsse sind die Definitionen (3.8) nicht mehr geeignet, da sie sich nicht zu einem Vierervektor ergänzen lassen. Es liegt nahe, Ansätze der folgenden Art zu versuchen:

$$I_\alpha^j = \varrho^j (v_\alpha^j - C^j v_\alpha), \quad (\alpha = 1, 2, 3), \qquad I_4^j = i c \, \varrho^j (1 - C^j). \tag{26.5}$$

Damit diese Größen Vierervektoren sind, muß $C^j \varrho^j / \varrho$ für jedes j eine Invariante sein; ϱ^j / ϱ selbst ist wegen der Diffusion, d.h. wegen $\boldsymbol{v}^j \neq \boldsymbol{v}$, keine Invariante. Ferner wird man für C^j eine Abhängigkeit nur von $\boldsymbol{v}$ und $\boldsymbol{v}^j$ anzunehmen haben. Die zweckmäßigste Wahl ist nach KLUITENBERG

$$C^j = \frac{c^2 - \boldsymbol{v}^j \cdot \boldsymbol{v}}{c^2 - \boldsymbol{v}^2}. \tag{26.6}$$

Mit ihr haben die relativen Materieflüsse folgende Eigenschaften: 1. Im nichtrelativistischen Grenzfall gehen sie in (3.8) über. 2. Es gilt in Erweiterung von (4.12)

$$\sum_{j=1}^{n} I_\alpha^j = 0 \quad (\alpha = 1, 2, 3, 4). \tag{26.7}$$

3. Die relativen Materieflüsse sind senkrecht zum Vierervektor des gesamten Materieflusses

$$\sum_{\alpha=1}^{4} m_\alpha I_\alpha^j = 0 \quad (j = 1, 2, \ldots, n). \tag{26.8}$$

---

[1] Die Vierervektoren und Vierertensoren der relativistischen Theorie werden im folgenden stets in Komponentenschreibweise dargestellt. Zur Erhöhung der Übersichtlichkeit werden daher die chemischen Komponenten durch hochgestellte Indices bezeichnet.

Statt m_α kann man in dieser Gleichung auch u_α schreiben, wobei

$$u_\alpha = \frac{v_\alpha}{\sqrt{c^2 - v^2}} \quad (\alpha = 1, 2, 3), \quad u_4 = \frac{ic}{\sqrt{c^2 - v^2}}. \tag{26.9}$$

Dieser Vierervektor kann, wenn man von einem Faktor absieht, als das vierdimensionale Analogon der baryzentrischen Geschwindigkeit gedeutet werden.

Für das folgende ist es zweckmäßig, noch einige Bezeichnungen und Definitionen einzuführen. Alle Größen werden, wenn sie im baryzentrischen Lorentz-System gemessen werden, mit einem Strich bezeichnet. Es gilt also beispielsweise

$$\varrho' = \varrho \sqrt{1 - \frac{v^2}{c^2}}. \tag{26.10}$$

Die substantielle Zeitableitung ist durch den Lorentz-invarianten Operator

$$D = c \sum_{\alpha=1}^{4} u_\alpha \frac{\partial}{\partial x_\alpha} = \frac{\partial}{\partial t'} \tag{26.11}$$

gegeben, wobei t' die Zeit im baryzentrischen Lorentz-System bedeutet. D ist das relativistische Analogon des in (4.5) und (4.8) eingeführten Operators D/Dt. Für die Umrechnung der Lorentz-kovarianten Gesetze auf den dreidimensionalen Raum ist die Einführung des symmetrischen Tensors

$$\Delta_{\alpha\beta} = \delta_{\alpha\beta} + u_\alpha u_\beta \quad (\alpha, \beta = 1, \ldots 4) \tag{26.12}$$

nützlich. Er spielt in der relativistischen Theorie die Rolle des dreidimensionalen Tensors $\delta_{\alpha\beta}(\alpha, \beta = 1, 2, 3)$ der nicht-relativistischen Theorie. Im baryzentrischen System lautet er

$$\Delta'_{\alpha\beta} = \begin{pmatrix} 1 & 0 & 0 & 0 \\ 0 & 1 & 0 & 0 \\ 0 & 0 & 1 & 0 \\ 0 & 0 & 0 & 0 \end{pmatrix}. \tag{26.13}$$

Mit Hilfe des Tensors $\Delta_{\alpha\beta}$ lassen sich die relativen Materieflüsse (26.5) besonders einfach darstellen.

$$I_\alpha^j = \sum_{\beta=1}^{4} \Delta_{\alpha\beta} m_\beta^j \quad (\alpha = 1, 2 \ldots 4; j = 1 \ldots n). \tag{26.14}$$

27. Die Erhaltungssätze und ihre Bilanzgleichungen. α) *Der Energie-Impulstensor der Materie.* Sei e_v die Energie pro Volumeneinheit — unter Einschluß der Ruhenergie — und $\boldsymbol{J_e}$ der entsprechende Energiefluß. Dem Energiefluß sei eine Impulsdichte $\boldsymbol{g}$ gemäß

$$\boldsymbol{J_e} = c^2 \boldsymbol{g} \tag{27.1}$$

zugeordnet. Der Energie-Impulstensor ist dann

$$\left. \begin{aligned} W_{\alpha\beta} &= t_{\alpha\beta} + g_\alpha v_\beta \quad (\alpha, \beta = 1, 2, 3), & W_{\alpha 4} &= i c g_\alpha \quad (\alpha = 1, 2, 3), \\ W_{4\alpha} &= \frac{i}{c} J_{e\alpha} \quad (\alpha = 1, 2, 3), & W_{44} &= -e_v. \end{aligned} \right\} \tag{27.2}$$

Die Komponenten $W_{\alpha\beta}(\alpha, \beta = 1, 2, 3)$ sind in einen konvektiven Impulstransport $g_\alpha v_\beta$ und einen Spannungstensor $t_{\alpha\beta}$ aufgespalten[1]. Es sei angenommen, daß allgemein

$$W_{\alpha\beta} = W_{\beta\alpha} \quad (\alpha, \beta = 1, 2, 3, 4). \tag{27.3}$$

---

[1] Siehe z.B. M. v. Laue: Die Relativitätstheorie, Bd. 1, S. 162. Braunschweig 1955.

Mittels der Identität

$$W_{\alpha\beta} = u_\alpha u_\beta \sum_\gamma \sum_\zeta u_\gamma W_{\gamma\zeta} u_\zeta - u_\beta \sum_\gamma \sum_\zeta \Delta_{\alpha\gamma} W_{\gamma\zeta} u_\zeta - \\ - u_\alpha \sum_\gamma \sum_\zeta \Delta_{\beta\gamma} W_{\zeta\gamma} u_\zeta + \sum_\gamma \sum_\zeta \Delta_{\alpha\gamma} W_{\gamma\zeta} \Delta_{\zeta\beta} \qquad (27.4)$$

kann man den Energie-Impulstensor (27.2) in folgender zweckmäßigen und leicht interpretierbaren Weise darstellen:

$$W_{\alpha\beta} = u_\alpha u_\beta e_v' + \frac{1}{c}(u_\alpha I_\beta^0 + u_\beta I_\alpha^0) + w_{\alpha\beta} \qquad (\alpha, \beta = 1, 2, 3, 4), \qquad (27.5)$$

wobei

$$e_v' = \sum_\gamma \sum_\zeta u_\gamma W_{\gamma\zeta} u_\zeta, \qquad (27.6)$$

$$I_\alpha^0 = c \sum_\gamma \sum_\zeta \Delta_{\alpha\gamma} W_{\gamma\zeta} u_\zeta, \qquad (27.7)$$

$$w_{\alpha\beta} = \sum_\gamma \sum_\zeta \Delta_{\alpha\gamma} W_{\gamma\zeta} \Delta_{\zeta\beta}. \qquad (27.8)$$

c_v' ist eine Invariante, I_0^α ein Vierervektor und $w_{\alpha\beta}$ ein Vierertensor. Die Bedeutung dieser Größen ergibt sich, wenn man (27.2) in (27.6) bis (27.8) einsetzt und ein baryzentrisches Lorentz-System mit $v = 0$ zugrundelegt. Wie schon die Bezeichnung andeutet, ist e_v' in (27.6) der Wert von $e_v' = -W_{44}$ im baryzentrischen System. Weiter ist

$$I_\alpha^{0'} = J_{e\alpha}' \qquad (\alpha = 1, 2, 3), \qquad I_4^{0'} = 0, \qquad (27.9)$$

$$w_{\alpha\beta}' = t_{\alpha\beta}' \quad (\alpha, \beta = 1, 2, 3), \qquad w_{\alpha4}' = w_{4\alpha}' = 0 \quad (\alpha = 1, 2, 3, 4). \qquad (27.10)$$

Der Vektor $\boldsymbol{J}_e'$ entspricht dem Fluß der inneren Energie $\boldsymbol{J}_u$. Man wird daher den Vierervektor I_α^0 als die relativistische Verallgemeinerung von $\boldsymbol{J}_u$ ansehen dürfen. Ebenso hat man den Vierertensor $w_{\alpha\beta}$ als relativistische Verallgemeinerung des Spannungstensors $t_{\alpha\beta}$ ($\alpha, \beta = 1, 2, 3$) zu deuten[1].

β) *Die Bilanz der Ruhmasse.* Die Bilanz (4.11) kann unmittelbar übernommen werden, wenn man $\varrho^j \boldsymbol{v}$ mittels (3.8) eliminiert. Unter Γ^j ist jetzt aber die Produktion von *Ruhmasse* der Komponente j zu verstehen. Mit (26.2) läßt sich (4.11) in

$$\sum_{\alpha=1}^{4} \frac{\partial m_\alpha^j}{\partial x_\alpha} = \Gamma^j \qquad (j = 1 \dots n) \qquad (27.11)$$

umformen. Da die linke Seite Lorentz-invariant ist, gilt dies auch für die Produktionsdichten Γ^j. Dieses bemerkenswerte Ergebnis folgt daraus, daß sich Längenkontraktion und Zeitdilatation in den Produktionsdichten kompensieren.

Wegen der Erhaltung der gesamten Ruhmasse, $\sum_{j=1}^{n} \Gamma^j = 0$ folgt

$$\sum_{\alpha=1}^{4} \frac{\partial m_\alpha}{\partial x_\alpha} = 0. \qquad (27.12)$$

γ) *Die Bilanz von Impuls und Energie.* Sei $\boldsymbol{F}^j$ die äußere Kraft auf die Einheit der Ruhmasse der Komponente j („spezifische äußere Kraft"). Sie sei von der Geschwindigkeit $\boldsymbol{v}^j$ unabhängig angenommen. Dann ist der dieser Kraft zugeordnete Vierervektor in bekannter Weise gegeben durch

$$K_\alpha^j = \frac{c}{\sqrt{c^2 - v^{j2}}} F_\alpha^j = \frac{\varrho^j}{\varrho} F_\alpha^j \qquad (\alpha = 1, 2, 3); \qquad K_4^j = i\frac{\varrho^j \boldsymbol{v}^j \cdot \boldsymbol{F}^j}{\varrho_0^j c}. \qquad (27.13)$$

---

[1] Vgl. M. v. Laue: Die Relativitätstheorie, Bd. 1, S. 160 ff.. Braunschweig 1955.

Mit ϱ_0^j ist die Dichte der Komponente j in einem Lorentz-System gemeint, in welchem $\boldsymbol{v}^j = 0$ ist. Da der Unterschied von $\boldsymbol{v}^j$ und $\boldsymbol{v}$ in der Regel klein gegen die Lichtgeschwindigkeit ist, kann man in guter Näherung $\varrho^{j''}$ mit ϱ_0^j gleichsetzen. Davon wird jedoch nicht Gebrauch gemacht.

Die Bilanzgleichungen für den Impuls und die Energie lauten

$$\frac{\partial g_\alpha}{\partial t} + \sum_{\beta=1}^{3} \frac{\partial g_\alpha v_\beta}{\partial x_\beta} = \sum_{j=1}^{n} \varrho^j F_\alpha^j - \sum_{\beta=1}^{3} \frac{\partial t_{\alpha\beta}}{\partial x_\beta} \qquad (\alpha = 1, 2, 3), \qquad (27.14)$$

$$\frac{\partial e_v}{\partial t} = -\operatorname{div} \boldsymbol{J}_u + \sum_{j=1}^{n} \varrho^j \boldsymbol{v}^j \cdot \boldsymbol{F}^j. \qquad (27.15)$$

Sie lassen sich in der Lorentz-kovarianten Gleichung

$$\sum_{\beta=1}^{4} \frac{\partial W_{\alpha\beta}}{\partial x_\beta} = \sum_{j=1}^{n} \varrho_0^j K_\alpha^j \qquad (\alpha = 1, 2, 3, 4) \qquad (27.16)$$

vereinigen.

$\delta)$ *Die Gibbssche Fundamentalgleichung.* Es seien auch hier, wie schon in Ziff. 3 ff. Bifluida (Helium II, Supraleiter, Plasmen mit schwach gekoppelten Komponenten) ausgeschlossen. Unter den auch in der nicht-relativistischen Theorie üblichen Voraussetzungen darf man die Gültigkeit der Gibbsschen Fundamentalgleichung wenigstens im baryzentrischen Lorentz-System voraussetzen. Sie lautet nach (3.4) bzw. (5.1)

$$T' \frac{\partial s'}{\partial t'} = \frac{\partial e'}{\partial t'} + p' \frac{\partial}{\partial t'} \left(\frac{1}{\varrho'}\right) - \sum_{j=1}^{n} \mu^{j'} \frac{\partial \gamma^{j'}}{\partial t'}. \qquad (27.17)$$

γ_j' ist die Konzentration im baryzentrischen Lorentz-System. Allgemein ist γ definiert durch

$$\gamma^j = \frac{\varrho^j}{\varrho}. \qquad (27.18)$$

e_v' hängt mit e' zusammen über die Gleichung

$$e_v' = \varrho'(e' + a) \qquad (27.19)$$

mit einer willkürlichen und aus den Endergebnissen herausfallenden Konstanten a, welche nur den Energienullpunkt festlegt. Wegen (26.11) ist (27.17) identisch mit

$$T' D s' = D e' + p' D \frac{1}{\varrho'} - \sum_{j=1}^{n} \mu^{j'} D \gamma^{j'}. \qquad (27.20)$$

Diese Gleichung gilt dann in einem beliebigen Lorentz-System, wenn die gestrichenen Größen sich zwar auf das baryzentrische System beziehen, aber in den Koordinaten $x_\alpha (\alpha = 1, \ldots 4)$ des beliebigen Lorentz-Systems ausgedrückt werden.

$\varepsilon)$ *Die Bilanzen der inneren Energie und der Entropie.* Die Bilanz (4.19) der inneren Energie ergab sich durch Subtraktion der Gln. (4.16) und (4.18). Dem entspricht hier die Multiplikation von (27.16) mit u_α und Summation über α:

$$\sum_{\alpha=1}^{4} \sum_{\beta=1}^{4} u_\alpha \frac{\partial W_{\alpha\beta}}{\partial x_\beta} = \sum_{j=1}^{n} \sum_{\alpha=1}^{4} \varrho_0^j u_\alpha K_\alpha^j. \qquad (27.21)$$

Einsetzen von (27.5) und einige längere Umformungen führen auf die Bilanz der inneren Energie:

$$\varrho'\left(De' + p'D\,\frac{1}{\varrho'}\right) = -\sum_{\beta=1}^{4}\left(\frac{\partial I_\beta^0}{\partial x_p} + \frac{1}{c}\,I_\beta^0 D u_p\right) \left.\begin{array}{c}\\[2mm] + \sum_{j=1}^{n}\sum_{\alpha=1}^{4}\omega^j I_\alpha^j K_\alpha^j + c\sum_{\alpha=1}^{4}\sum_{\beta=1}^{4}(p'\Delta_{\alpha\beta} - w_{\alpha\beta})\,\frac{\partial u_\beta}{\partial x_\alpha}\end{array}\right\} \quad (27.22)$$

mit der Abkürzung

$$\omega^j = \frac{\varrho_0^j}{\varrho^{j'}}. \tag{27.23}$$

Durch Kombination von (27.22) mit (27.20) erhält man die Entropiebilanz

$$\varrho'Ds' + \sum_{\alpha=1}^{4}\frac{\partial}{\partial x_\alpha}\left\{\frac{1}{T'}\left(I_\alpha^0 - \sum_{j=1}^{n}I_\alpha^j\mu^{j'}\right)\right\} \left.\begin{array}{c}\\[4mm] = \frac{1}{T'}\sum_{j=0}^{n}\sum_{\alpha=1}^{4}I_\alpha^j X_\alpha^j + \frac{c}{T'}\sum_{\alpha=1}^{4}\sum_{\beta=1}^{4}\widetilde{P}_{\alpha\beta}\,\frac{\partial u_\beta}{\partial x_\alpha} - \frac{1}{T'}\Pi\varrho'Dr' - \sum_{k=1}^{r}\frac{A^k}{T'}\Gamma^k\end{array}\right\} \quad (27.24)$$

mit

$$X_\alpha^0 = \frac{\partial}{\partial x_\alpha}\,\frac{1}{T'} - \frac{1}{cT'}\,Du_\alpha \quad (\alpha = 1,2,3,4), \tag{27.25}$$

$$X_\alpha^j = \frac{\omega^j}{T'}\,K_\alpha^j - \frac{\partial}{\partial x_\alpha}\,\frac{\mu^{j'}}{T'} \quad (\alpha = 1,\dots 4;\, j = 1\dots n). \tag{27.26}$$

Die Affinitäten A^k sind Invarianten und definiert wie in (11.14). Π ist der Unterschied zwischen dynamischem und statischem Druck

$$\Pi - p' - \frac{1}{3}\sum_{\alpha=1}^{4}w_{\alpha\alpha}. \tag{27.27}$$

$\widetilde{P}_{\alpha\beta}$ ist das Analogon zum Tensor der Schubspannungen

$$\widetilde{P}_{\alpha\beta} = -w_{\alpha\beta} + \frac{1}{3}\sum_{\gamma=1}^{4}w_{\gamma\gamma}\Delta_{\alpha\beta}. \tag{27.28}$$

Die Bilanz (27.22) der inneren Energie ist analog zu (4.19) gebaut mit Ausnahme des zweiten Terms der rechten Seite. Er wurde bereits von ECKART[1] gefunden. Auch die Entropiebilanz (27.24) entspricht der nicht-relativistischen Gl. (5.3). Sie kann auch analog interpretiert werden. Der zweite Term der linken Seite von (27.24) ist die Divergenz des Vierervektors des nicht-konvektiven Entropieflusses I_α^s. Die Terme der rechten Seite geben der Reihe nach die Entropieproduktion durch Wärmeleitung ($j=0$) und Diffusion ($j=1,\dots,n$) durch Schubviscosität, Volumenviscosität und durch die chemischen Reaktionen.

Die linke Seite von (27.24) läßt sich noch zusammenfassen zu $\sum_{\alpha=1}^{4}\frac{\partial}{\partial x_\alpha}(m_\alpha s' + I_\alpha^s)$. Der Vierervektor des gesamten Entropieflusses hat daher die Komponenten $S_\alpha = m_\alpha s + I_\alpha^s$ ($\alpha = 1,\dots 4$). Er vereinigt die Entropie und den dreidimensionalen Entropiefluß. Speziell im baryzentrischen Lorentz-System geben die ersten drei Komponenten das Analogon zum dreidimensionalen Entropiefluß, während seine vierte Komponente gleich $ic\varrho's'$, d.h. abgesehen vom Faktor ic die Entropiedichte ist.

---

[1] C. ECKART: Phys. Rev. **58**, 919 (1940).

Ein Vergleich mit (5.8) legt nahe, $i s_4/c$ allgemein als Entropiedichte ϱs zu interpretieren; daraus ergibt sich die spezifische Entropie s für ein beliebiges Lorentz-System nach der Gleichung

$$i c \varrho s = m_4 s' + I_4^s = i c \varrho s' + I_4^s. \tag{27.29}$$

Danach ist also die spezifische Entropie nur dann Lorentz-invariant, wenn weder Diffusion noch Wärmeleitung stattfindet. Für Zustände, wie sie hier betrachtet werden — kleine Abweichungen vom thermodynamischen Gleichgewicht — ist $|v^{j'}| \ll c$ $(j = 1, 2, \dots n)$ und $|S'_\alpha| \ll c \varrho' s'$ anzunehmen. Dann ergibt sich $s \approx s'$ für beliebige v, d. h. die spezifische Entropie ist fast Lorentz-invariant.

ζ) *Die linearen phänomenologischen Gleichungen und die O.C.R.B.* Die Entropiebilanz (27.24) ist wieder der Ausgangspunkt für die Aufstellung der linearen phänomenologischen Gleichungen bei kleinen Abweichungen vom thermodynamischen Gleichgewicht. Für isotrope Medien gilt das Curiesche Gesetz in seiner vierdimensionalen Formulierung; danach hängen die Kräfte von einem gewissen Tensorcharakter nur von den Flüssen desselben Tensorcharakters ab. Es ergeben sich so zunächst die Wärmeleitungs- und Diffusionsgleichungen

$$I_\alpha^j = \sum_{k=0}^{n} \sum_{\beta=1}^{4} L_{\alpha\beta}^{jk} X_\beta^k \qquad (\alpha = 1 \dots 4; \, j = 0, 1 \dots n). \tag{27.30}$$

Die Eigenschaft der Isotropie legt den phänomenologischen Koeffizienten $L_{\alpha\beta}^{jk}$, die einen Sitz von $(n+1)^2$ Vierertensoren bilden, noch erhebliche Einschränkungen auf. Es ist jedoch zu beachten, daß die Isotropie nur im baryzentrischen Lorentz-System gilt. In jedem anderen Lorentz-System gibt es ja im Material eine ausgezeichnete Richtung, nämlich die von v. Im baryzentrischen Lorentz-System müssen sich also die Gln. (27.30) infolge der Isotropie auf

$$I_\alpha^{j'} = \sum_{k=0}^{n} \sum_{\beta=1}^{3} L_{\alpha\beta}^{jk'} X_\beta^{k'} \qquad (\alpha = 1 \dots 3), \qquad I_4^{j'} = 0 \tag{27.31}$$

mit
$$L_{\alpha\beta}^{jk'} = L^{jk} \delta_{\alpha\beta} \qquad (\alpha, \beta = 1, 2, 3)$$

reduzieren. Da $X_4^{k'}$ in der Regel nicht verschwindet, so folgt daraus weiter, daß $L_{\alpha\beta}^{jk'} = 0$, wenn α oder β oder beide den Wert 4 annehmen. Man kann daher allgemein setzen

$$L_{\alpha\beta}^{jk'} = L^{jk} \Delta'_{\alpha\beta} \tag{27.32}$$

und damit in jedem Lorentz-System

$$L_{\alpha\beta}^{jk} = L^{jk} \Delta_{\alpha\beta} \qquad (\alpha, \beta = 1, 4; \, j, k = 0, 1 \dots n). \tag{27.33}$$

Die L^{jk} sind die phänomenologischen Koeffizienten im baryzentrischen System und damit die der nicht-relativistischen Theorie. Für sie gelten daher die O.C.R.B. $L^{jk} = L^{kj}$ $(k, j = 0, 1, \dots n)$. Damit folgt dann

$$L_{\alpha\beta}^{jk} = L_{\beta\alpha}^{jk} = L_{\alpha\beta}^{kj} = L_{\beta\alpha}^{kj} \qquad (\alpha, \beta = 1 \dots 4; \, j, k = 1 \dots n). \tag{27.34}$$

Die O.C.R.B. gelten daher auch in der relativistischen Theorie, und sie sind invariant gegen Lorentz-Transformationen.

In ähnlicher Weise findet man die weiteren phänomenologischen Gleichungen

$$\tilde{P}_{\alpha\beta} = \eta c \sum_{\gamma,\zeta=2}^{4} \left\{ \Delta_{\alpha\beta} \Delta_{\beta\gamma} \left(\frac{\partial u_\gamma}{\partial x_\zeta} + \frac{\partial u_\zeta}{\partial x_\gamma} \right) - \frac{2}{3} \Delta_{\alpha\beta} \Delta_{\gamma\zeta} \frac{\partial u_\gamma}{\partial x_\zeta} \right\} \qquad (\alpha, \beta = 1, \dots 4).$$

Die letzten beiden Gleichungen sind praktisch identisch mit den früheren Gln. (11.16) und (11.17).

28. Systeme mit elektromagnetischen Feldern. $\alpha)$ *Die Maxwellschen Gleichungen.* Im folgenden benutzen wir die Maxwellschen Gleichungen in der Gestalt

$$\operatorname{rot} \boldsymbol{E} = -\dot{\boldsymbol{B}}, \qquad \operatorname{div} \boldsymbol{D} = \varrho_{\mathrm{el}}, \atop \operatorname{rot} \boldsymbol{H} = \dot{\boldsymbol{D}} + \boldsymbol{j}, \qquad \operatorname{div} \boldsymbol{B} = 0. \tag{28.1}$$

$\boldsymbol{j}$ ist die Stromdichte, ϱ_{el} die elektrische Ladungsdichte. Diese Größen sind gegeben durch

$$\varrho_{\mathrm{el}} = \sum_{j=1}^{n} e^{j} \varrho^{j}, \qquad \boldsymbol{j} = \sum_{j=1}^{n} e^{j} \varrho^{j} \boldsymbol{v}^{j}, \tag{28.2}$$

wenn e^{j} die Ladung pro Ruhmasseneinheit der chemischen Komponente j ist. Die Materialgleichungen seien

$$\boldsymbol{D}' = \varepsilon \boldsymbol{E}', \qquad \boldsymbol{B}' = \mu \boldsymbol{H}' \tag{28.3}$$

mit Werten von ε und μ, die nur vom Zustand, aber nicht von der Vorgeschichte — oder damit gleichwertig, von der Frequenz — abhängen sollen. ε_0, μ_0 sind die Vakuumwerte von ε, μ. $(\varepsilon_0 \mu_0)^{-\frac{1}{2}}$ ist gleich der Lichtgeschwindigkeit c im Vakuum. Die Maxwellschen Gleichungen sind damit als Größengleichungen mit vier unabhängigen dimensionierten Größen geschrieben. Bei Verwendung des Giorgischen Einheitensystems sind sie in dieser Schreibweise gleichzeitig Zahlenwertgleichungen. Die Striche an $\boldsymbol{D}, \boldsymbol{E}$ usw. deuten wieder an, daß die Gln. (28.3) im baryzentrischen Lorentz-System gelten. Ferner seien die Polarisation und die Magnetisierung durch

$$\boldsymbol{P} = \boldsymbol{D} - \varepsilon_0 \boldsymbol{E}, \qquad \boldsymbol{M} = \frac{1}{\mu_0} \boldsymbol{B} - \boldsymbol{H} \tag{28.4}$$

eingeführt.

$\beta)$ *Die Maxwellschen Gleichungen in Lorentz-kovarianter Schreibweise.* Die Lorentz-kovariante Formulierung der Maxwellschen Gleichungen geschieht mit Hilfe der antisymmetrischen Vierertensoren

$$B_{\alpha\beta} = \begin{pmatrix} 0 & B_3 & -B_2 & -iE_1/c \\ -B_3 & 0 & B_1 & -iE_2/c \\ B_2 & -B_1 & 0 & -iE_3/c \\ iE_1/c & iE_2/c & iE_3/c & 0 \end{pmatrix}, \tag{28.5}$$

$$H_{\alpha\beta} = \begin{pmatrix} 0 & H_3 & -H_2 & -icD_1 \\ -H_3 & 0 & H_1 & -icD_2 \\ H_2 & -H_1 & 0 & -icD_3 \\ icD_1 & icD_2 & icD_3 & 0 \end{pmatrix}, \tag{28.6}$$

$$M_{\alpha\beta} = \begin{pmatrix} 0 & M_3 & -M_2 & icP_1 \\ -M_3 & 0 & M_1 & icP_2 \\ M_2 & -M_1 & 0 & icP_3 \\ -icP_1 & -icP_2 & -icP_3 & 0 \end{pmatrix}. \tag{28.7}$$

Dann lauten die Maxwellschen Gleichungen

$$\sum_{\beta=1}^{4} \frac{\partial H_{\alpha\beta}}{\partial x_\beta} = \sum_{j=1}^{n} e^{j} m^{j}_{\alpha} \quad (\alpha = 1 \ldots 4), \tag{28.8}$$

$$\frac{\partial B_{\alpha\beta}}{\partial x_\gamma} + \frac{\partial B_{\beta\gamma}}{\partial x_\alpha} + \frac{\partial B_{\gamma\alpha}}{\partial x_\beta} = 0 \quad (\alpha, \beta, \gamma = 1 \ldots 4). \tag{28.9}$$

Hierzu treten die Materialgleichungen, ebenfalls Lorentz-kovariant geschrieben:

$$\sum_{\gamma,\zeta=1}^{4} \Delta_{\alpha\gamma}(B_{\gamma\zeta} - \mu H_{\gamma\zeta})\Delta_{\zeta\beta} = 0 \quad (\alpha,\beta = 1,\dots 4) \tag{28.10}$$

und

$$\sum_{\gamma=1}^{4}\left(B_{\alpha\gamma} - \frac{1}{c^2\varepsilon}H_{\alpha\gamma}\right)u_\gamma = 0 \quad (\alpha = 1,\dots 4). \tag{28.11}$$

$\gamma)$ *Der Energieimpulstensor des elektromagnetischen Feldes.* Ein naheliegender Weg zur Berücksichtigung des elektromagnetischen Feldes geht von (27.16) aus. Die Wirkung des Feldes wird durch die von ihm auf die Materie ausgeübte Kraft ausgedrückt. Es treten demnach zu eventuell vorhandenen äußeren Kräften die Lorentz-Kräfte, die auf die Ladungen der einzelnen Komponenten wirken und die ponderomotorischen Kräfte, die auf der Polarisation und der Magnetisierung der Materie beruhen. Damit erweitert sich (27.16) zu

$$\sum_{\beta=1}^{4}\frac{\partial W_{\alpha\beta}}{\partial x_\beta} = \sum_{j=1}^{n}\varrho_0^j K_{A\alpha}^j + \sum_{j=1}^{n}\varrho_0^j K_{L\alpha} + k_{p\alpha}. \tag{28.12}$$

$K_{A\alpha}^j$ ist der Vierervektor der äußeren Kräfte (Gravitation) auf die Komponente j, $K_{L\alpha}^j$ ist der Vierervektor der Lorentz-Kraft auf die Komponente j und $k_{p\alpha}$ ist der Vierervektor der ponderomotorischen Kraftdichte. Das Hauptproblem besteht nun darin, geeignete Ansätze für die elektromagnetischen Kräfte zu finden. Für Ladungen im leeren Raum würde man einfach die Lorentz-Kraft gemäß

$$c\,\varrho_0^j K_{L\alpha}^j = e^j \sum_{\beta=1}^{4} B_{\alpha\beta}m_\beta^j \quad (\alpha = 1,\dots 4; j = 1,\dots n) \tag{28.13}$$

anzusetzen haben. Für Ladungen in der kontinuierlichen Materie sind aber gewisse lokale elektrische und magnetische Felder an Stelle von $B_{\alpha\beta}$ zu wählen. Um die Viererkovarianz und die lineare Abhängigkeit vom Feld zu wahren, muß man sich auf Ansätze der Art

$$\lambda_1^j B_{\alpha\beta} + \lambda_2^j \sum_{\gamma=1}^{4}(u_\alpha B_{\beta\gamma} - u_\beta B_{\alpha\gamma})u_\gamma + \lambda_3^j M_{\alpha\beta} + \lambda_4^j \sum_{\gamma=1}^{4}(u_\alpha M_{\beta\gamma} - u_\beta M_{\alpha\gamma})u_\gamma \tag{28.14}$$

an Stelle von $B_{\alpha\beta}$ in (28.13) mit Invarianten λ_1^j bis λ_4^j beschränken. Wie Kluitenberg zeigt, läßt sich schon mit dem einfacheren Ansatz (28.13) eine konsistente Theorie entwickeln. Indessen erscheint es nicht uninteressant, auch die Möglichkeiten des allgemeineren Ansatzes (28.14) zu untersuchen.

Problematischer noch ist der Ansatz für die ponderomotorische Kraft $k_{p\alpha}$. Es ist daher angezeigt, einen anderen Weg zur Berücksichtigung des elektromagnetischen Feldes einzuschlagen. Man postuliert dazu die Existenz eines für isotrope Medien symmetrischen und in den elektromagnetischen Feldstärken quadratischen Energie-Impulstensors $W_{f\alpha\beta}$ des elektromagnetischen Feldes von der Art, daß

$$\sum_{\beta=1}^{4}\frac{\partial}{\partial x_\beta}(W_{\alpha\beta} + W_{f\alpha\beta}) = \sum_{j=1}^{n}\varrho_0^j K_{A\alpha}^j. \tag{28.15}$$

Dann ergeben sich die Lorentz-Kräfte auf die chemischen Komponenten und die ponderomotorischen Kräfte aus

$$\sum_{j=1}^{n}\varrho_0^j K_{L\alpha}^j + k_{p\alpha} = -\sum_{\beta=1}^{4}\frac{\partial}{\partial x_\beta}W_{f\alpha\beta}. \tag{28.16}$$

Da $W_{f\alpha\beta}$ ein Vierertensor sein muß, bleibt verhältnismäßig wenig Spielraum für einen Ansatz. Um ihn aufzubauen, stehen nur die Vierertensoren $B_{\alpha\beta}$, $H_{\alpha\beta}$ ($M_{\alpha\beta}$ ist ja von diesen beiden linear abhängig) und der Vierervektor u_α zur Verfügung. KLUITENBERG wählt speziell den Ansatz

$$W_{f\alpha\beta} = -\sum_{\gamma=1}^{4} B_{\alpha\beta} H_{\gamma\beta} - \frac{1}{4\mu_0}\delta_{\alpha\beta}\sum_{\gamma,\zeta=1}^{4} B_{\gamma\zeta} B_{\zeta\gamma} + \\ + \sum_{\gamma,\zeta=1}^{4} u_\zeta (B_{\zeta\gamma} H_{\gamma\alpha} - H_{\zeta\gamma} B_{\gamma\alpha}) u_\beta \quad (\alpha,\beta = 1,\dots 4). \tag{28.17}$$

Hieraus gewinnt man mit (28.16) die Lorentz-Kräfte (28.13) und die ponderomotorische Kraft

$$k_{p\alpha} = \frac{1}{2}\sum_{\beta,\gamma=1}^{4} M_{\beta\gamma}\frac{\partial B_{\beta\gamma}}{\partial x_\alpha} + \frac{\varrho'}{c}D\left\{\frac{1}{\varrho'}\sum_{\beta,\gamma=1}^{4} u_\beta (B_{\beta\gamma} M_{\gamma\alpha} - M_{\beta\gamma} B_{\gamma\alpha})\right\} \\ (\alpha = 1,\dots 4). \tag{28.18}$$

δ) *Die Bilanzgleichungen für Ruhmasse, Impuls, Energie und innere Energie.* Die Bilanzgleichungen (27.11) und (27.12) für die Ruhmassen bleiben unverändert bestehen. Auch die Bilanzgleichungen (27.16) und (27.22) für Impuls und Energie und für die innere Energie können ohne weiteres übernommen werden, wenn man für die rechte Seite von (27.16) die Gesamtkraft nach (28.12) schreibt. Es ist jedoch zweckmäßig, den dritten Term rechts in (27.22) umzuformen in

$$\sum_{j=1}^{n}\sum_{\alpha=1}^{4} I_\alpha^j\left(\omega^j K_{A\alpha}^j + e^j\sum_{\beta=1}^{4} B_{\alpha\beta} u_\beta\right) - c\sum_{\alpha=1}^{4} u_\alpha k_{p\alpha}. \tag{28.19}$$

ε) *Die Gibbssche Fundamentalgleichung.* Ebenso wie die Ansätze für Lorentz-Kraft und ponderomotorische Kraft sowie für den Energieimpulstensor ist der Ansatz für die Gibbssche Fundamentalgleichung mit einer gewissen Willkür behaftet. Diese Schwierigkeit ist bereits aus der nicht-relativistischen Theorie bekannt. Man benutzt als Zusatzterme infolge des elektrischen Feldes Ausdrücke wie $E\,d(D/\varrho)$, $D\,d(E/\varrho)$, $(D/\varrho)\,dE$, $E\,d(P/\varrho)$ usw. Analoge Möglichkeiten bestehen für die durch ein Magnetfeld bedingten Zusatzterme. Allen diesen Ansätzen ist gemeinsam, daß zu beiden Seiten des Differentials Feldvektoren stehen. Für die relativistische Formulierung der Gibbsschen Fundamentalgleichung ist daher ein Ansatz folgender Art

$$T'Ds' = De' + p'D\frac{1}{\varrho'} + \frac{1}{2}\sum_{\alpha,\beta=1}^{4} G_{\alpha\beta}'DZ_{\alpha\beta}' - \sum_{j=1} \mu^{j'}D\gamma_j' \tag{28.20}$$

nahegelegt, in welchem $G_{\alpha\beta}'$ und $Z_{\alpha\beta}'$ antisymmetrische Vierertensoren sind. Beide wird man analog zu (28.14) darstellen mit acht unbestimmten Koeffizienten, von denen entweder die ersten vier oder die letzten vier den Faktor $1/\varrho'$ enthalten.

Die Lorentz-Invarianz von (28.20) ist nicht unmittelbar ersichtlich. Sie beruht auf der Existenz der Materialgleichungen (28.3) oder (28.10) und (28.11), mit deren Hilfe man den Zusatzterm in (28.20) in

$$\tfrac{1}{2}\sum_{\alpha,\beta=1}^{4} G_{\alpha\beta}DZ_{\alpha\beta} + \sum_{\alpha,\beta,\gamma=1}^{4} u_\alpha (G_{\alpha\gamma} Z_{\gamma\beta} - Z_{\alpha\gamma} G_{\gamma\beta}) D u_\beta \tag{28.21}$$

umformen kann.

Die Entropiebilanz (27.24) ändert sich nun insofern, als statt (27.26) zu schreiben ist

$$\frac{1}{T'}X_\alpha^j = \frac{\omega^j}{T'} K_{A\alpha}^j + \frac{1}{T'}e^j\sum_{\beta=1}^{4} B_{\alpha\beta} u_\beta - \frac{\partial}{\partial x_\alpha}\frac{\mu^{j'}}{T'} \quad (\alpha=1,\dots 4, j=1\dots n) \tag{28.22}$$

und als zur rechten Seite von (27.24) folgender Term hinzutritt

$$-\frac{c}{T'}\sum_{\alpha=1}^{4}u_\alpha\left[k_{p\alpha}-\frac{1}{2}\varrho'\sum_{\beta,\gamma=1}^{4}G_{\beta\gamma}\frac{\partial Z_{\beta\gamma}}{\partial x_\alpha}-\varrho'\sum_{\beta,\gamma,\zeta=1}^{4}u_\beta\,(G_{\beta\gamma}Z_{\gamma\zeta}-Z_{\beta\gamma}G_{\gamma\zeta})\frac{\partial u_\zeta}{\partial x_\alpha}\right].\quad(28.23)$$

Er enthält nichts, was auf Diffusion, Wärmeleitung und chemische Reaktionen Bezug hat. Wohl könnte man den letzten Anteil mit der Entropieproduktion durch Viskosität zusammenfassen und so zu einer abgeänderten Definition des Reibungsdrucktensors gelangen. Die einfachste Möglichkeit ist jedoch, den Ausdruck (28.23) insgesamt gleich Null zu setzen. Dies ist in der Tat mit dem Energie-Impulstensor (28.17) und der daraus abgeleiteten ponderomotorischen Kraft (28.18) verträglich, wenn man

$$G_{\alpha\beta}=\frac{1}{\varrho'}\,M_{\alpha\beta},\qquad Z_{\alpha\beta}=H_{\alpha\beta}\quad(\alpha,\beta=1,\dots4)\qquad(28.24)$$

wählt.

ζ) *Transformationseigenschaften.* Die speziellen Annahmen (28.17) für den Energie-Impulstensor und (28.24) für die Gibbssche Fundamentalgleichung werfen natürlich die Frage auf, ob damit dem Formalismus zuviel Zwang auferlegt ist, d.h. ob nicht andere Annahmen zu inhaltlich wesentlich verschiedenen Formulierungen der relativistischen (und damit auch der nicht-relativistischen) Thermodynamik der irreversiblen Prozesse mit elektromagnetischen Feldern führen. Diese Frage scheint bisher nicht erschöpfend behandelt zu sein. Man kann jedoch so viel sagen, daß viele andere Annahmen tatsächlich inhaltsgleiche Formulierungen zur Folge haben. Ein interessantes Beispiel ist der Abrahamsche symmetrische Energie-Impulstensor

$$\left.\begin{aligned}W^A_{f\alpha\beta}=&-\sum_{\gamma=1}^{4}B_{\alpha\gamma}H_{\gamma\beta}-\tfrac{1}{4}\delta_{\alpha\beta}\sum_{\gamma,\zeta=1}^{4}B_{\gamma\zeta}H_{\gamma\zeta}+\\&+\sum_{\gamma,\zeta=1}^{4}u_\zeta\,(B_{\zeta\gamma}H_{\gamma\alpha}-H_{\zeta\gamma}B_{\gamma\alpha})\,u_\beta\quad(\alpha,\beta=1,\dots4).\end{aligned}\right\}\quad(28.25)$$

Er unterscheidet sich von (28.17) nur in den Diagonaltermen und hat gegenüber (28.17) den formalen Vorzug, daß er konsequent Komponenten des Vierertensors $B_{\alpha\beta}$ mit solchen des Vierertensors $H_{\alpha\beta}$ multipliziert. Die Folge der neuen Wahl des Energie-Impulstensors ist, daß die Energiedichte und der Druck des elektromagnetischen Feldes andere Werte annehmen; dies korrigiert sich jedoch dadurch, daß der statische Druck und die spezifische innere Energie der Materie diese Änderungen kompensieren. So sind statischer Druck und spezifische innere Energie bei Zugrundelegung des Abrahamschen Energie-Impulstensors durch

$$p^{A'}=p'-\tfrac{1}{4}\sum_{\gamma,\zeta=1}^{4}B_{\gamma\zeta}M_{\gamma\zeta}$$

und

$$e^{A'}=e'+\frac{1}{4\varrho'}\sum_{\gamma,\zeta=1}^{4}B_{\gamma\zeta}M_{\gamma\zeta}$$

gegeben.

Natürlich hat für das Verhalten des Systems nur die Summe der Energie-Impulstensoren $W_{\alpha\beta}+W_{f\alpha\beta}$ Bedeutung. Die Theorie wird inhaltlich nicht dadurch abgeändert, daß man von $W_{f\alpha\beta}$ einen Teil abspaltet und nach $W_{\alpha\beta}$ hinüberschiebt. Ob sich aber die Gesamtheit der möglichen Ansätze für $W_{f\alpha\beta}$, $G_{\alpha\beta}$ und $Z_{\alpha\beta}$ auf diese Weise ineinander überführen läßt, scheint noch nicht hinreichend geklärt zu sein.

η) *Bemerkungen.* Aus der relativistischen Fassung der Thermodynamik der irreversiblen Prozesse gelangt man zur nicht-relativistischen Fassung, indem man alle Terme von der Ordnung v/c gegen 1 vernachlässigt. Neben dieser nicht-relativistischen Fassung ist aber auch noch die relativistische Fassung in drei-dimensionaler Formulierung interessant. Sie läßt insbesondere deutlich die relativistischen Korrekturen erkennen. Wegen ihrer Herleitung sei jedoch nochmals auf die ausführliche Darstellung von KLUITENBERG verwiesen.

Eine Ergänzung der relativistischen Theorie scheint noch nötig für Systeme, deren Teilchen innere Freiheitsgrade haben. Hier ist insbesondere an die ver-zögerte Einstellung von elektrischen und magnetischen Dipolmomenten bei Feldänderungen zu denken. Solche Systeme sind mit der vorhergehenden Behandlung nicht erfaßt, da diese ausdrücklich Materialkonstanten voraussetzen, die nur vom Zustand des Materials aber nicht von seiner Vorgeschichte abhängen.

F. Zur statistischen Theorie der irreversiblen Prozesse.

29. Die statistische Begründung der O.C.R.B. Die O.C.R.B. sind in vielen Einzelfällen empirisch bestätigt. Sie haben auch für verschiedene spezielle Modelle Begründungen aus der kinetischen Theorie erhalten. Man kann sich daher fragen, ob den kinetischen Begründungen ein allgemeines Prinzip zugrunde-liegt, das von den speziellen Modellvorstellungen unabhängig ist.

Der erste Schritt in dieser Richtung stammt von BOHR[1]. Er hat die Ableitung der zweiten Thomsonschen Beziehung für die thermoelektrischen Effekte mit Hilfe der klassischen Elektronentheorie analysiert und bemerkt, daß sie letzten Endes darauf beruht, daß die Bewegungsgleichung der Elektronen in bezug auf Zeitumkehr, $t \to -t$, invariant sind. Dieselbe Wurzel hat der in (14.7) ausgedrückte Zusammenhang zwischen Thermodiffusion und Diffusionsthermik, wie man an seiner Herleitung aus der kinetischen Gastheorie erkennen kann.

Das besondere Verdienst ONSAGERs[2] ist es, die Invarianz der Bewegungs-gleichungen der atomaren Bestandteile eines Systems gegen Zeitumkehr, das sog. Prinzip der mikroskopischen Reversibilität, für die Schwankungstheorie nutzbar gemacht und diese in Beziehung zu den irreversiblen Prozessen gesetzt zu haben. CASIMIR[3] hat die O.C.R.B. dann auf den Fall erweitert, daß sowohl gerade als auch ungerade Variable zur Beschreibung des irreversiblen Ablaufs benötigt werden.

Wir bringen in dieser Ziffer den Beweis der O.C.R.B. für skalare, gerade oder ungerade Variable. In der darauffolgenden Ziff. 30 wird auf Modifikationen des Beweises durch CALLEN und GREENE[4] und KAPLAN[5] eingegangen, und es wird die Verallgemeinerung auf vektorielle und tensorielle Variable am Beispiel der Wärmeleitung auseinandergesetzt.

α) *Schwankungstheorie.* Wir betrachten zunächst homogene Systeme und schließen Transporterscheinungen aus. Das System sei adiabatisch isoliert, d.h. nur über Arbeitsvariable $\alpha_1, \alpha_2, \ldots, \alpha_f$ im Kontakt mit seiner Umgebung. Innere Zustandsänderungen werden durch extensive Variable $\alpha_{f+1}, \ldots, \alpha_n$ beschrieben. Ohne Beschränkung der Allgemeinheit nehmen wir an, daß alle Variablen α_i im thermodynamischen Gleichgewicht den Wert Null haben. Wegen der adia-batischen Isolierung ist die innere Energie keine unabhängige Variable und daher

---

[1] N. BOHR: Studier over Metallernes Elektronteori. Kopenhagen 1913.

[2] L. ONSAGER: Phys. Rev. **37**, 405 (1931); **38**, 2265 (1931).

[3] H.B.G. CASIMIR: Rev. Mod. Phys. **17**, 343 (1945).

[4] H.B. CALLEN u. R.F. GREENE: Phys. Rev. **86**, 702 (1952). — R.F. GREENE u. H.B. CALLEN: Phys. Rev. **88**, 1387 (1952).

[5] T.A. KAPLAN: Phys. Rev. **102**, 1447 (1956).

unter den α_i nicht eingeschlossen. Bei kleinen Abweichungen vom thermodynamischen Gleichgewicht, die mit der adiabatischen Isolierung verträglich sind, kann man die Entropie in erster Näherung durch

$$S(\alpha_i) = S_0 - \tfrac{1}{2} \sum_{i,k=1}^{n} S_{ik}\, \alpha_i \alpha_k \qquad (29.1)$$

darstellen. Die Matrix S_{ik} kann symmetrisch angenommen werden; sie hat eine positiv definite quadratische Form.

Über die Gibbssche Fundamentalgleichung

$$T\, dS = dU - \sum_{i=1}^{n} \zeta_i\, d\alpha_i \qquad (29.2)$$

sind den Variablen α_i thermodynamisch konjugierte Variable $\zeta_i(\alpha_j)$ zugeordnet. Sie mögen im thermodynamischen Gleichgewicht mit $\alpha_i = 0$ die Werte $\zeta_i(0) = \zeta_i^+$ haben. Die Abweichungen $\zeta_i(\alpha_j) - \zeta_i^+$ seien mit $\Delta\zeta_i(\alpha_j)$ bezeichnet. Im Gleichgewicht mit der Umgebung sind die intensiven Parameter $\zeta_i(\alpha_j)$ (Druck, Affinitäten usw.) des Systems gleich den intensiven Parametern der Umgebung. Bei Abweichungen vom Gleichgewicht ist zwischen den $\zeta_i(\alpha_j)$ und den ζ_i der Umgebung zu unterscheiden. Wir wählen als Werte dieser Variablen für die Umgebung die Werte ζ_i^+ des Bezugszustandes und halten diese konstant. Adiabatische Isolierung bedeutet daher $dU - \sum_{i=1}^{n} \zeta_i^+ d\alpha_i = 0$. Dann lassen sich die Koeffizientenmatrix S_{ik} in (29.1) und ihre reziproke Matrix S^{ik} in folgender Weise ausdrücken:

$$S_{ik} = \frac{1}{T}\frac{\partial \zeta_i}{\partial \alpha_k}, \qquad S^{ik} = T\frac{\partial \alpha_i}{\partial \zeta_k}. \qquad (29.3)$$

Diese Differentiationen sind so auszuführen, daß die übrigen α_j bzw. ζ_j konstant gehalten werden und die Bedingung der adiabatischen Isolierung gewahrt bleibt. Die Zustandsgleichungen für kleine Abweichungen vom thermodynamischen Gleichgewicht lauten damit

$$\Delta\zeta_i = T\sum_{k=1}^{n} S_{ik}\, \alpha_k \qquad (i = 1, 2, \ldots, n). \qquad (29.3\,\mathrm{a})$$

Auf Grund der Schwankungen können kleine Abweichungen der extensiven Parameter α_i von ihren Gleichgewichtswerten $\alpha_i = 0$ bei festgehaltenen intensiven Parametern ζ_k^+ auch spontan auftreten. Die Variablen α_i sind danach schwankende Funktionen $\alpha_i(t)$ der Zeit. Die Wahrscheinlichkeit von Werten α_i in den Intervallen $d\alpha_1, \ldots, d\alpha_n$ ist proportional zu $\exp\dfrac{S - S_0}{k}\, d\alpha_1, \ldots, d\alpha_n{}^1$, k ist die Boltzmann-Konstante. Dies ist eine Gaußsche Wahrscheinlichkeitsverteilung mit den Momenten

$$\langle \alpha_i(t) \rangle = 0, \quad \langle \alpha_i(t)\, \alpha_k(t) \rangle = k\, S^{ik}. \qquad (29.4)$$

$\beta)$ *Statistische Funktionen.* Man nimmt an, daß die schwankenden Funktionen $\alpha_i(t)$ statistische Funktionen (random functions) im Sinne der Wahrscheinlichkeitstheorie sind. Damit ist folgendes gemeint: Man betrachte eine Gesamtheit von sehr vielen gleichartigen Systemen. Jedes System hat einen anderen Zeitablauf der Schwankungen $\alpha_i(t)$. Es lassen sich aber Häufigkeits- bzw. Wahrscheinlichkeitsaussagen über die Gesamtheit der Funktionensätze $\alpha_i(t)$ machen. Wählt

---

[1] Zur Theorie der Schwankungen bei adiabatischer Isolierung vergleiche die oben zitierten Arbeiten von Callen und Greene.

man Zeitpunkte $t_1, t_2, \ldots, t_s$, so gibt es eine Wahrscheinlichkeit

$$W_s \begin{pmatrix} \alpha_{i1}, & \alpha_{i2}, & \ldots & \alpha_{is} \\ t_1, & t_2, & \ldots & t_s \end{pmatrix} \prod_{i,k} d\alpha_{ik} \tag{29.5}$$

dafür, daß die Variablen α_i zur Zeit t_1 Werte zwischen α_{i1} und $\alpha_{i1} + d\alpha_{i1}$, zur Zeit t_2 Werte zwischen α_{i2} und $\alpha_{i2} + d\alpha_{i2}$ usw. annehmen. Die Schwankungsfunktionen $\alpha_i(t)$ haben noch die weitere Eigenschaft, stationäre statistische Funktionen zu sein; d.h. die Wahrscheinlichkeiten sind gegen eine Zeittranslation invariant und damit nur von den Zeitdifferenzen $t_2 - t_1, t_3 - t_1, \ldots, t_s - t_1$ abhängig. Wir benötigen im folgenden nur die Wahrscheinlichkeiten W_s für $s = 1, 2$. W_1 hängt also nicht von der Zeit, W_2 nur von $\tau = t_2 - t_1$ ab:

$$W_1 = W_1(\alpha_i), \qquad W_2 = W_2(\alpha_{i1}, \tau, \alpha_{i2}). \tag{29.6}$$

W_1 ist die schon oben erwähnte Gauß-Verteilung mit den Momenten (29.4).

Wir gehen vorübergehend zu einer kompakten Bezeichnung über und fassen die α_i zu einem Vektor $\boldsymbol{\alpha}$ zusammen. Unter $d(\boldsymbol{\alpha})$ verstehen wir $d\alpha_1 \ldots d\alpha_n$.

Sind zu einer Zeit t die Werte $\boldsymbol{\alpha}_1$ festgestellt, so kann man eine Wahrscheinlichkeit dafür angeben, daß man zur Zeit $t + \tau$ Werte des Vektors $\boldsymbol{\alpha}$ zwischen $\boldsymbol{\alpha}_2$ und $\boldsymbol{\alpha}_2 + d\boldsymbol{\alpha}_2$ findet. Man nennt sie *bedingte Wahrscheinlichkeit*. Sie ist gegeben durch

$$P_2(\boldsymbol{\alpha}_1 \mid \tau, \boldsymbol{\alpha}_2)\, d(\boldsymbol{\alpha}_2) = \frac{W_2(\boldsymbol{\alpha}_1, \tau, \boldsymbol{\alpha}_2)\, d(\boldsymbol{\alpha}_2)}{W_1(\boldsymbol{\alpha}_1)} . \tag{29.7}$$

Den Mittelwert von $\boldsymbol{\alpha}_2$ bezüglich dieser Wahrscheinlichkeitsverteilung bezeichnet man als *mittlere Schwankungsregression*

$$\langle \boldsymbol{\alpha}_1 \mid \tau, \boldsymbol{\alpha}_2 \rangle = \int \boldsymbol{\alpha}_2\, P_2(\boldsymbol{\alpha}_1 \mid \tau, \boldsymbol{\alpha}_2)\, d(\boldsymbol{\alpha}_2). \tag{29.8}$$

Sie gibt den mittleren Wert des Vektors $\boldsymbol{\alpha}_2$ zur Zeit $t + \tau$ wenn man zur Zeit t den Wert $\boldsymbol{\alpha}_1$ beobachtet hat. Diese Mittelwerte streben für $\tau \to \infty$ gegen Null.

Von Interesse sind ferner die Momente zweiter Ordnung, deren Gesamtheit durch die sog. Korrelationsmatrix

$$\mathsf{R}(\tau) = \langle \boldsymbol{\alpha}(t)\, \tilde{\boldsymbol{\alpha}}(t + \tau) \rangle \tag{29.9}$$

dargestellt wird. Ihre Elemente lassen sich auch in folgender Weise ausdrücken:

$$\left. \begin{aligned} R_{ik}(\tau) &= \langle \alpha_i(t)\, \alpha_k(t + \tau) \rangle \\ &= \int \alpha_{i1} \alpha_{k2} W_2(\boldsymbol{\alpha}_1, \tau, \boldsymbol{\alpha}_2)\, d(\boldsymbol{\alpha}_1)\, d(\boldsymbol{\alpha}_2) \\ &= \int \alpha_{i1} W_1(\boldsymbol{\alpha}_1) \langle \alpha_{i1} \mid \tau, \alpha_{k2} \rangle\, d(\boldsymbol{\alpha}_1). \end{aligned} \right\} \tag{29.10}$$

Insbesondere ist nach (29.4)

$$R_{ik}(0) = k\, S^{ik}. \tag{29.10a}$$

γ) *Mikroskopische Reversibilität.* Eine weitere wichtige Annahme über die statistischen Funktionen $\boldsymbol{\alpha}(t)$ ist ihr Verhalten bei Zeitumkehr. Sind die $\boldsymbol{\alpha}$ gerade Variable, d.h. gerade Funktionen der Geschwindigkeit der atomaren Bestandteile (z.B. Volumen, Ladung, Konzentrationen), so soll die Wahrscheinlichkeit W_2 eine gerade Funktion von τ sein. Das Paar von Ereignissen $\boldsymbol{\alpha}_1$ und $\boldsymbol{\alpha}_2$ soll also dieselbe Wahrscheinlichkeit haben, gleichgültig ob t_1 um τ früher oder um τ später als t_2 liegt. Dieselbe Aussage gilt dann für die bedingte Wahrscheinlichkeit $P_2(\boldsymbol{\alpha}_1 \mid \tau, \boldsymbol{\alpha}_2)$ und für die mittlere Schwankungsregression $\langle \boldsymbol{\alpha}_1 \mid \tau, \boldsymbol{\alpha}_2 \rangle$.

Anders ist es, wenn die $\boldsymbol{\alpha}(t)$ sowohl gerade als ungerade Variable enthalten. Man kann diesen Fall allerdings leicht auf den Fall gerader Variabler zurückführen, wenn man die ungeraden Variablen durch ihre Zeitintegrale ersetzt, wie dies schon in Ziff. 23 geschah. Auf dasselbe läuft es hinaus, wenn man setzt

$$W_2(\boldsymbol{\alpha}_1, \tau, \boldsymbol{\alpha}_2) = W_2(\varepsilon\,\boldsymbol{\alpha}_1, -\tau, \varepsilon\,\boldsymbol{\alpha}_2). \tag{29.11}$$

Um gerade und ungerade Variable nicht explizit unterscheiden zu müssen, ist in (29.11) die Diagonalmatrix ε eingeführt worden. Ihre Diagonalelemente ε_i sind gleich $+1$, wenn α_i eine gerade Variable ist, und gleich -1, wenn α_i eine ungerade Variable ist. Aus (29.11) folgt durch Integration über α_2

$$W_1(\boldsymbol{\alpha}) = W_1(\varepsilon\,\boldsymbol{\alpha}). \tag{29.12}$$

Da $W_1(\boldsymbol{\alpha})$ eine Gaußsche Verteilung ist, so folgt aus (29.12), daß ihr Exponent keine gemischten Produkte $\alpha_i \alpha_k$ von geraden und ungeraden Variablen enthalten kann. Dies ist gleichwertig der Aussage, daß $S_{ik} = 0$, wenn i und k sich auf eine gerade und eine ungerade Variable beziehen. Das bedeutet weiter wegen (29.1), daß die Entropie gerade in bezug auf Zeitumkehr ist. Weiter folgt aus (29.11) und (29.12)

$$P_2(\boldsymbol{\alpha}_1 \,|\, \tau, \boldsymbol{\alpha}_2) = P_2(\varepsilon\,\boldsymbol{\alpha}_1 \,|\, -\tau, \varepsilon\,\boldsymbol{\alpha}_2) \tag{29.13}$$

und

$$\langle \boldsymbol{\alpha}_1 \,|\, \tau, \boldsymbol{\alpha}_2 \rangle = \langle \varepsilon\,\boldsymbol{\alpha}_1 \,|\, -\tau, \varepsilon\,\boldsymbol{\alpha}_2 \rangle. \tag{29.14}$$

Mit (29.11) und den Folgerungen (29.12), (29.13) ist das Prinzip der mikroskopischen Reversibilität formuliert. Für die Momente zweiter Ordnung ergibt sich daraus mit (29.9)

$$\langle \boldsymbol{\alpha}(t)\,\tilde{\boldsymbol{\alpha}}(t+\tau)\rangle = \langle \varepsilon\,\boldsymbol{\alpha}(t)\,\widetilde{\varepsilon\,\boldsymbol{\alpha}}(t-\tau)\rangle. \tag{29.15}$$

Eine Modifikation tritt ein, wenn ungerade Parameter im System vorhanden sind; z.B. die Winkelgeschwindigkeit eines rotierenden Systems oder ein Magnetfeld mit der magnetischen Induktion $\boldsymbol{B}$. Im zweiten Beispiel ist links in (29.11) das Argument $\boldsymbol{B}$, rechts das Argument $-\boldsymbol{B}$ hinzuzufügen und es ergibt sich statt (29.15)

$$\langle \boldsymbol{\alpha}(t, \boldsymbol{B})\,\tilde{\boldsymbol{\alpha}}(t+\tau, \boldsymbol{B})\rangle = \langle \varepsilon\,\boldsymbol{\alpha}(t, -\boldsymbol{B})\,\widetilde{\varepsilon\,\boldsymbol{\alpha}}(t-\tau, -\boldsymbol{B})\rangle. \tag{29.16}$$

Analog ist die Situation bei anderen ungeraden Parametern.

$\delta)$ *Mittlere Schwankungsregression und irreversibler Prozeß.* Es handelt sich nun darum, die mittlere Schwankungsregression mit dem Ablauf von irreversiblen Prozessen in Verbindung zu bringen. Man legt die Idee zugrunde, daß es für den Ablauf eines irreversiblen Prozesses gleichgültig sein sollte, wie der Anfangszustand zur Zeit t_0 hergestellt wurde. Man kann entweder einen Schwankungszustand zur Zeit t_0 mit Werten $\boldsymbol{\alpha}_1$ herausgreifen, oder man kann der Umgebung und dem System für $t \leqq t_0$ solche Werte der intensiven Variablen geben, daß zur Zeit $t = t_0$ gerade die Werte $\boldsymbol{\alpha}_1$ der extensiven Variablen realisiert sind, und dann die Werte der intensiven Variablen der Umgebung plötzlich auf $\boldsymbol{\zeta}^+$ abändern. Im ersten Fall wäre der Ablauf des irreversiblen Prozesses sinngemäß durch die mittlere Schwankungsregression, im zweiten Fall durch die phänomenologischen Gleichungen zu beschreiben. In der Annahme der Äquivalenz beider Beschreibungen liegt die entscheidende Voraussetzung für den folgenden Beweis.

Wenn man die phänomenologischen Gleichungen in der Gestalt

$$\dot{\boldsymbol{\alpha}} = \mathsf{L}\,\varDelta\boldsymbol{\zeta}, \qquad \varDelta\boldsymbol{\zeta} = -\,T\mathsf{S}\,\boldsymbol{\alpha} \tag{29.17}$$

schreibt, so lautet also diese Voraussetzung wegen (29.3)

$$\frac{d}{d\tau}\langle \boldsymbol{\alpha}_1 \,|\, \tau, \boldsymbol{\alpha}_2 \rangle = -\,T\mathsf{L}\mathsf{S}\,\langle \boldsymbol{\alpha}_1 \,|\, \tau, \boldsymbol{\alpha}_2 \rangle, \tag{29.18}$$

gültig für jeden Anfangszustand $\boldsymbol{\alpha}_1$. Multipliziert man diese Gleichung mit $\boldsymbol{\alpha}_1 W_1(\boldsymbol{\alpha}_1)$ und integriert über die $\boldsymbol{\alpha}_1$, so ergibt sich nach (29.10)

$$\frac{d}{d\tau}\langle \boldsymbol{\alpha}(t)\,\tilde{\boldsymbol{\alpha}}(t+\tau)\rangle = -\,T\langle \boldsymbol{\alpha}(t)\,\tilde{\boldsymbol{\alpha}}(t+\tau)\rangle\, \mathsf{S}\,\tilde{\mathsf{L}}. \tag{29.19}$$

Nun ist nach (29.16), wenn wir wieder zur Komponentenschreibweise übergehen,

$$\begin{aligned} R_{ij}(\tau) &= \langle \alpha_i(t)\,\alpha_j(t+\tau)\rangle = \langle \alpha_i(t-\tau)\,\alpha_j(t)\rangle \\ &= \varepsilon_i\,\varepsilon_j\langle \alpha_j(t)\,\alpha_i(t+\tau)\rangle = \varepsilon_i\,R_{ji}(\tau)\,\varepsilon_j, \end{aligned} \tag{29.20}$$

da eine Ersetzung von t durch $t-\tau$ wegen des stationären Charakters der $\alpha_i(t)$ nichts an den Mittelwerten ändert. Vertauscht man i und j in (29.19) und macht von (29.20) Gebrauch, so folgt für $\tau > 0$

$$\frac{d}{d\tau}\langle \alpha_i(t)\,\alpha_j(t+\tau)\rangle = -\,T\sum_{k,l} L_{ik}\,S_{kl}\langle \alpha_l(t)\,\alpha_j(t+\tau)\rangle\,\varepsilon_i\,\varepsilon_l. \tag{29.21}$$

Die rechten Seiten von (29.19) und (29.21) sind einander gleich für alle τ und auch im Grenzfall $\tau = 0$. Für $\tau = 0$ kann man aber die Mittelwerte nach (29.9) und (29.10a) ausdrücken und erhält zunächst

$$\sum_{k,l} L_{jk}\,S_{kl}\,S^{il} = \sum_{k,l} L_{ik}\,S_{kl}\,S^{jl}\,\varepsilon_i\,\varepsilon_l. \tag{29.22}$$

Da die Matrizen S_{kl} und S^{il} reziprok sind, so gibt die linke Seite L_{ji}. Da S_{jl} und $S^{il}=0$ wenn $\varepsilon_j\,\varepsilon_l = -1$, so kann man $S^{il}\,\varepsilon_l$ durch $\varepsilon_j\,S^{il}$ ersetzen. Die rechte Seite von (29.22) gibt dann, wieder unter Berücksichtigung der Reziprozität von S_{kl} und S^{il}, $\varepsilon_i\,\varepsilon_j\,L_{ij}$. Somit ist

$$L_{ji} = \varepsilon_i\,\varepsilon_j\,L_{ij}. \tag{29.23}$$

Damit sind die O.C.R.B. unter der erwähnten Annahme bewiesen.

30. Ergänzungen und Verallgemeinerungen. $\alpha)$ *Allgemeine Bemerkungen.* Die Annahme äquivalenter Gesetze für die mittlere Schwankungsregression und für den makroskopischen irreversiblen Prozeß bedarf noch der Erläuterung. Mit der Postulierung des Differentialgleichungssystems (29.18) für die mittlere Schwankungsregression wird dem statistischen Prozeß $\alpha_i(t)$ unterstellt, daß er ein Markoff-Prozeß ist. Eine solche Eigenschaft kann jedoch nicht begründet werden. Indessen ist sie bei geeigneter Wahl der Variablen α_i (s. dazu unten) wenigstens näherungsweise erfüllt, wenn man die Werte $\alpha_i(t)$ und $\alpha_i(t+\tau)$ in nicht zu kleinen Zeitabständen ($\tau \geq t_1 > 0$) vergleicht. Dann ist es aber nicht mehr erlaubt, in (29.21) den Differentialquotienten nach τ für $\tau = 0$ zu bilden. Indessen läßt sich die Argumentation in Ziff. 30 aufrechterhalten, wenn man die Differentialquotienten nach der Zeit in (29.18), (29.19) und (29.21) als Differenzenquotienten für ein endliches Zeitintervall der Größenordnung t_1 erklärt.

Damit scheinen die O.C.R.B. (29.23) allerdings den Charakter von nur näherungsweise gültigen Aussagen anzunehmen, die um so ungenauer sind, je größer das Zeitintervall t_1 gewählt werden muß.

Diese Schwierigkeit läßt sich in einfacher Weise beheben, wenn man beachtet, daß schon die Gln. (29.17) für den makroskopischen irreversiblen Prozeß problematisch und im selben Umfang ungenau sind wie die O.C.R.B. Denn jede Beschreibung eines irreversiblen Prozesses durch einen endlichen Satz von Variablen ist im Prinzip unvollständig und daher ungenau. Sie kann sich jeweils nur auf entsprechend gealterte Systeme beziehen. In diesem Sinne gelten auch die Gln. (29.17) nur in einem Zeitintervall $0 < t_1 < \tau < t_2$. Dabei ist t_2 die Zeit, nach der der irreversible Prozeß praktisch abgelaufen ist — sie ist von der Größenordnung der größten Relaxationszeit des Gleichungssystems (29.17) — und t_1 ist die Zeit, nach der die im Variablensatz $\alpha_1, \alpha_2, \ldots, \alpha_n$ nicht berücksichtigten Variablen $\alpha_{n+1}, \alpha_{n+2}, \ldots$ entweder Gleichgewichtswerte angenommen haben oder von $\alpha_1, \ldots, \alpha_n$ abhängig geworden sind. Welcher Spielraum zwischen t_1 und t_2 besteht, hängt von der Natur des Systems und der getroffenen Auswahl der Variablen α_i ab. [Es ist z. B. nicht zweckmäßig beim idealen Gas mit zweiatomigen Molekülen die Besetzungszahlen der Rotationszustände unter die α_i aufzunehmen, ohne daß man die Besetzungszahlen der Schwingungszustände mitnimmt; denn die ersteren stellen sich sehr viel schneller als die letzteren auf die Boltzmann-Verteilung ein und die Gültigkeitsdauer für die Gln. (29.17) wäre viel kleiner als die Relaxationszeit der Schwingung.]

Man wird dieser Situation gerecht, indem man die Existenz eines vollständigen Satzes von Variablen α_i annimmt mit der Eigenschaft, daß der irreversible Prozeß durch seine Anfangswerte $\alpha_i(0)$ bestimmt ist, unabhängig davon, wie diese Anfangswerte entstanden sind. Dem entspricht beim statistischen Prozeß $\alpha_i(t)$, daß die bedingte Wahrscheinlichkeit, Werte $\alpha_i(t)$ zu einer Zeit $t > 0$ zu beobachten, wenn zur Zeit $t = 0$ Werte $\alpha_i(0)$ vorgelegen haben, unabhängig von der Kenntnis früherer Werte der $\alpha_i(t)$ (also für Zeitpunkte $t < 0$) ist. Der Prozeß $\alpha_i(t)$ ist dann ein Markoffscher Prozeß und genügt damit einem Differentialgleichungssystem der Gestalt (29.18). Die Äquivalenz der Gesetze des makroskopischen irreversiblen Prozesses und der mittleren Schwankungsregression bedeutet damit nur noch eine Feststellung über die Gleichheit der Koeffizienten in (29.17) und (29.18), zwingt aber der mittleren Schwankungsregression nicht Differentialgleichungen von einem Typ auf, der nicht schon anderweitig begründet ist.

Der Beweis der O.C.R.B. in Ziff. 29 gilt somit zunächst nur für einen vollständigen Satz von Variablen. Von hier aus schließt man jedoch leicht auf die O.C.R.B. für geeignete Teilmengen dieser Variablen, die mehr oder weniger gealterte Systeme beschreiben. Die dazu erforderlichen Überlegungen können in ähnlicher Weise wie in einer Arbeit von Meixner[1] durchgeführt werden.

Problematisch bleibt zwar noch die Existenz eines vollständigen Systems von Variablen. Diese Frage findet sich bei Green[2] und Ludwig[3] angeschnitten. Immerhin sieht man, daß man durch Erweiterung des Variablensystems die Größe der Zeit t_1 herunterdrücken kann.

$\beta)$ *Das Schwankungs-Dissipationstheorem.* Einen anderen Weg zur Vermeidung der Schwierigkeiten, die auf der Verwendung eines unvollständigen Variablensatzes beruhen, haben Callen-Greene[4] und Kaplan[5] eingeschlagen. Sie nutzen die Tatsache aus, daß die betrachteten Systeme bei kleinen Abweichungen vom thermodynamischen Gleichgewicht lineare Systeme (s. Ziff. 23) sind, daß für sie also das Superpositionsprinzip gilt. Variieren insbesondere die $\Delta\zeta_k$ harmonisch mit der Zeit, so gilt dasselbe für die α_i, und es besteht eine lineare Abhängigkeit

$$\alpha_i(w) = \sum_k \frac{Y_{ik}(\omega)}{i\,\omega}\,\Delta\zeta_k(\omega)\,. \tag{30.1}$$

---

[1] J. Meixner: Z. Physik **131**, 456 (1952).

[2] M. S. Green: J. Chem. Phys. **22**, 398 (1954).

[3] G. Ludwig: Z. Naturforsch. **12**a, 662 (1957).

[4] H. B. Callen u. R. F. Greene: Phys. Rev. **86**, 702 (1952). — R. F. Greene u. H. B. Callen: Phys. Rev. **88**, 1387 (1952).

[5] T. A. Kaplan: Phys. Rev. **102**, 1447 (1956).

Die Matrix $Y_{ik}(\omega)$ wird *Admittanzmatrix* genannt. $\sum\limits_{i,k} Y_{ik}(ip)\,x_i x_k$ ist, als Funktion der komplexen Variablen p betrachtet, für beliebige reelle x_i eine positive Funktion im Sinne der Netzwerktheorie (s. Ziff. 23), und es gilt

$$\lim_{\omega \to 0} \frac{Y_{ik}(\omega)}{i\,\omega} = \frac{1}{T}\,S^{ik}. \tag{30.2}$$

da im thermodynamischen Gleichgewicht bei vorgegebenen konstanten Werten der ζ_k (30.1) mit (29.3) äquivalent sein muß. Hieraus folgt für irgendeinen vorgeschriebenen Zeitverlauf $\zeta_k(t)$ durch Fourier-Zerlegung

$$\Delta\zeta_k(t) = \int\limits_{-\infty}^{+\infty} e^{i\omega t}\,\Delta\zeta_k(\omega)\,d\omega, \quad \Delta\zeta_k(\omega) = \frac{1}{2\pi}\int\limits_{-\infty}^{+\infty} e^{-i\omega t}\,\Delta\zeta_k(t)\,dt \tag{30.3}$$

der Zeitverlauf

$$\alpha_i(t) = \sum_k \int\limits_{-\infty}^{+\infty} e^{i\omega t}\,\frac{Y_{ik}(\omega)}{i\,\omega}\,\Delta\zeta_k(\omega)\,d\omega. \tag{30.4}$$

Wir betrachten nun einen speziellen Zeitverlauf $\Delta\zeta_k(t) = \Delta\zeta_k^0$ für $t<0$ und $\Delta\zeta_k(t) = 0$ für $t>0$. Die Fourier-Koeffizienten dieser Sprungfunktionen sind

$$\Delta\zeta_k(\omega) = \left[-\frac{1}{2\pi i\,\omega} + \frac{1}{2}\,\delta(\omega)\right]\Delta\zeta_k^0. \tag{30.5}$$

$\delta(\omega)$ ist eine δ-Funktion; das Fourier-Integral (30.3) für $\Delta\zeta_k(t)$ ist als Cauchyscher Hauptwert zu verstehen. Man kann übrigens δ-Funktion und Cauchyschen Hauptwert vermeiden, wenn man die Funktionen $\Delta\zeta_k(t)$ nur in einem endlichen Intervall konstant und von Null verschieden und sonst gleich Null annimmt und das Superpositionsprinzip anwendet. Die getroffene Wahl bringt jedoch formale Vereinfachungen. Aus (30.4) und (30.5) folgt dann für $\alpha_i(t)$ der spezielle Zeitverlauf

$$\alpha_i(t) = \sum_k \int\limits_{-\infty}^{+\infty} e^{i\omega t}\,\frac{Y_{ik}(\omega)}{i\,\omega}\left[-\frac{1}{2\pi i\,\omega} + \frac{1}{2}\,\delta(\omega)\right]d\omega\,\Delta\zeta_k^0. \tag{30.6}$$

Unter Verwendung von (30.2) kann man das Integral über den zweiten Term ausführen und erhält

$$\alpha_i(t) = \sum_k \left[\frac{1}{2\pi}\int\limits_{-\infty}^{+\infty} e^{i\omega t}\,\frac{Y_{ik}(\omega)}{\omega^2}\,d\omega + \frac{1}{2T}\,S^{ik}\right]\Delta\zeta_k^0, \tag{30.7}$$

eine für alle t gültige Beziehung. Für $t<0$ nehmen die $\alpha_i(t)$ für den vorgegebenen Zeitverlauf der $\Delta\zeta_k(t)$ die Gleichgewichtswerte

$$\alpha_i(t) = \frac{1}{T}\sum_k S^{ik}\Delta\zeta_k^0 \quad (t<0) \tag{30.8}$$

an. Durch Vergleich von (30.7) und (30.8) folgt somit

$$\frac{1}{2T}\,S^{ik} = \frac{1}{2\pi}\int\limits_{-\infty}^{+\infty} e^{i\omega t}\,\frac{Y_{ik}(\omega)}{\omega^2}\,d\omega \quad (t<0). \tag{30.9}$$

Die Matrix S^{ik} hat wegen des Fehlens von Produkten gerader und ungerader Variablen in (29.1) die Eigenschaft

$$\varepsilon_i\,S^{ik}\,\varepsilon_k = S^{ik} = S^{ki}. \tag{30.10}$$

Damit schließt man aus (30.9) für $t < 0$ auf

$$\int_{-\infty}^{+\infty} e^{i\omega t}\, \frac{\varepsilon_i Y_{ik}(\omega)\, \varepsilon_k}{\omega^2}\, d\omega = \int_{-\infty}^{+\infty} e^{i\omega t}\, \frac{Y_{ki}(\omega)}{\omega^2}\, d\omega. \tag{30.11}$$

Wir zeigen nun, daß diese Beziehung auch für $t > 0$ richtig ist. Dazu betrachten wir die stationären statistischen Funktionen $\alpha_i(t)$ und fragen nach der mittleren Schwankungsregression bei Ausgangswerten

$$\alpha_i(0) = \frac{1}{T} \sum_k S^{ik}\, \Delta \zeta_k^0.$$

Sie ist gegeben durch die Mittelwerte $\langle \alpha_k | t, \alpha_i \rangle$. Von ihr wird postuliert, daß sie mit $\alpha_i(t)$ aus (30.6) für $t > 0$ übereinstimmt. Auf Grund der Linearität des stationären statistischen Prozesses $\alpha_i(t)$ folgt aus (29.8) und (29.10) für die Korrelationsmatrix R_{ik}

$$\sum_{k,l} R_{ki}(\tau)\, R^{lk}(0)\, \alpha_l(0) = \langle \alpha_{l1} | \tau, \alpha_{i2} \rangle \tag{30.12}$$

oder

$$-\frac{1}{kT} \sum_k R_{ki}(\tau)\, \Delta\zeta_k^0 = \langle \alpha_{l1} | \tau, \alpha_{i2} \rangle. \tag{30.13}$$

Somit entsteht nach (30.7) für $t > 0$

$$\frac{1}{2\pi} \int_{-\infty}^{+\infty} e^{i\omega t}\, \frac{Y_{ik}(\omega)}{\omega^2}\, d\omega + \frac{1}{2T}\, S^{ik} = -\frac{1}{kT}\, R_{ki}(t). \tag{30.14}$$

Wegen (30.10) und (29.20) folgt hieraus die Richtigkeit von (30.11) auch für $t > 0$, d.h. für alle Zeiten. Dies ist aber gleichwertig mit

$$\varepsilon_i Y_{ik}(\omega)\, \varepsilon_k = Y_{ki}(\omega). \tag{30.15}$$

Diese Beziehungen haben die O.C.R.B. zur Folge. Dies sieht man, wenn man für $\Delta\zeta_k(t)$ den oben angenommenen Verlauf wählt und die Zeitableitung von $\alpha_i(t)$ zur Zeit $t = 0$ als Differenzenquotient für ein hinreichend kleines Zeitintervall t_1 (s. Ziff. 30α) berechnet. Es ergibt sich aus (30.6)

$$\frac{\alpha_i(t_1) - \alpha_i(0)}{t_1} = \frac{1}{2\pi t_1} \sum_k \int_{-\infty}^{+\infty} (e^{i\omega t} - 1)\, \frac{Y_{ik}(\omega)}{\omega^2}\, d\omega\, \Delta\zeta_k^0.$$

Die Koeffizienten der $\Delta\zeta_k^0$ sind mit den phänomenologischen Koeffizienten in (29.17) zu identifizieren, und damit erhält man die O.C.R.B.

Schließlich ergibt sich noch, wenn man S^{ik} durch Gleichsetzen der rechten Seiten von (30.7) und (30.8) berechnet, t durch $-t$ ersetzt und dann (29.10a) in (30.14) einsetzt

$$\left.\begin{aligned}
R_{ki}(t) &= -\frac{kT}{\pi} \int_{-\infty}^{+\infty} \cos\omega t\, \frac{Y_{ik}(\omega)}{\omega^2}\, d\omega \\
&= -\frac{kT}{2\pi} \int_{-\infty}^{+\infty} e^{i\omega t}\, \frac{\operatorname{Re} Y_{ik}(\omega)}{\omega^2}\, d\omega
\end{aligned}\right\} \quad (0 < t < +\infty). \tag{30.16}$$

Die letzte Beziehung verknüpft die Korrelationsmatrix der Schwankungs-vorgänge mit der Admittanzmatrix des linearen dissipativen Systems. Sie wird daher auch *Schwankungs-Dissipationstheorem* genannt. Dieses ist eine Verall-gemeinerung der Nyquistschen Formel[1] für das elektrische Rauschen.

γ) Die O.C.R.B. für vektorielle und tensorielle Phänomene. Der Beweis der O.C.R.B. wurde oben für skalare Variable durchgeführt. Mit den Ausführungen in Ziff. 25 ist jedoch bereits gezeigt, wie man die O.C.R.B. für vektorielle und tensorielle Phänomene auf die O.C.R.B. bei skalaren Variablen zurückführen kann. Man kann aber auch nach DE GROOT und MAZUR die Schwankungstheorie so erweitern, daß man ihre Ergebnisse unmittelbar auf die phänomenologischen Koeffizienten bei Transportphänomenen anwenden kann. Wir setzen dies wieder am Beispiel der Wärmeleitung im Magnetfeld auseinander und verweisen wegen der Behandlung anderer Probleme auf DE GROOT und Mitarbeiter[2].

Wie in Ziff. 24 nehmen wir an, daß im System nur Temperaturänderungen möglich sind und keine Bewegungen erfolgen, daß der Ausdehnungskoeffizient verschwindet und das System adiabatisch isoliert ist. Zunächst teilen wir das wärmeleitende System in kleine Volumenelemente gleicher Größe V auf. Ihre Lage sei durch die Ortsvektoren r_i gekennzeichnet. Die Temperatur im Volumen-element i ist $T(r_i) = T_0 + \Delta T(r_i)$ wobei $|\Delta T(r_i)| \ll T_0$ angenommen werden soll. Die innere Energie des Volumenelements V am Ort r_i ist $u_v(r_i) \cdot V = [u_v(T_0) + \Delta u_v(r_i)] V$. Bei einer Zustandsänderung $d\, \Delta u_v(r_i)$ ändert sich die gesamte Entropie S um

$$dS = \sum_i -\frac{\Delta T(r_i)}{T_0^2}\, d\, \Delta u_v(r_i)\, V. \qquad (30.17)$$

Man kann nun die Überlegungen in Ziff. 29 übertragen und die Momente der Schwankungen berechnen. An Stelle von $\alpha_i(t)$ haben wir jetzt $V \Delta u_v(r_i, t)$, für $\Delta \zeta_i$ steht $\Delta T(u_v(r_i, t))/T_0^2$. Aus (29.4) gewinnt man dann

$$V \langle \Delta u_v(r_i, t)\, \Delta T(u_v(r_k, t)) \rangle = k T_0^2\, \delta_{ik}. \qquad (30.18)$$

Sind die Volumenelemente hinreichend klein gewählt, so kann diese Beziehung durch

$$\langle \Delta u_v(r, t)\, \Delta T(u_v(r', t)) \rangle = k T_0^2\, \delta(r' - r) \qquad (30.19)$$

ersetzt werden. Man überzeugt sich davon indem man (30.19) über ein Volumen V nach r integriert. Aus Gründen der Allgemeinheit lassen wir zu, daß auch ein Magnetfeld mit der magnetischen Induktion B vorliegt. Ist $\Omega(r)$ ein linearer Operator

$$\Omega(r) = \Omega_0(r) + \sum_{\alpha=1}^{3} \Omega_\alpha(r)\, \frac{\partial}{\partial x_\alpha} + \sum_{\alpha=1}^{3} \sum_{\beta=1}^{3} \Omega_{\alpha\beta}(r)\, \frac{\partial^2}{\partial x_\alpha\, \partial x_\beta} + \cdots \qquad (30.20)$$

mit Funktionen $\Omega_0(r)$, $\Omega_\alpha(r)$, ..., so gilt ersichtlich auch

$$\langle \Delta u_v(r, t, B(r))\, \Omega(r')\, \Delta T(u_v(r', t, B(r'))) \rangle = k T^2 \Omega(r') \cdot \delta(r - r'). \qquad (30.21)$$

Aus dem Prinzip der mikroskopischen Reversibilität folgt weiter, da die $\Delta u_v(r)$ Variable gleichen Typs sind,

$$\left.\begin{aligned}
&\langle \Delta u_v(r, t, B(r))\, \Delta u_v(r', t + \tau, B(r')) \rangle \\
&= \langle \Delta u_v(r', t, -B(r'))\, \Delta u_v(r, t + \tau, -B(r)) \rangle
\end{aligned}\right\} \qquad (30.22)$$

---

[1] H. NYQUIST: Phys. Rev. **32**, 110 (1928).
[2] S.R. DE GROOT u. P. MAZUR: Phys. Rev. **94**, 218 (1954). — P. MAZUR u. S.R. DE GROOT: Phys. Rev. **94**, 224 (1954).

und durch Differentiation nach τ, wenn man hinterher $\tau = 0$ setzt,

$$\left.\begin{aligned}
&\left\langle \Delta u_v(\boldsymbol{r}, t, \boldsymbol{B}(r)) \frac{\partial}{\partial t} \Delta u_v(\boldsymbol{r}', t, \boldsymbol{B}(r')) \right\rangle \\
&= \left\langle \Delta u_v(\boldsymbol{r}', t, -\boldsymbol{B}(r')) \Delta u_v(\boldsymbol{r}, t, -\boldsymbol{B}(r)) \right\rangle.
\end{aligned}\right\} \tag{30.23}$$

Damit sind alle Vorbereitungen getroffen, um die O.C.R.B. für den Wärmeleitungstensor $\lambda_{\alpha\beta}$ in den phänomenologischen Gleichungen der Wärmeleitung, d.h. in

$$\frac{\partial}{\partial t} u_v(\boldsymbol{r}, t) = \sum_{\alpha,\beta=1}^{3} \frac{\partial}{\partial x_\alpha} \lambda_{\alpha\beta}(\boldsymbol{r}, \boldsymbol{B}) \frac{\partial}{\partial x_\beta} T(\boldsymbol{r}, t) \tag{30.24}$$

zu finden. Die Mittelwerte in (30.23) kann man in zwei Schritten gewinnen. Man mittelt erst den zweiten Faktor für gegebene Werte des ersten Faktors. Diese bedingten Mittelwerte genügen wegen der Äquivalenz von irreversiblem Prozeß und mittlerer Schwankungsregression der Gl. (30.24). Dann mittelt man über die Werte des ersten Faktors. Dies läuft also darauf hinaus, daß man (30.24) in (30.23) einsetzt. Berücksichtigt man dann (30.21), so folgt

$$\left.\begin{aligned}
&\sum_{\alpha,\beta=1}^{3} \frac{\partial}{\partial x_\alpha'} \lambda_{\alpha\beta}(\boldsymbol{r}', \boldsymbol{B}(r')) \frac{\partial}{\partial x_\beta'} \delta(\boldsymbol{r}' - \boldsymbol{r}) \\
&= \sum_{\alpha,\beta=1}^{3} \frac{\partial}{\partial x_\alpha} \lambda_{\alpha\beta}(\boldsymbol{r}, -\boldsymbol{B}(r)) \frac{\partial}{\partial x_\beta} \delta(\boldsymbol{r}' - \boldsymbol{r}).
\end{aligned}\right\} \tag{30.25}$$

Eine einfache Rechnung führt von hier auf die O.C.R.B. (25.15) und (25.16).

31. Das Prinzip von Onsager und Machlup. α*) Die Langevinschen Gleichungen.* Die in Ziff. 29 zum Beweis der O.C.R.B. angenommene Äquivalenz von mittlerer Schwankungsregression und irreversiblem Ablauf setzt einen sehr engen Zusammenhang zwischen Schwankungen und irreversiblen Prozessen voraus. Einen Versuch, beide Erscheinungen in einheitlicher Weise zu formulieren, stellen die Langevinschen Gleichungen dar; sie liefern die genannte Äquivalenz als Folgerung.

Die Langevinschen Gleichungen beschreiben die Wirkung der intensiven Parameter ζ_k^0 der Umgebung durch eine statistische Kraft $\zeta_k^0(t) + \delta_k(t)$, in welcher die $\delta_k(t)$ statistische Funktionen mit dem Mittelwert $\langle \delta_k(t) \rangle = 0$ und mit verschwindender Korrelationsfunktion $\langle \delta_k(t) \delta_k(t+\tau) \rangle$ für $\tau \neq 0$ (sog. weißes Spektrum der Schwankungen) sind. Die Langevinschen Gleichungen lauten also

$$\dot{\alpha}_i(t) = \sum_k L_{ik}\left(\Delta\zeta_k(\alpha_j) + \delta_k(t)\right), \quad \zeta_k - \zeta_k^0 \equiv \Delta\zeta_k = -T \sum_l S_{kl}\alpha_l(t). \tag{31.1}$$

Sie sind lineare stochastische Differentialgleichungen für die statistischen Funktionen $\alpha_i(t)$. Aus (31.1) ergeben sich die phänomenologischen Gleichungen (29.17) durch Mittelung über eine Gesamtheit von Systemen. Die Schwankungseigenschaften im Gleichgewicht ergeben sich, indem man jene Lösungen von (31.1) bei konstanten $\zeta_k^0 = \zeta_k^+$ sucht, welche stationäre statistische Funktionen sind. Der Ansatz (31.1) läßt sich auf kontinuierliche Sätze von Variablen, also etwa auf Kramerssche Gleichungen übertragen (Hashitsume[1]).

Die Langevinschen Gleichungen bilden den Ausgangspunkt der statistischen Theorie der irreversiblen Prozesse von Onsager und Machlup[2]. Wir stellen diese

---

[1] N. Hashitsume: Progr. Theor. Phys. **8**, 461 (1952); **15**, 369 (1956).
[2] L. Onsager u. S. Machlup: Phys. Rev. **91**, 1505 (1953). — S. Machlup u. L. Onsager: Phys. Rev. **91**, 1512 (1953).

Theorie im Anschluß an TISZA und MANNING[1] dar, welche die Langevinsche Gleichung nicht an die Spitze stellen, sondern als Folgerung gewinnen.

β) Gaußsche Markoff-Prozesse. Wir machen erst einige Angaben über stationäre statistische Prozesse, welche Markoff-Charakter und Gauß-Verteilungen haben. Die Variablen α_i sollen zunächst alle gerade sein. Dann kann man zeigen, daß es unabhängige Linearkombinationen der Variablen α_i gibt derart, daß sich die Gaußsche Wahrscheinlichkeitsverteilung $W_2(\alpha_1, \tau, \alpha_2)$ in ein Produkt von Gaußschen Verteilungen zerlegen läßt, so daß jeder Faktor nur von einer solchen Linearkombination abhängt[2]. Ohne Beschränkung der Allgemeinheit kann man sich daher im folgenden auf die Betrachtung einer einzigen Variablen α beschränken.

Der stationäre Markoff-Prozeß $\alpha(t)$ mit Gauß-Verteilung wird durch die Wahrscheinlichkeit

$$W_2(\alpha_1, \tau, \alpha_2) \sim \exp\left\{ -\tfrac{1}{2}\left(p_{11}(\tau)\,\alpha_1^2 - 2p_{12}(\tau)\,\alpha_1\alpha_2 + p_{22}(\tau)\,\alpha_2^2 \right) \right\} \qquad (31.2)$$

vollständig beschrieben. Der Proportionalitätsfaktor, der von α_1 und α_2 unabhängig ist, ergibt sich aus der Normierung der Wahrscheinlichkeit. Der Prozeß $\alpha(t)$ soll reversibel sein; daraus folgt nach (29.11) $p_{11}(\tau) = p_{22}(\tau)$. Durch Integration von (31.2) nach α_2 erhält man

$$W_1(\alpha) \sim \exp\left\{ -\frac{\alpha^2}{2\sigma^2} \right\} \qquad (31.3)$$

mit der zeitunabhängigen Größe

$$\frac{1}{\sigma^2} \equiv p_{11}(\tau) - \frac{p_{12}^2(\tau)}{p_{11}(\tau)}. \qquad (31.4)$$

Durch Division von (31.2) und (31.3) ergibt sich die bedingte Wahrscheinlichkeit

$$P_2(\alpha_1 \mid \tau, \alpha_2) \sim \exp\left\{ -\frac{1}{2} p_{11}(\tau)\left(\alpha_2 - \frac{p_{12}(\tau)}{p_{11}(\tau)}\,\alpha_1 \right)^2 \right\} \qquad (31.5)$$

und der bedingte Mittelwert

$$\langle \alpha_1 \mid \tau, \alpha_2 \rangle = \frac{p_{12}(\tau)}{p_{11}(\tau)}\,\alpha_1. \qquad (31.6)$$

Der Prozeß ist also linear. Die Markoff-Eigenschaft des Prozesses drückt sich in der Chapman-Kolmogoroff-Beziehung[3]

$$P_2(\alpha_1 \mid \tau + \tau', \alpha_3) = \int P_2(\alpha_1 \mid \tau, \alpha_2)\, d\alpha_2\, P_2(\alpha_2 \mid \tau', \alpha_3) \qquad (\tau, \tau' > 0) \qquad (31.7)$$

aus. Multipliziert man sie mit α_3 und integriert nach α_3, so folgt wegen (31.6) die Funktionalgleichung

$$\frac{p_{12}(\tau + \tau')}{p_{11}(\tau + \tau')} = \frac{p_{12}(\tau)}{p_{11}(\tau)} \cdot \frac{p_{12}(\tau')}{p_{11}(\tau')} \qquad (\tau, \tau' > 0). \qquad (31.8)$$

Sie hat neben der trivialen und hier nicht interessierenden Lösung $p_{12}(\tau) = 0$ für $\tau > 0$ die Lösung

$$\frac{p_{12}(\tau)}{p_{11}(\tau)} = e^{-\tau/\tau_0} \equiv \varrho(\tau) \qquad (31.9)$$

mit $\tau_0 > 0$, damit die Korrelation zwischen α_1 und α_2 für $\tau \to \infty$ verschwindet. Aus (31.4) und (31.9) folgt somit

$$p_{11}(\tau) = \frac{1}{\sigma^2\,[1 - \varrho(2\tau)]}, \qquad p_{12} = \frac{\varrho(\tau)}{\sigma^2\,[1 - \varrho(2\tau)]} \qquad (\tau \geqq 0). \qquad (31.10)$$

---

[1] L. TISZA u. I. MANNING: Phys. Rev. **105**, 1695 (1957).
[2] Siehe Fußnote 1, S. 514.
[3] A. KOLMOGOROFF: Math. Ann. **104**, 415 (1931).

Der stationäre Markoff-Prozeß mit Gauß-Verteilung ist also durch die zwei Konstanten σ^2 und τ_0 vollständig bestimmt.

$\gamma)$ *Die Onsager-Machlup-Funktion.* Onsager und Machlup haben bemerkt, daß man die bedingte Wahrscheinlichkeit P_2 des stationären Gaußschen Markoff-Prozesses der thermodynamischen Variablen $\alpha(t)$ durch ein Stationaritätsprinzip

$$P_2\binom{\alpha_1\,|\,\alpha_2}{t_1\,|\,t_2} \sim \exp\left\{-\frac{1}{k}\operatorname{stat}\int_{t_1,\alpha_1}^{t_2,\alpha_2}\mathfrak{O}(\alpha,\dot\alpha)\,dt\right\} \tag{31.11}$$

ausdrücken kann. Die Onsager-Machlup-Funktion $\mathfrak{O}(\alpha,\dot\alpha)$ ist durch

$$\mathfrak{O}(\alpha,\dot\alpha) = \frac{1}{4LT}\left(\dot\alpha + \frac{1}{\tau_0}\alpha\right)^2 \tag{31.12}$$

gegeben. Die Bezeichnung stat in (31.11) bedeutet, daß der stationäre Wert des Integrals mit den Randbedingungen $\alpha(t_1)=\alpha_1$, $\alpha(t_2)=\alpha_2$ zu bilden ist. Er ergibt sich, indem man die durch diese Randbedingungen bestimmte Lösung der Euler-Lagrangeschen Differentialgleichung

$$\frac{\partial\mathfrak{O}}{\partial\alpha} - \frac{d}{dt}\frac{\partial\mathfrak{O}}{\partial\dot\alpha} = 0, \quad \text{d.h.} \quad \tau_0^2\ddot\alpha - \alpha = 0 \tag{31.13}$$

in $\mathfrak{O}(\alpha,\dot\alpha)$ in (31.11) einsetzt. Hieraus folgt in einfacher Weise durch Multiplikation von Ausdrücken der Form (31.11)

$$P_n\binom{\alpha_1\,|\,\alpha_2,\alpha_3,\ldots,\alpha_n}{t_1\,|\,t_2,\,t_3,\,\ldots,\,t_n} \sim \exp\left\{-\frac{1}{k}\operatorname{stat}\int_{(t_i,\alpha_i)}\mathfrak{O}(\alpha,\dot\alpha)\,dt\right\}, \tag{31.14}$$

wobei der Exponent ersichtlich als

$$-\frac{1}{k}\operatorname{stat}\int_{(t_i\alpha_i)}\mathfrak{O}(\alpha,\dot\alpha)\,dt = -\frac{1}{k}\sum_{i=1}^{n-1}\operatorname{stat}\int_{t_i,\alpha_i}^{t_{i+1},\alpha_{i+1}}\mathfrak{O}(\alpha,\dot\alpha)\,dt \tag{31.15}$$

zu verstehen ist. Die bedingte Wahrscheinlichkeit (31.14) gibt also die Wahrscheinlichkeit dafür, daß die statistische Funktion von einem Wert α_1, zur Zeit t_1 ausgeht und zu späteren Zeitpunkten $t_2, t_3, \ldots, t_n$ Werte $\alpha_2, \alpha_3, \ldots, \alpha_n$ in Intervallen $d\alpha_2, \ldots, d\alpha_n$ annimmt. Indem man bei festgehaltenem Zeitintervall (t_1, t_n) die Einteilung $t_2, \ldots, t_{n-1}$ immer feiner macht, kann man in der Grenze die Wahrscheinlichkeit für eine „Bahn" $\alpha(t)$ mit $\alpha(t_1)=\alpha_1$ angeben. Dieser Grenzübergang ist von Siegel[1] sorgfältig diskutiert worden; er hat insbesondere die dabei auftretenden Schwierigkeiten untersucht und gedeutet.

Es sei noch bemerkt, daß sich die Onsager-Machlupsche Funktion auch in der Gestalt

$$\mathfrak{O}(\alpha,\dot\alpha) = \tfrac{1}{2}\left[\Phi(\alpha) + \Psi(\zeta) - \sigma(\dot\alpha,\zeta)\right] \tag{31.16}$$

schreiben läßt, wenn man unter $\Phi(\alpha)$ und $\Psi(\zeta)$ die quadratische Form der phänomenologischen Dissipationsfunktion [s. Gl. (23.14)], in $\dot\alpha$ bzw. in ζ ausgedrückt, und unter $\sigma(\dot\alpha,\zeta)$ die bilineare Form der Entropieproduktion versteht.

$\delta)$ *Grundlegung der zeitabhängigen Schwankungstheorie nach Tisza und Manning.* Wir betrachten wieder nur den Fall einer thermodynamischen Variablen α bei konstanten Werten der intensiven Parameter der Umgebung und formulieren zunächst einige durch die früheren Ausführungen nahegelegte Postulate:

1. Die Variable $\alpha(t)$ beschreibt einen Markoffschen Prozeß.

2. Der Prozeß ist stationär.

3. Der Prozeß ist reversibel.

---

[1] A. Siegel: Phys. Rev. **102**, 953 (1956); **106**, 609 (1957).

Die Vereinigung der phänomenologischen und der statistischen Theorie zu einer einheitlichen Theorie der Schwankungen und der Dissipation läßt sich mit Hilfe von zwei weiteren Postulaten entwickeln, die in verwandter Weise formuliert werden können.

4. Die Wahrscheinlichkeit ist eine Funktion der Entropie

$$W_1(\alpha) \sim f\big(S(\alpha)\big) \tag{31.17}$$

d.h. sie hängt nur über $S(\alpha)$ von der Variablen α ab.

5. Die bedingte Wahrscheinlichkeit $P_n \begin{pmatrix} \alpha_1 & \alpha_2, \ldots \alpha_n \\ t_1 & t_2, \ldots t_n \end{pmatrix}$ ist eine Funktion von stat $\int \mathfrak{D}(\alpha, \dot\alpha)\, dt$, wobei dieser Ausdruck wie in (31.15) zu verstehen ist:

$$P_n \begin{pmatrix} \alpha_1 & \alpha_2, \ldots, \alpha_n \\ t_1 & t_2, \ldots, t_n \end{pmatrix} \sim g\left(\text{stat} \int \mathfrak{D}(\alpha, \dot\alpha)\, dt\right). \tag{31.18}$$

Das vierte Postulat ist die bekannte Grundlage der klassischen Schwankungstheorie. Es hat die Boltzmannsche Beziehung

$$W_1(\alpha) \sim \exp\left\{ -\frac{S(\alpha)}{k} \right\} \tag{31.19}$$

zur Folge, womit die Funktion $f(S)$ in (31.17) festgelegt ist.

Das fünfte Postulat gibt in entsprechender Weise die Grundlage für die Theorie der zeitabhängigen Schwankungen. Die explizite Gestalt der Funktion g ergibt sich aus dem Markoff-Charakter des Prozesses. Es ist ja

$$P_n \begin{pmatrix} \alpha_1 & \alpha_2, \ldots, \alpha_n \\ t_1 & t_2, \ldots, t_n \end{pmatrix} = P_{n-1} \begin{pmatrix} \alpha_1 & \alpha_2, \ldots, \alpha_{n-1} \\ t_1 & t_2, \ldots, t_{n-1} \end{pmatrix} P_2 \begin{pmatrix} \alpha_{n-1} & \alpha_n \\ t_{n-1} & t_n \end{pmatrix}. \tag{31.20}$$

Mit (31.18) entsteht hieraus für die Einteilung $t_1, t_2, \ldots, t_{n-1}, t_n$

$$g\left(\text{stat} \int_{t_1,\alpha_1}^{t_n,\alpha_n} \mathfrak{D}(\alpha, \dot\alpha)\, dt\right) \sim g\left(\text{stat} \int_{t_1,\alpha_1}^{t_{n-1},\alpha_{n-1}} \mathfrak{D}(\alpha, \dot\alpha)\, dt\right) \cdot g\left(\text{stat} \int_{t_{n-1},\alpha_{n-1}}^{t_n,\alpha_n} \mathfrak{D}(\alpha, \dot\alpha)\, dt\right) \tag{31.21}$$

mit einem von den α_i unabhängigen Proportionalitätsfaktor. Wegen der leicht einsehbaren Additivitätseigenschaft von stat $\int \mathfrak{D}(\alpha, \dot\alpha)\, dt$ ist diese Beziehung mathematisch äquivalent mit der Funktionalgleichung (31.8) und hat daher als einzige nicht singuläre und nicht identisch verschwindende Lösung die Funktion

$$g \sim \exp\left(a \,\text{stat} \int_{t_1}^{t_n} \mathfrak{D}(\alpha, \dot\alpha)\, dt\right). \tag{31.22}$$

Die Konstante a bestimmt sich daraus, daß für $t_2 \to \infty$

$$P_2 \begin{pmatrix} \alpha_1 & \alpha_2 \\ t_1 & t_2 \end{pmatrix} \to W_1(\alpha_1)$$

zu $a = k^{-1}$.

Der stationäre Wert des Integrals über die Onsager-Machlup-Funktion läßt sich als Integral dieser Funktion über den sog. exponentiell geglätteten Weg $\alpha_{\exp}(t)$ berechnen. Darunter versteht man jene stetige Funktion, die in jedem Intervall t_i, t_{i+1} die Gestalt

$$\alpha_{\exp}(t) = A_i \, e^{-\frac{t}{\tau_0}} + B_i \, e^{+\frac{t}{\tau_0}}, \quad t_i \leqq t \leqq t_{i+1}, \quad i = 1, 2, \ldots, n-1 \tag{31.23}$$

mit den Randbedingungen $\alpha_{\exp}(t_i) = \alpha_i$ $(i = 1, 2, \ldots, n)$ besitzt (s. Fig. 4). Wir nennen den ersten Summanden $\alpha_{\exp}^{-}(t)$, den zweiten $\alpha_{\exp}^{+}(t)$. Die Onsager-Machlup-Funktion (31.12) hat die Eigenschaft, nur den Beitrag von $\alpha_{\exp}^{+}(t)$ übrig zu lassen. Man sieht leicht an (31.16), daß

$$\mathfrak{D}\big(\alpha_{\exp}(t)\big) = -\,\sigma(\dot\alpha_{\exp}^{+}, \zeta)\,. \tag{31.24}$$

Daraus folgt

$$\int_{t_1}^{t_n} \mathfrak{D}\big(\alpha_{\exp}(t)\big)\,dt = -\sum_{i=1}^{n} \Delta S_i, \tag{31.25}$$

wobei ΔS_i die (negative) Entropie bedeutet, welche beim exponentiellen Anstieg $B_i\,e^{-\frac{t}{\tau_0}}$, von $\alpha_{\exp}^{+}(t_i)$ bis $\alpha_{\exp}^{+}(t_{i+1})$ produziert wird. Man kann also sagen:

Die Abweichung der Funktion $\alpha_{\exp}(t)$ vom phänomenologischen irreversiblen Ablauf ist um so größer, je größer das Integral in (31.25) ist. Andererseits ist die bedingte Wahrscheinlichkeit um so kleiner, je größer die Abweichung der Werte $\alpha(t_2)$, $\alpha(t_3)$, $\ldots$, $\alpha(t_n)$ von dem von $\alpha_1 = \alpha(t_1)$ ausgehenden phänomenologischen Ablauf ist. Diese Tatsachen legen nahe, einen funktionalen Zusammenhang zwischen der bedingten Wahrscheinlichkeit und dem Onsager-Machlupschen Integral anzunehmen. Mit dem fünften Postulat wird gerade dieser Sachverhalt ausgedrückt. Diese Überlegung läuft offensichtlich ganz analog zu der Überlegung, mit welcher die Beziehung (31.17), also das vierte Postulat begründet wird.

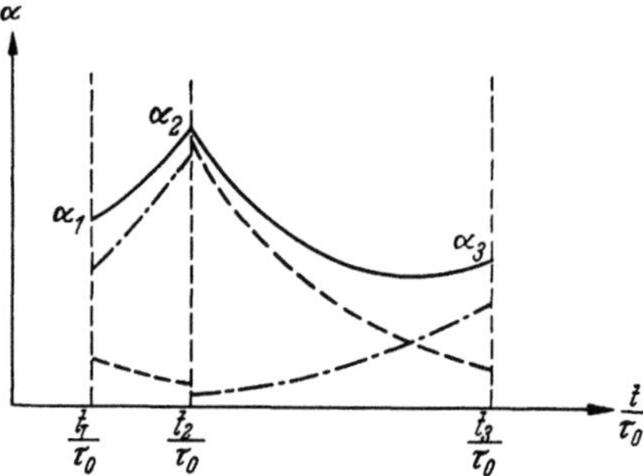

Fig. 4. Die Funktion $\alpha_{\exp}(t)$ für drei Zeitpunkte t_1, t_2, t_3. Die gestrichelten Kurven geben die exponentiell abfallenden Anteile e^{-t/τ_0}, die strichpunktierten die exponentiell anwachsenden Anteile e^{+t/τ_0}, aus denen sich $\alpha_{\exp}(t)$ zusammensetzt.

Die Langevinsche Gleichung folgt nun in einfacher Weise, indem man für den stationären Prozeß $\alpha(t)$ die Funktion

$$\vartheta(t) \equiv \tau_0\,\dot\alpha(t) + \alpha(t) \tag{31.26}$$

berechnet und nach ihren Eigenschaften fragt. Nach allgemeinen Sätzen ist $\vartheta(t)$ ebenfalls ein stationärer Prozeß mit $\langle\vartheta(t)\rangle = 0$. Man schließt aus (31.26) auf die beiden Gleichungen

$$\langle\vartheta(t)\,\vartheta(t+s)\rangle = \Big(\tau_0\frac{d}{ds} + 1\Big)\langle\vartheta(t)\,\alpha(t+s)\rangle,$$

$$\langle\alpha(t)\,\vartheta(t+s)\rangle = \Big(\tau_0\frac{d}{ds} + 1\Big)\langle\alpha(t)\,\alpha(t+s)\rangle = \langle\alpha(t-s)\,\vartheta(t)\rangle.$$

Hieraus folgt

$$\langle\vartheta(t)\,\vartheta(t+s)\rangle = \Big(1 - \tau_0^2\frac{d^2}{ds^2}\Big)\langle\alpha(t)\,\alpha(t+s)\rangle = 0 \quad \text{für} \quad s \neq 0.$$

$\vartheta(t)$ ist also ein stationärer Prozeß ohne Korrelation, da

$$\langle\alpha(t)\,\alpha(t+s)\rangle = \langle\alpha(t)\rangle^2\,e^{-\frac{s}{\tau_0}}$$

ist.

$\varepsilon)$ *Verallgemeinerungen.* Zwei wichtige Verallgemeinerungen sollen noch kurz angedeutet werden. Hat man gerade und ungerade Variable, so kann man die

Überlegungen nicht mehr auf die Betrachtung einer einzigen Variablen zurückführen. Indessen treten keine wesentlichen Änderungen ein, außer daß die Onsager-Machlup-Funktion nun in der verallgemeinerten Gestalt

$$\mathfrak{D}(\alpha, \dot{\alpha}, \ddot{\alpha}) = (\tilde{\ddot{\alpha}}\, m + \tilde{\dot{\alpha}}\, R + \tilde{\alpha}\, s)\, R^{-1} (m\, \ddot{\alpha} + R\, \dot{\alpha} + s\, \alpha)$$

zu schreiben ist, wenn die phänomenologischen Gleichungen in der Form $m\, \ddot{\alpha} + R\, \dot{\alpha} + s\, \alpha = 0$ zugrundegelegt werden, daß der exponentiell ausgeglichene Weg durch die Lösung der Differentialgleichung vierter Ordnung

$$\frac{\partial \mathfrak{D}}{\partial \alpha_i} - \frac{d}{dt}\frac{\partial \mathfrak{D}}{\partial \dot{\alpha}_i} + \frac{d^2}{dt^2}\frac{\partial \mathfrak{D}}{\partial \ddot{\alpha}_i} = 0$$

in jedem Intervall mit Randbedingungen für $\alpha_i, \dot{\alpha}_i$ an den Enden jedes Intervalls definiert ist, und daß in entsprechender Weise wie im Postulat 5 das Onsager-Machlupsche Integral in funktionale Beziehung zur bedingten Wahrscheinlichkeit

$$P_n\begin{pmatrix}\alpha_1\dot{\alpha}_1 & \alpha_2,\dot{\alpha}_2, \ldots, \alpha_n\dot{\alpha}_n \\ t_1 & t_2, \quad \ldots, \quad t_n\end{pmatrix}$$

zu setzen ist. Wegen Einzelheiten sei auf MACHLUP und ONSAGER[1] verwiesen.

Eine andere Verallgemeinerung hebt die Beschränkung auf natürlich ablaufende irreversible Prozesse auf und befaßt sich mit Systemen, bei welchen die intensiven Parameter der Umgebung vorgeschriebene Funktionen $\zeta_i(t)$ der Zeit sind. Eine mögliche Erweiterung ergibt sich dann mit einer verallgemeinerten Onsager-Machlup-Funktion. Für eine Variable $\alpha(t)$ würde sie lauten

$$\mathfrak{D}(\alpha, \dot{\alpha}) = \frac{1}{4LT}\left[\dot{\alpha} + \frac{1}{\tau_0}\alpha - L\zeta(t)\right]^2.$$

Man erhält aus ihr die bedingte Wahrscheinlichkeit

$$P_2\begin{pmatrix}\alpha_1 & \alpha_2 \\ t_1 & t_2\end{pmatrix} \sim \exp\left\{-\frac{1}{2k}(e^{2t_2/\tau_0} - e^{2t_1/\tau_0})^{-1} \times\right.$$
$$\left. \times \left[\alpha_2\, e^{t_2/\tau_0} - \alpha_1\, e^{t_1/\tau_0} - L\int_{t_1}^{t_2}\zeta(t')\, e^{t'/\tau_0}\, dt'\right]^2\right\}$$

und die einfache Wahrscheinlichkeit

$$W_1\begin{pmatrix}\alpha \\ t\end{pmatrix} \sim \exp\left\{-\frac{1}{2k}\left[\alpha - L\int_{t_1}^{t_2}\zeta(t')\, e^{\frac{t'-t}{\tau_0}}\, dt'\right]^2\right\}.$$

Dieses Ergebnis stimmt mit einem auf Grund allgemeiner quantenmechanischer Behandlung von KUBO[2] und CALLEN[3] erhaltenen Ergebnis im Spezialfall von Markoff-Prozessen überein.

32. Zum Problem der Irreversibilität. Die statistische Mechanik der irreversiblen Prozesse versucht die Gesetze der phänomenologischen Thermodynamik der irreversiblen Prozesse aus den mechanischen Eigenschaften eines Systems mit großer Teilchenzahl zu begründen. Die Entwicklung dieser Theorie ist noch im Flusse, doch zeichnet sich bereits ab, in welcher Weise der Übergang zu vollziehen ist.

---

[1] S. MACHLUP u. L. ONSAGER: Phys. Rev. **91**, 1512 (1954).
[2] R. KUBO: J. Phys. Soc. Japan **12**, 570 (1957).
[3] H. B. CALLEN: Phys. Rev. **111**, 367 (1958).

Das wesentliche Problem ist natürlich die Auflösung des Paradoxons, daß die Gesetze für die Bewegung der einzelnen atomaren Bausteine reversibel sind, während die makroskopische Materie ein typisch irreversibles Verhalten zeigt.

Wir besprechen nun einige Beispiele, die die Auflösung des Paradoxons versuchen und die Voraussetzungen, die im einzelnen gemacht wurden, um den Übergang von den mechanischen Gleichungen zu den Gesetzen der Thermodynamik irreversibler Prozesse zu vollziehen. Ein elementares Beispiel hat Temperley[1] diskutiert. Es handelt sich um die Bewegung von N Teilchen in M auf einem Kreis angeordneten gleichen Potentiallöchern. Die Häufigkeit der Übergänge zwischen benachbarten Plätzen wird durch Wahrscheinlichkeitsamplituden im Sinne der Quantenmechanik geregelt. Der Übergang von einer quantenmechanischen und reversiblen Beschreibung der Teilchengesamtheit zu einer Gleichung vom Typ der Diffusionsgleichung läßt sich in wenigstens drei getrennte Schritte zerlegen: Phasenmittelung, Übergang zu großen N, Übergang zu großen M. Der erste Schritt zerstört bereits die Symmetrie zwischen Vergangenheit und Zukunft. Im zweiten Schritt geht die Ergodizität des Systems verloren, während im dritten Schritt die Schwankungen verschwinden.

Ein anderes, vielleicht noch durchsichtigeres Beispiel geben Meixner und König[2]. Sie zeigen, daß man ein Netzwerk mit irreversiblem Verhalten, etwa bestehend aus einem Kondensator und einem Widerstand in Serienschaltung, durch ein Netzwerk bestehend aus Kondensatoren und Induktivitäten (CL-Netzwerk) ersetzen kann in dem Sinne, daß dieses zweite Netzwerk für alle Spannungskurven $u(t)$ mit $u(t)=0$ für $t<0$ (Einschaltvorgang) in einem gewissen Zeitintervall $(0<t<t_1)$ denselben Gesamtstrom $i(t)$ führt wie das erste Netzwerk. Durch geeignete Wahl des zweiten Netzwerkes kann man t_1 beliebig groß machen. Notwendig für große Werte von t_1 ist, daß die Zahl der Elemente des zweiten Netzwerkes ebenfalls sehr groß ist. Geht man wesentlich über t_1 hinaus, so geht der Charakter der Irreversibilität verloren, die Wiederkehreigenschaften machen sich geltend und das System verhält sich fastperiodisch und reversibel.

Umgekehrt kann man ein CL-Netzwerk mit vielen Elementen im oben genannten Sinne durch ein Netzwerk, welches auch Widerstände enthält, ersetzen, wenn das Spektrum seiner Eigenschwingungen gewisse Eigenschaften erfüllt; dazu gehört, daß die Eigenfrequenzen genügend dicht liegen.

Ersichtlich kommt es für die gegenseitige Ersetzbarkeit von reversiblen und irreversiblen Netzwerken darauf an, daß man den „Zweipol-Standpunkt" (allgemeiner $2n$-Polstandpunkt) einnimmt, d.h., daß man darauf verzichtet, auch die Vorgänge im Inneren der beiden Systeme zu vergleichen. Dieser „Zweipol-Standpunkt" ist aber gerade für die phänomenologische Behandlung thermodynamischer Systeme charakteristisch. Hierbei interessiert man sich ja auch nur für das Verhalten der makroskopischen Variablen und ihrer thermodynamisch konjugierten Variablen, ohne daß man in das mikroskopische Geschehen hineinzuschauen versucht.

Untersuchungen anderer Art sind von Prigogine und Mitarbeitern[3] durchgeführt worden. Sie beschäftigen sich mit einem System von schwach gekoppelten Elementen, etwa dem System der Normalschwingungen eines Kristalls mit schwach anharmonischer Wechselwirkung. Der Ausgangspunkt ist die Liouvillesche

---

[1] H. N. V. Temperley: Proc. Cambridge Phil. Soc. **52**, Nr. 4, 712 (1956).

[2] J. Meixner u. H. König: Im Druck.

[3] R. Brout u. I. Prigogine: Physica, Haag **22**, 621 (1956). — I. Prigogine u. R. Balescu: Physica, Haag **23**, 555 (1957). — I. Prigogine u. J. Philippot: Physica, Haag **23**, 569 (1957). — I. Prigogine u. P. Henin: Physica, Haag **23**, 585 (1957).

Gleichung des Gesamtsystems mit einer Hamilton-Funktion $H = \sum_{i=1}^{n} H_i(J_i) + \lambda V$.
Als Variable für die einzelne Normalschwingung werden die Wirkungsvariablen J_i und die Winkelvariablen α_i gewählt. λV stellt die anharmonische Wechselwirkung dar; V ist eine Funktion aller J_i und α_i. Die Lösung der Liouvilleschen Gleichung wird mit dem Ansatz

$$\varrho = \sum_{\{n\}} \varrho_{\{n\}} (J_1, J_2, \ldots, J_n, t)\, e^{i \Sigma n_i (\alpha_i - \omega_i t)}$$

nach Eigenfunktionen des ungestörten Liouville-Operators ($\lambda V = 0$) entwickelt. Mit $\{n\}$ ist ein Satz von ganzzahligen n_i-Werten bezeichnet. Die ω_i sind die Eigenfrequenzen $\omega_i = \partial H / \partial J_i$.

Das Hauptergebnis ist, daß jeder Koeffizient $\varrho_{\{n\}}$ für sich unter sehr allgemeinen Voraussetzungen einer partiellen Differentialgleichung erster Ordnung in der Zeit genügt, wenn das System sehr groß ist ($N \to \infty$) und wenn $\lambda \to 0$, $t \to \infty$, während $\lambda^2 t$ endlich bleibt. Insbesondere wird die zeitliche Entwicklung von $\varrho_{\{0\}}$ durch einen Markoff-Prozeß beschrieben.

Entsprechende Überlegungen auf quantenmechanischer Grundlage sind von van Hove[1] durchgeführt worden. Brout[2] hat auseinandergesetzt, daß die Ergebnisse von Brout und Prigogine sich aus der quantenmechanischen Theorie als Grenzfall für $\hbar \to 0$ ergeben.

In der Durchführung der Grenzübergänge in beiden Theorien scheinen noch Schwierigkeiten zu liegen (Ingraham[3]).

In den genannten Untersuchungen von Prigogine u. a. wird ein Schritt zum irreversiblen Verhalten hin untersucht, nämlich der Einfluß wachsender Größe des Systems. Ein weiterer Schritt müßte in der Reduktion der Zahl der Variablen auf die makroskopischen Variablen liegen. Dazu ist es notwendig, erst einmal zu klären, was man unter makroskopischen Variablen zu verstehen hat. Eine prinzipielle Definition der makroskopischen Variablen gibt Ludwig[4]. Es scheint aber nicht einfach zu sein, diese Definition auf reale Systeme anzuwenden.

Besonders anschaulich setzt van Kampen[5] auseinander, welchen Weg man gehen und welche heute noch nicht vollständig bewiesenen Annahmen man machen muß, um von der reversiblen Liouvilleschen Gleichung im Γ-Raum des Systems der elementaren Bestandteile zu den makroskopischen Gesetzen der Thermodynamik irreversibler Prozesse zu gelangen.

Van Kampen legt die Existenz eines Satzes von makroskopischen Variablen $\alpha_1, \alpha_2, \ldots, \alpha_n$ zugrunde, der ausreicht, um das makroskopische irreversible Verhalten in einem gewissen Umfang (bei harmonischen Vorgängen etwa bis zu einer gewissen maximalen Frequenz, s. auch Ziff. 23) zu beschreiben. Das Ziel sind die Gleichungen der Thermodynamik irreversibler Prozesse. In einer Zwischenstufe müssen auch noch die Schwankungen enthalten sein, und man wird im allgemeinen erwarten, daß diese durch einen Gaußschen Prozeß, also durch eine Fokker-Plancksche Gleichung beschrieben werden.

Der erste Schritt ist dann der Übergang von der Liouvilleschen Gleichung im Phasenraum aller elementaren Teilchen zur sog. *master equation* im Raum der makroskopischen Variablen α_i. Dazu wird der Phasenraum in Zellen j endlicher

---

[1] L. van Hove: Physica, Haag **21**, 517, 901 (1955); **22**, 343 (1956); **23**, 341 (1957).

[2] R. Brout: Physica, Haag **23**, 953 (1957).

[3] R. Ingraham: Nuovo Cim. **9**, 99 (1958).

[4] G. Ludwig: Z. Physik **152**, 98 (1958).

[5] N. G. van Kampen: Physica, Haag **20**, 603 (1954); **23**, 707 (1957). — Fortschr. Physik **4**, 405 (1956).

Größe aufgeteilt, in welchen die makroskopischen Variablen Werte α_i mit möglichen Fehlern $\Delta\alpha_i$ haben. Es handelt sich dabei um eine sog. grobkörnige Struktur, welche von der aus der quantenmechanischen Behandlung folgenden feinkörnigen Struktur wohl zu unterscheiden ist. In der Sprache der Quantenmechanik entspricht die grobkörnige Struktur der Zusammenfassung der Eigenzustände des Systems zu Gruppen j. Das Volumen der Zelle j sei G_j. Die Wahrscheinlichkeit, daß sich das System in der Zelle j mit den Werten $\alpha_i(j)$ der makroskopischen Variablen befindet, sei P_j. Unter gewissen plausiblen Annahmen läßt sich dann begründen, daß die P_j Gleichungen der Gestalt

$$\frac{dP_j}{dt} = \sum_{j'} W_{jj'}\, P_{j'} - \sum_{j'} W_{j'j}\, P_j \tag{32.1}$$

genügen, in denen die nicht-negativen Größen $W_{jj'}$ den Charakter von Übergangswahrscheinlichkeiten mit der Symmetrieeigenschaft

$$W_{jj'}\, G_{j'} = G_j W_{j'j} \tag{32.2}$$

besitzen. Dabei ist der Einfachheit halber angenommen, daß alle Variablen α_i gerade sind. Nimmt man an, daß sich sowohl die α_j als auch die Größen P_j, $W_{jj'}$, G_j nur langsam ändern, so geht die master equation (32.1) über in

$$\frac{\partial P(\alpha, t)}{\partial t} = \int W(\alpha, \alpha')\, P(\alpha', t)\, d\alpha' - P(\alpha) \int W(\alpha, \alpha')\, d\alpha \tag{32.3}$$

mit der Symmetriebedingung

$$W(\alpha, \alpha')\, G(\alpha') = G(\alpha)\, W(\alpha', \alpha). \tag{32.4}$$

Hier steht α als Abkürzung für den gesamten Variablensatz α_i.

Man muß eine ganze Hierarchie von Größenordnungen durchlaufen, um zu den makroskopischen Gesetzen der Thermodynamik der irreversiblen Prozesse zu gelangen. Die kleinste zu betrachtende Größe ist die Größe der Zelle $\Delta\alpha_i$ im Raum der Variablen α_i. Die Funktion $W(\alpha, \alpha')$ darf über eine solche Zelle nur wenig variieren, damit die master equation gilt. Andererseits ist es wohl zulässig und plausibel anzunehmen, daß sie für genügend große $\alpha_i - \alpha_i'$ praktisch verschwindet. Die Breite $\Delta_1\alpha_i$ des Bereiches, in welchem $W(\alpha, \alpha')$ wesentlich von Null verschieden ist, gibt die nächste Größenordnung, $\Delta_1\alpha_i \gg \Delta\alpha_i$. Damit nun für $P(\alpha)$ aus (32.3) eine Fokker-Plancksche Differentialgleichung hergeleitet werden kann, darf $P(\alpha)$ selbst wieder nur wenig über den Bereich $\Delta_1\alpha_i$ variieren. Da der Variationsbereich $\Delta_2\alpha_i$ von $P(\alpha)$ im wesentlichen durch die Schwankungen gegeben ist, muß also weiter $\Delta_2\alpha_i \gg \Delta_1\alpha_i$ sein. Schließlich muß die Beobachtungsungenauigkeit $\Delta_3\alpha_i$, wenn man die Schwankungen ausschließen und nur die Scharmittelwerte der α_i bestimmen will, groß gegen $\Delta_2\alpha_i$ sein.

Die master equation gibt insbesondere dann, wenn sie sich auf eine Fokker-Plancksche Gleichung reduzieren läßt, einen besonders einfachen Ausgangspunkt zur Begründung der O.C.R.B. Diese sind eine einfache Folge der Symmetrieeigenschaft (32.4), d.h. der mikroskopischen Reversibilität[1]. Der Fall ungerader Variabler bereitet keine Schwierigkeiten, wenn man nur die Symmetrieeigenschaft (32.4) entsprechend modifiziert[2]. Eine wichtige Eigenschaft der Fokker-Planckschen Gleichung ist die Persistenz der Gauß-Verteilung. Sie hat zur Folge,

---

[1] Wir vermerken hier, daß Thomson [Phys. Rev. **91**, 1263 (1953)] eine feine Unterscheidung zwischen diesem Prinzip und dem des detaillierten Gleichgewichts macht, als welches er die Eigenschaft $P_j W_{jj'} = W_{jj'} P_{j'}$ im Gleichgewicht bezeichnet. Hieraus und aus der Ergodenhypothese, daß im Gleichgewicht P_j proportional zu G_j ist, folgt das Prinzip der mikroskopischen Reversibilität in diesem Sinne.

[2] N. G. van Kampen: Physica, Haag **23**, 641 (1957).

daß eine über die makroskopischen Variablen α_i bewirkte Störung sich stets über Gauß-Verteilungen der Variablen α_i ausgleicht[1].

Der Weg von der Liouvilleschen Gleichung zur makroskopischen Thermodynamik irreversibler Prozesse kann auch auf andere Weise gegangen werden. So lehnen sich die Untersuchungen von KIRKWOOD[2] sowie BORN und GREEN[3] mehr an die traditionelle Herleitung der Boltzmannschen Fundamentalgleichung an. Ihr Ziel ist nämlich eine Verallgemeinerung der kinetischen Gastheorie, welche auch Mehrfachstöße berücksichtigt und damit eine Grundlage für die Theorie des kondensierten Zustandes geben kann.

Zunächst werden aus der Verteilungsfunktion $f_N(\boldsymbol{q}_1, \ldots, \boldsymbol{q}_N, \boldsymbol{p}_1, \ldots, \boldsymbol{p}_N; t)$ des ganzen Systems von N Teilchen mit den Koordinaten $\boldsymbol{q}_i$ und den Impulsen $\boldsymbol{p}_i$ (von inneren Freiheitsgraden sehen wir hier ab) reduzierte Verteilungsfunktionen $f_h(\boldsymbol{q}_{s1}, \ldots, \boldsymbol{q}_{sh}; \boldsymbol{p}_{s1}, \ldots, \boldsymbol{p}_{sh}; t)$ gebildet. Sie entstehen aus f_N durch Integration über die restlichen Koordinaten und Impulse. Diese Verteilungsfunktionen genügen Gleichungen, die man durch die entsprechenden Integrationsprozesse aus der Liouvilleschen Gleichung erhält. Die Änderung von f hängt nicht nur von f_h selbst, sondern auch von der Verteilungsfunktion f_{h+1} ab.

Man erhält so eine Hierarchie von Verteilungsfunktionen $f_N, f_{N-1}, \ldots, f_1$ und ebenso eine Hierarchie von Integrodifferentialgleichungen für diese. Zu ihrer Behandlung muß erst die Liouvillesche Gleichung für f_N gelöst werden; dann lassen sich sukzessive die Verteilungsfunktionen $f_{N-1}, \ldots, f_1$ berechnen. Ersichtlich ist in diesen Gleichungen noch keine Irreversibilität enthalten; denn sie geben nur an, wie man aus der Liouvilleschen Gleichung die Verteilungsfunktionen $f_{N-1}, \ldots, f_1$ zu berechnen hat.

Die Irreversibilität kommt durch ein Postulat zustande, das eine Verallgemeinerung der Hypothese des molekularen Chaos, wie sie in der kinetischen Gastheorie benutzt wird, bedeutet.

Es verlangt, daß man für ein gewisses n die Verteilungsfunktion f_{n+1} durch $f_1, f_2, \ldots, f_n$ ausdrücken kann. Dann sind die Gleichungen für die Änderung von $f_1, \ldots, f_n$ in sich verträglich und vollständig. Abgesehen von Symmetrieforderungen ist der Ansatz für f_{n+1} noch sehr willkürlich. Die traditionelle kinetische Gastheorie gewinnt man mit $n = 1$ und $f_2(\boldsymbol{q}_1, \boldsymbol{p}_1; \boldsymbol{q}_2, \boldsymbol{p}_2; t) = f_1(\boldsymbol{q}_1, \boldsymbol{p}_1, t)\, f_1(\boldsymbol{q}_2, \boldsymbol{p}_2, t)$. Für Flüssigkeiten hat KIRKWOOD als erste Näherung $n = 2$ und, in abgekürzter Schreibweise,

$$f_3(1, 2, 3) = \frac{f_2(1, 2)\, f_2(2, 3)\, f_2(3, 1)}{f_1(1)\, f_1(2)\, f_1(3)}$$

vorgeschlagen.

Das Gleichungssystem für $f_1, \ldots, f_n$ zeigt nun die Eigenschaft der Irreversibilität; dies ist vom Beispiel der kinetischen Gastheorie her wohlbekannt.

Das genannte Postulat ist häufig schon aus geometrischen Gründen plausibel. Eine ins einzelne gehende Rechtfertigung und Analyse steht jedoch noch aus. Für eine eingehendere Diskussion der Verhältnisse verweisen wir auf die Originalliteratur[2,3], sowie auf die Darstellungen von HIRSCHFELDER, CURTISS und BIRD[4] und auf EISENSCHITZ[5].

---

[1] E. I. TAKIZAWA u. J. MEIXNER: Z. Naturforsch. (im Druck).

[2] J. G. KIRKWOOD: J. Chem. Phys. **14**, 180 (1946); **15**, 72 (1946).

[3] M. BORN u. H. S. GREEN: Proc. Roy. Soc. Lond., Ser. A **188**, 10 (1946); **189**, 103 (1947); **190**, 455 (1947).

[4] J. O. HIRSCHFELDER, F. C. CURTISS u. R. BIRD: The Molekular Theory of Gases and Liquids. New York 1954.

[5] R. EISENSCHITZ: Statistical Theory of Irreversible Processes. Oxford 1958.

Probability and Stochastic Processes.

By

ALLADI RAMAKRISHNAN.

Prefatory note.

This essay needs an apology rather than a preface. It is an attempt to present to the physicist a physical approach to the theory of stochastic processes, a field till recently the close preserve of the mathematician. The only justification for the style and form adopted here lies in that the theory of stochastic processes as formulated in abstract mathematical treatises and papers is difficult to read even to those who are trained to a rigorous mathematical discipline. But in many problems where some random element is introduced, the physicist needs a knowledge of the results of stochastic theory and he would like to use them without being diverted by mathematical details or trammelled by the demands of rigour. Such examples are cited, but no pretence is made to completeness, and the emphasis is laid only on the methods used in applying a general theory to particular problems.

This article would have justified its purpose if it furnishes the physicist with such mathematical tools as are already available. It is hoped that incidentally it will persuade pure mathematicians who are inclined to look upon the subject of probability and stochastic processes as just a branch of measure theory to take a lively interest in the fields in which stochastic theory is applied.

Introduction.

The theory of stochastic processes is a natural extension of the theory of probability to dynamical problems, the word "dynamical" being used in its most general sense to denote the changes with respect to a parameter or a set of parameters. While probability theory and dynamics are old and well established branches of science, almost treated as classical, the theory of stochastic processes is of comparatively recent origin. The abstract mathematical formulation of the theory of stochastic processes has attracted the attention of many mathematicians. But somehow, the application to physical problems has been delayed. This delay can be ascribed not merely to the inadequate acquaintance of the physicists with the mathematical formalism but to a deeper and more fundamental reason —the difficulty of translating the physical conditions of the problems into a proper mathematical formulation. The object of the present contribution is merely to present stochastic theory in a serviceable form which can be applied to physical problems by a physicist. No attempt is made to intrude into the abstract field of pure mathematical theory which is for example found in the classic work of DOOB [1] and others. In presenting a physical approach we draw attention also to the correspondence with the abstract formalism. Incidentally we emphasise that in the very task of transcribing the physical conditions into a rigorous formulation, new concepts come to light which would otherwise escape attention if a purely logical and axiomatic approach is made.

The essay is divided into two parts. Part A is a brief and rapid survey of probability theory, a refresher course on a subject which is too well known to demand any repetitive discussions. But it is presented in a manner that leads directly to the concept of stochastic processes. In addition, the standard techniques used in probability theory are brought together in a compact form to serve as a ready reference. In this part of the essay, proofs or elaborate discussions are not given but the implications and scope of the fundamental theorems of probability are explained.

In Part B which deals with stochastic processes, we begin with a physical approach which is followed by a short chapter establishing the correspondence with the abstract formulation of DOOB. As many physical examples as the author could lay his hands on to illustrate stochastic processes are cited. Since modern physics is almost identified with quantum mechanics, the role of the theory of stochastic processes in physics is closely connected with that of quantum mechanics. Hence an attempt is made to point out the stochastic aspects of some processes in quantum mechanics. In this task, the author has been encouraged by the proceedings of a seminar course on probability problems in physics organised by LOEVE [2] and his associates and secondly by the scheme of subjects discussed by the probability seminar group at the M.I.T.

The first book which is an explicit attempt at the application of stochastic theory to physical problems is that of my teacher Professor BARTLETT [3] and this work is intended as a sequel to it. Hence the subjects dealt with in detail by him are not discussed here at length but referred to when necessary.

A. Probability.

I. Probability and measure theory.

Probability is an accepted branch of mathematics and it is but natural that mathematicians should have formulated it without any appeal to auxiliary concepts like experiments, trials, events and their occurrences. As DOOB [1] has emphasised, "such a formulation is necessary to avoid spurious simplifications of some parts of the subject and a genuine distortion of all of it". However, since the theory of probability and stochastic processes finds direct application throughout the field of physics and other branches of science, it is equally necessary to establish a clear correspondence between the physical nature of a problem and the mathematical abstraction of it. Thus we shall use intuitive concepts like experiments, events and trials, but take care to show how these can be carried over to the notions of the theory of sets and measure to which probability theory really belongs.

1. Events and probabilities. We assume that the outcome of an experiment $\mathscr{E}$ is one of a mutually exclusive set of n events, $E_1, E_2, E_3, \ldots, E_n$. For brevity the set is denoted by $\langle E_i \rangle$[1]. We shall consider n to be finite and later extend it to the enumerable and continuous infinite cases. To each event E_k we assign a number $P(E_k)$ representing the probability that event E_k is the outcome. Before we ascribe a meaning to this number, we assume that without loss of generality, it is possible to set

$$\sum_k P(E_k) = 1. \tag{1.1}$$

The decomposition of the outcome into n mutually exclusive events is not necessarily unique. An event E', say, may denote either the event E_k or E_l and in such a case, we write symbolically

$$E' = E_k + E_l \tag{1.2}$$

and to E' we assign a number $P(E')$ the probability that E' i.e. E_k or E_l is the outcome of the experiment.

$$P(E') = P(E_k) + P(E_l). \tag{1.3}$$

Thus in the aggregate of mutually exclusive events $E_1, E_2, \ldots, E_n$ we may remove E_k and E_l and replace them by E' and now we have another set of $n-1$ mutually

---

[1] We shall in this essay denote aggregates by the symbol $\langle \, \rangle$. The members of the aggregate can be inferred from the context through the symbol used inside the bracket.

exclusive events. The above law of composition can be extended to any number of mutually exclusive events. Thus if we define

$$E = E_1 + E_2 + \cdots + E_k$$

then

$$P(E) = P(E_1) + P(E_2) + \cdots + P(E_k). \tag{1.4}$$

In a similar manner if we define $E\{\text{total}\} = E_1 + E_2 + \cdots + E_n$

$$P(E\{\text{total}\}) = \sum_k P(E_k) = 1. \tag{1.5}$$

Since we have assumed that the aggregate $E_1, E_2, \ldots, E_n$ exhausts all possibilities, or in other words, one of the aggregate must be the outcome of the experiment, the event $E\{\text{total}\}$ must occur and since $P(E\{\text{total}\})$ is equal to 1, we may safely state that an event that must occur must be assigned the number unity.

If an event cannot be the outcome, then it is not a part of the aggregate and we can assign the number 0 to it and add it to our aggregate of events without any consequence to our algebra.

Though a certain event can be assigned a probability 1 and an impossible event a probability 0, the converse is not obvious, for when we say that an event has a probability 1, this number 1 might have been approached as a limit and on careful examination, it may be possible to show that it is the limit of $1 - \delta$ as δ tends to zero. The same considerations apply to an event with probability zero. We shall for the moment accept that an event with probability tending to 1, is "almost certain" to be the outcome of the experiment. This will be fitted into the concept of measure later. Accepting the intuitively satisfactory interpretation for probability zero and probability one, we shall give a meaning[1] for $P(E_k)$ the probability of an event E_k, when

$$1 > P(E_k) > 0. \tag{1.6}$$

This is done easily by noting that experiments can be composed to form new experiments. We shall assume that the experiment $\mathscr{E}$ can be repeated and the two $\mathscr{E}$ experiments can be treated together as one single experiment $\mathscr{E}^{(2)}$. Thus the outcome of $\mathscr{E}^{(2)}$ is given by the sequence of the outcomes of two experiments. That is, if $\langle E_k^{(2)} \rangle$ is the set of mutually exclusive events which are the outcomes of the experiment $\mathscr{E}^{(2)}$, any member $E_k^{(2)}$ of that set can be represented as a sequence of two events (E_l, E_m). This concept can be generalised to define an experiment $\mathscr{E}^{(n)}$ which consists of n repetitions of the experiment $\mathscr{E}$. If we assume that each experiment is unaffected by the previous one, then the outcome of $\mathscr{E}^{(n)}$ is the sequence of n events of the type E_k. To interpret $P(E_k)$ we shall adopt the following procedure. In the experiment $\mathscr{E}$ we shall group all the events other than E_k into one event called $\bar{E}_k$ (i.e. non-E_k). Then

$$P(\bar{E}_k) = 1 - P(E_k). \tag{1.7}$$

Now in an $\mathscr{E}^{(n)}$-experiment which consists of a sequence of n $\mathscr{E}$-experiments the outcome is a sequence of n events, each event being either E_k or $\bar{E}_k$. Instead of dealing with this complex group we shall define a new set of events called $\eta_1, \eta_2, \ldots, \eta_n$ as the outcome of $\mathscr{E}^{(n)}$ where η_i is the event denoting that E_k is the outcome of i out of the n experiments of $\mathscr{E}$-type.

---

[1] Professor L. JANOSSY seems to attach considerable significance to this interpretation. Starting with this as an axiom, he has attempted a somewhat generalised addition and multiplication rule for probabilities, but he admits that his generalised scheme does not lead to any results different from those obtained in the usual way. [Acta Phys. Acad. Sci. Hung. **4**, 333 (1955).]

Denoting f, the frequency of the outcome of the event $\boldsymbol{E}_k$ in n $\mathscr{E}$-experiments as the ratio of the number of times it is obtained as the outcome to n, we interpret $\eta_1, \eta_2, \ldots, \eta_n$ to mean events representing the frequencies $1/n, 2/n, \ldots, n/n$. Thus we can group all events representing a frequency f where

$$P(\boldsymbol{E}_k) + \delta > f > P(\boldsymbol{E}_k) - \delta \tag{1.8}$$

into a single event called $\eta(\delta)$. The outcome of $\mathscr{E}^{(n)}$ can be $\eta(\delta)$ or non $\eta(\delta)$. If we are able to show that the probability for $\eta(\delta)$ tends to 1 however small δ may be made, by suitably choosing n, then we would have successfully interpreted the probability of the original event $\boldsymbol{E}_k$ in the experiment $\mathscr{E}$. This is known as the frequency interpretation of the probability of the occurrence of the event $\boldsymbol{E}_k$ which in the opinion of the author amounts to an interpretation of the probability P (which lies between 0 and 1) in terms of the intuitively satisfactory concept of probability 0 and 1. The frequency interpretation is identical with the famous BERNOULLI's theorem or the *law of large numbers*.

Having discussed the intuitive basis of the probability of an event, we shall now draw the correspondence with the abstract formalism of measure theory.

2. Concept of the basic ω space[1]. We have associated a number with each event which is the probable outcome of an experiment. Let us now assume that any such event can be put into correspondence with any set of points which forms a sub-set of a total set of points called the basic ω-space. The entire aggregate of events corresponds to a set of points which defines the entire ω-space. Since our aggregate consists of mutually exclusive events, the sets of points corresponding to these events have no points in common. Each set, we may say, has a measure equal to the probability of the event to which this set corresponds. Once we have introduced the notion of the set of points we shall assume that this set is the fundamental one from which the set of events can be derived. In passing from the ω-space to the set of events, we are establishing a many-to-one correspondence. In principle, we can derive a set of events from this basic ω-space and hence the definition of events as the outcome is not unique, but depends on the nature of the sub-division of the basic ω-space. Any set of points can be put into correspondence with an event and the field of all sub-sets, their joins, their intersections, correspond to events, the events being mutually exclusive only if their corresponding sets have no common points. This field is known as the *Borel field*. The advantage in this abstract concept of the ω-space over the intuitive concept of events which is known as the sample space is that the formulation is brought into the frame work of measure theory and a rigorous approach is made possible in view of the analytical apparatus already available. If we accept the basic ω-space to be fundamental, we immediately notice that the original set of events with which we started need not be the only way of defining the possible outcomes of the experiment. Since, starting with the ω-space we may define a new set of events which correspond to the points in the ω-space and the relation between the new set and the original set of events may be established through their corresponding sets of points. In fact it is possible to define a set of events which bear a one-to-one correspondence with the points in the ω-space and if the ω-space is non-enumerable, we can conceive of a non-enumerable set of events to be the outcome and the original enumerable set which we have started with may be obtained from this non-enumerable set by suitable composition.

---

[1] We shall in this section assume familiarity on the part of the reader with the basic concepts of measure theory. The purpose of the section is merely to establish correspondence between events and point sets.

Thus the introduction of the concept of the ω-space directly leads us to the concept of the sets of non-enumerable events which will be discussed presently.

With the well known techniques of mathematical analysis, we may be able to define another ω'-space which may be put into correspondence with the ω-space. The transformations which are measure preserving play a fundamental role in probability. If this transformation is measure preserving, then ω' can equally be chosen as the basic space. We shall now illustrate the correspondence by particular examples. Consider the following case. The outcome of an experiment is, let us say, one of six events $E_1, E_2, ..., E_6$ with equal probability $\frac{1}{6}$, and let the experiment relate, for example, to the determination of the coordinate of a particle. If the coordinate of a particle lies between 0 and 1, let us postulate six mutually exclusive events as the possible outcome, the events being that the coordinate of the particle lies in the intervals $(0, \frac{1}{6})$, $(\frac{1}{6}, \frac{1}{3})$, ..., $(\frac{5}{6}, 1)$ which are the events $E_1, E_2, ..., E_6$. One of the basic ω-spaces corresponding to this linear coordinate space between 0 and 1 is the set of all points between 0 and 1. We can define ω' also to mean a volume instead of a linear coordinate space. If the measure of the volume is 1, and is divided into six equal parts, the events may be made to correspond to these. Thus we see that the law of composition of events and their probabilities can be immediately carried over to the laws of addition and computation of the measures of sets of points.

3. Random variable; physical approach. In accordance with the spirit of this article, the notion of the random variable will be introduced in two equivalent ways, (1) a physical and intuitive approach, (2) the abstract approach of DOOB.

We start with the assumption that the possible outcome of an experiment is a non-enumerable set of events. (The case of discrete and enumerable set will follow easily.) One of the easiest ways of defining a non-enumerable set of events is to denote each event by a number, the value assumed by a single continuous variable or by an aggregate of n numbers. If there is one variable, it is called a *random variable*. n variables may be treated either as an n-dimensional random variable or an aggregate of n random variables. The physicist is unaware of any other method of describing a non-enumerable set of events except through the values assumed by one continuous variable or an aggregate. Let us for simplicity take the case of a single random variable which can assume any value between 0 and 1 (inclusive of 0 and 1). Now let us discuss the problem of assigning probabilities to the events. The total probability measure is 1 and so two possibilities arise: (1) The probability of each of the events of the non-enumerable set may be zero. (2) Among this non-enumerable set, there may exist a finite or enumerable set of events each of which has a probability greater than zero and each of the rest of the events has a probability zero. In both cases it is suitable to group events which correspond to the values of the random variable lying within certain intervals. In particular it is found convenient to define the possible values of a single random variable from $-\infty$ to $+\infty$ and ask for the probability of events which correspond to the value of the random variable, x lying between $-\infty$ and λ (λ inclusive). Thus we speak of an event which corresponds to $x \leq \lambda$ and its probability $P(x \leq \lambda)$. It is clear that two events corresponding to $x \leq \lambda_1$ and $x \leq \lambda_2 (\lambda_2 > \lambda_1)$ are not mutually exclusive, since the second event always includes the first. We note that since λ is a variable the probability that $x \leq \lambda$ is a function of λ, say $F(\lambda)$. This function is known as the *distribution function* of the random variable x. It may be stated at this juncture that the non-enumerable set can be suitably grouped into an enumerable set of mutually exclusive events.

The essential problem in any physical situation is to obtain the probabilities of events. If this is done a priori, then there exists no problem at all to be solved. One of the ways of obtaining these probabilities is to put these events into correspondence with some basic ω-space, which in turn can be put into correspondence with another set of events which are more fundamental than the previous set and go on till we reach a set of events for which some a priori probability can be ascribed intuitively. The points of the ω-space itself may be put into direct correspondence with the events, but as we have already stated, any ω'-space obtained by a measure preserving transformation from ω is as good as the original and ω' can be treated as a more convenient starting point. Thus there is no limit to the tracing of the space in which the probability measure is a priori defined and this depends on the physical needs of the problem. This, according to the author, is connected with the tracing of the origin of randomness which is treated in Part B, Chap. X of this article.

4. Random variable; measure theoretical approach. In the physical approach we first introduced the intuitive idea of an event and the number denoting the event as the value of the random variable. Since any event can be put into correspondence with a set of points in the ω-space, so can the value of the random variable. Thus we immediately see that the *random variable is a function of the points of the ω-space* i.e., its value depends upon the points of the ω-space. Hence DOOB writes $x(\omega)$ for x. For completeness, we quote his definition.

"A real function $x(\omega)$ defined on a space of points ω will be called a real random variable if there is a probability measure defined on ω-sets and for every real number λ, the inequality $x(\omega) \leq \lambda$ delimits an ω-set whose probability is defined i.e. a measurable ω-set. Thus

$$F(\lambda) = P\big(x(\omega) \leq \lambda\big) \tag{4.1}$$

is defined for all real values of λ. In mathematical language, a real random variable is thus simply a real measurable function. A complex random variable is an ω-function whose real and imaginary parts are measurable. $F(\lambda)$ is monotone, non-decreasing, continuous on the right[1] and

$$\left.\begin{aligned} F(\lambda) &= P\{x(\omega) \leq \lambda\} \\ \lim_{\lambda \to -\infty} F(\lambda) &= 0, \quad \lim_{\lambda \to +\infty} F(\lambda) = +1. \end{aligned}\right\} \tag{4.2}$$

Any function F satisfying all these conditions will be called a distribution function."

5. Aggregate joint distributions. We have introduced the notion of the distribution function $F(\lambda)$ of a single random variable. When we have a collection of n random variables $\langle x_i(\omega) \rangle$, $i = 1, 2, \ldots, n$ it is obvious that this notion can be easily extended and we can define

$$F(\lambda_1, \lambda_2, \ldots, \lambda_n) = \text{probability} \begin{pmatrix} x_1(\omega) \leq \lambda_1 \\ x_2(\omega) \leq \lambda_2 \\ \cdots\cdots\cdots \\ x_n(\omega) \leq \lambda_n \end{pmatrix}. \tag{5.1}$$

This is known as the joint distribution function of the aggregate $\langle x_i(\omega) \rangle$.

---

[1] We can remove the restriction on continuity if we introduce the Dirac-delta function as will be explained later.

6. Marginal distributions. We now go back to the physical concept of events. Let us consider the case where an event can be represented by the values of n random variables $x_1, x_2, \ldots, x_n$. Suppose we are interested in only one of them, say, x_1 and ask for the probability that $x_1 < \lambda_1$. We know that the measure of the event defined by $x_1 < \lambda_1, x_2 < \lambda_2, \ldots, x_n < \lambda_n$ is the measure of the set of points in the ω-space to which these events correspond. If we are interested only in x_1 we have merely to replace each of the quantities $\lambda_2, \lambda_3, \ldots, \lambda_n$ in the joint distribution function $F(\lambda_1, \lambda_2, \ldots, \lambda_n)$ by ∞ and this obviously gives us the probability measure of the event $x_1 \leqq \lambda_1$ irrespective of other variables. This is defined by a function $F(\lambda_1)$[1] known as the marginal distribution of λ. If we are interested in a partial set $(\lambda_1, \lambda_2, \ldots, \lambda_k)$ we put each of the $\lambda_{k+1}, \lambda_{k+2}, \ldots, \lambda_n$ equal to ∞.

Suppose we are interested in the value of the random variables $x_2, x_3, \ldots, x_n$ only when the random variable x_1 assumes a value $\leqq \lambda_1$. Since we know that the event corresponding to $x_1 \leqq \lambda_1$ irrespective of the value of other events has a measure $F(\lambda_1)$ and that the set of points in the ω-space corresponding to the events $x_1 \leqq \lambda_1, x_2 \leqq \lambda_2, \ldots, x_n \leqq \lambda_n$ is $F(\lambda_1, \lambda_2, \ldots, \lambda_n)$ the ratio of the two measures,

$$\frac{F(\lambda_1, \lambda_2, \ldots, \lambda_n)}{F(\lambda_1)} \tag{6.1}$$

can be defined to be the probability of the event $x_2 \leqq \lambda_2, \ldots, x_n \leqq \lambda_n$ given that $x_1 \leqq \lambda_1$. We immediately recognise this ratio as a *conditional probability*.

II. Probability frequency functions[2].

7. Definition. In the definition of the distribution function of a single random variable we just ask for the probability that the random variable has a value less than or equal to λ. The random variable may assume either a continuous set of values from $-\infty$ to $+\infty$ or assume only a discrete set of values. When it can assume a discrete set of values, say $\lambda_1, \lambda_2, \ldots, \lambda_n$, n being at most enumerably infinite, then $F(\lambda)$ is obviously a step function i.e. is characterised by discontinuous jumps at $\lambda = \lambda_i$ and $F(\lambda)$ does not vary between two consecutive λ_i's. In such a case we can define $P(\lambda_i)$ as the probability that $\lambda = \lambda_i$ and

$$F(\lambda) = \sum_{i,\,(\lambda_i \leqq \lambda)} P(\lambda_i). \tag{7.1}$$

The summation is over all values of i consistent with the condition $\lambda_i \leqq \lambda$. $P(\lambda_i)$ is then a probability magnitude and it is found more convenient to work with $P(\lambda_i)$ rather than with $F(\lambda)$ since there is no event corresponding to values $\lambda \neq \lambda_i$.

When the random variable can assume a continuous set of values between $-\infty$ and $+\infty$ and $F(\lambda)$ is continuous in λ it can be expressed as a Riemann integral $F(\lambda) = \int_{-\infty}^{\lambda} f(x)\, dx$. In such a case $f(x)$ is called the probability frequency function of the random variable. There is no necessity that $f(x)$ should be continuous; it is well known that it can have finite discontinuities at an enumerable number of points. If however $F(\lambda)$ has finite discontinuities at a finite number of values of λ, then $f(x) = F'(x)$ becomes infinite at these points. Then it is customary to

---

[1] We use the same symbol F to denote any distribution function, the distinction between two different distributions being apparent from the context.

[2] Sometimes the word "distribution function" is loosely used instead of the "frequency function" and this will be clear from the context. In fact, many well-known frequency functions like the Poisson and the binomial are referred to as distribution functions.

write $F(\lambda)$ as a Stieltjes integral $\int\limits_{-\infty}^{\lambda} dF(x)$. This being a notation has only a formal advantage and it is found more convenient to work with frequency functions wherever possible. Even in the case when $F(\lambda)$ has finite discontinuities (this therefore includes the case of the discrete random variable also), the concept of frequency function can be retained if we introduce the Dirac delta function.

In the application of stochastic theory to physical problems, we shall see that the frequency function plays a fundamental role. *In fact in many problems relating to stochastic processes, the distribution function is obtained only as an integral of the frequency function.* To enable one to work directly with frequency functions even when $F(\lambda)$ is discontinuous at a finite number of values of λ it is very convenient to introduce δ-functions. There are many ways of defining this function which has been the subject of criticism from pure mathematicians. Since it has come to stay in mathematical physics and its use is justified by elaborate mathematical treatments, we shall avoid such discussions on rigour and summarise the useful results.

8. The Dirac delta function. There are many ways of introducing the delta function provided we understand the delta function to be an operational quantity which is meaningful only when it occurs under an integral. A bare delta function though analytically incomplete can also be suitably interpreted in probability theory.

The simplest way of defining a δ-function is to understand it to mean the differential coefficient of the Heaviside unit function

$$\left.\begin{array}{l} H(x) = 0 \quad \text{if} \quad x < 0, \\ H(x) = 1 \quad \text{if} \quad x > 0. \end{array}\right\} \tag{8.1}$$

Thus $H(x)$ is discontinuous at $x=0$. It is clear that $dH(x)/dx = 0$ for $x > 0$ or $x < 0$ and is infinite at $x=0$. Calling $dH(x)/dx$ as $\delta(x)$ we can write

$$\int\limits_{-\infty}^{+\infty} \delta(x)\,dx = \int\limits_{-b}^{a} \delta(x)\,dx = 1 \tag{8.2}$$

where a and b are chosen positive and arbitrary and can be as small as we please. The function $H(x)$ has a simple probabilistic interpretation. It can denote the distribution function of a random variable x. It is zero if $x < 0$ and is equal to 1 if $x > 0$. The set of events is non-enumerable and consists of all points on the real axis $-\infty$ to $+\infty$, the probability measure being unity. *The entire measure is concentrated at the point $x=0$.* Thus the outcome of the experiment will be $x=0$ with probability unity.

Thus in the above experiment the delta function can be looked upon as the frequency function of the random variable x. It has a singularity at $x=0$ and is zero everywhere else. It is a probability density and is infinite only because the density is the ratio of the probability measure of the interval to the linear measure of the interval. If this probability measure is finite even if the linear measure tends to zero, the density is obviously infinite though, from the point of view of events, it is clear that the event represented by the point $x=0$ has a finite probability equal to unity. As one of the immediate consequences of the definition of the δ-function we have

$$\int\limits_{-\infty}^{+\infty} f(x)\,\delta(x)\,dx = f(0) = \int\limits_{-b}^{a} f(x)\,\delta(x)\,dx. \tag{8.3}$$

The meaning of $\delta(x-\alpha)$ where α is a constant follows by defining a variable $x' = x - \alpha$.

Eq. (8.3) *can be taken as an adequate definition of the delta function.* On such a definition $\delta(x)$ is also represented as

$$\delta(x) = \frac{1}{2\pi} \lim_{k\to\infty} \int_{-k}^{+k} e^{ik'x}\,dk' = \lim_{k\to\infty} \frac{1}{\pi k}\int_{0}^{k} \cos k'x\,dk' = \lim_{k\to\infty} \frac{\sin kx}{k\pi}. \tag{8.4}$$

The limit $k\to\infty$ does not exist in itself. If however the right hand side of the above equation is multiplied by a function $f(x)$ which is regular at $x=0$ and is integrated with respect to x over any interval covering the point $x=0$, the limiting process can be carried out after the integration

$$\int_{-b}^{a} \delta(x)\,f(x)\,dx = \frac{1}{\pi}\lim_{k\to\infty}\int_{-bk}^{ak} f\left(\frac{y}{k}\right)\frac{\sin y}{y}\,dy = f(0). \tag{8.5}$$

$\delta(x)$ has no meaning when multiplied by a function $f(x)$ which is itself singular at $x=0$. $\delta(x)$ can also be expressed in many other ways. Every complete set of orthonormal functions $u_n(x)$ provides a representation of $\delta(x)$ through the completeness theorem

$$\sum_{n} u_n^*(x)\,u_n(x') = \delta(x - x') \tag{8.6}$$

where n be a discrete, continuous or mixed index.

It appears from Eq. (8.3) that $\delta(x)$ is an even function of x

$$\delta(-x) = \delta(x). \tag{8.7}$$

From the defining property of Eq. (8.3) we see that $\delta(x) = \delta(cx)/|c|$ and

$$\delta[(x - x_1)(x - x_2)] = \frac{\delta(x - x_1) + \delta(x - x_2)}{|x_2 - x_1|}. \tag{8.8}$$

This follows from the fact that the contributions to the integral arise from the points where the arguments of $\delta(x)$ vanish and from Eq. (8.7). In particular

$$\delta(x^2 - a^2) = \frac{1}{2a}\{\delta(x - a) + \delta(x + a)\}. \tag{8.9}$$

$\delta(x)$ can also be defined by contour integration in the complex plane. Let C be a closed path containing the point $x=0$, but no singularity of $f(x)$. Then $f(0) = \frac{1}{2\pi i}\oint_C \frac{f(x)}{x}\,dx$ and we can write

$$\delta(x) = \frac{1}{2\pi i}\frac{1}{x}\Big|_C. \tag{8.10}$$

The symbol $|_C$ means the the subsequent integration over x should be carried over C. C may be taken as a very small circle.

The derivative $\delta'(x)$ can be defined as

$$\delta'(x) = \frac{1}{\pi}\lim_{k\to\infty}\left(\frac{k\cos kx}{x} - \frac{\sin kx}{x^2}\right) \tag{8.11}$$

to be understood in the same sense as above. $\delta'(x)$ when multiplied by $f(x)$ has the simple property that follows from partial integration and the fact that $\delta(x)=0$ for $x\neq0$:

$$\int_{-b}^{a} \delta'(x)\,f(x)\,dx = -f'(0). \tag{8.12}$$

$\delta'(x)$ is obviously an odd function of x.

In many stochastic problems, we will be constrained to use delta functions of the form $\delta(f(x) - \alpha)$. To interpret this in terms of functions of the type $\delta(x - \alpha)$ we may introduce the transformation $f(x) = y$ and write

$$\int_{-\infty}^{+\infty} \varphi(x)\, \delta(f(x) - \alpha)\, dx = \int_{-\infty}^{+\infty} \psi(y)\, \delta(y - \alpha) \frac{1}{k(y)}\, dy \qquad (8.13)$$

where $\psi(y)$ and $k(y)$ are obtained by expressing x in terms of y in $\varphi(x)$ and $df(x)/dx$ respectively. This looks deceptively simple and care has to be exercised when $f(x) = \alpha$ has more than one solution. This gives rise to intriguing situations which have to be solved from the knowledge of the particular considerations of these problems. In employing the delta function in integral operations with respect to a function, there may occur a case when x cannot take negative values and where the delta function singularity occurs at zero. In such a case, tricky situations may arise by including or omitting the point zero.

It is now clear that if we wish to speak of probability densities in the case of a random variable capable of assuming only discrete values λ_i with probability $P(\lambda_i)$ we have to introduce delta functions. Since $F(\lambda)$ in such cases is given by (7.1), the frequency function is represented thus

$$\frac{dF(\lambda)}{d\lambda} = \sum_i P(\lambda_i)\, \delta(\lambda - \lambda_i). \qquad (8.14)$$

There is of course no specific advantage in using the language of continuous variable when the random variable takes only discrete values. In such a case we call $P(\lambda_i)$ itself the frequency function.

However in the case of a random variable capable of assuming continuous values, under certain physical circumstances a distribution function $F(\lambda)$ may have finite discontinuities at a finite number of points and is continuous elsewhere. In such a case the frequency function can be expressed as

$$f(\lambda) = f_1(\lambda) + \sum_i k(\lambda_i)\, \delta(\lambda - \lambda_i) \qquad (8.15)$$

where $f_1(\lambda)$ is a function with no delta function singularities, (i.e. may have only finite discontinuities) and the points λ_i are those at which the delta function singularities occur, $k(\lambda_i)$ being the corresponding coefficients. The above can be treated to be the frequency function of a random variable which is the super-position of a continuous variable with frequency distribution $f_1(\lambda)$ and a discrete variable x' assuming values λ_i with probability $k(\lambda_i)$. It is not advantageous then to treat the discrete variable separately since in stochastic processes where such distributions occur, the points of singularity may shift with respect to a certain parameter.

Thus when we speak of a continuous random variable we only mean that the random variable can assume a continuous set of values. It is not implied that its distribution function is continuous. It may exhibit finite discontinuities at a finite number of points but must be continuous elsewhere. At these points of discontinuities, the frequency function has delta function singularities. If elsewhere the frequency function is zero we are dealing with a discrete random variable and so we can work directly with the probabilities of the events corresponding to the discrete values assumed by the random variable.

The above discussion can be carried over to define the joint frequency function of n random variables. If $F(\lambda_1, \lambda_2, \ldots, \lambda_n)$ is the joint distribution function of n continuous random variables then

$$f(\lambda_1, \lambda_2, \ldots, \lambda_n) = \frac{\partial^n F}{\partial \lambda_1 \partial \lambda_2 \ldots \partial \lambda_n} \tag{8.16}$$

is the frequency function.

We shall now define the mean values of a function of a random variable. If $f(\lambda)$ is the frequency function of a random variable λ and $h(\lambda)$ is a function of the random variable, then the mean value of $h(\lambda)$ is defined as

$$\mathscr{E}\{h(\lambda)\} = \int_\lambda h(\lambda) f(\lambda) \, d\lambda = \int_{-\infty}^{+\infty} h(\lambda) \, dF(\lambda). \tag{8.17}*$$

In particular when

$$h(\lambda) = \lambda^n \tag{8.18}$$

$\mathscr{E}\{\lambda^n\}$ is called the n-th moment of the random variable or sometimes the n-th moment of the frequency function.

$$\sigma^2(\lambda) = \mathscr{E}\{\lambda^2\} - [\mathscr{E}\{\lambda\}]^2$$

is called the mean square deviation and considered an important characteristic of the distribution.

9. Independence and correlation of random variables. The definition of joint distribution and frequency functions leads us immediately to the concepts of independence and correlation of random variables.

α) *Independence.* If $F(\lambda_1, \lambda_2, \ldots, \lambda_n)$ be the joint distribution function of n random variables and F can be written in the form

$$F(\lambda_1, \lambda_2, \ldots, \lambda_n) = F_1(\lambda_1) F_2(\lambda_2) \ldots F_n(\lambda_n) \tag{9.1}$$

we say that the n random variables are independent of each other. This is equivalent to stating that the frequency function $f(\lambda_1, \lambda_2, \ldots, \lambda_n)$ can be expressed as

$$f(\lambda_1, \lambda_2, \ldots, \lambda_n) = f_1(\lambda_1) f_2(\lambda_2) \ldots f_n(\lambda_n). \tag{9.2}$$

β) *Correlation.* If x and y are two random variables and the joint frequency function is $f(x, y)$ we have the following mean values

$$\left.\begin{aligned}
\mathscr{E}\{x\} = m_1 &= \int_x \int_y f(x, y) \, x \, dx \, dy, \\
\mathscr{E}\{y\} = m_2 &= \int_x \int_y f(x, y) \, y \, dx \, dy, \\
\mathscr{E}\{xy\} &= \int_x \int_y xy f(x, y) \, dx \, dy, \\
\mathscr{E}\{xy\} - \mathscr{E}\{x\}\mathscr{E}\{y\} = \mathscr{E}\{(x - m_1)(y - m_2)\} &= \mu_{11}, \\
\mathscr{E}\{(x - m_1)^2\} = \sigma_1^2 &= \int_x \int_y (x - m_1)^2 f(x, y) \, dx \, dy, \\
\mathscr{E}\{(y - m_2)^2\} = \sigma_2^2 &= \int_x \int_y (y - m_2)^2 f(x, y) \, dx \, dy.
\end{aligned}\right\} \tag{9.3}$$

The coefficient of correlation ϱ is defined as

$$\varrho = \mu_{11}/\sigma_1 \sigma_2. \tag{9.4}$$

We just note that if the two variables are independent, by definition $f(x, y) = f(x) f(y)$ and hence the variables are uncorrelated. But two uncorrelated variables

---

* From now on, we shall use the following conventions: (i) Same symbols will be used for the random variable and the value it assumes. (ii) $\mathscr{E}$ denotes the mean or expectation value. (iii) Formally the domain of λ ranges from $-\infty$ to $+\infty$ but if $f(\lambda) = 0$ for certain ranges they can be omitted. This is implied in the symbol $\int_\lambda$.

(their $\varrho = 0$) are not necessarily independent. It merely means that independence is a much stronger condition than lack of correlation which is obvious since one is a restriction on the entire distribution and the other on μ_{11}. A simple example can be cited to illustrate this. Let us assume

$$f(x, y) = g\left(\sqrt{x^2 + y^2}\right) \tag{9.5}$$

where g is any function and f the joint frequency function of the random variables x and y. It can be easily verified that if the two are uncorrelated they need not be independent.

$g(x) = e^{-x}$ is one such function.

In fact from any two correlated variables ξ and η we can obtain two uncorrelated variables by the following transformation

$$\left.\begin{aligned}
x &= (\xi - m_1) \cos \varphi + (\eta - m_2) \sin \varphi, \\
y &= - (\xi - m_1) \sin \varphi + (\eta - m_2) \cos \varphi.
\end{aligned}\right\} \tag{9.6}$$

If m_1 and m_2 are their means and φ satisfies the equation $\tan \varphi = \dfrac{2\,\mu_{11}}{\sigma_2 - \sigma_1}$ then x and y are uncorrelated.

It would be worthwhile to conclude this section by writing the joint distribution of a random variable and a function of the same random variable. Let $\pi(x)$ be the probability frequency function of x. The other random variable y is a function of x, say $\varphi(x)$, i.e. it assumes a value $\varphi(a)$ if $x = a$. Therefore

$$\left.\begin{aligned}
f(x, y)\,dx\,dy &= \pi(x)\,\delta\{y - \varphi(x)\}\,dx\,dy \\
\mathscr{E}\{xy\} &= \int_x \int_y x\,y\,\pi(x)\,\delta\{y - \varphi(x)\}\,dx\,dy \\
&= \int_x x\,\varphi(x)\,\pi(x)\,dx.
\end{aligned}\right\} \tag{9.7}$$

When y is a function of x, we say it is correlated in a "gear-box" manner with x. It just means that the values of y are uniquely determined if x is known.

10. Composition of distribution functions of independent random variables. If $x_1, x_2, \ldots, x_n$ are n independent random variables, and if we define a new random variable $z = \sum\limits_{i=1}^{n} x_i$ then $\pi(z)$, the probability frequency function of z, is obtained in terms of the frequency function $\pi_i(x_i)$ of the variables x_i.

$$\left.\begin{aligned}
\pi(z) = \int_{x_1} \int_{x_2} \cdots \int_{x_{n-1}} \pi_1(x_1)\,\pi_2(x_2) \ldots \pi_{n-1}(x_{n-1}) \times \\
\times \pi_n(z - x_1 - x_2 - \cdots - x_{n-1})\,dx_1\,dx_2, \ldots dx_{n-1}.
\end{aligned}\right\} \tag{10.1}$$

One simple and useful corollary of the above equation is that the mean square deviation of z is the sum of the mean square deviations of the x_i's.

There is a very important theorem which connects the frequency function of the sum of a large number of independent variables with the mean and mean square derivations only of the individual distribution. It is not necessary to know the individual distribution functions in the limiting case when the number of variables is very large and certain analytical conditions are satisfied. The usefulness of this limit theorem is obvious since it dispenses with the exact knowledge of the individual distributions. Hence it plays a dominant role in statistics and recently has been used by KHINCHIN [4] to derive the fundamental law of classical statistical mechanics. It is stated in many forms and derived under many conditions and restrictions. We state here the central limit theorem without giving the proof for which the reader is referred to the book of KHINCHIN.

11. Central limit theorem. Suppose we have a sequence of mutually independent random variables with probability densities $u_k(x)$ $(k=1, 2, \ldots)$ and let

$$g_k(t) = \int_x e^{itx} u_k(x)\, dx \qquad k = 1, 2, \ldots \tag{11.1}$$

represent the characteristic functions[1] corresponding to these densities. Let us assume that

1. The functions $u_k(x)$ possess continuous derivatives and there exists such a constant A that

$$\int_x |u_k'(x)| \cdot dx < A \qquad (k = 1, 2, \ldots). \tag{11.2}$$

2. The functions $u_k(x)$ possess finite moments of the first five orders which we will denote by a_k, b_k, c_k, d_k and e_k. Without loss of generality we can put $a_k = 0$, $(k = 1, 2, \ldots)$. Then there exist positive constants α and β such that

$$0 < \alpha < b_k < \beta; \quad \bar{c}_k < \beta, \quad d_k < \beta, \quad \bar{e}_k < \beta \quad (k = 1, 2, \ldots) \tag{11.3}$$

where $\bar{c}_k$ and $\bar{e}_k$ represent the absolute moments of the third and fifth order of the functions $u_k(x)$.

3. There exist positive constants a and b such that for $|t| < a$,

$$|g_k(t)| > b \quad (k = 1, 2, \ldots). \tag{11.4}$$

4. For each interval (c_1, c_2), $c_1 c_2 > 0$ there exists a number $\varrho(c_1, c_2) < 1$ such that for any t within the interval (c_1, c_2) we have

$$|g_k(t)| < \varrho \quad (k = 1, 2, \ldots). \tag{11.5}$$

Let $u_n(x)$ be the probability density of the sum of the first n terms in the given sequence of random quantities. Then for $n \to \infty$ and for $|x| < 2(\log n)^2$ we have

$$u_n(x) = \frac{1}{(2\pi B_n)^{\frac{1}{2}}} \exp\left[-\frac{x^2}{2 B_n}\right] + \frac{s_n + t_n x}{B_n^{\frac{3}{2}}} + O\left[\frac{1 + |x^3|}{n^2}\right]. \tag{11.6}$$

For any arbitrary x we have

$$u_n(x) = \frac{1}{(2\pi B_n)^{\frac{1}{2}}} \exp\left[-\frac{x^2}{2 B_n}\right] + O\left(\frac{1}{n}\right) \tag{11.7}$$

where $B_n = \sum_1^n b_k$ and S_n and t_n are quantities independent of x increasing not faster than n.

12. Transformation of random variables. We have till now referred to the determination of the frequency functions of sums or products of independent variables the frequency function of each being known. We shall now refer to the more general problem relating to the transformation of random variables.

Let $x_1, x_2, \ldots, x_n$ be an aggregate of independent random variables the frequency functions of x_i being $f(x_i)$. If y is a function of the x_i's then y is a random variable and our object is to determine its frequency function. In the simple case when we have only one random variable x and

$$y = \varphi(x), \quad dy = \varphi'(x)\, dx \tag{12.1}$$

the frequency function $g(y)$ of y is determined as follows: If Δ is an arbitrary infinitesimal quantity

$$g(y)\, \Delta = f(x)\, \Delta / \varphi'(x). \tag{12.2}$$

On the right hand side we must express x in terms of y to get $g(y)$.

---

[1] A general discussion on the characteristic function follows later. We need here only the definition (11.1).

On the other hand, if we have an aggregate $\langle x_i \rangle$ and y is a function of all the x_i's the situation is far from simple. We shall content ourselves by outlining the general procedure to be adopted.

Let
$$y = \varphi(x_1, x_2, , \ldots, x_n). \tag{12.3}$$

We require the probability that φ lies between y and $y + \Delta$. Let us first fix all the x_i's except x_n. The change in φ is effected only through a change in x_n.

$$\Delta = \frac{\partial \varphi}{\partial x_n} d x_n \quad \text{or} \quad d x_n = \Delta \Big/ \frac{\partial \varphi}{\partial x_n} . \tag{12.4}$$

The x_i's except x_n can be varied arbitrarily and

$$g(y)\, \Delta = \Delta \int_{x_1} \int_{x_2} \cdots \int_{x_{n-1}} \frac{f(x_1)\, f(x_2) \ldots f(x_n)}{\partial \varphi / \partial x_n}\, d x_1\, d x_2 \ldots d x_{n-1}. \tag{12.5}$$

On the right hand side before integration we must substitute for x_n in terms of $x_1, \ldots x_{n-1}$ and y.

Written in this form, this method looks simple but the difficulties of expressing x_n in terms of the x_i's $(i \neq n)$ and y and subsequent integration sometimes render this method useless and devices have to be made to suit the particular cases*. The problem becomes much more complicated if we require the joint distribution of two or more functions of the x_i's.

Sometime we may impose constraints on the n variables, say l constraints, by requiring
$$k_j(x_1, x_2, \ldots, x_n) = 0 \quad (j = 1, 2, \ldots l) \tag{12.6}$$

and ask for the frequency function of $y = \varphi(x_1, x_2, \ldots x_n)$. Formally this is equivalent to defining random variables y and k_j and asking for the frequency function of y conditional upon $k_j = 0$.

One particular case of the above class of problems deserves mention. If we start with n independent random variables and introduce the transformation

$$y_i = \varphi_i(x_1, x_2, \ldots, x_n) \quad (i = 1, 2, \ldots n) \tag{12.7}$$

the joint frequency function of the y_i's can be expressed directly in terms of the frequency functions of x_i.

$$g(y_1, y_2, \ldots, y_n)\, \Delta_1 \Delta_2, \ldots \Delta_n = \frac{f(x_1)\, f(x_2) \ldots f(x_n)}{J(x_1, x_2, \ldots x_n)}\, \Delta_1 \Delta_2 \ldots \Delta_n \tag{12.8}$$

where $g(y_1, y_2, \ldots, y_n)\, \Delta_1 \Delta_2 \ldots \Delta_n$ is the joint probability that φ_i lies between y_i and $y_i + \Delta y_i$ $(i = 1, 2, \ldots n)$ and J is the Jacobian. On the right hand side, we must express all the x_i's in terms of the y_i's.

Situations as described above arise frequently in stochastic processes when we have to deal with transformations of dynamical variables.

13. Composition of frequency functions under constraint. We have just referred to the composition of independent random variables. By composition and transformation, we meant the determination of the frequency function of sums or

---

* For example in writing (12.2) and requiring x to be expressed in terms of y, we have made the simplifying assumptions that (1) x is a single valued function of y, (2) the interval $(y, y + \Delta)$ in the y-space corresponds to an interval $(x, x + \Delta')$ in the x-space, where $\Delta = \Delta'\, \varphi'(x)$. More generally the interval Δ may correspond to a set of points δs in the x-space with measure $m(\delta s)$. If s denotes the set in the x-space defined by $\varphi(x) < a$ the probability that $y < a$ is defined by the Lebesgue integral of $f(x)$ over the set s. In simple cases when Δ corresponds to a set of intervals in the x-space, it may be possible to obtain the probability that $y < a$ as a sum of Riemann integrals. Similar observations apply to (12.4) and (12.5).

products of the random variables, more generally, functions of the independent random variables. If the variables are not independent, the frequency function of a random variable which is a function of the set of random variables cannot be derived without recourse to the joint frequency function of the entire set of variables. There is, however, one important case where the non-independence of a set of n variables can be traced to the imposition of a constraint upon n independent variables. In such a case the composition of such non-independent variables can be made in terms of the independent variables.

Consider an aggregate of n independent continuous random variables $\langle x_i \rangle$, $i = 1, 2, \ldots n$, the frequency function of x_i being $f_i(x_i)$. Let $\varphi(x_1, x_2, \ldots, x_n)$ be a function of the x_i's. Hence it is a random variable. If we impose the condition that $\varphi = c$, the frequency function of x_i conditional on $\varphi = c$ is different from $f_i(x_i)$ and shall be denoted by $f_i(x_i | c)$. In other words, we may say we have a new random variable x_i' with frequency function $f_i(x_i' | c)$. But it is customary to call $f_i(x_i | c)$ the frequency function of x_i conditional upon $\varphi = c$. It is obvious that the x_i's under constraint are no longer independent. We shall now consider the simplest type of constraint

$$\varphi_n(x_1, x_2, \ldots, x_n) = x_1 + x_2 + \cdots + x_n = c. \tag{13.1}$$

If the frequency function of φ_n is denoted by $g_n(\varphi)$, simple probability arguments lead to the expression

$$g_n(\varphi) = \int_{x_1} \int_{x_2} \cdots \int_{x_{n-1}} f_n\left(\varphi - \sum_{i=1}^{n-1} x_i\right) \prod_{i=1}^{n-1} f_i(x_i) \, d x_1 \, d x_2 \ldots d x_{n-1} \tag{13.2}$$

i.e. g_n is the convolution of the f_i's. If $f(x_1, x_2, \ldots, x_{n-1} | c)$ is the joint frequency function of $x_1, x_2, \ldots, x_{n-1}$ conditional upon $\varphi = c$ (note that x_n is determined if all other x_i's are known)

$$f(x_1, x_2, \ldots, x_{n-1} | c) = f_n\left(c - \sum_{i=1}^{n-1} x_i\right) \prod_{i=1}^{n-1} f_i(x_i) / g_n(c). \tag{13.3}$$

The frequency function $f_i(x_i | c)$ of x_i conditional upon $\varphi_n = c$ can be obtained by integrating the above expression with respect to all the x_i's except $x_j = x_i$. If we write

$$\varphi_{n-1} = x_1 + x_2 + \cdots + x_{n-1} \tag{13.4}$$

and denote its frequency function by $g_{n-1}(\varphi)$ then

$$f_n(x_n | c) = f_n(x_n) \, g_{n-1}(c - x_n) / g_n(c). \tag{13.5}$$

This is precisely the formula on which KHINCHIN's work on statistical mechanics is based. In view of the considerable attention that his work has drawn from mathematicians[1] we shall discuss in Part B of this article how the above formula plays an important role in statistical mechanics.

III. Functions associated with the frequency functions[2].

It is found convenient in practice to work with functions which are associated with frequency functions rather than with frequency functions themselves. This is mainly a matter of mathematical convenience for analytical operations. We shall here give a summary of the various "transforms" associated with the frequency functions of a single random variable. The extension to the joint

---

[1] At the recent Berkeley symposium (1955) on probability and statistics, BLANC-LA-PIERRE and TORTRAT [5] have presented a critical estimate of KHINCHIN's work.

[2] For convenience, sometimes we will refer to the probability frequency function as p.f.f.

frequency functions of several variables is immediate since the "transforms" can be defined with respect to each variable. It is customary to treat the case when the random variable can assume discrete values and continuous values separately.

a) Discrete random variables.

14. Generating function. We shall take the case when the random variable can take only zero or positive integral values. Suitable adaptations have to be made when the random variable takes non-integral values or negative values.

Let $\pi(n)$ be the probability that the stochastic variable x takes a value n. We define $G(u)$ as the generating function corresponding to $\pi(n)$ in the following manner:

$$G(u) = \sum_{n=0}^{\infty} u^n \pi(n). \tag{14.1}$$

The inversion formula for this is obviously obtained by expanding G in the neighborhood of $u = 0$,

$$G(u) = G(0) + u\,G'(0) + \frac{u^2}{2!}\,G''(0) + \cdots \tag{14.2}$$

Comparing with this the definition of $G(u)$ we write

$$\pi(n) = \left[\frac{\partial^n G(u)}{\partial u^n}\,\frac{1}{n!}\right]_{u=0}. \tag{14.3}$$

The moments of x can be obtained from $G(u)$.

$$\mathscr{E}\{x^m\} = \left(u\,\frac{\partial}{\partial u}\right)^m G(u) \quad \text{at } u = 1. \tag{14.4}$$

One simple result can immediately be obtained. If $G(u)$ is an even function then the probability that x is odd is zero. If $G(u)$ is odd the probability that x is even is zero.

If we have two independent stochastic variables x_1 and x_2 with frequency functions $\pi_1(x_1)$ and $\pi_2(x_2)$ respectively and we require the distribution function of $\pi(z)$ where $z = x_1 + x_2$, then by definition

$$\pi(z) = \sum_{x_1=0}^{z} \pi_1(x_1)\,\pi_2(z - x_1) = \sum_{x_2=0}^{z} \pi_2(x_2)\,\pi_1(z - x_2). \tag{14.5}$$

If G_1, G_2 and G are the generating functions corresponding to π_1, π_2 and π we immediately observe that

$$G(u) = G_1(u)\,G_2(u). \tag{14.6}$$

This can easily be extended to the generating function of n independent stochastic variables. Denoting the independent stochastic variables by x_1, x_2, ..., x_n, and the generating functions by $G_1(u)$, $G_2(u)$, ..., $G_n(u)$ we obtain

$$G(u) = G_1(u)\,G_2(u) \dots G_n(u) \tag{14.7}$$

as the generating function corresponding to the p.f.f. of the variable

$$z = x_1 + x_2 + \cdots + x_n. \tag{14.8}$$

15. Partition function. In some cases $\pi(n)$ may be written as $\pi(n, \lambda)$ where λ is a continuous parameter. This parameter enters due to the physical conditions of the problem on the basis of which the distribution is derived. Let us deal with a class of p.f.f.'s which can be written in the form

$$\pi(n, \lambda) = \frac{a(n)}{Q(\lambda)}\,e^{n\lambda}. \tag{15.1}$$

This merely means that π can be decomposed into a product of functions of n and λ except for one factor which cannot be so separated but is of the exponential type. In such a case we note that

$$Q(\lambda) = \sum_n a(n)\, e^{n\lambda} \tag{15.2}$$

and

$$\mathscr{E}\{n^\nu\} = Q^\nu(\lambda)/Q(\lambda) \tag{15.3}$$

where $Q^\nu(\lambda)$ is the ν-th derivative of Q with respect to λ. $Q(\lambda)$ is called the partition function of the distribution and serves the same purpose as the generating function i.e. generating moments. It is the sum of the "unnormalised" probabilities $a(n)\, e^{n\lambda}$ and its use depends on its being a function of a continuous parameter λ.

For the definition of the partition function it is not necessary that n should be a discrete random variable. We have in such a case merely to replace the sum by an integral. *It is now easy to realise why the partition function plays such a dominant role in thermodynamics where the exponential term (with the energy variable E instead of n) is the keystone on which statistical mechanics rests.*

b) Continuous random variables.

In the case of continuous x any transform must be an integral transform. We shall consider a few familiar integral transforms.

16. The characteristic function[1]. The most familiar is the characteristic function. If $f(x)$ is the frequency function of the random variable x, the characteristic function $\varphi(t)$ is defined as

$$\varphi(t) = \int_{-\infty}^{+\infty} f(x)\, e^{itx}\, dx \tag{16.1}$$

where t is a real variable. The inversion formula for this is given by

$$f(x) = \frac{1}{2\pi} \int_{-\infty}^{+\infty} \varphi(t)\, e^{-itx}\, dt. \tag{16.2}$$

The moments can be obtained from the characteristic function:

$$\mathscr{E}\{x^n\} = \left(-i\frac{\partial}{\partial t}\right)^n \varphi(t) \quad \text{at } t = 0. \tag{16.3}$$

If we have two stochastic variables x and y with the distribution functions $f_1(x)$ and $f_2(y)$ and if the stochastic variable $z = x + y$ with z having a distribution function $f(z)$, then

$$\left. \begin{aligned} f(z) &= \int_x f_1(x)\, f_2(z-x)\, dx, \\ &= \int_y f_1(z-y)\, f_2(y)\, dy. \end{aligned} \right\} \tag{16.4}$$

There exists a simple relation between the characteristic functions as in the case of generating functions.

$$\varphi(t) = \varphi_1(t)\, \varphi_2(t). \tag{16.5}$$

The form of the integral (16.4) is called the *convolution* and the characteristic function of a convolution is the product of the characteristic functions of the

---

[1] In the following sections the most important formulae of the theory of integral transforms have been summarised as far as they concern us here. For more details the reader is referred to I. N. SNEDDON's article in Vol. II of this Encyclopedia.

terms in the integrand. A very interesting symmetry property between φ and f can be noticed if we interpret $\sqrt{2\pi}\,\varphi$ as the characteristic function with argument t corresponding to $f(x)$. Then $\sqrt{2\pi}\,f(x)$ can equally be interpreted as the characteristic function with argument x corresponding to $\varphi(t)$. It is interesting to write the characteristic function of the product $f_1(x)\,f_2(x)$ as

$$\varphi_{1,2}(t) = \int_{-\infty}^{+\infty} \varphi_1(\tau)\,\varphi_2(t-\tau)\,d\tau. \tag{16.6}$$

This follows the symmetry property of f and φ referred to above.

17. Fourier cosine and sine transforms. Since the characteristic function is a generalisation of the Fourier cosine and sine transforms we define them here for the sake of completeness.

α) Fourier sine transform

$$\bar{f}(p) = \int_0^\infty f(x)\sin p\,x\,d x. \tag{17.1a}$$

The inversion formula is given by

$$f(x) = \frac{2}{\pi}\int_0^\infty \bar{f}(p)\sin x\,p\,dp. \tag{17.1b}$$

β) Fourier cosine transform

$$\bar{f}(p) = \int_0^\infty f(x)\cos p\,x\,dx. \tag{17.2a}$$

The corresponding inversion formula is given by

$$f(x) = \frac{2}{\pi}\int_0^\infty \bar{f}(p)\cos x\,p\,dp. \tag{17.2b}$$

18. Laplace transform. A more general transform which includes the characteristic function is the Laplace transform. If s is a complex variable we define

$$\varphi(s) = \int_0^\infty f(x)\,e^{-s x}\,dx \tag{18.1a}$$

and its inversion formula is

$$2\pi i\,f(x) = \int_{\gamma-i\infty}^{\gamma+i\infty} \varphi(s)\,e^{s x}\,ds \tag{18.1b}$$

where γ is greater than the real parts of all the singularities of $\varphi(s)$. If $s=\gamma+it$ where γ and t are the real and imaginary parts, putting $\gamma=0$ and treating $\varphi(s)$ as a function of t, we get the characteristic function. The advantage of the Laplace transform over the characteristic function lies in that even if $\int f(x)\,dx$ is not convergent, we can obtain convergence of the Laplace transform by choosing the real part to be suitably large and negative. If $f(x)$ is a probability frequency function, its integral is always convergent and hence both the characteristic functions and Laplace transforms can be used equally well. The convolution theorem of Laplace transform is identical with that of the characteristic function i.e. if $\varphi_1(s)$ and $\varphi_2(s)$ be the Laplace transforms of $f_1(x)$ and $f_2(x)$ respectively, then defining a function

$$f(y) = \int_0^\infty f_1(y)\,f_2(x-y)\,dy \tag{18.2}$$

the Laplace transform $\varphi(s)$ of $f(y)$ is given by

$$\varphi(s) = \varphi_1(s)\,\varphi_2(s). \tag{18.3}$$

19. MELLIN's transform. The Mellin transform of a function $f(y)$ is defined as

$$\varphi(s) = \int\limits_0^\infty f(y)\, y^{s-1}\, dy,\tag{19.1a}$$

and its inverse is given by

$$f(y) = \frac{1}{2\pi i} \int\limits_{\gamma-i\infty}^{\gamma+i\infty} \varphi(s)\, y^{-s}\, ds.\tag{19.1b}$$

Defining a new variable x such that $x = -\log y$, we recognise the correspondence between the Laplace transform and the Mellin transform. The convolution theorem for the Mellin transform is given as follows:

If $z = xy$ and if $f(z) = \int\limits_x f_1(x)\, f_2\left(\frac{z}{x}\right) \frac{dx}{x}$ and $\varphi_1(s)$ and $\varphi_2(s)$ are the Mellin transforms of $f_1(x)$ and $f_2(y)$ then the Mellin transform $\varphi(s)$ of $f(z)$ is given by

$$\varphi(s) = \varphi_1(s)\, \varphi_2(s).\tag{19.2}$$

Since in the description of the frequency function of a continuous random variable we may introduce δ-functions, it will be worthwhile to mention the transform of the δ-function.

(i) The Laplace transform of a δ-function is defined as follows:

$$\varphi(s, a) = \int\limits_0^\infty \delta(x - a)\, e^{-sx} dx = e^{-sa},\tag{19.3}$$

$$\text{if}\quad a = 0,\quad \varphi(s) = 1.$$

(ii) The MELLIN's transform of $\delta(x)$ is

$$\varphi(s, a) = \int\limits_0^\infty \delta(x - a)\, x^{s-1}\, dx = a^{s-1}.\tag{19.4}$$

We are faced with a difficulty when $a = 0$ since the Mellin transform of zero is zero. However a^{s-1} as $a \to 0$ can be distinguished from the Mellin transform of zero since at $s = 1$ it is undefined. But it can be interpreted to be unity at $s = 1$ since $\int\limits_0^\infty \delta(x - a)\, dx = 1$. Hence if in any computation we get a function $\varphi(s)$ which is zero everywhere and is undefined at $s = 1$, we can interpret that function to be the Mellin transform of a δ-function if we are able to justify from the physical considerations that $\varphi(s) = 1$ at $s = 1$.

IV. Standard probability frequency functions.

Certain standard probability frequency functions occur very frequently in statistical problems. It will be convenient to give a short list of such functions and explain the circumstances under which they arise since in studying stochastic processes it has become customary to compare the frequency function occurring in any stochastic process with some standard function.

For convenience we divide these distribution functions into two classes,

a) functions of discrete random variables taking only positive integral values, and

b) functions of continuous random variables.

a) Discrete distributions.

20. The binomial distribution of Bernoulli. This arises when an experiment consists of n independent trials and in each trial we get a success with probability p and failure with probability $1-p$. Then, x, the number of successes in n trials is a stochastic variable and by elementary arguments it can be shown to have the following frequency function,

$$\varphi(x, n) = \binom{n}{x} p^x q^{n-x} \tag{20.1}$$

where $q = 1 - p$. The corresponding generating function is given by

$$G(u) = (pu + q)^n \tag{20.2}$$

and

$$\mathscr{E}\{x\} = np, \quad \sigma_B^2(x) = npq. \tag{20.3}$$

21. The binomial distribution of Poisson. The experiment consists of a sequence of n independent trials, the probability of success is p_j in the j-th trial. These p_j are different for different trials. The generating function in this case is

$$G(u) = (p_1 u + q_1)(p_2 u + q_2) \cdots (p_n u + q_n). \tag{21.1}$$

If x is the number of successes in n trials,

$$\mathscr{E}\{x\} = \sum_{i=1}^{n} p_i = np, \quad \sigma_P^2 = \sum p_i q_i. \tag{21.2}$$

22. The Lexian distribution. The experiment consists of k sets of trials, the number of trials in each set being n, the probability of success being the same in all trials belonging to the same set. p_j is the probability of success in a trial belonging to the j-th set. The generating function of the Lexian distribution is given by

$$G(u) = \frac{1}{k} \sum_j (1 + p_j u)^n. \tag{22.1}$$

The mean and variance of the number of successes x is given by

$$\mathscr{E}\{x\} = n \sum \frac{p_j}{k} = np, \tag{22.2}$$

$$\left. \begin{aligned} \sigma_L^2 &= npq + n(n-1) \sum \frac{(p_j - p)^2}{k} \\ &= \sigma_B^2 + n(n-1)\sigma_P^2. \end{aligned} \right\} \tag{22.3}$$

Thus the variance exceeds that of the Bernoullian variance.

23. Coolidge's extension of Lexian scheme. This distribution arises from Poisson's binomial sets (instead of Bernoullian) each with different set of probabilities in its constituent n trials. If p_{ij} is the probability of success in the j-th trial of the i-th set, then the moment generating function of the distribution is

$$G(u) = k^{-1} \sum_i \pi_j (1 + p_{ij} u). \tag{23.1}$$

If we put $\sum_j p_{ij} = p_{i0}$ and $\sum_i p_{i0} = kp$, the mean of the distribution is obviously np and also

$$\sigma_C^2 = npq + n(n-1) \sum_i \frac{(p_{i0} - p)^2}{k} - \sum_i \sum_j \frac{(p_{ij} - p_{i0})^2}{k}. \tag{23.2}$$

The three terms may be regarded as belonging to the Bernoullian, Lexian and Poisson types respectively. Hence

$$\sigma_C^2 = \sigma_B^2 + n^2\,\sigma_{P_i}^2 - n\,\sigma_P^2. \tag{23.3}$$

24. Hypergeometric distribution. Consider a population of N individuals of which $M = Np$ are of one kind A. Let n drawings be made without replacement. It is required to find the probability distribution of x, the number of individuals of kind A drawn. The desired probability is

$$\varphi(x) = \binom{n}{x} M^{(x)}(N - M)^{(n-x)}/N^{(n)} \tag{24.1}$$

where $M^{(x)} = M(M - 1) \ldots (M - x + 1)$ and so on. $\varphi(x)$ is a typical term of the hypergeometric series. The generating function is the hypergeometric series

$$\frac{M^{(n)}}{N^{(n)}} F(-M, -n, N - M - n + 1; u). \tag{24.2}$$

From this the mean number is obtained as M_n/N and the variance as

$$\sigma^2 = M(M - 1)\, n\,(n - 1)/N(N - 1). \tag{24.3}$$

25. The Poisson distribution. Here the probability frequency of a random variable n is given by

$$\pi(n) = \frac{e^{-\lambda}\lambda^n}{n!}. \tag{25.1}$$

The generating function of the distribution is

$$G(u) = e^{\lambda(u-1)} \tag{25.2}$$

where λ is the parameter of the distribution. The partition function of this is given by

$$Q(\lambda) = e^{e^\lambda}. \tag{25.3}$$

The mean and variance are given by

$$\left.\begin{array}{c} \mathscr{E}\{n\} = \lambda, \\ \sigma_P^2 = \lambda. \end{array}\right\} \tag{25.4}$$

The Poisson distribution is obtained as the limiting case of a Binomial when $N \to \infty$ and p tends to 0 such that $Np \to \lambda$. There is however a direct way of arriving at the Poisson distribution in a stochastic process. This is dealt with in the chapter on stochastic processes (Sect. 53).

26. Frequency functions derivable from a Poisson distribution. Let the random variable be n and consider $P(n)$, the frequency function of n defined by

$$P(n) = \int_0^\infty e^{-x}\frac{x^n}{n!}\,dV(x) \tag{26.1}$$

where $V(x)$ is the distribution function for $x \geq 0$. Consael [6] has proved the very elegant result that, if

$$G(u) = \sum_n P(n)\,u^n \tag{26.2}$$

and

$$\varphi(t) = \int_0^\infty e^{itx}\,dV(x) \tag{26.3}$$

then G and φ are connected by the relation

$$G(u) = \varphi\left(\frac{u-1}{i}\right). \tag{26.4}$$

We shall now consider particular cases of $V(x)$.

1. $$dV(x) = dH(x - \lambda) = \delta(x - \lambda)\, dx \tag{26.5}$$

where H is the Heaviside unit function defined as usual. This yields the Poisson distribution with parameter λ.

2. $$dV(x) = e^{-a}\, a^x/x!, \quad a > 0, \quad x = 0, 1, 2, \ldots \tag{26.6}[1]$$

$$P(n) = \sum_{x=0}^{\infty} e^{-x}\,\frac{x^n}{n!}\, e^{-a}\,\frac{a^x}{x!}. \tag{26.7}$$

The characteristic function $\varphi(t)$ is given by

$$\varphi(t) = e^{a(e^{it}-1)}, \tag{26.8}$$

$$G(u) = e^{a(e^{u-1}-1)}, \tag{26.9}$$

$$\mathscr{E}\{n\} = a \quad \text{and} \quad \sigma^2 = 2a. \tag{26.10}$$

3. $$dV(x) = \binom{N+x-1}{x} p^x q^N \tag{26.11}$$

where
$$N > 0, \quad 0 < p < 1 \quad \text{and} \quad p + q = 1, \quad x = 0, 1, 2, \ldots.$$

In this case

$$\varphi(t) = \left(\frac{q}{1 - p\, e^{it}}\right)^N, \tag{26.12}$$

$$G(u) = \left(\frac{q}{1 - p\, e^{u-1}}\right)^N, \tag{26.13}$$

$$\mathscr{E}\{n\} = \frac{Np}{q}; \quad \sigma^2 = \frac{Np}{q^2}(q+1). \tag{26.14}$$

4. Polya-Eggenberger distribution.

$$dV(x) = \frac{1}{x!}\,\frac{\left(\frac{h}{d} + x - 1\right)!}{\left(\frac{h}{d} - 1\right)!}\,\frac{d^x}{(1 + d)^{\frac{h}{d} + x}} \tag{26.15}$$

where
$$d > 0, \quad h > 0 \quad \text{and} \quad x = 0, 1, 2, \ldots.$$

Therefore

$$\varphi(t) = \left[1 + d(1 - e^{it})\right]^{-\frac{h}{d}} \tag{26.16}$$

and

$$G(u) = \left[1 + d(1 - e^{u-1})\right]^{-\frac{h}{d}} \tag{26.17}$$

$$\mathscr{E}\{n\} = h, \quad \sigma^2 = h(d + 2). \tag{26.18}$$

---

[1] In accordance with our convention as elaborated in Sect. 8, we should write $dV(x')$ as
$$\sum_x e^{-a}\, a^{x'}/x'!\; \delta(x' - x)\, dx'$$
but we refrain from doing so to be in conformity with Consael's notation which is of course defective. Similar observations apply to other discrete distributions obtained using his method.

5. Pearson's distribution of type III.

$$dV(x) = \frac{1}{\left(\frac{1}{b}-1\right)!}\, e^{-\frac{x}{b}}\left(\frac{x}{b}\right)^{\frac{1}{b}-1}\frac{dx}{b} \qquad x>0,\ b>0, \tag{26.19}$$

$$\varphi(t) = (1-i\,b\,t)^{-\frac{1}{b}}, \tag{26.20}$$

$$G(u) = \{1-b(u-1)\}^{-\frac{1}{b}}, \tag{26.21}$$

$$\mathscr{E}\{n\} = 1, \qquad \sigma^2 = 1+b. \tag{26.22}$$

We shall now consider the case when instead of $V(x)$ we use the m-fold convolution $V_m(x)$ of $V(x)$. Then defining the corresponding frequency function $P_m(n)$ as

$$P_m(n) = \int\limits_0^\infty e^{-x}\frac{x^n}{n!}\,dV_m(x). \tag{26.23}$$

CONSAEL has shown that $P_m(n)$ is the m-fold convolution of $P(n)$. The generating function of $P_m(n)$ is $[G(u)]^m$. If we now define a new frequency function

$$Q(n) = \sum_{m=0}^\infty e^{-\lambda}\frac{\lambda^m}{m!}\,P_m(n) \tag{26.24}$$

then $F(u)$, the generating function of $Q(n)$ is given by

$$F(u) = e^{\lambda[G(u)-1]} = e^{\lambda\left[\varphi\left(\frac{u-1}{i}\right)-1\right]}. \tag{26.25}$$

As an example consider

$$dV(x) = e^{-\frac{x}{b}}\frac{dx}{b}, \tag{26.26}$$

$$G(u) = \frac{1}{b}\,\frac{1}{1+\frac{1}{b}-u}. \tag{26.27}$$

So

$$F(u) = \exp\left[\frac{\lambda}{1+(1-u)\,b}-\lambda\right], \tag{26.28}$$

$$Q(n) = \sum_{k=0}^\infty e^{-\lambda}\frac{\lambda^k}{k!}\,\frac{1}{(1+b)^k}\left(\frac{b}{1+b}\right)^n\frac{(n+k-1)!}{n!\,(k-1)!}. \tag{26.29}$$

b) Continuous distributions.

27. The normal distribution. The normal frequency function is

$$\varphi(x) = \frac{1}{\sigma\sqrt{2\pi}}\, e^{-\frac{1}{2}(x-\mu)^2/\sigma^2} \tag{27.1}$$

with μ as the mean and σ^2 as the variance. If the mean is taken as the origin, this becomes

$$\varphi(x) = \frac{1}{\sigma\sqrt{2\pi}}\, e^{-\frac{x^2}{2\sigma^2}}. \tag{27.2}$$

The distribution is symmetrical about the mean position with a peak there which becomes sharper and sharper for smaller and smaller values of σ. It is customary to describe it as normal $(0, \sigma)$. The moments are

$$\left.\begin{aligned}\mu_{2r+1} &= 0, \\ \mu_{2r} &= 1 \cdot 3 \ldots (2r - 1)\, \sigma^{2r}.\end{aligned}\right\} \tag{27.3}$$

28. The χ^2 distribution. Let $\xi_1, \xi_2, \ldots, \xi_n$ be n random variables each of which is normal $(0, \sigma)$ and consider the variable

$$\chi^2 = \sum_{r=1}^{n} \xi_r^2. \tag{28.1}$$

Now the frequency function of $\xi_r^2 = x$ is obviously $\dfrac{1}{\sigma \sqrt{2\pi x}}\, \mathrm{e}^{-\frac{x}{2\sigma^2}}$ for $x > 0$ and the corresponding characteristic function is $(1 - 2\sigma^2 i t)^{-\frac{1}{2}}$. Hence the characteristic function of χ^2 is $(1 - 2\sigma^2 i t)^{-n/2}$. The corresponding frequency function is

$$\left.\begin{aligned}k_n(x) &= \frac{1}{\sigma^n\, 2^{n/2}\, \Gamma(n/2)}\; x^{\frac{n}{2}-1}\, \mathrm{e}^{-\frac{x}{2\sigma^2}}, \qquad x > 0, \\ &= 0, \qquad\qquad\qquad\qquad\qquad\;\; x \leqq 0.\end{aligned}\right\} \tag{28.2}$$

The parameter n is often referred to as the number of degrees of freedom. Since there is a term with exponential dependence on x, the partition function of the distribution is given by

$$Q(\lambda) = Q\left(-\frac{1}{2\sigma^2}\right) = \sigma^n\, 2^{n/2}\, \Gamma\left(\frac{n}{2}\right) = \Gamma\left(\frac{n}{2}\right) (-\lambda)^{-n/2}. \tag{28.3}$$

The moments of x are given by

$$\left.\begin{aligned}\mu_1' &= n\,\sigma^2, & \mu_r' &= n(n+2) \ldots (n + 2r - 2)\, \sigma^{2r}, \\ \mu_2' &= n(n+2)\, \sigma^4, & \sigma_x^2 &= 2n\, \sigma^4.\end{aligned}\right\} \tag{28.4}$$

The sum of two independent variables, each having a χ^2-distribution also has a χ^2-distribution. In practical applications the probability of getting a value greater than a specified χ_0^2 is often required. Other useful distributions derivable from the χ^2 distribution are given below [each x_i normal $(0, \sigma)$].

Variable	Frequency function
$\displaystyle\sum_1^n x_r^2$	$\dfrac{1}{\sigma^2}\, k_n\!\left(\dfrac{x}{\sigma^2}\right)$
$\dfrac{1}{n}\displaystyle\sum_1^n x_r^2$	$\dfrac{n}{\sigma^2}\, k_n\!\left(\dfrac{n x}{\sigma^2}\right)$
$\sqrt{\displaystyle\sum_1^n x_r^2}$	$\dfrac{2x}{\sigma^2}\, k_n\!\left(\dfrac{x^2}{\sigma^2}\right)$
$\sqrt{\dfrac{1}{n}\displaystyle\sum_1^n x_r^2}$	$\dfrac{2n x}{\sigma^2}\, k_n\!\left(\dfrac{n x^2}{\sigma^2}\right)$

29. Student's t-distribution. Let us consider the case of the aggregate of $n+1$ random variables, ξ and ξ_i, $i = 1, 2, \ldots, n$ which are normal $(0, \sigma)$ all of them being independent. Define a variable $t = \dfrac{\xi}{\sqrt{\dfrac{1}{n}\displaystyle\sum_1^n \xi_r^2}}$. The frequency function $S_n(t)$ of t is called the STUDENT's t-distribution given by

$$S_n(t) = \frac{1}{\sqrt{n\pi}} \frac{\Gamma\left(\frac{n+1}{2}\right)}{\Gamma\left(\frac{n}{2}\right)} \left(1 + \frac{t^2}{n}\right)^{-\frac{n+1}{2}} \tag{29.2}$$

and is seen to be independent of the σ value of the component distribution. The r-th moment is finite provided $r<n$. The distribution is symmetrical about $t=0$.

$$\left. \begin{aligned} \sigma^2 &= \frac{n}{n-2}, \\ \mu'_{2r} &= \frac{1.\,3.\ldots(2r-1)\,n^r}{(n-2)\,(n-4)\ldots(n-2r)}. \end{aligned} \right\} \tag{29.3}$$

For large n the distribution is asymptotically normal $(0, 1)$.

30. FISHER's Z distribution. Let the random variables $\xi_1, \xi_2, \ldots, \xi_m$ and $\eta_1, \eta_2, \ldots, \eta_n$ be independent and normal $(0, \sigma)$. Putting

$$\xi = \sum_1^m \xi_r^2, \qquad \eta = \sum_1^n \eta_r^2 \tag{30.1}$$

the distribution of $k=\xi/\eta$ is shown to be

$$f_{m,n}(k) = \frac{\Gamma\left(\frac{m+n}{2}\right)}{\Gamma\left(\frac{m}{2}\right)\Gamma\left(\frac{n}{2}\right)} \frac{k^{\frac{m}{2}-1}}{(k+1)^{\frac{m+n}{2}}}. \tag{30.2}$$

This is independent of σ. If we define $e^{2Z} = \frac{n}{m} k$ the mean and variance of the variable e^{2Z} are

$$\mu'_1 = \frac{n}{n-2} \quad \text{for} \quad n>2 \quad \text{and} \quad \sigma^2 = \frac{2n^2(m+n-2)}{n(n-2)^2(n-4)} \quad n>4. \tag{30.3}$$

31. The Pearsonian systems. Some generalised frequency curves due to PEARSON are dealt with here. These curves are widely different in their shapes but have a common origin in the sense that they are solutions of a single differential equation.

Let us consider a frequency function $y(x)$ satisfying the differential equation

$$\frac{dy}{dx} = \frac{(x+a)\,y}{c_0 + c_1 x + c_2 x^2}. \tag{31.1}$$

We shall be concerned here only with those values of x for which y takes positive values. We list below several frequency functions that arise by taking particular values of c_1 and c_2.

$$\text{Type I:} \quad y = y_0 \left(1 + \frac{x}{a_1}\right)^{m_1} \left(1 - \frac{x}{a_2}\right)^{m_2} \quad \text{for} \quad -a_1 < x < a_2 \tag{31.2}$$

where $\dfrac{m_1}{a_1} = \dfrac{m_2}{a_2}$.

$$\text{Type II:} \quad y = y_0 \left(1 - \frac{x^2}{a^2}\right)^{m} \quad \text{for} \quad -a < x < a. \tag{31.3}$$

$$\text{Type III:} \quad y = y_0\, e^{-rx} \left(1 + \frac{x}{a}\right)^{ra} \quad \text{for} \quad x > -a. \tag{31.4}$$

Type IV: $\quad y = y_0 \left(1 + \dfrac{x^2}{a^2}\right)^{-m} e^{-r \arctan \frac{x}{a}}$. $\hfill (31.5)$

Type V: $\quad y = y_0\, x^{-p}\, e^{-r/x}$. $\hfill (31.6)$

Type VI: $\quad y = y_0 (x - a)^{q_2}\, x^{-q_1}$. $\hfill (31.7)$

Type VII: $\quad y = y_0\, e^{-x^2/2\sigma^2}$. $\hfill (31.8)$

This is the normal curve obtained by putting $c_1 = 0 = c_2$.

Type VIII: $\quad y = y_0 \left(1 + \dfrac{x}{a}\right)^{-m} \quad$ for $\quad -a < x < 0$. $\hfill (31.9)$

This degenerates into an equilateral hyperbola when $m = 1$.

Type IX: $\quad y = y_0 \left(1 + \dfrac{x}{a}\right)^{m} \quad$ for $\quad -a < x < 0$. $\hfill (31.10)$

This degenerates into a straight line when $m = 1$.

Type X: $\quad y = \dfrac{n}{\sigma}\, e^{\pm \frac{x}{\sigma}} \quad$ for $\quad -\infty < x < +\infty$. $\hfill (31.11)$

This type is LAPLACE's first frequency curve while the normal curve is called his second frequency curve.

Type XI: $\quad y = y_0^{x-m}$. $\hfill (31.12)$

Type XII: $\quad y = y_0 \left(\dfrac{a_1 + x}{a_2 - x}\right)^{p} \quad$ for $\quad -a_1 < x < a_2$. $\hfill (31.13)$

These curves have a wide range of applications as probability frequency curves.

V. Product densities, Janossy densities and the Characteristic Functional.

32. Preliminary remarks. Till now we were dealing with a system that could occupy any one of a discrete number of states, may be as large as we please or a continuous infinite set. More complex situations may arise in the description of the system of states. For example, let us take the problem of defining the statistical distribution of a discrete number of particles in a continuous infinite space characterised by the parameter E which may be one or many dimensional. If the E-space were discrete and the states represented by $E_1, E_2, \ldots$, the statistical distribution can be described by the number of particles in each state, say $\nu_1, \nu_2, \ldots$ respectively. Thus the aggregate $\langle \nu_i \rangle$ will represent a typical state of the entire particle system. (The E-space may also be called the system of states for each particle.) Therefore it is sometimes customary to call the ν's the occupation numbers. The word "state" has been used in such a general sense that caution has to be taken by examining the meaning in each particular context. In the case of discrete E, we can define the joint probability $\pi(\nu_1, \nu_2, \ldots; E_1, E_2, \ldots)$ that there are ν_1 particles in E_1, ν_2 particles in $E_2 \ldots$ and ν_n in E_n. This can also be called the joint distribution of the aggregate of stochastic variables ν_i. If the E-space were continuous, it is clear that no such π function can be defined since we have a continuous infinity of stochastic variables, corresponding to a continuous infinite number of points in any interval in the E-space. In such a case we resort to the following technique to deal with the statistical distribution of a discrete number in a continuous infinity of states.

33. Product densities. Let $M(E)$ represent the stochastic variable denoting the number of particles with parametric values less than E. Then $dM(E)$ represents the stochastic variable denoting the number of particles in the energy range dE. We shall assume that the probability that there occurs one particle in dE is proportional to dE while the probability that there occurs more than one particle, say n, is of the order $(dE)^n$ and hence is vanishingly small compared to the probability for the occurrence of one particle. Thus it is possible to define a function $f_1(E)$ such that

$$f_1(E)\, dE = \mathscr{E}\{dM(E)\} \tag{33.1}$$

where $\mathscr{E}\{dM(E)\}$ represents the average number of particles in dE. If we say that the probability that n particles occur in dE is $p(n)$ where n is zero or a positive integer then

$$
\begin{aligned}
p(1) &= f_1(E)\, dE + O(dE^2) = \mathscr{E}\{dM(E)\} + O(dE^2), \\
p(0) &= 1 - f_1(E)\, dE - O(dE^2), \\
p(n) &= O(dE^n), \quad n > 1.
\end{aligned}
\tag{33.2}
$$

If we define

$$\sum n^r\, p(n) = \mathscr{E}\{n^r\} \tag{33.3}$$

then we have

$$\mathscr{E}\{n^r\} = \mathscr{E}\{[dM(E)]^r\} = \mathscr{E}\{n\} = \mathscr{E}\{dM(E)\}. \tag{33.4}$$

Thus all the moments of the stochastic variable $dM(E)$ are equal to the probability that the stochastic variable assumes the value 1. But it must be noted that while $f_1(E)\, dE$ is a probability magnitude its integral over E yields only the mean number of particles in the range of integration as the integration does not correspond to the addition of infinitesimal probabilities relating to mutually exclusive events. Thus

$$\mathscr{E}\{M(E_u) - M(E_l)\} = \int_{E_l}^{E_u} \mathscr{E}\{dM(E)\} = \int_{E_l}^{E_u} f_1(E)\, dE, \quad E_u > E_l. \tag{33.5}$$

We now introduce the concept of product density. We form the product of the stochastic variables $dM(E_1)$, $dM(E_2)$ and accordingly define

$$f_2(E_1, E_2)\, dE_1\, dE_2 = \mathscr{E}\{dM(E_1)\}\, \mathscr{E}\{dM(E_2)\} \tag{33.6}$$

which is also equal to the joint probability that a particle lies in dE_1 and a particle lies in dE_2 where dE_1 and dE_2 do not overlap; when they do a degeneracy occurs and

$$\mathscr{E}\{[dM(E_1)]^2\} = \mathscr{E}\{dM(E_1)\} = f_1(E)\, dE. \tag{33.7}$$

We call f_2 a product density of degree 2. In view of the degeneracy it directly follows that

$$
\begin{aligned}
\mathscr{E}\{[M(E_u) - M(E_l)]^2\} &= \int_{E_l}^{E_u} \int_{E_l}^{E_u} \mathscr{E}\{dM(E_1)\, dM(E_2)\} \\
&= \int_{E_l}^{E_u} f_1(E)\, dE + \int_{E_l}^{E_u} \int_{E_l}^{E_u} f_2(E_1, E_2)\, dE_1\, dE_2.
\end{aligned}
\tag{33.8}
$$

Similarly we can define product densities of any higher order, say n, where

$$f_n(E_1, E_2, \dots, E_n)\, dE_1\, dE_2 \dots dE_n = \mathscr{E}\{dM(E_1)\, dM(E_2) \dots dM(E_n)\} \tag{33.9}$$

represents the joint probability that there lies a particle in dE_1, one in dE_2, one in dE_n provided the intervals dE_1, dE_2, ..., dE_n do not overlap. It should be noted that in the above expression, no two of the dE_i's should overlap for if $dE_n = dE_{n-1}$ then

$$\mathscr{E}\left\{dM(E_1)\,dM(E_2)\ldots[dM(E_{n-1})]^2\right\} = \mathscr{E}\left\{dM(E_1)\,dM(E_2)\ldots dM(E_{n-1})\right\} \atop = f_{n-1}(E_1, E_2, \ldots, E_{n-1})\,dE_1\,dE_2\ldots dE_{n-1}. \qquad (33.10)$$

This consideration is of importance when we are concerned with the n-th order moments of $M(E_u) - M(E_l)$. Before proceeding to this problem we shall consider special cases of f_n which yield results useful for the problem of determining higher moments of $M(E_u) - M(E_l)$. Consider the case of N particles distributed in a continuum represented by the E-space when N is a constant and not a stochastic variable. But the N particles are distributed in the E-space according to probability laws. We can then write in this case

$$f_1(E)\,dE = N f_1^0(E)\,dE \qquad \left(\int\limits_{\text{whole range}} f_1^0(E)\,dE = 1\right) \qquad (33.11)$$

where $f_1^0(E)\,dE$ represents the probability that any one of the N particles selected at random lies in dE. By a simple argument it follows that

$$f_2(E_1, E_2)\,dE_1\,dE_2 = N(N-1)\,f_1^0(E_1)\,f_1^0(E_2)\,dE_1\,dE_2, \atop f_N(E_1, E_2, \ldots, E_N)\,dE_1\,dE_2\ldots dE_N = N!\,f_1^0(E_1)\,f_1^0(E_2)\ldots f_1^0(E_N)\,dE_1\,dE_2\ldots dE_N. \quad (33.12)$$

If we take a finite fragment of the E-space, say ΔE, and write

$$\int\limits_{\Delta E} f_1^0(E)\,dE = P_{\Delta E} \qquad (33.13)$$

we get the mean number of particles in ΔE as $NP_{\Delta E}$ and the mean square deviation of the number of particles in ΔE as $NP_{\Delta E}(1 - P_{\Delta E})$ in accordance with the well-known binomial distribution law. It is quite clear from our definitions that in this case product densities of order greater than N vanish and when $N = 1$, the density of degree 1 coincides with the ordinary concept of probability density. Now we are in a position to calculate the general (say r-th) moment of the number of particles in any finite range $\Delta E = E_u - E_l$ represented by

$$\mathscr{E}\left\{[M(E_u) - M(E_l)]^r\right\} = \mathscr{E}\left\{M_{\Delta E}^r\right\}. \qquad (33.14)$$

In view of the degeneracies mentioned we can write the r-th moment as

$$\sum_{s=1}^{r} C_s^r \int\limits_{E_l}^{E_u}\int\limits_{E_l}^{E_u}\ldots\int\limits_{E_l}^{E_u} f_s(E_1, E_2, \ldots, E_s)\,dE_1\,dE_2\ldots dE_s. \qquad (33.15)$$

The coefficients C_s^r are functions of r and s and in no way dependent on the function f_s. To calculate C_s^r we take the particular case of the distribution of N particles in the E-space, substitute the values for the various product densities in terms of $f_1^0(E)$ and integrate over the whole space. Then we get

$$N^r = \sum_{s=1}^{r} C_s^r N(N-1)\ldots(N-s+1). \qquad (33.16)$$

The coefficients are now obtained by taking $N = 1, 2, \ldots$. This yields

$$C_r^r = C_1^r = 1; \qquad C_2^r = \tfrac{1}{2}(2^r - 2). \qquad (33.17)$$

Thus we have obtained the r-th moment of the stochastic variable as the sum of the integrals of the product densities of order equal to and less than r. Now if we consider the case where

$$f_r(E_1, E_2, \ldots, E_r) = f_1(E_1)\, f_1(E_2) \ldots f_1(E_r) \tag{33.18}$$

i.e. where there is no correlation between the particles in two different energy ranges the r-th moment of the number of particles in a finite range ΔE is

$$\mathscr{E}\{N_{\Delta E}^r\} = C_1^r\, \mathscr{E}\{N_{\Delta E}\} + C_2^r[\mathscr{E}\{N_{\Delta E}\}]^2 + \cdots + C_r^r[\mathscr{E}\{N_{\Delta E}\}]^r \tag{33.19}$$

which can be shown to coincide with the moments of a variable with a Poisson distribution.

34. Janossy densities. To deal with the problem of distribution of a discrete number of particles in a continuous infinity of states, JANOSSY [7] introduced the following function which bears a close relationship to the product density function.

$$J_N(E_1, E_2, \ldots, E_N)\, dE_1\, dE_2 \ldots dE_N \tag{34.1}$$

represents the joint probability that there exists one particle in dE_1, one in $dE_2, \ldots$ and one in dE_N and *none elsewhere*. It is to be noted that

$$\sum_N \frac{1}{N!} \int_{E_1} \int_{E_2} \cdots \int_{E_N} J_N(E_1, E_2, \ldots, E_N)\, dE_1\, dE_2 \ldots dE_N = 1. \tag{34.2}$$

The $N!$ arises since we are integrating the functions with respect to all variables over the whole range and hence the event representing the N particles being found in the corresponding N intervals is counted $N!$ times. In view of this correcting factor we cannot call J_N a probability density though $J_N(E_1, E_2, \ldots, E_n)\, dE_1\, dE_2 \ldots dE_N$ is a probability magnitude. The product density $f_N(E_1, E_2, \ldots, E_n)$ is related to Janossy density in the following manner:

$$\begin{aligned}
f_n(E_1, &E_2, \ldots, E_n) \\
&= \sum_{N=n}^{\infty} \frac{1}{(N-n)!} \int_{E_{n+1}} \int_{E_{n+2}} \cdots \int_{E_N} J_N(E_1, E_2, \ldots, E_N)\, dE_{n+1} dE_{n+2} \ldots dE_N.
\end{aligned} \tag{34.3}$$

It may seem at first sight that in problems involving the distribution of particles in a continuous infinity of states the Janossy densities should suffice. However in practice the product densities are found to be more useful and easy to work with since their integrals are directly connected with factorial moments.

35. The characteristic functional. In a previous chapter, we defined the characteristic function corresponding to the frequency function of a random variable or an aggregate of a finite number of random variables. It is obvious that in the case of a continuous infinite set of random variables a suitable extension can be made. If the stochastic variable $N(E)$ represents the number of particles with energy below E, then symbolically $dN(E)$ is a stochastic variable which represents the number in dE. Then if

$$C[\vartheta(E)] = \mathscr{E}\left\{ e^{+i \int_E \vartheta(E)\, dN(E)} \right\} \tag{35.1}$$

where $\vartheta(E)$ is an arbitrarily chosen function of E, $C[\vartheta(E)]$ is called the characteristic functional and $\vartheta(E)$ the argument. "The characteristic functional when it can be determined contains in portmanteau form all the probability relationships associated with the state of the 'population' of particles in the energy space."

We have here used the symbol $dN(E)$ for the stochastic variable representing the number of particles in dE. This can assume a value n with probability $O(dE^n)$ and hence only $n=1$ will contribute to the integral. This is precisely the basis of the theory of product densities. The probability that $dN(E)$ assumes the value 1 is infinitesimal i.e. proportional to dE. Thus in the above class of problems, we deal with the composition of a continuous infinity of stochastic variables which can assume finite values (here unity) with infinitesimal probabilities. Later in the theory of stochastic integration we will be faced with the composition of a continuous infinity of stochastic variables which can assume infinitesimal values with finite probabilities. In such a case we also use the notation $d[x(t)]$ to denote a random variable but this can assume only infinitesimal values (proportional to dt) with finite probability. Hence caution has to be exercised in the use of such symbolic integrals[1].

VI. Combinatorial analysis.

We shall now describe one of the old and well known methods for deriving the distribution functions of a discrete random variable capable of assuming non-negative integral values, $0, 1, 2, \ldots$. This is the method of combinatorial analysis and one of its most important applications is the derivation of the Maxwell-Boltzmann, Fermi-Dirac and Bose-Einstein statistics in statistical mechanics. We shall first describe the general principle on which the derivations of the distributions from combinatorial analysis are based.

Let the outcome of an experiment be one of an enumerable set of events $\langle E_i \rangle$ ($i = 1, 2, \ldots$) and the probability of the event E_i be p_i. If we are interested in a random variable x, we assume that the value of the random variable x is the result of an experiment which is composed of the fundamental experiments whose outcome is to be one of the E_i's. The distribution of the random variable is thus given in terms of the p_i's. In many cases when the number of the fundamental events is finite, equal to N (say), equal probabilities may be assumed for these events on intuitive or apriori grounds. If the events are assumed to be the states of a physical system, this method is also known as the method of enumeration of states, the meaning of which will be clear when we consider the following examples.

36. Some elementary results of combinatorial analysis. (i) The number of modes in which N distinguishable objects can be arranged in order is $N!$. Hence the probability of realising a particular arrangement i.e. permutation is $1/N!$.

(ii) The number of different modes in which M objects may be selected from N distinguishable objects irrespective of the order of choice is

$$\binom{N}{M} = \frac{N!}{M!\,(N-M)!} \, . \tag{36.1}$$

If we identify the N objects with N trials and the M selected objects with M successes, the probability that we have M successes in N trials is

$$P(M, N) = \frac{1}{2^N} \binom{N}{M}; \quad \sum_{M=0}^{N} P(M, N) = 1 \, . \tag{36.2}$$

---

[1] Another interpretation can be given to the characteristic functional as follows. If the random points (particles) lie at $E_1, E_2, \ldots$, we can define a random variable $y = \sum_i \vartheta(E_i)$ associated with the random distribution of points. If $\varphi(t)$ be the Fourier transform of the requency function of y we note that $\varphi(1) = C[\vartheta(E)]$.

The random variable here is M. The normalising factor $1/2^N$ implies equal probability for success and failure in a single trial.

(iii) The multinomial coefficient, the coefficient of $\prod_i x_i^{m_i}$ in $(\sum x_i)^N$ is

$$\frac{N!}{\prod_i m_i!}. \tag{36.3}$$

(iv) The number of modes in which N objects may be placed in p piles, m_1 in the first pile, m_2 in the second, ... m_p in the p-th pile, with $\sum_1^p m_i = N$ is

$$\frac{N!}{\prod_i m_i!}. \tag{36.4}$$

The probability of realising a typical complexion $\langle m_i \rangle$ is

$$\frac{1}{p^N} \frac{N!}{\prod_i m_i!} = P(\langle m_i \rangle) \tag{36.5}$$

if for each mode we assume equal probability $1/p$. The total number of ways for all complexions is p^N. We have here an aggregate of random variables $\langle m_i \rangle$.

(v) The number of ways in which N distinguishable objects may be assorted into piles with m_1 piles of one object each, m_2 piles of two objects each, m_k piles of k objects each, such that

$$\sum k m_k = N, \qquad \sum m_k = p$$

is given by

$$\frac{N!}{\prod_k m_k! \, (k!)^{m_k}}. \tag{36.6}$$

The probability for such a complexion is obtained if we multiply the above by $1/p^N$.

(vi) The number of modes in which N distinguishable objects may be placed in C numbered boxes with no restrictions on the number per box is C^N. The probability of a typical mode is $1/C^N$.

(vii) The number of modes in which N distinguishable objects can be placed in C numbered boxes with no more than one object each to a box $[N \leq C]$ is $C!/(C-N)!$. The entire aggregate or modes depends on the particular problem, e.g. if we assume that N can take values 1 to C then the total number of modes is clearly

$$\sum_{N=1}^{C} \frac{C!}{(C-N)!} = T. \tag{36.7}$$

Assuming equal probability for each of these modes, the probability of realising the distribution of N particles in C numbered boxes is

$$\frac{1}{T} \frac{C!}{(C-N)!}. \tag{36.8}$$

This may be assumed to be the frequency function of the random variable N.

(viii) The number of modes in which N indistinguishable objects can be placed in C numbered boxes no more than one in a box $[N \leq C]$ is $\dfrac{C!}{(C-N)!\,N!}$.

(ix) The number of modes in which N indistinguishable objects can be placed in C numbered boxes with no restriction on the number per box is $\dfrac{(C+N-1)!}{(C-1)!\,N!}$.

In cases (viii) and (ix) if we assume N can be varied, we obtain the probabilities for the random variable N by dividing the expressions by the corresponding normalising factors as in case (vii). We thus find that combinatorial analysis helps us to arrive at the frequency functions of a random variable provided we assume some method of realisation of this random variable, i.e. the basic set of events for which we assign equal or known probabilities. That there are other ways of realising the random variable is obvious.

In case (ix), a little explanation may be necessary Let the C boxes and N objects be arranged in a row such that the objects belonging to any box are placed to the right of that box. Then one box should be placed at the beginning of the row and the different arrangements are obtained by rearranging the $(C+N-1)$ members (boxes and objects) of the row. There are $(C+N-1)!$ ways of doing it of which any permutation of the $C-1$ boxes within themselves in $(C-1)!$ ways produce no new complexion and the permutation of N objects in $N!$ ways also does not give new complexions. Hence, they should be divided out of $(C+N-1)!$ leading to the result $\dfrac{(C+N-1)!}{(C-1)!\,N!}$.

It is worthwhile considering the following example which illustrates how a probability distribution of a random variable depends entirely upon the definition of the basic set of events, a suitable composition of which yields the value of the random variable. Let us be interested in the probability that a random variable x assumes the value N, N being a positive integer. The determination will depend upon how we conceive the random variable is realised. For example, we can think of two ways in which it is realised.

(A) Let this number be the sum of k quantities $q_1, q_2, \dots q_k$. We assume that each of the q's is an integer between 1 and N, with equal probability $1/N$, i.e. each is the result of an experiment the outcome of which is the number between 1 and N with equal probability. If we now take a sequence of k experiments we will get the sequence of k numbers, the sum of which is the value of our random variable x. The number of possible sequences is clearly N^k and to find out the probability frequency function of x, we merely enumerate the number of sequences v_x that give the number x. Then v_x/N_k is the probability of getting x. In this case it is clear that a permutation of a sequence gives a new sequence and is taken into account in the enumeration.

(B) On the other hand, we can conceive of our random variable being the sum of k integers, the probability of any typical sequence being equal to any other but two sequences being considered identical if one is the permutation of the other. Thus the total number of sequences is not N^k but less and v_x will also be different. This way of obtaining the probabilities of the number is as natural as the previous one. Thus in the method of enumeration of states the computation of probabilities depends upon the assignment of apriori probabilities to the basic set of states.

A simple example illustrates the above point. In the case when x is a sum of three integers being 1, 2 or 3, the number of possible sequences is 27. In case A, the probability of getting $x=3$ is $\frac{1}{27}$ since there is only one mode of realising it. In case B, the number of modes is nine and the probability of getting $q_1=q_2=q_3=1$ is $\frac{1}{9}$.

37. The three fundamental distributions in statistical mechanics. The most important application of combinatorial analysis is the derivation of the three well known distributions in statistical mechanics.

α) Boltzmann distribution. We shall assume that we have energy cells, each denoted by energies $E_1, E_2, \dots, E_j$. We distribute the N particles in these cells and impose the additional condition that

$$\sum_j N_j E_j = E_0 = N\overline{E}$$

where E_0 is the total energy of the N particles. We now assume that each cell E_j consists of C_j sub-cells. Hence the total number of sub-cells is $\sum_j C_j$. We consider the distribution of N particles in $\sum_j C_j$ sub-cells. The number of ways in which N particles can be divided into groups of N_j each, such that $\sum_j N_j = N$ is $\dfrac{N!}{\prod_j N_j!}$; since the particles are distinguishable and the number of ways in which N_j independent particles can be arranged in C_j cells is $C_j^{N_j}$ the number of ways consistent with the distribution is given by

$$\Omega_D = N! \prod_j \frac{C_j^{N_j}}{N_j!}. \tag{37.1}$$

Let us now consider the asymptotic case of physical interest when N and C_j are very large. We impose the conditions that

$$N = \sum_j N_j, \qquad E_0 = \sum_j N_j E_j, \tag{37.2}$$

$$\log \Omega_D = \log N! + \sum_j [N_j \log C_j - N_j \log N_j + N_j]. \tag{37.3}$$

Omitting the first term, we have to calculate those values of N_j for which Ω_D is a maximum subject to the conditions above. Using the usual Lagrangian method, we obtain

$$\frac{\partial}{\partial N_j}\left[\log \Omega_D - \alpha \sum_j N_j - \beta \sum_j E_j N_j\right] = 0 \tag{37.4}$$

where α and β are constants characteristic of the constraint. This yields

$$\log C_j - \log N_j - \alpha - \beta E_j = 0; \qquad N_j = C_j e^{-\alpha - \beta E_j}, \tag{37.5}$$

which is the Boltzmann distribution.

β) *Bose-Einstein distribution.* In this case, N_j identical particles have to be distributed in the C_j subcells without restricting the number in each sub-cell. The number of ways in which this can be done has been shown to be

$$\frac{(C_j + N_j - 1)!}{(C_j - 1)! N_j!}. \tag{37.6}$$

The total number of states is the product of this expression for all energies.

$$\left.\begin{aligned}
\Omega_D &= \prod_j \frac{(C_j + N_j - 1)!}{(C_j - 1)! N_j!}, \\
\log \Omega_D &= \sum_j \left[C_j \log\left(1 + \frac{N_j}{C_j}\right) + N_j \log\left(1 + \frac{C_j}{N_j}\right)\right].
\end{aligned}\right\} \tag{37.7}$$

Applying the restrictive conditions (37.2), using the Lagrangian method and maximising $\log \Omega_D$, we get

$$\log(C_j + N_j) - \log N_j - \alpha - \beta E_j = 0; \qquad \frac{N_j}{C_j} = \frac{1}{e^{\alpha + \beta E_j} - 1}. \tag{37.8}$$

This is the Bose-Einstein distribution.

γ) Fermi-Dirac distribution. If we require that there cannot be more than one particle in a cell, N_j must be equal to or less than C_j. The number of ways in which N_j indistinguishable particles can be distributed in C_j cells, not more than one in a cell, has to be calculated. This amounts to a calculation of the number of ways in which C_j cells can be divided into two groups of N_j occupied ones and $(C_j - N_j)$ empty ones. This number is given by

$$\frac{C_j!}{N_j!(C_j - N_j)!} \tag{37.9}$$

and the total number of states is the product of this expression over all energies.

$$\left. \begin{aligned} \Omega_D &= \prod_j \frac{C_j!}{(C_j - N_j)!\, N_j!}\,; \\ \log \Omega_D &= \sum_j \left[-C_j \log\left(1 - \frac{N_j}{C_j}\right) + N_j \log\left(\frac{C_j}{N_j} - 1\right) \right]. \end{aligned} \right\} \tag{37.10}$$

Maximising $\log \Omega_D$ under the restrictive conditions (37.2) we obtain

$$\frac{N_j}{C_j} = \frac{1}{e^{\alpha + \beta E_j} + 1} \tag{37.11}$$

which is the Fermi-Dirac distribution.

It is easily seen that both (37.8) and (37.11) reduce to the Boltzmann distribution if e^α is much greater than unity.

In the above derivations, we have to concede that the procedure for deriving the Maxwell-Boltzmann distribution is slightly different from that adopted in the other cases. For the Maxwell-Boltzmann distribution we first assume the totality of N particles and the distribution of N_j's was conditional upon the totality. Such a restriction was not placed in deriving the number of ways corresponding to the N_j distribution for the Bose-Einstein and Fermi-Dirac particles. This is anyhow imposed as a constraint along with the total energy restriction later when the method of Lagrangian multipliers is used. It may seem paradoxical that if we do not make any assumption upon the total number there is no way of arriving at the term $C_j^{N_j}/N_j!$ in considering the Maxwell-Boltzmann distribution since the $N_j!$ comes from the multinomial coefficient $\dfrac{N!}{\prod_j N_j!}$ and not from indistinguishability arguments as in the case of the Bose-Einstein and Fermi-Dirac particles.

Some attempts have been made to give a unified method of deriving the distribution functions and one of them is the method of Brillouin statistics. The author has however devised a simple way of arriving at these functions from a composition of basic distributions. The three distributions are obtained by making three different assumptions regarding the apriori distribution of particles in the sub-cells when there is no constraint on the total number and total energy.

1. In any sub-cell the probability that there are k particles is given by a Poisson distribution

$$e^{-\lambda}\frac{\lambda^k}{k!} \tag{37.12}$$

λ being an arbitrary parameter.

This is true for any sub-cell belonging to any cell. Thus in the cell j consisting of C_j sub-cells, the probability there are N_j particles is obtained by composing C_j Poisson distributions as

$$\frac{e^{-\lambda_j C_j}(\lambda_j C_j)^{N_j}}{N!} \tag{37.13}$$

and so the joint probability distribution of the complexion $\langle N_j \rangle$ is proportional to

$$\prod_j \frac{C_j^{N_j}}{N_j!} \quad (j = 1, 2, \ldots) \tag{37.14}$$

which is the actual result aimed at in the Boltzmann distribution.

2. We assume that in each sub-cell we have 0 or 1 with probability $(1 - \lambda)$ and λ respectively. The probability that we have N_j particles in C_j sub-cells is given by

$$\frac{C_j! \, \lambda^{N_j} (1 - \lambda)^{C_j - N_j}}{N_j! \, (C_j - N_j)!} \tag{37.15}$$

and the joint probability distribution of the complexion $\langle N_j \rangle$ is proportional to

$$\prod_j \frac{C_j!}{N_j! \, (C_j - N_j)!} \tag{37.16}$$

which gives the Fermi-Dirac distribution.

3. In the case of Bose-Einstein statistics a slightly novel assumption is to be made. In any sub-cell there is equal probability of there being any number of particles. At first sight, it may seem that if this probability is non-zero, the total probability of getting any number is infinite. However dealing with *unnormalised* probabilities we can assume the generating function for this distribution in the sub-cells as $\frac{1}{1-u}$. The corresponding generating function for the distribution in the cell j is $G_j(u) = (1-u)^{-C_j}$. Note if $u = 1$ this is infinite, though for normalised probabilities this should be equal to 1. The corresponding unnormalised probability for having N_j particles in C_j sub-cells is the coefficient of u^{N_j} in $(1-u)^{-C_j}$. This is

$$\frac{(N_j + C_j - 1)!}{N_j! \, (C_j - 1)!} \tag{37.17}$$

which is just the number of ways obtained for a single cell by the standard arguments in case of Bose-Einstein statistics.

B. Stochastic processes.

I. A physical approach to stochastic processes.

38. Introduction. Till now, we were concerned with a random variable or an aggregate of a finite number of random variables and their distribution functions. It is clear that in physics, the distribution function of the random variable will be determined by the physical conditions characteristic of the problem. We shall now see how the notion of sequences of random variables directly leads us to the concept of random functions and stochastic processes.

In many physical problems, we deal with "processes"—a convenient term to denote the variation of the state of a system with respect to certain parameters giving the most general interpretation to the words "state", "systems" and "parameters". Dynamics deals with such processes and the quantities, the magnitudes of which are studied, are known as the dynamical variables. In classical dynamics, these variables, say x_i $(i = 1, 2, \ldots)$, are supposed to be functions of a certain parameter and if that parameter is one-dimensional (like time) and can be ordered, we say the process "evolves" with time and the dynamical variables are deterministic functions of time and certain other parameters which enter into the physical problem. Since the process is said to "evolve" only if t

varies while the other parameters are unaltered, we speak of the dynamical variable as a function of t, the parameter with respect to which the process evolves. Without any reference to the other variables, t can in general be many-dimensional.

We can now see how probability theory can enter into dynamics. Instead of the dynamical variables x_i being deterministic functions of t, they can be conceived to be random functions of t. In simpler terms this means that we cannot say that these variables assume definite values but can only speak of their probability distributions. Considering for simplicity a one-dimensional parameter t, with each value of t we can associate an aggregate of random variables $\langle x_i(t) \rangle$ characteristic of the dynamical system and we can speak of the joint distribution of the aggregate. Since t can vary we have a family of aggregates corresponding to all the points in any interval of t. If t is continuous, we have a continuous infinite set of aggregates of random variables and, if t is discrete, we have a finite or enumerable set of aggregates. In order to avoid confusion between the aggregate of variables corresponding to one value of t and the family generated by varying t we shall assume that at one value of t we have only a single random variable $x(t)$. The extension to an aggregate $\langle x_i(t) \rangle$ corresponding to each t is immediate. *If t is one dimensional, this set of random variables corresponding to various values of t can be arranged as a sequence and when the study of the random variables at t can be made without any reference to these variables corresponding to $t' > t$ we call the process evolutionary.* If t is many-dimensional, ordering of t is not possible and so we have to consider the entire set $\langle x(t) \rangle$ in toto in any interval of the many dimensional parameter t. *This aggregate characterises a stochastic process.* While it is possible to define a stochastic process in abstract mathematical language as we shall do later in terms of families of random variables, we shall in conformity with the spirit of the article deal first with the physical approach in the case of a stochastic process evolving with respect to t[1].

39. General phenomenological description. We shall now describe, from a purely phenomenological point of view, the principal features of a stochastic process in comparison with a deterministic dynamical system progressing with respect to a one-dimensional parameter t. A deterministic dynamical system is assumed to be capable of existing at any instant t in any one of an aggregate of states called, for obvious reasons, *occupation states* which may be finite, or enumerably infinite or continuous infinite in number. The state of the system at t is defined by specifying the occupation state in which the system is at that time. If we assume t to be a continuous parameter, we wish to study the variation of the state of the system with t. Since we have postulated a deterministic system, the state of the system at t is a function of t. If an occupation state is defined by a number, this number is a function of t and if we change t to $t + \Delta$, the number must change by an infinitesimal quantity proportional to Δ at almost all t except at a finite or enumerably infinite number of points. If the set of states were discrete and t continuous, we shall show that either the system can change only at a finite number of points in a finite interval, the system remaining static between any two such changes, or we are forced to the conclusion that the system changes an infinite number of times in an infinitesimal interval. The first possibility is trivial while the second we exclude due to physical considerations. Thus for a deterministic system if we ignore the first possibility, a discrete set of states is incompatible with continuous t. This can be illustrated by a simple example. If we have only two possible occupation states defined by two colours "blue"

---

[1] When t is discrete, the aggregate is sometimes referred to in literature as a *chain*.

and "red", is it possible to conceive of the colour as a function of t? For, if at t we have "blue" then at $t+\Delta$ if there should be a change, it should be to "red" and thus we are led to the conclusion that in a finite period of time which can be made as small as we please there will be infinite changes, not taking into account that we get into insoluble conceptual difficulties in dividing Δ into sub-intervals. The only possibility therefore left for us is that any interval $(0, t)$ is to be divided into finite intervals and in each interval the state of the system continues to be the same. This would mean that except at a finite or enumerable number of points the system does not change.

The normal problem of dynamics stated in the most general terms is to study the variation of the dynamical system, in a finite interval if we know the mode of variation in an infinitesimal interval. This is expressible in the form of differential equations. As soon as we write a differential equation with differential coefficients referring to the variation of a quantity defining the occupation states with t, we have conceded that the state at $t+\Delta$ is completely defined by the state at t and the mode of transition is defined by the physical conditions of the problem. If we are given the state at $t=0$ and we wish to obtain the state at t, we integrate the differential equation.

In a stochastic process, at t we cannot say whether the system is in any particular occupation state. The system can occupy any one of the occupation states with different probabilities and we therefore speak of the probability that the system is in a state S at t. This is defined by the probability frequency function at t. If an observation were made it will be found in one of the occupation states which we call the realised state at t. The most natural question is to ask why at any t we can only speak of the probability that the system is in any state. There is one obvious way in which such randomness occurs. We have stated that the evolution of the system evolving in a deterministic manner with respect to t may depend on certain parameters—for example the initial conditions of the problem. Given these parameters, the state of the system is a deterministic function of time. But if these parameters are treated as random variables then the state of the system at any t is not a deterministic function of t and we can only speak of the probability that the system is in a particular state at t. To obtain this probability distribution we first fix the parameters and express the state of the system as a function of t and the parameters. Then treating the parameters as random variables we should ask for the probability distribution in a particular state. This amounts obviously to the problem of transformation of variables for which no general method can be devised. In other words in many cases this is equivalent to solving partial differential equations with random parameters. A reference to this class of problems is made later in this article.

We shall turn our attention to cases where the randomness is not just due to some parameters which do not depend on t, but when the randomness enters at every stage of the evolution of the system with t. To this purpose we must study the variation of the system with t. The probability distribution in the various occupation states of course changes with t and one of the objects of the theory of stochastic processes is to predict the probability distribution in the occupation states at t given the initial distribution at $t=0$. A more detailed study can be made by determining the joint probability distribution of the system in the occupation states at a finite number of points on the t-axis, the number being made as large as we please. It is obvious that it is almost an intractable problem (except in some simple cases) to obtain the joint distribution at all points in the interval $(0, t)$ since they form a non-enumerable set. However if we plot in any particular experiment the realised state of the system at τ as τ varies from 0 to t

we obtain the *realised curve of the trajectory of the stochastic process*. It is difficult to ascribe a measure to this trajectory except in some simple cases since we meet with the pathological problem of the joint distribution at a continuous infinite number of points.

Turning to the simpler task of determining the probability distribution at a particular t, by analogy with the dynamical system we should know the variation of this distribution for an infinitesimal change of t. Here we meet with a difficulty which has no parallel in deterministic systems. The change in the probability frequency function (p.f.f.) of the system as t varies from t to $t+\Delta$ need not be completely specified by the state at t. This complication may be explained in clear terms if we state that at $t=0$, the system has been observed to be in a particular state. So our object is to determine the p.f.f. at t. In many cases the specification of initial conditions may not be sufficient to derive this p.f.f. as the development of the process between 0 and t may depend upon the "history" of the process prior to $t=0$, i.e. we may need information about the realised states of the system prior to $t=0$. In other words if a system was observed to be, say, in the state S' at $t=-a$ and S at $t=0$, it may evolve in a different manner from a system which was observed to be in S'' at $t=-a$ and S at $t=0$. If the development of a process from $t=0$ is independent of its history prior to $t=0$, we call it *Markovian*. All other processes are non-Markovian and it is clear that all types of non-Markovian processes cannot be enumerated and they can only be conceived of as a residuary class. The distinction between Markovian and non-Markovian processes does not arise in deterministic systems since in those cases the occupation state in which the system lies is completely determined by t.

Throughout this article we shall confine ourselves to Markovian processes. This does not mean that non-Markovian processes are completely neglected. From the previous discussion it is clear that the p.f.f. of the system refers to the occupation states. By redefining a proper set of occupation states a process which is non-Markovian when we consider the first set, may be viewed to be Markovian if we consider the system to exist in the suitably defined set of occupation states. This technique is well known and cannot be cast into any general form but is easily illustrated by taking particular examples.

In a Markovian process, we are interested in determining $\pi(S\,|\,S_0;\,t,\,t_0)$ the probability that the system is in S at t given that it was in S_0 at $t_0\,(t_0<t)$. If in addition we assume the process to be homogeneous in t, π is a function only of $t-t_0$ and we can set $t_0=0$ and write $\pi(S\,|\,S_0;\,t)$ instead of $\pi(S\,|\,S_0;\,t,\,t_0)$. To determine π, our first task is to find out what happens in an infinitesimal interval to a system which is known to exist at S_0 at $t=0$. This amounts to knowing $\lim \pi(S\,|\,S_0;\,t)$ as $t\to\Delta\to0$. If we make an observation at $t=\Delta$, the system will be found to be in any one of the many possible states. This transition from S_0 to S occurs with probability $\lim \pi(S\,|\,S_0;\,t)$ as $t\to\Delta\to0$. The knowledge of this limit gives us a complete picture of what happens in the infinitesimal interval and it is therefore possible to express $\pi(S\,|\,S_0;\,t)$ for finite t in terms of this limit. This is expressed in the form of a differential equation with respect to t. If the occupation states are discrete in number, summations may occur with respect to them, and if they are continuous infinite in number and describable with the aid of variables in a continuous domain, integrals with respect to the variables will occur. This equation is known as the *forward differential equation* of the process.

Let us now consider the nature of $\lim \pi(S'\,|\,S;\,t)$ as $t\to\Delta\to0$ for two cases,

    (i) when we have a discrete set of occupation states,

(ii) when the set of occupation states is continuous infinite.

(i) If S belongs to a discrete set say S_1, S_2, S_3, ... we have ruled out by phenomenological arguments any deterministic change from any S to any other state in the interval Δ. On the other hand the system may jump from S_k to S_j in the interval with probability proportional to Δ. This condition of proportionality to Δ "safeguards" the process against infinite changes in a finite interval which can be made as small as we please. If an experiment were performed between 0 and t and the states plotted, the number of realised transitions will be finite and the probability of infinite number of transitions vanishes. Thus in a stochastic system "jumps" from one state to another are possible in an infinitesimal interval with infinitesimal probabilities.

In the case when we have a discrete set of states, let us write for convenience $\pi(k|l;t)$ for $\pi(S_k|S_l;t)$ and assume

$$\left.\begin{aligned}
\pi(k|l;t) &\to R(k|l)\,\Delta && \text{as } t \to \Delta \to 0 \text{ for } k \neq l, \\
\pi(k|l;t) &\to 1 - \Delta \sum_{j \neq l} R(j|l) && \text{as } t \to \Delta \to 0 \text{ for } k = l,
\end{aligned}\right\} \tag{39.1}$$

i.e. the probability of change in an infinitesimal interval Δ is proportional to Δ while the probability of "continuance" is $1 - O(\Delta)$. π can now be shown to satisfy the forward differential equation

$$\frac{\partial \pi(k|l;t)}{\partial t} = -\pi(k|l;t) \sum_{j,\,j \neq k} R(j|k) + \sum_{j,\,j \neq k} \pi(j|l;t)\,R(k|j). \tag{39.2}$$

Since $\pi(j|i;t)$ is a probability magnitude $\pi(j|i;t) \geq 0$ for all t, j, i; equally so $R(j|i) \geq 0$ for all $j, i\,(j \neq i)$. If $R(j|i) > 0$ note that $\pi(j|i;t)$ is $O(\Delta)$ as $t \to \Delta \to 0$, but it is finite and greater than zero for finite t. For convenience we divide the above type of stochastic processes into two groups.

1. When $R(j|i) > 0$ for all j, i, i.e. the system is "connected" for infinitesimal transformations. In this case, while $\pi(j|i;t)$ is $O(\Delta)$ if $t \to \Delta \to 0$ it is finite and greater than zero for finite t.

2. When $R(j|i) = 0$ for some j, i say j^+, i^+. Even in this case the author [8] has shown that $\pi(j^+|i^+;t) > 0$ provided for every pair of states j^+, i^+ for which $R(j^+|i^+) > 0$, there exists a sequence of states $S_{k_1}, S_{k_2}, ..., S_{k_n}$ such that each of the terms

$$R(j|k_n),\ R(k_n|k_{n-1}),\ ...,\ R(k_2|k_1),\ R(k_1|i) \tag{39.3}$$

is greater than zero. In such a case we say the system is completely connected for finite transformations and not for infinitesimal transformations.

(ii) If the occupation states form a continuous system describable by a one-dimensional[1] variable, say E, in a deterministic system E is a function of t. Infinitesimal changes of t will cause infinitesimal changes in E but in the case of a stochastic system the system in a state E_1 can jump to a state between E_2 and $E_2 + dE_2$, $(E_1 - E_2)$ being finite, with probability proportional to Δ in an interval Δ. A stochastic system may be characterised not only by these possible changes but also may involve deterministic changes. Thus in defining $\lim \pi(S'|S;t)$ as $t \to \Delta \to 0$ *we must first unravel the possible stochastic transitions from the deterministic changes.* This is necessary only when the S states form a continuous system, now defined by the one-dimensional variable E. Thus replacing the S's by the E's we can safely state that $\pi(E|E_0;t)$ tends to a limit $R(E|E_0)\Delta$ as

---

[1] Actually there is no restriction on the number of dimensions. The extension to a many-dimensional variable presents no difficulty in the description of the process.

$t \to \varDelta \to 0$ provided $E_0 - E$ is finite since it will refer only to stochastic transitions. If no stochastic transition occurs, then the change in E is determined by the mode of the deterministic change characteristic of the process.

Consider the infinitesimal changes in E for an infinitesimal change $\varDelta$ in t. It is physically reasonable that this change is proportional to $\varDelta$ since $\varDelta$ is an arbitrary increment and what is true of $\varDelta$ must be true of $\varDelta/2$, $\varDelta/3$, $\varDelta/4$ etc. But the factor of proportionality can be either (i) a deterministic function of E or t or (ii) a stochastic variable with a defined frequency function, (iii) or more generally a random function of t. We shall for the moment disregard the last two possibilities and consider only the first one. Even in that case we shall assume that the factor of proportionality is not a function of t for as soon as we concede that it is a function of t we tacitly assume that t has to be measured from a particular point and this amounts to a knowledge of the prehistory of the process, thus disturbing the Markovian property. For simplicity therefore we shall assume that the factor is a deterministic function of the realised value of E at t which we shall denote by $f(E)$. $f(E)$ need not be a continuous function of E but should be bounded, provided it is understood that whenever differential coefficients of π are involved we introduce suitable delta functions. We can now write the forward differential equation for the probability frequency function as

$$
\left.
\begin{aligned}
\frac{\partial \pi (E \mid E_0; t)}{\partial t} = &- \pi (E \mid E_0; t) \int\limits_{E'} R(E' \mid E)\, dE' \\
&+ \int\limits_{E'} \pi (E' \mid E_0; t)\, R(E \mid E')\, dE' \\
&- \frac{\partial}{\partial E} \{ \pi (E \mid E_0; t)\, f(E) \}
\end{aligned}
\right\}
\qquad (39.4)
$$

where $f(E)\, dt$ is the deterministic change in E if no stochastic transitions occur. Such a process we call a "blended" process. If $f(E) = 0$ the author has called the process a "basic random process".

In the above discussion we take into consideration either a discrete set of states $S_1, S_2, \ldots$ or a continuous set defined by a single variable E. It must be mentioned the state can be defined with the aid of as many discrete and continuous variables as are necessary for the description of a physical system and in this case the differential equation will naturally be more complicated than the one given above. But our point here is merely to mention that the differential equations in the case of the continuous set may consist of two sets of terms, one denoting stochastic and the other deterministic changes. A large freedom is allowed to define the basic occupation states and this usually depends upon the ingenuity in adapting the physical theory to actual problems.

40. A process defined by the forward differential equation. A typical trajectory of a process defined by (39.4) in the interval 0 to t will be characterised by finite jumps at a finite number of points (due to the realisation of stochastic transitions) and a continuous curve elsewhere. As mentioned earlier, it is possible to have discrete deterministic changes at a finite number of points on the t-axis but in this case since no changes occur between these points it is of no interest for our discussion and we therefore exclude this possibility. If in addition we do not postulate infinitesimal deterministic changes for infinitesimal changes in t, i.e. $f(E) \equiv 0$ the realised value of E will be constant between two transitions. Such a process has been called a *basic random process*. The measure of a typical trajectory of either a basic random process or a simple blended process with transitions at $t_1, t_2, \ldots$, is just the probability that the transitions take place at those points. The mathematical difficulties of assigning a measure to a typical trajectory of a random function do not occur in this simplified case. Thus we can say that provided we know f and $R(E'|E)$ all the characteristic features of the process can be studied.

Till now we considered t to be continuous. t can also be assumed to be discrete. If t is discrete and the set of occupation states continuous the process can be only stochastic. For if it were assumed to be deterministic this would be equivalent to stating that to each value of t there corresponds one occupation state (if the state is a single-valued function of t) or a finite number. Thus corresponding to all the values of t there can be at most an enumerable number since t is discrete. Thus the set of occupation states cannot be continuous. Denoting the discrete set of t by $t_1, t_2, \ldots$ and using the continuous parameter E to represent an occupation state, we define $R(E'|E; t_{n+1}, t_n)\, dE'$ as the probability that the system is between E' and $E'+dE'$ at t_{n+1} given that it was in E at t_n.

If the set of occupation states is discrete (say denoted by $S_1, S_2, \ldots$) it is obvious that the process can be either deterministic or stochastic. If the process is stochastic we define $R(j|i; t_{n+1}, t_n)$ as the probability that the system jumps to S_j at t_{n+1} given that it was in S_i at t_n.

Summarising our results we can classify processes from the point of view of the nature of occupation states and of the parameter t.

(A) Occupation states—discrete, t—discrete; a process can either be deterministic or stochastic or "blended".

(B) Occupation states—discrete, t—continuous; process can be only stochastic.

(C) Occupation states—continuous, t—discrete; process can be only stochastic.

(D) Occupation states—continuous, t—continuous; process can be either stochastic or deterministic or "blended".

41. Recurrence and first passage times[1]. A more limited purpose than studying the typical trajectory is to consider first passage and recurrence times for any particular state. We shall treat the problem from a phenomenological point of view and explain in physical terms the difficulties involved by considering the classes of stochastic processes listed above as A, B, C and D one by one.

Class A. We shall denote the states by S_j and the discrete t-axis by t_k where the t_k's are ordered. Given that at t_l the system is in S_i, we ask what is the probability $F(j|i; t_k, t_l)$ $(k > l)$ that the system enters S_j for the first time at t_k. In this case if a system is in a certain state at some t_j and continues to be in the same state at t_{j+1} we shall not treat it as continuance but treat it as a "fresh" entry into that state. F satisfies the equation

$$\left.\begin{aligned}
\pi(j|i; t_k, t_l) &= \sum_{n=l+1}^{k} F(j|i; t_n, t_l)\, \pi(j|j; t_k, t_n)\,, \\
\pi(j|i; t_{l+1}, t_l) &\equiv F(j|i; t_{l+1}, t_l)\,.
\end{aligned}\right\} \tag{41.1}$$

By successive substitution we obtain the probability of first passage at all values of t. Note that $F(j|i; t_{l+1}, t_l)$ corresponds to the R functions in the case of continuous t.

The probability of recurrence of a state S_l at t_k given that it occupied S_l at t_i is obtained merely by taking $j = l$ in the expression for first passage, $F(j|i;$

---

[1] Not much work from a physical point of view has been done on this aspect of stochastic theory and BARTLETT's [3] chapter on this problem deals with only the beginnings of this subject and does not advert to some of the difficulties in the case of complex stochastic processes. But he makes specific reference to this in his recent paper [9] (BARTLETT 1953).

t_k, t_i). Thus the problem of determining the distribution of first passage and recurrence times is straightforward.

Class B. In this case a system is in S_j at t and till $t+dt$ it continues with probability $1-O(dt)$. Continuance at $t+dt$ should not be treated as an arrival at this state in the interval dt. We have therefore to define $F(j|i;t)\,dt$ as the probability that S_j is entered for the first time in the interval dt given that the system was at S_i at $t=0$. It is clear that i should not be equal to j in the above definition. π will satisfy the equation

$$\pi(j|i;t) = \int_\tau F(j|i;\tau)\,\pi(j|j;t-\tau)\,d\tau. \tag{41.2}$$

F is determined if we know π and π satisfies the well-known stochastic equation (39.2).

To obtain the recurrence time we require the probability $\varrho(j|j;t)\,dt$ that S_j recurs in the interval dt. It means that the system has jumped off from the state S_j between t and $t+dt$ and is again entering S_j for the first time since its exit from that state somewhere between 0 and t.

$$\varrho(j|j;t)\,dt = dt \int_0^t C(j;\tau) \sum_k R(k|j)\,F(j|k;t-\tau)\,d\tau \tag{41.3}$$

where $R(k|j)$ has been defined earlier and $C(j;\tau)$ is the probability of continuing in the state S_j in the interval 0 to τ. This is given by

$$C(j;\tau) = \exp\left[-\sum_k R(k|j)\,\tau\right]. \tag{41.4}$$

ϱ is therefore completely determined, if we know R, π and F.

Class C. This is exactly similar to Class A, except that we have to define $F(E|E';t_k,t_l)\,dE(k>l)$ the probability that the system enters the occupation state between E and $E+dE$ at t_k given that it was at E' at t_l. The other arguments run on similar lines and so integrals suitably replace the summations in Class A.

Class D. This is the most interesting class, especially in view of the difficulties that occur in the case of "blended" processes. We therefore shall consider the problem in increasing order of complexities.

(i) Basic random process. Consider the system to be at E_0 at $t=0$; we wish to know the probability $F(E|E_0;t)\,dE\,dt\,(E\neq E_0)$ that a state between E and $E+dE$ is entered for the first time between t and $t+dt$. In this case we get the most remarkable result that the probability that the system enters the state between E and $E+dE$ between 0 and t then jumps off and re-enters a state between E and $E+dE$ between t and $t+dt$ is of the order $(dE)^2\,dt$ while the first passage between t and $t+dt$ is of the order of $dE\,dt$. Hence we can write

$$F(E|E_0;t) = \int_{E'} \pi(E'|E_0;t)\,R(E|E')\,dE'. \tag{41.5}$$

Compare this with Eqs. (41.1) and (41.2) where F occurred as a "kernel", i.e. as a term within the summation or integral signs. The integral equation for π corresponding to (41.2) can be written in the form

$$\left.\begin{aligned}\pi(E|E_0;t) &= \int_\tau F(E|E_0;\tau)\,C(E;t-\tau)\,d\tau, \\ C(E;t-\tau) &= e^{-(t-\tau)\int_{E'} R(E'|E)\,dE'}.\end{aligned}\right\} \tag{41.6}$$

$C(E; \tau)$ is the probability that a system which is found in E at $t = 0$ continues to be in the same state till $t = \tau$. Differentiating the above expression and substituting the value for F according to Eq. (41.5) we obtain

$$\frac{\partial \pi(E \mid E_0; t)}{\partial t} = - \pi(E \mid E_0; t) \int\limits_{E'} R(E' \mid E)\, dE' \left.\right\}$$
$$+ \int\limits_{E'} \pi(E' \mid E_0; t)\, R(E \mid E')\, dE' \left.\right\} \qquad (41.7)$$

which is the standard equation for a basic random process.

The problem of recurrence times does not occur in this case since we have shown that the probability of recurrence is of higher order of smallness, $O(dE^2 dt)$ as compared to the probability of first passage $O(dE\, dt)$ as is also evident from the derivation of the integral equation for π. Hence the probability of recurrence can be taken to be zero. This striking result we must note is due to the infinitesimal nature of dE and not that of dt. Hence if we take a group of states between E and $E + \Delta E$, ΔE being finite, the problem of their recurrences and first passage times into this group is not so simple as the one considered above. The computation of this leads to almost intractable difficulties as a division of the E space into finite intervals will render the process non-Markovian.

(ii) Simple blended processes. Here we have postulated an infinitesimal deterministic change of $f(E)\, dt$ for a change dt in t if the system is in a state E at t. Of course we assume the possibility of random transitions from E to an interval between E' and $E' + dE'$ $(E - E'$ being finite) in the interval dt with probability $R(E' \mid E)\, dt\, dE'$. A typical trajectory is characterised by a continuous curve representing deterministic changes except at a finite number of points where random transitions have been realised. Phenomenologically there is no meaning in speaking of first entry into a state between E and $E + dE$ between t and $t + dt$ because this state is entered in two ways either by a jump from some state E' to the interval between E and $E + dE$ in a random transition, or *by crossing the state E in the interval t and $t + dt$*. A slight reflection will show that the two concepts of crossing and jumping cannot be defined by the same function and hence the definition of first passage and recurrence breaks down unless in considering jumping we do not consider crossing and vice versa..

The same problem will arise if the finite random transitions are blended with infinitesimal random changes. If the finite random transitions are absent and only infinitesimal random changes are postulated we can define a function $F(E \mid E_0; t)\, dt$ giving the probability that the state E is crossed between t and $t + dt$. (Note the absence of dE.) This problem is dealt with presently in the following treatment on integrals of random functions.

42. Integrals of random functions. The possibility of infinitesimal random changes for infinitesimal changes of t leads us directly to a phenomenological definition of integrals of random functions. The concept of stochastic integration is shrouded so much in abstract mathematical language, that it is not possible for one who is interested in its physical interpretation to glean from such rigorous treatment the essential physical basis which is necessary in applying any theory to concrete problems. The author therefore felt it necessary to attempt this problem from the beginning, using only the concept of the realised curve or the trajectory of a stochastic process. As the phenomenological theory has been elaborated in a series of three papers by the author [10], we confine ourselves only to the fundamental definition of an integral of a random function and explain in general terms how it can be applied to physical problems. We now consider the postulate of infinitesimal random changes in infinitesimal periods of time. If $x^R(t)$ is the realised value of the random function at t and if we increase t to $t+dt$ our postulate requires that $x^R(t+dt) - x^R(t)$ is $O(dt)$ or more precisely $k\,dt$, k being the realised value of a random function of t, say $\psi(t)$. This random function of t can be symbolically represented as the differential coefficient of $x(t)$ i.e.

$$\psi(t) = \frac{d}{dt}\,x(t). \tag{42.1}$$

At first sight it would seem that it is the knowledge of the random function $x(t)$ that determines the differential coefficient. A closer examination will show that it is more logical to treat $x(t)$ as an integral of the random function $\psi(t)$ and a knowledge of $\psi(t)$ is necessary to study $x(t)$. This point can easily be illustrated by asking for the distribution function of $\psi(t)$, given the nature of $x(t)$ or vice versa.

We shall assume that the joint distribution of $x(t)$ at two points t_1 and t_2 is known and is specified by the function $\pi(x_1, x_2; t_1, t_2)$. We shall not assume that $x(t)$ is Markovian. Therefore the joint distribution cannot be expressed simply in terms of $\pi(x; t)$, the distribution function at a point t. The question of arriving at this distribution function will be discussed later. But for the moment we assume that we are provided with such a distribution function. Defining $x(t_2) - x(t_1)$ as a new random function $\psi(t_1, t_2)$ the joint frequency function of $x(t_1)$ and $x(t_2)$ immediately determines the joint frequency function of $x(t_1)$ and $\psi(t_1, t_2)$ and equally $x(t_1)$ and $\psi(t_1, t_2)/(t_2 - t_1)$. If we now let $t_2 \to t_1$ and postulate that the values which $\psi(t_1, t_2)$ can assume as $t_2 \to t_1$ are $O(t_2 - t_1)$, then we can obtain the joint distribution of $x(t_1)$ and $\psi(t_1, t_2)/(t_2 - t_1)$ as $t_2 \to t_1$. It is reasonable to denote $\lim \psi(t_1, t_2)/(t_2 - t_1)$ symbolically by the random variable $\dot{x}(t)$. Since we are now in possession of the joint distribution of $x(t)$ and $\dot{x}(t)$ by integrating this function over the values of x we obtain the frequency function of $\dot{x}(t)$.

Thus from the above discussion we obtain the lemma that the joint distribution of $x(t)$ at two points helps us to obtain the distribution of $\dot{x}(t)$. But the important question remains as to how the joint distribution of $x(t_1)$ and $x(t_2)$ can be obtained or derived or at least the distribution function of $x(t)$ given that of $\dot{x}(t)$ or that of $\dot{x}(t)$ given that of $x(t)$ can be obtained. In the opinion of the author it is more logical and phenomenologically more satisfying to conceive of $x(t)$ as an integral of $\psi(t)$ and show that a knowledge of $\psi(t)$ helps us to obtain a knowledge of $x(t)$. Accepting for the moment this point of view, if $\psi(t)$ is characterised by infinitesimal random changes in infinitesimal periods of time, we should view the process as an integral of another process and so on, till we reach a process which is a random process not characterised by infinitesimal random changes. It is reasonable to suppose in many cases of physical interest that it is a basic random process. Denoting it as $x(t)$ we can build a whole class of processes characterised by infinitesimal random changes in infinitesimal periods of time by defining symbolically that class as

$$y_n(t) = \varphi_n(t) \int_0^t \varphi_{n-1}(\tau_{n-1})\,d\tau_{n-1} \int_0^{\tau_{n-1}} \varphi_{n-2}(\tau_{n-2})\,d\tau_{n-2} \dots \int_0^{\tau_1} \varphi_0(\tau_0)\,x(\tau_0)\,d\tau_0 \tag{42.2}$$

where $\varphi_0, \varphi_1, \dots$ etc. are fully deterministic functions of t. The author is not suggesting that any random process characterised by infinitesimal random changes in infinitesimal periods of time can be represented by an integral like (42.2) but only suggests that an entire class can be defined by the above integral and therefore by deterministic functions of such integrals and simple combinations of such functions. It is clear that if we assume the basic random process to be Markovian the integral process $y_n(t)$ is non-Markovian since the development of $y_n(t)$ requires a knowledge of $y_{n-1}(t)$ and so on and only if we consider the process represented by the entire aggregate $y_n(t), y_{n-1}(t), \ldots y_1(t)$ and $x(t)$, we obtain a "quasi" Markovian process (the significance of the word "quasi" will be explained presently) where the development can be studied by standard methods. From the above considerations it is clear that, even if we require the distribution function of $y_n(t)$ alone, we must first obtain the joint distribution of the aggregate $y_n(t)$, $y_{n-1}(t), \ldots$ and then integrate over the intermediate variables. This procedure has been adopted successfully by MATHEWS and SRINIVASAN [11] (1956) in the case of a class of basic random processes.

There is however an alternative treatment available if we are interested only in the distribution of $y_n(t)$ in the special case when the basic random process is of a very simple type with certain symmetric properties. We shall not however go into that method in any detail here, but merely indicate one of the important results that was obtained in the course of the application of that method. It led to the concept of "equivalent" processes. A process is said to be equivalent to another if the distribution function of the two stochastic variables $y(t)$ and $y^*(t)$ under consideration are identical at all t. Note that we are not stating that the joint distribution of $y(t)$ at more than one point, say t_1 and t_2, is identical with the joint distribution of $y^*(t)$ at the same points t_1 and t_2. We only emphasise that the distribution function of $y(t)$ at t_1 is identical with the distribution function of $y^*(t)$ at t_1 as t_1 takes any value. This result is probably implied in many abstract treatments of the theory of stochastic processes. But the author has not seen any particular illustration of this rather remarkable feature of stochastic processes. This feature has no parallel in the case of static distributions. If we have two stochastic variables x and y and they have identical distributions, we say that they display the same statistical features. On the other hand in stochastic processes we do not have just a stochastic variable, but a random function $x(t)$ instead of x and $y(t)$ instead of y. These represent a continuous infinite set of stochastic variables and the distribution function deals with one of them "at a time" and hence any information only about the distribution function cannot give us a complete picture of the stochastic process. The author has succeeded in citing a number of examples of such equivalent processes and we refer the reader to the series of papers on integrals of random functions.

43. Quasi-Markovian processes. In referring to Markovian processes we also assumed homogeneity with respect to t. It would be interesting to investigate the meaning of a process which is Markovian but inhomogeneous with respect to t.

Suppose we know the state of a system at a point P on the t-axis. If a process is Markovian the development of the process from P onwards should depend only on the state of the system at P. If we now ask for a description of the state at Q and assume homogeneity it depends only upon the distance PQ and not upon the "coordinates" of P and Q separately on the t-axis. If the process is inhomogeneous we must know the coordinates of P besides knowing the state at P to study the development of the process from P onwards. The ascription of a coordinate to Q on the t-axis assumes a "knowledge" of a previous history consisting of, in some sense, the location of the point from which the coordinate Q is measured. Therefore in the opinion of the author, the term Markovian for processes inhomogeneous in t is not happy. It would be more appropriate to call them quasi-Markovian since forward differential equations for the probability distributions involved in such processes can be written in the same way as for Markovian processes.

44. The Chapman-Kolmogoroff equation. There is a very important equation of consistency satisfied by the probability frequency function in the case of Markovian and quasi-Markovian processes. We shall write down the equations corresponding to the four cases discussed earlier in this chapter.

(A) If $\pi(k\,|\,i;\,t_n,\,t_l)$ is the probability that the system is in state S_k at t_n given that it was in S_i at t_l $(t_n > t_l)$ then

$$\pi(k\,|\,i;\,t_n,\,t_l) = \sum_j \pi(k\,|\,j;\,t_n,\,t_m)\,\pi(j\,|\,i;\,t_m,\,t_l) \qquad (44.1)$$

for all $t_n > t_m > t_l$.

(B) The same equation is valid provided we understand $t_n,\,t_m,\,t_l$ are points on the continuous t-axis and $t_n > t_m > t_l$.

(C) Denoting the continuous infinite number of states by the parameter E we can write

$$\pi(E_k\,|\,E_i;\,t_n,\,t_l) = \int\limits_{E_j} \pi(E_k\,|\,E_j;\,t_n,\,t_m)\,\pi(E_j\,|\,E_i;\,t_m,\,t_l)\,dE_j \qquad (44.2)$$

for all $t_n > t_m > t_l$.

(D) The same equation as for (C) holds provided we understand $t_n,\,t_m,\,t_l$ are points on the continuous t-axis.

These equations are self-explanatory and are a direct consequence of the Markovian property. They express only a consistency relation i.e. they are satisfied by all π's which represent probability frequency functions of Markovian and quasi-Markovian processes. However if we make $t_n \to t_m$ in the case of continuous t and we assume that we know the limit of

$$\pi(j\,|\,k;\,t_n,\,t_m) \quad \text{as} \quad t_n \to t_m$$

then we get integro-differential equations for π in terms of these limits. This is precisely what we have done in writing down the forward differential equation of the process. In the case of discrete t we must assume the transition probability for adjacent "times", i.e.

$$\pi(j\,|\,k;\,t_n,\,t_{n-1}).$$

If we assume in addition that the process is homogeneous in t then

$$\pi(k\,|\,i;\,t_n,\,t_m) = \pi(k\,|\,i;\,t_n - t_m) \qquad (44.3)$$

and t_m can be set equal to zero without loss of generality.

If in the case of continuous t by making $t_m \to t_n$ we obtain the "forward" differential equation, are we not entitled to ask the question: what happens if $t_m \to t_l$? This leads us to what is known as the "backward" differential equation of the process. Instead of writing the corresponding backward differential equations for all the four cases we shall write it for case (B) when the process is homogeneous in t.

$$\frac{\partial \pi(j\,|\,i;\,t)}{\partial t} = -\pi(j\,|\,i;\,t) \sum_k R(k\,|\,i) + \sum_k \pi(j\,|\,k;\,t)\,R(k\,|\,i). \qquad (44.4)$$

We note that in the backward equation it is the argument referring to the "initial" condition in π that is varied in the sums (or integrals) on the right hand side while in the forward differential equation it is the "final" state in a π function that is varied. Or in other words, in writing the backward equation we ask for what happens in the first small interval Δ and in writing the forward equation we ask for what happens in the last small interval Δ. The two equations though equivalent are not identical and in many cases it may not be possible to go from one to the other except by identifying them as the limiting cases of the Chapman-Kolmogoroff equation. In the case of the discrete system of states we note that $\pi(j|i;t)$ are terms of a matrix as j and i vary. $\pi(j|i;t)$ as j varies is a column vector, the variation of which with t is studied by the forward equation while $\pi(j|i;t)$ as i varies is a row vector, the variation of which with t is represented by the backward equation. Such a simple connection is however not possible when we have a continuous system of states. Some times it is found easier to work with the forward equation, some times with the backward. The choice depends on the particular situation.

However the backward equation has an additional advantage over the forward since *for some non-Markovian processes it is possible to write backward equations though the general Chapman-Kolmogoroff equation is not valid.* The Chapman-Kolmogoroff equation and the forward equation can be written down only for Markovian or quasi-Markovian processes. The peculiar feature of the backward equation is discussed presently in the following section on non-Markovian processes.

45. Remarks on Markovian, non-Markovian and quasi-Markovian processes. We have referred to the integrals of random functions as non-Markovian processes and we have emphasised earlier that non-Markovian processes are a residuary class and cannot be exhausted by classification. Integration is one way of generating non-Markovian from Markovian processes. But these belong to that class of non-Markovian processes which by the introduction of additional parameters can be rendered quasi-Markovian. The difficulty of defining a general non-Markovian process can easily be illustrated in the following manner. Suppose $x(t)$ is non-Markovian and t is continuous. This means that if we know the value of x at some value of t it is not possible to predict what happens between t and $t+dt$ merely if we are in possession of the knowledge of the value of $x(t)$. What other information do we require? Let us postulate that we require information about the realised value of $x(\tau)$, $\tau < t$: If we require this information about all the states in a domain $\tau < t$, it leads to the mathematically difficult and possibly intractable problem of defining a function of a continuous infinite number of points or variables represented by the value of τ. Or to put it more precisely, we can speak of the function of a variable τ or of an enumerable number of variables $\tau_1, \tau_2, \ldots$ etc., but not of a continuous infinity of variables in an interval $t_0 < \tau < t$.

An interesting class of non-Markovian processes can be quoted easily which have a simpler dependence on previous history. Suppose we have the distribution of random points in a line represented by the t-axis and we postulate the following law of distribution. The probability that a point lies between τ_1 and $\tau_1 + d\tau_1$ is equal to $\psi(\tau_1 - \tau)\, d\tau_1$, given that the previous point has occurred at τ and *does not depend upon* the distribution of points prior to $t = \tau$. To put it picturesquely the memory of the process can be traced only up to the last realised point. Let a point be realised at $t = 0$ and let us require $\pi(n; t)$ the probability that there are n points in the interval 0 to t. $\pi(n; t)$ is non-Markovian since it is impossible to predict the probability whether a point will lie between t and $t+dt$, if we are just given that n points lie between 0 and t. We have to know where the n-th point has occurred. However it possesses a very interesting property which we have not adverted to till now. It is *"regenerative"* with respect to an occurrence of a point. Or in other words, when a point occurs the process "loses all memory" and starts "afresh". Regenerative processes first attracted the attention of BELLMAN and HARRIS [12] and have been applied with success to many important processes later on. BARTLETT in his book has given an interesting definition of a regenerative process in fairly general terms. If a system can occupy a discrete set of states $S_1, S_2, \ldots$ etc. and we are interested in the stochastic process which is described by defining the occupation state in which the system lies at parametric value t as t varies, we call it non-Markovian if, given that it is in a state S_j at t, we are not able to predict the development of the process between t and $t + dt$. However, if among the aggregate of occupation states there is a sub-aggregate $S_1', S_2', S_3', \ldots$ etc. and if the system is found in any one of those states at t, it is possible to predict what happens between t and $t + dt$, we will call the process regenerative with respect to these states. This definition is fairly general and has to be adapted to particular circumstances. In fact some fundamental difficulties occur in translating this general description into exact terms in physical problems. The regeneration process referred to above which is so delightfully simple becomes quite complicated from the view point described above. If we define the occupation states as the number of points 0, 1, 2, etc. the system is not regenerative with respect to anyone of the states, but we know that it is regenerative with respect to the occurrence of the point. So let us divide the interval 0 to t into a large number of small and equal intervals of length Δ and $N\Delta = t$. In every interval there can be 0, 1, 2, 3, etc. random points. As Δ becomes very small, let us postulate that the probability that there is one random point is $O(\Delta)$ and no random point is $1 - O(\Delta)$ while the probability of k random points is $O(\Delta^k)$ which is small compared to $O(\Delta)$. We therefore define two occupation states which are characterised by 0 and 1, the number of random points in the interval. The state at t is to be interpreted by taking a small interval of width Δ in the neighbourhood of t and asking the question whether there is a random point in it or not and the system is now regenerative with respect to the occupation state 1, representing the number in Δ.

It is quite obvious that a Markovian process is always a regenerative process, the converse of course not being true. It is possible to study regenerative processes by an elegant mathematical technique though they are not Markovian. The technique is suggested by the regenerative property itself. Suppose we are interested in the probability distribution function of the system at t, we first assume that the system is in a state S_j with respect to which the process is regenerative. We now study how the system for the first time enters another regenerative state S_k at some point between τ and $\tau + d\tau$. At that point τ the process "loses memory" and starts afresh with the new initial condition S_k and the duration of this freshly started process is $t - \tau$. Since S_k can represent any of the regenerative states and τ can be any point between 0 and t this approach leads to the integral equation

$$\pi(l \mid j^+; t) = \int\limits_0^t \sum_{k^+} F(k^+ \mid j^+; \tau)\, \pi(l \mid k^+; t - \tau)\, d\tau. \tag{45.1}$$

In some cases it may be possible to express $\pi(l \mid k^+; t)$ in terms of $\pi(l \mid j^+; t)$ and this makes the situation simpler. We must remember that $l, j, k, \ldots$ etc. refer to the states $S_l, S_j, S_k, \ldots$ which can be interpreted in the widest possible manner to suit the needs of the problem. Accordingly, the summation sign over k^+ must be interpreted as an integral or a sum. We also assume that the first transition from j^+ can be only to a regenerative state. *Differentiating this expression with respect to t yields the backward differential equation for π.*

We have now described the general "physical" features of a stochastic process evolving with respect to a one-dimensional parameter t, the Markovian processes being a simple class of such processes. If t were two- or many-dimensional, it will be difficult to define Markovian processes since t cannot be ordered and the meaning of "previous history" becomes obscure. In a similar manner a "regenerative" property cannot also be easily attributed to such processes. A brief discussion on this is given in Chap. XII (p. 640).

II. Stochastic process: Measure theoretical approach.

46. With the background of the physical approach to guide us we can now adopt the definition of a stochastic process as given by mathematicians, for example Doob [1].

A stochastic process is any family of random variables the family being generated by the variation of a parameter t. The family of random variables is denoted by $x(t), t \in T$. Doob in his book starts with the parameter t as one-dimensional, T representing the "time" range involved. The extension to many-dimensional t is immediate. He considers two cases, (a) t is a discrete parameter and T an infinite sequence, (b) T is an interval, then $x(t), t \in T$ becomes a continuous parameter family.

"This definition of a stochastic process is embarrassingly inclusive in that there are not many problems in probability that cannot be formulated as problems in families of random variables. Historically however the term stochastic process has been reserved for families of random variables with some simple relationships between the variables. A basic problem is to devise suitable relationships, that is to discover new types of stochastic processes that are useful or mathematically elegant or which conform otherwise to the investigator's criterion of importance." An additional comment may be added to these remarks of Doob. The author in his physical approach has given the meaning of the word "process" from a physical point of view. There must be some mechanism by which we change the parameter t and as we change it we get the sequence of "realised values" of $x(t)$. In such a case a process results. If t is a one-dimensional parameter which can be ordered we call it an *evolution*. If t is many-dimensional the concept of evolution breaks down and the process is not evolutionary. Nevertheless we can still speak of the family as constituting a process provided the physical conditions of the problem give us a mechanism of studying the family of random variables by moving from one parametric value to another.

Let $x(t), t \in T$ be a stochastic process. A function $x(t, \omega)$ obtained by fixing ω in the basic ω-space in which the entire set $x(t), t \in T$ is defined and letting t vary is called a *sample function* of the process. If T is finite or denumerably infinite the sample functions are of course sample sequences. *In Sect. 39, we called it a realisation function $x^R(t)$ corresponding to the stochastic prosess $x(t)$.* Let $t_1, t_2, \ldots, t_n$ be any finite set of parameter values of the process $x(t), t \in T$. The multivariate distribution of the random variables $x(t_1), x(t_2), \ldots, x(t_n)$ is called a finite-dimensional distribution of the process. In fact a random function $x(t)$ according to Kolmogoroff (see Doob [1]) is completely defined by giving the joint probability distribution for any set of values $t_1, t_2, \ldots, t_n$ as n takes a value as large as we please or by the corresponding joint characteristic function.

This description of a random function can be compared with that of an ordinary function $x(t)$; the latter may be considered as completely defined if we can give a rule for finding the values $x_1, x_2, \ldots, x_n$ of the dependent variable attached to every set $t_1, t_2, \ldots, t_n$ of the independent variable t. In the definition of a random function we correspondingly replace the exact knowledge of the x_i's by the knowledge of their joint probability distributions when t is continuous in the interval $0 \leq t \leq a$; t takes a continuous set of values and a comprehensive knowledge of the random function should in principle be obtained only if we know the joint distribution at this continuous infinite set of points. Though no function can be explicitly defined which represents this joint distribution function Kolmogoroff asserts that the function is completely defined if we know the joint distribution at a finite set of points. If however we wish to know the behaviour of the continuous set the physical approach of the previous chapter helps us to comprehend the difficulties of defining analytical operations like integration and differentiation relating to the random function and the meaning of such operations. We shall now describe in mathematical terms the contents of the previous Chapter. The extension to the continuous infinite sets has been given by Kolmogoroff himself, Doob, Pitt (see Moyal [13]) and other authors. The problem is essentially to define a measure for the trajectory corresponding to any interval. Kolmogoroff's equation is based on the distribution at a finite discrete set and is considered unsatisfactory by Moyal [13] because it leaves unmeasurable, sets of functions for which we require a measure. Pitt starts instead from the distribution function giving the joint probability that $a_i \leq x_i(t) \leq b_i$ for each t_i belonging to an interval τ_i and $\tau_1, \tau_2, \ldots, \tau_n$ are a finite set of non-overlapping intervals. Such a distribution function is a functional of the realised function of $x(t)$. In the opinion of Moyal it has a clear advantage that if one wants to deal with the whole functional field, it depends on the behaviour of $x(t)$ over the whole interval and not at discrete points.

Doob tackles this problem by introducing the notion of separable processes which are defined by the joint distribution for a finite set of points and establishing that to every process there corresponds a separable process under well defined conditions.

The advantage of dealing with finite-dimensional distributions is obvious since it is possible to ascribe measure in a Borel field of ω-sets on which the finite set of random variables is defined. If we take the entire set of random variables $x(t)$, $t \in T$ we have to define a non-enumerable set of random variables on a suitable ω-space and it may be difficult to define measure in a Borel field of sets in such an ω-space.

As long as we are dealing with joint distributions at a finite set of points the definitions for the mean, moments and characteristic functions are identical with those of the aggregate of n random variables.

III. Calculus of random functions[1].

47. Introduction. We have introduced the analytical process of integration of random functions and thereby that of differentiation by appealing to the physical concept of a realised trajectory. In doing so, we have assumed that the realised curve of the integrand was integrable in the ordinary Riemann sense. It is obvious that such a treatment is useful only if from the physical nature of the problem all the characteristic features of the trajectory of the integrand are known. The correlation functions associated with the integral process were also obtained on the basis of this knowledge. However it is possible to define analytical operations with respect to stochastic processes without appealing to the realised trajectories but taking recourse only to correlation functions if we use what is known as the concept of mean square convergence of sequences of random variable.

---

[1] I am deeply indebted to Professor C. T. Rajagopal in the preparation of this chapter.

We first realise that in a stochastic process we are dealing with sequences of random variables, and therefore our first task is to carry over the notions of ordinary convergence to that of random variables. For this purpose it is convenient to start with a discrete sequence of random variables $x_1, x_2, \ldots, x_n$ and then extend the notions to a continuous parametric sequence. What do we mean by saying that a sequence of random variables $x_1, x_2, \ldots, x_n$ converges to a random variable x? The simplest and most direct definition of convergence is based on the following interpretation: if we perform an experiment to determine the values of x and the sequence x_n, in each experiment the sequence of realised values of x_n converges to the realised value of x. Obviously this definition of convergence (hereinafter called *ordinary* convergence) is too restrictive. In the theory of probability, four other modes of convergence have been introduced.

48. Modes of convergence. α) *Convergence in distribution functions (d.f.).* Let the distribution function of x_n and x be $f(x_n)$ and $f(x)$ respectively. For d.f. convergence, we merely require that $f(x_n) \rightarrow f(x)$ as $n \rightarrow \infty$ at every point of continuity of $f(x)$. If we perform an experiment to determine x and x_n the probability that $|x_n - x| > \varepsilon$ can be finite for finite ε. This situation is easily conceivable since we know that even in the simple case when there can exist two random variables x and y with the same distribution function, $x - y$ is a random variable with its corresponding probability distribution.

β) *Convergence in probability (i.p.).* Here we require that

$$\lim_{n \to \infty} \text{prob} \{| x_n - x| > \varepsilon\} = 0$$

for every $\varepsilon > 0$. The meaning of this becomes clear if we restate it in the following manner. Given an ε and Δ which are positive and can be chosen as small as we please, it is possible to choose an n_0 such that for $n > n_0$

$$\text{prob} \{| x_n - x| < \varepsilon\} > 1 - \Delta. \tag{48.1}$$

γ) *Almost certain (a.c) convergence.* In the previous definition we required the probability that the difference between any random variable x_n and x is less than ε can be made greater than $1 - \Delta$, for $n > n_0$, n_0 being suitably chosen. We can impose a more stringent form of restriction if for every ε and Δ chosen as small as we please, it is possible to find an n_0 such that

$$\text{prob} \left[\{|x_{n_0+1} - x|, |x_{n_0+2} - x|, \ldots\} < \varepsilon \right] > 1 - \Delta. \tag{48.2}$$

We then say that the sequence is almost certain (a.c.) convergent. In defining the above probability we recognise that no member of the aggregate $x_{n_0+1}, x_{n_0+2}, \ldots$ (The aggregate in conformity with our previous notation can be written as $\langle x_n \rangle_{n > n_0}$) should differ from x by more than ε while in the previous case we consider only a pair x_n, x $(n > n_0)$ "at a time".

δ) *Mean square (m.s.) convergence.* If it is possible to choose an n_0 such that for $n > n_0$

$$\mathscr{E} \{|x_n - x|^2\} < \varepsilon \tag{48.3}$$

for any $\varepsilon > 0$ chosen as small as we please we say the sequence is mean square (m.s.) convergent.

It can be shown that γ and δ imply α and β. That γ implies β and α, is of course apparent from the definitions themselves. However γ and δ do not imply each other. The above notions can be understood through simple examples (Bartlett [3]). Let us assume that the random variables are independent.

(This is a convenient assumption if we take a discrete sequence, but leads to difficulties in the case of continuous parametric set.)

(i) Let x_n be a stochastic variate such that it can take only values n or 0 with probability $\frac{1}{n^2}$ and $1-\frac{1}{n^2}$ respectively. This sequence is obviously a.c. convergent. We require that

$$\text{prob}\,\{\langle x_n - 0\rangle < \varepsilon\} > 1 - \varDelta,$$

i.e., the infinite product

$$\prod_{m=1}^{\infty}\left[1 - \frac{1}{(n_0+m)^2}\right] \tag{48.4}$$

can be made to approach unity arbitrarily near by choosing a large enough n_0. The sequence is thus a.c. convergent while it is not m.s. convergent since $\lim_{n\to\infty}\mathscr{E}\,\{(x_n-0)^2\}=1$.

(ii) Consider the example when $x_n=1$ with probability $1/n$ and 0 with probability $1-\frac{1}{n}$. Then $\lim_{n\to\infty}\mathscr{E}\,\{(x_n-x_0)^2\}=0$ and hence the series is mean square convergent. However since

$$\text{prob}\,\{\langle x_n-0\rangle_{n>n_0} < \varepsilon\} = \prod_{m=1}^{\infty}\left[1 - \frac{1}{(n_0+m)^2}\right] \tag{48.5}$$

tends to zero the sequence is not a.c. convergent.

49. Convergence and random functions. The extension of the above concepts to the case of the continuous parametric systems is clear only in the case of i.p., m.s. and ordinary convergence but not in a.c. convergence.

$$\lim_{\varDelta\to 0}\mathscr{E}\,\{[x(t+\varDelta)-x(t)]^2\}=0 \tag{49.1}$$

is the criterion for mean square convergence and this is equivalent to the requirement

$$\lim_{\varDelta\to 0}\mathscr{E}\,\{x(t)\,x(t+\varDelta)\}=\mathscr{E}\,\{x^2(t)\}. \tag{49.2}$$

The definition for i.p. convergence runs on similar lines as for the discrete sequence

$$\lim_{\varDelta\to 0}\text{prob}\,\{[x(t+\varDelta)-x(t)] > \varepsilon\} = 0. \tag{49.3}$$

The concept of ordinary convergence is equally direct. If $x^R(t)$ is the realised trajectory (this is a sequence of realised values as t varies), we require that $x^R(t+\varDelta)$ converges to $x^R(t)$.

In the case of a.c. convergence we have to consider the aggregate in the continuous case corresponding to $\langle x_n\rangle_{n>n_0}$ in the discrete sequence. This aggregate is obviously the continuous infinite set of random variables $x(\tau)$ for $t+\varDelta \geq \tau \geq t$. A direct extension of the concept of almost certain convergence to the continuous case would mean that we can chose a suitable $\varDelta$ such that the probability that no member of the aggregate $x(\tau)$, $t+\varDelta \geq \tau \geq t$ differs from $x(t)$ by values greater than ε can be made greater than $1-\delta$, δ and ε being arbitrarily small.

In the case of continuous t instead of speaking of $x(t+\varDelta)$ converging to $x(t)$ we say that $x(t)$ is continuous at t. Ordinary continuity merely means that every realised trajectory is continuous and implies almost certain continuity.

On these definitions, the basic random process which we have considered in the previous sections is a.c. continuous at all points on the t-axis. We recall that the definition of a basic random process progressing with its parameter t,

$R(x'|x)\, dx'\, \Delta$ is the probability that $x(t+\Delta)$ lies between x' and $x'+dx'$, given that $x(t)=x\,(x\neq x')$. Therefore the probability that $x(t+\Delta)=x$ is given by $1-\Delta \int_{x'} R(x'|x)\, dx'$. Since $\Delta \int_{x'} R(x'|x)\, dx' \to 0$ as $\Delta \to 0$ it is obvious that

$$\lim_{\Delta \to 0} \operatorname{prob}\{[x(t+\Delta)-x(t)]>\varepsilon\}=0, \tag{49.4}$$

ε and Δ being arbitrarily small. Hence a basic random process is a.c. continuous at all points. To investigate whether it is m.s. continuous at any point, we note that

$$\mathscr{E}\{[x(t+\Delta)-x(t)]^2\}=\Delta \int_{x'} (x-x')^2 R(x'|x)\, dx'. \tag{49.5}$$

This tends to zero as $\Delta \to 0$ if the integral on the right-hand side is finite. But for the definition of a basic random process this is not necessary.

However it is obvious that a basic random process does not satisfy the require-ments of ordinary continuity since ordinary continuity would mean that every realised trajectory should be continuous and this is not the case with a basic random process.

The calculus of random functions can be developed on the concept of mean square convergence. In fact, this is the standard procedure adopted by mathe-maticians. For the sake of completeness, we summarise the analytical operations based on mean square convergence.

50. Stochastic differentiation. A random function is continuous in mean square at t if

$$\lim_{\Delta \to 0} \mathscr{E}\{[x(t+\Delta)-x(t)]^2\}=0. \tag{50.1}$$

A necessary and sufficient condition for mean square continuity is that

$$\lim_{t_1,t_2 \to t} \mathscr{E}\{x(t_1)\, x(t_2)\} = \lim_{t_1,t_2 \to t} \mu(t_1,t_2) = \mathscr{E}\{x^2(t)\}. \tag{50.2}$$

If this is true for all t in an interval, it is mean square continuous in that interval. We say that $x(t)$ is mean square differentiable if there exists a random function $\dot{x}(t)$ such that

$$\lim_{\Delta \to 0} \mathscr{E}\left\{\left[\frac{x(t+\Delta)-x(t)}{\Delta}-\dot{x}(t)\right]^2\right\}=0. \tag{50.3}$$

A necessary and sufficient condition for this is that the partial derivatives $\dfrac{\partial \mu}{\partial t_1}$, $\dfrac{\partial \mu}{\partial t_2}$ and $\dfrac{\partial^2 \mu}{\partial t_1 \partial t_2}$ should exist at $t_1=t_2=t$. If this condition is fulfilled on an in-terval, the mean square derivative exists in that interval.

Let us consider a basic random process. We know that in this case

$$\operatorname{prob}\{[x(t+\Delta)-x(t)]>\varepsilon\} \to k\,\Delta \tag{50.4}$$

Δ being infinitesimal and k being finite and determinate, and that is why the probability tends to zero as $\Delta \to 0$. Therefore $\operatorname{prob}\left\{\dfrac{x(t+\Delta)-x(t)}{\Delta}>\dfrac{\varepsilon}{\Delta}\right\}=k\,\Delta$ and hence as $\Delta \to 0$,

$$\mathscr{E}\left\{\left[\frac{x(t+\Delta)-x(t)}{\Delta}\right]^2\right\} \to k'\,\frac{1}{\Delta}, \tag{50.5}$$

k' being finite and determinate, and therefore tends to infinity as $\Delta \to 0$.

In the case when $\mu(t_1,t_2)$ is only a function of t_1-t_2, sufficient conditions for the mean square continuity and differentiability can be stated in the follow-ing manner also: if

$$\mu(t_1,t_2) \to k+k'\,\Delta \quad \text{as} \quad t_1-t_2 \to \Delta \to 0 \tag{50.6}$$

it is mean square continuous without being mean square differentiable. If

$$\mu(t_1, t_2) \to k + k' \Delta^2 \tag{50.7}$$

it is mean square continuous and differentiable also. The physical meaning of the statement is that the mean square value of the difference $|x(t+\Delta) - x(t)|$ should be proportional to Δ^2. Two simple ways of achieving this are given below.

(i) $\text{prob}\{|x(t+\Delta) - x(t)| > \varepsilon\} = k\Delta^2$, i.e. the probability of a finite change is proportional to Δ^2 [1]. (It must be remembered that in a basic random process this probability is proportional to Δ.)

(ii) $x(t+\Delta) - x(t)$ can assume only values proportional to Δ and the probability that the modulus of the coefficient of proportionality is above a finite value is finite.

On the basis of (50.6) and (50.7) we immediately see that the random functions with $\mu(t_1, t_2)$ of the type $\cos \omega(t_1 - t_2)$ and $\exp[-\alpha(t_1 - t_2)^2]$ are obviously mean square continuous and differentiable, while random functions with correlations of the type $\exp[-\alpha(t_1 - t_2)]$ are continuous without being differentiable in mean square. Correlation of type $\exp[-\alpha(t_1 - t_2)]$ is a characteristic feature of basic random processes and also occurs in a large class of physical processes.

Loeve [14] has given simple sufficient conditions for a.c. continuity and differentiability. By arguments similar to those given above, i.e. by studying the relationship between $x(t)$ and $x(t+\Delta)$ for infinitesimal Δ, we are led to the sufficient condition that $x(t)$ is a.c. continuous if it is differentiable once in mean square.

On the basis of this definition of differentiation, the following simple extensions can be made.

1. If we are dealing with aggregates of random functions, we can introduce the analytical process of partial differentiation.

2. The operations of m.s. differentiation and taking expectation values commute:

$$\mathscr{E}\{\dot{x}(t)\} = \frac{\partial}{\partial t} \mathscr{E}\{x(t)\}. \tag{50.8}$$

3. If $\mu(t_1, t_2, \ldots, t_n)$ is the product moment

$$\mathscr{E}\{x(t_1)\, x(t_2) \ldots x(t_n)\}$$

and if by $x^{(p)}(t)$, we mean the p-th mean square differential coefficient of $x(t)$, by virtue of the fact that the operation m.s. differentiation and stochastic averaging commute, we have

$$\mathscr{E}\{x^{(p_1)}(t_1)\, x^{(p_2)}(t_2) \ldots x^{(p_n)}(t_n)\} = \left(\frac{\partial}{\partial t_1}\right)^{p_1} \left(\frac{\partial}{\partial t_2}\right)^{p_2} \ldots \left(\frac{\partial}{\partial t_n}\right)^{p_n} \mu(t_1, t_2, \ldots, t_n) \tag{50.9}$$

and in particular $\mathscr{E}\{\dot{x}(t_1)\dot{x}(t_2)\} = \dfrac{\partial^2 \mu(t_1, t_2)}{\partial t_1 \partial t_2}$. It also follows that

$$\mathscr{E}\{x(t_1)\, \dot{x}^2(t_2)\} = \mathscr{E}\{x(t_1)\, \dot{x}(t_2)\, \dot{x}(t_3)\} \quad \text{when} \quad t_2 = t_3. \tag{50.10}$$

Similar "contractions" can be made when powers of random functions are involved.

---

[1] This possibility can be ruled out in evolutionary Markovian processes since it leads to the trivial result that no random transitions occur in a finite interval t.

51. Stochastic integration. Integration of a random function in the Riemann sense can be defined closely following that of ordinary integration. Let $x(t)$ be a random process. We wish to investigate the meaning of the integral

$$y(t) = \int_0^t x(\tau)\, d\tau. \tag{51.1}$$

As in ordinary integration, we divide the interval 0 to t into n intervals, $\Delta_1, \Delta_2, \ldots, \Delta_n$ and let $t_1, t_2, \ldots, t_n$ be the arbitrarily chosen points in the n intervals respectively. $U_n(t) = \sum_i x(t_i)\Delta_i$ is then a stochastic variable since each $x(t_i)$ is. $U_n(t)$ is a member of the sequence as we vary n. If

$$\lim_{n \to \infty} \mathscr{E}\{[U_n(t) - y(t)]^2\} = 0 \tag{51.2}$$

we say that $y(t)$ is the mean square Riemann integral of $x(t)$. The necessary and sufficient condition for this to exist is the existence of the ordinary double integral

$$\int_0^t \int_0^t \mu(t_1, t_2)\, dt_1\, dt_2.$$

The above notion can be extended to the Lebesgue integral (MOYAL [13]) of the random function $x(t)$. For this purpose it is necessary to introduce a measure $\eta(S)$, a non-negative additive set function in the space suitable for all Borel sets S of t. While in the Riemann case we divide 0 to t into n intervals, here we have to divide the t space into n arbitrary disjoint sets S_i $(i = 1, 2, \ldots, n)$. If x_i is the random variable attached to an arbitrary point in S_i we require the sum $\sum_i x_i \eta(S_i)$ should converge in mean square to the L integral of $x(t)$. A necessary and sufficient condition for this is that the repeated L integral of $\mu(t_1, t_2)$ should exist.

For both Riemann and Lebesgue integrals, the integral and averaging operations commute.

The above discussion is a brief summary of the standard definition of analytical operations given in mathematical papers on stochastic processes[1]. It should be noted that in the above treatment, no appeal was made to the concept of the trajectory of random functions. In the opinion of the author it is desirable that the connection between the concept of trajectory and the concept of convergence should be made clear if physical phenomena were to be represented through stochastic integrals or differential coefficients.

In our phenomenological definition of integration, we were concerned with the distribution functions of the integrals, given all information about the integrand—by all information we mean that the measure of any typical trajectory is known. To fix our ideas we started with a basic random process and studied its integrals. In the definition of integration we assumed that the realised values of the integrals were the integrals of the realised values of the integrands. A comparison of this definition based on physical approach with those given above will form the subject matter of a future contribution[2]. But in most physical examples dealing with evolutionary processes, the physical approach gives all answers and in fact the following general theorem connecting the moments of

---

¹ For a comprehensive list of references see MOYAL [13].
² To be published by ALLADI RAMAKRISHNAN and R. VASUDEVAN.

the integral with the correlation function of the integrand was established by the author [15] by physical arguments:

$$\mathscr{E}\left\{y^{n}(t)\right\} = \int_{0}^{t}\int_{0}^{t} \cdots \int_{0}^{t} \mathscr{E}\left\{x(t_1)\, x(t_2) \ldots x(t_n)\right\} dt_1\, dt_2, \ldots, dt_n. \tag{51.3}$$

IV. Physical examples of stochastic processes.

Class A: x discrete, t discrete.

52. Before the advent of the methods of stochastic theory it was customary to consider a process with continuous t as the limiting case of a process progressing with discrete t. Since in most of the physical processes the dynamical nature is associated with time which is a continuous parameter, not many examples with considerable physical interest can be cited with t discrete. However for the sake of completeness we shall formally write the stochastic equation for a Markovian process relating to a discrete variable and discrete time.

Let $t_1, t_2, \ldots, t_m$ denote the values which the discrete parameter can take, the number of values m being at most enumerably infinite and the random variable x can assume the values $x_1, x_2, \ldots, x_n$ where n can atmost be enumerably infinite. Defining $\pi(j; t_i)$ as the probability that $x = x_j$ at time $t = t_i$, then $\pi(j; t_i)$ satisfies the difference equation

$$\pi(j; t_i) = \sum_{k} \pi(k; t_{i-1})\, R(j\,|\,k; t_i, t_{i-1}) \tag{52.1}$$

where $R(j\,|\,k; t_i, t_{i-1})$ is the fundamental transition probability of jumping from x_k at t_{i-1} to x_j at t_i. If x is continuous, we have to replace the summation suitably by an integral. The simplest example of a process with x and t both discrete is the elementary one-dimensional random walk problem which can be described as follows. A particle suffers displacement along a straight line in a series of steps equal in length a, each step being taken forward with a probability p and backward with a probability $1 - p$. The number of steps is integral and non-negative and the displacement can take only a discrete set of values.

We can ask for the probability $W(n, m)$ that after n steps the particle has suffered displacement $m\,a$. $W(n, m)$ satisfies the difference equation

$$W(n, m) = p\, W(n - 1, m - 1) + (1 - p)\, W(n - 1, m + 1) \tag{52.2}$$

with the initial condition

$$W(0, 0) = 1, \qquad W(0, m) = 0 \tag{52.3}$$

for all $m \neq 0$. The solution of this is obviously the familiar binomial distribution

$$W(n, m) = \binom{n}{\dfrac{m+n}{2}} p^{\frac{n+m}{2}} (1 - p)^{\frac{n-m}{2}}. \tag{52.4}$$

Note that if n is even, m should be even. If n is odd, m should be odd. For very large n, say N, this tends to the well-known Gaussian formula when $p = \frac{1}{2}$

$$W(N, m) = \left(\frac{2}{\pi N}\right)^{\frac{1}{2}} \exp\left(-\frac{m^2}{2N}\right). \tag{52.5}$$

If we make a very small we can treat the variable $x = m\,a$ when N is large, as a continuous variable and ask for the probability that x lies between x and $x + \Delta x$, where Δx is large compared to a. Then defining $W(N, x)\, \Delta x$ as the probability that the displacement lies between x and $x + \Delta x$ after N steps, we have

$$W(N, x)\, \Delta x = W(N, m)\, \frac{\Delta x}{2a} \tag{52.6}$$

since m can take only even or odd values depending on whether N is even or odd and since $x = m\,a$

$$W(N, x) = \frac{1}{(2\pi N a^2)^{\frac{1}{2}}} \exp\left[-\frac{x^2}{2N a^2}\right]. \tag{52.7}$$

The above equation can be modified and used for slightly more complicated problems like random walk with reflecting and absorbing barriers. As it is dealt with in great detail in the classic article of CHANDRASEKHAR [16] on stochastic problems, no purpose will be served by repeating them here.

Class B: x discrete, t continuous.

53. Some simple examples. *α) The Poisson process.* Physical examples for class B are easily found if x takes zero or positive integer values. The simplest case is the Poisson distribution of events occurring along the t-axis. If we assume that $\lambda\, dt$ is the probability that an event occurs between t and $t + dt$ then the number of events $n(t)$ that occur between 0 and t is a stochastic variable. Defining $\pi(n, t)$ as the probability frequency function of $n(t)$ we obtain by simple probability arguments,

$$\frac{\partial \pi(n, t)}{\partial t} = -\lambda\, \pi(n, t) + \lambda\, \pi(n - 1, t) \tag{53.1}$$

with the initial condition that $\pi(n, 0) = 0$ if $n \neq 0$, $\pi(0, 0 = 1)$. In case when λ is a constant we have the solution

$$\pi(n, t) = \frac{e^{-\lambda t}(\lambda t)^n}{n!}. \tag{53.2}$$

If λ is a function of t, the process becomes quasi-Markovian in the sense of Sect. 45 and the equation is still valid. The solution is given by

$$\pi(n; t) = e^{-k}\, \frac{k^n}{n!} \tag{53.3}$$

where

$$k = \int_0^t \lambda(\tau)\, d\tau. \tag{53.4}$$

It can easily be verified that the m-th moment of $n(t)$ can be expressed in terms of the first moment k.

The generating function is given by

$$G(u) = \sum_n u^n\, \pi(n; t) = e^{-k\,(u-1)}. \tag{53.5}$$

The Poisson distribution is so familiar that it is not necessary to give a particular example. It is obtained also as a limiting case of the binomial distribution defined by

$$\pi(n, N) = \binom{N}{n} p^n (1 - p)^{N-n}, \tag{53.6}$$

where p is the probability of success in a single trial and $\pi(n, N)$ is the probability of getting n success in N trials. Making $N \to \infty$ and $p \to 0$ such that

$N p \rightarrow k$, a finite quantity, it can be shown that the binomial tends to the Poisson distribution.

In the Poisson distribution we have assumed that λ is not a function of n. *We call a process multiplicative if the probability of an event occurring between t and $t+dt$ is a function of n for the obvious reason that the number of events occurring between t and $t+dt$ depends upon the number between 0 and t which is essentially a feature of multiplication.*

β) *The Furry* [17] *process.* The simplest multiplicative process occurs if the events follow the law that the probability of the occurrence of an event in dt is $n \lambda\, dt$, if n events have occurred between 0 and t. The physical meaning of this is obvious if by events we mean occurrence of particles and each particle existing at t has a probability $\lambda\, dt$ of producing another particle in dt. Then the equation for $\pi(n, t)$ will read as follows

$$\frac{\partial \pi(n, t)}{\partial t} = (n - 1)\, \lambda\, \pi(n - 1, t) - n\, \lambda\, \pi(n, t) \tag{53.7}$$

with the initial condition

$$\pi(n, t) = 1 \quad \text{if} \quad n = 1 \quad \text{at} \quad t = 0.$$

This initial condition is necessary in view of the dependence of the probability of occurrence of the event between t and $t+dt$ on n. The solution of the above equation is the well known Furry distribution given by

$$\pi(n, t) = e^{-\lambda t}(1 - e^{-\lambda t})^{n-1}. \tag{53.8}$$

Its generating function is given by

$$G(u) = u\, e^{-\lambda t}[1 - u\,(1 - e^{-\lambda t})]^{-1}. \tag{53.9}$$

If we require $\pi(n \mid m; t)$ the probability that there are n particles between 0 and t given that there were m at $t=0$, we first recognise that since each of the m particles behave independently the generating function of $\pi(n \mid m; t)$ is $[G(u)]^m$ where $G(u)$ is the generating function of $\pi(n \mid 1; t)$, the Furry distribution. We note that

$$\lim \pi(n \mid m; t) \quad \text{as} \quad t \to \Delta \to 0 \\ \qquad = m\, \lambda\, \Delta \quad \text{if} \quad n = m + 1, \\ \qquad = 0 \quad \text{otherwise.} \tag{53.10}$$

More complicated limits for $\pi(n \mid m; t)$ can be assumed depending upon the nature of the process. If we introduce interaction, the limit will be a complicated function of n.

The above treatment can also be interpreted to give the law of distribution of random points on a line if we identify every random point with an event. Hence the name "point processes" to such a class.

γ) *A simple regenerative non-Markovian process.* We shall now consider another class of point processes defined by the following law. The probability that the n-th point occurs between t and $t+dt$ is a function of the distance from the occurrence of the previous point. Obviously this renders the process represented by the distribution function $\pi(n; t)$ non-Markovian, as it is not enough to know the number of points that have occurred between 0 and t to predict what happens between t and $t+dt$. But since the "memory" extends only up to the last point, this is just the property required for the regeneration point method. Thus if we define $\varphi(a)\,\Delta$ as the probability that a point occurs between t and $t+\Delta$ given that the previous point has occurred at $t-a$ then $\varphi(a)$ determines the whole process. Applying the regeneration point technique we have

$$\pi(n;t) = \int_0^t \pi(n-1, t-\tau)\, \varphi(\tau)\, d\tau. \tag{53.11}$$

We can recognise this to be the random walk problem from a new point of view. We ask for the number of steps that can be taken between 0 and t, if the distribution of the length of a single step is of the form $\varphi(a)$. If $\varphi(\tau) = e^{-\lambda\tau}\lambda$ then $\pi(n;t)$ reduces to the Poisson distribution. The solution for the above is easily obtained by considering the generating function with respect to n and then taking the Laplace transform with respect to t.

The immediate application of this equation is to "counter problems" where the occurrence of a recorded event is followed by a dead time and $\varphi(\tau) = 0$ if $\tau < a$ and $\varphi(\tau) = e^{-\lambda\tau}\lambda$ if $\tau > a$. (If the dead time a tends to 0, it is clear we obtain the Poisson distribution.) In this case, every event is followed by a dead time if registered. The total number of events obey the Poisson law $e^{-\lambda t}\,\dfrac{(\lambda t)^n}{n!}$ while the recorded events satisfy the equations for $\pi(n;t)$ given above. The above equation cannot be directly applied if every event recorded or unrecorded is followed by a dead time.

$\delta)$ *Birth, death and immigration processes.* We have introduced till now two modes of increase of the number of "particles" namely (1) increase per infinitesimal interval independent of the number (non-multiplicative), (2) increase depending on number (multiplicative).

We now introduce the additional random element of death or annihilation of particles, which may or may not depend on the number of "particles" present at t. This leads us to KENDALL's [18] birth, death and immigration process.

The simplest stochastic process involving the three random elements of birth, death and immigration is defined as follows:

1. The probability that a new individual ("particle") is created is proportional to n, the number of individuals i.e., $\lambda\,n\,dt$.

2. The probability of an increase of one individual by "influx" from outside is $k\,dt$.

3. The probability that one of the n individuals dies in the interval dt is $\mu\,n\,dt$.

Then $\pi(n;t)$ the probability that the population consists of n individuals at t, obeys the equation

$$\left.\begin{aligned}
\frac{\partial\pi(n;t)}{\partial t} = {}& \{\lambda(n-1)+k\}\,\pi(n-1;t) - \\
& - (\lambda n + k)\,\pi(n;t) + \\
& + \mu\{\pi(n+1;t) - \pi(n;t)\}.
\end{aligned}\right\} \tag{53.12}$$

The problem becomes more complicated if λ, μ, k are functions of t which renders the process quasi-Markovian in the sense of Sect. 45. The above equation can be solved by the standard method of generating functions, the initial conditions being known.

A still more complex situation will arise if we consider the age distribution of particles and λ and μ depend on age. Obviously this involves defining the statistical distribution of a discrete number of particles in a continuous space α, α representing age. It is then necessary to use the concept of product densities as was done by KENDALL.

ε) Simple multiple production. We can introduce another feature in the multiplicative nature of the process. We may assume that in any interval, k particles can be produced with probability $\varphi(k)\,dt$. Till now we have assumed that k can take a value equal to 1 only, i.e. λ is assumed to be $\varphi(1)$ in the above equation. If we introduce the above idea of multiple production, we have merely to replace the term

$$\lambda\{(n-1)\,\pi(n-1;t) - n\,\pi(n;t)\} \tag{53.13}$$

by the following term,

$$\sum_{k=1}^{N} \varphi(k)\,\{(n-k)\,\pi(n-k;t) - n\,\pi(n;t)\}. \tag{53.14}$$

We do not attempt a general solution of the equation. To realise the nature of the difficulty in obtaining the solution we write the alternative equation for $\pi(n;t)$ or equivalently its generating function using the regeneration point method,

$$\frac{\partial G(u;t)}{\partial t} = \{\psi(G)\,G - G\}\,\varphi \tag{53.15}$$

where

$$G(u;t) = G = \sum_{n=0}^{\infty} u^{n}\,\pi(n;t), \tag{53.16}$$

$$\varphi = \sum_{1}^{\infty} \varphi(n), \quad \psi(u) = \sum u^{n}\,\varphi(n). \tag{53.17}$$

This equation is obtained by asking for what happens in the first small interval $\varDelta$. If k particles are produced, these k particles in addition to the initial particles become independent primaries of a stochastic process of duration $t-\varDelta$.

ζ) Branching stochastic processes[1]. Till now we considered only a single stochastic variable. We can consider a process relating to the multiplication of different types of particles, i.e. we are interested in the function $\pi(n_1, n_2, \ldots, n_k; t)$, the probability that there are n_1 particles of type 1, n_2 particles of type 2, $\ldots, n_k$ of type k at time t. Following the arguments of the preceding section for the computation of π, we need the fundamental transition probability $R_i(m_1, m_2, \ldots, m_k)\,dt$ that in an interval dt, a particle i produces m_1 of the type 1, m_2 of the type 2, etc. Such a process is called a branching process. An equation can be written down for π; but no use will be served since no new principle is involved other than what has been outlined above. If we introduce additional parameters like energy, or other dynamical variables associated with these particles, we are again led to the use of the product density technique.

54. General remarks on Class B [2]. We shall now make some general observations regarding processes evolving with continuous t and involving one discrete random variable or an aggregate of discrete random variables. Instead of speaking of a random variable, it is more convenient to define our stochastic process as follows.

We shall assume a physical system is capable of occupying one of a discrete set of states $S_1, S_2, \ldots, S_n$ called occupation states. If we can ascribe numbers to these states, the number represents a random variable. However, we shall refrain temporarily from doing so, since a state may be described not merely by

---

[1] By a private communication, I hear that the mathematical theory of branching processes is under preparation by T. E. HARRIS of the Rand Corporation, California.

[2] I am grateful to N. R. RANGANATHAN for communicating his unpublished results on which this section is based.

one but by an aggregate of random variables and it will be more convenient to use a comprehensive term state in its widest sense. If we assume the process is Markovian in t, say time, we can speak of the probability $\pi(j\,|\,i;t_1,t_2)$ that the system is in a state S_j at t_2 given that it was in S_i at t_1. If in addition we assume it is homogeneous in time we can set without loss of generality $t_1=0$ and $t_2=t$. In Sect. 39, we obtained the differential equation of the process in terms of $R(j\,|\,k)$ where $R(j\,|\,k)\,\varDelta$ is obtained as the limit of π. In writing this equation we assumed that the R's are independent of t. However this restriction is not necessary and the equations are still valid even when R's can be functions of t. But it is clear that as soon as we say that R's are functions of time and since this time has to be measured from the point of reference the process is not homogeneous with respect to t but only with respect to the point of reference which may be anywhere on the t axis. Such processes have been called quasi-Markovian for reasons elaborated earlier. Our object now is to consider special cases of the **R** matrix with elements $R(j\,|\,k)$ which in the following section are referred to as R_{jk} for convenience.

Particular cases of the **R** *matrix*:

(i) **R** *is independent of time.*

(a) R_{ik}'s are completely connected for infinitesimal transformation i.e. $R_{ik}>0$ for all i,k $(i\neq k)$.

(b) Some R_{ik} may be zero, but for every such pair of states i^+ and k^+ there exists a sequence of states $l_1, l_2, \ldots, l_n$ such that $R_{l_1 i^+}, R_{l_2 l_1} \ldots R_{k^+ l_n}$ each is greater than 0. Such a system is said to be connected only for finite transformations.

(c) $R_{ik}=R_i$ for all values of k.
(d) $R_{ik}=R_k$ for all values of i.
(e) $R_{ik}=R_{ki}$.

(ii) **R** *is time dependent.*

We shall consider the features of the stochastic processes one by one without giving any proof for the results. FELLER's book [*19*] treats the case of discrete x. An exhaustive summary of these processes is given in BARTLETT's book with particular reference to asymptotic properties. Mention must be made of the very interesting mathematical papers by LEDERMANN [*20*] and others in this connection.

Let us now consider case (i). $\pi(j\,|\,k;t)>0$ for all j,k if the states are completely connected. Even if condition (b) alone is satisfied it can be shown that $\pi(j\,|\,k;t)>0$ for all j and k and that is why we speak of states as being connected for finite transformations. If condition (a) or (b) is satisfied the system is called *ergodic*.

If we observe the system as it evolves between 0 and t and plot its realisation, the system makes only a discrete number of transitions in the continuous interval 0 to t and in between these transitions it stays in any particular state. Thus we can speak of $\tau(j)$, the time spent in the state S_j between 0 and t, and of the fraction $x(j;t)=\tau(j)/t$. $\tau(j)$ and $x(j;t)$ are random variables and they have frequency functions. To find out their frequency functions we should know the measure of a typical trajectory of the process. This is easily done formally since the measure of the class of trajectories which start from the state S_{i_0} at $t=0$ and make transitions to the states $S_{i_1}, S_{i_2}, \ldots$ in the intervals $dt_1, dt_2, \ldots$ can be expressed in terms of π functions and R_{ji}'s. But to compute the time spent in any particular state from this is a formidable task. However if we are interested only in the moments of the distribution a simple lemma is available which helps us to compute this.

Let $d\tau(j;t)$ be a random variable $\{\tau(j;t+\Delta)-\tau(j;t)\}$ representing the time spent in state j in interval Δ. $d\tau(j;t)$ has simpler features than $\tau(j;t)$ since in a small interval Δ either the whole of the time is spent in j or not. Thus $\pi(j;t)\,\Delta$ is the mean value of the time spent in Δ in the state j and this can be immediately extended to give the following result,

$$
\left.
\begin{aligned}
&\mathscr{E}\{d\tau(j;t)\} = \pi(j\mid i)\,\Delta,\\
&\mathscr{E}\{d\tau(j;t_1)\,d\tau(j;t_2)\ldots d\tau(j;t_n)\}\\
&\qquad = \pi(j\mid i;t_1)\,\pi(j\mid j;t_2-t_1)\ldots\pi(j\mid j;t_n-t_{n-1})\,\Delta_1\Delta_2\ldots\Delta_n.
\end{aligned}
\right\}
\tag{54.1}
$$

To study the ergodic nature of any process, we have to let $t\to\infty$ and in such a case it is not $\tau(j;t)$ that is interesting but only $x(j;t)$. If we assume that as $t\to\infty$, $\pi(j\mid i;t)$ tends to $\pi(j)$ independent of i, we can show that

$$
\mathscr{E}\{x^n(j;t)\} \to [\pi(j)]^n,
\tag{54.2}
$$

i.e. the random variable $x(j)$ converges in probability to the number $\pi(j)$, i.e. the fraction of the time spent has a sharp peak at $\pi(j)$. Another interesting result relating to these systems is that if we define $K(j\mid i;t)$ as the probability that the system does not enter the state j in time interval $(0, t)$, $K(j\mid i;t)\to0$ as $t\to\infty$, i.e. every state is entered at least once with probability 1. This immediately implies that every state is entered infinitely many times as $t\to\infty$.

Let us consider cases (c) and (d).

(c) $R_{ij}=R_i$. The system reduces to

$$
\frac{\partial\pi(j;t)}{\partial t} = -\pi(j;t)\sum_i R_i + R_j.
\tag{54.3}
$$

The limit as $t\to\infty$ is obtained in this case very easily as

$$
\pi(j,\infty) = R_j\Big/\sum_i R_i.
\tag{54.4}
$$

(d) $R_{ij}=R_j$. The corresponding system of equations is

$$
\left.
\begin{aligned}
\frac{\partial\pi(j\mid i;t)}{\partial t} &= -(n-1)\,R_j\,\pi(j;t) + \sum_{i\neq j}\pi(i;t)\,R_i\\
&= -n\,R_j\,\pi(j;t) + \sum_i \pi(i;t)\,R_i.
\end{aligned}
\right\}
\tag{54.5}
$$

We observe that the second term $\sum_i \pi(i;t)\,R_i$ occurs in all the equations as we vary j and for convenience we call it $f(t)$. As $t\to\infty$, if we assume $\pi(j;t)$ tends to a limit, it is given by

$$
\pi(j;\infty) = f(\infty)/n\,R_j.
\tag{54.6}
$$

There is no need to determine $f(\infty)$ since we know that $\sum_j \pi(j;t)=1$ and therefore from the above equation we get

$$
\pi(j;t) = \frac{1/R_j}{\sum_i 1/R_i}.
\tag{54.7}
$$

(e) $R_{ij}=R_{ji}$. This is very important in statistical mechanics[1]. The reduced system of equations in this case is

$$
\frac{\partial\pi(j;t)}{\partial t} = -\pi(j;t)\sum_i R_{ij} + \sum_i \pi(i;t)\,R_{ji} = \sum_i [\pi(i;t)-\pi(j;t)]\,R_{ij}.
\tag{54.8}
$$

---

[1] In macroscopic theory of irreversibility, such R matrices occur, their elements being the ONSAGER [Phys. Rev. **37**, 405 (1931)] coefficients. The vector matrix equation in ONSAGER's theory reads $\dot{\boldsymbol{\alpha}} = [R]\,\boldsymbol{F}$ where the elements of $\boldsymbol{\alpha}$ are the Onsager coordinates α_i and of $\boldsymbol{F}$ are the thermodynamic forces related to the α_i's as $F_i = -\sum g_{ik}\,\alpha_k$.

As time $t \to \infty$ it is clear that $\pi(j; \infty) = \pi(i; \infty)$ for all j and i and hence each equals $1/N$ if there are, N states.

We now consider the case when the R matrix is time dependent. There exists a formal way of solving the equation

$$\frac{\partial \pi(t)}{\partial t} = R(t)\,\pi(t) \tag{54.9}$$

where $\pi(t)$ is the column vector with its typical term $\pi(j; t)$. We can rewrite the equation as an infinitesimal transformation of $\pi(t)$.

$$\pi(t + \Delta t) = [I + R\,dt]\,\pi(t) \tag{54.10}$$

where I is the unit matrix. If $\pi(0)$ is known, $\pi(t)$ can be written as an infinite product

$$\pi(t) = [I + R(n\Delta)\,\Delta]\,[I + R(\overline{n-1}\,\Delta)\,\Delta] \ldots [I + R(\Delta)\,\Delta]\,\pi(0). \tag{54.11}$$

This product as $n \to \infty$ can be written as a sum of integrals,

$$\pi(t) = \left[I + \int\limits_0^t R(\tau)\,d\tau + \int\limits_0^t d\tau_1 \int\limits_{\tau_1}^t R(\tau_2)\,R(\tau_1)\,d\tau_2 + \cdots + \right.$$
$$\left. + \int\limits_0^t d\tau_1 \int\limits_{\tau_1}^t d\tau_2 \ldots \int\limits_{\tau_{n-1}}^t R(\tau_n)\,R(\tau_{n-1}) \ldots R(\tau_1)\,d\tau_n + \cdots \right]\pi(0), \tag{54.12}$$

where $\tau_1, \tau_2, \ldots$ are ordered i.e. $\tau_1 < \tau_2 < \cdots$. If $R(\tau)$ can be written as $R\,\varphi(\tau)$ when R is a time independent matrix, the above expression can be written as

$$\pi(t) = \left[I + R \int\limits_0^t \varphi(\tau)\,d\tau + \frac{R^2}{2!} \int\limits_0^t \int\limits_0^t \varphi(\tau_1)\,\varphi(\tau_2)\,d\tau_1\,d\tau_2 + \right.$$
$$\left. + \cdots + \frac{R^n}{n!} \int\limits_0^t \int\limits_0^t \cdots \int\limits_0^t \varphi(\tau_1)\,\varphi(\tau_2) \ldots \varphi(\tau_n)\,d\tau_1\,d\tau_2 \ldots d\tau_n + \cdots \right]\pi(0). \tag{54.}$$

With this form we can symbolically write it as

$$\pi(t) = e^{R \int\limits_0^t \varphi(\tau)\,d\tau}\,\pi(0) \tag{54.14}$$

where the exponential matrix is equivalent to the above expression. However if $R(\tau)$ cannot be written in the simple form $R\,\varphi(\tau)$ then we must be content with the solution in the first form. But if we wish to retain the above form in this case also, we have to use what is known as the DYSON [21] operator,

$$\pi(t) = \left[I + \int\limits_0^t R(\tau)\,d\tau + \frac{1}{2!} \int\limits_0^t \int\limits_0^t \mathscr{P}\,[R(\tau_1)\,R(\tau_2)]\,d\tau_1\,d\tau_2 + \right.$$
$$\left. + \cdots + \frac{1}{n!} \int\limits_0^t \int\limits_0^t \cdots \int\limits_0^t \mathscr{P}\,[R(\tau_1)\,R(\tau_2) \ldots R(\tau_n)]\,d\tau_1\,d\tau_2 \ldots d\tau_n + \cdots \right]\pi(0) \tag{54.}$$

where $\mathscr{P}$ is the Dyson operator which has the property that operating on a product of t labelled operators it re-arranges them in the same order as the t-sequence of their labels the latest one in t occurring first in the product. We thus conclude that the exponential form is only a conventional way of writing this and it is clear that the inverse of the above matrix cannot be written as $e^{-\int_0^t R(\tau)\,d\tau}$. To find the inverse we go back to the original equation and write

$$\left.\begin{aligned}
\boldsymbol{\pi}(0) &= [I - R(\varDelta)\,\varDelta]\,[I - R(2\varDelta)\,\varDelta]\ldots[I - R(n\varDelta)\,\varDelta]\,\boldsymbol{\pi}(t)\\
&= \left[e^{+\int_0^t R(\tau)\,d\tau}\right]^{-1}\boldsymbol{\pi}(t).
\end{aligned}\right\} \quad (54.15')$$

Symbolically we can define the inverse to be

$$e^{-\int_0^t R(t-\tau)\,d\tau}. \tag{54.16}$$

Exponential matrices play a fundamental role in the invariant perturbation theory of the interaction representation of Dyson [21] and an ingenious operator calculus has been devised by Feynman [22] very recently. It starts with the basic idea that e^{A+B} where A and B are matrices has to be distinguished from $e^A\,e^B$ if B and A are non-commutative. The reader is referred to Feynman's paper.

Class C: x continuous, t discrete.

55. The most obvious example is obtained by introducing another random element in the problem of random walk discussed earlier (Class A: x discrete, t discrete) by treating the length of the step itself as a continuous stochastic variable with a probability frequency distribution $\varphi_j(\eta)\,d\eta$. In the language of this distribution, the case when the displacement is $\pm a$ with probability p and $1-p$ can be written as

$$\varphi(\eta) = \delta(\eta - a)\,p + \delta(\eta + a)\,(1 - p) \tag{55.1}$$

where δ is the Dirac delta function. Treating the general case of $\varphi_j(\eta)$ we note that x, the total displacement, is a continuous variable. The equation for $W(N, x)$ reads

$$W(N, x) = \int_{-\infty}^{+\infty} W(N - 1, x - \eta)\,\varphi_N(\eta)\,d\eta. \tag{55.2}$$

We immediately ecognise this to be the N-fold convolution of the functions $\varphi_j(\eta)$ and hence defining the characteristic function of $\varphi_j(\eta)$ as

$$\Phi_j(\varrho) = \int_{-\infty}^{+\infty} e^{i\varrho x}\,\varphi_j(x)\,dx \tag{55.3}$$

we obtain the characteristic function of $W(N, x)$ as

$$A(N, \varrho) = \prod_{j=1}^{N} \Phi_i(\varrho). \tag{55.4}$$

The above considerations can be immediately extended to the three-dimensional random walk. In such a case we define $\varphi_j(x, y, z)\,dx\,dy\,dz$ as the probability that in the j-th step the components of the displacement lie between x and $x + dx$, y and $y + dy$, z and $z + dz$. The equation now reads

$$W(N, X, Y, Z) = \iiint_{-\infty}^{+\infty} W(N - 1, X - x, Y - y, Z - z)\,\varphi_N(x, y, z)\,dx\,dy\,dz \tag{55.5}[1]$$

---

[1] We use the capitals $\boldsymbol{R}$, X, Y, Z to denote the total displacement and its components to distinguish from $\boldsymbol{r}$, x, y, z, the displacement and its components corresponding to a single step.

where $W(N, X, Y, Z)\, dX\, dY\, dZ$ is defined as the probability that the components of the displacement lie between X and $X + dX$, Y and $Y + dY$ and Z and $Z + dZ$ respectively.

We can define the three-dimensional characteristic function of

$$\varphi_j(x, y, z) \quad \text{or} \quad \varphi_j(\boldsymbol{r}) \quad \text{as} \quad \Phi_j(\varrho_x, \varrho_y, \varrho_z) \quad \text{or} \quad \Phi_j(\varrho):$$

$$\Phi_j(\varrho) = \int\!\!\!\int\!\!\!\int_{-\infty}^{+\infty} \exp(i\,\varrho \cdot \boldsymbol{r}_j)\, \varphi_j(\boldsymbol{r}_j)\, dx_j\, dy_j\, dz_j. \tag{55.6}$$

Then as before the characteristic function of $W(N, \boldsymbol{R})$ is given by

$$A(N, \varrho) = \prod_{j=1}^{N} \Phi_j(\varrho). \tag{55.7}$$

If all the φ_j's and hence the Φ_j's are identical and denoted by φ and Φ respectively

$$A(N, \varrho) = [\Phi(\varrho)]^N, \tag{55.8}$$

$$W(N, \boldsymbol{R}) = \frac{1}{8\pi^3} \int\!\!\!\int\!\!\!\int_{-\infty}^{+\infty} \exp(-i\,\varrho \cdot \boldsymbol{R})\, A(N, \varrho)\, d\varrho_x\, d\varrho_y\, d\varrho_z. \tag{55.9}$$

Three cases of particular interest have been considered by CHANDRASEKHAR.

Case I:

$$\varphi_j(\boldsymbol{r}_j) = \frac{1}{\left[\dfrac{2\pi}{3}\, \mathscr{E}\{l_j^2\}\right]^{\frac{3}{2}}} \exp\left[-3\,|\boldsymbol{r}_j|^2/2\, \mathscr{E}\{l_j^2\}\right] \tag{55.10}$$

where $\mathscr{E}\{l_j^2\}$ denotes the mean square displacement on the j-th occasion l_j being the length of the j-th step, i.e. $l_j = |\boldsymbol{r}_j|$. He arrives at the equation

$$W(N, \boldsymbol{R}) = \frac{1}{\left(\dfrac{2\pi}{3}\, N l^2\right)^{\frac{3}{2}}} \exp\left[-3\,|\boldsymbol{R}|^2/2N l^2\right] \tag{55.11}$$

where

$$l^2 = \frac{1}{N} \sum_j \mathscr{E}\{l_j^2\}. \tag{55.12}$$

This is an exact solution valid for any value of N.

Case II: Let the displacement on the j-th occasion be of length exactly equal to l_j in a random direction. Under this circumstance we define the distribution function φ_j as

$$\varphi_j(\boldsymbol{r}_j)\, dx_j\, dy_j\, dz_j = \frac{1}{4\pi l_j^2}\, \delta(|\boldsymbol{r}_j| - l_j)\, dx_j\, dy_j\, dz_j \tag{55.13}\,[1]$$

where δ is the Dirac delta function and $j = 1, 2, \ldots N$. This leads to an expression for $W(N, \boldsymbol{R})$ given by

$$W(N, \boldsymbol{R}) = \frac{1}{2\pi^2 |\boldsymbol{R}|} \int_0^{\infty} \sin(|\varrho|\,|\boldsymbol{R}|) \left\{\prod_{j=1}^{N} \frac{\sin|\varrho|\,l_j}{|\varrho|\,l_j}\, |\varrho|\, d|\varrho|\right\} \tag{55.14}$$

---

[1] In CHANDRASEKHAR's article the distribution function has been taken to be $\frac{1}{4\pi l_j^2}\, \delta(|\boldsymbol{r}_j|^2 - l_j^2)$. But the author is of opinion that (55.13) is the correct representation.

where ϱ occurs in the characteristic function

$$A(N, \varrho) = \prod_{j=1}^{N} \frac{1}{4\pi l_j^3} \int \exp(i\,\varrho \cdot r_j)\,\delta(|r_j| - l_j)\,d x_j\,d y_j\,d z_j. \qquad (55.15)$$

The above equations have been given by Lord RAYLEIGH [23] in one of his classic papers.

Case III: The most interesting case occurs when all the l_j's are equal, $l_j = l = $ a constant and φ_j is independent of j. In this case if $N \to \infty$

$$\lim \left[\frac{\sin(|\varrho|\,l)}{|\varrho|\,l}\right]^N = \lim_{N\to\infty}\left[1 - \frac{1}{6}|\varrho|^2\,l^2 + \cdots\right]^N = \exp(-N|\varrho|^2\,l^2/6). \qquad (55.16)$$

Here, writing $W(N, R)$ as $W(R)$ when $N \to \infty$ we obtain

$$W(R) = \frac{1}{\left(\frac{2\pi}{3}\,N l^2\right)^{\frac{3}{2}}} \exp\left[-3|R|^2/2N l^2\right]. \qquad (55.17)$$

Case IV: The solution to the general problem of random flights for $W(N, R)$.

The solution $W(N, R)$ for large values of N, with no special assumptions concerning the distribution of the different displacements except that all the $\varphi_j(r_j)$ represent the same function $\varphi(r_j)$, has been given by CHANDRASEKHAR. We define as usual

$$A(N, \varrho) = \left[\int\!\!\!\int\!\!\!\int_{-\infty}^{+\infty} \exp\left[i(\varrho_x x + \varrho_y y + \varrho_z z)\right]\varphi(x, y, z)\,d x\,d y\,d z\right]^N \qquad (55.18)$$

where ϱ_x, ϱ_y, and ϱ_z denote the components of ϱ in some fixed system of coordinates.

As $N \to \infty$, $A(N, \varrho)$ reduces to the form

$$A(N, \varrho) = \exp\left[iN\{\varrho_x\,\mathscr{E}\{x\} + \varrho_y\,\mathscr{E}\{y\} + \varrho_z\,\mathscr{E}\{z\}\} - \tfrac{1}{2}N\,Q(\varrho)\right] \qquad (55.19)$$

where

$$\left.\begin{aligned}Q(\varrho) = \varrho_x^2\,\mathscr{E}\{x^2\} + \varrho_y^2\,\mathscr{E}\{y^2\} + \varrho_z^2\,\mathscr{E}\{z^2\} + 2\varrho_x\varrho_y\,\mathscr{E}\{x y\} + \\ + 2\varrho_y\varrho_z\,\mathscr{E}\{y z\} + 2\varrho_z\varrho_x\,\mathscr{E}\{z x\},\end{aligned}\right\} \qquad (55.20)$$

the expectation values on the right hand side relating to $\varphi(x, y, z)$. To evaluate the integral, we can rotate the coordinate system to bring the quadratic form $Q(\varrho)$ to its diagonal form,

$$Q(\varrho) = \mathscr{E}\{h^2\}\,\varrho_h^2 + \mathscr{E}\{k^2\}\,\varrho_k^2 + \mathscr{E}\{l^2\}\,\varrho_l^2 \qquad (55.21)$$

where (h, k, l) denote the components of r in the new coordinate system and $\mathscr{E}\{h^2\}$, $\mathscr{E}\{k^2\}$, $\mathscr{E}\{l^2\}$ are the eigenvalues of the symmetric matrix formed by the second moments:

$$\left\{\begin{matrix} \mathscr{E}\{x^2\} & \mathscr{E}\{x y\} & \mathscr{E}\{x z\} \\ \mathscr{E}\{y x\} & \mathscr{E}\{y^2\} & \mathscr{E}\{y z\} \\ \mathscr{E}\{z x\} & \mathscr{E}\{z y\} & \mathscr{E}\{z^2\} \end{matrix}\right\}. \qquad (55.22)$$

$R = (H, K, L)$ is the vector in the new system of coordinates. The probability distribution $W(R)$ of the position of the particle after suffering a large number of displacements governed by the basic distribution function $\varphi(x, y, z)$ is an ellipsoidal distribution given by

$$
\left.
\begin{aligned}
W(\boldsymbol{R}) = {} & \frac{1}{[8\pi^3 N^3 \,\mathscr{E}\{h^2\}\,\mathscr{E}\{k^2\}\,\mathscr{E}\{l^2\}]^{\frac12}} \times \\
& \times \exp\left[-\frac{(H-N\,\mathscr{E}\{h\})^2}{2N\,\mathscr{E}\{h^2\}} - \frac{(K-N\,\mathscr{E}\{k\})^2}{2N\,\mathscr{E}\{k^2\}} - \frac{(L-N\,\mathscr{E}\{l\})^2}{2N\,\mathscr{E}\{l^2\}}\right].
\end{aligned}
\right\}
\tag{55.23}
$$

This is centred at $N\,\mathscr{E}\{h\}$, $N\,\mathscr{E}\{k\}$, $N\,\mathscr{E}\{l\}$. In other words, the particle suffers an average systematic net displacement of amount $N\,\mathscr{E}\{h\}$, $N\,\mathscr{E}\{k\}$, $N\,\mathscr{E}\{l\}$ and superposed on this is a general random distribution.

To extend the above theory to the case of random flights with respect to a continuous parameter t, say time, instead of discrete N, CHANDRASEKHAR assumed that the particle suffers n displacements per unit time and wrote the expression for $W(\boldsymbol{R})$ as

$$
W(\boldsymbol{R}, t) = \frac{1}{(8\pi t)^{\frac32}(D_1 D_2 D_3)^{\frac12}} \exp\left[-\frac{(X+\beta_1 t)^2}{4D_1 t} - \frac{(Y+\beta_2 t)^2}{4D_2 t} - \frac{(Z+\beta_3 t)^2}{4D_3 t}\right]
\tag{55.24}
$$

where

$$
\left.
\begin{aligned}
D_1 &= \tfrac12 n\,\mathscr{E}\{x^2\}, & D_2 &= \tfrac12 n\,\mathscr{E}\{y^2\}, & D_3 &= \tfrac12 n\,\mathscr{E}\{z^2\}, \\
\beta_1 &= -n\,\mathscr{E}\{x\}, & \beta_2 &= -n\,\mathscr{E}\{y\}, & \beta_3 &= -n\,\mathscr{E}\{z\},
\end{aligned}
\right\}
\tag{55.25}
$$

the coordinate system being so chosen that X, Y, Z are measured along the principal axes of the moment ellipsoid (55.20). This expression can be derived as the solution of a partial differential equation. There are direct methods for arriving at this equation, but for historical interest we outline here the derivation of CHANDRASEKHAR. We consider an interval Δt which is long enough for a particle to suffer a large number of individual displacements but short enough for the mean square displacement $\mathscr{E}\{(\Delta\boldsymbol{R})^2\}$ to be small. The probability that the particle suffers an increment $\Delta\boldsymbol{R}$ in an interval Δt is therefore governed by the distribution function

$$
\left.
\begin{aligned}
W(\Delta\boldsymbol{R}, \Delta t) = {} & \frac{1}{(8\pi\Delta t)^{\frac32}(D_1 D_2 D_3)^{\frac12}} \times \\
& \times \exp\left[-\frac{(\Delta X+\beta_1\Delta t)^2}{4D_1\Delta t} - \frac{(\Delta Y+\beta_2\Delta t)^2}{4D_2\Delta t} - \frac{(\Delta Z+\beta_3\Delta t)^2}{4D_3\Delta t}\right].
\end{aligned}
\right\}
\tag{55.26}
$$

Hence we write

$$
\left.
\begin{aligned}
W(\boldsymbol{R}, t+\Delta t) &= W(\boldsymbol{R}, t) + \frac{\partial W}{\partial t}\Delta t + O(\Delta t^2) \\
&= \int W(\boldsymbol{R}-\Delta\boldsymbol{R}, t)\, W(\Delta\boldsymbol{R}, \Delta t)\, d(\Delta\boldsymbol{R})
\end{aligned}
\right\}
\tag{55.27}
$$

and pass to the limit as $\Delta t \to 0$. Expanding $W(\boldsymbol{R}-\Delta\boldsymbol{R}, t)$ in Taylor series and integrating term by term, we obtain due to the following properties of the distribution

$$
\left.
\begin{aligned}
\mathscr{E}\{\Delta X\} &= -\beta_1\Delta t, \\
\mathscr{E}\{\Delta Y\} &= -\beta_2\Delta t, \\
\mathscr{E}\{\Delta Z\} &= -\beta_3\Delta t; \\
\mathscr{E}\{\Delta X^2\} &= 2D_1\Delta t + \beta_1(\Delta t)^2, \\
\mathscr{E}\{\Delta Y^2\} &= 2D_2\Delta t + \beta_2(\Delta t)^2, \\
\mathscr{E}\{\Delta Z^2\} &= 2D_3\Delta t + \beta_3(\Delta t)^2; \\
\mathscr{E}\{\Delta X\,\Delta Y\} &= \beta_1\beta_2(\Delta t)^2, \\
\mathscr{E}\{\Delta Y\,\Delta Z\} &= \beta_2\beta_3(\Delta t)^2, \\
\mathscr{E}\{\Delta Z\,\Delta X\} &= \beta_3\beta_1(\Delta t)^2.
\end{aligned}
\right\}
\tag{55.28}
$$

and the differential equation

$$\frac{\partial W}{\partial t} = \beta_1 \frac{\partial W}{\partial X} + \beta_2 \frac{\partial W}{\partial Y} + \beta_3 \frac{\partial W}{\partial Z} + D_1 \frac{\partial^2 W}{\partial X^2} + D_2 \frac{\partial^2 W}{\partial Y^2} + D_3 \frac{\partial^2 W}{\partial Z^2}. \tag{55.29}$$

According to this equation we can describe the phenomenon under discussion as a general process of diffusion in which the number of particles crossing elements of area normal to the X, Y and Z directions per unit area per unit time are given respectively by

$$-\beta_1 W - D_1 \frac{\partial W}{\partial X}, \qquad -\beta_2 W - D_2 \frac{\partial W}{\partial Y}, \qquad -\beta_3 W - D_3 \frac{\partial W}{\partial Z}. \tag{55.30}$$

Class D: Processes with x continuous and t continuous.

As mentioned in Sect. 39 if both x and t are continuous and the process is Markovian and homogeneous in t, we are interested in the function $\pi(x|x_0; t)\,dx$, the probability that the random variable $x(t)$ lies between x and $x+dx$ given that the random variable $x(0)=x_0$. If we do not postulate any deterministic changes and assume that $\pi(x|x_0; t) \to R(x|x_0)\Delta$ as $t\to\Delta\to 0$, π satisfies the forward differential equation,

$$\frac{\partial \pi(x|x_0; t)}{\partial t} = -\pi(x|x_0; t)\int_{x'} R(x'|x)\,dx' + \int_{x'} \pi(x'|x_0; t)\, R(x|x')\,dx'.$$

We called this a *Basic Random Process* (B.R.P.). We shall now consider some particular cases which are very useful in the application of stochastic theory to physical problems.

56. Additive process. $R(x|x')$ is a function only of $x-x'$ and can be written as $R(x-x')$. Consequently $\pi(x|x_0; t)$ is a function only of $x-x_0$ and hence can be written as $\pi(x-x_0; t)$. In such a case $\int^x R(x|x')\,dx = \varrho$ which is independent of x'. In addition we shall assume that $R(x|x') = 0$ for $x<x'$.

$$\frac{\partial \pi(x; t)}{\partial t} = -\pi(x; t)\,\varrho + \int_{x'} \pi(x', t)\, R(x-x')\,dx'. \tag{56.1}$$

Since the probability of transition in dt is $\varrho\,dt$, the total number of transitions n in the interval 0 to t has a Poisson distribution $e^{-\varrho t}\frac{(\varrho t)^n}{n!}$. If n transitions occur, the conditional distribution of x is an n-fold convolution of $R(x)$ i.e. if the Laplace transform of R is defined as

$$\left.\begin{aligned}
\int_0^\infty R(x)\, e^{-sx}\,dx &= \varrho(s), \\
\varrho(0) &= \varrho.
\end{aligned}\right\} \tag{56.2}$$

Then the transform of the n-fold convolution of $R(x)$ is $[\varrho(s)]^n$. Thus the Laplace transform of $\pi(x; t)$ is given by

$$\left.\begin{aligned}
\int_x \pi(x; t)\, e^{-sx}\,dx &= \sum_n e^{-\varrho t}\frac{(\varrho t)^n}{n!}\,[\varrho(s)]^n \\
&= e^{-\varrho t[1-\varrho(s)]}.
\end{aligned}\right\} \tag{56.3}$$

Hence

$$\pi(x; t) = \frac{1}{2\pi i}\int_{\sigma-i\infty}^{\sigma+i\infty} e^{-\varrho t[1-\varrho(s)]}\, e^{sx}\,ds. \tag{56.4}$$

The above solution could have been obtained by reducing the forward differential equation (56.1) by the Laplace transform. The same arguments apply if $R(x|x')=0$ for $x'<x$ and $R(x|x')$ is a function only of $x'-x$ if $x<x'$.

57. q-process. We shall now assume that $R(x|x')\,dx$ can be written as $R(q)\,dq$ where $q=x/x'$ and $R(q)=0$ for $q>1$. In such a case the equation for the B.R.P. reduces to

$$\frac{\partial \pi(x|x_0;t)}{\partial t} = -\pi(x|x_0;t)\int_0^1 R(q)\,dq + \int_0^1 \pi\left(\frac{x}{q}\Big|x_0;t\right)\frac{R(q)}{q}\,dq. \qquad (57.1)$$

Obviously this can be reduced by the Mellin transformation defined by

$$p(s;t) = \int_x \pi(x|x_0;t)\,x^{s-1}dx. \qquad (57.2)$$

The solution is obtained in the following form

$$p(s;t) = p(s;0)\,e^{-A_s t} \qquad (57.3)$$

where $p(s;0)$ is the Mellin transform of $\pi(x|x_0;0)$.

58. Fluctuating density field (F.D.F.). We now assume that $R(x|x')$ is a function only of x and can be written as $\varrho(x)$. We also note that

$$\int_x R(x|x')\,dx = \int_x \varrho(x)\,dx = a. \qquad (58.1)$$

In this case the forward differential equation (56.1) reduces to

$$\left.\begin{aligned}\frac{\partial \pi(x|x_0;t)}{\partial t} &= -\pi(x|x_0;t)\,a + \int_x \pi(x'|x_0;t)\,\varrho(x)\,dx' \\ &= -\pi(x|x_0;t)\,a + \varrho(x).\end{aligned}\right\} \qquad (58.2)$$

The solution is given by

$$\pi(x|x_0;t) = \frac{\varrho(x)}{a}(1-e^{-at}) + \delta(x-x_0)\,e^{-at} \qquad (58.3)$$

where δ is the Dirac delta function. As $t\to\infty$,

$$\pi(x|x_0;t) \to \frac{\varrho(x)}{a} = \psi(x). \qquad (58.4)$$

A case of particular interest is when a is very large, i.e. $\frac{1}{a}\ll 1$. To study the stochastic process in such a case, we note that

$$\mathscr{E}\{x(t_1)\,x(t_2)\} = \int_{x_1}\int_{x_2} x_1\,x_2\,\pi(x_1|x_0;t_1)\,\pi(x_2|x_1;t_2-t_1)\,dx_1\,dx_2. \qquad (58.5)$$

Writing t_2-t_1 as t and letting t_1 and $t_2\to\infty$, we get

$$\mathscr{E}\{x(t_1)\,x(t_2)\} - [\mathscr{E}\{x\}]^2 = \{\mathscr{E}\{x^2\} - [\mathscr{E}\{x\}]^2\}\,e^{-at} \qquad (58.6)$$

where

$$\mathscr{E}\{x^n\} = \int_0^\infty x^n\,\psi(x)\,dx. \qquad (58.7)$$

If a is very large, the quantity on the right-hand side of (58.6) $\to 0$ even for small t provided $t\gg\frac{1}{a}$ i.e. the stochastic variables $x(t_1)$, $x(t_2)$ are independent if $t_2-t_1\gg\frac{1}{a}$. The quantity $1/a$ is well known in turbulence as the "mixing length". If x represents the density of matter, we have a wildly fluctuating density field, the degree of wildness depending on the smallness of $1/a$. The concept of such a field was introduced by CHANDRASEKHAR and MUNCH [24] in connection with certain astrophysical problems. Perhaps the most interesting feature of the field is that it directly leads to what are known as Gaussian processes. Defining $M(t) = \int_0^t x(\tau)\, d\tau$ where $x(t)$ represents the F.D.F., the integral being given the phenomenological interpretation of Sect. 39, it can be shown after some calculation that

$$\mathscr{E}\{M^n(t)\} = [\mathscr{E}\{M(t)\}]^n + \alpha^2 \tau_0 \, n\,(n-1)\,[\mathscr{E}\{M(t)\}]^{n-1}, \qquad (58.8)$$

where

$$\alpha^2 = \frac{\mathscr{E}\{x^2\} - (\mathscr{E}\{x\})^2}{(\mathscr{E}\{x\})^2}$$

and

$$\tau_0 = \frac{\mathscr{E}\{x\}}{a}. \qquad (58.9)$$

In particular

$$\sigma^2 = \mathscr{E}\{M^2(t)\} - [\mathscr{E}\{M(t)\}]^2 = 2\alpha^2\,\tau_0\,\mathscr{E}\{M(t)\}, \qquad \mathscr{E}\{M(t)\} = t\,\mathscr{E}\{x(t)\}. \qquad (58.10)$$

Eq. (58.8) can be written as

$$\mathscr{E}\{M^n(t)\} = [\mathscr{E}\{M(t)\}]^n + \frac{\sigma^2 n\,(n-1)}{2}\,[\mathscr{E}\{M(t)\}]^{n-2}.$$

These expressions can easily be shown to be the moments of a "degenerate" Gaussian distribution. For, consider a stochastic variable y with a Gaussian distribution

$$\varphi(y) = \frac{1}{\sigma\sqrt{2\pi}}\,e^{-\frac{(y-\mu)^2}{2\sigma^2}}$$

where μ is the mean and σ^2 is the mean-square deviation. The moments of $y-\mu$ are given by

$$\left.\begin{aligned}\mathscr{E}\{(y-\mu)^{2r}\} &= \frac{(\tfrac{1}{2})^r \sigma^{2r}\,(2r)!}{r!}, \\ \mathscr{E}\{(y-\mu)^{2r+1}\} &= 0,\end{aligned}\right\} \qquad (58.11)$$

where r is an integer > 0. If we assume that $\sigma^2 \ll 1$, such that its powers can be neglected,

$$\mathscr{E}\{y^n\} = [\mathscr{E}\{y\}]^n + \sigma^2\,\frac{n\,(n-1)}{2}\,[\mathscr{E}\{y\}]^{n-2}. \qquad (58.12)$$

It is possible to calculate the correlation functions associated with $M(t)$. RANGANATHAN[1] has proved that if we define $F(t) = M(t) - \alpha$ and $\mathscr{E}\{F(t_1)\,F(t_2)\ldots F(t_n)\}$ as the expectation value of the product of $F(t_i)$'s then the above expression vanishes if n is odd. If n is even, i.e. $n = 2m$,

$$\mathscr{E}\{F(t_1)\,F(t_2)\ldots F(t_{2m})\} = \sum_{\text{all pairs}} \mathscr{E}\{F(t_i)\,F(t_j)\}\,\mathscr{E}\{F(t_k)\,F(t_l)\}\ldots. \qquad (58.13)$$

59. Gaussian approximation to the basic random process. In the above case we arrived at the Gaussian distribution by first obtaining the F.D.F. approximation of a B.R.P. and introducing the concept of integration. There is however another way of obtaining the Gaussian process from the general equation for a

---

[1] Private communication.

B.R.P. We assume that $R(x|x')$ rapidly diminishes as x differs from x' in such a way that

$$\int_x (x - x') R(x|x') \, dx = \beta$$

and

$$\int_x (x - x')^2 R(x|x') \, dx = \gamma$$

but

$$\int_x (x - x')^n R(x|x') \, dx \ll \beta \text{ or } \gamma$$

for all values of $n > 2$.

This merely means that transitions from x' occur to values in the immediate neighbourhood of x'. It may be perplexing why, if higher moments vanish, the second moment is of the same order as the first. The finiteness of the second moment would suggest that $R(x|x')$ is sufficiently large for $x \approx x'$ and rapidly tends to 0 as x differs from x'. Hence if the effective range of $x - x'$ is ε, $R(x|x')$ is of the order of $1/\varepsilon^2$ if $x - x' < \varepsilon$. This may suggest that the first moment is safeguarded since $x - x'$ can be positive and negative. In fact if $R(x|x')$ is symmetrical in x, x' then the mean is zero. Using the above approximation and expanding $\pi(x|x_0)$ within the integral, the differential equation yields the *Fokker-Planck equation*,

$$\frac{\partial \pi}{\partial t} = - \frac{\partial}{\partial x} \{\beta(x)\, \pi\} + \frac{1}{2} \frac{\partial^2}{\partial x^2} \{\gamma(x)\, \pi\}. \tag{59.1}$$

Another extreme approximation is to assume $\gamma = 0$; i.e., $R(x|x')$ is of the order $1/\varepsilon^3$ and the first moment is not zero. Then

$$\frac{\partial \pi}{\partial t} = - \frac{\partial}{\partial x} \{\beta(x)\, \pi\}. \tag{59.2}$$

The solution of this equation is a delta function and this corresponds to a deterministic loss of $\beta(x)\, dt$ in dt. Thus it is usually stated that in the Fokker-Planck equation β determines the deterministic loss and γ the fluctuating Gaussian part. This derivation can be compared with CHANDRASEKHAR's treatment of the Fokker-Planck equation.

The results obtained by CHANDRASEKHAR in his discussion on the Fokker-Planck equation can be fitted into the framework of the theory of stochastic integration. But in view of the attention his classic article has attracted in mathematical physics, we present a summary of his treatment as is found in the article.

60. CHANDRASEKHAR's approach. CHANDRASEKHAR starts with LANGEVIN's equation written in a symbolic form as

$$\frac{d\boldsymbol{u}}{dt} = - \beta\, \boldsymbol{u} + \boldsymbol{A}(t) \tag{60.1}$$

where $\boldsymbol{u}$ denotes the velocity of the particle. According to this equation the influence of the surrounding medium on the particle can be split up into two parts, first a systematic part $-\beta\boldsymbol{u}$ representing dynamical friction experienced by the particle and second a fluctuating part $\boldsymbol{A}(t)$ which is characteristic of Brownian motion. Regarding the frictional term $-\beta\boldsymbol{u}$ it is assumed that this is governed by STOKES' law which states that the frictional force decelerating a spherical particle of radius a and mass m is given by $6\pi a \eta \dfrac{\boldsymbol{u}}{m}$ where η is the coefficient of viscosity of the surrounding fluid and

$$\beta = 6\pi a \eta / m.$$

As for the fluctuating part $A(t)$, the following principal assumptions are made:

1. $A(t)$ is independent of u.

2. $A(t)$ varies extremely rapidly compared to variations of u.

The second assumption implies that time intervals of duration Δt exist such that during Δt, the variations in u that are to be expected are very small indeed while during the same interval, $A(t)$ may undergo several fluctuations. Alternatively we may say that though $u(t)$ and $u(t+\Delta)$ are expected to differ by negligible amounts, no correlation between $A(t)$ and $A(t+\Delta)$ exists.

Our problem is to solve the stochastic differential equation subject to the restrictions on $A(t)$ stated above. But "solving" a stochastic differential equation is not the same thing as solving an ordinary differential equation[1]. For one thing, Eq. (60.1) involves the function $A(t)$ which as we shall presently see has only statistically defined properties. Consequently "solving" the Langevin equation (60.1) has to be understood rather in the sense of specifying a probability distribution, $\pi(u|u_0; t)$ which governs the probability of occurrence of velocity u at time t, given that $u = u_0$ at $t = 0$. Of this function $\pi(u|u_0; t)$ we should clearly require that as $t \to 0$,

$$\pi(u|u_0; t) = \delta(u_x - u_{x0})\, \delta(u_y - u_{y0})\, \delta(u_z - u_{z0}) \tag{60.2}$$

where the δ's are the Dirac delta functions. Further the physical circumstances of the problem require that we demand of π that it tends to Maxwellian distribution for temperature T of the surrounding fluid, independently of u_0 as $t \to \infty$.

$$\lim_{t \to \infty} \pi(u|u_0; t) \to \left(\frac{m}{2\pi kT}\right)^{\frac{3}{2}} \exp\left[-m|u^2|/2kT\right]. \tag{60.3}$$

This last demand on π conversely requires that $A(t)$ satisfies certain statistical requirements. For according to LANGEVIN's equation we have the formal solution

$$u - u_0\, e^{-\beta t} = e^{-\beta t} \int_0^t e^{\beta \xi} A(\xi)\, d\xi. \tag{60.4}$$

Consequently the statistical properties of $(u - u_0\, e^{-\beta t})$ must be the same as that of

$$e^{-\beta t} \int_0^t e^{\beta \xi} A(\xi)\, d\xi. \tag{60.5}$$

As $t \to \infty$ the quantity $u - u_0\, e^{-\beta t}$ tends to u, hence the distribution of

$$\lim_{t \to \infty} \left\{ e^{-\beta t} \int_0^t e^{\beta \xi} A(\xi)\, d\xi \right\} \tag{60.6}$$

must be the Maxwellian distribution

$$\left(\frac{m}{2\pi kT}\right)^{\frac{3}{2}} \exp\left[-m|u^2|/2kT\right]. \tag{60.7}$$

Now one of our principal assumptions concerning $A(t)$ is that it varies extremely rapidly compared to any of the other quantities that enter into our discussion. Further the fluctuating acceleration experienced by the Brownian particles is statistical in character in the sense that Brownian particles having the same initial coordinates and velocities will suffer accelerations which will differ from particle to particle, both in magnitude and in their dependence on time. However, on account of the rapidity of these fluctuations, we can always divide an

---

[1] A full discussion on the "solution" of stochastic differential equations follows later in this article, Sect. 70.

interval of time which is long enough for any of the physical parameters like position and velocity of the Brownian particle to change appreciably into a very large number of subintervals of duration Δ such that during each of these subintervals we can treat all functions of time except $A(t)$ which enter into our formulae as constants. Thus the quantity $e^{-\beta t} \int_0^t e^{\beta \xi} A(\xi)\, d\xi$ may be written as

$$e^{-\beta t} \sum_j e^{\beta j\Delta} \int_{j\Delta}^{(j+1)\Delta} A(\xi)\, d\xi. \tag{60.8}$$

Let

$$B(\Delta) = \int_t^{t+\Delta} A(\xi)\, d\xi. \tag{60.9}$$

The physical meaning of $B(\Delta)$ is that it represents the net acceleration which a Brownian particle may suffer on a given occasion during an interval of time Δ. Eq. (60.4) becomes

$$u - u_0 e^{-\beta t} = \sum_j e^{\beta(j\Delta - t)} B(\Delta) \tag{60.10}$$

and we require that, as $t \to \infty$, the quantity on the right-hand side tends to a Maxwellian distribution. We now assert that this requires the probability of occurrence of different values for $B(\Delta)$ be governed by the distribution function

$$W\{B(\Delta)\} = \frac{1}{(4\pi q \Delta)^{\frac{3}{2}}} \exp\left[-|B(\Delta)|^2/4q\,\Delta\right] \tag{60.11}$$

where $q = \beta\, kT/m$. To prove this assertion we have to show that the distribution function $\pi(u\,|\,u_0;\,t)$ derived on the basis of Eqs. (60.10) and (60.11) does in fact tend to the Maxwellian distribution (60.7) as $t \to \infty$. We shall presently show that this is the case but may remark meanwhile on the formal similarity of Eq. (60.11), giving the probability distribution of the acceleration $B(\Delta)$ suffered by a Brownian particle in the time Δ, to the equation giving the probability distribution of the increment ΔR in the position of a particle describing random flights in time Δ:

$$\psi(\Delta R, \Delta) = \frac{1}{(4\pi D\Delta)^{\frac{3}{2}}} \exp\left[-|\Delta R|^2/4D\Delta\right]. \tag{60.12}$$

For the validity of this assertion, Δ must be long enough for a large number of individual displacements to occur; analogously our expression for $W\{B(\Delta)\}$ is valid only for times Δ large compared to the average period of a single fluctuation in $A(t)$. Now the period of fluctuation of $A(t)$ is clearly of the order of time between successive collisions of the Brownian particle with the molecules of the surrounding fluid; in a liquid this is generally of the order of 10^{-21} sec. Accordingly the similarity of the expressions for $W\{B(\Delta)\}$ with the equation for the theory of random flights leads us to interpret the acceleration $B(\Delta)$ suffered by a Brownian particle in time Δ large compared with the frequency of collisions with the surrounding particles, as the result of superposition of the large number of random accelerations caused by collisions with the individual molecules. This is of course reasonable but the reason why q in Eq. (60.11) is given by $q = \beta\, kT/m$ is due to our requirement that $\pi(u\,|\,u_0;\,t)$ tends to Maxwellian distribution as $t \to \infty$.

It can be proved that if

$$R = \int_0^t \psi(\xi)\, A(\xi)\, d\xi \tag{60.13}$$

subject to the conditions pertaining to this problem, the probability distribution of $\boldsymbol{R}$ can be shown to be

$$W(\boldsymbol{R}) = \frac{1}{\left[4\pi q \int_0^t \psi^2(\xi)\,d\xi\right]^{\frac{3}{2}}} \exp\left[-\,|\boldsymbol{R}|^2/4q \int_0^t \psi^2(\xi)\,d\xi\right]. \tag{60.14}$$

In our present case

$$\psi(\xi) = e^{\beta(\xi-t)}. \tag{60.15}$$

Then the probability distribution of $\boldsymbol{u} - \boldsymbol{u}_0\,e^{-\beta t}$ can be easily written down. Since

$$\int_0^t \psi^2(\xi)\,d\xi = \int_0^t e^{2\beta(\xi-t)}\,d\xi = \frac{1}{2\beta}\,(1 - e^{-2\beta t})$$

and

$$\frac{q}{\beta} = \frac{kT}{m},$$

we arrive at

$$\pi(\boldsymbol{u}\,|\,\boldsymbol{u}_0;\,t) = \left[\frac{m}{2\pi kT\,(1 - e^{-2\beta t})}\right]^{\frac{3}{2}} \exp\left[\frac{-\,m\,|\boldsymbol{u} - \boldsymbol{u}_0\,e^{-\beta t}|^2}{2kT(1 - e^{-2\beta t})}\right]. \tag{60.16}$$

We verify according to this equation

$$\lim_{t\to\infty} \pi(\boldsymbol{u}\,|\,\boldsymbol{u}_0;\,t) \to \left(\frac{m}{2\pi kT}\right)^{\frac{3}{2}} \exp\left[-\,m\,|\boldsymbol{u}^2|/2kT\right],$$

the Maxwellian distribution. This proves the assertion that with the statistical properties of $\boldsymbol{B}(\varDelta)$ implied in Eq. (60.11) we arrive at $\pi(\boldsymbol{u}\,|\,\boldsymbol{u}_0;\,t)$ which is Maxwellian, independent of $\boldsymbol{u}_0$ as $t\to\infty$.

We shall now show how with the assumption made concerning $\boldsymbol{B}(\varDelta)$ we further arrive at $W(\boldsymbol{r}\,|\,\boldsymbol{r}_0,\,\boldsymbol{u}_0;\,t)$, the distribution of the displacement $\boldsymbol{r}$ of a Brownian particle at time t, given that it is at $\boldsymbol{r}_0$ with velocity $\boldsymbol{u}_0$ at time $t=0$.

Since

$$\boldsymbol{r} - \boldsymbol{r}_0 = \int_0^t \boldsymbol{u}(t)\,dt \tag{60.17}$$

we have according to Eq. (60.4)

$$\boldsymbol{r} - \boldsymbol{r}_0 = \int_0^t d\eta \left\{\boldsymbol{u}_0\,e^{-\beta\eta} + e^{-\beta\eta} \int_0^\eta e^{\beta\xi}\,\boldsymbol{A}(\xi)\,d\xi\right\}$$

or

$$\boldsymbol{r} - \boldsymbol{r}_0 - \beta^{-1}\boldsymbol{u}_0(1 - e^{-\beta t}) = \int_0^t d\eta\,e^{-\beta\eta} \int_0^\eta d\xi\,e^{\beta\xi}\,\boldsymbol{A}(\xi). \tag{60.18}$$

Integrating the right-hand side by parts

$$\boldsymbol{r} - \boldsymbol{r}_0 - \beta^{-1}\boldsymbol{u}_0(1 - e^{-\beta t}) = -\,\frac{1}{\beta}\,e^{-\beta t} \int_0^t e^{\beta\xi}\,\boldsymbol{A}(\xi)\,d\xi + \frac{1}{\beta} \int_0^t \boldsymbol{A}(\xi)\,d\xi. \tag{60.19}$$

Again we reduce this equation to the form

$$\boldsymbol{r} - \boldsymbol{r}_0 - \beta^{-1}\boldsymbol{u}_0(1 - e^{-\beta t}) = \int_0^t \psi(\xi)\,\boldsymbol{A}(\xi)\,d\xi \tag{60.20}$$

by defining

$$\psi(\xi) = \frac{1}{\beta}\,[1 - e^{\beta(\xi-t)}]. \tag{60.21}$$

Thus, the probability distribution of $r - r_0 - \beta^{-1} u_0 (1 - e^{-\beta t})$, i.e. of r at time t for given r_0 and u_0 is given by (60.21). Since

$$\int_0^t \psi^2(\xi)\, d\xi = \frac{1}{\beta^2} \int_0^t \{1 - e^{\beta(\xi - t)}\}^2\, d\xi = \frac{1}{2\beta}\left[2\beta t - 3 + 4e^{-\beta t} - e^{-2\beta t}\right] \quad (60.22)$$

we have

$$\left. \begin{aligned} W(r|r_0, u_0; t) &= \left[\frac{m\beta^2}{2\pi\, kT(2\beta t - 3 + 4e^{-\beta t} - e^{-2\beta t})}\right]^{\frac{3}{2}} \times \\ &\quad \times \exp\left[-\frac{m\beta^2\, \overline{(r - r_0 - u_0\, 1 - e^{-\beta t}\, \beta^{-1})}}{2kT\,(2\beta t - 3 + 4e^{-\beta t} - e^{-2\beta t})}\right]. \end{aligned} \right\} \quad (60.23)$$

For intervals of time long compared with β^{-1} the foregoing simplifies considerably. For under these circumstances we can ignore the exponential and the constant terms as compared to $2\beta t$. Further we shall presently show $\mathscr{E}\{|r - r_0|^2\}$ is of the order t; hence we can also neglect $u_0(1 - e^{-\beta t})/\beta$ compared to $r - r_0$. This reduces Eq. (60.23) to

$$W(r|r_0, u_0; t) \approx \frac{1}{(4\pi D t)^{\frac{3}{2}}} \exp\left[-|r - r_0|^2/4Dt\right], \quad t \gg \beta^{-1} \quad (60.24)$$

where we have introduced the "diffusion coefficient" D defined by

$$D = \frac{kT}{m\beta} = \frac{kT}{6\pi a\eta}. \quad (60.25)$$

From Eq. (60.24) we obtain for the mean-square displacement along any given direction, say x-direction,

$$\left. \begin{aligned} \mathscr{E}\{(x - x_0)^2\} &= \tfrac{1}{3}\, \mathscr{E}\{(r - r_0)^2\} \\ &= 2Dt = \left(\frac{kT}{3\pi a\eta}\right) t. \end{aligned} \right\} \quad (60.26)$$

This is EINSTEIN's result. Eq. (60.26) has been verified by PERRIN to lead to consistent and satisfactory values for the Boltzmann constant k by observation of $\mathscr{E}\{(x - x_0)^2\}/t$ over wide ranges of T, η and a.

V. Cascade processes involving continuous parameters.

We shall now deal with processes which involve the statistical distribution of a discrete number of particles in a continuous infinity of states. This distribution changes with a continuous parameter t in a random manner characteristic of the process. Instead of dealing in a general manner with such a class, we shall formulate a particular problem from which the generalisation can easily be made.

61. Formulation of the cascade problem[1]. We formulate a cascade process involving n types of particles distributed in energy space. At $t = 0$, we have a particle of type i with energy equal to E_0. The particle passes through matter and then we shall assume that the following aggregate is produced by the particle of type i and energy E_0 with probability $R\, dt$ where R will be defined presently. The particle if it survives is included in the aggregate which consists of k_1 particles of type 1, k_2 of type 2, ... k_n of type n. The k_j particles of type j are distributed in energy as follows; one of them lies in the range between E_1^j and $E_1^j + dE_1^j$,

---

[1] For cosmic ray cascades cf. Vol. XLVI of this Encyclopedia.

one between E_2^i and $E_2^i + dE_2^i$, ... and one between $E_{k_j}^j$ and $E_{k_j}^j + dE_{k_j}^j$. As j runs over all the indices the energies of all the particles are defined. This aggregate will be denoted by $\langle E_l^j \rangle$. The probability of production of this aggregate by a particle of type i with energy E_0 in thickness dt is written for the sake of brevity as

$$R^i(\langle E_l^j \rangle | E_0)\, dt.$$

The total cross-section $R^i(E_0)$ is obtained by integration over all the energy variables and summing over all values of k_j and j and multiplying by $\dfrac{1}{\prod\limits_j k_j!}$. This factor arises since integration with respect to each of the energy variables over the whole range implies counting an event representing the occurrence of k_j particles $k_j!$ times. (A similar situation arises in Janossy densities.) R will define the development of the entire process. Given this probability, our object is to obtain the statistical distribution at t with the initial condition as specified above.

It is quite clear that in describing the distribution at t we run into difficulties adverted to in the earlier section. Therefore we are constrained to use the concept of product densities. If we wish to avoid this, we take recourse to defining the joint probability $\pi^i(\langle \nu_m \rangle | E_0, t)$ the probability that we have ν_m particles of type m with energy greater than $E_m (m = 1, 2, ...)$[1]. The joint probability that nothing happens to the initial particle of type i and energy E_0 till τ and that an aggregate $\langle E_l^j \rangle$ is produced in $d\tau$ is $R^i(\langle E_l^j \rangle | E_0) \exp(-R^i(E_0)\,\tau)\, d\tau$. Now adopting the regeneration point method we use the following argument. Somewhere between 0 and t in the interval $(\tau, \tau + d\tau)$ the initial particle of type i produces an aggregate and the members of this aggregate become independent primaries of cascades. This aggregate can be formed anywhere between 0 and t and therefore we obtain the integral equation

$$
\left.
\begin{aligned}
\pi^i(\langle \nu_m \rangle | E_0; t) = &\int\limits_0^t \int\limits_{\langle E_i \rangle} \mathrm{e}^{-R^i(E_0)\tau} \frac{R^i(\langle E_l^j \rangle | E_0; \tau)}{\prod\limits_j k_j!} \times \\
&\times \sum_j \sum_{l=1}^k \pi(\langle \nu_m \rangle | \langle E_l^j \rangle; t - \tau)\, d\langle E_l^j \rangle\, d\tau + \\
&+ \mathrm{e}^{-R^i(E_0)t} \prod_{j, j \neq i} \delta(\nu_j)\, \delta(\nu_i - 1)\, \delta(E_1^i - E_0)
\end{aligned}
\right\}
\quad (61.1)
$$

where $\delta(\nu_i)$ is the Kronecker delta function. The last term merely indicates that nothing happens between 0 and t and we have the initial particle with energy E_0. $\pi(\langle \nu_m \rangle | \langle E_l^j \rangle, t - \tau)$ is the probability that we have the aggregate $\langle \nu_m \rangle$ at t given that we started with the aggregate $\langle E_l^j \rangle$ at τ. This is a function of the duration $t - \tau$ and not of t and τ separately since the process is assumed to be homogeneous in t. The probability frequency function $\pi(\langle \nu_m \rangle | \langle E_l^j \rangle, t - \tau)$ can be decomposed in terms of the probability frequency function with simpler initial conditions i.e. the one-particle initial condition as in $\pi^i(\langle \nu_m \rangle | E_0, t)$ in the following manner if we assume that the particles of the aggregate $\langle E_l^j \rangle$ behave independently as primaries of cascade processes.

$$\pi(\langle \nu_m \rangle | \langle E_l^j \rangle, t - \tau) = \prod_{l, j} \pi^j(\langle \nu_m^{jl} \rangle | E_l^j; t - \tau) \quad \text{such that} \quad \sum_{l, j} \langle \nu_m^{jl} \rangle = \langle \nu_m \rangle, \quad (61.2)$$

<hr>

[1] Note the difference in definition of the aggregate $\langle \nu_m \rangle$ and $\langle E_l^j \rangle$. $\langle E_l^j \rangle$ is the most comprehensive definition of an aggregate since the energy of each of particle is defined.

where $\langle v_m^{jl} \rangle$ is the aggregate of v_m^{jl} particles with energy greater than E_m ($m = 1, 2, \ldots$) produced by a particle of type j and energy E_l^j.

Eq. (61.1) can be reduced to a differential equation by differentiating with respect to t. We then have a system of simultaneous integro-differential equations. These equations are amenable to the technique of generating functions. If appeal is made to generating functions, (61.2) can be simplified to be a product of generating functions by the convolution theorem. In spite of this simplification the equations are still too general and comprehensive to be of any use and no purpose is served by attempting a general solution of this problem. The equations when there are only two types of particles were first given by JANOSSY [25] and in view of the considerable interest that has been evinced in the stochastic problem of cosmic radiation, we discuss JANOSSY's equation in the next section.

New complications arise if we introduce various types of deterministic changes in the system, e.g. deterministic loss of energy. If for example we postulate that a particle of type i loses energy $\beta_i \Delta$ in passing through matter we have merely to replace $R^i(E)$ by an integral

$$\int_0^\tau R^i(E - \beta_i \tau')\, d\tau'$$

and

$$R^i(\langle v_j \rangle | E) \quad \text{by} \quad R^i(\langle v_j \rangle | E - \beta_i \tau). \tag{61.3}$$

If β_i is a function of energy also, the problem is much more complex but no new principle is involved in writing expressions using β as a function of energy.

62. Outline of JANOSSY's treatment of cosmic ray cascades. JANOSSY defined a function $\pi^{(i)}(n, E, E_0; t)$ representing the probability that there are n electrons above energy E at t, given that at $t = 0$ there was a primary of energy E_0. If the primary is an electron we put $i = 1$. If it is a photon we put $i = 2$. He visualised the stochastic process from the regeneration point method. Somewhere between 0 and t, say between τ and $\tau + d\tau$, the electron may create a photon and the newly created photon and electron become independent primaries of stochastic processes of duration $t - \tau$. If at $t = 0$ we have a photon, it may create a pair of electrons between τ and $\tau + d\tau$ and these electrons become independent primaries of stochastic processes of duration $t - \tau$. Viewing the development of the process thus, we find that $\pi^{(i)}(n, E, E_0; t)$ satisfies the following integral equations:

$$\pi^{(1)}(n, E, E_0; t) = \int_0^t e^{-\alpha_1 \tau}\, d\tau \int_E^{E_0} \left[\sum_{n_1 + n_2 = n} \pi^{(1)}(n_1, E, E'; t - \tau) \times \right.$$
$$\left. \times\, \pi^{(2)}(n_2, E, E_0 - E'; t - \tau) \right] R(E', E_0)\, dE' + \delta(n - 1)\, e^{-\alpha_1 t},$$

$$\pi^{(2)}(n, E, E_0; t) = \int_0^t e^{-\alpha_2 \tau}\, d\tau \int_E^{E_0} \left\{ \sum_{n_1 + n_2 = n} \pi^{(1)}(n_1, E, E'; t - \tau) \times \right.$$
$$\left. \times\, \pi^{(1)}(n_2, E, E_0 - E'; t - \tau) \right\} R'(E', E_0)\, dE' + \delta(n)\, e^{-\alpha_2 t},$$

$$\quad (62.1)$$

where

$$\alpha_1 = \int_0^E R(E', E_0)\, dE'$$

and

$$\alpha_2 = \int_0^E R'(E', E_0)\, dE'.$$

$R(E', E_0)$ represents the probability per unit t that an electron of energy E_0 drops to an interval of energy lying between E' and $E' + dE'$ radiating a photon of energy $E_0 - E'$. $R'(E', E_0)\, dE'$ represents the probability per unit t that a photon of energy E_0 splits up into a pair of electrons, one of which has an energy lying between E' and $E' + dE'$. These equations are self-explanatory. Differentiating these equations with respect to t we get

$$
\begin{aligned}
\frac{\partial \pi^{(1)}(n, E, E_0; t)}{\partial t} &= -\alpha_1\, \pi^{(1)}(n, E, E_0; t) + \int\limits_E^{E_0} \sum_{n_1+n_2=n} \pi^{(1)}(n_1, E, E'; t) \times \\
&\qquad \times \pi^{(2)}(n_2, E, E_0 - E'; t)\, R(E', E_0)\, dE', \\[2mm]
\frac{\partial \pi^{(2)}(n, E, E_0; t)}{\partial t} &= -\alpha_2\, \pi^{(2)}(n, E, E_0; t) + \int\limits_E^{E_0} \sum \pi^{(1)}(n_1, E, E'; t) \times \\
&\qquad \times \pi^{(1)}(n_2, E, E_0 - E'; t)\, R'(E', E_0)\, dE.
\end{aligned}
\tag{62.2}
$$

To solve these equations we use the method of generating functions. We also assume that $R(E', E)\, dE'$ and $R'(E', E)\, dE'$ can be expressed as functions of E'/E i.e. $R(\varepsilon)\, d\varepsilon$ and $R'(\varepsilon)\, d\varepsilon$ respectively.

Defining

$$
\begin{aligned}
G(u, \varepsilon; t) &= \sum_n \pi^{(1)}(n, \varepsilon, t)\, u^n, \\
F(u, \varepsilon; t) &= \sum_n \pi^{(2)}(n, \varepsilon, t)\, u^n
\end{aligned}
\qquad \varepsilon = E/E_0
\tag{62.3}
$$

we obtain

$$
\begin{aligned}
\frac{\partial G(u, \varepsilon; t)}{\partial t} &= -\alpha_1\, G(u, \varepsilon, t) + \int\limits_\varepsilon^1 G\!\left(u, \frac{\varepsilon}{\varepsilon'}, t\right) F\!\left(u, \frac{\varepsilon}{1-\varepsilon'}, t\right) R(\varepsilon')\, d\varepsilon', \\[2mm]
\frac{\partial F(u, \varepsilon; t)}{\partial t} &= -\alpha_2 F(u, \varepsilon, t) + \int\limits_\varepsilon^1 G\!\left(u, \frac{\varepsilon}{\varepsilon'}, t\right) G\!\left(u, \frac{\varepsilon}{1-\varepsilon'}, t\right) R'(\varepsilon')\, d\varepsilon'.
\end{aligned}
\tag{62.4}
$$

These are the equations obtained by Janossy. Analytical solutions have been obtained only for the first and second moments and considerable effort has been expended on developing techniques for the solution of the problem.

We shall now take a simplified special case of the stochastic problem described till now. We shall ask for the generating functions for the *total number* of electrons or photons. This only means that in the Janossy equations we must make ε tend to 0. Such a limiting process would result in exaggerating the multiplicative capacity of the low energy particles which is contrary to physical facts. But it is suggested that such a simplification is justified when almost all the particles are above the critical energy, say E_c, and this will be true for small thicknesses when the average number of particles, say n, is such that, if E_0 is the initial energy of the primary, $E_0/n > E_c$.

Putting

$$
\begin{aligned}
G^{(1)}(u, t) &= G, & G^{(2)}(u, t) &= F, \\
\pi^{(1)}(n, \varepsilon, t) &= \pi^{(1)}(n, t), & \pi^{(2)}(n, \varepsilon, t) &= \pi^{(2)}(n, t), \\
\alpha_1 &= \alpha, & \alpha_2 &= \beta
\end{aligned}
\tag{62.5}
$$

we get

$$
\begin{aligned}
\frac{\partial \pi^{(1)}(n, t)}{\partial t} &= \left\{ \sum_{n_1+n_2=n} \pi^{(1)}(n_1, t)\, \pi^{(2)}(n_2, t) - \pi^{(1)}(n, t) \right\} \alpha, \\
\frac{\partial \pi^{(2)}(n, t)}{\partial t} &= \left\{ \sum_{n_1+n_2=n} \pi^{(1)}(n_1, t)\, \pi^{(1)}(n_2, t) - \pi^{(2)}(n, t) \right\} \beta, \\
\frac{\partial G}{\partial t} &= (GF - G)\, \alpha, \qquad \frac{\partial F}{\partial t} = (G^2 - F)\, \beta.
\end{aligned}
\tag{62.6}
$$

Initial conditions are given by

$$
\left.\begin{array}{ll}
\pi^{(1)}(n, 0) = 0 & \text{for } n \neq 1, \\
\pi^{(2)}(n, 0) = 0 & \text{for } n \neq 0
\end{array}\right\} \tag{62.7}
$$

and

$$
\left.\begin{array}{l}
\pi^{(1)}(1, 0) = 1, \\
\pi^{(2)}(0, 0) = 1.
\end{array}\right\} \tag{62.8}
$$

But, for all t $\pi^{(1)}(0, t) = 0$. This means that for an electron-initiated shower there is at least one electron at all thicknesses since we do not have an absorption probability for electrons. It is difficult to solve for G and F if $\alpha \neq \beta$. To simplify our problem we assume $\alpha = \beta = 1$. The solutions for G and F are given as follows:

Electron distribution.

$$
\left.\begin{array}{l}
G = \sqrt{1-u^2}\, e^{-\alpha t}\, \text{Cosech} \left\{ (e^{-\alpha t} - 1)\sqrt{1-u^2} + \text{Ar Coth}\, \dfrac{1}{\sqrt{1-u^2}} \right\}, \\[3mm]
F = \sqrt{1-u^2}\, e^{-\alpha t}\, \text{Coth} \left\{ (e^{-\alpha t} - 1)\sqrt{1-u^2} + \text{Ar Coth}\, \dfrac{1}{\sqrt{1-u^2}} \right\}.
\end{array}\right\} \tag{62.9}
$$

To obtain the generating functions for the photon distribution corresponding to electron and photon initiated showers the initial conditions imposed on the G and F functions have to be suitably altered since they refer to photons. The following solutions are obtained.

Photon distribution.

$$
\left.\begin{array}{l}
G = \sqrt{u^2-1}\, e^{-\alpha t}\, \text{Cosech} \left\{ (e^{-\alpha t} - 1)\sqrt{u^2-1} + \text{Ar Coth}\, \dfrac{u}{\sqrt{1-u^2}} \right\}, \\[3mm]
F = \sqrt{u^2-1}\, e^{-\alpha t}\, \text{Coth} \left\{ (e^{-\alpha t} - 1)\sqrt{u^2-1} + \text{Ar Coth}\, \dfrac{u}{\sqrt{1-u^2}} \right\}.
\end{array}\right\} \tag{62.10}
$$

The mean number of electrons and photons for electron and photon initiated showers are given below.

Electron initiated shower.

$$
\left.\begin{array}{l}
\text{Mean number of electrons} = \tfrac{2}{3}\left(e^{\alpha t} + \tfrac{1}{2} e^{-2\alpha t}\right), \\[2mm]
\text{Mean number of photons} = \tfrac{1}{3}\left(e^{\alpha t} - e^{-2\alpha t}\right).
\end{array}\right\} \tag{62.11}
$$

Photon-initiated shower.

$$
\left.\begin{array}{l}
\text{Mean number of electrons} = \tfrac{2}{3}\left(e^{\alpha t} - e^{-2\alpha t}\right), \\[2mm]
\text{Mean number of photons} = \tfrac{1}{3}\left(e^{\alpha t} + 2\, e^{-2\alpha t}\right).
\end{array}\right\} \tag{62.12}
$$

It is to be noted that if we do not distinguish between photons and electrons the total number of particles is $e^{\alpha t}$ which is as it should be. Calculating the mean and mean-square deviation of the number of electrons and photons, under the assumption $e^{\alpha t} \gg e^{-\alpha t}$ we find the mean-square deviation of electrons $\simeq$ (mean number of electrons)^2 and the mean square deviation of photons $\simeq$ (mean number of photons)^2.

We conclude this chapter by giving the corresponding product density treatment of the stochastic problem of the following cascade process involving only one type of particle.

63. Cascades involving only one type of particle. The problem of nucleon cascades can be stated thus:

Given the initial energy spectrum of the nucleons at $t=0$ (t is the thickness of matter traversed) and that a nucleon of energy E drops to an energy between E_i and E_i+dE_i creating a recoil nucleon of energy between E_j and E_j+dE_j with a probability per unit distance given by $W(E_i, E_j|E)\, dE_i\, dE_j$, our object is to calculate the probability distribution of the number of nucleons at t above a certain energy. Since we cannot distinguish between recoil and incident nucleons we can define

$$\omega(E_i, E_j|E) = W(E_i, E_j|E) + W(E_j, E_i|E) \qquad (63.1)$$

where $\omega(E_i, E_j|E)\, dE_i\, dE_j$ is the probability per unit distance that a particle of energy E is replaced by a pair, one of energy E_i and another of energy E_j.

Let us write the diffusion equations for the product density of degree n for nucleons, defined by $f_n(E_1, E_2, \ldots, E_n)$. For this purpose we define

$$\left.\begin{aligned}
W'(E_k|E_i) &= \int_{E_j} W(E_k, E_j|E_i)\, dE_j, \\
W'(E) &= \int_{E_j}\int_{E_i} W(E_i, E_j|E)\, dE_i\, dE_j.
\end{aligned}\right\} \qquad (63.2)$$

We can write the diffusion equation for the product density of degree n:

$$\left.\begin{aligned}
\frac{\partial f_n(E_1, E_2, \ldots, E_n)}{\partial t} =\ & -f_n(E_1, E_2, \ldots, E_n) \sum_{1}^{n} W'(E_i) \\
& + \sum_{1}^{n}\int_E f_n(E_1', E_2', \ldots, E_{n-1}', E)\, W'(E_i|E)\, dE \\
& + \sum_{\substack{i=1 \\ }}^{n}\sum_{\substack{j=1 \\ i\neq j}}^{n}\int_E f_{n-1}(E_1'', E_2'', \ldots, E_{n-2}'', E)\, W(E_i, E_j|E)\, dE
\end{aligned}\right\} \quad (63.3)$$

where the notation E_k' with $k=1, 2, \ldots, n-1$ represents the values for E_k with $k=1, 2, \ldots, n$ omitting the value E_i, and the notation E_k'' with $k=1, 2, \ldots, n-2$ stands for values of E_k with $k=1, 2, \ldots, n$ with the omission of values E_i and E_j.

If it is assumed that $W(E_i, E_j|E)\, dE_i\, dE_j$ can be expressed in the form $W(\varepsilon_i, \varepsilon_j)\, d\varepsilon_i\, d\varepsilon_j$ where $\varepsilon_i = E_i/E$, $\varepsilon_j = E_j/E$, the above equations can be reduced to differential equations by MELLIN's transformation and formal transform solutions obtained[1].

VI. Stochastic problems in astrophysics.

Recently many statistical problems have been discussed in astrophysics by various groups of workers. Earlier, interest has been evinced in the statistical aspect of astrophysics[2] and to illustrate the nature of the new fields of application, we shall refer to three types of problems.

1. The problems discussed by CHANDRASEKHAR [16] dealing with the analysis of the statistical nature of the force acting on a star which is a member of a system.

---

[1] Over fifty papers have been published on cascades by MESSEL and other workers and for such literature, the reader is referred to a review article by the author in Progr. Theor. Phys. **11**, 95 (1954).

[2] See for example, TRUMPLER and WEAVER: Statistical Astronomy. University of California Press 1953.

2. Problems initiated by CHANDRASEKHAR and MUNCH [24] relating to the fluctuations in the brightness of the Milky Way and elaborated by other workers.

3. The statistical theory of clusters developed by NEYMAN and SCOTT [25]. As this has been discussed in great detail[1], we merely make a brief reference to it for the sake of completeness and fitting it into the general pattern of this article.

64. The random force problem. One of the principal problems in stellar dynamics is concerned with the analysis of the nature of the force acting on a star which is a member of the stellar system. The randomness of this force arises from the randomness of the distribution of the members of the stellar system. To make the problem tractable, CHANDRASEKHAR distinguishes broadly between the influence of the system as a whole and that of the immediate local neighbourhood. The former will be a smoothly varying function of position and time i.e. deterministic. The latter will be subject to rapid fluctuations, i.e. stochastic in nature. We shall summarise CHANDRASEKHAR's treatment, incidentally pointing out the relation to some of the general observations made in this article.

We first define the average density function, $n(\mathbf{r}, M, t)$ governing the spatial distribution of stars of different masses at time t, $\mathbf{r}$ representing a point in space and M the mass of the star. In our notation this represents the product density of degree 1 in the $\mathbf{r}$ and M space.

The gravitational potential on any star is given by

$$\eta(\mathbf{r}, t) = - G \iint\limits_{M\,\tau} \frac{M\,n(\mathbf{r}, M, t)}{|\mathbf{r} - \mathbf{r}_1|} \, dM \, d\tau \qquad (64.1)$$

where $d\tau$ is the elementary volume surrounding $\mathbf{r}_1$ and the integration is over the whole volume. The force on the star is given by

$$\mathbf{k} = - \nabla\eta. \qquad (64.2)$$

This force is a random function since it depends on the "complexion" of the stellar distribution or in CHANDRASEKHAR's approximation, on the "local" distribution. To elucidate the nature and origin of this fluctuation we surround the star under consideration by an element of volume σ small enough to contain only a relatively few stars. The number of stars in this volume can be assumed to be governed by a Poisson distribution. This is stationary but if we observe the number of stars in the volume σ as time varies, i.e. we plot the realised "trajectory" of the stochastic variable denoting the number of stars, the number will vary in time; so will also the positions of the stars. Hence the realised value of the force fluctuates as t varies. CHANDRASEKHAR introduces the notion of period of fluctuation to denote the interval of time that should elapse before the number of stars changes appreciably. This interval itself is a stochastic variable but the order of magnitude of its average value may be computed from simple arguments. We here see a similar situation as in the theory of "self consistent" fields. The change in the number in a small volume is a direct consequence of the varying influence of the state in the surrounding volume and it is precisely the same types of changes in the surrounding volume that cause the varying influence!

The average period of such a fluctuation is estimated as follows. It is of the order of time involved for two stars to separate by a distance equal to the average distance D between the stars. The period of fluctuation will be of the order of

$$\tau = \frac{D}{\sqrt{\mathscr{E}\{v^2\}}} \qquad (64.3)$$

---

[1] Cf. the contribution of NEYMAN and SCOTT in Vol. LIII of this Encyclopedia

where v denotes the relative velocity between two stars. If D is equal to three parsecs and $\sqrt{\mathscr{E}}\{v^2\} = 50$ km/sec, τ near the sun works out to be 6×10^4 years, which is fast enough compared to the period of galactic rotation, 2×10^8 years. The force on unit mass of the star can be written as $\mathfrak{F} = \boldsymbol{K}(\boldsymbol{r}, t) + \boldsymbol{F}(t)$ where $\boldsymbol{K}$ is the smoothed out distribution and $\boldsymbol{F}$ is the fluctuating force due to near neighbours. The fluctuating force is a random force depending on τ. If we now take Δt to denote an interval of time very large compared to τ but small compared to t, we can assume the correlation between $\boldsymbol{F}(t)$ and $\boldsymbol{F}(t+\Delta t)$ is very small indeed and we can write

$$\mathfrak{F}\,\Delta t = \boldsymbol{K}\Delta t + \boldsymbol{\delta}(t + \Delta t, t) \tag{64.4}$$

where $\boldsymbol{\delta}(t+\Delta t, t)$ is the integral

$$\int\limits_{t}^{t+\Delta t} \boldsymbol{F}(\xi)\, d\xi . \tag{64.5}$$

The changes in velocity $\boldsymbol{\delta}(t+\Delta t, t)$ and $\boldsymbol{\delta}(t+2\Delta t, t+\Delta t)$ are treated to be independent. Thus the change in velocity in an interval is the sum of two terms $\boldsymbol{K}\Delta t$ and the stochastic term $\boldsymbol{\delta}(t+\Delta t, t)$. Stated in this fashion, we immediately see the relevance of Brownian motion to stellar dynamics with this important difference that we have analysed the statistial properties of $\boldsymbol{F}(t)$ suitably for the present problem instead of appealing to apriori considerations as in Brownian motion.

Let us now consider the random force on a star. This is a random variable since it depends upon the positions and masses of the stars in space and these quantities themselves are random variables. Hence we are interested in $W(N, \boldsymbol{F})$, the frequency function of $\boldsymbol{F}$, the force on the star, per unit mass when there are N other stars in space. This force can be written as

$$\boldsymbol{F} = \sum_{i=1}^{N} \boldsymbol{F}_i \tag{64.6}$$

$$\boldsymbol{F}_i = G\,\frac{M_i\,\boldsymbol{r}_i}{|\boldsymbol{r}_i|^3} \tag{64.7}$$

where we assume the star to be situated at the origin and $\boldsymbol{F}_i$ is the force on it due to a star i of mass M_i at $\boldsymbol{r}_i$. The $\boldsymbol{F}_i$ are independent random variables and therefore the frequency function $W(N, \boldsymbol{F})$ is just the N-fold convolution of the frequency functions of $\boldsymbol{F}_i$. Thus the characteristic function $A(N, \boldsymbol{\varrho})$ corresponding to $W(N, \boldsymbol{F})$ is the product of the characteristic functions corresponding to the frequency functions of $\boldsymbol{F}_i$ which are the same for all the stars. To determine the latter, we just need the fundamental probability frequency function $\tau(\boldsymbol{r}_i, M_i)$ that the star i occurs at $\boldsymbol{r}_i$ and has a mass M_i. We shall also assume that all the N stars are confined in a sphere of radius R surrounding the origin and write

$$\tfrac{4}{3}\,\pi R^3 n = N; \quad n = \mathrm{const}. \tag{64.8}$$

The characteristic function $a(\boldsymbol{\varrho})$ corresponding to the frequency function of $\boldsymbol{F}_i$ is given by

$$a(\boldsymbol{\varrho}) = \int\limits_{0}^{\infty} dM_i \int\limits_{|\boldsymbol{r}_i|=0}^{R} \exp\,(i\,\boldsymbol{\varrho}\cdot\boldsymbol{F}_i)\,\tau(\boldsymbol{r}_i, M_i)\,d^3\boldsymbol{r}_i . \tag{64.9}$$

Note that there is no necessity to determine explicitly the frequency function of $\boldsymbol{F}_i$ to obtain its characteristic function since the integration in (64.9) is performed with respect to $\boldsymbol{r}_i$ and M_i, $\boldsymbol{F}_i$ being expressed in terms of $\boldsymbol{r}_i$ and M_i.

Therefore we obtain

$$A(N, \varrho) = [a(\varrho)]^N , \qquad (64.10)$$

$$W(N, \boldsymbol{F}) = \frac{1}{8\pi^3} \int\limits_{-\infty}^{+\infty} \exp\left(- i\,\varrho \cdot \boldsymbol{F}\right) A(N, \varrho)\, d^3\varrho . \qquad (64.11)$$

If we now suppose that the N stars are uniformly distributed in the sphere of radius R we can write

$$\tau(\boldsymbol{r}_i, M_i) = \frac{3}{4\pi R^3}\, \tau(M_i) \qquad (64.12)$$

where $\tau(M_i)$ now governs the frequency of different masses among the stars. This implies

$$\frac{3}{4\pi R^3} \int\limits_0^\infty dM \int\limits_{|\boldsymbol{r}|=0}^R \tau(M)\, d^3\boldsymbol{r} = 1 , \qquad (64.13)$$

The Eq. (64.10) reduces to

$$A(N, \varrho) = \left\{ \frac{3}{4\pi R^3} \int\limits_0^\infty dM \int\limits_0^\infty \exp\left(i\,\varrho \cdot \boldsymbol{\varphi}\right) \tau(M)\, d^3\boldsymbol{r} \right\}^N \qquad (64.14)$$

where we have written

$$\boldsymbol{\varphi} = \frac{G M \boldsymbol{r}}{|\boldsymbol{r}|^3} . \qquad (64.15)$$

We now let R and N tend to infinity keeping n finite in Eq. (64.8).

$$W(\boldsymbol{F}) = \frac{1}{8\pi^3} \int\limits_{-\infty}^{+\infty} \exp\left(- i\,\varrho \cdot \boldsymbol{F}\right) A(\varrho)\, d^3\varrho \qquad (64.16)$$

where

$$A(\varrho) = \lim_{N\to\infty} \left[\frac{3}{4\pi R^3} \int\limits_0^\infty dM \int\limits_0^R \exp\left(i\,\varrho \cdot \boldsymbol{\varphi}\right) \tau(M)\, d^3\boldsymbol{r} \right]^{\frac{4}{3}\pi R^3 n} . \qquad (64.17)$$

Defining the auxiliary function $C(\varrho)$ by the relation

$$A(\varrho) = \exp\left[- n\, C(\varrho)\right], \qquad (64.18)$$

CHANDRASEKHAR obtained the following expression for $C(\varrho)$ after considerable calculation

$$C(\varrho) = \tfrac{4}{15} (2\pi G)^{\frac{3}{2}} \mathscr{E}\{M^{\frac{3}{2}}\} |\varrho|^{\frac{3}{2}} . \qquad (64.19)$$

Therefore

$$W(\boldsymbol{F}) = \frac{1}{8\pi^3} \int\limits_{-\infty}^{+\infty} \exp\left(- i\,\varrho \cdot \boldsymbol{F} - a\,|\varrho|^{\frac{3}{2}}\right) d^3\varrho \qquad (64.20)$$

where

$$a = \tfrac{4}{15} (2\pi G)^{\frac{3}{2}} \mathscr{E}\{M^{\frac{3}{2}}\}\, n . \qquad (64.21)$$

Using the frame of reference in which one principal axis is in the direction of $\boldsymbol{F}$ and changing to polar coordinates, (64.20) can ultimately be reduced to

$$W(\boldsymbol{F}) = \frac{1}{2\pi^2 |\boldsymbol{F}|^3} \int\limits_0^\infty \exp\left(- a\, x^{\frac{3}{2}}/|\boldsymbol{F}|^{\frac{3}{2}}\right) x \sin x\, dx \qquad (64.22)$$

where

$$x = |\varrho|\,|\boldsymbol{F}| . \qquad (64.23)$$

If we now introduce the "normal field" Q_H defined by

$$Q_H = a^{\frac{3}{2}} = \left(\tfrac{4}{15}\right)^{\frac{3}{2}} (2\pi G) \left[(\mathscr{E}\,\{M^{\frac{3}{2}}\})\, n\right]^{\frac{2}{3}} \tag{64.24}$$

and expressing $|\boldsymbol{F}|$ in terms of this unit

$$|\boldsymbol{F}| = \beta\, Q_H, \tag{64.25}$$

we have

$$W(\boldsymbol{F}) = H(\beta)/(4\pi\, a^2\, \beta^2) \tag{64.26}$$

where

$$H(\beta) = \frac{2}{\pi\beta} \int\limits_0^\infty \exp\left\{-\left(\frac{x}{\beta}\right)^{\frac{3}{2}}\right\} x \sin x\, dx. \tag{64.27}$$

Since

$$W(|\boldsymbol{F}|) = 4\pi\, |\boldsymbol{F}|^2\, W(F) \tag{64.28}$$

we obtain

$$W(|\boldsymbol{F}|) = H(\beta)/Q_H. \tag{64.29}$$

$H(\beta)$ is the distribution function of $|\boldsymbol{F}|$ expressed in terms of Q_H. The asymptotic behaviour of the distribution is given by

$$W(|\boldsymbol{F}|) \simeq \left(\frac{4}{3\pi\, Q_H^3}\right) |\boldsymbol{F}|^2 \qquad (|\boldsymbol{F}| \to 0) \tag{64.30}$$

and

$$W(|\boldsymbol{F}|) \simeq 2\pi\, G^{\frac{3}{2}}\, \mathscr{E}\,\{M^{\frac{3}{2}}\}\, n\, |\boldsymbol{F}|^{-\frac{5}{2}} \quad (|\boldsymbol{F}| \to \infty). \tag{64.31}$$

Eq. (64.26) is an exact formula for the distribution of $\boldsymbol{F}$. We can however adopt an approximate treatment based on the assumption that the force acting on the star is entirely due to its nearest neighbour.

The law of distribution of the nearest neighbour is obtained as follows: Let $\omega(r)\, dr$ denote the probability that the nearest neighbour to a star occurs between r and $r + dr$; this probability must clearly be equal to the product of the probability that no stars exist interior to r and the probability that a star exists in the spherical shell between r and $r + dr$. Accordingly the function $\omega(r)$ must satisfy the relation

$$\omega(r) = \left[1 - \int\limits_0^r \omega(r)\, dr\right] 4\pi\, r^2\, n, \tag{64.32}$$

where n denotes the number of stars per unit volume. From (64.32) we derive

$$\frac{\partial}{\partial r}\left[\frac{\omega(r)}{4\pi\, r^2\, n}\right] = -\,\omega(r). \tag{64.33}$$

The solution is obtained as

$$\omega(r) = \exp\left(-\frac{4\pi\, r^2\, n}{3}\right) 4\pi\, r^2\, n. \tag{64.34}$$

On the assumption that the random force on the star is only due to its nearest neighbour, we obtain

$$W(|\boldsymbol{F}|)\, d|\boldsymbol{F}| = \exp\left[-4\pi\,(GM)^{\frac{3}{2}}\,\frac{n}{3}\,|\boldsymbol{F}|^{\frac{3}{2}}\right] \cdot 2\pi\,(GM)^{\frac{3}{2}}\, n\, |\boldsymbol{F}|^{-\frac{5}{2}}\, d|\boldsymbol{F}|,$$

and asymptotically

$$W(|\boldsymbol{F}|) \simeq 2\pi\,(G\,M)^{\frac{3}{2}}\, n\, |\boldsymbol{F}|^{-\frac{5}{2}}. \tag{64.35}$$

It may at first seem surprising that the asymptotic formula is in exact agreement with that derived from the Holtsmark distribution. The physical meaning of this agreement for $|\boldsymbol{F}| \to \infty$ is simply that the highest fields are in reality produced by the nearest neighbour.

65. Fluctuations in brightness of the Milky Way. Another series of problems were recently initiated by CHANDRASEKHAR and MUNCH. We give here the mathematical formulation of the problem and indicate the method of attack that was used, without going into the astrophysical details.

CHANDRASEKHAR and MUNCH considered the following problem:

Given (1) that there is a deterministic contribution of amount $\beta \, d\tau$ from the element of length $d\tau$ at $t=\tau$ (this is due to the stars occurring with a uniform distribution along the one-dimensional t-axis) to the intensity measured at $t=0$[1], (2) that clouds occur with a Poisson distribution $e^{-\lambda t}\dfrac{(\lambda t)^n}{n!}$ in any element of length t where λ is the probability per unit t that a cloud occurs in any interval, (3) that a cloud has a transparency factor q (i.e. it reduces the intensity of radiation of the stars immediately behind it by this factor) with a probability density $\psi(q)$ so that $\lambda \, \psi(q) \, dq$ is the probability per unit t that radiation of a given intensity u jumps to an interval between $u(q)$ and $u(q+dq)$.

What is the frequency function $g(u, t)$ governing the probability with which an observer at the origin, i.e. at $t=0$ will measure an intensity u when the system extends to a distance t?

If we use the Markovian property of the stochastic process defined by $g(u, t)$ and express $g(u, t+\Delta t)$ in terms of $g(u, t)$, we obtain the integro-differential equation

$$\frac{\partial g(u, t)}{\partial t} = -\lambda g(u, t) - \beta \frac{\partial g(u, t)}{\partial u} + \lambda \int_{u'} g(u', t)\, \psi\left(\frac{u}{u'}\right) \frac{du'}{u'}. \qquad (65.1)[2]$$

Without loss of generality λ and β can be chosen to be unity by suitably choosing the units of distance and intensity. Noting that

$$\int_u^t g(u', t)\, \psi\left(\frac{u}{u'}\right) \frac{du'}{u'} = \int_{u/t}^1 g\left(\frac{u}{q}, t\right) \frac{\psi(q)}{q} \, dq \qquad (65.2)$$

we obtain the equation in the form given by CHANDRASEKHAR and MUNCH:

$$\frac{\partial g(u, t)}{\partial t} = -g(u, t) - \frac{\partial g(u, t)}{\partial u} + \int_q g\left(\frac{u}{q}, t\right) \psi(q) \frac{dq}{q}. \qquad (65.3)$$

Defining the MELLIN's transform of $g(u, t)$ and $\psi(q)$ as

$$p(s, t) = \int_0^\infty g(u, t)\, u^s \, du, \qquad q_s = \int_0^\infty \psi(q)\, q^s \, dq = \int_0^1 \psi(q)\, q^s \, dq, \qquad (65.4)$$

where s is a complex variable, we obtain

$$\frac{\partial p(s, t)}{\partial t} = -p(s, t) + q_s\, p(s, t) + s\, p(s-1, t). \qquad (65.5)$$

---

[1] In astrophysical problems it is customary to assume that the observer is at $t=0$ and distances are measured from him.

[2] Note that this equation is a special case of (39.4).

The corresponding initial conditions are $g(u, 0) = \delta(u)$, i.e. $p(s, 0) = 0$ if $s \neq 0$, $p(0, 0) = 1$ (see Sect. 19 for definition of the Mellin transform of the δ function). The inverse corresponding to Eq. (65.4) is

$$g(u, t) = \frac{1}{2\pi i} \int_{\sigma - i\infty}^{\sigma + i\infty} p(s, t)\, u^{-s-1}\, ds. \tag{65.6}$$

Since s is complex, (65.5) cannot be solved by iteration. But if we put $s = n$, $n \geq 0$ being an integer, $p(n, t)$ represents the n-th moment of u. By iteration it will be found

$$p(n, t) = n! \sum_{k=0}^{n} \frac{e^{-(1-q_k)t}}{\prod\limits_{j=0, j \neq k}^{n} (q_j - q_k)}. \tag{65.7}$$

In the above discussion, the stochastic element is due to the fluctuations in the transparency factor $Q(\tau)$ of the clouds occurring in a distance 0 to τ. Between τ and $\tau + d\tau$, radiation of intensity of magnitude $\beta\, d\tau\,(\beta = 1)$ is produced, but, in its passage through a distance τ before reaching the observer is cut in intensity to $Q(\tau)\,\beta\, d\tau$. Hence $Q(\tau)$ is also a stochastic variate with a probability frequency distribution given by $P(Q, \tau)$ satisfying the integro-differential equation

$$\frac{\partial P(Q, \tau)}{\partial \tau} = -\lambda\, P(Q; \tau) + \int_{Q}^{1} P\left(\frac{Q}{q}, \tau\right) \frac{\psi(q)\, dq}{q} \tag{65.8}$$

with the initial condition $P(Q, 0) = \delta(Q - 1)$ where δ is the Dirac delta function and $\psi(q)$ is the probability frequency function of q, the transparency factor of a single cloud.

Another random element can be introduced by considering a discrete distribution of stars which corresponds to the actual case. In this case, we ask for a more detailed question regarding the statistical distribution of the observed intensity of stars. The problem becomes complicated if we introduce attenuation with distance of the intensity. We here give a brief description of the mode of approach.

A star can be seen by an observer only if the intensity reaching the observer is above a certain value i.e. below a certain magnitude. The stars are distributed at random in space and possess a random distribution in intensity. There is a deterministic attenuation of the intensity of the star with distance and clouds intercepting the light of the stars occur in the line of sight in a random manner with a random transparency factor. Taking into account all these factors, we want to find the mean and mean-square number of stars visible along our line of sight through a given distance per unit solid angle.

CHANDRASEKHAR and MUNCH [26] have found solutions to this problem giving the mean and mean square of $N(m)$, the number of stars below a certain apparent magnitude m and visible to us along a line of sight per unit solid angle, taking the occurrence of the clouds and their transparency factor as the only two random elements in the problem. In their treatment they assumed a "deterministic" distribution of stars in space. We can introduce another probabilistic element into the problem by taking the occurrence of stars and their distribution in intensity also to be stochastic in nature.

1. The occurrence of a star at τ per unit solid angle is governed by the function $\tau^2 f(\tau)\, \lambda\, d\tau$.

2. The distribution in intensity η of a star can be taken as $\varphi(\eta)$ and for a star to be seen with an intensity η' by the observer the intrinsic intensity of the star should be $\eta'\,\tau^2/Q(\tau)$ to allow for attenuation with distance and transparency factor $Q(\tau)$ of the clouds occurring between the observer and the star. The total transparency factor itself is a stochastic variate with distribution function $\pi(Q, \tau)$. Hence the probability of a star occurring between τ and $\tau+d\tau$ and being visible is given by

$$F_1(\tau)\,d\tau = \tau^2 f(\tau)\,d\tau \int_Q \psi\left(\frac{\eta'\,\tau^2}{Q}\right)\pi(Q, \tau)\,dQ\,, \qquad \psi(\eta) = \int_\eta^\infty \varphi(\eta')\,d\eta'. \qquad (65.9)$$

$F_1(\tau)$ is a product density of degree one. Hence the mean number of stars seen with apparent intensity greater than η' is given by

$$\mathscr{E}\{N(\eta')\} = \int_0^L F_1(\tau)\,d\tau \qquad (65.10)$$

where L is the total distance under observation.

$F_2(\tau_1, \tau_2)\,d\tau_1\,d\tau_2$ the joint probability that a star occurs in $d\tau_1$ and another at $d\tau_2$ and these are visible is given by

$$F_2(\tau_1, \tau_2)\,d\tau_1\,d\tau_2 = \tau_1^2\,\tau_2^2\,f(\tau_1)\,f(\tau_2)\,d\tau_1\,d\tau_2\times$$
$$\left.\times \int_{Q_1}\int_{Q_2} \pi(Q_1, Q_2; \tau_1, \tau_2)\,\psi\left(\frac{\eta'\,\tau_1^2}{Q_1}\right)\psi\left(\frac{\eta'\,\tau_2^2}{Q_2}\right)dQ_1\,dQ_2.\right\} \qquad (65.11)$$

$F_2(\tau_1, \tau_2)$ is a product density of degree two and the mean-square number of visible stars is given by

$$\mathscr{E}\{N^2(\eta')\} = \int_0^L F_1(\tau)\,d\tau + \int_0^L\int_0^L F_2(\tau_1, \tau_2)\,d\tau_1\,d\tau_2. \qquad (65.12)$$

This problem has been discussed in detail by RAMAKRISHNAN and VASUDEVAN [27].

66. Statistical problems relating to the spatial distribution of galaxies. Recently in a series of papers, NEYMAN and SCOTT have developed a probabilistic theory of spatial distribution of galaxies. In a general essay of this type it is not possible to give a detailed summary of their work which was a cooperative study extending over several years[1].

The essential idea behind their theory is based on the "clustering" of galaxies. Our purpose here is merely to explain the concept of clustering from a probabilistic point of view.

It is assumed that points known as cluster centres are distributed uniformly in a Poissonian manner in space, i.e. if we take a volume V the probability that there are N centres is given by

$$e^{-\lambda V}(\lambda V)^N/N!, \qquad (66.1)$$

i.e. the product density of degree n of cluster centres is given by

$$f_n(|r_1|, |r_2|, \ldots, |r_n|)\,d\tau_1\,d\tau_2 \ldots d\tau_n = \lambda^n\,d\tau_1\,d\tau_2 \ldots d\tau_n \qquad (66.2)$$

where $d\tau_1\,d\tau_2 \ldots d\tau_n$ are the volumes surrounding the points $r_1, r_2, \ldots, r_n$ respectively. With respect to each cluster centre, we associate the following distribution of galaxies. Assuming a centre to be situated at $r=0$ stars associated with this centre are distributed in a random manner described by the product densities $\varphi_n(r_1, r_2, \ldots, r_n)\,d\tau_1\,d\tau_2 \ldots d\tau_n$ where the r's are measured from the

---

[1] In a contribution to Vol. LIII of this Encyclopedia entitled "Large scale organisation of distribution of galaxies", NEYMAN and SCOTT have summarised their work.

origin of the cluster centre. We can assume a particular case that φ_n can be written in the form

$$\varphi_n(\boldsymbol{r}_1, \boldsymbol{r}_2, \ldots, \boldsymbol{r}_n) = \varphi_1(|\boldsymbol{r}_1|)\, \varphi_1(|\boldsymbol{r}_2|) \cdots \varphi_1(|\boldsymbol{r}_n|). \tag{66.3}$$

This means that we have a Poissonian distribution of stars with radial symmetry. But the parameter of the Poissonian distribution in any volume depends upon the coordinates of the volume with respect to the cluster centre. Each of these cluster centres has thus a statistical distribution of stars associated with it and the functions defining a statistical distribution of stars are the same for all centres. Our object is to determine the statistical distribution of stars without reference to the cluster centres. Instead of speaking in general terms, we shall derive the integral equations for the product densities of stars when we assume radial symmetry for the function.

α) *Product density of degree 1.* We ask for the probability $f_1(\boldsymbol{r})\, d\tau$ that a star occurs in $d\tau$ where $\boldsymbol{r}$ is measured from an arbitrary origin (not a cluster centre) and $d\tau$ the volume surrounding the point $\boldsymbol{r}$.

$$f_1(\boldsymbol{r})\, d\tau = \int_{\tau_1} \varphi(|\boldsymbol{r} - \boldsymbol{r}_1|)\, d\tau_1\, \lambda\, d\tau \tag{66.4}$$

This is obtained merely by arguing that a cluster centre is situated in $d\tau_1$ with probability $\lambda\, d\tau_1$ and the star at $\boldsymbol{r}$ belonging to that cluster occurs at $d\tau$.

β) *Product density of degree 2.* If $f(\boldsymbol{r}_1, \boldsymbol{r}_2)\, d\tau_1\, d\tau_2$ is the joint probability that a star occurs in $d\tau_1$ surrounding $\boldsymbol{r}_1$ and another in $d\tau_2$ surrounding $\boldsymbol{r}_2$ then we have

$$\left. \begin{aligned} f(\boldsymbol{r}_1, \boldsymbol{r}_2)\, d\tau_1\, d\tau_2 &= \iint \lambda^2\, d\tau_1'\, d\tau_2'\, \varphi(|\boldsymbol{r}_1 - \boldsymbol{r}_1'|)\, \varphi(|\boldsymbol{r}_2 - \boldsymbol{r}_2'|)\, d\tau_1\, d\tau_2 + \\ &\quad + \int \lambda\, d\tau'\varphi(|\boldsymbol{r}_1 - \boldsymbol{r}'|)\, \varphi(|\boldsymbol{r}_2 - \boldsymbol{r}'|)\, d\tau_1\, d\tau_2. \end{aligned} \right\} \tag{66.5}$$

The first term is obtained by arguing that $\lambda^2\, d\tau_1'\, d\tau_2'$ is the probability that a cluster centre occurs in $d\tau_1'$, another in $d\tau_2'$ and that a star belonging to the centre at $d\tau_1'$ occurs in $d\tau_1$ and a star belonging to the centre at $d\tau_2'$ occurs in the volume $d\tau_2$. The second term is interesting since it relates to the stars at $\boldsymbol{r}_1$ and $\boldsymbol{r}_2$ which belong to the same centre at $d\tau'$. If this term were not there, $f(\boldsymbol{r}_1, \boldsymbol{r}_2)\, d\tau_1\, d\tau_2$ can be written as a product of $f(\boldsymbol{r}_1)$ and $f(\boldsymbol{r}_2)$. It introduces the element of correlation in the problem and $f(\boldsymbol{r}_1, \boldsymbol{r}_2)$ is not equal to $f(\boldsymbol{r}_1)\, f(\boldsymbol{r}_2)$. Hence the distribution of stars without reference to cluster centres is non-Poissonian though the distribution of stars with respect to their corresponding cluster centres is Poissonian. This is intriguing and on this idea is based the entire work of NEYMAN and SCOTT.

VII. Statistical theory of the structure of simple fluids[1].

67. Introductory remarks. The theory of product densities has recently found applications in the theory of the structure of simple fluids. It is clear that the essential difference between a liquid and a gas from a statistical point of view is that we have to assume that in a liquid each molecule is simultaneously in interaction with its neighbours. But unlike the case of solids, there is no single "regular" periodic configuration of dominant probability which can be used as the starting point of development for small displacements as in the crystal. One is forced to a situation intermediate between that of the ideal gas (where we assume no interaction) and that of the crystal (where there exists almost a

---

[1] Details on the structure of fluids are given in GREEN's contribution to Vol. X of this Encyclopedia.

deterministic element, completely controlling the motion of each particle by the configuration of its neighbours.) This led BORN and GREEN [28] to define what are known as density functions which are identical with product densities. We can define

$$\varrho_m(\boldsymbol{r}_1, \boldsymbol{r}_2, \ldots, \boldsymbol{r}_m)\, d\tau_1\, d\tau_2 \ldots d\tau_m \tag{67.1}$$

as the joint probability that a particle is found in the volume element $d\tau_1$ surrounding the point $\boldsymbol{r}_1$, one in $d\tau_2$ surrounding $\boldsymbol{r}_2 \ldots$ and one in $d\tau_m$ surrounding $\boldsymbol{r}_m$ irrespective of the number elsewhere. ϱ_1 is thus the average density and $\varrho_2(\boldsymbol{r}_1, \boldsymbol{r}_2)$ the product density of degree two. This product density can be found as the Fourier transform of the scattering of X-ray intensity. The higher product densities are experimentally unobservable atleast at present, but are essential concepts in almost any attempt to treat the transport properties of the fluids such as viscosity or conductivity. It is quite clear that to determine these product densities from a rigorous stochastic point of view we are led to coupled integral equations. The three-dimensional nature of the volume leads to the difficulty of dealing with non-Markovian and non-regenerative processes. If we wish to take advantage of the simplicity of the regeneration property of a stochastic process, we shall be forced to consider a one-dimensional theory of fluids. But we have to admit that a one-dimensional theory cannot lead to an extension to three dimensions. However to gain familiarity with product density functions we briefly give the formulation of the one-dimensional model and later refer to the three-dimensional fluids. For this purpose, we first discuss the distribution of particles along a one-dimensional t-axis.

68. One-dimensional fluids. We assume the particles are randomly distributed on a one-dimensional t-axis in the following manner. The probability that a particle lies between τ_1 and $\tau_1 + d\tau_1$ given that the "previous" particle is situated at τ is given by $g(\tau_1 - \tau)$. g is called the basic function of the random distribution of particles. If $g(n, t)\, dt$ is the probability that the n-th particle lies between t and $t + dt$ given that a particle (uncounted) occurs at $t = 0$, then $g(n, t)$ satisfies the integral equation

$$\left.\begin{aligned} g(n, t) &= \int_0^t g(n - 1, t - \tau)\, g(\tau)\, d\tau, \qquad n > 1 \\[2mm] g(1, t) &\equiv g(t). \end{aligned}\right\} \tag{68.1}$$

It is clear that $g(n, t)$ is the n-fold convolution of $g(t)$. If $\varrho_1^F(t)$[1] is the product density of degree one of particles at t, by simple reasoning we note

$$\varrho_1^F(t) = \sum_{n=1}^{\infty} g(n, t). \tag{68.2}$$

If $\varrho_n^F(t_1, t_2, \ldots, t_n)$ is the product density of degree n of particles, it can be expressed in terms of ϱ_1^F as

$$\varrho_n^F(t_1, t_2, \ldots, t_n) = \varrho_1^F(t_1)\, \varrho_1^F(t_2 - t_1) \ldots \varrho_1^F(t_n - t_{n-1}). \tag{68.3}$$

We call the above distribution of particles on a line, a "free" distribution with basic function $g(t)$.

In considering a one-dimensional liquid we impose the condition that the $(N + 1)$-th particle is at L. In such a case the corresponding basic function

---

[1] The superscript F indicates a "free distribution" as contrasted with a constrained distribution to be described presently.

$g^c(n, t, L)$ and the product density $\varrho_n^c(t_1, t_2, \ldots, t_n)$ for the "constrained" distribution of particles are given by

$$g^c(n, t, L) = \frac{g(n, t)\, g(N+1-n, L-t)}{g(N+1, L)}, \tag{68.4}$$

$$\varrho_n^c(t_1, t_2, \ldots, t_n) = \sum_{m_i} \frac{g(m_1, t_1)\, g(m_2, t_2-t_1) \ldots g(m_n, t_n-t_{n-1})\, g(m_{n+1}, L-t_n)}{g(N+1, L)}. \tag{68.5}$$

In particular, as before we have

$$\varrho_1^c(t) = \sum_{n=1}^{\infty} g^c(n, t, L).$$

One of the interesting properties of the constrained distribution is as follows. If we replace $g(n, t)$ by a function $G(n, t)$ such that

$$G(n, t) = g(n, t)\, e^{\mu n + \lambda t} \tag{68.6}$$

we obtain the same expressions for the product densities for the constrained distribution. In other words, $G(n, t)\, dt$ is the unnormalized probability that the n-th particle occurs between t and $t+dt$. Since $\int_0^L g(n, t)\, dt = 1$ for all n, we have

$$e^{-\mu n} \int_0^{\infty} G(n, t)\, e^{-\lambda t}\, dt = \frac{e^{-\mu n}}{Q_n(\lambda)} \quad \text{i.e.} \quad Q_n(\lambda) = e^{-\mu n}. \tag{68.7}$$

In particular

$$e^{-\mu} = Q_1(\lambda).$$

We can recognise $Q_1(\lambda)$ to be the partition function corresponding to the distribution $g(t) = \dfrac{G(t)\, e^{-\lambda t}}{Q_1(\lambda)}$. It is customary in statistical mechanics to take $G(t_2-t_1) = e^{-\beta(t_2-t_1)\, u}$ where $u(t_2-t_1)$ is the potential between two particles. Thus in the one-dimensional fluid we consider potentials only between nearest neighbours.

Salzburg, Zwanzig and Kirkwood [29] have developed a one-dimensional theory of liquids by choosing proper functions for the potentials. It can be shown that in the case of simple potentials, the product densities of small orders have the same form as the free distribution but with λ being a specified function of the parameters characteristic of the constraint. To illustrate this statement if we choose the hardcore potential

$$\left.\begin{aligned} u(t) &= \infty \quad \text{for} \quad 0 < t < a \\ u(t) &= 0 \quad\ \ \text{for} \quad t > a \end{aligned}\right\} \tag{68.8}$$

the product density of degree one for the free distribution is given by

$$\varrho_1^F = \sum_{n=1}^{\infty} H(t - n a)\, \exp\left[-\lambda(t - n a)\right] \frac{\lambda^n (t - n a)^{n-1}}{(n-1)!} \tag{68.9}$$

where H is the Heaviside unit function. In the corresponding expression for the constraints, λ assumes the particular value $\dfrac{N}{L - N a}$.

In fact, this result is quite general and forms the basis of Khinchin's work on statistical mechanics as will be mentioned later (Sect. 76).

We now proceed to a short discussion of the problem of three-dimensional fluids. We cannot take advantage of the "regenerative" property by ordering the particles. We will have to consider the entire distribution of all the N particles by appealing to the definition of Janossy densities and then by a process of integration obtain product densities of small orders. In this we deal only with constrained distributions.

69. Three-dimensional fluids. The extension to the three-dimensional case of an actual fluid is immediate as long as we are concerned with a formal definition of product density functions. In the study of the structure of three-dimensional fluids, we define

$$\varrho_n(\boldsymbol{r}_1, \boldsymbol{r}_2, \ldots, \boldsymbol{r}_n)\, d\tau_1\, d\tau_2 \ldots d\tau_n = \varrho_n(\langle n \rangle)\, d\langle n \rangle \tag{69.1}$$

as the joint probability that there exists one molecule in the volume element $d\tau_1$ surrounding $\boldsymbol{r}_1$, one in $d\tau_2$ surrounding $\boldsymbol{r}_2 \ldots$ and one in $d\tau_n$ surrounding $\boldsymbol{r}_n$ irrespective of the number elsewhere. If $J_m(\boldsymbol{r}_1, \boldsymbol{r}_2, \ldots, \boldsymbol{r}_m)$ is the Janossy density of degree m then $J_m(\boldsymbol{r}_1, \boldsymbol{r}_2, \ldots, \boldsymbol{r}_m)\, d\tau_1\, d\tau_2 \ldots d\tau_m$ represents the joint probability of a particle being in $d\tau_1$, one in $d\tau_2$, $\ldots$ and one in $d\tau_m$ and *none elsewhere*. If the total number of particles of a liquid is fixed and is equal to N, then all Janossy densities of order $m \neq N$ vanish and

$$J_N(\boldsymbol{r}_1, \boldsymbol{r}_2, \ldots, \boldsymbol{r}_N) = \varrho_N(\boldsymbol{r}_1, \boldsymbol{r}_2, \ldots, \boldsymbol{r}_N). \tag{69.2}$$

$\varrho_1(\boldsymbol{r}_1)\, d\tau_1$ for a fluid is a trivial constant equal to the average number density. $\varrho_2(\boldsymbol{r}_1, \boldsymbol{r}_2)\, d\tau_1\, d\tau_2$ is a function only of $r_{12} = |\boldsymbol{r}_1 - \boldsymbol{r}_2|$ and is significant experimentally as the Fourier transform of scattered X-ray intensity.

MAYER defines a function G_N^0 as

$$G_N^0 = \exp\{- U_N \langle N \rangle / kT\} \tag{69.3}$$

with $U_N \langle N \rangle$ representing the mutual potential energy of the molecules which is obviously a function of all the coordinates and hence a function of $\dfrac{N(N-1)}{2}$ distances r_{ij} $(i, j = 1, 2, \ldots, N)$, since $U_N \langle N \rangle$ is a potential function

$$\left.\begin{aligned} U_N\langle N \rangle &\to 0 \quad \text{for all} \quad r_{ij} \to \infty, \\ U_N\langle N \rangle &\to \infty \quad \text{for any} \quad r_{ij} \to 0. \end{aligned}\right\} \tag{69.4}$$

A quantity Z is also defined as

$$Z = \exp\left[(\mu - \mu_0)/kT\right] \tag{69.5}$$

where μ is the chemical potential of GIBBS and μ_0 is so chosen that Z becomes equal to the number density $\varrho = N/V$, at the limit that both become zero

$$\lim_{Z \to 0}\left\{\frac{Z}{\varrho}\right\} = 1 \tag{69.6}$$

(determines μ_0), for which density the system always approaches a perfect gas in properties. He makes the assumption that

$$J_N(\boldsymbol{r}_1, \boldsymbol{r}_2, \ldots, \boldsymbol{r}_N) = Z^N G_N^0 C \tag{69.7}$$

where C has to be so chosen that J_N is a Janossy density. That is

$$\int J_N(\boldsymbol{r}_1, \boldsymbol{r}_2, \ldots, \boldsymbol{r}_N)\, d\tau_1\, d\tau_2 \ldots d\tau_N = N!. \tag{69.8}[1]$$

---

[1] Note that $J_N/N!$ corresponds to $g^c(N, L)$ and G_N^0 to G, the function in the earlier discussion on the one-dimensional fluid (Sect. 68).

Therefore C is given by

$$[C]^{-1} = \sum_N \frac{Z^N}{N!} \int\int\cdots\int G_N^0 \langle N\rangle \, d\langle N\rangle. \tag{69.9}$$

This constant has the physical significance that it is equal to $e^{-PV/kT}$. MAYER also defines the following function

$$G_n\langle n\rangle = \sum_{N\geq 0} \frac{Z^N}{N!} \int\int\cdots\int G_{N+n}^0 \left(\langle n\rangle + \langle N\rangle\right) d\langle N\rangle. \tag{69.10}$$

It is obviously connected with the product density ϱ_n in the following manner

$$G_n\langle n\rangle = e^{PV/kT} Z^{-n} \varrho_n\langle n\rangle. \tag{69.11}$$

Since P is expressed as a function of Z, T in Eq. (69.5), for $n=0$ the thermodynamic relations

$$\left.\begin{aligned}
Z\left[\frac{\partial}{\partial Z}(P/kT)\right] &= \varrho, \\[4pt]
\frac{\partial(P/kT)}{\partial T} &= H - H_0,
\end{aligned}\right\} \tag{69.12}$$

with H as the enthalpy and H_0 the enthalpy of a perfect gas, serve to give all the thermodynamic properties of the system from those of a perfect gas. The function $G_n\langle n\rangle$ for $n=1,\ 2,\ \ldots$ go beyond the thermodynamic specifications of the system and the physical meaning is

$$G_n\langle n\rangle = e^{PV/kT} Z^{-n} \varrho_n\langle n\rangle. \tag{69.13}$$

We have here adapted the discussion of MAYER [30] in a manner suitable to demonstrate the role of product densities. The starting point of any type of liquid is based upon the type of approximation used for the potential. For a preliminary acquaintance of the theory, we refer to the lucid article by J. E MAYER [30]. A trivial case arises when $U_N\langle N\rangle = 0$ for a perfect gas:

$$\left.\begin{aligned}
G_N^0 &= 1, \\[4pt]
G_0 &= e^{PV/kT} = \sum_N \frac{Z^N}{N!} V^N = e^{VZ}.
\end{aligned}\right\} \tag{69.14}$$

We get $\dfrac{PV}{NkT} = 1$ which is the perfect gas equation.

The essential difficulty in a comprehensive treatment of the three-dimensional fluid lies in performing the integrations in Eq. (69.10) since the term involving the potential is a function of all the $n(n-1)$ distances between the n particles. It is customary in such cases to adopt the following approximation. The potential between any two particles vanishes for all except small distances in which case it becomes very large. Hence writing $e^{-u(r_{ij})}$ as $1+v_{ij}$ we note that v_{ij} is equal to zero for all except small distances. If the integral is written in terms of v_{ij}'s it can be expressed as a sum of what was called by MAYER "cluster integrals" involving particles at small distances from each other[1].

VIII. Differential equations involving random functions of time.

70. General formulation. The phenomenological interpretation of random functions given in Chap. B I can be directly applied to the solution of linear

---

[1] Cf. J. E. MAYER's article in Vol. XII of this Encyclopedia.

differential equations involving random functions of time. Suppose we have a linear differential equation of the m-th order

$$\frac{d^m y}{dt^m} + a_1(t)\frac{d^{m-1}y}{dt^{m-1}} + \cdots + a_m(t)\, y = x(t). \tag{70.1}$$

If $x(t)$ is a random function of t representing a stochastic process, what is the meaning of the solution of this differential equation? Since $x(t)$ is a random function of t, $y(t)$ obviously is. And so let us understand the "solution" to mean the determination of the joint distribution of $y(t)$ and $x(t)$ at any k points on the t-axis, k being as large as we please, or at least the probability frequency function (p.f.f.) of $y(t)$ as t varies. Ignoring the random nature of $x(t)$, we can treat the above equation as symbolic and first write the solution of $y(t)$ as an m-th order iterated integral of $x(t)$.

$$y(t) = \varphi_{m+1}(t)\int_0^t \varphi_m(\tau_m)\, d\tau_m \int_0^{\tau_m}\cdots\int_0^{\tau_2}\varphi_1(\tau_1)\, x(\tau_1)\, d\tau_1 \tag{70.2}$$

where the φ_i's are determined through the differential equations connected with a_i's but not involving $x(t)$. It is enough if we consider the nature of

$$y_m(t) = \int_0^t \varphi_m(\tau_m)\int_0^{\tau_m}\cdots\int_0^{\tau_2}\varphi_1(\tau_1)\, x(\tau_1)\, d\tau_1 \tag{70.3}$$

i.e. omitting the term $\varphi_{m+1}(t)$. On this definition $y_0(t) = x(t)$.

According to the phenomenological interpretation, a "knowledge" about the random function of t is necessary to obtain the p.f.f. of $y_m(t)$. By "knowledge" we mean that the measure of a typical "trajectory" of $x(t)$ in any interval on the t-axis is known. If $x^R(t)$ and $y^R(t)$ are the realised curves, then they are connected by the same relation as above but, in this case, the integration is not merely symbolic but meaningful in the ordinary sense since we are dealing with actual realised curves. We can write

$$\left.\begin{aligned} y_m^R(t) &= \int_0^t \varphi_m(\tau)\, y_{m-1}^R(\tau)\, d\tau, \\ y_0^R(t) &= x^R(t). \end{aligned}\right\} \tag{70.4}$$

The standard procedure to determine the p.f.f. of $y_m(t)$ is to express the variation of the realised value of $y_m(t)$ corresponding to the variations in the realised values of $x(t)$. This procedure is cumbersome and amounts to finding the joint p.f.f. of the aggregate $\langle y_i(t)\rangle$, $i = 0, 1, 2, \ldots, m$. However a direct method can be devised to arrive at the joint p.f.f. of $y(t)$ and $x(t)$ without recourse to the joint p.f.f. of the aggregate $\langle y_i(t)\rangle$. To this end, the integral can be "telescoped" and written as

$$y^R(t) = \int_0^t F(t,\tau)\, x^R(\tau)\, d\tau \tag{70.5}$$

where $F(t,\tau)$ is a fully deterministic function.

$$F(t,\tau) = \varphi_1(\tau)\int_\tau^t \varphi_m(\tau_m)\, d\tau_m \ldots \int_\tau^{\tau_3}\varphi_2(\tau_2)\, d\tau_2. \tag{70.6}$$

Hence we are justified in writing the random function $y_m(t)$ as

$$y_m(t) = \int_0^t F(t,\tau)\, x(\tau)\, d\tau. \tag{70.7}$$

The replacement of the iterated integral by the single integral does not simplify the problem as far as the study of the trajectory of $y(t)$ is concerned since in computing the variation in the realised value of $y_m(t)$ in the interval $(t, t+dt)$ we are effectively computing the corresponding realised value of $y_1, y_2, \ldots, y_m$ and this is essentially the complication we wish to avoid. However the form of $y_m(t)$ as given above is suitable for evaluating the moments of $y_m(t)$ using a general theorem expressing the moments of the integral in terms of the correlation functions of the integrand and an examination of this theorem shows that the occurrence of t in the integrand which is the source of complication in the study of the trajectory of $y_m(t)$ does not affect the direct application of the theorem which yields

$$\mathscr{E}\{y_m^n(t)\} = \int\limits_0^t \int\limits_0^t \cdots \int\limits_0^t F(t, \tau_1)\, F(t, \tau_2) \ldots F(t, \tau_n) \times \left.\vphantom{\int}\right\} \tag{70.8}$$
$$\times\, \mathscr{E}\{x(\tau_1)\, x(\tau_2) \ldots x(\tau_n)\}\, d\tau_1\, d\tau_2 \ldots d\tau_n.$$

If $x(\tau)$ is Markovian and homogeneous with respect to t, writing out the formal expression for $\mathscr{E}\{x(\tau_1)\, x(\tau_2) \ldots x(\tau_n)\}$ is an easy task:

$$\mathscr{E}\{x(\tau_1)\, x(\tau_2) \ldots x(\tau_n)\} = \int\limits_{x_1}\int\limits_{x_2} \cdots \int\limits_{x_n} x_1\, x_2 \ldots x_n\, \pi(x_1 \mid x_0; \tau_1) \left.\vphantom{\int}\right\} \tag{70.9}$$
$$\pi(x_2 \mid x_1; \tau_2 - \tau_1) \ldots \pi(x_n \mid x_{n-1}; \tau_n - \tau_{n-1})\, dx_1\, dx_2 \ldots dx_n$$

where $\pi(x_i \mid x_{i-1}; \tau_i - \tau_{i-1})$ represents the p.f.f. of $x(\tau_i)$ given that $x(\tau_{i-1}) = x_{i-1}$. If x_0 is not specified but has a p.f.f., we have to integrate the above expression over this p.f.f.

As long as t occurs in the integrand, $y_n(t) = \int\limits_0^t F(t, \tau)\, x(\tau)\, d\tau$ does not represent a simplification of (70.3). To resolve this difficulty we adopt the following device which it is hoped can be applied to a great variety of problems. Defining the stochastic process represented by the random function

$$y_m(t, a) = \int\limits_0^t F(a, \tau)\, x(\tau)\, d\tau \tag{70.10}$$

[the suffix m being used merely to indicate the comparison with $y_m(t)$ defined by (70.2)] where a is a parameter independent of t, we find that the realised value of $y_m(t, a)$ is identical with the realised value of $y_m(t)$ at $t = a$ for a corresponding realisation of $x(t)$ in the interval $(0, t)$. At no other t are their realisation values the same. Hence the p.f.f.'s of $y_m(t, a)$ and $y_m(t)$ are identical at $t = a$ though different at any other value of t; in other words, we can obtain the p.f.f. of $y_m(t)$ by first obtaining the p.f.f. of $y_m(t, a)$ and in that expression just replacing a by t. This procedure is very simple since the trajectory of $y_m(t, a)$ can be studied in an easier manner than that of $y_m(t)$. For, defining $\pi(y_m, x; t, a)$ as the joint p.f.f. of $y_m(t, a)$ and $x(t)$ we have, by arguments now familiar in stochastic theory, the equation

$$\frac{\partial \pi(y_m, x; t, a)}{\partial t} = -\, \pi(y_m, x; t, a) \int\limits_{x'} R(x' \mid x)\, dx' + \left.\vphantom{\int}\right\}$$
$$+ \int\limits_{x'} \pi(y_m, x'; t, a)\, R(x \mid x')\, dx' \tag{70.11}$$
$$-\, x\, F(a, t)\, \frac{\partial \pi(y_m, x; t, a)}{\partial x}.$$

assuming that $R(x' \mid x)\, dx'\, dt$ is the probability of transition from the state x to the state $x' + dx'$ during the interval dt. We recognise that $\pi(y_m, x; t)$ the joint distribution of $y_m(t)$ and $x(t)$ is obtained on replacing a by t in $\pi(y_m, x; t, a)$ and therefore it satisfies a much more complicated equation than Eq. (70.11) given above.

We shall not attempt to obtain the solution for a general basic random process but concern ourselves with the simple case where $x(t)$ represents random points in the interval 0 to t. The distribution of the random points need not be Poissonian but we shall assume that it is described by product densities of various degrees, the definition of which we shall now recall.

If $f_n(t_1, t_2, \ldots, t_n)$ is the product density of degree n then

$$f_n(t_1, t_2, \ldots, t_n) \, dt_1 \, dt_2 \ldots dt_n$$

represents the probability that one random point lies in dt_1, one in dt_2 etc., irrespective of the distribution of random points elsewhere. In this case, it is not necessary to assume $x(t)$ to be Markovian as we did earlier in deriving (70.11) and we shall show that to obtain the first k moments of $y_m(t)$ it is enough if we know f_n for all values of $n \leq k$. The correlation functions of $x(t)$ are connected with the product densities in a complicated manner to be referred to presently.

If the distribution of $x(t)$ is Poissonian (in such a case it is Markovian) we shall speak of $n(t)$ instead of $x(t)$ i.e. the probability that $n(t) = n$ is given by $\pi(n, t) = e^{-\lambda t} \frac{(\lambda t)^n}{n!}$, where λ is the characteristic parameter of the Poisson distribution. In the language of product densities

$$f_1(t) = \lambda, \qquad f_n(t_1, t_2, \ldots, t_n) = \lambda^n. \tag{70.12}$$

A few more interesting results can be obtained. Let us assume we can write symbolically

$$\int_0^t F(t, \tau) \, n(\tau) \, d\tau = \int_0^t \varphi(t, \tau) \, \frac{dn(\tau)}{d\tau} \, d\tau \tag{70.13}$$

where

$$\varphi(t, \tau) = \int_\tau^t F(t, \tau') \, d\tau' \tag{70.14}$$

and $dn(\tau)/d\tau$ has the following interpretation: since $n(\tau)$ is a random variable so is $dn(\tau)/d\tau$. Its realisation corresponding to a realisation of n random points at $\tau_1, \tau_2, \ldots, \tau_n$ in the interval $(0, t)$ is represented by the curve

$$\frac{dn^R(\tau)}{d\tau} = \sum_i \delta(\tau - \tau_i). \tag{70.15}$$

Hence with every point occurring at τ_i in $(0, t)$ we associate the value $\varphi(t, \tau_i)$ and with the interval $(0, t)$ we associate the stochastic variable $\sum_i \varphi(t, \tau_i)$. The correlation function theorem (70.8) cannot be applied to the form

$$\int_0^t \varphi(t, \tau) \, \frac{dn(\tau)}{d\tau} \, d\tau \tag{70.16}$$

for the following reason.

For the validity of applying the theorem, $\frac{dn(\tau)}{d\tau} \, d\tau$ must be infinitesimal for all τ in $(0, t)$. This is not so when $n(\tau)$ represents, as we have assumed, the number of random points in the interval $(0, \tau)$ since $\dfrac{d n^R(\tau)}{d\tau}$ has delta function singularities at $\tau_1, \tau_2, \ldots, \tau_n$, the values of τ at which the points are realised. In such cases we note

$$\int_0^t \varphi(t, \tau)\, \frac{d n(\tau)}{d\tau}\, d\tau = \sum_i \varphi_i(t, \tau_i). \tag{70.17}$$

We now call into aid the meaning of product densities elaborated earlier. We here give the expression of the first, second and third moments of $y_m(t)$.

$$\left.\begin{aligned}
\mathscr{E}\{y_m(t)\} &= \int_0^t \varphi(t, \tau)\, f_1(\tau)\, d\tau, \\[4pt]
\mathscr{E}\{y_m^2(t)\} &= \int_0^t \int_0^t \varphi(t, \tau_1)\, \varphi(t, \tau_2)\, f_2(\tau_1, \tau_2)\, d\tau_1\, d\tau_2 + \\
&\quad + \int_0^t \varphi^2(t, \tau)\, f_1(\tau)\, d\tau, \\[4pt]
\mathscr{E}\{y_m^3(t)\} &= \int_0^t \int_0^t \int_0^t \varphi(t, \tau_1)\, \varphi(t, \tau_2)\, \varphi(t, \tau_3)\, f_3(\tau_1, \tau_2, \tau_3)\, d\tau_1\, d\tau_2\, d\tau_3 + \\
&\quad + 3 \int_0^t \int_0^t \varphi^2(t, \tau_1)\, \varphi(t, \tau_2)\, f_2(\tau_1, \tau_2)\, d\tau_1\, d\tau_2 + \\
&\quad + \int_0^t \varphi^3(t, \tau)\, f_1(\tau)\, d\tau.
\end{aligned}\right\} \tag{70.18}$$

71. A particular case of the p.f.f. of $\int_0^t F(t, \tau)\, n(\tau)\, d\tau$. We now consider the symbolic integral for $y(t) = \int_0^t F(t, \tau)\, n(\tau)\, d\tau$ where $n(\tau)$ has a Poisson distribution. We must then compute the variation in $\sum_i \varphi(t, \tau_i)$ for an infinitesimal change in t and this requires a knowledge of the positions of τ_i. This difficulty can be avoided if we adopt the device introduced before. We consider the stochastic variate

$$y(t, a) = \sum_i \varphi(a, \tau_i) \tag{71.1}$$

a being independent of t. This represents a process which unlike $y_m(t)$ has all the essential features of a basic random process except that it is quasi-Markovian. If a random point occurs between t and $t + dt$, it jumps up by a finite value $\varphi(a, t)$, otherwise it remains constant. Hence by a simple argument we have

$$\frac{\partial \pi(y; t, a)}{\partial t} = -\lambda\, \pi(y; t, a) + \lambda\, \pi\big(y - \varphi(a, t); t, a\big). \tag{71.2}$$

Defining the Laplace transforms of $\pi(y, t, a)$ as $p(s, t, a)$

$$p(s, t, a) = \int_0^\infty \pi(y; t, a)\, e^{-s y}\, dy \tag{71.3}$$

where s is a complex variable, we obtain

$$p(s; t, a) = \exp\left[-\int_0^t \lambda\{1 - e^{-s \varphi(a, \tau)}\}\, d\tau\right]. \tag{71.4}$$

This expression may be inverted in principle to obtain $\pi(y; t, a)$. $\pi(y; t)$ is obtained by replacing a by t in $\pi(y; t, a)$. Note that it is not possible to write a simple stochastic equation for $\pi(y; t)$ since its trajectory is continuous and the infinitesimal changes in $\pi(y; t)$ require a complete knowledge of the random distribution of points in $(0, t)$. The Laplace transform of $\pi(y; t)$ is given by $p(s, t, t)$ which for simplicity we write as

$$p(s, t) = \exp\left[-\lambda \int_0^t \{1 - e^{-s\varphi(t,\tau)}\}\, d\tau\right]. \tag{71.5}$$

To demonstrate the usefulness of the above technique we cite first and second order processes and give a few examples.

72. Linear differential equations of the first order. Let us consider a stochastic variable $y(t)$ defined by the symbolic equation

$$\frac{dy}{dt} + a(t)\, y = k(t)\, x(t) \tag{72.1}$$

where $a(t)$ and $k(t)$ are fully determinate functions of t and $x(t)$ is a random variable representing a process progressing with t. Our object is to obtain the p.f.f. of $y(t)$ or its moments, at least the first few, given the nature of the process $x(t)$. The first step consists in formally writing the solution of (72.1) as

$$y(t) - y_0\, e^{-b(t)} = y'(t) = \int_0^t e^{-\{b(t)-b(\tau)\}}\, k(\tau)\, x(\tau)\, d\tau \tag{72.2}$$

as if $x(t)$ were a determinate function; $b(t)$ is given by

$$b(t) = \int_0^t a(\tau)\, d\tau. \tag{72.3}$$

The initial condition is defined as $y(t) = y_0$ at $t = 0$. We shall consider a particular case of $x(t)$ as an example.

$x(t) = dn(t)/dt$ where $n(t)$ represents a Poisson process. It is well known that corresponding to a realisation of events at $t_1, t_2, \ldots, t_n$ along the t-axis the realised value of $dn(t)/dt$ is given by

$$\sum_i \delta(t - t_i)$$

where δ is the Dirac delta function. The distribution of $y'(t)$ has been studied and defining the Laplace transform $p(s; t)$ of $\pi(y'; t)$, the probability frequency function of $y'(t)$, as

$$p(s, t) = \int_0^\infty e^{-sy'}\, \pi(y'; t)\, dy', \tag{72.4}$$

the solution of $p(s, t)$ is obtained as

$$p(s, t) = \exp\left[-\lambda t + \lambda \int_0^t \exp\{-s\, k(\tau)\, e^{-[b(t)-b(\tau)]}\}\, d\tau\right]. \tag{72.5}$$

Moments of y' and hence those of y can be obtained from the above. An example of such a process is provided by the fluctuation of voltage at the anode of a thermionic valve due to the fluctuations in the number of electrons per unit time emitted by the cathode. If we assume the probability of emission of an electron per unit time to be constant, we may reasonably expect the number $n(t)$ of electrons emitted in the interval $(0, t)$ to be governed by a Poisson distribution. The fluctuating voltage $V(t)$ across R satisfies the equation

$$\frac{dV}{dt} + \frac{V}{RC} = \frac{e}{C}\frac{dn}{dt} \tag{72.6}$$

where we have assumed that the circuit between anode and earth is equivalent to a resistance R in parallel with capacity C. e is the charge of the electron. Substituting values of $a(t)$, $k(t)$, which are constants in this particular case, the Laplace transform solution of $\pi(V; t)$, the p.f.f. of V, is obtained as

$$p(s; t) = \exp\left[-\lambda t + \lambda \int_0^t \exp\left[\frac{se}{C}e^{-(t-\tau)/RC}\right]d\tau\right] \tag{72.7}$$

where λ is the mean number of electrons emitted per unit time. The moments are given by

$$\left.\begin{aligned}
&\mathscr{E}\{V(t) - V_0 e^{-t/RC}\} = -\lambda e R(1 - e^{-t/RC}),\\[4pt]
&\mathscr{E}\{[V(t) - V_0 e^{-t/RC}]^2\} = \lambda^2 e^2 R^2(1 - e^{-t/RC})^2 + \frac{\lambda e^2 R}{2C}(1 - e^{-2t/RC}),\\[4pt]
&\mathscr{E}\{[V(t) - V_0 e^{-t/RC}]^m\} = \lambda \sum_{r=0}^{m-1}\left(\frac{-e}{C}\right)^{m-r}\mathscr{E}\{[V(t) - V_0 e^{-t/RC}]^r\}\times\\[4pt]
&\qquad\times\frac{[1 - e^{-t(m-r)/RC}][RC\cdot(m-1)!]}{(m-r-1)!\,r!}\,.
\end{aligned}\right\} \tag{72.8}$$

73. Linear differential equations of second order. We now proceed to discuss the case where $y(t)$ satisfies the second order linear differential equation

$$\frac{d^2y}{dt^2} + a_1(t)\frac{dy}{dt} + a_2(t)\,y = k(t)\,x(t) \tag{73.1}$$

where $a_1(t)$, $a_2(t)$ and $k(t)$ are deterministic functions and $x(t)$ is a random function as before. The solution of the above can be put in the form

$$y = e^{-\frac{1}{2}b(t)}\int_0^t e^{-\{\alpha(t)-\alpha(\tau)\}}d\tau\int_0^\tau e^{\alpha(\tau)-\alpha(\tau')}k(\tau')\,x(\tau')\,d\tau' \tag{73.2}$$

under proper initial conditions where b and α are expressible in terms of a_1 and a_2.

Consider the case when $x(t) = dn(t)/dt$ where $n(t)$ is a random variable representing a Poisson process.

The Laplace transform of $\pi(y; t)$ is given by

$$p(s; t) = \exp\left\{-\lambda t + \lambda \int_0^t \exp\left[-sf(t, \tau)\,k(\tau)\right]d\tau\right\} \tag{73.3}$$

where

$$f(t, \tau) = \exp\left\{-\tfrac{1}{2}b(t) - \alpha(t) - \alpha(\tau)\right\}\int_0^t e^{2\alpha_2(\tau')}d\tau'. \tag{73.4}$$

The moments of y of any order can be obtained from the above equations.

IX. Stationary processes.

74. Stationarity in evolutionary stochastic processes. In a dynamical process "evolving" deterministically with respect to a one-dimensional parameter t (for example, time) the dynamical variables $\langle q_i \rangle$ $i = 1, 2, \ldots$ are functions of t and the simplest definition of stationarity is as follows. The process tends to stationarity if

$$\langle q_i(t) \rangle \rightarrow \langle q_i \rangle$$

i.e. functions independent of t. This looks trivial and is so in fact. A more general definition would be as follows: A process is stationary if as t tends to infinity the function becomes constant or periodic or quasi-periodic. To speak more loosely "if functions of t do not die down but remain finite as t tends to infinity" they may be called stationary. This definition is intuitively satisfactory. The periodicity may or may not depend on the initial conditions i.e. at $t = 0$. We have here assumed that the process starts at $t = 0$ under certain initial conditions and tends to stationarity. Here stationarity is approached as the result of "evolution". However, without any reference to evolution, it is sometimes possible to deal directly with "stationary" processes. In such a case, we start with the periodic or quasi-periodic process itself. In other words, since in a deterministic process we can trace the values of the dynamical variables either backwards or forwards with respect to t, if we start with a stationary process, it merely means that the periodicity extends from $-\infty$ to $+\infty$. In fact it is easier to arrive at periodicity directly rather than by evolution. Simple examples of these are obviously (i) any oscillation under restoring forces, (ii) motion under a central force. Periodic processes can be obtained as limits if a dynamical motion is considered as a "mixture" of an ab initio stationary process and a non-stationary process in which the dynamical variables tend to zero. The "trajectories" of such processes have been found by STÖRMER [31] after considerable calculation in dealing with the motion of charged particles in a magnetic field. These orbits were called by him orbits asymptotic to periodic orbits.

The extension of this concept to stochastic processes is immediate if we adopt the intuitive and physical approach though a rigorous formulation would involve concepts of convergence and distinction between types of stationarity. To serve as an introduction to the better understanding of such mathematical formulations, we shall first outline the physical features of stationarity by appealing to the concept of the realised trajectory.

Let us take the case of a realised trajectory of a Markovian stochastic process $x(t)$ in the interval $(0, t)$. In principle all the features of the stochastic process are supposed to be known if we are able to assign a measure to a typical trajectory. This measure, if it can be defined will in general depend upon the initial conditions. Let us now be interested in the measure of a class of trajectories which start at $t = t_0$ but which are identical between $t = t_1$ and $t = t_2$. This measure for finite values of t_1 and t_2 is a function of t_1 and t_2 or $t_1 - t_0$ and $t_2 - t_1$, if the process is homogeneous in t. If t_1 and t_2 tend to infinity in such a way that $t_2 - t_1$ remains finite and if the measure tends to a limit which is a function only of $t_2 - t_1$, we call the process stationary. However this definition is obviously too strong and a weaker definition can be given in terms of the distribution function by saying that the distribution function $\pi(x, t)$ tends to a limit $\pi(x)$ as t tends to ∞. Using a rigorous mathematical approach, it is necessary to introduce notions of convergence and various types of stationarity. These definitions do not take into account the evolutionary nature of the process and hence are wider in scope since it is always possible to conceive of stationarity without any reference to the asymptotic approach to stationarity through evolution. We shall not however deal with such definitions but confine ourselves with evolutionary processes that tend to stationarity. However, for the sake of completeness, a short section on spectral theory of stationary random functions is included.

75. Ergodic theory in statistical mechanics. *α) Preliminary remarks.* One of the most important problems in regard to stationary processes is the so-called ergodic problem which has played a fundamental role in statistical mechanics. This problem, originated by BOLTZMANN, was interpreted by physicists to mean a convenient justification for replacing time averages by phase (space) averages in statistical mechanical problems. The mathematical implications were neglected by investigators for a long time after some unsuccessful attempts. Only in 1931, the remarkable work of BIRKHOFF [32] attracted the attention of many investigators and since then this group of problems has never ceased to interest mathematicians.

We shall here outline the results of the ergodic theory for deterministic dynamical systems and indicate their extension to stochastic systems. In the case of simple stochastic systems, an alternative direct method using elementary principles of stochastic theory has been already given and the results will in this chapter be shown to be consistent with viewing the ergodic nature of the system through the Birkhoff approach. No proofs or elaborate mathematical discussions will be given. Before dealing with ergodic theory a few observations are necessary about the relative role of dynamics and probability in statistical mechanics.

Once the kinetic theory of gases was established and the gross thermodynamic properties of matter were associated with the motions and interactions of individual particles, the introduction of probabilistic methods was unavoidable. However since the number of particles was very large, it was clear that even if the nature of the interactions between the particles was clearly understood, it was impossible to expect to develop a theory of gross property by means of the apparatus of differential equations relating to these particles. Thus the standard methods of dynamics where we specify initial conditions and study the evolution of the process obviously fail and we have to invoke methods which deal with properties in the gross which are determined by these interactions but are insensitive to detailed motions. Hence it is that the limit theorems of probability calculus and its analytical apparatus can be used for the purpose though the *evolutionary approach of non-stationary stochastic processes will not be useful in studying these problems,* especially relating to stationary systems. The evolutionary approach is however finding applications in non-stationary phenomena as in the thermodynamics of irreversible process.

β) Geometry and kinematics of phase space[1]. The state of a mechanical system with s degrees of freedom is best described by the values of the Hamiltonian variables $q_1, q_2, \ldots, q_s, p_1, p_2, \ldots, p_s$. The equations of motion of the system then assume the "canonical form"

$$\frac{dq_i}{dt} = \frac{\partial H}{\partial p_i}, \quad \frac{dp_i}{dt} = -\frac{\partial H}{\partial q_i} \quad (1 \leq i \leq s), \tag{75.1}$$

---

[1] Sects. β to ε are reproduced from GAMOW's translation of the classic work of KHINCHIN (see bibliography).

where H is the Hamiltonian function of the $2s$ variables. The function $H(p_i q_i)$ is an integral of the system of equations since we have

$$
\begin{aligned}
\frac{dH}{dt} &= \sum_{i=1}^{s} \frac{\partial H}{\partial q_i} \frac{dq_i}{dt} + \sum_{i=1}^{s} \frac{\partial H}{\partial p_i} \frac{dp_i}{dt} \\
&= \sum_{i=1}^{s} \frac{\partial H}{\partial q_i} \frac{\partial H}{\partial p_i} - \sum_{i=1}^{s} \frac{\partial H}{\partial p_i} \frac{\partial H}{\partial q_i} = 0.
\end{aligned}
\tag{75.2}
$$

Since the system of Eqs. (75.1) contains only equations of the first order the values of the Hamiltonian variables $q_1, q_2, \ldots, q_s, p_1, p_2, \ldots, p_s$ given for some time $t = t_0$ determine their values at any other time. In a Euclidian space Γ of $2s$ dimensions whose points are determined by the $2s$ Cartesian coordinates q_i's and p_i's, the state of a system G can be specified by a point called its image point in the phase space Γ. Since the state of the system derermines uniquely the state at another time the motion of the image point is uniquely determined if we know its position at a particular time. Thus through every point in the phase space there passes a unique trajectory determined by the equations of motion. Consider the case when the image point is represented by M_0 in phase space at $t = 0$. At t it goes over to M_t. In a similar manner every point in phase space moves to a corresponding point. Or in other words, the space is transformed into itself in an one-to-one way. The two important theorems relating to phase space when the motion of the point on the phase space are given by the Hamiltonian equations are (1) the LIOUVILLE's theorem, (2) the BIRKHOFF's theorem. We shall state and explain them from a physical point of view.

γ) *Liouville's theorem.* Let M be any measurable set of points in phase space Γ. In the natural motion of the space the set M_0 goes to another set M_t; the LIOUVILLE's theorem states that the measure of the set M_t coincides with the measure of the set M_0. *In other words, the measure of measurable point sets is an invariant under natural motion of Γ.* A simple corollary may be obtained from LIOUVILLE's theorem. Let $f(P)$ be an arbitrary phase function of a point P in phase space. In time t, P goes over to a point P_t. Let us write

$$
f(P_t) = f(P, t). \tag{75.3}
$$

Every point P goes over to a point P_t, a volume dV goes over to a volume dV_t and a set M_0 to a set M_t. LIOUVILLE's theorem states that $dV_t = dV$ and that the measure of M_0 is equal to measure of M_t:

$$
\int_{M_t} f(P_t)\, dV_t = \int_{M_t} f(P, t)\, dV_t. \tag{75.4}
$$

But if $M_t = M_0$, i.e. the set M is invariant under the transformation, then we immediately recognize that

$$
\int_{M_t} f(P, t)\, dV_t = \int_{M_0} f(P)\, dV = \int_{M_0} f(P, t)\, dV. \tag{75.5}
$$

δ) *The theorem of Birkhoff.* If V is an invariant part of the phase space and $f(P)$ a phase function summable over V and determined at all points P belonging to V, then the theorem of Birkhoff states that

$$
\frac{1}{T} \int_{0}^{T} f(P, t)\, dt \tag{75.6}
$$

exists for all points P of the set V except at most of a certain set of measure 0, we interpret this quantity as the time average of the function $f(P)$ along the trajectory passing through P during the interval 0 to T. Such a terminology is suitable only if this limit is independent of the choice of P at $t=0$. We shall now discuss a special case of the Birkhoff theorem. Let V be some invariant part of the space Γ. We shall call this part metrically indecomposable if it cannot be represented in the form $V = V_1 + V_2$ where V_1 and V_2 are invariant parts of positive measure. The physical content of this motion is as follows. It is impossible to find a set of points with a finite measure through which the trajectories pass which are confined only to the volume V_1 or V_2. If this condition is satisfied, then

$$\lim_{T \to \infty} \frac{1}{T} \int_0^T f(P, t)\, dt = \frac{1}{V} \int_V f(P)\, dV. \tag{75.7}$$

That is, the time average is equal to the space average of the function $f(P)$. This is known as the *ergodic property* of the trajectory or dynamical process.

ε) *Extension to stochastic processes.* The language used in the above section seems at first sight to confine the application of BIRKHOFF's theorem to deterministic dynamical processes. This is because we have chosen the phase space as the space in which the points are moving with respect to time and the position of the point P_t at time t is determined uniquely by its function at $t=0$, i.e. P_0. Let us adapt the above results to stochastic processes. For simplicity, we shall consider Markovian stochastic processes and assume that a dynamical system can occupy at time t one of a discrete system of mutually exclusive states called $s_1, s_2, \ldots, s_j \ldots$. If we identify the system of states to be determined by the canonical coordinates and momenta of the system of particles, then of course we do not have a discrete set of states, the aggregate of states being the Γ-space of $2s$ dimensions, corresponding to the $2s$ dynamical coordinates, but we have a continuous infinite set of states, every state corresponding to a particular point in the Γ space. However to facilitate easier understanding we shall assume a discrete set of occupation states $s_1, s_2, \ldots, s_n$. *Thus the space of occupation states corresponds to the Γ space of the previous set.* If the system is stochastic, at any particular t we can speak only of the probability distribution in the various states and if an observation were made, the system will be found to be in some particular state. As t changes if we plot the state in which the system is found, against time we get a "trajectory" in the space of occupation states. However as we have explained in detail, starting with the same point in this space at $t=0$, if we perform another experiment we shall observe another trajectory. So unlike the trajectories in the phase space, through every point in the new space of occupation states there does not pass a unique trajectory. In such a case, how are we going to apply the results of the BIRKHOFF's theorem? This is achieved not by comparing the deterministic trajectories in the phase space with the realised trajectories in the "states" space but by the following method. *This is based on realising that the Birkhoff's theorem refers to any general space on which measure is defined and the ergodic property is applied to a metrically indecomposable volume.* We shall in addition assume the parameter t to be discrete. The results can be generalised to include processes relating to a discrete system of states and continuous t and further to the case of a continuous infinite system of states.

ζ) *Ergodic property of a Markovian stochastic process with x discrete*[1]. Consider a Markovian process with t discrete, assuming values $-N', -N'-1, \ldots, 0, 1, \ldots N''$ with N' and N'' tending to ∞. Let the process relate to a system which can

---

[1] I am indebted to T. E. HARRIS who helped me to write this section.

occupy any one of a mutually exclusive set of states $S_1, S_2, \ldots, S_n$. Let $\pi(j|i; l+m, l)$ be the probability that the system is in a state S_j at $t = l+m$ given that it was in a state S_i at $t = l$. In view of the homogeneity condition on t, it is a function only of m and we can write $\pi(j|i; m)$ for $\pi(j|i; l+m, l)$. We shall suppose that all states are completely connected in the sense of Sec. 54. If we assume that the process is stationary, we can define $\pi(i)$ as the probability that the system is in state i at $t = n$. This is independent of n.

$$\pi(i) = \sum_j \pi(j)\, \pi(i|j; n) \tag{75.8}$$

for all n.

Consider a realisation of the whole process as t takes values from $-\infty$ to $+\infty$. This gives a sequence which we shall denote by α when

$$\langle \alpha \rangle = \langle (\ldots x_{-2}, x_{-1}, x_0, x_1, x_2, \ldots) \rangle, \tag{75.9}$$

where we have used the following notation: each of the x_i is an integer 1 to n, and $x_i = j$ would mean that the system is found to be in S_j at $t = i$. Then $\langle \alpha \rangle$ represents the complete history of the process corresponding to one realisation. Let Ω be the space of all "points" $\langle \alpha \rangle$. Although $\langle \alpha \rangle$ is a complicated type of point, it is clearly possible to map $\langle \alpha \rangle$ on to the real numbers so that the representation is not a theoretical difficulty. In fact there is no particular reason to perform this mapping. When we specify a law for Markovian processes, we specify a corresponding measure on certain measurable subsets of Ω. As regards the definition of measure it is this Ω space that corresponds to the Γ space of the previous subsection though for a description of the realised state of the system it is the space of occupation states that corresponds to the Γ space. Thus for example if A is a subset defined by

$$A = \{\langle \alpha \rangle; \ x_0 = i, \ x_n = j, \ x_m = k\}, \tag{75.10}$$

then the measure of A is given by

$$P(A) = \pi(i)\, \pi(j|i; n)\, \pi(k|j; m - n). \tag{75.11}$$

The general theory of stochastic processes ensures that a suitably additive measure is established in this way on a certain Borel field defined on Ω.

Let T be the transformation (usually called "shift") defined as follows:
If

then
$$\left.\begin{aligned}
\langle \alpha \rangle &= (\ldots x_{-2}, x_{-1}, x_0, x_1, x_2, \ldots) \\
T\langle \alpha \rangle &= (\ldots x_{-1}, x_0, \ x_1, x_2, x_3 \ldots)
\end{aligned}\right\} \tag{75.12}$$

i.e. each coordinate is shifted one unit to the left. We can verify that for any stationary process (MARKOVIAN or not), $P(TA) = P(A)$ for all measurable sets A. Since we have a measure preserving transformation the Birkhoff theorem states that for any function $f(\langle \alpha \rangle)$ such that

$$\int_{\Omega} f(\langle \alpha \rangle)\, dP < \infty \tag{75.13}$$

$$\lim_{N \to \infty} \frac{f(\langle \alpha \rangle) + f(T\langle \alpha \rangle) + \cdots + f(T^{N-1}\langle \alpha \rangle)}{N} = g(\langle \alpha \rangle) \tag{75.14}$$

exists for almost all $\langle \alpha \rangle$, i.e. with probability 1. In general $g(\langle \alpha \rangle)$ would depend on $\langle \alpha \rangle$. However if t is metrically transitive and (in the present case, the fact that all states communicate ensures that it is) we have

$$g(\langle\alpha\rangle) = \text{constant} = \int_\Omega f(\langle\alpha\rangle)\,dP = \mathscr{E}\{f(\langle\alpha\rangle)\}. \tag{75.15}$$

Now let A be the set

$$A = \{\langle\alpha\rangle, x_0 = i\}, \tag{75.16}$$

then

$$TA = \{\langle\alpha\rangle, x_1 = i\} \ldots T^N A = \{\langle\alpha\rangle, x_N = i\}. \tag{75.17}$$

Let $f(\langle\alpha\rangle) = 1$ if $\alpha \in A$ and 0 if $\alpha \notin A$. Then

$$f(T^N\langle\alpha\rangle) = 1 \quad \text{if} \quad T^N\langle\alpha\rangle \in A, \quad \text{i.e.} \quad x_N = i. \tag{75.18}$$

Also observe that

$$\int f(\langle\alpha\rangle)\,dP = \mathscr{E}\{f(\langle\alpha\rangle)\} = \pi(i). \tag{75.19}$$

So we have

1. $\lim\limits_{N\to\infty}\left\{\text{number of times } \dfrac{x_m = i\,(m = 0, 1, \ldots, N-1)}{N}\right\} = \pi(i)$ with probability 1.

Let m_N be the numerator of the fraction in the above. Then according to the above equation.

2. $\lim \dfrac{m_N}{N} = \pi(i)$ with probability 1.

This is the "strong law" for Markovian chains.

To relate this to theorems about convergence of probabilities note that letting

$$\pi(i\,|\,j; 0) = \delta_{ij}$$

we have

$$\mathscr{E}\{m_N\,|\,x_0 = j\} = \pi(j\,|\,i, 0) + \pi(j\,|\,i, 1) + \cdots + \pi(j\,|\,i, n-1). \tag{75.20}$$

Now

$$\frac{\mathscr{E}\{m_N\,|\,x_0 = i\}}{N} = \frac{1}{\pi(j)}\int_{\Omega_j}\frac{m_N}{N}\,dP \tag{75.21}$$

(by the definition of conditional expectation) where Ω_j is the set of $\langle\alpha\rangle$ defined by

$$\Omega_j = \{\langle\alpha\rangle;\ x_0 = j\}. \tag{75.22}$$

Since $m_N/N \to \pi(i)$ almost everywhere in Ω and in particular almost everywhere in Ω_j and since m_N/N is bounded, being ≤ 1, we have (due to the Lebesgue convergence theorem),

$$\lim_{N\to\infty}\frac{\mathscr{E}\{m_N\,|\,x_0 = j\}}{N} = \lim_{N\to\infty}\{\pi(j\,|\,i, 0) + \pi(j\,|\,i, 1) + \ldots + \pi(j\,|\,i, n-1\}, \tag{75.23}$$

$$= \lim\int_{\Omega_j}\frac{1}{\pi(j)}\frac{m_N}{N}\,dP = \frac{1}{\pi(j)}\int_{\Omega_j}\left(\lim\frac{m_N}{N}\right)dP, \tag{75.24}$$

$$= \frac{1}{\pi(j)}\int_{\Omega_j}\pi(i)\,dP = \frac{\pi(i)}{\pi(j)}\,P(\Omega_j) = \pi(i). \tag{75.25}$$

A close examination of the above method shows the correspondence with the physical approach to the same problem in the case of a continuous time parameter discussed earlier.

76. Structure functions of KHINCHIN [4]. In considering ergodic theory, we used the most important property relating to the transformation of space into itself when each point P in the phase space moves to a corresponding point P_t according to the equations of motion. From the point of view of physics, the most important phase function of a mechanical system is the total energy E.

$$E = E(q_1, q_2, \ldots q_s, p_1, p_2, \ldots, p_s).\qquad(76.1)$$

If the system is isolated this function has a constant value and the point P representing the system in the phase space moves on a constant energy "surface".

Let our system consist of n particles. Suppose we are interested in the probability $f_1(E|E_0)$ that a particle picked at random has an energy between E and $E + dE$ given that the total energy of the n particles is E_0. To arrive at this, let us assume that there exists a probability for a particle to have an energy between E and $E + dE$ without any reference to the total energy. Let this frequency function be denoted by $f_1(E)dE$. This has no physical significance since there is no reason why an isolated particle should have an energy distribution. We shall find this fiction useful in computing the physically significant conditional probability. The probability that a system composed of n particles has a total energy E_0 is an n-fold convolution of the function $f_1(E)$,

$$f_n(E_0) = \underset{E_1\,E_2\quad E_{n-1}}{\int\!\!\int \cdots \int} f_1(E_1)\, f_1(E_2) \ldots f_1(E_0 - E_1 - E_2 - \cdots - E_{n-1}) \times \\ \times\, dE_1\, dE_2 \ldots dE_{n-1}. \qquad(76.2)$$

Then by simple probability arguments

$$f_1(E|E_0) = \frac{f_1(E)\, f_{n-1}(E_0 - E)}{f_n(E_0)}.\qquad(76.3)$$

Now the free distribution function $f_1(E)$ is used here to generate the conditional probability $f_1(E|E_0)$. From the form of $f_1(E|E_0)$ in Eq. (76.3) it will be clear that any $\omega_1(E) = A f_1(E)\, e^{+\lambda E}$ is good enough to generate the same $f_1(E|E_0)$. A is determined as follows[1]:

i.e.

$$\int\limits_E f_1(E)\, dE = \int\limits_E \frac{\omega_1(E)\, e^{-\lambda E}}{A}\, dE = 1,$$

$$A = \Phi(\lambda) = \int\limits_E \omega_1(E)\, e^{-\lambda E}\, dE \qquad(76.4)$$

and

$$f_1(E) = \frac{\omega_1(E)\, e^{-\lambda E}}{\varphi(\lambda)}.$$

We shall now give an interpretation to a properly chosen function $\omega(E)$. It is of course connected with the energy of the n particle system. KHINCHIN chooses $\omega_n(E)dE$ to represent the volume of phase space between the constant energy surfaces corresponding to E and $E + dE$ respectively. It is easy to convince oneself without recourse to abstract mathematical arguments that the measure of an "area" $d\Sigma$ on the energy surface is proportional to the volume standing above it between the two energy surfaces and since the total volume between E and $E + dE$ is $\omega(E)dE$, the probability that a representative point of the system will fall on a set M on the constant energy surface Σ is

$$\frac{1}{\omega_n(E)} \int\limits_M \frac{d\Sigma}{|\operatorname{grad} E|}, \qquad(76.5)$$

grad E being used in a general sense with respect to the constant energy "surface". The $\omega_n(E)$ is called the structure function of the system of n particles.

---

[1] Note that the procedure is similar to that adopted in the theory of the one-dimensional fluid f_1, ω, φ corresponding to g, G, Q.

KHINCHIN shows that the constrained distribution $f_1(E|E_0)$ can be obtained in the asymptotic case when n is very large and $f_{n-1}(E_0-E)$ and $f_n(E_0)$ in Eq. (76.3) are replaced by their corresponding limiting forms using the central limit theorem. If this is done $f_1(E|E_0)$ assumes the same form as $f_1(E)$ except that λ in the exponent is determined by the constraint that the total energy and number of particles are constant.

To determine $\omega(E)$ we must make (i) an apriori assumption regarding the distribution of a particle in phase space, (ii) express the energy in terms of the momentum and coordinates. As regards (i) it is customary to assume equal probability in phase space, i.e. the apriori probability that a particle lies in the elementary phase volume $dp_x\,dp_y\,dp_z\,dx\,dy\,dz$ is proportional to that volume. Concerning (ii), the energy can be written as $E = H(p, q)$. The dependence on q is due to an external potential. (We assume the potential to be external since in our arguments we have treated the particles as non-interacting. The theory of interacting particles will lead to an unnecessary digression on to the ensemble theory of GIBBS which we wish to avoid.)

$\omega(E)\,dE$ is the volume of phase space between the two surface $H(p, q) = E$ and $H(p, q) = E + dE$. If we assume that $E = \dfrac{p^2}{2m}$, $\omega(E)$ can be shown to be proportional to $\sqrt{E}$ and thus we get the Maxwell-Boltzmann formula for $f_1(E|E_0)$ with λ in the exponent given by $\dfrac{1}{\lambda} = \dfrac{1}{kT} = \dfrac{2E_0}{3n}$, k being the Boltzmann constant and T, the temperature. We recognise $\Phi(\lambda)$ to be the partition function of a single particle and on the assumption of equal probability on the phase space for the free distribution, this is given by

$$\Phi(\lambda) = \iint e^{-H(p,\,q)/kT}\,dp\,dq. \tag{76.6}$$

77. Power spectra and correlation functions. A theory for analysing "functions of time t which do not die down and which remain finite as t approaches infinity" has been gradually developed over the last sixty years. Its historical development together with an extensive bibliography is given in a fundamental paper by WIENER [33] on "Generalised harmonic analysis".

The possibility of application of this to random functions of time is obvious. The realised functions of an evolutionary stochastic process which reaches stationarity must extend from $t = 0$ to $t = \infty$. The realised function of a stationary stochastic process is assumed to extend from $-\infty$ to $+\infty$ when we need not consider the evolutionary nature of the process. We shall first summarise some results of generalised harmonic analysis with respect to a function $I(t)$ extending from $t = -\infty$ to $+\infty$ and then extend it to the case of random functions. $I(t)$ may be regarded as composed of a great number of sinusoidal components, whose frequencies range from 0 to ∞.

Let $\varphi(t)$ be a function of t which is 0 outside the interval $0 \leq t \leq T$ and is equal to $I(t)$ inside the interval. Its spectrum $S(f)$ is given by

$$S(f) = \int_0^T I(t)\,e^{-2\pi i f t}\,dt. \tag{77.1}$$

The spectrum of the power $\omega(f)$ is defined as

$$\omega(f) = \lim_{T \to \infty} \frac{2\,|s(f)|^2}{T} \tag{77.2}$$

where we consider only the values of f greater than 0 and assume that this limit exists. This is substantially the definition of $\omega(f)$ given by CARSON. The correlation function $\psi(\tau)$ of $I(t)$ is defined by the limit

$$\psi(\tau) = \lim_{T \to \infty} \frac{1}{T} \int_0^T I(t)\, I(t+\tau)\, dt \tag{77.3}$$

which is assumed to exist. $\psi(\tau)$ is closely related to the correlation coefficients used in stochastic theory to measure the correlation of random variables. The spectrum of the power $\omega(f)$ and the correlation function $\psi(\tau)$ are related by the equations

$$\left. \begin{aligned} \omega(f) &= 4 \int_0^\infty \psi(\tau) \cos 2\pi f \tau\, d\tau, \\ \psi(\tau) &= \int_0^\infty \omega(f) \cos 2\pi f \tau\, df. \end{aligned} \right\} \tag{77.4}$$

It is seen that $\psi(\tau)$ is an even function of τ. When either $\psi(\tau)$ or $\omega(f)$ is known the other may be obtained provided that the corresponding integral converges.

The extension to random functions is of course obvious if we suitably interpret ω and ψ. $I(t)$ is a random function representing a stochastic process progressing with respect to t. For any particular realisation $I^R(t)$ of $I(t)$ from $-\infty$ to $+\infty$ we have a definite power spectrum $\omega^R(f)$ and correlation function $\psi^R(\tau)$. However the principal interest is in those functions which give the average values of $\omega^R(f)$ and $\psi^R(\tau)$ for fixed f and τ. These functions are obtained by over-all realisations. Thus

$$\left. \begin{aligned} \mathscr{E}\{\omega(f)\} &= 4 \int_0^\infty \mathscr{E}\{\psi(\tau)\} \cos 2\pi f \tau\, d\tau, \\ \mathscr{E}\{\psi(\tau)\} &= \int_0^\infty \mathscr{E}\{\omega(f)\} \cos 2\pi f \tau\, df \end{aligned} \right\} \tag{77.5}$$

where the expectation values of $\omega(f)$ and $\psi(\tau)$ are obtained by keeping f and τ fixed respectively.

The corresponding definitions of ω and ψ in terms of s and I are given by

$$\left. \begin{aligned} \mathscr{E}\{\omega(f)\} &= \lim_{T \to \infty} \frac{2}{T} \mathscr{E}\{|s(f)|^2\}, \\ \mathscr{E}\{\psi(\tau)\} &= \lim_{T \to \infty} \frac{1}{T} \int_0^T \mathscr{E}\{I(t)\, I(t+\tau)\} dt. \end{aligned} \right\} \tag{77.6}$$

The values of t and T are held fixed while averaging over the parameters. $S^R(f)$ is a function of a typical realisation $I^R(t)$:

$$S^R(f) = \int_0^T I^R(t)\, e^{-2\pi i f t}\, dt. \tag{77.7}$$

Sometimes $\mathscr{E}\{I(t)\, I(t+\tau)\}$ in the definition of $\mathscr{E}\{\psi(\tau)\}$ is independent of time t. This enables us to perform the integration at once and obtain

$$\mathscr{E}\{\psi(\tau)\} = \mathscr{E}\{I(t)\, I(t+\tau)\}. \tag{77.8}$$

This introduces a considerable simplification and it appears that the simplest method of computing $\mathscr{E}\{\omega(f)\}$ for an $I(t)$ of this sort is first to compute $\mathscr{E}\{\psi(t)\}$ and then use the inversion formula given above. Since this article mainly deals with evolutionary stochastic processes and the theory of stationarity is merely introduced for the sake of completeness we give one example to illustrate the use of harmonic analysis in the solutions of differential equations of random functions when the process is stationary.

There is a way to deal directly with a stationary stochastic process. This is obtained by first extending the theory of Fourier transformations to random functions. To fix our ideas and establish correspondence with our previous treatment, we shall take the case of a second order differential equation with constant coefficients

$$\frac{d^2 y}{dt^2} + \lambda \frac{dy}{dt} + n^2 y = x(t).\tag{77.9}$$

We shall assume that the random function $x(t)$ is stationary and can be decomposed into random functions in which t occurs in the exponent; λ and n^2 are positive real constants. This is the usual equation for the displacement of a body, the free motion of which consists of damped simple harmonic one-dimensional oscillations with angular frequency $(n^2 - \frac{1}{4}\lambda^2)^{\frac{1}{2}}$ subject to a random external force proportional to $x(t)$. The variation of $y(t)$ will likewise be random and the object is to determine the statistical behaviour of $y(t)$ in terms of that of $x(t)$. When $x(t)$ is periodic in t we have the well known resonance solution for $y(t)$. Since the relation between $y(t)$ and $x(t)$ is linear it is clear that a stationary random variation of $x(t)$ will produce a similar variation of $y(t)$, and the solution lies in the resolution of $x(t)$ and $y(t)$ into Fourier components. We write

$$x(t) = \int e^{ift}\, dz(f), \qquad y(t) = \int e^{ift}\, dw(f)\tag{77.10}$$

where f is a frequency variable and increments in the random functions $z(f)$ and $w(f)$ have the properties

$$\left.\begin{aligned}
\lim_{df_1=df_2=df\to 0} \mathscr{E}\left\{\frac{dz^*(f_1)\, dz(f_2)}{df}\right\} &= 0,\\[4pt]
\lim_{df_1=df_2-df\to 0} \mathscr{E}\left\{\frac{dw^*(f_1)\, dw(f_2)}{df}\right\} &= 0,\\[4pt]
\lim_{df\to 0} \mathscr{E}\left\{\frac{dz^*(f)\, dz(f)}{df}\right\} &= \Phi(f),\\[4pt]
\lim_{df\to 0} \mathscr{E}\left\{\frac{dw^*(f)\, dw(f)}{df}\right\} &= \Psi(f)
\end{aligned}\right\}\tag{77.11}$$

where $\Phi(f)$ and $\Psi(f)$ are the spectral (density) functions of $x(t)$ and $y(t)$ respectively. On substituting into the original equations we find

$$(-f^2 + n^2 + if\lambda)\, dw(f) = dz(f).\tag{77.12}$$

The relation between the spectra of the displacement of the body and of the impressed force is therefore

$$\Psi(f) = \frac{1}{(n^2 - f^2)^2 + \lambda^2 f^2}\, \Phi(f)\tag{77.13}$$

as is to be expected from resonance theory. A quantity which is often more readily observable is the correlation between the displacements of the body at two different times. This correlation is simply the Fourier transform of the spectral function, that is

$$\left.\begin{aligned}
\mathscr{E}\left\{y(t)\, y(t+\tau)\right\} &= \int_{-\infty}^{+\infty} \Psi(f)\, e^{if\tau}\, df\\[4pt]
&= \int_{-\infty}^{+\infty} \frac{\Phi(f)}{(n^2 - f^2)^2 + f^2 \lambda^2}\, e^{if\tau}\, df.
\end{aligned}\right\}\tag{77.14}$$

We can make use of PARSEVAL's formula to relate this integral to the correlation between the values of $x(t)$ at two different times. Since

$$\int_{-\infty}^{+\infty} \frac{1}{(n^2-f^2)^2+f^2\lambda^2}\, e^{if\tau}\, df = \frac{\pi}{\lambda n'n}\, e^{-\frac{1}{2}\lambda|\tau|} \sin\left(n'\,|\tau| + \arcsin \frac{n'}{n}\right) \qquad (77.15)$$

where $n'=(n^2-\frac{1}{4}\lambda^2)^{\frac{1}{2}}$ is the damped angular frequency of the body and, since

$$\int_{-\infty}^{+\infty} \Phi(f)\, e^{if\tau}\, df = \mathscr{E}\,\{x(t)\, x(t+\tau)\} = R(\tau), \qquad (77.16)$$

we have

$$\mathscr{E}\,\{y(t)\, y(t+\tau)\} = \frac{1}{2\lambda n n'} \int_{-\infty}^{+\infty} R(\tau-\xi)\, e^{-\frac{1}{2}\lambda|\xi|} \sin\left(n'\,|\xi| + \arcsin \frac{n'}{n}\right) d\tau. \qquad (77.17)$$

These are the most important of the desired relations between the statistical parameters $y(t)$ and $x(t)$.

The result (77.17) has been obtained using the phenomenological interpretation of random functions relating to evolutionary stochastic processes by MATHEWS and SRINIVASAN [34].

X. On the computation of infinitesimal transition probabilities.

78. As has been repeatedly emphasised in the course of this article, the solution of any problem relating to evolutionary stochastic processes essentially consists in being able to describe the statistical nature of a system at a parametric value t given the state of the system at $t=0$. For this we need the transition probability function which defines what happens in infinitesimal intervals. The mathematical theory of stochastic processes is built *assuming the existence of such limits*. But in any physical problem we must take into consideration the particular special features characteristic of the physical process and arrive through logical or intuitive arguments at this infinitesimal probability. The ways in which we arrive at this infinitesimal probability can be broadly divided into three classes:

$\alpha)$ Apriori assumptions based on intuitive considerations.

$\beta)$ We assume the infinitesimal interval is sufficiently large for us to decompose "what happens in that interval" into elementary processes for which apriori probabilities can be assigned and combinatorial or similar probability arguments used to arrive at the infinitesimal transition probability.

$\gamma)$ The same as above except that the elementary processes are not subject to classical probability laws but to quantum mechanical laws. This leads us to the close connection between the quantum mechanical nature of the elementary processes occurring in Δ, the infinitesimal interval, to the stochastic nature in the gross, i.e. to the finite interval t. We shall deal with the above three techniques in detail.

$\alpha)$ *Apriori assumptions.* This is usually done in genetical and biological problems where it is not possible to decompose what happens in Δ into more elementary processes. Thus the probabilities for birth, death, immigration etc. in population problems must be assumed apriori based on intuitive arguments. For example the probability of the death of an individual can be assumed to be a function of the age and this function must be arrived at by some sort of intuitive argument or upon some statistical data available. Since it is not possible to classify intuitive arguments we shall not discuss these any further.

β) Decomposition of what happens in Δ into elementary events. In this case also, it is not possible to classify and we illustrate the technique by three examples in increasing order of complexity.

(i) Simple decomposition. The probability in the Furry process (p. 581) that a new particle is created in Δ is $n\lambda\Delta$. This result is arrived at follows: It is assumed that any particle produces a new one in Δ with probability $\lambda\Delta = p$. The probability that k particles are produced in Δ by the n particles at t is $\binom{n}{k} p^k (1-p)^{n-k}$ and since all the terms in this expression other than $np = n\lambda\Delta$ are of order greater than or equal to Δ^2 we obtain $n\lambda\Delta$ as the probability for birth.

(ii) Technique using the concept of density of states. For simplicity let us assume that the state of the system is characterised by the value of a single continuous parameter E and let us be interested in the infinitesimal transition probability $R(E'|E)\,dE'\,\Delta$ that the system makes a transition from E to a state between E' and $E'+dE'$. To arrive at this it is sometimes found convenient to take first E as a discrete variable capable of assuming values $E_1, E_2, \ldots$. These discrete values are assumed to be so close to each other that we can take the infinitesimal interval dE to be sufficiently large to contain $\varrho(E)\,dE$ discrete states and call $\varrho(E)$ the density of states. If on the discrete model we are able to arrive at R_{jk}, the probability that the system with energy E_k jumps to E_j, by elementary laws of composition of probability similar to the one used above, we can write

$$R(E_j|E_k)\,dE_j\,\Delta = \varrho(E_j)\,R_{jk}\,dE_j\,\Delta.$$

This technique is especially useful in quantum mechanical processes. The calculation of the density of states ϱ is illustrated by examples in Sect. 81.

(iii) A technique used by CHANDRASEKHAR. Here we assume Δ the infinitesimal interval to be large enough for a very large number of elementary processes to occur i.e. Δ itself is divided into sub-intervals in which there occur elementary processes for which apriori probabilities are assigned. The technique has been well illustrated in Sect. 55.

γ) Cases when the elementary events occurring in Δ are quantum mechanical in nature. The assignment of probabilities to these events has to be made according to the principles of quantum mechanics. We shall confine ourselves to electromagnetic processes since quantum electrodynamics is perhaps the best established part of quantum mechanics. Here the parameter with respect to which a stochastic process progresses is usually the thickness of matter traversed by an incident particle. Stochastic processes involving quantum mechanical events can be described in a general manner as follows: A free particle is incident on matter of thickness t. The study of the stochastic process consists in a specification or computation of the probabilities of events reckoned at t, the events corresponding to the numbers of the different types of particles that may emerge and the dynamical variables attached to them. This can be done by standard stochastic theory if we know the transition probabilities corresponding to small intervals Δ and these are known as "cross-sections"[1]. In a very small interval Δ, quantum mechanical collision processes can occur. The collision process is described as follows: the incident particle is assumed to be free and represented by the corresponding wave function of a free particle. It is perturbed by the Coulomb potential of a heavy nucleus of the matter traversed. After collision it passes

---

[1] In quantum mechanics the term "transition probability" is used with reference to time. The connection between transition probabilities per unit time and cross-sections will be established in the next chapter.

out of the range of the perturbing potential and travels as a free particle. During the collision new particles or radiation may be created. The interval Δ of matter traversed in which we require the transition probability is very large compared with the dimensions of the perturbing field of each nucleus and hence the collisions with the different nuclei occur independently. Thus in computing the cross-sections we need a clear picture of the collision process from a quantum mechanical point of view. At the risk of repetition of well-known ideas, we recall the well-established results of quantum theory in the next chapter in so far as they are necessary for the computation of transition probabilities. No reference will be made to field theory as this will lead to an unnecessary digression from the main purpose of this article.

XI. Applications to quantum mechanics[1].

We shall first rapidly outline the main ideas necessary when we postulate that, in a thickness Δ of matter, quantum mechanical processes occur.

79. The wave function. In quantum mechanics, any elementary particle is represented by a wave function ψ which is a complex function either of space and time or of momenta and time. In the first case it is called a configuration representation; in the other a momentum representation. This wave function consists of one or more components, i.e. more than one function may be necessary to represent the entire wave function according to the requirements of spin, relativity and other considerations imposed upon the particle. For the purpose of our discussion, we confine ourselves to a one-component wave function since the understanding of the arguments for the computation of cross-sections does not depend so much upon the existence of components as upon the connection between wave functions and probability concepts. In the configurational representation $\psi^* \psi \, dx \, dy \, dz$ is defined as the probability that the particle is found in the volume $dx \, dy \, dz$. In the momentum representation, $\psi^*(\boldsymbol{p}) \, \psi(\boldsymbol{p}) \, dp_x \, dp_y \, dp_z$ is the probability that the particle has momentum between $\boldsymbol{p}$ and $\boldsymbol{p} + d\boldsymbol{p}$. Without going into abstruse discussions regarding complementarity, we can state the fundamental principle of complementarity as follows: the state of a system can be described either in configuration space or in momentum space and not simultaneously in both. As regards the probability interpretation, this merely implies that we cannot arrive at a joint distribution in momentum and space coordinates but only at the marginal distribution $\psi^* \psi$, referring either to the configuration or momentum space. The main difference between a stochastic process and a quantum mechanical system is that in *the quantum mechanical system we study the variation of the ψ function, and the variation of $\psi^* \psi$ is only a subsidiary concept.* The variation of ψ is governed by the Hamiltonian characteristic of the system. If the Hamiltonian is denoted by H then the variation of ψ is given by the Schrödinger equation

$$-i\,\hbar\,\frac{\partial \psi}{\partial t} = H\,\psi. \tag{79.1}$$

If H is independent of time, we assume that ψ can be written as

$$\psi = u\,e^{i\,Et/\hbar} \tag{79.2}$$

i.e.

$$H\,u = E\,u. \tag{79.3}$$

---

[1] In this chapter, t denotes time but not the parameter with respect to which the stochastic process progresses. The relation between this parameter and time is established at the end of this chapter.

In other words the only time dependence in ψ lies in the exponent; $\psi^*\psi = u^*u$ is however independent of time. *Thus the probability density does not reveal the time dependence in the exponent of the wave function.* The probability distribution therefore is an inadequate description of the state of the system. No explicit equation exists involving only the probability density without recourse to the wave function, and the statistical behaviour of a quantum mechanical system is essentially different from that of a classical stochastic system due to the oscillatory or wave nature of ψ.

Eq. (79.3) has solutions only for specified values of $E_1, E_2, \ldots$, characteristic of the Hamiltonian. The solutions $u_1, u_2, \ldots$ corresponding to $E_1, E_2, \ldots$ are known as the eigen-functions. If a system is in an eigen-state at $t=0$, it continues to be in the same eigen-state if the Hamiltonian is unaltered.

80. Elementary perturbation theory. In a collision process, the Hamiltonian is that of a free particle and the collision consists in "switching on" a perturbation H' at $t=0$ and switching it off at t. During the period $(0, t)$ H' does not vary with time. The effect of this perturbation is as follows:

At t the wave function is represented by

$$\psi(t) = \sum_n a_n(t)\, u_n\, e^{i\,E_n t/\hbar} \tag{80.1}$$

where $a_n^*(t)\, a_n(t)$ is interpreted as the probability that the particle is in a state u_n at t. If we are given that at $t=0$, the system was in a state ψ_0, then it is a well-known result of perturbation theory that

$$|a_n(t)|^2 = \frac{2|H'_{n0}|^2}{(E_0 - E_n)^2}\left\{1 - \cos(E_0 - E_n)\frac{t}{\hbar}\right\} \tag{80.2}$$

where

$$H'_{n0} = \int \psi_n^*\, H'\, \psi_0\, dx\, dy\, dz = M. \tag{80.3}$$

$H'_{n0} = M$ is called the matrix element for the transition from the state ψ_0 to ψ_n. We shall not deal with the case $H'_{n0} = 0$.

In collision problems, we are not interested in transitions from a state ψ_0 to a particular state ψ_n but to a group of states with E_n belonging to the energy interval $(E, E + dE)$. For this purpose we need the number of states lying between E and $E + dE$. To compute this, it is not enough to consider only the particle in which we are interested. Since the incident particle has an energy E_0 and it emerges with energy E and since energy must be conserved, it is obvious that unless $E = E_0$, i.e. collision is elastic, there must be some other particle or particles which carry away the residue. The energy of the entire system of emergent particles i.e. after perturbation is equal to that of the system before perturbation. Thus if we have n emergent particles, which we shall consider distinguishable for convenience and which we require to be in the interval

$$(E_1, E_1 + dE_1), \ (E_2, E_2 + dE_2), \ \ldots, \ (E_n, E_n + dE_n),$$

the density of states of the entire system characterised by these energies is a function of $E_1, E_2, \ldots, E_n$ and E_0, $E_0 = E_1 + E_2 + \cdots + E_n$ and will be denoted by $\varrho(E_1, E_2, \ldots, E_{n-1}; E_0)$. If in the group of emergent particles there is more than one of the same type as the incident particle, there is no meaning in saying which of them corresponds to the incident particle. This leads to the complication of the computation of the matrix element itself and we do not now take such cases into consideration and in any process we are interested in asking the transition probability that there emerges a particle of a certain type in a particular energy state assuming the incidence of a particle. The density of final states i.e. states of the emergent system corresponding to our requirement that the energy of the particle of interest lies between E and $E + dE$ is obtained by putting $E_1 = E$ and integrating over all the energy variables except E_1. This yields $\varrho(E, E_0)$. We now multiply our matrix element by $\varrho(E) dE$ and integrate over a small energy interval ΔE. On performing this integration, the probability comes out proportional to t as it should, with a factor $2\pi/\hbar$. This leads to the definition of transition probability per unit time, that a particle of energy between E and $E + dE$ is produced given that there is an incident particle of energy E_0 as

$$\frac{2\pi}{\hbar} \varrho(E_0, E) |M|^2 dE.$$

The calculation of the density $\varrho(E)$ for one, two and three particle final states is elaborated in Sect. 81 below.

What we have now obtained is the transition probability per second of an incident particle with a certain energy colliding with a particle of matter and being in a state characterised by a certain momentum and energy after collision. This is usually obtained on the basis of the normalisation that the wave function of the incident particle corresponds to a density of one particle per unit volume. If the particle is moving with a velocity v this corresponds to a flux of v particles per second crossing unit area. Since we are interested in what happens to a given incident particle, we divide the transition probability per second by the flux per second per unit area, thus obtaining the cross-section σ which is of the dimensions of an area and which is numerically equal to the probability that an incident particle collides with the other.

Let us consider the passage of a fast particle through matter containing N particles per cm³. In a thickness Δ there are $N\Delta$ particles, assuming the cross-sectional area to be unity. The probability that the incident particle makes a collision in Δ is $N\sigma\Delta$. The unit area cross-section is chosen for the volume $N\Delta$ since σ was obtained on the basis of the normalisation obtained above.

We must remember that in computing a transition probability, the wave function was integrated over the whole volume from $-\infty$ to $+\infty$. This infinite limit is to be interpreted only with reference to the dimensions of the nuclear phenomenon and $N\Delta$ scattering centres perturb the incident wave independently. *Δ is infinitesimal for the stochastic treatment of the change in the state of the incident particle but is sufficiently large to accommodate independent quantum mechanical collisions.* $N\sigma$ is however a finite quantity. N is very large, of the order of the Avogadro number. σ is small so that $N\sigma$ is reasonably finite. Once we are in possession of $N\sigma$ we can apply the processes of analysis to the variations with respect to the thickness of matter traversed and consider Δ to be sufficiently small to say that $N\sigma\Delta$ is the probability that one collision takes place. Thus we have arrived at the transition probability per unit thickness of matter. *Once the cross-section of the process is given the development can be treated classically and quantum mechanical considerations do not enter at all.* They have already been embedded in the derivation of σ itself.

81. Density of states. α) *One-particle system.* The number of states available to a single particle shall be assumed to be uniformly distributed in the six-dimensional phase space. Therefore the number of states available for a particle of momentum between $\boldsymbol{k}$ and $\boldsymbol{k} + d\boldsymbol{k}$ is proportional to d^3k per unit volume. According to the fundamental principles of quantum mechanics, the number of states is given by d^3k/h^3. If we use the notation $h/2\pi = \hbar = 1$ the number of states available is $d^3\boldsymbol{k}/(2\pi)^3$. We note that this is the number of normal modes per unit volume derived in classical theory. The number of states available for a single particle with the magnitude of momentum between p and $p+dp$ emerging in a solid angle $d\Omega$ is

$$p^2 \frac{dp\, d\Omega}{(2\pi)^3}\,, \qquad p = |\boldsymbol{k}|. \tag{81.1}$$

Hence the density of states in $d\Omega$ per unit energy interval is given by[1]

$$p^2 \frac{dp}{dE} \frac{d\Omega}{(2\pi)^3}\,. \tag{81.2}$$

For a relativistic particle of energy E, $E^2 = p^2 + m^2$ so that

$$\frac{dp}{dE} = \frac{E}{p}\,. \tag{81.3}$$

Hence the density of states per unit volume for a single particle with energy E is given by

$$(2\pi)^{-3} E\, p\, d\Omega. \tag{81.4}$$

β) *Two-particle system.* If after a quantum mechanical collision, two particles emerge with masses m_1 and m_2, energies E_1 and E_2 and momenta p_1 and p_2 respectively, the following relations hold

$$\begin{aligned} E = E_{\text{final}} &= E_{\text{initial}} = E_1 + E_2, \\ E_1^2 = p_1^2 + m_1^2, \quad &E_2^2 = p_2^2 + m_2^2, \end{aligned} \right\} \tag{81.5}$$

$$\boldsymbol{p} = \boldsymbol{p_1} + \boldsymbol{p_2}. \tag{81.6}$$

$\boldsymbol{p}$ is the original momentum before the event. We now ask for the density of final states when one of the particles has a momentum of magnitude between p_1 and $p_1 + dp_1$ and emerges in a solid angle $d\Omega$. This is given by

$$(2\pi)^{-3} p_1^2 d\Omega \left(\frac{dp_1}{dE_{\text{total}}}\right)_\vartheta \tag{81.7}$$

where ϑ is the angle of emergence of the particle with momentum $\boldsymbol{p_1}$,

$$\left. \begin{aligned} \frac{dp_1}{dE} &= \frac{dp_1}{dE_1} \cdot \frac{dE_1}{dE} \\ &= \frac{E_1}{p_1}\left(\frac{dE_1}{dE}\right)_\vartheta = \frac{E_1}{p_1} \Big/ \left(\frac{dE}{dE_1}\right)_\vartheta, \\ \left(\frac{dE}{dE_1}\right) &= \frac{p_1}{E_1}\frac{dp_1}{dE_1} + \frac{1}{E_2}\left\{\frac{d}{dE_1}(p^2 + p_1^2 - 2p_1 p\cos\vartheta + m_2^2)\right\} \\ &= \frac{E p_1 - E_1 p\cos\vartheta}{E_2 p_1}, \\ E = E_1 + E_2 &= (p_1^2 + m_1^2)^{\frac{1}{2}} + (p_2^2 + m_2^2)^{\frac{1}{2}}. \end{aligned} \right\} \tag{81.8}$$

The density of final states is therefore given by

$$\frac{(2\pi)^{-3} E_1 E_2 p_1^2 d\Omega_1}{(E p_1 - E_1 p\cos\vartheta)} = \frac{(2\pi)^{-3} E_1 E_2 p_1^3 d\Omega_1}{E p_1^2 - E_1(\boldsymbol{p}\cdot\boldsymbol{p_1})}\,. \tag{81.9}$$

γ) *Three-particle system.* In this case, we ask for the density of final states per unit interval of total energy E and per unit interval of the energy of the

---

[1] In literature, the term "density of states" is used rather loosely either to denote the number of states in an infinitesimal interval dp or the number per unit energy range. We however wish to keep this distinction clear. To obtain the number from the density we multiply by the factor $\dfrac{p}{E}\,dp$.

particle of interest when the particles emerge with a momentum of magnitude between p_1 and $p_1 + dp_1$ in a solid angle $d\Omega_1$ and another with a momentum of magnitude between p_2 and $p_2 + dp_2$ in a solid angle $d\Omega_2$. Note that the momentum and the angle of emergence of the third particle is determined by the disposition of the other two particles. The density of states is given by

$$(2\pi)^{-6} \frac{E_1 E_2 E_3 p_2^3 p_1 \, d\Omega_1 \, d\Omega_2}{p_2^2 (E - E_1) - E_2 \boldsymbol{p}_2 \cdot (\boldsymbol{p} - \boldsymbol{p}_1)} . \tag{81.10}$$

This has been obtained by multiplying the density of states of a single particle with momentum $\boldsymbol{p}_1$ emerging in $d\Omega_1$ by the two-particle final states of the other two particles with total energy $E - E_1$ (and total momentum $\boldsymbol{p} - \boldsymbol{p}_1$), if one of them emerges in $d\Omega_2$ with momentum magnitude between p_2 and $p_2 + dp_2$. If one of the final particles has infinite mass $m_3 = \infty$, and hence energy $E_3 = \infty$, the density of the three-particle final states reduces to

$$(2\pi)^{-6} E_1 E_2 \, p_1 p_2 \, d\Omega_1 \, d\Omega_2 . \tag{81.11}$$

82. Many-particle systems. We have adverted to the probabilistic interpretation of the wave function of a single particle in quantum mechanics and emphasised that the wave function has a deeper meaning than is contained in the probabilistic interpretation. It is obvious that situations will arise in considering many particle systems which do not have a correspondence in the classical theory of stochastic processes. Since a detailed comparison would lead us to an unnecessary digression, we will just refer to the phenomenon of exchange energy in two-particle systems, a phenomenon which is purely quantum mechanical in nature and cannot be understood from simple probability arguments.

We shall deal with the problem of two electrons in a helium atom. The wave functions should contain two arguments (in the stationary state) $\boldsymbol{r}_1$ with components x_1, y_1, z_1 and $\boldsymbol{r}_2$ with components x_2, y_2, z_2 relating to the position of the two electrons and the Schrödinger equation should read

$$\nabla_1^2 \psi + \nabla_2^2 \psi + \frac{2m}{\hbar^2} \left(E - V_1 - V_2 - \frac{e^2}{r_{12}} \right) \psi = 0, \tag{82.1}$$

using the following abbreviations

$$\left. \begin{aligned}
&V_1 = V(x_1, y_1, z_1), \qquad V_2 = V(x_2, y_2, z_2), \\
&\nabla_1^2 = \frac{\partial^2}{\partial x_1^2} + \frac{\partial^2}{\partial y_1^2} + \frac{\partial^2}{\partial z_1^2}, \\
&\nabla_2^2 = \frac{\partial^2}{\partial x_2^2} + \frac{\partial^2}{\partial y_2^2} + \frac{\partial^2}{\partial z_2^2}, \\
&r_{12} = |\boldsymbol{r}_1 - \boldsymbol{r}_2| .
\end{aligned} \right\} \tag{82.2}$$

In the absence of the perturbing term e^2/r_{12} the solution for ψ can be written as

$$\psi_a(1)\, \psi_b(2) = \psi \tag{82.3}$$

where $\psi_a(1)$ is the solution of

$$\nabla_1^2 \psi_a(1) + \frac{2m}{\hbar^2} (E_a - V_1) \psi_a(1) = 0 \tag{82.4}$$

and $\psi_b(2)$ is the solution of

$$\nabla_2^2 \psi_b(2) + \frac{2m}{\hbar^2} (E_b - V_2) \psi_b(2) = 0 \tag{82.5}$$

where

$$E_a + E_b = E . \tag{82.6}$$

Physically however the two electrons are indistinguishable from each other. Equally well we could have written the solution as $\psi_a(2)\,\psi_b(1)$. Any linear combination of these two can also be chosen. For instance, we can choose the sum or the difference

$$\psi_+ = \psi_a(1)\,\psi_b(2) + \psi_a(2)\,\psi_b(1),\qquad(82.7)$$

$$\psi_- = \psi_a(1)\,\psi_b(2) - \psi_a(2)\,\psi_b(1).\qquad(82.8)$$

It is clear that when we are dealing with two-electron wave functions the condition that we impose is that the total energy is E and that the energy of one of the electrons is E_a and the other E_b and there is no meaning in distinguishing between the two electrons. Thus if ψ be any linear combination of the wave function $\psi_a(1)\,\psi_b(2)$ and $\psi_a(2)\,\psi_b(1)$ then $\psi^*\,\psi$ should be invariant under interchange of E_a and E_b. Thus the only linear combinations allowed are ψ_+ and ψ_-.

Let us now write down the expression for $\psi_\pm^*\,\psi_\pm = \psi_\pm^2$:

$$\psi_\pm^2 = \psi_a^2(1)\,\psi_b^2(2) + \psi_b^2(1)\,\psi_a^2(2) \pm 2\psi_a(1)\,\psi_b(1)\,\psi_a(2)\,\psi_b(2).\qquad(82.9)$$

Integrating with respect to $dx_1\,dy_1\,dz_1$ and $dx_2\,dy_2\,dz_2$, the last term vanishes on the right-hand side. But when we calculate the perturbation energy due to the term e^2/r_{12}, we note that

$$\Delta E = \frac{\int V\psi^2\,d\tau}{\int \psi^2\,d\tau} = C + A \qquad(82.10)$$

where

$$C = \int \psi_a^2(1)\,\psi_b^2(2)\,\frac{e^2}{r_{12}}\,d\tau_1\,d\tau_2 \qquad(82.11)$$

and

$$A = \int \psi_a(1)\,\psi_a(2)\,\psi_b(1)\,\psi_b(2)\,\frac{e^2}{r_{12}}\,d\tau_1\,d\tau_2. \qquad(82.12)$$

C can be simply interpreted as the potential energy of the mutual interaction. This is what we would have got even on the probability interpretation of $\psi^*\psi$. But A is not capable of a simple probability interpretation. It can be understood if we realise that in a quantum mechanical description an electron in r_1 is partly in a and partly in b states. This new feature is a typical result of quantum mechanics and A is the term due to exchange charge density.

A similar situation arises in the case of the two electrons of the hydrogen molecule where the additional feature is present that ψ_a and ψ_b are not orthogonal since the wave functions belong to different atoms.

Till now we were referring to the wave functions as representing either the spatial distribution of the electron in configuration space or in momentum space. There may be other degrees of freedom for the fundamental particle such as spin and isotopic spin. Wave functions are defined to describe the state of the particle in the spin and isotopic spin spaces, and by total wave function of a system of particles we mean the wave function which takes into account the extra parameters. The wave functions of fundamental particles obeying the Fermi-Dirac statistics obey also the Pauli principle which states that the total wave function of the n particles should be anti-symmetric under an exchange of two particles. We shall illustrate this by taking the case of the two electrons in the helium atom. If the two spin wave functions of an electron are denoted by α and β for spin up and down ($+\frac{1}{2}$ and $-\frac{1}{2}$) respectively, four spin wave functions are possible for the two-electron system.

$\alpha(1)\,\alpha(2)$ corresponds to the spin $+1$ of the total system,

$\alpha(1)\,\beta(2) + \alpha(2)\,\beta(1)$ corresponds to the spin 0 of the total system,

$\beta(1)\,\beta(2)$ corresponds to the spin -1 of the total system,

$\alpha(1)\,\beta(2) - \alpha(2)\,\beta(1)$ corresponds to the spin 0 of the total system.

If we now insist on the Pauli principle it is clear that we should choose the orbital wave function to be symmetric or antisymmetric according as the spin wave function to be antisymmetric or symmetric. We give here the table of the wave functions.

Orbital and spin wave functions.

Orbital	Spin	Total spin
$\psi_a(1)\,\psi_b(2) + \psi_a(2)\,\psi_b(1)$	$\alpha(1)\,\beta(2) - \alpha(2)\,\beta(1)$	0 singlet
$\psi_a(1)\,\psi_b(2) - \psi_a(2)\,\psi_b(1)$	$\left\{\begin{array}{l}\alpha(1)\,\alpha(2)\\ \beta(1)\,\beta(2)\\ \alpha(1)\,\beta(2) + \alpha 2()\,\beta(1)\end{array}\right\}$	1 triplet
$a = b \cdot \psi_a(1)\,\psi_a(2)$	$\alpha(1)\,\beta(2) - \alpha(2)\,\beta(1)$	0 singlet

The above discussion, it is hoped, illustrates the striking difference between the probability concepts that occur in quantum mechanics and those that occur in classical stochastic theory.

XII. Random functions of a many-dimensional parameter.

83. Most of the processes we have considered till now were represented an random functions of a one-dimensional parameter t. The one-dimensional nature of t permits the ordering of t which leads naturally to the concept of evolutions with respect to t. If the equations for the frequency functions of the random variables were to be written down, we assumed that the process was either Markovian or regenerative. Both these properties depend on the ordering of the parameter t.

If t were many-dimensional there is no meaning in ordering and hence we cannot appeal to the Markovian and regenerative property. To save a bad situation the approximation of "nearest neighbour" interactions is sometimes assumed and since these approximations cannot be brought into a general framework but are introduced to suit the necessities of a particular problem there is no point in referring to them in detail.

It is obvious why the random function of a three-dimensional parameter is called a "field". Denoting the three-dimensional parameter by r (with components x, y, z) and not by t (for obvious reasons) $X(r)$ defines the random field. We have then a continuous infinity of random variables corresponding to the points r and we should be content with the joint distribution of $X(r_i)$, $i = 1, 2, \ldots$ at n points. Since no Markovian or regenerative property can be attributed with respect to r, it is difficult to imagine how this joint distribution can be obtained. One way out of the difficulty is to assume that the statistical distribution of $X(r)$ changes with respect to a one-dimensional parameter t. With the one-dimensional t can be introduced the concept of evolution. The change of the statistical distribution of $X(r)$ with respect to t inevitably leads to concepts like random current and analytical processes like differentiation and integration with respect to t.

Such problems arise in the theory of turbulence where we consider velocity fields. The three-dimensional parameter describes ordinary space (x, y, z). With every point in space we associate the velocity components v_x, v_y, v_z. In the theory of turbulence, the velocity components are random functions of x, y, z. The central problem in turbulence is to investigate how this randomness arises. The theory of turbulence has now developed into one of the most important branches of fluid dynamics and we do not wish to advert to the mathematical problems of that rapidly expanding field. Suffice to say that one of the main objects of the theory should be to find the joint distribution of the velocity components at n points. This is an almost intractable problem and applied mathematicians have confined their attention to the correlation functions at two or more points in space.

Since the velocity field changes with time due to fluid motion, not only are we interested in the velocity correlations at different points in space at the same time but also at different times. For example, we may define

$$\mathscr{E}\left\{u_x(x_1, y_1, z_1, t_1)\, u_y(x_2, y_2, z_2, t_2)\right\}$$

as the mean product of the x-component of the velocity at x_1, y_1, z_1, t_1 and the y-component of the velocity at (x_2, y_2, z_2, t_2). It is natural to denote the above expectation value as a component of the velocity correlation tensor.

It is to be emphasised that in the case of random fields, with every point r in the three- (or many-) dimensional space, we associate a random variable $X(r)$ which can take a continuous range of values. This problem should be clearly distinguished from the statistical distribution of a discrete number of particles in a many-dimensional space. Here also we have a continuous infinity of stochastic variables corresponding to the numbers of particles in every element of volume in the space. In such a case, we define product densities and characteristic functionals as in Chap. IV and to derive these product densities we may have to assume variation or evolution with respect to a one-dimensional parameter.

A well known illustration of this is the famous Boltzmann transport equation of statistical mechanics. The discussion of this equation and its solution under various approximations has become almost a "mathematical recreation". In the simplest case when the molecules of a gas are considered as hard spheres of diameter δ which are allowed to exchange energy only through elastic collisions, the Boltzmann equation assumes the form

$$\left.\begin{aligned}
\frac{\partial}{\partial t} f(r, v, t) + v \cdot V_r f + X(r) \cdot V_v f &= \frac{\delta^2}{2} \int d^3w \int dl \times \\
\times \{f(r, v + (w - v) \cdot ll, t)\, f(r, w - (w - v) &\cdot ll, t) - \\
- f(r, v, t)\, f(r, w, t)\}\, &|(w - v) \cdot l|
\end{aligned}\right\} \qquad (83.1)$$

where $f(r, v, t)\, dr\, dv$ is the average number of molecules in $dr\, dv$ (r is the position, v the velocity) at r, v, $V_r f$ the gradient of f with respect to r, $V_v f$ the gradient with respect to v, l a unit vector and dl the surface element on the unit sphere. $X(r)$ is an outside force (for example gravity) acting on the particle at r. The Boltzmann equation can be extended to the case when the molecules are not hard spheres but considered as centres of force when

$$\frac{\delta^2}{2}\, |(w - v) \cdot l| \qquad (83.2)$$

has to be replaced by an expression depending on the nature of the force.

The entire equation is based on the assumption of randomness of the collision parameters derivable from elementary considerations of the collisions between two "balls of finite size". It is customary to demonstrate that in the stationary case the Maxwell distribution satisfies the Boltzmann equation.

Recently the derivation of the equation has been re-examined critically by KAC [35] in the hope of providing "sharp formulations and paving the way toward further work on this fascinating borderline between mathematics and physics".

XIII. Equations involving random parameters.

84. In the processes till now considered in detail, we assumed that the randomness occurs in every interval of time Δt and the "cumulative" effect was studied knowing the infinitesimal nature of the random changes in Δt. There is however another class of problems in which, given a set of parameters $\langle A \rangle$ which may be specified by the initial conditions at $t=0$, the process develops deterministically. But the parameters $\langle A \rangle$ are assumed to be random variables and consequently the dynamical variables associated with the process become random functions of time. However the statistical properties of $\langle A \rangle$ are assumed to be independent of time. If they themselves are random functions then we are dealing with random functions associated with other random functions and this is precisely the problem discussed in the previous chapters.

KAMPE DE FÉRIET [36] formulates the mathematical problem of equations involving random parameters as follows. He refers to partial differential equations.

Given (a) a partial differential equation

$$a(u) = 0 \tag{84.1}$$

and (b) an open domain D with boundary C (in an n-dimensional Euclidean space), our object is to find a function $u(P)$ satisfying (84.1) in D and taking values $f(A)$ on C,

$$u(P) \to f(A) \quad \text{if} \quad P \to A. \tag{84.2}$$

In each case one has to specify: (a) what conditions (continuity, boundedness, existence of derivatives, etc.) $u(P)$ must satisfy in order to be considered a regular solution; (b) what precise meaning the limit in (84.2) has (limit along given paths, limit in the mean, etc.); and (c) what class of functions $f(A)$ (continuous, integrable, etc.) is one allowed to use for the boundary conditions, in order to secure existence and uniqueness theorems.

When one looks at the physical facts which give support to this mathematical setting, one is readily convinced that, due to errors in measurements, to neglected fluctuations in the phenomena, etc., the boundary conditions cannot be expressed by only one well-determined function $f(A)$, but that a whole set of functions $f_\omega(A)$ must be considered. Furthermore, among these we can not identify the one that will actually materialize in an experiment; in general, we are only able to say that some functions in the set are more likely to be observed than some others. We can translate this idea in mathematical language by saying that, in the boundary conditions corresponding to the real physical problem, the function $f(A)$ must be replaced by a random function $f_\omega(A)$ where ω represents a point in a suitable probability space.

KAMPE DE FÉRIET [36] was led to this formulation by his researches on the statistical theory of turbulence. In his opinion, the key problem in this theory is to find random solutions of the Navier-Stokes equations of hydrodynamics corresponding to a given random velocity field at time $t=0$. Unfortunately, since the equations of fluid dynamics are non-linear almost nothing is known about existence and uniqueness of their solutions. KAMPE DE FÉRIET was attracted to this point of view that the randomness of the velocity field can be traced to the boundary conditions but had to satisfy himself by considering only linear partial differential equations in which the assumptions of randomness in boundary values make sense.

It is not our purpose here to go into the question of the nature of turbulence but to emphasise the fact that the central problem in any stochastic process is to "trace the randomness" to some more elementary "parts" imbedded in the "structure" of the stochastic process.

Speaking of randomness in parameters, it is relevant to refer to the solutions of algebraic equations whose coefficients are random variables. A polynomial of the n-th degree has atmost n solutions and these solutions are random variables if the coefficients are. HAMMERSLEY [37] has considered this problem in great detail and it is rather interesting to note how product densities and characteristic functionals enter into the problem. The n solutions are now random points on a line and it is natural to introduce the concept of product densities and characteristic functionals in relation to the distribution of a discrete set of points on a line which is a continuous parametric space.

XIV. Inverse probability.

85. A few observations will now be made on inverse probability. The subject of inverse probability has unfortunately been regarded with general suspicion and so not many attempts have been made to apply the theory even to cases where unambiguous answers can be given on a proper mathematical formulation of the problems.

Let the result of an experiment α be one of an aggregate of mutually exclusive events $\langle A_i \rangle$ with probabilty $P_\alpha(A_i)$, and of an experiment β be one of $\langle B_i \rangle$ with probability $P_\beta(B_i)$. Consider the compound experiment $\alpha + \beta$ the result of which is a pair of values (A_j, B_k) with probabilities $P(A_j, B_k)$. $\langle A_j B_k \rangle$ is the corresponding aggregate of events for the compound experiment. If the events A_j and B_k are independent

$$P(A_j, B_k) = P_\alpha(A_j) P_\beta(B_k). \tag{85.1}$$

If the events are not independent we cannot express $P(A_j, B_k)$ in terms of $P_\alpha(A_j)$ and $P_\beta(B_k)$ only. Let us examine the meaning of non-independence. If we postulate the conditional probability $P_\alpha(A_j | B_k)$ that A_j is the result of experiment α given that B_k was the result of β, it implies that it is possible to perform the experiment β without performing α. In such a case

$$P(A_j, B_k) = P_\beta(B_k) P_\alpha(A_j | B_k). \tag{85.2}$$

In a like manner, if we postulate the conditional probability $P_\beta(B_k | A_j)$,

$$P(A_j, B_k) = P_\alpha(A_j) P_\beta(B_k | A_j). \tag{85.3}$$

This assumes the possibility of performance of the experiment α without performing β.

On the other hand, for the validity of the following result,

$$\left. \begin{aligned} P_\alpha(A_j) &= \sum_k P(A_j, B_k), \\ P_\beta(B_j) &= \sum_k P(A_k, B_j), \end{aligned} \right\} \tag{85.4}$$

we need only admit the possibility of the simultaneous performance of experiments α and β.

The problem of inverse probability is as follows. If we know $P_\alpha(A_j)$, $P_\beta(B_k)$ and $P_\beta(B_k|A_j)$, are we right in assuming

$$P_\beta(B_k|A_j) = P_\alpha(A_j|B_k)\, P_\beta(B_k)/P_\alpha(A_j) \ ? \qquad (85.5)[1]$$

This is a direct consequence of (85.2) and (85.3). If both (85.2) and (85.3) are valid, we must assume α and β can be performed without reference to each other. Is this possible? No general answer can be given, but in the case of stochastic processes we can discuss this question unambiguously.

86. Markovian stochastic processes evolving with respect to a one-dimensional parameter. Let us consider a stochastic process representing the "evolution" of the random variable $X(t)$ in a Markovian manner with respect to the one-dimensional parameter t. Let our experiment α consist in the determination of the value of $x(t)$ at t_0 and experiment β, the determination of $x(t)$ at t_1. Let $t_1 > t_0$. In Chap. I we stated that when such a process progresses with t we can define $\pi(x_1|x_0; t_1, t_0)\, dx_1$ as the probability that $x(t_0) = x_0$ and $x(t_1) = x_1$. This is obviously a conditional probability and corresponds to $P_\beta(B_k|A_j)$ in the previous discussion, since A_j's correspond to the values of $x(t_0)$ and B_k's to those of $x(t_1)$. *This conditional probability is determined if we know the infinitesimal transition probabilities corresponding to* $(t_1 - t_0) \to \Delta \to 0$. We shall assume the process to be homogeneous in t and so $\pi(x|x_0; t_1, t_0)$ is a function only of the interval $t_1 - t_0$ and not of t_1, t_0, x separately. Though it is customary to set $t_0 = 0$ and write $t_1 - t_0$ as t_1, we shall deliberately refrain from doing so for reasons which will be clear presently. On the other hand we shall use the notation $\pi(x; t)$ to represent the probability frequency function (p.f.f.) of $x(t)$ when we do not specify the "initial" conditions i.e. the p.f.f. at some $t'(t' < t)$. Thus by definition we immediately write

$$\pi(x_1; t_1) = \int_{x_0} \pi(x_0; t_0)\, \pi(x_1|x_0; t_1, t_0)\, dx_0 \qquad (t_0 < t_1). \qquad (86.1)$$

When we deal with the distributions at two different values of t, say, t_0 and t_1 $(t_1 > t_0)$, we shall call the distribution at t_1 the emergent spectrum with respect to the injected spectrum at t_0 and $\pi(x_1|x_0; t_1, t_0)$ the fundamental conditional probability for the finite interval $t_1 - t_0$. We immediately note that $\pi(x; t)$ corresponds to $P_\beta(B_k)$ and $\pi(x_0; t_0)$ to $P_\alpha(A_j)$, the distributions corresponding to the experiments β and α respectively. In the general case of the two experiments α and β, the knowledge of $P_\alpha(A_j)$, $P_\beta(B_k)$ does not give us the conditional probability unless it is given apriori, or the joint probability $P(A_j, B_k)$ is given. But in the case of a stochastic process, $\pi(x_1; t_1)$ satisfies the same equation as $\pi(x_1|x_0; t_1, t_0)$ except that in the case of $\pi(x_1|x_0)$ the injection spectrum at t_0 is given by

$$\pi(x_1; t_0) = \delta(x - x_0) \qquad (86.2)$$

where δ is the Dirac delta function. $\pi(x; t_1)$ satisfies the Chapman-Kolmogoroff equation

$$\pi(x; t_1) = \int_{x'} \pi(x'; t_0 + \tau)\, \pi(x|x'; t_1, t_0 + \tau)\, dx' \qquad (86.3)$$

for all values of τ such that $t_0 < t_0 + \tau < t_1$ and in particular

$$\pi(x_1|x_0; t_1, t_0) = \int_{x'} \pi(x'|x_0; t_0 + \tau)\, \pi(x_1|x'; t_1, t_0 + \tau)\, dx'. \qquad (86.4)$$

---

[1] This is known in statistical literature as Bayes theorem and the validity of its application has been the subject of considerable discussion.

Thus in a stochastic process if we know the infinitesimal transition probability we can determine $P_\beta(B_k|A_j)$, also $P_\beta(B_k)$ given $P_\alpha(A_j)$. The problem of inverse probability is as follows: Is there a meaning in asking the following questions:

(i) Can we obtain $P_\alpha(A_j)$ i.e. $\pi(x_0; t_0)$ given $P_\beta(B_k)$?

(ii) Can we obtain $P_\alpha(A_j|B_k)$ i.e. $\pi(x_0|x_1; t_0, t_1)$, $t_0 < t_1$?

The answer to the first question is direct while the second leads to complications which we shall resolve presently. The forward differential equation of $\pi(x_1; t_1)$ is obtained by expressing $\pi(x_1; t_1 + \Delta)$ in terms of $\pi(x_1; t_1)$. In the derivation, Δ is assumed to be positive. The author has shown that this restriction is not necessary for π though the infinitesimal transition probability $R(x|x_0)\Delta$ obtained as the limit of $\pi(x|x_0; t_1, t_0)$ as $t_1 - t_0 \to \Delta \to 0$ is defined only for positive Δ. $\pi(x; t - \Delta)$ can be obtained by merely replacing Δ by $-\Delta$ in $\pi(x; t + \Delta)$. Thus $\pi(x_0; t_0)$ can be obtained by a process of integration from $\pi(x_1; t_1)$, $t_1 > t_0$. The author [38] proved by very simple arguments

$$\pi(x_0; t_0) = \int_{x_1} \pi(x_1; t_1) \hat{\pi}(x_0|x_1; t_1, t_0)\, d x_1 \qquad (t_1 > t_0) \tag{86.5}$$

when $\hat{\pi}(x_0|x_1; t_1, t_0)$ is a function obtained by replacing $t_1 - t_0$ by the negative quantity $t_0 - t_1$ in $\pi(x_0|x_1; t_0, t_1)$ which we remember is only a function of $t_1 - t_0$ and not t_1, t_0 separately. $\hat{\pi}$ unlike π has no probability significance and is only a functional operator. In Eq. (86.3), since $\pi(x; t)$ and $\pi(x_1|x_0; t, t_0)$ are both probability densities, they are positive, and so $\pi(x, t)$ exists for any $t_1 > t_0$, i.e. $t_1 - t_0$ is positive and finite and can be chosen as large as we please. If $\pi(x|x_0; t_1, t_0)$ tends to a stationary distribution as $t_1 - t_0 \to \infty$, so also does $\pi(x_1; t_1)$. On the contrary in (86.5) since $\hat{\pi}$ is not a p.f.f. $\pi(x; t_0)$ does not remain a p.f.f. for all values of $t_0 < t_1$. It can be proved that there exists a t_p such that $\pi(x_0, t_0)$ is a p.f.f. in the domain $t_p \leq t_0 \leq t_1$ and the process cannot be traced back to a point $t < t_p$. t_p is the primitive origin of the stochastic process. $t_1 - t_p$ is the "age" of the process at t_1. Thus tracing a process backward can be done in an exactly similar manner as tracing it forward with the only difference that in tracing backward, we reach a point t_p at which we must assume an apriori injection spectrum. In other words, the spectrum at t_p cannot be the spectrum of a process started earlier.

We have observed that the p.f.f. $\pi(x_1|x_0; t_1, t_0)$ $(t_1 > t_0)$ is a particular case of $\pi(x_1; t_1)$ with $\pi(x; t_0) = \delta(x - x_0)$. In this case, the process cannot be traced back further than t_0, i.e. to a point $t < t_0$. Since $\pi(x; t_0) = 0$ for $x \neq x_0$ $\pi(x; t_0 - \Delta)$ is negative (Δ positive) for $x \neq x_0$. Therefore a delta function can only be an injected spectrum and cannot be the emergent spectrum of a stochastic process. Thus $\hat{\pi}(x|x_0; t, t_0)$, $t > t_0$, has no probability interpretation.

Now we come to the second question. What does correspond to the conditional probability $P_\alpha(A_j, B_k)$ if $\pi(x_1|x_0; t_1, t_0)$ corresponds to $P_\alpha(A_j|B_k)$? We must note that $\hat{\pi}(x|x_0; t, t_0)$ does not correspond to $P_\alpha(A_j|B_k)$. We have to define a new function $P(x_0|x_1; t_0, t_1)$ as the conditional probability that $x(t_0)$ lies between x_0 and $x_0 + dx_0$ given that $x(t) = x_1$ at $t = t_1$. This has to be distinguished from $\hat{\pi}$. When in defining P we state that $x(t_1) = x_1$ we mean that the observed value of the emergent spectrum at t_1 is x_1 and not that the emergent spectrum is a delta function.

$$P(x_0|x_1; t_0, t_1) = \pi(x_0; t_0)\, \pi(x_1|x_0; t_1, t_0)/\pi(x_1; t_1). \tag{86.6}$$

If we know the infinitesimal transition probability $R(x_1|x_0)$, $\pi(x_1|x_0, t_1, t_0)$ can be determined. So can $\pi(x_1; t_1)$ given $\pi(x_0; t_0)$, or vice versa. Thus in the case of a stochastic process progressing with respect to t we can summarise our results as follows:

If the transition probability for infinitesimal $\varDelta$ is known,

(i) $P_\alpha(A_j)$ can be obtained if we know $P_\beta(B_k)$ and vice versa,

(ii) $P_\beta(B_k|A_j)$ can be determined directly without reference to $P_\alpha(A_j)$ or $P_\beta(B_k)$.

(iii) $P_\alpha(A_j|B_k)$ can be determined if one of the two $P_\alpha(A_j)$ and $P_\beta(B_k)$ is known since the other two can be determined from the transition probability.

87. Inverse probability in relation to random functions of t where t is many-dimensional. If t is many-dimensional and replaced by the symbol r the process cannot be considered evolutionary and hence we have to first obtain the joint distribution of the stochastic variables at two points r_1 and r_2. Earlier by considering the value of the stochastic variable at t_1, we connected the value at t_2 by evolution and obtained the distribution at t_2 conditional upon the value of $x(t_1)$. The joint distribution of $x(t_1)$ and $x(t_2)$ is derived therefrom. Here we first speak of joint distributions and marginal distributions and conditional probabilities derived therefrom.

If however we define another parameter t with respect to which the aggregate of stochastic variates corresponding to different values of r evolves, the problem reduces to the one discussed before, the experiments α and β now referring to different points on the t-axis and not in the r-space.

XV. Stochastic processes involving "back-scattering".

88. We conclude this essay by a few observations on a class of stochastic processes which are defined with respect to a single parameter but contain new features hitherto not discussed. The most important characteristic of Markovian stochastic processes is the "development" or "unfolding" as the one-dimensional parameter t is varied. The initial conditions are defined at $t=t_0$ and we ask for the state at $t=t_1(>t_0)$. In inverse probability we trace back the process for any $t<t_0$. The forward differential equation was obtained by expressing the state at $t+dt$ in terms of the state at t; the corresponding backward differential equation was obtained by considering the first event after "injection" or equivalently by considering the events in the small interval of time $(t_0, t_0+\varDelta)$. We shall now discuss a new type of stochastic process first through an example and then by a suitable generalisation.

Let us consider a process in which the state of a particle is characterised by its velocity which can take positive or negative values as the one-dimensional parameter l[1], the distance travelled by the particle, is varied. If the particle is assumed to move in the direction of $+l$ if the velocity is positive, it is necessarily implied that it moves in the direction of $-l$ if the velocity is negative. If we choose time as the one-dimensional parameter, the process can be supposed to "develop" with t but if l is chosen as the parameter of the process, the concept of "development" with l breaks down. However, transition probabilities can be defined with respect to l as follows.

We define a probability $R_1(v'|v)\,dv'\varDelta$ that a particle in a state characterised by velocity v (positive) at l makes a transition to a state characterised by a velocity lying between v' and $v'+dv'$ (v' positive or negative) when it moves to $l+\varDelta$ ($\varDelta$ positive); $R_2(v'|v)\,dv'\varDelta$ that a particle in a state characterized by

---

[1] We are using l in this chapter since t will be used when the one-dimensional parameter is time.

velocity v (negative) at l makes a transition to a state characterised by a velocity lying between v' and $v' + dv'$ (v' positive or negative) when it moves to $l - \Delta$ (Δ positive). It is also assumed that

$$R_1(v_1' | v_1) \equiv R_2(v_2' | v_2) \quad \text{if} \quad v_2' = -v_1' \quad \text{and} \quad v_2 = -v_1 \tag{88.1}$$

i.e. the transition probabilities are symmetric with respect to change of direction of both the velocities v_1' and v_1.

In this case, the question arises whether we can write differential equations with respect to l. The essential difficulty is due to "propagation" in the direction of $+l$ when v is positive and in the direction of $-l$ when v is negative. We should distinguish "propagation" in the direction of $-l$ from tracing a process backwards, for in tracing a process to $t - \Delta$ we replace Δ by $-\Delta$ as has been elaborated in the previous chapter. On the other hand, we have taken the positive quantity $R(v'|v)\, dv'\Delta$ for transition as we proceed backwards and so we call it "*backward propagation*". By "back scattering" we mean that the particle switches from positive to negative velocities or vice versa. If such "back-scattering" is possible at all values of l, no differential equation can be written with respect to l by any of the arguments envisaged in this essay, without the aid of an additional parameter t.

A way out of this difficulty is suggested by a method used recently by BELLMAN *et al.* [*39*], by "switching off" the backward-scattering probabilities outside an interval[1] (l_1, l_2) and considering the states only at the boundaries l_1 and l_2. To illustrate the method, we shall obtain the differential equations and their solutions in the very simple case when only one positive velocity $v = +a$ and ooe negative velocity $v = -a$ are envisaged.

Let the interval (l_1, l_2) be of length l. We shall define $p_1(l)$, $q_1(l)$ as the probabilities of the particle emerging out at l_1 (with the negative velocity) and at l_2 (with the positive velocity) respectively. Let $a\Delta$ be the probability of "backward scattering" in an interval of length Δ. Then, using the familiar arguments of the regeneration point method we have

$$p_1(l + \Delta) = (1 - a\Delta)\cdot p_1(l)\,(1 - a\Delta) + p_1(l)\cdot a\Delta\cdot p_1(l)\cdot(1 - a\Delta) + a\Delta \tag{88.2}$$

or

$$p_1'(l) = a\{p_1^2(l) - 2p_1(l) + 1\} \tag{88.3}$$

with the initial condition

$$p_1(0) = 0.$$

Similar arguments lead to

$$q_1'(l) = -a\,q_1(l)\,\{1 - p_1(l)\} \tag{88.4}$$

with

$$q_1(0) = 1.$$

These equations have very simple solutions with interesting physical interpretations:

$$p_1(l) = \frac{al}{1 + al}, \tag{88.5}$$

$$q_1(l) = \frac{1}{1 + al}, \tag{88.6}$$

i.e., $p_1(l) > q_1(l)$ if $l > \dfrac{1}{a}$ the mean free path for back-scattering and tends to unity as l approaches infinity.

---

[1] The language used by BELLMAN *et al.* is not the same but essentially equivalent to the above though in his article and its review by CHANDRASEKHAR [*41*] the stochastic significance of "switching off" was not stated in this form presumably because no comparison was made with the more general process.

In deriving the equation for $p_1(l)$, the "first" event is considered between l_1 and $l_1 + \Delta$ at which the emergent state is also defined: This is an interesting modification of the usual application of the regeneration point method in a stochastic process where there is no "back-scattering". If we had introduced the time parameter t, the states $p_1(l)$ and $q_1(l)$ correspond to $t = \infty$.

BELLMAN *et al.* have applied this method to the case when other processes like "fission", i.e. production of particles with velocities positive or negative, can occur in the interval (l_1, l_2). Without "back-scattering", these problems would just reduce to Markovian processes described before, but it is the possibility of such "back-scattering" that necessitates the distinction between Markovian processes and the "ambigenous" process described in this chapter.

XVI. Some concluding remarks.

89. We have attempted in this article to develop a physical approach to the theory of stochastic processes in a form intended to be useful to a physicist dealing with problems involving some random element in their structure. Attention has been drawn as to how the random element enters when quantum mechanical processes are concerned. The author is aware of the dangers in drawing the correspondence between stochastic processes and quantum mechanical phenomena.

To a fresh entrant to the theory of stochastic processes, the relation between the wave functions in quantum mechanics and their probabilistic interpretation may seem to point out some sort of correspondence between the theory of stochastic processes and quantum mechanical phenomena. *This correspondence is only superficial.* Any reference in this article to quantum mechanics has been made only to familiarise the worker on stochastic processes with the use of quantum mechanical cross-sections.

Earlier in Chap. V, a reference was made to branching stochastic processes which deal with phenomena where we envisage different types of particles whose numbers and dynamical variables characterising their states vary in a random manner. This may suggest some resemblance to the quantum theory of fields which deals with varying number of particles which are the quanta of the field. It may seem at first sight that the difference lies in the fact that we introduce wave functions and operators in field theory. The difference lies much deeper. "It consists in that the mathematics of stochastic processes deals with the solutions of clearly formulated problems whereas in field theory, one is using mathematics only as an imperfect guide in trying to formulate a problem which is not yet physically defined."

Despite this cautious and rather sceptical attitude of theoretical physicists to any discussion about probability concepts in quantum theory, attempts have been made by some to understand the relationship between probability and quantum mechanics. Special mention must be made of such an attempt by FEYNMAN [42], himself one of the creators of modern quantum electrodynamics. He draws particular attention to the fact that while the square of the modulus of the wave function (or complex probability amplitude) represents a probability density, the principle of superposition (additivity) applies only to wave functions (or complex probability amplitudes). It looks as though in the microscopic realm of elementary particles "nature with her infinite imagination has found another set of principles for determining probabilities, a set other than that of LAPLACE, which nevertheless does not lead to logical inconsistencies. The laws of probability which are conventionally applied are quite satisfactory in analyzing the behaviour of the roulette wheel but not the behaviour of a single electron or a photon of light".

Using the principle of superposition of infinitesimal complex probability amplitudes FEYNMAN "derived" the Schrödinger equation. Without going into a critical discussion of his derivation the author wishes to point out that FEYNMAN's methods bear close correspondence (almost one to one) with those used in this article in connection with the equation for an additive basic random process and the elaborate discussions on its Gaussian approximation (see Sects. 56, 59 and 50), keeping in mind of course that additivity applies in FEYNMAN's case to complex amplitudes. The transition probability amplitude for infinitesimal times is *assumed* so as to satisfy the requirements of the Lagrangian principle of least action. The remarks on trajectories of stochastic processes (see Sect. 42) can be translated directly into FEYNMAN's language provided we replace probability densities by amplitudes.

Acknowledgment.

I am grateful to Mr. S. K. SRINIVASAN, Mr. R. VASUDEVAN and Mr. N. R. RANGANATHAN for their valuable assistance and hearty co-operation throughout the preparation of this work. They belong to the group working in Madras on the applications of the theory of stochastic processes to physical problems and their work was the main stimulus for an article of this form and style with accepted emphasis on the physical approach. My thanks are due to Professors CHANDRASEKHAR and DOOB, for allowing me to quote abstracts from their work and to the Dover Publications Inc. for permission to reproduce some relevant passages from KHINCHIN's book.

Dr. M. VENKATARAMAN, my colleague in the Madras University acted as my mathematical guide and I cannot adequately express my gratitude to him for the valuable discussions we had in comparing the physical and the measure-theoretic formulations of the theory of stochastic processes.

References[1].

Introduction.

[1] DOOB, J.L.: Stochastic processes. New York 1953.
[2] LOEVE, MICHEL: Probability methods in Physics. Mimeographed notes of a Seminar. Berkeley: University of California 1948—49.
[3] BARTLETT, M.S.: Stochastic processes. Cambridge 1955.

Part A.

Chapter I.

[1] DOOB, J.L.: Stochastic processes. New York 1953.
USPENSKY, J.V.: Introduction to mathematical probability. New York 1937.

Chapter II.

[4] KHINCHIN, I.: Mathematical Foundations of Statistical Mechanics. Dover 1949.
[5] BLANC-LAPIERRE, ANDRÉ: Proceedings of the Third Berkeley Symposium on mathematical statistics and probability, Vol. III, p. 145, 1956.
CRAMER, H.: Methods of mathematical statistics. Princeton 1946.

Chapter III.

TRANTER, C.J.: Integral transforms in mathematical physics. London: Methuen & Co. 1951.

Chapter IV.

AITKEN, A.C.: Statistical mathematics. London: Oliver & Boyd 1947.
[6] CONSAEL, R.: Bulletin de la Classe des Sciences. Acad. Roy. Belg. **38**, 442 (1952).

---

[1] The list given here is representative but not exhaustive. A comprehensive list can be obtained from the papers and books cited. The numbered references have been actually cited in the article.

Chapter V.

Ramakrishnan, Alladi: Proc. Cambridge Phil. Soc. **46**, 595 (1950).
[7] Janossy, L.: Proc. Roy. Irish Acad. Sci. **53**, 181 (1950).
Bartlett, M. S., and D. G. Kendall: Proc. Cambridge Phil. Soc. **47**, 65 (1950).

Chapter VI.

Mayer, J. E., and M. G. Mayer: Statistical Mechanics. New York 1940.
Ter Haar, D.: Statistical Mechanics. New York 1954.

Part B.

Chapter I.

Ramakrishnan, Alladi: Proc. Ind. Acad. Sci. **44**, 428 (1956).
Bartlett, M. S.: Stochastic processes. Cambridge 1955.
[8] Ramakrishnan, Alladi: Z. angew. Math. Mech. **37**, 336 (1957).
[9] Bartlett, M. S.: Proc. Cambridge Phil. Soc. **49**, 263 (1953).
Feller, W.: An introduction to probability theory and its applications. New York 1950.
[10] Ramakrishnan, Alladi: Proc. Kon. Nederl. Akad. Wetensch. **58**, 470, 634 (1955); **59**, 120 (1956).
[11] Mathews, P. M., and S. K. Srinivasan: Proc. Nat. Inst. Sci. **22**, 369 (1956).
[12] Bellman and Harris: Proc. Nat. Akad. Sci. U.S.A. **34**, 601 (1948).

Chapter II.

[1] Doob, J. L.: Stochastic processes. New York 1953.
Kolmogoroff: Grundbegriffe der Wahrscheinlichkeitsrechnung. Ergebn. Math. 2, Nr. 3 (1933).
[13] Moyal, J. E.: J. Roy. Statist. Soc. B **11**, 150 (1949).

Chapter III.

[14] Loeve, M.: Rev. Sci. Paris **83**, 297 (1945).
[15] Ramakrishnan, Alladi: Astrophys. J. **119**, 443 (1954).
[13] Moyal, J. E.: J. Roy. Statist. Soc. B **11**, 150 (1949).

Chapter IV.

[16] Chandrasekhar, S.: Rev. Mod. Phys. **15**, 1 (1943).
[17] Furry, W. H.: Phys. Rev. **52**, 569 (1937).
[18] Kendall, D. G.: Biometrika **35**, 6 (1948).
[19] Feller, W.: An introduction to probability theory and its applications. New York 1950.
[20] Ledermann, W.: Proc. Cambridge Phil. Soc. **46**, 581 (1950).
[21] Dyson, F. J.: Phys. Rev. **75**, 486, 1736 (1949).
[22] Feynman, R. P.: Phys. Rev. **84**, 108 (1951).
[23] Rayleigh, Lord: Collected papers, Vol. 6, p. 604.
[24] Chandrasekhar, S., and G. Munch: Astrophys. J. **115**, 103 (1952).

Chapter V.

[25] Janossy, L.: Proc. Phys. Soc. Lond. A **63**, 1101 (1950).
Ramakrishnan, Alladi: J. Roy. Statist. Soc. B **13**, 131 (1951).
Ramakrishnan, Alladi: Proc. Cambridge Phil. Soc. **148**, 451 (1952).

Chapter VI.

[16] Chandrasekhar, S.: Rev. Mod. Phys. **15**, 1 (1943).
[24] Chandrasekhar, S., and G. Munch: Astrophys. J. **112**, 380, 393 (1950).
[25] Neyman, J., and E. L. Scott: Proceedings of the Third Berkeley Symposium on mathematical statistics and probability, Vol. III, p. 75, 1956.
[26] Chandrasekhar, S., and G. Munch: Astrophys. J. **112**, 380, 393 (1950).
[27] Ramakrishnan, Alladi, and R. Vasudevan: Astrophys. J. **126**, 573 (1957).

Chapter VII.

[28] Born, M., and H. S. Green: A general kinetic theory of liquids. Oxford 1949.
Ramakrishnan, Alladi: Phil. Mag. **45**, 401, 1053 (1954).
[29] Salzburg, Z. W., R. W. Zwanzig and J. G. Kirkwood: J. Chem. Phys. **21**, 1098 (1953).
[30] Mayer, J. E.: Bull. Amer. Math. Soc. **62**, 352 (1956).

Chapter VIII.

Mathews, P. M., and S. K. Srinivasan: Proc. Ind. Acad. Sci. **43**, 4 (1956).

Chapter IX.

[31] Störmer: Z. Astrophys. **1**, 237 (1930).
[32] Birkhoff, G.D.: Proc. Nat. Acad. **17**, 650, 656 (1931).
 [4] Khinchin, I.: Mathematical Foundations of statistical mechanics. Dover 1949.
[33] Wiener, N.: Acta math. **55**, 117—258 (1930).
[34] Mathews, P.M., and S.K. Srinivasan: Proc. Ind. Acad. Sci. **43**, 4 (1956).

Chapter X.

Ramakrishnan, Alladi and R. Vasudevan: Proc. Ind. Acad. Sci. **1958** (to be published).

Chapter XI.

Heitler, W.: Quantum theory of radiation. 3rd edition. Oxford 1954.
Heitler, W.: Elementary wave mechanics with applications. Oxford 1956.

Chapter XII.

[35] Kac, M.: Proceedings of the Third Berkeley Symposium on mathematical statistics and probability, Vol. III, 1956.
Bass, J.: Space time correlation in a turbulent fluid. Berkeley: University of California, Publications in Statistics 1955.
Batchelor, G.K.: Theory of homogeneous turbulence. Cambridge 1953.

Chapter XIII.

[36] Kampe de Fériet: Proceedings of the Third Berkeley Symposium on mathematical statistics and probability, Vol. III, 1956.
[37] Hammersley, J.M.: Proceedings of the Third Berkeley Symposium on mathematical statistics and probability, Vol. III, 1956.

Chapter XIV.

[38] Ramakrishnan, Alladi: Proc. Ind. Acad. Sci. A **41**, 145 (1955).

Chapter XV.

[39] Bellman, R., R.E. Kalaba and G.M. Wing: Proc. Nat. Acad. Sci. USA. **43**, 517 (1957).
[40] Bellman, R., R.E. Kalaba and G.M. Wing: J. Math. Mech. **7**, 149 (1958).
[41] Chandrasekhar, S.: Math. Reviews **19**, 506 (1958).

Chapter XVI.

[42] Feynman, R.P.: Rev. Mod. Phys. **20**, 267 (1948).

Sachverzeichnis.

(Deutsch-Englisch.)

Bei gleicher Schreibweise in beiden Sprachen sind die Stichwörter nur einmal aufgeführt.

Subject Index.

(English-German.)

Where English and German spelling of a word is identical the German version is omitted.